One-Stop Internet Resources

Log on to msmath2.net

Online Book

- Complete Student Edition
- Links to Online Study Tools

Online Study Tools

- Extra Examples
- Self-Check Quizzes
- Vocabulary Review
- Chapter Test Practice
- Standardized Test Practice

Online Activities

- WebQuest Projects
- Game Zone Activities
- Career Links
- Data Updates

Parent & Student Study Guide Workbook

- Printable Worksheets

Graphing Calculator Keystrokes

- Calculator Keystrokes for other calculators

GLENCOE MATHEMATICS

Mathematics

Applications and Concepts

Course 2

Bailey
Day
Frey
Howard
Hutchens
McClain
Moore-Harris
Ott
Pelfrey
Price
Vielhaber
Willard

McGraw Hill Glencoe

New York, New York Columbus, Ohio Chicago, Illinois Peoria, Illinois Woodland Hills, California

The McGraw·Hill Companies

The Standardized Test Practice features in this book were aligned and verified by The Princeton Review, the nation's leader in test preparation. Through its association with McGraw-Hill, The Princeton Review offers the best way to help students excel on standardized assessments.

The Princeton Review is not affiliated with Princeton University or Educational Testing Service.

Send all inquiries to:
Glencoe/McGraw-Hill
8787 Orion Place
Columbus, OH 43240-4027

ISBN: 0-07-829633-1

9 10 043/027 10 09 08 07 06

Contents in Brief

Authors

Rhonda Bailey
Mathematics Consultant
Mathematics by Design
DeSoto, Texas

Roger Day, Ph.D.
Associate Professor
Illinois State University
Normal, Illinois

Patricia Frey
Director of Staffing and Retention
Buffalo City Schools
Buffalo, New York

Arthur C. Howard
Mathematics Teacher
Houston Christian High School
Houston, Texas

Deborah T. Hutchens, Ed.D.
Assistant Principal
Great Bridge Middle School
Chesapeake, Virginia

Kay McClain, Ed.D.
Assistant Professor
Vanderbilt University
Nashville, Tennessee

Beatrice Moore-Harris
Mathematics Consultant
League City, Texas

Jack M. Ott, Ph.D.
Distinguished Professor of Secondary Education Emeritus
University of South Carolina
Columbia, South Carolina

Ronald Pelfrey, Ed.D.
Mathematics Specialist
Appalachian Rural Systemic Initiative
Lexington, Kentucky

Jack Price, Ed.D.
Professor Emeritus
California State Polytechnic University
Pomona, California

Kathleen Vielhaber
Mathematics Specialist
Parkway School District
St. Louis, Missouri

Teri Willard, Ed.D.
Assistant Professor of Mathematics Education
Central Washington University
Ellensburg, Washington

Contributing Authors

USA TODAY
The USA TODAY Snapshots®, created by USA TODAY®, help students make the connection between real life and mathematics.

Dinah Zike
Educational Consultant
Dinah-Might Activities, Inc.
San Antonio, Texas

Aaron Haupt

Content Consultants

Each of the Content Consultants reviewed every chapter and gave suggestions for improving the effectiveness of the mathematics instruction.

Mathematics Consultants

L. Harvey Almarode
Curriculum Supervisor, Mathematics K–12
Augusta County Public Schools
Fishersville, VA

Claudia Carter, MA, NBCT
Mathematics Teacher
Mississippi School for Mathematics and Science
Columbus, MS

Carol E. Malloy, Ph.D.
Associate Professor, Curriculum Instruction, Secondary Mathematics
The University of North Carolina at Chapel Hill
Chapel Hill, NC

Melissa McClure, Ph.D.
Mathematics Instructor
University of Phoenix On-Line
Fort Worth, TX

Robyn R. Silbey
School-Based Mathematics Specialist
Montgomery County Public Schools
Rockville, MD

Leon L. "Butch" Sloan, Ed.D.
Secondary Mathematics Coordinator
Garland ISD
Garland, TX

Barbara Smith
Mathematics Instructor
Delaware County Community College
Media, PA

Reading Consultant

Lynn T. Havens
Director
Project CRISS
Kalispell, MT

ELL Consultants

Idania Dorta
Mathematics Educational Specialist
Miami–Dade County Public Schools
Miami, FL

Frank de Varona, Ed.S.
Visiting Associate Professor
Florida International University College of Education
Miami, FL

Teacher Reviewers

Each Teacher Reviewer reviewed at least two chapters of the Student Edition, giving feedback and suggestions for improving the effectiveness of the mathematics instruction.

Royallee Allen
Teacher, Math Department Head
Eisenhower Middle School
San Antonio, TX

Dennis Baker
Mathematics Department Chair
Desert Shadows Middle School
Scottsdale, AZ

Rosie L. Barnes
Teacher
Fairway Middle School–KISD
Killeen, TX

Charlie Bialowas
Math Curriculum Specialist
Anaheim Union High School District
Anaheim, CA

Stephanie R. Boudreaux
Teacher
Fontainebleau Jr. High School
Mandeville, LA

Dianne G. Bounds
Teacher
Nettleton Junior High School
Jonesboro, AR

Susan Peavy Brooks
Math Teacher
Louis Pizitz Middle School
Vestavia Hills, AL

Karen Sykes Brown
Mathematics Educator
Riverview Middle School
Grundy, VA

Kay E. Brown
Teacher, 7th Grade
North Johnston Middle School
Micro, NC

Renee Burgdorf
Middle Grades Math Teacher
Morgan Co. Middle
Madison, GA

Kelley Summers Calloway
Teacher
Baldwin Middle School
Montgomery, AL

Carolyn M. Catto
Teacher
Harney Middle School
Las Vegas, NV

Claudia M. Cazanas
Math Department Chair
Fairmont Junior High
Pasadena, TX

David J. Chamberlain
Secondary Math Resource Teacher
Capistrano Unified School District
San Juan Capistrano, CA

David M. Chioda
Supervisor Math/Science
Marlboro Township Public Schools
Marlboro, NJ

Carrie Coate
7th Grade Math Teacher
Spanish Fort School
Spanish Fort, AL

Toinette Thomas Coleman
Secondary Mathematics Teacher
Caddo Middle Career & Technology School
Shreveport, LA

Linda M. Cordes
Math Department Chairperson
Paul Robeson Middle School
Kansas City, MO

Polly Crabtree
Teacher
Hendersonville Middle School
Hendersonville, NC

Dr. Michael T. Crane
Chairman Mathematics
B.M.C. Durfee High School
Fall River, MA

Tricia Creech, Ph.D.
Curriculum Facilitator
Southeast Guilford Middle School
Greensboro, NC

Lyn Crowell
Math Department Chair
Chisholm Trail Middle School
Round Rock, TX

B. Cummins
Teacher
Crestdale Middle School
Matthews, NC

Debbie Davis
8th Grade Math Teacher
Max Bruner, Jr. Middle School
Ft. Walton Beach, FL

Diane Yendell Day
Math Teacher
Moore Square Museums Magnet Middle School
Raleigh, NC

Wendysue Dodrill
Teacher
Barboursville Middle School
Barboursville, WV

Judith F. Duke
Math Teacher
Cranford Burns Middle School
Mobile, AL

Carol Fatta
Math/Computer Instructor
Chester Jr. Sr. M.S.
Chester, NY

Cynthia Fielder
Mathematics Consultant
Atlanta, GA

Georganne Fitzgerald
Mathematics Chair
Crittenden Middle School
Mt. View, CA

Jason M. Fountain
7th Grade Mathematics Teacher
Bay Minette Middle School
Bay Minette, AL

Sandra Gavin
Teacher
Highland Junior High School
Cowiche, WA

Ronald Gohn
8th Grade Mathematics
Dover Intermediate School
Dover, PA

Larry J. Gonzales
Math Department Chairperson
Desert Ridge Middle School
Albuquerque, NM

Shirley Gonzales
Math Teacher
Desert Ridge Middle School
Albuquerque, NM

Paul N. Hartley, Jr.
Mathematics Instructor
Loudoun County Public Schools
Leesburg, VA

Deborah L. Hewitt
Math Teacher
Chester High School
Chester, NY

Steven J. Huesch
Mathematics Teacher/Department Chair
Cortney Jr. High
Las Vegas, NV

Sherry Jarvis
8th Grade Math/Algebra 1 Teacher
Flat Rock Middle School
East Flat Rock, NC

Mary H. Jones
Math Curriculum Coordinator
Grand Rapids Public Schools
Grand Rapids, MI

Vincent D.R. Kole
Math Teacher
Eisenhower Middle School
Albuquerque, NM

Ladine Kunnanz
Middle School Math Teacher
Sequoyah Middle School
Edmond, OK

Barbara B. Larson
Math Teacher/Department Head
Andersen Middle School
Omaha, NE

Judith Lecocq
7th Grade Teacher
Murphysboro Middle School
Murphysboro, IL

Paula C. Lichiello
7th Grade Math and Pre-Algebra Teacher
Forest Middle School
Forest, VA

Michelle Mercier Maher
Teacher
Glasgow Middle School
Baton Rouge, LA

Jeri Manthei
Math Teacher
Millard North Middle School
Omaha, NE

Albert H. Mauthe, Ed.D.
Supervisor of Mathematics (Retired)
Norristown Area School District
Norristown, PA

Karen M. McClellan
Teacher & Math Department Chair
Harper Park Middle
Leesburg, VA

Ken Montgomery
Mathematics Teacher
Tri-Cities High School
East Point, GA

Helen M. O'Connor
Secondary Math Specialist
Harrison School District Two
Colorado Springs, CO

Cindy Ostrander
8th Grade Math Teacher
Edwardsville Middle School
Edwardsville, IL

Michael H. Perlin
8th Grade Mathematics Teacher
John Jay Middle School
Cross River, NY

Denise Pico
Mathematics Teacher
Jack Lund Schofield Middle School
Las Vegas, NV

Ann C. Raymond
Teacher
Oak Ave. Intermediate School
Temple City, CA

M.J. Richards
Middle School Math Teacher
Davis Middle School
Dublin, OH

Linda Lou Rohleder
Math Teacher, Grades 7 & 8
Jasper Middle School
Jasper, IN

Dana Schaefer
Pre-Algebra & Algebra I Teacher
Coachman Fundamental Middle School
Clearwater, FL

Donald W. Scheuer, Jr.
Coordinator of Mathematics
Abington School District
Abington, PA

Angela Hardee Slate
Teacher, 7th Grade Math/Algebra
Martin Middle School
Raleigh, NC

Mary Ferrington Soto
7th Grade Math
Calhoun Middle School-Ouachita Parish Schools
Calhoun, LA

Diane Stilwell
Mathematics Teacher/Technology Coordinator
South Middle School
Morgantown, WV

Pamela Ann Summers
K–12 Mathematics Coordinator
Lubbock ISD–Central Office
Lubbock, TX

Marnita L. Taylor
Mathematics Teacher/Department Chairperson
Tolleston Middle School
Gary, IN

Susan Troutman
Teacher
Dulles Middle School
Sugar Land, TX

Barbara C. VanDenBerg
Math Coordinator, K–8
Clifton Board of Education
Clifton, NJ

Mollie VanVeckhoven-Boeving
7th Grade Math and Algebra Teacher
White Hall Jr. High School
White Hall, AR

Mary A. Voss
7th Grade Math Teacher
Andersen Middle School
Omaha, NE

Christine Waddell
Teacher Specialist
Jordan School District
Sandy, UT

E. Jean Ware
Supervisor
Caddo Parish School Board
Shreveport, LA

Karen Y. Watts
9th Grade Math Teacher
Douglas High School
Douglas, AL

Lu Wiggs
Supervisor
I.S. 195
New York, NY

Teacher Advisory Board and Field Test Schools

Teacher Advisory Board

Glencoe/McGraw-Hill wishes to thank the following teachers for their feedback on *Mathematics: Applications and Concepts*. They were instrumental in providing valuable input toward the development of this program.

Katie Davidson
Legg Middle School
Coldwater, MI

Lynanne Gabriel
Bradley Middle School
Huntersville, NC

Kathleen M. Johnson
New Albany-Plain Local Middle School
New Albany, OH

Ronald C. Myer
Indian Springs Middle School
Columbia City, IN

Mike Perlin
John Jay Middle School
Cross River, NY

Reema Rahaman
Brentwood Middle School
Brentwood, MO

Diane T. Scheuber
Elizabeth Middle School
Elizabeth, CO

Deborah Sykora
Hubert H. Humphrey Middle School
Bolingbrook, IL

DeLynn Woodside
Roosevelt Middle School,
Oklahoma City Public Schools
Oklahoma City, OK

Field Test Schools

Glencoe/McGraw-Hill wishes to thank the following schools that field-tested pre-publication manuscript during the 2002–2003 school year. They were instrumental in providing feedback and verifying the effectiveness of this program.

Knox Community Middle School
Knox, IN

Roosevelt Middle School
Oklahoma City, OK

Brentwood Middle School
Brentwood, MO

Elizabeth Middle School
Elizabeth, CO

Legg Middle School
Coldwater, MI

Great Hollow Middle School
Nesconset, NY

Student Advisory Board

The Student Advisory Board gave the authors, editorial staff, and design team feedback on the lesson design, content, and covers of the Student Editions. We thank these students for their hard work and creative suggestions in making *Mathematics: Applications and Concepts* more student friendly.

Front Row: Joey Snyder, Tiffany Pickenpaugh, Craig Hammerstein

Back Row: Brittany Yokum, Alex Johnson, Cimeone Starling, Kristina Smith, Kate Holt, Ben Ball

UNIT 1

Decimals, Algebra, and Statistics 2

Chapter 1 Decimal Patterns and Algebra 4

INTERDISCIPLINARY PROJECT
• Math and Science **3**

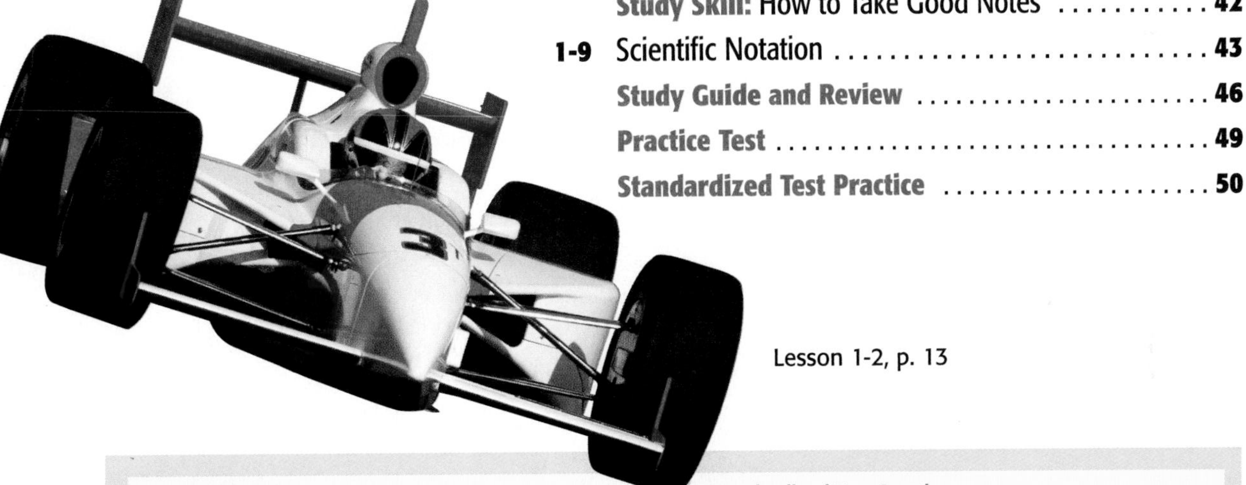

Lesson 1-2, p. 13

Chapter 2 Statistics: Analyzing Data 52

Lesson 2-2a, p. 59

UNIT 2

Integers and Algebra

Lesson 3-5, p. 129

UNIT 2

Chapter 4 Algebra: Linear Equations and Functions

Lesson 4-3, p. 163

UNIT 3

Lesson 5-4, p. 210

UNIT 3

Chapter 6 Applying Fractions

Lesson 6-2, p. 245

UNIT 4

Lesson 7-8, p. 323

UNIT 4

Chapter 8 Applying Percent

Lesson 8-1, p. 334

UNIT 4

Chapter 9 Probability

Lesson 9-4, p. 382

UNIT 5

Lesson 10-9, p. 458

UNIT 5

Chapter 11 Geometry: Measuring Two-Dimensional Figures 468

Lesson 11-5, p. 490

Chapter 12 Geometry: Measuring Three-Dimensional Figures 510

Lesson 12-2, p. 522

HOW TO...

Use Your Math Book

Why do I need my math book?

Have you ever been in class and not understood all of what was presented? Or, you understood everything in class, but at home, got stuck on how to solve a couple of problems? Maybe you just wondered when you were ever going to use this stuff?

These next few pages are designed to help you understand everything your math book can be used for . . . besides homework problems!

BEFORE YOU READ

Have a Goal

- What information are you trying to find?
- Why is this information important to you?
- How will you use the information?

Have a Plan

- Read *What You'll Learn* at the beginning of the lesson.
- Look over photos, tables, graphs, and opening activities.
- Locate boldfaced words and read their definitions.
- Find Key Concept and Concept Summary boxes for a preview of what's important.
- Skim the example problems.

Have an Opinion

- Is this information what you were looking for?
- Do you understand what you have read?
- How does this information fit with what you already know?

IN CLASS

During class is the opportunity to learn as much as possible about that day's lesson. Take advantage of this time! Ask questions about things that you don't understand, and take notes to help you remember important information.

To help keep your notes in order, try making a Foldables Study Organizer. It's as easy as 1-2-3! Here's a Foldable you can use to keep track of the rules for addition, subtraction, multiplication, and division.

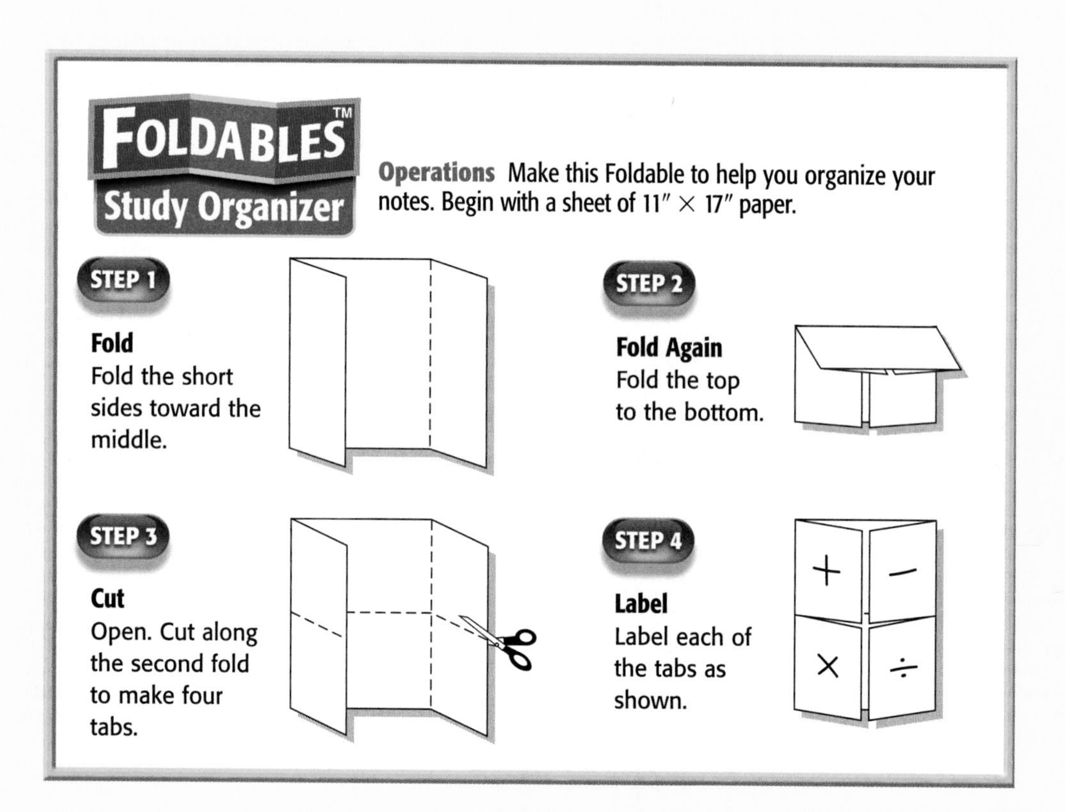

DOING YOUR HOMEWORK

Regardless of how well you paid attention in class, by the time you arrive at home, your notes may no longer make any sense and your homework seems impossible. It's during these times that your book can be most useful.

- Each lesson has example problems, solved step-by-step, so you can review the day's lesson material.
- A Web site has extra examples to coach you through solving those difficult problems.
- Each exercise set has Homework Help boxes that show you which examples may help with your homework problems.
- Answers to the odd-numbered problems are in the back of the book. Use them to see if you are solving the problems correctly. If you have difficulty on an even problem, do the odd problem next to it. That should give you a hint about how to proceed with the even problem.

BEFORE A TEST

Admit it! You think there is no way to study for a math test! However, there *are* ways to review before a test. Your book offers help with this also.

- Review all of the new vocabulary words and be sure you understand their definitions. These can be found on the first page of each lesson or highlighted in yellow in the text.
- Review the notes you've taken on your Foldable and write down any questions that you still need answered.
- Practice all of the concepts presented in the chapter by using the chapter Study Guide and Review. It has additional problems for you to try as well as more examples to help you understand. You can also take the Chapter Practice Test.
- Take the self-check quizzes from the Web site.

$3n - 8 = 7$

$y = x + 0.5$

LOOK FOR...

The Web site with self-check quizzes on these pages in Chapter 1: 9, 13, 16, 21, 27, 33, 36, 38, and 45.

The Study Guide and Review for Chapter 1 on page 46.

LET'S GET STARTED

To help you find the information you need quickly, use the Scavenger Hunt below to learn where things are located in each chapter.

1. What is the title of Chapter 1?
2. How can you tell what you'll learn in Lesson 1-1?
3. Sometimes you may ask, "When am I ever going to use this?" Name a situation that uses the concepts from Lesson 1-2.
4. What is the key concept presented in Lesson 1-3?
5. How many Examples are presented in Lesson 1-3?
6. What is the web address where you could find extra examples?
7. In Lesson 1-4, there is a paragraph that mentions it is common to use the first letter of a word as a variable. What is the main heading above that paragraph?
8. There is a Real-Life Career mentioned in Lesson 1-4. What is it?
9. List the new vocabulary words that are presented in Lesson 1-5.
10. List the math symbols that are presented in Lesson 1-5.
11. Suppose you're doing your homework on page 32 and you get stuck on Exercise 12. Where could you find help?
12. What is the web address that would allow you to take a self-check quiz to be sure you understand the lesson?
13. On what pages will you find the Study Guide and Review?
14. Suppose you can't figure out how to do Exercise 25 in the Study Guide on page 47. Where could you find help?
15. You complete the Practice Test on page 49 to study for your chapter test. Where could you find another test for more practice?

UNIT 1

Decimals, Algebra and Statistics

Chapter 1

Decimal Patterns and Algebra

Chapter 2

Statistics: Analyzing Data

Your study of math requires analysis of data, much of which will be in decimal form. In this unit, you will learn how to evaluate algebraic expressions involving decimals, as well as analyze and graph data.

WebQuest
INTERDISCIPLINARY PROJECT
MATH and SCIENCE
LIONS, TIGERS, AND BEARS
Are you ready to join a team of animal experts? As part of your application to be a Zoo's new coordinator, you must complete several challenging tasks. You'll make decisions about what animals to purchase for the zoo based on financial information provided to you. You'll gather specific data about the animals you choose, including their weight and expected lifespan. Finally, you'll present your findings to the hiring committee. So pack up your gear and don't forget your algebra tool kit. This adventure is going to be wild!
Log on to msmath2.net/webquest to begin your WebQuest.

Decimal Patterns and Algebra

CHAPTER 1

"What does exercising have to do with math?"

A 13-year-old has a minimum training heart rate of about 124 beats per minute. You can use the expression **0.6(220 − *a*)**, where *a* stands for a person's **age** to find a person's minimum training heart rate. Other real-life situations can also be modeled using expressions and variables.

You will solve problems about your health in Lesson 1-4.

GETTING STARTED

Take this quiz to see whether you are ready to begin Chapter 1. Refer to the lesson or page number in parentheses if you need more review.

Vocabulary Review

State whether each sentence is *true* or *false*. If *false*, replace the underlined word to make a true sentence.

1. A decimal point separates the ones place and the <u>tenths</u> place. (Page 555)
2. When <u>multiplying</u> or subtracting decimals, you must first line up the decimal points. (Page 555)

Prerequisite Skills

Add. (Page 559)

3. 89.3 + 16.5
4. 7.9 + 32.45
5. 54.25 + 6.39
6. 10.8 + 2.6

Subtract. (Page 559)

7. 24.6 − 13.3
8. 9.1 − 6.6
9. 30.55 − 2.86
10. 17.4 − 11.2

Multiply. (Page 560)

11. 4×7.7
12. 9.8×3
13. 2.7×6.3
14. 8.5×1.2

Divide. (Page 562)

15. $32.6 \div 4$
16. $10.6 \div 2$
17. $5.5 \div 5$
18. $17.84 \div 4$

Multiply or divide. (Page 562)

19. $2.45 \times 1{,}000$
20. $87.3 \div 100$
21. 0.61×100
22. $10 \div 1{,}000$

Decimal Patterns and Algebra Make this Foldable to help you organize your notes. Begin with ten sheets of notebook paper.

Staple
Staple the ten sheets together to form a booklet.

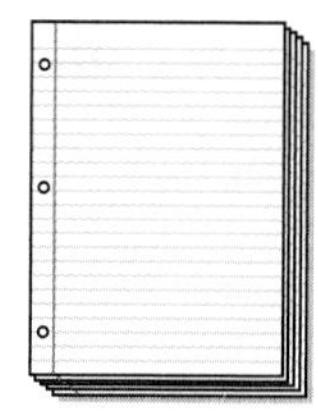

STEP 2 **Cut Tabs**
On the second page, make the top tab the width of the white space. On the third page, make the tab 2 lines longer, and so on.

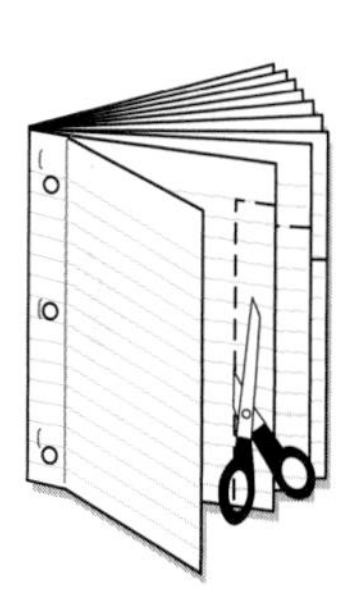

STEP 3 **Label**
Write the chapter title on the cover and label each tab with the lesson number.

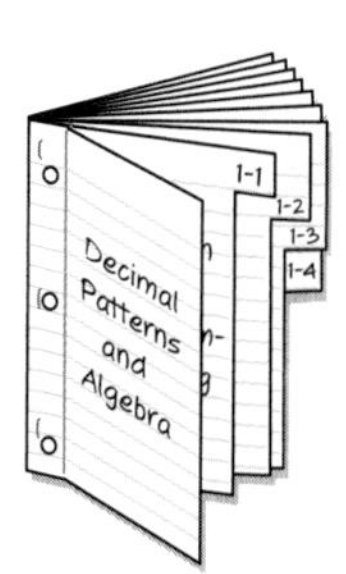

Reading and Writing As you read and study each lesson, write examples on the corresponding page.

1-1 A Plan for Problem Solving

WHEN am I ever going to use this?

What You'll LEARN

Solve problems using the four-step plan.

ANIMALS The graphic shows the top five turkey-producing states in a recent year. What is the total number of turkeys produced in the top five states?

1. Do you have all of the information necessary to solve this problem?
2. Explain how you would solve this problem. Then solve it.
3. Does your answer make sense? Explain.
4. What can you do if your first attempt at solving the problem does not work?

In mathematics, there is a *four-step plan* you can use to help you solve math problems.

1. Explore
- Determine what information is given and what you need to find.
- Do you have all the information to solve the problem?
- Is there any information given that is not needed?

2. Plan
- Select a strategy for solving the problem. There may be several that you can use.
- It is usually helpful to estimate the answer.

3. Solve
- Carry out your plan.
- If your plan doesn't work, keep trying until you find one that does work.
- Make sure your solution contains appropriate units or labels.

4. Examine
- Does your answer make sense given the facts in the problem?
- Is your answer reasonable compared to your estimate?
- If your answer is not correct, make a new plan and start again.

EXAMPLE Use the Four-Step Plan

1 **TELEVISION** **Color TV sets were first mass-produced in 1954. In that year, 5,000 color TVs were produced. How many years passed before at least 100 times as many were produced?**

Year	Color TVs Produced	Year	Color TVs Produced
1954	5,000	1960	120,000
1955	20,000	1961	140,000
1956	90,000	1962	450,000
1957	85,000	1963	750,000
1958	80,000	1964	1,500,000
1959	90,000		

Source: *Sams Photofact Book on Color TV*

Explore *What are you trying to find?*
The number of years since 1954 that 100 times as many TVs were produced.

What information do you need to solve the problem?
From the table, you know the number of TVs produced each year from 1954 to 1964.

Plan You can estimate the number of years that passed before at least 100 times 5,000 TVs, or 500,000 TVs were produced. This happened between 1962 and 1963. So, approximately 10 years had passed.

Solve $100 \times 5{,}000$ TVs $= 500{,}000$ TVs

In 1963, 750,000 TVs were produced, which is more than 500,000 TVs. Since $1963 - 1954 = 9$, nine years passed before production was at least 100 times greater than the number of sets produced in 1954.

Examine *Is your answer reasonable?*
Look at the data in the table and count the number of years from 1954 to 1963. Also, compare your answer to the estimate.

Your Turn **Solve using the four-step plan.**

a. How many times more TVs were produced in 1960 than in 1954?

b. Between which two years did TV production increase the most?

REAL-LIFE MATH

TELEVISION The average American watches more than 4 hours of TV per day.
Source: A.C. Nielsen Company

Problems can be solved using different operations or strategies.

Key Concept **Problem-Solving Strategies**

guess and check	use a graph
look for a pattern	work backward
make an organized list	eliminate possibilities
draw a diagram	estimate reasonable answers
act it out	use logical reasoning
solve a simpler problem	make a model

STUDY TIP

Four-Step Plan No matter what strategy you use, you can always use the four-step plan to solve a problem.

EXAMPLE Use a Strategy in the Four-Step Plan

2 **SPORTS In a basketball tournament, 128 teams play in the first round. Then there are 64 teams left to play in the second round, 32 teams left to play in the third round, and so on. How many rounds does it take to determine a winner?**

Explore You know the number of teams at the start of the tournament and how many teams play in each of the first three rounds.

Plan You can look for a pattern by organizing the information in a table. Then continue the pattern until there is one team left.

Solve

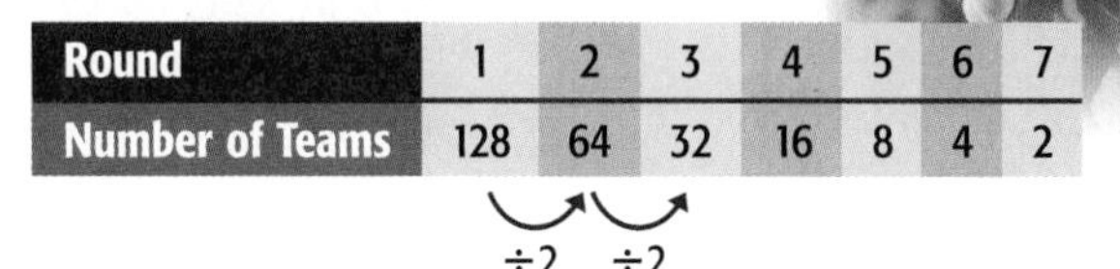

Round	1	2	3	4	5	6	7
Number of Teams	128	64	32	16	8	4	2

So, a winner is determined in 7 rounds.

Examine You could complete the diagram below to show that the answer is reasonable.

Round 1 → 128 teams: 64 teams win, 64 teams lose

Round 2 → 64 teams: 32 teams win, 32 teams lose

Round 3 → 32 teams: 16 teams win, 16 teams lose

Your Turn

c. How many rounds would it take to determine a winner if 512 teams participated in the tournament?

Skill and Concept Check

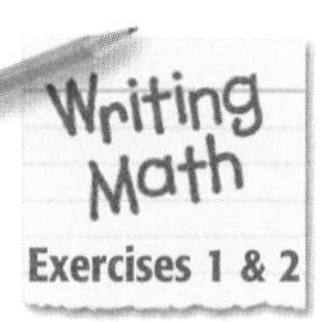

1. **Explain** why it is important to plan before solving a problem.
2. **Describe** what to do if your solution to a problem seems unrealistic.
3. **OPEN ENDED** Write a real-life problem that can be solved by adding 79 and 42.

GUIDED PRACTICE

Use the four-step plan to solve each problem.

4. **MONEY** To attend the class picnic, each student will have to pay $5.00 for transportation and $6.50 for food. If there are 300 students in the class, how much money will be collected for the picnic?
5. **PATTERNS** If Isaac receives an E-mail every 20 minutes during the workday, how many E-mails would he expect to receive between 8:00 A.M. and noon?

Practice and Applications

HOMEWORK HELP	
For Exercises	See Examples
6–9	1
10–11	2

Extra Practice
See pages 564, 596.

Use the four-step plan to solve each problem.

6. **VIDEO RENTALS** A video store took in $5,400 in video rentals during July. January sales are expected to be double that amount. If videos rent for $4, how many video rentals are expected in January?

7. **RACING** The Indianapolis Motor Speedway has 3,200,000 bricks that lie beneath the 2.5 miles of track. Find how many bricks lie under one mile of track at the race track.

8. **COOKING** Mr. Sanchez is serving Cajun fried turkey at 3:00 P.M. The 15-pound turkey has to remain in the fryer 3 minutes for every pound and must cool for at least 45 minutes before it is carved. What is the latest time he can start frying?

9. **GEOMETRY** Draw the next two figures in the pattern.

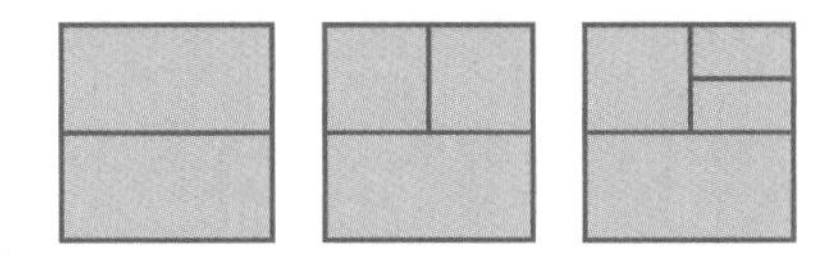

10. **MULTI STEP** Kishi wants to buy a DVD player that costs $250 with tax. So far, she has saved $145. If she saves $5 every week, in how many weeks will she be able to purchase the DVD player?

11. **COLLECTIONS** In 1999, Dustin started collecting 6 die-cast cars every year. In 2001, his brother Logan started collecting 9 cars per year. In what year did Dustin and Logan have the same number of cars?

12. **CRITICAL THINKING** Use the digits 1, 2, 3, 4, and 5 to form a two-digit and a three-digit number so that their product is the least product possible. Use each digit only once.

Standardized Test Practice and Mixed Review

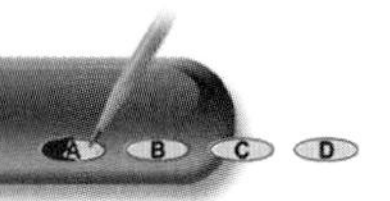

13. **SHORT RESPONSE** Jeannie has $86 to spend on birthday gifts for her mother. Which three items from the table could Jeannie purchase for her mother, not including tax?

Item	Cost
sweater	$29.50
gloves	$22.75
purse	$32.00
chocolates	$15.00
movie passes	$27.75

14. **MULTIPLE CHOICE** In a field near Duns, Scotland, a reproduction of the painting *Sunflowers* was created with 250,000 plants and flowers. The "painting" covered a 46,000-square foot area. What is a reasonable number of plants and flowers per square foot?

   Ⓐ 5   Ⓑ 50   Ⓒ 100   Ⓓ 500

GETTING READY FOR THE NEXT LESSON

BASIC SKILL Multiply.

15. 10×10 16. $3 \times 3 \times 3$ 17. $5 \times 5 \times 5 \times 5$ 18. $2 \times 2 \times 2 \times 2 \times 2$

1-2 Powers and Exponents

WHEN am I ever going to use this?

What You'll LEARN

Use powers and exponents.

NEW Vocabulary

factors
exponent
base
powers
squared
cubed
evaluate
standard form
exponential form

TECHNOLOGY Computer engineer Gordon E. Moore made this observation in 1964. Moore's Law says the amount of available storage space on a computer chip doubles every year.

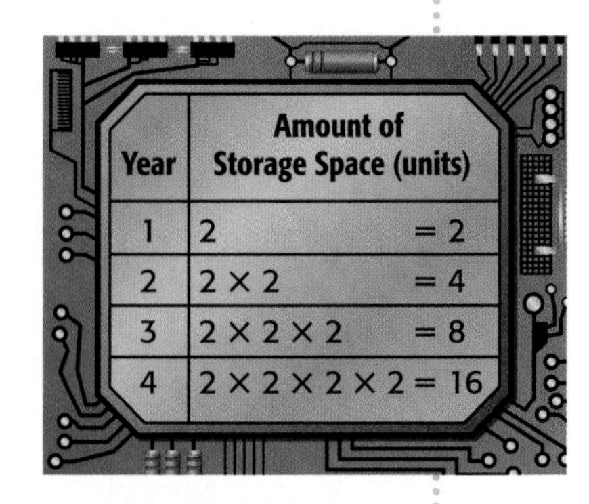

Year	Amount of Storage Space (units)
1	2 = 2
2	2 × 2 = 4
3	2 × 2 × 2 = 8
4	2 × 2 × 2 × 2 = 16

1. How is doubling shown in the table?
2. If the pattern continued, how much storage space would be available by year 6?
3. What is the relationship between the number of 2s and the year?

Two or more numbers that are multiplied together to form a product are called **factors**. The amount of storage space on a computer chip in year 4 can be written using only factors of 2. When the same factor is used, you may use an exponent to simplify the notation.

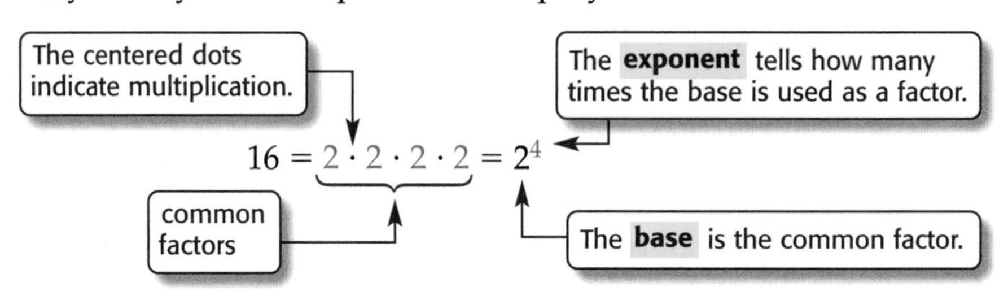

Numbers expressed using exponents are called **powers**. You can say that 2^4 is a power of 2. This and the powers 5^2 and 4^3 are read as follows.

Symbols	Words
5^2	five to the second power or five **squared**
4^3	four to the third power or four **cubed**
2^4	two to the fourth power

READING Math

First Power When a number does not have an exponent, it is understood to be 1. For example, $8 = 8^1$.

EXAMPLES Write Powers as Products

Write each power as a product of the same factor.

1 7^5

The base is 7. The exponent 5 means that 7 is used as a factor five times.

$7^5 = 7 \cdot 7 \cdot 7 \cdot 7 \cdot 7$

2 3^2

The base is 3. The exponent 2 means the 3 is used as a factor twice.

$3^2 = 3 \cdot 3$

You can **evaluate**, or find the value of, powers by multiplying the factors. Numbers written without exponents are in **standard form**.

EXAMPLES Write Powers in Standard Form

Evaluate each expression.

3 2^5

$2^5 = 2 \cdot 2 \cdot 2 \cdot 2 \cdot 2$

$= 32$

4 4^3

$4^3 = 4 \cdot 4 \cdot 4$

$= 64$

Your Turn **Evaluate each expression.**

a. 10^2 b. 7^3 c. 5^4

STUDY TIP

Technology You can use a calculator to compute large exponents. To find 4^5, press 4 [^] 5 [ENTER]. The display shows **1024**.

Numbers written with exponents are in **exponential form**.

EXAMPLE Write Numbers in Exponential Form

5 **Write $3 \cdot 3 \cdot 3 \cdot 3$ in exponential form.**

3 is the base. It is used as a factor 4 times. So, the exponent is 4.

$3 \cdot 3 \cdot 3 \cdot 3 = 3^4$

Your Turn **Write each product in exponential form.**

d. $5 \cdot 5 \cdot 5$ e. $12 \cdot 12 \cdot 12 \cdot 12 \cdot 12 \cdot 12$

Skill and Concept Check

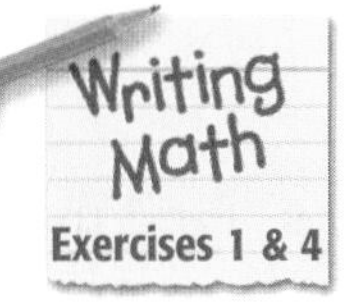

1. **Explain** what *five to the fourth power* means.
2. **Write** 7^5 in words.
3. **OPEN ENDED** Write one number in exponential form and another number in standard form.
4. **Which One Doesn't Belong?** Identify the number that cannot be written as a power with an exponent greater than 1. Explain your reasoning.

4	9	16	50

GUIDED PRACTICE

Write each power as a product of the same factor.

5. 6^2 6. 4^4 7. 8^5

Evaluate each expression.

8. 3^4 9. 5^5 10. 10^3

Write each product in exponential form.

11. $5 \cdot 5 \cdot 5 \cdot 5 \cdot 5 \cdot 5$ 12. $1 \cdot 1 \cdot 1 \cdot 1$

13. Evaluate *eleven to the third power.*

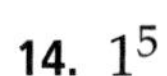

Practice and Applications

For Exercises	See Examples
14–19	1, 2
20–31, 38, 42–47, 54	3, 4
32–37, 39, 57, 59, 61	5

HOMEWORK HELP

Extra Practice
See pages 564, 596.

Write each power as a product of the same factor.

14. 1^5 **15.** 4^2 **16.** 3^8

17. 9^3 **18.** 10^4 **19.** 11^3

Evaluate each expression.

20. 2^3 **21.** 2^6 **22.** 4^3 **23.** 5^4

24. 1^{10} **25.** 10^1 **26.** 3^8 **27.** 4^6

28. 10^5 **29.** 12^4 **30.** 20^4 **31.** 50^3

Write each product in exponential form.

32. $3 \cdot 3$ **33.** $7 \cdot 7 \cdot 7 \cdot 7$ **34.** $1 \cdot 1 \cdot 1$

35. $6 \cdot 6 \cdot 6 \cdot 6 \cdot 6$ **36.** $4 \cdot 4 \cdot 4 \cdot 4 \cdot 4 \cdot 4 \cdot 4$ **37.** $13 \cdot 13 \cdot 13 \cdot 13 \cdot 13$

38. Find the value of *6 cubed.*

39. Express 256 in exponential form as a power of 2.

GEOMETRY For Exercises 40 and 41, use the figures below.

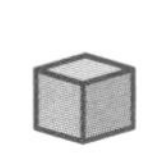
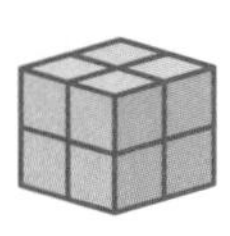

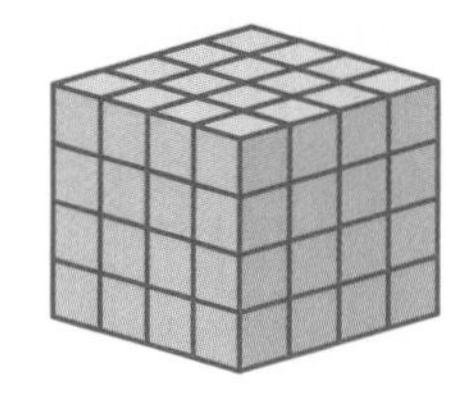
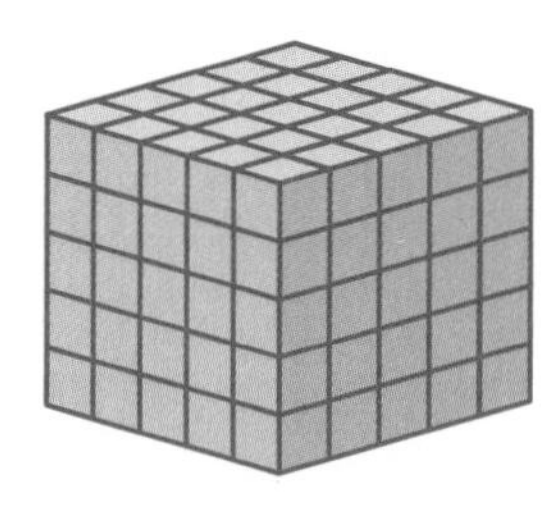

40. Find the number of unit cubes that make up each large cube. Write your answers using exponents.

41. Why do you think the expression 2^3 is sometimes read as *2 cubed*?

Use a calculator to determine whether each sentence is *true* or *false.*

42. $3^7 > 7^3$ **43.** $14^4 = 19^2$ **44.** $927 > 9^3$

45. $3^2 = 2^3$ **46.** $6^3 < 4^4$ **47.** $3^4 < 9^2$

Order the following numbers from least to greatest.

48. $6^5, 1^{14}, 4^{10}, 17^3$ **49.** $2^8, 15^2, 6^3, 3^5$ **50.** $5^3, 4^6, 2^{11}, 7^2$

51. $6^4, 3^6, 5^5, 8^3$ **52.** $3^7, 10^3, 8^2, 2^{10}$ **53.** $6^5, 9^4, 11^3, 5^5$

54. POPULATION There are approximately 7^{10} people living in the United States. About how many people is this?

55. Explain why a *quadrillion,* 10^{15}, is usually written in exponential form instead of standard form.

56. Tell whether the following statement is *true* or *false.* Explain your reasoning. *The number 64 can be written in exponential form in only one way.*

57. Use exponents to write $5 \cdot 5 \cdot 5 \cdot 5 \cdot 4 \cdot 4 \cdot 4$ in the shortest form possible.

58. **MULTI STEP** Use a calculator to express 1,679,616 in exponential form as a power of 6.

59. **POPULATION** The population of Fort Worth, Texas, is approximately 534,000. Find a number less than 534,000 and a number greater than 534,000 that can be expressed in exponential form.

60. **RESEARCH** Our numbering system is based on the powers of 10. The ancient Mayan system is based on another power. Use the Internet or another source to write a paragraph about the Mayan numbering system.

61. **PHYSICAL SCIENCE** The speed of sound varies as it travels through different substances. At ordinary temperatures, sound travels through water at a rate of 5,000 feet per second. Write 5,000 using exponents and the factors 5 and 10.

62. **CRITICAL THINKING** Based on the number pattern shown at the right, write a convincing argument that any number, except 0, raised to the 0 power equals 1.

$2^4 = 16$
$2^3 = 8$
$2^2 = 4$
$2^1 = 2$
$2^0 = ?$

Standardized Test Practice and Mixed Review

63. **MULTIPLE CHOICE** Which is *two to the sixth power* written in standard form?

Ⓐ 26　　Ⓑ $6 \cdot 6$

Ⓒ 64　　Ⓓ $2 \cdot 2 \cdot 2 \cdot 2 \cdot 2 \cdot 2$

64. **GRID IN** How many zeros does the value of 10^8 have?

65. **RACING** The graph shows which numbered-cars have the most wins at the Indianapolis 500. How many more wins did the number 3 car have than the number 14 car? (Lesson 1-1)

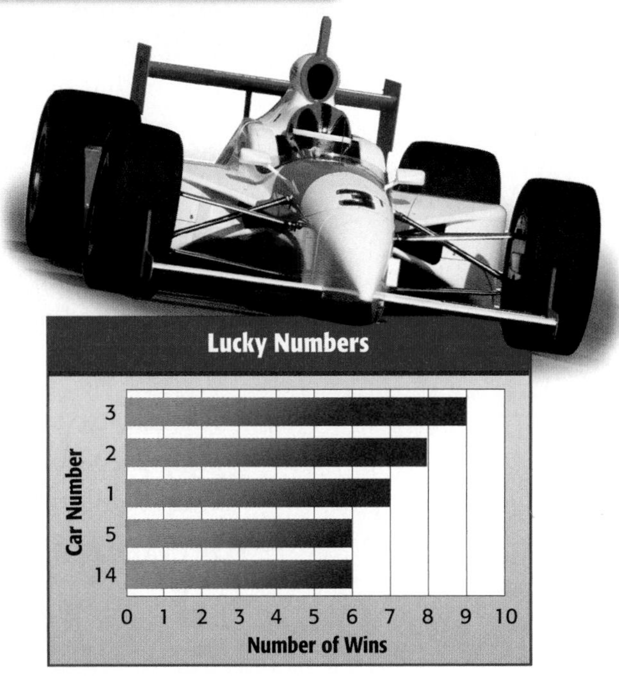

Source: *Indianapolis 500 Record Book*

66. **PRODUCTION** A machine on a production line fills 8 soft drink cans per minute. How many cans does it fill in 8 hours? (Lesson 1-1)

67. *True* or *False*? In the four-step plan, the *Solve* step comes last. (Lesson 1-1)

GETTING READY FOR THE NEXT LESSON

PREREQUISITE SKILL Add, subtract, multiply, or divide. (Pages 559, 560, 562)

68. $13.4 + 8.7$　69. $13.6 - 10.2$　70. 1.5×6.8　71. $8.6 - 5.53$

72. $36 \div 3$　73. $1.7 + 32.28$　74. 3.2×6　75. $24.5 \div 7$

1-3 Order of Operations

WHEN am I ever going to use this?

What You'll LEARN

Evaluate expressions using the order of operations.

NEW Vocabulary

numerical expression
order of operations

GAMES Kaitlyn and Percy are each writing as many different expressions as possible using the numbers rolled on a number cube. In the first round, 3, 5, 4, and 6 were rolled in order.

Kaitlyn	Percy
$3 \cdot 5 - (4 + 6) = 3 \cdot 5 - 10$ $= 15 - 10$ or 5	$3 \cdot (5 - 4) + 6 = 3 \cdot 1 + 6$ $= 3 + 6$ or 9

1. List the similarities and differences between the two expressions.
2. What was Kaitlyn's first step in simplifying her expression? Percy's first step?
3. **Make a conjecture** about what should be the first step in simplifying $(3 \cdot 5) - 4 + 6$.

The expressions $3 \cdot 5 - (4 + 6)$ and $3 \cdot (5 - 4) + 6$ are **numerical expressions**. Mathematicians have agreed upon steps called the **order of operations** to find the values of such expressions. These rules ensure that numerical expressions have only one value.

Key Concept — **Order of Operations**

1. Do all operations within grouping symbols first.
2. Evaluate all powers before other operations.
3. Multiply and divide in order from left to right.
4. Add and subtract in order from left to right.

You can use the order of operations to evaluate numerical expressions.

EXAMPLES Evaluate Expressions

1 Evaluate $5 + (12 - 3)$.

$5 + (12 - 3) = 5 + 9$ Subtract first since $12 - 3$ is in parentheses.
$= 14$ Add 5 and 9.

2 Evaluate $8 - 3 \cdot 2 + 7$.

$8 - 3 \cdot 2 + 7 = 8 - 6 + 7$ Multiply 3 and 2.
$= 2 + 7$ Subtract 6 from 8.
$= 9$ Add 2 and 7.

Your Turn **Evaluate each expression.**

a. $39 \div (9 + 4)$

b. $10 + 8 \div 2 - 6$

STUDY TIP

Powers If an exponent lies outside a grouping symbol, complete the operations within the grouping symbol before applying the power.

EXAMPLES Evaluate Expressions with Powers

3 Evaluate 15×10^3.

$15 \times 10^3 = 15 \times 1{,}000$ Find the value of 10^3.

$= 15{,}000$ Multiply.

4 Evaluate $36 \div (1 + 2)^2$.

$36 \div (1 + 2)^2 = 36 \div 3^2$ Add 1 and 2 inside the parentheses.

$= 36 \div 9$ Find the value of 3^2.

$= 4$ Divide.

Your Turn **Evaluate each expression.**

c. 3×10^4

d. $(5 - 1)^3 \div 4$

In addition to using the symbols × and ·, multiplication can be indicated by using parentheses.

$2(3 + 5)$ means $2 \times (3 + 5)$ or $2 \cdot (3 + 5)$.

$(4 - 2)3$ means $(4 - 2) \times 3$ or $(4 - 2) \cdot 3$.

EXAMPLE Evaluate an Expression

5 Evaluate $14 + 3(7 - 2) - 2 \cdot 5$.

$14 + 3(7 - 2) - 2 \cdot 5$

$= 14 + 3(5) - 2 \cdot 5$ Subtract 2 from 7.

$= 14 + 15 - 2 \cdot 5$ Multiply from left to right, $3 \cdot 5 = 15$.

$= 14 + 15 - 10$ Multiply from left to right, $2 \cdot 5 = 10$.

$= 29 - 10$ Add from left to right, $14 + 15 = 29$.

$= 19$ Subtract 10 from 29.

REAL-LIFE MATH

VIDEO GAMES A team of 500 engineers and technicians built the first video-game console in 1966.

Source: www.geekcomix.com

EXAMPLE Use an Expression to Solve a Problem

6 VIDEO GAMES Evita is buying a video game station, three extra controllers, and four new video games. What is the total cost?

Item	Quantity	Unit Cost
game station	1	$180.00
controller	3	$24.95
game	4	$35.99

cost of game station + number of controllers × cost of controller + number of games × cost of game

$180 + 3 × $24.95 + 4 × $35.99

$= 180 + 74.85 + 143.96$ Multiply from left to right.

$= 398.81$ Add.

So, the total cost is $398.81.

Check Check the reasonableness of the answer by estimating. The cost is about $180 + (25 \times 3) + (40 \times 4) = 180 + 75 + 160$, or $415. So, the solution of $398.81 is reasonable.

Skill and Concept Check

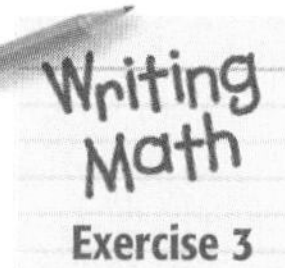

1. **Identify** the operation that should be done first in each expression.
   a. $9 \div 3 + (14 - 7)$
   b. $3 + 24 \div 3 \cdot 4$
2. **OPEN ENDED** Write an expression containing five numbers that is evaluated by first multiplying.
3. **FIND THE ERROR** Yutaka and Cynthia are evaluating $16 - 24 \div 6 \cdot 2$. Who is correct? Explain your reasoning.

Yutaka
$16 - 24 \div 6 \cdot 2$
$= 16 - 24 \div 12$
$= 16 - 2$ or 14

Cynthia
$16 - 24 \div 6 \cdot 2$
$= 16 - 4 \cdot 2$
$= 16 - 8$ or 8

GUIDED PRACTICE

Evaluate each expression.

4. $11 - (3 \cdot 2)$
5. $25 \div (9 - 4)$
6. $8 - 4 + 3.7$
7. $2 + 5 \cdot 5$
8. $14 \div 2 \cdot 6$
9. $8 \cdot 5 - 4 \cdot 3$
10. 4×10^2
11. $3.5 \times 5 + 6^2$
12. $7 + 4(5.6 - 2) - 9$
13. Evaluate $(16 \div 4)^3 - 6$.
14. Find the value of $(6 + 8) \div (10 - 8)$.

Practice and Applications

HOMEWORK HELP	
For Exercises	See Examples
15–28	1, 2
29–36, 43–45	3, 4
37–40, 46–48	5
41–42, 50	6

Extra Practice
See pages 564, 596.

Evaluate each expression.

15. $(1 + 8) \times 3$
16. $10 - (3 + 4)$
17. $(25 \div 5) + 8$
18. $(11 - 2) \div 9$
19. $3 \cdot 2 + 7$
20. $15 \div 3 + 4$
21. $12 + 6.6 \div 3$
22. $18 - 3 \cdot 6$
23. $8 - 7.2 + 5$
24. $28 \div 7(5)$
25. $(17 + 3) \div (4 + 1)$
26. $(6 + 5) \cdot (8 - 6)$
27. $21 \div 3 \times 2 - 4$
28. $35 \div 5 + 56 \div 7$
29. $2 \times 9 - 4^2$
30. $24 \div 3 + 5^3$
31. $7 + (8 - 7 + 2)^4$
32. $(2 + 10)^2 \div 4$
33. 6×10^2
34. 18×10^3
35. 1.95×10^2
36. 3.7×10^4
37. $6 + 2(9.4 - 1)$
38. $3(4.5 + 7.2) - 5 \cdot 4$
39. $72 \div 3 - 5(8.8 - 6) + 9$
40. $9 \div 3 \cdot 14(10 - 8) - 60$

41. **GEOMETRY** The distance around a geometric figure is called its *perimeter.* Write a numerical expression to find the perimeter of the figure at the right. Then evaluate the expression.

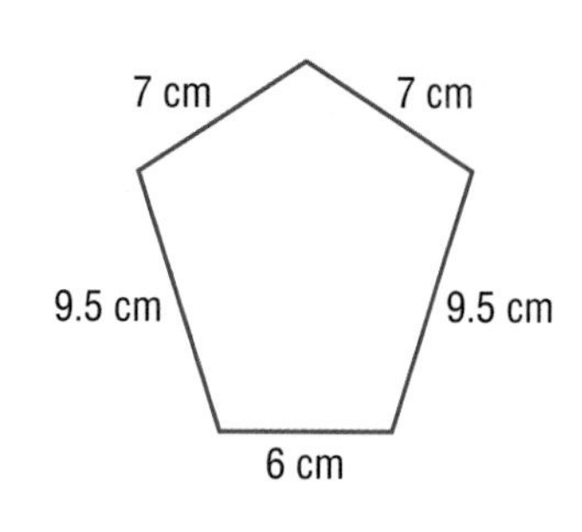

42. **MARATHONS** On Mondays, Wednesdays, and Thursdays, Jacob trains for a marathon for 3.5 hours. On Tuesdays and Fridays, he trains for 2 hours, and on Saturdays, he trains for 4.5 hours. How many hours does Jacob train per week?

Find the value of each expression.

43. 4.5×10^2
44. 3.08×10^4
45. 2.965×10^5
46. $3 \cdot 4(5 - 3.8) + 2.7$
47. $7.1 \times 9 - (4 + 3) + 1$
48. $(6 + 1)^2 - 2.5(3)$

49. Insert parentheses to make $72 \div 9 + 27 - 2 = 0$ true.

50. **MONEY MATTERS** Luke is ordering nine reams of paper, four boxes of pens, and two rolls of tape. Use the table to find the cost of his order.

Item	Cost
ream of paper	$3.95
box of pens	$7.49
roll of tape	$1.29

CRITICAL THINKING Some calculators are programmed to follow the order of operations.

51. Ruby evaluates $12 \div 2 + 2 \cdot 5$ and gets 15. When she checks using her calculator, she gets 16. Which is correct? Explain.

52. Explain how you could tell whether your calculator follows the order of operations.

Standardized Test Practice and Mixed Review

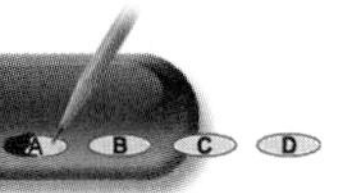

53. **MULTIPLE CHOICE** Which is the first step in evaluating $10 - 3 \cdot 2 + 4(15 - 5)$?

(A) $10 - 3$ (B) $3 \cdot 2$ (C) $2 + 4$ (D) $15 - 5$

54. **MULTIPLE CHOICE** Anthony has 2 boxes of cookies containing 24 cookies each and 3 packages of brownies containing 15 brownies each. Which expression *cannot* be used to find the total number of items he has?

(F) $3 \times 15 + 2 \times 24$ (G) $5 \times (24 + 15)$

(H) $2(24) + 3(15)$ (I) $15 + 15 + 15 + 24 + 24$

55. **TESTING** Use the graphic to estimate how much longer educators spend preparing students in grades 6–8 for tests than students in grades 9–12. (Lesson 1-1)

USA TODAY

USA TODAY Snapshots®

Time spent preparing for testing

Average class time educators say they spend annually preparing their students for state and/or national assessment tests:

Grades K-2: 30.2 days
Grades 3-5: 39.4 days
Grades 6-8: 44.1 days
Grades 9-12: 19.3 days

Source: Market Data Retrieval and Education Market Research

By Cindy Hall and Bob Laird, USA TODAY

Evaluate each expression. (Lesson 1-2)

56. 2^4
57. 11^2
58. 3^6
59. 4^5

Write each power as a product of the same factor. (Lesson 1-2)

60. 3^4
61. 8^5
62. 6^2
63. 4^6
64. 1^4
65. 9^1

GETTING READY FOR THE NEXT LESSON

BASIC SKILL **Divide.**

66. $45 \div 15$
67. $90 \div 5$
68. $110 \div 10$
69. $52 \div 4$

1-4 Algebra: Variables and Expressions

What You'll LEARN

Evaluate simple algebraic expressions.

NEW Vocabulary

variable
algebraic expression
algebra
term
coefficient
constant

REVIEW Vocabulary

evaluate: find the value **(Lesson 1-2)**

HANDS-ON Mini Lab

Materials

- isometric dot paper

Work with a partner.

The pattern below is made up of triangles, each with side lengths of 1.

1. Draw the next three figures in the pattern.
2. Find the perimeter of each figure and record your data in a table like the one shown below. The first three are completed for you.

Number of Triangles	1	2	3	4	5	6
Perimeter	3	4	5	?	?	?

3. Without drawing the figure, determine the perimeter of a figure made up of 10 triangles. Check by making a drawing.
4. Find a relationship between the number of triangles and the perimeter at each stage of the pattern.

In the Mini Lab, you found that the perimeter of the figure is two more than the number of triangles. You can use a placeholder, or **variable**, to represent the number of triangles.

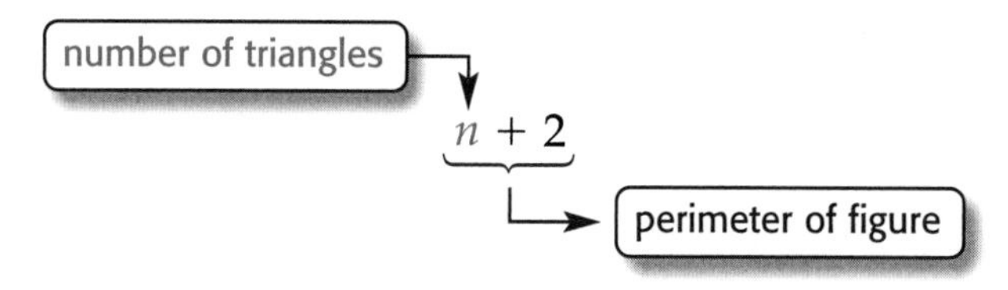

The expression $n + 2$ is called an **algebraic expression** because it contains variables, numbers, and at least one operation. The branch of mathematics that involves expressions with variables is called **algebra**.

READING Math

Variables The letter x is often used as a variable. It is also common to use the first letter of a word as the variable. For example, use t for *time* or n for *number*.

EXAMPLE Evaluate an Expression

1 Evaluate $n + 3$ if $n = 4$.

$n + 3 = 4 + 3$ Replace n with 4.

$= 7$

Your Turn **Evaluate each expression if $c = 8$.**

a. $c + 2$ b. $c - 3$ c. $15 - c$

The following symbols are used for multiplication and division with variables. Note that the multiplication sign is usually omitted.

$6d$	means	$6 \times d$ or $6 \cdot d$	$9st$	means	$9 \times s \times t$
ef	means	$e \times f$	$\frac{c}{d}$	means	$c \div d$

When plus or minus signs separate an algebraic expression into parts, each part is a **term**. The numerical factor of a term that contains a variable is called a **coefficient**. A term that does not contain a variable is called a **constant**.

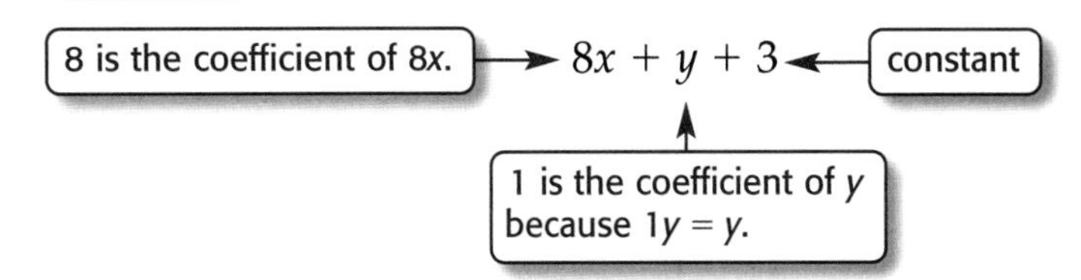

READING Math

Multiplication In Example 2, the $8w$ means 8 times w, but 85 does not mean 8 times 5. So, there are parentheses around the 5 in 8(5).

EXAMPLES Evaluate Expressions

2 Evaluate $8w - 2v$ if $w = 5$ and $v = 3$.

$8w - 2v = 8(5) - 2(3)$ Replace w with 5 and v with 3.

$= 40 - 6$ Use the order of operations.

$= 34$ Subtract 6 from 40.

3 Evaluate $\frac{mn}{6}$ if $m = 8$ and $n = 12$.

$\frac{mn}{6} = \frac{(8)(12)}{6}$ Replace m with 8 and n with 12.

$= \frac{96}{6}$ The fraction bar is like a grouping symbol.

$= 16$ Divide.

4 Evaluate $y^2 + 2$ if $y = 3$.

$y^2 + 2 = 3^2 + 2$ Replace y with 3.

$= 9 + 2$ or 11 Use the order of operations.

Your Turn **Evaluate each expression if $a = 4$ and $b = 3$.**

d. $9a - 6b$ **e.** $\frac{ab}{2}$ **f.** $a^2 + 5$

REAL-LIFE CAREERS

How Does an Athletic Trainer Use Math?

Athletic trainers use the formula $(220 - a) \times 0.6$, where a is a person's age, to find their minimum training heart rate.

Online Research
For information about a career as an athletic trainer, visit: msmath2.net/careers

EXAMPLE Use an Expression to Solve a Problem

5 HEALTH Use the formula at the left to find Kaylee's minimum training heart rate if she is 17 years old.

$(220 - a) \times 0.6 = (220 - 17) \times 0.6$ Replace a with 17.

$= 203 \times 0.6$ or 121.8 Use the order of operations.

Kaylee's minimum training heart rate of 121.8 beats per minute is rounded up to 122 because her heart cannot beat 0.8 times.

Skill and Concept Check

Writing Math
Exercise 3

1. **Identify** the coefficient and constant in the expression $2x - 6$.
2. **OPEN ENDED** Write an algebraic expression that involves multiplication and addition.
3. **Which One Doesn't Belong?** Identify the expression that does not have the same characteristic as the other three. Explain your reasoning.

$4x$	ab	$r \cdot 8$	$\frac{x}{y}$

GUIDED PRACTICE

Evaluate each expression if $a = 3$, $b = 5$, and $c = 4$.

4. $a + 7$
5. $4b$
6. $b + c$
7. $6c - a$
8. $\frac{c}{4}$
9. $\frac{5c + b}{5}$
10. $4b^2$
11. $\frac{3c^2}{a}$
12. What operations are used in the algebraic expression $\frac{xyz}{2}$?
13. Evaluate $n^2 + 1$ if $n = 2$.

Practice and Applications

Evaluate each expression if $d = 8$, $e = 3$, $f = 4$, and $g = 1$.

14. $d + 9$
15. $10 - e$
16. $4f$
17. $8g$
18. $f - e$
19. $d + f$
20. $10g - 6$
21. $8 + 5d$
22. $7e + 2d$
23. $\frac{d}{5}$
24. $\frac{16}{f}$
25. $\frac{5d - 25}{5}$
26. e^2
27. $\frac{d^2}{2}$
28. $6f^2$
29. $11.9d - 10e$
30. $\frac{2f}{g^2}$
31. $\frac{2.5f + 5}{3}$
32. $7e^2 + 5$
33. $\frac{(5 + g)^2}{2}$
34. Evaluate $9h + h^2$ if $h = 3.5$.
35. For what value of x is the expression x^2 equal to 9?

HOMEWORK HELP

For Exercises	See Examples
14–19	1
20–22	2
23–25	3
26–35	4
38–40, 43–45	5

Extra Practice
See pages 565, 596.

Determine whether each statement is *sometimes, always,* or *never* true. Explain.

36. The expressions $x - 3$ and $y - 3$ represent the same value.
37. The expression ab means the same as the expression $a \times b$.

BASEBALL For Exercises 38–40, use the following information.
Charlie earns $16 per game for umpiring Little League baseball games.

38. Make a table that shows how much money Charlie will earn for umpiring 1, 2, 3, 4, and 5 baseball games.
39. Write an expression to find how much money Charlie will earn for any number of games. Let n be the number of games he umpires.
40. Charlie umpires 18 games during the summer. How much will he earn?

Complete each table.

41.

Quarts (q)	Gallons $\left(\frac{q}{4}\right)$
4	1
6	1.5
8	?
10	?
12	?
14	?

42.

Dollars (d)	Quarters ($4d$)
1	4
2	8
3	?
4	?
5	?
6	?

43. **HEALTH** The expression $110 + \frac{A}{2}$, where A stands for a person's age, is used to estimate a person's normal systolic blood pressure. Estimate the normal systolic blood pressure for a 16-year old.

BOATING For Exercises 44 and 45, use the information below and at the right.
The Lerman family is vacationing at the lake and wants to rent a boat.

44. Write an expression showing the cost of renting a boat for h hours.

45. How much would it cost if they rented the boat for 2, 3, or 4 hours?

46. **WRITE A PROBLEM** Write an algebraic expression with the variable x that has a value of 3 when evaluated.

47. **CRITICAL THINKING** Find values of x and y so that the value of $7x + 2$ is greater than the value of $3y + 23$.

Standardized Test Practice and Mixed Review

48. **MULTIPLE CHOICE** Which expression could be used to find the cost of buying b baseball bats at \$75 each and g baseball gloves at \$98 each?

Ⓐ $75b + 98g$ Ⓑ $b + g$ Ⓒ $75 + 98$ Ⓓ $75b \div 98g$

49. **SHORT RESPONSE** Juan opened a bank account one month with \$40 and deposited \$15 each month after that. He was able to save \$100 by December. In which month did he open the bank account?

Evaluate each expression. (Lesson 1-3)

50. $6(5) - 2$ 51. $9 + 9 \div 3$ 52. $4 \cdot 2(8 - 1)$ 53. $(17 + 3) \div 5$

54. Write 2^5 as a product of the same factor. (Lesson 1-2)

GETTING READY FOR THE NEXT LESSON

PREREQUISITE SKILL Determine whether each sentence is *true* or *false*. (Lesson 1-3)

55. $15 - 2(3) = 9$ 56. $20 \div 5 \times 4 = 1$ 57. $4^2 + 6 \cdot 7 = 154$ 58. $24 \div (7 - 3) = 6$

1-5a Problem-Solving Strategy

A Preview of Lesson 1-5

What You'll Learn
Solve problems using the guess and check strategy.

Guess and Check

The total ticket sales for the school play were $263.50. Amber, how many adult and student tickets were sold?

Adult tickets are $7.50 and student tickets are $4.00. Also, twice as many students bought tickets as adults. Let's **guess and check** to find out.

Explore	We know adult tickets are $7.50, student tickets are $4.00, and twice as many students bought tickets than did adults.		
Plan	Let's make a guess and check if it is correct. Then adjust the guess until we get the correct answer.		
Solve	Make a guess.		
	10 adults, 20 students	$7.5(10) + 4(20) = \$155$	*too low*
	Adjust the guess upward.		
	20 adults, 40 students	$7.5(20) + 4(40) = \$310$	*too high*
	Adjust the guess downward slightly.		
	18 adults, 36 students	$7.5(18) + 4(36) = \$279$	*still too high*
	Adjust the guess downward again.		
	17 adults, 34 students	$7.5(17) + 4(34) = \$263.50$	*correct*
	There were 17 adult and 34 student tickets sold.		
Examine	Seventeen adult tickets cost $127.50, and 34 student tickets cost $136. Since $\$127.50 + \$136 = \$263.50$ and 34 tickets are twice as many as 17 tickets, the guess is correct.		

Analyze the Strategy

1. **Explain** why you must keep a careful record of each of your guesses and their results in the *solve* step of the problem-solving plan.
2. **Write** a problem that could be solved by guess and check. Then write the steps you would take to find the solution to your problem.

Apply the Strategy

Solve. Use the guess and check strategy.

3. **CAR WASH** The band sponsored a car wash to help pay for uniforms. They charged $4 for a car and $6 for an SUV. During the first hour, they washed 16 vehicles and earned $78. How many of each type of vehicle did they wash?

4. **SOUVENIRS** The Pike's Peak souvenir shop sells standard-sized postcards in packages of 5 and large-sized postcards in packages of 3. If Juan bought 16 postcards, how many packages of each did he buy?

Mixed Problem Solving

Solve. Use any strategy.

5. **MUSIC** Flor is burning a CD for her friend Derek. The CD will hold 30 minutes of music. Which songs should she select from the list at the right to record the maximum time on the CD without going over?

SONG	TIME
A	3 min 30 s
B	6 min 0 s
C	4 min 45 s
D	5 min 30 s
E	5 min 6 s
F	4 min 15 s
G	4 min 30 s
H	3 min 48 s

6. **NUMBER THEORY** A number is squared, and the result is 256. Find the number.

7. **PATTERNS** Copy the table below. Extend the pattern to complete the table.

Input	1	2	3	4	5	6	7
Output	1	8	27	?	?	?	?

8. **MULTI STEP** The Wicked Twister roller coaster can accommodate about 1,000 passengers per hour. The coaster has 8 vehicles that each carry 4 passengers. Approximately how many runs are made in 1 hour?

9. **PHYSICAL SCIENCE** Telephone calls travel through optical fibers at the speed of light, which is 186,000 miles per second. A millisecond is 0.001 of a second. How far can your voice travel over an optical line in 1 millisecond?

10. **MONEY MATTERS** Syreeta needs to buy four markers to make posters for a science project. She has $4. Does she have enough money if the cost of each marker, including tax, is 89¢? Explain.

SNOWFALL For Exercises 11 and 12, use the graph below.

Source: www.usatoday.com

11. What is the total average snowfall for October through April?

12. Between which two consecutive months is the difference in snowfall the least?

13. **STANDARDIZED TEST PRACTICE**
Kelly has $2.80 in change in her purse. If the change is made up of an equal number of nickels, dimes, and quarters, how many of each does she have?

(A) 4 (B) 5 (C) 6 (D) 7

You will use the guess and check strategy in the next lesson.

1-5 Algebra: Equations

What You'll LEARN

Solve equations using mental math.

NEW Vocabulary

equation
solution
solving an equation
defining the variable

MATH Symbols

= is equal to
≠ is not equal to

WHEN am I ever going to use this?

BASKETBALL The table shows the number of wins for six WNBA Eastern Division teams after playing 32 games each.

1. How many losses did each team have?
2. Write a rule to describe how you found the number of losses.
3. Let w represent the number of wins and ℓ represent the number of losses. Rewrite your rule using numbers, variables, and an equals sign.

Women's National Basketball Association, 2001

Team	Wins	Losses
Cleveland	22	?
New York	21	?
Miami	20	?
Charlotte	18	?
Orlando	13	?
Indiana	10	?

Source: cnn.si.com

An **equation** is a sentence in mathematics that contains an equals sign. The equals sign tells you that the expression on the left is equivalent to the expression on the right.

$4 + 3 = 8 - 1$ $\qquad$ $3(4) = 24 \div 2$ $\qquad$ $17 = 13 + 2 + 2$

An equation that contains a variable is neither true nor false until the variable is replaced with a number. The **solution** of an equation is a number that makes the sentence true.

The process of finding a solution is called **solving an equation**. Some equations are easy to solve using mental mathematics.

EXAMPLE Solve an Equation Mentally

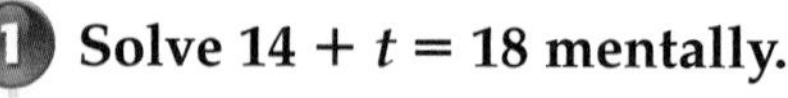

1 Solve $14 + t = 18$ mentally.

$14 + t = 18$ Write the equation.

$14 + 4 = 18$ You know that 14 + 4 is 18.

$18 = 18$ Simplify.

The solution is 4.

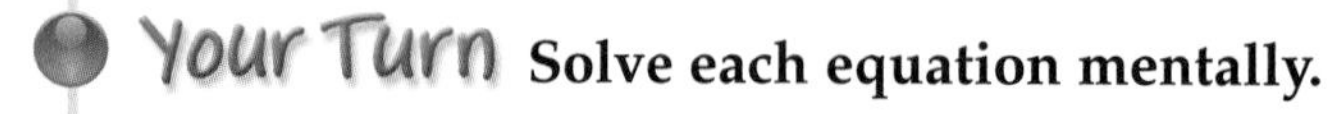

Your Turn Solve each equation mentally.

a. $p - 5 = 20$ $\qquad$ b. $8 = y \div 3$ $\qquad$ c. $7h = 56$

EXAMPLE Graph the Solution of an Equation

2 **Graph the solution of the equation in Example 1.**

Locate the point named by the solution on a number line. Then draw a dot at the solution, 4.

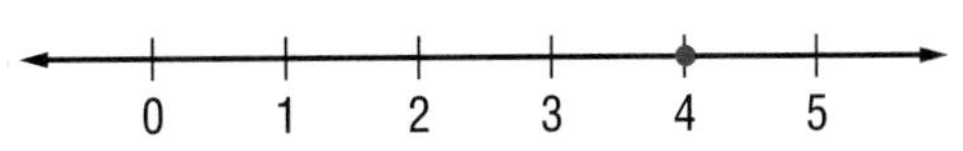

When you write an equation that represents a real-life problem, you are *modeling* the problem. First, choose a variable to represent one of the unknowns. This is called **defining the variable**.

EXAMPLE Write an Equation to Solve a Problem

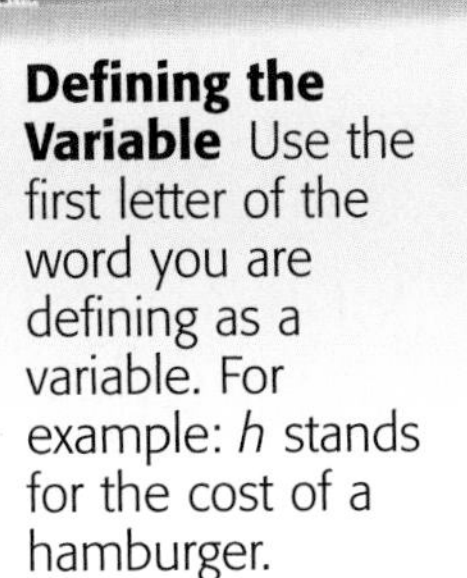

STUDY TIP

Defining the Variable Use the first letter of the word you are defining as a variable. For example: h stands for the cost of a hamburger.

3 **FOOD The total cost of a hamburger, fries, and soft drink is $5.50. If the fries and drink cost $2.50 together, what is the cost of the hamburger?**

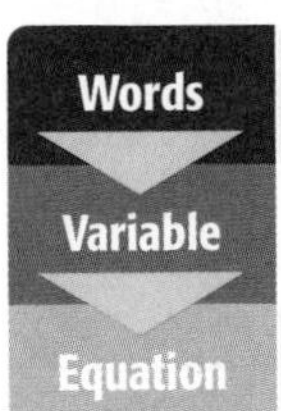

Words	The cost of a hamburger, fries, and soft drink is $5.50.
Variable	Let h represent the cost of the hamburger.
Equation	$h + 2.50 = 5.50$

$h + 2.50 = 5.50$ Write the equation.

$3 + 2.50 = 5.50$ Replace h with 3 to make the equation true.

$5.50 = 5.50$ Simplify.

The number 3 is the solution of the equation. So, the cost of the hamburger is $3.00.

EXAMPLE Find a Solution of an Equation

Standardized Test Practice

4 **MULTIPLE-CHOICE TEST ITEM**

What value of x is a solution of $x + 14.6 = 30.2$?

Ⓐ 14.9 Ⓑ 15.2 Ⓒ 15.6 Ⓓ 16.1

Test-Taking Tip (The Princeton Review)

Solutions Since an equation with one variable has only one solution, the value of 16.1, or choice D, does not need to be checked.

Read the Test Item

Substitute each value for x to determine which makes the left side of the equation equivalent to the right side.

Solve the Test Item

Replace x with 14.9.	Replace x with 15.2.	Replace x with 15.6.
$x + 14.6 = 30.2$	$x + 14.6 = 30.2$	$x + 14.6 = 30.2$
$14.9 + 14.6 \stackrel{?}{=} 30.2$	$15.2 + 14.6 \stackrel{?}{=} 30.2$	$15.6 + 14.6 \stackrel{?}{=} 30.2$
$29.5 \neq 30.2$ false	$29.8 \neq 30.2$ false	$30.2 = 30.2$ true ✔

The value 15.6 makes the equation true. So, the answer is C.

Skill and Concept Check

1. **Explain** what it means to solve an equation.

2. **OPEN ENDED** Write an example of an equation that is easily solved using mental mathematics and an equation that is not as easily solved using mental mathematics.

3. **FIND THE ERROR** Joshua and Ivan had a contest to see who could solve $w - 25 = 50$ the fastest. Their solutions are shown below. Who is correct? Explain.

Joshua	Ivan
$w = 25$	$w = 75$

GUIDED PRACTICE

Solve each equation mentally.

4. $y - 18 = 20$
5. $5 + a = 22$
6. $w \div 4 = 7$
7. $\frac{r}{9} = 6$
8. $75 = w + 72$
9. $69 = 3f$

10. Graph the solution of $x + 5 = 11$.

Name the number that is the solution of the given equation.

11. $x + 15 = 19$; 4, 5, 6
12. $13k = 80.6$; 5.5, 5.8, 6.2

13. **BASKETBALL** Jason scored 14 more points than David in the last basketball game. If David scored 9 points, how many did Jason score?

Practice and Applications

HOMEWORK HELP	
For Exercises	See Examples
14–31	1
32–35	2
40–46	3
36–39	4

Extra Practice
See pages 565, 596.

Solve each equation mentally.

14. $b + 7 = 18$
15. $8 + x = 15$
16. $y - 15 = 71$
17. $a - 18 = 20$
18. $25 - 19 = n$
19. $24 + 39 = m$
20. $12 \cdot 5 = s$
21. $n = \frac{30}{6}$
22. $b - 42 = 84$
23. $7t = 77$
24. $3d = 99$
25. $22 + h = 42$
26. $20 = y \div 5$
27. $16 = \frac{u}{4}$
28. $\frac{z}{7} = 12$
29. $18j = 360$
30. $m - 36 = 123$
31. $108 = 9b$

Graph the solution of each equation on a number line.

32. $w + 3 = 10$
33. $x - 7 = 11$
34. $12y = 48$
35. $11 = 132 \div z$

Name the number that is the solution of the given equation.

36. $13.4 \cdot 9 = h$; 117.8, 118.7, 120.6
37. $n \div 10 = 4$; 20, 30, 40
38. $9.9 + r = 24.2$; 12.7, 14.3, 16.3
39. $c - 8 = 17$; 15, 25, 35

40. **ORNAMENTS** Last year, Desiree had 25 glass ornaments. This year, she has 38 ornaments. Solve $25 + n = 38$ to find how many new ornaments Desiree collected.

41. **MONEY** Maria was paid \$9 per hour and earned \$67.50. How many hours did she work? Use the equation $67.50 = 9h$, where h is hours worked.

WHALES For Exercises 42 and 43, use the information below.
Each winter, Humpback whales migrate about 1,500 miles north to the Indian Ocean. However, scientists tracked one whale that migrated 5,000 miles in one season.

42. Write an equation that can be used to find how many more miles the whale migrated compared to the usual migration.

43. How much farther did the whale travel than the usual migration?

44. **MULTI STEP** A long-running TV show aired 179 episodes from October 1951 to September 1961. About how many episodes were filmed each year?

45. **PENGUINS** On average, how much taller are Emperor penguins than Adelie penguins? Write an equation and solve.

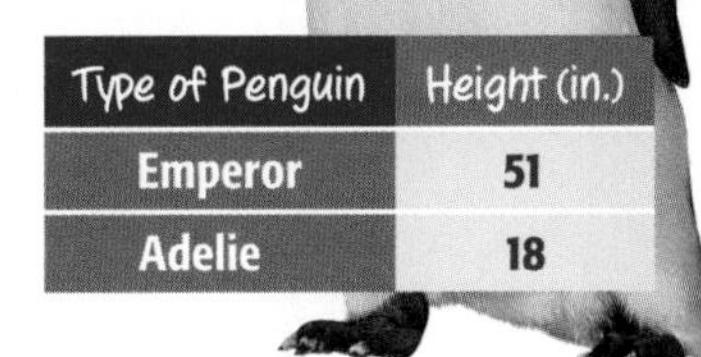

Type of Penguin	Height (in.)
Emperor	51
Adelie	18

Data Update How do the heights of other types of penguins compare to Adelie penguins? Visit msmath2.net/data_update to learn more.

46. **BASKETBALL** Michael Jordan returned to the NBA as a member of the Washington Wizards in 2001. During one game, he scored 51 of the team's 107 points. How many points did the rest of the team score?

47. **CRITICAL THINKING** Consider the equation $0 \times a = b$. Find the values of a and b.

Standardized Test Practice and Mixed Review

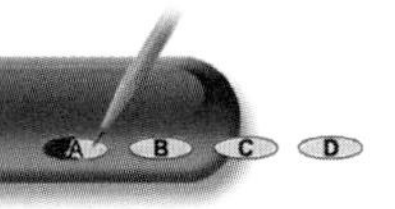

48. **MULTIPLE CHOICE** Choose the solution of the equation $\frac{48}{c} = 4 + 2$.

Ⓐ 2 Ⓑ 4 Ⓒ 6 Ⓓ 8

49. **GRID IN** Use the table to determine how much farther in air miles Phoenix is from New Orleans than Memphis.

From	To	Distance (air mi)
Memphis, TN	New Orleans, LA	358
Phoenix, AZ	New Orleans, LA	1,316

50. **ALGEBRA** Evaluate $3a + b^2$ if $a = 2$ and $b = 3$. (Lesson 1-4)

Evaluate each expression. (Lesson 1-3)

51. $11 \cdot 6 \div 3 + 9$
52. $5 \cdot 13 - 62$
53. $1 + 2(8 - 5)^2$

54. **DINING** Four hundred sixty people are scheduled to attend a banquet. If each table seats 8 people, how many tables are needed? (Lesson 1-1)

GETTING READY FOR THE NEXT LESSON

BASIC SKILL Multiply.

55. $2 \cdot (4 + 10)$
56. $(9 \cdot 1) \cdot 8$
57. $(5 \cdot 3)(5 \cdot 2)$
58. $(6 + 8) \cdot 12$

Mid-Chapter Practice Test

Vocabulary and Concepts

1. **List** the steps in the four-step problem-solving plan. (Lesson 1-1)
2. **OPEN ENDED** Write an algebraic expression and identify any coefficients, variables, and constants. (Lesson 1-4)

Skills and Applications

3. **SPORTS** A baseball stadium holds 20,000 people. If 3,650 people can be seated in the bleachers, how many seats are available in the rest of the stadium? (Lesson 1-1)
4. Write $2 \cdot 2 \cdot 2 \cdot 2 \cdot 2 \cdot 2 \cdot 2$ in exponential form. (Lesson 1-2)

Evaluate each expression. (Lessons 1-2 and 1-3)

5. 4^5
6. 9^6
7. 3^{10}
8. $25 - (3 + 2 \cdot 5)$
9. $(2^3 + 4) + 3^2$
10. $\frac{2(7-3)}{2^2}$

Evaluate each expression if $x = 12$, $y = 4$, and $z = 8$. (Lesson 1-4)

11. $x - 5$
12. $3y + 10z$
13. $\frac{yz}{2}$
14. $\frac{(y+8)^2}{x}$

Solve each equation mentally. (Lesson 1-5)

15. $x + 16 = 23$
16. $36 = 3y$
17. $\frac{65}{z} = 5$

18. **GEOMETRY** The perimeter of a rectangle is given by the expression $2(\ell + w)$ where ℓ = length and w = width. Find the perimeter of the rectangle. (Lesson 1-4)

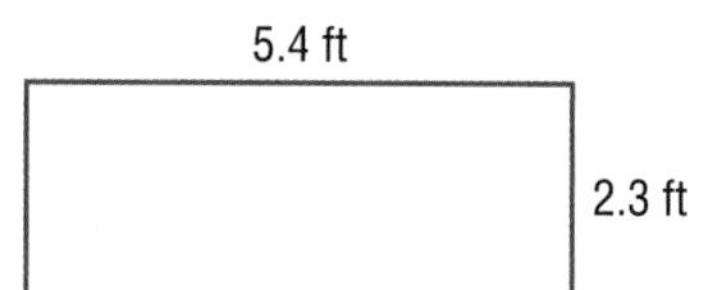

Standardized Test Practice

19. **MULTIPLE CHOICE** A cycling club is planning a 1,800-mile trip. They can average 15 miles per hour for 6 hours each day. How many days will it take them to complete the trip? (Lesson 1-1)

    (A) 20 days (B) 21 days
    (C) 22 days (D) 23 days

20. **GRID IN** The expression $\frac{w}{30}$, where w is a person's weight in pounds, is used to find the approximate number of quarts of blood in the person's body. How much blood does a 120-pound person have? (Lesson 1-4)

The Game Zone

A Place To Practice Your Math Skills

Math Skill

Evaluating Expressions

GET READY!

Players: two, three, or four

Materials: scissors, 18 index cards cut in half, one number cube

GET SET!

- On each card, write a different expression containing only one variable. The coefficients and constants are to be values less than or equal to ten and any exponents should be less than four. Four examples are shown at the right.

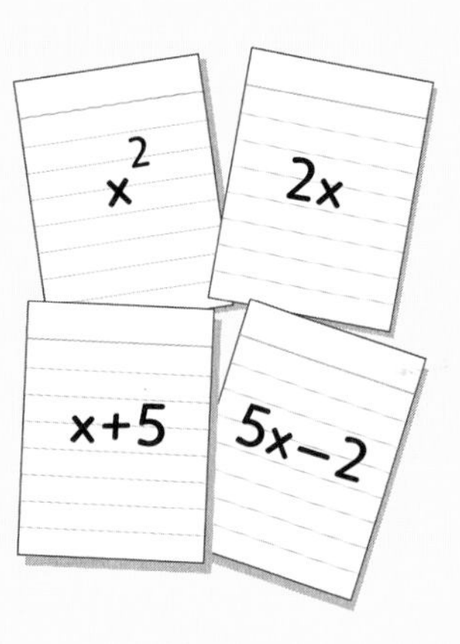

GO!

- Deal all cards to the players. The dealer then rolls the number cube. The number rolled is the value of the variable for the first round.
- The player to the left of the dealer puts a card of his or her choice faceup on the table, evaluates the expression on the card, and announces its value. Play continues until all players have placed one card on the table. This is the end of the first round. The person whose card has the greatest value wins all of the cards for that round.
- The player to the left of the dealer rolls the number cube. This is the value of the variable for the next round. Play continues until all the cards are played.
- **Who wins?** The person who has the most cards at the end of the game is the winner.

1-6 Algebra: Properties

WHEN am I ever going to use this?

RESTAURANTS Land-Ho! Fish Market is having a Friday night special.

FISH-BAKE DINNER
-2 FISH FILLETS
-GARLIC MASHED POTATOES
-SIDE OF MIXED GREENS
-BEVERAGE $ 8.95
ADD A SLICE OF CHEESECAKE FOR ONLY $2.15

1. Find the total cost for a 5-member family, without tax and tip, if each one orders a fish-bake dinner and cheesecake.
2. Describe the method used to find the total cost.
3. Is there more than one way to find the total cost?

What You'll LEARN

Use addition and multiplication properties to solve problems.

NEW Vocabulary

equivalent expressions
properties

Link to READING

Everyday Meaning of Distribute: to divide among several, as in distribute a deck of cards

Here are two ways to find the total cost of the dinner:

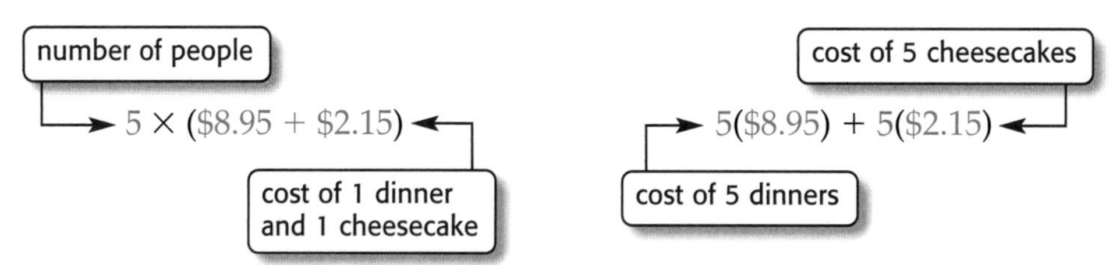

The expressions 5($8.95 + $2.15) and 5($8.95) + 5($2.15) are **equivalent expressions** because they have the same value, $55.50. This shows how the **Distributive Property** combines addition and multiplication.

Key Concept — **Distributive Property**

Words To multiply a sum by a number, multiply each addend of the sum by the number outside the parentheses.

Symbols

Arithmetic	Algebra
$3(4 + 6) = 3(4) + 3(6)$	$a(b + c) = a(b) + a(c)$

EXAMPLES Use the Distributive Property

Use the Distributive Property to write each expression as an equivalent expression. Then evaluate the expression.

1 **5(3 + 2)**

$5(3 + 2) = 5 \cdot 3 + 5 \cdot 2$
$= 15 + 10$ Multiply.
$= 25$ Add.

2 **(7 + 4)3**

$(7 + 4)3 = 7 \cdot 3 + 4 \cdot 3$
$= 21 + 12$ Multiply.
$= 33$ Add.

REAL-LIFE MATH

HISTORY Each rider on the Pony Express would travel 100 miles and change horses every 10–20 miles.

Source: http://members.tripod.com

EXAMPLE Use the Distributive Property

3 HISTORY The riders for the Pony Express mail service carried mail from St. Joseph, Missouri, to Sacramento, California, in eight days. On average, the riders covered 250 miles each day. About how far did the Pony Express riders travel over the eight-day period?

You can find how many miles the riders traveled over the eight-day period by finding 8×250. You can use the Distributive Property to multiply mentally.

$8(250) = 8(200 + 50)$	Rewrite 250 as 200 + 50.
$= 8(200) + 8(50)$	Distributive Property
$= 1{,}600 + 400$	Multiply.
$= 2{,}000$	Add.

So, the riders on the Pony Express traveled about 2,000 miles over an eight-day period.

In algebra, **properties** are statements that are true for any number or variable.

STUDY TIP

Properties The *Commutative Property* shows how the terms move and change places.

The *Associative Property* shows how the parentheses move and change places.

Key Concept — Properties

Commutative Property	The order in which two numbers are added or multiplied does not change their sum or product.
Associative Property	The way in which three numbers are grouped when they are added or multiplied does not change their sum or product.
Identity Property	The sum of an addend and zero is the addend. The product of a factor and one is the factor.

EXAMPLES Identify Properties

Name the property shown by each statement.

4 $6 + (2 + 7) = (6 + 2) + 7$ Associative Property of Addition

5 $15 \times 10 = 10 \times 15$ Commutative Property of Multiplication

6 $4 \times 1 = 4$ Identity Property of Multiplication

7 $4(6 + 8) = 4(6) + 4(8)$ Distributive Property

Your Turn **Name the property shown by each statement.**

a. $1 \times (3 \times 4) = (1 \times 3) \times 4$ **b.** $a + 0 = a$

Skill and Concept Check

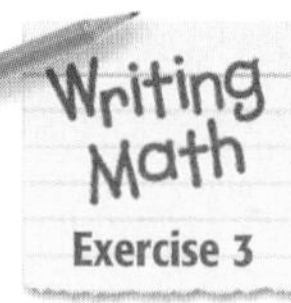

1. **Rewrite** $p(q) + p(r)$ using the Distributive Property.
2. **OPEN ENDED** Write an equation that illustrates the Associative Property of Addition.
3. **NUMBER SENSE** *True* or *False*? Explain your answer.
   $(11 + 18) \times 5 = 11 + 18 \times 5$

GUIDED PRACTICE

Use the Distributive Property to write each expression as an equivalent expression. Then evaluate the expression.

4. $7(4 + 3)$
5. $(10 + 8)2$
6. $5(6 - 2)$

Name the property shown by each statement.

7. $5 + 4 = 4 + 5$
8. $(2 \times 3) \times 7 = 2 \times (3 \times 7)$
9. $6(9 + 3) = 6(9) + 6(3)$
10. $5 \times 1 = 5$

Practice and Applications

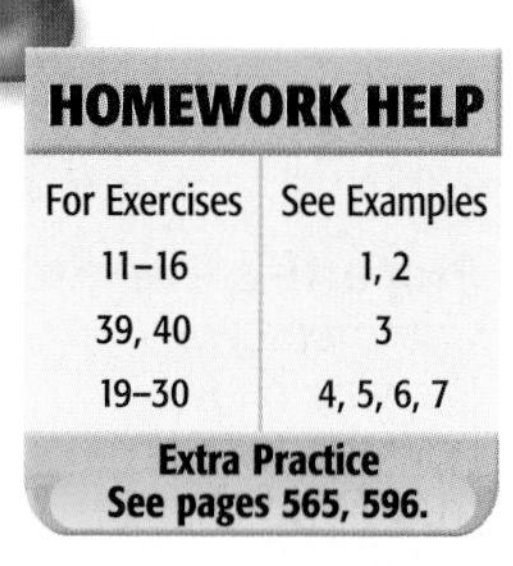

HOMEWORK HELP

For Exercises	See Examples
11–16	1, 2
39, 40	3
19–30	4, 5, 6, 7

Extra Practice
See pages 565, 596.

Use the Distributive Property to write each expression as an equivalent expression. Then evaluate the expression.

11. $2(6 + 7)$
12. $(3 + 8)4$
13. $(11 + 3)8$
14. $2(5 - 4)$
15. $7(8 - 6)$
16. $6(12 + 5)$
17. Which expression is easier to evaluate mentally: 3(452) or 3(400 + 50 + 2)? Explain.
18. Is $5 + 15 = 5(1 + 3)$ a true equation? Explain your reasoning.

Name the property shown by each statement.

19. $1 \times (5 \times 9) = (1 \times 5) \times 9$
20. $4 + b = b + 4$
21. $m \times n = n \times m$
22. $8 + 0 = 8$
23. $(1 + r) + s = 1 + (r + s)$
24. $(2.5 + 7)e = 2.5e + 7e$
25. $x(y + z) = xy + xz$
26. $19 \times 1 = 19$

TRAVEL For Exercises 27 and 28, use the table at the right.

27. Write a sentence comparing the mileage from Jacksonville to Charleston to Norfolk and the mileage from Norfolk to Charleston to Jacksonville.
28. Name the property that is shown in the sentence.

From	To	Driving Distance (mi)
Jacksonville, FL	Charleston, SC	239
Charleston, SC	Norfolk, VA	531

COIN COLLECTING Mrs. Jackson has collected 152 rare coins.

29. If she does not add any coins to her collection, write a sentence that represents this situation.
30. Name the property that is illustrated.

Use properties to rewrite each expression as an equivalent expression without parentheses.

31. $(y + 1) + 4$ **32.** $2 + (x + 4)$ **33.** $4(8b)$ **34.** $(3a)2$

35. $2(x + 3)$ **36.** $4(2 + b)$ **37.** $6(c + 1) - c$ **38.** $3(f + 4) + 2f$

39. **MENTAL MATH** Use one of the properties to find 6(48) mentally. Which property did you use?

40. **BASKETBALL** Use mental math to find the total number of NBA draftees that have come from the schools in the graph.

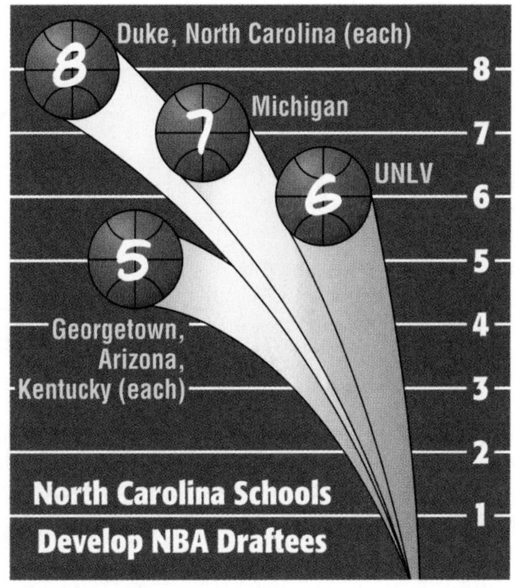

Source: nbcsports.com

41. **WRITE A PROBLEM** Write about a real-life situation that can be solved using the Distributive, Commutative, Associative, or Identity Properties. Then use the property to solve the problem.

42. **CRITICAL THINKING** A *counterexample* is an example showing that a statement is not true. Provide a counterexample to the following statement. *Division of whole numbers is associative.*

Standardized Test Practice and Mixed Review

43. **MULTIPLE CHOICE** Rewrite $p \times (q \times r)$ using the Associative Property.

Ⓐ $(p \times q) \times r$ Ⓑ $p \times q \times r$

Ⓒ $(p + q) + r$ Ⓓ $p \times q \times r \times 1 = p \times q \times r$

44. **SHORT RESPONSE** Rewrite $6(9 + 8)$ using the Distributive Property.

Name the number that is the solution of the given equation. (Lesson 1-5)

45. $x + 12 = 20$; 8, 9, 10 **46.** $7.3 = t - 4$; 10.3, 11.3, 12.3

47. $35.5 = 5n$; 5.1, 7.1, 9.1 **48.** $z \div 6 = 18$; 88, 98, 108

LIFE SCIENCE For Exercises 49 and 50, use the information below.
It is believed that a dog ages 7 human years for every calendar year. (Lesson 1-4)

49. Write an expression for determining a dog's age in human years. Let y represent the number of calendar years the dog has lived.

50. Find the human age of a dog that has lived for 12 calendar years.

51. Evaluate $(14 - 9)^4$. (Lesson 1-3)

GETTING READY FOR THE NEXT LESSON

BASIC SKILL Find the next number in each pattern.

52. 2, 4, 6, 8, ? **53.** 10, 21, 32, 43, ? **54.** 1.4, 2.2, 3.0, 3.8, ?

55. 4, 8, 12, 16, ? **56.** 64, 32, 16, 8, ? **57.** 0.4, 1.2, 3.6, 10.8, ?

1-7 Sequences

What You'll LEARN

Recognize and extend patterns for sequences.

NEW Vocabulary

sequence
term
arithmetic sequence
geometric sequence

Link to READING

Everyday Meaning of Sequence: a connected series, as in a sequence of events

HANDS-ON Mini Lab

Materials

- straightedge

Work with a partner.

A *diagonal* connects two nonconsecutive vertices in a figure, as shown at the right.

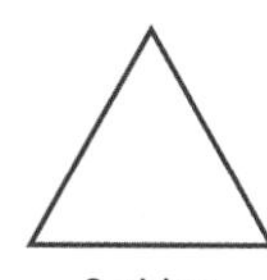
3 sides
0 diagonals

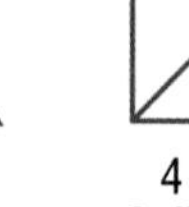
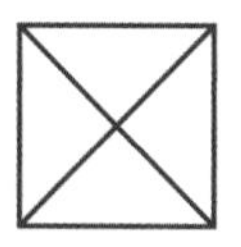
4 sides
2 diagonals

5 sides
5 diagonals

1. Draw a six-sided figure and all of its diagonals.
2. How many diagonals does the figure have?
3. Describe the pattern formed by the number of diagonals: 0, 2, 5, 9, … .
4. **Make a prediction** about how many diagonals a 7-sided, an 8-sided, and a 9-sided figure would have.

A **sequence** is an ordered list of numbers. Each number in a sequence is called a **term**.

8, 11, 14, 17, 20, …
+3 +3 +3 +3

In an **arithmetic sequence**, each term is found by *adding* the same number to the previous term.

3, 6, 12, 24, 48, …
×2 ×2 ×2 ×2

In a **geometric sequence**, each term is found by *multiplying* the previous term by the same number.

EXAMPLES Describe Patterns in Sequences

Describe the pattern in each sequence and identify the sequence as *arithmetic, geometric,* or *neither.*

1 **8, 13, 18, 23, …**

8, 13, 18, 23, …

+5 +5 +5

Each term is found by adding 5 to the previous term. This sequence is arithmetic.

2 **1, 4, 16, 64, …**

1, 4, 16, 64, …

×4 ×4 ×4

Each term is found by multiplying the previous term by 4. This sequence is geometric.

Your Turn **Describe the pattern in each sequence and identify the sequence as *arithmetic, geometric,* or *neither.***

a. 4, 12, 36, 108, …

b. 1, 2, 4, 7, 11, …

If you know the pattern of a sequence, you can use it to determine the terms.

EXAMPLES Determine Terms in Sequences

Write the next three terms of each sequence.

3 **0, 13, 26, 39, …**

0, 13, 26, 39, … Each term is 13 greater than the previous term.
+13 +13 +13

Continue the pattern to find the next three terms.

39 + 13 = 52 52 + 13 = 65 65 + 13 = 78

The next three terms are 52, 65, and 78.

4 **0.16, 0.8, 4, 20, 100, …**

0.16, 0.8, 4, 20, … Each term is 5 times the previous term.
×5 ×5 ×5

Continue the pattern to find the next three terms.

20 × 5 = 100 100 × 5 = 500 500 × 5 = 2,500

The next three terms are 100, 500, and 2,500.

Your Turn **Write the next three terms of each sequence.**

c. 0.5, 1.5, 2.5, 3.5, … d. 0.5, 1, 2, 4, …

Skill and Concept Check

Writing Math
Exercises 1–3

1. **Compare and contrast** arithmetic and geometric sequences.
2. **Explain** how you would find the next term in the sequence 14, 22, 30, 38, 46, … .
3. **OPEN ENDED** Write five terms of an arithmetic sequence and describe the rule for finding the terms.

GUIDED PRACTICE

Describe the pattern in each sequence and identify the sequence as ***arithmetic, geometric,*** **or** ***neither.***

4. 0, 9, 18, 27, … 5. 200, 202, 206, 212 6. 36, 144, 576, 2,304, …

Write the next three terms of each sequence.

7. 4, 12, 36, 108, … 8. 22, 33, 44, 55, … 9. 5, 5.4, 5.8, 6.2, …

10. Find the missing terms in the sequence 26, 33, 40, __?__, 54, 61, __?__, …

11. Create a sequence of four terms in which the first term is 70 and the next terms are found by multiplying 0.1.

Practice and Applications

HOMEWORK HELP	
For Exercises	See Examples
12–17, 26	1, 2
18–25	3, 4

Extra Practice
See pages 566, 596.

Describe the pattern in each sequence and identify the sequence as *arithmetic, geometric,* or *neither.*

12. 1, 6, 36, 216, …

13. 19, 31, 43, 55, …

14. 2, 14, 98, 686, …

15. 3.5, 10.5, 31.5, 94.5, …

16. 2.0, 3.1, 4.2, 5.3, …

17. 1, 2, 6, 24, 120, …

Write the next three terms of each sequence.

18. 33, 38, 43, 48, …

19. 4.6, 8.6, 12.6, 16.6, …

20. 0.05, 0.4, 3.2, 25.6, …

21. 0.1, 0.4, 0.7, 1.0, …

22. 3, 12, 48, 192, …

23. 125, 25, 5, 1, …

24. **INSECTS** Every 17 years, billions of insects called *cicada* emerge from hibernation. The last time they emerged was in 1996. Find the next three years in which they will come out of hibernation.

GEOMETRY **For Exercises 25 and 26, use the following information.** Numbers that can be represented by a triangular arrangement of dots are called *triangular numbers.* The first five triangular numbers are shown below.

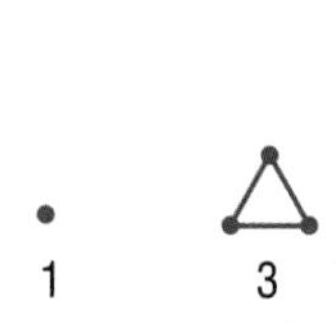

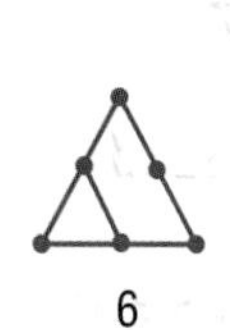

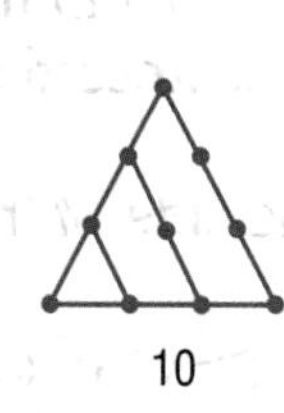

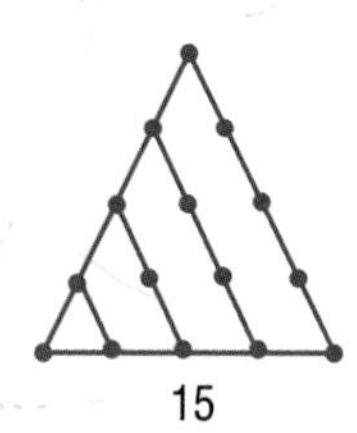

1 3 6 10 15

25. Write a sequence formed by the first eight triangular numbers.

26. Write a rule for generating the sequence.

27. **CRITICAL THINKING** You can find any term in an arithmetic sequence by using the expression $a + (n - 1)d$, where a is the first term, d is the difference between each pair of consecutive terms, and n is the position in the sequence. Find the eleventh term in the sequence 6, 13, 20, 27, 34, … .

Standardized Test Practice and Mixed Review

28. **SHORT ANSWER** The table shows the height of a bamboo plant after each number of hours. How tall will the plant be after 6 hours?

Time (h)	Height (in.)
0	5
1	27
2	49
3	71

29. **GRID IN** Find the next term in the sequence 44, 58, 72, 86, … .

Name the property shown by each statement. (Lesson 1-6)

30. $8 + 4.1 = 4.1 + 8$

31. $3(n + 6) = 3n + 3 \cdot 6$

32. Solve $30 + y = 50$ mentally. (Lesson 1-5)

GETTING READY FOR THE NEXT LESSON

PREREQUISITE SKILL **Multiply or divide.** (Pages 560, 562)

33. 52×10

34. $400 \div 1{,}000$

35. $32 \div 10$

36. 0.31×100

1-7b HANDS-ON LAB

A Follow-Up of Lesson 1-7

Exploring Sequences

What You'll LEARN

Explore patterns in sequences using paper folding.

Materials

- calculator
- piece of paper
- recording table
- colored pencils

INVESTIGATE *Work with a partner.*

Number of Folds	Layers of Paper	Fraction of Paper *Not* Shaded
1	2	$\frac{1}{2}$
2	4	$\frac{1}{4}$
3	8	
4		
⋮		

STEP 1 Fold a piece of paper in half. Make a table like the one at the right and record the number of layers of paper.

STEP 2 Shade one side of the folded paper.

STEP 3 Open the piece of paper. Record the fractional part of the paper that is *not* shaded. Refold the paper.

STEP 4 Fold your paper in half again so that the unshaded side is on the outside. Record the number of layers of paper.

STEP 5 Shade one side of the folded paper.

STEP 6 Open the piece of paper. Record the fractional part of the paper that is *not* shaded. Completely refold the paper.

STEP 7 Continue this folding, shading, and recording process for five folds.

Writing Math

Work with a partner.

1. **Examine** the sequence of numbers in the "Layers" column of your table. Is the sequence arithmetic or geometric? Then write a rule to find the next three terms.
2. **Examine** the sequence of numbers in the "Fraction" column of your table. Is the sequence arithmetic or geometric? Then write a rule to find the next three terms.
3. Assume you could continue the paper-folding process indefinitely. Suppose your unfolded piece of paper is 0.002 inch thick. Add a column to your table and find the thickness of the paper for the first five folds.
4. How many folds would it take until the paper is as tall as you?
5. **Explain** the relationship between the number of layers and the fraction of the shaded region.

1-8 Measurement: The Metric System

What You'll LEARN

Change metric units of length, capacity, and mass.

NEW Vocabulary

meter
metric system
gram
kilogram
liter

HANDS-ON Mini Lab

Materials

- centimeter ruler
- computer diskette
- CD case
- meterstick

Work with a partner.

The widths of two objects are shown below.

Object	Width (millimeters)	Width (centimeters)
computer diskette	?	8.9
CD case	?	14.4

1. Find three other objects. Use a ruler to find and record the width of all five objects to the nearest millimeter and tenth of a centimeter. Add your measurements to the table.
2. Compare the measurements of each object, and write a rule that describes how to convert from millimeters to centimeters.
3. Use a meterstick to measure the length of your classroom in meters. **Make a conjecture** about which operation you would use to convert this measure to centimeters. Explain.

The **meter** (m) is the base unit of length in the **metric system**. A meter is about the distance from the floor to a doorknob, or a little more than a yard. One kilometer is equivalent to 1,000 meters, and 1 meter is equivalent to 100 centimeters and also to 1,000 millimeters.

Prefix	Meaning in Words	Meaning in Numbers
kilo-	thousands	1,000
centi-	hundredths	0.01
milli-	thousandths	0.001

Metric prefixes indicate the decimal place-value position of the measurement.

MULTIPLY to convert from larger units to smaller units.

DIVIDE to convert from smaller units to larger units.

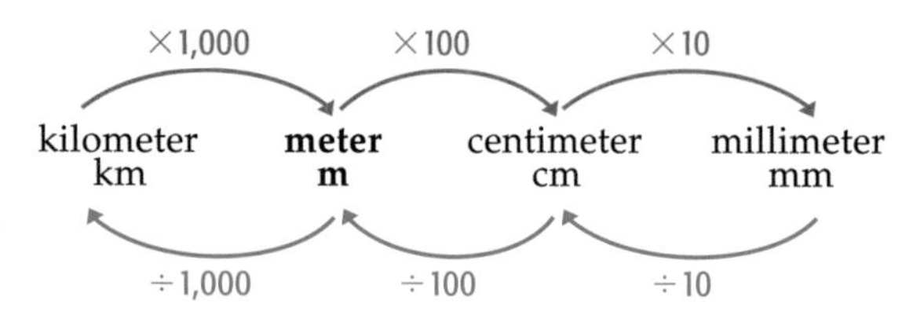

STUDY TIP

Mental Math To multiply or divide by a power of ten, you can move the decimal point right or left.

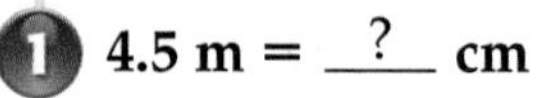

EXAMPLES Convert Units of Length

Complete.

1 **4.5 m = __?__ cm**

To convert from meters to centimeters, multiply by 100.

$4.5 \times 100 = 450$

4.5 m = 450 cm

2 **2,600 m = __?__ km**

To convert from meters to kilometers, divide by 1,000.

$2{,}600 \div 1{,}000 = 2.6$

2,600 m = 2.6 km

READING Math

Prefixes The kilogram is the only base unit in the metric system that has a prefix as part of its name.

The **gram** (g) measures *mass*, the amount of matter in an object. A paper clip has a mass of about one gram. The base unit of mass in the metric system is the **kilogram**, which is equivalent to 1,000 grams. One gram is equivalent to 1,000 milligrams.

EXAMPLES Convert Units of Mass

Complete.

3 **8.43 kg = ___?___ g**

To convert from kilograms to grams, multiply by 1,000.

$8.43 \times 1{,}000 = 8{,}430$

8.43 kg = 8,430 g

4 **500 mg = ___?___ g**

To convert from milligrams to grams, divide by 1,000.

$500 \div 1{,}000 = 0.5$

500 mg = 0.5 g

Your Turn **Complete.**

a. 23.5 g = ___?___ mg

b. 9,014 g = ___?___ kg

The **liter** (L) is widely used and accepted as the metric measure for capacity. *Capacity* is the amount of dry or liquid material an object can hold. Some soft drinks are sold in one-liter bottles. Kiloliter, liter, and milliliter are related in a manner similar to kilometer, meter, and millimeter.

EXAMPLES Convert Units of Capacity

Complete.

5 **3.2 L = ___?___ mL**

To convert from liters to milliliters, multiply by 1,000.

$3.2 \times 1{,}000 = 3{,}200$

3.2 L = 3,200 mL

6 **1,640 L = ___?___ kL**

To convert from liters to kiloliters, divide by 1,000.

$1{,}640 \div 1{,}000 = 1.64$

1,640 L = 1.64 kL

Your Turn **Complete.**

c. 0.75 kL = ___?___ L

d. 8,800 mL = ___?___ L

REAL-LIFE MATH

METRIC SYSTEM The Metric Act of 1866 made the use of the metric system legal in the United States.

Source: lamar.colostate.edu

The following table is a summary of how to convert measures in the metric system.

Concept Summary — **Converting Units**

	Larger Units → Smaller Units (multiply by a power of 10)	Smaller Units → Larger Units (divide by a power of 10)
Units of Length	1 km = 1,000 m 1 m = 100 cm 1 cm = 10 mm	1 mm = 0.1 cm 1 cm = 0.01 m 1 m = 0.001 km
Units of Mass	1 kg = 1,000 g 1 g = 1,000 mg	1 mg = 0.001 g 1 g = 0.001 kg
Units of Capacity	1 kL = 1,000 L 1 L = 1,000 mL	1 mL = 0.001 L 1 L = 0.001 kL

Skill and Concept Check

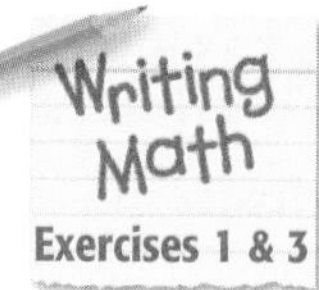

1. **Explain** why it makes sense to multiply when changing from a larger unit to a smaller unit.
2. **OPEN ENDED** List a set of measurement references, made up of everyday objects, that will help you remember the basic units in the metric system.
3. **FIND THE ERROR** Hunter and Arturo are converting 45.7 milligrams to grams. Who is correct? Explain.

Hunter	Arturo
45.7 mg = 0.0457 g	45.7 mg = 45,700 g

GUIDED PRACTICE

Complete.

4. 550 m = __?__ km
5. 45 g = __?__ mg
6. 1,460 mL = __?__ L
7. 3.7 m = __?__ cm
8. 3,800 mg = __?__ g
9. 2.34 kL = __?__ L
10. How many milliliters are in 0.04 liter?

Practice and Applications

HOMEWORK HELP	
For Exercises	See Examples
11–41	1–6

Extra Practice See pages 566, 596.

Complete.

11. 720 cm = __?__ m
12. 983 mm = __?__ m
13. 64 kg = __?__ g
14. 7.5 g = __?__ mg
15. 925 mg = __?__ g
16. 345 mL = __?__ L
17. 0.85 cm = __?__ mm
18. 3.2 m = __?__ cm
19. 9.1 L = __?__ mL
20. 73.2 g = __?__ mg
21. 43 L = __?__ kL
22. 130.5 kL = __?__ L
23. 997 g = __?__ kg
24. 0.046 kL = __?__ L
25. 82.1 g = __?__ kg
26. 0.03 m = __?__ mm
27. 57,000 mL = __?__ L
28. 0.22 mm = __?__ cm

MONUMENTS For Exercises 29 and 30, use the information about the Washington Monument at the right.

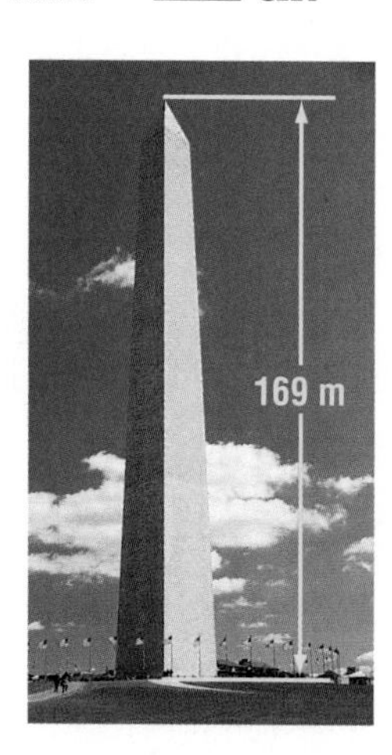

29. What is the height of the monument in kilometers?
30. What is the height in centimeters? in millimeters?
31. Order 0.02 km, 50 m, and 3,000 cm from least to greatest length.
32. Order 0.05 kg, 32 g, and 430,000 mg from greatest to least mass.
33. Order 660 mL, 0.06 L, and 6.6 kL from least to greatest capacity.

Complete.

34. 0.05 km = __?__ mm
35. 0.93 km = __?__ cm
36. 23,000 mg = __?__ kg
37. 40,000 mL = __?__ kL

38. **TRACK AND FIELD** Refer to the table. Convert the distance of each event to kilometers.

Event	Number of Kilometers
10,000-meter run	?
110-meter hurdles	?
1,600-meter relay	?
3,000-meter steeplechase	?

39. **MULTI STEP** Heather needs a 2.5-meter pole for a birdfeeder that she is building. How many centimeters will she need to cut off a 3-meter pole in order to use it for the birdfeeder?

40. **MULTI STEP** Margarita is running the school concession stand. The fruit punch machine holds 15 liters. Each drink cup holds 500 milliliters of punch. How many cups of punch can Margarita sell before the fountain is empty?

41. **BAKING** Evan is making peach cobblers for the fair. The recipe requires 900 grams of peaches per cobbler. If Evan has 4.5 kilograms of peaches, how many cobblers can he make?

42. **RESEARCH** Use the Internet or another source to investigate relationships between metric units of mass and capacity. Explain the relationships and give examples of how they are used in real life.

CRITICAL THINKING The metric prefix *giga* refers to something that is one billion times larger than the base unit. The metric prefix *nano* refers to something that is one billion times smaller than the base unit.

43. How many meters are in 3.9 gigameters?

44. How many meters are in 3.9 nanometers?

Standardized Test Practice and Mixed Review

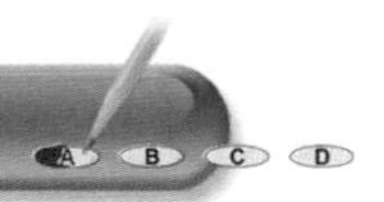

45. **MULTIPLE CHOICE** Choose the best estimate for the mass of a golf ball.

(A) 4.5 mg (B) 4.5 g (C) 45 g (D) 45 kg

46. **GRID IN** What is the mass in grams of a 200-milligram dosage of medication?

Describe the pattern in each sequence and identify the sequence as *arithmetic, geometric,* or *neither.* (Lesson 1-7)

47. 27, 36, 45, 54, …
48. 6, 36, 216, 1,296, …
49. 1.1, 2.2, 3.3, …
50. 1, 4, 9, 16, …

Use the Distributive Property to write each expression as an equivalent expression. Then evaluate the expression. (Lesson 1-6)

51. $5(9 + 7)$
52. $(12 + 4)4$
53. $8(7 - 2)$
54. $(6 - 5)10$

GETTING READY FOR THE NEXT LESSON

PREREQUISITE SKILL Multiply. (Lesson 1-3)

55. $8 \cdot 10^2$
56. $5 \cdot 10^3$
57. $4 \cdot 10^5$
58. $2.7 \cdot 10^3$
59. $65 \cdot 10^2$
60. $1.35 \cdot 10^4$

Study Skill

HOW TO...

Take Good Notes

POWER NOTES

Do you ever have trouble organizing your notes? Try using power notes. They can help you organize the main ideas and details in a lesson.

Power notes are similar to lesson outlines, but they are simpler to organize. Power notes use the numbers 1, 2, 3, and so on.

Power 1: This is the main idea.
  Power 2: This provides details about the main idea.
    Power 3: This provides details about Power 2.
      and so on....

Here's a sample of power notes from Lesson 1-8. Notice that you can even add drawings or examples to your power notes.

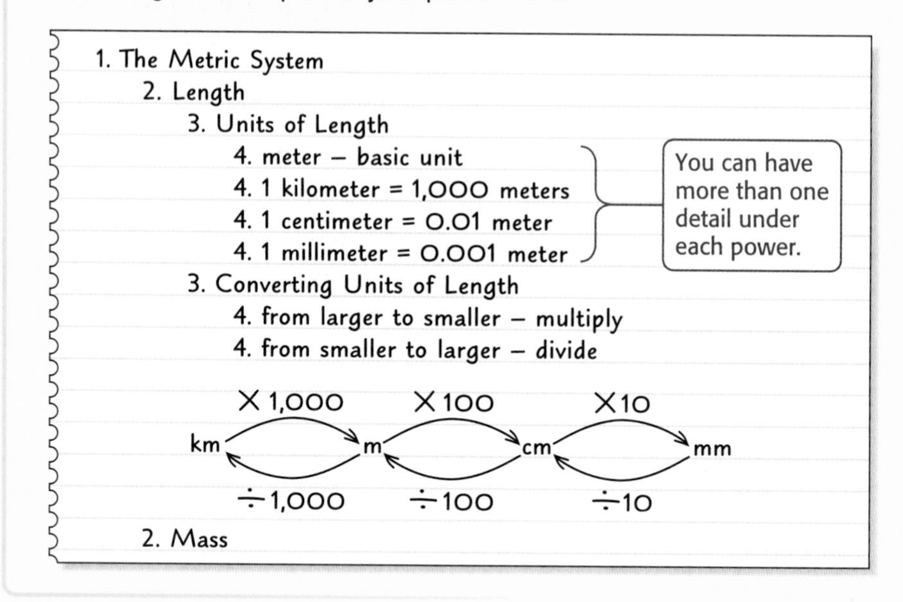

SKILL PRACTICE

1. **Explain** why the four-step plan in Lesson 1-1 is a form of power notes.
2. Complete the power notes above for units of mass and capacity.

Use power notes to make an outline for each concept.

3. order of operations (*Hint:* See Lesson 1-3.)
4. addition and multiplication properties (*Hint:* See Lesson 1-6.)
5. algebraic and geometric sequences (*Hint:* See Lesson 1-7.)

1-9 Scientific Notation

WHEN am I ever going to use this?

What You'll LEARN

Write numbers greater than 100 in scientific notation and in standard form.

NEW Vocabulary

scientific notation

REVIEW Vocabulary

standard form: numbers written without exponents **(Lesson 1-2)**

ENTERTAINMENT The Mall of America in Bloomington, Minnesota, is the largest shopping and entertainment center in the United States. The table shows two features of the mall.

	Standard Form	Abbreviated Form
Number of Parking Spaces	?	13 thousand
Area of Mall (square feet)	4,200,000	?

1. Complete the table.
2. Write both numbers as the product of a number and a power of ten.
3. Rewrite both powers of ten using exponents.

When you work with very large numbers, it is difficult to keep track of the place value. You can write numbers like 4.2 million in **scientific notation** by using a power of ten.

$$4.2 \text{ million} = 4.2 \times 1{,}000{,}000$$
$$= 4.2 \times 10^6 \qquad 10^6 = 1{,}000{,}000$$

Key Concept — **Scientific Notation**

Words A number is expressed in scientific notation when it is written as the product of a number and a power of ten. The number must be greater than or equal to 1 and less than 10.

Examples $60{,}000 = 6.0 \times 10^4$ $\qquad$ $3.5 \text{ billion} = 3.5 \times 10^9$

In Lesson 1-3, you used the order of operations to evaluate expressions like 5.2×10^5. You can also use the order of operations to write a number in standard form.

EXAMPLE Write a Number in Standard Form

1 Write 5.2×10^5 in standard form.

$5.2 \times 10^5 = 5.2 \times 100{,}000$ $\qquad$ $10^5 = 100{,}000$

$= 520000$ $\qquad$ Move the decimal point 5 places to the right.

$= 520{,}000$

Your Turn **Write each number in standard form.**

a. 7.2×10^4 $\qquad$ b. 3.8×10^9 $\qquad$ c. 1.0×10^8

To write a number in scientific notation, move the decimal to the right of the first nonzero number. Then find the power of ten.

Move the decimal point to the right of the first nonzero digit.

$13,000 = 1.3 \times 10^4$

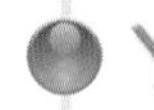

The decimal point was moved 4 places.

EXAMPLE Write a Number in Scientific Notation

2 Write 23,500,000 in scientific notation.

$23,500,000 = 2.35 \times 10,000,000$ Move the decimal point 7 places to find a number between 1 and 10.

$= 2.35 \times 10^7$

Your Turn Write each number in scientific notation.

d. 38,000,000 **e.** 19,000 **f.** 86.1 billion

When you use a calculator to compute large numbers, the numbers are often displayed in scientific notation.

STUDY TIP

Technology To enter 3.42×10^{10} on a calculator, press 3.42 [EE] 10 [ENTER]. The screen would look like this: **3.42**$_{\times 10^{10}}$

EXAMPLE Compute with Large Numbers

3 RESERVOIRS Lake Francis Case in South Dakota contains 5.7 billion cubic meters of water. Lake Mead in Nevada has about 6 times as much water. Find the approximate amount of water in Lake Mead.

To find the amount of water in Lake Mead, multiply 5.7 billion by 6.

5.7 billion = 5,700,000,000

Enter 5700000000 [×] 6 [ENTER].

The number **3.42**$_{\times 10^{10}}$ on the calculator represents 3.42×10^{10}.

So, Lake Mead has about 3.42×10^{10} cubic meters of water.

Skill and Concept Check

1. **Explain** why scientific notation is used with large numbers.
2. **OPEN ENDED** Write a number in standard form and then in scientific notation.
3. **NUMBER SENSE** Which number describes the population of Chicago, 3.0×10^6 or 3.0×10^3? Explain your reasoning.

GUIDED PRACTICE

Write each number in standard form.

4. 5.1×10^4
5. 1.0×10^5
6. 8.05×10^6

Write each number in scientific notation.

7. 70,000
8. 510,000
9. 17,500,000

Practice and Applications

HOMEWORK HELP

For Exercises	See Examples
10–21	1
24–35, 42	2
22, 43	3

Extra Practice
See pages 566, 596.

Write each number in standard form.

10. 4.2×10^3
11. 7.3×10^4
12. 6.9×10^6
13. 3.38×10^5
14. 1.1×10^8
15. 8.98×10^6
16. 5.16×10^4
17. 7.98×10^5
18. 1.05×10^7
19. 9.90×10^8
20. 5.33×10^{10}
21. 8.06×10^7

22. Triple the number 435,000 and write it in scientific notation.

23. Order the numbers 53,400, 4.98×10^6, 5.03×10^5, and 4,980,100 from least to greatest.

Write each number in scientific notation.

24. 46,000
25. 19,800
26. 169,000
27. 9,970,000
28. 3,720,000
29. 505,000
30. 607,000
31. 23,600,000
32. 7 million
33. 1.1 million
34. 40 billion
35. 60.8 billion

NUMBER NAMES **For Exercises 36 and 37, use the table.**

Power of 10	Name
10^6	million
10^9	billion
10^{12}	trillion
10^{15}	quadrillion
10^{18}	quintillion

36. Write 13 quintillion in scientific notation.

37. Write the name of the number 5.43×10^{16}.

Replace each ● with $<$, $>$, or $=$ to make a true sentence.

38. $5{,}000$ ● 5.0×10^3
39. 600 ● 6.0×10^1
40. $81{,}900$ ● 8.19×10^5
41. 1 million ● 1.1×10^6

42. **GEOGRAPHY** The Mariana Trench is the deepest part of the Pacific Ocean with a depth of about 36,000 feet. Write this depth in scientific notation.

43. **EARTH SCIENCE** A light-year, the distance light travels in one year, is about 5,880,000,000,000 miles. How many miles is 50 light years?

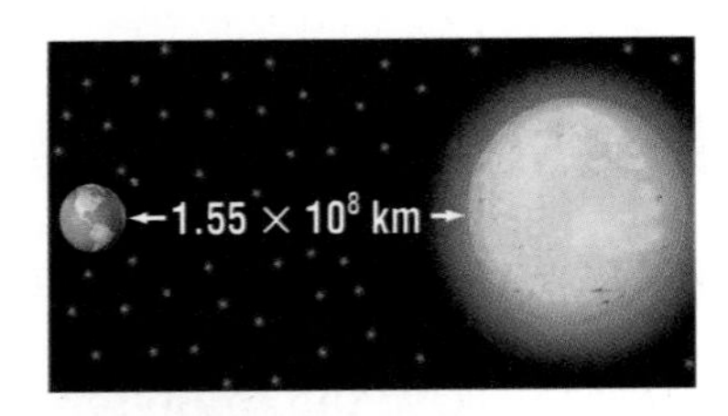

44. **CRITICAL THINKING** The distance from Earth to the Sun is 1.55×10^8 kilometers. If the speed of light is 3×10^5 kilometers per second, about how long does it take for light to travel from the Sun to Earth? (*Hint*: distance = rate × time)

Standardized Test Practice and Mixed Review

45. **MULTIPLE CHOICE** Asia has an area of 44,579,000 square kilometers. Write 44,579,000 in scientific notation.

Ⓐ 4.4579×10^6 Ⓑ 4.4579×10^7 Ⓒ 4.4579×10^8 Ⓓ 4.4579×10^9

46. **SHORT RESPONSE** Write the standard form of 5.56×10^{12}.

Complete. (Lesson 1-8)

47. 1.01 kg = __?__ g
48. 36 mL = __?__ L
49. 2.33 km = __?__ m

50. **SOCCER** The World Cup is played every four years. If the World Cup was played in 2002, when are the next three years that it will be played? (Lesson 1-7)

msmath2.net/self_check_quiz

Study Guide and Review

Vocabulary and Concept Check

algebra (p. 18)
algebraic expression (p. 18)
arithmetic sequence (p. 34)
base (p. 10)
coefficient (p. 19)
constant (p. 19)
cubed (p. 10)
defining the variable (p. 25)
equation (p. 24)
equivalent expressions (p. 30)
evaluate (p. 11)
exponent (p. 10)
exponential form (p. 11)
factors (p. 10)
geometric sequence (p. 34)
gram (p. 39)
kilogram (p. 39)
liter (p. 39)
meter (p. 38)
metric system (p. 38)
numerical expression (p. 14)
order of operations (p. 14)
powers (p. 10)
properties (p. 31)
scientific notation (p. 43)
sequence (p. 34)
solution (p. 24)
solving an equation (p. 24)
squared (p. 10)
standard form (p. 11)
term (pp. 19, 34)
variable (p. 18)

Choose the letter of the term that best matches each phrase.

1. a number that makes an equation true
2. numbers expressed using exponents
3. two or more numbers that are multiplied together to form a product
4. rules used to ensure that numerical expressions have only one value
5. tells how many times the base is used as a factor

a. algebra
b. solution
c. order of operations
d. factors
e. powers
f. exponent

Lesson-by-Lesson Exercises and Examples

A Plan for Problem Solving (pp. 6–9)

Use the four-step plan to solve each problem.

6. **CELLULAR PHONES** Sherita's service charges a monthly fee of \$20.00 plus \$0.15 per minute. One monthly bill is \$31.25. How many minutes did Sherita use during the month?
7. **CAR RENTAL** ABC Car Rental charges \$25 per day to rent a mid-sized car plus \$0.20 per mile driven. Mr. Ruiz rents a mid-sized car for 3 days and drives a total of 72 miles. Find the amount of Mr. Ruiz's bill.

Example 1 **Use the four-step plan to solve the problem.**
GARDENING A bag of mulch covers 25 square feet of garden space. Taylor uses 7 bags of mulch to cover her garden. How large is Taylor's garden?

Explore Taylor used 7 bags of mulch, each covering 25 square feet.
Plan Multiply 25 by 7.
Solve $25 \cdot 7 = 175$
Taylor's garden is 175 square feet.
Examine Add 25 seven times.

msmath2.net/vocabulary_review

1-2 Powers and Exponents (pp. 10–13)

Evaluate each expression.

8. 3^5
9. 7^9
10. 2^8
11. 18^2
12. 9^4
13. 10^4
14. 20^3
15. 100^1
16. Write $15 \cdot 15 \cdot 15$ in exponential form.

Example 2 **Evaluate 4^5.**

The base is 4. The exponent 5 means that 4 is used as a factor 5 times.

$4^5 = 4 \cdot 4 \cdot 4 \cdot 4 \cdot 4$

$= 1{,}024$

1-3 Order of Operations (pp. 14–17)

Evaluate each expression.

17. $24 - 8 + 3^2$
18. $48 \div 6 + 2 \cdot 5$
19. $5 \cdot 9 - (2 + 3)^2$
20. $(14 - 4) \div 2$
21. $9 + 3(7 - 5)^3$
22. 12×10^4
23. $42 \div 6 - 63 \div 9$
24. $15 + 9 \div 3 - 7$

Example 3 **Evaluate $24 - (8 \div 4)^4$.**

$24 - (8 \div 4)^4 = 24 - 2^4$ Divide first since $8 \div 4$ is in parentheses.

$= 24 - 16$ Find the value of 2^4.

$= 8$ Subtract.

1-4 Algebra: Variables and Expressions (pp. 18–21)

Evaluate each expression if $a = 10$, $b = 4$, and $c = 8$.

25. $b + 7$
26. $7b$
27. $(a - b)^2$
28. $ab \div c$
29. $3b^2 + c$
30. $\frac{(b + c)^2}{3}$
31. **PRODUCTION** The cost of producing T-shirts is given by the expression $350 + 0.82x$, where x is the number of T-shirts produced. Find the cost of producing 750 T-shirts.

Example 4 **Evaluate $2m^2 - 5n$ if $m = 4$ and $n = 3$.**

$2m^2 - 5n = 2(4)^2 - 5(3)$ Replace m with 4 and n with 3.

$= 2(16) - 5(3)$ Find the value of 4^2.

$= 32 - 15$ Multiply.

$= 17$ Subtract.

1-5 Algebra: Equations (pp. 24–27)

Solve each equation mentally.

32. $h + 9 = 17$
33. $5 = c - 3$
34. $32 - 19 = w$
35. $8a = 56$
36. $31 - y = 8$
37. $\frac{t}{9} = 12$
38. $100 = 20g$
39. $p - 49 = 18$

Example 5 **Solve $14 = 5 + x$ mentally.**

$14 = 5 + x$ Write the equation.

$14 = 5 + 9$ You know that $5 + 9 = 14$.

$14 = 14$ Simplify.

The solution is 9.

1-6 Algebra: Properties (pp. 30–33)

Name the property shown by each statement.

40. $3 + 8 = 8 + 3$

41. $3 \cdot (2 \cdot 7) = (3 \cdot 2) \cdot 7$

42. $6(1 + 4) = 6(1) + 6(4)$

43. $29 \cdot 1 = 29$

Example 6 **Name the property shown by the statement $12 + (8 + 2) = (12 + 8) + 2$.**

This statement shows that the way in which three numbers are grouped when they are added does not change their sum. This is an example of the Associative Property of Addition.

1-7 Sequences (pp. 34–36)

Describe the pattern in each sequence and identify the sequence as *arithmetic, geometric,* or *neither.* Then find the next three terms.

44. 5, 15, 45, 135, …

45. 2.0, 3.2, 4.4, 5.6, …

46. 1, 3, 7, 13, …

47. **CONTESTS** The prizes for first, second, and third places in a spelling bee are $1,000, $500, and $250, respectively. If this pattern continues, what would the fourth, fifth, and sixth prizes be?

Example 7 **Describe the pattern in the sequence 2, 4, 8, 16, 32, …, and identify the sequence as *arithmetic, geometric,* or *neither.* Then find the next three terms.**

This is a geometric sequence created by multiplying the previous term by 2.

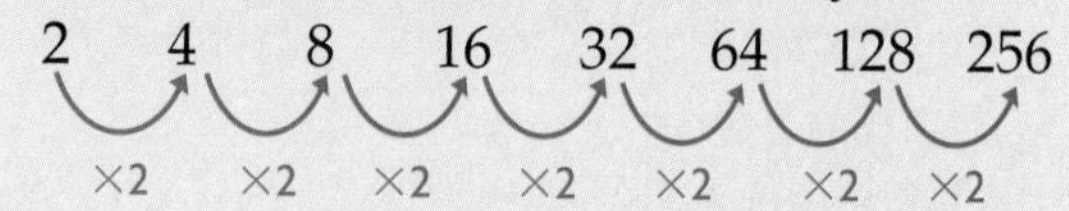

The next three terms are 64, 128, and 256.

1-8 Measurement: The Metric System (pp. 38–41)

Complete.

48. 3.6 km = __?__ m

49. 29 L = __?__ mL

50. 237 mg = __?__ g

51. 7 mL = __?__ L

52. 3,200 cm = __?__ m

Example 8 **Complete.**
6,750 m = __?__ km

Divide 6,750 by 1,000.

$6{,}750 \div 1{,}000 = 6.75$

So, 6,750 meters is equivalent to 6.75 kilometers.

1-9 Scientific Notation (pp. 43–45)

Write each number in scientific notation.

53. 59,000

54. 84,600,000

55. 324,000

56. 10,000,000,000

57. 1,030,000

58. 333,000

59. 9.1 million

60. 6 billion

Example 9 **Write 23,100,000 in scientific notation.**

Move the decimal point 7 places to find a number between 1 and 10.

$23{,}100{,}000 = 2.31 \times 10^7$

Practice Test

Vocabulary and Concepts

1. **List** the order of operations used to find the value of an expression.
2. **Explain** the difference between an arithmetic sequence and a geometric sequence.

Skills and Applications

Evaluate each expression.

3. 3^5
4. 15^4
5. $18 - 3 + 5$
6. $8 + (12 \div 3)^3$

Evaluate each expression if $x = 12$, $y = 5$, and $z = 3$.

7. $7 - x \div z$
8. $yz + 23$
9. $(y - z)^5$
10. $\frac{xz}{y + 13}$

Solve each equation mentally.

11. $9 + m = 16$
12. $d - 14 = 37$
13. $23 = \frac{92}{t}$
14. $6x = 126$

Name the property shown by each statement.

15. $2 + (3 + 7) = (2 + 3) + 7$
16. $5(w + 2) = 5w + 10$

Complete.

17. 5.45 m = __?__ cm
18. 27 mL = __?__ L
19. 8,200 g = __?__ kg

20. **CAPACITY** The table lists the capacity of three different containers. Determine which container has the largest capacity.

Container	Capacity
#1	5400 mL
#2	0.008 kL
#3	4.78 L

21. **ANIMALS** Irene left both her dog and cat at a kennel for 3 nights. The kennel charges \$8 per night for the dog and \$5 per night for the cat. Find Irene's bill.

Write each number in scientific notation.

22. 42,300
23. 3,780,000
24. 502,000,000

Standardized Test Practice

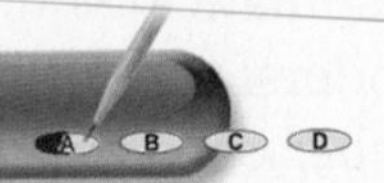

25. Ms. Carter runs a pizza parlor. Her daily costs are \$125 for rent, employee wages, and utilities, plus \$2 for every pizza. If n represents the number of pizzas the pizza parlor makes one day, which expression represents Ms. Carter's total cost for that day?

   (A) $\$125 + 2n$ (B) $125n + \$2$ (C) $\$125n + 2n$ (D) \$127

msmath2.net/chapter_test

CHAPTER 1

Standardized Test Practice

PART 1 Multiple Choice

Record your answers on the answer sheet provided by your teacher or on a sheet of paper.

1. The strategy of guess and check would be *most* useful in solving which problem? (Lesson 1-1)

   (A) About how long does it take to travel 62 miles at 58 miles per hour?

   (B) What is the area of a rectangle if it has a length of 4.5 centimeters and a width of 9 centimeters?

   (C) It costs $3 to cross the town bridge to enter town, but it is free of charge to cross it to leave town. How much does a person who makes 150 trips across the bridge into town and 50 trips across the bridge to leave town spend in bridge tolls?

   (D) The product of two consecutive even numbers is 728. Find the two numbers.

2. Ciro earns $8 per hour plus tips cleaning rooms at Sandman's Inn. Which expression represents Ciro's earnings for a week in which he works 36 hours and collects $25 in tips? (Lesson 1-3)

   (F) $(36 \times 8) + 25$

   (G) $(36 + 25) \times 8$

   (H) $(8 \times 25) + 36$

   (I) $(36 \div 8) + 25$

3. To estimate the number of miles a thunderstorm is from you, count the number of seconds between the thunder and the lightning. Then divide the number of seconds by 5. Which expression can be used to estimate the distance? (Lesson 1-4)

   (A) $5 \div s$ (B) $s - 5$

   (C) $5s$ (D) $s \div 5$

4. Which value of p makes $64 \div p = 8$ a true sentence? (Lesson 1-5)

   (F) 2 (G) 2^2

   (H) 2^3 (I) 3^2

5. Which is the graph of the solution of $x - 15 = 33$? (Lesson 1-5)

   (A) 46 47 48 49 50

   (B) 13 14 15 16 17

   (C) 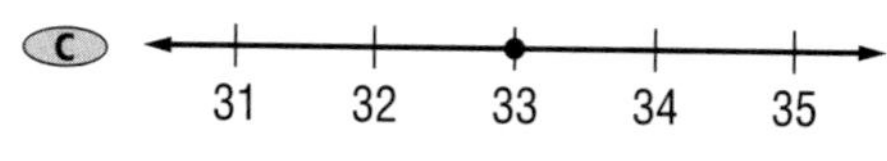

   (D) 16 17 18 19 20

6. Identify the property shown by the sentence $16 + (4 + 7) = (16 + 4) + 7$. (Lesson 1-6)

   (F) Commutative Property of Addition

   (G) Associative Property of Addition

   (H) Associative Property of Multiplication

   (I) Identity Property of Multiplication

7. Which is the next term of the sequence 6, 12, 18, 24, ... ? (Lesson 1-7)

   (A) 28 (B) 30

   (C) 36 (D) 48

8. Every morning, Morgan swims 0.5 kilometer. How many meters does she swim? (Lesson 1-8)

   (F) 5 m (G) 50 m

   (H) 100 m (I) 500 m

9. Which is 5,300,000 written in scientific notation? (Lesson 1-9)

   (A) 0.53×10^6 (B) 53×10^5

   (C) 5.3×10^6 (D) 5.3×10^7

PART 2 Short Response/Grid In

Record your answers on the answer sheet provided by your teacher or on a sheet of paper.

10. What value is represented by $(1 \times 10^4) + (7 \times 10^3) + (3 \times 10^2) + (0 \times 10^1)$? (Prerequisite Skill, p. 561)

11. The graph shows the number of World Series Championships for selected teams from 1903–2001. How many total championships have been won by these teams? (Lesson 1-1)

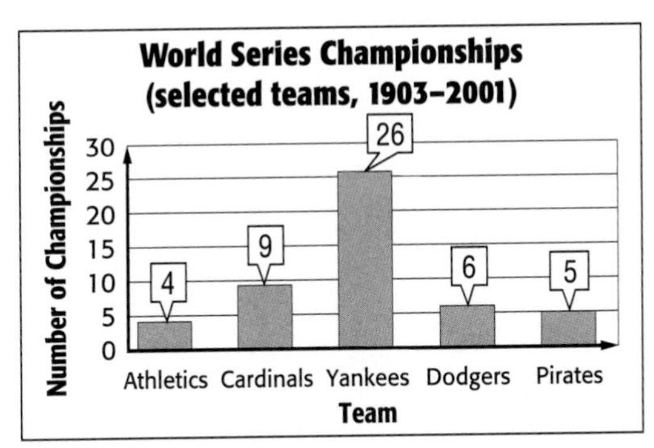

Source: World Almanac, 2001

12. Write x^6 as a product of the same factor. (Lesson 1-2)

13. What is the value of $2^2 + 4^3$? (Lesson 1-3)

14. Tyler works at the library 3 times as many hours as Lugo and Lydia combined. Write an algebraic expression that represents the number of hours Tyler works per week if h represents the number of hours Lugo and Lydia work combined. (Lesson 1-4)

15. The town of Holden's water tank can hold 10 kiloliters. The town of Jasper's water tank can hold half the amount of Holden's tank. How many liters of water can Jasper's water tank hold? (Lesson 1-8)

TEST-TAKING TIP

Question 13 Fill in the answer bubbles below each numeral completely on the answer grid-in sheet. That way your answer will be clear.

16. Four countries and their approximate populations in 2001 are shown in the table. Which country has the least population? (Lesson 1-9)

Country	Population, 2001
Honduras	6.406×10^6
Australia	1.936×10^7
Croatia	4.334×10^6
Denmark	5.353×10^6

Source: www.factmonster.com

PART 3 Extended Response

Record your answers on a sheet of paper. Show your work.

17. Figures 1, 2, and 3 form a pattern made of equal-sized squares. Each side of the square in Figure 1 is 2 units in length. (Lesson 1-7)

1 2 3

a. Extend the pattern to find the perimeters of Figures 3–8.

Figure	Perimeter
1	8
2	12
3	?
4	?
5	?
6	?
7	?
8	?

b. Describe the sequence of perimeters and identify the sequence as *arithmetic*, *geometric*, or *neither*.

c. The perimeter of a term above is given by the expression $4f + 4$, where f is the figure number in the pattern. What is the perimeter of figure 26?

18. Marlene made her special punch by mixing 4 liters of ginger ale and 2,650 milliliters of cranberry juice. (Lesson 1-8)

a. How many liters are in the punch bowl when the punch is complete?

b. About how many 200-mL servings of punch can be served?

Statistics: Analyzing Data

"What do bridges have to do with math?"

Twenty-two of the longest suspension bridges in North America are 1,500 feet to 4,260 feet long. You can use a box-and-whisker plot to summarize the data about bridges. In mathematics, you will use tables, graphs, and diagrams to organize lists of numbers that describe real-life events.

You will solve problems by organizing data in box-and-whisker plots in Lesson 2-6.

GETTING STARTED

Take this quiz to see whether you are ready to begin Chapter 2. Refer to the page number in parentheses if you need more review.

Vocabulary Review

Choose the correct term to complete each sentence.

1. 173.02 and 37.125 have the same digit in the (hundreds, hundredths, tenths) place. (Page 555)
2. The number 0.02 is (less than, equal to, greater than) 0.002. (Page 556)
3. The numbers 10.88, 10.089, and 1.08 are in order from (least to greatest, greatest to least). (Page 556)

Prerequisite Skills

Name the place value of each underlined digit. (Page 555)

4. $7.\underline{9}5$
5. $\underline{2},064$
6. $48.0\underline{5}$
7. $16\underline{3}.3$

Order each set of decimals from least to greatest. (Page 556)

8. 96.2, 96.02, 95.89
9. 5.61, 5.062, 5.16
10. 22.02, 22, 22.012
11. 3.98, 39.8, 0.398

Order each set of decimals from greatest to least. (Page 556)

12. 74.65, 74.67, 74.7
13. 1.26, 1.026, 10.26
14. 3.304, 3.04, 3.340
15. 68.9, 69.79, 68.99

Add or divide. (Pages 559 and 562)

16. 88.21 + 4.75 + 62.4 + 39.60
17. 215 ÷ 2.5
18. 5.9 + 28.1 + 9.2
19. 333 ÷ 0.6
20. 51 ÷ 1.7

Statistics Make this Foldable to help you organize information about analyzing data. Begin with eight sheets of notebook paper.

STEP 1 Fold
Fold 8 sheets of paper in half along the width.

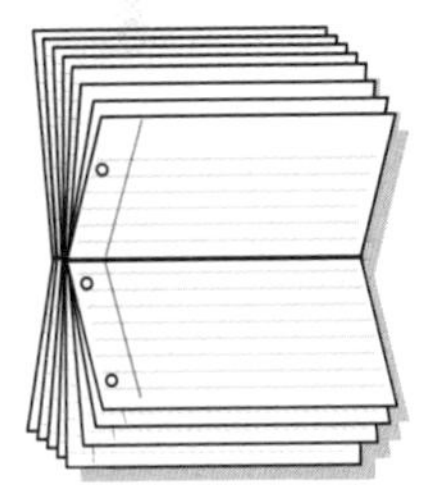

STEP 2 Cut
Cut a 1″ tab along the left edge through one thickness.

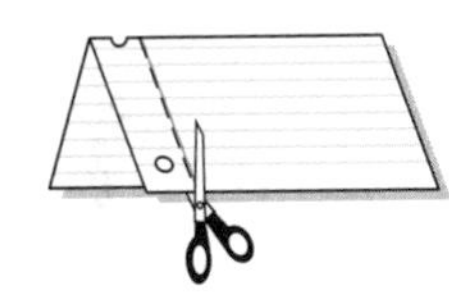

STEP 3 Glue and Label
Glue the 1″ tab down. Write the lesson number and title on the front tab.

STEP 4 Repeat and Staple
Repeat Steps 2 and 3 for the remaining sheets. Staple them together on the glued tabs to form a booklet.

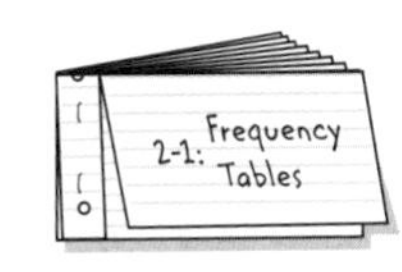

Reading and Writing As you read and study the chapter, fill the booklet with notes and examples of statistical graphs.

2-1 Frequency Tables

What You'll LEARN

Organize and interpret data in a frequency table.

NEW Vocabulary

statistics
data
frequency table
scale
interval

WHEN am I ever going to use this?

SCOOTERS Riding a scooter is a fun way to get exercise. The table shows prices for 35 types of nonmotorized scooters found at an online store.

1. What is the cost of the least expensive scooter? the most expensive?
2. How many scooters cost $51 to $75?
3. How could you reorganize the prices so that they are easier to find and read?

Prices of Scooters ($)				
60	25	29	25	40
35	50	80	30	70
80	90	80	55	70
40	100	52	45	60
99	60	100	89	11
99	100	49	35	90
92	20	49	80	50

Statistics deals with collecting, organizing, and interpreting data. **Data** are pieces of information, which are often numerical. Large amounts of data can be organized in a frequency table. A **frequency table** shows the number of pieces of data that fall within given intervals.

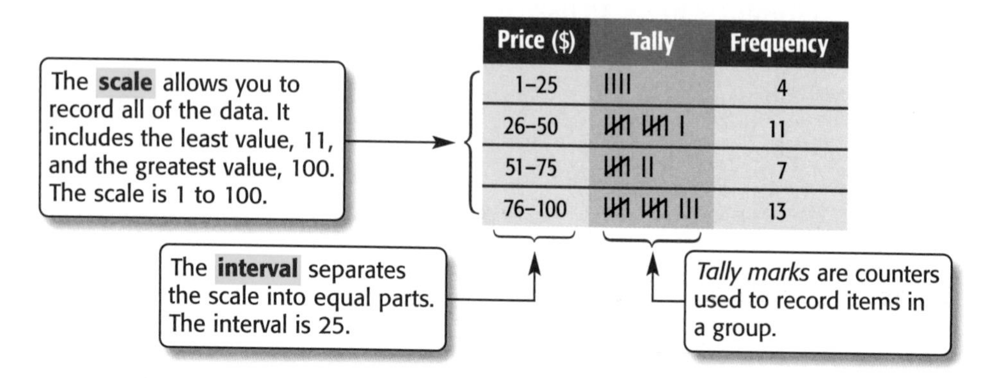

Price ($)	Tally	Frequency
1–25	IIII	4
26–50	~~IIII~~ ~~IIII~~ I	11
51–75	~~IIII~~ II	7
76–100	~~IIII~~ ~~IIII~~ III	13

From the frequency table, you can see that just over half the scooters cost between $51 and $100.

EXAMPLE Make a Frequency Table

1 RACING ***Cyclo-cross*** **is cycle racing that takes place on short off-road courses. The table shows the total points earned by the top 20 racers in the cyclo-cross World Cup events. Make a frequency table of the data.**

Cyclo-Cross Points			
310	179	116	91
225	173	107	76
223	150	105	55
218	149	100	52
190	123	100	50

Source: www.cyclo-cross.com

Step 1 Choose an appropriate interval and scale for the data. The scale should include the least value, 50, and the greatest value, 310.

interval: 50
scale: 50 to 349

} The scale includes all of the data, and the interval separates it into equal parts.

Step 2 Draw a table with three columns and label the columns *Points, Tally,* and *Frequency.*

Step 3 Complete the table.

Points	Tally	Frequency
50–99	𝍸	5
100–149	𝍸 II	7
150–199	IIII	4
200–249	III	3
250–299		0
300–349	I	1

Frequency tables can also have categories without scales and intervals.

EXAMPLE Make and Use a Frequency Table

2 OLYMPICS Enrique asked his classmates to name their favorite winter Olympic event. The results are shown at the right. Make a frequency table of the data. Then determine the overall favorite event.

Favorite Winter Olympic Events					
S	F	F	S	B	F
F	S	H	J	B	S
J	F	J	F	H	F
F	B	S	F	S	H
S	B	H	F	J	J

B = bobsleigh, F = figure skating, H = hockey, J = ski jumping, S = snowboarding

Draw a table with three columns. In the first column, list the events. Then complete the rest of the table.

Event	Tally	Frequency
bobsleigh	IIII	4
figure skating	𝍸 𝍸	10
hockey	IIII	4
ski jumping	𝍸	5
snowboarding	𝍸 II	7

Figure skating, with 10 tallies, was the favorite event. Snowboarding was the second favorite, with 7 tallies.

REAL-LIFE MATH

OLYMPICS In 2002, Vonetta Flowers became the first African-American to win a gold medal in the Winter Olympics. She and Jill Bakken won the first ever 2-woman bobsleigh event with a two-mile time of 1 minute 37.76 seconds.

Source: www.olympic-usa.org

Frequency tables are useful for interpreting data.

EXAMPLE Interpret Data

3 HISTORY The frequency table shows the reigns of the last forty rulers of England. How many of them ruled England for more than 30 years?

There are three categories with more than 30 years.

31–45 years = 6 rulers

46–60 years = 4 rulers

61–75 years = 1 ruler

So, 6 + 4 + 1 or 11 of the last forty rulers each ruled England for more than 30 years.

Rulers of England		
Reign (years)	Tally	Frequency
1–15	𝍸 𝍸 𝍸 III	18
16–30	𝍸 𝍸 I	11
31–45	𝍸 I	6
46–60	IIII	4
61–75	I	1

Source: www.scotlandroyalty.org/kings.html

Skill and Concept Check

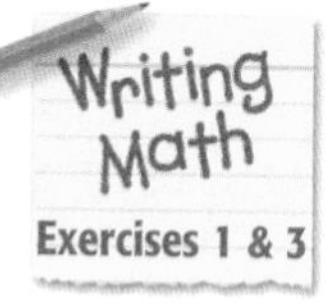

1. **Describe** an advantage and a disadvantage of using a frequency table to display data.

2. **OPEN ENDED** Make a frequency table that displays thirty data items and has an interval of 5.

3. **DATA SENSE** Explain why the frequency table in Example 1 does not have 300–310 as its last category when the greatest data value is 310.

GUIDED PRACTICE

For Exercises 4 and 5, use the table below.

Number of Movie Rentals During the Month							
10	9	2	17	12	9	11	10
4	14	3	15	0	6	5	9
20	10	11	8	3	1	13	4

4. Choose an appropriate interval and scale to make a frequency table.
5. Make a frequency table of the data.

For Exercises 6 and 7, use the table below.

Favorite Breakfast Foods					
e	c	e	p	c	c
c	p	p	e	c	e
c	e	c	p	p	e

c = cereal, e = eggs, p = pancakes

6. Make a frequency table of the data.
7. What is the favorite breakfast food?

Practice and Applications

HOMEWORK HELP	
For Exercises	See Examples
8–12	1
13–18	2, 3

Extra Practice
See pages 567, 597.

Make a frequency table of each set of data.

8.

Ages of Students in Ms. Warren's Class				
12	13	13	14	14
14	14	13	13	14
13	14	13	13	12
14	13	14	12	13

9.

Years of Major Hurricanes				
1900	1975	1985	1995	1996
1938	1944	1957	1980	1992
1961	1935	1989	1972	1928
1955	1969	1979	1915	1964
1970	1960	1954	1926	1967
1999	1998			

Source: *The World Almanac*

10.

How Do You Listen to the Most Music?					
I	C	M	I	R	M
R	M	I	M	C	R
R	C	C	R	I	C
M	C	C	R	C	M
R	I	R	R	R	C

C = CDs, R = radio, I = Internet, M = music videos

11.

Which Chore Do Students Like Least?						
D	L	Y	L	V	R	L
D	L	R	D	R	D	Y
V	D	V	R	R	Y	R
R	R	Y	L	D	V	R
D	R	V	R	R	D	L

D = wash dishes, V = vacuum, R = clean room, L = do laundry, Y = do yard work

12. **Determine** whether the categories 0–100, 100–200, and 200–300 could be used in a frequency table displaying data. Explain.

MOVIES For Exercises 13–15, use the table at the right. It shows when the 20 top-grossing movies of all time were released as of 2002.

Years of Top-Grossing Movies			
1997	1993	1983	1990
1977	1994	1996	2001
1982	2001	2002	1975
1999	1994	1999	2000
2002	2001	1980	2001

Source: www.movieweb.com

13. Make a frequency table of the data. Use the intervals 1970–1979, 1980–1989, 1990–1999, and 2000–2009.

14. In which years were the greatest number of top-grossing movies released?

15. **RESEARCH** Use the Internet or another source to find the current 20 top-grossing movies. Make a frequency table of the data and compare it to the table you made in Exercise 13.

CEREAL For Exercises 16–18, use the table at the right. It shows the number of Calories contained in one serving of students' favorite cereals.

Number of Calories per Serving of Dry Cereal				
70	120	100	100	140
70	120	100	100	100
110	110	110	120	100
110	100	100	110	150
90	110	120	110	100
140	130	100	80	100
100	110	80	90	110

16. Make a frequency table of the data.

17. How many Calories are in one serving of most of the cereals?

18. If a student eats two bowls of his favorite cereal, what is the greatest and least number of Calories possible that he consumes?

19. **CRITICAL THINKING** Make a frequency table of the scooter data at the beginning of this lesson using a different scale and interval. Analyze the data in the new table and describe an advantage and a disadvantage of using this scale and interval.

Standardized Test Practice and Mixed Review

20. **MULTIPLE CHOICE** The following are scores of the Duke University women's basketball team in the 2001–2002 conference games: 107, 76, 90, 91, 81, 73, 86, 102, 102, 86, 66, 83, 77, 81, 88, 90. Choose the best scale for making a frequency table of the scores.

    (A) 66–105 (B) 68–102 (C) 66–110 (D) 70–108

21. **MULTIPLE CHOICE** If there are 24 students in math class, how many received a score on the exam in the interval 81–90?

    (F) 7 (G) 6 (H) 5 (I) 4

Score	Frequency
71–80	8
81–90	?
91–100	9

Write each number in standard form. (Lesson 1-9)

22. 1.8×10^3
23. 5.33×10^5
24. 2.4×10^4
25. 6.0×10^7

26. **MEASUREMENT** How many centimeters are in 8.05 meters? (Lesson 1-8)

GETTING READY FOR THE NEXT LESSON

PREREQUISITE SKILL

27. At 5:00 P.M. the outside temperature was 81°F; at 6:00 P.M. it was 80°F; at 7:00 P.M. it was 79°F. If the pattern continues, predict the outside temperature at 8:00 P.M. (Lesson 1-7)

2-2a Problem-Solving Strategy

A Preview of Lesson 2-2

What You'll Learn
Solve problems by using graphs.

Use a Graph

How many home runs do you think the Pittsburgh Pirates will get in 2004?

By just looking at the table, I can't really tell. Let's **use a graph** to get a better picture.

Pittsburgh Pirates Home Runs												
Year	'91	'92	'93	'94	'95	'96	'97	'98	'99	'00	'01	'02
Home Runs	126	106	110	80	125	138	129	107	171	168	161	142

Source: www.baseball-almanac.com

Explore	We know the number of home runs from 1991 to 2002. We need to predict how many there will be in 2004.
Plan	If we organize the information in a graph, we may be able to see if there are any trends in the number of home runs.
Solve	We can see from the graph that generally, as the years progress, the number of home runs increases. We can predict that the number of home runs in 2004 will be about 190.
Examine	Draw a line through the points that is close to as many of the points as possible, as shown above. The estimate is close to the line, so the prediction is reasonable.

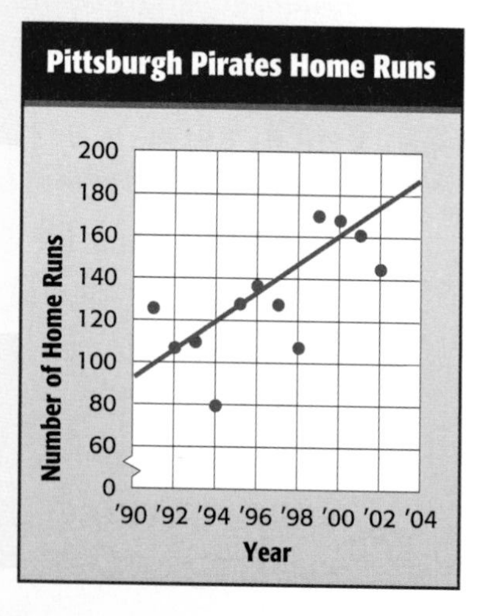

Analyze the Strategy

1. **Explain** why analyzing a graph is a useful way to quickly make conclusions about a set of data.
2. The last step of the four-step plan asks you to *examine* the solution. **Write** a problem in which using a graph would be a useful way to examine the solution.

Apply the Strategy

Solve. Use a graph.

EARTH SCIENCE For Exercises 3 and 4, refer to the graph at the right. It shows the average high and low temperatures each month from January to June in Memphis, Tennessee. Suppose the trends continue through July.

3. Predict the average high temperature for the month of July.
4. Predict the average low temperature for the month of July.

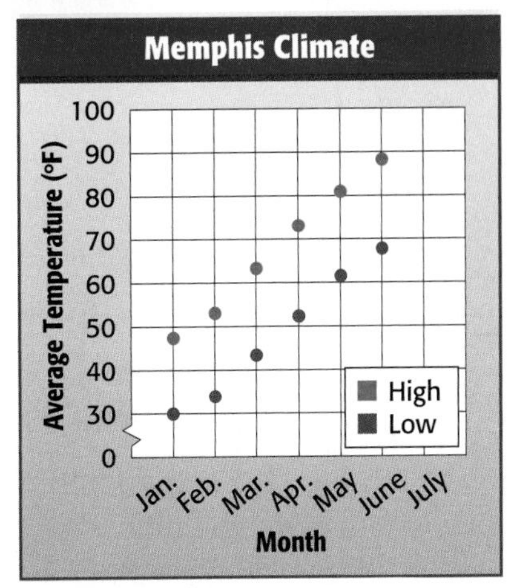

Source: www.weather.com

Mixed Problem Solving

Solve. Use any strategy.

5. **READING** Maya read 10 pages of a 150-page book on Monday and plans to read twice as many pages each day than she did the previous day. On what day will she finish the book?

6. **SCHOOL** The graph shows the number of energy bars sold in a school's cafeteria during one week.

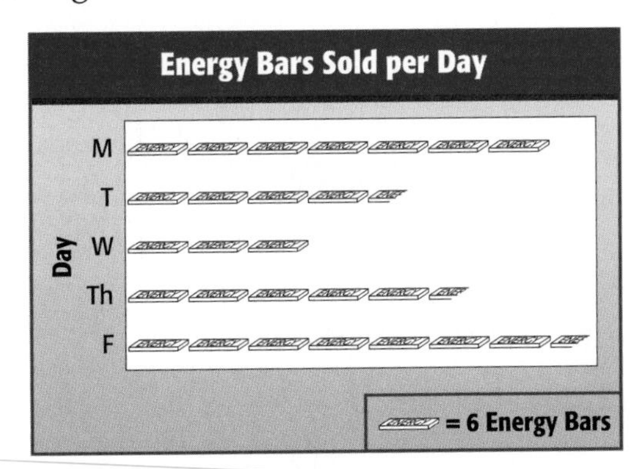

According to the graph, were more energy bars sold on Monday or Tuesday?

7. **SAFETY** An elevator sign reads *Do not exceed 2,500 pounds.* How many people, each weighing about 150 pounds, can be in the elevator at the same time?

8. **NUMBER THEORY** Find two numbers whose sum is 56 and whose product is 783.

9. **BOWLING** Tariq and three of his friends are going bowling, and they have a total of $70 to spend. Suppose they buy a large pizza, four drinks, and each rent bowling shoes. How many games can they bowl if they all bowl the same number of games?

Item	Price
large pizza	$13.75
beverage	$1.50
shoe rental	$3.50
game	$4.00

10. **STANDARDIZED TEST PRACTICE** The following sequence forms a pattern.

4, 8, 12, 16, …

If this pattern continues, which expression can be used to find the *sixth* term?

(A) $6 + 4$ (B) $6 - 4$

(C) $6 \div 4$ (D) 6×4

You will use a graph to make predictions in the next lesson.

2-2 Making Predictions

HANDS-ON Mini Lab

What You'll LEARN

Make predictions from graphs.

NEW Vocabulary

line graph
scatter plot

Materials

- measuring cup
- drinking glass
- metric ruler
- marbles
- grid paper

Work with a partner.

- Pour 1 cup of water into the drinking glass.
- Measure the height of the water and record it in a table like the one below.
- Place five marbles in the glass. Measure the height of the water and record it in your table.
- Continue adding marbles, five at a time, until there are twenty marbles in the glass. After each time, measure and record the height of the water.

Number of Marbles	Height of Water (cm)
0	
5	
10	
15	
20	

1. By how much did the height of the water change after each addition of five marbles?
2. Predict the height of the water when thirty marbles are in the drinking glass. Explain how you made your prediction.
3. Test your prediction by placing ten more marbles in the glass.
4. Draw a graph of the data that you recorded in the table.

You created a line graph in the Mini Lab. **Line graphs** can be useful in predicting future events when they show trends over time.

EXAMPLE Use a Line Graph to Predict

1 STATUE OF LIBERTY The table shows the time it takes Kyle to climb the Statue of Liberty. Make a line graph and predict the total time it will take him to climb 354 steps to the top.

Time (min)	0	2	4	6	8	10
Number of Steps	0	51	114	163	210	275

Climbing the Statue of Liberty

The data values go from 0 to 275. You want to predict the time when 354 steps are climbed. So, make the scale go from 0 to 400 with an interval of 100.

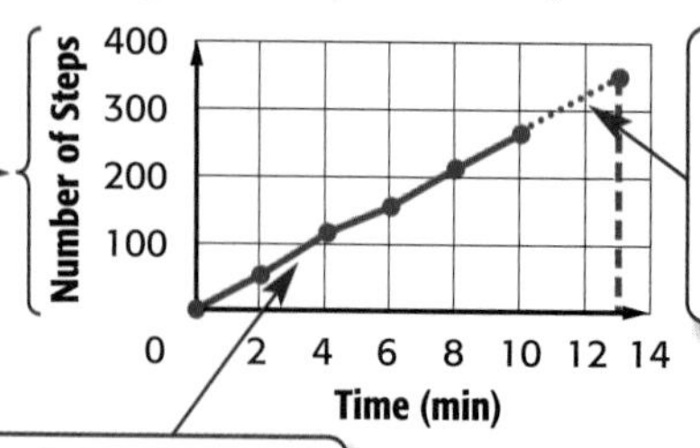

Continue the graph with a dotted line in the same direction until you reach a horizontal position of 354 steps.

Graph the data and connect the points.

It will take Kyle about 13 minutes to climb 354 steps.

READING Math

Scatter Plots Another name for scatter plot is *scattergram*.

A **scatter plot** displays two sets of data on the same graph. Like line graphs, scatter plots are useful for making predictions because they show trends in data. If the points on a scatter plot come close to lying on a straight line, the two sets of data are related.

EXAMPLE Use a Scatter Plot to Predict

2 SWIMMING The scatter plot shows the winning times for the women's 100-meter butterfly event at the Summer Olympics from 1968 to 2000. Use it to predict a winning time for this event at the 2012 Olympics.

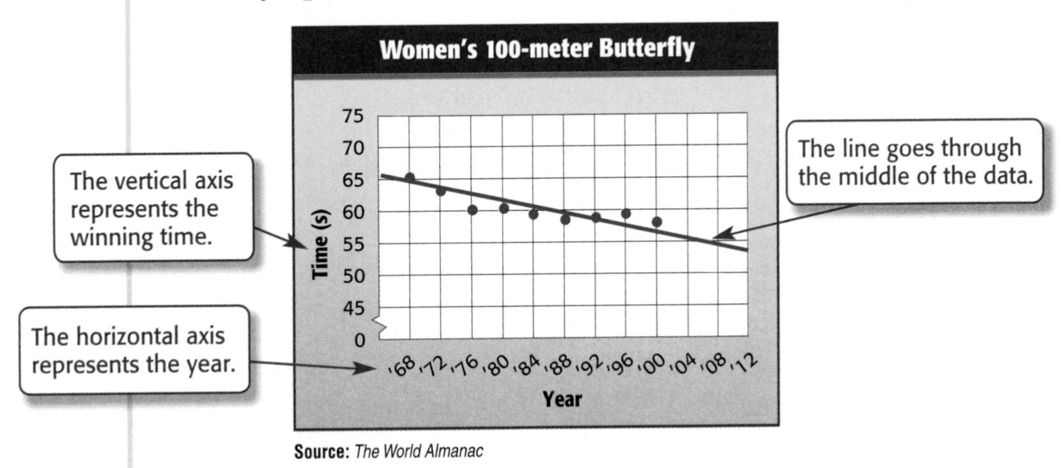

Source: *The World Almanac*

By looking at the pattern in the graph, we can predict that the winning time at the 2012 Olympics will be about 54 seconds.

Skill and Concept Check

Writing Math Exercises 1 & 3

1. **Explain** how a graph can be used to make predictions.
2. **OPEN ENDED** Name two sets of data that can be graphed on a scatter plot.
3. **Which One Doesn't Belong?** Identify the term that does not have the same characteristic as the other three. Explain your reasoning.

line graph	tally mark	frequency table	scatter plot

GUIDED PRACTICE

4. **SNACKS** The graph shows the sales, in billions of dollars, of salty snacks like potato chips and pretzels for several years. If the trend continues, about what will the sales be in 2006?

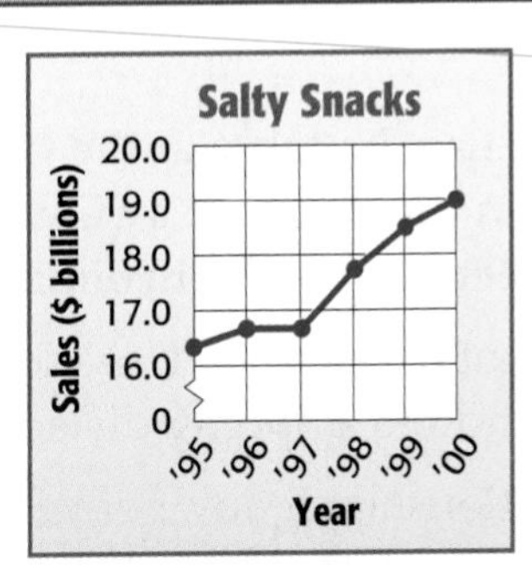

Source: *American Demographics*

Practice and Applications

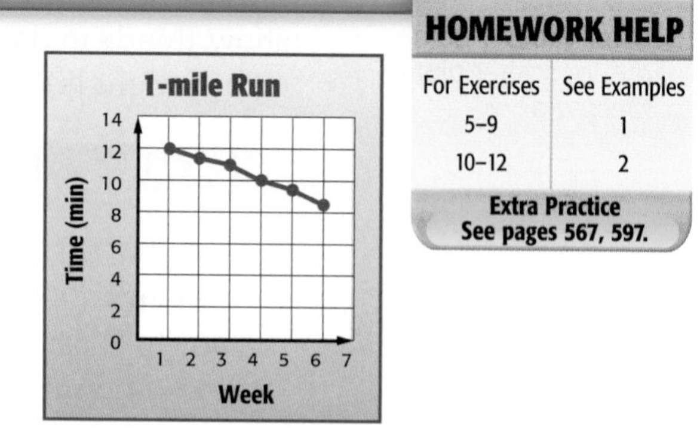

HOMEWORK HELP

For Exercises	See Examples
5–9	1
10–12	2

Extra Practice
See pages 567, 597.

5. **RUNNING** Mahkah has been training for several weeks for cross-country tryouts. His progress is shown in the graph at the right. To make the team, he must be able to run 1 mile in less than 8 minutes. If the trend continues, will Mahkah make the team? Explain.

6. **SCHOOLS** The line graph shows public school teachers' salaries for the past few years. If the trend continues, about what will the average annual salary be in 2005?

7. **RESEARCH** Use the Internet or another source to find a real-life example of a line graph. Write a description of what the graph displays and extend the graph to show where the data will be in the future.

USA TODAY

USA TODAY Snapshots®

Pay rises for public school teachers

How the average annual salaries of public school classroom teachers have risen nationally:

$44,047

$23,600

'85 '89 '93 '97 '01

Source: National Center for Education Statistics (The 2001 average is an estimate.)

By Lori Joseph and Adrienne Lewis, USA TODAY

OLYMPICS For Exercises 8 and 9, use the table below. It shows the number of nations that participated in the Summer Olympics from 1948 to 2000.

Year	Number of Nations	Year	Number of Nations
1948	59	1976	92
1952	69	1980	80
1956	72	1984	140
1960	83	1988	159
1964	93	1992	169
1968	112	1996	197
1972	121	2000	199

Source: www.factmonster.com

8. Make a line graph of the data.

9. Predict the number of nations that will participate in the 2008 Summer Olympics.

For Exercises 10–12, use the table at the right. It shows the relationship between the speed of a vehicle and the distance required to stop in certain weather conditions.

Speed (mph)	Stopping Distance (ft)
55	273
65	355
75	447

10. Make a scatter plot of the data. Use the speed on the horizontal axis and the stopping distance on the vertical axis.

11. Describe the relationship, if any, between the two sets of data.

12. Predict the stopping distance for a car traveling 45 miles per hour.

13. **CRITICAL THINKING** What can you conclude about the relationship between students' birth month and height shown in the scatter plot at the right?

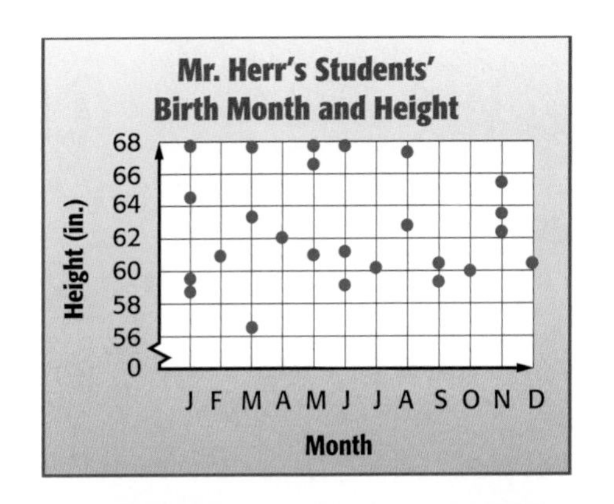

14. **EXTENDING THE LESSON** The graph at the right is called a *multiple-line graph* because it contains more than one line graph.

The multiple-line graph shows the population of Miami-Dade and Broward Counties in Florida from 1950 to 2000. Do you think that the population of Broward County will catch up to the population of Miami-Dade County in the next census of 2010? Explain.

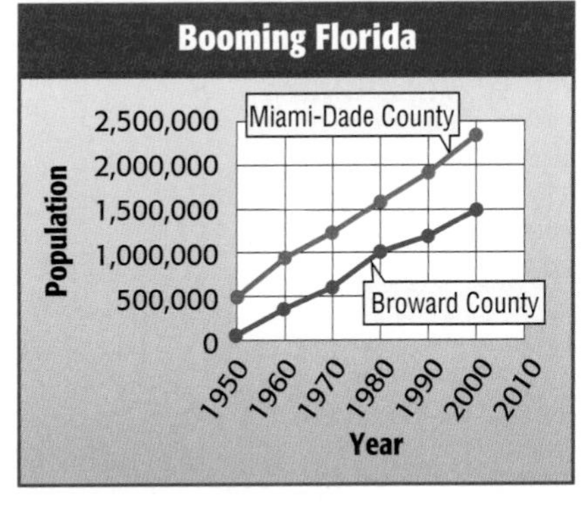

Source: U.S. Census Bureau

Standardized Test Practice and Mixed Review

For Exercises 15 and 16, refer to the graph at the right that shows the number of customers at Clothes Palace at different times during the day.

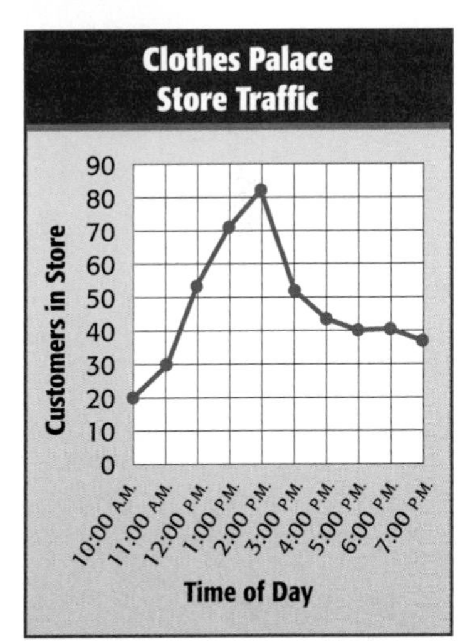

15. **MULTIPLE CHOICE** Extra workers are needed when the number of customers in the store exceeds 50. Between which hours is extra help needed?

(A) 12:00 P.M.–3:00 P.M. (B) 11:00 A.M.–4:00 P.M.

(C) 1:00 P.M.–4:00 P.M. (D) 1:00 P.M.–3:00 P.M.

16. **SHORT RESPONSE** Predict about how many customers are in the store at 8:00 P.M.

17. **STATISTICS** Choose an appropriate interval and scale for the following set of data: 35, 42, 18, 25, 32, 47, 34. (Lesson 2-1)

18. Write 72,000 in scientific notation. (Lesson 1-9)

Write each product in exponential form. (Lesson 1-2)

19. $9 \cdot 9$ 20. $14 \cdot 14 \cdot 14 \cdot 14$ 21. $7 \cdot 7 \cdot 7$ 22. $5 \cdot 5 \cdot 5 \cdot 5 \cdot 5$

GETTING READY FOR THE NEXT LESSON

PREREQUISITE SKILL The following data are the heights of the players, in feet, on the Miami Heat basketball team: 6.8, 7.0, 6.2, 6.7, 6.8, 6.4, 6.7, 6.1, 6.5, 6.2, 6.5, 6.8, 6.8, 7.0, 6.2. (Page 556)

23. Order the heights from least to greatest.

2-3 Line Plots

What You'll LEARN

Construct and interpret line plots.

NEW Vocabulary

line plot
cluster
outlier
range

Link to READING

Everyday Meaning of Range: in music, all the notes between the high and low notes, as in a singer with a wide range

WHEN am I ever going to use this?

BUILDINGS The table shows the number of stories in the 32 tallest buildings in Los Angeles.

1. Do any of the values seem much greater or much less than the other data values?
2. Do some of the buildings have the same number of stories? Is this easy to see? Explain.
3. What better ways can you suggest for organizing these data?

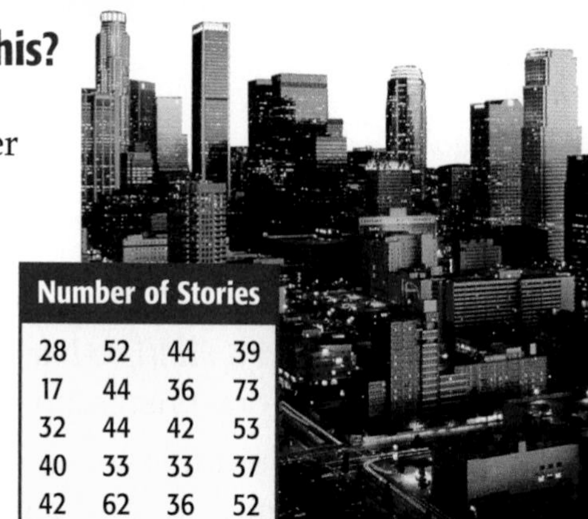

Number of Stories			
28	52	44	39
17	44	36	73
32	44	42	53
40	33	33	37
42	62	36	52
34	55	34	28
31	48	41	52
52	54	52	34

Source: *The World Almanac*

One way to show how data are spread out is to use a line plot. A **line plot** is a diagram that shows the frequency of data on a number line. An "×" is placed above a number on a number line each time that data value occurs.

EXAMPLE Make a Line Plot

1 **BUILDINGS** **Make a line plot of the data shown above.**

Step 1 Draw a number line. Because the shortest building in the table has 17 stories, and the tallest has 73, you can use a scale of 15 to 75 and an interval of 5. Other scales and intervals could also be used.

Step 2 Put an  above the number that represents the number of stories in each building.

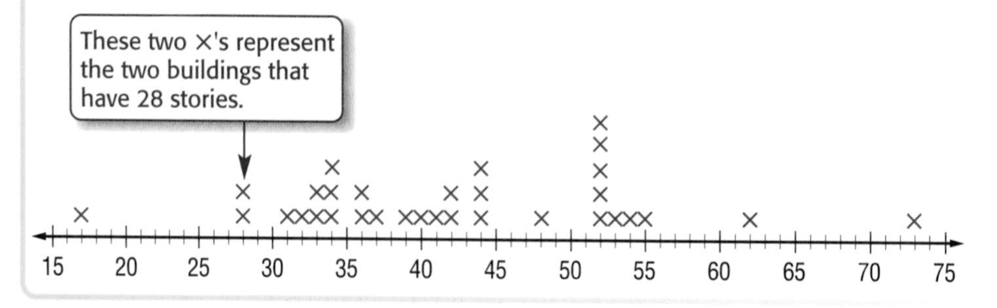

You can see on the line plot how many buildings have the same number of stories. This information is not easily found using the table above.

You can make some observations about the *distribution* of data, or how data are grouped together or spread out, using the line plot in Example 1.

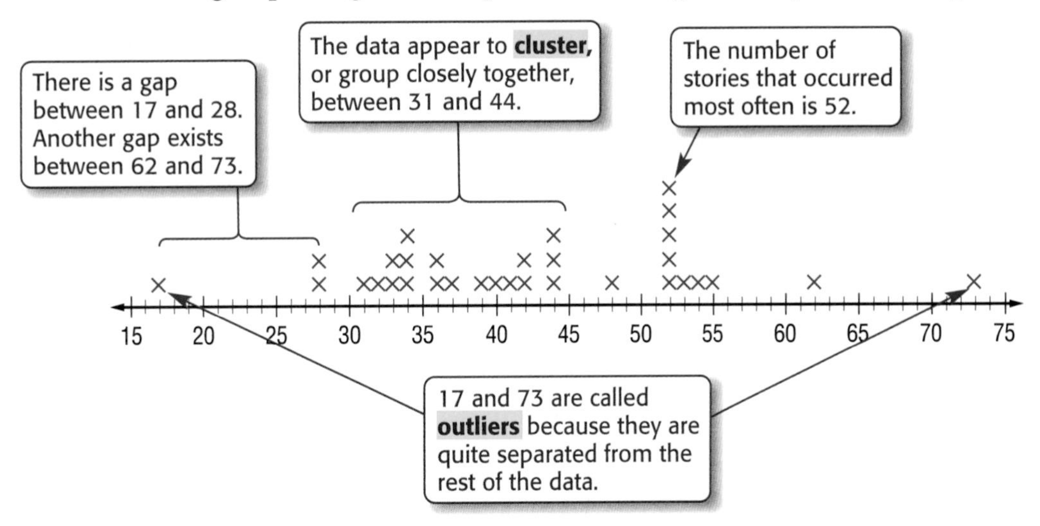

In a line plot, you can also easily find the **range** of the data, or the difference between the greatest and least numbers in the data set. This value is helpful in seeing how spread out the data are.

EXAMPLES Use a Line Plot to Analyze Data

2 SHOES The line plot below shows prices for different basketball shoes. What is the range of the data?

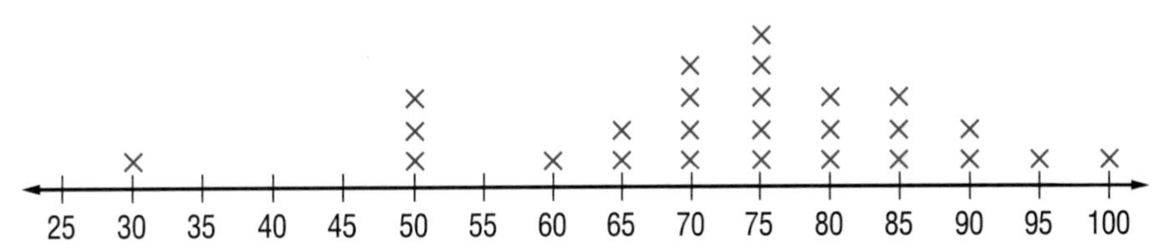

The greatest price is \$100, and the lowest price is \$30.

range: $100 - 30$ or 70

The range of the prices is \$70.

3 Identify any clusters, gaps, and outliers in the data in Example 2 and explain what they mean.

Many of the data cluster around \$75. You could say that most of the shoes cost from \$70 to \$85.

There is a gap from \$30 to \$50, so there were no shoes in this price range.

The number 30 appears removed from the rest of the data, so it could be considered an outlier. This means that \$30 is an extremely low price and not representative of the whole data set.

STUDY TIP

Clusters You can describe a cluster by stating a range of values or by giving a single value around which the data appear to be grouped.

Your Turn

a. Identify the range of the data in Example 1.

Skill and Concept Check

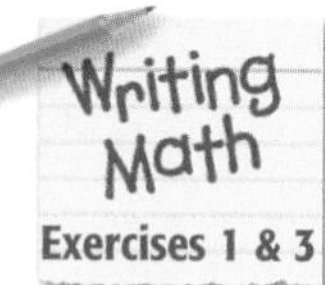

1. **List** the information you can identify from a line plot that is not obvious in an unorganized list of data.

2. **OPEN ENDED** Draw a line plot that displays fifteen pieces of data. Arrange the data so that there is a cluster and one outlier.

3. **FIND THE ERROR** Ryan and Darnell are analyzing the data shown in the line plot at the right. Who is correct? Explain.

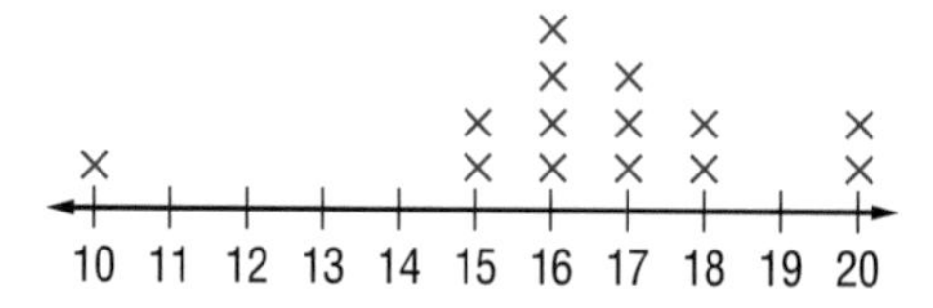

Ryan	Darnell
greatest data value: 16	greatest data value: 20
least data value: 10	least data value: 10

4. **DATA SENSE** Refer to the line plot in Exercise 3. Find the range and identify any clusters, gaps, or outliers.

GUIDED PRACTICE

For Exercises 5–7, use the table at the right. It shows the number of music CDs owned by students in a classroom.

5. Make a line plot of the data.
6. Identify any clusters, gaps, or outliers.
7. Summarize the data in a sentence or two.

Number of Music CDs				
12	9	11	8	12
0	7	10	20	11
6	4	10	10	8
10	9	8	11	6

WEATHER For Exercises 8–11, use the line plot below. It shows the record high temperatures recorded by weather stations in each of the fifty states.

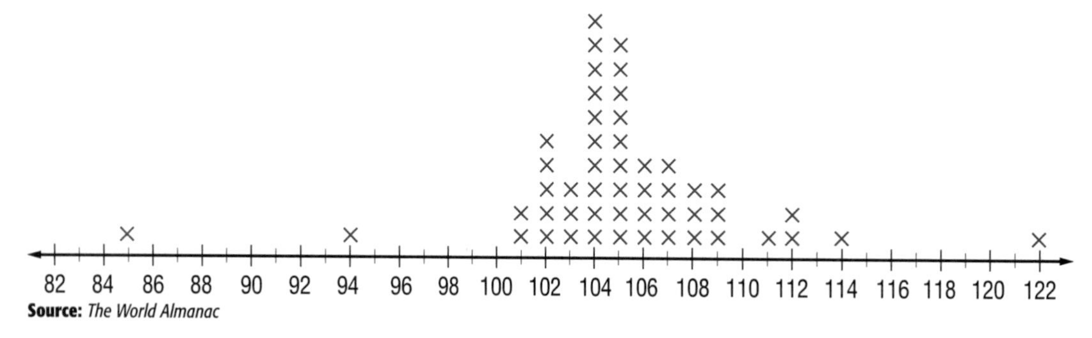

Source: *The World Almanac*

8. What is the range of the data?
9. What temperature occurred most often?
10. Identify any clusters, gaps, or outliers and analyze the data by describing what these values represent.
11. Describe how the range of the data would change if 122 were not part of the data set.

Data Update What are the record high temperatures for some of the cities in your state? Visit msmath2.net/data_update to learn more.

Practice and Applications

HOMEWORK HELP

For Exercises	See Examples
12–13, 19–22	1–3
14–18	2, 3

Extra Practice
See pages 567, 597.

Make a line plot for each set of data. Identify any clusters, gaps, or outliers.

12.

Heights of Desert Cacti (ft)				
30	10	1	15	10
10	10	10	2	10
20	3	2	15	5

Source: ag.arizona.edu/pubs/garden/mg/cacti

13.

Test Scores					
98	90	97	85	86	92
92	93	95	79	91	92
91	94	88	90	93	92

Determine whether each statement is *sometimes, always,* or *never* true. Explain.

14. You can determine the number of items in a data set from a line plot.
15. If there is a cluster, it will appear in the center of the line plot.

SURVEY For Exercises 16–18, use the line plot at the right and the information below.
Jamie asked her classmates how many cans of soda they drink on a typical day. The results are shown on the line plot.

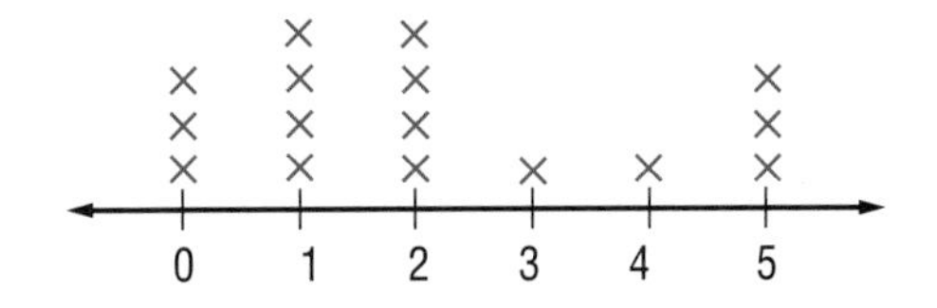

16. What was the most frequent response?
17. What is the range?
18. Summarize the data in a sentence or two.

ANIMALS For Exercises 19–22, use the table below.

Average Life Spans					
Animal	Span (yr)	Animal	Span (yr)	Animal	Span (yr)
black bear	18	giraffe	10	lion	15
camel	12	gorilla	20	pig	10
cat	12	grizzly bear	25	polar bear	20
chipmunk	6	horse	20	rhesus monkey	15
dog	12	kangaroo	7	squirrel	10
elephant	40	leopard	12	white rhinoceros	20

Source: *The World Almanac for Kids*

19. Make a line plot of the data.
20. Find the range and determine any clusters or outliers.
21. Summarize the data on the line plot in a sentence or two.
22. The following data are the *maximum* life spans, in order, of the animals in the table: 36, 50, 28, 10, 20, 77, 36, 54, 50, 50, 24, 23, 30, 27, 45, 37, 24, 50. Make a line plot of the data and compare it to the line plot you made in Exercise 19. Include a discussion about clusters, outliers, range, and gaps in data.

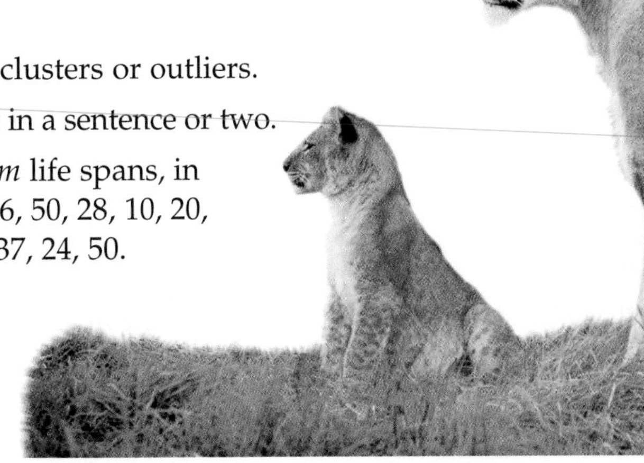

23. **DATA SENSE** The last fourteen bowling scores of two people are shown on the line plots below. Describe which person is more consistent and explain how you know.

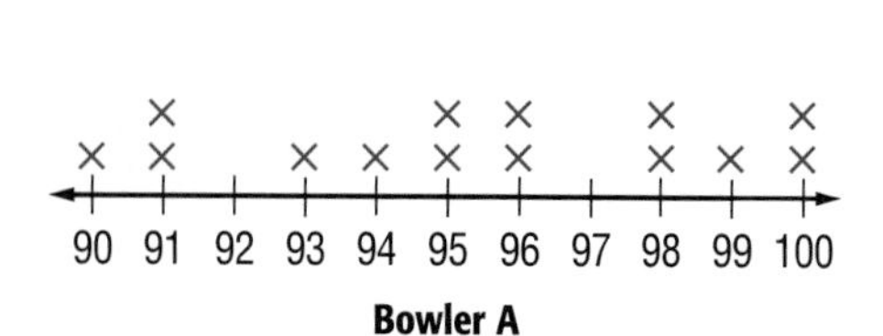

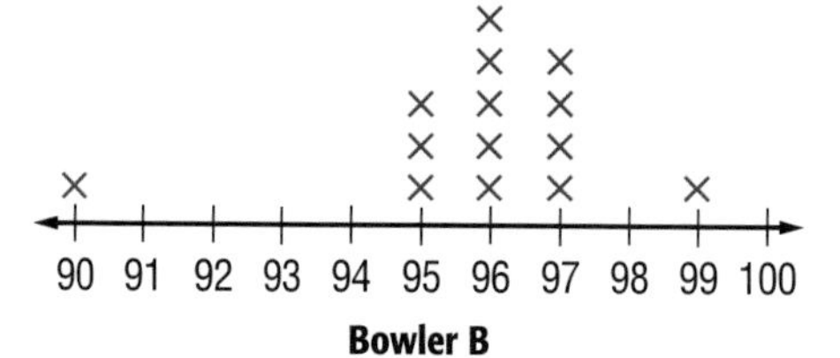

24. **CRITICAL THINKING** Compare and contrast line plots and frequency tables. Include a discussion about what they show and when it is better to use each one.

Standardized Test Practice and Mixed Review

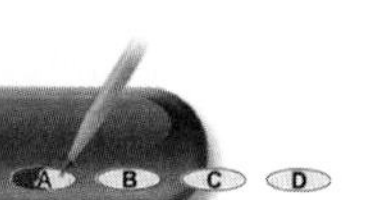

For Exercises 25 and 26, refer to the line plot shown below.

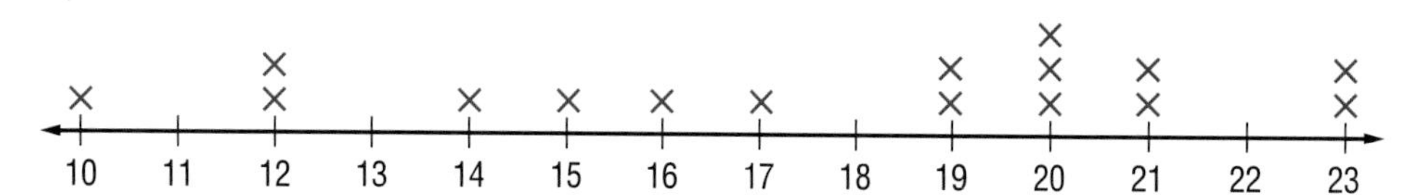

25. **MULTIPLE CHOICE** Which of the following describes a cluster of data?

Ⓐ 20–23 Ⓑ 14–17 Ⓒ 12–14 Ⓓ 19–21

26. **MULTIPLE CHOICE** Determine which sentence is *not* true.

Ⓕ The number that occurs most often is 20.

Ⓖ There are 23 numbers in the data set.

Ⓗ The majority of the data is greater than 18.

Ⓘ The range is 13.

27. **RAIN** The scatter plot shows the height of grass on several lawns and the amount of rain that has fallen on each lawn. Write a conclusion based on the graph. (Lesson 2-2)

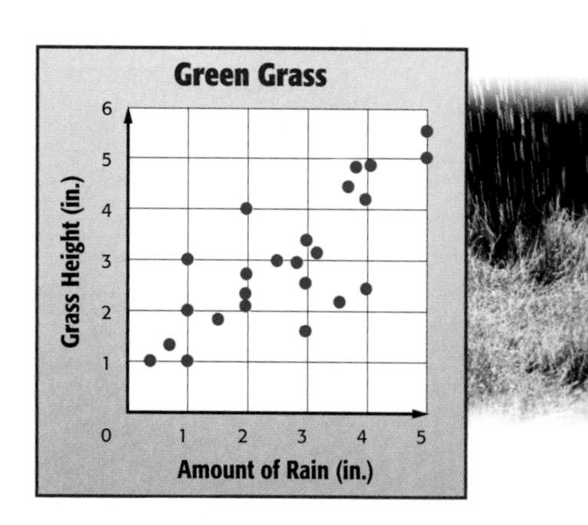

28. *True* or *False*? A frequency table is a good way to organize large amounts of data. (Lesson 2-1)

Solve each equation mentally. (Lesson 1-5)

29. $\frac{t}{5} = 15$

30. $m + 18 = 33$

31. $25 = a - 4$

32. $12x = 120$

GETTING READY FOR THE NEXT LESSON

BASIC SKILL Add or divide. Round to the nearest tenth if necessary.

33. $16 + 14 + 17$

34. $32 + 34 + 36$

35. $4.6 + 2.5 + 9$

36. $77 + 71 + 74$

37. $\frac{202}{16}$

38. $\frac{56}{3}$

39. $\frac{255}{7}$

40. $\frac{155.3}{30}$

2-4 Mean, Median, and Mode

HANDS-ON Mini Lab

Materials

- centimeter cubes

Work with a partner.

In five days, it snowed 4 inches, 3 inches, 5 inches, 1 inch, and 2 inches.

- Make a stack of cubes to represent the snowfall for each day, as shown at the right.
- Move the cubes from stack to stack until each stack has the same number of cubes.

4 in. 3 in. 5 in. 1 in. 2 in.

1. How many cubes are in each stack?
2. In the past five days, it snowed an average of __?__ inches per day.
3. Suppose on the sixth day it snowed 9 inches. If you moved the cubes again, how many cubes would be in each stack?

What You'll LEARN

Find the mean, median, and mode of a set of data.

NEW Vocabulary

measures of central tendency
mean
median
mode

Link to READING

Everyday Meaning of Median: the middle paved or planted section of a highway, as in median strip

If you had to choose one number to represent the data set 1, 2, 3, 4, 5, a good choice would be the middle number, 3. When analyzing data, it is often helpful to use a single number to describe the whole set.

Numbers called **measures of central tendency** can be used to describe the *center* of data. The most common of these measures are mean, median, and mode.

Key Concept — Mean

Words The **mean** of a set of data is the sum of the data divided by the number of items in the data set.

Example data set: 1, 1, 5, 2, 2, 4, 2, 5

mean: $\frac{1 + 1 + 5 + 2 + 2 + 4 + 2 + 5}{8}$ or 2.75

EXAMPLE Find the Mean

1 **NUTRITION The table shows the grams of sugar in twenty different breakfast bars. Find the mean.**

Grams of Sugar in Breakfast Bars				
14	11	7	14	15
22	18	18	21	10
10	10	11	9	11
11	9	12	8	12

Source: www.cspinet.org

$$\text{mean} = \frac{14 + 11 + \ldots + 12}{20} \quad \begin{array}{l}\leftarrow \text{sum of data} \\ \leftarrow \text{number of data items}\end{array}$$

$$= \frac{253}{20} \text{ or } 12.65$$

The mean amount of sugar for the breakfast bars is 12.65 grams.

Key Concept — Median

Words The **median** of a set of data is the middle number of the ordered data, or the mean of the middle two numbers.

Example data set: 7, 11, 15, 17, 20, 20

median: $\frac{15 + 17}{2}$ or 16 The median divides the data in half.

Mode

Words The **mode** or modes of a set of data is the number or numbers that occur most often.

Example data set: 50, 45, 45, 52, 49, 56, 56

modes: 45 and 56

REAL-LIFE MATH

SPACE The International Space Station measures 356 feet by 290 feet, and contains almost an acre of solar panels.

Source: *The World Almanac*

EXAMPLE Find the Mean, Median, and Mode

2 **SPACE Twenty-seven countries have sent people into space. The table shows the number of individuals from each country. Find the mean, median, and mode of the data.**

People in Space								
267	1	9	8	1	1	1	1	1
97	1	1	1	3	1	1	2	1
11	2	1	1	5	1	1	1	1

Source: *The World Almanac*

mean: sum of data divided by 27, or 15.6
median: 14th number of the ordered data, or 1
mode: number appearing most often, or 1

The mean of the data in Example 2 does not accurately describe the data set because it is greater than most of the data items. This happens when one or more of the data is an outlier.

EXAMPLE Analyze Data

3 **TESTS The line plot shows the test scores of the students in Mrs. Hiroshi's math class. Would the mean, median, or mode best represent the test scores?**

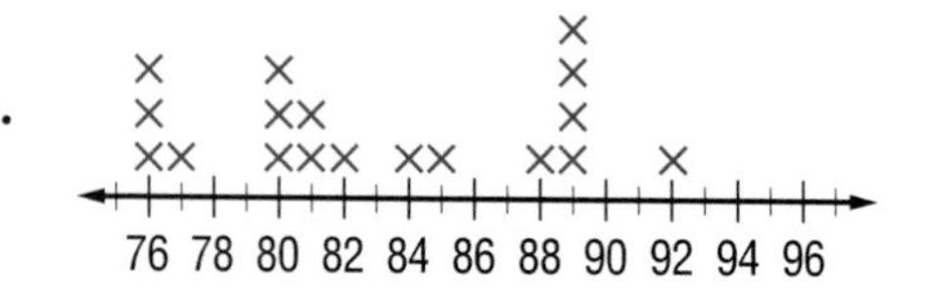

mean: $\frac{76 + 76 + 76 + \ldots + 92}{18}$ or 83

median: $\frac{\text{9th term} + \text{10th term}}{2} = \frac{81 + 82}{2}$ or 81.5

mode: 89

The mode of 89 misrepresents the scores as being higher than they really were. Either the mean of 83 or the median of 81.5 could be used to represent the scores.

STUDY TIP

Median When there is an odd number of data, the median is the middle number of the ordered set. When there is an even number of data, the median is the mean of the two middle numbers.

You can find measures of central tendency from charts, tables, graphs, and line plots. Below are some general guidelines for using these measures.

Concept Summary — **Mean, Median, and Mode**

Measure	Most Useful When . . .
Mean	• data set has no outliers
Median	• data set has outliers • there are no big gaps in the middle of the data
Mode	• data set has many identical numbers

Skill and Concept Check

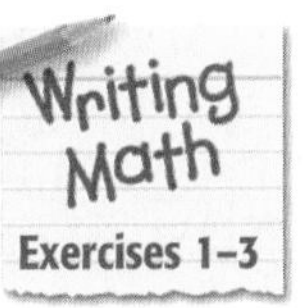

1. **Determine** whether the median is *sometimes*, *always*, or *never* part of the data set. Explain.
2. **OPEN ENDED** Write a problem in which the mean is not the best representation of the data set. Explain why not.
3. **FIND THE ERROR** Jared and Marissa are analyzing the data in the frequency table at the right. Who is correct? Explain.

Jared
median: 2 min
mode: 3 min

Marissa
median: 2.5 min
mode: 9 min

Time (min)	Tally	Frequency
1	卌	5
2	卌 I	6
3	卌 IIII	9
4	I	1

GUIDED PRACTICE

Find the mean, median, and mode for each set of data. Round to the nearest tenth if necessary.

4. 14, 5, 16, 7, 4, 15, 9, 7, 2, 14, 6
5. 29, 14, 80, 59, 78, 30, 59, 69, 55, 50
6. 40 41 42 43 44 45 46 47 48 49 50
7.

Number of Wins	Tally	Frequency
7	卌 卌 II	12
8	IIII	4
9	II	2
10	卌 II	7

SCHOOL For Exercises 8 and 9, use the line plot. It shows the number of times per day that students drink from the water fountain at school.

8. Find the mean, median, and mode.
9. Which measure best describes the data? Explain.

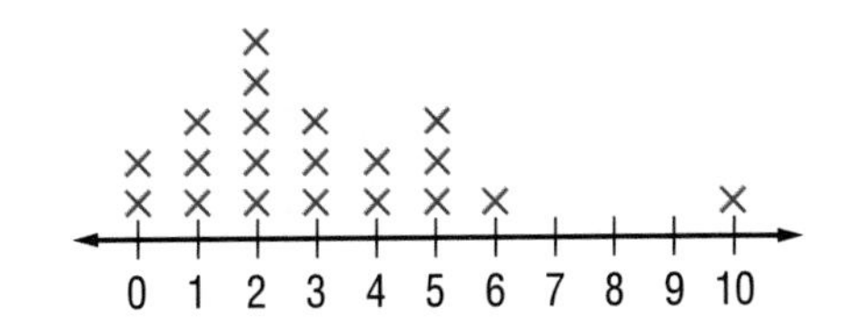

Practice and Applications

HOMEWORK HELP	
For Exercises	See Examples
10–16	1, 2
17, 18	3

Extra Practice
See pages 568, 597.

Find the mean, median, and mode for each set of data. Round to the nearest tenth if necessary.

10. 65, 56, 57, 75, 76, 66, 64
11. 9.3, 8.4, 8.0, 9.1, 9.4, 9.0, 7.8, 9.3, 8.0
12. 95, 90, 92, 94, 91, 90, 98, 88, 89, 100
13. \$10, \$18, \$11, \$6, \$6, \$5, \$10, \$11, \$46, \$7, \$6, \$8

14.

22	23	24	25	26	27	28	29	30	31	32
	× × ×	× ×	× × ×	×	× ×		× ×		× ×	×

15.

Price	Tally	Frequency
\$7.50	卌 III	8
\$8.00	III	3
\$8.50	卌 I	6

MUSIC For Exercises 16 and 17, use the data at the right. It shows the number of weeks that songs have been on the *Top 20 Hot Country Singles & Tracks* chart.

Number of Weeks in Top 20				
8	15	14	24	21
28	12	20	19	18
12	7	11	30	25
15	21	22	18	31

Source: www.billboard.com

16. Find the mean, median, and mode of the data.
17. Determine which measure best represents the data. Explain.

Data Update What is the mean, median, and mode of the number of weeks of the current songs that are on the *Top 20 Hot Country Singles & Tracks* chart? Visit msmath2.net/data_update to learn more.

18. **DATA SENSE** According to the U.S. Census Bureau, the average number of family members per household in Texas is 3.28. State whether this average is a mean or mode. Explain how you know.

19. **CRITICAL THINKING** Without calculating, would the mean, median, or mode be most affected by eliminating 1,000 from the list below? Which would be the least affected? Explain.

    50, 100, 75, 60, 75, 1,000, 90, 100

Standardized Test Practice and Mixed Review

20. **MULTIPLE CHOICE** Which measure describes the center of a set of data?

    (A) mean (B) median (C) mode (D) all of the above

21. **GRID IN** The Cardinals baseball team played 8 games and scored a total of 96 runs. What was the mean number of runs scored per game?

22. **NUTRITION** The grams of fiber in one serving of fifteen different cereals are 5, 5, 4, 3, 3, 3, 1, 1, 1, 2, 1, 1, 1, 1, and 0. Make a line plot of the data. (Lesson 2-3)

23. What type of graph often shows trends over time? (Lesson 2-2)

GETTING READY FOR THE NEXT LESSON

BASIC SKILL Name the place value of the highlighted digit.

24. 581
25. 6,295
26. 4,369
27. 284

msmath2.net/self_check_quiz

2-4b HANDS-ON LAB

A Follow-Up of Lesson 2-4

Are You Average?

What You'll LEARN

Use mean, median, mode, and range to describe a set of data.

Materials

- markers
- ruler

INVESTIGATE *Work in groups of four.*

What do you think of when you hear the word "average"? How would you describe an "average" student in your school? In this lab, you will investigate what an average student in your math class is like.

STEP 1 Your group will create a survey of ten appropriate questions to ask your classmates that will help you describe the "average" student in your class. For example:

- What is your height in inches?
- How many hours per week do you spend doing homework?
- What is your favorite subject in school?
- What is your favorite extracurricular activity?

STEP 2 After your teacher approves your list of questions, have your classmates complete your survey.

STEP 3 Compile the data in a frequency table. Find the mean, median, mode, and range of the data for each question if appropriate. Which measure best describes each set of data? Justify your answer.

STEP 4 Create appropriate graphs to display your data from all ten questions. Each group will present their graphs with a description of what the "average" student in your classroom is like.

Writing Math

1. **Explain** why results may vary if you survey another class in your school.
2. **Explain** why your group selected the graph that you did to display your data. Could you have used another type of graph? Explain.
3. The word *bias* means to influence. **Describe** any factors that could have unfairly influenced the responses given by your classmates. Is there a way to limit bias in a survey?

Mid-Chapter Practice Test

Vocabulary and Concepts

1. **Explain** what a frequency table shows. (Lesson 2-1)
2. **Describe** a situation in which a line graph would be useful. (Lesson 2-2)
3. **List** the information that you can identify from a line plot that is not obvious in an unorganized list of data. (Lesson 2-3)
4. **Define** *mean*. (Lesson 2-4)

Skills and Applications

For Exercises 5–10, use the data below that shows the ages of twenty people when they got their driver's licenses. (Lessons 2-1, 2-3, and 2-4)

16 17 16 16 18 21 16 16 18 18 17 25 16 17 17 17 17 16 20 16

5. Make a frequency table of the data.
6. Make a line plot of the data.
7. Identify any clusters or gaps.
8. Identify any outliers.
9. Find the mean, median, and mode.
10. What is the range of the data?
11. **HEATING** Use the scatter plot to make a conclusion about the average monthly outside temperature and heating costs of a home. (Lesson 2-2)

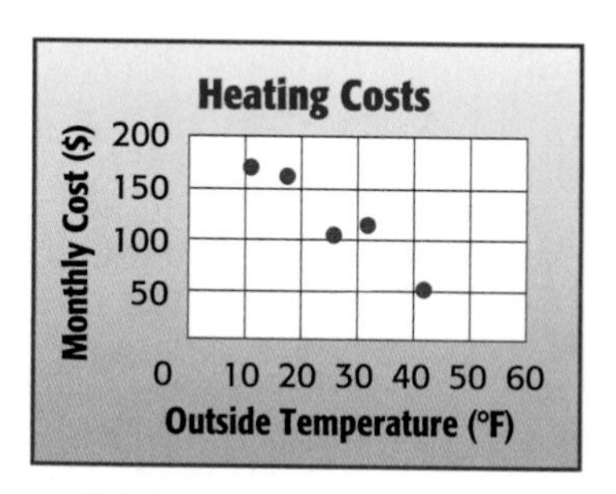

Find the mean, median, and mode for each set of data. (Lesson 2-4)

12. 27, 32, 18, 24, 32, 39, 41
13. 7.6, 6.8, 6.5, 7, 7.9, 6.8, 7, 7, 6.8, 8.1

Standardized Test Practice

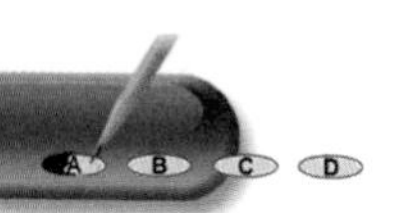

14. **GRID IN** Below are the quiz scores of a science class.

    89 92 67 75 95 89 82 92 89
    89 92 96 85 94 90 79 94 87

    What is the median quiz score? (Lesson 2-4)

15. **MULTIPLE CHOICE** Which may *not* show individual pieces of data? (Lessons 2-1, 2-2, and 2-3)

    (A) scatter plot
    (B) frequency table
    (C) line plot
    (D) an ordered list

The Game Zone

A Place To Practice Your Math Skills

Math Skill
Finding The Mean

Can You Guess?

GET READY!

Players: the entire class divided into three-person teams
Materials: cup, coins, water dropper, self-adhesive notes

GET SET!

- Draw a number line on the chalkboard with a scale from 10 to 25.
- Each team should fill a cup with water.

GO!

- Use the water dropper and find how many drops of water your team can get to stay on the head of a dime. Write the results on a sticky note. Post it on the chalkboard above the corresponding number on the number line. Find the mean of the numbers.

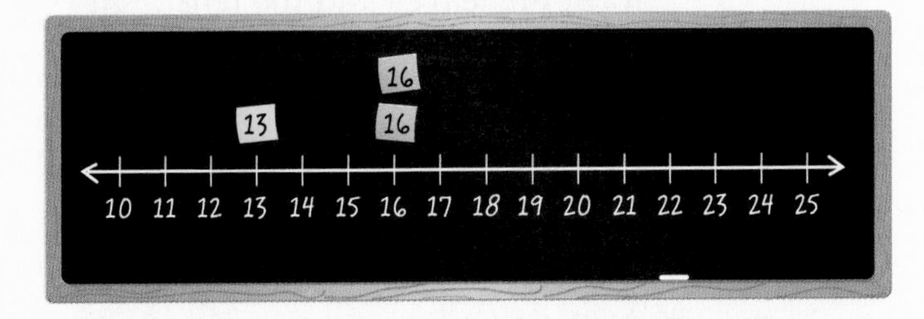

- Next, use the mean to predict how many drops of water will fit on the head of a penny. Record your team's prediction on a piece of paper.
- Find how many drops of water will stay on the head of a penny. Record the result on a sticky note and post it on the chalkboard. Find the mean of all of the results for the pennies. If your prediction is within three drops of the mean, your team gets to stay in the game.
- Continue predicting and testing using a nickel and a quarter.
- **Who Wins?** The teams left at the end of the game are the winners.

2-5 Stem-and-Leaf Plots

WHEN am I ever going to use this?

What You'll LEARN

Construct and interpret stem-and-leaf plots.

NEW Vocabulary

stem-and-leaf plot
leaf
stem

HISTORY Benjamin Franklin was the oldest signer of the Declaration of Independence. The table shows the ages in years of all the people who signed the Declaration of Independence.

1. What was the age of the youngest signer?
2. What methods do you already have for showing the distribution of ages?

Ages of Signers of Declaration of Independence

40	70	69	34	46	53	31
38	50	30	45	46	39	38
35	50	37	48	41	41	52
39	50	65	46	29	34	45
33	41	44	63	60	26	42
42	52	37	35	45	35	42
47	26	55	57	45	33	60
62	35	33	53	49	50	46

Source: *The World Almanac*

A useful way to organize data as you collect it is a stem-and-leaf plot. In a **stem-and-leaf plot**, the data are organized from least to greatest. The digits of the least place value usually form the **leaves**, and the next place value digits form the **stems**.

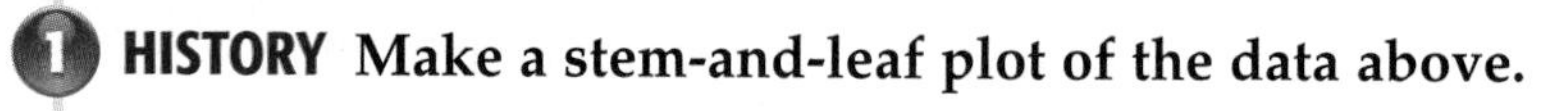

EXAMPLE Construct a Stem-and-Leaf Plot

1 HISTORY Make a stem-and-leaf plot of the data above.

Step 1 The digits in the least place value will form the leaves and the remaining digits will form the stems. In this data, 26 is the least value, and 70 is the greatest. So, the ones digits will form the leaves and the tens digits will form the stems.

Step 2 List the stems 2 to 7 in order from least to greatest in the *Stem* column. Write the leaves, the ones digits of the ages, to the right of the corresponding stems. The ages in the first column of the chart are shown at the right.

Stem	Leaf
2	
3	8 5 9 3
4	0 2 7
5	
6	2
7	

Step 3 Order the leaves and write a *key* that explains how to read the stems and leaves.

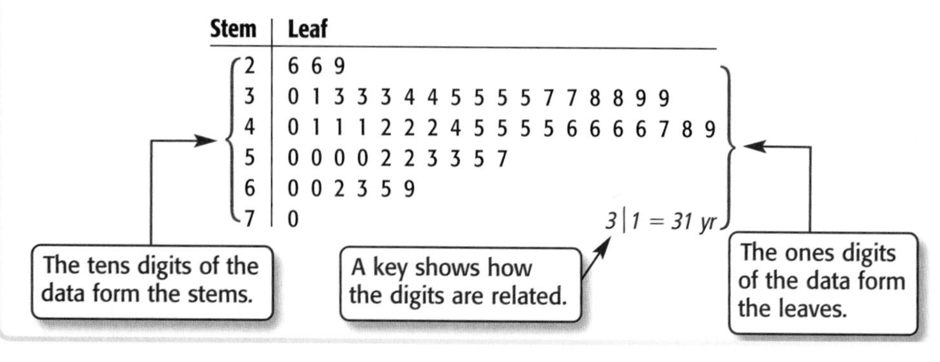

Stem	Leaf
2	6 6 9
3	0 1 3 3 3 4 4 5 5 5 5 7 7 8 8 9 9
4	0 1 1 1 2 2 2 4 5 5 5 5 6 6 6 6 7 8 9
5	0 0 0 0 2 2 3 3 5 7
6	0 0 2 3 5 9
7	0

$3|1 = 31$ *yr*

Stem-and-leaf plots are useful in analyzing data because you can see all the data values, including the greatest, least, mode, and median value.

EXAMPLE Analyze Data

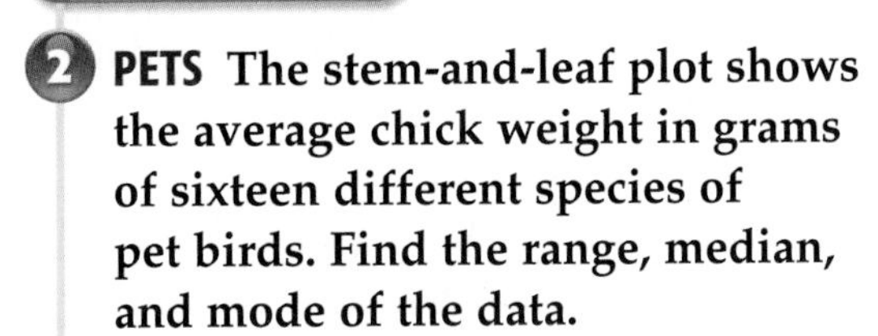

2 PETS The stem-and-leaf plot shows the average chick weight in grams of sixteen different species of pet birds. Find the range, median, and mode of the data.

Stem	Leaf
0	5 6 7
1	0 0 1 1 2 2 2 3 8 9
2	0 1 1 5

1|2 = 12 g

Source: www.upatsix.com

range: greatest weight − least weight = 25 − 5 or 20 grams

median: middle value, or 12 grams

mode: most frequent value, or 12 grams

REAL-LIFE CAREERS

How Does a Nutritionist Use Math?

Nutritionists use statistical graphs to summarize and compare the nutritional values of various foods.

Online Research
For information about a career as a nutritionist, visit:
msmath2.net/careers

Stem-and-leaf plots are also useful for making conclusions about data. Most of the data in the plot above are clustered in the 10–19 interval. So, you could say that most of those birds weigh 10 to 19 grams as chicks.

EXAMPLE Make Conclusions About Data

3 NUTRITION The table shows the average amounts of pasta that people in different countries consume each year. Make a stem-and-leaf plot of the data. Then use it to describe how the data are spread out.

Country	Pasta per Person (lb)	Country	Pasta per Person (lb)
Argentina	15	Netherlands	9
Australia	5	Portugal	15
Canada	14	Russia	15
Egypt	15	Spain	9
France	15	Sweden	8
Germany	10	Switzerland	20
Greece	19	Turkey	11
Italy	59	United Kingdom	4
Japan	3	United States	20
Mexico	8	Venezuela	28

Source: National Pasta & National Restaurant Associations

The least value is 3, and the greatest value is 59. So, the tens digits form the stems, and the ones digits form the leaves.

Stem	Leaf
0	3 4 5 8 8 9 9
1	0 1 4 5 5 5 5 5 9
2	0 0 8
3	
4	
5	9

2|8 = 28 lb

The data that do not have a digit in the tens place have 0 for a stem.

It is easy to see in the stem-and-leaf plot that people in most of the countries consume less than 20 pounds of pasta per year. You can also see that the mode is 15 and that 59 is an outlier, with a significant gap between that value and the next smaller value.

Skill and Concept Check

Writing Math Exercises 1 & 4

1. **Compare and contrast** a stem-and-leaf plot and a frequency table.
2. **OPEN ENDED** Make a stem-and-leaf plot in which the median of the data is 25.
3. **Write** the stems that would be used in a stem-and-leaf plot for the following set of data: 21, 8, 12, 7, 18, 9, 22, 31, 1, 14, 19, 11, 2.
4. **FIND THE ERROR** Selena and Lauren are analyzing the data in the stem-and-leaf plot at the right. Who is correct? Explain.

Stem	Leaf
2	6 6 9
3	
4	6
5	3 6 *2|6 = 26 in.*

Selena
None of the data have 3 in the tens place.

Lauren
Thirty is one of the data values.

GUIDED PRACTICE

Make a stem-and-leaf plot for each set of data.

5. 42, 47, 19, 16, 21, 23, 25, 25, 29, 31, 33, 34, 35, 39, 48
6. 225, 227, 230, 229, 246, 243, 269, 269, 267, 278, 278, 278

CHESS For Exercises 7–10, use the stem-and-leaf plot at the right. It shows the number of matches won by each member of the Avery Middle School Chess Team.

Stem	Leaf
0	8 8 9
1	9
2	0 0 2 4 4 8 9
3	1 1 2 4 5 5 6 6 7 7 8
4	0 0 0 3 8 9
5	2 4
6	1 *3|2 = 32 wins*

7. How many members are on the chess team?
8. What is the range of the number of matches won?
9. Find the mean, median, and mode for these data.
10. Describe how the data are spread out.

Practice and Applications

HOMEWORK HELP

For Exercises	See Examples
11–14, 19	1
15–17, 20	2
18, 21	3

Extra Practice
See pages 568, 597.

Make a stem-and-leaf plot for each set of data.

11. 15, 13, 28, 32, 38, 30, 31, 13, 36, 35, 38, 32, 38, 24
12. 56, 70, 96, 62, 80, 80, 69, 93, 66, 55, 95, 63, 90, 93, 60, 91, 67, 60
13. 451, 458, 439, 403, 411, 434, 433, 454, 457, 442, 449, 459, 454
14. 8.2, 9.3, 10, 9.9, 8, 9.2, 8.7, 8, 8.2, 9, 9.9, 8.7, 8.5, 8.1, 8.8, 9.3

SHOPPING For Exercises 15–18, use the stem-and-leaf plot at the right. It shows the prices of portable CD players found on an Internet shopping site.

Stem	Leaf
7	0
8	2 5 5
9	9 9
10	0 0 2 5 6 8
11	0 0 5 5 5 9 9
12	5 7 7
13	
14	
15	0 0 0 5 *11|5 = $115*

15. How many CD players are represented on the stem-and-leaf plot?
16. What is the range of the prices?
17. Find the median and mode for these data.
18. Write a sentence or two to describe how the prices of the portable CD players are spread out.

Practice and Applications

HOMEWORK HELP	
For Exercises	See Examples
7, 8, 12	1
9, 13	2
10, 11	3
Extra Practice	See pages 568, 597.

SPORTS For Exercises 7–9, refer to the table at the right. It shows the regular season games won by each men's professional basketball team in a recent season.

Number of Wins					
25	36	46	15	30	53
40	32	17	45	41	31
56	50	52	47	26	48
43	56	51	50	55	58
44	47	53	23	19	

7. Find the lower extreme, LQ, median, UQ, and upper extreme.
8. Draw a box-and-whisker plot of the data.
9. What fraction of the data are between 45 and 51.5?

BRIDGES For Exercises 10–13, refer to table at the right. It shows the lengths of 22 of the longest suspension bridges in North America.

Length of Bridges (ft)			
4,260	2,310	1,800	1,595
4,200	2,300	1,750	1,550
3,800	2,190	1,632	1,500
3,500	2,150	1,600	1,500
2,800	2,000	1,600	
2,800	1,850	1,600	

10. Determine the interquartile range.
11. What are the limits on outliers? Are there any outliers?
12. Draw a box-and-whisker plot of the data.
13. What fraction of the bridges are between 1,600 and 2,800 feet long?
14. Compare and contrast the data represented in the box-and-whisker plots at the right.

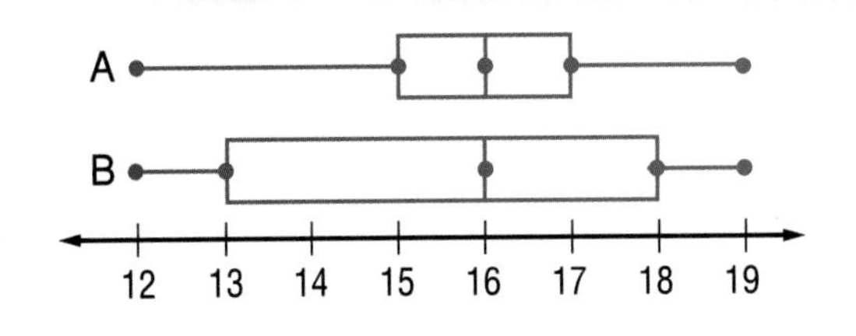

15. **CRITICAL THINKING** Describe a set of data in which there is only one whisker in its box-and-whisker plot.

Standardized Test Practice and Mixed Review

16. **MULTIPLE CHOICE** What is the lower quartile of the data in the box-and-whisker plot at the right?

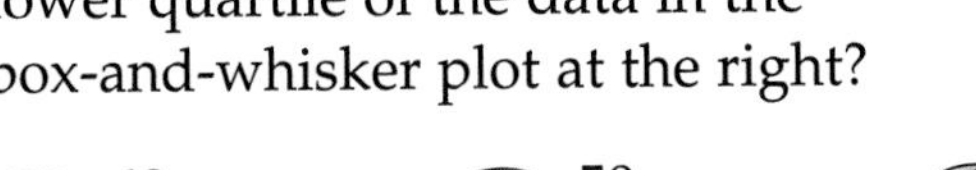

Ⓐ 60 Ⓑ 50 Ⓒ 30 Ⓓ 20

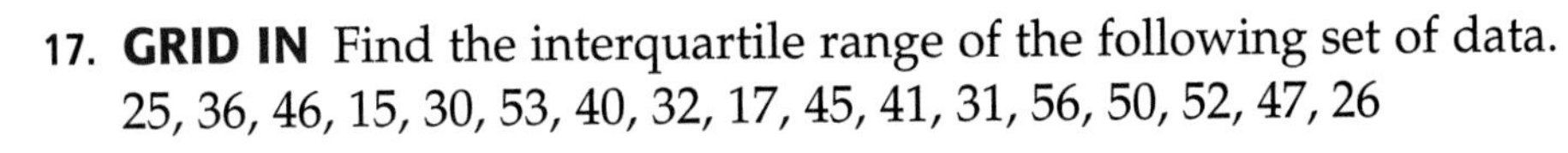

17. **GRID IN** Find the interquartile range of the following set of data. 25, 36, 46, 15, 30, 53, 40, 32, 17, 45, 41, 31, 56, 50, 52, 47, 26

RAINFALL For Exercises 18 and 19, refer to the table at the right.

Days of Rain per Year		
17	26	29
30	32	37
42	36	35

18. Make a stem-and-leaf plot of these data. (Lesson 2-5)
19. Find the mean, median, and mode. Round to the nearest tenth if necessary. (Lesson 2-4)

GETTING READY FOR THE NEXT LESSON

PREREQUISITE SKILL Choose an appropriate interval and scale to make a frequency table for each set of data. (Lesson 2-1)

20. 9, 0, 18, 19, 2, 9, 8, 13, 4
21. 30, 20, 60, 80, 90, 110, 40

msmath2.net/self_check_quiz

2-6b Graphing Calculator Investigation

A Follow-Up of Lesson 2-6

What You'll LEARN

Use a graphing calculator to make box-and-whisker plots.

Box-and-Whisker Plots

You can create a box-and-whisker plot using a TI-83 Plus graphing calculator.

ACTIVITY

Make a box-and-whisker plot of the data at the right. It shows the grades on Miss Romero's last Math 7 test.

Miss Romero's Math 7 Test							
78	94	85	92	72	56	89	92
90	84	98	82	75	100	94	87
92	85	94	70	78	95	70	80

STEP 1 Enter the data.
Begin by clearing any existing data in the first list, L1.
Press STAT ENTER ▲ CLEAR ENTER.

Then enter the data into L1. Input each number and press ENTER.

STEP 2 Choose the type of graph.
Press 2nd [STAT PLOT] ENTER to choose the first plot. Highlight On, the modified box-and-whisker plot for the type, L1 for the Xlist, and 1 as the frequency.

STEP 3 Choose the display window.
Press WINDOW and choose appropriate range settings for the x values. The window 50 to 110 with a scale of 5 includes all of this data.

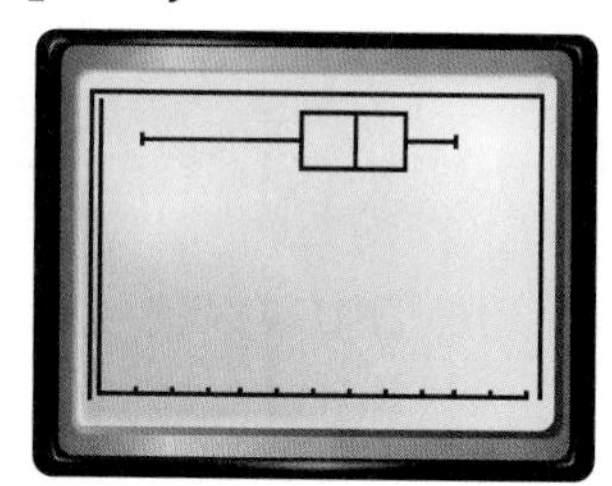

STEP 4 Display the graph.
Press GRAPH. Press TRACE and the arrow keys to determine the five key data points of your graph.

EXERCISES

1. What are the values of the five key data points of the graph? What do they represent?
2. What percent of the test scores are below 78?
3. What percent of the test scores are above the median? What percent of the test scores are below the median?
4. What percent of the scores are between 56 and 86?
5. Suppose you earned a grade of 80. Describe what percent of students scored higher and what percent scored lower than you.

msmath2.net/other_calculator_keystrokes

2-7 Bar Graphs and Histograms

WHEN am I ever going to use this?

What You'll LEARN

Construct and interpret bar graphs and histograms.

NEW Vocabulary

bar graph
histogram

REVIEW Vocabulary

frequency table: a table that shows how often an item occurs **(Lesson 2-1)**

ANIMALS The cheetah is the fastest known land animal. The table shows its fastest speed and the top speeds of four other land animals.

Animal	Speed (mph)
cheetah	70
wildebeest	50
lion	50
elk	45
zebra	40

Source: *The World Almanac*

1. What are the fastest and slowest speeds recorded in the table?
2. How can you create a visual representation to summarize the data?
3. Do any of these representations show both the animal name and its speed?

A **bar graph** is one method of comparing data by using solid bars to represent quantities.

EXAMPLE Construct a Bar Graph

1 **Make a bar graph to display the data in the table above.**

Step 1 Draw a horizontal axis and a vertical axis. Label the axes as shown. In this case, the scale on the vertical axis is chosen so that it includes all the speeds. Add a title.

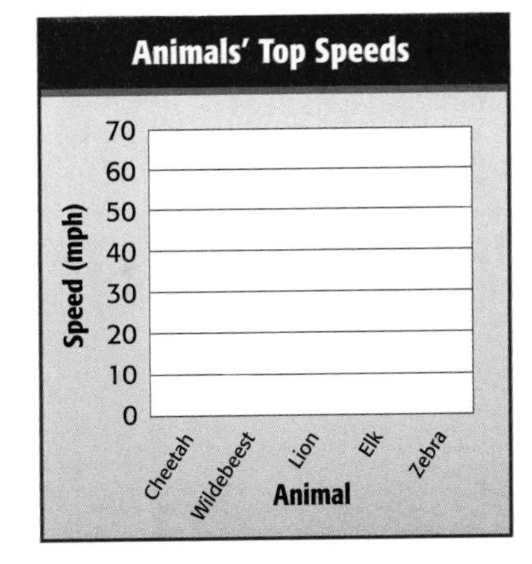

Step 2 Draw a bar to represent each category. In this case, a bar is used to represent the speed of each animal.

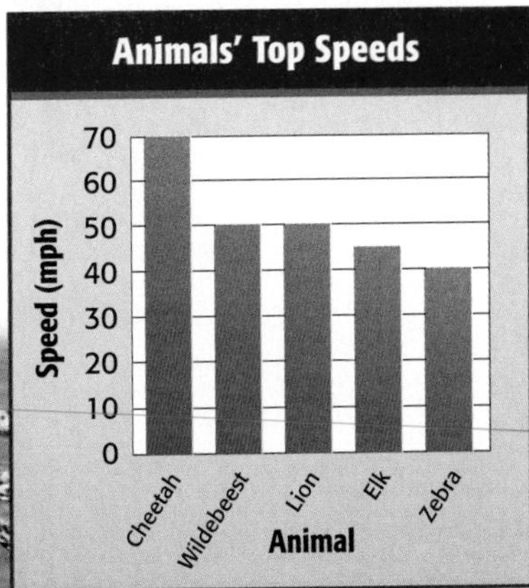

A special kind of bar graph, called a **histogram**, uses bars to represent the frequency of numerical data that have been organized in intervals.

EXAMPLE Construct a Histogram

2 **ANIMALS** **The top speeds of thirty different land animals have been organized into a frequency table. Make a histogram of the data.**

Maximum Speed (mph)	Frequency
11–20	5
21–30	5
31–40	10
41–50	8
51–60	0
61–70	2

Source: *The World Almanac*

STUDY TIP

Histograms
Because the intervals are equal, all of the bars have the same width, with no space between them. The space at 51–60 indicates that there are no data values on that interval.

Step 1 Draw and label horizontal and vertical axes. Add a title.

Step 2 Draw a bar to represent the frequency of each interval.

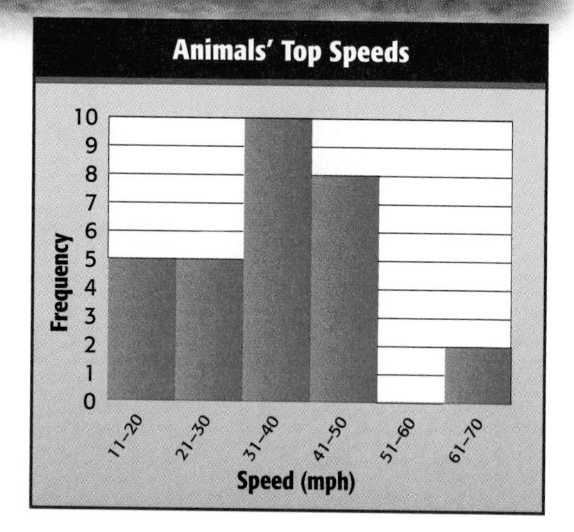

The two highest bars represent a majority of the data. From the graph, you can easily see that most of the animals have a top speed of 31–50 miles per hour.

EXAMPLE Compare Bar Graphs and Histograms

3 **TEXTBOOKS** **Refer to the graphs below.**

Graph A

Textbook Size
Pages: 0, 200, 400, 600, 800
Textbook: Health, Math, Science, History

Graph B

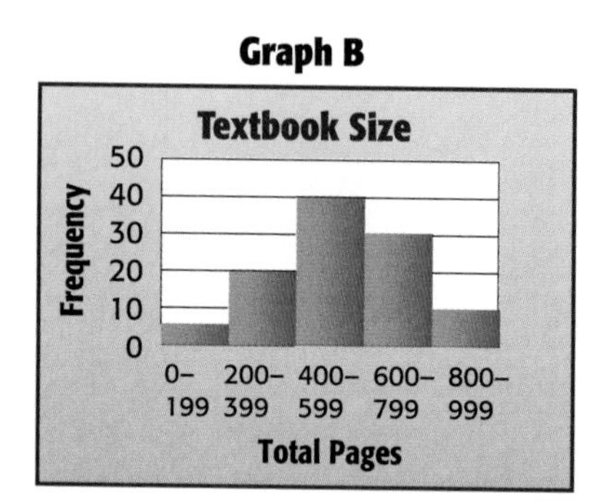

Which graph would you use to determine how many textbooks have less than 600 pages?

Graph B

Which graph would you use to compare the size of a science book and a history book?

Graph A

Standardized Test Practice

EXAMPLE Use Graphs to Solve a Problem

4 **MULTIPLE-CHOICE TEST ITEM** Which conclusion *cannot* be made about the data in Graph B in Example 3?

(A) There are 105 books in the data set.

(B) Twenty books have between 200 and 399 pages.

(C) Health books have the fewest pages.

(D) The interval with the fewest books is 0–199.

Test-Taking Tip The Princeton Review

Read a Graph
When analyzing data in a histogram, ask yourself questions such as, "What does each bar represent?"

Read the Test Item

Determine which of the four statements is *not* correct.

Solve the Test Item

A is correct; 5 books have 0–199 pages, 20 books have 200–399 pages, and so on, for a total of 105 books.

B is correct; the bar representing 200–399 has a value of 20.

C is incorrect; the graph does not describe individual books.

D is correct; only 5 books have 0 to 199 pages, which is less than the other intervals. The answer is C.

Skill and Concept Check

Writing Math Exercise 2

1. **Describe** how to determine the number of values in a data set using a histogram.

2. **OPEN ENDED** Can any data set be displayed using a histogram? If yes, explain why. If no, give a counterexample and explain why not.

GUIDED PRACTICE

For Exercises 3–6, use the table at the right.

3. Describe the data in the table.

4. Make a bar graph of the data.

5. What do the horizontal and vertical axes represent?

6. Which country won the most titles? the least titles?

Men's World Cup Soccer Titles, 1930–2002

Country	Titles Won
Uruguay	2
Italy	3
Germany	3
Brazil	5
England	1
Argentina	2
France	1

Source: fifaworldcup.yahoo.com/en

Practice and Applications

For Exercises 7–10, use the table below.

7. Describe the data in the table.
8. What does each interval represent?
9. Make a histogram of the data.
10. Describe general patterns in the histogram.

Candies in Snack-Size Packages	
Number of Candies	Frequency
9–11	8
12–14	15
15–17	24
18–20	3

HOMEWORK HELP

For Exercises	See Examples
7–9	1, 2
10–16	3

Extra Practice
See pages 569, 597.

TECHNOLOGY **For Exercises 11–13, use the bar graph below.**

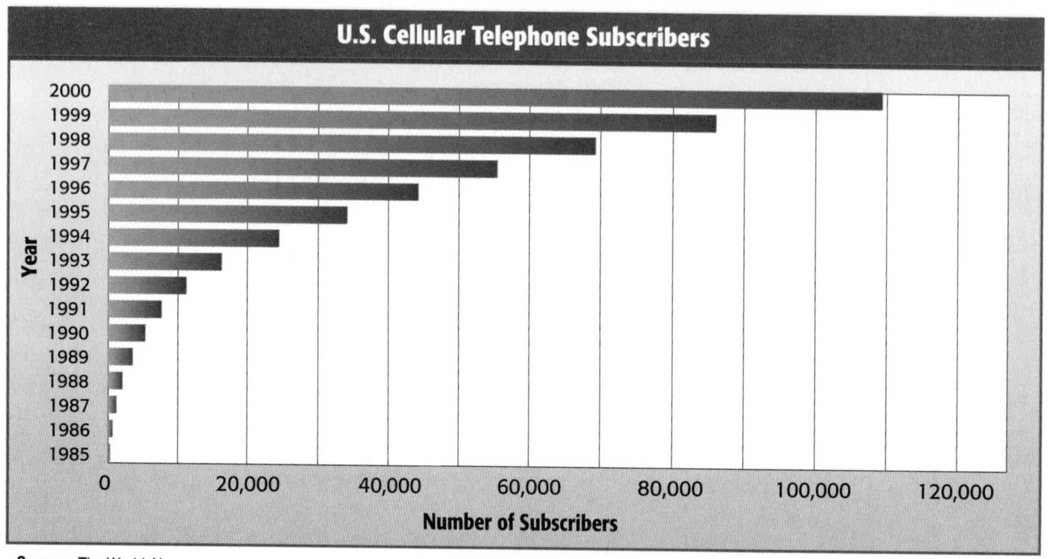

Source: *The World Almanac*

11. How is this bar graph different from the other bar graphs in the lesson?
12. Describe how the number of cell phone subscribers grew over the 15-year period.
13. Use the graph to predict the number of cell phone subscribers in 2006.

CIVICS **For Exercises 14–16, use the histogram at the right and the information below.**

Congress is made up of senators and representatives. Each state has 2 senators, and the number of representatives per state is determined by the state's population.

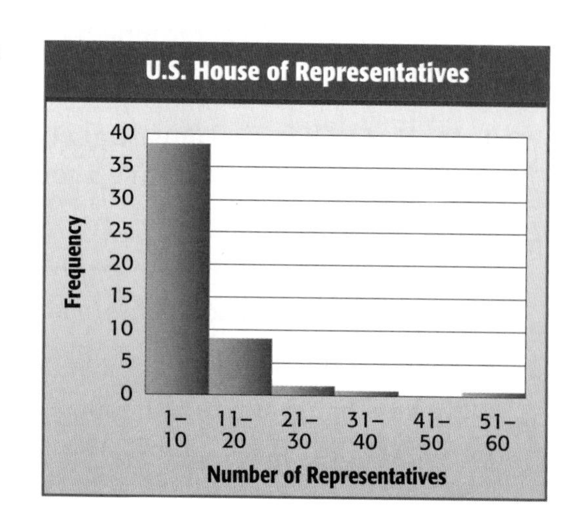

14. What does each interval represent?
15. Describe general patterns in the histogram.
16. There are 435 members in the House of Representatives. Determine the mean number of representatives per state. Compare this value to the value of the median number, according to the histogram. Discuss the significance of 38 states having 10 or fewer representatives.

17. **CRITICAL THINKING** The histograms show players' salaries to the nearest million dollars for the two major league baseball teams that played in the 2001 World Series. Compare the salary distributions of the two teams.

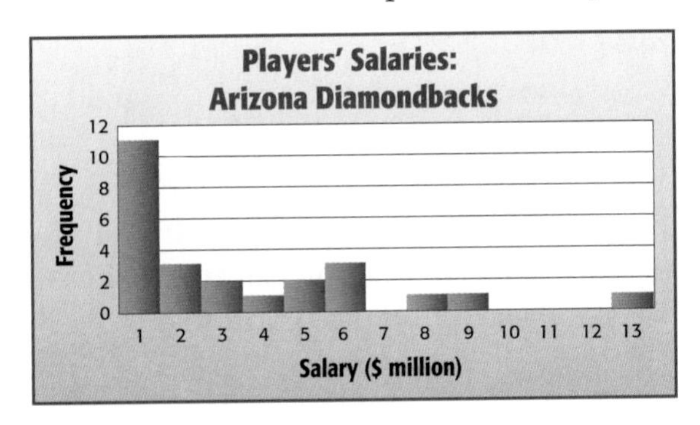

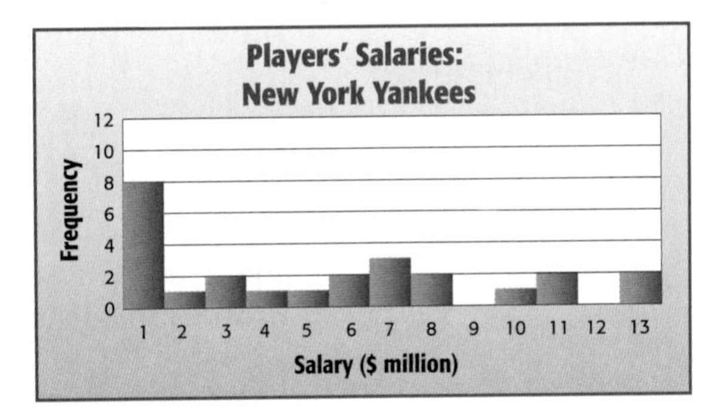

EXTENDING THE LESSON

A *multiple-bar graph* shows two or more sets of data on the same graph. The multiple-bar graph compares the annual costs of using old model appliances and using equivalent new model appliances.

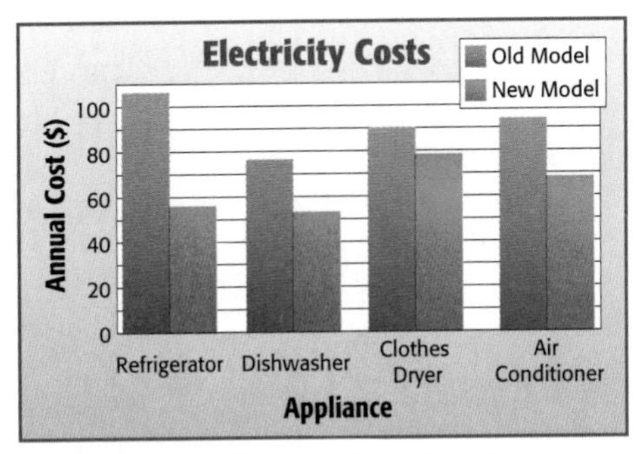

Source: Association of Home Appliance Manufacturers

18. For which appliance is the difference in electricity costs between the old and new model the greatest? Explain.

19. Describe an advantage of using a multiple-bar graph rather than two separate graphs to compare data.

Standardized Test Practice and Mixed Review

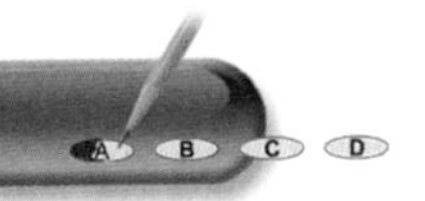

20. **MULTIPLE CHOICE** The histogram at the right shows the number of bordering states for each of the fifty states. Which is *not* a true statement?

    (A) The mode of the data is 4.

    (B) The greatest number of bordering states is 8.

    (C) Ten states have five bordering states.

    (D) No states have only 1 bordering state.

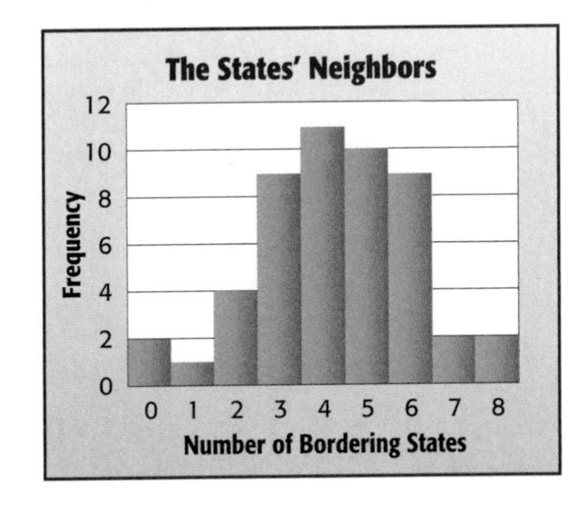

21. **SHORT RESPONSE** Which type of graph can be drawn using the data in a frequency table?

For Exercises 22 and 23, use the data 12, 17, 23, 5, 9, 25, 13, 16, 2, 25, 31, and 10.

22. Make a box-and-whisker plot of the data. (Lesson 2-6)

23. Make a stem-and-leaf plot of the data. (Lesson 2-5)

GETTING READY FOR THE NEXT LESSON

PREREQUISITE SKILL Determine whether each statement is *true* or *false*. (Lesson 2-2)

24. The vertical scale on a line graph must have equal intervals.

25. You do not need to label the axes of a line graph.

2-7b Spreadsheet Investigation

A Follow-Up of Lesson 2-2

What You'll LEARN

Use a spreadsheet to make a multiple-line graph and a multiple-bar graph.

Multiple-Line and Multiple-Bar Graphs

In Lessons 2-2 and 2-7, you interpreted data in a multiple-line graph and in a multiple-bar graph, respectively. You can use a Microsoft Excel® spreadsheet to make these two types of graphs.

ACTIVITY 1

The stopping distances on dry pavement and on wet pavement are shown in the table at the right.

Speed (mph)	Stopping Distance (ft)	
	Dry Pavement	Wet Pavement
50	200	250
60	271	333
70	342	430
80	422	532

Source: Continental Teves

Set up a spreadsheet like the one shown below.

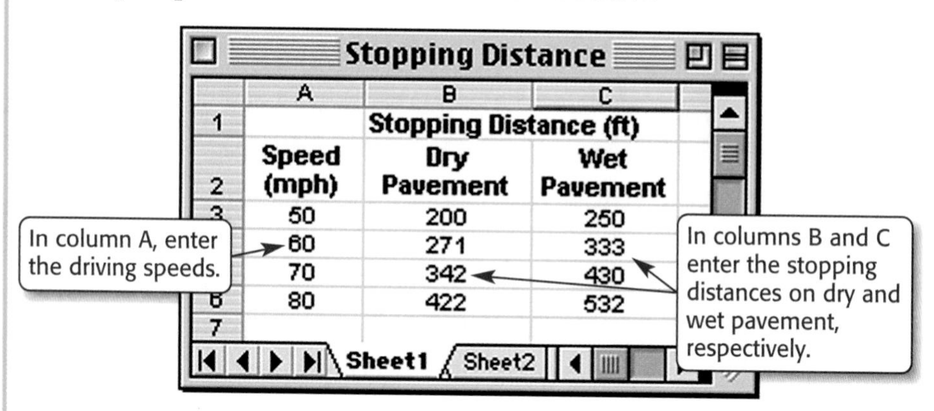

The next step is to "tell" the spreadsheet to make a double-line graph for the data.

1. Highlight the data in columns B and C, from B2 through C6.

   This tells the spreadsheet to read the data in columns B and C.

2. Click on the Chart Wizard icon.
3. Choose the line graph and click Next.
4. To set the x-axis, choose the Series tab and press the icon next to the Category (X) axis labels.
5. On the spreadsheet, highlight the data in column A, from A3 through A6.
6. Press the icon on the bottom of the Chart Wizard box to automatically paste the information.
7. Click Next and enter the chart title and labels for the x- and y-axes.
8. Click Next and then Finish.

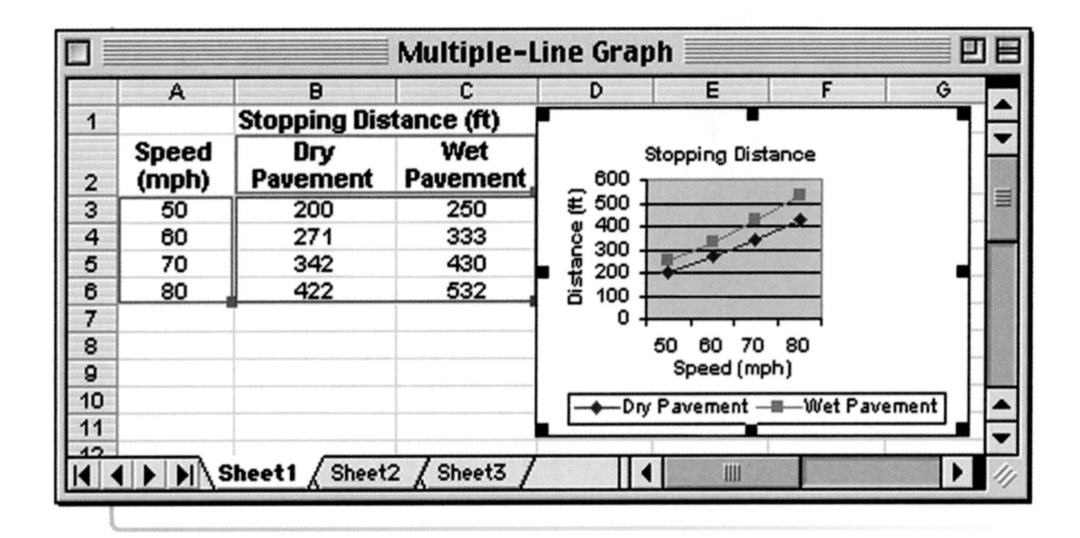

Multiple-Line Graph

Speed (mph)	Stopping Distance (ft): Dry Pavement	Stopping Distance (ft): Wet Pavement
50	200	250
60	271	333
70	342	430
80	422	532

ACTIVITY 2

Use the same data to make a multiple-bar graph. Set up the spreadsheet shown in Activity 1. The next step is to "tell" the spreadsheet to make a bar graph for the data.

- Highlight the data in columns B and C, from B2 through C6.
- Click on the Chart Wizard icon.
- Click on Column and Next to choose the vertical bar graph.
- Complete steps 4–8 from Activity 1.

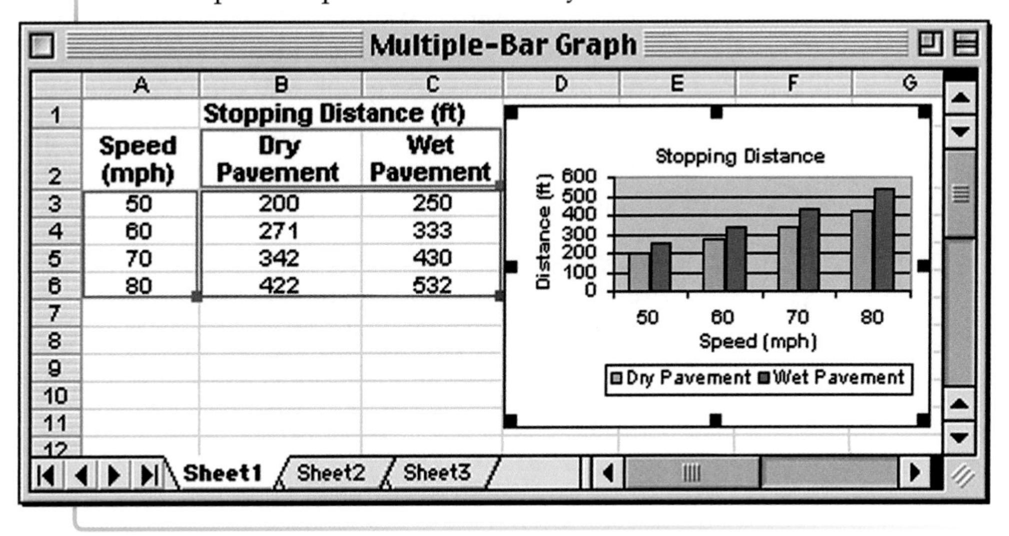

Multiple-Bar Graph

Speed (mph)	Stopping Distance (ft): Dry Pavement	Stopping Distance (ft): Wet Pavement
50	200	250
60	271	333
70	342	430
80	422	532

EXERCISES

1. **Explain** the steps you would take to make a multiple-line graph of the stopping distances that include the speeds 55, 65, and 75.
2. **OPEN ENDED** Collect some data that can be displayed in a multiple-line graph or a multiple-bar graph. Record the data in a spreadsheet. Then make a graph to display the data.
3. **Discuss** the advantages and disadvantages of using a spreadsheet to make a multiple-bar graph.

2-8 Misleading Statistics

WHEN am I ever going to use this?

What You'll LEARN

Recognize when statistics and graphs are misleading.

TRANSPORTATION A graph like the one at the right appeared in a brochure describing various modes of transportation.

Methods of Travel

Trains
40-foot Buses
Cars
2 4 6 8 10 12 14 16 18 20+
Passenger Capacity Per Lane/Per Hour (thousands)

1. Approximately how many more passengers per lane can a 40-foot bus transport in an hour than a car can transport?
2. Is the bus on the graph about twice as large as the car? Explain.
3. Do you think the graph appeared in a brochure for a train/bus transit company or for a car company? What makes you think so?

The graph above shows that a car can transport about 5,000 people per lane, per hour. A train can transport about 4 times that many. However, the train on the graph is much more than 4 times larger than the car. For this reason, the graph could be considered misleading.

Other ways that graphs could be misleading are if there is no title, there are no labels on the axes or scales, or the intervals on a scale are not equal.

EXAMPLE Misleading Graphs

1 BUSINESS The line graphs below show monthly profits of a company from October to March. Which graph is misleading? Explain.

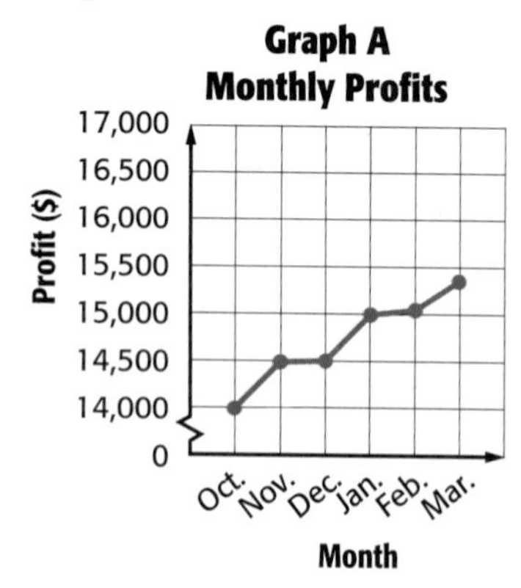

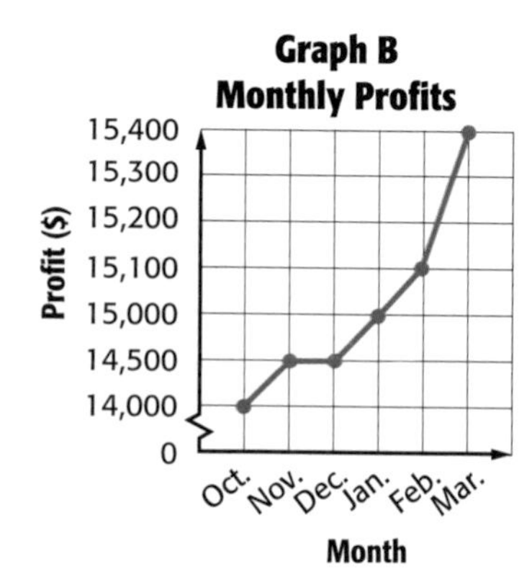

Graph B is misleading because the vertical scale has intervals of both $500 and $100. The increase in profits are exaggerated.

Statistical averages used to describe data may also be misleading. For example, whenever there are outliers, the mean may not be a good way to describe a data set because it is affected the most by outliers.

EXAMPLE Misleading Statistics

2 SALARY ABC Corporation claims the average salary for its employees is more than $60,000, while the average salary at XYZ Incorporated is only $25,000. Use the table to explain their reasoning and determine where you would prefer to work.

Position	Salary ($)	
	ABC Corp.	XYZ Inc.
President	500,000	120,000
1st Vice President	400,000	85,000
2nd Vice President	240,000	75,000
Sales Staff (5)	20,000	40,000
Supporting Staff (8)	15,000	25,000
Catalog Staff (7)	9,000	22,500

	ABC Corp.	XYZ Inc.
mean salary	$61,870	$36,413
median salary	$15,000	$25,000

ABC Corporation used the mean to represent its average salary. They used the median to represent the average salary at XYZ Inc.

Unless you could be a president or vice president, it would be better to work for XYZ Inc. because its 20 lowest-paid employees are all better paid than any of the 20 lowest-paid employees at ABC Corporation.

Skill and Concept Check

1. **Describe** two ways in which the presentation of data can be misleading.
2. **OPEN ENDED** Find an example of a graph in a newspaper or magazine that could be considered misleading.

GUIDED PRACTICE

For Exercises 3–6, use the graph at the right. It shows the prices for natural gas charged by an Illinois natural gas supplier.

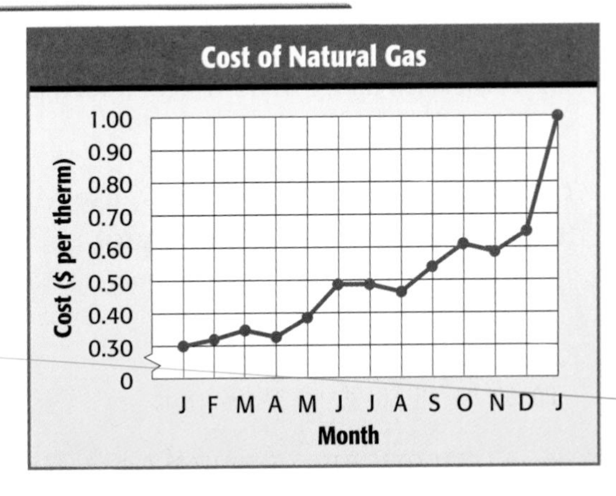

3. What trend seems to be revealed in the line graph?
4. What problems might there be in using this graph to predict future prices of natural gas?
5. Which measure of central tendency might the gas supplier use to advertise its prices? Explain.
6. Which measure of central tendency could competitors use to discourage customers from choosing that natural gas supplier? Explain.

Practice and Applications

HOMEWORK HELP	
For Exercises	See Examples
7–8	1
9–13	2

Extra Practice See pages 569, 597.

7. **ENTERTAINMENT** Both bar graphs show the percent of viewers that watch network television. Which graph makes it appear that the number of viewers is decreasing rapidly? Explain.

Graph A

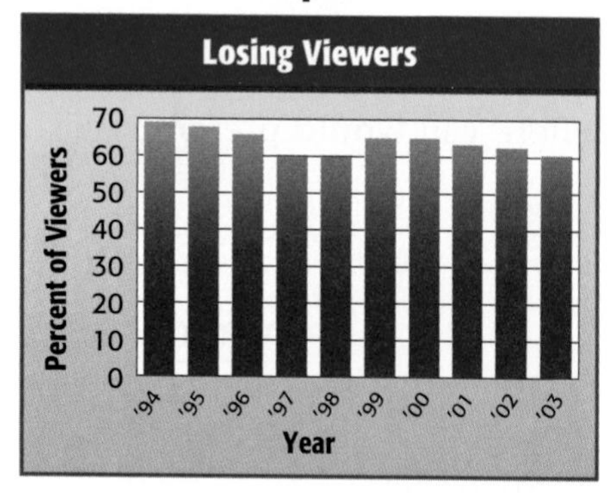

Graph B

Losing Viewers

Percent of Viewers: 70, 68, 66, 64, 62, 60, 58, 0

'94 '95 '96 '97 '98 '99 '00 '01 '02 '03

Year

8. **UTILITIES** The line graph shows the monthly electric bill for the condominium that Toshiko is interested in renting. Why is the graph misleading?

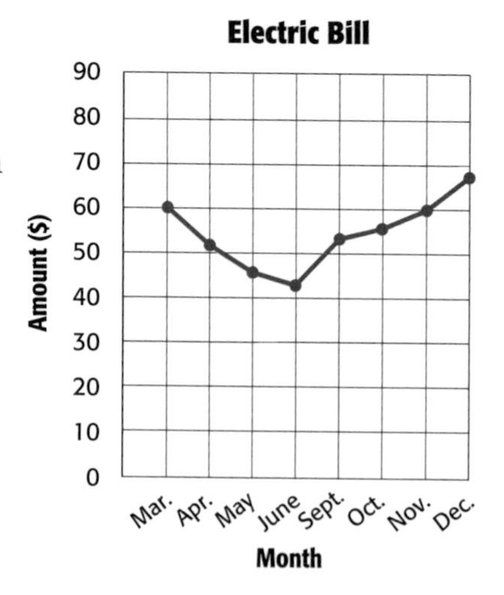

GEOGRAPHY For Exercises 9–11, use the table at the right.

9. Find the mean, mode, and median of the data.
10. Which measure of central tendency would be misleading in describing the size of the islands? Explain.
11. Which measure would most accurately describe the data?

Caribbean Islands

Island	Area (sq mi)	Island	Area (sq mi)
Antigua	108	Martinique	425
Aruba	75	Puerto Rico	3,339
Barbados	166	Tobago	116
Curacao	171	Virgin Islands, UL	59
Dominica	290	Virgin Islands U.S.	134

Source: U.S. Census Bureau

TESTS For Exercises 12 and 13, use Miguel's test scores shown in the line plot.

80 82 84 86 88 90 92 94 96 98 100

12. Which measure of central tendency might he use to emphasize the high scores?
13. Would this number be misleading? Explain.

14. **CRITICAL THINKING** Do values that are much greater or much less than the other values affect the median of a set of data? Give an example to support your answer.

Standardized Test Practice and Mixed Review

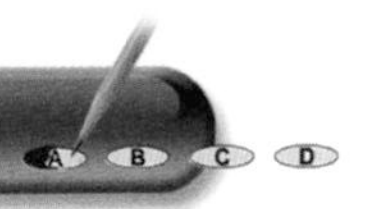

15. **MULTIPLE CHOICE** The table below shows the population of ten counties in Arizona from the 2000 Census. Which measure of central tendency, if any, is misleading in representing the data?

County	Population	County	Population
Apache	69,423	La Paz	19,715
Cochise	117,755	Maricopa	3,072,149
Coconino	116,320	Mohave	155,032
Gila	51,335	Navajo	97,470
Graham	33,489	Santa Cruz	38,381

Source: U.S. Census Bureau

Ⓐ mode Ⓑ mean Ⓒ median Ⓓ none of them

16. **SHORT RESPONSE** Describe one way that a graph could be misleading.

17. **SKATING** Describe what the height of the fourth bar in the graph represents. (Lesson 2-7)

18. Make a box-and-whisker plot of the following data. 12, 15, 20, 12, 13, 14, 15, 14, 17, 16, 16, 18, 15 (Lesson 2-6)

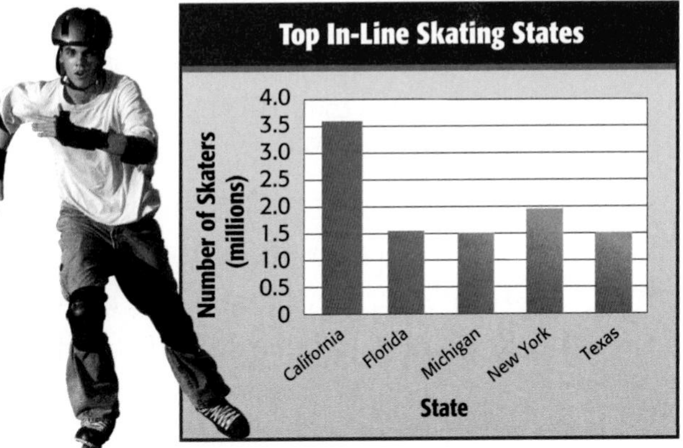

Source: National Sporting Goods Association

MOVIES For Exercises 19 and 20, use the movie ticket prices below.

$7.50, $6.50, $7.50, $8.00, $8.50, $8.00, $7.00

19. Find the mean, median, and mode. Round to the nearest tenth if necessary. (Lesson 2-4)

20. Find the range of the data. (Lesson 2-3)

Complete. (Lesson 1-8)

21. 8.5 g = ___?___ mg

22. 19 cm = ___?___ mm

23. 643 mg = ___?___ g

24. 5,375 mL = ___?___ L

WebQuest Interdisciplinary Project

Lions, Tigers, and Bears

It's time to complete your project. Use the information and data you have gathered about the animals you selected to prepare a video or poster. Be sure to include a spreadsheet and all the necessary graphs with your project.

msmath2.net/webquest

Study Guide and Review

Vocabulary and Concept Check

bar graph (p. 85)
box-and-whisker plot (p. 80)
cluster (p. 65)
data (p. 54)
frequency table (p. 54)
histogram (p. 86)
interquartile range (p. 81)
interval (p. 54)
leaf (p. 76)
line graph (p. 60)
line plot (p. 64)
lower extreme (p. 80)
lower quartile (p. 80)
mean (p. 69)
measures of central tendency (p. 69)
median (p. 70)
mode (p. 70)
outlier (p. 65)
range (p. 65)
scale (p. 54)
scatter plot (p. 61)
statistics (p. 54)
stem (p. 76)
stem-and-leaf plot (p. 76)
upper extreme (p. 80)
upper quartile (p. 80)

State whether each sentence is *true* or *false*. If *false*, replace the underlined word to make a true sentence.

1. The range is the difference between the greatest and the least values in a set of data.
2. The mode divides a set of data in half.
3. A graph that uses bars to make comparisons is a bar graph.
4. A scatter plot shows two sets of related data on the same graph.
5. The median is a data value that is quite separated from the rest of the data.
6. The mean is the arithmetic average of a set of data.
7. The number or item that appears most often in a set of data is the mode.
8. The lower quartile is the middle number in a set of data when the data are arranged in numerical order.

Lesson-by-Lesson Exercises and Examples

2-1 Frequency Tables (pp. 54–57)

Make a frequency table of each set of data.

9.

Hours of Sleep Per Night							
9	6	7	8	9	5	8	7
6	7	6	7	7	8	9	8

10.

Favorite Sport									
B	F	H	S	K	F	S	B	B	F
B	F	K	S	F	F	B	K	F	S

B = baseball, F = football, H = hockey, S = soccer, K = basketball

Example 1 Make a frequency table of the data at the right.

Age of Pet (yr)					
4	2	2	3	8	9
5	9	8	7	10	4

interval: 4
scale: 0–11

Age (yr)	Tally	Frequency
0–3	III	3
4–7	IIII	4
8–11	~~IIII~~	5

msmath2.net/vocabulary_review

2-2 Making Predictions (pp. 60–63)

11. **ENTERTAINMENT** The graph shows the attendance at a theme park. If the trend continues, predict the attendence in 2005.

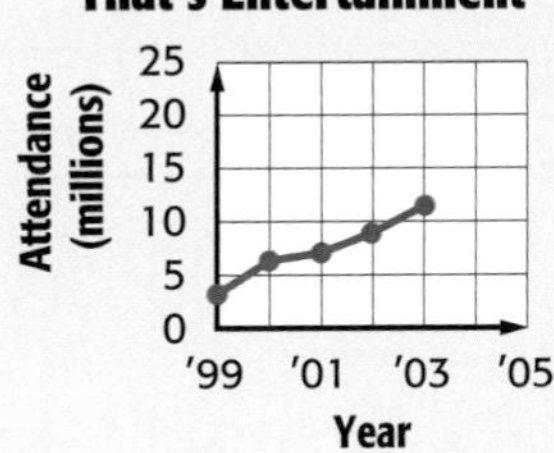

Example 2

SPORTS The graph shows the 400-meter winning times in the Summer Olympics. Predict the winning time in 2004.

By extending the graph, it appears that the winning time in 2004 will be about 42.1 seconds.

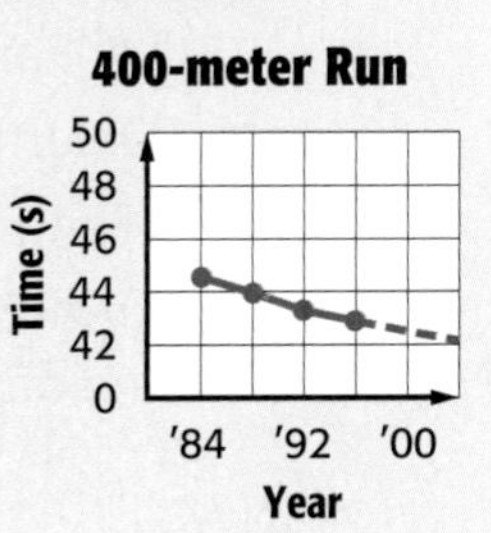

2-3 Line Plots (pp. 64–68)

Make a line plot for each set of data. Identify any clusters, gaps, or outliers.

12. 10, 12, 10, 8, 13, 10, 8, 12
13. 7, 8, 8, 9, 14, 9, 8, 7
14. 43, 41, 42, 45, 43, 42, 43, 46, 44, 44
15. 21, 23, 19, 21, 22, 28, 20, 22, 23, 19

Example 3 **Make a line plot for the data set 72, 75, 73, 72, 74, 68, 73, 74, 74, 75, and 73. Identify any clusters, gaps, or outliers.**

68 69 70 71 72 73 74 75

cluster: 72–75, gap: 68–72, outlier: 68

2-4 Mean, Median, and Mode (pp. 69–72)

Find the mean, median, and mode for each set of data.

16. 2, 3, 4, 3, 4, 3, 8, 7, 2
17. 31, 24, 26, 18, 23, 31, 18
18. 89, 76, 93, 100, 72, 86, 74
19. 125, 108, 172, 136, 121, 112, 148, 136

Example 4 **Find the mean, median, and mode for the data set 23, 22, 19, 19, and 20.**

mean: $\frac{23 + 22 + 19 + 19 + 20}{5}$ or 20.6

median: 20 middle value of the ordered set

mode: 19 value that occurs most often

2-5 Stem-and-Leaf Plots (pp. 76–79)

Make a stem-and-leaf plot for each set of data.

20. 29, 54, 31, 26, 38, 46, 23, 21, 32, 37
21. 75, 83, 78, 85, 87, 92, 78, 53, 87, 89, 91
22. 9, 5, 12, 21, 18, 7, 16, 24, 11, 10, 3, 14
23. 234, 218, 229, 204, 221, 219, 201, 225

Example 5 **Make a stem-and-leaf plot for the data set 12, 15, 17, 20, 22, 22, 23, 25, 27, and 45.**

The tens digits form the stems, and the ones digits form the leaves.

Stem	Leaf
1	2 5 7
2	0 2 2 3 5 7
3	
4	5 $2\|3 = 23$

2-6 Box-and-Whisker Plots (pp. 80–83)

GRADES For Exercises 24–27, use the following math test scores.

83, 92, 75, 59, 94, 82, 88, 79, 92, 90, 85, 96

24. Find the lower extreme, LQ, median, UQ, and upper extreme.
25. Determine the interquartile range.
26. Identify any outliers.
27. Draw a box-and-whisker plot of the data.

Example 6 Make a box-and-whisker plot of the data set 12, 12, 13, 14, 14, 15, 15, 15, 16, 16, 17, 18, and 20.

lower extreme = 12 upper extreme = 20
median = 15 LQ = 13.5 UQ = 16.5

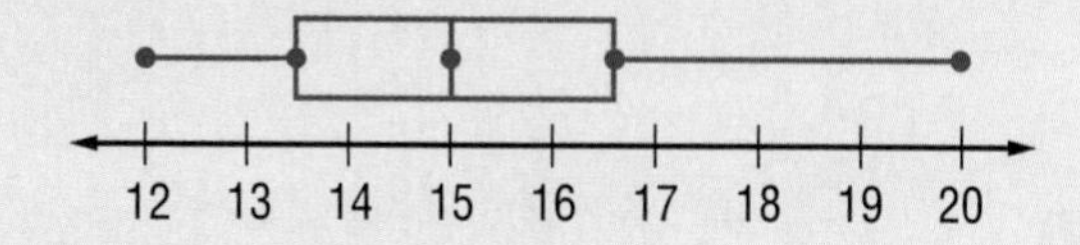

2-7 Bar Graphs and Histograms (pp. 85–89)

For Exercises 28 and 29, refer to the graph at the right.

28. How many students are 50 to 59 inches tall?
29. How many students are at least 60 inches tall?

Example 7 The graph shows the heights to the nearest inch of students in a classroom. Is this a bar graph or a histogram? Explain.

This is a histogram because it shows the number of students having the heights described by each interval.

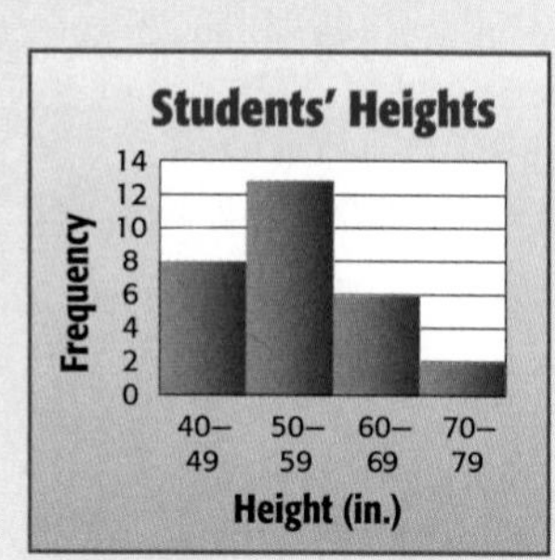

2-8 Misleading Statistics (pp. 92–95)

30. **SALES** The graph below shows the monthly CD sales at the Music Barn. Why is the graph misleading?

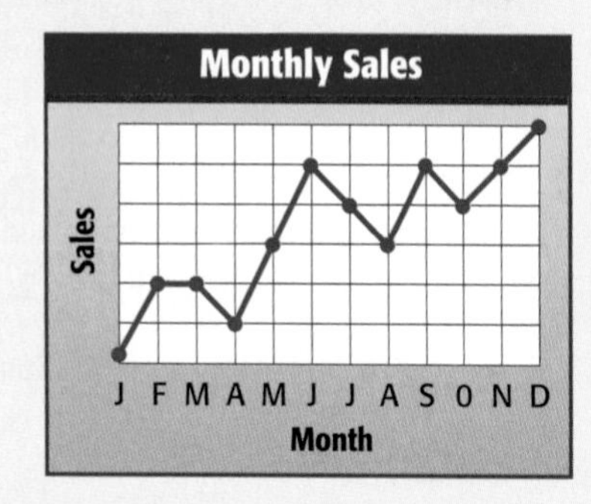

Example 8 The graph shows the pounds of aluminum cans that students recycled. Why is this graph misleading?

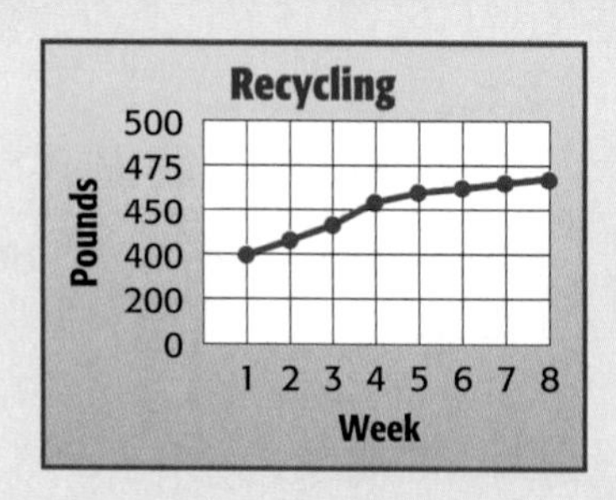

The vertical scale does not have equal intervals. It has intervals of 200 and 50.

CHAPTER 2

Practice Test

Vocabulary and Concepts

1. **Explain** how the range is different from the mean, median, and mode.
2. **Describe** how quartiles separate a set of data.

Skills and Applications

TESTS For Exercises 3–5, use the French test grades shown below.

95, 76, 82, 90, 83, 76, 79, 82, 95, 85, 93, 81, 63

3. Make a frequency table of the data.
4. Make a line plot of the data.
5. Identify any clusters, gaps, or outliers.
6. **EMPLOYMENT** The line graph shows the percent of women who had jobs outside the home from 1975 to 2000. Use the graph to predict the percent of women who will have jobs outside the home in 2005.

Women in Jobs

Percent (%): 0, 10, 20, 30, 40, 50

Year: '75, '80, '85, '90, '95, '00

Find the mean, median, and mode for each set of data. Round to the nearest tenth if necessary.

7. 4, 6, 11, 7, 4, 11, 4
8. 12.4, 17.9, 16.5, 10.2

Make a stem-and-leaf plot for each set of data.

9. 37, 59, 26, 42, 57, 53, 31, 58
10. 461, 422, 430, 425, 425, 467, 429

ALLOWANCES For Exercises 11 and 12, use the monthly allowances below.

\$25, \$32, \$40, \$10, \$18, \$15, \$30, \$40, \$25, \$36, \$20

11. Find the lower extreme, LQ, median, UQ, and upper extreme.
12. Make a box-and-whisker plot of the data.
13. **SALES** The line graph shows ship sales at Marvin's Marina in thousands of dollars. Why is the graph misleading?
14. Make a histogram for the French test grades given for Exercises 3–5.

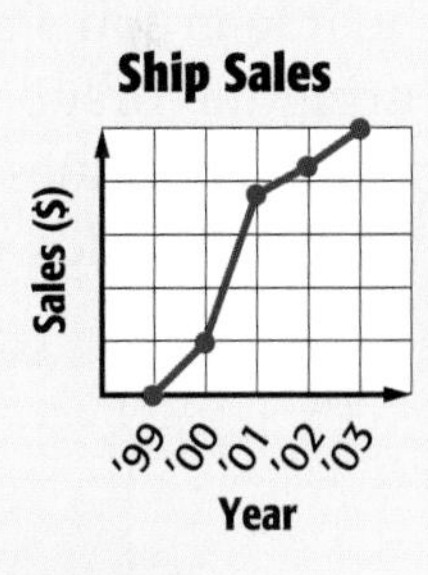

Standardized Test Practice

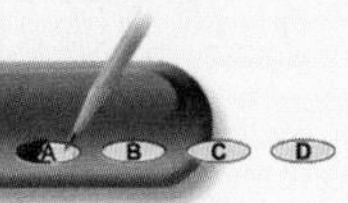

15. **MULTIPLE CHOICE** Choose the statement that is *not* correct.

    (A) The bars on a histogram represent the frequency of data in intervals.

    (B) The bars on a bar graph represent the number of data items in each interval.

    (C) The bars on a bar graph have space between them.

    (D) There is no space between the bars on a histogram.

Standardized Test Practice

CHAPTER 2

PART 1 Multiple Choice

Record your answers on the answer sheet provided by your teacher or on a sheet of paper.

1. Which is greater than 6.8? (Prerequisite Skill, p. 556)

   (A) 6.08 (B) 6.79

   (C) 6.80 (D) 6.9

2. What is the value of the expression $5 + 2 \times 4^2 - 9$? (Lesson 1-3)

   (F) 4 (G) 19

   (H) 28 (I) 103

3. Which is 430,000 written in scientific notation? (Lesson 1-9)

   (A) 4.3×10^5

   (B) 4.3×10^6

   (C) 43×10^5

   (D) 43×10^6

4. The line graph shows a baby's length during the first 8 months. If the pattern of growth continues, predict the baby's length at 9 months old. (Lesson 2-2)

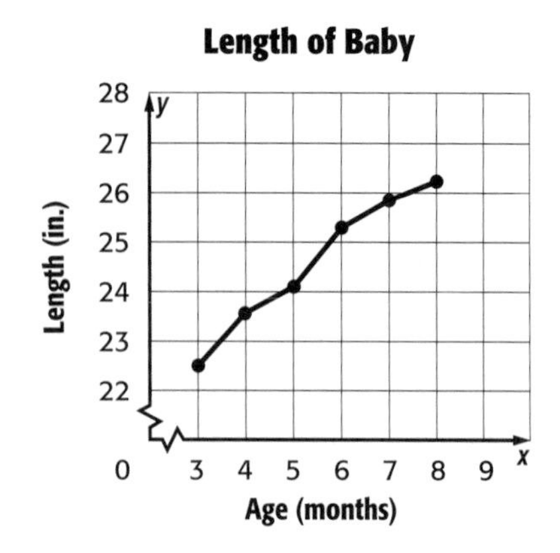

   (F) 26 in. (G) 27 in.

   (H) 28 in. (I) 29 in.

5. The line plot shows the number of minutes ten different students spent on homework last night. What is the range of the data? (Lesson 2-3)

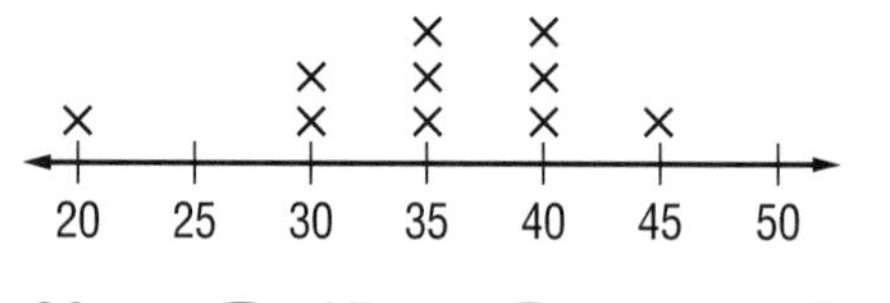

   (A) 20 (B) 25 (C) 30 (D) 50

6. The table shows the length of the phone calls made in one household on a given day. What is the median? (Lesson 2-4)

   (F) 2 min

   (G) 6 min

   (H) 7 min

   (I) 8 min

Length of Phone Calls (min)					
2	3	10	2	8	12
6	7	25	7	6	7

7. The stem-and-leaf plot shows the average monthly temperature in one city for a year. What is the mode? (Lesson 2-5)

Stem	Leaf
4	2 2 2 5
5	5 6
6	4
7	0 8 8
8	2 5

$5|6 = 56°F$

   (A) 42 (B) 55 (C) 65 (D) 78

8. The box-and-whisker plot below summarizes the ages of Leila's cousins. Which statement is *not* true? (Lesson 2-6)

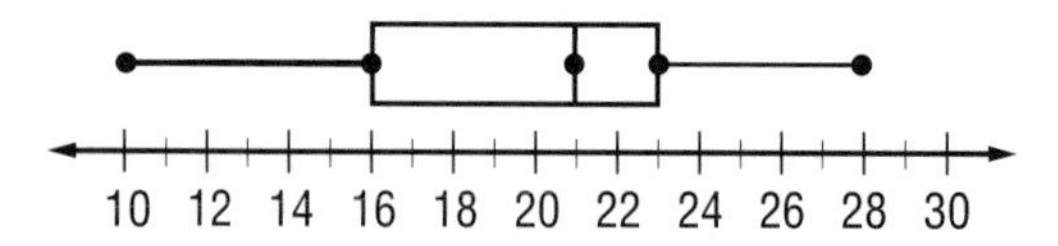

   (F) One fourth of Leila's cousins are between 10 and 16 years old.

   (G) Half of Leila's cousins are between 16 and 21 years old.

   (H) The range of ages is 18 years.

   (I) Half of Leila's cousins are older than 21.

PART 2 Short Response/Grid In

Record your answers on the answer sheet provided by your teacher or on a sheet of paper.

9. What is the value of $14 + x^2$ if $x = 3$? (Lesson 1-4)

10. Find the missing number in the sequence below. (Lesson 1-7)

    24.7, 25.0, __?__, 25.6, 25.9, …

11. The frequency table summarizes the number of miles Dino ran daily to train for a race. What interval shows the number of miles that he ran most frequently? (Lesson 2-1)

Number of Miles	Tally	Frequency
0–4	~~IIII~~ IIII	9
5–9	~~IIII~~ ~~IIII~~ III	13
10–14	IIII	4
15–19	I	1

12. If the number 18 was taken out of the number data set 7, 18, 46, 24, 2, 44, 31, 11, would the median or the mean change the most? (Lesson 2-4)

13. What is the upper extreme of the box-and-whisker plot below? (Lesson 2-6)

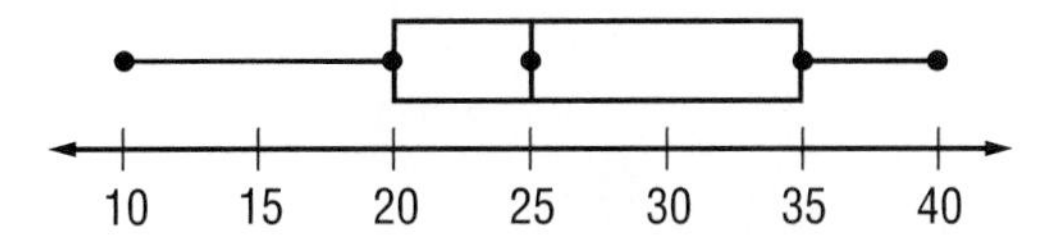

14. The graph below shows the sales of compact discs during a five-day sale. How many more compact discs were sold on Saturday than on Wednesday? (Lesson 2-7)

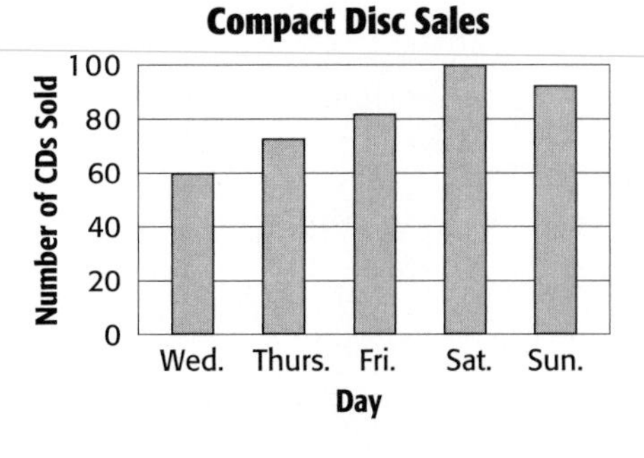

TEST-TAKING TIP

Question 15 Make sure you have answered every part of an open-ended question. If you find that you cannot answer the question entirely, do as much as you can, because you may earn partial credit.

PART 3 Extended Response

Record your answers on a sheet of paper. Show your work.

15. The table shows how the value of a painting increased over ten years. (Lesson 2-2)

Year	Value	Year	Value
1993	$350	1998	$1,851
1994	$650	1999	$2,151
1995	$950	2000	$2,451
1996	$1,250	2001	$2,752
1997	$1,551	2002	$3,052

    a. Make a line graph of the data.

    b. Use the graph to predict what the value of the painting will be in 2006.

16. Dominique had swimming practice before school one week and after school the following week. Her timed results are in the table below. (Lesson 2-7)

Day	Time Before School (s)	Time After School (s)
Monday	45.2	41.5
Tuesday	41.3	38.7
Wednesday	40.6	37.4
Thursday	42.9	35.2
Friday	40.7	38.9

    a. Make a double-line graph of the data. Let the x-axis represent the days and let the y-axis represent the times. Title the graph and label each part.

    b. Write a sentence comparing the data.

UNIT 2
Integers and Algebra

Chapter 3

Algebra: Integers

Chapter 4

Algebra: Linear Equations and Functions

Your study of math includes more than just whole numbers and decimals. In this unit, you will use negative numbers to describe many real-life situations and you will solve and graph equations that represent them.

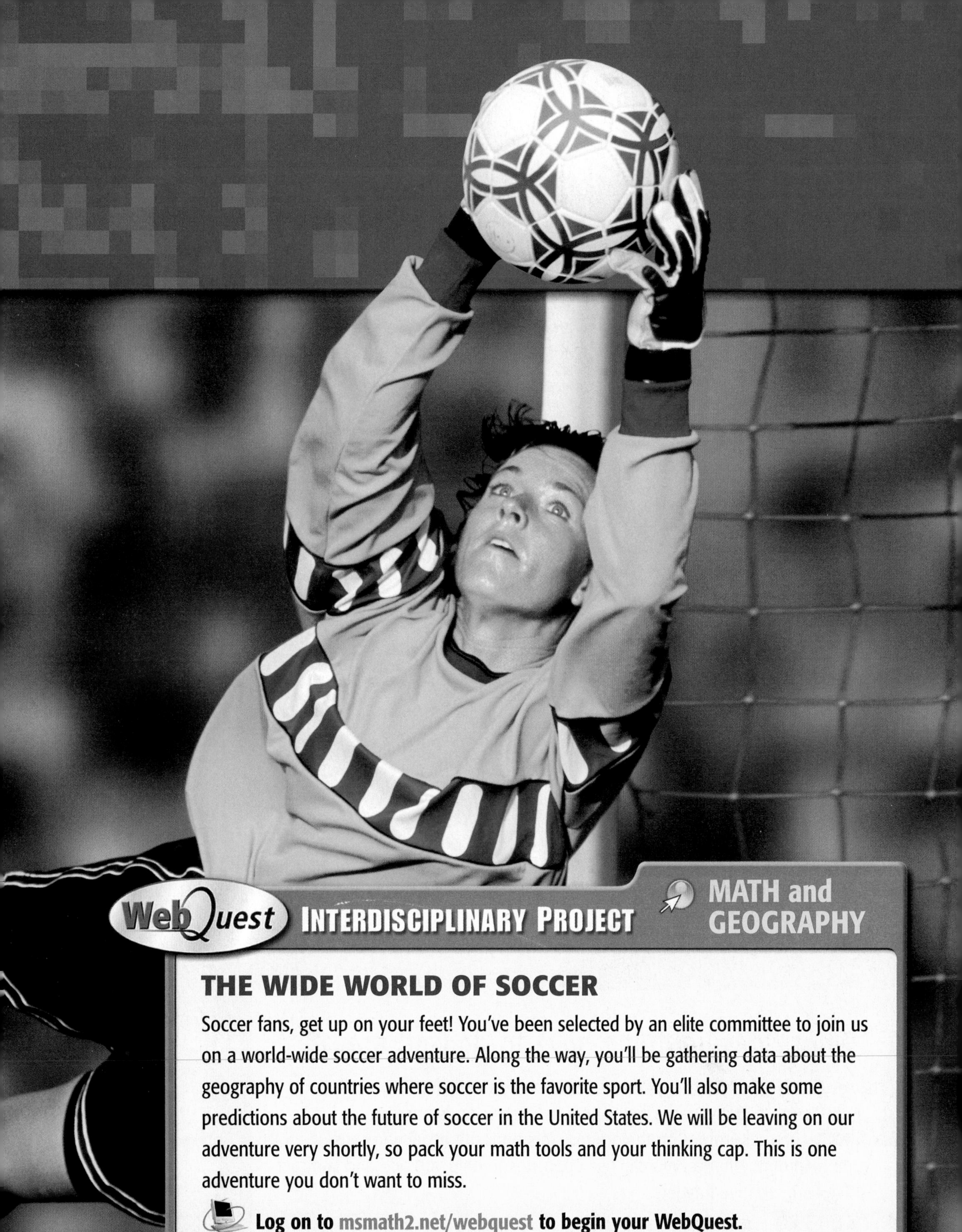

WebQuest INTERDISCIPLINARY PROJECT

MATH and GEOGRAPHY

THE WIDE WORLD OF SOCCER

Soccer fans, get up on your feet! You've been selected by an elite committee to join us on a world-wide soccer adventure. Along the way, you'll be gathering data about the geography of countries where soccer is the favorite sport. You'll also make some predictions about the future of soccer in the United States. We will be leaving on our adventure very shortly, so pack your math tools and your thinking cap. This is one adventure you don't want to miss.

Log on to msmath2.net/webquest to begin your WebQuest.

Algebra: Integers

"What does lightning have to do with math?"

Have you heard of the phrase "opposites attract"? During thunderstorms, negatively-charged electrons in the clouds are attracted to positively-charged protons on the ground. This opposite attraction causes lightning. In mathematics, you can use opposites to help you add and subtract positive and negative integers.

You will solve problems about thunderstorms in Lesson 3-4.

GETTING STARTED

Take this quiz to see whether you are ready to begin Chapter 3. Refer to the lesson or page number in parentheses if you need more review.

Vocabulary Review

State whether each sentence is *true* or *false*. If *false*, replace the underlined word to make a true sentence.

1. The mean of 1, 3, and 6 is 3. (Lesson 2-4)
2. The difference between the greatest number and the least number in a set of data is called the range. (Lesson 2-3)

Prerequisite Skills

Replace each ● with < or > to make a true sentence. (Page 556)

3. 1,458 ● 1,548
4. 36 ● 34
5. 1.02 ● 1.20
6. 76.7 ● 77.6

Add.

7. 84 + 39
8. 198 + 289
9. 826 + 904
10. 3,068 + 5,294

Multiply.

11. $2 \cdot 5 \cdot 3$
12. $18 \cdot 9$
13. $15 \cdot 6$
14. $10 \cdot 4 \cdot 7$

Divide.

15. $63 \div 9$
16. $96 \div 12$
17. $125 \div 5$
18. $187 \div 17$

Find the mean and range for each set of data.
(Lessons 2-3 and 2-4)

19. 12, 8, 25, 16, 9
20. 34, 57, 60, 45
21. 93, 88, 122, 97
22. 7, 10, 2, 6, 13

Integers Make this Foldable to help you organize information about integers. Begin with two sheets of $8\frac{1}{2}'' \times 11''$ paper.

STEP 1 Fold and Cut One Sheet
Fold in half from top to bottom. Cut along fold from edges to margin.

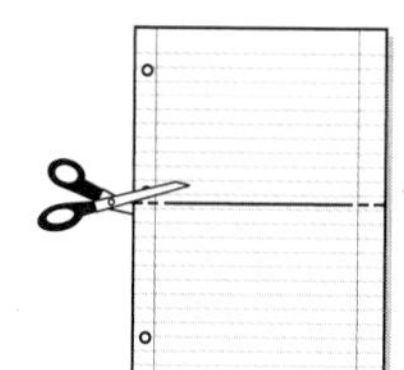

STEP 2 Fold and Cut the Other Sheet
Fold in half from top to bottom. Cut along fold between margins.

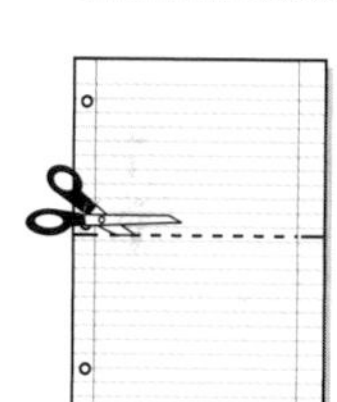

STEP 3 Fold
Insert first sheet through second sheet and align folds.

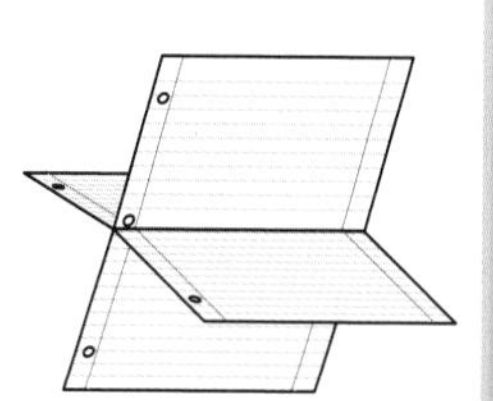

STEP 4 Label
Label each page with a lesson number and title.

3-1
Integers and Absolute Value

Reading and Writing As you read and study the chapter, fill the journal with notes, diagrams, and examples of integers.

3-1 Integers and Absolute Value

What You'll LEARN

Read and write integers, and find the absolute value of an integer.

NEW Vocabulary

integer
graph
positive integer
negative integer
absolute value

MATH Symbols

+3 positive three

−3 negative three

|3| absolute value of three

WHEN am I ever going to use this?

FOOTBALL The graph shows the number of yards the Bears gained or lost on the first four downs. A value of −3 represents a 3-yard loss.

1. What does a value of −2 represent?
2. On which down did they lose the most yards?
3. How can you represent a gain of 9 yards?

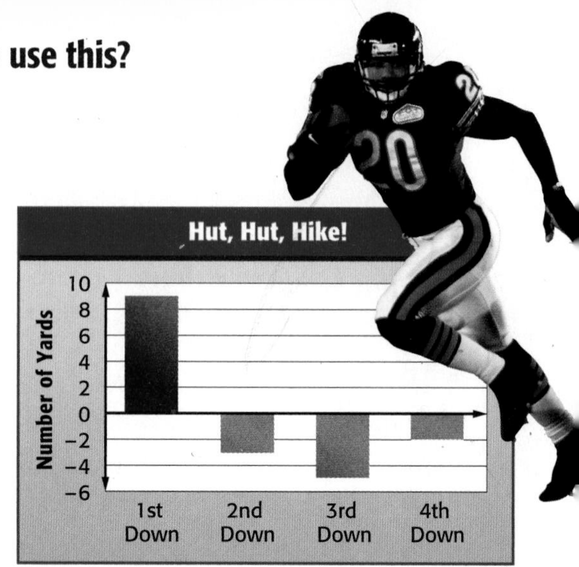

Numbers like 9 and −2 are called integers. An **integer** is any number from the set {..., −4, −3, −2, −1, 0, 1, 2, 3, 4, ...}. Integers can be graphed on a number line. To **graph** a point on the number line, draw a point on the line at its location.

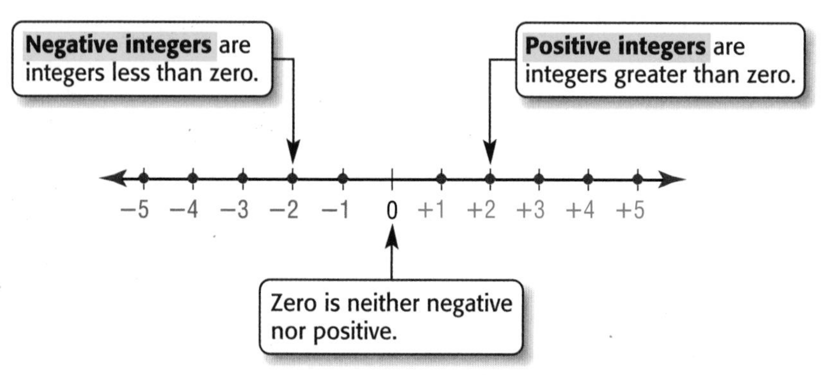

EXAMPLES Write Integers for Real-Life Situations

WEATHER Write an integer for each situation.

1 The average temperature in Tennessee for May was 5 degrees below normal.

Because it represents *below* normal, the integer is −5.

2 The average rainfall in Virginia for November was 5 inches above normal.

Because it represents *above* normal, the integer is +5 or 5.

Your Turn Write an integer for each situation.

a. 6 degrees above normal

b. 2 inches below normal

Integers Positive integers are usually written without the + sign.

READING Math

Set Theory The number 5 is an *element*, or member, of the set of integers.

The set $\{-5, 5\}$ is a *subset* of the set of integers.

You can represent the integers from Examples 1 and 2 on a number line.

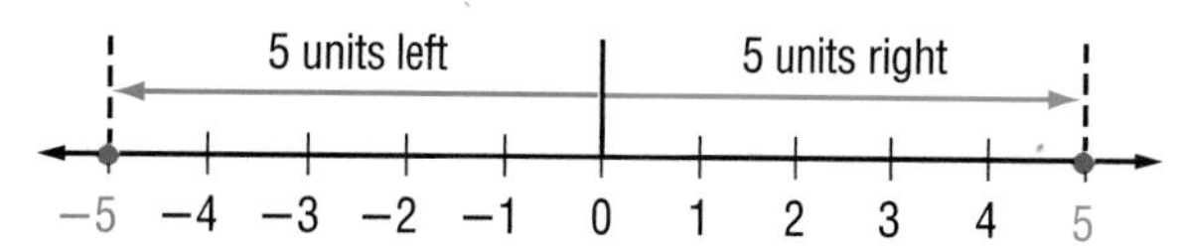

The numbers -5 and 5 are the same distance from 0, but on opposite sides of 0. So, -5 and 5 have the same **absolute value**.

Key Concept — Absolute Value

Words The absolute value of an integer is the distance the number is from zero on a number line.

Examples $|6| = 6$ $|-6| = 6$

EXAMPLES Evaluate Expressions

Evaluate each expression.

3 $|-4|$

On the number line, the graph of -4 is 4 units from 0.

So, $|-4| = 4$.

4 $|-5| - |2|$

$|-5| - |2| = 5 - 2$ $|-5| = 5, |2| = 2$

$= 3$ Subtract.

Skill and Concept Check

Writing Math Exercises 1 & 3

1. **Describe** a situation in everyday life where negative numbers are used.
2. **OPEN ENDED** On a number line, graph two different points that have the same absolute value.
3. **Which One Doesn't Belong?** Identify the expression that does not have the same value as the other three. Explain your reasoning.

$|-3|$	-3	$|3|$	3

GUIDED PRACTICE

Write an integer for each situation.

4. 6°F below 0
5. a loss of 11 yards
6. a deposit of $16

Evaluate each expression.

7. $|7|$
8. $|-4|$
9. $|-7| - |1|$
10. **STOCK MARKET** The price of a company's stock fell 21 points in two days. Write an integer to represent the amount the stock price fell.

msmath2.net/extra_examples

Practice and Applications

HOMEWORK HELP	
For Exercises	See Examples
11–18, 29	1, 2
19–22, 27, 28	3
23–26	4

Extra Practice
See pages 569, 598.

Write an integer for each situation.

11. a profit of $9
12. 53°C below 0
13. no gain on first down
14. an elevator goes up 12 floors
15. 2008 A.D.
16. 160 feet above sea level
17. a bank withdrawal of $50
18. 1000 B.C.

Evaluate each expression.

19. $|6|$
20. $|-12|$
21. $|-9|$
22. $|21|$
23. $|12| - |-8|$
24. $|-10| - 5$
25. $|-9| + |5|$
26. $|26| + |-4|$
27. What is the absolute value of 0?
28. Find $|x|$ if $x = -6$.
29. **STATIC ELECTRICITY** Electrical charges are made up of positively-charged protons and negatively-charged electrons. Suppose you rub a balloon through your hair to make the balloon stick to a wall. There are 2 protons on the wall and 5 electrons on the balloon. Write an integer for each charge.

Graph each set of integers on a number line.

30. $\{0, 1, -3\}$
31. $\{-4, 5, 4\}$
32. $\{-5, -1, 10, -9\}$
33. $\{-2, -4, -6, -8\}$
34. **WEATHER** A meteorologist reports a 20° change in the temperature from yesterday to today. Describe what this could mean.

CRITICAL THINKING Determine whether each statement is *true* or *false*. If *false*, give a counterexample.

35. Every integer has an absolute value.
36. The absolute value of every integer is positive.

Standardized Test Practice and Mixed Review

37. **MULTIPLE CHOICE** Identify the point that represents −5.

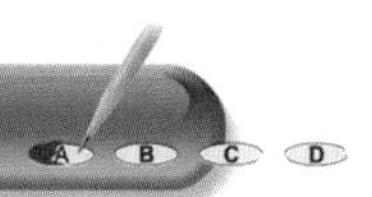

Ⓐ *A* Ⓑ *B* Ⓒ *C* Ⓓ *D*

38. **SHORT RESPONSE** Write an integer for 23°F below 0.

STATISTICS For Exercises 39 and 40, use the following information.
The mean income for a group of accountants is $36,266.67. Their incomes are $27,500, $36,100, $29,800, $33,400, $31,300, and $59,500.

39. In what way is the mean misleading? (Lesson 2-8)
40. Draw a bar graph of the data. (Lesson 2-7)

GETTING READY FOR THE NEXT LESSON

PREREQUISITE SKILL Replace each ● with < or > to make a true sentence. (Page 556)

41. 16 ● 6
42. 2.3 ● 3.2
43. 101 ● 111
44. 87.3 ● 83.7
45. 1,051 ● 1,015

msmath2.net/self_check_quiz

3-2 Comparing and Ordering Integers

What You'll LEARN

Compare and order integers.

REVIEW Vocabulary

median: the middle number in an ordered data set **(Lesson 2-4)**

MATH Symbols

< is less than

> is greater than

WHEN am I ever going to use this?

WEATHER The Wind Chill Temperature Index table shows how cold air feels on human skin.

WIND CHILL					
Wind (mph)	Temperature (°F)				
	15	10	5	0	−5
5	7	1	−5	−11	−16
10	3	−4	−10	−16	−22
15	0	−7	−13	−19	−26
20	−2	−9	−15	−22	−29

1. What is the wind chill if there is a wind at 20 miles per hour and the temperature is 5°?
2. Which is colder, a temperature of 15° with a 20 mile-per-hour wind or a temperature of 10° with a 10 mile-per-hour wind?
3. Graph both wind chills found in Exercise 2 on a number line.

When two numbers are graphed on a number line, the number to the left is always less than the number to the right. The number to the right is always greater than the number to the left.

Key Concept **Comparing Integers**

Model

−5 −4 −3 −2 −1 0 1

Words −4 is less than −2. −2 is greater than −4.

Symbols $-4 < -2$ $-2 > -4$

The symbol points to the lesser number.

EXAMPLE Compare Integers

1 Replace the ● with < or > to make −5 ● −3 a true sentence.

Graph each integer on a number line.

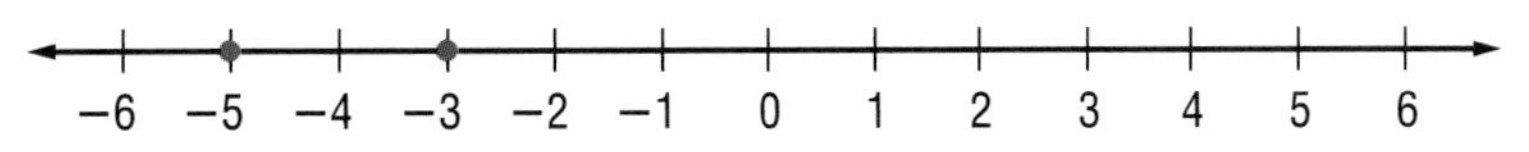

Since −5 is to the left of −3, $-5 < -3$.

Your Turn Replace each ● with < or > to make a true sentence.

a. −8 ● −4 b. 5 ● −1 c. −10 ● −13

Integers are used to compare numbers in many real-life situations.

Standardized Test Practice

EXAMPLE Order Integers

State	Record Low Temperature (°F)
Alaska	−80
Florida	−2
Hawaii	12
Montana	−70

Source: *The World Almanac and Book of Facts*

2 **MULTIPLE-CHOICE TEST ITEM The lowest temperatures in Alaska, Florida, Hawaii, and Montana are listed in the table. Order the temperatures from least to greatest.**

Ⓐ −80, −70, 12, −2

Ⓑ −80, −70, −2, 12

Ⓒ 12, −2, −70, −80

Ⓓ −2, 12, −70, −80

Read the Test Item To order the integers, graph them on a number line.

Solve the Test Item

A M F H

−90 −80 −70 −60 −50 −40 −30 −20 −10 0 10 20

Order the integers from least to greatest by reading from left to right: −80, −70, −2, 12. So, the answer is B.

Test-Taking Tip
The Princeton Review

Eliminating Answer Choices
If you are unsure of the correct answer, eliminate the choices you know are incorrect. Then consider the remaining choices. You can eliminate choice C since the list begins with a positive number.

Skill and Concept Check

Writing Math
Exercise 2

1. **Draw** a number line to show that −5 is less than −1.
2. **OPEN ENDED** Write an integer that is less than −9. Explain.
3. **NUMBER SENSE** Complete the sentence: −7 is greater than −12 because −7 lies to the __?__ of −12 on a number line.
4. **Name** the greatest negative integer.

GUIDED PRACTICE

Replace each ● with < or > to make a true sentence.

5. −5 ● −6
6. −2 ● 8
7. 0 ● −10

8. Order 51, −63, 49, −24, −38, and 38 from least to greatest.

MONEY For Exercises 9 and 10, use the information below and at the right.
Marva is saving money for a new bike and has already saved $21. She begins a log to keep track of her money.

9. Write each entry as an integer.
10. Order the integers from least to greatest.

Practice and Applications

HOMEWORK HELP	
For Exercises	See Examples
11–28	1
29, 31–32	2

Extra Practice
See pages 570, 598.

Replace each ● with < or > to make a true sentence.

11. $-17 \bullet -20$
12. $-21 \bullet -12$
13. $3 \bullet -10$
14. $-5 \bullet 17$
15. $4 \bullet -4$
16. $-25 \bullet -20$
17. $-52 \bullet -72$
18. $100 \bullet -10$
19. $|-8| \bullet 0$
20. $-13 \bullet |-14|$
21. $|36| \bullet -37$
22. $|-29| \bullet |92|$

Determine whether each sentence is *true* or *false*. If *false*, change one number to make the sentence true.

23. $-8 > 5$
24. $-7 < 0$
25. $|-9| = 9$
26. $|5| < -6$
27. $10 > |-8|$
28. $7 > |-7|$

WEATHER **For Exercises 29 and 30, use the information in the table. It shows the record low temperatures in Indianapolis, Indiana, for March 1–7 of a recent year.**

Day	Temperature (°F)
Mar. 1	−4
Mar. 2	−7
Mar. 3	1
Mar. 4	3
Mar. 5	−1
Mar. 6	−6
Mar. 7	6

Source: www.weather.com

29. Arrange the dates from the coldest temperature to the warmest.
30. Find the median temperature.
31. Order −7, 5, −6, −4, 1, and 3 from least to greatest.
32. Order $|51|$, −53, $|-52|$, 55, −56, and −57 from greatest to least.
33. **CRITICAL THINKING** If 0 is the greatest integer in a set of five integers, what can you conclude about the other four integers?

Standardized Test Practice and Mixed Review

34. **MULTIPLE CHOICE** The table shows the inventions of several toys. Order the inventions from earliest to most recent.
    - (A) chess, yo-yo, teddy bear, checkers
    - (B) checkers, yo-yo, chess, teddy bear
    - (C) yo-yo, teddy bear, checkers, chess
    - (D) chess, teddy bear, yo-yo, checkers

Toy	Year
Yo-Yo	1000 B.C.
Teddy Bear	1902 A.D.
Chess	600 A.D.
Checkers	2000 B.C.

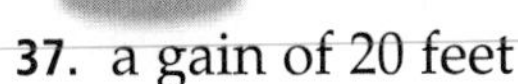

35. **GRID IN** Which is greater, −12 or 7?

Write an integer for each situation. (Lesson 3-1)

36. 9°C below 0
37. a gain of 20 feet

38. **PROFITS** The daily profits of T-shirts sold last week were \$55, \$35, \$25, \$30, and \$55. Which average might be misleading: the mode, the median, or the mean? Explain. (Lesson 2-8)

GETTING READY FOR THE NEXT LESSON

PREREQUISITE SKILL **Graph the solution of each equation on a number line.** (Lesson 1-5)

39. $x + 3 = 5$
40. $x - 4 = 8$
41. $3x = 9$
42. $5x = 30$

3-3 The Coordinate Plane

WHEN am I ever going to use this?

MAPS A map of Terrell's neighborhood is shown.

1. Suppose Terrell starts at the corner of Russel and Main and walks 1 block north and 2 blocks east. Name the intersection of his location.
2. Using the words *north*, *south*, *west*, and *east*, write directions to go from the corner of School and Highland to the corner of Main and Oak.

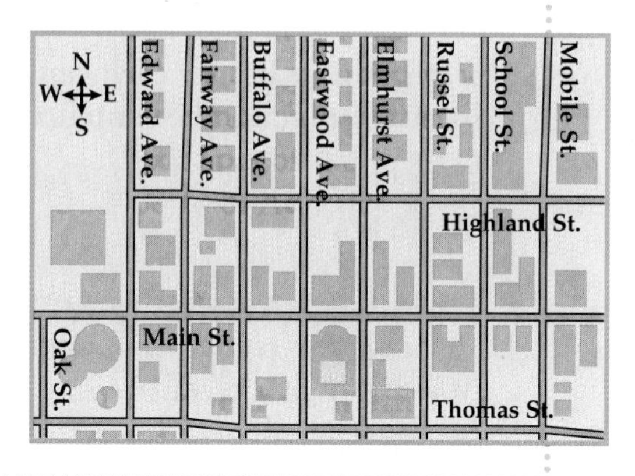

What You'll LEARN

Graph points on a coordinate plane.

NEW Vocabulary

coordinate plane
coordinate grid
***x*-axis**
***y*-axis**
origin
ordered pair
***x*-coordinate**
***y*-coordinate**
quadrant

A **coordinate plane** is used to locate points. It is a plane in which a horizontal number line and a vertical number line intersect at their zero points. A coordinate plane is also called a **coordinate grid**.

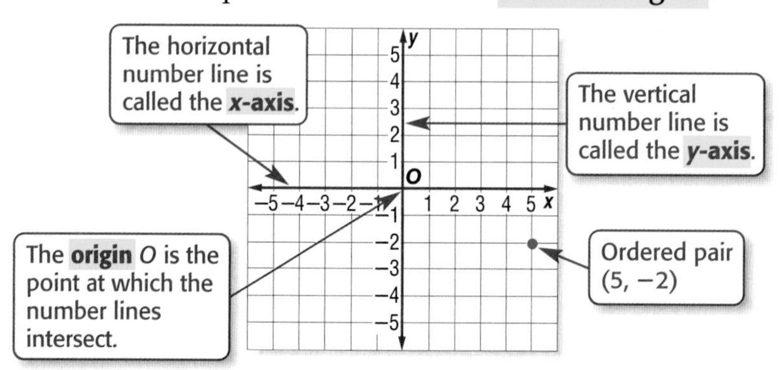

An **ordered pair** is a pair of numbers such as (5, −2) used to locate a point in the coordinate plane.

The ***x*-coordinate** is the first number, and it corresponds to a number on the *x*-axis.

(5, −2)

The ***y*-coordinate** is the second number, and it corresponds to a number on the *y*-axis.

EXAMPLE Name an Ordered Pair

1 **Name the ordered pair for point *P*.**

- Start at the origin.
- Move left to find the *x*-coordinate of point *P*, which is −4.
- Move up to find the *y*-coordinate, which is 2.

So, the ordered pair for point *P* is (−4, 2).

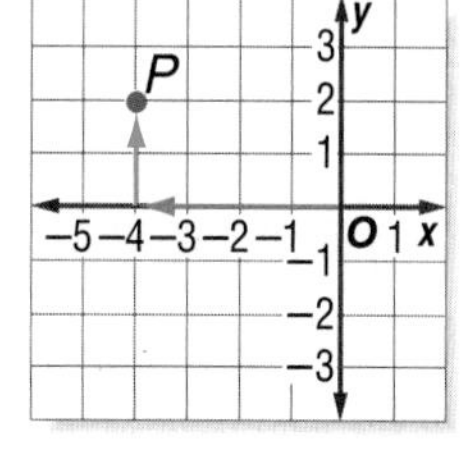

STUDY TIP

Graphing Moving right or up on a coordinate plane is in the *positive direction.* Moving left or down is in the *negative direction.*

READING Math

Scale When no numbers are shown on the *x*- or *y*-axis, you can assume that each grid is 1 unit long on each side.

EXAMPLE Graph an Ordered Pair

2 **Graph and label the point $Q(2, -5)$.**

- Draw a coordinate plane.
- Move 2 units to the right. Then move 5 units down.
- Draw a dot and label it $Q(2, -5)$.

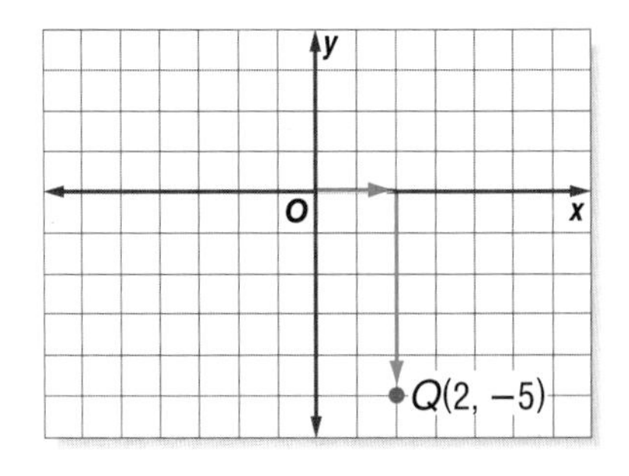

Your Turn **Graph each point.**

a. $A(6, 0)$

b. $B(-5, -3)$

The coordinate plane is separated into four sections called **quadrants**.

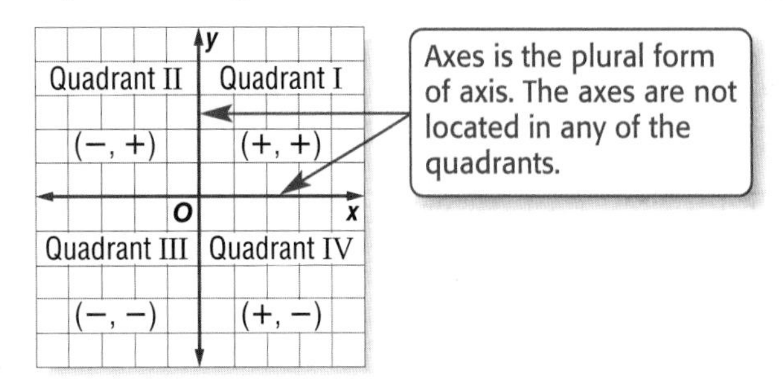

REAL-LIFE CAREERS

How Does a Cartographer Use Math?
Cartographers use coordinates to prepare geographic, political, and cultural maps of large areas.

Online Research
For more information about a career as a cartographer, visit: msmath2.net/careers

EXAMPLES Identify Quadrants

3 **GEOGRAPHY** **The world map can be divided into a coordinate grid where (x, y) represents (degrees longitude, degrees latitude). In which quadrant is the United States located?**

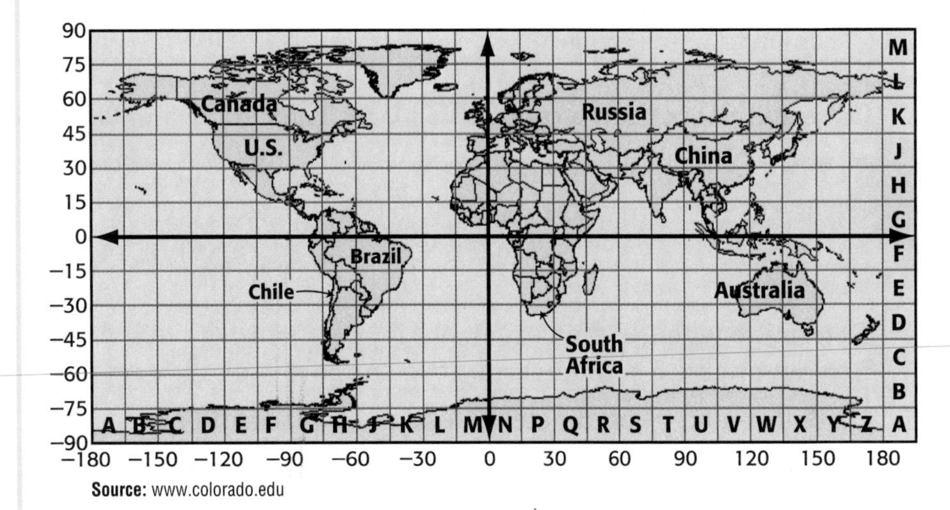

Source: www.colorado.edu

The United States is located in the upper left quadrant, quadrant II.

4 **Name a country from the map that is located in quadrant III.**

Quadrant III is the bottom-left quadrant. So, Chile is in quadrant III.

Skill and Concept Check

Writing Math
Exercise 1

1. **Explain** why point $A(1, -2)$ is different from point $B(-2, 1)$.

2. **OPEN ENDED** Name and graph a point in quadrant IV.

GUIDED PRACTICE

Name the ordered pair for each point graphed at the right. Then identify the quadrant in which each point lies.

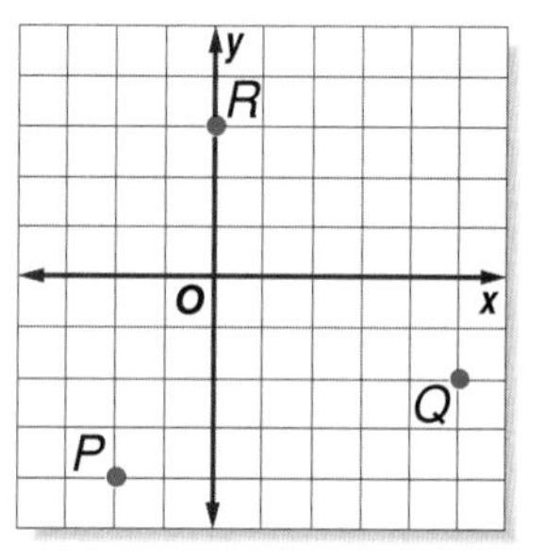

3. P 4. Q 5. R

On graph paper, draw a coordinate plane. Then graph and label each point.

6. $S(2, 3)$ 7. $T(-4, 6)$ 8. $U(-5, 0)$

Practice and Applications

Name the ordered pair for each point graphed at the right. Then identify the quadrant in which each point lies.

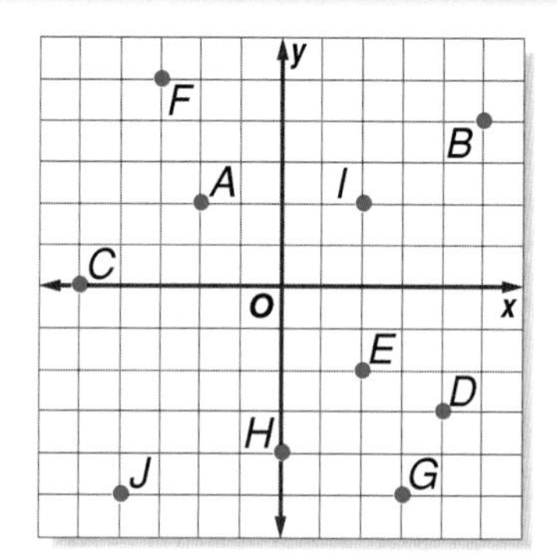

9. A 10. B
11. C 12. D
13. E 14. F
15. G 16. H
17. I 18. J

HOMEWORK HELP

For Examples	See Examples
9–18	1, 3
19–31	2
35–42	3, 4

Extra Practice
See pages 570, 598.

19. Write the ordered pair for the point that lies on the y-axis and is 32 units down from the origin.

On graph paper, draw a coordinate plane. Then graph and label each point.

20. $M(5, 6)$ 21. $N(-2, 10)$ 22. $P(7, -8)$ 23. $Q(3, 0)$
24. $R(-1, -7)$ 25. $S(0, 6)$ 26. $T(-3, 7)$ 27. $U(5, -2)$
28. $V(8, 1)$ 29. $W(-5, -7)$ 30. $X(1.5, -3)$ 31. $Y(-6.5, 6.5)$

Determine whether each statement is *sometimes*, *always*, or *never* true. Explain or give a counterexample to support your answer.

32. Both x- and y-coordinates of a point in quadrant I are negative.
33. The x-coordinate of a point that lies on the x-axis is negative.
34. The y-coordinate of a point in quadrant IV is negative.

GEOGRAPHY For Exercises 35 and 36, use the map in Example 3.

35. In what country is the point (105° longitude, 30° latitude) located?
36. Find an ordered pair that can represent the location of California.

For Exercises 37–41, use the map of the Brookfield Zoo.

Brookfield Zoo, Brookfield, IL

Hoofed Animals, Camels, North Gate, Dragonfly Marsh, Habitat Africa! The Forest, Australia House, Sea Lions, Habitat Africa! The Savannah, The Fragile Kingdom, Salt Creek Wilderness, Seven Seas, Pachyderm House, Roosevelt Fountain, Ibex Mountain, Tropic World: A Primate's Journey, Bear Grottos, Wolf Woods, The Living Coast, Baboon Island, The Swamp, Children's Zoo, Perching Bird House, South Gate

Source: www.brookfieldzoo.org

37. What exhibit is located at $(4, -2)$?
38. In which quadrant is the Dragonfly Marsh exhibit located?
39. Find the ordered pair that represents the location of Baboon Island.
40. What is located at the origin?
41. Describe how you would walk from the entrance of the Pachyderm House at $(-2, 2)$ to the entrance of The Swamp at $(-1, -2)$.
42. **GEOMETRY** Graph the points $A(-3, 2)$, $B(2, 2)$, $C(2, -4)$, and $D(-3, -4)$ on the same coordinate plane. Connect the points from A to B, B to C, C to D, and D to A. Name the figure.
43. **CRITICAL THINKING** Find the possible locations for any ordered pair whose x- and y-coordinates are always the same integer. Explain.

Standardized Test Practice and Mixed Review

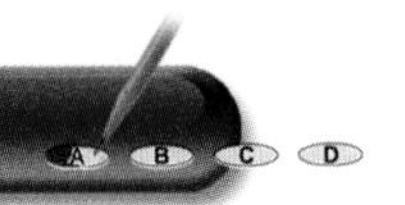

For Exercises 44–46, use the coordinate plane at the right.

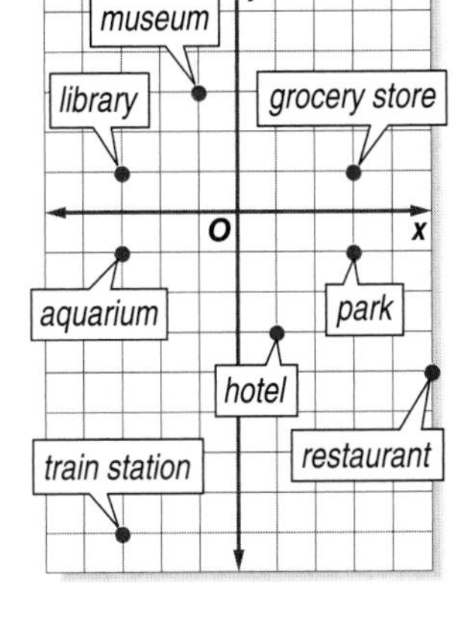

44. **MULTIPLE CHOICE** Which building has the coordinates of $(-3, -1)$?

   Ⓐ museum   Ⓑ park   Ⓒ library   Ⓓ aquarium

45. **MULTIPLE CHOICE** What are the coordinates of the point that shows the location of the hotel?

   Ⓕ $(-1, -3)$ Ⓖ $(1, -3)$ Ⓗ $(-1, 3)$ Ⓘ $(1, 3)$

46. **SHORT RESPONSE** In which quadrant is the grocery store located?

Replace each ● with < or > to make a true sentence. (Lesson 3-2)

47. $14 \bullet |-15|$ 48. $-8 \bullet -3$ 49. $26 \bullet -30$ 50. $-40 \bullet |40|$

51. Find the absolute value of -101. (Lesson 3-1)

52. **SPORTS** A triathlon competition consists of swimming 3 miles, running 10 miles, and bicycling 35 miles. How many miles does an athlete travel during the competition? (Lesson 1-1)

GETTING READY FOR THE NEXT LESSON

BASIC SKILL **Add.**

53. $138 + 246$ 54. $814 + 512$ 55. $2,653 + 4,817$ 56. $6,003 + 5,734$

Mid-Chapter Practice Test

Vocabulary and Concepts

1. **Define** *absolute value.* (Lesson 3-1)
2. **Write** the ordered pair which identifies a point 4 units to the left of the y-axis and three units above the x-axis. (Lesson 3-3)
3. **Draw** a coordinate plane, and label the quadrants. (Lesson 3-3)

Skills and Applications

Write an integer for each situation. (Lesson 3-1)

4. 45 feet below sea level
5. a deposit of $100
6. a gain of 8 yards
7. lost a $5 bill

Replace each ● with < or > to make a true sentence. (Lesson 3-2)

8. $-12 \bullet -9$
9. $-4 \bullet 4$
10. $|-14| \bullet |3|$

11. **FOOTBALL** The Tigers have recorded the following yardage on the past six plays: 9, −2, 5, 0, 12, and −7. Order the integers from least to greatest. (Lesson 3-2)

On graph paper, draw a coordinate plane. Then graph and label each point. (Lesson 3-3)

12. $D(4, -3)$
13. $E(1, 3)$
14. $F(0, -5)$

Standardized Test Practice

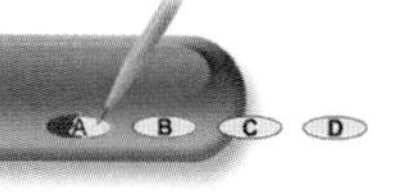

15. **MULTIPLE CHOICE** Which of the following points represents a number and its absolute value? (Lesson 3-1)

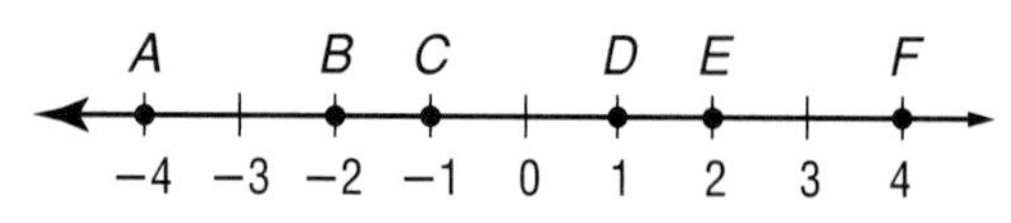

Ⓐ B and E Ⓑ C and F

Ⓒ B and D Ⓓ A and E

16. **SHORT RESPONSE** The table shows the number of inches of monthly precipitation above or below normal for a midwestern city in a recent year. Find the median monthly precipitation above or below normal. (Lesson 3-2)

Monthly Precipitation Above or Below Normal

J	F	M	A	M	J
4	−1	6	−2	−3	1
J	**A**	**S**	**O**	**N**	**D**
−2	−1	2	−3	1	−3

The Game Zone

A Place To Practice Your Math Skills

Math Skill
Graphing Points on a Coordinate Plane

Tic-Tac-Toe

GET READY!

Players: two
Materials: grid paper

GET SET!

- Draw a coordinate plane on grid paper.
- This game is similar to tic-tac-toe, except players must get four Xs or four Os in a row.

GO!

- Player 1 chooses two numbers: the first number is the *x*-coordinate of an ordered pair, and the second number is the *y*-coordinate. Each number must be between -5 and 5. Then Player 1 announces the ordered pair and plots the X or O on the coordinate plane.
- Player 2 then chooses his or her numbers, announces them, and plots the points.
- An ordered pair cannot be changed after it has been announced.
- If a player announces an ordered pair that has already been used or graphs an ordered pair incorrectly, the player loses a turn.
- **Who Wins?** The first player to get four Xs or Os in a row is the winner.

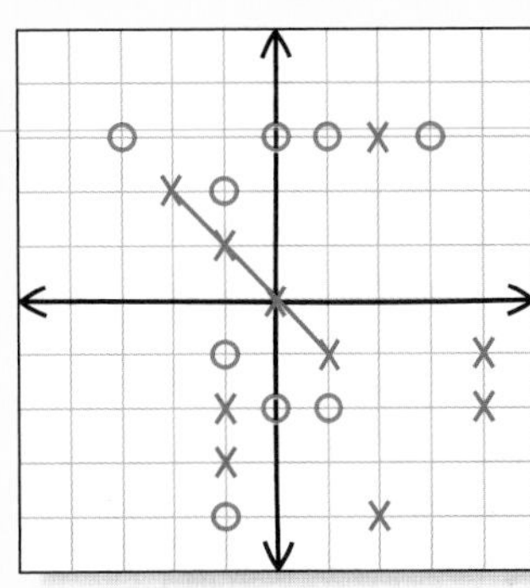

3-4a HANDS-ON LAB

A Preview of Lesson 3-4

Adding Integers

What You'll LEARN

Use counters to model the addition of integers.

Materials

- counters
- integer mat

You can use positive and negative counters to model the addition of integers. The counter ⊕ represents 1, and the counter ⊖ represents −1. Remember that addition means *combining* two sets.

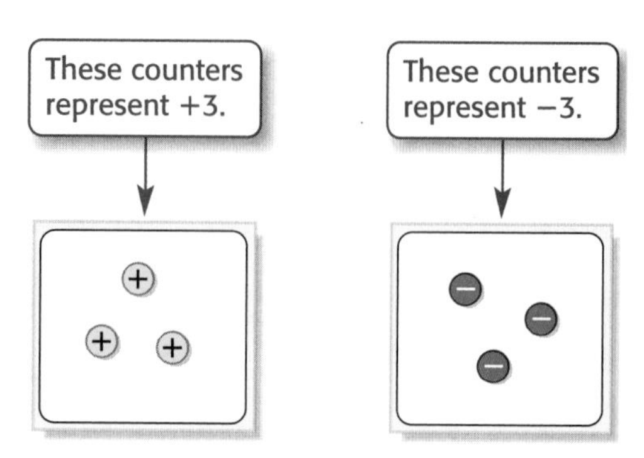

ACTIVITY *Work with a partner.*

1 **Use counters to find −2 + (−4).**

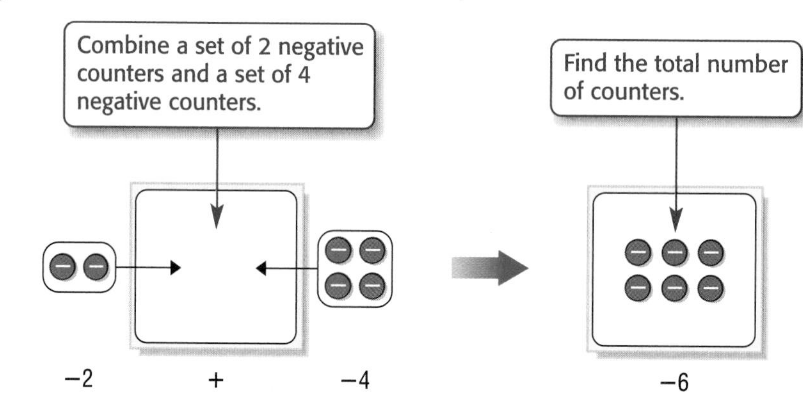

So, $-2 + (-4) = -6$.

Your Turn **Use counters to find each sum.**

a. $5 + 6$ b. $-3 + (-5)$ c. $-5 + (-4)$

d. $7 + 3$ e. $-2 + (-5)$ f. $-8 + (-6)$

READING Math

Addends and Sums The numbers you add are called *addends*. The result is called the *sum*.

The following two properties are important when modeling operations with integers.

- When one positive counter is paired with one negative counter, the result is called a **zero pair**. The value of a zero pair is 0.
- You can add or remove zero pairs from a mat because adding or removing zero does not change the value of the counters on the mat.

You will use zero pairs in Activity 2 and Activity 3.

ACTIVITIES *Work with a partner.*

Use counters to find each sum.

2 $-5 + 3$

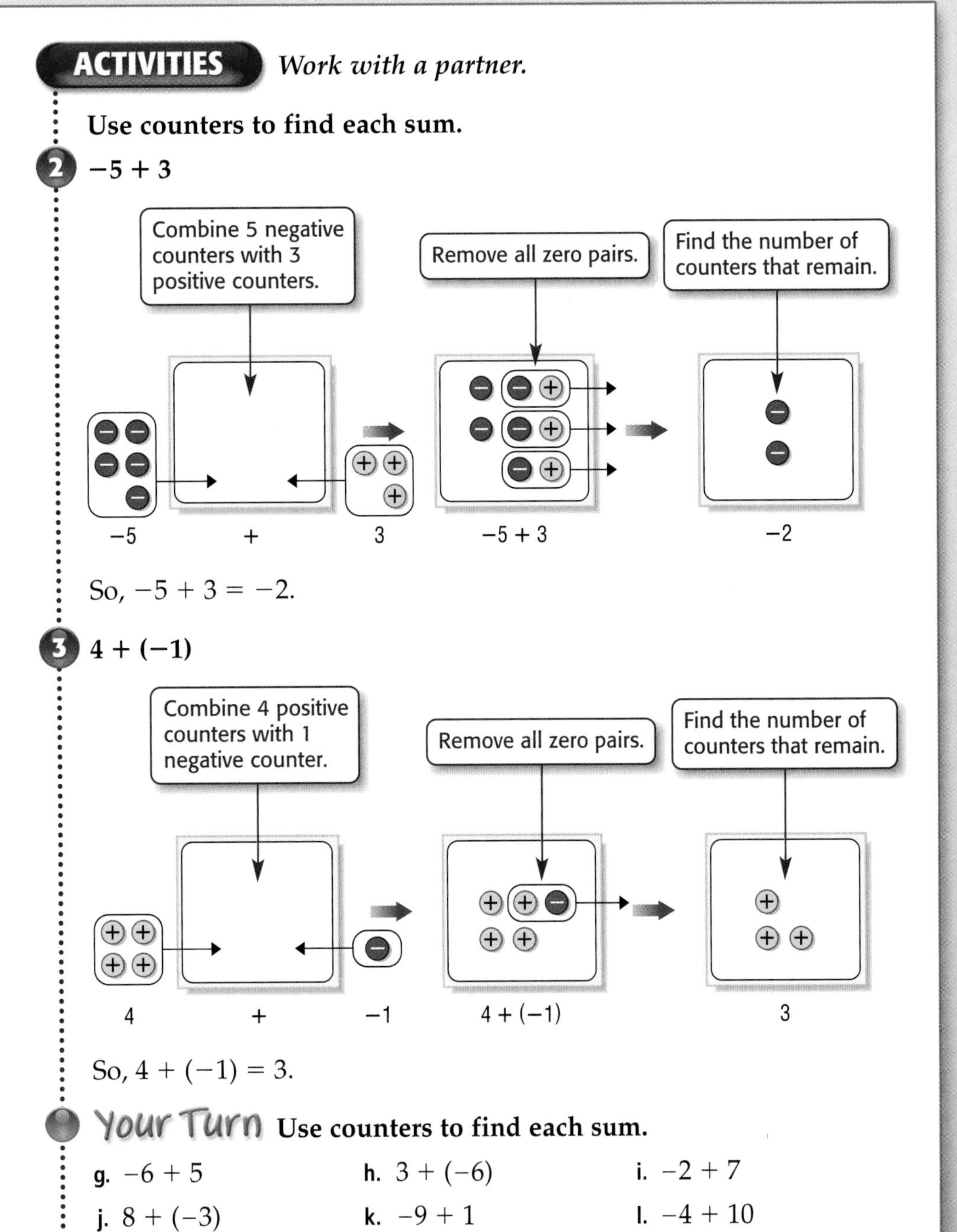

So, $-5 + 3 = -2$.

3 $4 + (-1)$

So, $4 + (-1) = 3$.

Your Turn **Use counters to find each sum.**

g. $-6 + 5$ **h.** $3 + (-6)$ **i.** $-2 + 7$

j. $8 + (-3)$ **k.** $-9 + 1$ **l.** $-4 + 10$

Writing Math

1. **Write** two addition sentences where the sum is positive. In each sentence, one addend should be positive and the other negative.
2. **Write** two addition sentences where the sum is negative. In each sentence, one addend should be positive and the other negative.
3. **MAKE A CONJECTURE** Write a rule that will help you determine the sign when finding the sum of integers.

3-4 Adding Integers

What You'll LEARN

Add integers.

NEW Vocabulary

opposites
additive inverse

Link to READING

Everyday Meaning of Opposite: something that is across from or is facing the other way, as in running the opposite way

WHEN am I ever going to use this?

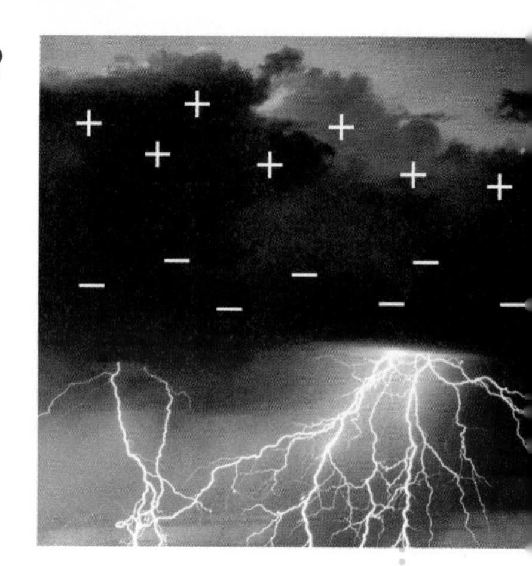

EARTH SCIENCE Thunderstorms are made of both positive and negative electrical charges. The negative charges (electrons) are at the bottom of a thundercloud and positive charges (protons) are at the top.

1. What is the charge at the top of a cloud where there are more protons than electrons?
2. What is the charge at the bottom of a cloud where there are more electrons than protons?

Combining positive and negative electrical charges in a thunderstorm is similar to adding integers.

EXAMPLE Add Integers with the Same Sign

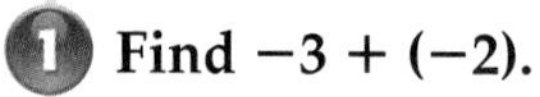

1 Find $-3 + (-2)$.

Use a number line.

- Start at 0.
- Move 3 units left to show -3.
- From there, move 2 units left to show -2.

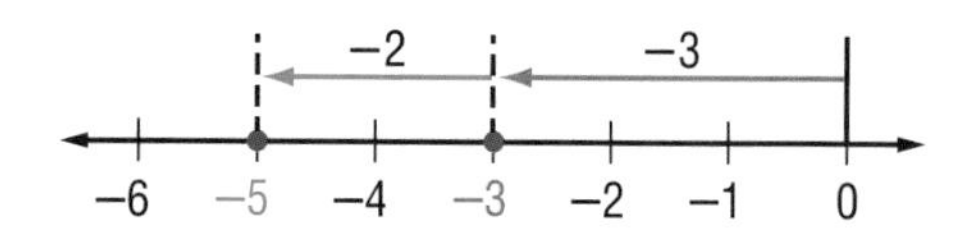

So, $-3 + (-2) = -5$.

Key Concept — Adding Integers with the Same Sign

Words The sum of two positive integers is positive.
The sum of two negative integers is negative.

Examples $7 + 4 = 11$ $\quad -7 + (-4) = -11$

STUDY TIP

Technology To enter a negative integer on a calculator, use the [(−)] key. For example, to enter −5, press [(−)] 5.

EXAMPLE Add Integers with the Same Sign

2 Find $-26 + (-17)$.

$-26 + (-17) = -43$ The sum of two negative integers is negative.

Your Turn Add.

a. $-14 + (-16)$ b. $23 + 38$ c. $-35 + (-49)$

The integers 43 and -43 are called **opposites** of each other because they are the same distance from 0, but on opposite sides of 0. Two integers that are opposites are also called **additive inverses**.

Key Concept — **Additive Inverse Property**

Words The sum of any number and its additive inverse is 0.

Example $5 + (-5) = 0$

EXAMPLES Add Integers with Different Signs

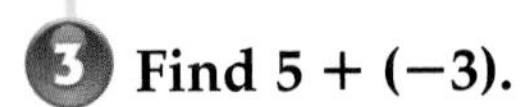

3 Find $5 + (-3)$.

Use counters.

Remove all zero pairs.

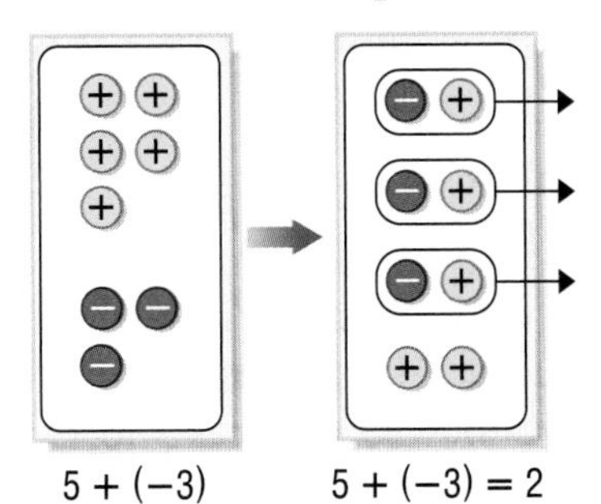

$5 + (-3)$ $\quad$ $5 + (-3) = 2$

So, $5 + (-3) = 2$.

4 Find $-3 + 2$.

Use a number line.

- Start at 0.
- Move 3 units left.
- Then move 2 units right.

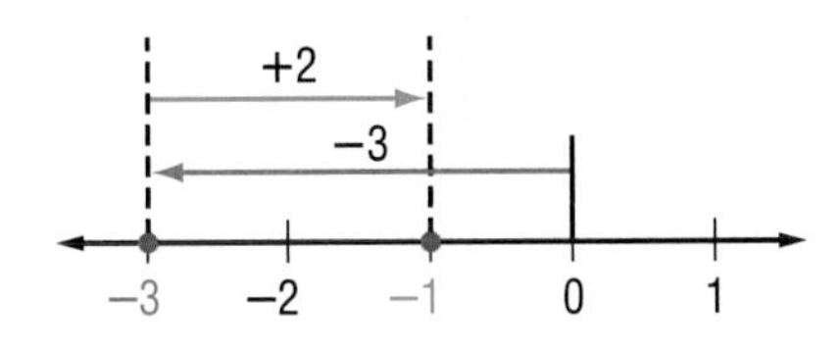

So, $-3 + 2 = -1$.

STUDY TIP

Look Back You can review **absolute value** in Lesson 3-1.

Key Concept — **Adding Integers with Different Signs**

Words To add integers with different signs, subtract their absolute values. The sum is:

- positive if the positive integer has the greater absolute value.
- negative if the negative integer has the greater absolute value.

Examples $9 + (-4) = 5$ $\qquad$ $-9 + 4 = -5$

EXAMPLES Add Integers with Different Signs

5 Find $7 + (-1)$.

$7 + (-1) = 6$ Subtract. Keep the sign of the integer with the greater absolute value.

6 Find $-8 + 4$.

$-8 + 4 = -4$ Subtract. Keep the sign of the integer with the greater absolute value.

Your Turn **Add.**

d. $10 + (-12)$ $\qquad$ **e.** $-16 + 9$ $\qquad$ **f.** $-13 + 18$

EXAMPLE Simplify an Expression with Integers

7 **ALGEBRA Simplify $12 + x + (-20)$.**

$12 + x + (-20) = 12 + (-20) + x$ Commutative Property of Addition

$= -8 + x$ Add.

Your Turn **Simplify each expression.**

g. $1 + y + (-5)$ **h.** $z + (-2) + 9$

REAL-LIFE MATH

OCEANOGRAPHY The Great Barrier Reef off the coast of Australia is home to 1,500 species of fish and 215 species of birds.

Source: www.ozramp.net.au

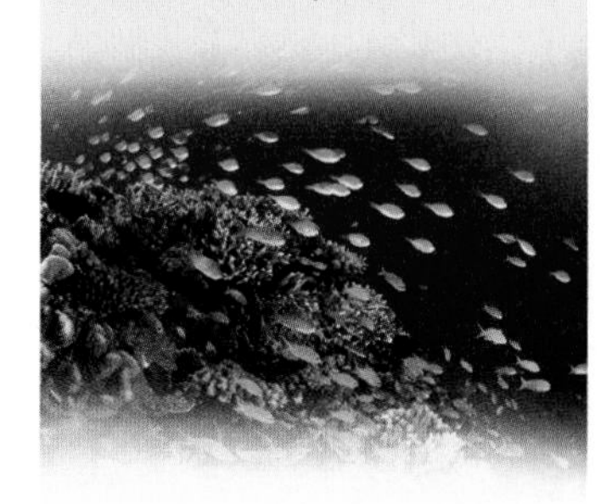

EXAMPLE Use Integers to Solve a Problem

8 **OCEANOGRAPHY Anna was scuba diving near the Great Barrier Reef 16 meters below the surface of the water. She saw a dolphin swim by 7 meters above her. What was the depth of the dolphin?**

Anna is 16 meters underwater, and the dolphin is 7 meters above her. So, the depth of the dolphin can be represented by the expression $-16 + 7$, or -9.

The dolphin is 9 meters below the surface of the water.

Skill and Concept Check

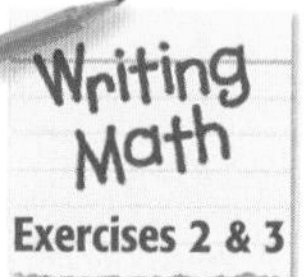

1. **Draw** a model to show $2 + (-7)$.

2. **OPEN ENDED** Give an example of integers that are additive inverses.

3. **FIND THE ERROR** Brooke and Javier are finding $-12 + 13$. Who is correct? Explain.

Brooke	Javier
$-12 + 13 = 1$	$-12 + 13 = -1$

4. **NUMBER SENSE Tell whether each sum is *positive*, *negative*, or *zero* without adding.**

   a. $-6 + (-7)$ b. $-8 + 10$ c. $-14 + 14$

GUIDED PRACTICE

Add.

5. $-6 + (-8)$ 6. $-3 + 10$ 7. $7 + (-11)$ 8. $9 + (-9)$

9. **MONEY** You pay your brother $42 that you owe him. The same week you earn $35 dog-sitting for the neighbors. Do you have more or less money than at the beginning of the week?

10. **ALGEBRA** Simplify $12 + y + (-8)$.

Practice and Applications

HOMEWORK HELP	
For Exercises	See Examples
11–42	1–6
46–57	7
43–45, 58	8

Extra Practice
See pages 570, 598.

Add.

11. $-8 + 8$ **12.** $-9 + 11$ **13.** $13 + (-19)$ **14.** $6 + 10$

15. $-10 + (-15)$ **16.** $-12 + 10$ **17.** $-30 + 16$ **18.** $18 + (-5)$

19. $21 + (-21)$ **20.** $18 + (-20)$ **21.** $-22 + (-16)$ **22.** $-24 + 19$

23. $-11 + 13 + 6$ **24.** $-16 + (-21) + 15$

25. $12 + (-17) + (-25)$ **26.** $20 + (-30) + (-40)$

Write an addition expression to describe each situation. Then find each sum.

27. **WEATHER** The temperature outside is −3°F. The temperature drops 6°.

28. **SUBMARINE** A submarine dives 106 feet below the water. Then, it rises 63 feet.

29. **SKATEBOARDING** Hakeem starts at the bottom of a half pipe 6 feet below street level. He rises 14 feet at the top of his kickturn.

30. **MONEY MATTERS** Stephanie has $43 in the bank. She withdraws $35.

ALGEBRA **Evaluate each expression if $x = -10$, $y = 7$, and $z = -8$.**

31. $x + 14$ **32.** $6 + y$ **33.** $z + (-5)$ **34.** $-17 + y$

35. $20 + z$ **36.** $-10 + x$ **37.** $z + 8$ **38.** $15 + x$

39. $x + y$ **40.** $y + z$ **41.** $x + z$ **42.** $x + y + z$

GOLF **For Exercises 43–45, use the information below.**

Scores over par in a golf tournament are recorded as positive integers. Scores under par are recorded as negative integers. Even par is recorded as 0. The person with the lowest total score wins. The table shows the top two finishers in the 2001 LPGA Championship.

43. Find Laura Diaz's final score.

44. Find Karrie Webb's final score.

45. Who had the better score? Explain.

	Round 1	Round 2	Round 3	Round 4
Laura Diaz	−4	0	−5	−3
Karrie Webb	−4	−7	−1	−2

Source: www.lpga.com

Data Update What were the four-round scores of the latest winners of the LPGA Championship? Visit msmath2.net/data_update to learn more.

ALGEBRA **Simplify.**

46. $x + (-5) + 1$ **47.** $4 + y + (-2)$ **48.** $-9 + m + (-6)$

49. $8 + (-8) + n$ **50.** $-1 + a + 7$ **51.** $f + (-19) + 11$

Explain how the Commutative and Associative Properties of Addition can help you find each sum mentally. Then find each sum.

52. $7 + (-2) + (-7)$
53. $-6 + 9 + (-4)$
54. $-5 + (-6) + (-3)$
55. $8 + 10 + (-8)$
56. $-5 + (-7) + (-10)$
57. $8 + (-9) + 9$

58. **STOCK MARKET** The members of the Investment Club purchased a stock for $50. The next day the price of the stock dropped $18. On the second and third days, the price dropped another $16 and then rose $21. How much was the stock worth at the end of the third day?

59. **WRITE A PROBLEM** Write about a real-life problem using the addition sentence $-8 + 11 = t$. Then solve the equation and explain what the solution represents.

CRITICAL THINKING For Exercises 60 and 61, use a number line to find each sum. Does the order of the addends make a difference? Explain.

60. $3 + (-8)$ and $-8 + 3$
61. $[7 + (-3)] + (-6)$ and $7 + [-3 + (-6)]$

Standardized Test Practice and Mixed Review

62. **MULTIPLE CHOICE** In a game with a standard deck of cards and the scoring system at the right, three cards are dealt and added together to get a final score. Dylan is dealt the 4 of hearts, the king of spades, and the 3 of diamonds. What is his final score?

(A) −11 (B) −3 (C) 3 (D) 9

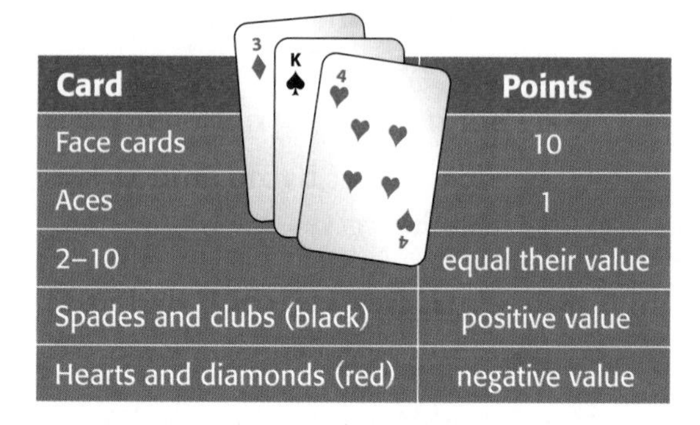

Card	Points
Face cards	10
Aces	1
2–10	equal their value
Spades and clubs (black)	positive value
Hearts and diamonds (red)	negative value

63. **SHORT RESPONSE** Jeremy owes his sister $5. Then he borrows $6 more from her. Write the total amount he owes as an integer.

Name the ordered pair for each point graphed at the right. Then identify the quadrant in which each point lies. (Lesson 3-3)

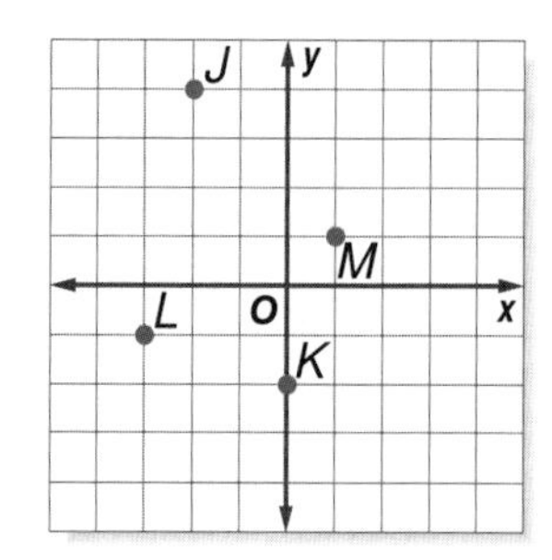

64. J
65. K
66. L
67. M

68. Order 6, −3, 0, 4, −8, 1, and −4 from least to greatest. (Lesson 3-2)

69. **STATISTICS** Construct a line plot for the following test scores: 81, 83, 75, 81, 82, 81, 75, 82, 82, 86, 83, 81, and 79. (Lesson 2-3)

GETTING READY FOR THE NEXT LESSON

PREREQUISITE SKILL Find the range for each set of data. (Lesson 2-3)

70. 13, 7, 6, 22, 21
71. 54, 32, 43, 49, 30
72. 62, 59, 85, 74, 82

Study Skill

HOW TO...

Take Good Notes

FLOWCHARTS

Have you ever tried to solve a math problem and then realized you left out an important step? Try using a flowchart when you take notes to map out the steps you should follow.

A flowchart is like a map that tells you how to get from the beginning of a problem to the end.

	Flowchart Symbols
(diamond)	A diamond contains a question. You need to stop and make a decision.
(rectangle)	A rectangle tells you what to do.
(oval)	An oval indicates the beginning or end.

Here's a flowchart for adding two integers. Just follow the arrows.

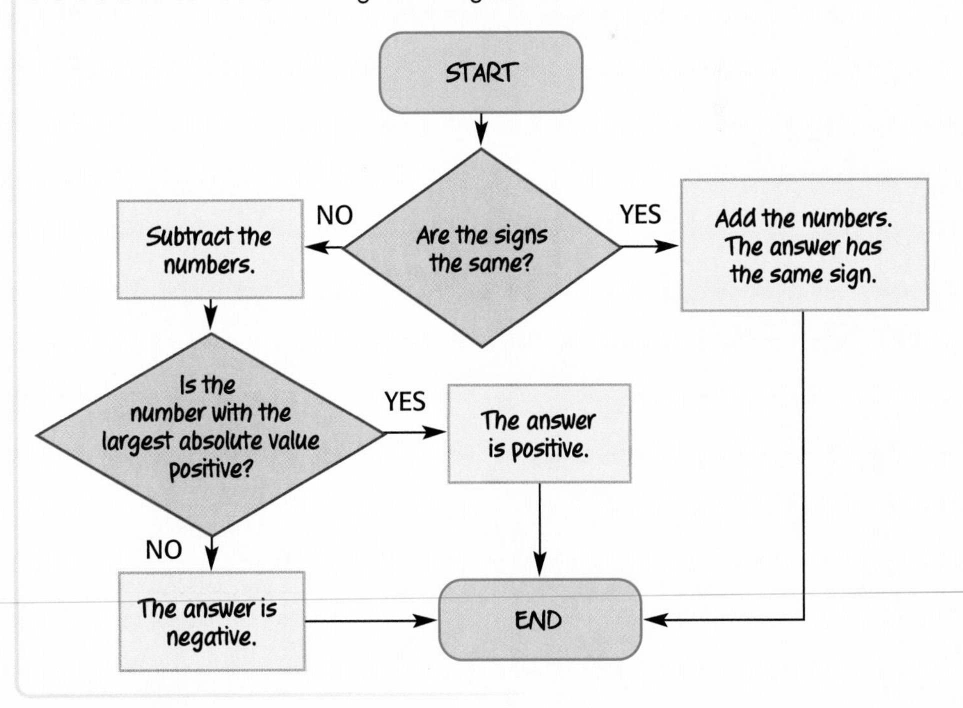

SKILL PRACTICE

Make a flowchart for each kind of problem.

1. rounding a decimal to a given place (See page 557.)
2. evaluating an expression using order of operations (See Lesson 1-3.)

Subtracting Integers

What You'll LEARN

Use counters to model the subtraction of integers.

Materials

- counters
- integer mat

You can also use counters to model subtraction of integers. Remember one meaning of subtraction is to *take away*.

ACTIVITIES *Work with a partner.*

Use counters to find each difference.

1 $6 - 4$

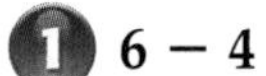

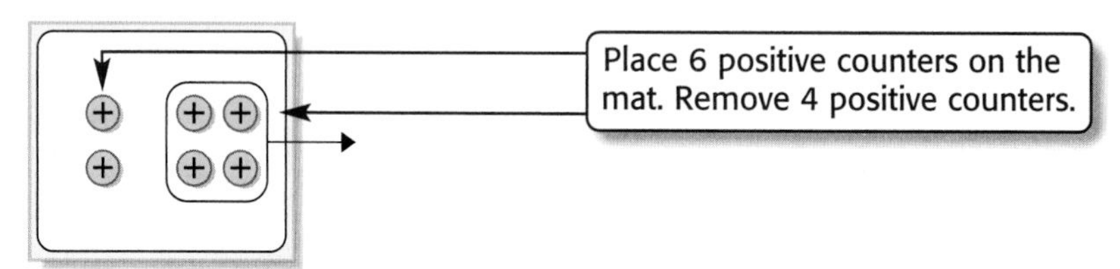

So, $6 - 4 = 2$.

2 $3 - (-2)$

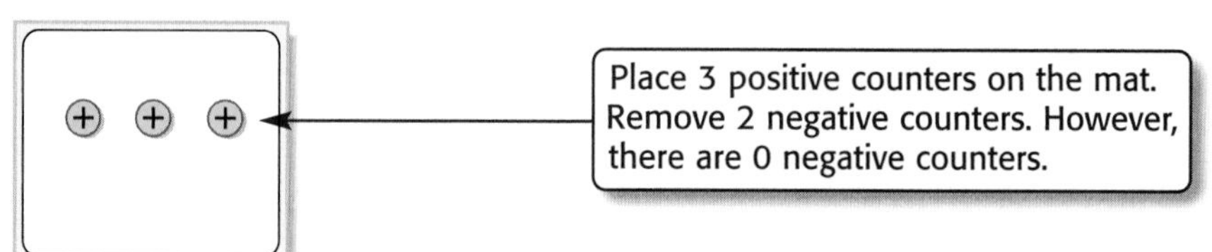

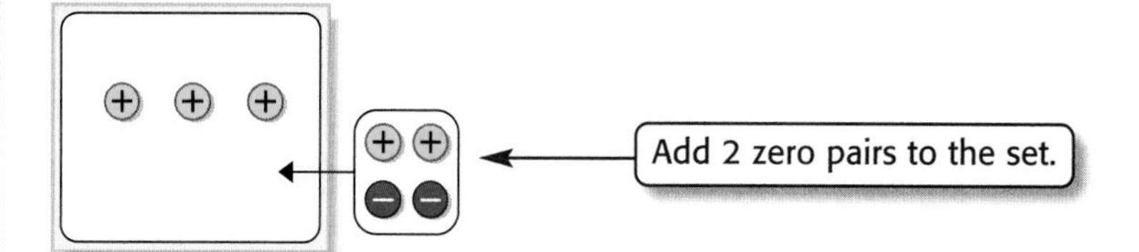

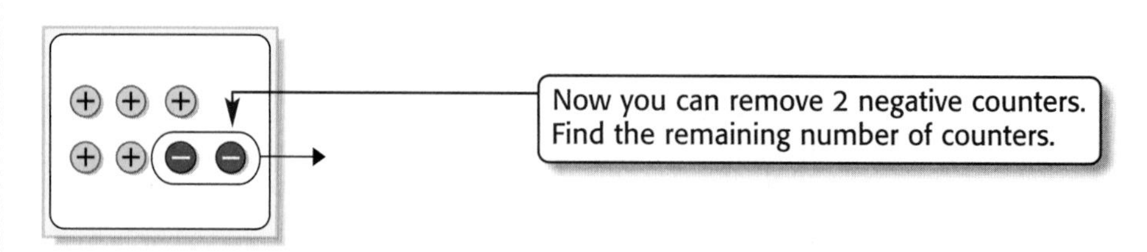

So, $3 - (-2) = 5$.

Your Turn **Use counters to find each difference.**

a. $7 - 6$ **b.** $5 - (-3)$ **c.** $6 - (-3)$ **d.** $5 - 8$

ACTIVITIES *Work with a partner.*

Use counters to find each difference.

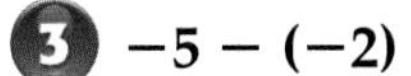

3 $-5 - (-2)$

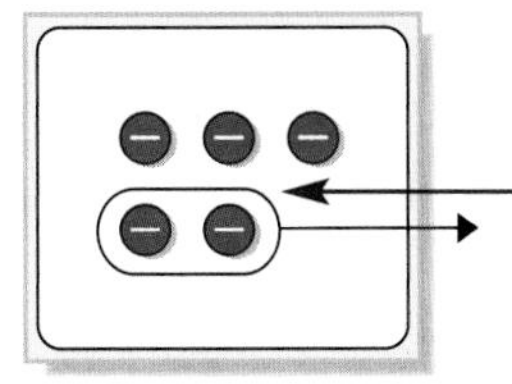

So, $-5 - (-2) = -3$.

4 $-4 - 3$

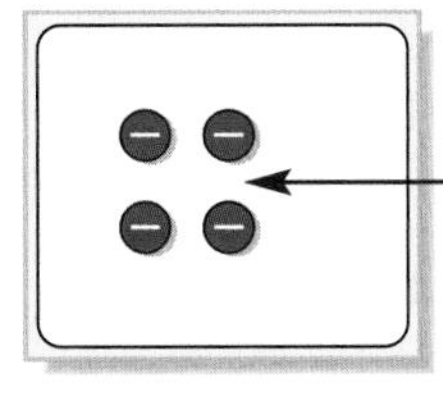

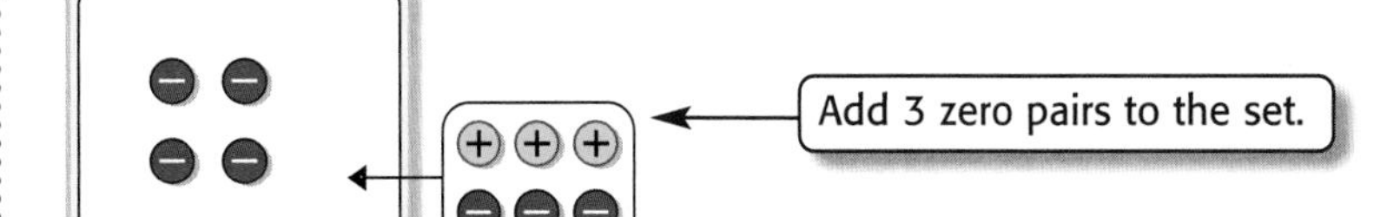

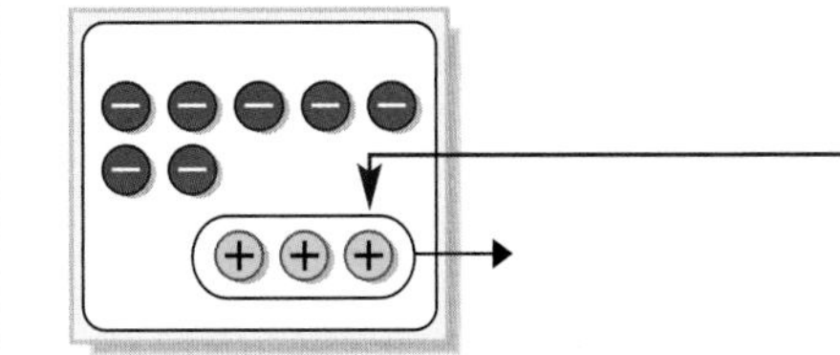

So, $-4 - 3 = -7$.

Your Turn **Use counters to find each difference.**

e. $-6 - (-3)$ **f.** $-7 - 3$ **g.** $-5 - (-7)$

READING Math

Minuends, Subtrahends, and Differences In the subtraction sentence $-4 - 3 = -7$, -4 is the *minuend,* 3 is the *subtrahend,* and -7 is the *difference*.

Writing Math

Work with a partner.

1. **Write** two subtraction sentences where the difference is positive. Make sure you use a combination of positive and negative integers.
2. **Write** two subtraction sentences where the difference is negative. Make sure you use a combination of positive and negative integers.
3. **MAKE A CONJECTURE** Write a rule that will help you determine the sign of the difference of two integers.

3-5 Subtracting Integers

What You'll LEARN

Subtract integers.

REVIEW Vocabulary

range: the difference between the greatest number and the least number in a set of data **(Lesson 2-3)**

HANDS-ON Mini Lab

Materials
- graph paper

Work with a partner.

The subtraction problems below are modeled on number lines.

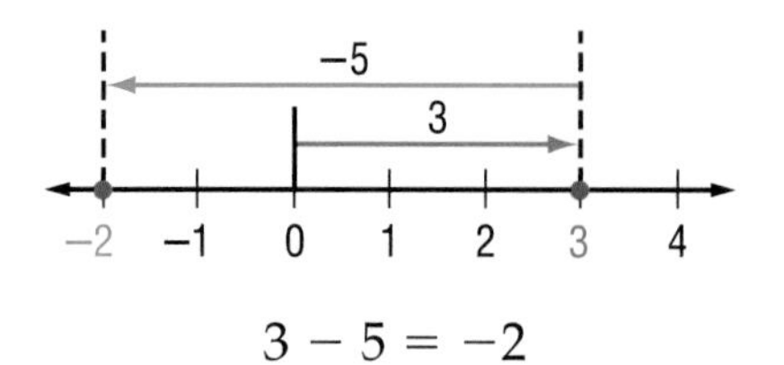

$3 - 5 = -2$

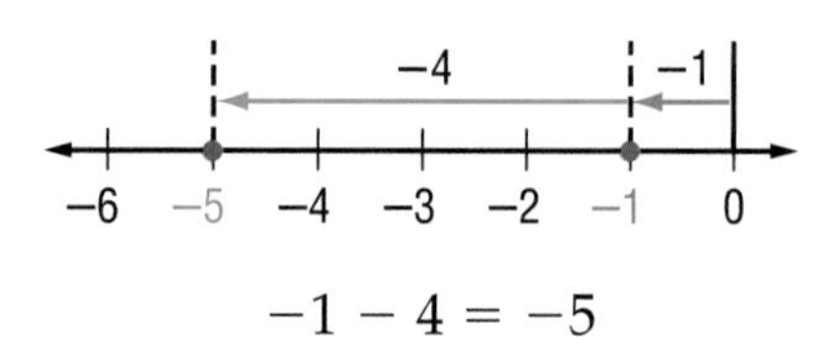

$-1 - 4 = -5$

1. Write a related addition sentence for each subtraction sentence.

Use a number line to find each difference. Write an equivalent addition sentence for each.

2. $1 - 5$
3. $-2 - 1$
4. $-3 - 4$
5. $0 - 5$
6. **Compare and contrast** subtraction sentences with their related addition sentences.

When you subtract 5, as shown in the Mini Lab, the result is the same as adding −5. When you subtract 4, the result is the same as adding −4.

opposite

$3 - 5 = -2 \quad 3 + (-5) = -2$

same result

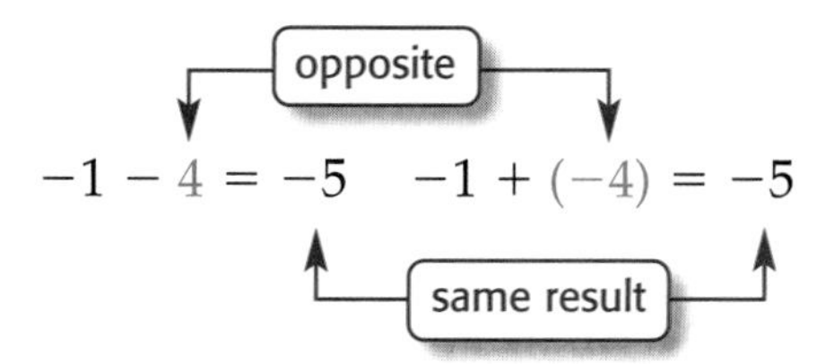

Key Concept — **Subtracting Integers**

Words To subtract an integer, add its opposite.

Example $4 - 9 = 4 + (-9) = -5$

EXAMPLES Subtract Positive Integers

Subtract.

1 $8 - 13$

$8 - 13 = 8 + (-13)$ To subtract 13, add −13.

$= -5$ Simplify.

2 $-10 - 7$

$-10 - 7 = -10 + (-7)$ To subtract 7, add −7.

$= -17$ Simplify.

EXAMPLES Subtract Negative Integers

3 **Find $1 - (-2)$.**

$1 - (-2) = 1 + 2$ — To subtract −2, add 2.

$= 3$ — Simplify.

4 **Find $-10 - (-7)$.**

$-10 - (-7) = -10 + 7$ — To subtract −7, add 7.

$= -3$ — Simplify.

STUDY TIP

Technology Use the [−] key to subtract a number. Use the [(−)] key to enter a negative number.

Your Turn **Subtract.**

a. $4 - (-12)$ b. $-15 - (-5)$ c. $18 - (-6)$

EXAMPLE Evaluate an Expression

5 **ALGEBRA** **Evaluate $x - y$ if $x = -6$ and $y = 5$.**

$x - y = -6 - 5$ — Replace x with −6 and y with 5.

$= -6 + (-5)$ — To subtract 5, add −5.

$= -11$ — Simplify.

Your Turn **Evaluate each expression if $a = 5$, $b = -8$, and $c = -9$.**

d. $b - 10$ e. $a - b$ f. $c - a$

EXAMPLE Use Integers to Solve a Problem

REAL-LIFE MATH

EARTH SCIENCE Yellowfin tuna can be found in waters that are between 17.8°C and 26.7°C.

Source: www.sstcharts.com

6 **EARTH SCIENCE** **The legend on the sea-surface temperature map shows the minimum temperature at −2°C and the maximum temperature at 31°C. What is the range of temperatures on the map?**

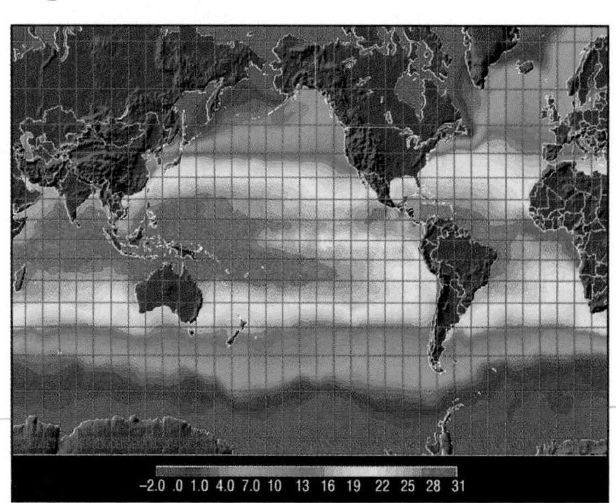

To find the range, or difference in temperatures, subtract the lowest temperature from the highest temperature.

$31 - (-2) = 31 + 2$ — To subtract −2, add 2.

$= 33$ — Simplify.

So, the range of temperatures is 33°C.

Skill and Concept Check

Writing Math
Exercises 1 & 3

1. **Explain** how additive inverses are used in subtraction.

2. **OPEN ENDED** Write a subtraction sentence using integers. Then, write the equivalent addition sentence and find the sum.

3. **FIND THE ERROR** Bradley and Mitsu are finding $-16 - (-19)$. Who is correct? Explain.

Bradley	Mitsu
$-16 - (-19) = -16 + (-19)$	$-16 - (-19) = -16 + (19)$
$= -35$	$= 3$

GUIDED PRACTICE

Subtract.

4. $-4 - 8$
5. $14 - 17$
6. $14 - (-10)$
7. $-3 - (-1)$

ALGEBRA Evaluate each expression if $p = 8$, $q = -14$, and $r = -6$.

8. $p - q$
9. $q - r$
10. $r - p$

11. **METEOROLOGY** The highest temperature ever recorded on Earth was 136°F in Libya. The lowest temperature was −129°F in Antarctica. What is the range of the highest and lowest temperatures on Earth?

Practice and Applications

HOMEWORK HELP

For Exercises	See Examples
12–31	1–4
32–44	5
45–48	6

Extra Practice
See pages 571, 598.

Subtract.

12. $-9 - 5$
13. $0 - 10$
14. $-8 - 9$
15. $17 - 13$
16. $27 - (-8)$
17. $-25 - (-5)$
18. $12 - 26$
19. $4 - (-19)$
20. $-11 - 42$
21. $15 - (-14)$
22. $-27 - (-19)$
23. $-18 - (-20)$
24. $31 - 48$
25. $-33 - (-27)$
26. $52 - (-52)$
27. $-44 - (-41)$
28. $-2 - 9 + 7$
29. $6 + (-1) - 4$
30. What is -3 minus 4?
31. Find $-23 - (-19)$.

ALGEBRA Evaluate each expression if $f = -6$, $g = 7$, and $h = 9$.

32. $5 - f$
33. $h - (-9)$
34. $f - g$
35. $g - 7$
36. $h - f$
37. $f - 6$
38. $g - h$
39. $4 - (-g)$
40. $-h - 10$
41. $-f - h$
42. $f - g - h$
43. $h - g - f$

44. **ALGEBRA** Find $|a - b|$ when $a = -7$ and $b = 11$.

45. **GEOGRAPHY** The Dead Sea's deepest part is 799 meters below sea level. A plateau to the east of the Dead Sea rises to about 1,340 meters above sea level. What is the difference between the deepest part of the Dead Sea and the top of the plateau?

HISTORY For Exercises 46–48, use the timeline that shows the lives of three rulers of Rome.

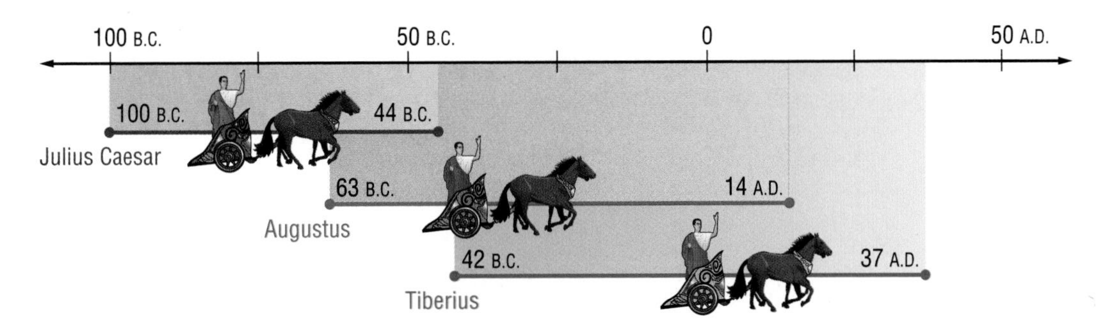

46. How old was Augustus when he died?
47. Who lived the longest? How old was he when he died?
48. How many years were there between when Julius Caesar was born and when Tiberius died?

Determine whether each statement is *sometimes, always,* or *never* true. Give an example or counterexample for each answer.

49. negative − positive = negative
50. negative − negative = positive
51. positive − positive = positive
52. positive − negative = negative

53. **CRITICAL THINKING**
True or *False*? When n is a negative integer, $n - n = 0$.

Standardized Test Practice and Mixed Review

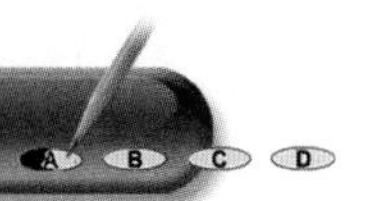

54. **MULTIPLE CHOICE** Find the correct subtraction sentence shown in the model.

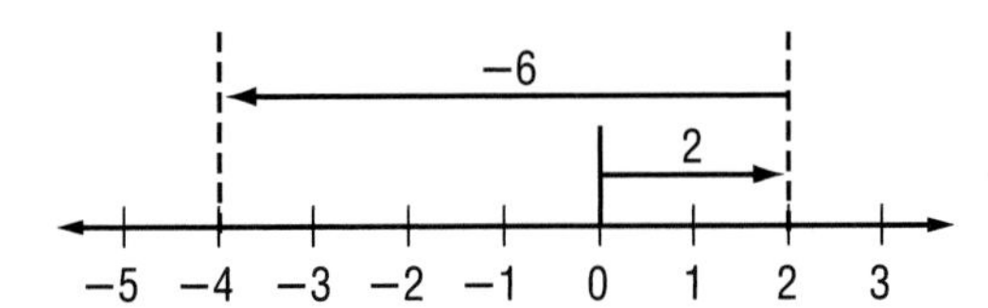

Ⓐ $2 - (-6) = -4$
Ⓑ $-6 - 2 = -4$
Ⓒ $-2 - 4 = -6$
Ⓓ $2 - 6 = -4$

55. **SHORT RESPONSE** The temperatures on the moon vary from −173°C to 127°C. Find the range of temperatures.

Add. (Lesson 3-4)

56. $10 + (-3)$
57. $-2 + (-9)$
58. $-7 + (-6)$
59. $-18 + 4$

60. In which quadrant do ordered pairs with a positive x-coordinate and a negative y-coordinate lie? (Lesson 3-3)

GETTING READY FOR THE NEXT LESSON

BASIC SKILL Multiply.

61. $14 \cdot 5$
62. $9 \cdot 16$
63. $6 \cdot 8 \cdot 4$
64. $11 \cdot 7 \cdot 7$

3-6a Problem-Solving Strategy

A Preview of Lesson 3-6

What You'll Learn
Solve problems using the look for a pattern strategy.

Look for a Pattern

We've already saved $155 in four months. If we keep saving our money at the same rate, how long do you think it will take to save enough money to buy a DVD player that costs $330?

I found the table where we listed our savings each month. Let's **look for a pattern** to figure it out.

Explore	We began with $50 and added more money to our savings every month. We need to find the number of months when we will have $330 to buy the DVD player.
Plan	Let's look for a pattern or rule that increases the balance each month. Then use the rule to extend the pattern and find the solution.
Solve	After the initial $50, we saved $35 per month. To extend the pattern, add $35 to each monthly balance until the balance equals $330. We will have enough money saved after 9 months.
Examine	We saved about 2 × $155, or $310 in 8 months. So, 9 months is a reasonable answer.

Month	Balance
1	$50
2	$85
3	$120
4	$155

Month	Balance
1	$50
2	$85
3	$120
4	$155
5	$190
6	$225
7	$260
8	$295
9	$330

Analyze the Strategy

1. **Explain** when you would use the look for a pattern strategy to solve a problem.
2. **Describe** how to solve a problem using the look for a pattern method as a problem-solving strategy.
3. **Write** a problem that could be solved by looking for a pattern. Explain your answer.

Apply the Strategy

Solve. Use the look for a pattern strategy.

4. **LIFE SCIENCE** The table shows about how many times a firefly flashes at different temperatures. Estimate how many times a firefly will flash when the temperature is 36°C.

Outside Temperature (°C)	Flashes per Minute
16	8
20	9
24	11
28	14

5. **CEREAL** A display of cereal boxes is stacked in the shape of a pyramid. There are 4 boxes in the top row, 6 boxes in the second row, 8 boxes in the next row, and so on. The display contains 7 rows of boxes. How many boxes are in the display?

Mixed Problem Solving

Solve. Use any strategy.

LIFE SCIENCE For Exercises 6–8, use the information and the graph.

6. What does 0 on this graph represent?

7. Write an integer to represent the rainfall for each month.

Summer Rainfall (compared to normal)

Rain (in.): 3, 2, 1, 0, −1, −2, −3, −4, −5

Month: June, July, August, September

8. Write a sentence that summarizes the message this graph conveys about this summer's rainfall.

9. **EARTH SCIENCE** Hydrothermal vents are similar to geysers, but are found on the ocean floor. A hydrothermal vent chimney can grow at an average rate of 9 meters in 18 months. What is the average rate of growth per month?

10. **MULTI STEP** Francisco is on vacation and is planning to send postcards and letters to his friends. He has $3.04 to spend on postage. A stamp for a letter costs 37¢, and a stamp for a postcard costs 23¢. If he is going to spend the entire $3.04 on postage, how many postcards and letters can he send?

11. **BASKETBALL** Laura makes 3 free throws out of every 5 she attempts. Find the number of free throws she will make after 15, 20, and 30 attempts.

12. **COINS** Olivia has seven coins that total $1.32. What are the coins?

13. **FOOD** The school cafeteria added a breakfast special to their menu. The table shows the foods that are part of the special and the number of Calories. Estimate how many Calories there are in the special.

Food	Calories
whole-wheat bagel	156
skim milk	90
nonfat strawberry yogurt	183
fresh fruit salad	68

14. **STANDARDIZED TEST PRACTICE** The total land area of Alabama is about 50,744 square miles. In 2000, about 87.6 persons per square mile were living in Alabama. What was the approximate population of Alabama in 2000?

(A) 45,000 (B) 450,000
(C) 4,500,000 (D) 45,000,000

You will use the look for a pattern strategy in the next lesson.

3-6 Multiplying Integers

HANDS-ON Mini Lab

What You'll LEARN

Multiply integers.

Materials

- counters
- integer mat

Work with a partner.

Counters can be used to multiply positive and negative integers.

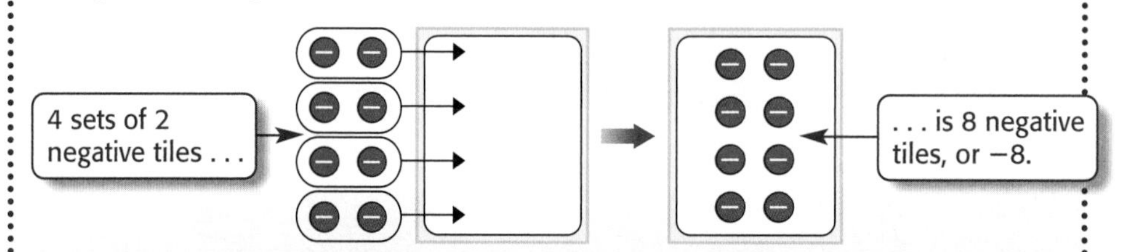

1. Write a multiplication sentence that describes the model above.

Find each product using counters.

2. 3(−2)
3. 4(−3)
4. 1(−7)
5. 5(−2)
6. Write a rule for finding the sign of the product of a positive and negative integer.

Remember that multiplication is the same as repeated addition. The multiplication expression 4(−2) in the Mini Lab means that −2 is used as an addend four times.

$$4(-2) = (-2) + (-2) + (-2) + (-2)$$
$$= -8$$

−2 −2 −2 −2

−8 −7 −6 −5 −4 −3 −2 −1 0 1

By the Commutative Property of Multiplication, 4(−2) = −2(4). When two integers have different signs, the following rule applies.

Key Concept **Multiplying Integers with Different Signs**

Words The product of two integers with different signs is negative.

Examples 6(−4) = −24 −5(7) = −35

EXAMPLES Multiply Integers with Different Signs

Multiply.

1 **3(−5)**

3(−5) = −15 The integers have different signs. The product is negative.

2 **−6(8)**

−6(8) = −48 The integers have different signs. The product is negative.

STUDY TIP

Zero The Multiplicative Property of Zero states that when any number is multiplied by zero, the product is zero.

The product of two positive integers is positive. You can use a pattern to find the sign of the product of two negative integers.

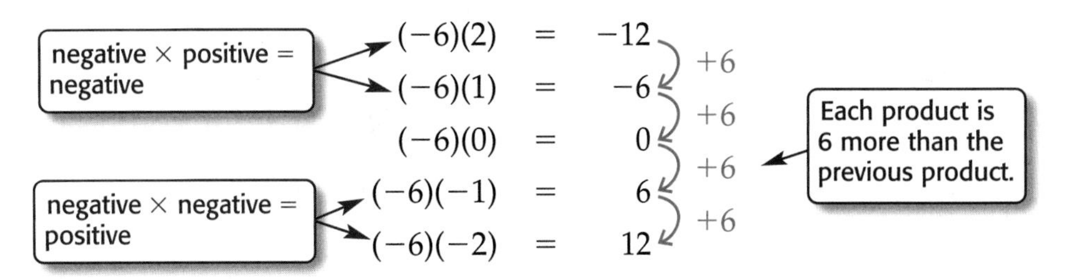

When two integers have the same sign, the following rule applies.

Key Concept **Multiplying Integers with the Same Sign**

Words The product of two integers with the same sign is positive.

Examples $2(6) = 12$ $\quad -10(-6) = 60$

EXAMPLES Multiply Integers with the Same Sign

Multiply.

3 $\mathbf{-11(-9)}$

$-11(-9) = 99$ The integers have the same sign. The product is positive.

STUDY TIP

Look Back You can review **exponents** in Lesson 1-2.

4 $\mathbf{(-4)^2}$

$(-4)^2 = (-4)(-4)$ There are two factors of −4.

$= 16$ The product is positive.

Your Turn Multiply.

a. $-12(-4)$ b. $(-5)^2$ c. $(-2)^3$

EXAMPLES Simplify and Evaluate Expressions

5 **ALGEBRA Simplify $-2(3x)$.**

$-2(3x) = (-2 \cdot 3)x$ Associative Property of Multiplication

$= -6x$ Simplify.

6 **ALGEBRA Evaluate pqr if $p = -3$, $q = 4$, and $r = -1$.**

$pqr = (-3)(4)(-1)$ Replace p with −3, q with 4, and r with −1.

$= (-12)(-1)$ Multiply −3 and 4.

$= 12$ Multiply −12 and −1.

Your Turn

d. Simplify $-5(2y)$.

e. Evaluate xyz if $x = -7$, $y = -4$, and $z = 2$.

Skill and Concept Check

Writing Math
Exercise 3

1. **Model** the product of 2 and -3 using counters. Then write the multiplication sentence.

2. **OPEN ENDED** Name two integers whose product is negative.

3. **NUMBER SENSE** What is the sign of the product of three negative integers? Give an example.

GUIDED PRACTICE

Multiply.

4. $6(-10)$
5. $-15(-3)$
6. $(-2)^2$

ALGEBRA Simplify each expression.

7. $-5(2a)$
8. $3(-6b)$
9. $-5(-9c)$

ALGEBRA Evaluate each expression if $f = -1$, $g = 7$, and $h = -10$.

10. $5f$
11. fgh
12. $-h^2$

13. **SUBMARINES** A submarine is diving from the surface of the water at a rate of 125 feet per minute. What is the depth of the submarine after 7 minutes?

Practice and Applications

HOMEWORK HELP

For Exercises	See Examples
14–27, 44	1–4
28–35	5
36–43	6

Extra Practice
See pages 571, 598.

Multiply.

14. $8(-13)$
15. $-16(-5)$
16. $(-9)^2$
17. $-10(-17)$
18. $-7(16)$
19. $(-6)^2$
20. $-20(-8)$
21. $-15(30)$
22. $-31(-5)$
23. $11(-20)$
24. $-(7^2)$
25. $(-4)^3$
26. Find the product of -13 and 13.
27. Find -7 squared.

ALGEBRA Simplify each expression.

28. $-3(6c)$
29. $-7(10d)$
30. $5(-4e)$
31. $9(-8f)$
32. $-2(-3g)$
33. $-6(-4h)$
34. $(2x)(-3y)$
35. $(-5r)(2s)$

ALGEBRA Evaluate each expression if $w = 7$, $x = -8$, $y = 5$, and $z = 10$.

36. $-4w$
37. xy
38. $-2xz$
39. xyz
40. $-7wy$
41. $-3z^2$
42. $12x^2$
43. $-wz^2$

44. **VOLUNTEERING** The Volunteer Club raked leaves at several senior citizens' homes in the neighborhood. If each group of three students could remove 8 cubic meters of leaves in one hour, find an integer to represent the number of cubic meters of leaves 12 students could remove in five hours.

45. **PATTERNS** Find the next two numbers in the pattern 1, -2, 4, -8, 16, … . Then describe the pattern.

GEOMETRY **For Exercises 46–48, use the graph at the right.**

46. Name the ordered pairs for P, Q, and R. Multiply each x- and y-coordinate by -1 to get three new ordered pairs.

47. Graph the ordered pairs and connect them to form a new triangle. Describe its position with respect to the original triangle.

48. In which quadrant does a triangle lie if only the y-coordinates of the original triangle are multiplied by -2?

49. **CRITICAL THINKING** For what values of n is $(-2)^n$ positive?

Standardized Test Practice and Mixed Review

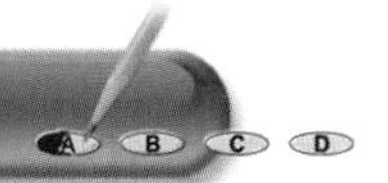

50. **MULTIPLE CHOICE** An oil rig is drilling into the ground at a rate of 7 feet per minute. What integer represents the position of the oil rig after 42 minutes?

    (A) -294 ft (B) -35 ft (C) 35 ft (D) 294 ft

51. **MULTIPLE CHOICE** Monifa has 100 shares of stock each worth \$15. If the price drops to \$8, what integer represents the change in Monifa's current investment?

    (F) $-\$700$ (G) $-\$7$ (H) \$7 (I) \$700

52. **GRID IN** Evaluate $-6[-2(3) + 0(-5)] + (-4)^2$.

53. Find $-25 - (-33)$. (Lesson 3-5)

ALGEBRA **Evaluate each expression if $x = -4$, $y = 6$, and $z = 1$.** (Lesson 3-4)

54. $x + (-2)$
55. $-1 + z$
56. $-15 + y$
57. $x + y$

58. **EARTH SCIENCE** The low temperatures in degrees Fahrenheit for ten cities on January 23 were -3, 27, 13, -6, -14, 36, 47, 52, -2, and 0. Order these temperatures from greatest to least. (Lesson 3-2)

MILITARY **For Exercises 59 and 60, use the double-bar graph at the right.** (Lesson 2-7)

59. In which age groups are there more members in the Navy than members in the Air Force?

60. About how many 36–40 year olds are in the Air Force?

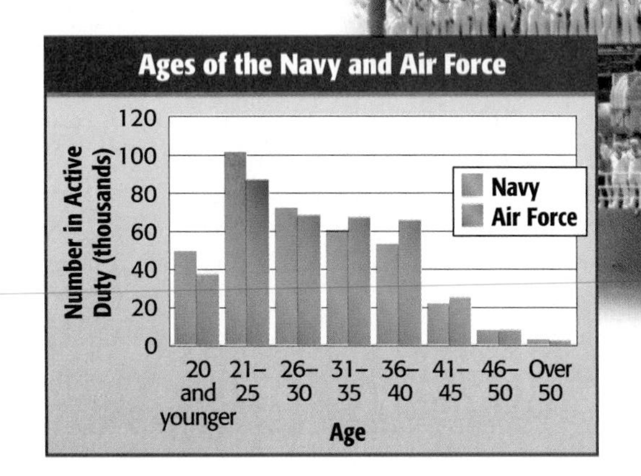

GETTING READY FOR THE NEXT LESSON

BASIC SKILL **Divide.**

61. $72 \div 9$
62. $120 \div 6$
63. $84 \div 21$
64. $215 \div 43$

3-7 Dividing Integers

HANDS-ON Mini Lab

What You'll LEARN

Divide integers.

Materials
- counters
- integer mat

Work with a partner.

You can use counters to model division of integers. Follow these steps to find $-10 \div 5$.

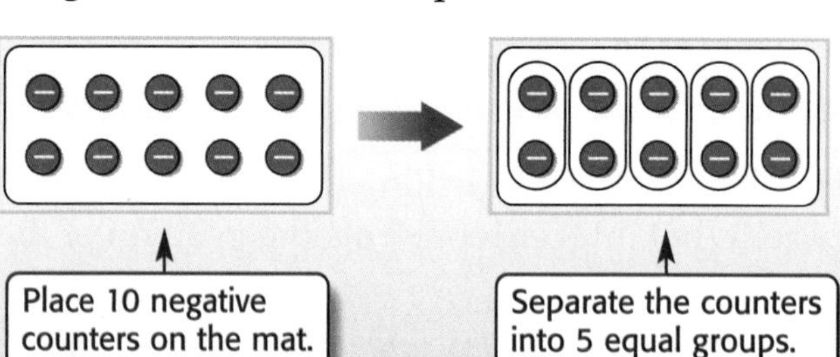

There are 2 negative counters in each group. So, $-10 \div 5 = -2$.

Find each quotient using counters.

1. $-6 \div 2$
2. $-12 \div 3$

Division of integers is related to multiplication. When finding the quotient of two integers, you can use a related multiplication sentence.

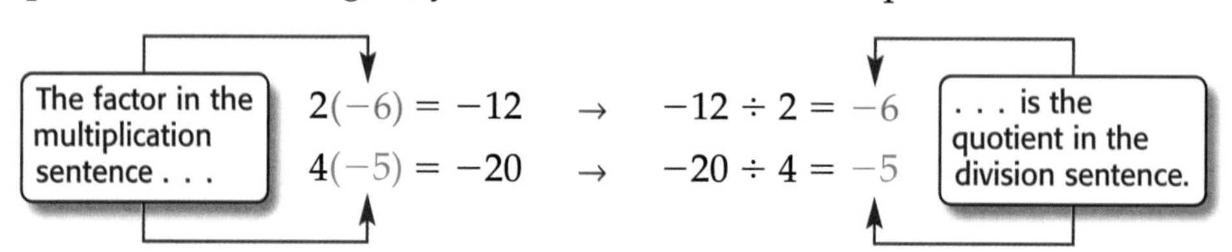

Since multiplication and division sentences are related, you can use them to find the quotient of integers with different signs.

different signs

$8(-9) = -72 \rightarrow -72 \div 8 = -9$

$-8(-9) = 72 \rightarrow 72 \div (-8) = -9$

negative quotient

different signs

These related sentences lead to the following rule.

Key Concept — **Dividing Integers with Different Signs**

Words The quotient of two integers with different signs is negative.

Examples $33 \div (-11) = -3$ $\quad -64 \div 8 = -8$

READING Math

Division Sentence
In the division sentence $80 \div (-10) = -8$, 80 is called the *dividend*, -10 is called the *divisor*, and -8 is called the *quotient*.

EXAMPLES Dividing Integers with Different Signs

Divide.

1 $80 \div (-10)$ — The integers have different signs.

$80 \div (-10) = -8$ — The quotient is negative.

2 $\frac{-55}{11}$ — The integers have different signs.

$\frac{-55}{11} = -5$ — The quotient is negative.

Your Turn Divide.

a. $20 \div (-4)$ b. $\frac{-81}{9}$ c. $-45 \div 9$

The quotient of two integers with the same sign is positive.

Key Concept — Dividing Integers with the Same Sign

Words The quotient of two integers with the same sign is positive.

Examples $15 \div 5 = 3$ $-21 \div (-3) = 7$

EXAMPLE Dividing Integers with the Same Sign

Divide.

3 $-14 \div (-7)$ — The integers have the same sign.

$-14 \div (-7) = 2$ — The quotient is positive.

Your Turn Divide.

d. $-24 \div (-4)$ e. $-9 \div (-3)$ f. $\frac{-28}{-7}$

REAL-LIFE MATH

ASTRONOMY The spacecraft *Mars Odyssey* was launched in April 2001 to study mineralogy and radiation of the surface of Mars.

Source: *The World Almanac for Kids*

EXAMPLE Evaluate an Expression

4 **ASTRONOMY The average surface temperature on Mars is -81°F. Use the expression $\frac{5(F-32)}{9}$, where F represents the number of degrees Fahrenheit, to find the temperature on Mars in degrees Celsius.**

$\frac{5(F-32)}{9} = \frac{5(-81-32)}{9}$ — Replace F with -81.

$= \frac{5(-113)}{9}$ — Subtract 32 from -81.

$= \frac{-565}{9}$ — Multiply 5 and -113.

≈ -62.8 — Divide.

The average temperature on the surface of Mars is about -63°C.

Skill and Concept Check

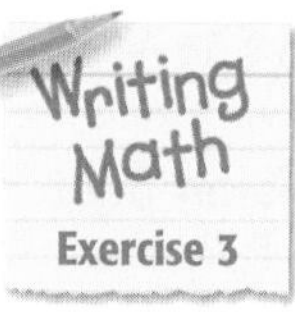

1. **Write** two division sentences related to the multiplication sentence $-6 \cdot 7 = -42$.

2. **OPEN ENDED** Write a division sentence. Then draw a model to show how the quotient can be found.

3. **Which One Doesn't Belong?** Identify the division expression whose quotient does not have the same sign as the other three. Explain your reasoning.

-24 ÷ 6	-18 ÷ (-9)	28 ÷ (-7)	-22 ÷ 11

GUIDED PRACTICE

Divide.

4. $32 \div (-8)$
5. $-16 \div 2$
6. $-60 \div (-5)$
7. $\frac{-6}{6}$

ALGEBRA Evaluate each expression if $d = -9$, $e = 36$, and $f = -6$.

8. $-108 \div f$
9. $e \div d$
10. $\frac{e-f}{f}$

Practice and Applications

Divide.

11. $-18 \div 9$
12. $50 \div (-5)$
13. $-15 \div (-3)$
14. $\frac{21}{-7}$
15. $56 \div (-8)$
16. $\frac{0}{-5}$
17. $-52 \div (-13)$
18. $-34 \div 2$
19. $\frac{90}{6}$
20. $-300 \div 25$
21. $99 \div (-99)$
22. $-184 \div (-23)$

HOMEWORK HELP	
For Exercises	See Examples
11–24	1–3
25–37	4

Extra Practice
See pages 571, 598.

23. Find the quotient of -65 and 13.

24. Divide 200 by -100.

ALGEBRA Evaluate each expression if $r = 12$, $s = -4$, and $t = -6$.

25. $-12 \div r$
26. $72 \div t$
27. $r \div s$
28. $rs \div 16$
29. $\frac{-r}{t}$
30. $\frac{16-(-r)}{-s}$
31. $t^2 \div r$
32. $\frac{r^2}{s^2}$

33. **FOOTBALL** During the fourth quarter, the Colts were penalized 3 times for the same amount for a total of 45 yards. Write a division sentence to represent this situation. Then find the number of yards for each penalty.

34. **EARTH SCIENCE** Use the expression $\frac{5(F-32)}{9}$, where F represents the number of degrees Fahrenheit, to convert 5°F to degrees Celsius.

35. **PATTERNS** Find the next two numbers in the pattern 729, -243, 81, -27, 9, … . Explain your reasoning.

36. **SALES** The graph shows five magazines that had losses in a recent year. The numbers represent the profit the magazines made in 2002 compared to 2001. What is the mean of the losses for these five magazines?

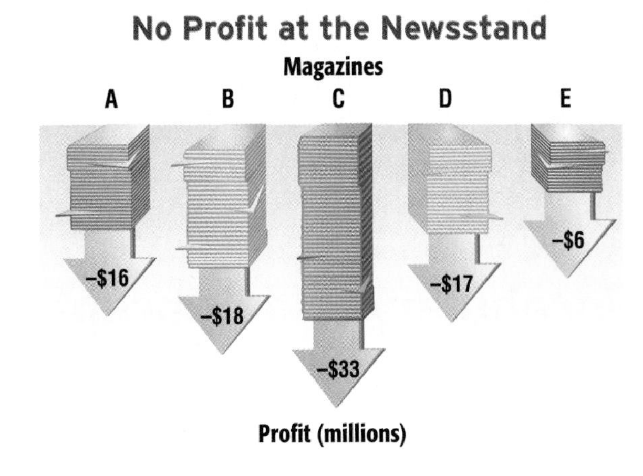

37. **MULTI STEP** The sea otter population is decreasing. There were 2,377 sea otters in 1995. The population dropped to 2,161 in 2001. Find the average rate of change for the sea otter population between 1995 and 2001.

38. **WRITE A PROBLEM** Write about a situation in your life in which you used positive and negative integers. Create a problem and solve it using any of the four operations.

39. **CRITICAL THINKING** List all of the numbers by which -20 is divisible.

Standardized Test Practice and Mixed Review

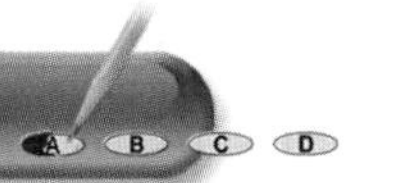

40. **MULTIPLE CHOICE** Shenequa is driving cross-country. If she expects to drive between 350 and 450 miles per day, which number of days is reasonable for her to drive 3,800 miles?

    (A) fewer than 6 days
    (B) between 6 and 8 days
    (C) between 9 and 11 days
    (D) more than 11 days

41. **MULTIPLE CHOICE** The width of a beach at 8:00 P.M. is 107 feet. At 3:00 A.M., the width of the beach narrows to 23 feet due to the high tide. What is the average rate of change per hour of the beach width?

    (F) -15 ft/h
    (G) -13 ft/h
    (H) -12 ft/h
    (I) -10 ft/h

42. **SHORT RESPONSE** Find the mean of the following low temperatures for a 5-day period: -2°F, -3°F, 1°F, 4°F, and -5°F.

Multiply. (Lesson 3-6)

43. $14(-2)$
44. $(-9)^2$
45. $-20(-3)$
46. $-5(7)$

47. Find $6 - (-12)$. (Lesson 3-5)

48. **DIVING** Mark jumped into 12 feet of water and touched the bottom of the pool before he surfaced. Which integer describes where Mark was in relation to the surface of the water? (Lesson 3-1)

49. **PHYSICAL SCIENCE** A chemistry experiment requires 3 milligrams of potassium chloride. How many grams of potassium chloride are needed? (Lesson 1-8)

Study Guide and Review

Vocabulary and Concept Check

absolute value (p. 107)
additive inverse (p. 121)
coordinate grid (p. 112)
coordinate plane (p. 112)
graph (p. 106)
integer (p. 106)
negative integer (p. 106)
opposite (p. 121)
ordered pair (p. 112)
origin (p. 112)
positive integer (p. 106)
quadrant (p. 113)
x-axis (p. 112)
x-coordinate (p. 112)
y-axis (p. 112)
y-coordinate (p. 112)

Choose the correct term or number to complete each sentence.

1. Integers less than zero are (positive, negative) integers.
2. Two numbers represented by points that are the same distance from 0 are (opposites, integers).
3. The absolute value of 7 is $(7, -7)$.
4. The opposite of $(-12, 12)$ is -12.
5. The (coordinate plane, origin) is the point where the horizontal and vertical number lines intersect.
6. The x-axis and the y-axis separate the plane into four (quadrants, coordinates).
7. The first number in an ordered pair is the (x-coordinate, y-coordinate).
8. The sum of two (positive, negative) integers is negative.
9. The product of a positive and a negative integer is (positive, negative).
10. The quotient of a negative integer and a (positive, negative) integer is negative.

Lesson-by-Lesson Exercises and Examples

Integers and Absolute Value (pp. 106–108)

Write an integer for each situation.

11. a loss of $150
12. 350 feet above sea level
13. a gain of 8 yards
14. 12°F below 0

Evaluate each expression.

15. $|-11|$
16. $|100|$
17. $|5|$
18. $|-32|$
19. $|-16| + |9|$

Example 1 **Write an integer for 8 feet below sea level.**

Since this situation represents an elevation *below* sea level, -8 represents the situation.

Example 2 **Evaluate $|-10|$.**

On the number line, the graph of -10 is 10 units from 0.

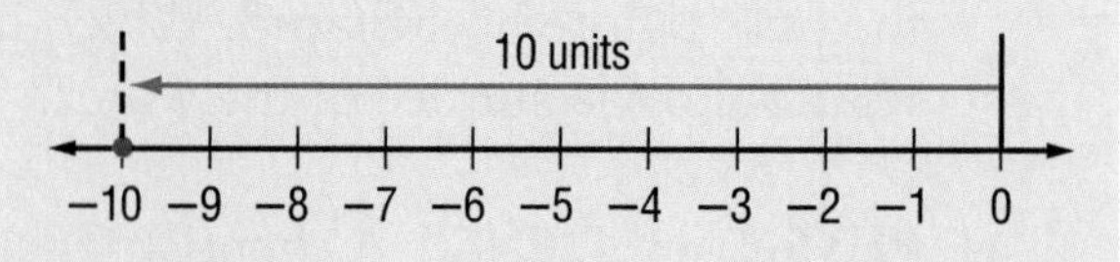

So, $|-10| = 10$.

msmath2.net/vocabulary_review

3-2 Comparing and Ordering Integers (pp. 109–111)

Replace each ● with < or > to make a true sentence.

20. -3 ● -9

21. 8 ● -12

22. -3 ● 3

23. $|-10|$ ● $|-13|$

24. 25 ● $|8|$

25. 0 ● $|-4|$

Order each set of integers from least to greatest.

26. $-3, 8, -10, 0, 5, -12, 9$

27. $-21, 19, -23, 14, -32, 25$

28. $|-17|, -18, 18, |15|, -16, |16|$

29. EARTH SCIENCE The predicted low temperatures for Monday through Friday are 3°, −1°, −2°, 0°, and 1°. Order the temperatures from greatest to least.

Example 3 **Replace ● with < or > to make −4 ● −7 a true sentence.**

Graph each integer on a number line.

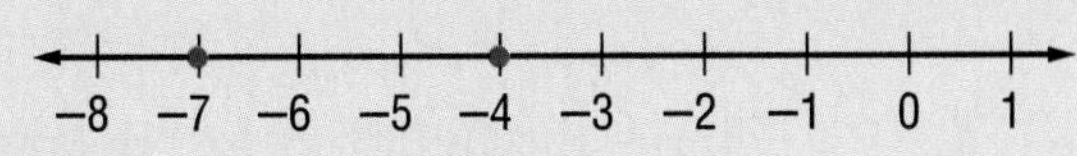

Since −4 is to the right of −7, $-4 > -7$.

Example 4 **Order the integers −4, −3, 5, 3, 0, −2 from least to greatest.**

Graph the integers on a number line.

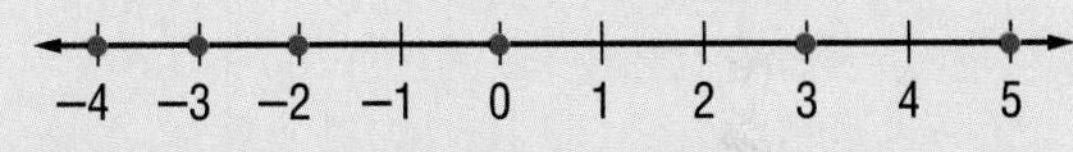

Order the integers by reading from left to right: −4, −3, −2, 0, 3, 5.

3-3 The Coordinate Plane (pp. 112–115)

Name the ordered pair for each point graphed at the right. Then identify the quadrant in which each point lies.

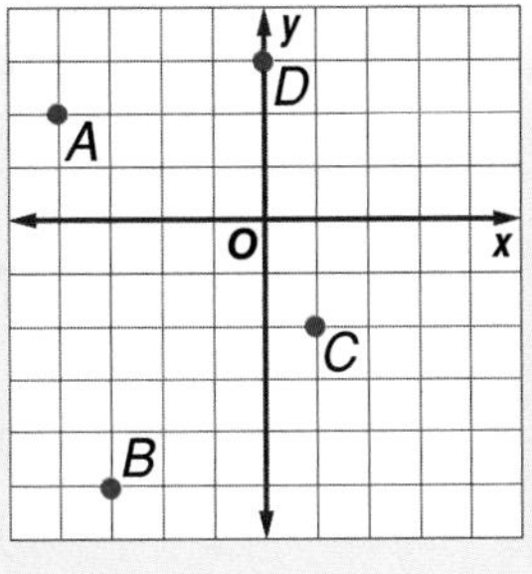

30. *A*

31. *B*

32. *C*

33. *D*

On graph paper, draw a coordinate plane. Then graph and label each point.

34. $E(1, -4)$

35. $F(-5, 2)$

36. $G(-2, -3)$

37. $H(4, 0)$

Example 5 **Name the ordered pair for point *W* graphed at the right. Then identify the quadrant in which point *W* lies.**

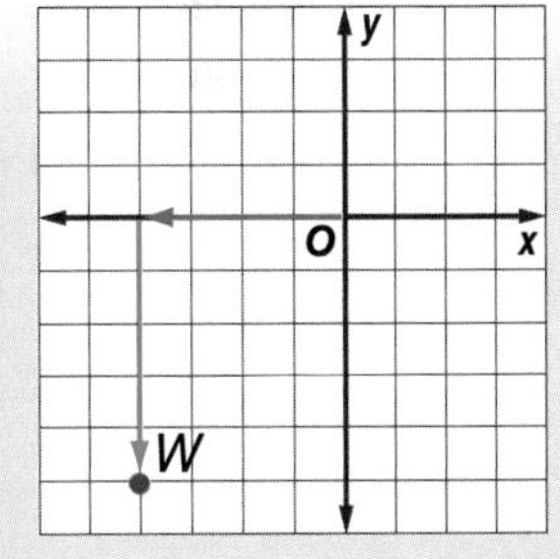

The ordered pair is (−4, −5). Point *W* is in quadrant III.

Example 6 **Graph and label the point *S*(3, −1).**

Draw a coordinate plane. Move 3 units to the right. Then move 1 unit down. Draw a dot and label it *S*(3, −1).

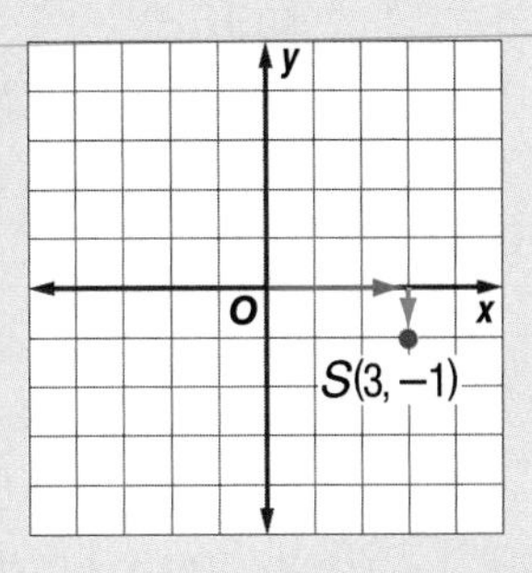

3-4 Adding Integers (pp. 120–124)

Add.

38. $-6 + 8$
39. $-4 + (-9)$
40. $7 + (-12)$
41. $-18 + 18$
42. **FOOTBALL** On the first play of the game, the Bulldogs lost 8 yards. On the second and third plays, they gained 5 yards and then lost 2 yards. Find the result of the first three plays.

Example 7 Find $-4 + 3$.

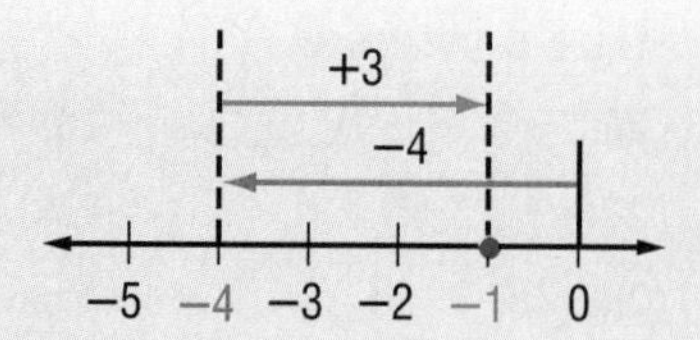

So, $-4 + 3 = -1$.

3-5 Subtracting Integers (pp. 128–131)

Subtract.

43. $-5 - 8$
44. $3 - 6$
45. $5 - (-2)$
46. $-4 - (-8)$

Example 8 Find $-3 - 9$.

$-3 - 9 = -3 + (-9)$ To subtract 9, add −9.

$= -12$ Simplify.

3-6 Multiplying Integers (pp. 134–137)

Multiply.

47. $-4(3)$
48. $8(-6)$
49. $-5(-7)$
50. $-2(40)$

ALGEBRA **Evaluate each expression if $a = -4$, $b = -7$, and $c = 5$.**

51. ab
52. $-3c$
53. bc
54. abc

Example 9 Find $-4(3)$.

$-4(3) = -12$ The integers have different signs. The product is negative.

Example 10 Evaluate xyz if $x = -6$, $y = 11$, and $z = -10$.

$xyz = (-6)(11)(-10)$ $x = -6, y = 11, z = -10$

$= (-66)(-10)$ Multiply −6 and 11.

$= 660$ Multiply −66 and −10.

3-7 Dividing Integers (pp. 138–141)

Divide.

55. $-45 \div (-9)$
56. $36 \div (-12)$
57. $-12 \div 6$
58. $-81 \div (-9)$
59. **HIKING** Marta started a hike at sea level and ended the hike 6 hours later at 300 feet below sea level. If Marta hiked at the same pace during the trip, how far did she travel each hour?

Example 11 Find $-72 \div (-9)$.

$-72 \div (-9) = 8$ The integers have the same sign. The quotient is positive.

CHAPTER 3 Practice Test

Vocabulary and Concepts

1. **Explain** what it means for two numbers to be opposites.
2. **Name** the rule for dividing integers with different signs.

Skills and Applications

Write an integer for each situation.

3. a stock increased by \$5
4. 1000 B.C.
5. an elevator goes down 11 floors

Replace each ● with $<$ or $>$ to make a true sentence.

6. -3 ● -9
7. $|9|$ ● $|-12|$
8. $|-7|$ ● 9

9. **WEATHER** The local weather service records the following changes in temperature during the last week: 4, −7, −3, 2, 9, −8, 1. Order these temperature changes from greatest to least.

Name the ordered pair for each point graphed at the right. Then identify the quadrant in which each point lies.

10. P
11. Q
12. R

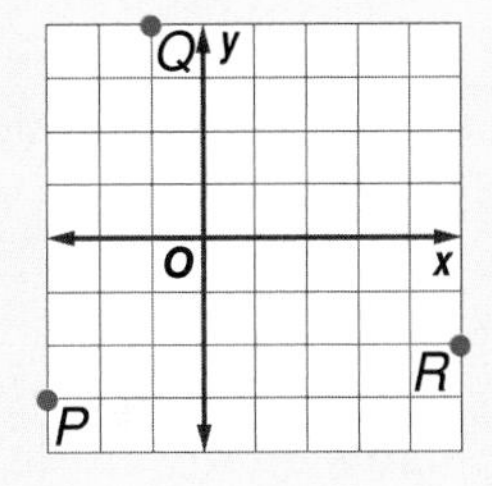

Add, subtract, multiply, or divide.

13. $-3 + 8$
14. $12 + (-19)$
15. $-3 - 8$
16. $-7 - (-20)$
17. $-7(-3)$
18. $5(-11)$
19. $-24 \div 8$
20. $-36 \div (-9)$

ALGEBRA Evaluate each expression if $a = -5$, $b = 4$, and $c = -12$.

21. $c - a$
22. ab
23. $ac \div b$

24. **STOCK MARKET** The value of a stock went down \$3 each week for a period of seven weeks. Describe the change in the value of the stock at the end of the seven week period.

Standardized Test Practice

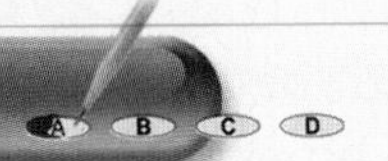

25. **MULTIPLE CHOICE** Choose the graph that shows the ordered pair (2, −1).

Ⓐ
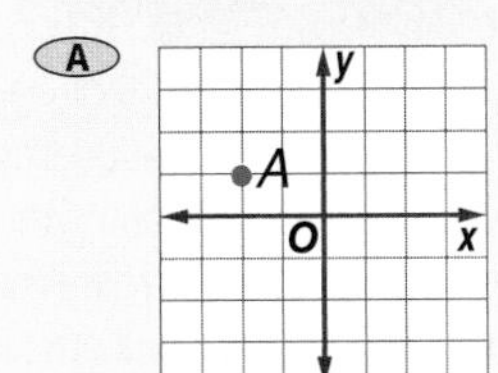

Ⓑ
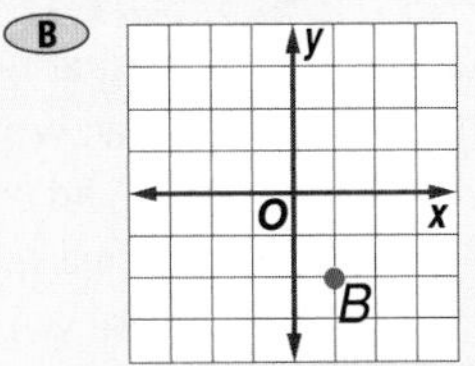

Ⓒ
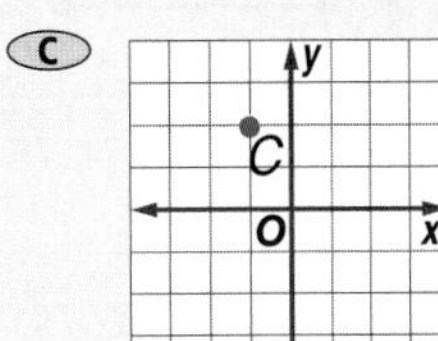

Ⓓ
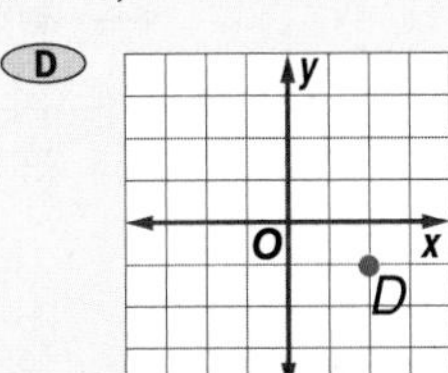

CHAPTER 3 Standardized Test Practice

PART 1 Multiple Choice

Record your answers on the answer sheet provided by your teacher or on a sheet of paper.

1. The table shows the five most common languages spoken in the United States other than English. How many more people speak Chinese than French? (Lesson 1-1)

Language	Number of People
Spanish	26,745,067
Chinese	1,976,564
French	1,914,918
German	1,224,213
Tagalog	1,184,939

Source: U.S. Census Bureau

A 61,646
B 1,914,918
C 1,976,564
D 3,891,482

2. Which is equivalent to 5^4? (Lesson 1-2)

F 20
G 125
H $5 \cdot 5 \cdot 5 \cdot 5$
I $4 \cdot 4 \cdot 4 \cdot 4 \cdot 4$

3. How many millimeters are in 13 centimeters? (Lesson 1-8)

A 0.13
B 1.3
C 13
D 130

4. Salvador recorded the number of minutes it took him to drive to work each day for a week. Find the mean for the following times: 12, 23, 10, 14, and 11. (Lesson 2-4)

F 12 min
G 13 min
H 14 min
I 15 min

5. Find the interquartile range of the data in the box-and-whisker plot. (Lesson 2-6)

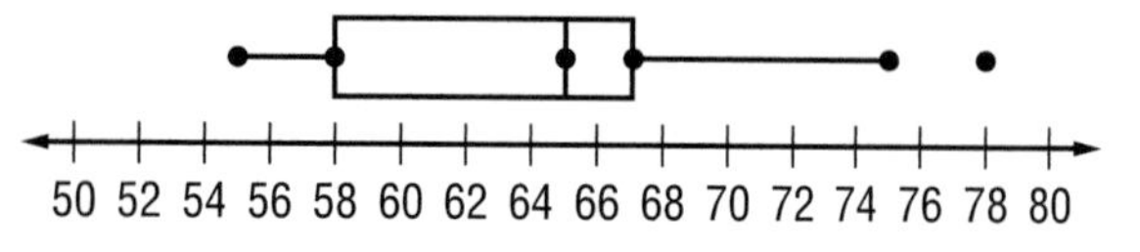

A 9
B 23
C 58
D 65

6. What is the value of $|-2|$? (Lesson 3-1)

F −2
G −1
H 0
I 2

7. Which of these is the correct order of the integers from greatest to least? (Lesson 3-2)

A 0, 1, −2, −5
B 1, 0, −2, −5
C −5, −2, 0, 1
D 1, −2, −5, 0

8. Which of these are the coordinates of point Z? (Lesson 3-3)

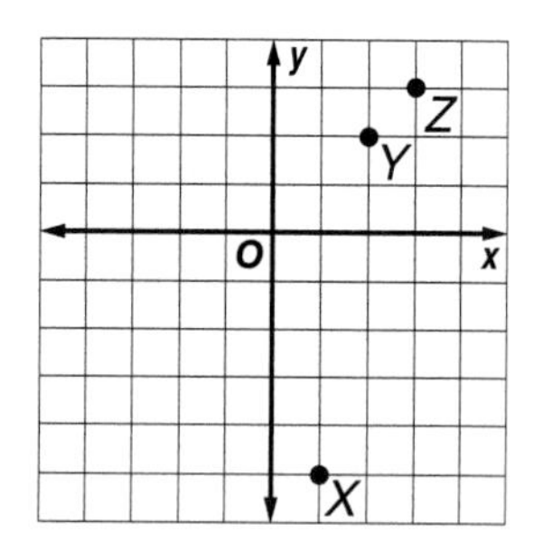

F (2, 2)
G (3, 3)
H (−2, −2)
I (−3, −3)

9. If you graph and connect the following points on a coordinate plane, what shape would you make? (Lesson 3-3)

(0, −3), (0, 4), (4, 4), (4, −3)

A rectangle
B triangle
C circle
D parallelogram

10. Which of the following points on the number line represents opposites? (Lesson 3-4)

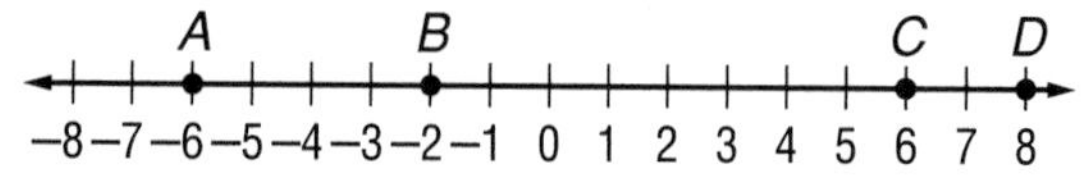

F C and D
G B and D
H A and D
I A and C

TEST-TAKING TIP

Question 10 Always be sure to check every answer choice of a multiple-choice question. Start with answer choice F. Each time you find an incorrect answer choice, cross it off so you remember that you've eliminated it.

PART 2 Short Response/Grid In

Record your answers on the answer sheet provided by your teacher or on a sheet of paper.

11. Wallpaper costs $16 per roll, and border costs $9 per roll. If 12 rolls of wallpaper and 6 rolls of border are needed for one room, find the total cost of the wallpaper and border. (Lesson 1-3)

12. Each triangle in the figure below is made from toothpicks. Extend the pattern. Find the number of toothpicks in the fifth figure. (Lesson 1-7)

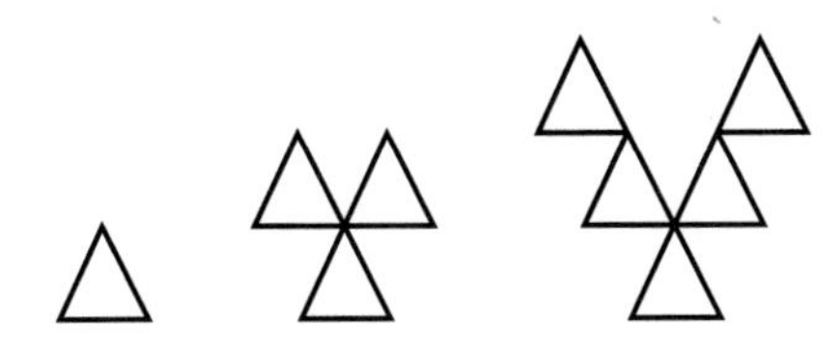

13. Write 63.5 million in scientific notation. (Lesson 1-9)

For Questions 14 and 15, use the information below.
The graph shows the number of orders taken each hour one day at a fast food restaurant. (Lesson 2-2)

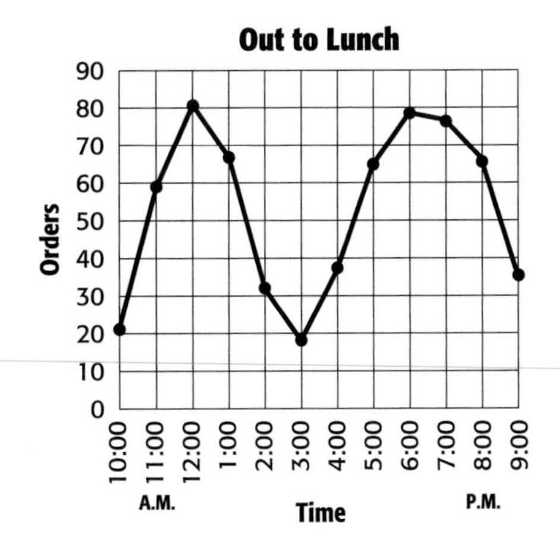

14. About how many orders were taken at 11:00 A.M.?

15. What time of the day appears to be the least busy at this restaurant?

For Questions 16 and 17, use the information below.
The temperature of the liquid in Connor's beaker changed drastically. The temperatures he recorded were −28°F, 59°F, 1°F, −16°F, 24°F, and 8°F. (Lesson 3-2)

16. Order the temperatures from least to greatest.

17. Find the mean temperature.

18. When Elena went hiking at 8 A.M., she started at an elevation of 16 meters below sea level. But at 2 P.M., she was 12 meters higher than when she started. What was Elena's elevation at 2 P.M.? (Lesson 3-5)

19. Find $\frac{-12 + 8(-3)}{-2 - 7}$. (Lesson 3-7)

PART 3 Extended Response

Record your answers on a sheet of paper. Show your work.

20. Use the table of ordered pairs at the right to answer the questions below. (Lesson 3-3)

x	y
−2	1
3	2
−1	−4

a. Plot the ordered pairs in the table on a coordinate plane.

b. In which quadrant is a point on the graph not represented?

c. Connect the points on the graph. What shape do they form?

d. How would you double the size of the figure you drew above?

e. Add four more columns to the table. Label the columns $x + y$, $x - y$, xy, and $x \div y$. Complete the table.

CHAPTER

4 Algebra: Linear Equations and Functions

"What do horses have to do with math?"

A horse can gallop at a speed of 43 miles per hour. You can use the linear equation $d = 43t$ to find the **distance d** a horse gallops in a certain **time t**. In algebra, you will use variables and equations to describe many real-life situations.

You will solve problems about distance, rate, and time in Lesson 4-3.

GETTING STARTED

Take this quiz to see whether you are ready to begin Chapter 4. Refer to the lesson or page number in parentheses if you need more review.

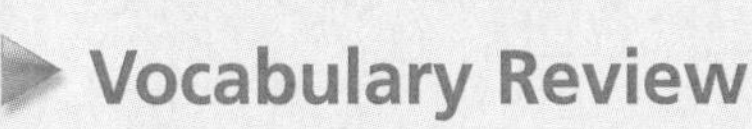

Vocabulary Review

Complete each sentence.

1. When you replace the variable with a number that makes an equation true, you have __?__ the equation. (Lesson 1-5)
2. The first number in an ordered pair is the __?__, and the second number is the __?__. (Lesson 3-3)

Prerequisite Skills

Name the number that is the solution of the given equation. (Lesson 1-5)

3. $a + 15 = 19; 4, 5, 6$
4. $11a = 77; 6, 7, 8$
5. $x + 9 = -2; 7, -11, 11$
6. $2x - 8 = 10; 1, 9, 18$

Graph each point on a coordinate plane. (Lesson 3-3)

7. $(-4, 3)$
8. $(-2, -1)$

Add. (Lesson 3-4)

9. $-3 + (-5)$
10. $-8 + 3$
11. $9 + (-5)$
12. $-10 + 15$

Subtract. (Lesson 3-5)

13. $-5 - 6$
14. $8 - 10$
15. $8 - (-6)$
16. $-3 - (-1)$

Divide. (Lesson 3-7)

17. $-6 \div (-3)$
18. $-12 \div 3$
19. $10 \div (-5)$
20. $-24 \div (-4)$

Solving Equations Make this Foldable to help you organize your notes. Begin with a sheet of $8\frac{1}{2}'' \times 11''$ paper.

STEP 1 Fold Fold the short sides toward the middle.

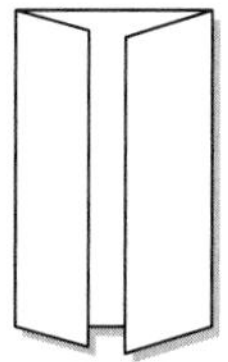

STEP 2 Fold Again Fold the top to the bottom.

STEP 3 Cut Open. Cut along the second fold to make four tabs.

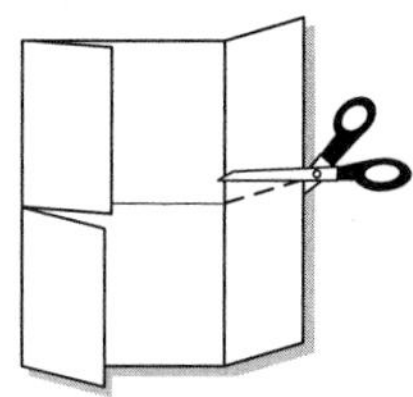

STEP 4 Label Label each of the tabs as shown.

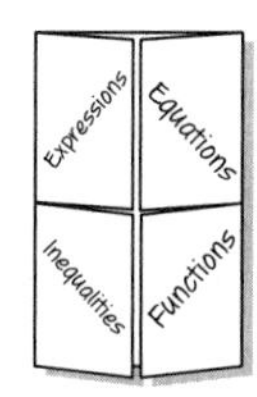

Reading and Writing As you read and study the chapter, write examples under each tab.

Writing Expressions and Equations

What You'll LEARN

Write verbal phrases and sentences as simple algebraic expressions and equations.

REVIEW Vocabulary

expression: a combination of variables, numbers, and at least one operation **(Lesson 1-4)**

WHEN am I ever going to use this?

COMICS Even these characters from *Peanuts* are using algebra.

1. Suppose the daughter is 12 years old. How old is the son?
2. What operation did you use to find the son's age? Explain.
3. Suppose the comic said that *the son is twice as old as the daughter.* If the daughter is 12 years old, how old is the son?
4. What operation did you use to find the son's age? Explain.

Words and phrases in problems often suggest addition, subtraction, multiplication, and division. Here are some examples.

Addition and Subtraction		Multiplication and Division	
plus	minus	times	divided
sum	difference	product	quotient
more than	less than	multiplied	per
increased by	less	twice	separate
in all	decreased by	each	

EXAMPLE Write a Phrase as an Expression

1 Write the phrase *five dollars less than Jennifer earned* as an algebraic expression.

Words	five dollars less than Jennifer earned
Variable	Let d represent the number of dollars Jennifer earned.
Expression	$d - 5$

Your Turn Write each phrase as an algebraic expression.

a. twice as many tomatoes as last year

b. 3 more runs than the Pirates scored

Remember, an equation is a sentence in mathematics that contains an equals sign. When you write a verbal sentence as an equation, you can use the equals sign (=) for the words *equals* or *is*.

REAL-LIFE MATH

FOOD In addition to potato chips and tortilla chips, other snacks, in pounds, consumed during the 2002 Super Bowl were:

popcorn: 4.2 million
pretzels: 3.8 million
nuts: 3.1 million

Source: Snack Food Association

EXAMPLES Write Sentences as Equations

Write each sentence as an algebraic equation.

Sentence	Equation
2 Five more than a number is 20.	$n + 5 = 20$
3 Three times Jack's age equals 12.	$3a = 12$

4 **FOOD It is estimated that 12.4 million pounds of potato chips were consumed during the 2002 Super Bowl. This was 3.1 million pounds more than the number of pounds of tortilla chips consumed. Write an equation that models this situation.**

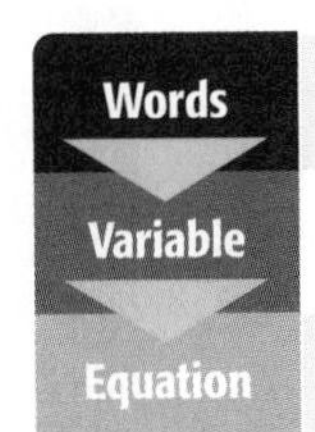

Words	Potato chips were 3.1 million more than tortilla chips.
Variable	Let t = number of millions of pounds of tortilla chips.
Equation	$12.4 = 3.1 + t$

The equation is $12.4 = 3.1 + t$.

Skill and Concept Check

Writing Math
Exercise 2

1. **OPEN ENDED** Write a verbal sentence that translates into the equation $n + 5 = 8$.

2. **FIND THE ERROR** Antonio and Julia are writing an algebraic expression for the phrase *2 less than a number.* Who is correct? Explain.

Antonio	Julia
2 − n	n − 2

GUIDED PRACTICE

Write each phrase as an algebraic expression.

3. eight more than x
4. nine less than t
5. ten times as many hours
6. −15 divided by some number

Write each sentence as an algebraic equation.

7. The sum of a number and four is −9.
8. Twice a number equals −10.

9. **POPULATION** The median age of people living in the United States was 35.3 years in 2000. This is 2.4 years older than the median age in 1990. Write an equation that models this situation.

msmath2.net/extra_examples

Practice and Applications

HOMEWORK HELP	
For Exercises	See Examples
10–17, 26–27	1
18–25, 28	2–4

Extra Practice
See pages 572, 599.

Write each phrase as an algebraic expression.

10. fifteen increased by t
11. five years older than Luis
12. nine dollars less than j
13. a number less six
14. the product of r and 8
15. twice as many oranges
16. Emily's age divided by 3
17. a number divided by -12

Write each sentence as an algebraic equation.

18. The sum of a number and four is -8.
19. Two more than the number of cookies is 4.
20. The product of a number and five is -20.
21. Ten times the number of students is 280.
22. Ten inches less than her height equals 26.
23. Five less than a number equals 31.
24. Seven more than twice his age is 51.
25. Three more than twice a number is 15.

MONEY For Exercises 26 and 27, use the table. The table shows the average lifespan of several kinds of paper currency in the United States. Let y represent the average lifespan of a $5 bill.

U.S. Currency

Kind	Lifespan (years)
$1	1.5
$5	2
$10	3
$20	4
$50	9
$100	9

Source: Federal Reserve System

26. Which lifespan can be represented by $2y$?
27. Write an expression to represent the lifespan of a $50 bill.
28. **TOURISM** The Washington Monument is 555 feet tall. It is 75 feet shorter than the Gateway to the West Arch. Write an equation that models this situation.
29. **CRITICAL THINKING** If x is an odd number, how would you represent the odd number immediately following it? preceding it?

Standardized Test Practice and Mixed Review

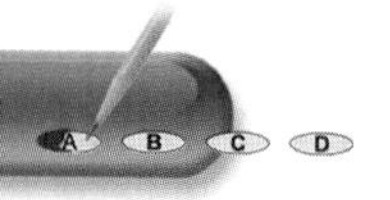

30. **MULTIPLE CHOICE** A mechanic charges a $35 initial fee and $32.50 for each hour he works. Which equation could be used to find the cost c of a repair job that lasts h hours?

    Ⓐ $c = 32.5 + 35h$
    Ⓑ $c = 35 + 32.5h$
    Ⓒ $c = 32.5 - 35h$
    Ⓓ $c = 32.5(35 - h)$

31. **MULTIPLE CHOICE** Translate *12 more than d* into an algebraic expression.

    Ⓕ $12d$
    Ⓖ $d - 12$
    Ⓗ $d + 12$
    Ⓘ $12 - d$

Divide. (Lesson 3-7)

32. $-18 \div 3$
33. $25 \div (-5)$
34. $-14 \div (-7)$
35. $72 \div (-9)$

GETTING READY FOR THE NEXT LESSON

PREREQUISITE SKILL Add. (Lesson 3-4)

36. $-8 + (-3)$
37. $-10 + 9$
38. $12 + (-20)$
39. $-15 + 15$

Study Skill

HOW TO...

Read Math Problems

SIMPLIFY THE PROBLEM

Have you ever tried to solve a long word problem and didn't know where to start? Try to rewrite the problem using fewer and fewer words. Then translate the problem into an equation.

▶**READ** the problem.

> Shopping networks on television are a popular way to shop. In addition to the cost of the items, you usually pay a shipping fee. Kylie wants to order several pairs of running shorts that cost $12 each. The total shipping fee is $7. How many shorts can she order with $55?

▶**REWRITE** the problem to make it simpler. Keep all of the important information but use fewer words.

Kylie has $55 to spend on some shorts that cost $12 each plus a shipping fee of $7. How many can she buy?

▶**REWRITE** the problem using even fewer words. Write a variable for the unknown.

The total cost of x shorts at $12 each plus $7 is $55.

▶**TRANSLATE** the words into an equation.

$12x + 7 = 55$

SKILL PRACTICE

Use the method above to write an equation for each word problem.

1. **FLYING** Orville and Wilbur Wright flew their airplane called Flyer I in Kitty Hawk, North Carolina, on December 17, 1903. Wilbur's flight was 364 feet. This was 120 feet longer than Orville's flight. How far was Orville's flight?

2. **ANIMALS** The cougars that are found in the colder parts of North and South America are about 75 inches long. They are about 1.5 times longer than the cougars that are found in the tropical jungles of Central America. Find the length of the tropical cougar.

3. **MONEY** Akira is saving money to buy a scooter that costs $125. He has already saved $80 and plans to save an additional $5 each week. In how many weeks will he have enough money for the scooter?

4. **TRAVEL** A taxi company charges $1.50 per mile plus a $10 fee. Suppose Olivia can afford to spend $19 for a taxi ride from her apartment to the mall. How far can she travel by taxi with $19?

4-2a HANDS-ON LAB

A Preview of Lesson 4-2

What You'll LEARN

Solve equations using models.

Materials

- cups and counters
- equation mat

REVIEW Vocabulary

equation: a mathematical sentence that contains an equals sign **(Lesson 1-5)**

Solving Equations Using Models

INVESTIGATE The scale at the right is balanced, and the bag contains a certain number of blocks.

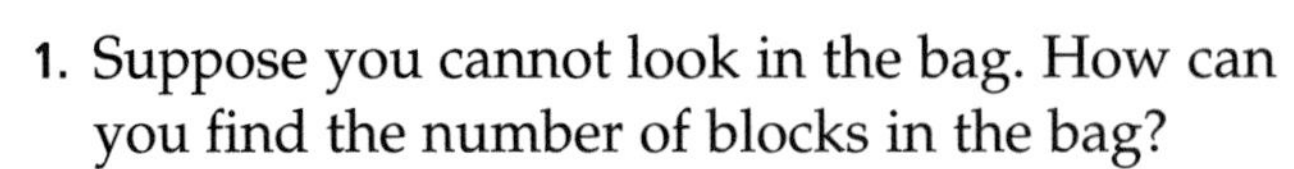

1. Suppose you cannot look in the bag. How can you find the number of blocks in the bag?
2. In what way is a balanced scale like an equation?
3. What does it mean to *solve an equation*?

To solve an equation using models, you can use these steps.

- You can add or subtract the same number of counters from each side of the mat.
- You can add or subtract zero from each side of the mat.

ACTIVITY *Work with a partner.*

1 Solve $x + 2 = 5$ using models.

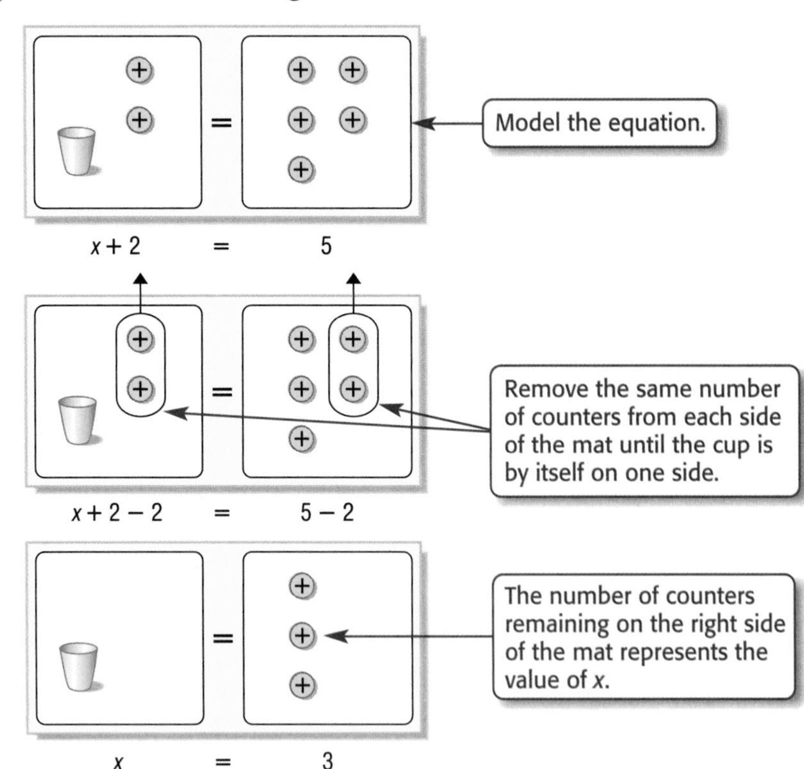

Therefore, $x = 3$. Since $3 + 2 = 5$, the solution is correct.

Your Turn **Solve each equation using models.**

a. $x + 1 = 3$ b. $x + 3 = 7$ c. $x + 4 = 4$

Look Back You can review **zero pairs** in Lesson 3-4.

Some equations are solved by using **zero pairs**. You can add or subtract a zero pair from either side of an equation without changing its value, because the value of a zero pair is zero.

ACTIVITY *Work with a partner.*

2 **Solve $x + 2 = -1$ using models.**

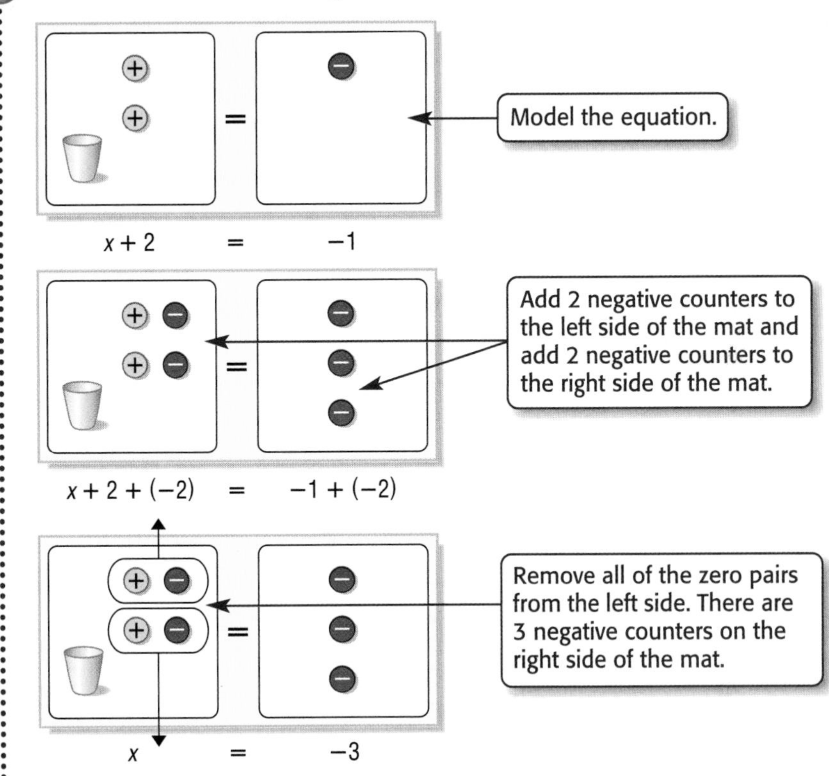

Therefore, $x = -3$. Since $-3 + 2 = -1$, the solution is correct.

Your Turn **Solve each equation using models.**

d. $x + 3 = -2$ **e.** $x + 4 = 1$ **f.** $-2 = x + 1$

g. $x - 3 = -2$ **h.** $x - 1 = -3$ **i.** $4 = x - 2$

Writing Math

1. How is solving an equation similar to keeping a scale in balance?
2. For any equation, how can you determine how many counters to add or subtract from each side?
3. **Identify** the property of numbers that is illustrated by a zero pair.
4. **Identify** the property of numbers that allows you to add or subtract zero without changing the value of a number.
5. **MAKE A CONJECTURE** Write a rule that you can use to solve an equation like $x + 3 = 2$ without using models.

4-2 Solving Addition and Subtraction Equations

What You'll LEARN

Solve addition and subtraction equations.

NEW Vocabulary

inverse operations

REVIEW Vocabulary

solve: finding a number that makes the equation true **(Lesson 1-5)**

WHEN am I ever going to use this?

SCIENCE Bottle-nosed dolphins and killer whales are the best-known species of the dolphin family. A killer whale, which can grow to a length of 9 meters, is 4 meters longer than a bottle-nosed dolphin.

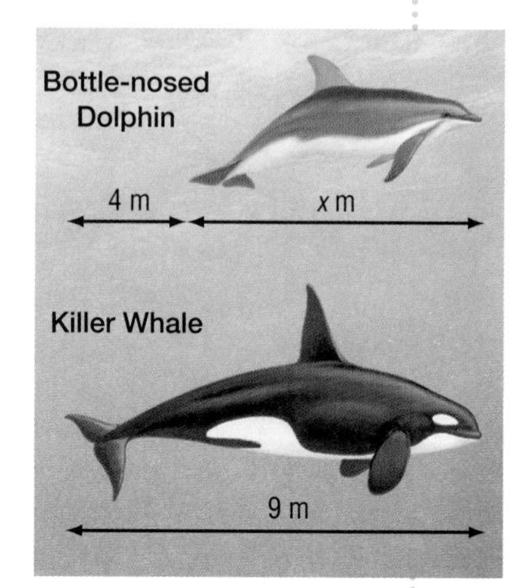

1. What does x represent in the figure?
2. Write an expression to represent *4 meters longer than a dolphin.*
3. Write an addition equation you could use to find the length of a dolphin.

You can use inverse operations to solve the equation $x + 4 = 9$. **Inverse operations** "undo" each other. To undo the addition of 4 in the equation $x + 4 = 9$, you would subtract 4 from each side of the equation.

EXAMPLE Solve an Addition Equation

1 **Solve $x + 4 = 9$.**

Method 1 Use symbols.

$x + 4 = 9$ Write the equation.

$$\begin{array}{rcl} x + 4 &=& 9 \\ -4 &=& -4 \end{array}$$ Subtract 4 from each side.

$$\begin{array}{rcl} x + 4 &=& 9 \\ -4 &=& -4 \\ \hline x &=& 5 \end{array}$$ Simplify.

The solution is 5.

Method 2 Use models.

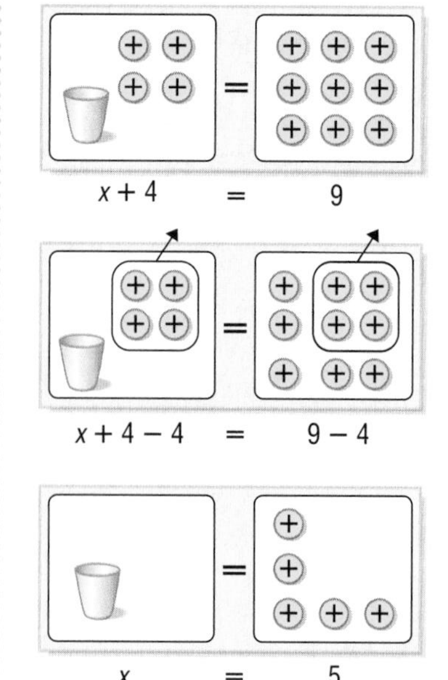

Your Turn Solve each equation.

a. $y + 6 = 9$ b. $x + 3 = 1$ c. $-3 = a + 4$

You can use the **Subtraction Property of Equality** and the **Addition Property of Equality** to solve equations like $x + 4 = 9$ and $x - 2 = 1$.

Key Concept — **Subtraction Property of Equality**

Words If you subtract the same number from each side of an equation, the two sides remain equal.

Symbols

Arithmetic	Algebra
$\begin{array}{r} 5 = 5 \\ \underline{-3 = -3} \\ 2 = 2 \end{array}$	$\begin{array}{r} x + 2 = 3 \\ \underline{-2 = -2} \\ x = 1 \end{array}$

Key Concept — **Addition Property of Equality**

Words If you add the same number to each side of an equation, the two sides remain equal.

Symbols

Arithmetic	Algebra
$\begin{array}{r} 5 = 5 \\ \underline{+3 = +3} \\ 8 = 8 \end{array}$	$\begin{array}{r} x - 2 = 4 \\ \underline{+2 = +2} \\ x = 6 \end{array}$

EXAMPLE Solve a Subtraction Equation

2 **Solve $x - 2 = 1$. Check your solution.**

Method 1 Use symbols.

$x - 2 = 1$ Write the equation.

$\begin{array}{r} x - 2 = 1 \\ \underline{+2 = +2} \end{array}$ Add 2 to each side.

$\begin{array}{r} x - 2 = 1 \\ \underline{+2 = +2} \\ x = 3 \end{array}$ Simplify.

Check $x - 2 = 1$ Write the original equation.

$3 - 2 \stackrel{?}{=} 1$ Replace x with 3.

$1 = 1$ ✔ This sentence is true.

The solution is 3.

Method 2 Use models.

$x - 2 = 1$

$x - 2 + 2 = 1 + 2$

$x = 3$

STUDY TIP

Checking Solutions It is always wise to check your solution. You can often use arithmetic facts to check the solutions of simple equations.

Your Turn **Solve each equation.**

d. $y - 3 = 4$ e. $r - 4 = -2$ f. $-9 = q - 8$

EXAMPLE Use an Equation to Solve a Problem

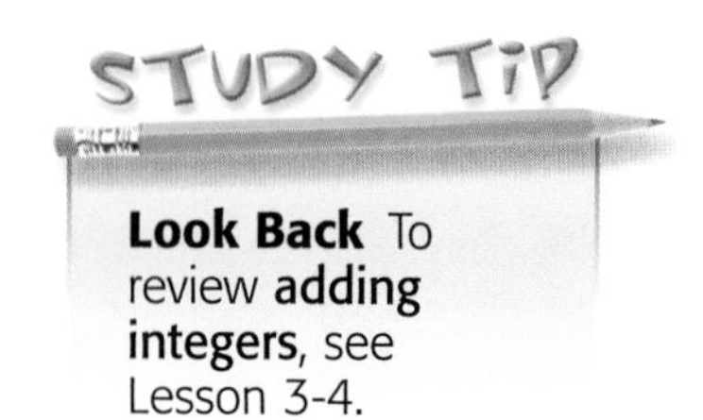

STUDY TIP

Look Back To review **adding integers**, see Lesson 3-4.

3 **SPORTS** **Tiger Woods won the 2002 Masters Championship with a final score of 12 under par, or -12. His scores for the first three rounds were -2, -3, and -6. What was his score for the fourth round?**

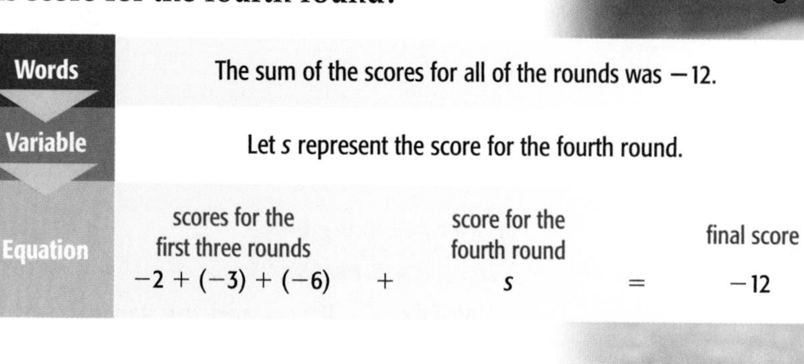

Words	The sum of the scores for all of the rounds was -12.				
Variable	Let s represent the score for the fourth round.				
Equation	scores for the first three rounds $-2 + (-3) + (-6)$	$+$	score for the fourth round s	$=$	final score -12

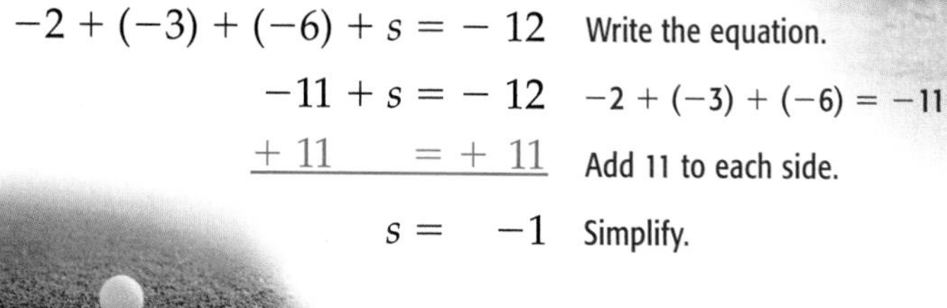

$$-2 + (-3) + (-6) + s = -12 \quad \text{Write the equation.}$$

$$-11 + s = -12 \quad -2 + (-3) + (-6) = -11$$

$$\underline{+11 \qquad = +11} \quad \text{Add 11 to each side.}$$

$$s = -1 \quad \text{Simplify.}$$

Check You can check the solution by adding.
$-2 + (-3) + (-6) + (-1) = -12$ ✔

Skill and Concept Check

Writing Math
Exercise 3

1. **Tell** what property you would use to solve $x - 4 = -2$.

2. **OPEN ENDED** Write two different equations that have -2 as a solution.

3. **Which One Doesn't Belong?** Identify the equation that does not have the same solution as the other three. Explain your reasoning.

$x + 3 = 4$	$b + 5 = 4$	$7 + y = 8$	$8 + a = 9$

GUIDED PRACTICE

Solve each equation. Check your solution.

4. $n + 6 = 9$
5. $m + 5 = 3$
6. $-2 = a + 6$
7. $x - 5 = 6$
8. $-1 = c - 6$
9. $s - 4 = -7$

10. **GEOMETRY** The sum of the measures of the angles of a triangle is 180°. Find the missing measure.

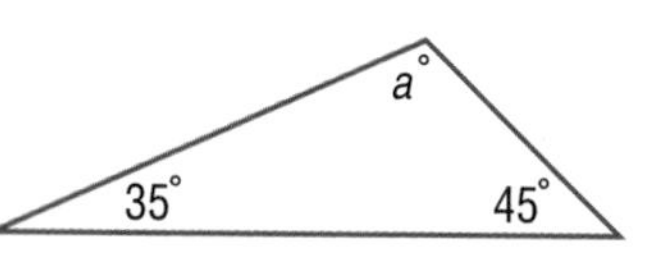

Practice and Applications

HOMEWORK HELP	
For Exercises	See Examples
11–16	1
17–22	2
29–30, 37–38	3

Extra Practice
See pages 572, 599.

Solve each equation. Check your solution.

11. $a + 3 = 10$
12. $y + 5 = 11$
13. $x + 8 = 5$
14. $y + 15 = 11$
15. $r + 6 = -3$
16. $k + 3 = -9$
17. $s - 8 = 9$
18. $w - 7 = 11$
19. $q - 8 = -1$
20. $p - 13 = -2$
21. $x - 5 = -9$
22. $w - 9 = -12$
23. $34 + r = 95$
24. $64 + y = 84$
25. $-23 = x - 18$
26. $-59 = m - 11$
27. $-18 + c = -30$
28. $-34 = t + 9$

29. The sum of a number and 3 is -2. Find the number.

30. If you decrease a number by 4, the result is -5. Find the number.

Solve each equation. Check your solution.

31. $a - 3.5 = 14.9$
32. $x - 2.8 = 9.5$
33. $r - 8.5 = -2.1$
34. $z - 9.4 = -3.6$
35. $n + 1.4 = 0.72$
36. $b + 2.25 = 1$

ROLLER COASTERS For Exercises 37 and 38, use the table.

37. *Superman The Escape* is 105 feet taller than *Millennium Force*. Write and solve an addition equation that you could use to find the height of *Millennium Force*.

38. The difference in the speeds of *Superman The Escape* and *Millennium Force* is 7 miles per hour. If *Superman The Escape* has the greater speed, write and solve a subtraction equation to find its speed.

39. CRITICAL THINKING Suppose $x + y = 10$ and the value of x increases by 3. What must happen to the value of y so that $x + y = 10$ is still a true sentence?

Tallest Steel Roller Coasters in North America	Height (feet)	Speed (mph)
Superman The Escape	415	?
Millenium Force	?	93
Titan	245	85
Goliath	235	85
Nitro	230	80

Source: www.rcdb.com

Standardized Test Practice and Mixed Review

40. MULTIPLE CHOICE The Sears Tower in Chicago is 1,454 feet tall. It is 204 feet taller than the Empire State Building in New York City. Use the equation $e + 204 = 1{,}454$ to find the height of the Empire State Building.

Ⓐ 1,250 ft Ⓑ 1,350 ft Ⓒ 1,450 ft Ⓓ 1,650 ft

41. GRID IN If $x + 3 = 2$, what is the value of $2x + 5$?

42. ALGEBRA Write an expression for 6 inches less than w. (Lesson 4-1)

Divide. (Lesson 3-7)

43. $-15 \div 3$
44. $36 \div (-9)$
45. $-63 \div (-7)$
46. $-27 \div 9$

GETTING READY FOR THE NEXT LESSON

PREREQUISITE SKILL Divide. (Page 562)

47. $15.6 \div 13$
48. $8.84 \div 3.4$
49. $75.25 \div 0.25$
50. $0.76 \div 0.5$

Solving Multiplication Equations

What You'll LEARN

Solve multiplication equations.

REVIEW Vocabulary

coefficient: the numerical factor of a multiplication expression **(Lesson 1-4)**

HANDS-ON Mini Lab

Materials

- cups and counters
- equation mat

Work with a partner.

Equations like $2x = -6$ are called multiplication equations because the expression $2x$ means *2 times the value of x.* Follow these steps to solve $2x = -6$.

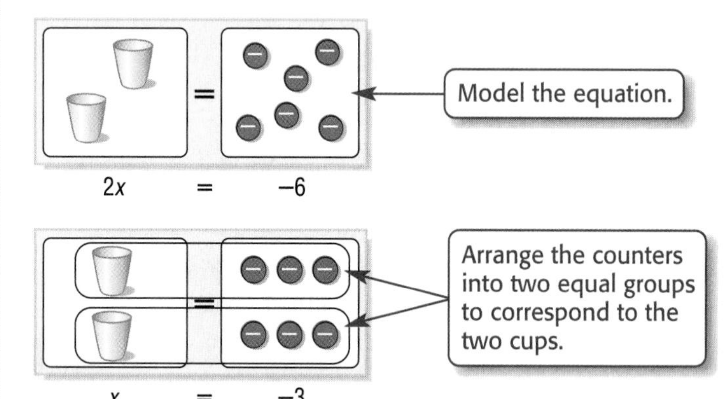

Each cup is matched with 3 negative counters. So, $x = -3$.

Solve each equation using models.

1. $4x = 8$
2. $-8 = 2x$
3. $3x = 3$
4. What operation did you use to find each solution?
5. How can you use the coefficient of x to solve $4x = 12$?
6. How can you use the coefficient of x to solve $-5x = 10$ without using cups and counters?

In the Mini Lab above, you matched each cup with an equal number of counters. This suggests the operation of division. You can use the **Division Property of Equality** to solve equations like $2x = -6$.

Key Concept — **Division Property of Equality**

Words If you divide each side of an equation by the same nonzero number, the two sides remain equal.

Symbols

Arithmetic	Algebra
$8 = 8$	$2x = -6$
$\frac{8}{2} = \frac{8}{2}$	$\frac{2x}{2} = \frac{-6}{2}$
$4 = 4$	$x = -3$

Notice that the division expression $8 \div 2$ can be written as the fraction $\frac{8}{2}$.

EXAMPLES Solve Multiplication Equations

1 **Solve $20 = 4x$. Check your solution.**

$20 = 4x$	Write the equation.
$\frac{20}{4} = \frac{4x}{4}$	Divide each side of the equation by 4.
$5 = x$	$20 \div 4 = 5$

Check

$20 = 4x$	Write the original equation.
$20 \stackrel{?}{=} 4(5)$	Replace x with 5. Is this sentence true?
$20 = 20$ ✓	

The solution is 5.

2 **Solve $-8y = 24$. Check your solution.**

$-8y = 24$	Write the equation.
$\frac{-8y}{-8} = \frac{24}{-8}$	Divide each side by −8.
$y = -3$	$24 \div (-8) = -3$

Check

$-8y = 24$	Write the original equation.
$-8(-3) \stackrel{?}{=} 24$	Replace y with −3. Is this sentence true?
$24 = 24$ ✓	

The solution is −3.

Your Turn **Solve each equation.**

a. $30 = 6x$ b. $-6a = 36$ c. $-9d = -72$

REAL-LIFE MATH

ANIMALS The tortoise is one of the slowest land animals, reaching a top speed of about 0.25 mile per hour. Dragonflies, the fastest insect, can fly at a speed of about 18 miles per hour.

Source: *Incredible Comparisons*

The equation $d = rt$ shows the relationship between the variables d (distance), r (rate or speed), and t (time).

EXAMPLE Use an Equation to Solve a Problem

3 **ANIMALS Refer to the information at the left. At this speed, how long will it take a tortoise to travel 1.5 miles?**

Words	Distance is equal to the rate times the time.				
Variables	d	$=$	r	$\cdot$	t
Equation	1.5	$=$		$0.25t$	

$1.5 = 0.25t$	Write the equation.
$\frac{1.5}{0.25} = \frac{0.25t}{0.25}$	Divide each side by 0.25.
$6 = t$	$1.5 \div 0.25 = 6$

At this speed, it would take a tortoise 6 hours to travel 1.5 miles. Check this solution.

Skill and Concept Check

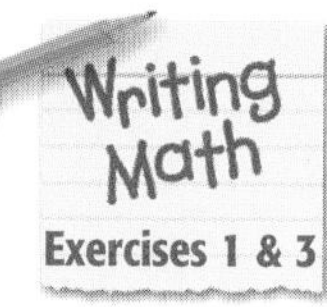

1. **Tell** whether -4 is a solution of $-3x = -12$. Explain.

2. **OPEN ENDED** Write two different multiplication equations that have a negative integer as a solution.

3. **FIND THE ERROR** Jesse and Haley are solving $-5x = 30$. Who is correct? Explain.

Jesse	Haley
$-5x = 30$	$-5x = 30$
$\frac{-5x}{-5} = \frac{30}{-5}$	$\frac{-5x}{5} = \frac{30}{5}$
$x = -6$	$x = -6$

GUIDED PRACTICE

Solve each equation. Check your solution.

4. $6c = 18$
5. $10y = 20$
6. $-6s = 24$
7. $-9r = 36$
8. $-8z = -40$
9. $-11r = -77$
10. $15 = 5z$
11. $72 = 12r$

12. The product of a number and -4 is 64. Find the number.

13. If you multiply a number by 3, the result is -21. What is the number?

Practice and Applications

Solve each equation. Check your solution.

14. $7a = 49$
15. $9e = 27$
16. $35 = 5v$
17. $112 = 8p$
18. $4j = -36$
19. $12y = -60$
20. $48 = -6r$
21. $266 = -2t$
22. $-3w = -36$
23. $-10g = -100$
24. $-28 = -7f$
25. $-275 = -5s$

HOMEWORK HELP

For Exercises	See Examples
14–19	1
20–25	2
26–31, 38–39	3

Extra Practice
See pages 572, 599.

26. When a number is multiplied by -12, the result is -168. Find the number.

27. The product of a number and 25 is 1,000. What is the number?

28. **BABY-SITTING** Gracia earns \$5 per hour when she baby-sits. How many hours does she need to work to earn \$75?

29. **TRAVEL** A Boeing 747 aircraft has a cruising speed of about 600 miles per hour. At that speed, how long will it take to travel 1,500 miles? Use the formula $d = rt$.

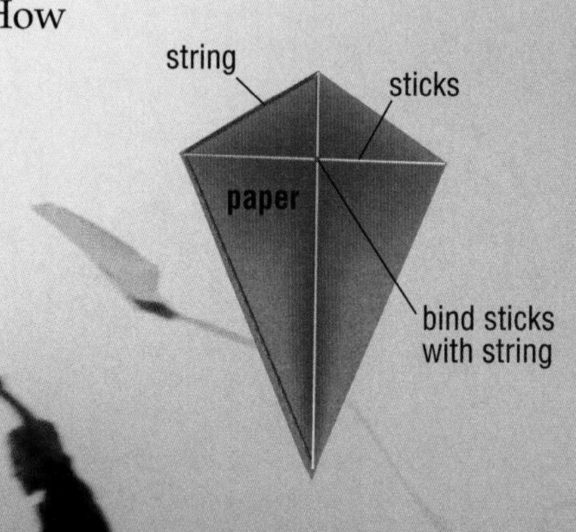

KITES For Exercises 30 and 31, use the following information.
In a simple kite, the length of the longer stick should be 1.5 times the length of the shorter stick.

30. Suppose the length of the longer stick is 36 inches. Write a multiplication equation to find the length of the shorter stick.

31. Solve the equation.

Solve each equation. Check your solution.

32. $4x = 9.2$
33. $9y = 13.5$
34. $5.4 = 0.3p$
35. $9.72 = 1.8a$
36. $3.9y = 18.33$
37. $2.6b = 2.08$

EARTHQUAKES For Exercises 38 and 39, use the following information. Scientists determine the epicenter of an earthquake by measuring the time it takes for surface waves to travel between two places. Surface waves travel about 6 kilometers per second through Earth's crust.

38. The distance from Los Angeles, California, to Phoenix, Arizona, is 600 kilometers. Write a multiplication equation to find how long it would take surface waves to travel from Los Angeles to Phoenix.
39. Solve the equation.

ROLLER COASTERS For Exercises 40 and 41 use the table.

40. Without calculating, explain whether the *Blue Streak* or *Magnum* has the greater speed.
41. Find the speed of each roller coaster in feet per second.

Name	Track Length (ft)	Time of Ride
Blue Streak	2,558	1 min 45 s
Corkscrew	2,050	2 min
Magnum	5,106	2 min
Mean Streak	5,427	2 min 45 s

Source: www.rcdb.com

42. **CRITICAL THINKING** Solve $3|x| = 12$.

Standardized Test Practice and Mixed Review

43. **MULTIPLE CHOICE** A car is traveling at a speed of 75 feet per second. How many yards will the car travel in 90 seconds if it maintains the same speed?

   A 75 yd B 270 yd C 2,250 yd D 6,750 yd

44. **MULTIPLE CHOICE** Use the formula $A = bh$ to find the height of a parallelogram with a base of 34 millimeters and an area of 612 square millimeters.

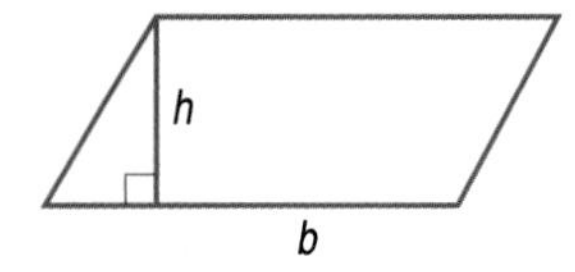

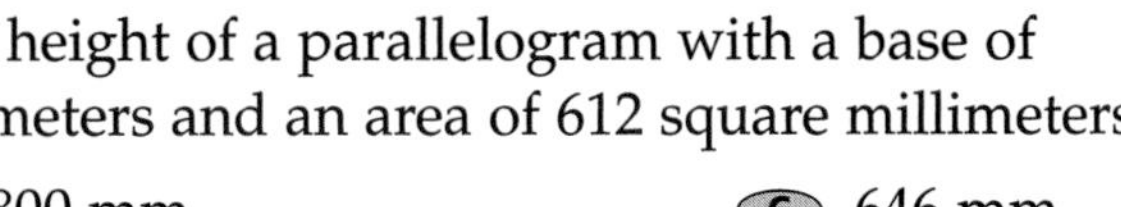

   F 20,800 mm G 646 mm
   H 578 mm I 18 mm

ALGEBRA Solve each equation. (Lesson 4-2)

45. $y + 8 = -2$
46. $x - 7 = -2$
47. $20 = z + 23$

ALGEBRA Write an algebraic expression for each phrase. (Lesson 4-1)

48. the product of –3 and y
49. 5 years older than Rafael
50. 10 fewer students than last year
51. twice as many runs as the Marlins scored

GETTING READY FOR THE NEXT LESSON

PREREQUISITE SKILL Subtract. (Lesson 3-5)

52. $8 - (-2)$
53. $-7 - 7$
54. $-3 - (-9)$
55. $-3 - 18$

4-4a Problem-Solving Strategy

A Preview of Lesson 4-4

What You'll Learn
Solve problems using the work backward strategy.

Work Backward

Alicia, how much money did we take to the movies today?

Well, we spent $9 on movie tickets. Then we spent $5 on candy, and one half of what was left on popcorn. We have $3 left. Let's **work backward** to figure it out.

Explore	We know we have $3 left and the amounts spent. We need to find the initial amount we had.	
Plan	Let's start with the end result and work backward.	
Solve	We have $3 left. **Undo** the half spent on popcorn. **Undo** the $5 spent on candy. **Undo** the $9 spent on movie tickets. So, we initially had $20.	3 × 2 6 + 5 11 + 9 20
Examine	Assume that we started with $20. After buying movie tickets, we had $20 − $9 or $11. We spent $5 on candy, so we had $11 − $5 or $6. Then we spent one half of the remaining money on popcorn, so we had $6 ÷ 2 or $3. So, our answer of $20 is correct.	

Analyze the Strategy

1. **Explain** when you would use the work backward strategy to solve a problem.
2. **Describe** how to solve a problem by working backward.
3. **Write** a problem that can be solved by working backward. Then write the steps you would take to find the solution to your problem.

Apply the Strategy

Solve. Use the work backward strategy.

4. **MONEY** Mia spent $4.50 at the bakery and then spent four times that amount at the grocery. She had $12.80 left. How much money did she have initially?

5. **NUMBER THEORY** A number is multiplied by 4. Then 6 is subtracted from the product. After adding 5, the result is 15. What is the number?

Mixed Problem Solving

Solve. Use any strategy.

6. **PATTERNS** Draw the sixth figure in the pattern shown.

7. **FOOD** Mateo goes to the grocery store and buys a ham for $24.98 and a vegetable tray for $17.49. There is no tax. He gives the cashier one bill and receives less than $10 in change. What was the denomination of the bill Mateo gave the cashier?

8. **TIME** A shuttle bus schedule is shown. What is the earliest time after noon when the bus departs?

Departs	Arrives
8:55 A.M.	9:20 A.M.
9:43 A.M.	10:08 A.M.
10:31 A.M.	10:56 A.M.
11:19 A.M.	11:44 A.M.

9. **GEOGRAPHY** The land area of Texas is 267,277 square miles. This is about 5 times the land area of Arkansas. Estimate the land area of Arkansas.

10. **NUMBER THEORY** How many different two-digit numbers can you make using the digits 2, 4, and 9 if no digit is repeated within a number?

11. **AGE** Maya is two years older than her sister Jenna. Jenna is 5 years older than their brother Trevor, who is 9 years younger than their brother Trent. Trent is 17 years old. How old is Maya?

FOOD For Exercises 12 and 13, use the graph below.

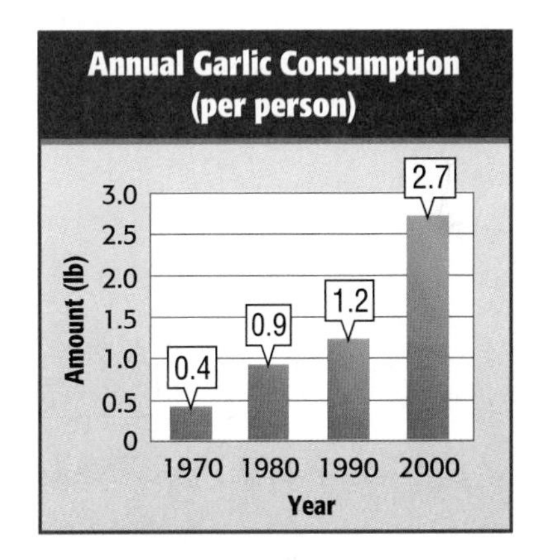

12. How much more garlic was eaten in 2000 than in 1990?

13. Find the average monthly consumption of garlic in 2000.

14. **VIDEOS** Carlos rented 2 times as many DVDs as Ashley last month. Ashley rented 4 fewer than Greg, but four more than Grace. Greg rented 9 DVDs. How many DVDs did each person rent?

15. **STANDARDIZED TEST PRACTICE** Which of the following is the most reasonable total amount for the items purchased?

Item	Price
Pens	$ 2.09
Ruler	$ 0.99
Paper	$ 1.49
Book	$14.99
Candy	$ 0.49
Glue	$ 0.89
Folder	$ 1.19
Erasers	$ 1.99
Pencils	$ 1.87

A $17 B $20 C $26 D $30

You will use the work backward strategy in the next lesson.

4-4 Solving Two-Step Equations

HANDS-ON Mini Lab

What You'll LEARN

Solve two-step equations.

NEW Vocabulary

two-step equation

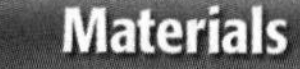

- cups and counters
- equation mat

Work with a partner.

A **two-step equation** has two different operations. Follow these steps to solve $2x - 3 = 1$.

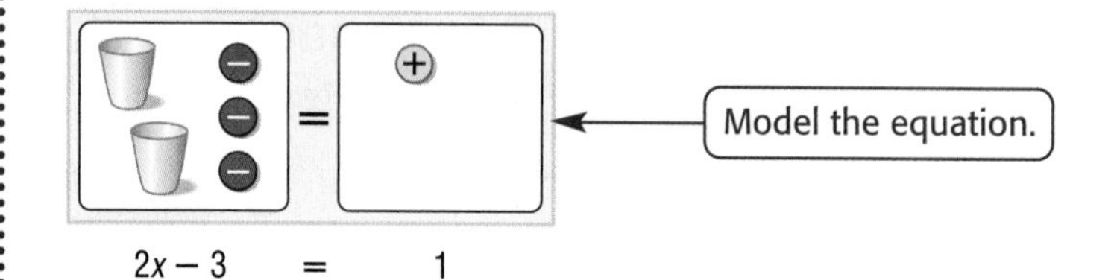

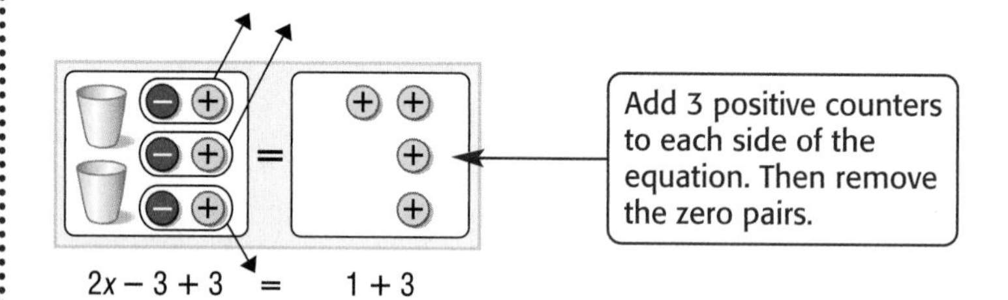

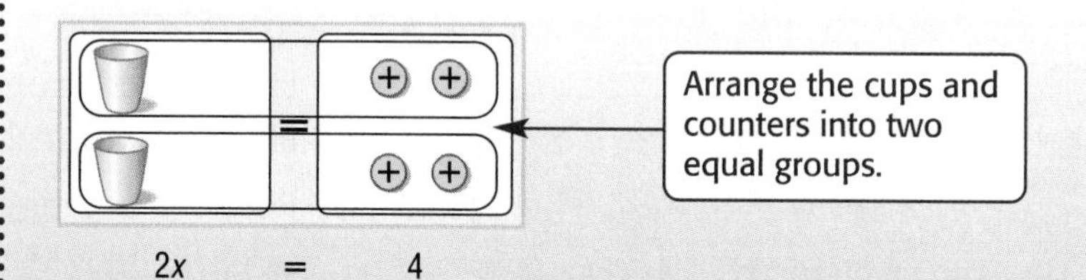

Each cup is matched with 2 positive counters. So, $x = 2$.

Solve each equation using models.

1. $3x + 1 = 7$ **2.** $2x - 4 = 2$ **3.** $2x + 3 = -3$

To solve two-step equations, "undo" the operations in reverse order of the order of operations. You are using the *work backward* strategy.

EXAMPLE Solve a Two-Step Equation

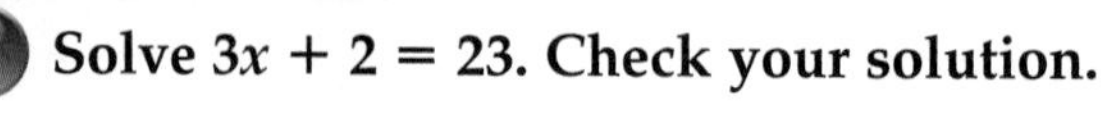

1 Solve $3x + 2 = 23$. Check your solution.

$3x + 2 = 23$	Write the equation.
$-2 = -2$	Subtract 2 from each side.
$3x = 21$	Simplify.
$\frac{3x}{3} = \frac{21}{3}$	Divide each side by 3.
$x = 7$	Simplify.

The solution is 7. Check the solution.

EXAMPLES Solve Two-Step Equations

STUDY TIP

Solving Equations When you solve a simple equation like $-2y = 10$, you can mentally divide each side by -2.

Solve each equation. Check your solution.

2 $-2y - 7 = 3$

$-2y - 7 = 3$	Write the equation.
$+7 = +7$	Add 7 to each side.
$-2y = 10$	Simplify.
$\frac{-2y}{-2} = \frac{10}{-2}$	Divide each side by -2.
$y = -5$	Simplify.

The solution is -5. Check the solution.

3 $4 + 5r = -11$

$4 + 5r = -11$	Write the equation.
$-4 = -4$	Subtract 4 from each side.
$5r = -15$	Simplify.
$\frac{5r}{5} = \frac{-15}{5}$	Divide each side by 5.
$r = -3$	Simplify.

The solution is -3. Check the solution.

Your Turn Solve each equation.

a. $4x + 5 = 13$ b. $-3n - 8 = 7$ c. $1 + 2y = -3$

REAL-LIFE MATH

BOWLING *Handicapping* allows bowlers with varying skills to score about the same. Your handicap h is added to your score s, so your final score is $h + s$.

Source: www.bowl.com

Some problems start with a given amount and increase at a certain rate.

EXAMPLE Use an Equation to Solve a Problem

4 **BOWLING** Suppose you spend \$6 to rent bowling shoes at The Bowling Alley. Each game costs \$3.50. How many games can you bowl if you have \$20 to spend?

Variable	Let x = the number of games.
Words	Cost of the shoes plus the cost of the games equals \$20.
Equation	$6 + 3.50x = 20$

$6 + 3.50x = 20$	Write the equation.
$-6 = -6$	Subtract 6 from each side.
$3.50x = 14$	Simplify.
$\frac{3.50x}{3.50} = \frac{14}{3.50}$	Divide each side by 3.50.
$x = 4$	$14 \div 3.50 = 4$

You can bowl 4 games. Is this answer reasonable?

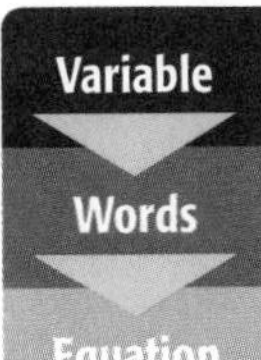

Skill and Concept Check

Writing Math
Exercise 2

1. **OPEN ENDED** Write a two-step equation that has -2 as the solution.
2. **FIND THE ERROR** Jackson and Michele are solving $4 + 2x = -8$. Who is correct? Explain.

Jackson	Michele
$4 + 2x = -8$	$4 + 2x = -8$
$4 + \frac{2x}{2} = \frac{-8}{2}$	$-4 \quad = -4$
$4 + x = -4$	$2x = -12$
$x = -8$	$x = -6$

GUIDED PRACTICE

Solve each equation. Check your solution.

3. $3x + 1 = 7$
4. $4h - 6 = 22$
5. $-6r + 1 = -17$
6. $13 = 1 + 4s$
7. Five more than three times a number is 23. Find the number.

Practice and Applications

HOMEWORK HELP	
For Exercises	See Examples
8–11	1
12–15	2
16–19	3
26–29, 38–41	4

Extra Practice
See pages 573, 599.

Solve each equation. Check your solution.

8. $3x + 1 = 10$
9. $5x + 4 = 19$
10. $2t + 7 = -1$
11. $6m + 1 = -23$
12. $-4w - 4 = 8$
13. $-7y + 3 = -25$
14. $-8s + 1 = 33$
15. $-2x + 5 = -13$
16. $3 + 8n = -5$
17. $5 + 4d = 37$
18. $14 + 2p = 8$
19. $25 + 2y = 47$
20. $2 = 3t - 13$
21. $57 = -8x - 7$
22. $18 = 9d - 18$
23. $4 = 4 + 7f$
24. $21 + 11x = -1$
25. $15x + 4 = 49$
26. Three more than the product of a number and 4 is 15. Find the number.
27. Five less than three times a number is 1. Find the number.
28. The product of 2 and a number is increased by 9. The result is -17. Find the number.
29. If you subtract 3 from twice a number, the result is 25. Find the number.

Solve each equation. Check your solution.

30. $2r - 3.1 = 1.7$
31. $4t + 3.5 = 12.5$
32. $16b - 6.5 = 9.5$
33. $5w + 9.2 = 19.7$
34. $16 = 0.5r - 8$
35. $0.2n + 3 = 8.6$
36. $7.5s - 2 = 28$
37. $1.5v - 16 = 8$
38. **MONEY MATTERS** Joshua has saved \$74 toward a new sound system that costs \$149. He plans on saving an additional \$15 each week. How many weeks will it take Joshua to save enough money to buy the sound system?
39. **WEATHER** The temperature is 20°F. It is expected to rise at a rate of 4° each hour for the next several hours. In how many hours will the temperature be 32°?

TEMPERATURE For Exercises 40 and 41, use the following information and the graph.

Temperature is usually measured on the Fahrenheit scale (°F) or the Celsius scale (°C). Use the formula $F = 1.8C + 32$ to convert from one scale to the other.

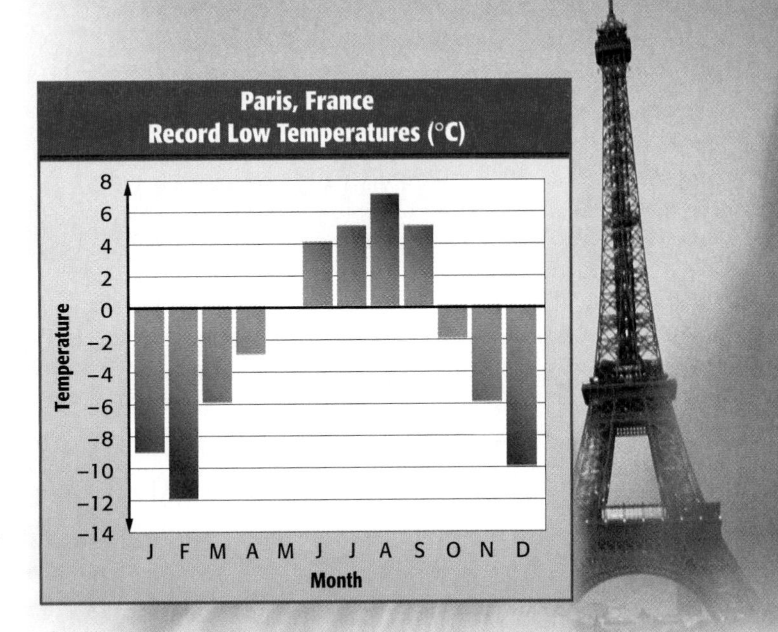

40. The highest temperature ever recorded in Virginia Beach, Virginia, was 104°F. Find this temperature in degrees Celsius.

41. **MULTI STEP** The lowest temperature ever recorded in Virginia Beach was −3°F. Is this temperature greater or less than the lowest temperature ever recorded in Paris, France?

Data Update What are the record high and low temperatures for a large city near you? Visit msmath2.net/data_update to learn more.

42. **CRITICAL THINKING** Is there a temperature at which the number of Celsius degrees is the same as the number of Fahrenheit degrees? If so, find it. If not, explain why not.

Standardized Test Practice and Mixed Review

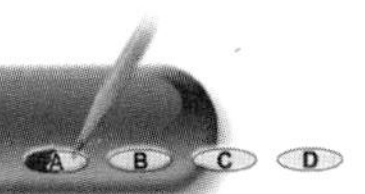

43. **MULTIPLE CHOICE** A taxi driver charges \$2.00 plus \$0.80 for each mile traveled. Which expression could be used to find the cost of a taxi ride if m miles are traveled?

Ⓐ $2.80m$ Ⓑ $2m + 0.80$ Ⓒ $2 + 0.80m$ Ⓓ $2m + 0.80m$

44. **SHORT RESPONSE** In 2002, the first-class postage rates were \$0.37 for the first ounce plus an additional cost for each ounce.

Weight (oz)	1	2	3	4	5	6
Postage (dollars)	0.37	0.60	0.83	1.06	1.29	1.52

What is the cost, in dollars, for a 12-ounce letter?

ALGEBRA Solve each equation. Check your solution. (Lessons 4-2 and 4-3)

45. $4f = 28$

46. $6p = -72$

47. $p - 14 = 27$

48. $26 = n + 2$

49. **DIVING** Find the distance between two divers if one diver is 27 feet below sea level and the other diver is 13 feet below sea level. (Lesson 3-5)

GETTING READY FOR THE NEXT LESSON

PREREQUISITE SKILL Replace each ● with < or > to make a true sentence. (Lesson 3-2)

50. −3 ● −12

51. 1 ● −1

52. 7 ● −18

53. −120 ● −30

Mid-Chapter Practice Test

Vocabulary and Concepts

1. **State** the property of equality used to solve $a - 7 = -2$. (Lesson 4-2)
2. **Write** a two-step equation. Then solve your equation. (Lesson 4-4)
3. **Define** *inverse operation* and give an example. (Lesson 4-2)

Skills and Applications

Write each sentence as an algebraic equation. (Lesson 4-1)

4. The product of a number and 3 is −16.
5. 10 less than a number is 45.

Solve each equation. Check your solution. (Lessons 4-2, 4-3, and 4-4)

6. $21 + m = 33$
7. $a - 5 = -12$
8. $7y = 63$
9. $5f = -75$
10. $-28 = -2d$
11. $-1.6w = 4.8$
12. $3z - 7 = 17$
13. $2g - 9 = -5$
14. $-4c - 1 = 11$

15. **FLYING** An airplane is flying at an altitude of t feet before it increases its altitude by 1,000 feet. Write an expression for its new altitude. (Lesson 4-1)

16. **GEOMETRY** The sum of the measures of the angles of a triangle is 180°. Find the missing measure. (Lesson 4-2)

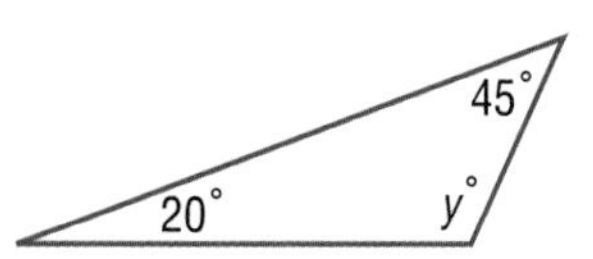

LAWN SERVICE For Exercises 17 and 18, use the following information.
Alex earned $326 this summer mowing lawns. The total was 4 times more than what he earned last summer. (Lesson 4-3)

17. Write a multiplication equation to find how much Alex earned last summer.
18. Solve the equation.

Standardized Test Practice

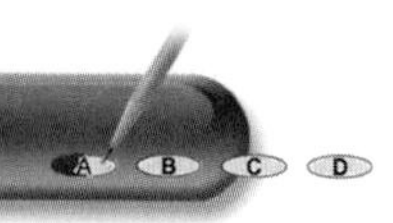

19. **MULTIPLE CHOICE** Kim's time for the 5K race was four minutes less than Tanya's time. If Tanya's time is t, which expression represents Kim's time? (Lesson 4-1)

    (A) $4 - t$
    (B) $t - 4$
    (C) $t + 4$
    (D) $4t$

20. **GRID IN** Last baseball season, Ryan had four less than twice the number of hits that Marcus had. Ryan had 48 hits. How many hits did Marcus have last season? (Lesson 4-4)

The Game Zone

A Place To Practice Your Math Skills

Math Skill
Solving Equations

Players: two, three, or four
Materials: scissors, 26 index cards, and 4 different colored markers

- Cut each index card in half, making 52 cards.
- To make a set of four cards, use the markers to put a different-colored stripe at the top of each card.
- Then write a different equation on each card. The solution of each equation should be 1.
- Continue to make sets of four cards having equations with solutions of 2, 3, 4, 5, 6, 0, −1, −2, −3, −4, and −5.
- Mark the remaining set of four cards "Wild."

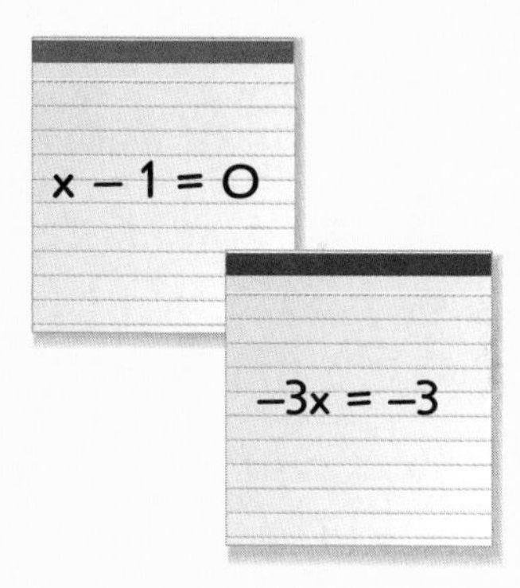

GO!

- The dealer shuffles the cards and deals five to each person. The remaining cards are placed in a pile facedown in the middle of the table. The dealer turns the top card faceup.
- The player to the left of the dealer plays a card with the same color or solution as the faceup card. Wild cards can be played any time. If the player cannot play a card, he or she takes a card from the pile and plays it, if possible. If it is not possible to play, the player places the card in his or her hand, and it is the next player's turn.
- **Who Wins?** The first person to play all cards in his or her hand is the winner.

4-5 Inequalities

WHEN am I ever going to use this?

What You'll LEARN

Solve inequalities.

NEW Vocabulary

inequality

MATH Symbols

$<$ is less than

$>$ is greater than

$\leq$ is less than or equal to

$\geq$ is greater than or equal to

BREAKFAST The table shows the nutrition requirements for a healthy breakfast cereal with milk.

1. Suppose your favorite cereal has 2 grams of fat, 7 grams of protein, 4 grams of fiber, and 3 grams of sugar. Is it a healthy cereal? Explain.
2. Is a cereal with 3 grams of fiber considered healthy? Explain.
3. Is a cereal with 5 grams of sugar considered healthy? Explain.

Healthy Breafast Cereals (per serving)	
Fat	less than 3 g
Protein	more than 5 g
Fiber	at least 3 g
Sugar	at most 5 g

An **inequality** is a mathematical sentence that contains the symbols $<, >, \leq$, or $\geq$.

Inequalities				
Words	is less than	is greater than	is less than or equal to is at most	is greater than or equal to is at least
Symbols	$<$	$>$	$\leq$	$\geq$

Any number that makes the inequality true is a solution of the inequality. Inequalities may have many solutions. The solutions are shown by shading a number line.

EXAMPLES Graph Solutions of Inequalities

Graph each inequality on a number line.

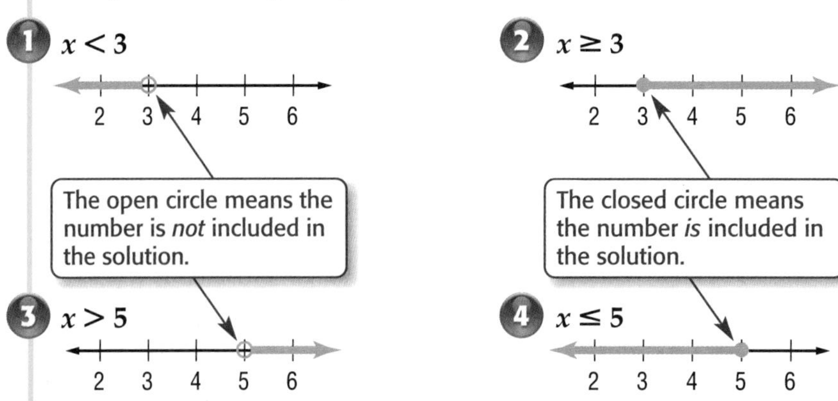

Your Turn Graph each inequality on a number line.

a. $x < -1$ b. $x \geq -2$ c. $x > 0$ d. $x \leq 3$

To solve an inequality, follow the same steps you use to solve an equation.

STUDY TIP

Checking Solutions You may also want to check $x = 5$. The solution of the equation is the *boundary* between numbers that are solutions and numbers that are not solutions.

EXAMPLES Solve One-Step Inequalities

5 Solve $x + 4 > 9$. Check your solution. Then graph the solution.

$x + 4 > 9$	Write the inequality.
$-4 = -4$	Subtract 4 from each side.
$x > 5$	Simplify.

Check Try 6, a number greater than 5.

$x + 4 > 9$	Write the inequality.
$6 + 4 \overset{?}{>} 9$	Replace x with 6. Is this sentence true?
$10 > 9$ ✓	

The solution is all numbers greater than 5.

6 Solve $4y \leq 8$. Graph the solution.

$4y \leq 8$	Write the inequality.
$\frac{4y}{4} \leq \frac{8}{4}$	Divide each side by 4.
$y \leq 2$	Check this solution.

The solution is all numbers less than or equal to 2.

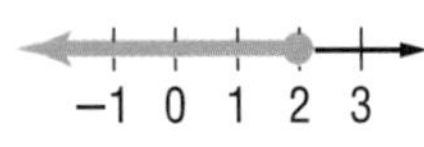

Your Turn **Solve each inequality. Graph the solution.**

e. $x + 6 > 8$ f. $x - 4 \geq -7$ g. $5x < 25$

REAL-LIFE CAREERS

How Does a Nurse Use Math?

Nurses use inequalities when they record test results like a person's blood pressure.

Online Research
For information about a career as a nurse, visit: msmath2.net/careers

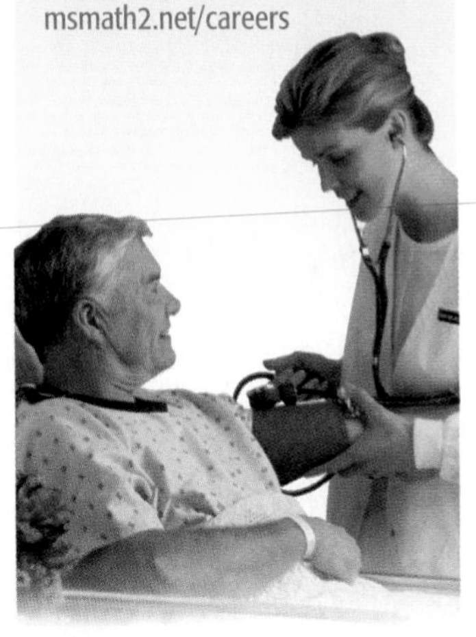

EXAMPLE Use an Inequality to Solve a Problem

7 HEALTH The formula $110 + 0.5A < P$ is used to estimate whether a person has high blood pressure. In the formula, A is the person's age and P is the blood pressure. For which ages is a blood pressure of 120 considered high?

$110 + 0.5A < P$	Write the formula.
$110 + 0.5A < 120$	Replace P with 120.
$-110 = -110$	Subtract 110 from each side.
$0.5A < 10$	Simplify.
$\frac{0.5A}{0.5} < \frac{10}{0.5}$	Divide each side by 0.5.
$A < 20$	$10 \div 0.5 = 20$

A blood pressure of 120 is considered high for a person who is younger than 20 years old.

Skill and Concept Check

Writing Math
Exercises 2 & 4

1. **Draw** a number line that shows all numbers less than or equal to -4.
2. Explain the difference between $x > 3$ and $x \geq 3$.
3. **OPEN ENDED** Write two different inequalities whose solution is $x < 2$.
4. **FIND THE ERROR** Courtney and Diego are writing an inequality for the expression *at least 2 hours of homework.* Who is correct? Explain.

Courtney	Diego
$h \leq 2$	$h \geq 2$

GUIDED PRACTICE

Graph each inequality on a number line.

5. $y > -3$
6. $x \geq 0$
7. $d < 7$

Solve each inequality.

8. $x + 3 > -4$
9. $6d \geq 24$
10. $3x + 8 < 15$

Practice and Applications

HOMEWORK HELP	
For Exercises	See Examples
11–19	1–4
20–25, 32–33	5–6
26–27, 40–41	7

Extra Practice
See pages 573, 599.

Graph each inequality on a number line.

11. $s > -4$
12. $y > 3$
13. $t \geq -4$
14. $a \geq 3$
15. $h < 2$
16. $g < -5$
17. $z \leq -1$
18. $p \leq 0$
19. $w \leq -6$

Solve each inequality.

20. $y + 5 < 14$
21. $x + 6 < 0$
22. $a - 3 \geq -5$
23. $g - 5 > 2$
24. $3r \leq 18$
25. $6u \geq 36$
26. $2x + 8 < 24$
27. $3y + 1 \leq 5$

Write the inequality for each graph.

28. 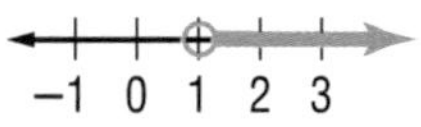

29. 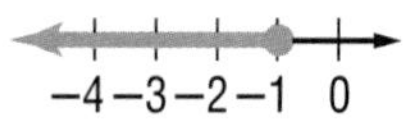

30. 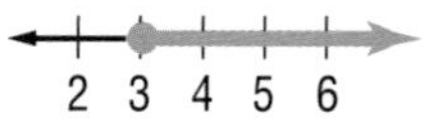

31. 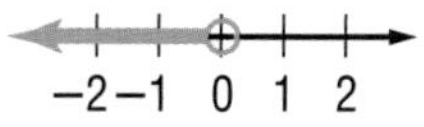

Write an inequality for each sentence. Then solve the inequality.

32. Five times a number is greater than 25.
33. The sum of a number and 1 is at least 5.
34. **WEATHER** A tropical depression has maximum sustained winds of less than 39 miles per hour. Write an inequality showing the wind speeds.
35. **DRIVING** In Ohio, you can get a driver's license if you are at least 16 years old. Write an inequality showing the age of all drivers in Ohio.

Data Update What are the minimum ages for getting a driver's license in other states? Visit msmath2.net/data_update to learn more.

SPORTS For Exercises 36–39, use the graphic.

36. In which sport(s) are more than 400,000 children hurt?
37. In which sport(s) are at least 185,000 children hurt?
38. Of the sports listed, which have less than 300,000 injuries?
39. Write an inequality that describes the number of children hurt each year playing sports.

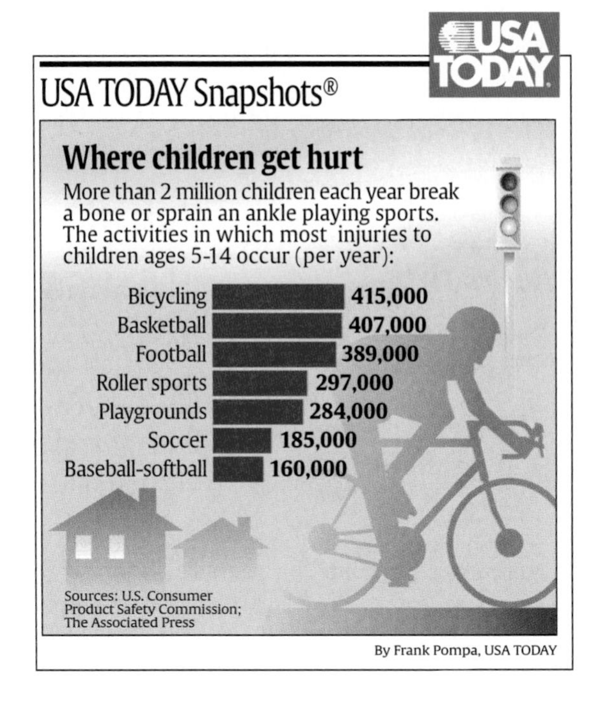

SHOPPING For Exercises 40 and 41, use the following information.
Suppose a pair of jeans costs \$29 and a necklace costs \$8. You have \$70 to spend on both.

40. Write an inequality to find how many pairs of jeans you can buy along with one necklace.
41. Solve the inequality.

42. **CRITICAL THINKING** A *compound inequality* is formed by two inequalities connected by the words *and* or *or*.

Type	Example	Solution	Graph
Intersection	$x > -3$ *and* $x < 4$	Both inequalities are true. $\{x \mid -3 < x < 4\}$	−5 −4 −3 −2 −1 0 1 2 3 4 5
Union	$x < 2$ *or* $x > 5$	Either inequality is true. $\{x \mid x < 2$ *or* $x > 5\}$	−5 −4 −3 −2 −1 0 1 2 3 4 5

Identify each compound inequality as an intersection or union. Then find and graph the solution.

a. $x > 2$ and $x < 7$ b. $x < -4$ or $x > 4$ c. $x \geq -3$ or $x < -6$ d. $x \leq 10$ and $x > 0$

Standardized Test Practice and Mixed Review

43. **MULTIPLE CHOICE** Which inequality is graphed on the number line?

Ⓐ $x < 3{,}000$ Ⓑ $x \leq 3{,}000$
Ⓒ $x > 3{,}000$ Ⓓ $x \geq 3{,}000$

0 1,000 2,000 3,000

44. **MULTIPLE CHOICE** Ayano has \$26 to spend. A hat costs \$8. Which inequality shows how much he can spend for a T-shirt?

Ⓕ $s + 8 < 26$ Ⓖ $s + 8 > 26$ Ⓗ $s + 8 \leq 26$ Ⓘ $s + 8 \geq 26$

ALGEBRA Solve each equation. Check your solution. (Lessons 4-2, 4-3, and 4-4)

45. $13 = s + 5$ 46. $18x = -54$ 47. $2q + 6 = -20$ 48. $-7q + 4.6 = -0.3$

GETTING READY FOR THE NEXT LESSON

PREREQUISITE SKILL Graph each point on a coordinate plane. (Lesson 3-3)

49. $(-4, 2)$ 50. $(3, -1)$ 51. $(-3, -4)$ 52. $(2, 0)$

4-6a HANDS-ON LAB

A Preview of Lesson 4-6

Functions and Graphs

What You'll LEARN

Graph a function on a scatter plot.

Materials

- stopwatch
- grid paper
- uncooked spaghetti

INVESTIGATE *Work as a class.*

Have you ever been at a sporting event when the crowd does the "wave"? In this Lab, you will investigate how long it would take the students at your school to complete the "wave."

STEP 1 Begin with five students sitting in a row.

STEP 2 At the timer's signal, the first student stands up, waves his or her arms overhead, and sits down. Each student repeats the wave in order.

STEP 3 When the last student sits down, the timer records the time in seconds.

STEP 4 Repeat for 6, 7, 8, and so on, up to 25 students.

Writing Math

Work with a partner.

1. **Graph** the ordered pairs (number of students, time) on a coordinate grid like the one at the right.

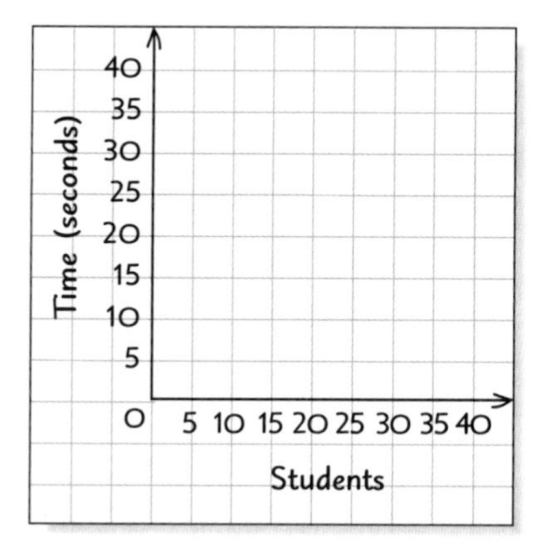

2. **Describe** how the points appear on your graph.
3. Place one piece of uncooked spaghetti on your graph so that it covers as many of the points as possible. **Predict** how long it would take 30 students to complete the "wave." Make a prediction for 50 students.
4. **Find a pattern** in the data and use the pattern to predict how long it would take the students in your school to complete the "wave." Explain your reasoning.
5. A *function* describes the relationship between two quantities. In a function, one quantity *depends on* the other. **Complete the sentence:** The time it takes to do the "wave" depends on __?__.

Functions and Linear Equations

What You'll LEARN

Graph linear equations.

NEW Vocabulary

function
function table
domain
range
linear equation

REVIEW Vocabulary

ordered pair: a pair of numbers used to locate a point in a coordinate system **(Lesson 3-3)**

WHEN am I ever going to use this?

FAST FOOD Suppose you can buy hamburgers for $2 each.

1. Copy and complete the table to find the cost of 2, 3, and 4 hamburgers.
2. On grid paper, graph the ordered pairs (number, cost). Describe how the points appear on the grid.
3. What happens to the cost as the number of hamburgers increases?

Hamburgers		
Number	Multiply by 2	Cost ($)
1	2 × 1	2
2		
3		
4		

The total cost of the hamburgers depends on the number of hamburgers. A relationship where one thing depends on another is called a **function**. In a function, you start with an *input* number, perform one or more operations on it, and get an *output* number.

You can organize the input numbers, output numbers, and the function rule in a **function table**.

EXAMPLE Make a Function Table

1 MONEY MATTERS Suppose you earn $5 each week. Make a function table that shows your total earnings after 1, 2, 3, and 4 weeks.

Input	Function Rule	Output
Number of Weeks	Multiply by 5	Total Earnings ($)
1	5 × 1	5
2	5 × 2	10
3	5 × 3	15
4	5 × 4	20

The set of input values is called the **domain**, and the set of output values is called the **range**. In Example 1, the domain is {1, 2, 3, 4}, and the range is {5, 10, 15, 20}.

Functions are often written as equations with two variables—one to represent the input and one to represent the output. Here's an equation for the situation in Example 1.

STUDY TIP

Input and Output When x and y are used in an equation, x usually represents the input, and y usually represents the output.

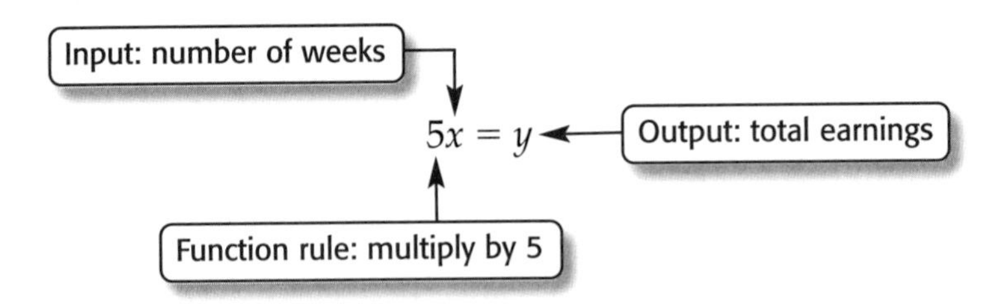

The solution of an equation with two variables consists of two numbers, one for each variable, that make the equation true. The solution is usually written as an ordered pair (x, y), which can be graphed.

EXAMPLE Graph Solutions of Linear Equations

2 Graph $y = 2x + 1$.

Select any four values for the input x. We chose 2, 1, 0, and -1. Substitute these values for x to find the output y.

x	$2x + 1$	y	(x, y)
2	$2(2) + 1$	5	(2, 5)
1	$2(1) + 1$	3	(1, 3)
0	$2(0) + 1$	1	(0, 1)
-1	$2(-1) + 1$	-1	$(-1, -1)$

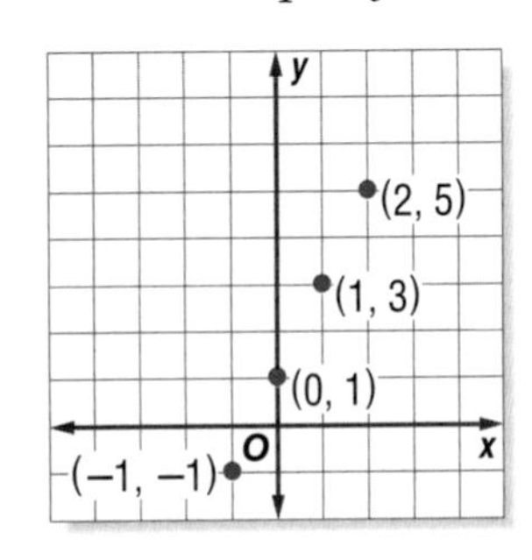

Four solutions are (2, 5), (1, 3), (0, 1), and $(-1, -1)$. The graph is shown above at the right.

Your Turn **Graph each equation.**

a. $y = x - 3$ b. $y = -3x$ c. $y = -3x + 2$

STUDY TIP

Graphing Equations Only two points are needed to graph the line. However, graph more points to check accuracy.

Notice that all four points in the graph lie on a line. Draw a line through the points to graph *all* solutions of the equation $y = 2x + 1$. The graph of (3, 7) is also on the line.

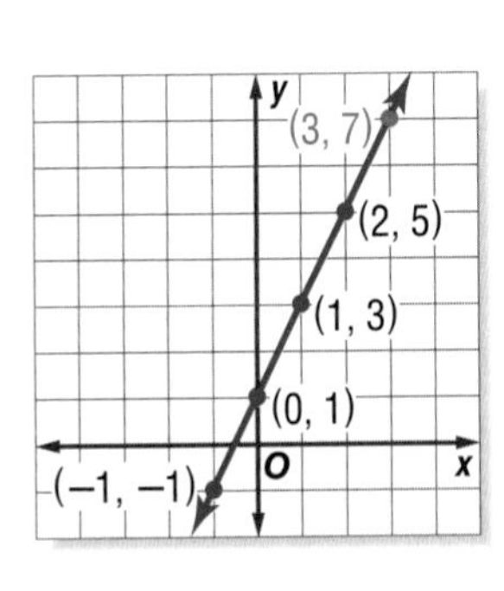

$y = 2x + 1$ Write the equation.

$7 \stackrel{?}{=} 2(3) + 1$ Replace x with 3 and y with 7.

$7 = 7$ ✔ This sentence is true.

So, (3, 7) is also a solution of $y = 2x + 1$.

An equation like $y = 2x + 1$ is called a **linear equation** because its graph is a straight line.

EXAMPLE Represent Real-World Functions

REAL-LIFE MATH

SKATEBOARDING About one fourth of all skateboarding purchases are made by teenagers from 14 to 17 years old.

Source: *Statistical Abstract of the United States*

3 SKATEBOARDING The top speed reached by a standing skateboarder is 80 feet per second. The equation $d = 80t$ describes the distance d that a skateboarder can travel in time t. Represent the function with a graph.

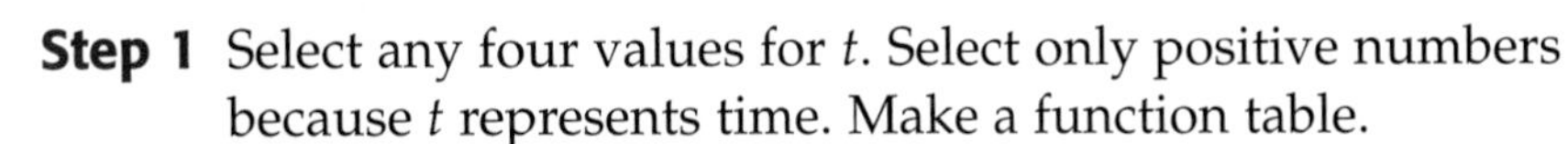

Step 1 Select any four values for t. Select only positive numbers because t represents time. Make a function table.

t	$80t$	d	(t, d)
1	80(1)	80	(1, 80)
2	80(2)	160	(2, 160)
3	80(3)	240	(3, 240)
4	80(4)	320	(4, 320)

Step 2 Graph the ordered pairs and draw a line through the points.

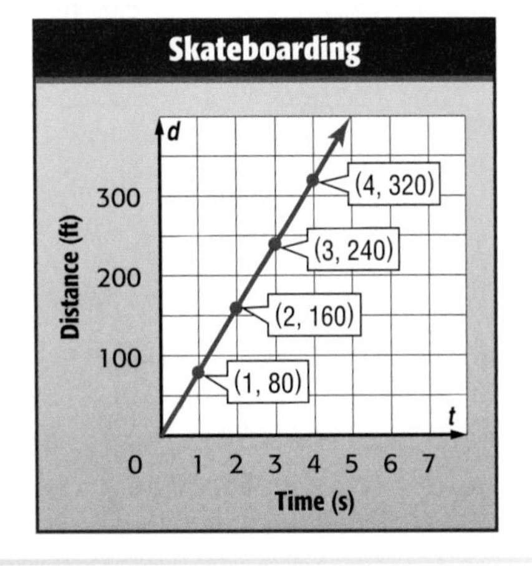

Skill and Concept Check

Writing Math
Exercise 2

1. **OPEN ENDED** Write an equation that has (1, 2) as a solution.

2. **Explain** the relationship among *input*, *output*, and *function rule.*

GUIDED PRACTICE

Copy and complete each function table. Identify the domain and range.

3. $y = x - 2$

x	$x - 2$	y
1		
2		
3		
4		

4. $y = 4x$

x	$4x$	y
−1		
0		
1		
2		

Graph each equation.

5. $y = x - 1$

6. $y = -1x$

7. $y = -2x + 3$

Practice and Applications

HOMEWORK HELP	
For Example	See Examples
8–13	1
14–25	2
29–35	3

Extra Practice
See pages 573, 599.

Copy and complete each function table. Identify the domain and range.

8. $y = x - 4$

x	$x - 4$	y
1		
2		
3		
4		

9. $y = x + 5$

x	$x + 5$	y
1		
2		
3		
4		

10. $y = 2x$

x	$2x$	y
−1		
0		
1		
2		

11. $y = -6x$

x	$-6x$	y
−1		
0		
1		
2		

12. $y = 2x - 1$

x	$2x - 1$	y
1		
2		
3		
4		

13. $y = -2x - 2$

x	$-2x - 2$	y
−1		
0		
1		
2		

Graph each equation.

14. $y = x + 1$
15. $y = x + 3$
16. $y = x$
17. $y = -2x$
18. $y = 2x + 3$
19. $y = 3x - 1$
20. $y = 4x - 2$
21. $y = 2x + 5$
22. $y = x + 0.5$
23. $y = 0.25x$
24. $y = 0.5x - 1$
25. $y = 2x - 1.5$

Make a function table for each sentence. Then write an equation using x to represent the first number and y to represent the second number.

26. The second number is three more than the first number.
27. The second number is five less than the first number.
28. The second number is ten times the first number.

SPENDING For Exercises 29–31, use the graph.

29. Make a function table that shows the total average defense spending per person for 1, 2, 3, and 4 days.
30. Write an equation in which x represents the days and y represents the total spending.
31. Graph the equation.

USA TODAY Snapshots®

Health research spending small

About 4.6 cents of each U.S. health-care dollar is spent on medical/health research. By comparison, federal spending per person per day on:

Source: Research America

By Lori Joseph, Keith Simmons and Marcy E. Mullins, USA TODAY

INTERNET For Exercises 32–34, use the following information.
An Internet provider charges \$20 each month.

32. Make a function table that shows the total charge for 1, 2, 3, and 4 months of service.
33. Write an equation in which x represents months and y represents the total charge.
34. Graph the equation.

35. **GEOMETRY** The formula for the area A of a rectangle whose length is 5 units is $A = 5w$, where w is the width. Graph the function.

36. **WRITE A PROBLEM** Write about a real-life situation that can be represented by the equation $y = 3x$.

CRITICAL THINKING For Exercises 37–40, write an equation for the function shown in each function table.

37.

x	y
1	3
2	4
3	5
4	6

38.

x	y
2	6
4	12
6	18
8	24

39.

x	y
−1	4
0	0
1	−4
2	−8

40.

x	y
1	3
2	5
3	7
4	9

EXTENDING THE LESSON Not all equations have graphs that are straight lines. The graphs at the right show two *nonlinear equations.*

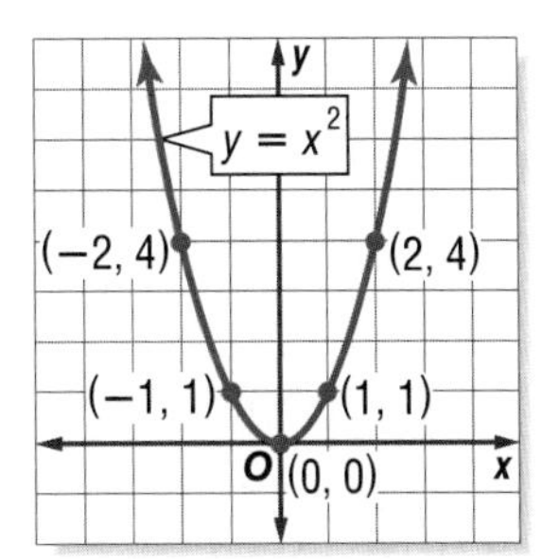

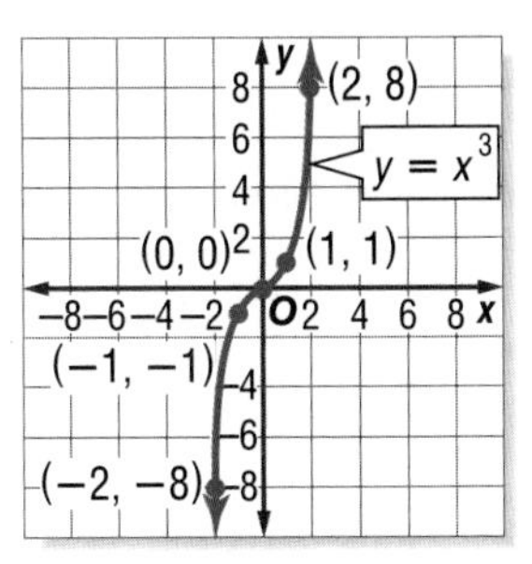

Graph each equation. Determine whether it is *linear* or *nonlinear*.

41. $y = x^2 + 1$

42. $y = x + 1$

43. $y = 2x$

44. $xy = 12$

45. $y = 2x^2$

46. $y = x^3 - 1$

Standardized Test Practice and Mixed Review

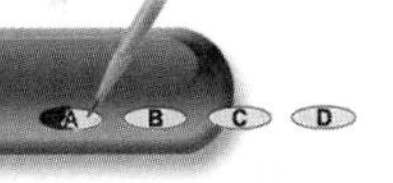

47. **MULTIPLE CHOICE** Choose the equation that is graphed at the right.

(A) $y = 2x$ (B) $y = x$

(C) $y = x - 1$ (D) $y = 2x - 2$

y
(2, 2)
(3, 4)
O
(1, 0)
x
(0, −2)

48. **MULTIPLE CHOICE** Which ordered pairs are solutions of the equation $y = 2x - 4$?

(F) (−2, −3), (0, 2) (G) (−2, −1), (2, − 3)

(H) (1, −2), (3, 2) (I) (−3, 2), (0, − 4)

49. **CIVICS** To serve as a U.S. Representative, a person must be at least 25 years old and a citizen of the United States for at least 7 years. Write an inequality showing the age of a person who may be a U.S. Representative. (Lesson 4-5)

ALGEBRA Solve each equation. Check your solution. (Lesson 4-4)

50. $8 = 3h - 1$

51. $2q + 6 = -20$

52. $32 = -4 + 9m$

GETTING READY FOR THE NEXT LESSON

PREREQUISITE SKILL Divide. (Lesson 3-7)

53. $-4 \div 2$

54. $10 \div (-5)$

55. $-12 \div (-4)$

56. $-16 \div 16$

4-7 Lines and Slope

What You'll LEARN

Find the slope of a line.

NEW Vocabulary

slope

Link to READING

Everyday Meaning of Slope: ground that forms an incline, as in a ski slope

WHEN am I ever going to use this?

COST OF GASOLINE In recent years, the cost of one gallon of gasoline has varied from a low of about \$1 per gallon to a high of about \$3 per gallon. The equations $y = 1x$ and $y = 3x$ are graphed.

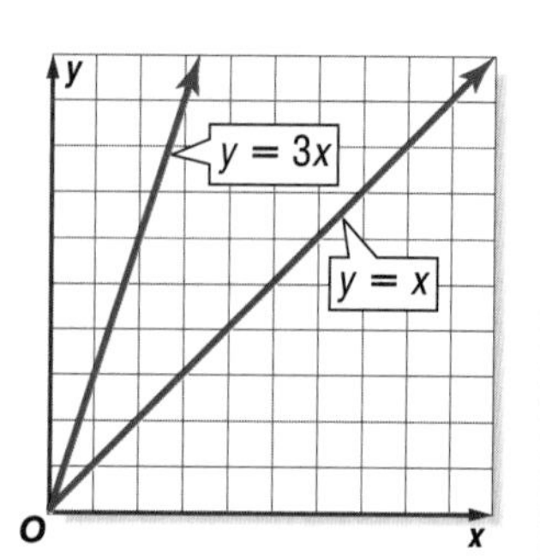

1. Which line is steeper?
2. What causes one line to be steeper?
3. **Make a conjecture** about where the line showing a cost of \$2 per gallon would be graphed. Explain.

The function table shows the total cost y of x gallons of gasoline at \$2 per gallon. The equation $y = 2x$ is graphed below.

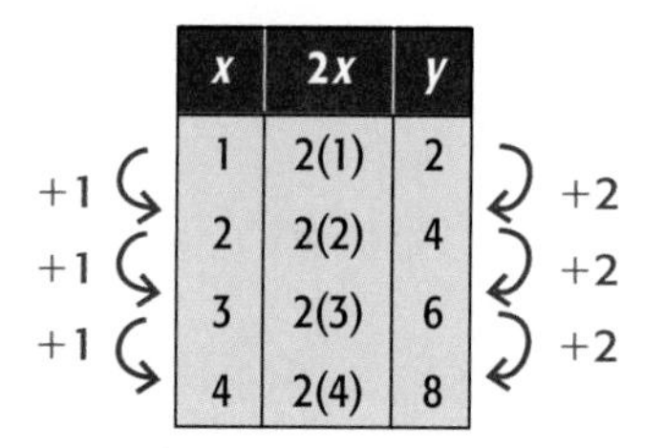

x	$2x$	y
1	2(1)	2
2	2(2)	4
3	2(3)	6
4	2(4)	8

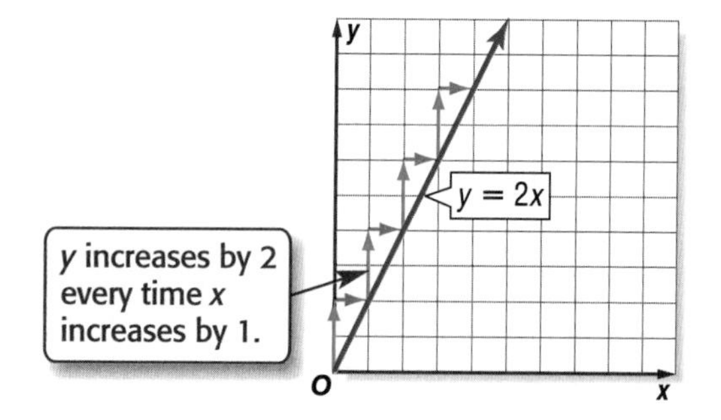

The change in y with respect to the change in x is called the **slope** of a line. Slope is a number that tells how steep the line is.

$$\text{slope} = \frac{\text{change in } y}{\text{change in } x} \quad \begin{array}{l}\leftarrow \text{vertical change} \\ \leftarrow \text{horizontal change}\end{array}$$

$$= \frac{2}{1} \text{ or } 2$$

The slope is the same for any two points on a straight line.

EXAMPLE Positive Slope

1 Find the slope of the line.

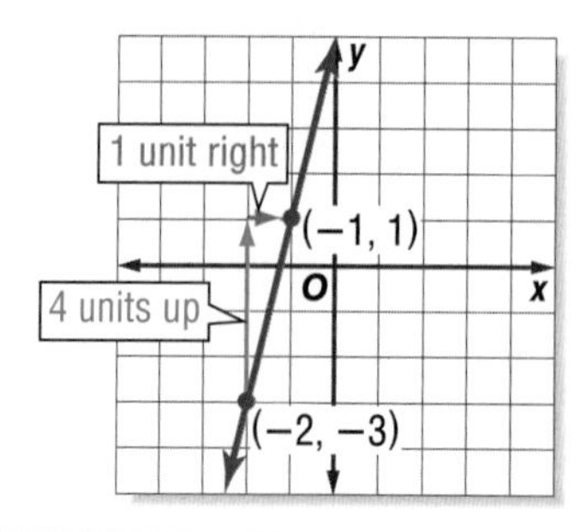

$$\text{slope} = \frac{\text{change in } y}{\text{change in } x}$$

$$= \frac{4}{1} \text{ or } 4$$

The slope of the line is 4.

When the value of y decreases as the value of x increases, the slope is a negative number. The slope of a line can also be a fraction.

> **STUDY TIP**
>
> **Slope** A line with a *positive slope* rises to the right. A line with a *negative slope* falls to the right.

EXAMPLES Negative Slope

2 **Find the slope of each line.**

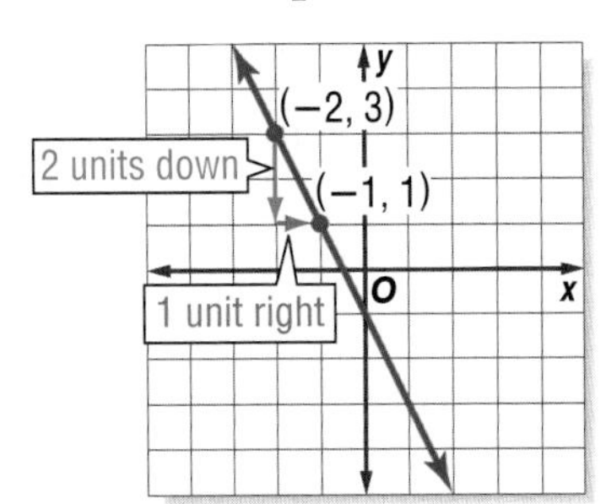

$$\text{slope} = \frac{\text{change in } y}{\text{change in } x}$$

$$= \frac{-2}{1} \text{ or } -2$$

The slope of the line is −2.

3

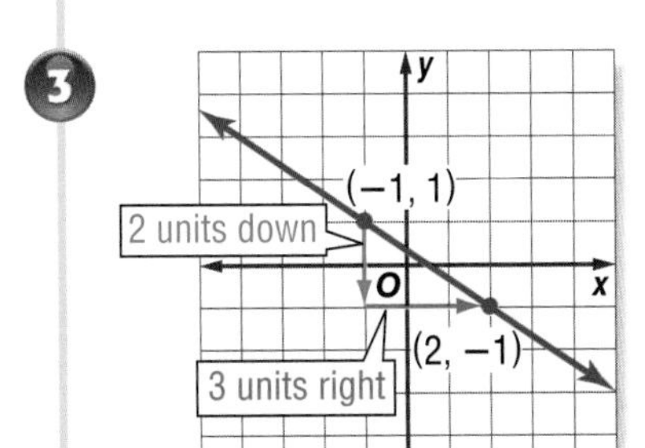

$$\text{slope} = \frac{\text{change in } y}{\text{change in } x}$$

$$= \frac{-2}{3} \text{ or } -\frac{2}{3}$$

The slope of the line is $-\frac{2}{3}$.

Your Turn **Find the slope of the line that passes through each pair of points.**

a. (−2, −4), (1, 5) **b.** (0, 3), (4, −1) **c.** (2, 2), (5, 3)

EXAMPLE Compare Slopes

Standardized Test Practice

4 **MULTIPLE-CHOICE TEST ITEM** The table shows information about the rise and run of three ski slopes in Pennsylvania. Which has the steepest slope?

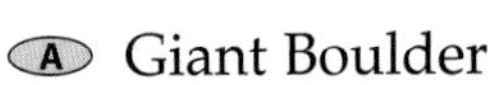

Ⓐ Giant Boulder Ⓑ Giant Steps

Ⓒ Gunner Ⓓ They have the same slope.

Ski Slope	Rise (ft)	Run (ft)
Giant Boulder	750	4,000
Giant Steps	750	3,200
Gunner	750	3,360

> **Test-Taking Tip** *The Princeton Review*
>
> **Make a Drawing** Whenever possible, make a drawing of the problem. Then use the drawing to estimate the answer.

Read the Test Item

The *rise* corresponds to the vertical change, or change in y. The *run* corresponds to the horizontal change, or change in x.

Solve the Test Item

Giant Boulder: $\frac{750}{4{,}000}$ Giant Steps: $\frac{750}{3{,}200}$ Gunner: $\frac{750}{3{,}360}$

All three fractions have the same rise. But Giant Steps has the shortest run. Therefore, its slope is greater. The answer is B.

Skill and Concept Check

Writing Math Exercises 1 & 2

1. **Define** *slope*.
2. **Explain** how a line can have a negative slope.
3. **OPEN ENDED** On a coordinate plane, draw a line with a slope of $\frac{1}{2}$.

GUIDED PRACTICE

Find the slope of the line that passes through each pair of points.

4. 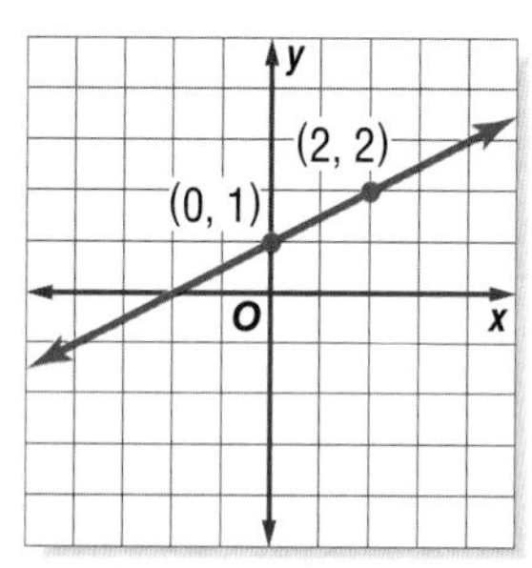

5. 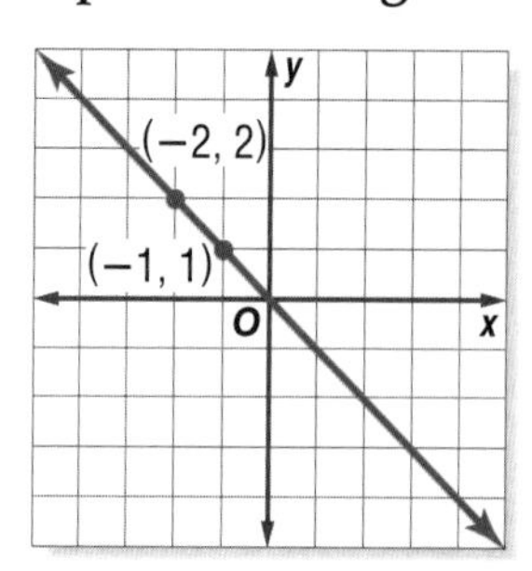

6. (0, 0), (1, 3)

Practice and Applications

HOMEWORK HELP	
For Exercises	See Examples
7–16	1–3
17–20	4

Extra Practice
See pages 574, 599.

Find the slope of the line that passes through each pair of points.

7. 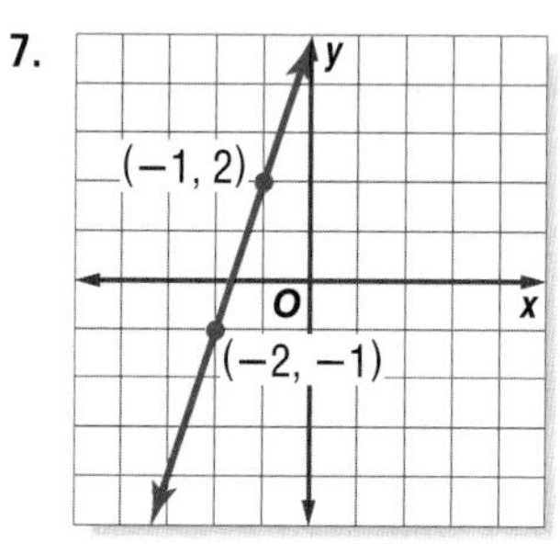

8. 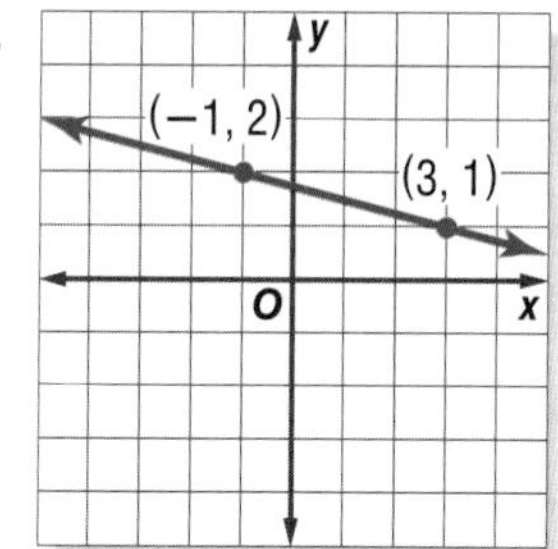

9. (−2, 0), (1, 3)
10. (3, 4), (4, 6)
11. (1, 0), (2, −3)
12. (−2, 2), (−1, −2)
13. (1, 1), (4, 2)
14. (−2, 1), (3, 3)
15. (1, −1), (3, −2)
16. (0, 0), (3, −2)

EARNINGS **For Exercises 17–19, use the table at the right.**

Hours Worked	Earnings ($)	
	Greg	Monica
1	4	5
2	8	10
3	12	15
4	16	20

17. Suppose each of the functions in the table was graphed on a coordinate plane. Which line is steeper? Explain.
18. Find the slope of each line.
19. What does the slope of each line represent?

20. **SKIING** Aerial skiers launch themselves into the air from a ramp like the one shown at the right. Is the slope of the ramp greater than one or less than one?

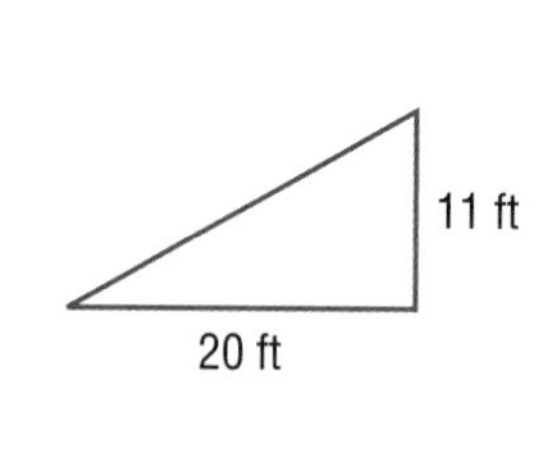

CAMPING **For Exercises 21–23, use the graph.**

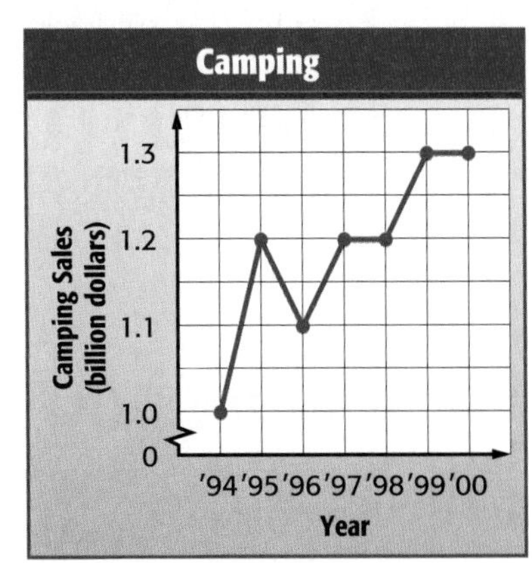

Source: National Sporting Goods Association

21. Which section of the graph shows the greatest increase in sales of camping gear? Describe the slope of this part of the graph.
22. What happened to sales between 1995 and 1996? Describe the slope of this part of the graph.
23. What happened to sales between 1997 and 1998? Describe the slope of this part of the graph.

CRITICAL THINKING **A linear function has a constant slope. Determine whether each function is *linear* or *nonlinear*.**

24.

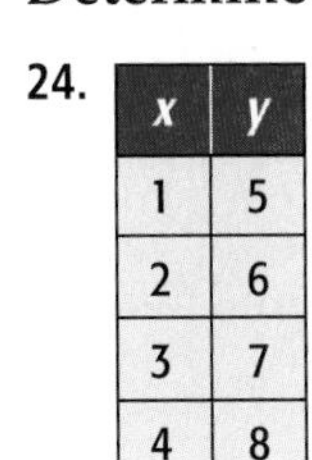

x	y
1	5
2	6
3	7
4	8

25.

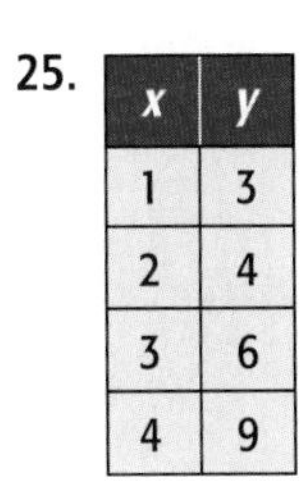

x	y
1	3
2	4
3	6
4	9

26.

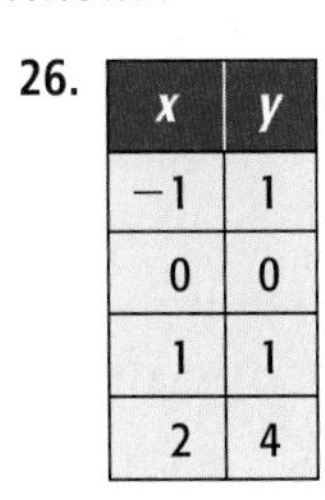

x	y
−1	1
0	0
1	1
2	4

EXTENDING THE LESSON You can graph a line if you know the slope and the coordinates of a point on the line. The figure at the right shows how to graph a line with slope 2 that passes through (1, −1).

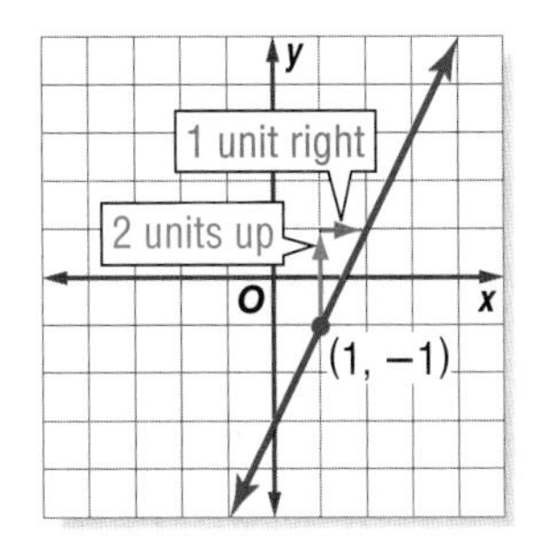

Graph each line with the given slope that passes through the given point.

27. slope = 3; (2, 3)
28. slope = −1; (−3, 2)
29. slope = −2; (−4, −1)
30. slope = 5; (0, −4)

Standardized Test Practice and Mixed Review

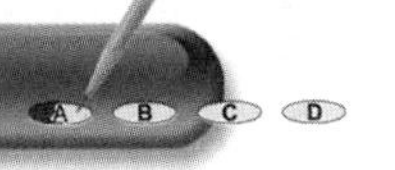

31. **MULTIPLE CHOICE** What is the slope of the line in the graph?

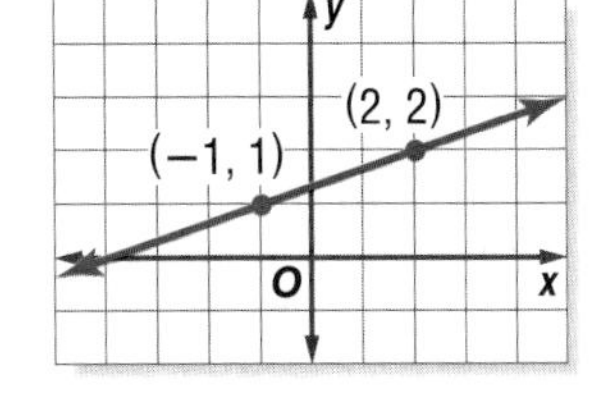

Ⓐ 3 Ⓑ −3 Ⓒ $\frac{1}{3}$ Ⓓ $-\frac{1}{3}$

32. **GRID IN** The graph of a straight line contains the points (0, −3), (1, 2), and (2, y). What is the value of y?

Graph each equation. (Lesson 4-6)

33. $y = 2x - 1$
34. $y = -3x$
35. $y = x + 4$

36. Solve $2x + 6 < 8$. (Lesson 4-5)

WebQuest Interdisciplinary Project

The Wide World of Soccer

It is time to complete your project. Use the information and data you have gathered about countries where soccer is a favorite sport to prepare a Web page or poster. Be sure to include a graph with your project.

msmath2.net/webquest

Study Guide and Review

Vocabulary and Concept Check

domain (p. 177)	inverse operations (p. 156)	slope (p. 182)
function (p. 177)	linear equation (p. 178)	two-step equation (p. 166)
function table (p. 177)	range (p. 177)	work backward strategy (p. 164)
inequality (p. 172)		

State whether each sentence is *true* or *false*. If *false*, replace the underlined word or number to make a true sentence.

1. Inverse operations "undo" each other.
2. The words "more than" sometimes suggest the operation of multiplication.
3. An inequality is a mathematical sentence that contains the symbols $<$, $>$, $\leq$, or $\geq$.
4. When graphing $t < 2$ on a number line, an open circle should be used to show that 2 is not included in the solution.
5. Slope is a number that tells how steep a line is.
6. An equation is called a linear equation if its graph is a point.
7. The solution of $m + 5 = 12$ is 17.
8. The solution of $g - 4 = 18$ is 22.
9. Six is the solution of $-8w = 48$.
10. The solution of $2y \leq 18$ is $y \leq 16$.

Lesson-by-Lesson Exercises and Examples

4-1 Writing Expressions and Equations (pp. 150–152)

Write each phrase as an algebraic expression.

11. the sum of a number and 5
12. six inches less than her height
13. twice as many apples
14. four times the number of dollars

Write each sentence as an algebraic equation.

15. Ten years older than Theresa's age is 23.
16. Four less than a number is 19.
17. The quotient of 56 and a number is 14.
18. Eight times the number of students is 64.

Example 1 **Write the phrase as an algebraic expression.**

four times the price

Let p represent the price.

The algebraic expression is $4p$.

Example 2 **Write the sentence as an algebraic equation.**

Six less than the number of cookies is 24.

Let c represent the number of cookies.

The equation is $c - 6 = 24$.

msmath2.net/vocabulary_review

4-2 Solving Addition and Subtraction Equations (pp. 156–159)

Solve each equation. Check your solution.

19. $x + 5 = 8$
20. $y + 4 = 12$
21. $a + 6 = 5$
22. $r + 8 = 2$
23. $p + 9 = -4$
24. $d + 14 = 23$
25. $s - 8 = 15$
26. $t - 6 = 7$
27. $n - 1 = -3$
28. $w - 9 = 28$

Example 3 **Solve $x + 6 = 4$.**

$$\begin{aligned} x + 6 &= 4 \\ -6 &= -6 \quad \text{Subtract 6 from each side.} \\ x &= -2 \end{aligned}$$

Example 4 **Solve $y - 3 = -2$.**

$$\begin{aligned} y - 3 &= -2 \\ +3 &= +3 \quad \text{Add 3 to each side.} \\ y &= 1 \end{aligned}$$

4-3 Solving Multiplication Equations (pp. 160–163)

Solve each equation. Check your solution.

29. $7c = 28$
30. $9y = 45$
31. $-2h = 24$
32. $-8w = 72$
33. $10y = -90$
34. $6q = -18$
35. $-11f = -121$
36. $-12r = -36$

Example 5 **Solve $-4b = 32$.**

$$-4b = 32$$

$$\frac{-4b}{-4} = \frac{32}{-4} \quad \text{Divide each side by } -4.$$

$$b = -8$$

4-4 Solving Two-Step Equations (pp. 166–169)

Solve each equation. Check your solution.

37. $3y - 12 = 6$
38. $6x - 4 = 20$
39. $2x + 5 = 3$
40. $5m + 6 = -4$
41. $10c - 8 = 90$
42. $3r - 20 = -5$
43. Ten more than five times a number is 25. Find the number.

Example 6 **Solve $3p - 4 = 8$.**

$$\begin{aligned} 3p - 4 &= 8 \\ +4 &= +4 \quad \text{Add 4 to each side.} \\ 3p &= 12 \end{aligned}$$

$$\frac{3p}{3} = \frac{12}{3} \quad \text{Divide each side by 3.}$$

$$p = 4$$

4-5 Inequalities (pp. 172–175)

Solve each inequality. Graph the solution.

44. $x + 3 < 8$
45. $y + 2 > 5$
46. $a + 4 \geq 10$
47. $d + 1 \leq 6$
48. $h - 5 \geq 7$
49. $s - 2 \leq 9$
50. $y + 2 < -3$
51. $m - 7 > -10$
52. $b + 9 \leq -11$
53. $t - 10 \geq -8$

Example 7 **Solve $g + 8 \leq 10$. Graph your solution.**

$$\begin{aligned} g + 8 &\leq 10 \\ -8 &= -8 \quad \text{Subtract 8 from each side.} \\ g &\leq 2 \end{aligned}$$

−5 −4 −3 −2 −1 0 1 2 3 4 5

4-6 Functions and Linear Equations (pp. 177–181)

Graph each equation.

54. $y = x + 5$
55. $y = x - 4$
56. $y = 2x$
57. $y = -1x$
58. $y = 3x + 2$
59. $y = -2x + 3$

MONEY MATTERS For Exercises 60–62, use the following infomation.

Angel earns $6 per hour working at the Ice Cream Shop.

60. Make a table that shows her total earnings for working 3, 5, 7, and 9 hours.
61. Write an equation in which x represents the number of hours and y represents Angel's total earnings.
62. Graph the equation.

Example 8 **Graph $y = x + 3$.**

Select four values for x. Substitute these values for x to find values for y.

x	$x + 3$	y
−1	−1 + 3	2
0	0 + 3	3
1	1 + 3	4
2	2 + 3	5

Four solutions are (−1, 2), (0, 3), (1, 4), and (2, 5). The graph is shown below.

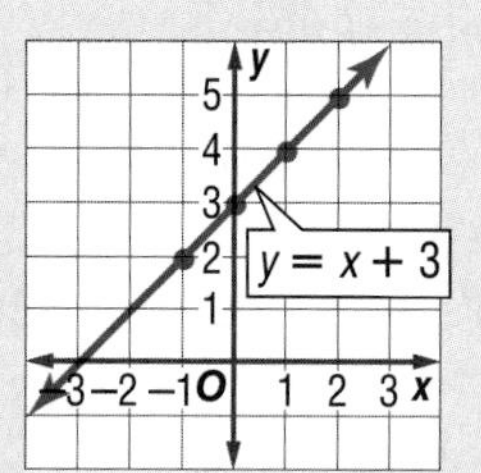

4-7 Lines and Slope (pp. 182–185)

Find the slope of the line that passes through each pair of points.

63.

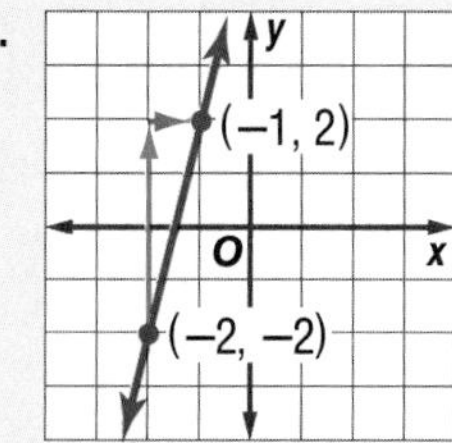

64.

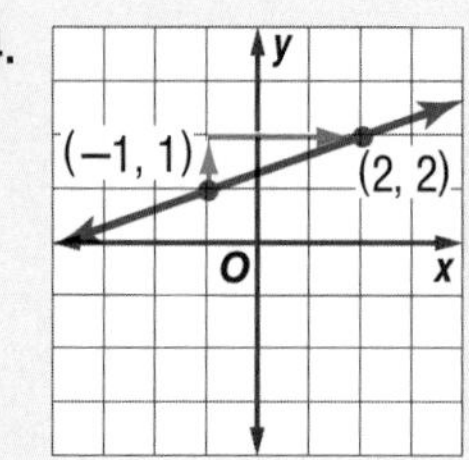

65. (1, 1), (2, 0)
66. (2, 3), (4, −1)

Example 9 **Find the slope of the line.**

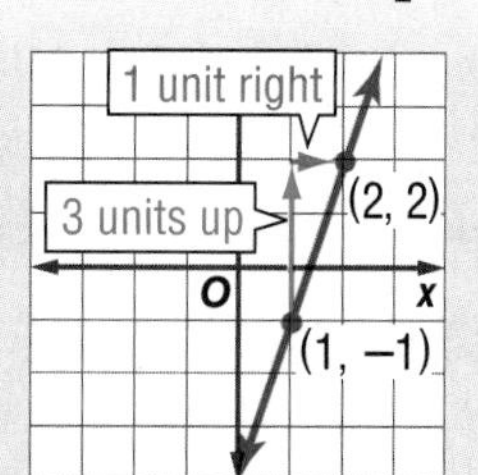

$$\text{slope} = \frac{\text{change in } y}{\text{change in } x}$$

$$= \frac{3}{1} \text{ or } 3$$

The slope of the line is 3.

CHAPTER 4 Practice Test

Vocabulary and Concepts

1. **Explain** the difference between an equation and an inequality.
2. **Define** *function* and give an example.

Skills and Applications

Write each phrase as an algebraic expression.

3. \$5 less than Matt has
4. 4 years older than Hana

Solve each equation. Check your solution.

5. $x + 5 = -8$
6. $y - 11 = 15$
7. $9z = -81$
8. $-6k - 4 = 38$

Solve each inequality. Graph the solution.

9. $p - 4 \geq -3$
10. $j + 5 > 2$

SHOPPING For Exercises 11 and 12, use the following information.
Suppose you want to buy 3 CDs and a new CD case that costs \$7. Each CD costs the same amount.

11. If you spend \$46, write an equation to find the cost of each CD.
12. Solve the equation.

Graph each equation.

13. $y = 3x - 2$
14. $y = -2x + 4$
15. $y = 0.5x$

MOVIES For Exercises 16–18, use the following information.
A student ticket to the movies costs \$3.

16. Make a table that shows the total cost of 2, 4, and 6 tickets.
17. Write an equation in which x represents the number of tickets, and y represents the total cost.
18. Graph the equation.

19. Find the slope of the line that passes through $(-2, 3)$ and $(-1, 2)$.

Standardized Test Practice

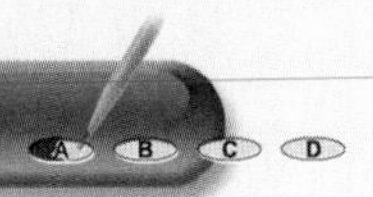

20. **MULTIPLE CHOICE** Which line has a slope of 2?

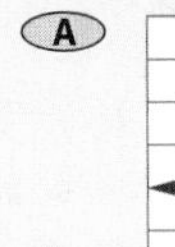

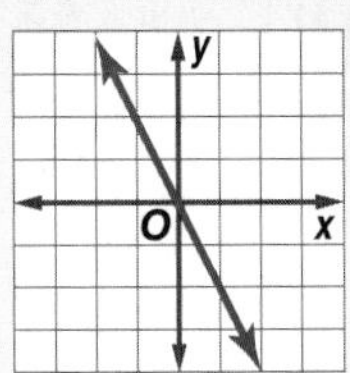

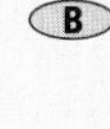

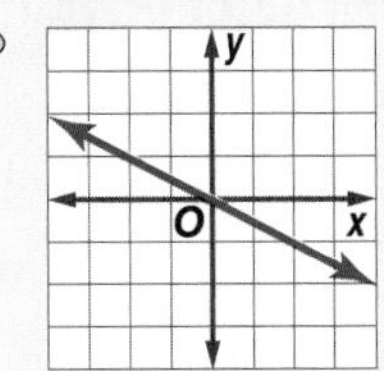

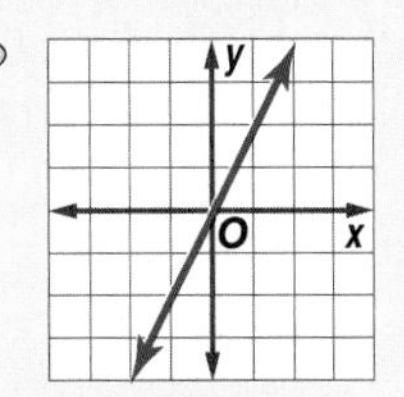

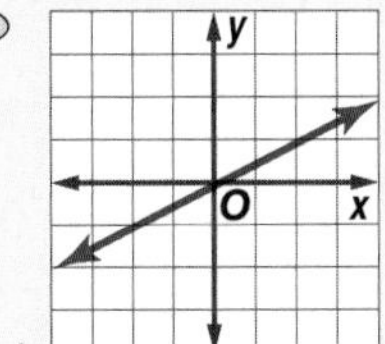

msmath2.net/chapter_test

CHAPTER 4 Standardized Test Practice

PART 1 Multiple Choice

Record your answers on the answer sheet provided by your teacher or on a sheet of paper.

1. Felicia's family wants to buy her a gift for \$220. Her parents will pay half. Her older sister will pay \$50. If her three other siblings split the remaining cost, how much will each pay? (Lesson 1-1)

   (A) \$20 (B) \$60 (C) \$130 (D) \$170

2. Which is equivalent to 3^6? (Lesson 1-2)

   (F) 18 (G) 36

   (H) $6 \cdot 6 \cdot 6$ (I) $3 \cdot 3 \cdot 3 \cdot 3 \cdot 3 \cdot 3$

3. Suppose you need 0.65 liter of water for a science experiment, but the container is measured in milliliters. How many milliliters of water do you need? (Lesson 1-5)

   (A) 0.0065 mL (B) 0.65 mL

   (C) 6.5 mL (D) 650 mL

4. If 18 is added to the data set below, which statement is true? (Lesson 2-4)

16, 14, 22, 16, 16, 18, 15, 25

   (F) The mode increases.

   (G) The mean decreases.

   (H) The mean increases.

   (I) The median increases.

5. The stem-and-leaf plot shows the number of points scored by the Bears in each of their basketball games this season. In how many games did they score at least 30 points? (Lesson 2-5)

Stem	Leaf
1	8 9
2	0 2 3 3 6 8 8 9
3	0 1 4 4 5 6 8 9
4	0 1 2

   1 | 8 = 18 points

   (A) 8 (B) 9 (C) 11 (D) 20

6. Suppose points given by (x, y) in the table are graphed. Which statement is true about the graphs? (Lesson 3-3)

x	y
−4	5
−5	10
−2	6

   (F) The graphs of the points are located in Quadrant I.

   (G) The graphs of the points are located in Quadrant II.

   (H) The graphs of the points are located in Quadrant III.

   (I) The graphs of the points are located in Quadrant IV.

7. The temperature at 6:00 A.M. was −5°F. What was the temperature at 8:00 A.M. if it had risen 7 degrees? (Lesson 3-4)

   (A) −12°F (B) −2°F

   (C) 2°F (D) 12°F

8. The Tigers scored four more runs than the Giants scored. Which expression represents the number of runs the Giants scored if the Tigers scored n runs? (Lesson 4-1)

   (F) $n + 4$ (G) $n - 4$

   (H) $4 - n$ (I) $4n$

9. Which is the graph of the equation $y = 3x - 2$? (Lesson 4-6)

   (A)
   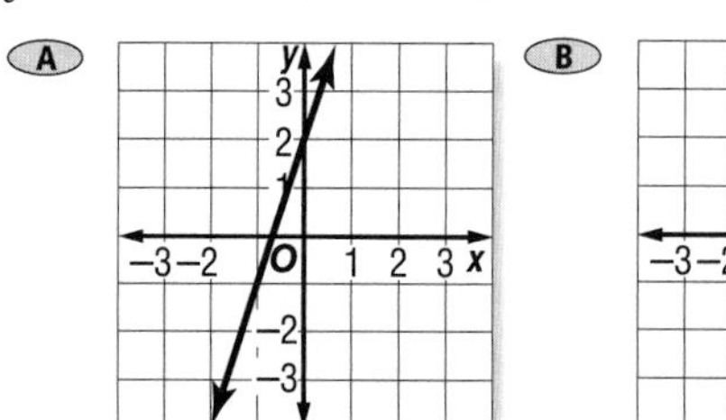

   (B)
   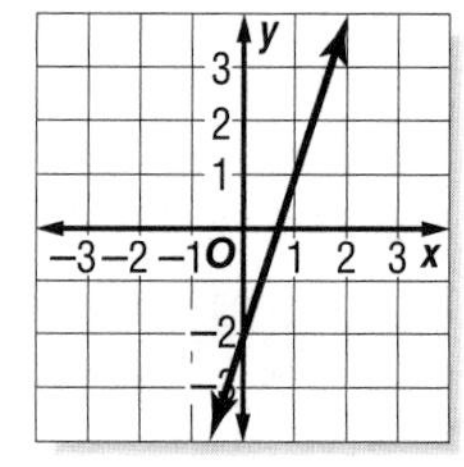

   (C)
   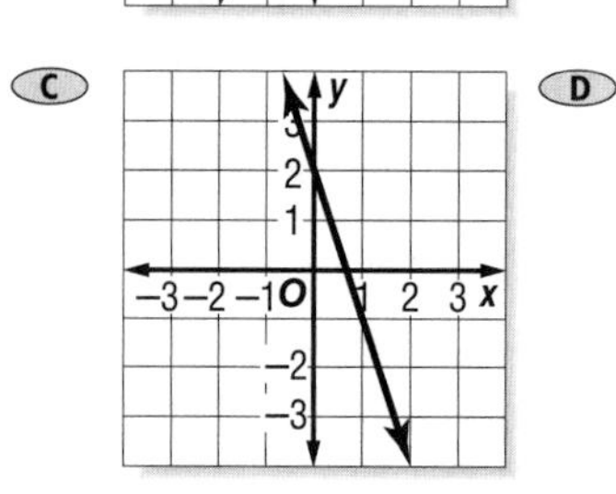

   (D)
   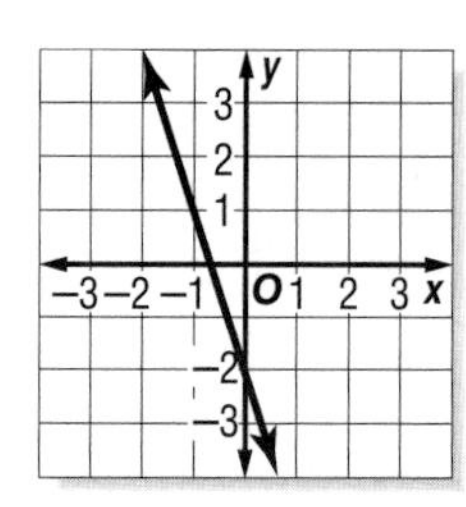

PART 2 Short Response/Grid In

Record your answers on the answer sheet provided by your teacher or on a sheet of paper.

10. The charge to enter a park is a flat amount per vehicle plus a fee for each person in the vehicle. The table shows the charge for vehicles holding up to 4 people.

Number of people	Charge (dollars)
1	2.00
2	2.50
3	3.00
4	3.50

What is the charge, in dollars, for a vehicle holding 8 people? (Lesson 1-1)

11. Evaluate $2(8 + 5^2)$. (Lesson 1-3)

12. The line plot shows how far in kilometers some students live from the school. How many students are represented in the plot? (Lesson 2-3)

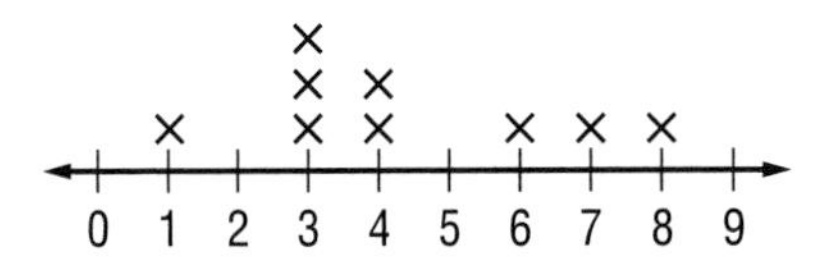

13. The ordered pairs (1, 2), (6, 2), and (1, −5) are coordinates of three of the vertices of a rectangle. What is the *y*-coordinate of the ordered pair that represents the fourth vertex? (Lesson 3-3)

14. What number should replace *y* in the table? (Lesson 3-6)

x	$2x - 5$
1	−3
2	y
4	3
8	11
16	27

TEST-TAKING TIP

Question 13 This question does not include a drawing. Make one. It can help you quickly see how to solve the problem.

15. What value of x makes $x - 2 = -4$ a true statement? (Lesson 4-2)

16. Twice a number plus 7 is 35. Find the number. (Lesson 4-4)

17. Winston earns $7 an hour landscaping. He wants to buy a DVD player that costs $140. Write an inequality for the least number of hours *h* he needs to work to reach his goal. (Lesson 4-5)

18. The graph of a line contains (1, 2) and (3, *y*). What is the value of *y* if the slope of the line is 2? (Lesson 4-7)

PART 3 Extended Response

Record your answers on a sheet of paper. Show your work.

19. Pete's Paints charges a $20 fee plus $7.50 per hour to rent a paint sprayer.
    a. Write an equation that could be used to determine the cost *c* of renting a paint sprayer for *h* hours.
    b. Suppose you want to spend no more than $50 to rent the paint sprayer. Write an inequality for this situation.
    c. Solve the inequality in part **b**. Explain the meaning of the solution.

20. The distances traveled by a bicycle rider are given in the table. (Lesson 4-6)
    a. Graph the ordered pairs.
    b. Write an equation that relates the time *t* to the distance *d*.
    c. Use your equation to predict the distance traveled in 3.5 hours.

Time (hours)	Distance (miles)
1	10
2	20
3	30
4	40
5	50

UNIT 3
Fractions

Chapter 5

Fractions, Decimals, and Percents

Chapter 6

Applying Fractions

Fractions, decimals, and percents are different ways of representing the same values. In this unit, you will describe real-life situations, perform basic operations, and solve equations with fractions.

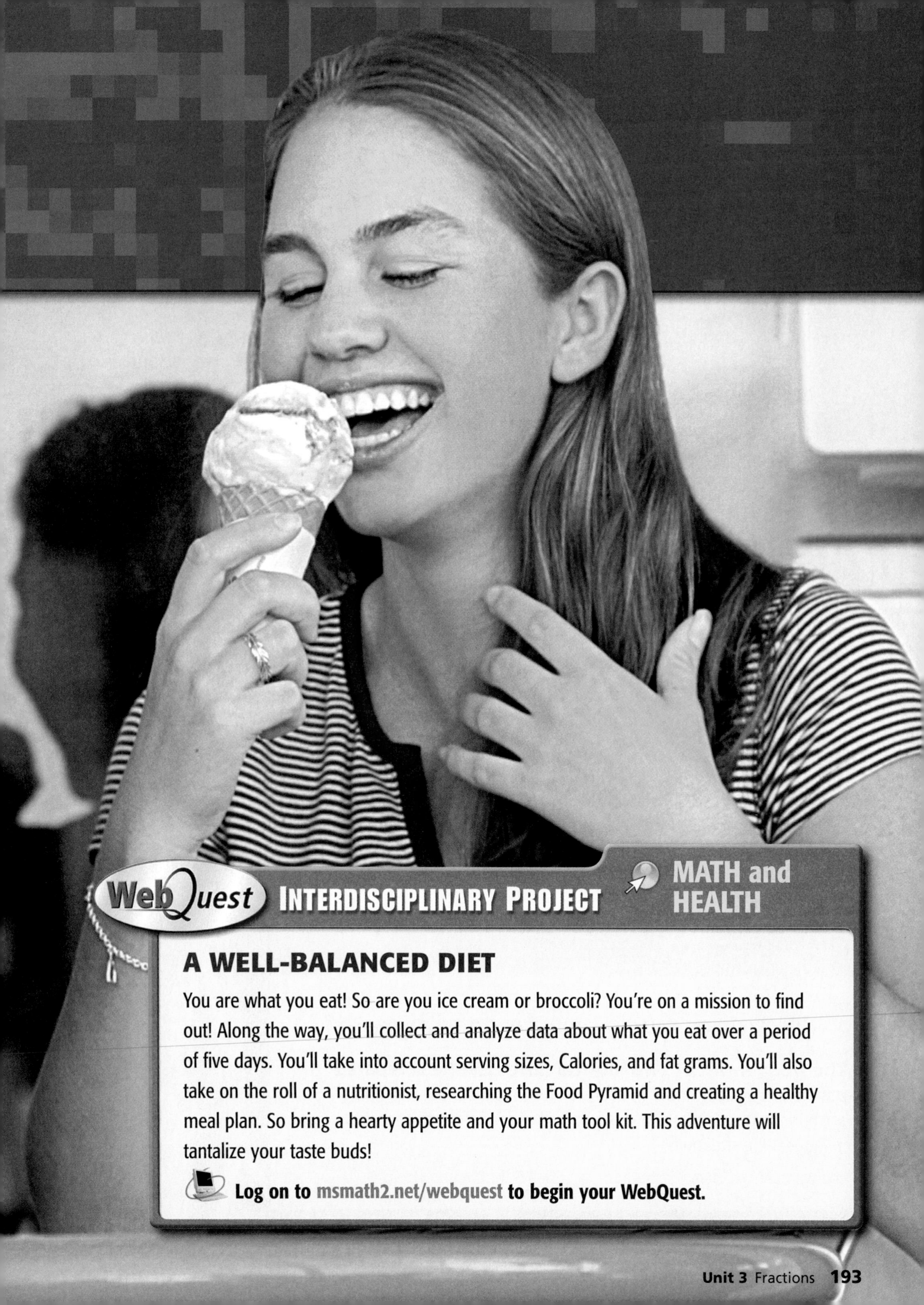

WebQuest **INTERDISCIPLINARY PROJECT** MATH and HEALTH

A WELL-BALANCED DIET

You are what you eat! So are you ice cream or broccoli? You're on a mission to find out! Along the way, you'll collect and analyze data about what you eat over a period of five days. You'll take into account serving sizes, Calories, and fat grams. You'll also take on the roll of a nutritionist, researching the Food Pyramid and creating a healthy meal plan. So bring a hearty appetite and your math tool kit. This adventure will tantalize your taste buds!

Log on to msmath2.net/webquest to begin your WebQuest.

CHAPTER 5

Fractions, Decimals, and Percents

"What does music have to do with math?"

The *frequencies* of two musical notes can be compared using a fraction. If the fraction can be simplified, the two notes are harmonious. In mathematics, you will use fractions to describe many real-life situations.

You will solve problems about music in Lesson 5-3.

GETTING STARTED

Take this quiz to see whether you are ready to begin Chapter 5. Refer to the lesson or page number in parentheses if you need more review.

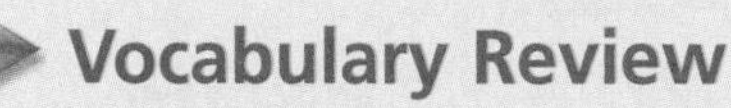

Vocabulary Review

Complete each sentence.

1. After comparing the decimals 0.41 and 0.04, you find that 0.41 is __?__ than 0.04. (Page 556)
2. Two or more numbers that are multiplied together to form a product are called __?__. (Lesson 1-2)

Prerequisite Skills

State which decimal is greater. (Page 556)

3. 0.6, 0.61
4. 1.25, 1.52
5. 0.33, 0.13
6. 1.08, 10.8

Use divisibility rules to determine whether each number is divisible by 2, 3, 5, 6, or 10. (Page 554)

7. 125
8. 78
9. 37
10. 214

Divide. (Page 562)

11. $5\overline{)2.0}$
12. $4\overline{)3.00}$

Write each power as a product of the same factor. (Lesson 1-2)

13. 2^3
14. 5^5
15. 7^2
16. 6^4

Write each product in exponential form. (Lesson 1-2)

17. $4 \times 4 \times 4$
18. 8×8
19. $1 \times 1 \times 1 \times 1 \times 1$
20. $7 \times 7 \times 7 \times 7$

Fractions, Decimals, and Percents Begin with a sheet of $8\frac{1}{2}''$ by 11″ construction paper and two sheets of notebook paper.

STEP 1 Fold and Label
Fold the construction paper in half lengthwise. Label the chapter title on the outside.

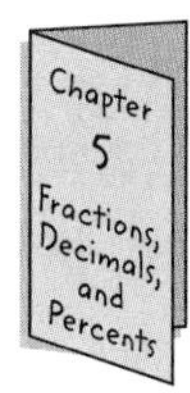

STEP 2 Fold
Fold the sheets of notebook paper in half lengthwise. Then fold top to bottom twice.

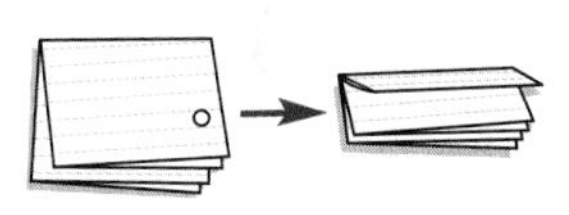

STEP 3 Cut
Open the notebook paper. Cut along the second folds to make four tabs.

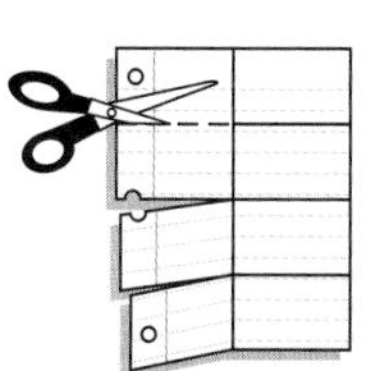

STEP 4 Glue and Label
Glue the uncut halves of notebook paper side by side onto the construction paper. Label each tab with the lesson number.

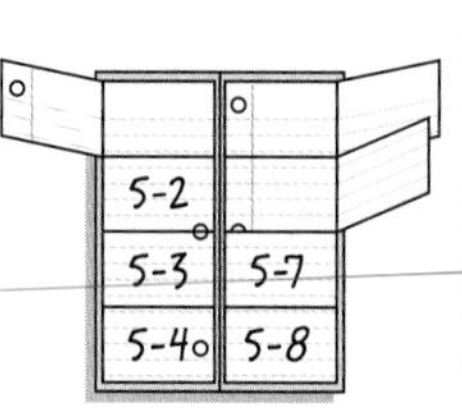

Reading and Writing As you read and study the chapter, write notes and examples for each lesson under the appropriate tab.

5-1a HANDS-ON LAB

A Preview of Lesson 5-1

Exploring Factors

What You'll LEARN

Discover factors of whole numbers.

Materials

- 15 index cards cut in half
- markers

INVESTIGATE *Work as a class.*

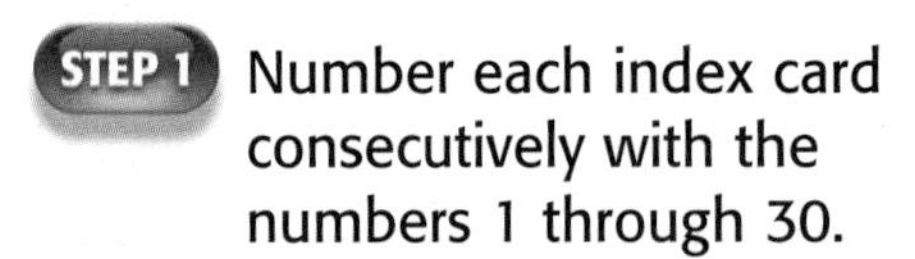

STEP 1 Number each index card consecutively with the numbers 1 through 30.

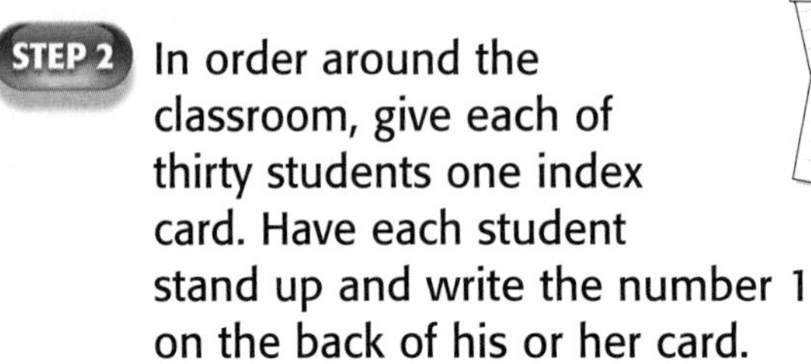

STEP 2 In order around the classroom, give each of thirty students one index card. Have each student stand up and write the number 1 on the back of his or her card.

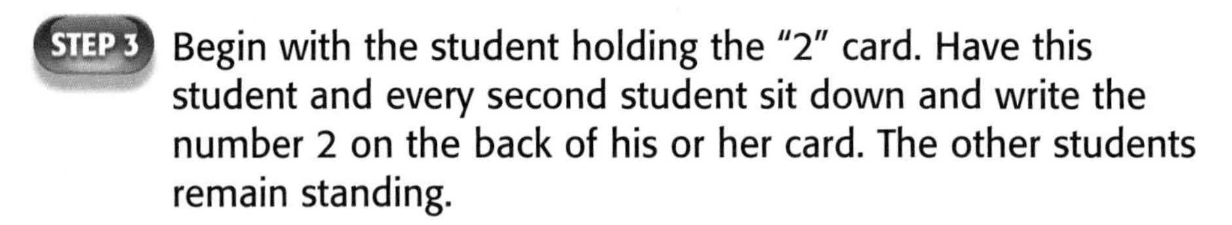

STEP 3 Begin with the student holding the "2" card. Have this student and every second student sit down and write the number 2 on the back of his or her card. The other students remain standing.

STEP 4 Next, start with the student holding the "3" card. Have this student and every third student stand up or sit down (depending on whether the student is already sitting or standing) and write the number 3 on the back of his or her card.

STEP 5 Continue this process for each of the remaining numbers up to 30. The thirtieth student ends the activity by standing or sitting, and writing the number 30 on the back of his or her card.

Writing Math

Work as a class.

1. How many students are standing at the end of the activity? Which cards are they holding?
2. **LOOK FOR A PATTERN** Suppose there were 100 students holding index cards. Extend the pattern in Exercise 1 to predict the numbers that would be held by students standing at the end of the activity.
3. **Explain** the relationship between the numbers on the front and the back of the cards.
4. Separate the cards into two groups: one group with exactly two numbers on the back of the card and one group with more than two numbers. **Describe** any special characteristics of each group.

5-1 Prime Factorization

HANDS-ON Mini Lab

Materials
- grid paper

Work with a partner.

There is only one way that 2 can be expressed as the product of whole numbers. The figure shows that there is only one way that two squares can form a rectangle.

1. Using your grid paper, draw as many different rectangles as possible containing 3, 4, 5, 6, 7, 8, 9, and 10 squares.
2. Which numbers of squares can be drawn in only one rectangle? In more than one rectangle?

What You'll LEARN

Find the prime factorization of a composite number.

NEW Vocabulary

prime number
composite number
prime factorization
factor tree

REVIEW Vocabulary

factors: two or more numbers that are multiplied together to form a product **(Lesson 1-2)**

The numbers of squares that can only be drawn in exactly one rectangle are prime numbers. The numbers that can be drawn in more than one rectangle are composite numbers.

A **prime number** is a whole number greater than 1 that has exactly two factors, 1 and itself.

A **composite number** is a whole number greater than 1 that has more than two factors.

The numbers 0 and 1 are neither prime nor composite.

Whole Numbers	Factors
2	1, 2
3	1, 3
5	1, 5
7	1, 7
4	1, 2, 4
6	1, 2, 3, 6
8	1, 2, 4, 8
9	1, 3, 9
10	1, 2, 5, 10
0	many
1	1

EXAMPLES Identify Numbers as Prime or Composite

Determine whether each number is *prime* or *composite*.

1 17

The number 17 has only two factors, 1 and 17, so it is prime.

2 12

The number 12 has six factors: 1, 2, 3, 4, 6, and 12. So, it is composite.

Your Turn **Determine whether each number is *prime* or *composite*.**

a. 11 **b.** 15 **c.** 24

msmath2.net/extra_examples

STUDY TIP

Commutative Property
Multiplication is commutative, so the order of the factors does not matter.

Every composite number can be written as a product of prime numbers in exactly one way. The product is called the **prime factorization** of the number. You can use a **factor tree** to find the prime factorization.

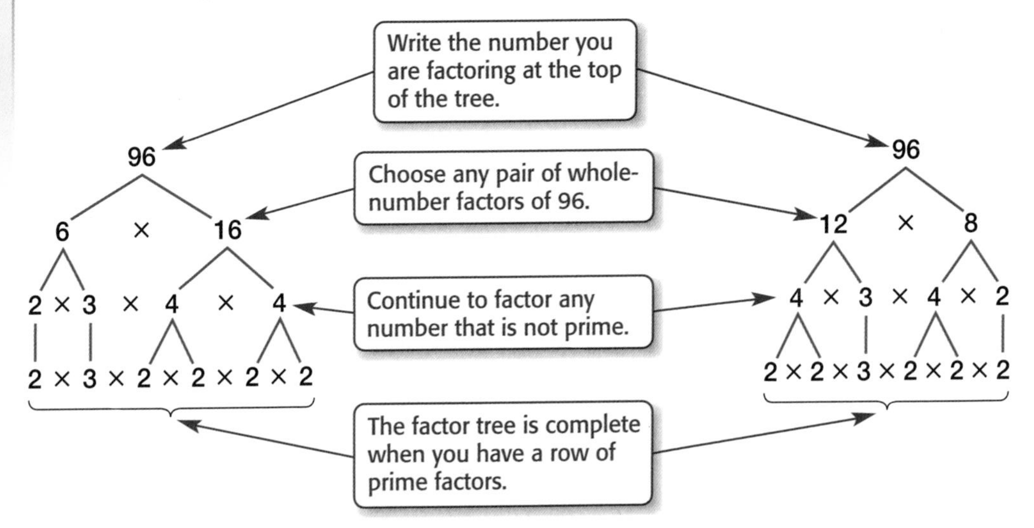

STUDY TIP

Mental Math
You can use divisibility rules to find prime factors. See page 554.

Both trees give the same prime factorization of 96, but in different orders. The prime factorization of 96 is $2 \times 2 \times 2 \times 2 \times 2 \times 3$, or $2^5 \times 3$ using exponents.

EXAMPLE Find the Prime Factorization

3 Find the prime factorization of 24.

Method 1 Use a factor tree.

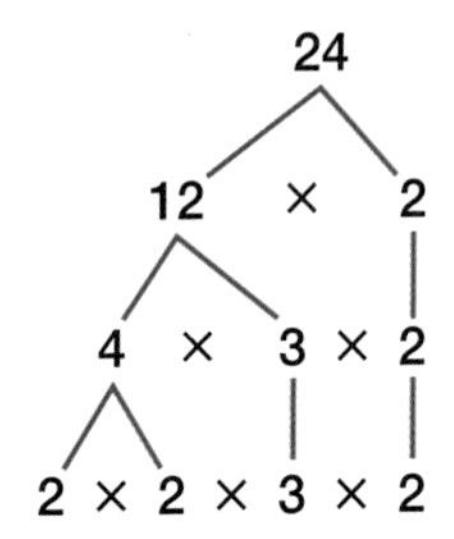

Method 2 Divide by prime numbers.

$$\begin{array}{r} 2 \\ 2\overline{)4} \\ 3\overline{)12} \\ 2\overline{)24} \end{array}$$

← Start here.

The divisors are 2, 2, 3, and 2.

The prime factorization of 24 is $2 \times 2 \times 3 \times 2$ or $2^3 \times 3$.

Your Turn **Find the prime factorization of each number.**

d. 18 **e.** 28 **f.** 16

You can factor an expression like $6ab$ as the product of prime numbers and variables.

EXAMPLE Factor an Algebraic Expression

4 ALGEBRA Factor $6ab$.

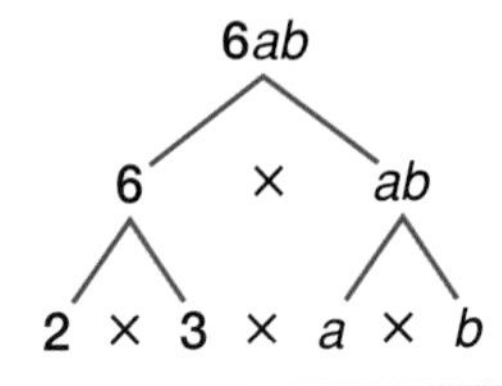

$6ab = 2 \cdot 3 \cdot a \cdot b$ Recall that a raised dot means multiplication.

Skill and Concept Check

1. **Compare and contrast** prime and composite numbers.
2. **OPEN ENDED** Draw two different factor trees to find the prime factorization of 36.
3. **Explain** why $2^3 \times 9$ is *not* the prime factorization of 72.
4. **Which One Doesn't Belong?** Identify the number that is not prime.

11	37	17	51

GUIDED PRACTICE

Determine whether each number is *prime* or *composite*.

5. 7 6. 50 7. 67

Find the prime factorization of each number.

8. 34 9. 30 10. 12

Factor each expression.

11. $10ac$ 12. $16x^2$

Practice and Applications

HOMEWORK HELP

For Exercises	See Examples
13–20, 29	1, 2
21–28, 36–38, 45–47	3
30–35	4

Extra Practice
See pages 574, 600.

Determine whether each number is *prime* or *composite*.

13. 22 14. 44 15. 13 16. 39
17. 81 18. 31 19. 97 20. 43

Find the prime factorization of each number.

21. 36 22. 42 23. 99 24. 64
25. 210 26. 180 27. 126 28. 375

29. **ALGEBRA** Is the value of $3a + 6b$ *prime* or *composite* if $a = 1$ and $b = 5$?

Factor each expression.

30. $15mn$ 31. $20pq$ 32. $34jkl$
33. $49y^2$ 34. $52gh^2$ 35. $48a^2b^2$

PACKAGING For Exercises 36–38, use the following information.
A juice company is designing a package that holds 36 juice cans. For example, a package could be 1 can high, 6 cans wide, and 6 cans long. Such a package would contain $1 \times 6 \times 6$, or 36 cans.

36. List two other ways to arrange 36 cans.
37. How many different ways can you arrange 36 cans?
38. In which arrangement should the 36 cans be packaged? Explain your reasoning.

Replace each ■ with prime factors to make a true sentence.

39. $2^3 \times ■ \times 11 = 616$
40. $2 \times ■ \times 5^2 = 450$
41. $3 \times 2^4 \times ■ = 1{,}200$

42. **NUMBER THEORY** *True* or *False*? All prime numbers are odd. If false, give a counterexample.

43. **NUMBER THEORY** Primes that differ by two are called *twin primes*. For example, 59 and 61 are twin primes. Give three examples of twin primes that are less than 50.

LANDSCAPING For Exercises 44–47, use the following information.
Mrs. Franks is building a patio that covers an area measuring 8 feet by 15 feet. She will use rectangular concrete tiles for the patio.

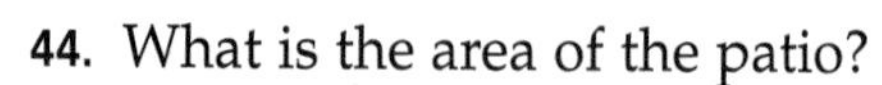

44. What is the area of the patio?

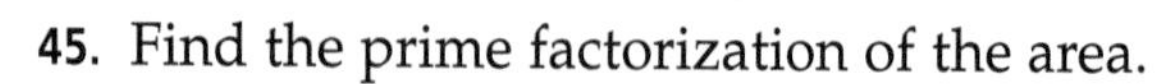

45. Find the prime factorization of the area.

46. If she can only buy tiles with dimensions that are prime numbers, name the dimensions of the concrete tiles available for her patio.

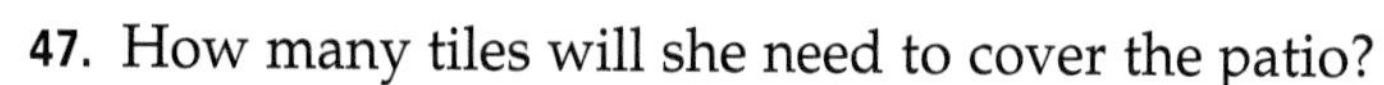

47. How many tiles will she need to cover the patio?

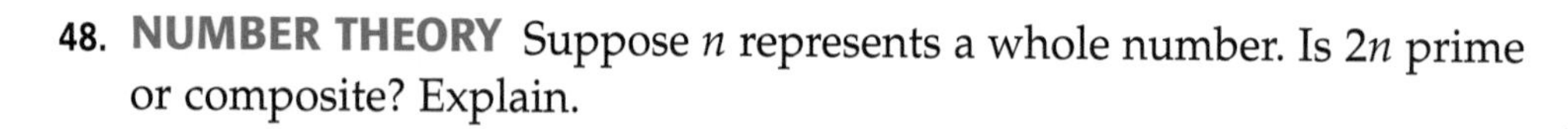

48. **NUMBER THEORY** Suppose n represents a whole number. Is $2n$ prime or composite? Explain.

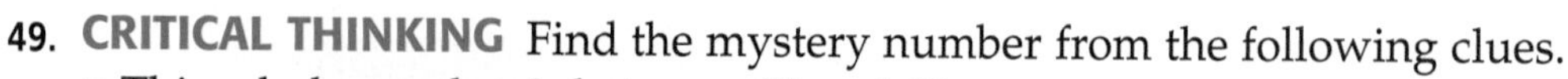

49. **CRITICAL THINKING** Find the mystery number from the following clues.
    - This whole number is between 30 and 40.
    - It has only two prime factors.
    - The sum of its prime factors is 5.

Standardized Test Practice and Mixed Review

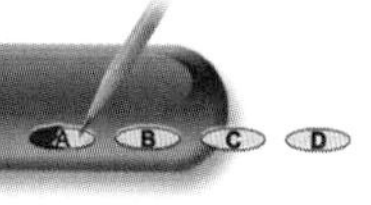

50. **MULTIPLE CHOICE** Which expression represents the prime factorization of 126?

    Ⓐ $2^2 \times 3 \times 7$ Ⓑ $2 \times 3^2 \times 7$ Ⓒ $2 \times 3 \times 7^2$ Ⓓ $2^2 \times 3^2 \times 7^2$

51. **SHORT RESPONSE** Determine whether the value of the expression $x^2 - 3x + 11$ is prime or composite when $x = 4$.

52. **ALGEBRA** What is the slope of the line that passes through points at $(-2, 5)$ and $(3, -1)$? (Lesson 4-7)

53. **ALGEBRA** Graph $y = 3x + 1$. (Lesson 4-6)

Tell whether each sum is *positive, negative,* or *zero*. (Lesson 3-4)

54. $6 + (-4)$
55. $-13 + 9$
56. $25 + (-26)$
57. $-5 + 5$

GETTING READY FOR THE NEXT LESSON

PREREQUISITE SKILL State whether each number is divisible by 2, 3, 5, 6, 9, or 10. (Page 554)

58. 24
59. 70
60. 120
61. 99
62. 125

5-2a Problem-Solving Strategy

A Preview of Lesson 5-2

What You'll Learn
Solve problems by making an organized list.

Make an Organized List

I wonder how many different ways a woodwind trio can be made if either a bass clarinet or a bassoon fills the first position and either a clarinet, oboe, or flute fills the other two positions?

Let's **make an organized list** of all of the possibilities.

Explore We know that either a bass clarinet or a bassoon must fill the first position. The other two positions are to be filled with two out of the three remaining instruments. We have to find all possible ways the trio can be made.

Plan Let's make an organized list.

Solve

	Trio 1	Trio 2	Trio 3	Trio 4	Trio 5	Trio 6
Position 1	bass clarinet	bass clarinet	bass clarinet	bassoon	bassoon	bassoon
Position 2	clarinet	oboe	flute	clarinet	oboe	flute
Position 3	oboe	flute	clarinet	oboe	flute	clarinet

There are six possibilities.

Examine We can draw a tree diagram to check our result.

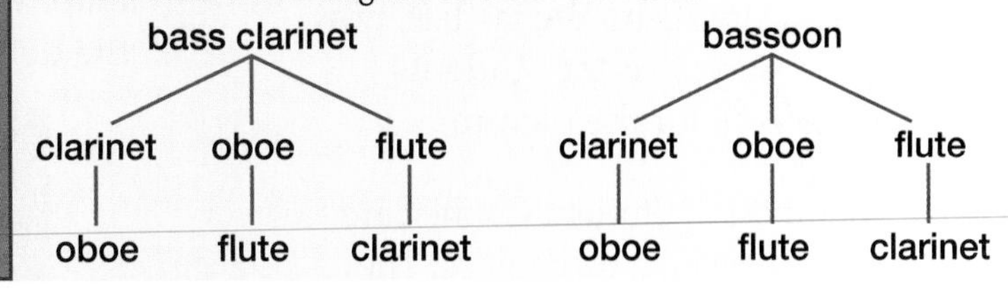

Analyze the Strategy

1. **Explain** why making an organized list was a useful strategy in solving this problem.
2. **Explain** why a tree diagram is another method for making an organized list.
3. **Write** a problem that could be solved by making an organized list. Explain.

Apply the Strategy

Solve. Use the make an organized list strategy.

4. **FOOD** Daniel is making a peanut butter and jelly sandwich. His choices are creamy or crunchy peanut butter, white or wheat bread, and grape, apple, or strawberry jelly. How many different types of sandwiches can Daniel make?

5. **GAMES** On the game board, you plan to move two spaces away from square A. You can either move horizontally, vertically, or diagonally. How many different moves can you make from square A? List them.

Mixed Problem Solving

Solve. Use any strategy.

6. **SAVINGS** The graph below shows deposits and withdrawals. During which week was the difference between deposits and withdrawals the greatest?

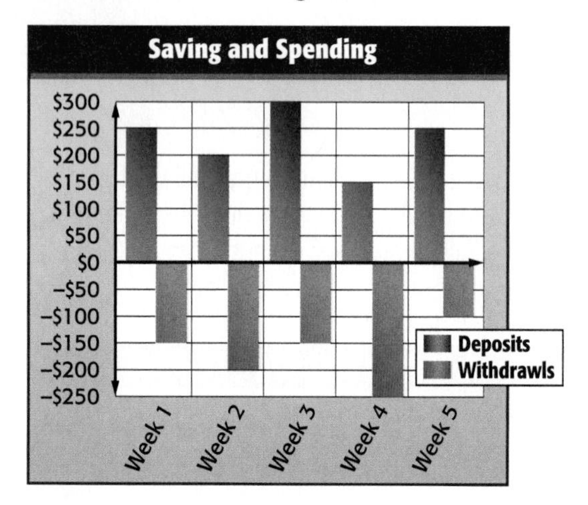

7. **SCIENCE FAIR** Mrs. Cassidy is taking a picture with the first, second, and third place winners of the science fair. If Mrs. Cassidy always stands on the far left, how many different ways can the students arrange themselves for the picture?

8. **SCHOOL** Members of Student Council are raising money to attend a conference. The total costs of the conference are shown in the table. If 24 students are attending the conference, how much does each student have to raise?

Transportation	$1,200
Registration	$288
Luncheon	$360

9. **NUMBER THEORY** Find two consecutive odd numbers that when added equal 56, and when multiplied equal 783.

10. **MULTI STEP** Terrez took a bag of cookies to play rehearsal. Half were given to the musicians and five to the director of the play. Terrez was left with 15 cookies. How many cookies did he take to rehearsal?

11. **EARTH SCIENCE** Giant kelp seaweed is found in the Pacific Ocean. One plant grows 3 feet the first two days. If it continues to grow at the same rate, what would be the length of the seaweed at the end of 80 days?

12. **STANDARDIZED TEST PRACTICE**
Mary has to make deliveries to three neighbors. She lives at house b on the map. Which is the shortest route to make the deliveries and return home?

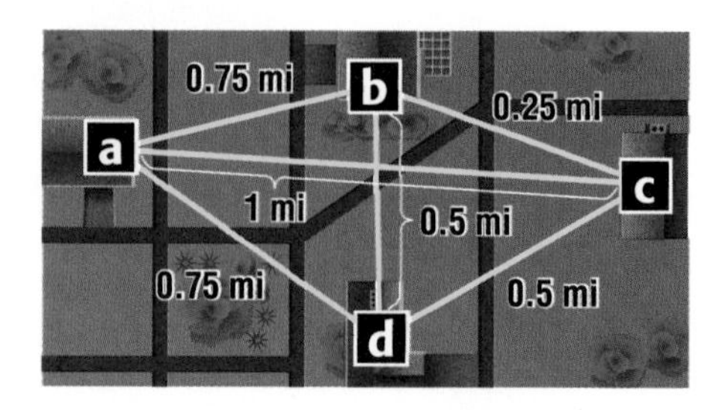

Ⓐ b-a-d-c-b　　Ⓑ b-a-c-d-b
Ⓒ b-c-a-d-b　　Ⓓ b-d-a-c-b

You will use the make an organized list strategy in the next lesson.

5-2 Greatest Common Factor

WHEN am I ever going to use this?

What You'll LEARN

Find the greatest common factor of two or more numbers.

NEW Vocabulary

Venn diagram
greatest common factor (GCF)

INTERNET A group of friends spent time in two Internet chat rooms. The diagram shows the chat rooms Angel, Sydney, Ian, Candace, and Christine visited. The friends were able to stay in one chat room or go to the other one.

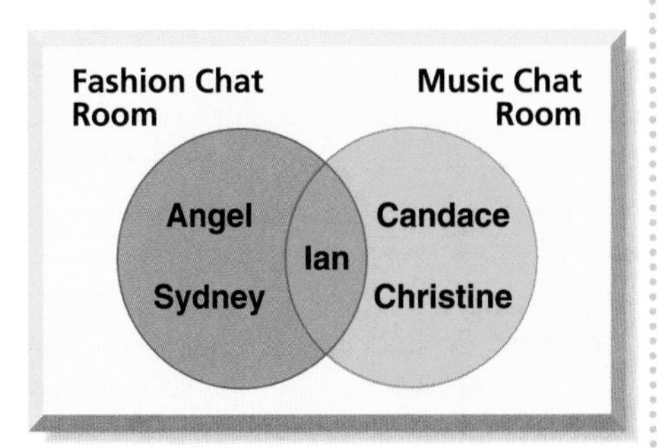

1. Who visited the Fashion Chat Room?
2. Who visited the Music Chat Room?
3. Who visited both chat rooms?

The diagram above is called a **Venn diagram**. It uses circles to show how elements among sets of numbers or objects are related. The region where circles overlap represents items that are common to two or more sets. It shows that Ian visited both chat rooms.

Venn diagrams can also show factors that are common to two or more numbers. The greatest of these common factors is called the **greatest common factor (GCF)**. The GCF of prime numbers is 1.

EXAMPLE Find the GCF by Listing Factors

1 Find the GCF of 20 and 24.

First, list the factors of 20 and 24.

factors of 20: 1, 2, 4, 5, 10, 20
factors of 24: 1, 2, 3, 4, 6, 8, 12, 24

common factors: 1, 2, 4

Notice that 1, 2, and 4 are common factors of 20 and 24. So, the GCF is 4.

Check You can draw a Venn diagram to check your answer.

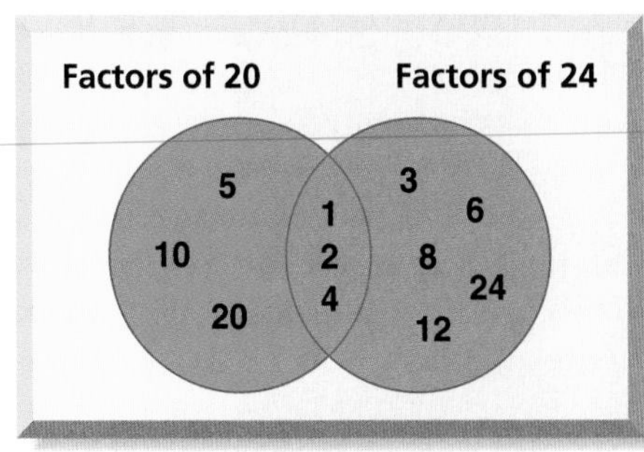

Your Turn **Find the GCF of each pair of numbers.**

a. 8 and 10 b. 6 and 12 c. 10, 17

EXAMPLES Find the GCF Using Prime Factors

Find the GCF of each set of numbers.

2 **18, 48**

Method 1 Write the prime factorization.

$18 = 2 \times 3 \times 3$

$48 = 2 \times 2 \times 2 \times 2 \times 3$

Method 2 Divide by prime numbers.

Divide both 18 and 48 by 2. Then divide the quotients by 3.

$$\begin{array}{r|rr} 2 & 18 & 48 \\ 3 & 9 & 24 \\ & 3 & 8 \end{array}$$

Start here.

The common prime factors are 2 and 3.
So, the GCF of 18 and 48 is 2×3, or 6.

3 **12, 24, 60**

$12 = 2 \times 2 \times 3$

$24 = 2 \times 2 \times 2 \times 3$

$60 = 2 \times 2 \times 3 \times 5$

Circle the common factors.

The common prime factors are 2, 2, and 3.
So, the GCF is $2 \times 2 \times 3$, or 12.

Your Turn **Find the GCF of each set of numbers.**

d. 36, 48 e. 10, 35 f. 30, 45, 75

STUDY TIP

Choosing a Method To find the GCF of two numbers, it is easier to:

- list factors for small numbers, and
- find the prime factorization or use division for large numbers.

EXAMPLE Find the GCF of an Algebraic Expression

4 **ALGEBRA Find the GCF of $10a$ and $15a^2$.**

Factor each expression.

$10a = 2 \cdot 5 \cdot a$

$15a^2 = 3 \cdot 5 \cdot a \cdot a$

Circle the common factors.

The GCF is $5 \cdot a$, or $5a$.

REAL-LIFE MATH

PARADES Forty pounds of dye is used to turn the Chicago River green for the annual St. Patrick's Day Parade.

Source: www.chicagostpatsparade.com

EXAMPLE Use the GCF to Solve a Problem

5 **PARADES In a parade, 64 eighth-graders are to march in front of 88 seventh-graders. Both groups should have the same number of students in each row. Find the greatest number of students in each row.**

The greatest number of students in each row is the GCF of the number of students in each group.

$64 = 2 \times 2 \times 2 \times 2 \times 2 \times 2 = 2^6$

$88 = 2 \times 2 \times 2 \times 11 = 2^3 \times 11$

The GCF of 64 and 88 is 2^3, or 8.

So, there should be 8 students in each row.

Skill and Concept Check

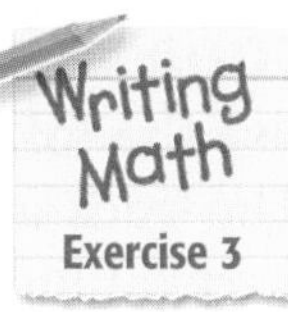

1. **OPEN ENDED** Find two numbers whose GCF is 16.

2. **Draw** a Venn diagram showing the factors of 30 and 42. Name the GCF.

3. **FIND THE ERROR** Charles and Tiffany both found the GCF of 12, 18, and 27. Who is correct? Explain.

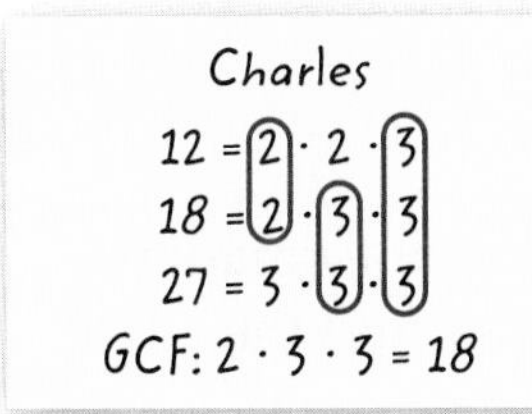

Tiffany
12 = 2 • 2 • 3
18 = 2 • 3 • 3
27 = 3 • 3 • 3
GCF: 3

GUIDED PRACTICE

Find the GCF of each set of numbers.

4. 18, 30
5. 45, 60
6. 6, 8, 12
7. 18, 42, 60

8. Find the GCF of $14xy$ and $7x^2$.

Practice and Applications

HOMEWORK HELP	
For Exercises	See Examples
9–16, 21–22	1, 2
17–20	3
26–29	4
30–33	5

Extra Practice
See pages 574, 600.

Find the GCF of each set of numbers.

9. 12, 78
10. 40, 50
11. 20, 45
12. 32, 48
13. 24, 48
14. 45, 75
15. 56, 96
16. 40, 125
17. 18, 24, 30
18. 36, 60, 84
19. 35, 49, 84
20. 36, 50, 130
21. What is the GCF of $2^4 \times 5$ and $5^2 \cdot 7$?
22. Find the GCF of $2^3 \times 3^2 \times 7$ and $2^2 \times 3 \times 11$.

Determine whether each statement is *sometimes, always,* or *never* true.

23. The GCF of two numbers is greater than both numbers.
24. If two numbers have no common prime factors, the GCF is 1.
25. The GCF of two numbers is one of the numbers.

Find the GCF of each set of algebraic expressions.

26. $24a$, $6a$
27. $27b^2$, $36b$
28. $16r^3$, $28r$
29. $20x^2$, $50xy^2$

INDUSTRIAL TECHNOLOGY For Exercises 30 and 31, use the following information.

Kibbe is building scale models of high-rise buildings for his class project. He designs the models using one-inch cubes. He is planning to build three buildings, the first with 108 blue cubes, the second with 270 red cubes, and the third with 225 yellow cubes. All of the buildings must be the same height, but not necessarily the same length and width.

30. What is the maximum height of each high-rise Kibbe can build?
31. What are the dimensions of all three buildings?

FOOD **For Exercises 32 and 33, use the following information.** The track-and-field coaches threw a party at the end of the season and bought a 32-inch, a 20-inch, and a 12-inch submarine sandwich. Suppose the sandwiches are cut into equal-sized pieces.

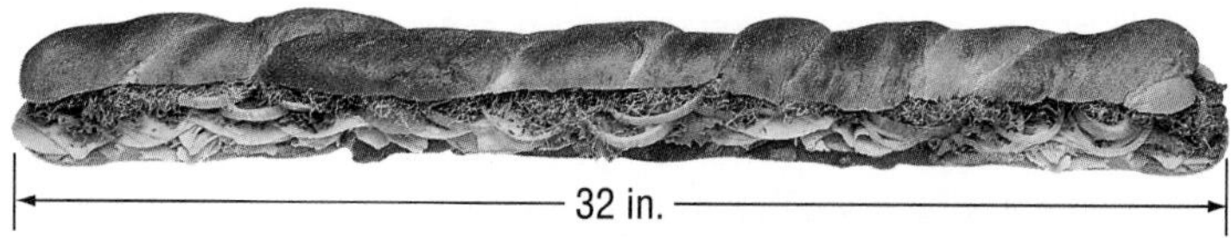
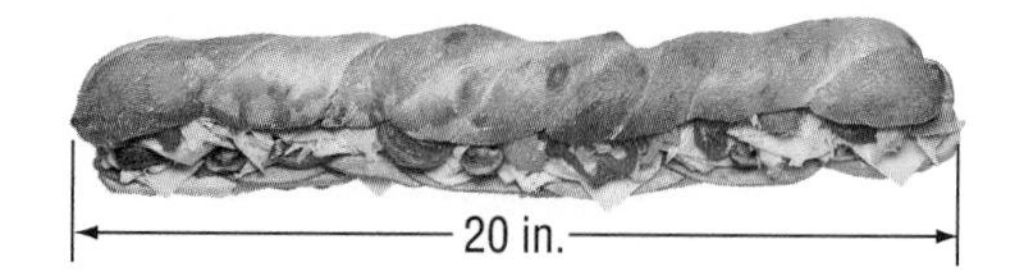

32. What is the length of the longest piece that can be cut?
33. If there are 18 people on the team, will everyone receive a sandwich? Explain.

34. **RESEARCH** Use the Internet or another source to find information about the Sieve of Eratosthenes. Explain how you can find the GCF of 18 and 54 using the Sieve.

CRITICAL THINKING **For Exercises 35–37, use the following information.** Numbers that have a GCF of 1 are *relatively prime*. Use this definition to determine whether each statement is true or false. Explain your answer.

35. Any two prime numbers are relatively prime.
36. If two numbers are relatively prime, one of them must be prime.
37. Two consecutive numbers are always relatively prime.

Standardized Test Practice and Mixed Review

38. **MULTIPLE CHOICE** Felisa is making shelves to store her books. How many shelves measuring 8 inches by 12 inches can be cut from a 36-inch-by-56-inch piece of plywood so that there is no waste?

    Ⓐ 17 Ⓑ 18 Ⓒ 20 Ⓓ 21

39. **MULTIPLE CHOICE** Which set of numbers does *not* have a GCF of 6?

    Ⓕ 42, 150 Ⓖ 54, 18 Ⓗ 24, 6 Ⓘ 18, 30

40. **SHORT RESPONSE** Find the GCF of $20st^2$ and $50s^2$.

Determine whether each number is *prime* or *composite*. (Lesson 5-1)

41. 21
42. 31
43. 65
44. 129

45. **ALGEBRA** Find the slope of the line that passes through the pair of points at the right. (Lesson 4-7)

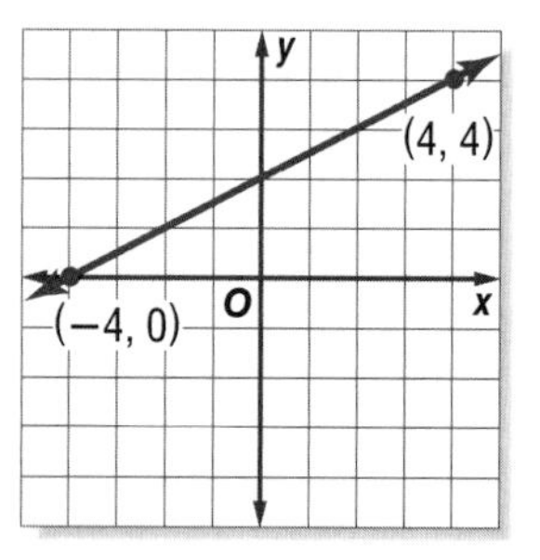

46. **GEOMETRY** Write the ordered pair that is 16 units up from and 11 units to the left of the origin. (Lesson 3-3)

47. **MEASUREMENT** Twenty milliliters is equal to how many liters? (Lesson 1-8)

GETTING READY FOR THE NEXT LESSON

BASIC SKILL **Divide.**

48. $27 \div 3$
49. $48 \div 2$
50. $160 \div 40$
51. $96 \div 6$

5-3 Simplifying Fractions

HANDS-ON Mini Lab

What You'll LEARN

Write fractions in simplest form.

NEW Vocabulary

equivalent fractions
simplest form

Materials

- grid paper
- colored pencils

Work with a partner.

On grid paper, draw the two figures at the right. Both figures should be the same size. Shade 8 out of the 20 squares in one figure. Shade 2 out of the 5 rectangles in the other.

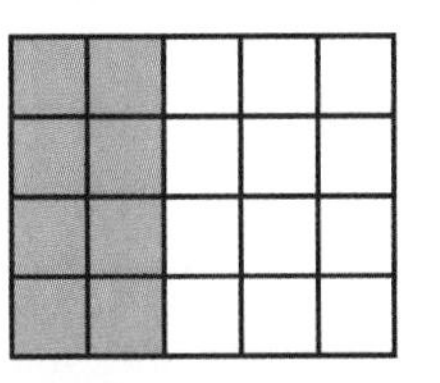

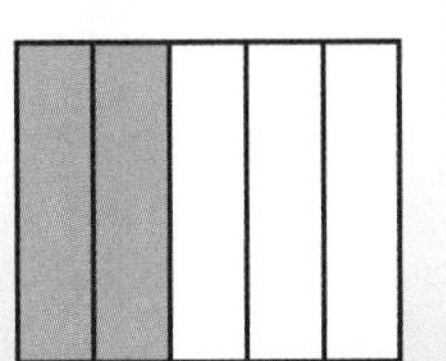

1. Write a fraction to describe each figure: $\frac{\text{number of shaded parts}}{\text{total number of parts}}$.
2. Which figure has a greater portion of its parts shaded?
3. What can you conclude about the fractions you wrote above?

The fractions $\frac{8}{20}$ and $\frac{2}{5}$ are **equivalent fractions** because they have the same value. The fraction $\frac{2}{5}$ is in **simplest form** because the GCF of the numerator and denominator is 1.

EXAMPLES Write Fractions in Simplest Form

Write each fraction in simplest form.

1 $\frac{8}{20}$

First, find the GCF of the numerator and denominator.

factors of 8: 1, 2, 4, 8
factors of 20: 1, 2, 4, 5, 10, 20

The GCF of 8 and 20 is 4.

Then, divide the numerator and the denominator by the GCF.

$\frac{8}{20} = \frac{8 \div 4}{20 \div 4} = \frac{2}{5}$ So, $\frac{8}{20}$ written in simplest form is $\frac{2}{5}$.

Check Multiply the numerator and denominator of the answer by the GCF. The result should be the original fraction.

2 $\frac{18}{30}$

$18 = 2 \cdot 3 \cdot 3$

$30 = 2 \cdot 3 \cdot 5$

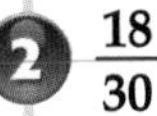

GCF: $2 \cdot 3 = 6$

$\frac{18}{30} = \frac{18 \div 6}{30 \div 6} = \frac{3}{5}$

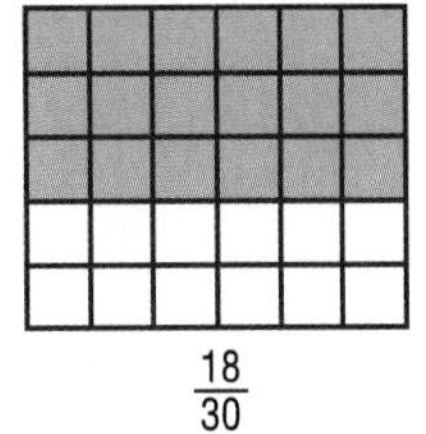

=

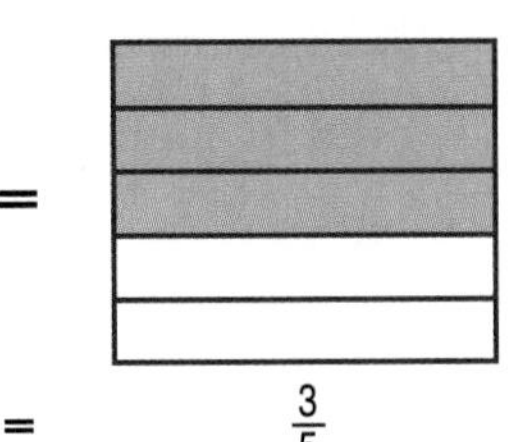

$\frac{18}{30}$ = $\frac{3}{5}$

So, $\frac{18}{30}$ written in simplest form is $\frac{3}{5}$.

STUDY TIP

Simplest Form A fraction is in simplest form when the GCF of the numerator and denominator is 1.

EXAMPLE Use Fractions to Solve a Problem

3 **MUSIC** **If the fraction of the frequencies of two notes can be simplified, the two notes are harmonious. Use the graphic below to find the simplified fraction of the frequency of notes C and E.**

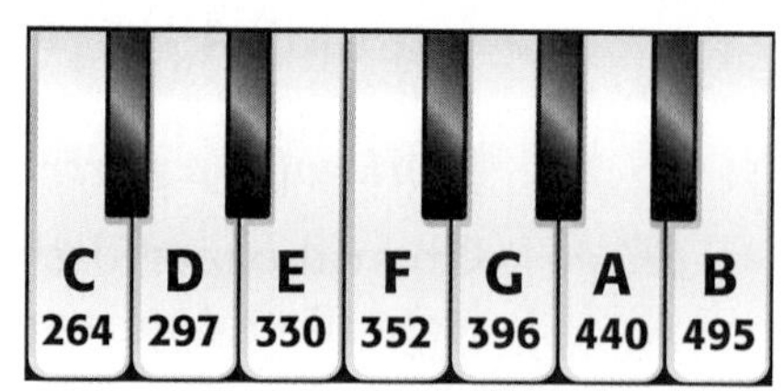

Frequency Chart (Hertz)

$$\frac{\text{frequency of note C}}{\text{frequency of note E}} = \frac{264}{330}$$

$$= \frac{\overset{1}{\cancel{2}} \times 2 \times 2 \times \overset{1}{\cancel{3}} \times \overset{1}{\cancel{11}}}{\underset{1}{\cancel{2}} \times \underset{1}{\cancel{3}} \times 5 \times \underset{1}{\cancel{11}}}$$

$$= \frac{4}{5}$$

The slashes mean that part of the numerator and part of the denominator are both divided by the same number. For example, $3 \div 3 = 1$.

The fraction of the frequency of notes C and E is $\frac{4}{5}$.

Your Turn **Find the simplified fraction of the frequency of each pair of notes.**

a. D and G

b. E and F

REAL-LIFE CAREERS

How Does a Music Composer Use Math? Music composers write notes that can vary in length. There are whole notes, half notes, quarter notes, eighth notes, and sixteenth notes.

Online Research For information about a career as a music composer, visit: msmath2.net/careers

Skill and Concept Check

Writing Math Exercises 1 & 3

1. **Explain** how to determine whether a fraction is in simplest form.

2. **OPEN ENDED** Write a fraction that is *not* in simplest form. Then, simplify it.

3. **FIND THE ERROR** Seki and Luther both wrote $\frac{16}{36}$ in simplest form. Who is correct? Explain.

Seki

$$\frac{16}{36} = \frac{16 \div 4}{36 \div 4} = \frac{4}{9}$$

Luther

$$\frac{16}{36} = \frac{16 \div 2}{36 \div 2} = \frac{8}{18}$$

GUIDED PRACTICE

Write each fraction in simplest form.

4. $\frac{3}{9}$

5. $\frac{18}{30}$

6. $\frac{81}{90}$

7. **AGRICULTURE** George Washington Carver produced about 100 products from sweet potatoes and about 325 from peanuts. Write the number of sweet potato products as a fraction of the number of peanut products in simplest form.

Practice and Applications

HOMEWORK HELP	
For Exercises	See Examples
8–23	1, 2
24–28	3

Extra Practice
See pages 575, 600.

Write each fraction in simplest form.

8. $\frac{9}{12}$ **9.** $\frac{25}{35}$ **10.** $\frac{16}{32}$ **11.** $\frac{14}{20}$

12. $\frac{10}{20}$ **13.** $\frac{12}{21}$ **14.** $\frac{15}{25}$ **15.** $\frac{24}{28}$

16. $\frac{48}{64}$ **17.** $\frac{32}{32}$ **18.** $\frac{50}{300}$ **19.** $\frac{80}{96}$

Write two fractions that are equivalent to each fraction.

20. $\frac{1}{2}$ **21.** $\frac{3}{5}$ **22.** $\frac{6}{7}$ **23.** $\frac{5}{8}$

24. PARKING New York holds the number one spot when it comes to the cost for parking your car. Use the graphic to express the cost of parking a car in Chicago as a fraction of the cost of parking a car in Boston in simplest form.

25. MUSIC Find the simplified fraction of the frequency of notes C and A in Example 3 on page 208.

Write each fraction in simplest form.

26. Fifteen minutes is what part of one hour?

27. Nine inches is what part of one foot?

28. Four days is what part of the month of April?

29. CRITICAL THINKING Both the numerator and denominator of a fraction are even. Is the fraction in simplest form? Explain.

Standardized Test Practice and Mixed Review

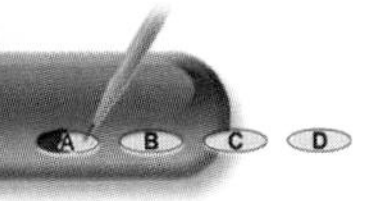

30. MULTIPLE CHOICE The drive to the football game took 56 minutes. Twenty-four minutes of the time was spent stopped in traffic. What fraction of the drive was spent stopped in traffic?

Ⓐ $\frac{3}{7}$ Ⓑ $\frac{6}{14}$ Ⓒ $\frac{15}{35}$ Ⓓ $\frac{1}{3}$

31. SHORT RESPONSE Draw a model that shows $\frac{15}{25} = \frac{3}{5}$.

32. Find the GCF of 345, 253, and 115. (Lesson 5-2)

33. Find the prime factorization of 630. (Lesson 5-1)

GETTING READY FOR THE NEXT LESSON

PREREQUISITE SKILL Divide. (Page 562)

34. $2\overline{)1.0}$ **35.** $4\overline{)1.00}$ **36.** $10\overline{)7.0}$ **37.** $8\overline{)3.000}$

5-4 Fractions and Decimals

WHEN am I ever going to use this?

What You'll LEARN

Write fractions as terminating or repeating decimals and write decimals as fractions.

NEW Vocabulary

terminating decimals
repeating decimals
bar notation

Link to READING

Everyday Meaning of terminate: coming to an end, as in terminate a game

SOFTBALL The United States Women's softball team won the gold medal in the 2000 Summer Olympic Games. The table shows the results of the games they played in Sydney, Australia. The number of runs scored by the United States is listed first.

Opponent	Score
Japan	2–1
Australia	1–0
China	3–0
Italy	6–0
New Zealand	2–0
Australia	1–2
China	0–2
Japan	1–2
Cuba	3–0
Canada	6–0

Source: www.usasoftball.com

1. How many games did the USA softball team win? How many did they play?
2. Write a fraction comparing the number of times the team won to the total number of games played.

The softball team won $\frac{7}{10}$ of the games played. When the denominator is a power of 10, you can use place value to write the fraction as a decimal.

Words	Model	Fraction	Decimal
seven tenths	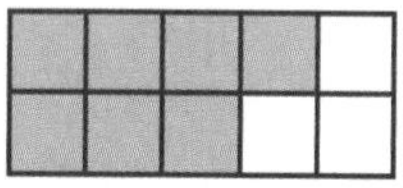	$\frac{7}{10}$	0.7

If the denominator is *not* a power of 10, you can write the fraction as a decimal using division.

EXAMPLES Write Fractions as Decimals

Write each fraction or mixed number as a decimal.

1 $\frac{1}{4}$

The fraction $\frac{1}{4}$ indicates $1 \div 4$.

Method 1 Use paper and pencil.

$$\begin{array}{r} 0.25 \\ 4\overline{)1.00} \\ \underline{8} \\ 20 \\ \underline{20} \\ 0 \end{array}$$

Division ends when the remainder is 0.

Method 2 Use a calculator.

1 [÷] 4 [ENTER] **0.25**

So, $\frac{1}{4} = 0.25$.

2 $5\frac{3}{4}$

The mixed number $5\frac{3}{4}$ is $5 + \frac{3}{4}$.

Method 1 Use paper and pencil.

$5\frac{3}{4} = 5 + \frac{3}{4}$ — Write as a sum.

$= 5 + 0.75$ — Write $\frac{3}{4}$ as 0.75.

$= 5.75$ — Add.

Method 2 Use a calculator.

3 [÷] 4 [+] 5 [ENTER] **5.75**

So, $5\frac{3}{4} = 5.75$.

In Example 1, the division ends, or terminates, when the remainder is zero. So, 0.25 is called a **terminating decimal**.

Repeating decimals have a pattern in the digits that repeats forever.

Consider $\frac{1}{3}$.

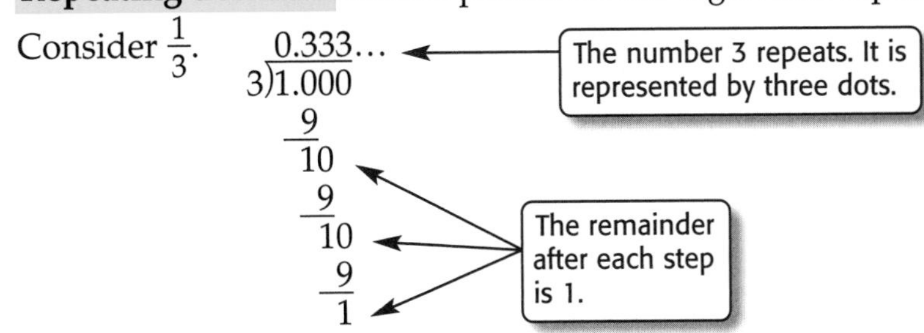

You can use **bar notation** to indicate that a number repeats forever. A bar is written over the digits that repeat.

$0.33333... = 0.\overline{3}$ $\quad 0.121212... = 0.\overline{12}$ $\quad 11.3858585... = 11.3\overline{85}$

EXAMPLES Write Fractions as Repeating Decimals

Write each fraction or mixed number as a decimal.

3 $\frac{7}{9}$

Method 1 Use paper and pencil.

$$\begin{array}{r} 0.77... \\ 9\overline{)7.00} \\ \underline{6\,3} \\ 70 \\ \underline{63} \\ 7 \end{array}$$

Method 2 Use a calculator.

7 [÷] 9 [ENTER] 0.7777...

So, $\frac{7}{9} = 0.\overline{7}$.

4 $8\frac{1}{3}$

Method 1 Use paper and pencil.

$8\frac{1}{3} = 8 + \frac{1}{3}$ Write as a sum.

$= 8 + 0.\overline{3}$ Write $\frac{1}{3}$ as $0.\overline{3}$.

$= 8.\overline{3}$ Add.

Method 2 Use a calculator.

1 [÷] 3 [+] 8 [ENTER] 8.3333...

So, $8\frac{1}{3} = 8.\overline{3}$.

Your Turn **Write each fraction or mixed number as a decimal. Use bar notation if the decimal is a repeating decimal.**

a. $\frac{5}{8}$ b. $\frac{2}{3}$ c. $1\frac{3}{11}$ d. $6\frac{2}{5}$ e. $4\frac{5}{6}$

STUDY TIP

Mental Math It will be helpful to memorize the following fraction-decimal equivalencies.

$\frac{1}{2} = 0.5$

$\frac{1}{3} = 0.\overline{3}$ $\quad \frac{2}{3} = 0.\overline{6}$

$\frac{1}{4} = 0.25$ $\quad \frac{3}{4} = 0.75$

$\frac{1}{5} = 0.2$ $\quad \frac{1}{10} = 0.1$

$\frac{1}{8} = 0.125$

You can use a power of 10 to write a decimal as a fraction. Use the place value of the final digit as the denominator.

EXAMPLE Write Decimals as Fractions

5 **Write 0.48 as a fraction in simplest form.**

$0.48 = \frac{48}{100}$ The 8 is in the hundredths place.

$= \frac{12}{25}$ Simplify.

Your Turn **Write each number as a fraction in simplest form.**

f. 0.56 g. 0.3 h. 2.8

Skill and Concept Check

Writing Math
Exercise 2

1. **OPEN ENDED** Write a fraction that is equivalent to a terminating decimal and one that is equivalent to a repeating decimal.

2. **NUMBER SENSE** Compare 1 out of 3 and 3 out of 10. Are these values equal? Explain your reasoning.

GUIDED PRACTICE

Write each repeating decimal using bar notation.

3. 0.6333...
4. 5.313131...
5. 12.470470470...

Write each fraction or mixed number as a decimal. Use bar notation if the decimal is a repeating decimal.

6. $\frac{2}{5}$
7. $3\frac{5}{8}$
8. $\frac{5}{9}$
9. $1\frac{5}{6}$

Write each decimal as a fraction in simplest form.

10. 0.22
11. 0.1
12. 4.6

13. **SOCCER** Cirilo surveyed his physical education class and found that 12 out of 23 students chose soccer as their favorite sport. Write the fraction of students who chose soccer as a decimal to the nearest thousandth.

Practice and Applications

HOMEWORK HELP

For Exercises	See Examples
20–21, 24–27, 32, 43	1, 2
14–19, 22–23, 28–31, 44	3, 4
33–42	5

Extra Practice
See pages 575, 600.

Write each repeating decimal using bar notation.

14. 0.999...
15. 5.92111...
16. 2.010101...
17. 13.1464646...
18. 26.993993...
19. 30.6841841841...

Write each fraction or mixed number as a decimal. Use bar notation if the decimal is a repeating decimal.

20. $\frac{3}{8}$
21. $\frac{4}{5}$
22. $\frac{4}{9}$
23. $\frac{8}{11}$
24. $\frac{5}{32}$
25. $\frac{3}{20}$
26. $1\frac{7}{20}$
27. $6\frac{17}{40}$
28. $\frac{13}{24}$
29. $\frac{5}{7}$
30. $8\frac{1}{33}$
31. $4\frac{203}{999}$

32. **DOGS** There are an estimated 53 billion dogs in the U.S. The table shows the results of a survey taken recently of dog owners. Write the fraction of people who pet and hug their dogs daily as a decimal.

Activity	Fraction of People
Play with your dog daily	$\frac{23}{25}$
Take your dog on vacation	$\frac{9}{20}$
Pet and hug your dog daily	$\frac{19}{20}$
Celebrate your dog's birthday	$\frac{43}{100}$

Source: Ralston Purina

Write each decimal as a fraction in simplest form.

33. 0.75
34. 0.34
35. 0.2
36. 0.9
37. 5.96
38. 2.66

BASKETBALL **Use the graph to write the number of rebounds and blocks per game for each player as a mixed number in simplest form.**

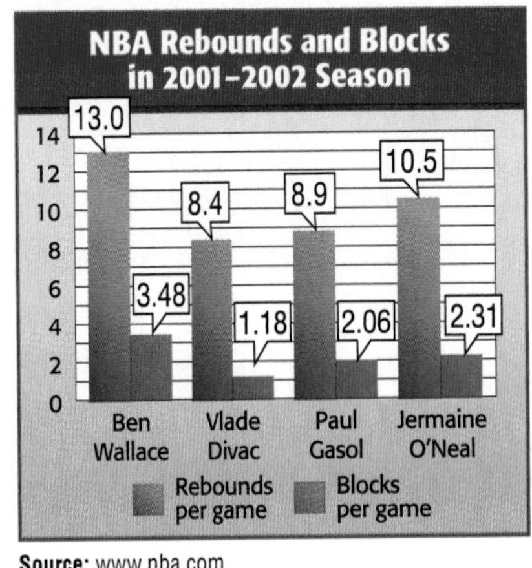

Source: www.nba.com

39. Wallace
40. Divac
41. Gasol
42. O'Neal

43. **MULTI STEP** Kelsey practiced playing the cello for 2 hours and 18 minutes. Write the time Kelsey spent practicing in hours as a decimal.

MATH HISTORY **For Exercises 44 and 45, use the following information.**

The value of pi (π) is 3.1415927… . Pi is a nonrepeating, nonterminating decimal. Mathematicians have used many methods to find the value of π.

44. Archimedes believed that π was between $3\frac{1}{7}$ and $3\frac{10}{71}$. Write each fraction as a decimal rounded to the nearest thousandth. Was Archimedes correct?
45. The Rhind Papyrus states that the Egyptians used $\frac{256}{81}$ for π. Write the fraction as a decimal rounded to the nearest thousandth. Which is closer to the actual value of π, Archimedes' value or the Egyptians' value?

46. **CRITICAL THINKING** Fractions with denominators of 2, 4, 8, 16, and 32 produce terminating decimals. Fractions with denominators of 6, 12, 18, and 24 produce repeating decimals. What do you think causes the difference? Explain.

Standardized Test Practice and Mixed Review

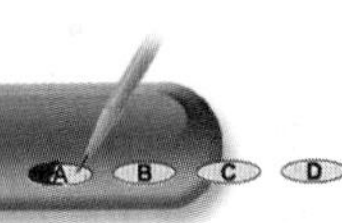

47. **MULTIPLE CHOICE** Express $\frac{9}{25}$ as a decimal.

Ⓐ 0.036 Ⓑ 0.36 Ⓒ $0.\overline{36}$ Ⓓ 3.6

48. **SHORT REPONSE** In 2001, Ichiro Suzuki had 242 hits in 692 at-bats. The batting average of a baseball player is the number of hits divided by the number of at-bats. Find Suzuki's batting average to the nearest thousandth.

Write each fraction in simplest form. (Lesson 5-3)

49. $\frac{10}{24}$
50. $\frac{39}{81}$
51. $\frac{28}{98}$
52. $\frac{51}{68}$

53. Find the GCF of 36 and 48. (Lesson 5-2)

GETTING READY FOR THE NEXT LESSON

BASIC SKILL **Write a fraction for the number of squares shaded to the total number of squares.**

54. 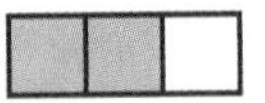

55. 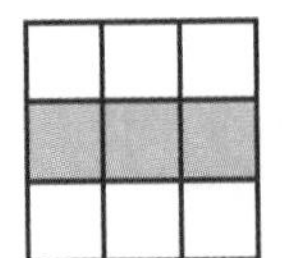

56. 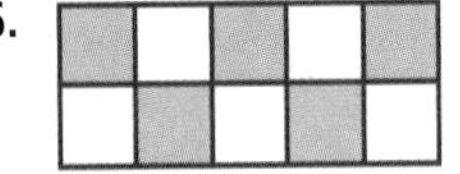

57. 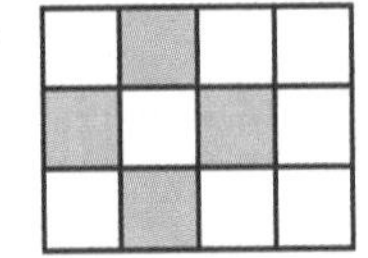

Mid-Chapter Practice Test

Vocabulary and Concepts

1. **Define** *prime number* and give an example. (Lesson 5-1)
2. **Describe** how you know when a fraction is in simplest form. (Lesson 5-3)

Skills and Applications

3. Determine whether 24 is *prime* or *composite*. (Lesson 5-1)

Factor each expression. (Lesson 5-1)

4. 30
5. 120
6. $14x^2y$
7. $50mn$

Find the GCF of each set of numbers or algebraic expressions. (Lesson 5-2)

8. 16, 40
9. 65, 100
10. $12x$, $20x^3$
11. $45a^2b^2$, $81ab^3$

12. **CYCLING** The *gear ratio* of a bicycle is the comparison of the number of teeth on a chainwheel to the number on a freewheel. If the gear ratio for Alexis' 10-speed bike is $\frac{52}{16}$, write this fraction in simplest form. (Lesson 5-3)

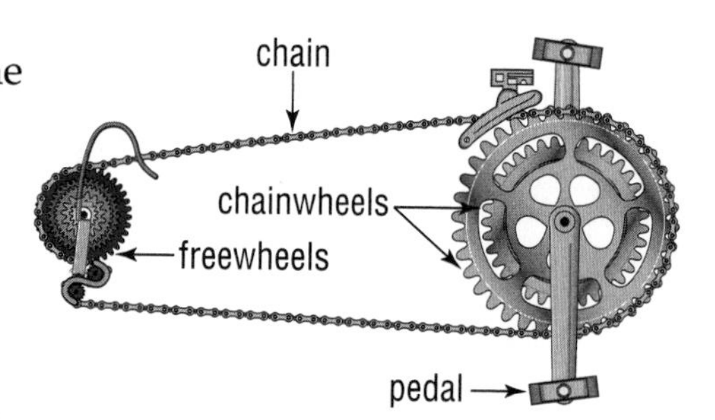

Write each fraction or mixed number as a decimal.
Use bar notation if the decimal is a repeating decimal. (Lesson 5-4)

13. $\frac{7}{8}$
14. $\frac{2}{9}$
15. $3\frac{13}{20}$

Write each decimal as a fraction in simplest form. (Lesson 5-4)

16. 0.6
17. 0.48
18. 7.02

Standardized Test Practice

19. **MULTIPLE CHOICE** What is the greatest number of crayons in each row of an 8-, a 64-, and a 96-crayon box if all rows have the same number? (Lesson 5-2)

Ⓐ 2 Ⓑ 4 Ⓒ 8 Ⓓ 12

20. **GRID IN** Amy spent 35 minutes in the afternoon practicing the piano. What part of one hour did she spend practicing? Write as a fraction in simplest form. (Lesson 5-3)

The Game Zone

A Place To Practice Your Math Skills

Math Skill

Finding Factors

GET READY!

Players: two teams
Materials: 36 index cards, tape

- Number the index cards 1 through 36.
- Tape the index cards in order on the chalkboard.
- Divide the class into two teams.

GO!

- Team A chooses one of the index cards, such as the 8 card, and takes it off the chalkboard. Team A gets 8 points. Team B gets all the cards that are factors of 8 that have not yet been taken. The sum of the factors are the points they receive. For example, Team B would receive 1 + 2 + 4, or 7 points.

Team A	Team B
8	1 2 + 4 7

- Team B then chooses a card and gets those points.
- A team loses a turn if it selects an illegal number. A number is considered illegal if it does not have at least one factor available.
- Teams continue to take turns until there are no legal plays remaining.
- **Who Wins?** The team with the most points is the winner.

5-5 Fractions and Percents

What You'll LEARN

Write fractions as percents and percents as fractions.

NEW Vocabulary

ratio
percent

REVIEW Vocabulary

simplest form: a fraction whose numerator and denominator have a GCF of 1 **(Lesson 5-3)**

WHEN am I ever going to use this?

EDUCATION The table shows the results of a survey in which students were asked to choose methods that make learning new subjects more interesting.

1. Shade a 10×10 grid that represents the number of students that chose each method.
2. What fraction of the students chose the Internet as the method that makes learning more interesting?

Method	Number of Students
Internet	34 out of 100
Teacher	29 out of 100
TV Program	24 out of 100
Textbook	12 out of 100

Source: Opinion Research Corporation

A **ratio** is a comparison of two numbers by division. Ratios, like the ones above, can be written in several different ways.

34 out of 100   34:100 $\frac{34}{100}$

When a ratio compares a number to 100, it can be written as a **percent**.

Key Concept — Percent

Words A percent is a ratio that compares a number to 100.

Symbols $\frac{n}{100} = n\%$

Ratio 34 to 100

Model

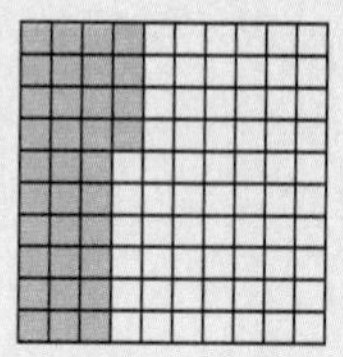

EXAMPLES Write Ratios as Percents

Write each ratio as a percent.

1 Annie answered 90 out of 100 questions correctly.

You can represent 90 out of 100 with a model.

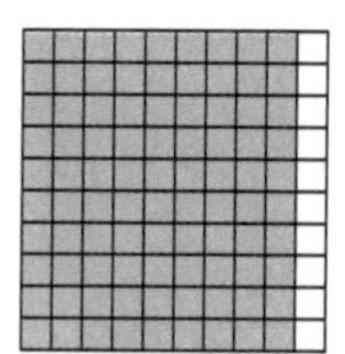

$\frac{90}{100} = 90\%$

2 On average, 50.5 out of 100 students own a pet.

$\frac{50.5}{100} = 50.5\%$

Your Turn Write each ratio as a percent.

a. $\frac{45}{100}$ b. 19.2 out of 100   c. \$3.30:\$100

READING Math

Percent *Percent* means *per hundred* or *hundredths*. The symbol % means percent.

Fractions and percents are ratios that can represent the same number. You can write a fraction as a percent by finding an equivalent fraction with a denominator of 100.

EXAMPLE Write a Fraction as a Percent

3 Write $\frac{3}{20}$ as a percent.

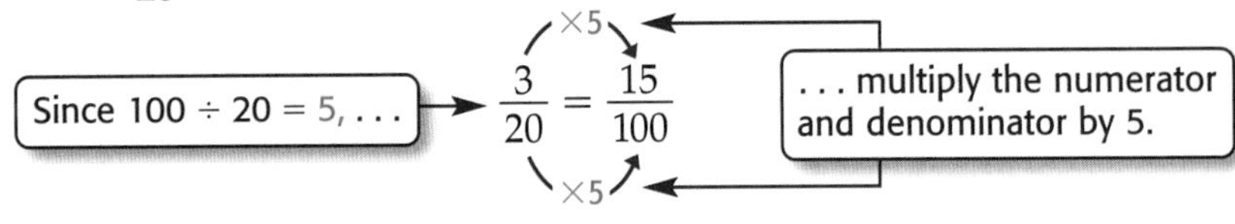

$\frac{15}{100} = 15\%$

So, $\frac{3}{20} = 15\%$.

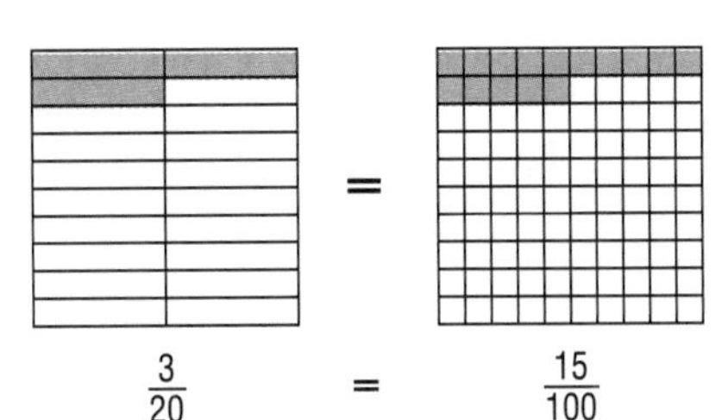

Your Turn **Write each fraction as a percent.**

d. $\frac{17}{20}$

e. $\frac{3}{5}$

f. $\frac{2}{25}$

You can also use the meaning of percent to write percents as fractions.

EXAMPLE Write a Percent as a Fraction

4 Write 48% as a fraction in simplest form.

$48\% = \frac{48}{100}$ Definition of percent

$= \frac{12}{25}$ Simplify.

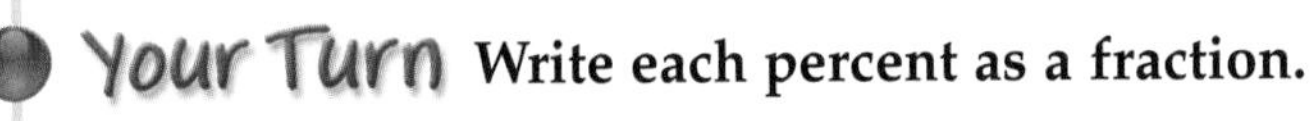

Your Turn **Write each percent as a fraction.**

g. 40%

h. 6%

i. 24%

REAL-LIFE MATH

COINS Each new quarter has a diameter of 24.26 millimeters, a thickness of 1.75 millimeters, and weighs 5.670 grams.

Source: www.usmint.gov

EXAMPLE Use Percent to Solve a Problem

5 COINS The 50 State Quarters® Program allows the United States Mint to release a new quarter every ten weeks from 1999 through 2008 commemorating the 50 states. By the end of 2004, 30 state coins will have been released. What percent of the coins is this?

By the end of 2004, 30 out of 50 state coins will have been released.

$\frac{30}{50} = \frac{60}{100}$ Write an equivalent fraction with a denominator of 100.

$= 60\%$ $\frac{60}{100} = 60\%$

So, 60% of the state coins will have been released by the end of 2004.

Skill and Concept Check

1. **Write** a fraction in simplest form and a percent to represent the shaded area at the right.

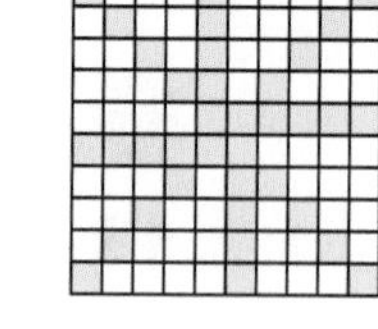

2. **OPEN ENDED** Draw a model that shows 25%.

3. **Which One Doesn't Belong?** Identify the number that does not have the same value as the other three. Explain your reasoning.

$\frac{10}{100}$	1%	$\frac{1}{10}$	100 out of 1,000

GUIDED PRACTICE

Write each ratio as a percent.

4. $\frac{57}{100}$
5. \$29.20 per \$100
6. 11 teams out of 100

Write each fraction as a percent.

7. $\frac{1}{4}$
8. $\frac{6}{10}$
9. $\frac{4}{5}$
10. $\frac{17}{20}$

Write each percent as a fraction in simplest form.

11. 90%
12. 80%
13. 75%
14. 22%

15. **ELECTRICITY** Electrical power is measured in a unit called the watt. Write a percent that compares the amount of power given off by a 75-watt bulb to the power given off by a 100-watt bulb.

Practice and Applications

HOMEWORK HELP

For Exercises	See Examples
16–23	1, 2
24–35	3
39–50	4
36–38	5

Extra Practice
See pages 575, 600.

Write each ratio as a percent.

16. $\frac{87}{100}$
17. $\frac{12.2}{100}$
18. 42 per 100
19. $11\frac{2}{7}$ out of 100
20. $66\frac{2}{3}$:100
21. 99.9:100
22. Fifty-one out of every 100 households have at least one computer.
23. Twenty out of every 100 adults use the Internet to check on news, weather, and sports.

Write each fraction as a percent.

24. $\frac{7}{10}$
25. $\frac{16}{20}$
26. $\frac{15}{25}$
27. $\frac{37}{50}$
28. $\frac{73}{100}$
29. $\frac{13}{50}$
30. $\frac{1}{5}$
31. $\frac{19}{20}$
32. $\frac{3}{5}$
33. $\frac{3}{25}$
34. $\frac{10}{10}$
35. $\frac{19}{25}$

36. **SPECIAL OLYMPICS** Mila and Randi are volunteers for their school district's Special Olympics competition. They are expecting 100 participants, but only 67 have registered so far. What percent of the Special Olympics participants still need to register before the competition begins?

EXAMPLES Write Percents as Decimals

3 Write 23% as a decimal.

$23\% = 23$ Divide by 100.

$= 0.23$ Remove the %.

So, 23% = 0.23.

4 Write $3\frac{1}{4}\%$ as a decimal.

$3\frac{1}{4}\% = 3.25\%$ Write $\frac{1}{4}$ as 0.25.

$= 03.25$ Divide by 100.

$= 0.0325$ Remove the %.

So, $3\frac{1}{4}\% = 0.0325$.

Your Turn Write each percent as a decimal.

a. 76% b. 8.5% c. $92\frac{1}{2}\%$

REAL-LIFE MATH

GEOGRAPHY A large Antarctic iceberg can weigh 400 million tons, rise 10 stories above the surface of the water, and supply water for 3 million people for a year.

Source: www.chennaionline.com

You can also use fractions to help you write decimals as percents.

EXAMPLES Write Decimals as Percents

5 GEOGRAPHY Nearly 0.02 of the world's fresh water comes from the Antarctic Icecap. Write 0.02 as a percent.

$0.02 = \frac{2}{100}$ Definition of decimal

$= 2\%$ Definition of percent

So, 0.02 = 2%.

6 Write 0.347 as a percent.

$0.347 = \frac{347}{1,000}$ Definition of decimal

$= \frac{34.7}{100}$ Divide both numerator and denominator by 10.

$= 34.7\%$ Definition of percent

So, 0.347 = 34.7%.

Study the pattern in the decimals and the equivalent percents in Examples 5 and 6. Notice that you can write the decimal as a percent by multiplying the number by 100 and adding the percent symbol.

Key Concept **Writing Decimals as Percents**

Words To write a decimal as a percent, multiply the percent by 100 and add a percent symbol.

Symbols 0.58 = 0.58 = 58%

STUDY TIP

Mental Math

- To write a percent as a decimal, move the decimal point two places to the left.
- To write a decimal as a percent, move the decimal point two places to the right.

EXAMPLES Write Decimals as Percents

7 Write 0.64 as a percent.

$0.64 = 0.64$ Multiply by 100.

$= 64\%$ Add the %.

So, 0.64 = 64%.

8 Write 0.875 as a percent.

$0.875 = 0.875$ Multiply by 100.

$= 87.5\%$ Add the %.

So, 0.875 = 87.5%.

Skill and Concept Check

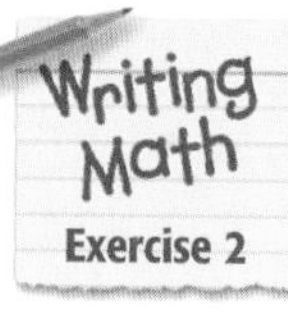

1. **OPEN ENDED** Choose a decimal. Then write it as a fraction in simplest form and as a percent.

2. **FIND THE ERROR** Jessica and Gregorio both wrote 0.881 as a percent. Who is correct? Explain.

Jessica	Gregorio
0.881 = 0.881	0.881 = 000.881
= 88.1%	= 0.00881%

GUIDED PRACTICE

Write each percent as a decimal.

3. 68% 4. 3% 5. 27.6% 6. $45\frac{1}{2}\%$

Write each decimal as a percent.

7. 0.56 8. 0.12 9. 0.08 10. 0.399

11. **GEOGRAPHY** Pennsylvania makes up 1.2% of the landmass of the United States. Write this percent as a decimal.

Practice and Applications

Write each percent as a decimal.

12. 27% 13. 70% 14. 6% 15. 1%

16. 18.5% 17. 2.2% 18. $15\frac{1}{2}\%$ 19. $30\frac{1}{4}\%$

HOMEWORK HELP	
For Exercises	See Examples
12–19, 34–37	1, 2, 3, 4
20–33	5, 6, 7, 8

Extra Practice
See pages 576, 600.

Write each decimal as a percent.

20. 0.95 21. 0.08

22. 0.17 23. 0.6

24. 0.675 25. 0.145

26. 0.012 27. 0.7025

TENNIS For Exercises 28–32, use the graphic at the right. Write each decimal as a percent for the following players.

28. Monica Seles
29. Martina Hingis
30. Venus Williams
31. Jennifer Capriati
32. Based on her winning percentage, which player wins 413 out of every 500 games played?

USA TODAY

USA TODAY Snapshots®

Seles soars at Australian Open

Monica Seles has a record at the Australian Open, which begins today, that any player would envy. Active female players with the best match winning percentage at the Australian open:

Source: International Tennis Federation

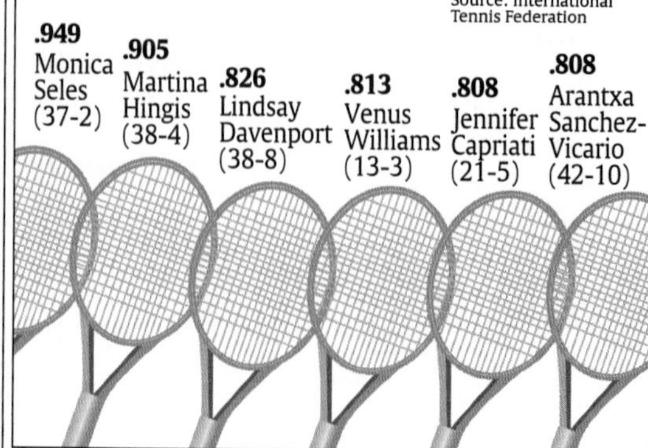

By Ellen J. Horrow and Quin Tian, USA TODAY

33. **BIRDS** Birds shown in a field guide are 0.10 to 0.67 their actual size. What percent of their actual size are they?

MODEL TRAINS For Exercises 34–37, use the table at the right.

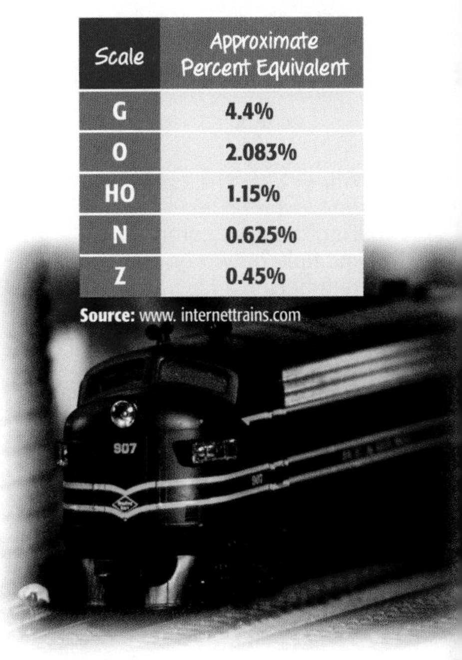

Scale	Approximate Percent Equivalent
G	4.4%
O	2.083%
HO	1.15%
N	0.625%
Z	0.45%

Source: www. internettrains.com

34. The five scales at the right can be used to build a model train based on the life-size original. For example, a G-scale train is about 4% of the size of the original. So, the model of a 75-foot-long locomotive measures 40 inches. Find the decimal equivalents for each scale.
35. Which scale is the smallest?
36. Find the approximate length of a 75-foot train in the N scale.
37. Name the scale used when a model of a 75-foot train measures approximately 18.75 inches.

CRITICAL THINKING For Exercises 38–43, choose the greater number in each pair.

38. $1\frac{1}{4}\%$, 0.125
39. 0.76, 76.5%
40. 42%, 4.2
41. 99%, 0.985
42. $\frac{13}{40}$, 0.30
43. 0.56, $\frac{45}{80}$

Standardized Test Practice and Mixed Review

44. **MULTIPLE CHOICE** Which is 0.08 written as a percent?

Ⓐ 0.08% Ⓑ 0.8% Ⓒ 8% Ⓓ 80%

45. **MULTIPLE CHOICE** Which is 96.3% written as a decimal?

Ⓕ 0.0963 Ⓖ 0.963 Ⓗ 9.63 Ⓘ 963

46. **GRID IN** Write $78\frac{1}{2}\%$ as a decimal.

Write each ratio as a percent. (Lesson 5-5)

47. $\frac{72}{100}$
48. 90.9:100
49. 18 per 100
50. 23.1 out of 100

51. Write $9\frac{3}{8}$ as a decimal. (Lesson 5-4)

52. **BALLOONS** Write an integer to represent a hot-air balloon descending 83 feet. (Lesson 3-1)

TESTS For Exercises 53 and 54, use the stem-and-leaf plot of Chapter 5 test scores at the right.

Stem	Leaf
10	0
9	2 3 3 3 4 5 9
8	0 1 1 2 3 5 9 9
7	0 4 6 8 8 9
6	4

8|1 = 81 points

53. How many students took the test? (Lesson 2-5)
54. What was the median score? (Lesson 2-4)

GETTING READY FOR THE NEXT LESSON

PREREQUISITE SKILL Use exponents to write the prime factorization of each number. (Lesson 5-1)

55. 50
56. 32
57. 76
58. 105

5-7 Least Common Multiple

What You'll LEARN

Find the least common multiple of two or more numbers.

NEW Vocabulary

multiple
least common multiple (LCM)

HANDS-ON Mini Lab

Materials

- centimeter cubes

Work with a partner.

Use 9 centimeter cubes to model the first floor of Building 1 and 12 centimeter cubes to model the first floor of Building 2, as shown.

Building 1

Building 2

1. Add a second floor to each building. Record the total number of cubes used in a table like the one shown below.

Number of Floors	1	2	3	4	5
Number of Cubes in Building 1	9	?	?	?	?
Number of Cubes in Building 2	12	?	?	?	?

2. Continue adding floors until each building has five floors. Record your results.
3. Describe two buildings that have the same number of cubes.
4. If you keep adding floors, will the two buildings have the same number of cubes again? Explain.

In the Mini Lab, you listed multiples of 9 and 12. A **multiple** is the product of a number and any whole number. The **least common multiple**, or **(LCM)**, of two or more numbers is the least of their common multiples, excluding zero. So, the LCM of 9 and 12 is 36.

You can use several strategies to find the LCM of two or more numbers.

EXAMPLE Find the LCM by Listing Multiples

1 Find the LCM of 6 and 10.

First, list the multiples of 6 and 10.

multiples of 6: 6, 12, 18, 24, 30, 36, 42, 48, 54, 60, …

multiples of 10: 10, 20, 30, 40, 50, 60, 70, 80, …

Notice that 30, 60, …, are common multiples. So, the LCM of 6 and 10 is 30.

Your Turn **Find the LCM of each pair of numbers.**

a. 3, 12 b. 10, 12 c. 4, 5

You can also use prime factorization to find the LCM. The LCM is the smallest product that contains the prime factors of each number.

EXAMPLES Find the LCM Using Prime Factors

2 **Find the LCM of 24 and 36.**

Method 1 Write the prime factorization.

$24 = 2 \times 2 \times 2 \times 3 = 2^3 \times 3$

$36 = 2 \times 2 \times 3 \times 3 = 2^2 \times 3^2$

The prime factors of 24 and 36 are 2 and 3. Multiply the greatest power of both 2 and 3.

Method 2 Divide by prime numbers.

```
    2     3
3)6      9
2)12    18
2)24    36  <—— Start here.
```

Start dividing by prime factors until both numbers cannot be divided by the same divisor. Then multiply the divisors and quotients to find the LCM.

The LCM of 24 and 36 is $2^3 \times 3^2$, or 72.

STUDY TIP

Choosing a Method To find the LCM of two or more numbers, it is easier to:
- list factors for small numbers, and
- use prime factorization or division for large numbers.

3 **Find the LCM of 4, 10, and 15.**

$4 = 2^2$

$10 = 2 \times 5$

$15 = 3 \times 5$

LCM: $2^2 \times 3 \times 5 = 60$

So, the LCM of 4, 10, and 15 is 60.

Your Turn **Find the LCM of each set of numbers.**

d. 9, 15 **e.** 10, 20, 30 **f.** 6, 17, 34

Skill and Concept Check

1. **Describe** the relationship between 12, 4, and 3 using the words *multiple* and *factor*.

2. **OPEN ENDED** Find three numbers that each have a multiple of 30.

3. **NUMBER SENSE** Determine whether the LCM for all pairs of odd numbers is *sometimes*, *always*, or *never* their product. Explain.

GUIDED PRACTICE

Find the LCM of each set of numbers.

4. 4, 10 5. 6, 7 6. 3, 5, 12

7. **PICNIC** Flavio is having a cookout for his class and serving hot dogs. Hot dogs come in packages of 10. Hot dog buns come in packages of 8. If Flavio wants to have the same number of hot dogs and buns, what is the least number of each that he will have to buy?

Practice and Applications

HOMEWORK HELP

For Exercises	See Examples
8–13, 20, 24	1, 2
14–19, 22–23	3

Extra Practice
See pages 576, 600.

Find the LCM of each set of numbers.

8. 6, 8 **9.** 12, 16 **10.** 30, 45 **11.** 8, 18

12. 11, 12 **13.** 45, 63 **14.** 2, 3, 5 **15.** 6, 8, 9

16. 8, 12, 16 **17.** 12, 15, 28 **18.** 22, 33, 44 **19.** 12, 16, 36

20. Find the LCM of 2×3^2 and $2^3 \times 3$.

21. PATTERNS Write a rule that describes the common multiples of 9 and 15 that are greater than 180.

PLANETS For Exercises 22 and 23, use the following information and the table at the right.

The nine planets in our solar system revolve around the Sun at different orbital speeds.

Planet	Orbital Speed (mi/s)
Earth	18
Mercury	30
Pluto	3

Source: infoplease.com

22. When will the number of miles revolved by Earth, Mercury, and Pluto be the same?

23. In how many seconds does Earth revolve the number of miles in Exercise 22? Mercury? Pluto?

24. NUMBER THEORY The LCM of two consecutive numbers is greater than 200 and is a multiple of 7. Name the numbers.

25. CRITICAL THINKING Two numbers have an LCM of $2^2 \times 3 \times 5^2$. Their GCF is 2×5. If one of the numbers is $2 \times 3 \times 5$, what is the other number?

Standardized Test Practice and Mixed Review

26. MULTIPLE CHOICE Find the LCM of 27 and 30.

Ⓐ 810 Ⓑ 300 Ⓒ 270

Ⓓ 3

27. SHORT RESPONSE Presidents are elected every four years. Senators are elected every six years. If a senator was elected in the presidential election year 2000, in what year will she campaign again during a presidential election year?

28. Write 55.7% as a decimal to the nearest hundredth. (Lesson 5-6)

DIAMONDS The graph shows the percent of diamond engagement rings that are different shapes. Write each percent as a fraction in simplest form. (Lesson 5-5)

29. round **30.** marquise **31.** others

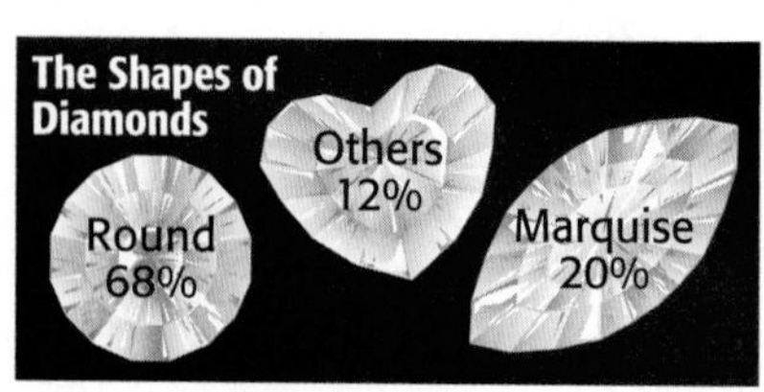

Source: Diamond Information Center

GETTING READY FOR THE NEXT LESSON

PREREQUISITE SKILL **Replace each ● with < or > to make a true sentence.** (Page 556)

32. 6.85 ● 5.68 **33.** 2.34 ● 2.43 **34.** 6.9 ● 5.99 **35.** 1.001 ● 1.1

5-8 Comparing and Ordering Rational Numbers

What You'll LEARN

Compare and order fractions, decimals, and percents.

NEW Vocabulary

common denominator
least common denominator (LCD)
rational numbers

WHEN am I ever going to use this?

SOFTBALL The table shows how Latanya and Patrick did during the first month of the softball season.

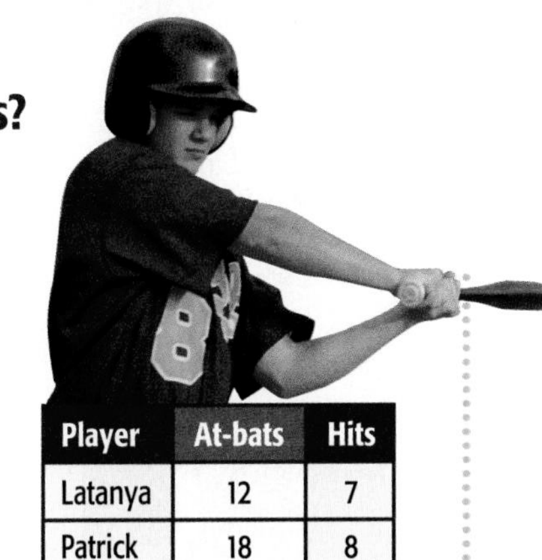

Player	At-bats	Hits
Latanya	12	7
Patrick	18	8

1. A batting average is the ratio of hits to at-bats. Write each player's batting average as a fraction.
2. Estimate which fraction is greater than $\frac{1}{2}$. Which is less than $\frac{1}{2}$?
3. Which player has the better batting average?

To compare and order fractions like the ones above, you can rewrite each fraction with a common denominator and then compare the numerators.

A **common denominator** is a common multiple of the denominators of two or more fractions. The **least common denominator (LCD)** is the LCM of the denominators.

EXAMPLE Compare Fractions

1 Who has the better batting average, Latanya or Patrick?

Method 1 Rename using the LCD. Then compare numerators.

Latanya: $\frac{7}{12} = \frac{7 \times 3}{12 \times 3} = \frac{21}{36}$

Patrick: $\frac{8}{18} = \frac{8 \times 2}{18 \times 2} = \frac{16}{36}$

The LCD is 36.

Since $\frac{21}{36} > \frac{16}{36}$, then $\frac{7}{12} > \frac{8}{18}$. Latanya has the better average.

Method 2 Write each fraction as a decimal. Then compare decimals.

Latanya: $\frac{7}{12} = 0.58\overline{3}$

Patrick: $\frac{8}{18} = 0.\overline{4}$

Since $0.58\overline{3} > 0.\overline{4}$, then $\frac{7}{12} > \frac{8}{18}$. Latanya has the better average.

Your Turn **Replace each ● with <, >, or = to make a true sentence.**

a. $\frac{5}{6}$ ● $\frac{7}{9}$

b. $\frac{1}{5}$ ● $\frac{7}{50}$

c. $\frac{9}{16}$ ● $\frac{7}{10}$

EXAMPLE Compare Ratios

REAL-LIFE MATH

MOUNTAIN BIKING Approximately 6.3 million people participated in off-road mountain biking in 2001.

Source: National Sporting Goods Association

2 **MOUNTAIN BIKING In Coach Ito's first period class, 19 out of 32 students owned a mountain bike. In his seventh period class, 16 out of 28 students owned a mountain bike. Did a greater fraction of students own a mountain bike in the first period class or the seventh period class?**

Estimate Both fractions are slightly greater than $\frac{1}{2}$.

Since the denominators are large, write $\frac{19}{32}$ and $\frac{16}{28}$ as decimals and then compare.

$19 \div 32 \approx 0.5938$ $16 \div 28 \approx 0.5714$ Use a calculator.

Since $0.5938 > 0.5714$, then $\frac{19}{32} > \frac{16}{28}$. So, a greater fraction of students in the first period class owned a mountain bike.

Some fractions, decimals, and percents are used more frequently than others. So, it is a good idea to be familiar with these common fraction-decimal-percent equivalents.

Concept Summary — **Fraction-Decimal-Percent Equivalents**

$\frac{1}{5} = 0.2 = 20\%$	$\frac{1}{10} = 0.1 = 10\%$	$\frac{1}{4} = 0.25 = 25\%$
$\frac{2}{5} = 0.4 = 40\%$	$\frac{3}{10} = 0.3 = 30\%$	$\frac{1}{2} = 0.5 = 50\%$
$\frac{3}{5} = 0.6 = 60\%$	$\frac{7}{10} = 0.7 = 70\%$	$\frac{3}{4} = 0.75 = 75\%$
$\frac{4}{5} = 0.8 = 80\%$	$\frac{9}{10} = 0.9 = 90\%$	

EXAMPLE Order Ratios

3 **Order 0.6, 48%, and $\frac{1}{2}$ from least to greatest.**

Use estimation to find that 48% is the least. Then, compare $\frac{1}{2}$ and 0.6 using decimals.

$\frac{1}{2} = 0.5$ and $0.5 < 0.6$ So, $48\% < \frac{1}{2} < 0.6$.

Check You can change 48% and 0.6 to fractions, then compare all three fractions using the LCD.

$0.6 = \frac{60}{100}$ $48\% = \frac{48}{100}$ $\frac{1}{2} = \frac{50}{100}$

Since $\frac{48}{100} < \frac{50}{100} < \frac{60}{100}$, $48\% < \frac{1}{2} < 0.6$.

Your Turn **Order each set of ratios from least to greatest.**

d. $22\%, 0.3, \frac{2}{10}$ e. $\frac{1}{5}, 2\%, 0.18$ f. $0.74, \frac{3}{4}, 70\%$

Fractions, terminating and repeating decimals, and integers are called **rational numbers** because they can be written as fractions.

Key Concept — Rational Numbers

Words Rational numbers are numbers that can be written as fractions.

Symbols $\frac{a}{b}$, where a and b are integers and $b \neq 0$

Model

Rational Numbers: 0.8, $\frac{1}{2}$, $5\frac{2}{3}$, $5.\overline{2}$, −1.4444...

Integers: −6, −10

Whole Numbers: 8, 13, 1

EXAMPLE Identify Rational Numbers

Standardized Test Practice

4 MULTIPLE-CHOICE TEST ITEM

Find the number that is *not* rational.

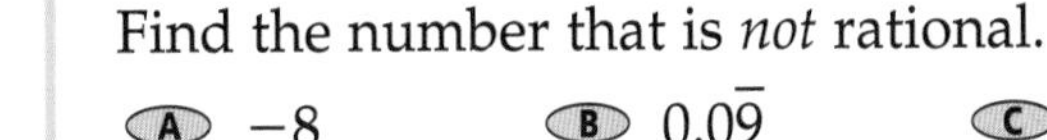
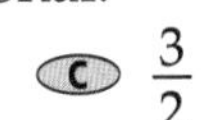

(A) -8 (B) $0.0\overline{9}$ (C) $\frac{3}{2}$ (D) 2.010010001…

Read the Test Item To find the number that is not rational, identify three rational numbers.

Solve the Test Item -8, $0.0\overline{9}$, and $\frac{3}{2}$ can be expressed as fractions. So, 2.010010001… is not rational, and the answer is D.

Test-Taking Tip (The Princeton Review)

Eliminate Possibilities Eliminate the possibilities that you know are incorrect. Then consider the choices that are left.

Skill and Concept Check

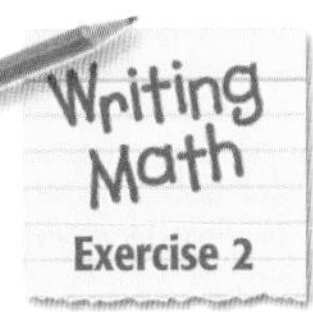

1. **OPEN ENDED** Write two fractions whose LCD is 30.
2. **Which One Doesn't Belong?** Identify the ratio that does not have the same value as the other three. Explain your reasoning.

9 out of 15	0.06	60%	$\frac{3}{5}$

GUIDED PRACTICE

Find the LCD for each pair of fractions.

3. $\frac{4}{5}, \frac{3}{9}$
4. $\frac{3}{8}, \frac{11}{32}$

Replace each ● with <, >, or = to make a true sentence.

5. $\frac{6}{15}$ ● $\frac{3}{8}$
6. $\frac{1}{6}$ ● $\frac{3}{18}$
7. $\frac{5}{4}$ ● $\frac{8}{7}$
8. $\frac{1}{8}$ ● 0.2

Order each set of ratios from least to greatest.

9. 60%, $\frac{31}{50}$, 0.59
10. $\frac{3}{20}$, 0.02, 16%

Determine whether each number is rational. Write *yes* or *no*.

11. $\frac{1}{11}$
12. 1.121121112…

Practice and Applications

HOMEWORK HELP	
For Exercises	See Examples
13–30, 36, 51–53	1
31–35, 43, 50, 54–55	2
37–42	3
44–49	4

Extra Practice
See pages 576, 600.

Find the LCD for each pair of fractions.

13. $\frac{1}{4}, \frac{3}{10}$
14. $\frac{5}{6}, \frac{7}{15}$
15. $\frac{9}{14}, \frac{2}{5}$
16. $\frac{1}{6}, \frac{3}{8}$
17. $\frac{7}{12}, \frac{13}{36}$
18. $\frac{13}{17}, \frac{3}{4}$

Replace each ● with <, >, or = to make a true sentence.

19. $\frac{7}{10}$ ● $\frac{2}{3}$
20. $\frac{5}{14}$ ● $\frac{3}{7}$
21. $\frac{4}{7}$ ● $\frac{5}{8}$
22. $\frac{2}{3}$ ● $\frac{10}{15}$
23. $\frac{16}{20}$ ● $\frac{40}{50}$
24. $\frac{11}{15}$ ● $\frac{5}{9}$
25. $\frac{9}{13}$ ● $\frac{14}{20}$
26. $\frac{17}{24}$ ● $\frac{11}{12}$
27. $\frac{7}{5}$ ● $\frac{14}{11}$
28. $2\frac{3}{4}$ ● $2\frac{2}{3}$
29. $\frac{11}{16}$ ● $\frac{7}{8}$
30. $\frac{3}{2}$ ● $1\frac{1}{2}$
31. 0.82 ● $\frac{5}{6}$
32. $\frac{9}{20}$ ● 0.45
33. $\frac{3}{5}$ ● 59%
34. 40% ● $\frac{11}{25}$

35. Which average is better for a soccer goalie: 3 saves out of 4 or 7 saves out of 11? Explain.

36. Order the fractions from greatest to least: $\frac{7}{8}, \frac{4}{5}, \frac{22}{25}$.

Order each set of ratios from least to greatest.

37. 0.23, 19%, $\frac{1}{5}$
38. $\frac{8}{10}$, 81%, 0.805
39. 0.615, $\frac{5}{8}$, 62%
40. 0.4, $\frac{1}{25}$, 0.25
41. 0.49, $\frac{49}{50}$, 0.5
42. $\frac{4}{7}$, 44%, 0.47

43. **DARTS** Bianca and Christopher were playing darts. Bianca hit the bull's-eye 5 out of 18 times. Christopher missed the bull's-eye 4 out of 15 times. Who hit the bull's-eye a greater fraction of the time?

Determine whether each number is rational. Write *yes* or *no*.

44. $\frac{1}{9}$
45. 1.141141114...
46. 1.2345...
47. π
48. $-3\frac{4}{5}$
49. $5.\overline{23}$

50. **WEATHER** Refer to the table at the right that shows about how much rain falls in Albuquerque and Denver. Which city has the greater fraction of inches of rain per day? Explain.

City	Amount of Rain (in.)	Number of Days
Albuquerque, NM	9	60
Denver, CO	15	90

Source: www.weather.com

51. **BAKING** Ofelia needs $2\frac{3}{4}$ cups of water for a cake recipe. If her measuring cup only shows decimals in increments of tenths, how will Ofelia measure out the water for the recipe?

52. **MULTI STEP** Find one terminating decimal and one repeating decimal between $\frac{2}{3}$ and $\frac{3}{4}$.

53. **NUMBER SENSE** Is $1\frac{15}{16}$, $\frac{17}{8}$, or $\frac{63}{32}$ nearest to 2? Explain.

BASEBALL For Exercises 54–56, use the following information and the table below.
Mark McGwire of the St. Louis Cardinals and Sammy Sosa of the Chicago Cubs competed in a home run race during the 1998 baseball season.

Player	Games	At-bats	Home Runs
McGwire	155	509	70
Sosa	159	643	66

54. Write the ratio of home runs to games as a decimal for both players. Who had a better chance of hitting a home run during a game?

55. Write the ratio of home runs to at-bats as a decimal for both players. Who had a better chance of hitting a home run during an at-bat?

56. **WRITE A PROBLEM** In 2001, Barry Bonds of the San Francisco Giants set the single-season home run record. He had 476 at-bats and hit 73 home runs in 153 games. Use this information and the information in Exercises 54 and 55 to write and then solve a problem involving ratios.

Data Update Has anyone surpassed Bonds' single-season mark of 73 home runs? Visit msmath2.net/data_update to learn more.

57. **CRITICAL THINKING** How do you order fractions that have the same numerator? Explain.

Standardized Test Practice and Mixed Review

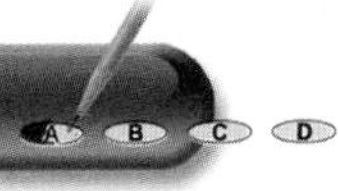

58. **MULTIPLE CHOICE** Find the least fraction.

Ⓐ $\frac{3}{4}$ Ⓑ $\frac{7}{9}$ Ⓒ $\frac{13}{18}$ Ⓓ $\frac{11}{15}$

59. **SHORT RESPONSE** Order the ratios from least to greatest: $47.\overline{4}$, $\frac{4}{7}$, $47\frac{2}{5}$, 47.41, 47%.

60. Find the LCM of 14 and 21. (Lesson 5-7)

FOOD The graph shows what percent of Americans chose the given pizza toppings as their favorite. Write the percent of people who chose each topping as a decimal. (Lesson 5-6)

61. onions

62. vegetables

63. mushrooms

64. **ALGEBRA** Translate the following sentence into an algebraic equation. *Four less than two times a number is 30.* (Lesson 4-1)

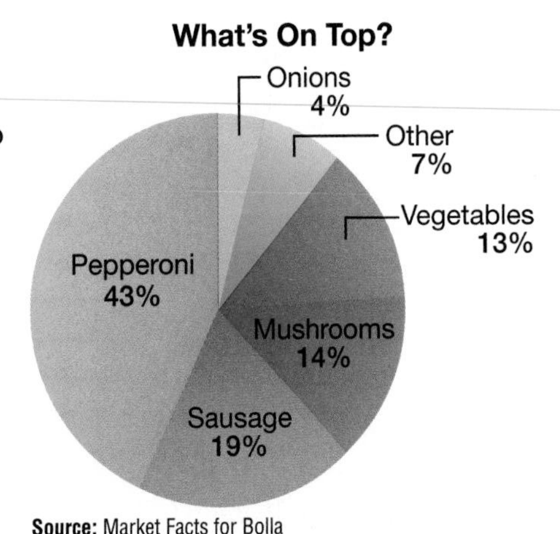

Study Guide and Review

Vocabulary and Concept Check

bar notation (p. 211)
common denominator (p. 227)
composite number (p. 197)
equivalent fractions (p. 207)
factor tree (p. 198)
greatest common factor (GCF) (p. 203)
least common denominator (LCD) (p. 227)
least common multiple (LCM) (p. 224)
multiple (p. 224)
percent (p. 216)
prime factorization (p. 198)
prime number (p. 197)
ratio (p. 216)
rational numbers (p. 229)
repeating decimals (p. 211)
simplest form (p. 207)
terminating decimals (p. 211)
Venn diagram (p. 203)

Choose the letter of the term or number that best matches each phrase.

1. a common multiple of the denominators of two or more fractions
2. a fraction in which the GCF of the numerator and denominator is 1
3. a ratio that compares a number to 100
4. a whole number greater than 1 that has exactly two factors, 1 and itself
5. a comparison of two numbers by division

a. ratio
b. simplest form
c. common denominator
d. percent
e. prime number
f. multiple
g. bar notation

Lesson-by-Lesson Exercises and Examples

Prime Factorization (pp. 197–200)

Determine whether each number is *prime* or *composite*.

6. 45
7. 37
8. 91
9. 118

Find the prime factorization of each number.

10. 54
11. 128
12. 68
13. 95
14. **ALGEBRA** Factor the expression $36x^2yz^3$.

Example 1 **Determine whether 21 is *prime* or *composite*.**

The number 21 has factors 1, 3, 7, and 21. So, 21 is a composite number.

Example 2 **Find the prime factorization of 18.**

Use a factor tree.

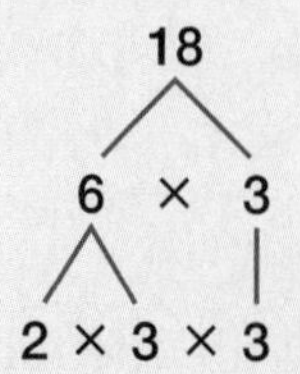

The prime factorization of 18 is 2×3^2.

5-2 Greatest Common Factor (pp. 203–206)

Find the GCF of each set of numbers.

15. 18, 27 **16.** 30, 72

17. 28, 70, 98 **18.** 42, 63, 105

Find the GCF of each set of algebraic expressions.

19. $18w, 54w^2y$ **20.** $21de^2, 35d^3e^3$

Example 3 **Find the GCF of 24 and 56.**

First, make a list of all of the factors of 24 and 56.

factors of 24: 1, 2, 3, 4, 6, 8, 12, 24
factors of 56: 1, 2, 4, 7, 8, 14, 28, 56
common factors: 1, 2, 4, 8

The GCF of 24 and 56 is 8.

5-3 Simplifying Fractions (pp. 207–209)

Write each fraction in simplest form.

21. $\frac{12}{15}$ **22.** $\frac{35}{60}$ **23.** $\frac{11}{121}$

24. $\frac{14}{63}$ **25.** $\frac{37}{45}$ **26.** $\frac{55}{110}$

27. **LIFE SCIENCE** The human body has 60,000 miles of blood vessels. The distance across the United States from east to west is 3,000 miles. Write a fraction in simplest form that compares the distance across the United States to the miles of blood vessels.

Example 4 **Write $\frac{24}{32}$ in simplest form.**

Find the GCF of the numerator and denominator.

$24 = 2 \times 2 \times 2 \times 3$
$32 = 2 \times 2 \times 2 \times 2 \times 2$

GCF: $2 \times 2 \times 2 = 8$

$\frac{24}{32} = \frac{24 \div 8}{32 \div 8} = \frac{3}{4}$ Divide the numerator and denominator by the GCF.

So, $\frac{24}{32}$ in simplest form is $\frac{3}{4}$.

5-4 Fractions and Decimals (pp. 210–213)

Write each fraction or mixed number as a decimal. Use bar notation if the decimal is a repeating decimal.

28. $\frac{3}{4}$ **29.** $\frac{7}{8}$

30. $\frac{5}{9}$ **31.** $4\frac{1}{3}$

32. $6\frac{2}{5}$ **33.** $1\frac{6}{7}$

Write each decimal as a fraction in simplest form.

34. 0.7 **35.** 0.44

36. 0.05 **37.** 0.18

38. 0.54 **39.** 0.08

Example 5 **Write $\frac{3}{8}$ as a decimal.**

$$\begin{array}{r} 0.375 \\ 8\overline{)3.000} \\ \underline{24} \\ 60 \\ \underline{56} \\ 40 \\ \underline{40} \\ 0 \end{array}$$

So, $\frac{3}{8} = 0.375$.

Example 6 **Write 0.64 as a fraction in simplest form.**

$0.64 = \frac{64}{100}$ Write as a fraction with a denominator of 100.

$= \frac{16}{25}$ Simplify.

5-5 Fractions and Percents (pp. 216–219)

Write each fraction as a percent.

40. $\frac{32}{100}$ **41.** $\frac{11}{25}$ **42.** $\frac{47}{50}$ **43.** $\frac{8}{20}$

Write each percent as a fraction in simplest form.

44. 68% **45.** 95% **46.** 42% **47.** 16%

48. FOOD A recent survey showed that 58% of school children chose peanut butter and jelly sandwiches as their favorite lunch food. Write this percent as a fraction in simplest form.

Example 7 **Write $\frac{27}{50}$ as a percent.**

$\frac{27}{50} = \frac{54}{100}$ Write an equivalent fraction with a denominator of 100.

$= 54\%$ Definition of percent

Example 8 **Write 96% as a fraction in simplest form.**

$96\% = \frac{96}{100}$ Definition of percent

$= \frac{24}{25}$ Simplify.

5-6 Percents and Decimals (pp. 220–223)

Write each percent as a decimal.

49. 48% **50.** 7% **51.** 12.5% **52.** $75\frac{1}{4}\%$

Write each decimal as a percent.

53. 0.61 **54.** 0.055 **55.** 0.19 **56.** 0.999

57. WEATHER Twenty-three percent of the days last month were rainy. Write this percent as a decimal.

Example 9 **Write 35% as a decimal.**

$35\% = \frac{35}{100}$ Write the percent as a fraction.

$= 0.35$ Write the fraction as a decimal.

Example 10 **Write 0.625 as a percent.**

$0.625 = 0.625$ Multiply by 100.

$= 62.5\%$ Add the %.

5-7 Least Common Multiple (pp. 224–226)

Find the LCM of each set of numbers.

58. 9, 15 **59.** 4, 8

60. 16, 24 **61.** 3, 8, 12

62. 4, 9, 12 **63.** 15, 24, 30

Example 11 **Find the LCM of 8 and 36.**

Write each prime factorization.

$8 = 2^3$ $36 = 2^2 \times 3^2$

LCM: $2^3 \times 3^2 = 72$

The LCM of 8 and 36 is 72.

5-8 Comparing and Ordering Rational Numbers (pp. 227–231)

Replace each ● with <, >, or = to make a true sentence.

64. $\frac{3}{8}$ ● $\frac{2}{3}$ **65.** 0.45 ● $\frac{9}{20}$

66. $\frac{8}{9}$ ● 85% **67.** $3\frac{3}{4}$ ● $3\frac{5}{8}$

68. 72% ● $\frac{8}{11}$ **69.** $\frac{5}{7}$ ● $\frac{60}{84}$

Example 12 **Replace ● with <, >, or = to make $\frac{3}{5}$ ● $\frac{5}{8}$ a true sentence.**

Find equivalent fractions. The LCD is 40.

$\frac{3}{5} = \frac{3 \times 8}{5 \times 8} = \frac{24}{40}$ $\frac{5}{8} = \frac{5 \times 5}{8 \times 5} = \frac{25}{40}$

Since $\frac{24}{40} < \frac{25}{40}$, then $\frac{3}{5} < \frac{5}{8}$.

CHAPTER 5

Practice Test

Vocabulary and Concepts

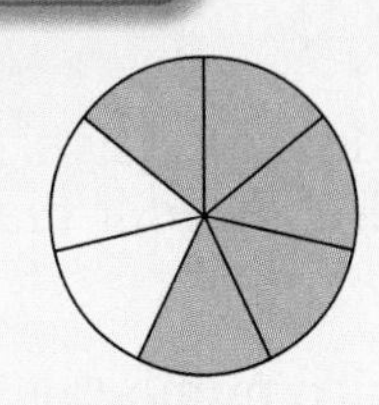

1. Which of the fractions modeled at the right is greater? Explain.
2. **Explain** how to write a percent as a decimal.

Skills and Applications

3. Find the prime factorization of 72.
4. Find the GCF of 24 and 40.

Write each fraction in simplest form.

5. $\frac{24}{60}$
6. $\frac{64}{72}$

Write each fraction, mixed number, or percent as a decimal. Use bar notation if the decimal is a repeating decimal.

7. $\frac{7}{9}$
8. $4\frac{5}{8}$
9. 91%

Write each decimal or percent as a fraction in simplest form.

10. 0.84
11. 0.006
12. 34%

Write each fraction or decimal as a percent.

13. $\frac{15}{25}$
14. 0.26
15. 0.135

16. **PAINTING** Suppose you are painting a wall that measures 10 feet by 10 feet. You have already painted 46 square feet. Write a percent for the portion of the wall that remains unpainted.

Replace each ● with $<$, $>$, or $=$ to make a true sentence.

17. $\frac{3}{5}$ ● $\frac{5}{9}$
18. $\frac{7}{12}$ ● $\frac{6}{8}$
19. $\frac{13}{20}$ ● 65%

Standardized Test Practice

20. **MULTIPLE CHOICE** One type of cicada emerges from hibernation every 17 years. Another type emerges every 13 years. If both types come out of hibernation one year, in how many years would this happen again?

(A) 30 yr (B) 68 yr (C) 120 yr (D) 221 yr

CHAPTER 5 Standardized Test Practice

PART 1 Multiple Choice

Record your answers on the answer sheet provided by your teacher or on a sheet of paper.

1. Which number is divisible by 2, 3, 5, 6, 9, and 10? (Prerequisite Skill, p. 554)

   (A) 90 (B) 120

   (C) 150 (D) 200

2. Students at Lincoln High School have collected money this year for charity. According to the graph, how much did all four classes collect this year? (Lesson 1-1)

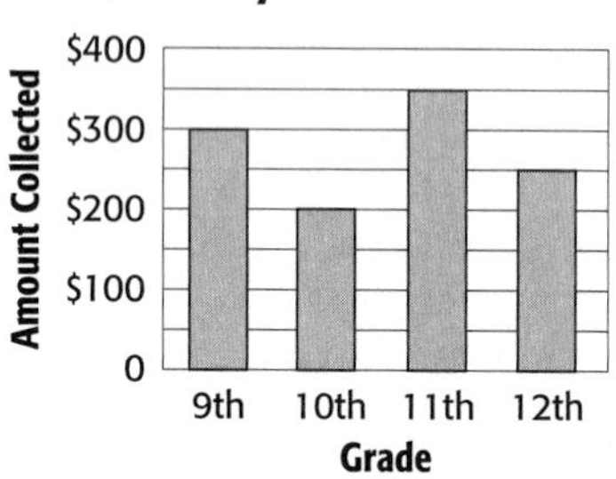

   (F) $950 (G) $1,000

   (H) $1,050 (I) $1,100

3. Which expression is the same as k^6? (Lesson 1-2)

   (A) $k + k + k + k + k + k$

   (B) $6k$

   (C) $k^3 + k^3$

   (D) $k \times k \times k \times k \times k \times k$

TEST-TAKING TIP

Question 2 When an item includes a graph, scan the graph to see what kind of information it includes and how the information is organized. Read each answer choice and compare it with the graph to see if the information in the answer choice is correct. Eliminate any wrong answer choices.

4. Which sentence shows the Commutative Property of Multiplication? (Lesson 1-6)

   (F) $f \cdot (g \cdot h) = (f \cdot g) \cdot h$

   (G) $q \cdot (r + s) = q \cdot r + q \cdot s$

   (H) $x \cdot y = y \cdot x$

   (I) $a + b = b + a$

5. In which quadrant is $(-5, -3)$ located? (Lesson 3-3)

   (A) quadrant I (B) quadrant II

   (C) quadrant III (D) quadrant IV

6. Write an addition sentence modeled by the number line. (Lesson 3-4)

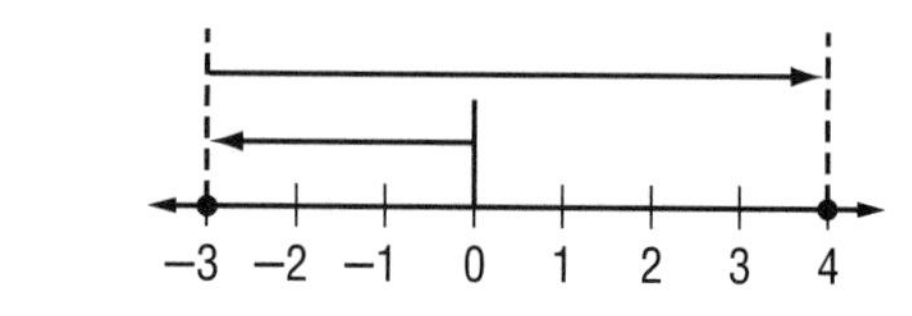

   (F) $-3 + 7 = 4$ (G) $3 + 4 = 7$

   (H) $-7 + 3 = -4$ (I) $3 + (-7) = -4$

7. Which of the ordered pairs is *not* a solution of $y = -3x + 7$? (Lesson 4-6)

   (A) $(-1, 10)$ (B) $(1, 4)$

   (C) $(2, 1)$ (D) $(3, 2)$

8. Find the percent that represents the shaded area. (Lesson 5-5)

   (F) 75% (G) 60%

   (H) 50% (I) 40%

9. Mrs. Benitez hired a contractor for a kitchen renovation project. The contractor renovated 60% of the kitchen the first week. What part of the project is left to do? (Lesson 5-6)

   (A) 0.04 (B) 0.4

   (C) 0.44 (D) 4.00

PART 2 Short Response/Grid In

Record your answers on the answer sheet provided by your teacher or on a sheet of paper.

10. Find the eighth term in the following sequence. 1, 3, 7, 13, 21, … (Lesson 1-7)

For Questions 11 and 12, use the information below.

The table gives weekly water-level readings of Snail Creek Lake over an 8-week period. The readings indicate the number of feet above or below flood level.

Week	1	2	3	4	5	6	7	8
Water Level	−8	−8	−6	−5	−2	−5	0	3

11. Find the median of the readings. (Lesson 3-2)

12. Find the range of readings. (Lesson 3-5)

13. The equation $4.5h + 75 = 165$, where h is the number of hours, describes how long Hannah must work to earn \$165. How many hours must she work to earn \$165? (Lesson 4-4)

14. Write an equation for the function shown in the table. (Lesson 4-6)

x	−1	0	1	2
y	3	1	−1	−3

15. Aaron invited 30 friends to a party, but only 22 of them were able to attend. What fraction of the friends could not attend? (Lesson 5-5)

16. A string of red, green, and blue lights blink at different rates. The red lights blink once every 4 seconds, the green lights blink once every 7 seconds, and the blue lights blink once every 9 seconds. How many seconds will go by between times when all three colors will blink at the same time? (Lesson 5-7)

17. Tino painted a mural on the side of his shed. He used $\frac{1}{7}$ can of yellow paint, $\frac{1}{4}$ can of red paint, and $\frac{2}{9}$ can of white paint. Of which color did he use the least? (Lesson 5-8)

PART 3 Extended Response

Record your answers on a sheet of paper. Show your work.

18. A soft drink company sells its beverages in packages of 12 and 24 plastic bottles. It added a package with 30 bottles. (Lessons 5-2 and 5-7)

a. Suppose the company designed boxes so that all three packages have the same number of bottles in each row. What is the greatest number of bottles in each row?

b. The time it takes to seal a package is equal to 1 second per bottle. If three assembly lines containing the 12, 24, and 30 packages begin at the same time, when is the first time that all three packages are sealed at the same time?

19. The table shows the number of students who volunteered at the food bank. (Lesson 5-8)

Classroom	Number of Students Volunteering	Total Number of Students
A	7	24
B	3	8
C	6	18

a. Model each ratio of the number of students volunteering to the total number of students on grid paper.

b. Which classroom had the greatest number of students volunteering?

c. Which classroom had the greatest fraction of students volunteering?

d. Write each ratio as a fraction in simplest form, as a decimal, and as a percent.

Chapter 6 Applying Fractions

"What do Ferris wheels have to do with math?"

A Ferris wheel follows a circular path. To find the distance that you travel when you go one time around a Ferris wheel, you can use the formula $C = \pi d$, where π is approximately $\frac{22}{7}$, or 3.14. In geometry, you will use formulas to solve many real-life problems.

You will solve problems about Ferris wheels in Lesson 6-9.

GETTING STARTED

Take this quiz to see whether you are ready to begin Chapter 6. Refer to the lesson or page number in parentheses if you need more review.

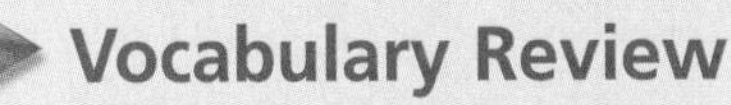

Vocabulary Review

Choose the correct term to complete each sentence.

1. The Division (Identity, Property) of Equality states that if you divide each side of an equation by the same nonzero number, the two sides remain equal. (Lesson 4-3)
2. The (GCF, LCD) of 12 and 16 is 4. (Lessons 5-2 and 5-7)

Prerequisite Skills

Find the LCD of each pair of fractions. (Lesson 5-8)

3. $\frac{5}{7}, \frac{3}{5}$
4. $\frac{1}{2}, \frac{4}{9}$
5. $\frac{8}{15}, \frac{1}{6}$
6. $\frac{3}{4}, \frac{7}{10}$

Multiply or divide. (Pages 560, 562)

7. 1.8×12
8. $99 \div 12$
9. $83 \div 100$
10. 4.6×0.3
11. 15×0.5
12. $216 \div 16$

Complete to show equivalent mixed numbers. (Page 563)

13. $3\frac{1}{5} = 2\frac{■}{5}$
14. $9\frac{2}{3} = ■\frac{5}{3}$
15. $6\frac{1}{4} = 5\frac{■}{4}$
16. $8\frac{6}{7} = 7\frac{■}{7}$

Write each mixed number as an improper fraction. (Page 563)

17. $10\frac{3}{4}$
18. $1\frac{7}{8}$
19. $4\frac{2}{5}$
20. $7\frac{2}{9}$

Fractions Make this Foldable to help you organize your notes. Begin with a sheet of $8\frac{1}{2}''$ by 11″ paper, four index cards, and glue.

Fold
Fold the paper in half widthwise.

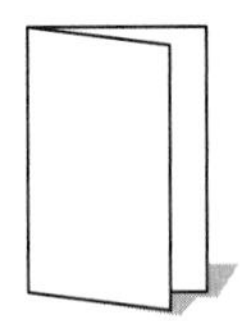

STEP 2 Open and Fold Again
Open and fold along the length about $2\frac{1}{2}''$ from the bottom.

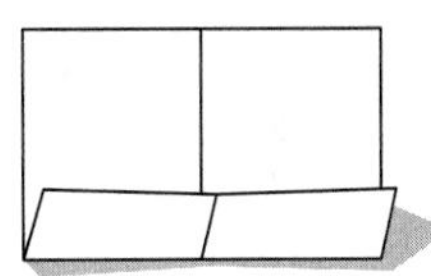

STEP 3 Glue
Glue the edges on either side to form two pockets.

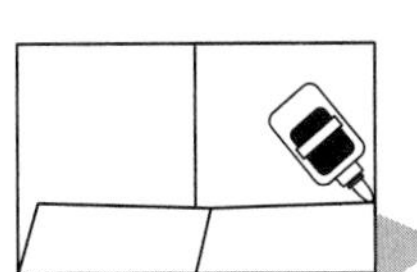

STEP 4 Label
Label the pockets *Fractions* and *Mixed Numbers*, respectively. Place two index cards in each pocket.

Reading and Writing As you read and study the chapter, write the rules for performing operations on fractions and mixed numbers. Show the rules by giving real-life examples.

6-1 Estimating with Fractions

What You'll LEARN

Estimate sums, differences, products, and quotients of fractions and mixed numbers.

NEW Vocabulary

compatible numbers

MATH Symbols

$\approx$ is approximately equal to

WHEN am I ever going to use this?

KITES For a kite to have balance while flying, the left and right sides of the horizontal support must each be $\frac{2}{3}$ as long as the bottom of the vertical support. Also, the top must be $\frac{1}{3}$ as long as the bottom portion.

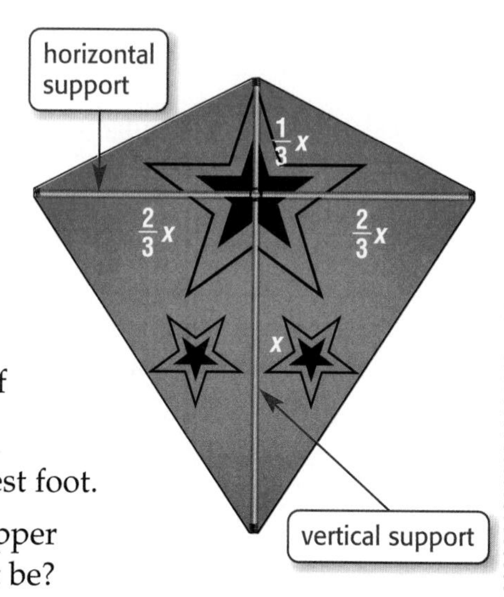

1. Suppose the bottom portion of the vertical support is $2\frac{3}{4}$ feet. Round this length to the nearest foot.
2. About how long should the upper portion of the vertical support be?
3. About how long should the left and right sides of the horizontal support be?

To estimate the sum, difference, product, or quotient of mixed numbers, round the mixed numbers to the nearest whole number.

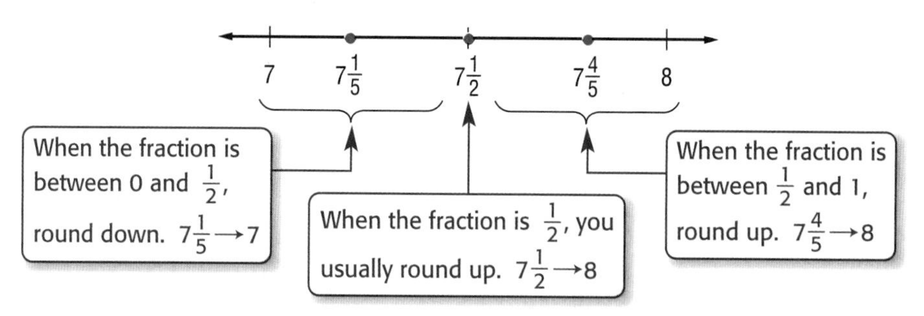

STUDY TIP

Front-End Estimation Add the integers, estimate the sum of the fractions, and find the total sum.

$3\frac{2}{3} + 5\frac{1}{6}$
$\approx 8 + 1$
≈ 9

EXAMPLES Estimate with Mixed Numbers

Estimate.

1 $3\frac{2}{3} + 5\frac{1}{6}$

$3\frac{2}{3} + 5\frac{1}{6} \rightarrow 4 + 5 = 9$

The sum is *about* 9.

2 $6\frac{2}{5} \times 1\frac{7}{8}$

$6\frac{2}{5} \times 1\frac{7}{8} \rightarrow 6 \times 2 = 12$

The product is *about* 12.

Your Turn Estimate.

a. $2\frac{1}{5} + 3\frac{1}{2}$

b. $4\frac{3}{8} \times 5\frac{1}{4}$

c. $8\frac{7}{9} \div 2\frac{3}{4}$

To estimate the sum, difference, product, or quotient of fractions, round each fraction to 0, $\frac{1}{2}$, or 1, whichever is closest. Number lines and fraction models can help you decide how to round.

Fractions Close to 0	Fractions Close to $\frac{1}{2}$	Fractions Close to 1
0 $\frac{1}{6}$ 1	0 $\frac{1}{2}$ $\frac{5}{8}$ 1	0 $\frac{9}{10}$ 1
$\frac{1}{7}$	$\frac{4}{9}$	$\frac{5}{6}$

EXAMPLES Estimate with Fractions

Estimate.

3 $\frac{1}{8} + \frac{2}{3}$

$\frac{1}{8}$ is about 0.

$\frac{2}{3}$ is about $\frac{1}{2}$.

$\frac{1}{8} + \frac{2}{3} \rightarrow 0 + \frac{1}{2} = \frac{1}{2}$ The sum is *about* $\frac{1}{2}$.

STUDY TIP

Fractions When the numerator and the denominator of a fraction are very close in value, such as $\frac{6}{7}$ or $\frac{8}{9}$, the fraction is close to 1.

4 $\frac{6}{7} - \frac{7}{10}$

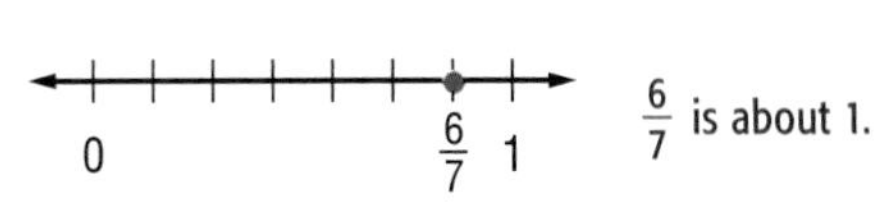

$\frac{6}{7}$ is about 1.

0 $\frac{1}{2}$ $\frac{7}{10}$ 1

$\frac{7}{10}$ is about $\frac{1}{2}$.

$\frac{6}{7} - \frac{7}{10} \rightarrow 1 - \frac{1}{2} = \frac{1}{2}$ The difference is *about* $\frac{1}{2}$.

5 $\frac{8}{9} \div \frac{5}{6}$

$\frac{8}{9} \div \frac{5}{6} \rightarrow 1 \div 1 = 1$ $\frac{8}{9}$ is about 1, and $\frac{5}{6}$ is about 1.

The quotient is *about* 1.

Your Turn Estimate.

d. $\frac{4}{5} + \frac{2}{7}$

e. $\frac{5}{8} - \frac{3}{7}$

f. $\frac{3}{5} \times \frac{11}{12}$

Sometimes it makes sense to round fractions to the nearest $\frac{1}{2}$, or mixed numbers to the nearest integer. Other times, it is useful to use **compatible numbers**, or numbers that are easy to compute mentally.

EXAMPLES Use Compatible Numbers

Estimate.

6 $\frac{1}{3} \times 14$ **THINK** What is $\frac{1}{3}$ of 14?

$\frac{1}{3} \times 14 \rightarrow \frac{1}{3} \times 15 = 5$ Round 14 to 15, since 15 is divisible by 3.

The product is *about* 5.

STUDY TIP

Compatible Numbers When dividing mixed numbers, round so that the dividend is a multiple of the divisor.

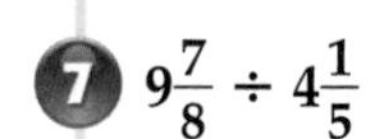

7 $9\frac{7}{8} \div 4\frac{1}{5}$

$9\frac{7}{8} \div 4\frac{1}{5} \rightarrow 10 \div 4\frac{1}{5}$ Round $9\frac{7}{8}$ to 10.

$\rightarrow 10 \div 5 = 2$ Round $4\frac{1}{5}$ to 5, since 10 is divisible by 5.

The quotient is *about* 2.

Your Turn **Estimate.**

g. $\frac{1}{4} \cdot 21$ h. $\frac{1}{2} \times 17$ i. $12 \div 6\frac{2}{3}$

Skill and Concept Check

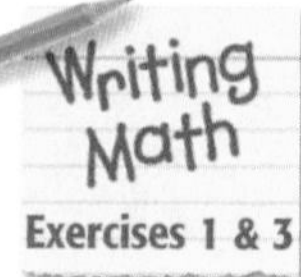

1. **Explain** how models are useful when estimating with fractions.
2. **OPEN ENDED** Describe when estimation is a better method for solving a problem rather than using pencil and paper, a calculator, or a computer. Then give a real-life example.
3. **NUMBER SENSE** Determine which of the following has a sum that is greater than 1. Write *yes* or *no* and explain.

a. $\frac{1}{2} + \frac{4}{7}$ b. $\frac{3}{4} + \frac{5}{8}$ c. $\frac{2}{5} + \frac{1}{6}$

GUIDED PRACTICE

Estimate.

4. $8\frac{3}{8} + 1\frac{4}{5}$
5. $5\frac{5}{7} \times 2\frac{7}{8}$
6. $\frac{1}{6} + \frac{2}{5}$
7. $\frac{6}{7} - \frac{1}{5}$
8. $\frac{1}{4} \cdot 15$
9. $21\frac{5}{6} \div 9\frac{3}{4}$
10. **CONSTRUCTION** About how many bookcase shelves shown at the right can a carpenter cut from a board that is 1 foot wide and 12 feet long?

1 ft

$3\frac{1}{2}$ ft

Practice and Applications

HOMEWORK HELP	
For Exercises	See Examples
11–16, 37–38	1, 2
17–24	3–5
25–28	6–7

Extra Practice
See pages 577, 601.

Estimate.

11. $3\frac{3}{4} + 4\frac{5}{6}$
12. $1\frac{1}{8} + 5\frac{11}{12}$
13. $5\frac{1}{3} - 3\frac{1}{6}$
14. $4\frac{2}{5} - 1\frac{1}{2}$
15. $2\frac{2}{3} \cdot 6\frac{1}{3}$
16. $6\frac{1}{8} \div 1\frac{2}{3}$
17. $\frac{3}{4} + \frac{3}{8}$
18. $\frac{5}{8} + \frac{3}{7}$
19. $\frac{5}{9} - \frac{1}{6}$
20. $\frac{3}{4} - \frac{3}{5}$
21. $\frac{1}{8} \times \frac{3}{4}$
22. $\frac{4}{9} \cdot \frac{11}{12}$
23. $\frac{4}{5} \div \frac{7}{8}$
24. $\frac{1}{10} \div \frac{5}{6}$
25. $\frac{1}{2} \times 13$
26. $\frac{1}{4} \times 39$
27. $25\frac{3}{10} \div 5\frac{2}{3}$
28. $27\frac{5}{8} \div 6\frac{1}{5}$
29. $12\frac{2}{7} \div 4\frac{1}{3}$
30. $5\frac{9}{10} \cdot 4\frac{1}{6}$

31. Estimate $23\frac{2}{9}$ divided by 3.

32. Estimate the sum of $4\frac{1}{8}$, $1\frac{5}{6}$, and $\frac{7}{9}$.

Estimate.

33. $-3\frac{2}{7} \times 8\frac{3}{4}$
34. $\frac{4}{5}\left(-\frac{5}{8}\right)$
35. $-\frac{1}{6} \times (-65)$
36. $12\frac{1}{4} \cdot 2\frac{7}{9}$

37. **BAKING** Kayla wants to make the bread recipe shown at the right, but she has only $1\frac{1}{3}$ cups of flour. About how much more flour does she need?

38. **SEWING** A skirt that is $15\frac{7}{8}$ inches long has a hem of $1\frac{1}{8}$ inches. Approximately how long will the skirt be if the hem is let down?

39. **CRITICAL THINKING** If a number being divided is rounded up and the divisor is rounded down, what is the effect on the quotient?

Standardized Test Practice and Mixed Review

40. **MULTIPLE CHOICE** Choose the best estimate for $2\frac{1}{5} + 3\frac{3}{4}$.

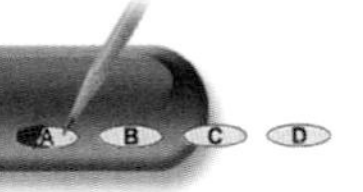

(A) 6 (B) 5 (C) 4 (D) 2

41. **MULTIPLE CHOICE** If Lucas's car gets $23\frac{1}{3}$ miles per gallon, about how many miles can he drive on $1\frac{3}{4}$ gallons?

(F) 0.46 mi (G) 4.6 mi (H) 46 mi (I) 460 mi

Replace each ● with <, >, or = to make a true sentence. (Lesson 5-8)

42. $\frac{7}{8}$ ● 0.75
43. $\frac{4}{5}$ ● $\frac{5}{7}$
44. $2\frac{1}{3}$ ● $\frac{7}{3}$
45. $\frac{6}{11}$ ● $\frac{9}{14}$

46. Find the LCM of 9 and 12. (Lesson 5-7)

GETTING READY FOR THE NEXT LESSON

PREREQUISITE SKILL Find the LCD of each pair of fractions. (Lesson 5-8)

47. $\frac{3}{4}, \frac{5}{12}$
48. $\frac{1}{2}, \frac{7}{10}$
49. $\frac{1}{6}, \frac{1}{8}$
50. $\frac{4}{5}, \frac{2}{3}$

6-2 Adding and Subtracting Fractions

What You'll LEARN

Add and subtract fractions.

REVIEW Vocabulary

LCD: the least common multiple of the denominators of two or more fractions **(Lesson 5-8)**

HANDS-ON Mini Lab

Materials

- ruler

Work with a partner.

1. Find $\frac{3}{8}$ inch on a ruler. From that point, add $\frac{4}{8}$ inch. What is the result?

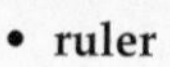

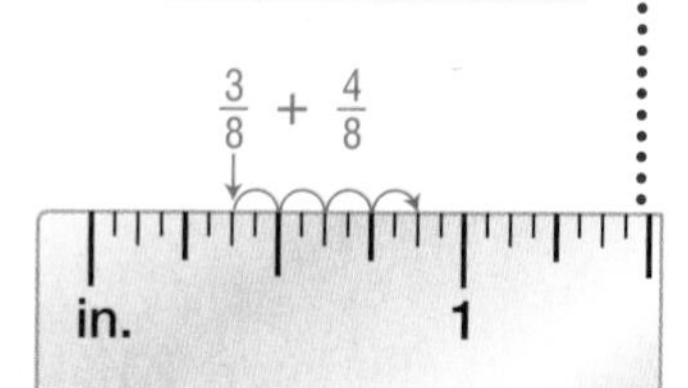

2. Use a ruler to add $\frac{1}{4}$ inch and $\frac{2}{4}$ inch.
3. **Make a conjecture** about how to find each sum. Check using a ruler.

a. $\frac{5}{8} + \frac{7}{8}$ b. $\frac{3}{16} + \frac{1}{16}$ c. $\frac{1}{2} + \frac{3}{4}$

Fractions that have the same denominators are called *like fractions.*

Key Concept **Adding and Subtracting Like Fractions**

Words To add or subtract like fractions, add or subtract the numerators and write the result over the denominator. Simplify if necessary.

Symbols

Arithmetic	Algebra
$\frac{1}{5} + \frac{3}{5} = \frac{4}{5}$	$\frac{a}{c} + \frac{b}{c} = \frac{a+b}{c}$, where $c \neq 0$
$\frac{11}{12} - \frac{7}{12} = \frac{4}{12}$ or $\frac{1}{3}$	$\frac{a}{c} - \frac{b}{c} = \frac{a-b}{c}$, where $c \neq 0$

EXAMPLES Add and Subtract Like Fractions

Add or subtract. Write in simplest form.

1 $\frac{5}{9} + \frac{2}{9}$

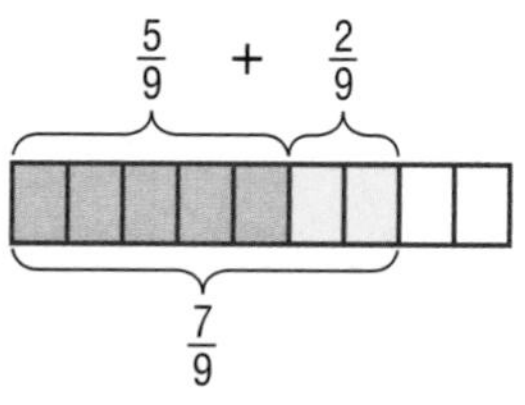

$\frac{5}{9} + \frac{2}{9} = \frac{5+2}{9}$ Add the numerators.

$= \frac{7}{9}$ Write the sum over the denominator.

2 $\frac{9}{10} - \frac{1}{10}$

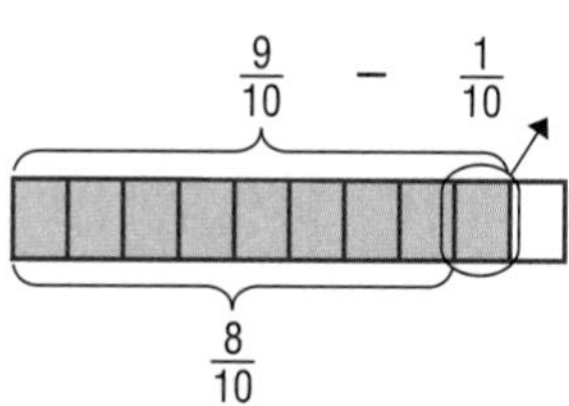

$\frac{9}{10} - \frac{1}{10} = \frac{9-1}{10}$ Subtract the numerators.

$= \frac{8}{10}$ Write the difference over the denominator.

$= \frac{4}{5}$ Simplify.

To add or subtract *unlike fractions*, or fractions with different denominators, rename the fractions using the LCD. Then add or subtract as with like fractions.

EXAMPLES Add and Subtract Unlike Fractions

Add or subtract. Write in simplest form.

3 $\frac{1}{2} + \frac{1}{6}$ **Estimate** $\frac{1}{2} + 0 = \frac{1}{2}$

The least common denominator of 2 and 6 is 6.

$\frac{1}{2} = \frac{1 \times 3}{2 \times 3} = \frac{3}{6}$ Rename $\frac{1}{2}$ using the LCD.

$$\begin{array}{rcr} \frac{1}{2} & \rightarrow & \frac{3}{6} \\ +\frac{1}{6} & \rightarrow & +\frac{1}{6} \\ \hline & & \frac{4}{6} \text{ or } \frac{2}{3} \end{array}$$

$\frac{1}{2}$ + $\frac{1}{6}$ → $\frac{3}{6} + \frac{1}{6}$ = $\frac{4}{6}$

$\frac{2}{3}$ is close to the estimate. So, $\frac{1}{2} + \frac{1}{6} = \frac{2}{3}$.

4 $\frac{7}{8} - \frac{5}{6}$ **Estimate** $1 - 1 = 0$

The LCD of 8 and 6 is 24.

$$\begin{array}{rcccr} \frac{7}{8} & \rightarrow & \frac{7 \times 3}{8 \times 3} & \rightarrow & \frac{21}{24} \\ -\frac{5}{6} & \rightarrow & \frac{5 \times 4}{6 \times 4} & \rightarrow & -\frac{20}{24} \\ \hline & & & & \frac{1}{24} \end{array}$$

So, $\frac{7}{8} - \frac{5}{6} = \frac{1}{24}$. Compare to the estimate.

REAL-LIFE MATH

WEATHER The average precipitation for November and December in Seattle, Washington, is $5\frac{7}{10}$ inches and 6 inches, respectively.

Source: *The World Almanac*

5 **WEATHER The average precipitation for November and December in Grand Junction, Colorado, is $\frac{7}{10}$ inch and $\frac{3}{5}$ inch, respectively. What is the total of the average precipitation for these two months?**

Estimate $\frac{1}{2} + \frac{1}{2} = 1$

$\frac{7}{10} + \frac{3}{5} = \frac{7}{10} + \frac{3 \times 2}{5 \times 2}$ The LCD of 10 and 5 is 10.

$= \frac{7}{10} + \frac{6}{10}$ Rename the fractions with the LCD.

$= \frac{13}{10}$ Add the numerators.

$= 1\frac{3}{10}$ Rename $\frac{13}{10}$ as a mixed number.

The total amount of precipitation is $1\frac{3}{10}$ inches.

Your Turn **Add or subtract. Write in simplest form.**

a. $\frac{8}{9} - \frac{2}{9}$ b. $\frac{5}{6} - \frac{3}{8}$ c. $\frac{7}{8} + \frac{3}{4}$ d. $\frac{2}{3} - \frac{1}{6}$

Skill and Concept Check

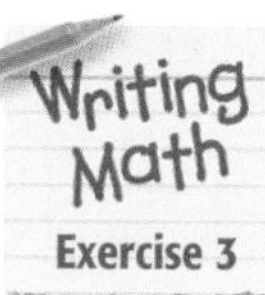

1. **Draw** a model to show $\frac{3}{8} + \frac{1}{4}$.
2. **OPEN ENDED** Write a subtraction problem with fractions in which the difference is $\frac{2}{7}$.
3. **FIND THE ERROR** Marissa and Jacinta are finding $\frac{5}{9} + \frac{1}{6}$. Who is correct? Explain.

Marissa
$\frac{5}{9} + \frac{1}{6} = \frac{5+9}{9+6}$

Jacinta
$\frac{5}{9} + \frac{1}{6} = \frac{5 \times 2}{9 \times 2} + \frac{1 \times 3}{6 \times 3}$

GUIDED PRACTICE

Add or subtract. Write in simplest form.

4. $\frac{4}{5} - \frac{2}{5}$
5. $\frac{6}{7} + \frac{3}{7}$
6. $\frac{1}{6} + \frac{3}{8}$
7. $\frac{5}{6} - \frac{7}{12}$
8. $\frac{4}{9} + \frac{2}{9}$
9. $\frac{3}{8} - \frac{1}{8}$
10. $\frac{2}{3} + \frac{5}{6}$
11. $\frac{5}{6} + \frac{4}{9}$
12. $\frac{3}{4} - \frac{1}{3}$
13. $\frac{7}{10} - \frac{1}{6}$
14. **PRESIDENTS** As of 2002, $\frac{5}{14}$ of the United States presidents were born in either Virginia or Ohio, and $\frac{1}{6}$ were born in either Massachusetts or New York. What fraction were born in one of these four states?

Practice and Applications

HOMEWORK HELP

For Exercises	See Examples
15–18, 23–26	1, 2
19–22, 27–41	3–5

Extra Practice
See pages 577, 601.

Add or subtract. Write in simplest form.

15. $\frac{3}{7} + \frac{1}{7}$
16. $\frac{5}{8} + \frac{7}{8}$
17. $\frac{5}{6} - \frac{1}{6}$
18. $\frac{7}{10} - \frac{3}{10}$
19. $\frac{1}{15} + \frac{3}{5}$
20. $\frac{7}{9} - \frac{1}{3}$
21. $\frac{4}{5} - \frac{1}{6}$
22. $\frac{7}{12} + \frac{7}{10}$
23. $\frac{4}{5} - \frac{3}{5}$
24. $\frac{8}{9} - \frac{5}{9}$
25. $\frac{3}{8} + \frac{7}{8}$
26. $\frac{5}{6} + \frac{5}{6}$
27. $\frac{5}{8} - \frac{7}{12}$
28. $\frac{2}{15} + \frac{4}{9}$
29. $\frac{5}{8} + \frac{11}{12}$
30. $\frac{3}{8} - \frac{1}{12}$
31. $\frac{3}{10} - \frac{1}{4}$
32. $\frac{9}{10} + \frac{4}{15}$
33. $\frac{9}{10} - \frac{1}{6}$
34. $\frac{7}{9} + \frac{5}{6}$

ALGEBRA Evaluate each expression if $a = \frac{3}{4}$ and $b = \frac{5}{6}$.

35. $\frac{1}{12} + a$
36. $b - \frac{7}{10}$
37. $b - a$
38. $a + b$

39. **MULTI STEP** After 1 hour, Jon had finished $\frac{5}{6}$ of a long-distance race, and Ling had finished $\frac{7}{9}$ of it. At that time, who had finished a greater fraction of the race, and by how much?

MONEY For Exercises 40 and 41, use the following information and the table at the right.
Sierra and Jacob each receive an equal allowance. The table shows the fraction of their allowance that they each deposit into their savings account and the fraction they each spend at the mall.

Where Money Goes	Fraction of Allowance	
	Sierra	Jacob
savings account	$\frac{1}{2}$	$\frac{1}{3}$
spend at mall	$\frac{1}{4}$	$\frac{3}{5}$
left over	?	?

40. What fraction of Jacob's allowance goes into his savings account or is spent at the mall?
41. Who has more money left over? Explain.

42. **CRITICAL THINKING** Does $\frac{1}{3} + \frac{5}{9} - \frac{5}{12} = \frac{5}{9} + \frac{5}{12} - \frac{1}{3}$? Explain.

Standardized Test Practice and Mixed Review

43. **MULTIPLE CHOICE** Makayla uses $\frac{1}{5}$ pound of ham and $\frac{1}{8}$ pound of turkey for her sandwich. How much meat does she use in all?

Ⓐ $\frac{1}{13}$ lb Ⓑ $\frac{2}{13}$ lb Ⓒ $\frac{13}{40}$ lb Ⓓ $\frac{7}{20}$ lb

44. **MULTIPLE CHOICE** Jamal used a bucket that was $\frac{7}{9}$ full with soapy water to wash his mother's car. After washing the car, the bucket was only $\frac{1}{6}$ full. What part of the bucket of soapy water did Jamal use?

Ⓕ $\frac{1}{9}$ Ⓖ $\frac{8}{15}$ Ⓗ $\frac{11}{18}$ Ⓘ $\frac{17}{18}$

Estimate. (Lesson 6-1)

45. $\frac{6}{7} - \frac{5}{12}$ 46. $4\frac{1}{9} + 3\frac{3}{4}$ 47. $16\frac{2}{3} \div 8\frac{1}{5}$ 48. $5\frac{4}{5} \cdot 3\frac{1}{3}$

PETS For Exercises 49 and 50, refer to the table at the right. It shows where pet owners get their pets. (Lesson 5-8)

49. Where do the greatest number of people get their pets?
50. Of the sources listed, where do the fewest people get their pets?

Pet Source	Portion of Pet Owners
animal shelter	$\frac{3}{20}$
friend/family	$\frac{21}{50}$
pet store	0.07
find as stray	0.14

Source: Yankelovich Partners

GETTING READY FOR THE NEXT LESSON

BASIC SKILL Complete.

Example: $8\frac{1}{2} = 7\frac{3}{2}$

51. $5\frac{2}{3} = 4\frac{■}{3}$ 52. $7\frac{8}{9} = 6\frac{■}{9}$ 53. $12\frac{1}{5} = ■\frac{6}{5}$ 54. $4\frac{3}{8} = ■\frac{11}{8}$

msmath2.net/self_check_quiz

6-3 Adding and Subtracting Mixed Numbers

What You'll LEARN

Add and subtract mixed numbers.

WHEN am I ever going to use this?

ASTRONOMY Astronomers use *astronomical units* (AU) to represent large distances in space. One AU is the average distance from Earth to the Sun. Mercury is about $\frac{2}{5}$ AU from the Sun.

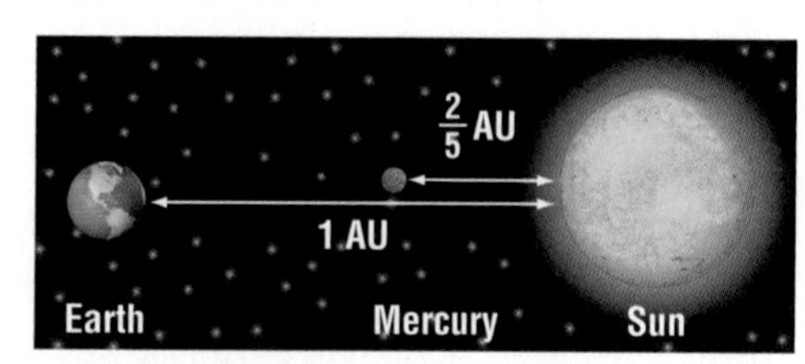

Diagram is not drawn to scale.

1. Jupiter is $5\frac{2}{5}$ AU from the Sun and Saturn's distance is $9\frac{1}{2}$ AU. Write an expression to find how much closer to the Sun Jupiter is than Saturn.
2. Find the difference of the fractional parts of the mixed numbers.
3. Find the difference of the whole numbers.
4. **Make a conjecture** about how to find $9\frac{1}{2} - 5\frac{2}{5}$. Then use your conjecture to find the difference.

To add or subtract mixed numbers, first add or subtract the fractions. If necessary, rename them using the LCD. Then add or subtract the whole numbers and simplify if necessary.

EXAMPLES Add and Subtract Mixed Numbers

Add or subtract. Write in simplest form.

1 $7\frac{4}{9} + 10\frac{2}{9}$

Estimate $7 + 10 = 17$

$$\begin{array}{r} 7\frac{4}{9} \\ +\ 10\frac{2}{9} \\ \hline 17\frac{6}{9} \text{ or } 17\frac{2}{3} \end{array}$$

Add the whole numbers and fractions separately.

Simplify.

Compare each sum to its estimate.

2 $8\frac{5}{6} - 2\frac{1}{3}$

Estimate $9 - 2 = 7$

$$\begin{array}{rcr} 8\frac{5}{6} & \rightarrow & 8\frac{5}{6} \\ -\ 2\frac{1}{3} & \rightarrow & -\ 2\frac{2}{6} \\ \hline & & 6\frac{3}{6} \text{ or } 6\frac{1}{2} \end{array}$$

Rename the fraction using the LCD.

Simplify.

Your Turn **Add or subtract. Write in simplest form.**

a. $6\frac{1}{8} + 2\frac{5}{8}$

b. $13\frac{7}{8} - 9\frac{3}{4}$

c. $1\frac{5}{9} + 4\frac{1}{6}$

EXAMPLE Use Mixed Numbers to Solve a Problem

REAL-LIFE CAREERS

How Does a Plumber Use Math?
Plumbers add and subtract mixed numbers when calculating the dimensions for installing house fixtures such as shower stalls and sinks.

Online Research
For information about a career as a plumber, visit: msmath2.net/careers

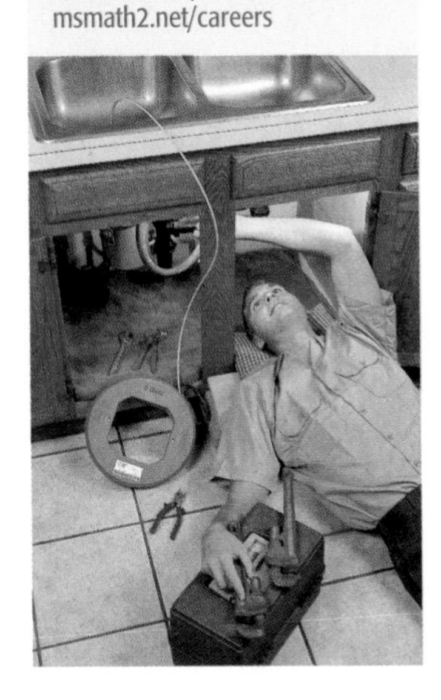

3 **BUILDING** There should be $2\frac{2}{3}$ feet of clearance space in front of a bathtub. What is the total width of the bathtub and clearance space shown at the right?

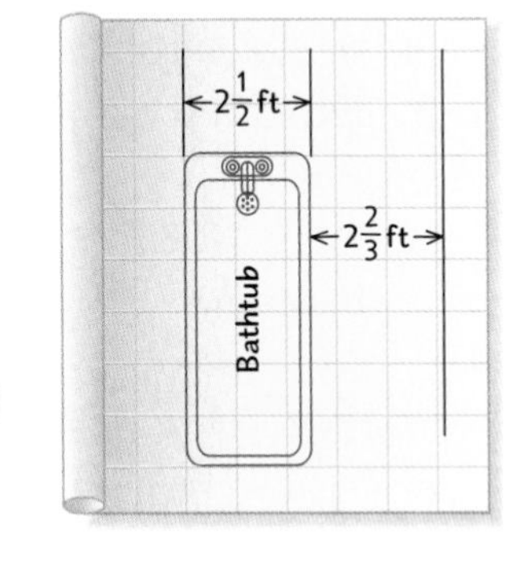

$2\frac{1}{2} + 2\frac{2}{3} = 2\frac{3}{6} + 2\frac{4}{6}$ Rename the fractions.

$= 4 + \frac{7}{6}$ Add the whole numbers and add the fractions.

$= 4 + 1\frac{1}{6}$ Rename $\frac{7}{6}$ as $1\frac{1}{6}$.

$= 5\frac{1}{6}$ Simplify.

The total width is $5\frac{1}{6}$ feet.

Sometimes when you subtract mixed numbers, the fraction in the first mixed number is less than the fraction in the second.

EXAMPLES Rename Mixed Numbers to Subtract

4 **Find $2\frac{1}{3} - 1\frac{2}{3}$.**

Rename $2\frac{1}{3}$ before subtracting.

1, 1, $\frac{1}{3}$ → 1, $\frac{3}{3}$, $\frac{1}{3}$ ← Change 1 to $\frac{3}{3}$.

$2\frac{1}{3} = 1\frac{3}{3} + \frac{1}{3}$ or $1\frac{4}{3}$

$2\frac{1}{3} \rightarrow 1\frac{4}{3}$ Rename $2\frac{1}{3}$ as $1\frac{4}{3}$.

$-1\frac{2}{3} \rightarrow -1\frac{2}{3}$ First subtract the whole numbers and then the fractions.

$\frac{2}{3}$

So, $2\frac{1}{3} - 1\frac{2}{3} = \frac{2}{3}$.

5 **Find $8\frac{7}{12} - 3\frac{3}{4}$.** **Estimate** $9 - 4 = 5$

$8\frac{7}{12} \rightarrow 8\frac{7}{12} \rightarrow 7\frac{19}{12}$ $\frac{7}{12} < \frac{9}{12}$, so rename $8\frac{7}{12}$ as $7\frac{19}{12}$.

$-3\frac{3}{4} \rightarrow -3\frac{9}{12} \rightarrow -3\frac{9}{12}$ Subtract.

$4\frac{10}{12}$ or $4\frac{5}{6}$ The difference is close to the estimate.

Your Turn **Subtract. Write in simplest form.**

d. $11\frac{2}{5} - 2\frac{3}{5}$ e. $5\frac{3}{8} - 4\frac{11}{12}$ f. $7 - 1\frac{1}{2}$

Skill and Concept Check

Writing Math
Exercises 2 & 3

1. **OPEN ENDED** Write a problem that can be solved by finding $8\frac{1}{2} + 2\frac{1}{3}$. Explain what the numbers represent and find the sum.

2. **FIND THE ERROR** Evan and José are finding $3\frac{3}{4} - 1\frac{7}{8}$. Who is correct? Explain.

Evan	José
$3\frac{3}{4} - 1\frac{7}{8} = 3\frac{6}{8} - 1\frac{7}{8}$	$3\frac{3}{4} - 1\frac{7}{8} = 3\frac{6}{8} - 1\frac{7}{8}$
$= 2\frac{1}{8}$	$= 2\frac{14}{8} - 1\frac{7}{8}$ or $1\frac{7}{8}$

3. **NUMBER SENSE** Use estimation to determine whether $6\frac{3}{4} + \frac{4}{5}$ is *greater than, less than,* or *equal to* $2\frac{1}{9} + 6\frac{7}{8}$. Explain.

GUIDED PRACTICE

Add or subtract. Write in simplest form.

4. $1\frac{5}{7} + 8\frac{1}{7}$
5. $7\frac{5}{6} - 3\frac{1}{6}$
6. $8\frac{1}{2} + 3\frac{4}{5}$
7. $6\frac{3}{4} + 2\frac{9}{10}$
8. $9\frac{4}{5} - 2\frac{3}{5}$
9. $2\frac{3}{8} + 5\frac{7}{8}$
10. $4\frac{3}{10} + 4\frac{9}{10}$
11. $7\frac{5}{6} + 9\frac{3}{8}$
12. $3\frac{1}{4} - 1\frac{3}{4}$
13. $11 - 6\frac{3}{8}$

14. **AQUARIUMS** A fish tank that holds $18\frac{2}{3}$ gallons of water has $10\frac{1}{2}$ gallons so far. How much more water can be added?

Practice and Applications

HOMEWORK HELP

For Exercises	See Examples
15–25, 28–30	1, 2
35–38	3
26–27, 31–34	4, 5

Extra Practice
See pages 577, 601.

Add or subtract. Write in simplest form.

15. $2\frac{1}{9} + 7\frac{4}{9}$
16. $10\frac{4}{5} - 2\frac{1}{5}$
17. $3\frac{1}{6} + 5\frac{1}{6}$
18. $6\frac{5}{7} + 8\frac{6}{7}$
19. $11\frac{3}{4} - 4\frac{1}{3}$
20. $4\frac{3}{8} + 10\frac{5}{12}$
21. $9\frac{4}{5} - 2\frac{3}{10}$
22. $6\frac{5}{6} - 2\frac{1}{3}$
23. $2\frac{3}{8} + 5\frac{7}{8}$
24. $8\frac{3}{7} + 1\frac{4}{7}$
25. $6\frac{2}{3} - 1\frac{3}{5}$
26. $4\frac{3}{10} - 1\frac{3}{4}$
27. $14\frac{1}{6} - 7\frac{1}{3}$
28. $3\frac{7}{9} + 3\frac{5}{9}$
29. $8\frac{5}{12} + 11\frac{1}{4}$
30. $7\frac{7}{8} + 10\frac{5}{6}$
31. $9\frac{1}{5} - 2\frac{3}{5}$
32. $12\frac{1}{2} - 6\frac{5}{8}$
33. $8 - 3\frac{2}{3}$
34. $13 - 5\frac{5}{6}$

STOCK MARKET For Exercises 35–37, use the following information.
Until several years ago, stock prices were listed as mixed numbers. Find the difference between the high and low price of each restaurant chain stock shown in the table.

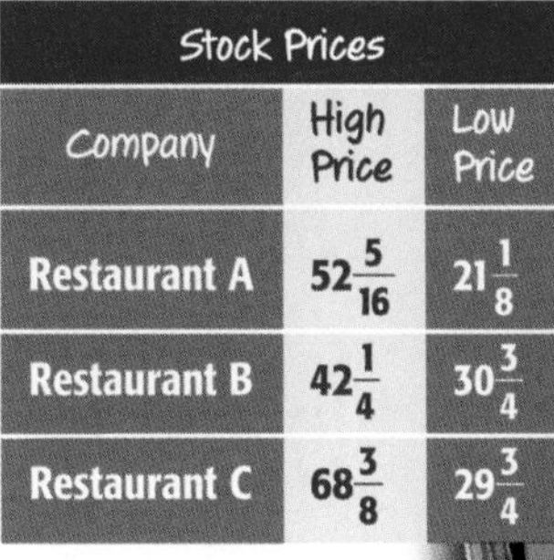

Stock Prices		
Company	High Price	Low Price
Restaurant A	$52\frac{5}{16}$	$21\frac{1}{8}$
Restaurant B	$42\frac{1}{4}$	$30\frac{3}{4}$
Restaurant C	$68\frac{3}{8}$	$29\frac{3}{4}$

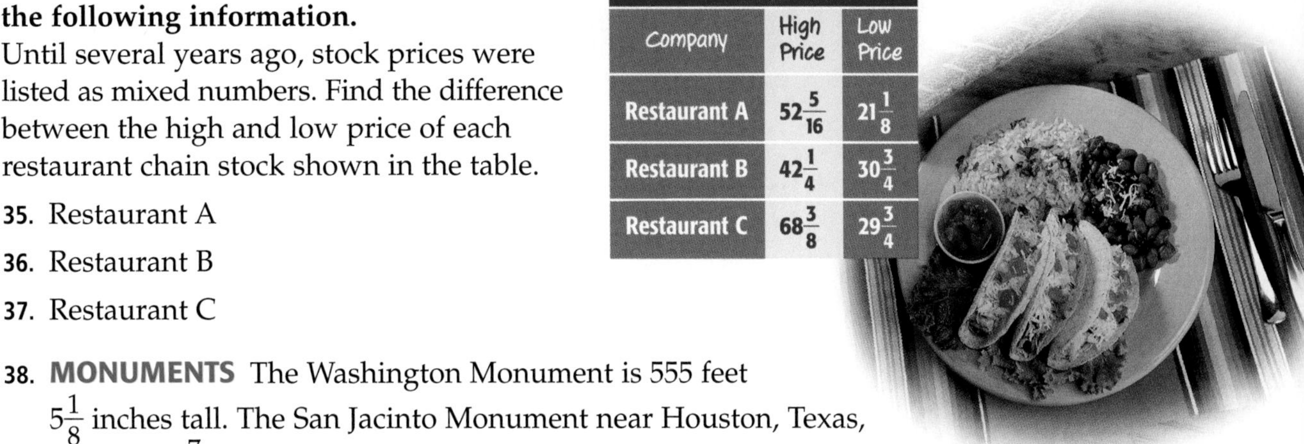

35. Restaurant A
36. Restaurant B
37. Restaurant C

38. **MONUMENTS** The Washington Monument is 555 feet $5\frac{1}{8}$ inches tall. The San Jacinto Monument near Houston, Texas, is 14 feet $6\frac{7}{8}$ inches taller. How tall is the San Jacinto Monument?

Add or subtract. Write in simplest form.

39. $10 - 3\frac{5}{11}$
40. $24 - 8\frac{3}{4}$
41. $6\frac{1}{6} + 1\frac{2}{3} + 5\frac{5}{9}$
42. $3\frac{1}{4} + 2\frac{5}{6} - 4\frac{1}{3}$

43. **CRITICAL THINKING** A string is cut in half, and one of the halves is used to bundle newspapers. Then one fifth of the remaining string is cut off. The piece left is 8 feet long. How long was the string originally?

Standardized Test Practice and Mixed Review

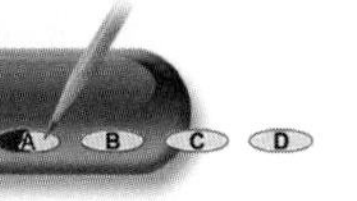

44. **MULTIPLE CHOICE** What is the sum of $7\frac{1}{2}$ and $2\frac{1}{6}$?

Ⓐ $5\frac{1}{3}$ Ⓑ $9\frac{1}{2}$ Ⓒ $9\frac{2}{3}$ Ⓓ $10\frac{2}{3}$

45. **MULTIPLE CHOICE** Melanie had $4\frac{2}{3}$ pounds of chopped walnuts. She used $1\frac{1}{4}$ pounds in a recipe. How many pounds of chopped walnuts did she have left?

Ⓕ $2\frac{1}{3}$ lb Ⓖ $2\frac{5}{12}$ lb Ⓗ $3\frac{5}{12}$ lb Ⓘ $3\frac{1}{2}$ lb

46. Find $\frac{7}{10} - \frac{1}{3}$. Write in simplest form. (Lesson 6-2)

Estimate. (Lesson 6-1)

47. $\frac{8}{9} \div \frac{9}{10}$
48. $3\frac{1}{2} + 6\frac{2}{3}$
49. $8\frac{4}{5} \times 7\frac{1}{9}$
50. $4\frac{2}{9} - 1\frac{1}{4}$

Replace each ● with <, >, or = to make a true sentence. (Lesson 5-8)

51. $\frac{4}{5}$ ● $\frac{7}{9}$
52. $\frac{2}{3}$ ● $\frac{5}{6}$
53. $\frac{1}{8}$ ● 0.15
54. $\frac{3}{7}$ ● 0.4

GETTING READY FOR THE NEXT LESSON

PREREQUISITE SKILL Write each mixed number as an improper fraction. (Page 563)

55. $2\frac{3}{8}$
56. $1\frac{2}{7}$
57. $5\frac{1}{10}$
58. $6\frac{4}{5}$

6-3b Problem-Solving Strategy

A Follow-Up of Lesson 6-3

What You'll Learn
Solve problems by eliminating possibilities.

Eliminate Possibilities

I recorded $3\frac{1}{4}$ hours of a miniseries on a videotape that can record 6 hours of programming. What is the most that I can record on the rest of the same tape—2 hours, $2\frac{1}{2}$ hours, or 3 hours?

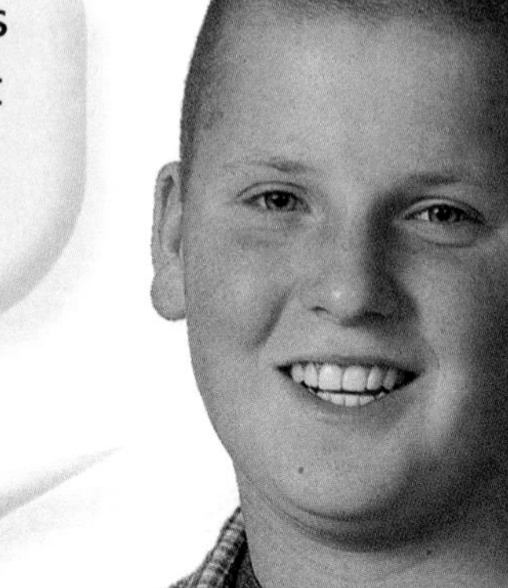

Well, we can **eliminate** some **possibilities** by estimating.

Explore	We know the combined hours of programming must be less than or equal to 6 hours.
Plan	Let's eliminate answers that are not reasonable.
Solve	You couldn't record 3 more hours on the tape because $3\frac{1}{4} + 3 = 6\frac{1}{4}$. So, we can eliminate that choice. Now let's check the choice of $2\frac{1}{2}$ hours. $3\frac{1}{4} + 2\frac{1}{2} = 5\frac{3}{4}$ Since this is less than 6 hours, this choice is correct. You could record $2\frac{1}{2}$ more hours on the tape.
Examine	Recording 2 more hours would give $3\frac{1}{4} + 2$ or $5\frac{1}{4}$ hours. This is less than the 6-hour maximum, but not the most that you could record.

Analyze the Strategy

1. **Describe** different ways that you can eliminate possibilities when solving problems.
2. **Explain** how the strategy of eliminating possibilities is useful for taking multiple choice tests.
3. **Write** a problem that could be solved by eliminating the possibilities. Explain your answer.

Apply the Strategy

Solve. Use the eliminate possibilities strategy.

4. **JUICE** Lauren has a 3-gallon cooler with $1\frac{3}{4}$ gallons of juice in it. If she wants the cooler full for her soccer game, how much juice does she need to add?

   (A) 4 gal (B) $3\frac{1}{4}$ gal
   (C) $1\frac{1}{4}$ gal (D) $\frac{1}{4}$ gal

5. **ELEPHANTS** An elephant in a zoo eats 58 cabbages in a week. About how many cabbages does an elephant eat in one year?

   (F) 7 (G) 700
   (H) 1,500 (I) 3,000

Mixed Problem Solving

Solve. Use any strategy.

6. **RAIN FOREST** In some areas of the rain forest, 325 inches of rain may fall in a year. Which is the *best* estimate for the average rainfall per day in such an area?

   (A) $\frac{1}{3}$ in. (B) 1 in.
   (C) 5 in. (D) 33 in.

7. **GRADES** Explain why the graph showing a student's science grades is misleading.

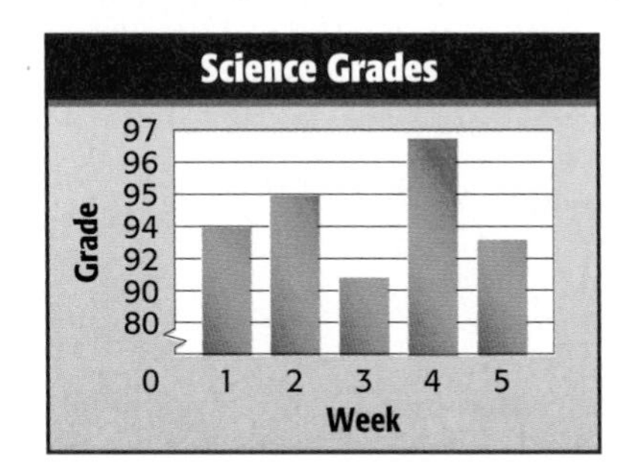

8. **ELEVATORS** An elevator can hold a maximum weight of 3,500 pounds. Which is the *best* estimate for the number of adults that the elevator can hold?

   (F) 10 adults (G) 20 adults
   (H) 35 adults (I) 80 adults

9. **SUPPLIES** Vanessa has $55 to buy school supplies. She bought a backpack that costs $23.50, a combination lock that costs $6.25, and 4 binders that are $3.99 each. If mechanical pencils are $2.50 per pack, how many packs can she buy?

10. **SHOPPING** Abby bought the items at the right for a party. Which is the *best* estimate of the total cost of the items, not including tax?

Item	Price
balloons	$2.95
cups and plates	$9.30
streamers	$4.50

   (A) less than $15
   (B) between $15 and $25
   (C) between $25 and $35
   (D) more than $35

11. **TRAVEL** Mr. Rollins drove 780 miles on a five-day trip. He rented a car for $23 per day plus $0.15 per mile after 500 free miles. About how much did the rental car cost?

   (F) $100 (G) $130 (H) $160 (I) $180

12. **STANDARDIZED TEST PRACTICE** If the trend in the graph continues, which is the best estimate for the number of acres of wheat harvested in the United States in 2004?

   (A) 60 million
   (B) 55 million
   (C) 45 million
   (D) 40 million

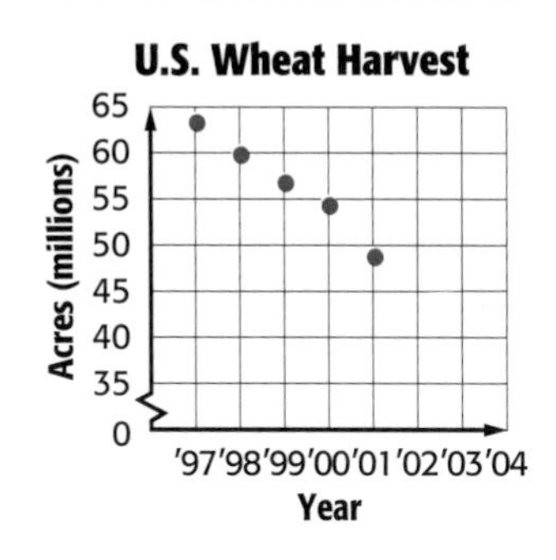

6-4 Multiplying Fractions and Mixed Numbers

What You'll LEARN

Multiply fractions and mixed numbers.

REVIEW Vocabulary

GCF: the greatest of the common factors of two or more numbers **(Lesson 5-2)**

WHEN am I ever going to use this?

EARTH SCIENCE About $\frac{1}{3}$ of the land in the United States is forests. About $\frac{2}{5}$ of U.S. forests are publicly owned.

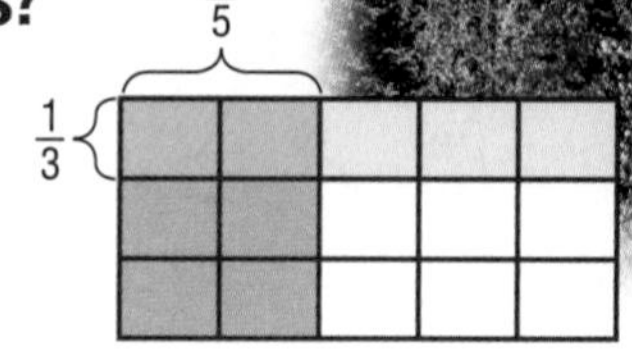

1. What part of the rectangle represents $\frac{1}{3}$?
2. What part of the rectangle represents $\frac{2}{5}$ of $\frac{1}{3}$?
3. **Make a conjecture** about what fraction of U.S. land is publicly owned forests.

You can multiply fractions by using the rule below.

Key Concept — **Multiplying Fractions**

Words To multiply fractions, multiply the numerators and multiply the denominators.

Symbols

Arithmetic: $\frac{1}{3} \times \frac{2}{5} = \frac{1 \times 2}{3 \times 5} = \frac{2}{15}$

Algebra: $\frac{a}{b} \times \frac{c}{d} = \frac{a \times c}{b \times d} = \frac{ac}{bd}$

EXAMPLES Multiply Fractions

Multiply. Write in simplest form.

1 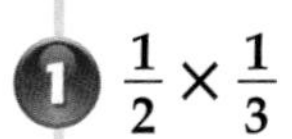 $\frac{1}{2} \times \frac{1}{3}$

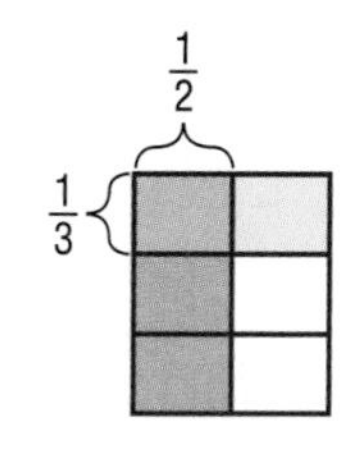

$\frac{1}{2} \times \frac{1}{3} = \frac{1 \times 1}{2 \times 3}$ ← Multiply the numerators. ← Multiply the denominators.

$= \frac{1}{6}$ Simplify.

2 $2 \times \frac{3}{4}$

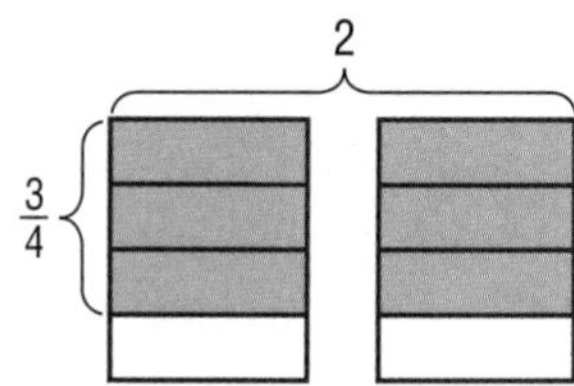

$2 \times \frac{3}{4} = \frac{2}{1} \times \frac{3}{4}$ Write 2 as $\frac{2}{1}$.

$= \frac{2 \times 3}{1 \times 4}$ Multiply the numerators and multiply the denominators.

$= \frac{6}{4}$ or $1\frac{1}{2}$ Simplify.

Your Turn **Multiply. Write in simplest form.**

a. $\frac{3}{5} \times \frac{1}{2}$ b. $\frac{1}{3} \times \frac{3}{4}$ c. $\frac{2}{3} \times 4$

If the numerator and denominator of either fraction have common factors, you can simplify before multiplying.

STUDY TIP

Simplifying You can simplify by dividing a numerator and a denominator by the same factor. Doing this step before you multiply is easier than doing it after you multiply.

EXAMPLE Simplify Before Multiplying

3 **Find $\frac{2}{7} \times \frac{3}{8}$. Write in simplest form.**

$\frac{2}{7} \times \frac{3}{8} = \frac{\overset{1}{\cancel{2}}}{7} \times \frac{3}{\underset{4}{\cancel{8}}}$ Divide 2 and 8 by their GCF, 2.

$= \frac{1 \times 3}{7 \times 4}$ Multiply the numerators and multiply the denominators.

$= \frac{3}{28}$ Simplify.

Your Turn **Multiply. Write in simplest form.**

d. $\frac{1}{3} \times \frac{3}{7}$ e. $\frac{4}{9} \times \frac{1}{8}$ f. $\frac{5}{6} \times \frac{3}{5}$

EXAMPLE Multiply Mixed Numbers

4 **Find $\frac{1}{2} \times 4\frac{2}{5}$. Write in simplest form.** **Estimate** $\frac{1}{2} \times 4 = 2$

Method 1 Rename the mixed number.

$\frac{1}{2} \times 4\frac{2}{5} = \frac{1}{\underset{1}{\cancel{2}}} \times \frac{\overset{11}{\cancel{22}}}{5}$ Rename $4\frac{2}{5}$ as an improper fraction, $\frac{22}{5}$.

$= \frac{1 \times 11}{1 \times 5}$ Multiply.

$= \frac{11}{5}$ or $2\frac{1}{5}$ Simplify.

The product is close to the estimate.

Method 2 Use mental math.

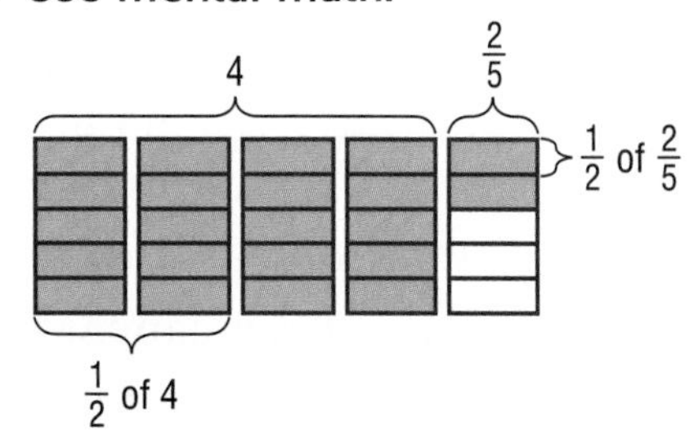

$\frac{1}{2} \times 4\frac{2}{5} = \frac{1}{2} \times \left(4 + \frac{2}{5}\right)$ Write $4\frac{2}{5}$ as a sum of its parts.

$= \left(\frac{1}{2} \times 4\right) + \left(\frac{1}{2} \times \frac{2}{5}\right)$ Distributive Property

$= 2 + \frac{1}{5}$ or $2\frac{1}{5}$ Multiply.

Compare this product to the model shown above.

STUDY TIP

Mental Math When you see a problem like $\frac{1}{2} \times 4\frac{2}{5}$, you can use the Distributive Property. Think, "What is $\frac{1}{2}$ of 4 and what is $\frac{1}{2}$ of $\frac{2}{5}$?" This is equal to $\frac{1}{2}\left(4 + \frac{2}{5}\right)$.

Your Turn **Multiply. Write in simplest form.**

g. $\frac{1}{4} \times 8\frac{4}{9}$ h. $3 \times 5\frac{1}{3}$ i. $1\frac{7}{8} \times 2\frac{2}{5}$

msmath2.net/extra_examples

Skill and Concept Check

1. **OPEN ENDED** Write a pair of fractions whose product is $\frac{8}{15}$.

2. **Which One Doesn't Belong?** Identify the expression that does not have the same value as the other three. Explain your reasoning.

$\frac{1}{5}$ of 25	$25 \times \frac{1}{5}$	$5 \div 25$	$\frac{1}{5} \cdot \frac{25}{1}$

3. **NUMBER SENSE** Is $18 \times \frac{4}{5}$ *greater than, less than,* or *equal to* 18? Explain.

GUIDED PRACTICE

Multiply. Write in simplest form.

4. $\frac{2}{3} \times \frac{1}{3}$
5. $\frac{1}{2} \times \frac{2}{5}$
6. $\frac{2}{3} \times \frac{3}{8}$
7. $\frac{1}{6} \times 4$
8. $2\frac{1}{4} \times \frac{2}{3}$
9. $1\frac{5}{6} \times 3\frac{3}{5}$

10. **FOOD** An average slice of American cheese is about $\frac{1}{8}$ inch thick. What is the height of a package containing 20 slices?

Practice and Applications

HOMEWORK HELP

For Exercises	See Examples
11–12	1
21–24, 34	2
13–20, 33	3
25–32, 35–36	4

Extra Practice
See pages 578, 601.

Multiply. Write in simplest form.

11. $\frac{3}{4} \times \frac{1}{8}$
12. $\frac{2}{5} \times \frac{2}{3}$
13. $\frac{1}{5} \times \frac{5}{6}$
14. $\frac{4}{9} \times \frac{1}{4}$
15. $\frac{2}{3} \times \frac{1}{4}$
16. $\frac{1}{12} \times \frac{3}{5}$
17. $\frac{4}{7} \times \frac{7}{8}$
18. $\frac{2}{5} \times \frac{15}{16}$
19. $\frac{3}{8} \times \frac{10}{27}$
20. $\frac{9}{10} \times \frac{5}{6}$
21. $9 \times \frac{1}{2}$
22. $\frac{4}{5} \times 6$

23. **ELECTIONS** In an election in which 4,500 votes were cast, one candidate received $\frac{3}{5}$ of the votes. How many votes did the candidate receive?

24. **PACKAGING** The plastic cases used to store compact disks and DVDs are about $\frac{1}{5}$-inch thick. A company wants to sell 10 of these cases in plastic wrapping. What is the height of 10 cases?

Multiply. Write in simplest form.

25. $4\frac{2}{3} \times \frac{4}{7}$
26. $\frac{5}{8} \times 2\frac{1}{2}$
27. $14 \times 1\frac{1}{7}$
28. $3\frac{3}{4} \times 8$
29. $9 \times 4\frac{2}{3}$
30. $4 \times 7\frac{5}{6}$
31. $3\frac{1}{4} \times 2\frac{2}{3}$
32. $5\frac{1}{3} \times 3\frac{3}{4}$

33. **TELEVISION** A media research survey showed that one evening, $\frac{2}{3}$ of all U.S. households had their TVs on, and $\frac{3}{8}$ of them were watching a World Series baseball game. What fraction of U.S. households was watching the game?

34. **ANIMALS** Komodo dragons are the largest lizards in the world. A 250-pound komodo dragon can eat enough at one time to increase its weight by $\frac{3}{4}$. Find $\frac{3}{4} \times 250$ to determine how much weight a 250-pound komodo dragon could gain after eating.

35. **TURTLES** A giant tortoise can travel about one tenth of a kilometer in an hour. At this rate, how far can it travel in $1\frac{3}{4}$ hours?

36. **FLAGS** By law, the length of an official United States flag must be $1\frac{9}{10}$ times its width. What is the length of the flag shown at the right?

37. **CRITICAL THINKING** Two positive fractions which are *not* improper are multiplied. Is the product *sometimes*, *always*, or *never* less than 1? Explain.

Standardized Test Practice and Mixed Review

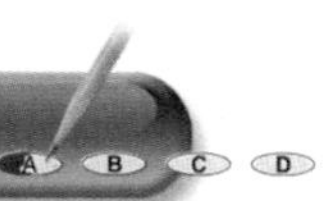

38. **MULTIPLE CHOICE** A box of books weighs $8\frac{2}{3}$ pounds. How much do $4\frac{1}{2}$ boxes weigh?

(A) 19 lb (B) $27\frac{2}{3}$ lb (C) $32\frac{1}{2}$ lb (D) 39 lb

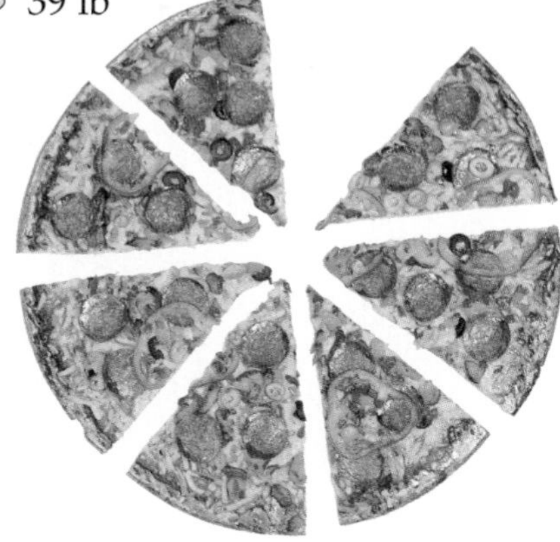

39. **GRID IN** Jeanette and Vanesa are each taking half of the leftover pizza shown at the right. What fraction of the whole pizza does each person take?

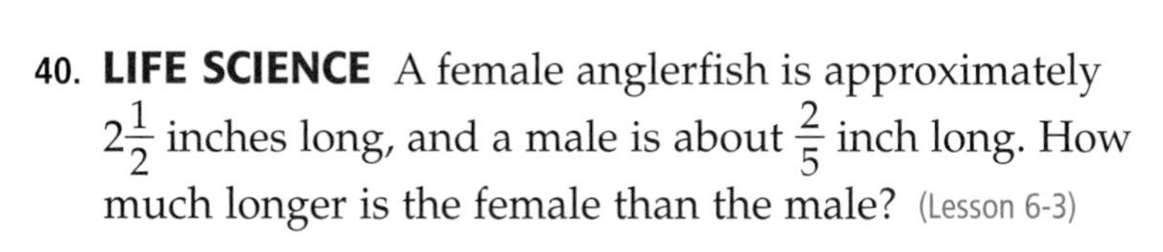

40. **LIFE SCIENCE** A female anglerfish is approximately $2\frac{1}{2}$ inches long, and a male is about $\frac{2}{5}$ inch long. How much longer is the female than the male? (Lesson 6-3)

Add or subtract. Write in simplest form. (Lesson 6-2)

41. $\frac{2}{7} + \frac{4}{7}$
42. $\frac{1}{2} - \frac{1}{8}$
43. $\frac{5}{9} + \frac{5}{6}$
44. $\frac{3}{4} - \frac{1}{6}$

Find the greatest common factor of each set of numbers. (Lesson 5-2)

45. 44, 60
46. 25, 75
47. 6, 15, 27
48. 12, 30, 48

49. **ALGEBRA** On a January day in Buffalo, New York, the temperature dropped to -20.8°C. Find this temperature in degrees Fahrenheit by using the formula $C = 0.6F - 17.8$. (Lesson 4-4)

GETTING READY FOR THE NEXT LESSON

PREREQUISITE SKILL **Multiply.** (Page 560)

50. $2.8 \cdot 5$
51. $1.9 \cdot 33$
52. $7 \cdot 12.5$
53. $3.6 \cdot 0.8$

6-5 Algebra: Solving Equations

HANDS-ON Mini Lab

Materials

- grid paper
- colored pencils

Work with a partner.

The model below shows $\frac{1}{2} \cdot 2$.

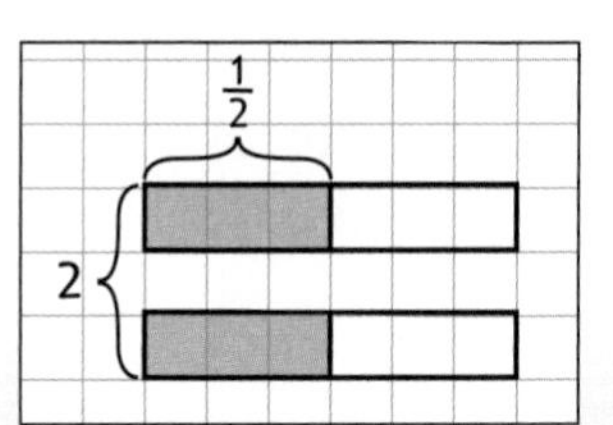

1. What is the product?
2. Use grid paper to model $\frac{1}{3} \cdot 3$. What is the product?
3. Copy and complete the table below.

$\frac{1}{2} \cdot 2 = \underline{\ ?\ }$	$\frac{2}{3} \cdot \frac{3}{2} = \underline{\ ?\ }$	$\frac{5}{6} \cdot \underline{\ ?\ } = 1$
$\frac{1}{3} \cdot 3 = \underline{\ ?\ }$	$\frac{3}{5} \cdot \frac{5}{3} = \underline{\ ?\ }$	$\frac{9}{20} \cdot \underline{\ ?\ } = 1$
$\frac{1}{4} \cdot \frac{4}{1} = \underline{\ ?\ }$	$\frac{11}{12} \cdot \frac{12}{11} = \underline{\ ?\ }$	$\underline{\ ?\ } \cdot \frac{17}{8} = 1$

What You'll LEARN

Solve equations with rational number solutions.

NEW Vocabulary

multiplicative inverse
reciprocal

Link to READING

Everyday Meaning of Inverse: opposite in order, as in an inverse statement in logic

Two numbers whose product is 1 are called **multiplicative inverses**, or **reciprocals**.

Key Concept — Multiplicative Inverse Property

Words The product of a number and its multiplicative inverse is 1.

Symbols

Arithmetic: $\frac{3}{4} \times \frac{4}{3} = 1$

Algebra: $\frac{a}{b} \times \frac{b}{a} = 1$, for $a, b \neq 0$

EXAMPLES Find Multiplicative Inverses

Find the multiplicative inverse of each number.

1 $\frac{2}{5}$

$\frac{2}{5} \cdot \frac{5}{2} = 1$ Multiply $\frac{2}{5}$ by $\frac{5}{2}$ to get the product 1.

The multiplicative inverse of $\frac{2}{5}$ is $\frac{5}{2}$, or $2\frac{1}{2}$.

2 $2\frac{1}{3}$

$2\frac{1}{3} = \frac{7}{3}$ Rename the mixed number as an improper fraction.

$\frac{7}{3} \cdot \frac{3}{7} = 1$ Multiply $\frac{7}{3}$ by $\frac{3}{7}$ to get the product 1.

The multiplicative inverse of $2\frac{1}{3}$ is $\frac{3}{7}$.

Your Turn **Find the multiplicative inverse of each number.**

a. $\frac{5}{6}$ b. $1\frac{1}{2}$ c. 8 d. $-\frac{4}{3}$

Look Back You can review the other **properties of equality** in Lessons 4-2 and 4-3.

In Chapter 4, you learned to solve equations using the Addition, Subtraction, and Division Properties of Equality. You can also solve equations by multiplying each side by the same number. This is called the **Multiplication Property of Equality**.

Key Concept — **Multiplication Property of Equality**

Words If you multiply each side of an equation by the same nonzero number, the two sides remain equal.

Symbols

Arithmetic	Algebra	
$5 = 5$	$\frac{x}{2} = -3$	$\frac{2}{3}x = 4$
$5 \cdot 2 = 5 \cdot 2$	$\frac{x}{2}(2) = -3(2)$	$\frac{3}{2} \cdot \frac{2}{3}x = \frac{3}{2} \cdot 4$
$10 = 10$	$x = -6$	$x = 6$

EXAMPLE Solve a Division Equation

3 Solve $\frac{n}{4} = 7$. Check your solution.

$\frac{n}{4} = 7$ Write the equation.

$\frac{n}{4} \cdot 4 = 7 \cdot 4$ Multiply each side of the equation by 4.

$n = 28$ Simplify.

The solution is 28.

When the coefficient of x is a fraction, multiply each side of the equation by its reciprocal.

EXAMPLE Use a Reciprocal to Solve an Equation

Standardized Test Practice

4 MULTIPLE-CHOICE TEST ITEM

What value of x makes $\frac{2}{3}x = -9$ a true sentence?

Ⓐ $\frac{3}{2}$ Ⓑ $-\frac{18}{27}$ Ⓒ $-\frac{9}{1}$ Ⓓ $-13\frac{1}{2}$

Test-Taking Tip *The Princeton Review*

Backsolving
Sometimes it may be easier to substitute each of the choices into the given equation until you get a true sentence.

Read the Test Item To find the value of x, solve the equation.

Solve the Test Item

$\frac{2}{3}x = -9$ Write the equation.

$\left(\frac{3}{2}\right)\frac{2}{3}x = \left(\frac{3}{2}\right)(-9)$ Multiply each side by the reciprocal of $\frac{2}{3}$, $\frac{3}{2}$.

$x = -\frac{27}{2}$ or $-13\frac{1}{2}$ Simplify.

The answer is D.

Your Turn **Solve each equation. Check your solution.**

e. $\frac{a}{6} = 2$ f. $6 = \frac{3}{8}d$ g. $\frac{x}{2.5} = -3$

Skill and Concept Check

1. **Tell** whether 8 is a solution of $\frac{n}{3} = 24$. Explain.

2. **OPEN ENDED** Write a division equation that can be solved by multiplying each side by $\frac{9}{4}$.

3. **Which One Doesn't Belong?** Identify the pair of numbers that does not have the same relationship as the other three. Explain your reasoning.

$\frac{7}{8}, \frac{8}{7}$	$5, \frac{1}{5}$	$\frac{2}{3}, 3$	$\frac{10}{3}, \frac{3}{10}$

GUIDED PRACTICE

Find the multiplicative inverse of each number.

4. $\frac{8}{5}$
5. $\frac{2}{9}$
6. -9
7. $5\frac{4}{5}$

Solve each equation. Check your solution.

8. $\frac{k}{16} = 2$
9. $-4 = \frac{y}{3}$
10. $6 = \frac{4}{7}u$
11. $\frac{1}{4}t = \frac{3}{8}$
12. $\frac{5}{7}y = -1.5$
13. $\frac{b}{8.2} = 2.5$

14. **MEASUREMENT** The weight in pounds p of an object with a mass m of 25 kilograms is given by the equation $\frac{p}{m} = 2.2$. How many pounds does the object weigh?

Practice and Applications

HOMEWORK HELP

For Exercises	See Examples
15–22	1, 2
23–36	3, 4

Extra Practice
See pages 578, 601.

Find the multiplicative inverse of each number.

15. $\frac{11}{2}$
16. $-\frac{9}{5}$
17. $-\frac{3}{8}$
18. $\frac{1}{6}$
19. 3
20. -14
21. $4\frac{2}{5}$
22. $6\frac{2}{3}$

Solve each equation. Check your solution.

23. $\frac{x}{12} = 3$
24. $\frac{d}{4} = 28$
25. $-\frac{2}{5}t = -12$
26. $-24 = \frac{3}{4}a$
27. $\frac{7}{8}k = -21$
28. $14 = \frac{8}{3}b$
29. $\frac{1}{2}z = -\frac{2}{5}$
30. $\frac{3}{5} = \frac{3}{7}r$
31. $35.1 = \frac{5}{6}m$
32. $-\frac{a}{3.2} = 5$
33. $0.8 = \frac{h}{3.6}$
34. $\frac{m}{4.6} = 2.8$

35. **VACATION** The distance Katie travels in her car while driving 55 miles per hour for 2.5 hours is given by the equation $\frac{d}{2.5} = 55$. How far did she travel?

36. **MONEY** Based on recent exchange rates, the equation $d = \frac{31}{50}c$ shows the value in U.S. dollars d for an amount of Canadian dollars c. To the nearest cent, find the value in Canadian currency for \$250 in U.S. dollars.

Data Update What is the value in Canadian currency for \$250 in U.S. dollars today? Visit msmath2.net/data_update to learn more.

37. **CRITICAL THINKING** In Lesson 11-5, you will learn that the area of a triangle A is given by the equation $A = \frac{1}{2}bh$, where b is the base of the triangle and h is the height. Explain how you can use the properties of equality to find the value of b in terms of A and h. Then solve for b.

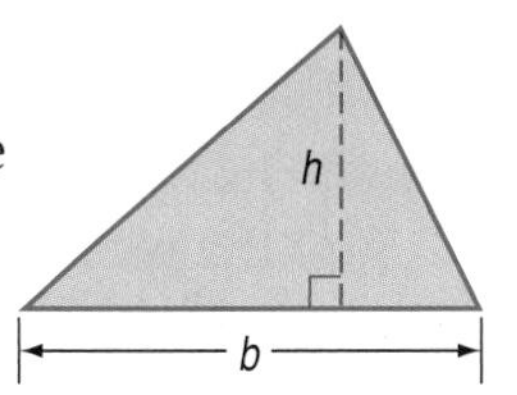

Standardized Test Practice and Mixed Review

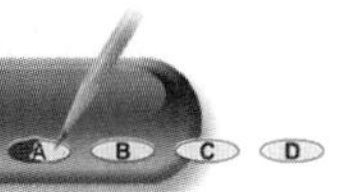

38. **MULTIPLE CHOICE** What is the reciprocal of $2\frac{1}{5}$?

Ⓐ $2\frac{5}{1}$ Ⓑ $\frac{11}{5}$ Ⓒ $\frac{5}{11}$ Ⓓ $\frac{1}{10}$

39. **GRID IN** Aaron is serving a 12-pound turkey at a dinner party. As a rule, you should allow about $\frac{3}{4}$ of a pound of meat per person. Use $\frac{3}{4}p = 12$ to find the number of people p that can be served at the dinner party.

Multiply. Write in simplest form. (Lesson 6-4)

40. $\frac{3}{8} \times \frac{4}{9}$

41. $1\frac{1}{2} \times 6$

42. $2\frac{2}{5} \times \frac{1}{6}$

43. $1\frac{1}{2} \times 1\frac{7}{9}$

44. Find $7\frac{1}{3} - 3\frac{5}{9}$. (Lesson 6-3)

For Exercises 45–47, use the graphic at the right. Write a fraction that compares the number of women champions to the total number of champions for each college. Write in simplest form. (Lesson 5-3)

45. Villanova

46. Texas–El Paso

47. Texas

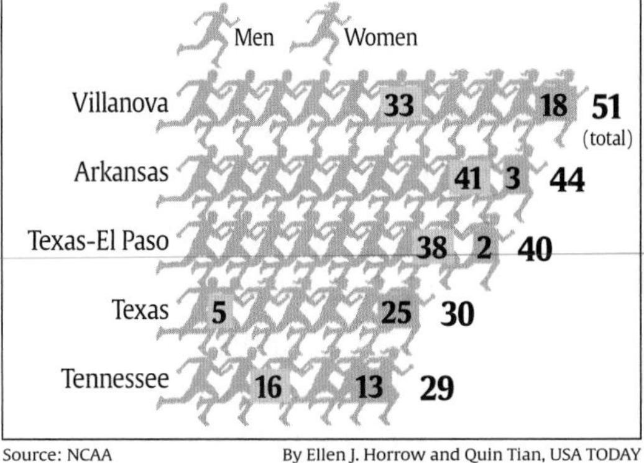

GETTING READY FOR THE NEXT LESSON

PREREQUISITE SKILL Estimate. (Lesson 6-1)

48. $18\frac{1}{6} \div 3$

49. $24\frac{3}{8} \div 11\frac{7}{9}$

50. $\frac{2}{11} \div \frac{11}{12}$

51. $\frac{9}{10} \div \frac{6}{7}$

Mid-Chapter Practice Test

Vocabulary and Concepts

1. **Write** an addition expression involving fractions shown by the model at the right. Then find the sum. Write in simplest form. (Lesson 6-2)

2. **Define** reciprocals. (Lesson 6-5)

Skills and Applications

Estimate. (Lesson 6-1)

3. $7\frac{1}{9} + 1\frac{1}{6}$
4. $13\frac{1}{2} \div 7\frac{2}{9}$
5. $\frac{11}{20} - \frac{5}{8}$

6. **SAVINGS** Jessica saves $\frac{1}{3}$ of the money she earns baby-sitting. If she earns $25 one evening, estimate the amount she saves. (Lesson 6-1)

Add, subtract, or multiply. Write in simplest form. (Lessons 6-2, 6-3, and 6-4)

7. $\frac{11}{15} - \frac{1}{15}$
8. $\frac{4}{7} - \frac{3}{14}$
9. $\frac{5}{8} + \frac{3}{4}$
10. $5\frac{1}{6} - 1\frac{1}{3}$
11. $\frac{7}{12} \times \frac{4}{9}$
12. $2\frac{3}{5} + 6\frac{13}{15}$
13. $2\frac{3}{4} \times 12$
14. $4\frac{2}{7} \times 5\frac{5}{6}$

15. **AIRPLANES** The aircraft *Voyager* weighed 2,000 pounds. In 1986, it carried about $3\frac{1}{2}$ times its weight in fuel to fly nonstop around the world. How many pounds of fuel did *Voyager* carry? (Lesson 6-4)

Solve each equation. Check your solution. (Lesson 6-5)

16. $\frac{t}{5} = -11$
17. $2 = \frac{3}{8}y$
18. $16.2 = \frac{3}{4}k$

Standardized Test Practice

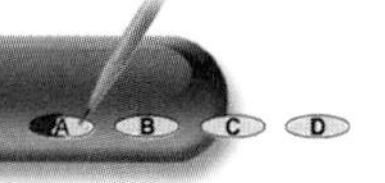

19. **MULTIPLE CHOICE** One batch of cookies uses $2\frac{1}{2}$ cups of flour and $1\frac{2}{3}$ cups of sugar. Which is the best estimate of the total amount of flour and sugar used in eight batches of cookies? (Lesson 6-1)
    - A less than 30 c
    - B between 30 c and 45 c
    - C between 45 c and 55 c
    - D more than 55 c

20. **MULTIPLE CHOICE** How much does a $12\frac{3}{4}$-pound package weigh after a $3\frac{5}{8}$-pound book is taken out of it? (Lesson 6-3)
    - F $8\frac{1}{8}$ lb
    - G 9 lb
    - H $9\frac{1}{8}$ lb
    - I 15 lb

The Game Zone

A Place To Practice Your Math Skills

Math Skill
Multiplying Fractions

Totally Mental

GET READY!

Players: two
Materials: 2 index cards, spinner with the digits 1 through 9

- Each player should make a game sheet on an index card like the one shown at the right.

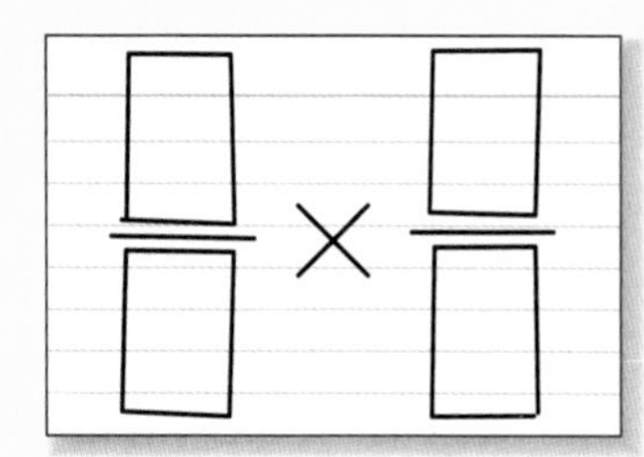

GO!

- One player spins the spinner. The number that is spun should be written in one of the four boxes on his or her game sheet.

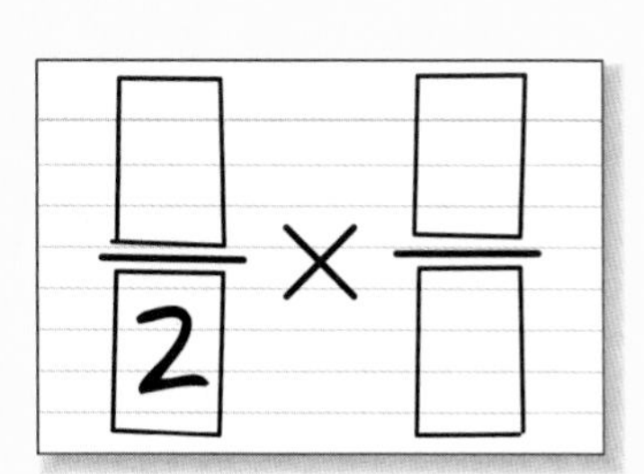

- The second player spins the spinner and writes the number from the spinner in one of the boxes on his or her game sheet.
- Continue until each person has spun the spinner four times.
- Find the product of the fractions on your game sheet.
- **Who Wins?** After four spins, the player with the greatest product is the winner.

6-6 Dividing Fractions and Mixed Numbers

What You'll LEARN

Divide fractions and mixed numbers.

HANDS-ON Mini Lab

Materials

- grid paper

Work with a partner.

The model at the right shows 2 units divided into thirds, or $2 \div \frac{1}{3}$.

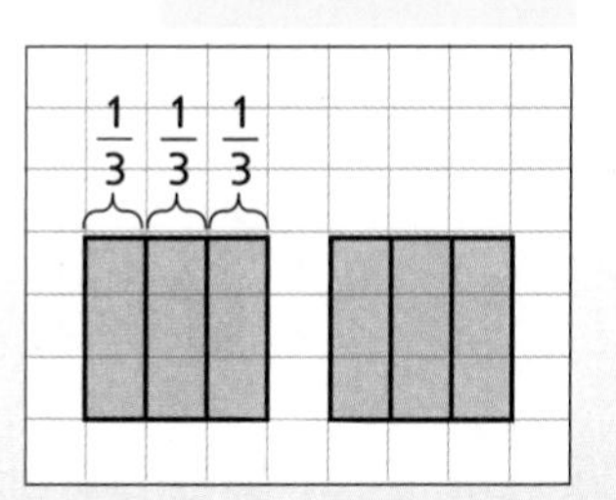

1. How many thirds are in 2 units?
2. What is $2 \div \frac{1}{3}$?
3. Draw a model to show $3 \div \frac{1}{2}$.
4. What is $3 \div \frac{1}{2}$?

Dividing 8 by 2 gives the same result as multiplying 8 by the reciprocal of 2, or $\frac{1}{2}$.

reciprocals

$8 \div 2 = 4 \qquad 8 \cdot \frac{1}{2} = 4$

same result

In the same way, dividing 4 by $\frac{1}{3}$ is the same as multiplying 4 by the reciprocal of $\frac{1}{3}$, or 3. This pattern is true for any rational number.

Key Concept — **Division by Fractions**

Words To divide by a fraction, multiply by its multiplicative inverse, or reciprocal.

Symbols

Arithmetic	Algebra
$\frac{7}{8} \div \frac{3}{4} = \frac{7}{8} \cdot \frac{4}{3}$	$\frac{a}{b} \div \frac{c}{d} = \frac{a}{b} \cdot \frac{d}{c}$, where $b, c, d \neq 0$

EXAMPLE Divide by a Fraction

1 Find $\frac{3}{4} \div \frac{1}{2}$. Write in simplest form.

$\frac{3}{4} \div \frac{1}{2} = \frac{3}{4} \cdot \frac{2}{1}$ Multiply by the reciprocal of $\frac{1}{2}$, which is $\frac{2}{1}$.

$= \frac{3}{\cancel{4}_2} \cdot \frac{\cancel{2}^1}{1}$ Divide by the GCF, 2.

$= \frac{3}{2}$ or $1\frac{1}{2}$ Multiply and simplify.

Your Turn **Divide. Write in simplest form.**

a. $\frac{3}{4} \div \frac{1}{4}$ b. $\frac{4}{5} \div \frac{8}{9}$ c. $\frac{5}{6} \div \frac{2}{3}$

To divide by a mixed number, rename the mixed number as an improper fraction.

EXAMPLES Divide by Mixed Numbers

2 **SURVEYING Fifteen acres of land are to be divided into $1\frac{1}{2}$-acre lots. How many lots will there be?**

Estimate $16 \div 2 = 8$

$15 \div 1\frac{1}{2} = 15 \div \frac{3}{2}$ Rename $1\frac{1}{2}$ as an improper fraction.

$= 15 \cdot \frac{2}{3}$ Multiply by the reciprocal of $\frac{3}{2}$, which is $\frac{2}{3}$.

$= \frac{\overset{5}{\cancel{15}}}{1} \cdot \frac{2}{\underset{1}{\cancel{3}}}$ Divide out common factors.

$= 10$ Multiply.

There will be 10 lots.

REAL-LIFE MATH

SURVEYING A surveyor can survey a line for nine miles before having to use spherical geometry and trigonometry to correct for Earth's curved surface.

Source: Kansas Society of Land Surveyors

3 **Find $\frac{2}{3} \div 3\frac{1}{3}$. Write in simplest form.**

Estimate $\frac{1}{2} \div 3 = \frac{1}{2} \times \frac{1}{3}$ or $\frac{1}{6}$

$\frac{2}{3} \div 3\frac{1}{3} = \frac{2}{3} \div \frac{10}{3}$ Rename $3\frac{1}{3}$ as an improper fraction.

$= \frac{2}{3} \cdot \frac{3}{10}$ Multiply by the reciprocal of $\frac{10}{3}$, which is $\frac{3}{10}$.

$= \frac{\overset{1}{\cancel{2}}}{\underset{1}{\cancel{3}}} \cdot \frac{\overset{1}{\cancel{3}}}{\underset{5}{\cancel{10}}}$ Divide out common factors.

$= \frac{1}{5}$ Multiply.

The quotient is close to the estimate.

4 **Find $6\frac{1}{2} \div 3\frac{5}{7}$.**

Estimate $6 \div 3 = 2 \leftarrow$ compatible numbers

$6\frac{1}{2} \div 3\frac{5}{7} = \frac{13}{2} \div \frac{26}{7}$ Rename the mixed numbers as improper fractions.

$= \frac{13}{2} \cdot \frac{7}{26}$ Multiply by the reciprocal of $\frac{26}{7}$, which is $\frac{7}{26}$.

$= \frac{\overset{1}{\cancel{13}}}{2} \cdot \frac{7}{\underset{2}{\cancel{26}}}$ Divide out common factors.

$= \frac{7}{4}$ Multiply.

$= 1\frac{3}{4}$ Simplify.

The quotient, $1\frac{3}{4}$, is close to the estimate.

Your Turn **Divide. Write in simplest form.**

d. $5 \div 1\frac{1}{3}$ **e.** $\frac{3}{4} \div 1\frac{1}{2}$ **f.** $2\frac{1}{3} \div 5\frac{5}{6}$

Skill and Concept Check

Writing Math
Exercises 1 & 2

1. **OPEN ENDED** Write a problem that is solved by finding $10 \div \frac{1}{4}$.

2. **Describe** the steps you would take to find *six divided by three-fourths.*

GUIDED PRACTICE

Divide. Write in simplest form.

3. $\frac{3}{5} \div \frac{1}{4}$
4. $\frac{3}{4} \div 6$
5. $\frac{1}{2} \div 7\frac{1}{2}$
6. $5\frac{3}{5} \div 4\frac{2}{3}$

7. **FOOD** How many $\frac{1}{8}$-pound boxes of mints can be made with 3 pounds?

Practice and Applications

HOMEWORK HELP

For Exercises	See Examples
8–14	1
15–17, 20	2, 3
18–19	4

Extra Practice
See pags 579, 601.

Divide. Write in simplest form.

8. $\frac{3}{8} \div \frac{6}{7}$
9. $\frac{5}{9} \div \frac{5}{6}$
10. $\frac{2}{3} \div \frac{1}{2}$
11. $\frac{7}{8} \div \frac{3}{4}$
12. $6 \div \frac{1}{2}$
13. $\frac{4}{9} \div 2$
14. $2\frac{2}{3} \div 4$
15. $5 \div 1\frac{1}{3}$
16. $\frac{2}{3} \div 2\frac{1}{2}$
17. $\frac{8}{9} \div 5\frac{1}{3}$
18. $4\frac{1}{2} \div 6\frac{3}{4}$
19. $5\frac{2}{7} \div 2\frac{1}{7}$

20. **CRAFTS** Jared is making bookmarks like the one shown at the right. How many bookmarks can he make from a 15-yard spool of ribbon?

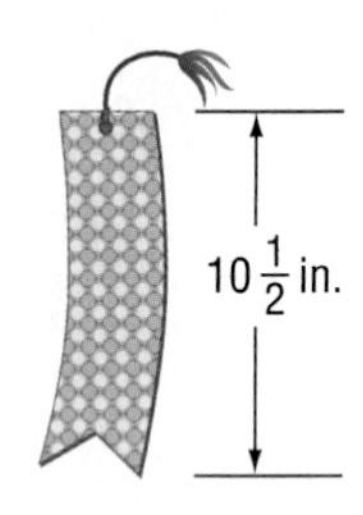

21. **CRITICAL THINKING** Will the quotient $7\frac{1}{6} \div 3\frac{2}{3}$ be a fraction less than 1 or greater than 1? Explain.

22. **EXTENDING THE LESSON** If you add any two rational numbers, the sum is always a rational number. So, the set of rational numbers is *closed* under addition. Is the set of rational numbers closed under subtraction, multiplication, and division also? Explain.

Standardized Test Practice and Mixed Review

A B C D

23. **SHORT RESPONSE** Leticia is dividing $2\frac{3}{4}$ pounds of trail mix equally among each of her four friends. How much does each receive?

24. **GRID IN** What is $\frac{8}{9}$ divided by $\frac{2}{9}$?

Find the multiplicative inverse of each number. (Lesson 6-5)

25. $\frac{6}{7}$
26. $\frac{4}{13}$
27. 8
28. $5\frac{1}{4}$

29. Find $\frac{1}{10} \times \frac{5}{8}$. Write in simplest form. (Lesson 6-4)

GETTING READY FOR THE NEXT LESSON

PREREQUISITE SKILL Multiply or divide. (Pages 560, 562)

30. 2.5×20
31. 3.5×4
32. $4{,}200 \div 2.1$
33. $104 \div 6.5$

6-7 Measurement: Changing Customary Units

What You'll LEARN

Change units in the customary system.

NEW Vocabulary

pound
ounce
ton
cup
pint
quart
gallon

WHEN am I ever going to use this?

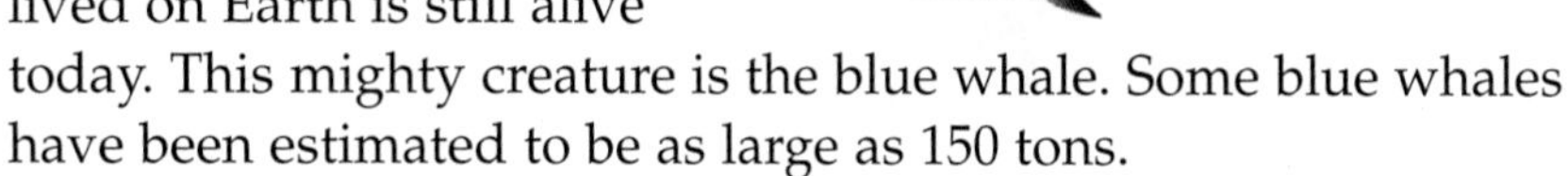

ANIMALS The largest creature that has ever lived on Earth is still alive today. This mighty creature is the blue whale. Some blue whales have been estimated to be as large as 150 tons.

1. There are 2,000 pounds in 1 ton. How many pounds are in 150 tons?
2. What operation did you use to find the weight in pounds? Explain.

The relationships among customary units of length, weight, and capacity are shown in the table at the right.

To change units, use the following rules.

- To convert from larger units to smaller units, multiply.
- To convert from smaller units to larger units, divide.

Customary Units of Length
1 foot (ft) = 12 inches (in.)
1 yard (yd) = 3 feet
1 mile (mi) = 5,280 feet
Customary Units of Weight
1 **pound** (lb) = 16 **ounces** (oz)
1 **ton** (T) = 2,000 pounds
Customary Units of Capacity
1 **cup** (c) = 8 fluid ounces (fl oz)
1 **pint** (pt) = 2 cups
1 **quart** (qt) = 2 pints
1 **gallon** (gal) = 4 quarts

larger units

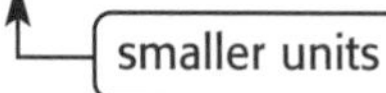

EXAMPLES Convert Larger Units to Smaller Units

Complete.

1 **20 ft = __?__ in.**

Since 1 foot = 12 inches, multiply by 12.

$20 \times 12 = 240$

20 feet = 240 inches

2 **$3\frac{1}{2}$ lb = __?__ oz**

Since 1 pound = 16 ounces, multiply by 16.

$3\frac{1}{2} \times 16 = 56$

$3\frac{1}{2}$ pounds = 56 ounces

Your Turn **Complete.**

a. 36 yd = __?__ ft

b. $2\frac{3}{4}$ T = __?__ lb

c. $1\frac{1}{2}$ c = __?__ fl oz

EXAMPLE Convert Units to Solve a Problem

3 **SKIING Speed skiing takes place on a course that is about two thirds of a mile long. How many feet long is the course?**

$\frac{2}{3} \times 5{,}280 = 3{,}520$ Multiply by 5,280 since there are 5,280 feet in 1 mile.

So, the course is about 3,520 feet long.

REAL-LIFE MATH

SKIING Ross Anderson is an American Indian speed skier. He has been clocked on skis going as fast as 137 miles per hour.

Source: *Native Peoples*

To convert from smaller units to larger units, divide.

EXAMPLES Convert Smaller Units to Larger Units

Complete.

4 **750 lb = ___?___ T**

Since 2,000 pounds are in 1 ton, divide by 2,000.

$750 \div 2{,}000 = \frac{3}{8}$

$750 \text{ pounds} = \frac{3}{8} \text{ ton}$

5 **10 c = ___?___ pt**

Since 2 cups are in 1 pint, divide by 2.

$10 \div 2 = 5$

10 cups = 5 pints

Your Turn **Complete.**

d. 2,640 ft = ___?___ mi e. 100 oz = ___?___ lb f. 14 pt = ___?___ qt

Skill and Concept Check

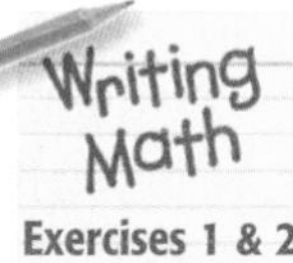

1. **OPEN ENDED** Write a problem in which you would need to convert pints to cups.

2. **Which One Doesn't Belong?** Identify the unit of measure that does not have the same characteristic as the other three. Explain your reasoning.

gallon	pint	fluid ounce	pound

GUIDED PRACTICE

Complete.

3. 48 oz = ___?___ lb
4. $5\frac{1}{3}$ yd = ___?___ ft
5. 12 qt = ___?___ gal
6. 28 in. = ___?___ ft
7. $\frac{1}{4}$ T = ___?___ lb
8. 15 pt = ___?___ qt

9. **DINOSAURS** The average weight of the dinosaur *Argentinosaurus* was estimated to be 200,000 pounds. How many tons did it weigh?

10. **HISTORY** Liquid products such as oil and vinegar were once shipped in huge containers called *hogsheads*. A hogshead contained 63 gallons of liquid. How many quarts did it contain?

Practice and Applications

For Exercises	See Examples
11–30	1, 2, 4, 5
31–32	3

HOMEWORK HELP

Extra Practice
See pages 579, 601.

Complete.

11. 18 ft = __?__ yd **12.** 2 lb = __?__ oz **13.** 4 gal = __?__ qt

14. 5,000 lb = __?__ T **15.** $4\frac{1}{2}$ pt = __?__ c **16.** 72 oz = __?__ lb

17. 2 mi = __?__ ft **18.** $1\frac{1}{4}$ mi = __?__ ft **19.** 9 c = __?__ pt

20. 3 c = __?__ fl oz **21.** $2\frac{3}{4}$ qt = __?__ pt **22.** 120 ft = __?__ yd

23. 7,040 ft = __?__ mi **24.** $3\frac{3}{8}$ T = __?__ lb **25.** 172 oz = __?__ lb

26. If 4 cups = 1 quart, then 9 cups = __?__ quarts.

27. If 36 inches = 1 yard, then 2.3 yards = __?__ inches.

Complete.

28. $1\frac{1}{4}$ gal = __?__ c **29.** 880 yd = __?__ mi **30.** 24 fl oz = __?__ qt

31. MULTI STEP Suppose a car repair company changes the oil of 50 cars and they recover an average of $3\frac{1}{2}$ quarts of oil from each car. How many gallons of oil did they recover?

32. MULTI STEP A window-washing solution can be made by mixing $1\frac{1}{3}$ cups of ammonia and $1\frac{1}{2}$ cups of vinegar with baking soda and water. Will the solution fit in a $\frac{1}{2}$-quart pan? Explain.

33. CRITICAL THINKING Make a table that shows the number of ounces in 1, 2, 3, and 4 pounds. Graph the ordered pairs (pounds, ounces) on a coordinate plane and connect the points. Describe the graph.

Standardized Test Practice and Mixed Review

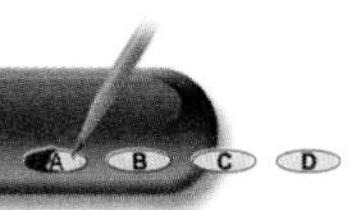

34. SHORT RESPONSE How many cups of milk are shown at the right?

35. MULTIPLE CHOICE A can of orange juice concentrate makes 48 fluid ounces of orange juice. How many pints is this?

(A) 3 pt (B) 4 pt (C) 6 pt (D) 12 pt

36. Find $1\frac{4}{7} \div 1\frac{5}{6}$. Write in simplest form. (Lesson 6-6)

Solve each equation. Check your solution. (Lesson 6-5)

37. $\frac{y}{4} = 7$ **38.** $\frac{1}{3}x = \frac{5}{9}$ **39.** $-4 = \frac{p}{2.7}$ **40.** $6n = -15$

GETTING READY FOR THE NEXT LESSON

PREREQUISITE SKILL **Evaluate each expression.** (Lesson 1-3)

41. $2 \cdot 8 + 2 \cdot 9$ **42.** $3(7) + 4(2)$ **43.** $2(6.5 + 3)$ **44.** $5 \cdot 2 + 5 \cdot 8.4$

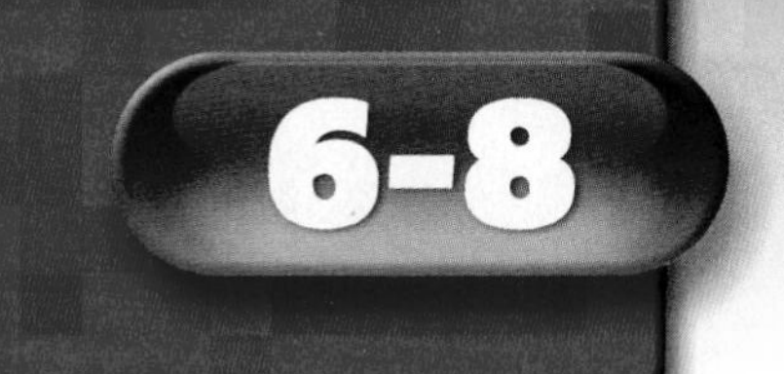

6-8 Geometry: Perimeter and Area

What You'll LEARN

Find the perimeters and areas of figures.

NEW Vocabulary

perimeter
formula
area

WHEN am I ever going to use this?

PARKS Central Park in New York City contains a running track, walking paths, playgrounds, and even a carousel.

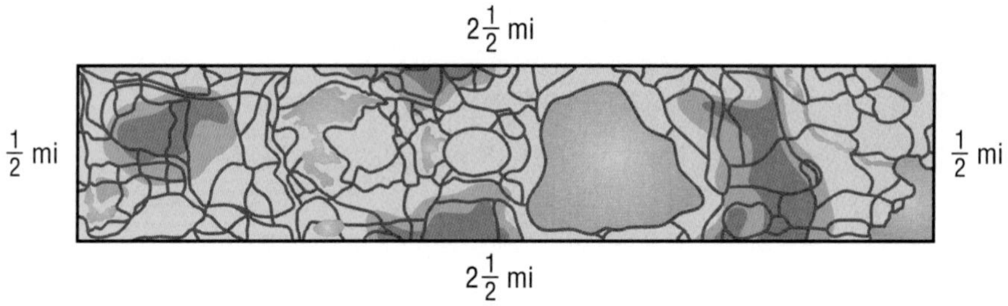

1. If you walked around the outer edge of the entire park, how far would you walk?
2. Describe how you found the distance.
3. Explain how you can use both multiplication and addition to find the distance.

The distance around a geometric figure is called the **perimeter**. To find the perimeter P of a rectangle, add the measures of the four sides.

Key Concept — Perimeter of a Rectangle

Words The perimeter P of a rectangle is twice the sum of the length ℓ and width w.

Symbols $P = \ell + \ell + w + w$

$P = 2\ell + 2w$ or $2(\ell + w)$

Model

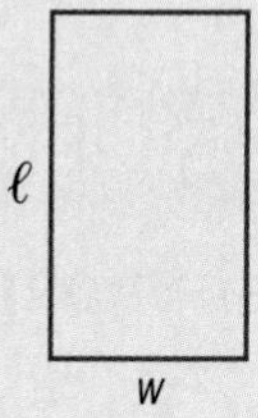

EXAMPLE Find the Perimeter of a Rectangle

1 **Find the perimeter of the rectangle shown at the right.**

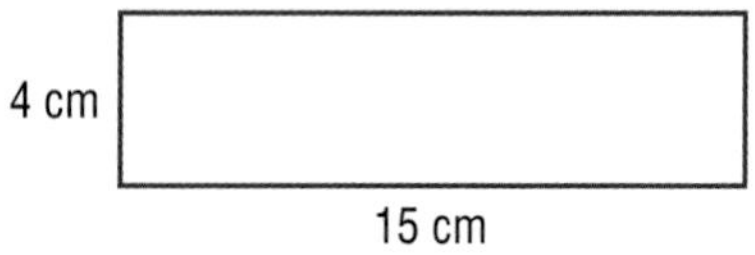

$P = 2\ell + 2w$	Perimeter of a rectangle
$P = 2(15) + 2(4)$	Replace ℓ with 15 and w with 4.
$P = 30 + 8$	Multiply.
$P = 38$	Add.

The perimeter is 38 centimeters.

You can find the perimeter of irregular figures by adding the lengths of the sides.

EXAMPLE Find the Perimeter of an Irregular Figure

2 Find the perimeter of the figure.

Estimate 3 + 5 + 7 + 3 = 18 in.

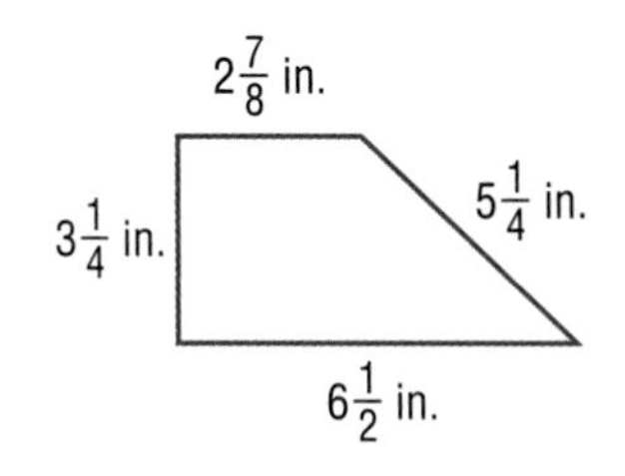

$P = 2\frac{7}{8} + 5\frac{1}{4} + 6\frac{1}{2} + 3\frac{1}{4}$

$P = 2\frac{7}{8} + 5\frac{2}{8} + 6\frac{4}{8} + 3\frac{2}{8}$

$P = 16\frac{15}{8}$ or $17\frac{7}{8}$

The perimeter is $17\frac{7}{8}$ inches. This is close to the estimate.

Your Turn

a. Find the perimeter of the figure.

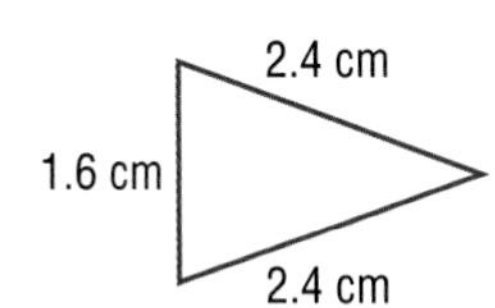

The equation $P = 2\ell + 2w$ is called a **formula** because it shows a relationship among quantities. The formula for the **area** of a rectangle, or the measure of the surface enclosed by a figure, is shown below.

Key Concept — Area of a Rectangle

Words The area A of a rectangle is the product of the length ℓ and width w.

Symbol $A = \ell \cdot w$

Model

w

ℓ

EXAMPLE Find the Area of a Rectangle

3 VOLLEYBALL Find the area of the volleyball court shown at the right.

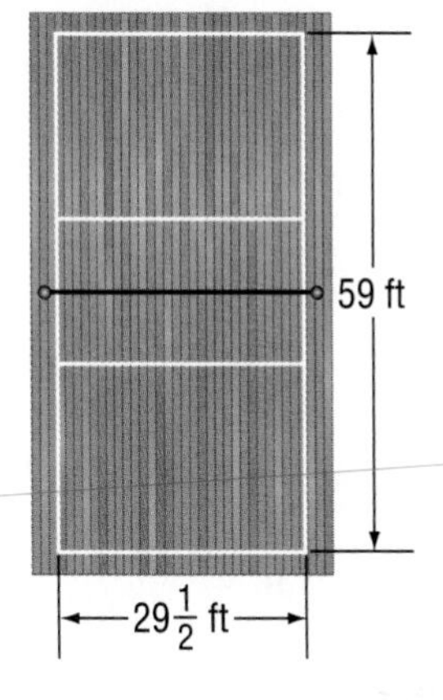

$A = \ell \cdot w$ Area of a rectangle

$A = 59 \cdot 29\frac{1}{2}$ Replace ℓ with 59 and w with $29\frac{1}{2}$.

$A = \frac{59}{1} \cdot \frac{59}{2}$ Rename 59 and $29\frac{1}{2}$.

$A = 1{,}740\frac{1}{2}$ Multiply and simplify.

The area is $1{,}740\frac{1}{2}$ square feet.

STUDY TIP

Area Units When finding area, the units are also multiplied. So, area is given in *square units*. Consider a rectangle 2 ft by 3 ft.

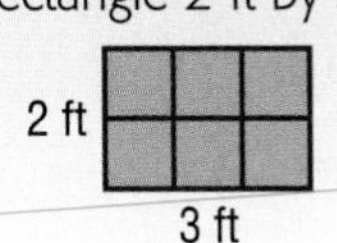

$A = 2 \text{ ft} \times 3 \text{ ft}$

$A = (2 \times 3)(\text{ft} \times \text{ft})$

$A = 6 \text{ ft}^2$

Your Turn Find the perimeter and area of each rectangle.

b.

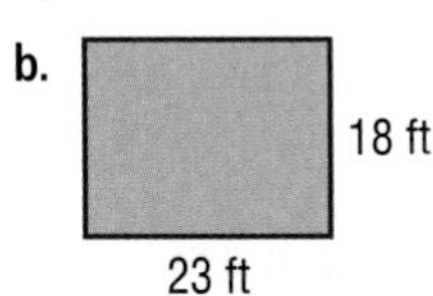

c. length = 13.2 mm
width = 8 mm

Skill and Concept Check

1. **Explain** why perimeter is given in units and area is given in square units.

2. **OPEN ENDED** Draw and label a rectangle that has an area of 24 square centimeters. What is the perimeter of your rectangle?

GUIDED PRACTICE

Find the perimeter and area of each rectangle.

3. 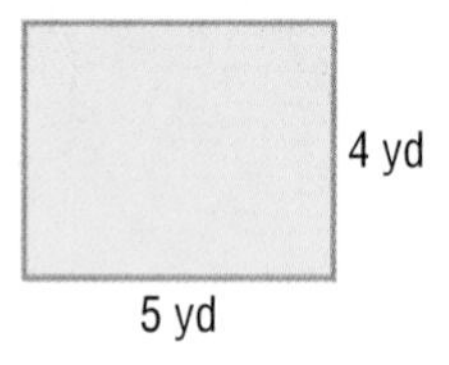

4. 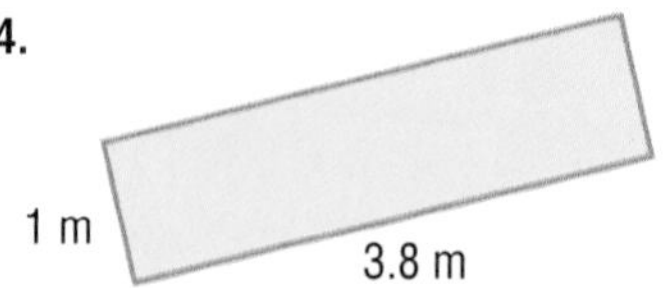

5. $\ell = 7$ cm, $w = 6$ cm

6. $\ell = 5\frac{1}{2}$ in., $w = 3$ in.

7. Find the perimeter of the figure at the right.

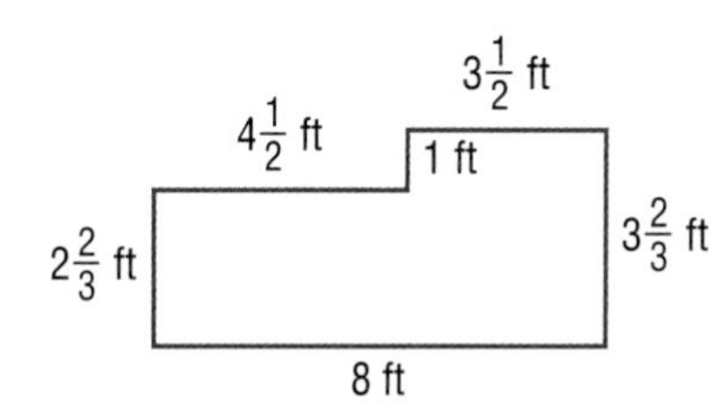

Practice and Applications

Find the perimeter and area of each rectangle.

HOMEWORK HELP	
For Exercises	See Examples
8–22, 25–28	1, 3
23–24	2
Extra Practice See pages 579, 601.	

8. 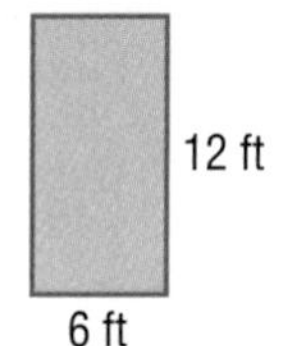

9. 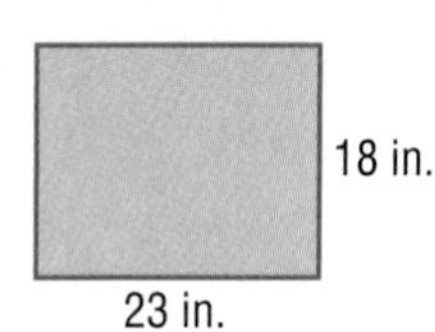

10. 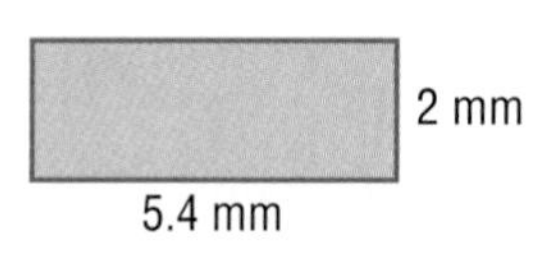

11. 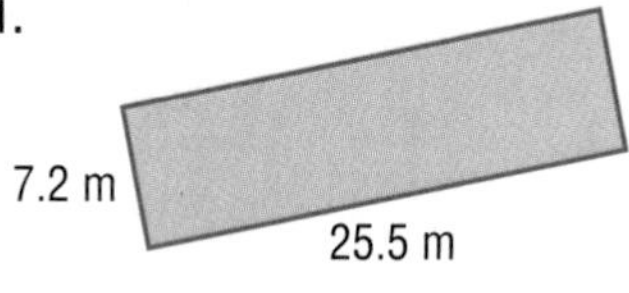

12. 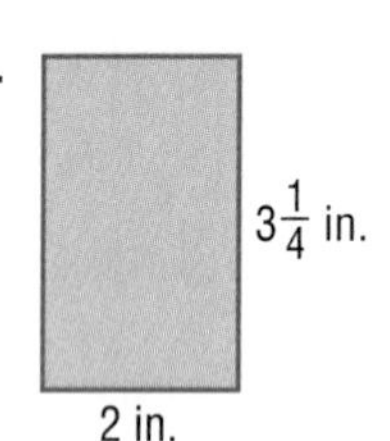

13. 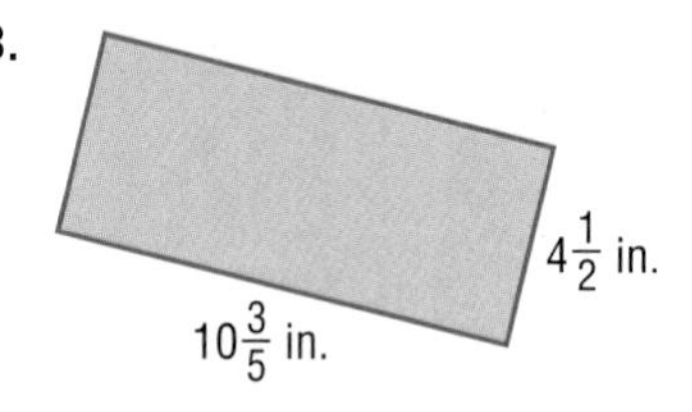

14. $\ell = 5$ ft, $w = 1$ ft

15. $\ell = 18$ cm, $w = 12$ cm

16. $\ell = 6.5$ m, $w = 4$ m

17. $\ell = 4\frac{3}{8}$ in., $w = 3\frac{1}{5}$ in.

18. $\ell = 2$ ft, $w = 18$ in.

19. $\ell = 35$ ft, $w = 7$ yd

For Exercises 20 and 21, use the square at the right.

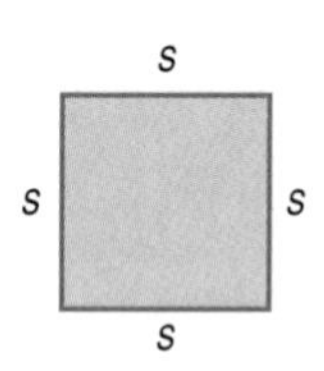

20. Write formulas for the perimeter P and area A of the square.

21. If the side length is doubled, what happens to the perimeter and area?

22. Find the width of a rectangle with an area of 30 square inches and a length of 5 inches.

Find the perimeter of each figure.

23.

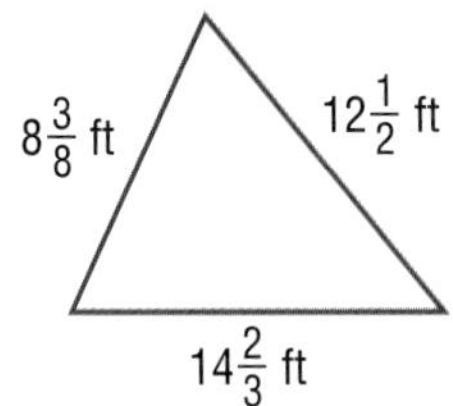

24.

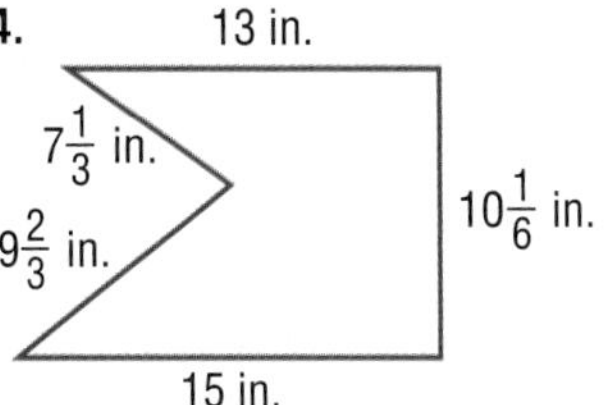

25. **FLOORING** Jasmine plans to use 1-foot square tiles to tile her kitchen floor, which measures 18 feet by 14 feet. If there are 40 tiles per box, how many boxes must she buy?

FOOTBALL For Exercises 26 and 27, use the table at the right.

Football Field	Length (ft)	Width (ft)
American	300	160
Canadian	330	197

26. How much greater is the area of a Canadian football field than an American football field?

27. An *acre* equals 43,560 square feet. How many acres is a Canadian football field? Round to the nearest tenth of an acre and explain your method.

28. A rectangle is made with exactly 9 feet of string. One side is $2\frac{5}{16}$ feet long. What is the length of the other side?

29. **CRITICAL THINKING** Compare and contrast the perimeters and areas of rectangles that have the following dimensions: 1 by 9, 2 by 8, 3 by 7, and 4 by 6.

EXTENDING THE LESSON For Exercises 30–32, refer to the figures at the right.

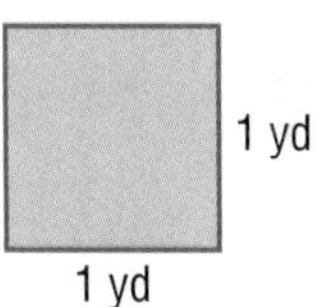

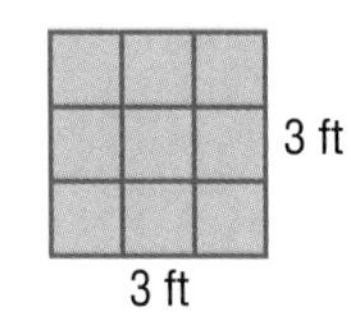

30. 1 yd^2 = __?__ ft^2 31. 4 yd^2 = __?__ ft^2 32. 1 ft^2 = __?__ in^2

Standardized Test Practice and Mixed Review

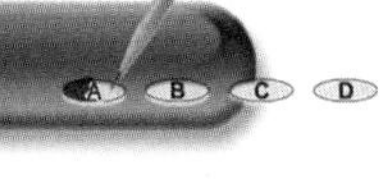

33. **MULTIPLE CHOICE** The perimeter of the rectangle shown at the right is $41\frac{1}{2}$ feet. What is the value of x?

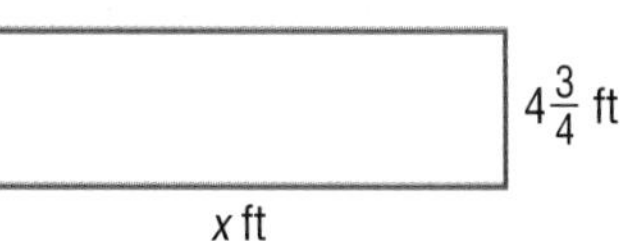

Ⓐ 16 Ⓑ 32 Ⓒ $36\frac{3}{4}$ Ⓓ 64

34. **SHORT RESPONSE** Determine the area of a rectangle that is 3 centimeters wide and 7 centimeters long.

Complete. (Lesson 6-7)

35. $5\frac{1}{4}$ T = __?__ lb 36. 8 yd = __?__ ft 37. 15 pt = __?__ qt 38. 72 in. = __?__ ft

39. Find $22 \div \frac{2}{3}$. Write in simplest form. (Lesson 6-6)

GETTING READY FOR THE NEXT LESSON

PREREQUISITE SKILL Multiply. Write in simplest form. (Lesson 6-4)

40. $\frac{9}{8} \cdot 16$ 41. $\frac{22}{7} \cdot 14$ 42. $2 \cdot \frac{3}{7} \cdot 35$ 43. $\frac{22}{7} \cdot 1\frac{1}{2}$

6-9a HANDS-ON LAB

A Preview of Lesson 6-9

Circumference

What You'll LEARN

Find a relationship between circumference and diameter.

Materials

- ruler
- measuring tape
- circular objects

INVESTIGATE *Work with a partner.*

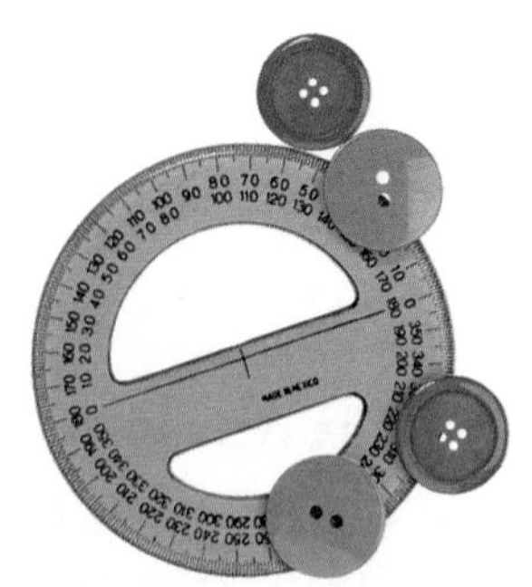

In this lab, you will investigate how *circumference*, or the distance around a circle, is related to its *diameter*, or the distance across a circle.

STEP 1 Use a ruler to measure the diameter of a circular object. Record the measure in a table like the one shown below.

Object	Diameter (cm)	Circumference (cm)

STEP 2 Make a small mark at the edge of the circular object. The measuring tape should be on a flat surface. Place the mark at the beginning of the measuring tape. Roll the object along the tape for one revolution, until you reach the mark again.

STEP 3 Record the length in the table. This is the circumference.

STEP 4 Repeat this activity with circular objects of various sizes.

Writing Math

Work with a partner.

1. For each object, divide the circumference by the diameter. Add another column to your table and record the results. Round to the nearest tenth if necessary.
2. What do you notice about the ratios?
3. **Graph** the ordered pair (diameter, circumference) on a coordinate plane for each object. What do you find?
4. Select two points on the graph and find the slope between them. Select two different points and find the slope. What do you observe about the slopes?
5. Use the graph to predict the circumference of a circular object that has a diameter of 18 centimeters.
6. **Write** a rule describing how you would find the circumference C of a circle if you know the diameter d.

STUDY TIP

Look Back You can review **slope** in Lesson 4-7.

6-9 Geometry: Circles and Circumference

What You'll LEARN

Find the circumference of circles.

NEW Vocabulary

circle
center
diameter
radius
circumference

MATH Symbols

π (pi)

WHEN am I ever going to use this?

FERRIS WHEELS The London Eye Ferris wheel measures 450 feet across.

1. Which point appears to be the center of the Ferris wheel?
2. Is the distance from G to F greater than, less than, or equal to the distance from G to J?

A **circle** is the set of all points in a plane that are the same distance from a given point, called the **center**.

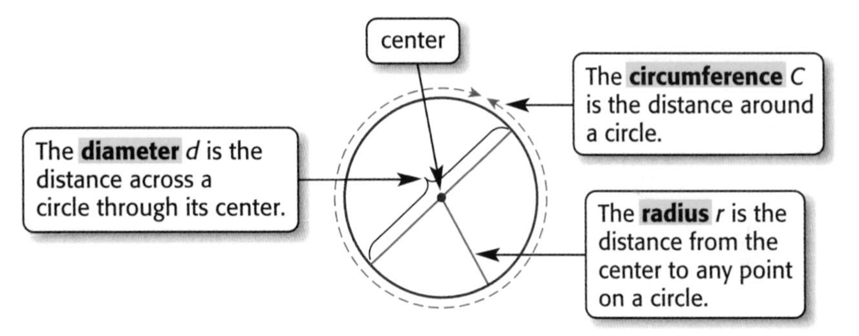

The diameter of a circle is 2 times the radius, or $d = 2r$. Another relationship that is true of all circles is $\frac{C}{d} = 3.1415926\ldots$. This nonterminating and nonrepeating number is represented by the Greek letter **π (pi)**. An approximation often used for π is 3.14.

Key Concept — Circumference of a Circle

Words The circumference C of a circle is equal to its diameter d times π, or 2 times its radius r times π.

Symbols $C = \pi d$ or $C = 2\pi r$

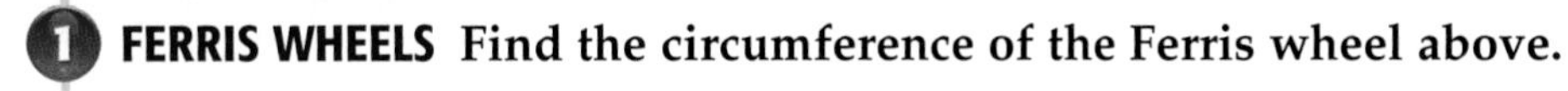

EXAMPLE Find Circumference

1 FERRIS WHEELS Find the circumference of the Ferris wheel above.

$C = \pi d$	Circumference of a circle
$C \approx 3.14(450)$	Replace π with 3.14 and d with 450.
$C \approx 1{,}413$	Multiply.

So, the distance around the Ferris wheel is about 1,413 feet.

Another approximation for π is $\frac{22}{7}$. Use this value when the radius or diameter of a circle is a multiple of 7 or has a multiple of 7 in its numerator.

Technology You can use a calculator to find the circumference. To find $2\pi(21)$, press 2 [×] [π] [×] 21 [ENTER]. The circumference is about 131.9468915.

EXAMPLE Find Circumference

2 Find the circumference of a circle with a radius of 21 inches.

Since 21 is a multiple of 7, use $\frac{22}{7}$ for π.

$C = 2\pi r$ — Circumference of a circle

$C \approx 2 \cdot \frac{22}{7} \cdot 21$ — Replace π with $\frac{22}{7}$ and r with 21.

$C \approx 2 \cdot \frac{22}{\cancel{7}_1} \cdot \frac{\cancel{21}^3}{1}$ — Divide by the GCF, 7.

$C \approx 132$ — Simplify.

21 in.

The circumference of the circle is about 132 inches.

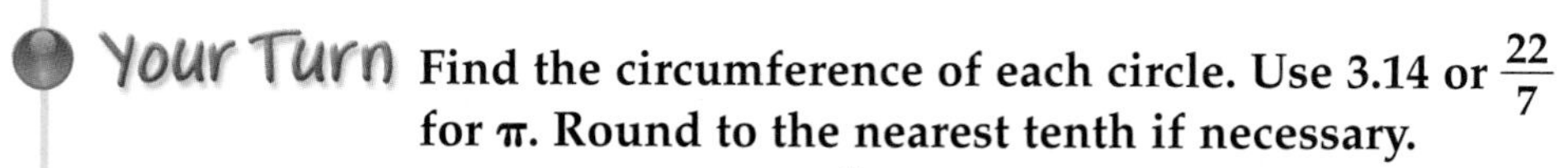

Your Turn **Find the circumference of each circle. Use 3.14 or $\frac{22}{7}$ for π. Round to the nearest tenth if necessary.**

a. diameter = 4.5 cm b. radius = $\frac{7}{8}$ ft c. radius = 35 in.

Skill and Concept Check

1. **Explain** how circumference is affected by an increase in radius.

2. **OPEN ENDED** Describe a real-life situation in which finding the circumference of a circle would be useful.

3. **FIND THE ERROR** Aidan and Mya are finding the circumference of a circle with a radius of 5 inches. Who is correct? Explain.

Aidan	Mya
$C = \pi \cdot 5 = 5\pi$	$C = 2 \cdot \pi \cdot 5 = 10\pi$

GUIDED PRACTICE

Find the circumference of each circle. Use 3.14 or $\frac{22}{7}$ for π. Round to the nearest tenth if necessary.

4. 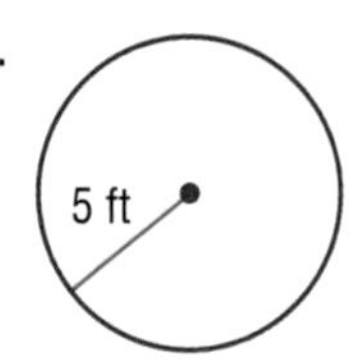

5. 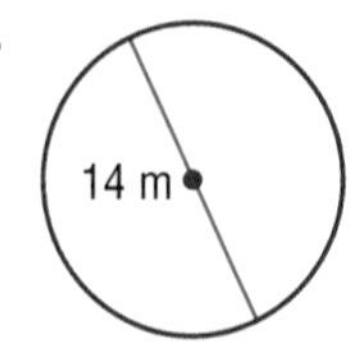

6. 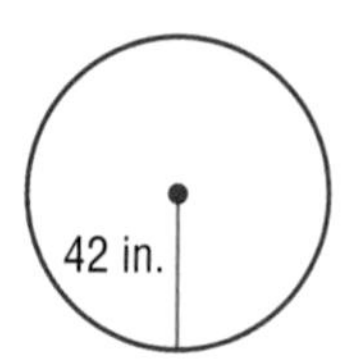

7. radius = 11.7 cm

8. radius = 28 ft

9. diameter = $3\frac{1}{2}$ yd

10. **MUSIC** Purdue University's marching band has a drum with a diameter of 8 feet. What is its circumference to the nearest tenth?

Practice and Applications

HOMEWORK HELP	
For Exercises	See Examples
11–23	1, 2

Extra Practice
See pages 579, 601.

Find the circumference of each circle. Use 3.14 or $\frac{22}{7}$ for π. Round to the nearest tenth if necessary.

11. 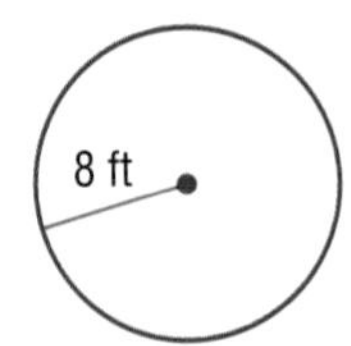

12. 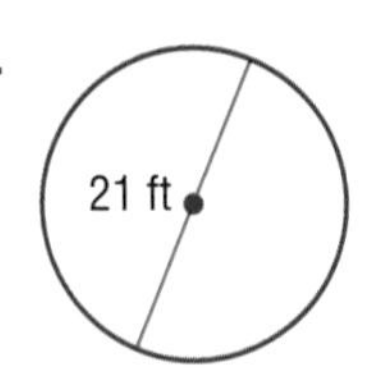

13. 7 yd

14. radius = 38.4 cm
15. diameter = 15.1 m
16. radius = $1\frac{3}{4}$ in.
17. radius = 2 km
18. diameter = 10 ft
19. radius = 45.5 m
20. radius = 56 cm
21. radius = $2\frac{5}{8}$ in.
22. diameter = $10\frac{1}{2}$ in.

23. **CROPS** The mysterious crop circle shown at the right was created in England in a single night by an unknown source. What is the circumference of the circle?

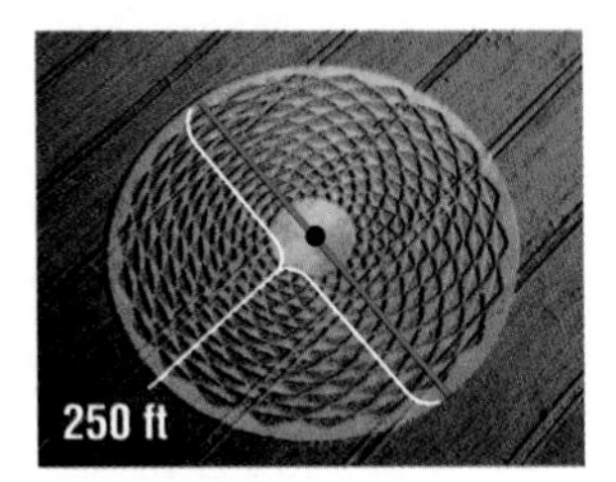

24. **RESEARCH** Use the Internet or another source to find three other parts of circles: *arcs*, *central angles*, and *inscribed angles*. Draw a circle and label these parts.

CRITICAL THINKING For Exercises 25 and 26, refer to the figure at the right.

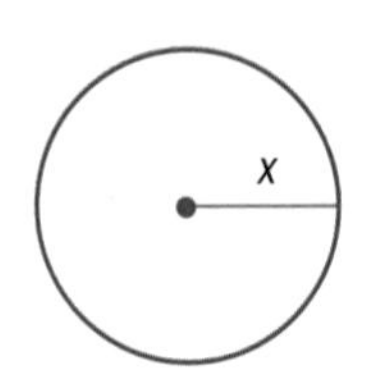

25. How many lengths x will fit on the circumference of the circle?
26. If the value of x is doubled, what effect will this have on the diameter? on the circumference?

Standardized Test Practice and Mixed Review

27. **SHORT RESPONSE** What is the circumference of the tree trunk whose cross section is shown at the right? Use 3.14 for π.

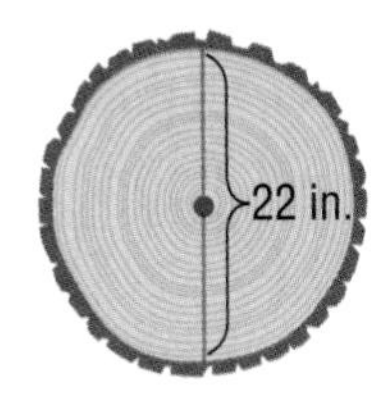

28. **MULTIPLE CHOICE** About how far does a bicycle wheel travel in 150 revolutions of the wheel if its diameter is 11 inches? Use 3.14 for π.

Ⓐ 431.75 ft   Ⓑ 863.5 ft   Ⓒ 5,181 ft   Ⓓ 10,362 ft

Find the perimeter and area of each rectangle. (Lesson 6-8)

29. $\ell = 5$ cm, $w = 3.8$ cm
30. $\ell = 2\frac{1}{4}$ ft, $w = 11$ ft

31. Twenty-four feet equals how many yards? (Lesson 6-7)

WebQuest Interdisciplinary Project

A Well-Balanced Diet

It's time to complete your project. Use the data you have gathered about the food you eat and about the Food Pyramid to prepare a Web page or poster. Be sure to include a chart and calculations with your project.

msmath2.net/webquest

CHAPTER 6 Study Guide and Review

Vocabulary and Concept Check

area (p. 271)	diameter (p. 275)	pint (p. 267)
center (p. 275)	formula (p. 271)	pound (p. 267)
circle (p. 275)	gallon (p. 267)	quart (p. 267)
circumference (p. 275)	multiplicative inverse (p. 258)	radius (p. 275)
compatible numbers (p. 242)	ounce (p. 267)	reciprocal (p. 258)
cup (p. 267)	perimeter (p. 270)	ton (p. 267)

Choose the correct term or number to complete each sentence.

1. The (radius, diameter) is the distance across a circle through its center.
2. To add like fractions, add the (numerators, denominators).
3. (Perimeter, Area) is the measure of the surface enclosed by a figure.
4. Fractions with different denominators are called (like, unlike) fractions.
5. When dividing by a fraction, multiply by its (value, reciprocal).
6. One cup is equivalent to (8, 12) fluid ounces.

Lesson-by-Lesson Exercises and Examples

6-1 Estimating with Fractions (pp. 240–243)

Estimate.

7. $2\frac{9}{10} \div 1\frac{1}{8}$
8. $6\frac{2}{9} - 5\frac{1}{7}$
9. $\frac{13}{15} \times \frac{1}{5}$
10. $\frac{1}{2} + \frac{3}{8}$
11. $\frac{1}{2} \cdot 25$
12. $15\frac{6}{7} \div 7\frac{1}{3}$

Example 1 Estimate $5\frac{1}{12} + 2\frac{5}{6}$.

$5\frac{1}{12} + 2\frac{5}{6} \rightarrow 5 + 3 = 8$

Example 2 Estimate $\frac{7}{8} - \frac{4}{7}$.

$\frac{7}{8} - \frac{4}{7} \rightarrow 1 - \frac{1}{2} = \frac{1}{2}$

6-2 Adding and Subtracting Fractions (pp. 244–247)

Add or subtract. Write in simplest form.

13. $\frac{2}{6} - \frac{1}{6}$
14. $\frac{3}{7} + \frac{9}{14}$
15. $\frac{5}{6} - \frac{3}{4}$
16. $\frac{1}{9} + \frac{5}{9}$
17. $\frac{4}{5} + \frac{4}{5}$
18. $\frac{9}{10} - \frac{3}{10}$
19. $\frac{5}{8} - \frac{5}{12}$
20. $\frac{11}{12} - \frac{1}{6}$
21. $\frac{3}{4} + \frac{7}{20}$

Example 3 Find $\frac{1}{8} + \frac{3}{8}$.

$\frac{1}{8} + \frac{3}{8} = \frac{1+3}{8}$ Add the numerators.

$= \frac{4}{8}$ or $\frac{1}{2}$ Simplify.

Example 4 Find $\frac{3}{10} - \frac{1}{4}$.

$\frac{3}{10} - \frac{1}{4} = \frac{6}{20} - \frac{5}{20}$ or $\frac{1}{20}$

msmath2.net/vocabulary_review

6-3 Adding and Subtracting Mixed Numbers (pp. 248–251)

Add or subtract. Write in simplest form.

22. $3\frac{2}{15} + 6\frac{9}{15}$
23. $9\frac{4}{5} - 2\frac{1}{5}$
24. $4\frac{1}{3} - 2\frac{2}{3}$
25. $8\frac{2}{7} + 1\frac{6}{7}$
26. $7\frac{11}{12} - 4\frac{3}{12}$
27. $7\frac{3}{5} - 5\frac{1}{3}$
28. $5\frac{3}{4} + 1\frac{1}{6}$
29. $3\frac{5}{8} + 11\frac{1}{2}$
30. $4\frac{3}{10} - 2\frac{4}{5}$

Example 5 Find $5\frac{2}{3} + 3\frac{1}{2}$.

$5\frac{2}{3} + 3\frac{1}{2} = 5\frac{4}{6} + 3\frac{3}{6}$ Rename the fractions.

$= 8\frac{7}{6}$ or $9\frac{1}{6}$ Add the whole numbers and add the fractions.

Example 6 Find $4\frac{1}{5} - 2\frac{3}{5}$.

$4\frac{1}{5} - 2\frac{3}{5} = 3\frac{6}{5} - 2\frac{3}{5}$ Rename $4\frac{1}{5}$ as $3\frac{6}{5}$.

$= 1\frac{3}{5}$ Subtract the whole numbers and subtract the fractions.

6-4 Multiplying Fractions and Mixed Numbers (pp. 254–257)

Multiply. Write in simplest form.

31. $\frac{3}{5} \times \frac{2}{7}$
32. $\frac{5}{12} \times \frac{4}{9}$
33. $\frac{3}{5} \times \frac{10}{21}$
34. $4 \times \frac{13}{20}$
35. $2\frac{1}{3} \times \frac{3}{4}$
36. $4\frac{1}{2} \times 2\frac{1}{12}$
37. **TRACK AND FIELD** One lap around the high school track is $\frac{3}{8}$ of a mile. If Matthew runs $4\frac{1}{2}$ laps, how far does he run?

Example 7 Find $\frac{5}{9} \times \frac{2}{3}$.

$\frac{5}{9} \times \frac{2}{3} = \frac{5 \times 2}{9 \times 3}$ Multiply the numerators and multiply the denominators.

$= \frac{10}{27}$ Simplify.

Example 8 Find $3\frac{1}{2} \times 2\frac{3}{4}$.

$3\frac{1}{2} \times 2\frac{3}{4} = \frac{7}{2} \times \frac{11}{4}$ Rename $3\frac{1}{2}$ and $2\frac{3}{4}$.

$= \frac{7 \times 11}{2 \times 4}$ Multiply the numerators and multiply the denominators.

$= \frac{77}{8}$ or $9\frac{5}{8}$ Simplify.

6-5 Algebra: Solving Equations (pp. 258–261)

Find the multiplicative inverse of each number.

38. $\frac{7}{12}$
39. 5
40. $3\frac{1}{3}$

Solve each equation. Check your solution.

41. $8 = \frac{w}{2}$
42. $\frac{4}{5}b = 12$
43. $-7.6 = \frac{n}{3}$
44. **EARTH SCIENCE** In 1996, a new planet was discovered. Earth's diameter, 7,970 miles, is only $\frac{5}{86}$ the size of this planet's diameter. Solve $\frac{5}{86}d = 7{,}970$ to find d, the diameter of the new planet in miles.

Example 9 Find the multiplicative inverse of $\frac{9}{5}$.

$\frac{9}{5} \cdot \frac{5}{9} = 1$ The product of $\frac{9}{5}$ and $\frac{5}{9}$ is 1.

The multiplicative inverse of $\frac{9}{5}$ is $\frac{5}{9}$.

Example 10 Solve $\frac{3}{4}g = 2$.

$\frac{3}{4}g = 2$ Write the equation.

$\frac{4}{3} \cdot \frac{3}{4}g = \frac{4}{3} \cdot 2$ Multiply each side by the reciprocal of $\frac{3}{4}$.

$g = \frac{8}{3}$ or $2\frac{2}{3}$ Simplify.

6-6 Dividing Fractions and Mixed Numbers (pp. 264–266)

Divide. Write in simplest form.

45. $\frac{3}{5} \div \frac{6}{7}$
46. $\frac{1}{2} \div \frac{1}{3}$
47. $5 \div \frac{10}{13}$
48. $4 \div \frac{2}{3}$
49. $2\frac{3}{4} \div \frac{5}{6}$
50. $\frac{2}{5} \div 3$
51. $\frac{6}{11} \div 4$
52. $4\frac{3}{10} \div 2\frac{1}{5}$
53. $\frac{2}{7} \div \frac{8}{21}$

Example 11 **Find $2\frac{4}{5} \div \frac{7}{10}$.**

$2\frac{4}{5} \div \frac{7}{10} = \frac{14}{5} \div \frac{7}{10}$ Rename $2\frac{4}{5}$.

$= \frac{\overset{2}{\cancel{14}}}{\underset{1}{\cancel{5}}} \cdot \frac{\overset{2}{\cancel{10}}}{\underset{1}{\cancel{7}}}$ Multiply by the reciprocal of $\frac{7}{10}$.

$= \frac{4}{1}$ or 4 Simplify.

6-7 Measurement: Changing Customary Units (pp. 267–269)

Complete.

54. 4 qt = __?__ pt
55. 6 gal = __?__ qt
56. 48 oz = __?__ lb
57. 8,000 lb = __?__ T
58. 9 c = __?__ pt
59. 36 in. = __?__ ft

Example 12 **Complete: 32 qt = __?__ gal**

Since 4 quarts are in 1 gallon, divide by 4.

$32 \div 4 = 8$

32 quarts = 8 gallons

6-8 Geometry: Area and Perimeter (pp. 270–273)

Find the perimeter and area of each rectangle.

60.

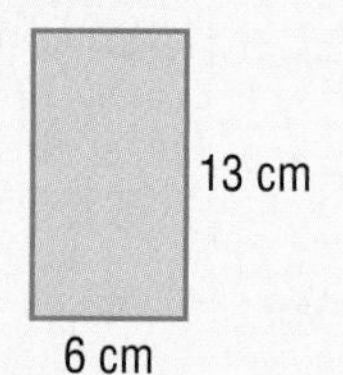

61.

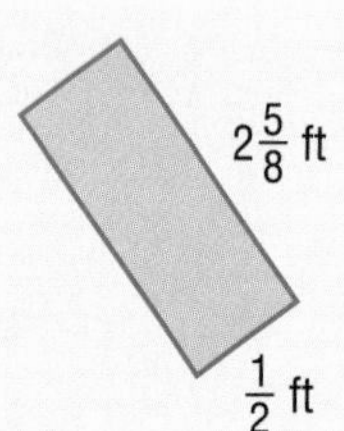

62. $\ell = 9$ cm, $w = 4$ cm
63. $\ell = 5$ in., $w = \frac{1}{2}$ in.
64. $\ell = 3.2$ m, $w = 6$ m
65. $\ell = 4\frac{1}{2}$ ft, $w = 2\frac{1}{3}$ ft

Example 13 **Find the perimeter and area of the rectangle.**

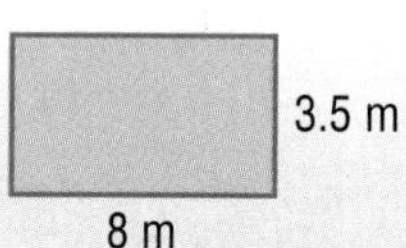

$P = 2\ell + 2w$ Perimeter of a rectangle

$P = 2(8) + 2(3.5)$ Substitution

$P = 23$ Simplify.

The perimeter is 23 meters.

$A = \ell \cdot w$ Area of a rectangle

$A = 8 \cdot 3.5$ Replace ℓ with 8 and w with 3.5.

$A = 28$ Multiply.

The area is 28 square meters.

6-9 Geometry: Circles and Circumference (pp. 275–277)

Find the circumference of each circle. Use 3.14 or $\frac{22}{7}$ for π. Round to the nearest tenth if necessary.

66. $r = 4.2$ cm
67. $d = 8$ yd
68. $r = \frac{7}{11}$ ft
69. $d = 8\frac{2}{5}$ ft

Example 14 **Find the circumference of a circle with a diameter of 12.2 meters. Round to the nearest tenth.**

$C = \pi d$ Circumference of a circle

$C \approx 3.14(12.2)$ $\pi \approx 3.14$ and $d = 12.2$

$C \approx 38.308$ Multiply.

The circumference is about 38.3 meters.

CHAPTER 6 Practice Test

Vocabulary and Concepts

1. **Explain** how to add unlike fractions.
2. **Define** perimeter.

Skills and Applications

Estimate.

3. $5\frac{7}{9} - 1\frac{2}{13}$
4. $3\frac{1}{12} + 6\frac{5}{7}$
5. $\frac{3}{7} \times \frac{13}{15}$

Add, subtract, multiply, or divide. Write in simplest form.

6. $\frac{4}{15} + \frac{8}{15}$
7. $\frac{7}{10} - \frac{1}{6}$
8. $\frac{5}{8} \times \frac{2}{5}$
9. $6 \times \frac{8}{21}$
10. $4\frac{5}{12} - 2\frac{1}{12}$
11. $6\frac{7}{9} + 3\frac{5}{12}$
12. $8\frac{2}{7} - 1\frac{5}{14}$
13. $\frac{5}{6} \div \frac{2}{3}$
14. $\frac{8}{9} \div 5\frac{1}{3}$

15. **COOKING** Taylor wants to make $2\frac{1}{2}$ times the quantity given in a recipe. The recipe calls for $1\frac{3}{4}$ cups of flour. How much flour will Taylor need?

16. **FLAG DAY** A giant cake decorated as an American flag measured 60 feet by 90 feet. What was the perimeter of the cake?

Solve each equation. Check your solution.

17. $\frac{y}{3} = 8$
18. $-6 = \frac{2}{5}m$
19. $\frac{3}{4} = \frac{5}{8}x$

Complete.

20. 42 ft = __?__ yd
21. 9 qt = __?__ pt
22. 7,600 lb = __?__ T

23. Find the perimeter and area of the rectangle.

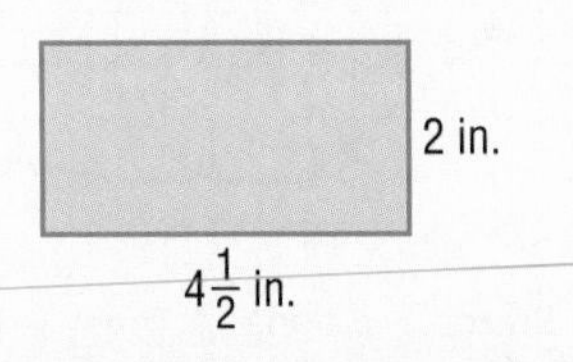

24. Find the circumference of a circle with a radius of 5 meters. Round to the nearest tenth.

Standardized Test Practice

25. **MULTIPLE CHOICE** In the 1999–2000 school year, the average backpack weighed $7\frac{1}{2}$ pounds. In the 2001–2002 school year, the average backpack weighed $7\frac{1}{5}$ pounds. By how much did the average backpack weight decrease?

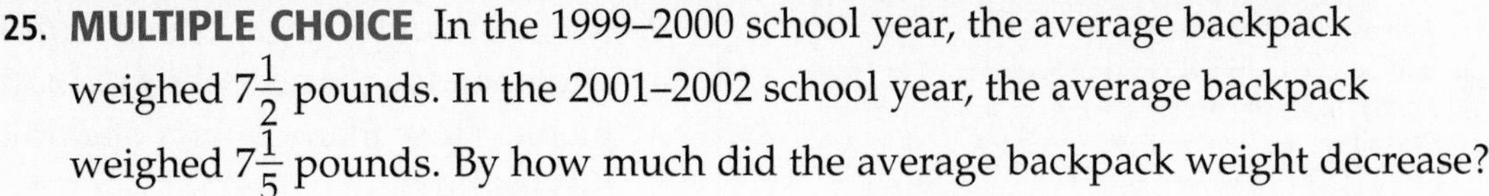

Ⓐ $\frac{1}{5}$ lb Ⓑ $\frac{3}{10}$ lb Ⓒ $\frac{1}{2}$ lb Ⓓ $\frac{7}{10}$ lb

CHAPTER 6 Standardized Test Practice

PART 1 Multiple Choice

Record your answers on the answer sheet provided by your teacher or on a sheet of paper.

1. The table shows four major rivers that run through Texas. Which is the most appropriate way to display this information? (Lessons 2-2 and 2-7)

River	Length (mi)
Brazos	950
Sabine	380
Trinity	360
Washita	500

Source: *The World Almanac*

(A)
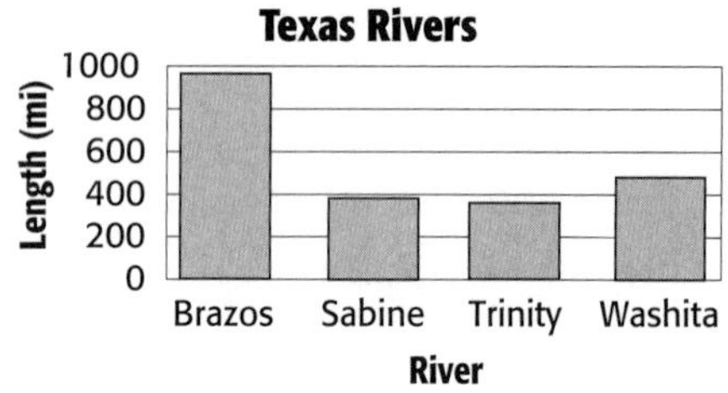

(B)
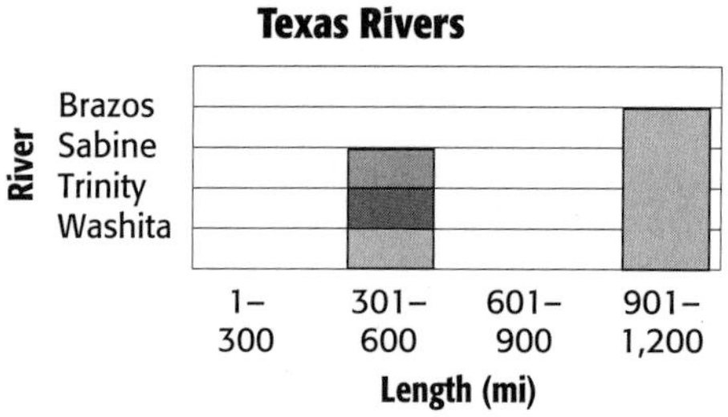

(C)
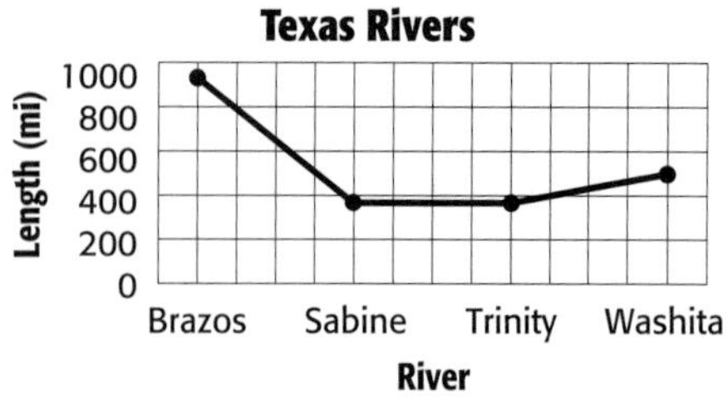

(D)
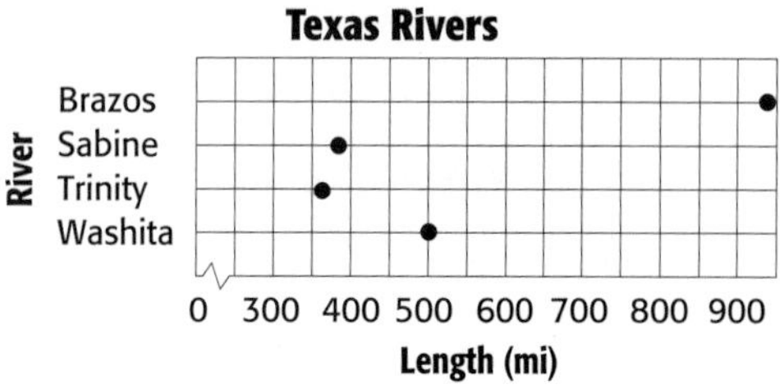

2. Which is the solution of $2x - 5 > 13$? (Lesson 4-5)

(F)
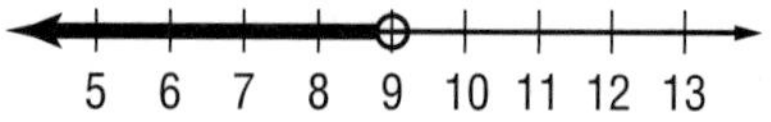

(G)
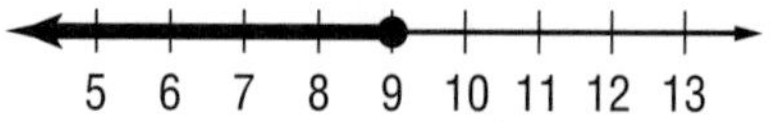

(H)
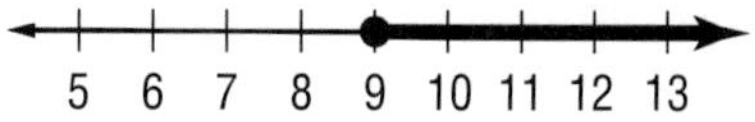

(I)
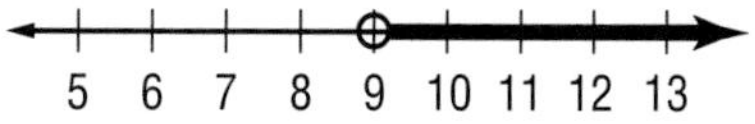

3. A jellyfish's body is made up of 95% water. What is 95% written as a decimal? (Lesson 5-6)

(A) 0.095 (B) 0.95

(C) 9.5 (D) 95

4. Dimitri has $6\frac{1}{8}$ ounces of orange juice and $10\frac{2}{3}$ ounces of cranberry juice. What is the best estimate of the total ounces of juice that he has? (Lesson 6-1)

(F) 15 oz (G) 16 oz

(H) 17 oz (I) 18 oz

5. Cole had $\frac{7}{8}$ of a tank of gas in the lawn mower. After mowing the grass, he had $\frac{1}{4}$ of a tank. What fraction of a tank did Cole use mowing the lawn? (Lesson 6-2)

(A) $\frac{1}{8}$ (B) $\frac{3}{8}$ (C) $\frac{5}{8}$ (D) $\frac{3}{4}$

6. What is the solution of $27 = \frac{3}{4}t$? (Lesson 6-5)

(F) 108 (G) 36

(H) 20.25 (I) 9

7. Kaylee bought 16 gallons of fruit juices and soda for the school's graduation picnic. How many quarts of drinks did Kaylee buy? (Lesson 6-7)

(A) 4 qt (B) 16 qt (C) 18 qt (D) 64 qt

PART 2 Short Response/Grid In

Record your answers on the answer sheet provided by your teacher or on a sheet of paper.

8. If 4 computers are needed for every 7 students in a grade, how many computers are needed for 280 students? (Lesson 1-1)

9. What are the coordinates of point K? (Lesson 3-3)

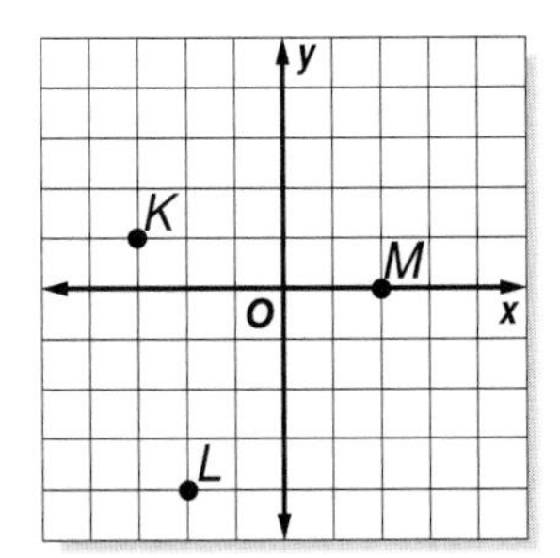

10. The graph shows the average number of greeting cards purchased yearly by the average person in the United States.

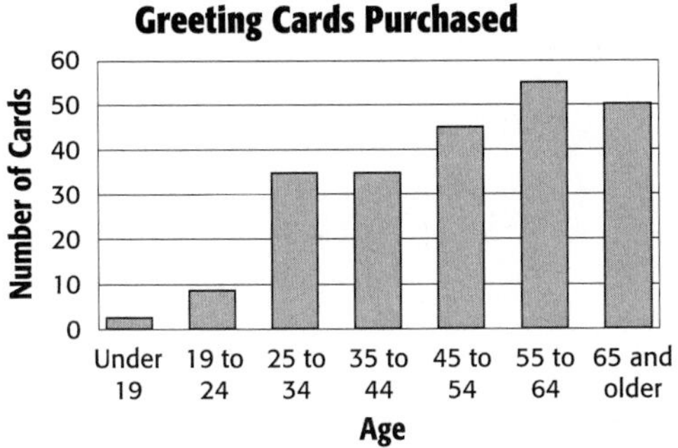

Source: American Greetings Corporation

Let c represent the number of cards purchased by Americans ages 35 to 44. Write an expression using c to represent the number of cards purchased by Americans ages 45 to 54. (Lesson 4-1)

11. Write a percent to represent the shaded area. (Lesson 5-5)

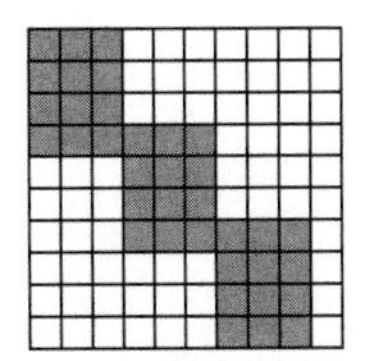

12. Find $7\frac{2}{7} + 3\frac{1}{4}$. (Lesson 6-3)

13. To make one batch of cookies, you need $\frac{3}{4}$ cup of butter. How much butter would you need to make $2\frac{1}{2}$ batches of cookies? (Lesson 6-4)

14. A box of laundry detergent contains 35 cups. If you use $1\frac{1}{4}$ cups per load of laundry, how many loads can you wash with 1 box? (Lesson 6-6)

15. What is the area of the rectangle? (Lesson 6-8)

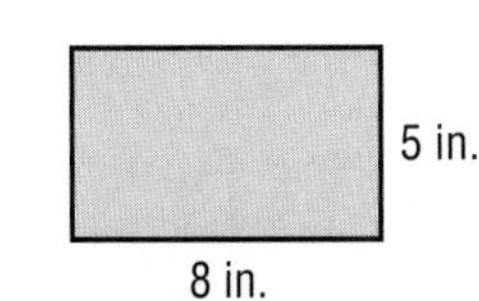

PART 3 Extended Response

Record your answers on a sheet of paper. Show your work.

16. Suppose you want to build a 4-foot wide deck around a circular swimming pool that has a radius of 66 inches. You also want to put a fence around the deck. (Lesson 6-9)

    a. Make a drawing of the problem. Include labels.

    b. About how much fencing will you need to the nearest foot?

    c. Suppose if you put a fence around just the pool instead of the deck, you save \$375. If this is the case, about how much does one foot of fencing cost? Explain.

TEST-TAKING TIP

Question 16 Many standardized tests include any necessary formulas in the test booklet. It helps to be familiar with formulas such as the area of a rectangle and the circumference of a circle, but use any formulas that are given to you with the test.

UNIT 4
Proportional Reasoning

Chapter 7
Ratios and Proportions

Chapter 8
Applying Percent

Chapter 9
Probability

In Unit 3, you learned about percents. In this unit, you will solve real-life problems involving proportions and percents. You will also discover how percents are used to describe probabilities.

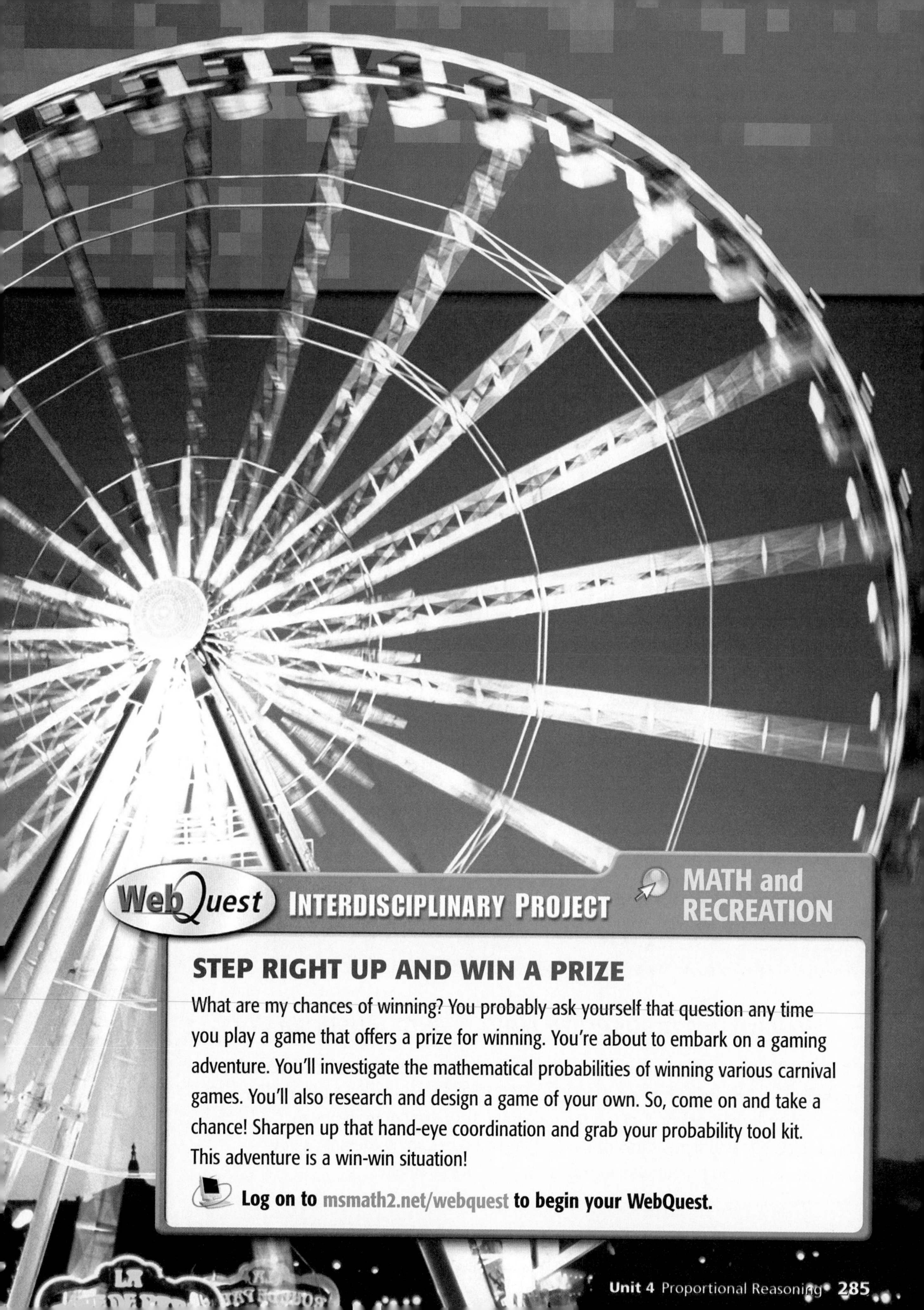

STEP RIGHT UP AND WIN A PRIZE

What are my chances of winning? You probably ask yourself that question any time you play a game that offers a prize for winning. You're about to embark on a gaming adventure. You'll investigate the mathematical probabilities of winning various carnival games. You'll also research and design a game of your own. So, come on and take a chance! Sharpen up that hand-eye coordination and grab your probability tool kit. This adventure is a win-win situation!

Log on to msmath2.net/webquest to begin your WebQuest.

CHAPTER 7 Ratios and Proportions

"How was math used to build Mount Rushmore?"

To build Mount Rushmore, carvers used large models of the former presidents' heads that had a ratio of 1 : 12. So, one inch on the models equaled 12 inches, or one foot, on the mountain. They used proportions to determine the actual measures of the sculptured figures. In mathematics, you will use ratios and proportions to solve many real-life problems.

You will solve problems about scale models in Lesson 7-4.

GETTING STARTED

Take this quiz to see whether you are ready to begin Chapter 7. Refer to the lesson or page number in parentheses if you need more review.

Vocabulary Review

Complete each sentence.

1. A comparison of two numbers by division is a ____. (Lesson 5-5)
2. A ratio that compares a number to 100 is called a ____. (Lesson 5-5)

Prerequisite Skills

Evaluate each expression. Round to the nearest tenth if necessary. (Lesson 1-3)

3. $100 \times 25 \div 52$
4. $10 \div 4 \times 31$
5. $\frac{63 \times 4}{34}$
6. $\frac{2 \times 100}{68}$

Write each fraction in simplest form. (Lesson 5-3)

7. $\frac{9}{45}$
8. $\frac{16}{24}$
9. $\frac{38}{46}$
10. $\frac{21}{35}$

Write each decimal as a fraction in simplest form. (Lesson 5-4)

11. 0.78
12. 0.320
13. 0.06
14. 0.90

Solve each equation. (Lesson 6-5)

15. $\frac{1}{4}x = 6$
16. $\frac{a}{5} = 8$
17. $4r = 22$
18. $2 = \frac{3}{5}z$

Complete. (Lesson 6-7)

19. 2 yd = __?__ ft
20. 48 oz = __?__ lb
21. $4\frac{1}{2}$ ft = __?__ yd
22. $3\frac{1}{4}$ h = __?__ min

Ratios and Proportions Make this Foldable to help you organize your notes. Begin with a sheet of notebook paper.

Fold Fold lengthwise to the holes.

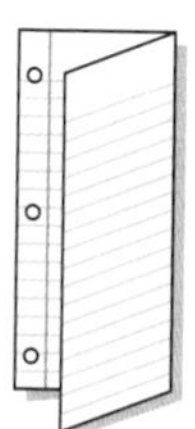

Cut Cut along the top line and then make equal cuts to form 10 tabs.

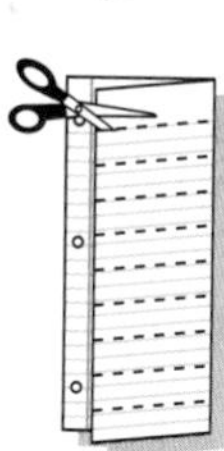

STEP 3 **Label** Label the major topics as shown.

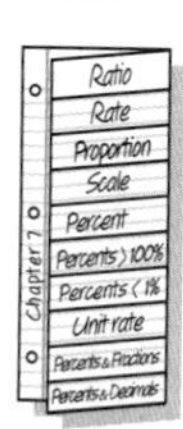

Reading and Writing As you read and study the lessons in this chapter, you can write notes under the corresponding tabs.

7-1 Ratios

What You'll LEARN

Write ratios as fractions and determine whether two ratios are equivalent.

NEW Vocabulary

equivalent ratios

REVIEW Vocabulary

ratio: a comparison of two numbers by division **(Lesson 5-5)**

WHEN am I ever going to use this?

GEARS Gears are used in objects that have spinning parts, such as bicycles and pendulum clocks. Suppose you have a larger gear that has 80 teeth being turned by a smaller gear that has 20 teeth. The comparison of the size of the larger gear to the size of the smaller gear is called a *gear ratio*.

1. Express the gear ratio as a fraction. Then write it as a fraction with a denominator of 1.
2. How many times does the smaller gear turn for every turn of the larger gear?
3. Describe some of the possible sizes of two gears that have a gear ratio of 3 to 1.

In Lesson 5-5, you learned that ratios can be written in several different ways. Ratios can also be simplified as shown below.

$$80 \text{ teeth} : 20 \text{ teeth} = \frac{\overset{4}{\cancel{80}} \text{ teeth}}{\underset{1}{\cancel{20}} \text{ teeth}} = \frac{4}{1}$$

EXAMPLES Write Ratios in Simplest Form

Write each ratio as a fraction in simplest form.

1 **25 to 10**

$25 \text{ to } 10 = \frac{25}{10}$ Write the ratio as a fraction.

$= \frac{5}{2}$ Simplify.

Written as a fraction in simplest form, the ratio 25 to 10 is $\frac{5}{2}$.

For ratios, leave the fraction as an improper fraction.

2 **6:18**

$6:18 = \frac{6}{18}$ Write the ratio as a fraction.

$= \frac{1}{3}$ Simplify.

The ratio 6:18 is $\frac{1}{3}$ in simplest form.

Your Turn **Write each ratio as a fraction in simplest form.**

a. 9 to 12 b. 27:15 c. 8 to 56

When writing ratios comparing units of length, units of time, units of weight, and so on, both measures should have the same unit.

EXAMPLE Write a Ratio by Converting Units

STUDY TIP

Canceling Units You can cancel units that are the same in the numerator and in the denominator.

3 **Write the ratio 2 feet to 10 inches as a fraction in simplest form.**

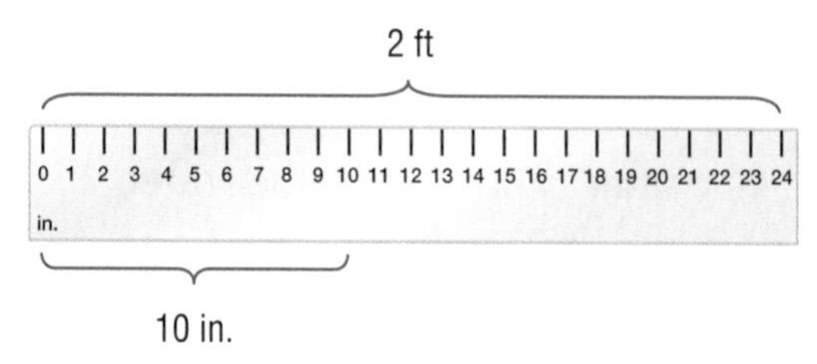

$\frac{2 \text{ feet}}{10 \text{ inches}} = \frac{24 \text{ inches}}{10 \text{ inches}}$ Convert 2 feet to inches.

$= \frac{\overset{12}{\cancel{24} \text{ inches}}}{\underset{5}{\cancel{10} \text{ inches}}}$ Divide by the GCF, 2 inches.

$= \frac{12}{5}$ Simplify.

So, the ratio 2 feet to 10 inches can be written as $\frac{12}{5}$.

Your Turn **Write each ratio as a fraction in simplest form.**

d. 4 feet : 4 yards

e. 15 ounces to 3 pounds

Two ratios that have the same value are **equivalent ratios**.

EXAMPLES Compare Ratios

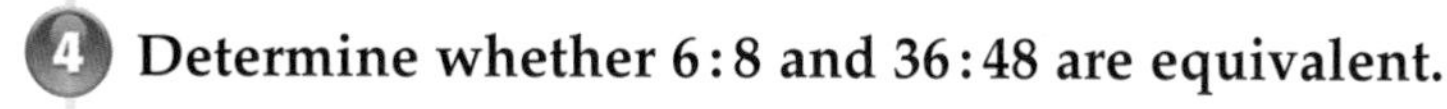

4 **Determine whether 6 : 8 and 36 : 48 are equivalent.**

Write each ratio as a fraction in simplest form.

$6:8 = \frac{6 \div 2}{8 \div 2}$ or $\frac{3}{4}$ The GCF of 6 and 8 is 2.

$36:48 = \frac{36 \div 12}{48 \div 12}$ or $\frac{3}{4}$ The GCF of 36 and 48 is 12.

The ratios in simplest form both equal $\frac{3}{4}$. So, 6 : 8 and 36 : 48 are equivalent ratios.

REAL-LIFE MATH

PONDS The recommended number of water lilies for a 9-square-foot pond is 1.

5 **PONDS It is recommended that for every 9 square feet of surface, a pond should have 2 fish. A pond that has a surface of 45 square feet contains 6 fish. Is this the correct number of fish based on the above recommendation?**

9 to 2 or $\frac{9}{2}$ Recommended ratio

45 to 6 or $\frac{45}{6} = \frac{15}{2}$ Actual ratio

Since $\frac{9}{2} \neq \frac{15}{2}$, the ratios are not equivalent.

So, the number of fish is not correct for the pond.

Your Turn **Determine whether the ratios are equivalent.**

f. $\frac{3}{8}$ and $\frac{6}{12}$

g. 35 students : 5 adults and 14 students : 2 adults

Skill and Concept Check

Writing Math
Exercises 3 & 4

1. **Write** the ratio 8 feet out of 15 feet in three different ways.

2. **OPEN ENDED** Write two different ratios that are equivalent to 3:2.

3. **Which One Doesn't Belong?** Identify the item that does not have the same ratio as the other three. Explain your reasoning.

$\frac{30}{45}$	4 to 6	26:39	3 to 2

4. **FIND THE ERROR** Marcus and Nicole are writing the ratio 2 days to 18 hours as a fraction in simplest form. Who is correct? Explain.

Marcus
$2:18 = \frac{2}{18}$
$= \frac{1}{9}$

Nicole
$48:18 = \frac{48}{18}$
$= \frac{8}{3}$

GUIDED PRACTICE

Write each ratio as a fraction in simplest form.

5. 8:30
6. 27 to 36
7. 11 to 4
8. 5 meters:1 meter
9. 28 feet:7 yards
10. 9 hours to 3 days

Determine whether the ratios are equivalent. Explain.

11. $\frac{3}{5}$ and $\frac{9}{15}$
12. 2:3 and 3:9

13. **HEIGHTS** Megan is 5 feet 3 inches tall, and her brother Troy is 5 feet 9 inches tall. Write a ratio in simplest form that compares Megan's height to Troy's height in inches.

Practice and Applications

Write each ratio as a fraction in simplest form.

14. 24 to 9
15. 14 to 70
16. 18:19
17. 45:21
18. 120:30
19. 33 to 90
20. 66 inches:72 inches
21. 12 pounds to 64 pounds
22. 4 pounds to 12 ounces
23. $8\frac{1}{3}$ yards:5 feet
24. 18 weeks:1 year
25. 45 minutes to 2 hours

HOMEWORK HELP

For Exercises	See Examples
14–19	1, 2
20–25	3
26–31, 33–37	4, 5

Extra Practice
See pages 580, 602.

Determine whether the ratios are equivalent. Explain.

26. $\frac{6}{9}$ and $\frac{2}{3}$
27. $\frac{9}{15}$ and $\frac{6}{10}$
28. 4:7 and 16:49
29. 8:21 and 16:42
30. 14 in.:9 ft and 16 ft:12 yd
31. 6 lb:72 oz and 2 lb:24 oz

32. **WRITE A PROBLEM** Write a problem that could be represented by using a ratio.

33. **MOVIES** One week in 2002, a movie about the mathematician John Nash earned 6 million dollars at the box office. Total revenues for all movies that week were 78 million dollars. Write a ratio to compare the money earned by that movie to the money earned by all the movies.

PHOTOGRAPHS The *aspect ratio* of a photograph is a ratio comparing the length and width. A 35 mm negative has an aspect ratio of 1.5:1. Photo sizes with the same aspect ratio can be printed full frame without cropping. Determine which size photos can be printed full frame from a 35 mm negative.

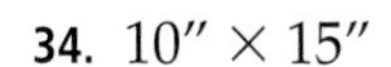

34. 10″ × 15″ 35. 8″ × 10″ 36. 5″ × 7.5″

37. **MONEY** In simplest form, write the ratio of the cost in cents to time in years for each type of currency shown in the table. Are they equivalent? Explain.

Currency	Cost to Make	Time in Use
Sacagawea dollar coin	12¢	30 years
Dollar bill	3.5¢	18 months

Source: www.sciencenews.org

38. **FIND A PATTERN** In decimal form, write the ratios of the first eight consecutive square numbers $1^2:2^2$, $2^2:3^2$, $3^2:4^2$, … . What do you notice about the ratios?

39. **CRITICAL THINKING** Find the next number in the pattern 480, 240, 80, 20, __?__. Explain your reasoning. (*Hint*: Look at the ratios of successive numbers.)

Standardized Test Practice and Mixed Review

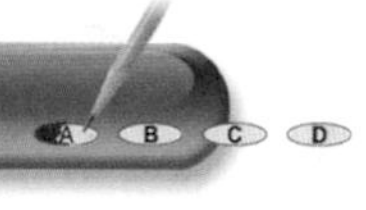

40. **MULTIPLE CHOICE** Express the ratio of shaded squares to total squares as a fraction in simplest form.

Ⓐ $\frac{3}{4}$ Ⓑ $\frac{15}{20}$ Ⓒ $\frac{7}{10}$ Ⓓ $\frac{3}{5}$

41. **MULTIPLE CHOICE** In the seventh grade, 16 of the 120 students are left-handed. Write the ratio of left-handed students to right-handed students as a fraction in simplest form.

Ⓕ $\frac{2}{15}$ Ⓖ $\frac{2}{13}$ Ⓗ $\frac{13}{2}$ Ⓘ $\frac{15}{2}$

42. **TREES** A giant sequoia tree has a diameter of about 30 feet. What is the circumference? Round to the nearest foot. (Lesson 6-9)

Find the perimeter and area of each rectangle. (Lesson 6-8)

43. $\ell = 9$ ft, $w = 4$ ft
44. $\ell = 12$ cm, $w = 10.6$ cm
45. $\ell = 7.5$ cm, $w = 2$ cm
46. $\ell = 1\frac{2}{5}$ in., $w = \frac{1}{2}$ in.

GETTING READY FOR THE NEXT LESSON

PREREQUISITE SKILL Divide. Round to the nearest hundredth if necessary. (Page 562)

47. 9.8 ÷ 2 48. \$4.30 ÷ 5 49. \$12.49 ÷ 40 50. 27.36 ÷ 3.2

7-2 Rates

HANDS-ON Mini Lab

What You'll LEARN

Determine unit rates.

NEW Vocabulary

rate
unit rate

Materials

- textbook
- computer
- stopwatch

Work with a partner.

Choose a page in a textbook and take turns typing as much as possible in 2 minutes.

1. Count the number of words that each of you typed.
2. Write the ratio *number of words to number of minutes* as a fraction.
3. Simplify the fractions by dividing the numerator and the denominator by 2.

A ratio that compares two quantities with different kinds of units is called a **rate**.

$\frac{\text{100 words}}{\text{2 minutes}}$ The units *words* and *minutes* are different.

When a rate is simplified so that it has a denominator of 1 unit, it is called a **unit rate**.

$\frac{\text{50 words}}{\text{1 minute}}$ The denominator is 1 unit.

The unit rate $\frac{\text{50 words}}{\text{1 minute}}$ can be read as *50 words per minute*. The table below shows some common unit rates used in everyday life.

Ratio	Unit Rate	Abbreviation	Name
$\frac{\text{number of miles}}{\text{1 hour}}$	miles per hour	mi/h or mph	speed
$\frac{\text{number of miles}}{\text{1 gallon}}$	miles per gallon	mi/gal or mpg	gas mileage
$\frac{\text{number of dollars}}{\text{1 pound}}$	price per pound	dollars/lb	unit price
$\frac{\text{number of dollars}}{\text{1 hour}}$	dollars per hour	dollars/h	hourly wage

EXAMPLE Find a Unit Rate

1 RUNNING Alethia ran 24 miles in 3 hours. What is her average speed in miles per hour?

Write the rate as a fraction. Then find an equivalent rate with a denominator of 1.

$$\text{24 miles in 3 hours} = \frac{24 \text{ mi}}{3 \text{ h}}$$ Write the rate as a fraction.

$$= \frac{24 \text{ mi} \div 3}{3 \text{ h} \div 3}$$ Divide the numerator and the denominator by 3.

$$= \frac{8 \text{ mi}}{1 \text{ h}}$$ Simplify.

The average speed, or unit rate, is 8 miles per hour.

Standardized Test Practice

EXAMPLES Find Unit Rates

2 **GRID-IN TEST ITEM** **Write *228 feet in 24 seconds* as a unit rate in feet per second.**

Read the Test Item

Write the ratio as a fraction. Then divide to get a denominator of 1.

Solve the Test Item

228 feet in 24 seconds

$= \frac{228 \text{ ft}}{24 \text{ s}}$ Write the rate as a fraction.

$= \frac{228 \text{ ft} \div 24}{24 \text{ s} \div 24}$ Divide the numerator and the denominator by 24.

$= \frac{9.5 \text{ ft}}{1 \text{ s}}$ Simplify.

The unit rate is 9.5 feet per second.

Fill in the Grid

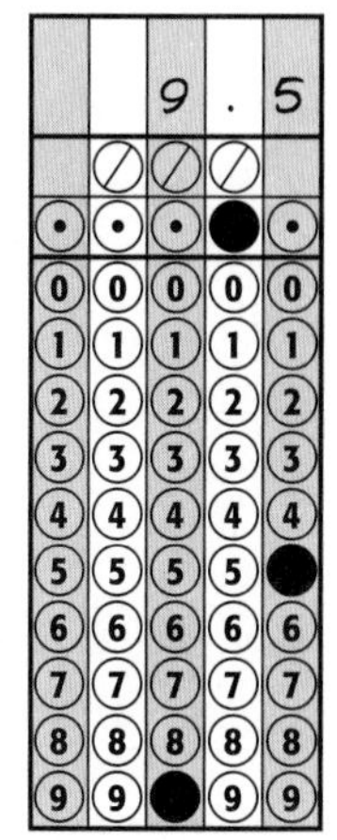

Test-Taking Tip (The Princeton Review)

Grid-In Answers To fill in the grid, first shade the appropriate bubbles. Then write the answer in the spaces above. You can align at the right as shown, or start in the left column.

3 **Find the unit price per orange if it costs $2 for six oranges. Round to the nearest hundredth if necessary.**

$2 for six oranges $= \frac{\$2}{6 \text{ oranges}}$ Write the rate as a fraction.

$= \frac{\$2 \div 6}{6 \text{ oranges} \div 6}$ Divide the numerator and the denominator by 6.

$\approx \frac{\$0.33}{1 \text{ orange}}$ Simplify.

The unit price is about $0.33 per orange.

Your Turn **Find each unit rate. Round to the nearest hundredth if necessary.**

a. 300 tickets in 6 days **b.** 220 miles in 8 gallons

In Example 3, you found a special kind of unit rate, called the *unit price*. This is the price per unit and is useful when you want to compare the cost of an item that comes in different sizes.

EXAMPLE Choose the Best Buy

4 **The costs of different sizes of peanut butter are shown at the right. Which jar costs the least per ounce?**

Size	Price
12 oz	$2.49
40 oz	$5.30
80 oz	$10.89

Find the unit price, or the cost per ounce, of each jar. Divide the price by the number of ounces.

12-ounce jar $2.49 ÷ 12 ounces ≈ $0.21 per ounce

40-ounce jar $5.30 ÷ 40 ounces ≈ $0.13 per ounce

80-ounce jar $10.89 ÷ 80 ounces ≈ $0.14 per ounce

The 40-ounce jar costs the least per ounce.

STUDY TIP

Estimation You can use compatible numbers to estimate the unit prices. For example, $2.49 ÷ 12 oz ≈ $2.40 ÷ 12 oz, or $0.20/oz.

Skill and Concept Check

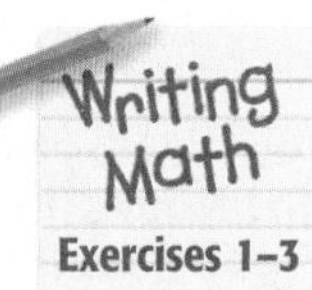

1. **OPEN ENDED** Write a rate and then convert it to a unit rate.

2. **FIND THE ERROR** Mikasi and Julie are determining which size sports drink is the better buy per ounce: a 16-ounce bottle for $1.95 or a 36-ounce bottle for $3.05. Who is correct? Explain.

Mikasi	Julie
16-oz bottle: 12.2¢ per ounce 36-oz bottle: 8.5¢ per ounce	16-oz bottle: 8.2¢ per ounce 36-oz bottle: 11.8¢ per ounce

3. **NUMBER SENSE** In which situation will the rate $\frac{x \text{ ft}}{y \text{ min}}$ increase? Give an example to explain your reasoning.
   a. x increases, y is unchanged
   b. x is unchanged, y increases

GUIDED PRACTICE

Find each unit rate. Round to the nearest hundredth if necessary.

4. 410 miles in 16 gallons
5. 1,500 words in 25 minutes
6. 5 pounds for $2.49
7. 3 cans of juice for $2.95

8. Which has the better unit price: a 6-pack of soda for $2.99 or a 12-pack for $4.50?

9. **MUSIC** Ethan can buy 4 CDs for $71.96 at CD Express or 9 CDs for $134.55 at Music Rox. Which has the better unit cost? Explain.

Practice and Applications

HOMEWORK HELP	
For Exercises	See Examples
10–17, 21–23	1–3
18–20	4

Extra Practice
See pages 580, 602.

Find each unit rate. Round to the nearest hundredth if necessary.

10. 360 miles in 6 hours
11. 6,840 customers in 45 days
12. 150 people for 5 classes
13. $7.40 for 5 pounds
14. $1.12 for 8.2 ounces
15. 810 Calories in 3 servings
16. 40 meters in 13 seconds
17. 144 miles in 4.5 gallons

Choose the best unit price.

18. $3.99 for a 16-ounce bag of candy or $2.99 for a 12-ounce bag
19. aspirin sold in bottles of 50 for $5.49, 100 for $8.29, or 150 for $11.99

20. **NUTRITION** Which soft drink described in the table has a lower amount of sodium per ounce? Explain.

Soft Drink	Serving Size (oz)	Sodium (mg)
A	12	70
B	8	35

21. **FOUNTAINS** The Prometheus Fountain in New York City pumps 60,000 gallons of water every 15 minutes. What is the unit rate?

22. **WHALES** A humpback whale can migrate 3,000 miles in 30 days. Find a unit rate to describe the average speed per day during migration.

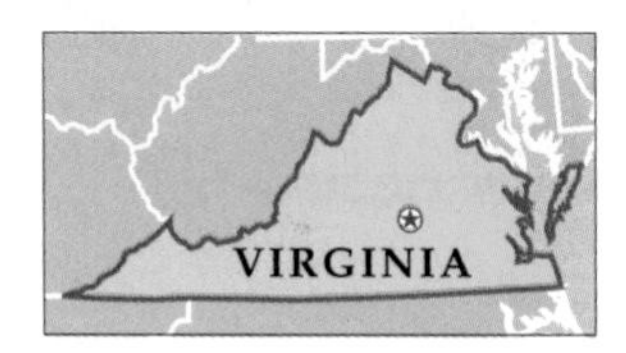

23. **POPULATION** The population of Virginia is approximately 7.1 million people, and its land area is approximately 39,600 square miles. Find the *population density*, or the population per square mile.

24. **RESEARCH** Use the Internet or another source to find the population and land area of your state. Determine the population density.

Determine whether the following statements are *sometimes*, *always*, or *never* true. Explain by giving an example or a counterexample.

25. A ratio is a rate.

26. A rate is a ratio.

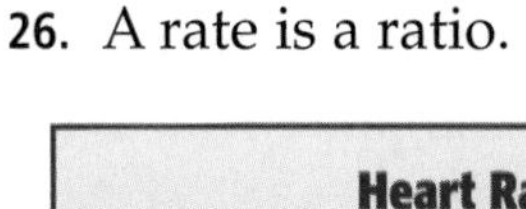

LIFE SCIENCE For Exercises 27 and 28, use the graph. It shows the average number of heartbeats for an adult elephant and an adult human.

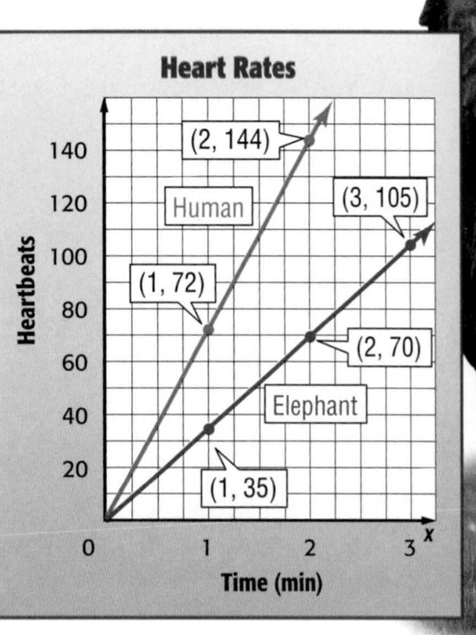

27. Whose heart rate is greater? Explain.

28. Whose heart beats more times in one hour?

29. **CRITICAL THINKING** If 8 tickets are sold in 15 minutes, find the rate in tickets per hour.

Standardized Test Practice and Mixed Review

30. **MULTIPLE CHOICE** Which of the following has the lowest unit price?

(A) 18 oz: \$5.40 (B) 16 oz: \$4.64 (C) 12 oz: \$3.72 (D) 10 oz: \$3.30

31. **MULTIPLE CHOICE** Which rate is the fastest?

(F) 589 miles in 11 hours (G) 360 miles in 7 hours

(H) 111 miles in 2 hours (I) 283 miles in 5 hours

Write each ratio as a fraction in simplest form. (Lesson 7-1)

32. 11:12 33. 24:4 34. 15 to 25 35. 20 to 14

36. A serving plate has a radius of 6 inches. Find the circumference to the nearest tenth. (Lesson 6-9)

GETTING READY FOR THE NEXT LESSON

PREREQUISITE SKILL Solve each equation. (Lesson 6-5)

37. $\frac{n}{8} = 7$ 38. $\frac{2}{3}x = 5$ 39. $9t = 12$ 40. $\frac{r}{5.2} = 4$

7-2b HANDS-ON LAB

A Follow-Up of Lesson 7-2

Rate of Change

What You'll LEARN

Investigate rate of change.

Materials

- square tiles
- grid paper

INVESTIGATE *Work in groups of three.*

A *rate of change* is a ratio that shows a change in one quantity with respect to a change in another quantity. In this lab, you will use tables and graphs to represent *constant* rates of change.

STEP 1 Use tiles to build the models shown below. Then continue the pattern to build the fourth and fifth models.

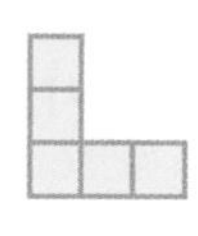

STEP 2 For each model, record the number of tiles and the perimeter in a table like the one shown at the right. The first two are done for you.

Model	Number of Tiles (x)	Perimeter (y)
1	1	4
2	3	8
3		
4		
5		

STEP 3 Draw a coordinate plane on grid paper and graph the ordered pairs (x, y).

Writing Math

1. What do you notice about the points?
2. Find the ratio $\frac{\text{change in perimeter}}{\text{change in tiles}}$ between the second and third points, the third and fourth points, and the fourth and fifth points. Each ratio is a rate of change. **Describe** what you observe.
3. **Complete:** As the number of tiles increases by 2 units, the perimeter of the models increases by __?__ units.
4. Refer to the table at the right that appeared in Lesson 4-7. Find the ratio $\frac{\text{change in earnings}}{\text{change in hours worked}}$ for Greg and Monica. How do these values compare to the slopes that you found in Lesson 4-7?

Hours Worked	Earnings ($)	
	Greg	Monica
1	4	5
2	8	10
3	12	15
4	16	20

5. **Make a conjecture** about the relationship between rate of change and a graph of the two quantities.

STUDY TIP

Look Back You can review **slope** in Lesson 4–7.

7-3 Solving Proportions

What You'll LEARN

Solve proportions.

NEW Vocabulary

proportion
cross product

WHEN am I ever going to use this?

NUTRITION As part of a healthy diet, teens need 60 milligrams of vitamin C each day.

Vitamin C (mg)	Amount of Cereal (c)
15	0.5
60	2

1. Write the ratio $\frac{\text{vitamin C}}{\text{amount of cereal}}$ for a half-cup serving of cereal.
2. Rewrite the ratio in Exercise 1 to find the unit rate in milligrams per cup.
3. Simplify $\frac{60 \text{ mg}}{2 \text{ c}}$ to find the unit rate.
4. **Make a conjecture** about the rates $\frac{15 \text{ mg}}{0.5 \text{ c}}$ and $\frac{60 \text{ mg}}{2 \text{ c}}$.

When two ratios are equivalent, they form a **proportion**. Since rates are types of ratios, they can also form proportions.

Key Concept — Proportion

Words A proportion is an equation stating that two ratios are equivalent.

Symbols

Arithmetic: $\frac{1}{2} = \frac{3}{6}$, $\frac{8 \text{ ft}}{10 \text{ s}} = \frac{4 \text{ ft}}{5 \text{ s}}$

Algebra: $\frac{a}{b} = \frac{c}{d}$, where $b, d \neq 0$

In a proportion, a **cross product** is the product of the numerator of one ratio and the denominator of the other ratio.

$\frac{1}{2} = \frac{3}{6}$

1 × 6 is a cross product.

2 × 3 is a cross product.

$1 \times 6 = 2 \times 3$

$6 = 6$ — The cross products of a proportion are equal.

EXAMPLE Identify a Proportion

1 Determine whether $\frac{6}{10}$ and $\frac{3}{5}$ form a proportion.

$\frac{6}{10} \stackrel{?}{=} \frac{3}{5}$ Write a proportion.

$6 \times 5 \stackrel{?}{=} 10 \times 3$ Find the cross products.

$30 = 30$ ✓ Multiply.

The cross products are equal, so the ratios form a proportion.

EXAMPLE Identify a Proportion

2 **Determine whether $\frac{\$4}{9\text{ oz}}$ and $\frac{\$21}{45\text{ oz}}$ form a proportion.**

$\frac{4}{9} \stackrel{?}{=} \frac{21}{45}$ Write a proportion.

$4 \times 45 \stackrel{?}{=} 9 \times 21$ Find the cross products.

$180 \neq 189$ Multiply.

The cross products are not equal, so the ratios do not form a proportion.

Your Turn **Determine whether each pair of ratios forms a proportion.**

a. $\frac{4}{9}$ and $\frac{2}{3}$ b. $\frac{15}{9}$ and $\frac{10}{6}$ c. $\frac{8\text{ in.}}{21\text{ s}}$ and $\frac{12\text{ in.}}{31.5\text{ s}}$

You can use cross products to find a missing term in a proportion. This is known as *solving the proportion*. Solving a proportion is similar to solving an equation.

EXAMPLES Solve Proportions

3 **Solve $\frac{2.6}{13} = \frac{8}{n}$.**

$\frac{2.6}{13} = \frac{8}{n}$ Write the proportion.

$2.6 \cdot n = 13 \cdot 8$ Find the cross products.

$2.6n = 104$ Multiply.

$\frac{2.6n}{2.6} = \frac{104}{2.6}$ Divide each side by 2.6.

$n = 40$ Simplify.

The solution is 40.

How Does a Paleontologist Use Math?
In 2001, paleontologists discovered a 6-foot skull of a prehistoric crocodile called *SuperCroc*. By using proportions, they estimated its total length as 40 feet.

Source: *National Geographic*

Online Research
For information about a career as a paleontologist, visit: msmath2.net/careers

4 **LIFE SCIENCE** **The skull length of a crocodile is proportional to its body length. In one species, a 6-foot crocodile has a 2-foot skull. Estimate the length of a crocodile of that species that has a 3.5-foot skull.**

body length → / skull length → $\frac{6\text{ ft}}{2\text{ ft}} = \frac{c\text{ ft}}{3.5\text{ ft}}$ Write a proportion.

$6 \cdot 3.5 = 2 \cdot c$ Find the cross products.

$21 = 2c$ Multiply.

$\frac{21}{2} = \frac{2c}{2}$ Divide each side by 2.

$10.5 = c$ Simplify.

So, a crocodile with a 3.5-foot skull is about 10.5 feet long.

Your Turn **Solve each proportion.**

d. $\frac{1}{5} = \frac{x}{60}$ e. $\frac{10}{k} = \frac{2.5}{4}$ f. $\frac{2\text{ ft}}{6\text{ min}} = \frac{5\text{ ft}}{t\text{ min}}$

Skill and Concept Check

1. **Explain** how you can determine whether two ratios are equivalent.
2. **OPEN ENDED** Write a proportion that has $\frac{12}{15}$ as one of the ratios.
3. **Which One Doesn't Belong?** Identify the ratio that is not equivalent to the other three. Explain your reasoning.

$\frac{2}{3}$ $\quad$ $\frac{6}{9}$ $\quad$ $\frac{4}{6}$ $\quad$ $\frac{2}{9}$

GUIDED PRACTICE

Determine whether each pair of ratios forms a proportion.

4. $\frac{2}{3}$ and $\frac{6}{9}$
5. $\frac{16}{30}$ and $\frac{8}{10.5}$
6. $\frac{5 \text{ ft}}{7 \text{ s}}$ and $\frac{25 \text{ ft}}{49 \text{ s}}$

Solve each proportion.

7. $\frac{24}{6} = \frac{8}{x}$
8. $\frac{n}{5} = \frac{20}{25}$
9. $\frac{7}{t} = \frac{3}{9}$
10. $\frac{8}{12} = \frac{n}{3}$
11. $\frac{a}{6} = \frac{12}{8}$
12. $\frac{13}{25} = \frac{39}{r}$
13. $\frac{21}{5} = \frac{c}{7}$
14. $\frac{7}{x} = \frac{3.5}{9}$
15. **NUTRITION** An 8-ounce glass of orange juice contains 72 milligrams of vitamin C. How much juice contains 60 milligrams of vitamin C?

Practice and Applications

HOMEWORK HELP

For Exercises	See Examples
16–21	1, 2
22–46	3, 4

Extra Practice
See pages 580, 602.

Determine whether each pair of ratios forms a proportion.

16. $\frac{1}{6}$ and $\frac{4}{24}$
17. $\frac{3}{9}$ and $\frac{2}{15}$
18. $\frac{27}{6}$ and $\frac{9}{2}$
19. $\frac{1}{4}$ and $\frac{4}{1}$
20. $\frac{\$8}{20 \text{ lb}}$ and $\frac{\$24}{60 \text{ lb}}$
21. $\frac{21 \text{ mi}}{6 \text{ h}}$ and $\frac{14.3 \text{ mi}}{3 \text{ h}}$
22. If there are 12 wheels, how many bicycles are there?
23. If there are 32 paws, how many kittens are there?

Solve each proportion.

24. $\frac{5}{9} = \frac{10}{x}$
25. $\frac{2}{12} = \frac{a}{36}$
26. $\frac{16}{t} = \frac{2}{3}$
27. $\frac{n}{8} = \frac{3}{4}$
28. $\frac{c}{7} = \frac{18}{42}$
29. $\frac{30}{w} = \frac{8}{20}$
30. $\frac{9}{15} = \frac{b}{10}$
31. $\frac{3}{n} = \frac{27}{18}$
32. $\frac{x}{12} = \frac{12}{4}$
33. $\frac{45}{5} = \frac{t}{7}$
34. $\frac{6}{3} = \frac{5}{w}$
35. $\frac{y}{36} = \frac{15}{24}$
36. $\frac{350}{a} = \frac{2}{10}$
37. $\frac{380}{520} = \frac{760}{n}$
38. $\frac{3}{5} = \frac{0.2}{d}$
39. $\frac{2.5}{4.5} = \frac{7.5}{x}$
40. **MODELS** The toy car shown at the right is modeled after a real car. If the real car is 4.79 meters long, how wide is it? Write the width to the nearest hundredth of a meter.

41. **TESTS** Suppose you received a score of 13 out of 15 on a quiz. Write a proportion to find the percent. Then solve. Round to the nearest percent.

COOKING For Exercises 42–44, refer to the table.

Vegetable Equivalents
$\frac{3}{4}$ lb broccoli = 3 c florets
1 lb carrots = 3 c slices
1 stalk celery = $\frac{1}{2}$ c slices
2 peppers = $2\frac{1}{2}$ c rings
1 lb potatoes = $2\frac{3}{4}$ c cubes

42. How many stalks of celery would you need to get $3\frac{1}{2}$ cups of celery slices?
43. How many cups of broccoli florets would 1 pound of broccoli produce?
44. **WRITE A PROBLEM** Write a real-life problem that can be solved using a proportion and one of the vegetable equivalents. Then write a proportion and solve.
45. **MULTI STEP** Amanda can word process 7 words in 6 seconds. At this rate, how many words can she word process in 3 minutes?
46. **MONEY** In 2002, France adopted the *Euro* as its currency, replacing the French franc. At that time, one Euro was worth \$0.90 in American currency. In Paris, France, a quart of milk cost 3 Euros. In Los Angeles, a quart of milk cost \$1.65. In which city was the quart of milk more expensive?

Data Update What is the current U.S. value of one Euro? Visit msmath2.net/data_update to learn more.

47. **CRITICAL THINKING** In the proportion $\frac{3}{6} = \frac{6}{12}$, the number 6 appears in two of the diagonal positions. This repeated number is called the *geometric mean* of the other two numbers. Find a pair of numbers other than 3 and 12 for which 6 is the geometric mean.

Standardized Test Practice and Mixed Review

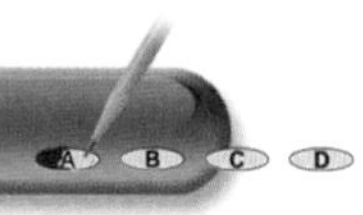

48. **MULTIPLE CHOICE** Ming is paid an hourly rate. One week he earned \$157.50 by working 30 hours. If he works 35 hours the next week, how much will he earn?

    (A) \$135.00 (B) \$160.50 (C) \$183.75 (D) \$210.25

49. **GRID IN** The ratio of boys to girls in the seventh grade is 7 to 8. If there are 120 students in the seventh grade, how many of them are girls?
50. Choose the better unit price: 5 pounds of onions for \$2.99 or 3 pounds of onions for \$1.29. (Lesson 7-2)

Write each ratio as a fraction in simplest form. (Lesson 7-1)

51. 45:81
52. 49 to 14
53. 14 ounces to 5 pounds

MEASUREMENT Complete. (Lesson 6-7)

54. 5 qt = __?__ pt
55. $3\frac{1}{2}$ lb = __?__ oz
56. 28 c = __?__ qt

GETTING READY FOR THE NEXT LESSON

PREREQUISITE SKILL Multiply. Write in simplest form. (Lesson 6-4)

57. $\frac{1}{5} \cdot 8$
58. $1\frac{3}{4} \cdot 5$
59. $16 \cdot 3\frac{1}{4}$
60. $24 \cdot 1\frac{3}{8}$

A Follow-Up of Lesson 7-3

Wildlife Sampling

What You'll LEARN

Use proportions to estimate.

Materials

- small bowl
- dried beans
- paper cup
- markers

INVESTIGATE *Work in groups of three.*

Naturalists can estimate the population in a wildlife preserve by using the *capture-recapture* technique. In this lab, you will model this technique using dried beans in a bowl to represent deer in a forest.

STEP 1 Fill a small bowl with dried beans.

STEP 2 Use the paper cup to scoop out some of the beans. These represent the original *captured* deer. Record the number in a table like the one shown at the right. Mark each bean with an × on both sides. Then return these beans to the bowl and mix well.

Original Number Captured ____			
Trial	Sample	Recaptured	P
A			
B			
C			
⋮			
J			
Total			

STEP 3 Scoop another cup of beans from the bowl and count them. This is the *sample* for Trial A. Count the beans with the ×'s. These are the *recaptured* deer. Record both numbers.

STEP 4 Use the proportion below to estimate the total number of beans in the bowl. This represents the total population P. Record the value of P in the table.

$$\frac{\text{captured}}{\text{total population } (P)} = \frac{\text{recaptured}}{\text{sample}}$$

STEP 5 Return all of the beans to the bowl.

STEP 6 Repeat Steps 3–5 nine times.

Writing Math

Work in groups of three.

1. Find the average of the estimates in column P. Is this a good estimate of the number of beans in the bowl? **Explain** your reasoning.
2. **Count** the actual number of beans in the bowl. How does this number compare to your estimate?

7-4a Problem-Solving Strategy

A Preview of Lesson 7-4

What You'll Learn
Solve problems by drawing a diagram.

Draw a Diagram

Mrs. Dixon said that we've gone about 90 miles, or $\frac{2}{3}$ of the way to the campsite.

So, how much farther do we have to go? I bet if we **draw a diagram**, we could figure it out.

Explore We know that 90 miles is about $\frac{2}{3}$ of the total distance.

Plan Let's draw a diagram showing the distance that we've already gone and the fractional part that it represents.

Solve

$\frac{1}{3}$ $\frac{2}{3}$ 1

school 90 miles campsite

If $\frac{2}{3}$ of the distance is 90 miles, then $\frac{1}{3}$ of the distance would be 45 miles. So, the missing third must be another 45 miles.

$\frac{1}{3}$ $\frac{2}{3}$ 1

school 45 miles 90 miles campsite

+ 45 + 45 + 45

The total distance is $90 + 45$ or 135 miles.

Examine Since $\frac{2}{3}$ of the total distance equals 90 miles, the equation $\frac{2}{3}x = 90$ represents this problem. Solving, we get $x = 135$ miles. So, the solution checks.

Analyze the Strategy

1. **Explain** why drawing a diagram can be a useful problem-solving strategy.
2. **Determine** how far the trip would have been if 90 miles was only $\frac{1}{3}$ of the total distance. Draw a new diagram for this situation.
3. **Write** a problem that could be solved by drawing a diagram. Exchange your problem with a classmate and solve.

Apply the Strategy

Solve. Use the draw a diagram strategy.

4. **PHYSICAL SCIENCE** A ball is dropped from 10 feet above the ground. It hits the ground and bounces up half as high as it fell. This is true for each successive bounce. What height does the ball reach on the fourth bounce?

5. **FAMILY** At Nate's family reunion, 80% of the people are 18 years of age or older. Half of the remaining people are under 12 years old. If 20 children are under 12 years old, how many people are at the reunion?

Mixed Problem Solving

Solve. Use any strategy.

6. **TESTS** The scores on a social studies test are found by adding or subtracting points as shown at the right.

Answer	Points
Correct	+8
Incorrect	−4
No answer	−2

If Mario's score on a 15-question test was 86 points, how many of his answers were correct, incorrect, and blank?

7. **GAMES** Six members of a video game club are having a tournament. In the first round, every player will play a video game against every other player. How many games will be in the first round of the tournament?

8. **BUSINESS** The graph shows the annual spending by five industries.

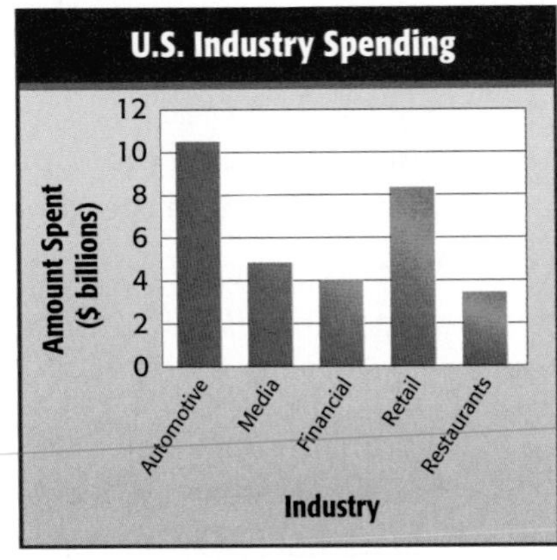

Source: *The Top Ten of Everything*

Estimate how much more the automotive industry spends per year than the retail industry. Write in scientific notation.

9. **ENVIRONMENT** What fraction of landfills is paper and plastic packaging? Write in simplest form.

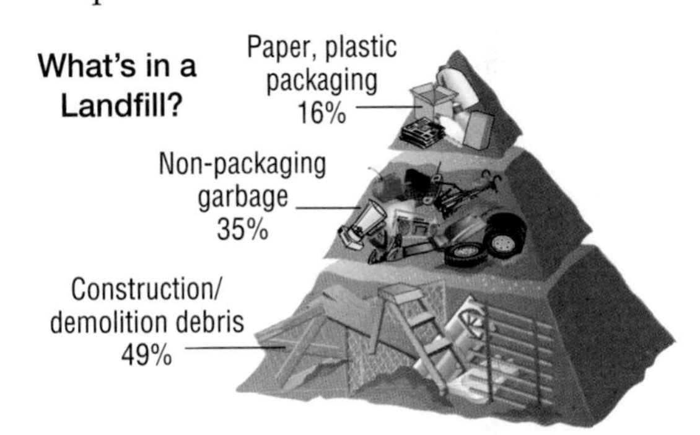

Source: *ULS Report*

10. **MONEY** Mr. Li has $240 in his checking account after writing checks for $15.70, $43.20, and $18. What was his balance before he wrote the three checks?

11. **STANDARDIZED TEST PRACTICE** Danielle is adding 3 feet to the length and width of her rectangular garden, as shown at the right. Which expression represents the area of land that will be added to the garden?

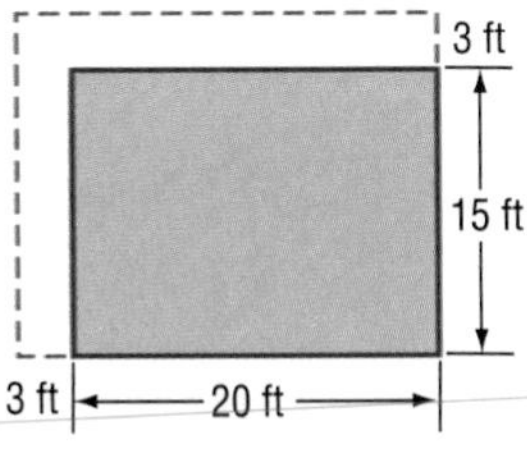

Ⓐ (15 + 3)(20 + 3)

Ⓑ (15 + 3)(20 + 3) − (15)(20)

Ⓒ (15 − 3)(20 − 3)

Ⓓ (15 − 3)(20 − 3) − (15)(20)

You will use the draw a diagram strategy in the next lesson.

7-4 Scale Drawings

HANDS-ON Mini Lab

What You'll LEARN

Solve problems involving scale drawings.

NEW Vocabulary

scale drawing
scale
scale factor
scale model

Materials

- measuring tape
- straightedge
- $\frac{1}{4}$-inch grid paper

Work with a partner.

- Measure the length of each wall, door, window, and chalkboard in your classroom.
- Record each length to the nearest $\frac{1}{2}$ foot.

1. Let 1 unit on the grid paper represent 1 foot. So, 6 units = 6 feet. Convert all of your measurements to units.
2. On grid paper, make a drawing of your classroom like the one shown at the right.

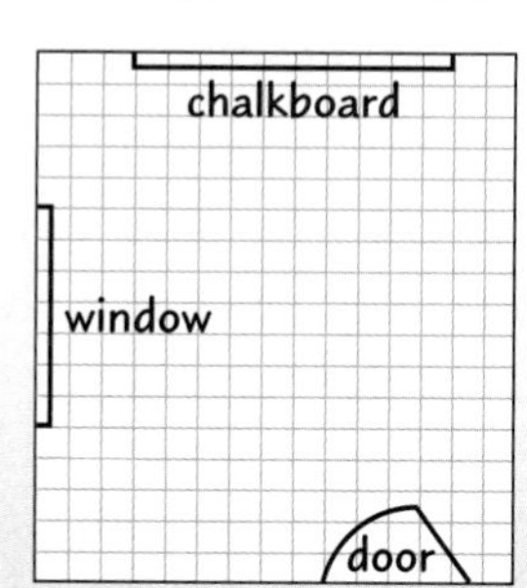

A **scale drawing** represents something that is too large or too small to be drawn at actual size.

The map is a scale drawing. Its **scale** gives the relationship between the distance on the map and the actual distance.

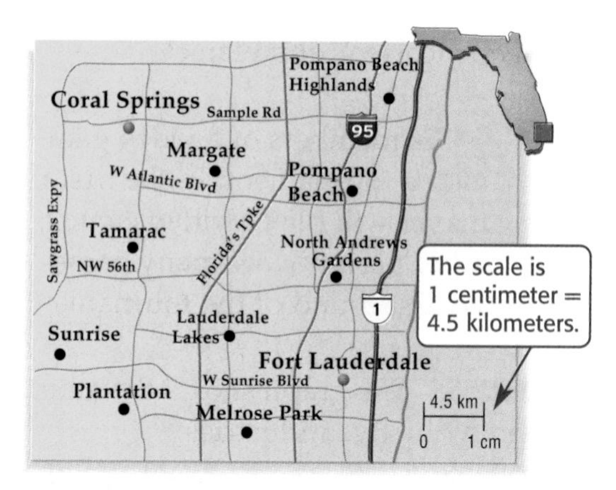

EXAMPLE Use a Scale Drawing

1 **MAPS On the map above, the distance between Coral Springs and Fort Lauderdale is about 4.1 centimeters. What is the actual distance? Round to the nearest tenth.**

Let d = the actual distance between the cities. Write and solve a proportion. Use the scale written as a fraction.

	Scale		Coral Springs to Ft. Lauderdale	
map → / actual →	$\frac{1 \text{ centimeter}}{4.5 \text{ kilometers}}$	=	$\frac{4.1 \text{ centimeters}}{d \text{ kilometers}}$	← map / ← actual

$1 \cdot d = 4.5 \cdot 4.1$ Cross products

$d = 18.45$ Simplify.

The distance between Coral Springs and Fort Lauderdale is about 18.5 kilometers.

A blueprint is another example of a scale drawing.

EXAMPLE Read a Scale Drawing

2 POOLS On the blueprint of the pool, each square has a side length of $\frac{1}{4}$ inch. What is the actual width of the pool?

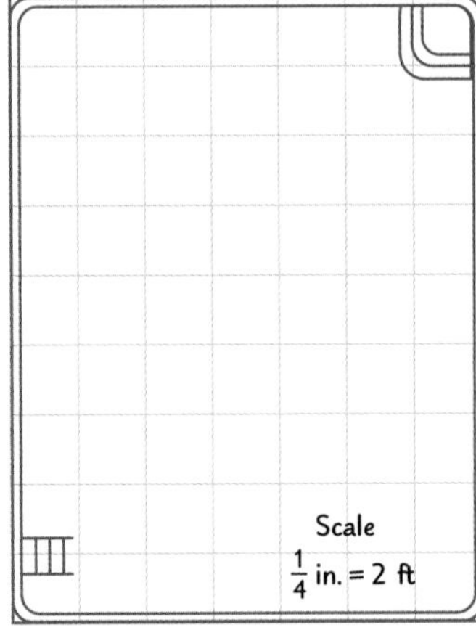

The pool on the blueprint is $1\frac{3}{4}$ inches wide. Let w = the actual width of the pool. Write and solve a proportion using the scale.

	Scale		Width of Pool	
blueprint →	$\frac{1}{4}$ inch		$1\frac{3}{4}$ inches	← blueprint
actual →	2 feet	=	w feet	← actual

$$\frac{\frac{1}{4}\text{ inch}}{2\text{ feet}} = \frac{1\frac{3}{4}\text{ inches}}{w\text{ feet}}$$

$\frac{1}{4} \cdot w = 2 \cdot 1\frac{3}{4}$ Cross products

$\frac{1}{4}w = \frac{14}{4}$ Multiply.

$w = 14$ Simplify. Multiply each side by 4.

The width of the pool is 14 feet.

In Lesson 7-1, you simplified ratios by converting units. You can use the same method to simplify the scale in Example 2. A scale written as a ratio in simplest form is called the **scale factor**.

scale → $\frac{\frac{1}{4}\text{ inch}}{2\text{ feet}} = \frac{\frac{1}{4}\text{ inch}}{24\text{ inches}}$ Convert 2 feet to inches.

$= \frac{4}{4} \cdot \frac{\frac{1}{4}\text{ inch}}{24\text{ inches}}$ Multiply by $\frac{4}{4}$ to eliminate the fraction in the numerator.

$= \frac{1}{96}$ ← scale factor

EXAMPLE Find the Scale Factor

3 Find the scale factor of a blueprint if the scale is 1 inch = 8 feet.

Write the ratio of 1 inch to 8 feet in simplest form.

$\frac{1\text{ inch}}{8\text{ feet}} = \frac{1\text{ inch}}{96\text{ inches}}$ Convert 8 feet to inches.

$= \frac{1}{96}$ Cancel the units.

The scale factor is $\frac{1}{96}$. That is, each measure on the blueprint is $\frac{1}{96}$ the actual measure.

STUDY TIP

Equivalent Scales The scales below are equivalent because their scale factors are equal, $\frac{1}{96}$.
- 1 inch = 8 feet
- $\frac{1}{4}$ inch = 2 feet

Your Turn

a. On a scale drawing of a new classroom, the scale is 1 centimeter = 2.5 meters. What is the scale factor?

If you know the actual length of an object and the scale, you can build a scale model. A **scale model** can be used to represent something that is too large or too small for an actual-size model.

EXAMPLE Make a Scale Model

STUDY TIP

Scale The scale is the ratio of the drawing/model measure to the actual measure. It is not always the ratio of a smaller measure to a larger measure.

4 COMPUTERS Designers are creating a larger model of the computer memory board to use in design work. If they use a scale of 20 inches = 1 inch, what is the length of the model?

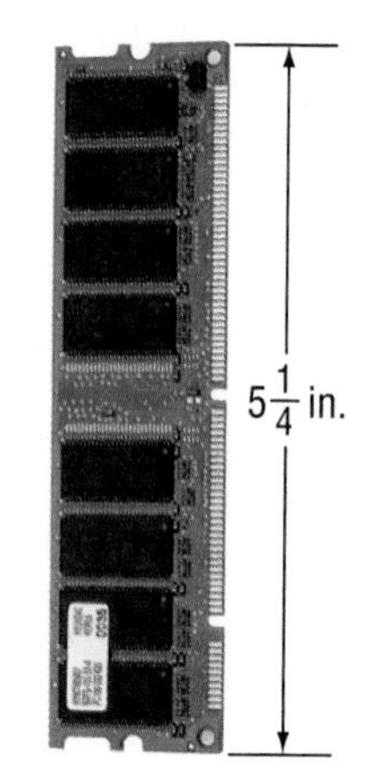

Write a proportion using the scale.

	Scale		Length of Model	
model →	20 inches	=	m inches	← model
actual →	1 inch		$5\frac{1}{4}$ inches	← actual

$$\frac{20 \text{ inches}}{1 \text{ inch}} = \frac{m \text{ inches}}{5\frac{1}{4} \text{ inches}}$$

$20 \cdot 5\frac{1}{4} = 1 \cdot m$ Find the cross products.

$105 = m$ Multiply.

The scale model is 105 inches long.

Skill and Concept Check

Writing Math Exercise 2

1. **OPEN ENDED** On grid paper, make a scale drawing of a room in your home. Include the scale that you used.

2. **NUMBER SENSE** Roberto built a model racecar that is 3 inches long and has a scale of 0.5 inch = 2 feet. He determined that the actual racecar must be $3 \cdot 2$ or 6 feet long. Is this correct? Explain.

GUIDED PRACTICE

Find the actual distance between each pair of cities. Round to the nearest tenth if necessary.

	Cities	Map Distance	Scale
3.	Kokomo, Indiana, and Chicago, Illinois	8 cm	1 cm = 25 km
4.	Raleigh and Charlotte, North Carolina	$3\frac{1}{4}$ in.	1 in. = 40 mi

BLUEPRINTS For Exercises 5–7, use the blueprint at the right. Each square has a side length of 0.5 centimeter.

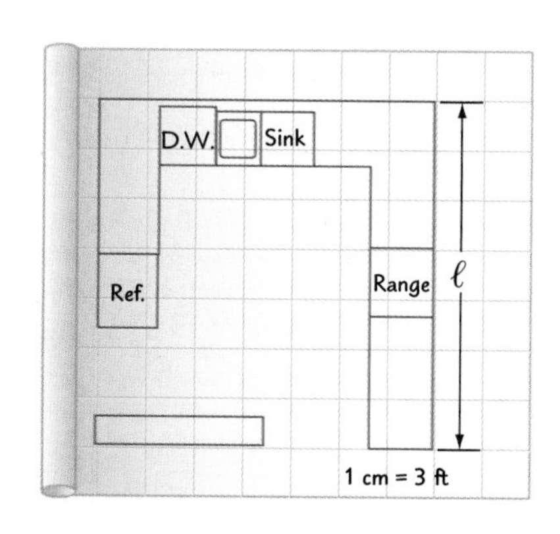

5. What is the length ℓ of the kitchen on the blueprint?
6. Find the actual length of the kitchen.
7. Find the scale factor. (*Hint*: 1 ft ≈ 30 cm)

8. **BRIDGES** A bridge is 28 meters long. Find the length of a scale model if the scale is 1 centimeter = 5.5 meters. Round to the nearest tenth.

Practice and Applications

HOMEWORK HELP

For Exercises	See Examples
9–12, 23	1
17–20	2
13–16, 21, 25	3
22, 24, 26	4

Extra Practice
See pages 581, 602.

Find the actual distance between each pair of cities. Round to the nearest tenth if necessary.

	Cities	Map Distance	Scale
9.	Virginia Beach, Virginia, and Washington, D.C.	8.5 cm	1 cm = 30 km
10.	Mobile and Huntsville, Alabama	$14\frac{1}{2}$ in.	1 in. = 20 mi
11.	Knoxville, Tennessee, and Choctaw, Oklahoma	19.28 cm	2 cm = 125 km
12.	Quebec, Canada, and Paris, France	$4\frac{5}{8}$ in.	$\frac{1}{2}$ in. = 355 mi

Suppose you are making a scale drawing. Find the length of each object on the scale drawing with the given scale. Then find the scale factor.

13. a back yard 120 feet deep; 1 inch : 20 feet
14. an airplane 87 feet long; 2 inches = 15 feet
15. an amusement park ride 36 meters high; 0.5 centimeter = 1.5 meters
16. a surgical instrument $5\frac{7}{8}$ inches long; 1 inch : $\frac{1}{2}$ inch

For Exercises 17–21, refer to the blueprint of the garden shown below. Each square has a side length of $\frac{1}{4}$ inch.

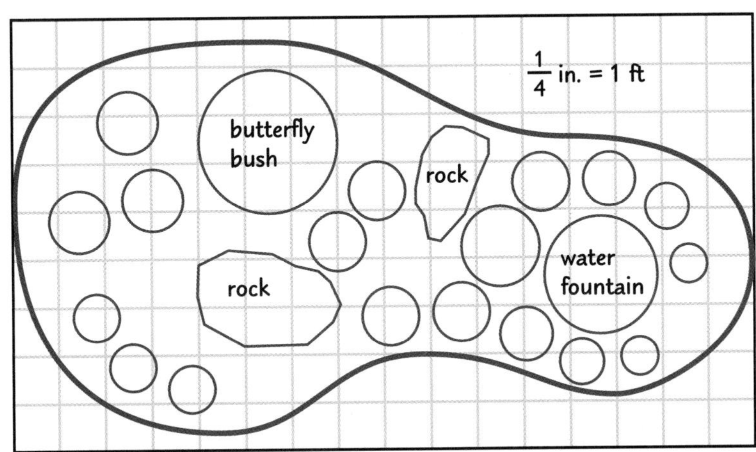

17. What is the length of the garden on the blueprint?
18. Find the actual length of the garden.
19. What is the diameter of the butterfly bush on the blueprint?
20. What is the actual circumference of the butterfly bush? Use 3.14 for π and round to the nearest tenth.
21. Find the scale factor.

22. **BRIDGES** The Natchez Trace Bridge in Franklin, Tennessee, is 1,500 feet long. Suppose you build a model of the bridge using a scale of 1 inch : 25 feet. How long is the model?

23. **GEOGRAPHY** A map of Brasilia has a scale of 1 inch to 5 miles. If the city is $2\frac{7}{16}$ inches across on the map, what is the distance across the actual city? Use estimation to check your answer.

ART For Exercises 24 and 25, refer to the figure at the right and the information below.

Mount Rushmore is a sculpture that was carved using a model with a scale of 1 inch:1 foot, as shown at the right.

24. How high is George Washington's face on the actual sculpture of Mount Rushmore?

25. What is the scale factor?

26. **MODEL** The diameter of an igloo is 10 feet. If you were to make a model of the igloo with sugar cubes, which might be an appropriate scale: 1 inch = 1 foot or 1 inch = 50 feet? Explain.

27. **CRITICAL THINKING** The distance between Atlanta and Savannah, Georgia, is 245 miles. The distance on a map is $3\frac{1}{2}$ inches. What is the scale of the map?

Standardized Test Practice and Mixed Review

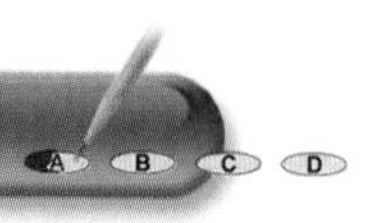

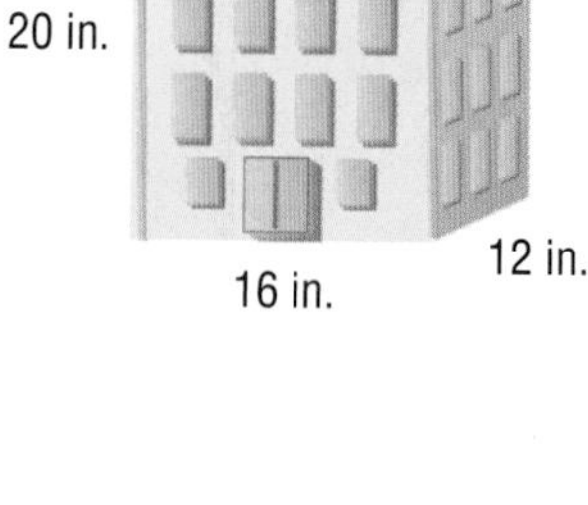

28. **SHORT RESPONSE** A scale model of a new office building is shown at the right. If the scale is 5 inches = 18 feet, how tall is the actual building?

29. **MULTIPLE CHOICE** Which of the following scales does *not* have a scale factor of $\frac{1}{60}$?

    (A) 1 in. = 5 ft
    (B) 1 cm = 60 cm
    (C) 1 ft = 20 yd
    (D) 1 in. = 60 ft

30. **BIRDS** A binocular has a *magnification factor* of 35 to 1; that is, objects that are actually 35 feet away appear to be only 1 foot away using the binoculars. If a goldfinch is 368 feet away, what will the distance appear to be in the binoculars? Round to the nearest tenth. (Lesson 7-3)

Find each unit rate. (Lesson 7-2)

31. 200 miles in 5 hours
32. 99¢ for 30 ounces
33. $6.20 for 4 pounds
34. 150 meters in 12 seconds

GETTING READY FOR THE NEXT LESSON

PREREQUISITE SKILL Divide. Write in simplest form. (Lesson 6-6)

35. $2\frac{3}{4} \div 10$
36. $4\frac{1}{3} \div 10$
37. $30\frac{2}{3} \div 100$
38. $87\frac{1}{2} \div 100$

7-4b Spreadsheet Investigation

A Follow-Up of Lesson 7-4

What You'll LEARN

Use a spreadsheet to calculate measurements for scale drawings.

Scale Drawings

A computer spreadsheet is a useful tool for calculating measures for scale drawings. You can change the scale factors and the dimensions, and the spreadsheet will automatically calculate the new values.

ACTIVITY

Suppose you want to make a scale drawing of your school. Set up a spreadsheet like the one shown below. In this spreadsheet, the actual measures are in feet, and the scale drawing measures are in inches.

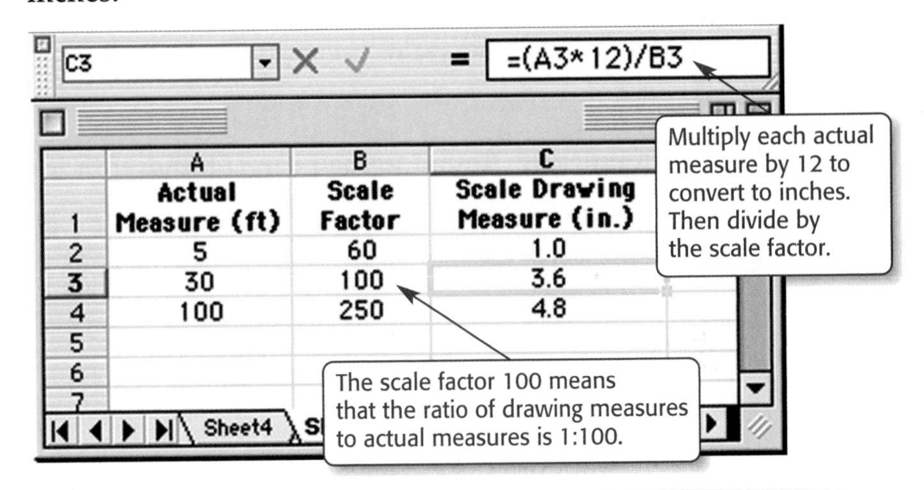

EXERCISES

1. The length of one side of the school building is 100 feet. If you use a scale factor of 1:250, what is the length on your scale drawing?
2. The length of a classroom is 30 feet. What is the scale factor if the length of the classroom on a scale drawing is 3.6 inches?
3. Calculate the length of a 30-foot classroom on a scale drawing if the scale factor is 1:10.
4. Suppose the actual measures of your school are given in meters. Describe how you could use a spreadsheet to calculate the scale drawing measures in centimeters using a scale factor of 1:50.
5. Choose three rooms in your home and use a spreadsheet to make scale drawings. First, choose an appropriate scale and calculate the scale factor. Include a sketch of the furniture in each room.

Mid-Chapter Practice Test

Vocabulary and Concepts

1. **Explain** the difference between a rate and a unit rate and give an example of each. (Lesson 7-2)
2. **Define** cross product. (Lesson 7-3)

Skills and Applications

Write each ratio as a fraction in simplest form. (Lesson 7-1)

3. 18 to 12
4. 7 to 49
5. 20 meters : 44 meters

Determine whether the ratios are equivalent. Explain. (Lesson 7-1)

6. $\frac{6}{9}$ and $\frac{2}{3}$
7. 150 : 15 and 3 : 1
8. 24 : 4 and 72 : 8

Find each unit rate. Round to the nearest hundredth if necessary. (Lesson 7-2)

9. 200 miles in 5 hours
10. $6.20 for 5 pounds

11. **SHOPPING** Which box of cereal shown in the table costs the least per ounce? (Lesson 7-2)

Weight (oz)	Cost ($)
12	2.50
18	3.69
24	4.95

12. Determine whether the ratios $\frac{5}{8}$ and $\frac{25}{40}$ form a proportion. (Lesson 7-3)

Solve each proportion. (Lesson 7-3)

13. $\frac{3}{d} = \frac{12}{20}$
14. $\frac{7}{8} = \frac{m}{48}$
15. $\frac{w}{8} = \frac{1}{3}$

16. **TRUCKS** A truck is 26 feet long. Find the length on a scale model if 1 inch represents $5\frac{1}{2}$ feet. Round to the nearest tenth if necessary. (Lesson 7-4)

Suppose you are making a scale drawing. Find the length of each object on the scale drawing with the given scale. Then find the scale factor. (Lesson 7-4)

17. a bedroom 12 feet long; 1 in. = 3 ft
18. a park area 38 meters wide; 0.5 cm = 1 m

Standardized Test Practice

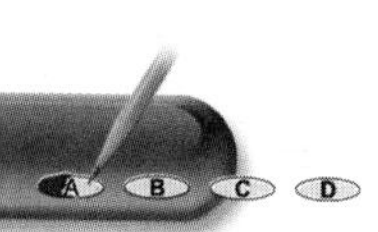

19. **GRID IN** A train travels 146 miles in 2 hours. At this rate, how many miles will it travel in 3.5 hours? (Lesson 7-3)

20. **SHORT RESPONSE** On a scale drawing, a dog pen is $4\frac{1}{2}$ inches long. If 1 inch = 3 feet, what is the actual length of the dog pen? (Lesson 7-4)

Math Skill

Solving Proportions

The Game Zone

A Place To Practice Your Math Skills

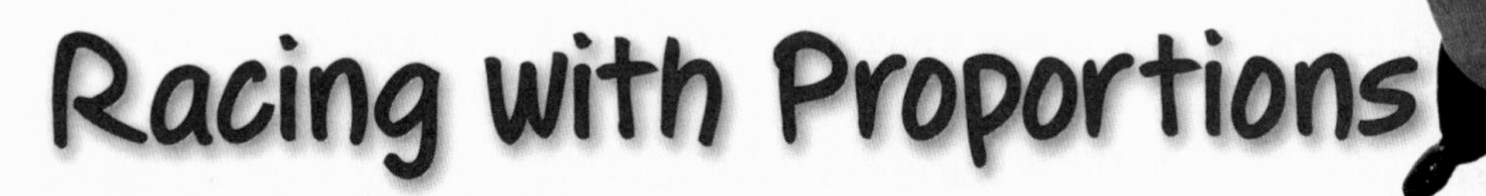

GET READY!

Players: two
Materials: centimeter grid paper, spinner, centimeter cubes

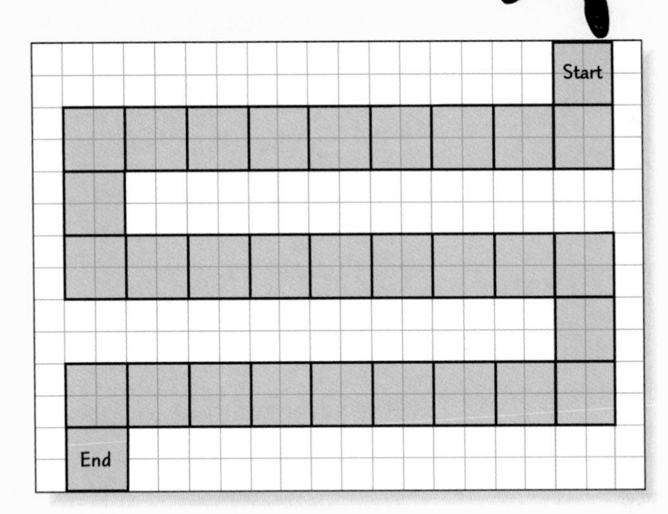

GET SET!

- Copy the game board at the right onto grid paper.
- Label equal sections of a spinner with 10, 12, 15, 20, 24, 30, 40, and 60.

GO!

- Each player places a cube on the *Start* square. (Note that both cubes will not fit on the same square. Place them side by side.)
- Each player spins the spinner and substitutes the number on the spinner for x in the proportion $\frac{y}{15} = \frac{8}{x}$. The player solves the proportion and moves his or her centimeter cube y spaces.
- Then each player spins the spinner and substitutes the number on the spinner for y in the given proportion. He or she solves the proportion and moves his or her centimeter cube x spaces.
- Continue to spin the spinner and solve the proportion for y and x.
- **Who Wins?** The first person to reach the *End* square wins the round. It is not necessary to land on *End* with an exact roll.

7-5 Fractions, Decimals, and Percents

What You'll LEARN

Write percents as fractions, and vice versa.

REVIEW Vocabulary

percent: a ratio that compares a number to 100 **(Lesson 5-5)**

WHEN am I ever going to use this?

SURVEYS The graph shows the results of a survey in which teens were asked to name the most important invention of the 20th century.

1. What percent of the teens said that the personal computer was the most important invention?
2. How is this percent written as a ratio?
3. Simplify the ratio.

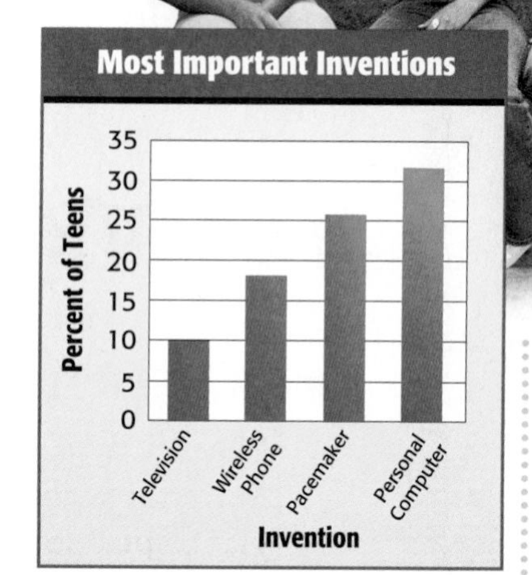

Source: Lemelson-MIT Program

In Lesson 5-5, you wrote percents like 32% as fractions by writing fractions with denominators of 100 and then simplifying. You can use the same method to write percents like 16.8% and $8\frac{1}{3}\%$ as fractions.

EXAMPLES Percents as Fractions

1 HOCKEY In men's college hockey, 16.8% of the players are from Ontario, Canada. What fraction is this? Write in simplest form.

$16.8\% = \frac{16.8}{100}$ Write a fraction with a denominator of 100.

$= \frac{16.8}{100} \cdot \frac{10}{10}$ Multiply by $\frac{10}{10}$ to eliminate the decimal in the numerator.

$= \frac{168}{1{,}000}$ or $\frac{21}{125}$ Simplify.

So, $\frac{21}{125}$ of the players are from Ontario, Canada.

2 Write $8\frac{1}{3}\%$ as a fraction in simplest form.

$8\frac{1}{3}\% = \frac{8\frac{1}{3}}{100}$ Write a fraction.

$= 8\frac{1}{3} \div 100$ Divide.

$= \frac{25}{3} \div 100$ Write $8\frac{1}{3}$ as an improper fraction.

$= \frac{25}{3} \times \frac{1}{100}$ Multiply by the reciprocal of 100, which is $\frac{1}{100}$.

$= \frac{25}{300}$ or $\frac{1}{12}$ Simplify.

To write a fraction like $\frac{8}{25}$ as a percent, multiply the numerator and the denominator by a number so that the denominator is 100. If the denominator is not a factor of 100, you can write fractions as percents by using a proportion.

EXAMPLES Fractions as Percents

3 TESTS On a math test, Mary got 7 questions correct out of 8. Find her grade as a percent.

To find her grade, write $\frac{7}{8}$ as a percent. **Estimate** $\frac{7}{8}$ is greater than $\frac{6}{8} = \frac{3}{4}$, or greater than 75%.

$\frac{7}{8} = \frac{n}{100}$ Write a proportion using $\frac{n}{100}$.

$700 = 8n$ Find the cross products.

$\frac{700}{8} = \frac{8n}{8}$ Divide each side by 8.

$87\frac{1}{2} = n$ Simplify.

So, $\frac{7}{8} = 87\frac{1}{2}\%$ or 87.5%.

This is greater than 75%, which was the estimate.

STUDY TIP

Choose the Method To write a fraction as a percent,
- use *multiplication* when a fraction has a denominator that is a factor of 100,
- use a *proportion* for any type of fraction.

4 Write $\frac{4}{15}$ as a percent. Round to the nearest hundredth.

Estimate $\frac{4}{15}$ is about $\frac{4}{16}$, which equals $\frac{1}{4}$ or 25%.

$\frac{4}{15} = \frac{n}{100}$ Write a proportion using $\frac{n}{100}$.

$400 = 15n$ Find the cross products.

400 [÷] 15 [ENTER] **26.66666667** Use a calculator to simplify.

So, $\frac{4}{15}$ is about 26.67%. This result is close to the estimate.

Your Turn Write each fraction as a percent. Round to the nearest hundredth if necessary.

a. $\frac{2}{15}$ b. $\frac{7}{16}$ c. $\frac{17}{24}$

In this lesson, you have written percents as fractions and fractions as percents. In Chapter 5, you wrote percents and fractions as decimals. You can also write a fraction as a percent by first writing the fraction as a decimal and then writing the decimal as a percent.

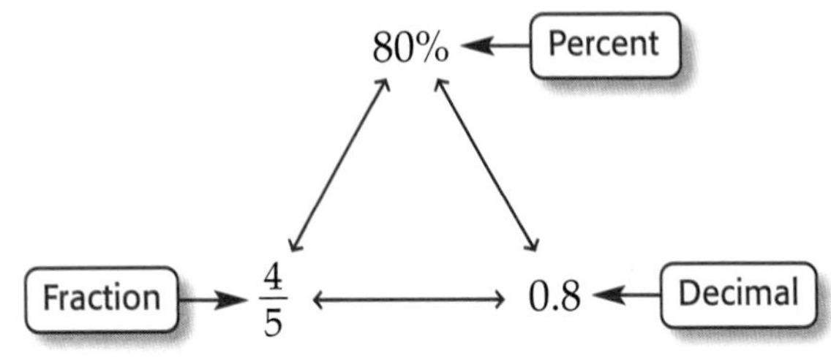

Percents, fractions, and decimals are all different names that represent the same number.

STUDY TIP

Look Back You can review **writing fractions as decimals** in Lesson 5-4.

EXAMPLES Fractions as Percents

5 **Write $\frac{1}{8}$ as a percent.**

$\frac{1}{8} = 0.125$ Write $\frac{1}{8}$ as a decimal.

$= 12.5\%$ Multiply by 100 and add the %.

6 **Write $\frac{5}{6}$ as a percent. Round to the nearest hundredth.**

$\frac{5}{6} = 0.833333333\ldots$ Write $\frac{5}{6}$ as a decimal.

$\approx 83.33\%$ Multiply by 100 and add the %.

Your Turn **Write each fraction as a percent. Round to the nearest hundredth if necessary.**

d. $\frac{5}{16}$ **e.** $\frac{7}{12}$ **f.** $\frac{2}{9}$

Some fractions whose denominators are not factors of 100 are used often in everyday situations. It is helpful to be familiar with these fractions and their equivalent decimals and percents.

Key Concept **Common Fraction/Decimal/Percent Equivalents**

Fraction	Decimal	Percent	Fraction	Decimal	Percent
$\frac{1}{3}$	$0.\overline{3}$	$33\frac{1}{3}\%$	$\frac{3}{8}$	0.375	$37\frac{1}{2}\%$
$\frac{2}{3}$	$0.\overline{6}$	$66\frac{2}{3}\%$	$\frac{5}{8}$	0.625	$62\frac{1}{2}\%$
$\frac{1}{8}$	0.125	$12\frac{1}{2}\%$	$\frac{7}{8}$	0.875	$87\frac{1}{2}\%$

Skill and Concept Check

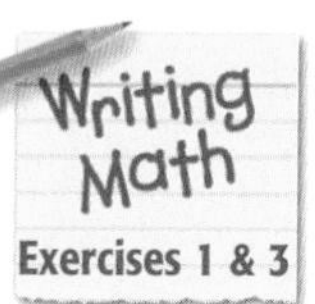

1. **Describe** two methods for writing a fraction as a percent.

2. **OPEN ENDED** Write a percent that is between 15.2% and 15.8%. Then write it as a fraction in simplest form.

3. **Explain** why percents are used in statistical graphs rather than fractions.

GUIDED PRACTICE

Write each percent as a fraction in simplest form.

4. 13.5% 5. 18.75% 6. $66\frac{2}{3}\%$ 7. $57\frac{1}{2}\%$

Write each fraction as a percent. Round to the nearest hundredth if necessary.

8. $\frac{4}{25}$ 9. $\frac{5}{8}$ 10. $\frac{9}{40}$ 11. $\frac{4}{11}$

12. **LANGUAGES** In Virginia, 1 person out of every 40 speaks an Asian language at home. What percent is this?

Practice and Applications

HOMEWORK HELP	
For Exercises	See Examples
13–24, 38	1, 2
25–37, 39	3–6

Extra Practice
See pages 581, 602.

Write each percent as a fraction in simplest form.

13. 17.5% **14.** 62.5% **15.** 6.2% **16.** 34.5%

17. 28.75% **18.** 56.25% **19.** $33\frac{1}{3}\%$ **20.** $16\frac{2}{3}\%$

21. $81\frac{1}{4}\%$ **22.** $93\frac{3}{4}\%$ **23.** $78\frac{3}{4}\%$ **24.** $23\frac{1}{3}\%$

Write each fraction as a percent. Round to the nearest hundredth if necessary.

25. $\frac{11}{20}$ **26.** $\frac{18}{25}$ **27.** $\frac{3}{8}$ **28.** $\frac{21}{40}$ **29.** $\frac{29}{30}$ **30.** $\frac{8}{9}$

31. $\frac{5}{7}$ **32.** $\frac{1}{16}$ **33.** $\frac{1}{80}$ **34.** $\frac{57}{200}$ **35.** $\frac{5}{12}$ **36.** $\frac{7}{15}$

37. What percent is equivalent to $\frac{60}{125}$?

CARS For Exercises 38 and 39, use the table at the right. It shows the percent of people who keep the listed items in their car.

38. What fraction of people keep a first-aid kit in their car?

39. Approximately 26 out of 125 people surveyed keep a hairbrush in their car. Is this greater or less than the percent who keep sports equipment? Explain.

40. CRITICAL THINKING For what value of x is $\frac{1}{x} = x\%$ a true sentence?

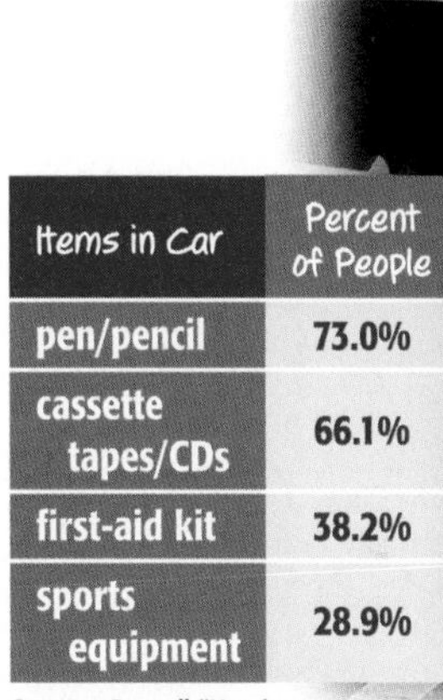

Items in Car	Percent of People
pen/pencil	73.0%
cassette tapes/CDs	66.1%
first-aid kit	38.2%
sports equipment	28.9%

Source: Penzoil "How's your Automotive Attitude?" survey

Standardized Test Practice and Mixed Review

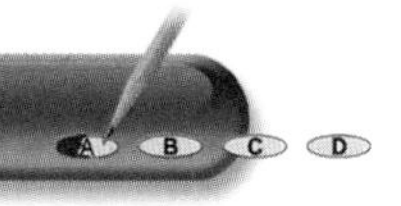

41. MULTIPLE CHOICE Two-thirds of Kodiak Island's 2 million acres has been set aside as a wildlife preserve for the Kodiak brown bear. What percent of the island is a preserve?

(A) $2\frac{2}{3}\%$ (B) 23% (C) $33\frac{1}{3}\%$ (D) $66\frac{2}{3}\%$

42. SHORT RESPONSE What is $\frac{7}{16}$ written as a percent?

43. MODELS On a scale model of a building, 2 inches represents 15 feet. If the model is 9 inches tall, how tall is the actual building? (Lesson 7-4)

Solve each proportion. (Lesson 7-3)

44. $\frac{6}{9} = \frac{4}{x}$ **45.** $\frac{3}{t} = \frac{9}{5}$ **46.** $\frac{n}{5} = \frac{2.8}{7}$ **47.** $\frac{7}{12} = \frac{r}{30}$

GETTING READY FOR THE NEXT LESSON

PREREQUISITE SKILL Write each decimal as a fraction with a denominator of 100. (Lesson 5-4)

48. 0.23 **49.** 0.09 **50.** 0.368 **51.** 0.425

7-6 Percents Greater Than 100% and Percents Less Than 1%

What You'll LEARN

Write percents greater than 100% and percents less than 1% as fractions and as decimals, and vice versa.

HANDS-ON Mini Lab

Materials
- grid paper
- colored pencils

Work with a partner.

- Draw three 10×10 squares on a piece of grid paper. Each large square represents 100%, and each small square represents 1%.
- For the first model, use two grids to shade 120 small squares.
- For the second model, shade half of one small square on a grid. Use a different color than on the first model.

1. Which model represents a percent greater than 100%? What is the percent?
2. Which model represents a percent less than 1%? What is the percent?
3. Shade grids to represent each percent.

   a. 150% b. 215% c. $\frac{1}{4}\%$

Percents greater than 100% or less than 1% can be written as decimals using the methods you learned in Chapter 5. They can also be written as mixed numbers or fractions.

$$250\% = \frac{250}{100} = 2.5 \text{ or } 2\frac{1}{2}$$

A percent greater than 100% equals a number greater than 1.

$$0.5\% = \frac{0.5}{100} = 0.005 \text{ or } \frac{1}{200}$$

A percent less than 1% equals a number less than 0.01 or $\frac{1}{100}$.

STUDY TIP

Alternate Method To divide by 100, move the decimal point 2 places to the left.

410% = 410 or 4.1

0.2% = 00.2 or 0.002

EXAMPLES Percents as Decimals or Fractions

Write each percent as a decimal and as a mixed number or fraction in simplest form.

1 410%

$410\% = \frac{410}{100}$ Definition of percent

$= 4.1 \text{ or } 4\frac{1}{10}$

2 0.2%

$0.2\% = \frac{0.2}{100}$ Definition of percent

$= 0.002 \text{ or } \frac{1}{500}$

msmath2.net/extra_examples

EXAMPLE Real-Life Percents as Decimals

READING Math

Percents In the example at the right, the solution 3 means *3 times as much.* So, before hibernation bears need 3 times as many Calories as usual.

3 BEARS Before bears hibernate, their need for Calories increases 300%. Write 300% as a decimal.

$300\% = 3.00$ Divide by 100.

$= 3.0$ or 3

You can write decimals less than 0.01 and decimals greater than 1 as percents by multiplying by 100 and adding a %.

EXAMPLES Decimals as Percents

Write each decimal as a percent.

4 1.68

$1.68 = 1.68$ Multiply by 100.

$= 168\%$

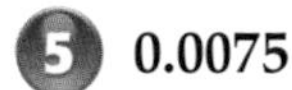

5 0.0075

$0.0075 = 0.0075$ Multiply by 100.

$= 0.75\%$

Your Turn **Write each decimal as a percent.**

a. 4.5 b. 0.002 c. 0.0016

Skill and Concept Check

Writing Math
Exercise 3

1. **Write** the fraction of squares that are shaded as a percent and as a decimal.

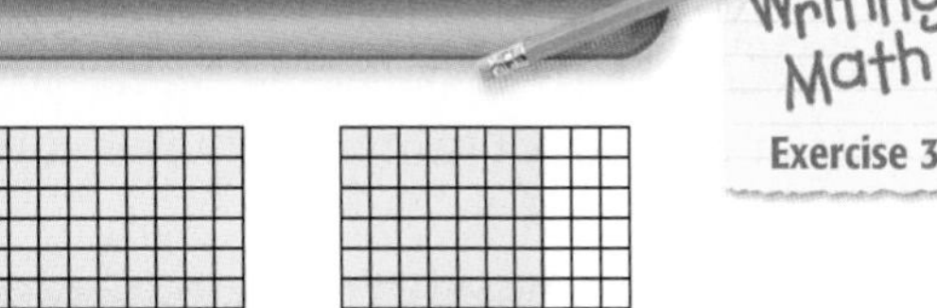

2. **OPEN ENDED** Write a number between 6 and 7 as a percent.

3. **Which One Doesn't Belong?** Identify the number that does not have the same value as the other three. Explain your reasoning.

63.5%	0.635	$63\frac{1}{2}$	$63\frac{1}{2}\%$

GUIDED PRACTICE

Write each percent as a decimal and as a mixed number or fraction in simplest form.

4. 325% 5. 200% 6. 0.7% 7. 0.15%

Write each decimal as a percent.

8. 1.8 9. 0.0015 10. 2.75 11. 0.0005

12. **PHOTOGRAPHY** A photograph developed from 35 mm film is 428% larger than the negative. Express this percent as a decimal.

Practice and Applications

HOMEWORK HELP	
For Exercises	See Examples
13–24, 33–37	1, 2
39–40	3
25–32, 38	4, 5

Extra Practice
See pages 581, 602.

Write each percent as a decimal and as a mixed number or fraction in simplest form.

13. 350% **14.** 475% **15.** 600%

16. 400% **17.** 0.6% **18.** 0.1%

19. 0.55% **20.** 0.48% **21.** 260%

22. 195% **23.** 0.05% **24.** 0.04%

Write each decimal as a percent.

25. 8.5 **26.** 35 **27.** 0.009 **28.** 0.003

29. 2.64 **30.** 1.07 **31.** 0.0034 **32.** 0.0077

Write each number as a percent.

33. $3\frac{1}{2}$ **34.** $9\frac{3}{4}$ **35.** $\frac{2}{500}$ **36.** $\frac{1}{400}$

37. Write $1\frac{5}{8}$ as a percent. **38.** Write $\frac{1}{8}$% as a decimal.

39. POPULATION In 2000, about 0.3% of the population in the United States was Japanese. Write this percent as a decimal and as a fraction.

40. RESEARCH Use the Internet or another source to find the percent of people in the United States that come from three other countries. Write the percents as decimals.

41. MULTI STEP One hour is what percent of one week?

42. CRITICAL THINKING Some North African ostriches are 150% of the height of a 6-foot human. What is the height of some ostriches?

Standardized Test Practice and Mixed Review

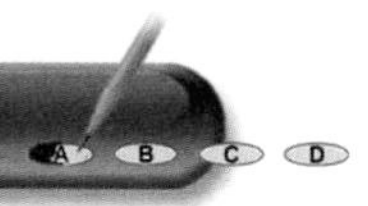

43. MULTIPLE CHOICE A certain stock increased its value by 467% over 10 years. What is this percent written as a decimal?

Ⓐ 4.67 Ⓑ 46.7 Ⓒ 0.467 Ⓓ 0.00467

44. MULTIPLE CHOICE Choose the best estimate for 100.5%.

Ⓕ 0 Ⓖ 1 Ⓗ 100 Ⓘ 1,000

Write each percent as a fraction in simplest form. (Lesson 7-5)

45. 7.5% **46.** 1.2% **47.** $6\frac{1}{4}$% **48.** $92\frac{1}{2}$%

49. SCALE DRAWING A garage is 18 feet wide. In a scale drawing, 1 inch = 3 feet. What is the width of the garage on the drawing? (Lesson 7-4)

GETTING READY FOR THE NEXT LESSON

PREREQUISITE SKILL Write each percent as a decimal. (Lesson 5-6)

50. 85% **51.** 6.5% **52.** 36.9% **53.** 12.3%

msmath2.net/self_check_quiz

7-7 Percent of a Number

What You'll LEARN

Find the percent of a number.

WHEN am I ever going to use this?

SURVEYS The graph at the right shows the results of a survey in which 1,016 people were asked to name the things they feared.

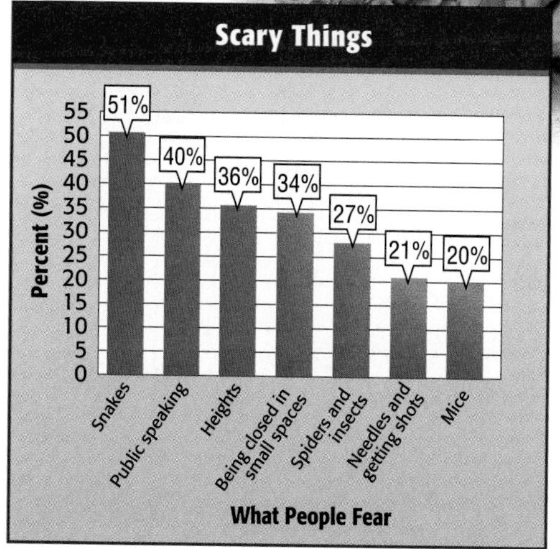

Source: Gallup Poll

1. About how many of them said they were afraid of snakes?
2. How did you estimate?
3. Suppose you pass 50 people at a mall. Based on the results of the survey, do you think more or less than 25 of them are afraid of mice? Explain.

One way to find the percent of a number is to use a proportion.

EXAMPLE Use a Proportion to Find a Percent

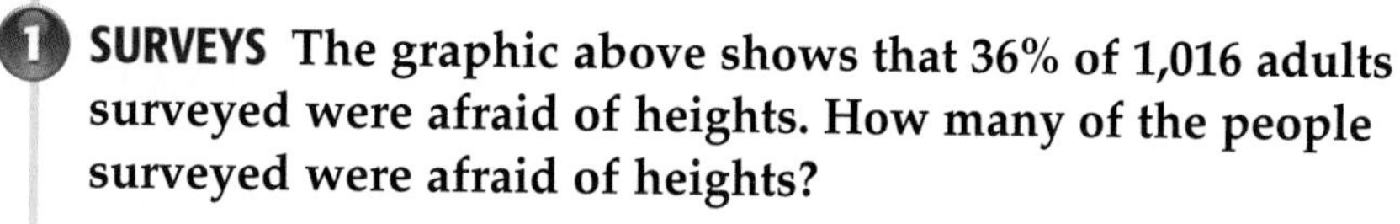

1 SURVEYS The graphic above shows that 36% of 1,016 adults surveyed were afraid of heights. How many of the people surveyed were afraid of heights?

36% means that 36 out of 100 people were afraid of heights. Find an equivalent ratio x out of 1,016 and write a proportion.

number afraid of heights → / total number in survey → $\frac{x}{1{,}016} = \frac{36}{100}$ } percent of people afraid of heights

Now solve the proportion.

$\frac{x}{1{,}016} = \frac{36}{100}$ Write the proportion.

$x \cdot 100 = 1{,}016(36)$ Find the cross products.

$100x = 36{,}576$ Multiply.

$\frac{100x}{100} = \frac{36{,}576}{100}$ Divide each side by 100.

$x = 365.76$ Simplify.

So, about 366 of the 1,016 people surveyed were afraid of heights.

Another method for finding the percent of a number is to use multiplication.

EXAMPLES Use Multiplication to Find a Percent

What number is 140% of 32?

140% of 32 = 140% × 32   Write a multiplication expression.

= 1.4 × 32   Write 140% as a decimal.

= 44.8   Multiply.

So, 140% of 32 is 44.8.

3 Find 75% of $800.

75% of $800 = 75% × $800   Write a multiplication expression.

= 0.75 × 800   Write 75% as a decimal.

= 600   Multiply.

So, 75% of $800 is $600.

> **STUDY TIP**
>
> **Mental Math** You can find 75% of 800 mentally by using the fraction $\frac{3}{4}$. $\frac{3}{4}$ of 800 is 600.

Your Turn Find each number.

a. Fifty percent of 140 is what number?

b. Find 37.5% of 64.

Skill and Concept Check

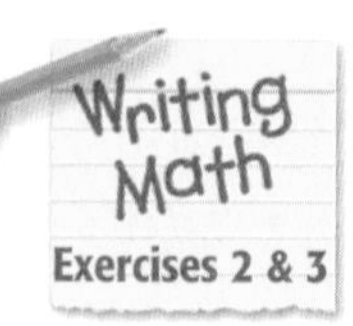

Exercises 2 & 3

1. **Write** a proportion that can be used to find 0.5% of 22.
2. **NUMBER SENSE** Use mental math to determine whether 30% of $150 is $80. Explain.
3. **OPEN ENDED** Write a problem that can be solved using the graphic at the right. Then solve the problem.

USA TODAY Snapshots®

Low-tech tools rule at meetings

1,500 white-collar workers were asked what tools they most commonly used to record information during a meeting.

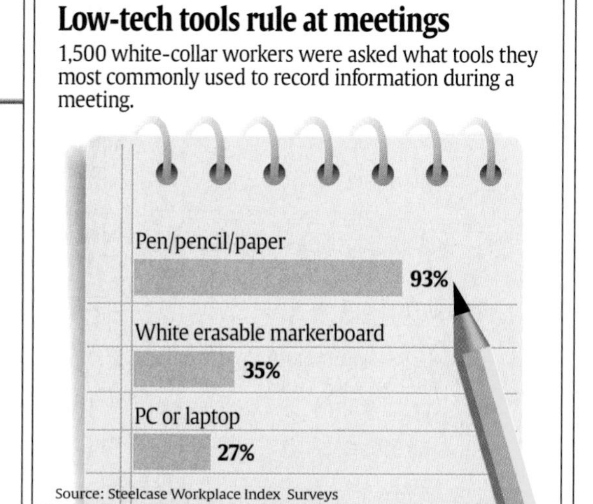

Source: Steelcase Workplace Index Surveys

By Darryl Haralson and Keith Simmons, USA TODAY

GUIDED PRACTICE

Find each number. Round to the nearest tenth if necessary.

4. Find 95% of $40.
5. Forty-two percent of 263 is what number?
6. What number is 200% of 75?
7. Find 0.25% of 40.
8. **COMMISSION** Ms. Sierra earns a 2% *commission*, or fee paid based on a percent of her sales, on every vacation package that she sells. One day, she sold vacation packages worth $6,500. What was her commission?

Practice and Applications

HOMEWORK HELP	
For Exercises	See Examples
9–23	1–3

Extra Practice
See pages 582, 602.

Find each number. Round to the nearest tenth if necessary.

9. What is 65% of 100?
10. Find 45% of $400.
11. Find 25% of $640.
12. What number is 18.5% of 500?
13. 67.5% of 76 is what?
14. 64% of 88 is what number?
15. What is 125% of 10?
16. Find 300% of 14.
17. Find 0.15% of 250
18. What number is 0.5% of 80?
19. Find $20\frac{1}{4}$% of 3.
20. 250.5% of 25 is what number?

SNACKING **For Exercises 21 and 22, use the graphic at the right. It shows the results of a poll of 1,746 college students.**

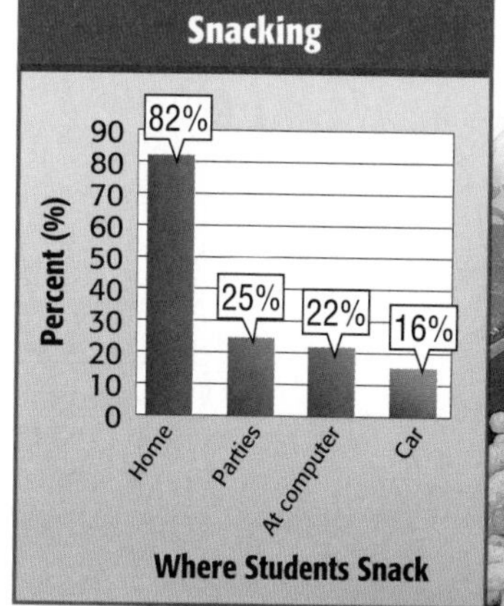

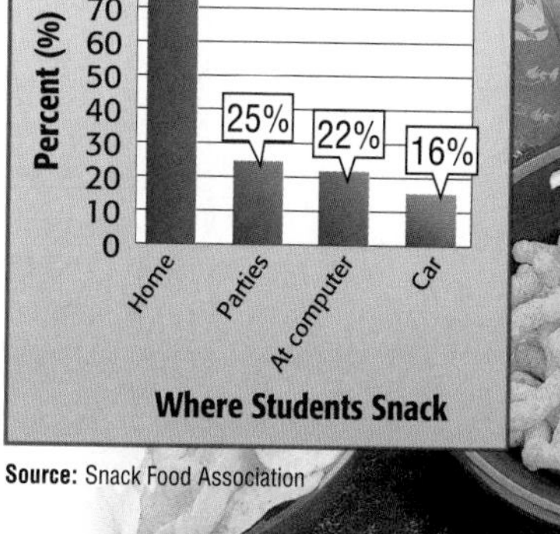

Source: Snack Food Association

21. Write a proportion that can be used to find how many students like to snack at their computer. Then solve.
22. How many students surveyed like to snack at home?
23. **DELIVERY** A mountain bike costs $288 plus a 4.5% delivery charge. What is the cost of the bike including the delivery charge?
24. **CRITICAL THINKING** Suppose you add 10% of a number to the number. Then you subtract 10% of the total. Is the result *greater than, less than*, or *equal to* the original number? Explain your reasoning.

Standardized Test Practice and Mixed Review

25. **MULTIPLE CHOICE** A football player has made 80% of the field goals he has attempted in his career. If he attempts 5 field goals in a game, how many would he be expected to make?

   (A) 5 (B) 4 (C) 3 (D) 2

26. **SHORT RESPONSE** What number is 12% of 16.5?

Write each decimal as a percent. (Lesson 7-6)

27. 7.5
28. 9
29. 0.0004
30. 0.0018

31. Write $\frac{11}{15}$ as a percent. Round to the nearest hundredth. (Lesson 7-5)

GETTING READY FOR THE NEXT LESSON

PREREQUISITE SKILL **Evaluate each expression. Round to the nearest tenth if necessary.** (Lesson 1-3)

32. $\frac{36 \cdot 2}{80}$
33. $\frac{14(5)}{3.5}$
34. $\frac{9 \times 100}{56}$
35. $22 \cdot 100 \div 490$

msmath2.net/self_check_quiz

7-8a HANDS-ON LAB

A Preview of Lesson 7-8

Using a Percent Model

You can find the percent of a number or the *part* of the whole by using a model.

What You'll LEARN

Use a percent model to find a part.

Materials

- grid paper

ACTIVITY *Work with a partner.*

Suppose you and a friend eat dinner at the Pizza Palace and your total bill is about $25. If you want to leave a 20% tip, how much should you leave? Use a model to find the amount.

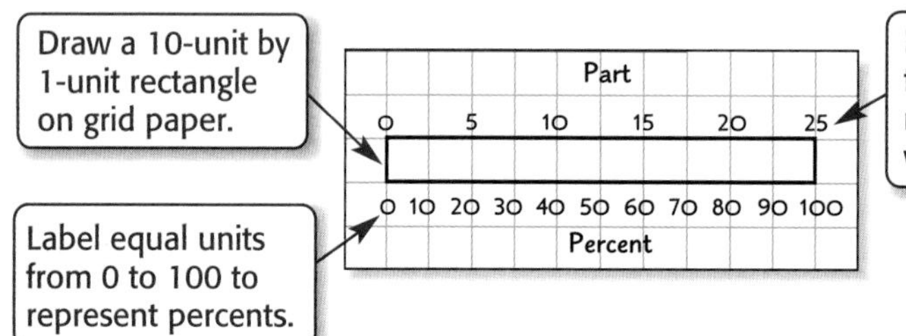

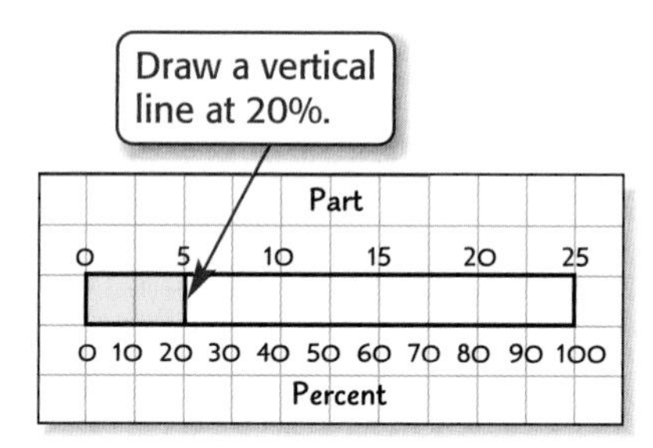

So, you should leave $5 as a tip.

Your Turn **Draw a model to find each part.**

a. 70% of 50 b. 45% of 20 c. 20% of 75

Writing Math

1. Suppose your whole family had dinner with you and the total bill was $50. How could you change your model to find the tip?
2. **Write** a sentence that describes what is represented by the model at the right.
3. **Write** a percent problem that can be represented and solved using a model.
4. **Make a conjecture** about how you could use a model to estimate the percent represented by the ratio 8 out of 50.

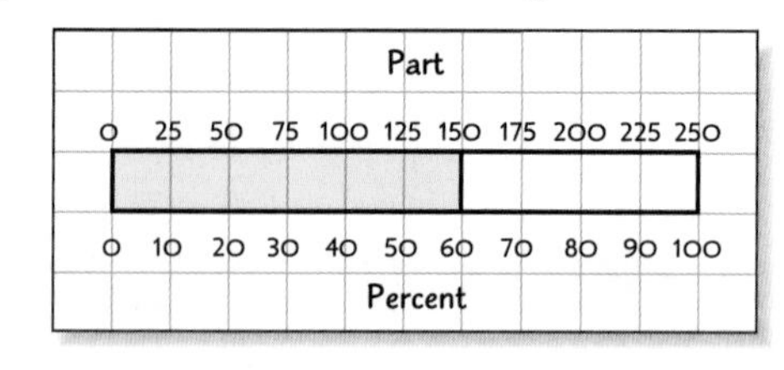

7-8 The Percent Proportion

What You'll LEARN

Solve problems using the percent proportion.

NEW Vocabulary

percent proportion
part
base

WHEN am I ever going to use this?

SPACE The engine on the space shuttle weighs approximately 19,700 pounds. The entire space shuttle weighs 178,000 pounds.

1. Write the ratio of engine weight to total weight as a fraction.
2. Use a calculator to write the fraction as a decimal to the nearest hundredth.
3. What percent of the space shuttle's weight is the engine?

A **percent proportion** compares **part** of a quantity to the whole quantity, called the **base**, using a percent. You can use a percent proportion to find what percent of a space shuttle's weight is the engine.

part: engine weight → $\frac{19{,}700}{178{,}000} = \frac{p}{100}$ ← base: total weight

The percent is written as a fraction with a denominator of 100.

Key Concept — Percent Proportion

Words The percent proportion is $\frac{\text{part}}{\text{base}} = \frac{\text{percent}}{100}$.

Symbols $\frac{a}{b} = \frac{p}{100}$, where a is the part, b is the base, and p is the percent.

To find the percent, write the percent proportion and solve for p.

EXAMPLE Find the Percent

1 What percent of $15 is $9?

9 is the part, and 15 is the base. You need to find the percent p.

$\frac{a}{b} = \frac{p}{100}$	Percent proportion
$\frac{9}{15} = \frac{p}{100}$	Replace a with 9 and b with 15.
$9 \cdot 100 = 15 \cdot p$	Find the cross products.
$900 = 15p$	Simplify.
$\frac{900}{15} = \frac{15p}{15}$	Divide each side by 15.
$60 = p$	So, 60% of $15 is $9. Compare the answer to the model.

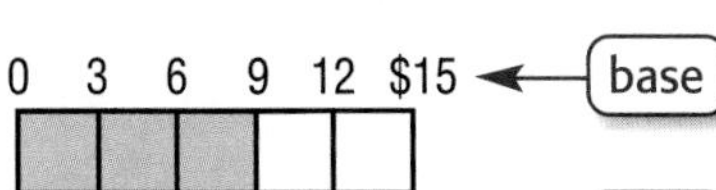

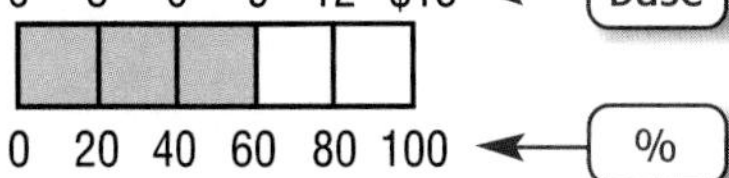

STUDY TIP

Finding the Base In a percent problem, the base follows the word *of.*

You can also use the percent proportion to find a missing part or base.

Concept Summary — **Types of Percent Problems**

Type	Example	Proportion
Find the Percent	What percent of 6 is 3?	$\frac{3}{6} = \frac{p}{100}$
Find the Part	What number is 50% of 6?	$\frac{a}{6} = \frac{50}{100}$
Find the Base	3 is 50% of what number?	$\frac{3}{b} = \frac{50}{100}$

To find the part, write the percent proportion and solve for a. To find the base, write the percent proportion and solve for b.

EXAMPLE Find the Part

2 What number is 40% of 120?

40 is the percent and 120 is the base. You need to find the part.

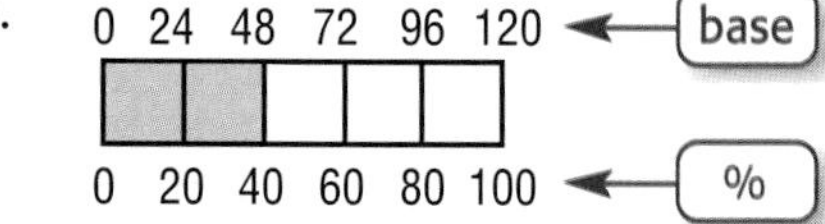

$\frac{a}{b} = \frac{p}{100}$ Percent proportion

$\frac{a}{120} = \frac{40}{100}$ Replace b with 120 and p with 40.

$a \cdot 100 = 120 \cdot 40$ Find the cross products.

$100a = 4{,}800$ Simplify.

$\frac{100a}{100} = \frac{4{,}800}{100}$ Divide each side by 100.

$a = 48$ So, 40% of 120 is 48. Compare to the model.

EXAMPLE Find the Base

3 18 is 25% of what number?

18 is the part and 25 is the percent. You need to find the base, or the whole quantity.

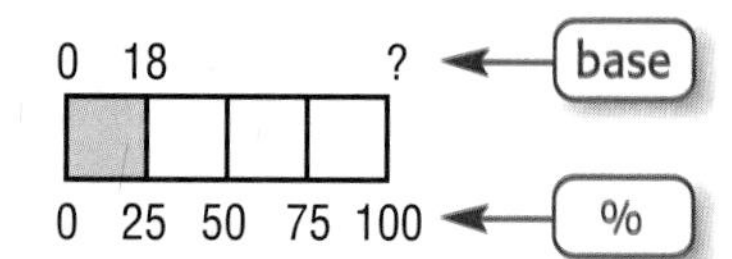

$\frac{a}{b} = \frac{p}{100}$ Percent proportion

$\frac{18}{b} = \frac{25}{100}$ Replace a with 18 and p with 25.

$18 \cdot 100 = b \cdot 25$ Find the cross products.

$1{,}800 = 25b$ Simplify.

$\frac{1{,}800}{25} = \frac{25b}{25}$ Divide each side by 25.

$72 = b$ So, 18 is 25% of 72.

Check In the model above, the whole quantity is $18 \cdot 4$ or 72.

Your Turn **Find each number. Round to the nearest tenth if necessary.**

a. What number is 5% of 60?

b. 80 is 75% of what number?

Skill and Concept Check

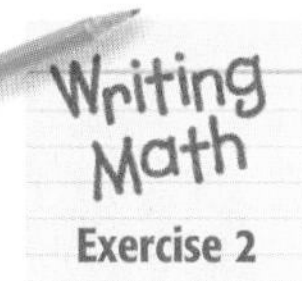

1. **Choose** the problem that can be solved by using $\frac{8}{b} = \frac{16}{100}$.
   a. What number is 16% of 8?
   b. 8 is 16% of what number?

2. **OPEN ENDED** Write a proportion that can be used to find the part if you know the base and the percent.

GUIDED PRACTICE

Find each number. Round to the nearest tenth if necessary.

3. What percent of $50 is $18?
4. What number is 25% of 180?
5. What number is 2% of 35?
6. 62 is 90.5% of what number?

Practice and Applications

HOMEWORK HELP

For Exercises	See Examples
7–22	1–3

Extra Practice
See pages 582, 602.

Find each number. Round to the nearest tenth if necessary.

7. What percent of 60 is 15?
8. 9 is 45% of what number?
9. What number is 12% of 72?
10. What percent of $12 is $9?
11. 75 is 20% of what number?
12. What number is 40% of 80?
13. $3 is what percent of $40?
14. 12.5% of what number is 24?
15. 8.2% of 50 is what number?
16. What percent of 300 is 0.6?
17. 45 is 150% of what number?
18. What percent of 25 is 30?
19. What number is 0.5% of 8?
20. 6% of what number is $10\frac{1}{2}$?

21. **SHOPPING** A paperback book originally priced at $12.50 is on sale for $7.50. What percent of the original cost is the sale price?

22. **MOVIES** Sarah and Monique spent $14, or 35% of their money, on movie tickets. How much money did they have to start with?

23. **CRITICAL THINKING** Without calculating, order the following from greatest to least value. Explain your reasoning.

    20% of 100, 20% of 500, 5% of 100

Standardized Test Practice and Mixed Review

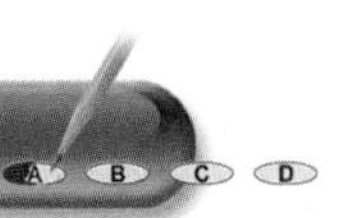

24. **MULTIPLE CHOICE** There are about 53,500 hotels in the United States. Use the table to find how many are located along highways.

    Ⓐ 22,577 Ⓑ 22,470 Ⓒ 17,976 Ⓓ 5,457

U.S. Hotels

Location	Percent
highways	42.2%
suburbs	33.6%
city	10.2%

25. **SHORT RESPONSE** 95 of 273 students volunteered. About what percent of the students did *not* volunteer?

Find each number. Round to the nearest tenth if necessary. (Lesson 7-7)

26. What is 25% of 120?
27. Find 45% of 70.

28. Write 1.2 as a percent. (Lesson 7-6)

msmath2.net/self_check_quiz

Study Guide and Review

Vocabulary and Concept Check

base (p. 323)
cross product (p. 297)
equivalent ratios (p. 289)
part (p. 323)
percent proportion (p. 323)
proportion (p. 297)
rate (p. 292)
scale (p. 304)
scale drawing (p. 304)
scale factor (p. 305)
scale model (p. 306)
unit rate (p. 292)

Choose the letter of the term that best matches each phrase.

1. a comparison of two numbers by division
2. two ratios that have the same value
3. a ratio of two measurements with different units
4. an equation that shows that two ratios are equivalent
5. used to represent something that is too large or too small for an actual-size drawing
6. the ratio of the distance on a map to the actual distance
7. the number that is compared to the whole quantity in a percent proportion
8. product of the numerator of one ratio in a proportion and the denominator of the other ratio
9. a rate that is simplified so that it has a denominator of 1
10. the number that represents the whole quantity in a percent proportion

a. rate
b. base
c. scale drawing
d. unit rate
e. ratio
f. scale
g. part
h. proportion
i. equivalent ratios
j. cross products

Lesson-by-Lesson Exercises and Examples

Ratios (pp. 288–291)

Write each ratio as a fraction in simplest form.

11. 16:12
12. 5 to 25
13. 50 cm to 75 cm
14. 6 ft:12 in.

Determine whether the ratios are equivalent. Explain.

15. $\frac{3}{5}$ and $\frac{21}{35}$
16. $\frac{18}{24}$ and $\frac{5}{20}$
17. 27:15 and 9:5
18. 4:21 and 2:7

Example 1 **Write the ratio 32 to 18 as a fraction in simplest form.**

$32 \text{ to } 18 = \frac{32}{18}$ Write the ratio as a fraction.

$= \frac{16}{9}$ Simplify.

Example 2 **Determine whether 5:6 and 15:18 are equivalent.**

$5:6 = \frac{5}{6}$ $\quad 15:18 = \frac{15}{18}$ or $\frac{5}{6}$

The ratios in simplest form both equal $\frac{5}{6}$. So, 5:6 and 15:18 are equivalent.

msmath2.net/vocabulary_review

7-2 Rates (pp. 292–295)

Find each unit rate. Round to the nearest hundredth if necessary.

19. $23.75 for 5 pounds
20. 810 miles in 9 days
21. $38 in 4 hours
22. 24 gerbils in 3 cages
23. 14 laps in 4 minutes
24. **SHOPPING** Which bottle of laundry detergent shown at the right costs the least per ounce?

Bottle	Price
12 oz	$2.16
32 oz	$5.60
64 oz	$10.24

Example 3 **Find the unit price per ounce of a 16-ounce box of pasta that is on sale for 96¢.**

To find the cost per ounce, write the rate as a fraction and then simplify.

$$\text{16-oz box for } 96¢ = \frac{96¢}{16 \text{ ounces}} = \frac{96¢ \div 16}{16 \text{ ounces} \div 16} = \frac{6¢}{1 \text{ ounce}}$$

The unit price is 6¢ per ounce.

7-3 Solving Proportions (pp. 297–300)

Solve each proportion.

25. $\frac{x}{10} = \frac{3}{5}$
26. $\frac{4}{9} = \frac{24}{m}$
27. $\frac{2}{t} = \frac{8}{50}$
28. $\frac{15}{w} = \frac{35}{21}$
29. $\frac{12}{8} = \frac{a}{6}$
30. $\frac{7}{18} = \frac{d}{6}$
31. **MODELS** A Boeing 747 jet is 70.5 meters long and has a wingspan of 60 meters. A model of the 747 has a wingspan of 80 centimeters. What is its length?

Example 4 **Solve $\frac{6}{9} = \frac{n}{12}$.**

$\frac{6}{9} = \frac{n}{12}$ Write the proportion.

$6 \times 12 = 9 \times n$ Find the cross products.

$72 = 9n$ Multiply.

$\frac{72}{9} = \frac{9n}{9}$ Divide each side by 9.

$8 = n$ Simplify.

The solution is 8.

7-4 Scale Drawings (pp. 304–308)

Suppose you are making a scale drawing. Find the length of each object on the drawing with the given scale.

32. a minivan 10 feet long; 2 inches = 1 foot
33. a rose garden 9 meters wide; 3 centimeters = 1 meter
34. a Ferris wheel with a diameter of 42 feet; 2 inches = 3 feet
35. a library 10.4 meters wide; 1 centimeter = 2 meters

Example 5 **On a map, the distance between Chicago, Illinois, and Mexico City, Mexico, is 10.9 centimeters. If the scale is 1 centimeter = 250 kilometers, what is the actual distance?**

Scale — **Chicago to Mexico City**

$\frac{1 \text{ cm}}{250 \text{ km}} = \frac{10.9 \text{ cm}}{n \text{ km}}$ Write a proportion.

$1 \times n = 250 \times 10.9$ Find the cross products.

$n = 2{,}725$ Simplify.

The actual distance is 2,725 kilometers.

7-5 Fractions, Decimals, and Percents (pp. 312–315)

Write each percent as a fraction in simplest form.

36. 27.5% **37.** 5.4% **38.** $45\frac{1}{4}\%$

Write each fraction as a percent. Round to the nearest hundredth if necessary.

39. $\frac{1}{8}$ **40.** $\frac{5}{6}$ **41.** $\frac{7}{40}$

Example 6 **Write 82.5% as a fraction in simplest form.**

$82.5\% = \frac{82.5}{100}$ Write a fraction with a denominator of 100.

$= \frac{825}{1,000}$ Multiply 82.5 and 100 by 10 to eliminate the decimal.

$= \frac{33}{40}$ Simplify.

7-6 Percents Greater Than 100% and Percents Less Than 1% (pp. 316–318)

Write each percent as a decimal and as a mixed number or fraction in simplest form.

42. 125% **43.** 0.75%

44. 563% **45.** 0.5%

Write each decimal as a percent.

46. 0.002 **47.** 4.75

48. 7.5 **49.** 0.0095

Example 7 **Write 235% as a decimal and as a mixed number in simplest form.**

235% = 2.35 Definition of percent

$= 2.35$ or $2\frac{7}{20}$

Example 8 **Write 0.008 as a percent.**

0.008 = 0.008 Multiply by 100.

= 0.8%

7-7 Percent of a Number (pp. 319–321)

Find each number. Round to the nearest tenth if necessary.

50. Find 78% of 50.

51. 45.5% of 75 is what number?

52. What is 225% of 60?

53. 0.75% of 80 is what number?

Example 9 **Find 24% of 200.**

24% of 200

$= 24\% \times 200$ Write a multiplication expression.

$= 0.24 \times 200$ Write 24% as a decimal.

$= 48$ Multiply.

So, 24% of 200 is 48.

7-8 The Percent Proportion (pp. 323–325)

Find each number. Round to the nearest tenth if necessary.

54. 6 is what percent of 120?

55. Find 0.8% of 35.

56. What percent of 375 is 40?

57. Find 310% of 42.

Example 10 **What percent of 90 is 18?**

$\frac{a}{b} = \frac{p}{100}$ Percent proportion

$\frac{18}{90} = \frac{p}{100}$ Replace *a* with 18 and *b* with 90.

$18 \cdot 100 = 90 \cdot p$ Find the cross products.

$20 = p$ So, 18 is 20% of 90.

Practice Test

Vocabulary and Concepts

1. **Define** *equivalent ratios.*
2. **Write** a proportion that can be used to find 76% of 512.

Skills and Applications

Write each ratio as a fraction in simplest form.

3. 45 to 18
4. 24:88
5. 15 minutes to 2 hours

Find each unit rate. Round to the nearest hundredth if necessary.

6. 24 cards for $4.80
7. 330 miles on 15 gallons of gasoline

Solve each proportion.

8. $\frac{2}{3} = \frac{x}{42}$
9. $\frac{t}{21} = \frac{15}{14}$
10. $\frac{9}{m} = \frac{12}{36}$

11. **BLUEPRINTS** The dimensions of a rectangular room in a new home are shown in the blueprint at the right. If the scale is 1 centimeter = 2.5 meters, find the length ℓ of the room on the scale drawing. Then find the scale factor.

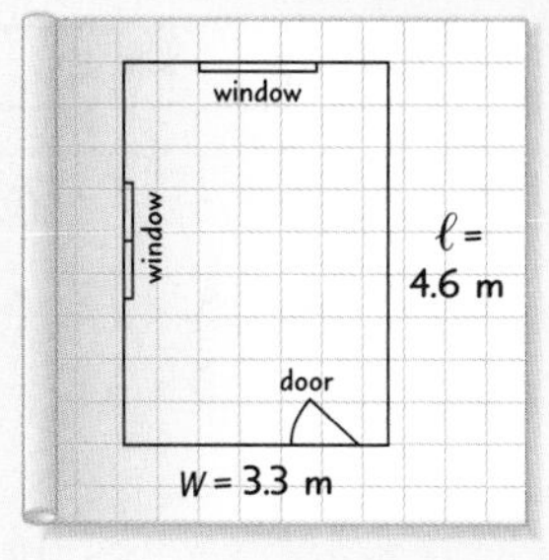

12. **SWIMMING** Alyssa swims 3 laps in 12 minutes. At this same rate, how many laps will she swim in 30 minutes?

Write each percent as a fraction in simplest form.

13. $56\frac{1}{4}\%$
14. $82\frac{1}{4}\%$
15. $83\frac{1}{3}\%$

Write each fraction as a percent. Round to the nearest hundredth if necessary.

16. $\frac{5}{8}$
17. $\frac{7}{15}$
18. $\frac{33}{40}$

19. Write 0.45% as a decimal and as a fraction in simplest form.
20. Write 8.25 as a percent.

Find each number. Round to the nearest tenth if necessary.

21. What is 18% of 30?
22. What number is 162.2% of 50?
23. What percent of 64 is 12.8?
24. 458 is 105% of what number?

Standardized Test Practice

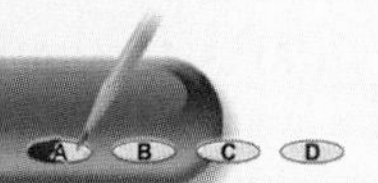

25. **MULTIPLE CHOICE** The purchase price of a bicycle is $140. The state tax rate is $6\frac{1}{2}\%$ of the purchase price. What is the tax rate expressed as a decimal?

Ⓐ 6.5 Ⓑ 0.65 Ⓒ 0.065 Ⓓ 0.0065

Standardized Test Practice

PART 1 Multiple Choice

Record your answers on the answer sheet provided by your teacher or on a sheet of paper.

1. Andre charges $4 per day for taking care of the neighbor's dog. Which expression represents the amount of money he earns in n days? (Lesson 1-4)

A $4n$ B $4 + n$

C $4 - n$ D n^4

2. The box-and-whisker plot shows the prices of single-scoop ice cream cones. What is the median price? (Lesson 2-6)

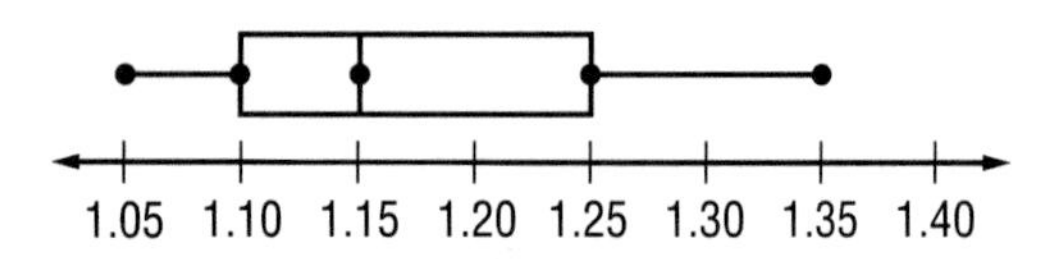

F $1.05 G $1.15

H $1.25 I $1.35

3. The graph shows the relationship between which two units of measure? (Lessons 4-6 and 6-7)

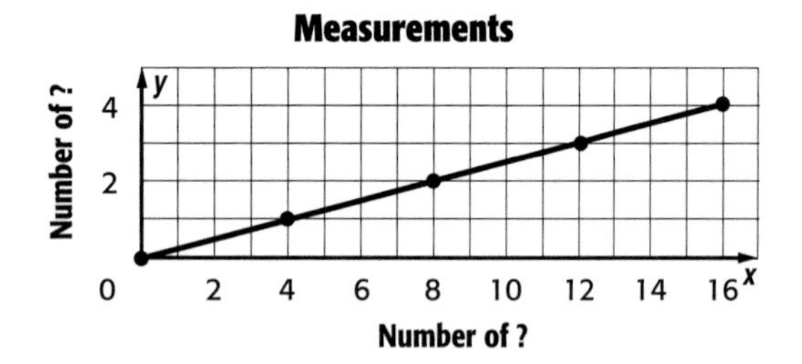

A feet and inches

B yards and feet

C pounds and ounces

D quarts and gallons

4. At Igoe's Campsite, 5 of the 15 sites were rented by campers. Which decimal equals the fraction of sites that were rented? (Lesson 5-4)

F $0.\overline{33}$ G 0.30 H 3.30 I 33.30

5. A survey showed that the ratio of tourists at the Grand Canyon who had cameras to those who did not was 8 : 10. A similar survey was conducted at the Statue of Liberty. If the ratios are equivalent, what could be the ratio at the Statue of Liberty? (Lesson 7-1)

A 32:30 B 30:24

C 30:32 D 24:30

6. Tre can play 18 holes of golf in 180 minutes. What is his average rate in number of minutes per hole? (Lesson 7-2)

F 1.8 G 3

H 6 I 10

7. Which of the following is the value of p in $\frac{p}{7} = \frac{21}{49}$? (Lesson 7-3)

A 3 B 6

C 11 D 16

8. A scale model of a computer circuit board is 1 centimeter = 0.2 centimeter. The actual circuit board is 4 centimeters wide. What is the width of the model? (Lesson 7-4)

F 0.2 cm G 2 cm

H 20 cm I 200 cm

9. The table shows the percent of Internet users in each region of the United States who are 3–17 years old. What fraction of Internet users in the South region are 3–17 years old? (Lesson 7-5)

Internet Users 3–17 Years Old	
Region	**Percent**
Midwest	32.1
Northeast	35.5
South	27.6
West	29.0

Source: U.S. Census Bureau

A $\frac{69}{500}$ B $\frac{69}{250}$

C $\frac{7}{25}$ D $\frac{7}{50}$

PART 2 Short Response/Grid In

Record your answers on the answer sheet provided by your teacher or on a sheet of paper.

10. The graph shows the number of school days per year in four countries. How many more days per year do Japanese students attend school than students in the United States? (Lesson 2-7)

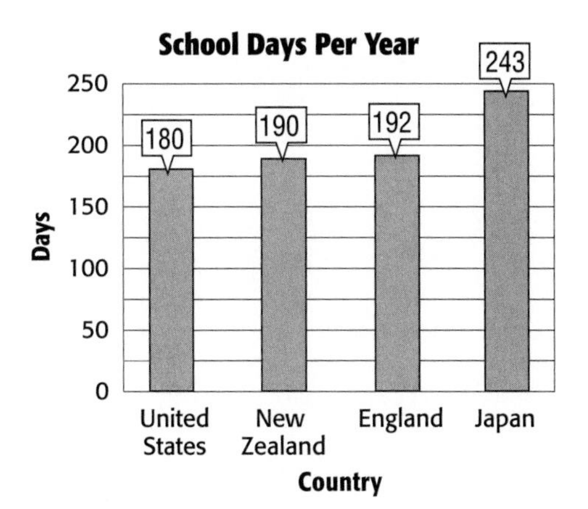

11. A train goes 65 miles per hour and travels 320 miles. How many hours will it take for the train to reach its destination? Round to the nearest tenth. (Lesson 3-7)

12. The table at the right represents a function. What is the missing x-coordinate? (Lesson 4-6)

x	y
3	9
4	12
7	21
?	27

13. Mellia has written $\frac{3}{5}$ of a research paper on fine art and French literature. What percent of the paper has she written? (Lesson 5-5)

14. The area of the rectangle is 18 square inches. What is the width w? (Lesson 6-8)

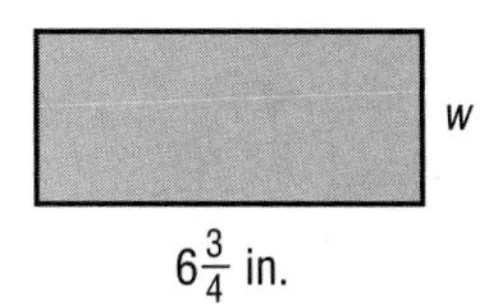

15. Ron's camera has a shutter speed of $\frac{1}{1{,}000}$. What is this shutter speed expressed as a percent? (Lesson 7-6)

16. Jean practices her headstand in gymnastics class every day. She has been able to stay in the headstand pose without tipping for 45% of the last 60 classes. How many classes is this? (Lesson 7-7)

17. The table shows the nutrition amounts in one serving of pretzels and the percent of the recommended daily amounts.

Nutrition Facts of Pretzels		
Item	**Amount**	**Percent***
fat	1 g	1.5%
sodium	486 mg	20%
carbohydrate	22 g	7.5%

Source: www.diet-data.com
*based on a 2,000-Calorie diet

How many daily milligrams of sodium are recommended? (Lesson 7-8)

PART 3 Extended Response

Record your answers on a sheet of paper. Show your work.

18. Jessie built a model of her town using items from her family's recycling bin.

a. One centimeter in Jessie's model equals 10 kilometers in the real town. If Lansdowne Road is 12 centimeters long in the model, how long is the real Lansdowne Road? (Lesson 7-4)

b. Jessie has cut up milk cartons to construct $\frac{4}{5}$ of her model town. What percent is this? (Lesson 7-5)

c. Jessie used a total of 70 recycled items to build her model town. Explain how to find the number of items that were milk cartons. Then solve. (Lesson 7-8)

TEST-TAKING TIP

Question 18 When a question involves an answer from a previous question, make sure to check that answer before you move on.

CHAPTER 8 Applying Percent

"What do snowboards have to do with math?"

How much money would you save buying a snowboard that is 35% off its original price of $169? You could use the equation $d = 0.35 \cdot 169$ to find the **discount** d. Equations are useful in finding the total cost of an item including the discount and the sales tax.

You will solve problems involving discount and sales tax in Lesson 8-5.

GETTING STARTED

Take this quiz to see whether you are ready to begin Chapter 8. Refer to the lesson number in parentheses if you need more review.

Vocabulary Review

State whether each sentence is *true* or *false*. If *false*, replace the underlined word to make a true sentence.

1. <u>Data</u> refers to pieces of information that are often numerical. (Lesson 2-1)
2. The percent <u>equation</u> is written as $\frac{\text{part}}{\text{base}} = \frac{\text{percent}}{100}$. (Lesson 7-8)

Prerequisite Skills

Multiply. (Page 560)

3. $300 \times 0.02 \times 8$
4. $85 \times 0.25 \times 3$
5. $560 \times 0.6 \times 4.5$
6. $154 \times 0.12 \times 5$

Simplify. Write as a decimal. (Lesson 1-3)

7. $\frac{22 - 8}{8}$
8. $\frac{50 - 33}{50}$
9. $\frac{35 - 7}{35}$
10. $\frac{63 - 14}{14}$

Solve. Round to the nearest tenth if necessary. (Lesson 4-3)

11. $0.4m = 52$
12. $21 = 0.28a$
13. $13 = 0.06s$
14. $0.95z = 37$

Write each percent as a decimal. (Lesson 7-6)

15. 40%
16. 3.25%
17. 7.5%
18. 2%

Use the percent proportion to find each number. Round to the nearest tenth if necessary. (Lesson 7-8)

19. What percent of 86 is 34?
20. 20% of what number is 55?

Percents Make this Foldable to help you organize information about percents. Begin with a piece of 11″ by 17″ paper.

Fold
Fold a 2″ tab along the long side of the paper. Then fold the rest in half.

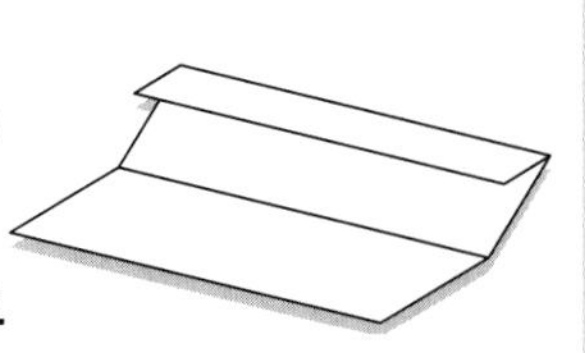

STEP 2 **Open and Fold**
Open the paper and fold in half widthwise 3 times to make 8 columns.

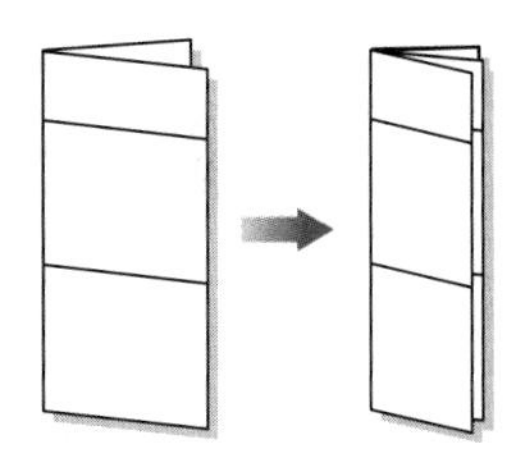

STEP 3 **Open and Label**
Draw lines along the folds and label as shown.

Ch 8	8-1	8-2	8-3	8-4	8-5	8-6	Other Facts
Main Ideas							
Examples							

Reading and Writing As you read and study each lesson, write the main ideas and some examples in the corresponding columns. In the last column, write other important facts that may apply to multiple lessons.

8-1 Percent and Estimation

What You'll LEARN

Estimate percents by using fractions and decimals.

WHEN am I ever going to use this?

GRILLING The graph shows the results of a survey in which people were asked which holiday was their favorite for grilling outdoors. Suppose 80 people were surveyed.

Source: Market Facts for Butterball Turkey

1. What fraction of people surveyed chose Labor Day as their favorite grilling day? How many of the 80 people surveyed is this?
2. About 50% of the people surveyed chose the Fourth of July as their favorite grilling day. Explain how you could use a fraction to estimate the number of people who chose this day. Then estimate.
3. Use a fraction to estimate the number of people surveyed who chose Memorial Day as their favorite grilling day.

Sometimes an exact answer is not needed when using percents. In these cases, you can estimate. One way to estimate the percent of a number is to use a fraction.

EXAMPLES Use Fractions to Estimate

1 Estimate 48% of 60.

48% is about 50% or $\frac{1}{2}$.

$48\% \text{ of } 60 \approx \frac{1}{2} \cdot 60$ Use $\frac{1}{2}$ to estimate.

≈ 30 Multiply.

So, 48% of 60 is about 30.

2 Estimate 82% of 195.

82% is about 80%, which is $\frac{8}{10}$ or $\frac{4}{5}$.

$82\% \text{ of } 195 \approx \frac{4}{5} \cdot 200$ Use $\frac{4}{5}$ to estimate and round 195 to 200.

≈ 160 Multiply.

So, 82% of 195 is about 160.

Your Turn **Estimate by using a fraction.**

a. 26% of 80 b. 75% of 23 c. 62% of 507

Another method for estimating the percent of a number is to first find 10% of the number and then multiply. For example, 70% = $7 \cdot 10\%$. So, 70% of a number equals 7 times 10% of the number.

STUDY TIP

Percents To use decimals in estimating, first round percents to the nearest 10%. To find 10% of a number, move the decimal point one place to the left.

EXAMPLE Estimate by Using 10%

3 Estimate 71% of 300.

Step 1 Find 10% of the number.

$$10\% \text{ of } 300 = 0.1 \cdot 300$$ To multiply by 10%, move the decimal point one place to the left.

$$= 30$$

Step 2 Multiply.

71% is about 70%.

70% of 300 is 7 times 10% of 300.

$7 \cdot 30 = 210$

So, 71% of 300 is about 210.

Your Turn **Estimate by using 10%.**

d. 19% of 40 **e.** 30% of 217 **f.** 63% of 91

You can also estimate percents of numbers when the percent is greater than 100 or the percent is less than 1.

EXAMPLES Percents Greater Than 100 or Less Than 1

4 Estimate 122% of 50.

122% is more than 100%, so 122% of 50 is greater than 50.

122% is about 120%.

$$120\% \text{ of } 50 = (100\% \text{ of } 50) + (20\% \text{ of } 50)$$ 120% = 100% + 20%

$$= (1 \cdot 50) + \left(\frac{1}{5} \cdot 50\right)$$ 100% = 1 and 20% = $\frac{1}{5}$

$$= 50 + 10 \text{ or } 60$$ Simplify.

So, 122% of 50 is about 60.

5 Estimate $\frac{1}{4}$% of 589.

$\frac{1}{4}$% is one fourth of 1%. 589 is about 600.

$$1\% \text{ of } 600 = 0.01 \cdot 600$$ To multiply by 1%, move the decimal point two places to the left.

$$= 6$$

One fourth of 1% is $\frac{1}{4} \cdot 6$ or 1.5.

So, $\frac{1}{4}$% of 589 is about 1.5.

Your Turn **Estimate.**

g. 174% of 200 **h.** 298% of 45 **i.** 0.25% of 789

EXAMPLE 6 Estimate Percent to Solve a Problem

POPULATION In 2000, about 0.5% of the people in Montana were of Asian descent. If Montana had a population of 902,195, estimate the number of people who were of Asian descent.

0.5% is half of 1%.

1% of 900,000 = $0.01 \cdot 900{,}000$ 902,195 is about 900,000.

= 9,000

So, 0.5% of 902,195 is about $\frac{1}{2}$ of 9,000 or 4,500.

So, about 4,500 people in Montana were of Asian descent.

Skill and Concept Check

Writing Math
Exercises 3 & 4

1. **Describe** two different ways to estimate 22% of 136.

2. **OPEN ENDED** Write a problem in which the answer can be found by estimating 12% of 50.

3. **FIND THE ERROR** Ian and Mandy are estimating 1.5% of 420. Who is correct? Explain.

Ian
1.5% of 420
$\approx$ 1% of 400 + 0.5% of 400
= $0.01 \cdot 400 + \frac{1}{2}(0.01 \cdot 400)$
= 4 + 2 or 6

Mandy
1.5% of 420
= $1 \cdot 400 + 0.5 \cdot 400$
= 400 + 200
= 600

4. **NUMBER SENSE** Explain whether an estimate for the percent of a number is *always*, *sometimes*, or *never* greater than the actual percent of the number. Give an example or a counterexample to support your answer.

GUIDED PRACTICE

Estimate by using fractions.

5. 52% of 160
6. 30% of 79
7. 77% of 22

Estimate by using 10%.

8. 40% of 62
9. 23% of 400
10. 89% of 98

Estimate.

11. 151% of 70
12. 305% of 6
13. $\frac{1}{2}$% of 82

14. **LIFE SCIENCE** The 639 muscles in your body make up about 40% of your total weight. If a person weighs 120 pounds, about how much of the weight is muscle?

Practice and Applications

HOMEWORK HELP	
For Exercises	See Examples
15–20	1, 2
21–26	3
27–34	4–6

Extra Practice
See pages 582, 603.

Estimate by using fractions.

15. 25% of 408
16. 80% of 37
17. 76% of 280
18. 39% of 20
19. 67% of 15.2
20. 10.5% of 238

Estimate by using 10%.

21. 60% of 39
22. 20% of 132
23. 76% of 80
24. 37% of 250
25. 28% of 121
26. 88% of 207

Estimate.

27. 132% of 54
28. 224% of 320
29. 410% of 12
30. 198% of 33
31. 0.4% of 400
32. 0.9% of 74

GEOLOGY **For Exercises 33 and 34, use the following information.**
Granite, a stone found in New Hampshire and Vermont, is 0.8% water.

33. About how many pounds of water are there in 3,000 pounds of granite?

34. About how much water is contained in a 15-pound piece of granite?

35. **CRITICAL THINKING** Explain how you could find $\frac{3}{8}$% of a number.

Standardized Test Practice and Mixed Review

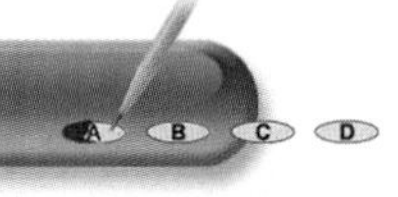

36. **MULTIPLE CHOICE** Estimate 15% of 61.

(A) 30 (B) 18 (C) 15 (D) 9

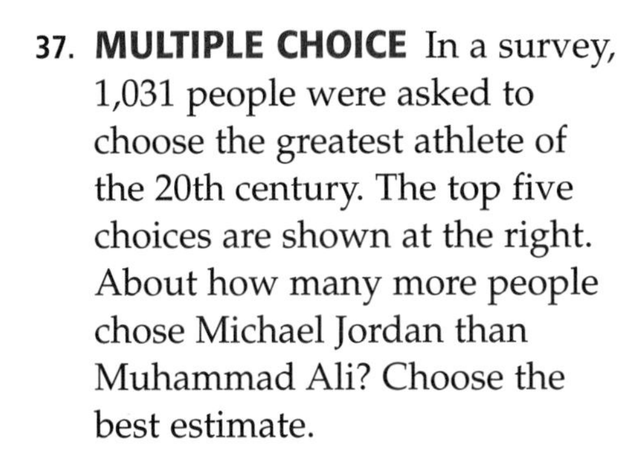

37. **MULTIPLE CHOICE** In a survey, 1,031 people were asked to choose the greatest athlete of the 20th century. The top five choices are shown at the right. About how many more people chose Michael Jordan than Muhammad Ali? Choose the best estimate.

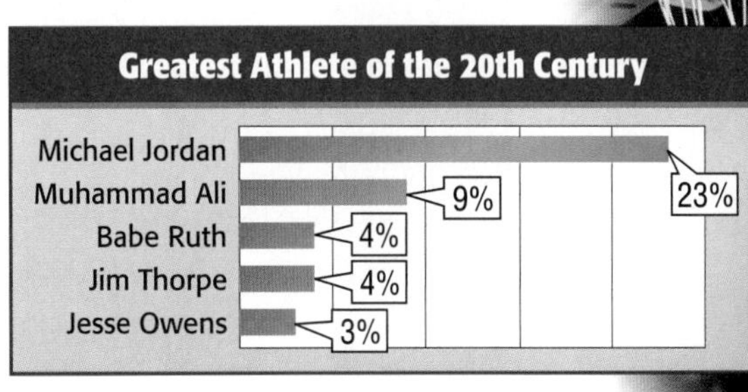

Source: Gallup Poll

(F) 90 people (G) 100 people (H) 150 people (I) 200 people

Find each number. Round to the nearest tenth if necessary. (Lesson 7-8)

38. 6 is what percent of 15?
39. Find 72% of 90.
40. What number is 120% of 60?
41. 35% of what number is 55?

42. Find 22% of 85. (Lesson 7-7)

GETTING READY FOR THE NEXT LESSON

PREREQUISITE SKILL **Solve. Round to the nearest tenth if necessary.** (Lesson 4-3)

43. $40 = 0.8x$
44. $10r = 61$
45. $25 = 0.07t$
46. $56 = 0.32n$

msmath2.net/self_check_quiz

8-1b Problem-Solving Strategy

A Follow-Up of Lesson 8-1

What You'll Learn
Solve problems by determining reasonable answers.

Reasonable Answers

The meals for our group cost a total of $38.95. I think we should leave a 15% tip, which would be about $4.

I think that the tip should be more than that. Let's estimate to find a **reasonable answer**.

Explore	We know that the total bill is $38.95 and we want to leave a 15% tip.
Plan	We can round $38.95 to $40 and then use mental math to find 15% of 40.
Solve	10% of $40 = 0.1 \cdot 40$ or 4 **10% = 0.1** 5% of $40 = \frac{1}{2} \cdot 4$ or 2 15% of $40 = (10\%$ of $40) + (5\%$ of $40)$ $= 4 + 2$ or 6 So, $6 would be a better amount to leave for a tip.
Examine	Use a calculator to check. .15 [×] 38.95 [ENTER] **5.8425** 5.8425 is close to 6, so the answer is reasonable.

Analyze the Strategy

1. The last step of the four-step plan for problem solving asks you to examine your solution. **Explain** how you can use estimation with decimals to help you examine a solution.

2. **Write** a problem that has an unreasonable answer and ask a classmate to explain why they think the answer is unreasonable.

3. **Describe** other problem-solving strategies that you could use to determine whether answers are reasonable.

Apply the Strategy

Solve. Use the reasonable answer strategy.

4. **COMMUNICATION** Sandra makes a long distance phone call to her grandparents and talks for 45 minutes. The phone company charges a rate of $0.20 per minute. How much does the call cost?

5. **SHOPPING** Suppose you are buying an entertainment system for $1,301.90 and the speakers are 57.6% of the total cost. What is a reasonable estimate for the cost of the speakers? Explain.

Mixed Problem Solving

Solve. Use any strategy.

6. **CHORES** Cameron is using a 2.5-liter container to fill a tank that holds 24 liters of water. How many times will he need to fill the container?

7. **MUSIC** A survey showed that 73% of teens who use computers listen to music at the same time. Suppose there are 410 teens in your school who use computers. Estimate how many of them listen to music while on the computer.

8. **VACATION** The graph below shows how the Mason family spent their time during their summer vacation. What percent of the time shown was spent touring historic sites?

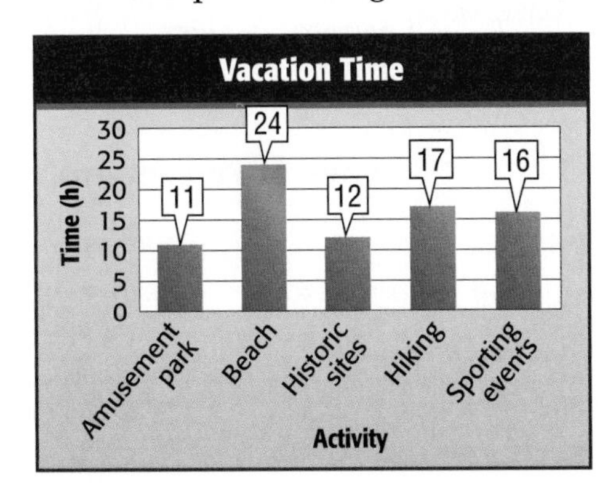

9. **BASKETBALL** Keisha made 18 points in one basketball game. How many possible shot combinations of 2- and 3-pointers could she have made? List the combinations in a table.

10. **POPULATION** About 9.4% of the people in Texas live in Houston. If the population of Texas is about 20,852,000, estimate the population of Houston.

11. **SEAT BELTS** The graph shows the percent of drivers that wore seat belts. Predict the percent of drivers that will wear seat belts in 2006. Explain why your answer is reasonable.

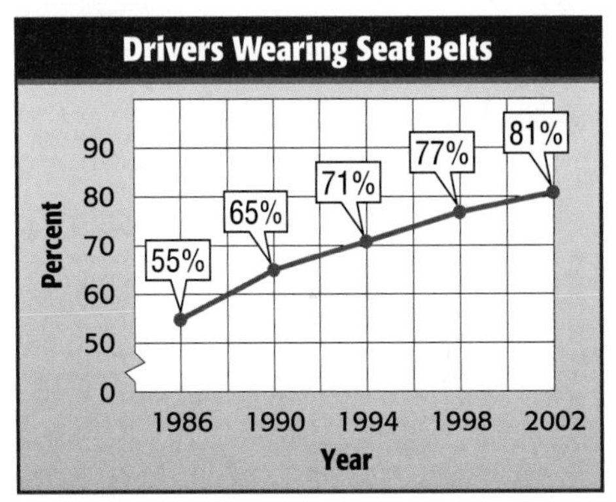

Source: Harris Interactive

12. **GEOMETRY** A rectangle has a length of $4\frac{1}{4}$ inches and a width of $3\frac{2}{5}$ inches. Is between 7 and 8 inches a reasonable estimate for the perimeter? Explain.

13. **STANDARDIZED TEST PRACTICE** Mr. Camacho is purchasing carpet for the two rooms described in the table. Which expression shows the number of square yards of carpet that he needs?

Room	Dimensions
living room	15 ft by 18 ft
TV room	18 ft by 20 ft

Ⓐ $[(15 \times 18) + (18 \times 20)] \div (3 \times 3)$

Ⓑ $[(15 \times 18(2) \times 20)] \div 3$

Ⓒ $[(15 \times 18) + (18 \times 20)] \div 3$

Ⓓ $(15 \times 18 \times 20) \div (3 \times 3)$

8-2 The Percent Equation

WHEN am I ever going to use this?

What You'll LEARN

Solve problems by using the percent equation.

NEW Vocabulary

percent equation

Link to READING

Everyday Meaning of Base: the bottom of something considered as its support, as in the base of a column

PHYSICAL SCIENCE The graph shows three main elements that make up most of the human body.

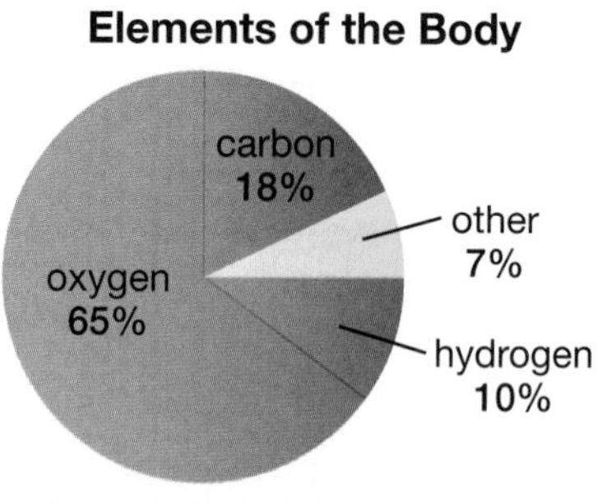

Source: *The New York Times Almanac*

1. Suppose a person weighs 120 pounds. Use the percent proportion to find the number of pounds of oxygen, carbon, and hydrogen.
2. Express the percent of each element as a decimal.
3. Multiply each decimal by 120. Record your results.
4. Compare the answers to Exercises 1 and 3.

In Lesson 7-8, you used the percent proportion $\frac{\text{part}}{\text{base}} = \frac{\text{percent}}{100}$ to find the missing part, percent, or base. You can also use an equation.

$$\frac{\text{part}}{\text{base}} = \text{percent}$$ The percent is written as a decimal.

$$\frac{\text{part}}{\text{base}} \cdot \text{base} = \text{percent} \cdot \text{base}$$ Multiply each side by the base.

$$\text{part} = \text{percent} \cdot \text{base}$$ ← This form is called the **percent equation**.

Concept Summary — Types of Percent Problems

Type	Example	Equation
Find the Part	What number is 50% of 6?	$n = 0.5 \cdot 6$
Find the Percent	3 is what percent of 6?	$3 = n \cdot 6$
Find the Base	3 is 50% of what number?	$3 = 0.5 \cdot n$

STUDY TIP

Percents When using the percent equation, always write the percent as a decimal or fraction before computing.

EXAMPLE Find the Part

1 What number is 12% of 350? **Estimate** $0.1 \cdot 350 = 35$

12% or 0.12 is the percent and 350 is the base. Let n represent the part.

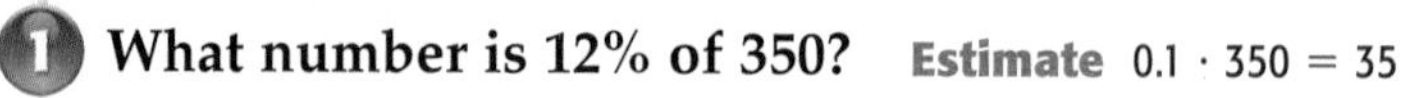

part = percent · base

$n = 0.12 \cdot 350$ Write an equation.

$n = 42$ Multiply. The part is 42.

So, 42 is 12% of 350. This is close to the estimate.

EXAMPLE Find the Percent

2 **21 is what percent of 40?** **Estimate** $\frac{21}{40} \approx \frac{1}{2}$ or 50%

Let n represent the percent.

$$\underbrace{\text{part}} = \underbrace{\text{percent}} \cdot \underbrace{\text{base}}$$

$21 = n \cdot 40$ Write an equation.

$\frac{21}{40} = \frac{40n}{40}$ Divide each side by 40.

$0.525 = n$ Simplify.

$52.5\% = n$ Write 0.525 as a percent.

So, 21 is 52.5% of 40. This is close to the estimate.

STUDY TIP

Percent Remember to write the decimal as a percent in your final answer.

EXAMPLE Find the Base

3 **13 is 26% of what number?** **Estimate** 13 is 25% or $\frac{1}{4}$ of 52.

Let n represent the base.

$$\underbrace{\text{part}} = \underbrace{\text{percent}} \cdot \underbrace{\text{base}}$$

$13 = 0.26 \cdot n$ Write an equation.

$\frac{13}{0.26} = \frac{0.26n}{0.26}$ Divide each side by 0.26.

$50 = n$ The base is 50.

So, 13 is 26% of 50. Compare to the estimate.

Your Turn **Write an equation for each problem. Then solve. Round to the nearest tenth if necessary.**

a. What percent of 125 is 75?

b. 39 is 84% of what number?

REAL-LIFE MATH

TECHNOLOGY The following cities in the United States have the highest percent of cell phone users.

Anchorage, AK	56%
Chicago, IL	55%
Honolulu, HI	54%
Atlanta, GA	52%
Detroit, MI	52%

Source: Polk's Research

EXAMPLE Apply the Percent Equation

4 **TECHNOLOGY Finland has the highest percent of people who have cell phones, 67.8%. If there are about 3,499,000 people with cell phones, what is the population of Finland?**

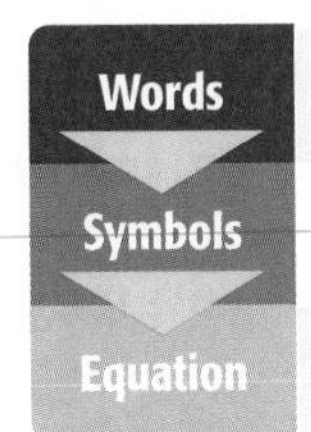

Words 3,499,000 is 67.8% of what number?

Symbols Let n represent the base.

Equation $3{,}499{,}000 = 0.678 \cdot n$

67.8% is written as the decimal 0.678.

$3{,}499{,}000 = 0.678 \cdot n$ Write the equation.

$\frac{3{,}499{,}000}{0.678} = \frac{0.678n}{0.678}$ Divide each side by 0.678. Use a calculator.

$5{,}160{,}767 \approx n$ Simplify.

The population of Finland is about 5,160,767.

Skill and Concept Check

Writing Math
Exercise 2

1. **State** whether the following problem represents a missing part, a missing percent, or a missing base.
17 is 10% of what number?

2. **OPEN ENDED** Write a real-life problem that can be solved by using the percent equation. State whether you need to find the part, the percent, or the base, and then solve.

GUIDED PRACTICE

Write an equation for each problem. Then solve. Round to the nearest tenth if necessary.

3. What number is 88% of 300?
4. 75 is what percent of 150?
5. 3 is 12% of what number?
6. 84 is 60% of what number?

7. **BASEBALL** In 2002, Derek Jeter had 191 hits in 644 times at bat. What was his *batting average*, or the percent of times at bat that were hits? Round the percent to the nearest tenth. Then write as a decimal.

Data Update How do current batting averages of other baseball players compare to Derek Jeter's average in 2002? Visit msmath2.net/data_update to learn more.

Practice and Applications

Write an equation for each problem. Then solve. Round to the nearest tenth if necessary.

8. What number is 65% of 98?
9. Find 53% of 470.
10. 9 is what percent of 45?
11. 26 is what percent of 96?
12. 84 is 75% of what number?
13. 17 is 40% of what number?
14. Find 24% of 25.
15. What percent of 64 is 30?
16. 98 is what percent of 392?
17. 33% of what number is 1.45?
18. Find 13.5% of 520.
19. What number is 75.2% of 600?
20. What number is 4% of 82.1?
21. 14 is 2.8% of what number?
22. Find 135% of 64.
23. What percent of 200 is 230?

HOMEWORK HELP

For Exercises	See Examples
8–23	1–3
24–28	4

Extra Practice
See pages 583, 603.

24. **SALES** Ms. Allon received a $325 *commission*, which is a fee paid based on a percent of her sales. If her sales totaled $8,125, what is the percent that she earns?

NUMBER THEORY **For Exercises 25 and 26, consider the whole numbers 1 through 100.**

25. For what percent of these numbers will the digits add to 5?
26. For what percent of these numbers will the digits add to an even number? an odd number?

CROPS For Exercises 27–29, use the graphic at the right. It shows the acres of cotton planted in the top five states. Round to the nearest percent.

27. About what percent of the cotton planted in the United States is planted in Texas?
28. About what percent of U.S. cotton is planted in North Carolina?
29. What percent of U.S. cotton is planted in states other than those listed on the graphic?

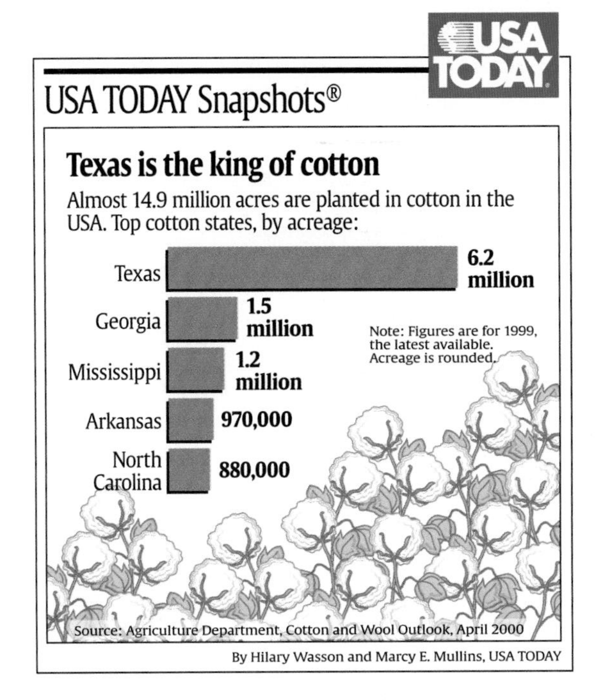

30. **MULTI STEP** A cargo of 18,000 tons of grain is 14% water due to moisture. Another cargo of 26,000 tons of grain is 12.2% water. Which contains more water? How much more?
31. **CRITICAL THINKING** If you need to find the percent of a number, explain how you can predict whether the part will be less than, greater than, or equal to the number.

Standardized Test Practice and Mixed Review

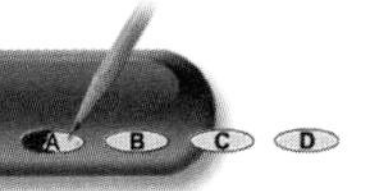

32. **MULTIPLE CHOICE** 23 is what percent of 64? Round to the nearest tenth.

    Ⓐ 2.8% Ⓑ 33.4% Ⓒ 34.0% Ⓓ 35.9%

33. **MULTIPLE CHOICE** 56 is 16% of what number? Round to the nearest tenth.

    Ⓕ 300.5 Ⓖ 305.0 Ⓗ 305.5 Ⓘ 350.0

Estimate by using fractions. (Lesson 8-1)

34. 11% of 79
35. 30.5% of 50
36. 48% of 311

37. **NUTRITION** A cup of fruit yogurt has approximately 315 milligrams of calcium. This is about 24% of the recommended daily allowance for people 9 to 18 years old. How many milligrams of calcium is recommended per day for people 9 to 18 years old? (Lesson 7-8)
38. **GEOMETRY** Find the circumference of the circle. Use $\frac{22}{7}$ for π. (Lesson 6-7)

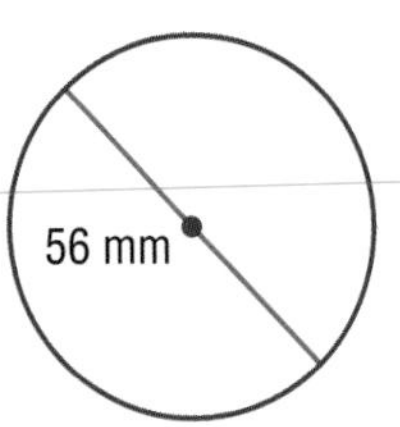

39. Find the greatest common factor of 72 and 270. (Lesson 5-2)

GETTING READY FOR THE NEXT LESSON

PREREQUISITE SKILL Use the percent proportion to find each number. Round to the nearest tenth if necessary. (Lesson 7-9)

40. What number is 45% of 60?
41. What percent of 180 is 30?
42. 20% of what number is 18?
43. 50% of what number is 15.8?

8-3a HANDS-ON LAB

A Preview of Lesson 8-3

Sampling

What You'll LEARN

Use sampling to collect data.

INVESTIGATE *Work as a class.*

To determine the favorite lunches of students at your school, you could ask each student. A more practical method is to survey a representative group of students and use that data to make a conclusion. This method is called **sampling**.

Term	Definition
population	total group of people or items in which the survey is interested
sample	part of the population
sample size	number of people or items in the survey

It is important to obtain a sample that is **unbiased**, that is, not limited or not favoring a particular outcome. Suppose you put the ID numbers of all the students in a box and randomly choose 65 numbers. This sample is unbiased because it is:

- a *random sample*. Each member of the population, or students at your school, has an equal chance of being selected.
- large enough to provide accurate data.

Writing Math

Work in groups of three.

State whether each sample is random. Explain.

1. To determine the favorite spectator sport for women over 25 years old, 350 women over 25 are surveyed at a professional basketball game.
2. To collect data about the study habits of middle school students in the Franklin School District, the name of every middle school student in the district is placed in a bag and 250 names are randomly selected.

For Exercises 3–7, refer to the information below.
Suppose you are to survey students in your school.

3. Formulate a hypothesis about students' activities.
4. Design and conduct a survey. Describe the technique that you used to get a random sample.
5. Organize and display the results of your survey in a table or graph.
6. Analyze ways in which the wording of your questions might have changed the outcome of the survey.
7. Use the results of your survey to evaluate your hypothesis.

Unbiased An unbiased sample is representative of the population.

8-3 Using Statistics to Predict

What You'll LEARN

Predict actions of a larger group by using a sample.

NEW Vocabulary

survey
population
random sample

REVIEW Vocabulary

data: pieces of information that are often numerical **(Lesson 2-1)**

WHEN am I ever going to use this?

TELEVISION The graphic shows the results of a survey in which children were asked whether they had rules about what they watched on television.

1. What ages of children are represented?
2. Can you tell how many were surveyed? Explain.
3. Describe how you could use the graphic to predict how many students in your school have no restrictions on their television viewing.

A **survey** is a question or set of questions designed to collect data about a specific group of people, called the **population**. If a survey uses a **random sample** of a population, or a sample chosen without preference, you can assume that the sample represents the population. Then you can use the results to make predictions about the population.

EXAMPLE Predict Using Percent Proportion

1 TELEVISION Refer to the survey results in the graphic above. Predict how many of the 1,250 Gallatin Middle School students have some restrictions on the television that they watch.

You can use the percent proportion and the survey results to predict what part of the population of students have TV restrictions.

part of the population →	$\frac{a}{b} = \frac{p}{100}$	Percent proportion
entire population →	$\frac{a}{1{,}250} = \frac{54}{100}$	Survey results: $54\% = \frac{54}{100}$
	$100a = 1{,}250(54)$	Cross products
	$a = 675$	Simplify.

So, about 675 of the Gallatin Middle School students have some television restrictions.

REAL-LIFE MATH

COMMUNICATION The survey at the right also found that 81% of people 12 to 17 years old use the Internet to send E-mails.

You can also use the percent equation to make predictions using survey results.

EXAMPLE Predict Using Percent Equation

2 **COMMUNICATION A survey showed that 70% of people 12 to 17 years old use the Internet to send instant messages. Predict how many of the 2,640 students at Washington Middle School send instant messages.**

You need to predict how many of the 2,640 students send instant messages.

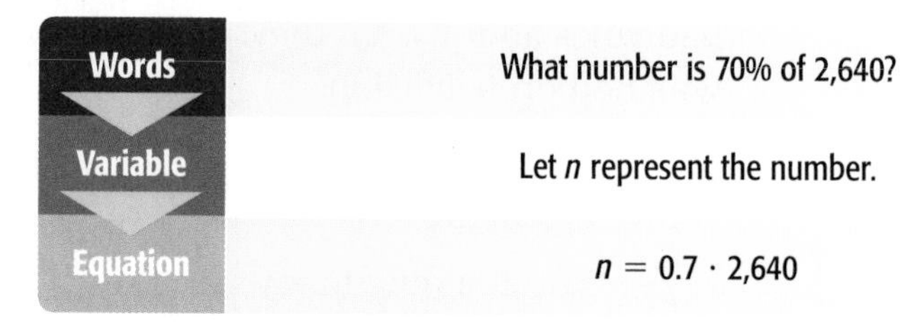

$n = 0.7 \cdot 2{,}640$ Write the equation.

$n = 1{,}848$ Multiply.

So, you could predict that about 1,848 students at Washington Middle School use the Internet to send instant messages.

Skill and Concept Check

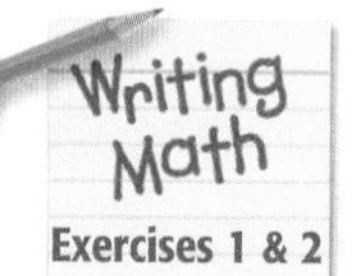

1. **Explain** how to use a sample to predict what a group of people prefer.

2. **OPEN ENDED** Select a newspaper or magazine article that contains a table or graph. Identify the population and explain how you think the results were found.

GUIDED PRACTICE

3. **SPENDING** The circle graph shows the results of a poll to which 60,000 teens responded. Predict how many of the approximately 28 million teens in the United States would buy a music CD if they were given $20.

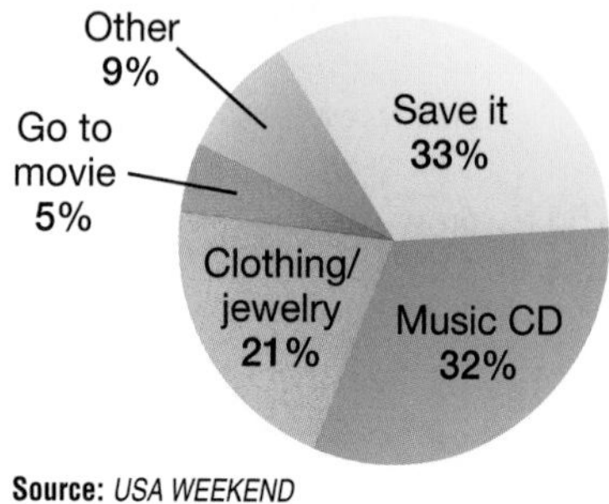

Source: *USA WEEKEND*

4. **TRANSPORTATION** The graph shows the results of a survey in which working adults in America were asked how they get to work. Use the data to predict how many working adults in America out of 143 million walk or ride a bicycle to work.

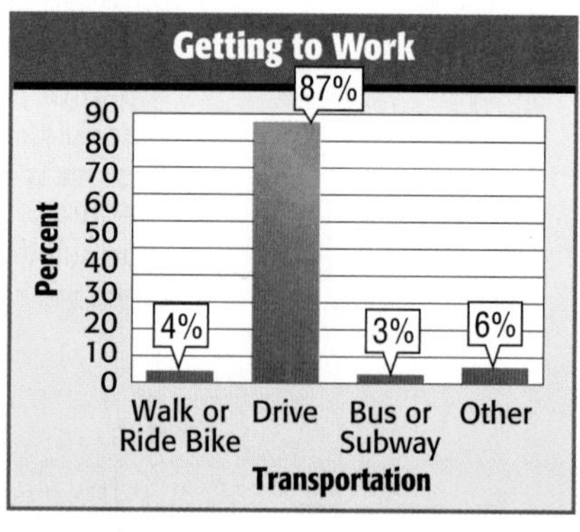

Source: Gallup Poll

Practice and Applications

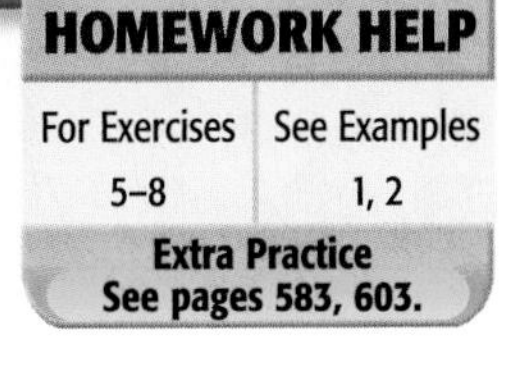
HOMEWORK HELP

For Exercises	See Examples
5–8	1, 2

Extra Practice
See pages 583, 603.

5. **COMPUTERS** The table shows the results of a survey in which students were asked how they use a personal computer at home. Use the results to predict how many of the 1,745 Allegheny Valley Middle School students use a PC for homework.

PC Use	Percent
educational programs	93%
homework	80%
games	70%

Source: U.S. Census Bureau

6. **CAMERAS** In a survey, 14% of teens said they own a digital camera. Predict how many of the 420,000 teens in Arizona own digital cameras.

CATS For Exercises 7 and 8, use the graphic at the right. It shows the percent of cat owners who train their cats to prevent each problem.

7. Out of 255 cat owners, how many would you predict have trained their cat not to climb on furniture?
8. Predict how many cat owners in a group of 316 have trained their cat not to claw furniture.

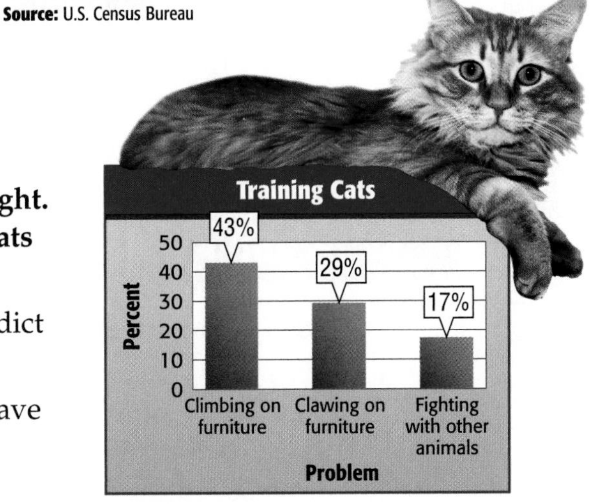

Source: Purina Cat Chow

9. **CRITICAL THINKING** A survey found that 80% of teens enjoy going to the movies in their free time. This was the response of 5,200 teens that were surveyed. What was the total number of teens surveyed?

Standardized Test Practice and Mixed Review

10. **MULTIPLE CHOICE** A survey of 80 seventh graders at Lincoln Middle School was taken to find how they get to school each day. The results are shown in the table. Of the 423 seventh graders in the school, predict how many walk to school.

Getting to School	Percent
take a bus	33%
walk	29%
adult drives	18%
other	20%

Ⓐ 23  Ⓑ 64 Ⓒ 123 Ⓓ 394

11. **SHORT RESPONSE** A survey showed that 16% of the people in Tennessee over the age of 16 belong to a fitness center. Predict how many of the 5 million people in Tennessee over the age of 16 belong to a fitness center.

Write an equation for each problem. Then solve. Round to the nearest tenth if necessary. (Lesson 8-2)

12. What number is 12% of 60?
13. 54 is 72% of what number?
14. Estimate 30% of 149. (Lesson 8-1)

GETTING READY FOR THE NEXT LESSON

PREREQUISITE SKILL Simplify. Write as a decimal. (Lesson 1-3)

15. $\frac{10 - 7}{10}$
16. $\frac{50 - 18}{50}$
17. $\frac{22 - 4}{4}$
18. $\frac{39 - 15}{15}$

Mid-Chapter Practice Test

Vocabulary and Concepts

1. **Describe** two methods that can be used to estimate the percent of a number. (Lesson 8-1)
2. **Explain** what the *population* is in a survey. (Lesson 8-3)

Skills and Applications

Estimate by using fractions. (Lesson 8-1)

3. 20% of 392
4. 78% of 112
5. 52% of 295

Estimate by using 10%. (Lesson 8-1)

6. 30% of 42
7. 79% of 88
8. 41.5% of 212

Write an equation for each problem. Then solve. Round to the nearest tenth if necessary. (Lesson 8-2)

9. What number is 35% of 72?
10. 16.1 is what percent of 70?
11. 27.2 is 68% of what number?
12. 16% of 32 is what number?

ENTERTAINMENT **For Exercises 13 and 14, refer to the graph at the right. It shows the results of a survey of students' favorite TV programs at Morgan Middle School.** (Lesson 8-3)

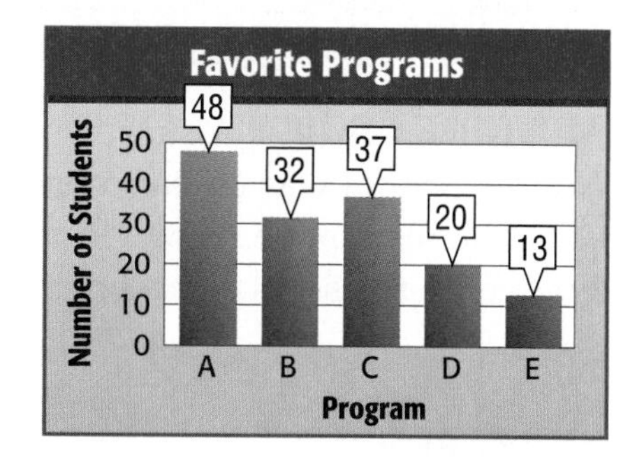

13. What percent of the students surveyed preferred Program A?
14. How many of the 925 students in the school would you expect to choose Program A as their favorite?

Standardized Test Practice

15. **MULTIPLE CHOICE** Miyoki has read 82% of a book that has 214 pages. Which is the best estimate of the number of pages that she has read? (Lesson 8-1)

    Ⓐ 16 pages Ⓑ 50 pages
    Ⓒ 80 pages Ⓓ 160 pages

16. **MULTIPLE CHOICE** A cookie company received 1,600 E-mails in one week. Of those E-mails, 12.5% were people requesting catalogues. How many people requested catalogues that week? (Lesson 8-2)

    Ⓕ 2,000 people Ⓖ 1,280 people
    Ⓗ 200 people Ⓘ 128 people

The Game Zone

A Place To Practice Your Math Skills

Math Skill

Percent of a Number

Spinning for Percents

GET READY!

Players: two or three
Materials: 6 index cards, 2 spinners

GET SET!

- Copy the numbers 11, 28, 45, 62, 84, and 98 onto index cards, one number per card.
- Shuffle the cards and place them facedown in a pile.
- Label equal sections of two spinners with the digits 0 through 9.
- Decide which spinner will stand for digits in the tens place and which spinner will stand for the ones place. The number formed by spinning both spinners is the percent.

These spinners represent 62%.

GO!

- One player selects the top card from the pile.
- The player spins both spinners and records the percent that is formed. He or she then finds that percent of the number that is on the index card. Round to the nearest whole number if necessary. This is the player's score for this turn.
- Continue in this way, taking turns selecting a card, until no cards remain in the pile.
- **Who Wins?** The player with the greatest total score wins.

8-4 Percent of Change

HANDS-ON Mini Lab

What You'll LEARN

Find the percent of increase or decrease.

NEW Vocabulary

percent of change
percent of increase
percent of decrease

Materials

- paper strips
- scissors
- tape

Work with a partner.

You can use paper strips to model a 50% increase.

- Begin with two paper strips. On each strip, label 0% on the left side and 100% on the right side.

0%	100%

- Fold one of the paper strips in half. Mark 50% in the center.

0%	50%	100%

- Cut the second strip at the 50% mark and tape the piece onto the end of the first strip. The new longer strip represents a 50% increase or 150%.

0%	100%	50%

Model each percent of change.

1. 25% increase
2. 75% increase
3. 30% increase
4. **Describe** a model that represents a 100% increase, a 200% increase, and a 300% increase.
5. **Describe** how this process would change to show percent of decrease.

One way to describe a change in quantities is to use percent of change.

Key Concept — **Percent of Change**

Words A **percent of change** is a ratio that compares the change in quantity to the original amount.

Equation $\text{percent of change} = \frac{\text{amount of change}}{\text{original amount}}$

If the original quantity is increased, then it is called a **percent of increase**. If the original quantity is decreased, then it is called a **percent of decrease**.

$$\text{percent of increase} = \frac{\text{amount of increase}}{\text{original amount}}$$

amount of increase: new amount − original amount

$$\text{percent of decrease} = \frac{\text{amount of decrease}}{\text{original amount}}$$

amount of decrease: original amount − new amount

EXAMPLE Find Percent of Increase

1 TREES Find the percent of change in tree height from year 1 to year 10. Round to the nearest whole percent if necessary.

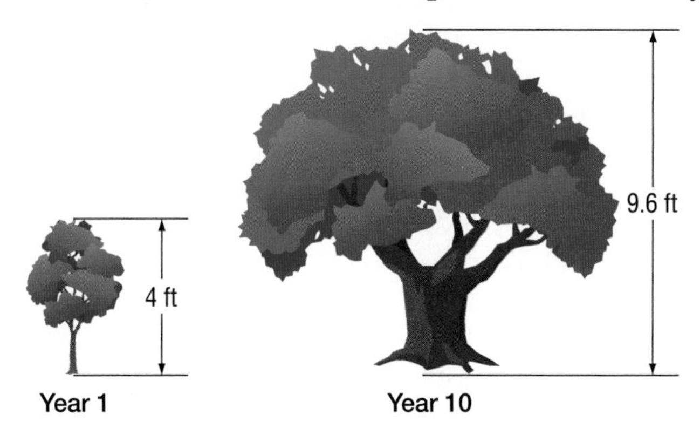

Since the new height is greater than the original height, this is a percent of increase. The amount of increase is 9.6 − 4 or 5.6 feet.

percent of increase $= \frac{\text{amount of increase}}{\text{original amount}}$

$= \frac{5.6}{4}$ Substitution

$= 1.4$ Simplify.

$= 140\%$ Write 1.4 as a percent.

The percent of increase in the tree height is 140%.

REAL-LIFE CAREERS

How Does a Stockbroker Use Math?
A stockbroker must be able to compare percents of change of different stock prices.

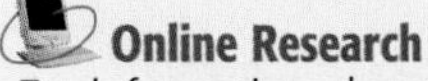

Online Research
For information about a career as a stockbroker, visit: msmath2.net/careers

EXAMPLE Find Percent of Decrease

2 STOCKS Find the percent of change if the original price of a stock was \$75 and the new price is \$60. Round to the nearest whole percent if necessary.

Stock	Price
original	\$75
new	\$60

Since the new price is less than the original price, this is a percent of decrease. The amount of decrease is 75 − 60 or \$15.

percent of decrease $= \frac{\text{amount of decrease}}{\text{original amount}}$

$= \frac{15}{75}$ Substitution

$= 0.2$ Simplify.

$= 20\%$ Write 0.2 as a percent.

The percent of decrease of the stock is 20%.

Your Turn **Find each percent of change. Round to the nearest whole percent if necessary. State whether the percent of change is an *increase* or *decrease*.**

a. original: 10
new: 13

b. original: 20
new: 15

Skill and Concept Check

Writing Math
Exercises 1–3

1. **OPEN ENDED** Write a percent of change problem using 14 and 25. State whether there is a percent of increase or decrease. Then solve.

2. **NUMBER SENSE** The costs of two different shelf stereo systems are decreased by $10. The original costs were $90 and $60, respectively. Without calculating, which had greater percent of decrease? Explain.

3. **FIND THE ERROR** Jada and Miranda are finding the percent of change from 46 to 130. Who is correct? Explain.

Jada

$\frac{130 - 46}{46} \approx 1.83$ or 183%

Miranda

$\frac{130 - 46}{130} \approx 0.65$ or 65%

GUIDED PRACTICE

Find each percent of change. Round to the nearest whole percent if necessary. State whether the percent of change is an *increase* or *decrease*.

4. original: 30
   new: 24
5. original: $126
   new: $150
6. original: 624
   new: 702
7. original: $75.80
   new: $94.75
8. original: 1.6
   new: 0.95
9. original: 20.5
   new: 35.5

10. **SPORTS** The table shows the number of people ages 7 to 17 who played soccer. What was the percent of increase in soccer players from 1990 to 2000? Round to the nearest whole percent.

Playing Soccer	
Year	Number (millions)
1990	10.9
2000	12.9

Source: National Sporting Goods Association

Practice and Applications

HOMEWORK HELP	
For Exercises	See Examples
11–31	1, 2

Extra Practice
See pages 583, 603.

Find each percent of change. Round to the nearest whole percent if necessary. State whether the percent of change is an *increase* or *decrease*.

11. original: 15
    new: 18
12. original: 100
    new: 140
13. original: $12
    new: $6
14. original: $240
    new: $320
15. original: 48
    new: 14
16. original: 360
    new: 120
17. original: 125
    new: 87.5
18. original: $15.60
    new: $11.70
19. original: $89.50
    new: $105.20
20. original: 132
    new: 125.4
21. original: 782
    new: 789.82
22. original: 12
    new: 60

GEOMETRY For Exercises 23 and 24, refer to the rectangle at the right. Suppose the side lengths are doubled.

8 cm

3 cm

23. Find the percent of increase in the perimeter.
24. Find the percent of increase in the area.

25. **SALES** Use the graphic at the right to find the percent of change in ketchup sales from 2000 to 2001.

Source: Heinz North America

26. **EDUCATION** A person with a high school diploma earns an average of $16,053 per year. A person with some college earns an average of $25,686 per year. What is the percent of increase?

27. **POPULATION** On July 1, 2001, the U.S. had an estimated 284,796,887 residents. This is 3,374,981 more than on April 1, 2000. To the nearest tenth, what was the percent of change in population?

28. **ART** In 2001, the sales of fine art were $40.8 billion. Art sales are projected to increase by 25% from 2001 to 2005. What is the projected amount of art sales in 2005?

ALLOWANCES For Exercises 29–31, refer to the table at the right.

Age	Average Weekly Allowance
12	$9.58
13	$9.52
14	$13.47
15	$15.57
16	$17.84
17	$30.66

Source: www.kidsmoney.org

29. Find the percent of increase of allowance from age 13 to 14. Round to the nearest whole percent.

30. State two consecutive ages where the change in allowance is a percent of decrease. Then find the percent of decrease. Round to the nearest tenth.

31. Between which two consecutive years is the percent of increase the greatest? What is the percent of increase? Round to the nearest whole percent.

32. **CRITICAL THINKING** If a quantity increases by 10% and then decreases by 10%, will the result be the original quantity? Explain.

Standardized Test Practice and Mixed Review

33. **MULTIPLE CHOICE** The table shows the average number of hours that Americans worked in 1990 and in 2000. Find the percent of increase to the nearest tenth.

Year	Time (h)
1990	1,943
2000	1,979

Source: International Labor Organization

(A) 19.0% (B) 2.0% (C) 1.9% (D) 0.2%

34. **SHORT RESPONSE** Find the percent of decrease from 85 to 68.

35. **FOOD** In a survey of 150 students at Kennedy Middle School, 48% said that their favorite type of pizza crust is thick crust. Predict how many of the 1,375 students in the school prefer thick crust pizza. (Lesson 8-3)

Write an equation for each problem. Then solve. Round to the nearest tenth if necessary. (Lesson 8-2)

36. 30% of what number is 17?

37. What is 21% of 62?

GETTING READY FOR THE NEXT LESSON

PREREQUISITE SKILL Write each percent as a decimal. (Lesson 7-5)

38. 6.5%

39. $5\frac{1}{2}\%$

40. $8\frac{1}{4}\%$

41. $6\frac{3}{4}\%$

8-5 Sales Tax and Discount

What You'll LEARN

Solve problems involving sales tax and discount.

NEW Vocabulary

sales tax
discount

WHEN am I ever going to use this?

COMPUTERS Julie Ann plans to buy a new computer with a flat screen that costs $1,299. She lives in Florida where there is a sales tax of 6%.

1. Calculate the sales tax by finding 6% of $1,299.
2. What will be the total cost including the sales tax?
3. Use a calculator to multiply 1.06 and 1,299. How does the result compare to your answer in Exercise 2?

One everyday use of percent is sales tax. **Sales tax** is an additional amount of money charged on items that people buy. The local, state, or federal government receives this money. The total cost of an item is the regular price plus the sales tax.

EXAMPLE Find the Total Cost

1 TECHNOLOGY A graphing calculator costs $90, and the sales tax is 4.25%. What is the total cost of the calculator?

First, find the sales tax.

4.25% of $90 = $0.0425 \cdot 90$

≈ 3.83 The sales tax is $3.83.

Next, add the sales tax to the regular price.

$3.83 + 90 = 93.83$

The total cost of the calculator is $93.83.

Another way to find the cost of an item with sales tax is to add the percent of tax to 100%.

EXAMPLE Find the Total Cost

2 CLOTHES What is the total cost of a sweatshirt if the regular price is $42 and the sales tax is $5\frac{1}{2}$%?

$100\% + 5\frac{1}{2}\% = 105\frac{1}{2}\%$ Add the percent of tax to 100%.

The total cost is $105\frac{1}{2}$% of the regular price.

$105\frac{1}{2}$% of $42 = $1.055 \cdot 42$ Use a calculator.

$= 44.31$

The total cost of the sweatshirt is $44.31.

Discount is the amount by which the regular price of an item is reduced. The sale price is the regular price minus the discount.

EXAMPLE Find the Sale Price

Standardized Test Practice

3 **MULTIPLE-CHOICE TEST ITEM** Alan wants to buy a snowboard that has a regular price of \$169. This week, the snowboard is on sale at a 35% discount. What is the sale price of the snowboard?

Ⓐ \$59.15 Ⓑ \$109.85 Ⓒ \$134.00 Ⓓ \$228.15

Test-Taking Tip The Princeton Review

Elimination
The sale price of an item is less than the regular price. Eliminate any choices in which the sale price given is greater than the regular price.

Read the Test Item

The sale price is \$169 minus the discount.

Solve the Test Item

Method 1 First, find the amount of the discount d.

$\underbrace{d}_{\text{part}} = \underbrace{0.35}_{\text{percent}} \cdot \underbrace{169}_{\text{base}}$ Use the percent equation.

$d = 59.15$ The discount is \$59.15.

So, the sale price is \$169 − \$59.15 or \$109.85.

Method 2 First, subtract the percent of discount from 100%.

$100\% - 35\% = 65\%$

So, the sale price s is 65% of the regular price.

$s = 0.65 \cdot 169$ Use the percent equation.

$s = 109.85$ The sale price is \$109.85.

So, the sale price of the snowboard is \$109.85. The answer is B.

EXAMPLE Find the Percent of Discount

4 **MUSIC An electric guitar is on sale as shown at the right. What is the percent of discount?**

Original Price: ~~\$299.95~~
Now: \$179.99

First, find the *amount* of discount.

$\$299.95 - \$179.99 = \$119.96$

Next, use the percent equation to find the percent discount.

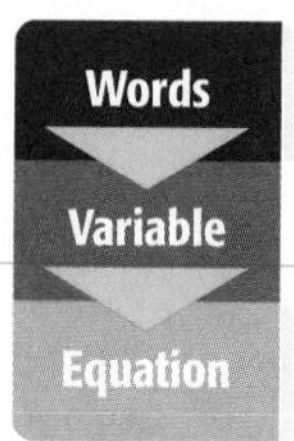

Words	\$119.96 is what percent of \$299.95?
Variable	Let n represent the percent.
Equation	$119.96 = n \cdot 299.95$

$119.96 = n \cdot 299.95$ Write the equation.

$0.40 \approx n$ Divide each side by 299.95 and simplify.

The percent of discount is about 40%.

STUDY TIP

Discount You also could have used the percent proportion to find what percent \$119.96 is of \$299.95.

Your Turn

a. Find the percent of discount if the sale price of the guitar is \$224.96.

Skill and Concept Check

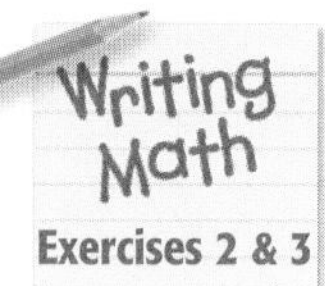

1. **Find** the sales tax of a $98 chair if the tax rate is 7%.

2. **Describe** two methods for finding the sale price of an item that is discounted 30%. Which method do you prefer? Explain.

3. **OPEN ENDED** Give an example of the regular price of an item and the total cost including sales tax if the tax rate is 5.75%.

GUIDED PRACTICE

Find the total cost or sale price to the nearest cent.

4. $2.95 notebook; 5% tax
5. $1,575 computer; 15% discount
6. $119.50 skateboard; 20% off
7. $46 shoes; 2.9% tax

Find the percent of discount to the nearest percent.

8. lotion: regular price, $4.50
   sale price, $2.25
9. in-line skates: regular price, $99
   sale price, $90

10. **VIDEO GAMES** What is the sales tax of a $178.90 video game system if the tax rate is 3.75%?

Practice and Applications

HOMEWORK HELP

For Exercises	See Examples
11–19, 25–31	1–3
20–24	4

Extra Practice
See pages 584, 603.

Find the total cost or sale price to the nearest cent.

11. $58 ski lift ticket; 20% discount
12. $1,500 computer; 7% tax
13. $99 CD player; 5% tax
14. $12.25 pen set; 60% discount
15. $4.30 notebook; 40% discount
16. $7.50 meal; 6.5% tax
17. $39.60 sweater; 33% discount
18. $89.75 scooter; $7\frac{1}{4}$% sales tax

19. **BOOKS** A book about candle making is $24.95. Find the total cost of the book including the 4.5% sales tax.

20. **TICKETS** A local movie theater is selling movie tickets for $5.25 during the first week of May. The regular price of a ticket is $6.75. What is the percent of discount to the nearest percent?

Find the percent of discount to the nearest percent.

21. calendar: regular price, $9
    sale price, $2.25
22. telescope: regular price, $180
    sale price, $126
23. concert tickets: regular price, $44
    sale price, $34
24. TV: regular price, $625
    sale price, $562.50

25. **MULTI STEP** A sound system has a regular price of $249. Find the total cost if it is on sale for 50% off and the sales tax is 5.75%.

26. **MULTI STEP** Suppose your restaurant bill comes to $28.35. Find your total cost if the tax is 6.25% and you leave a 20% tip on the amount before tax.

MOUNTAIN BIKES For Exercises 27–30, use the information below and at the right.

State	2002 Sales Tax Rate
Alabama	4%
Florida	6%
Indiana	5%
North Carolina	4.5%

Source: www.salestaxinstitute.com

A mountain bike has a regular price of $575.

27. Suppose Antonio lives in North Carolina. What is the total cost of the bike including tax?

28. If the mountain bike is on sale with a 25% discount, how much will Antonio pay for the bike, including tax?

29. Before 8 A.M., the bike will be discounted an additional 15% off the already discounted price. What will be the sale price, not including tax?

30. **RESEARCH** Use the Internet or another source to find the current tax rates of the states listed in the table or of other states. Find the cost of the bike including tax in one of these states.

31. **PROFIT** To make a profit, stores sell items for more than they paid. The increase in price is called the *markup*. Suppose Sports Galore purchases tennis racquets for $45 each. Find the markup price if the racquets are sold for 28% over the price paid for them.

32. **CRITICAL THINKING** Find the total percent of change on the price of an item if it is 15% off and the sales tax is 5%. Does it matter in which order the discount and the sales tax are applied? Explain.

Standardized Test Practice and Mixed Review

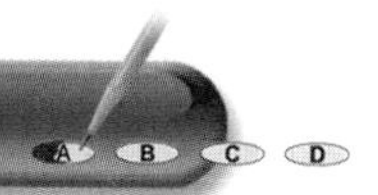

33. **MULTIPLE CHOICE** A T-shirt at the mall costs $14.95. It is on sale for 30% off. What is the sale price to the nearest cent?

    Ⓐ $4.49 Ⓑ $10.47 Ⓒ $10.50 Ⓓ $10.65

34. **SHORT RESPONSE** All of a department store's jackets are 20% off. To the nearest cent, what is the total cost of a jacket if the original price is $74.99 and the sales tax is $8\frac{1}{2}$%?

Find each percent of change. Round to the nearest whole percent if necessary. State whether the percent of change is an *increase* or *decrease*. (Lesson 8-4)

35. original: 4
    new: 6

36. original: $556
    new: $500

37. original: 20.5
    new: 35.5

38. **FOOD** The graph shows the results of a survey in which magazine readers were asked to name their favorite take-out foods. Predict how many out of 5,000 readers would choose Chinese food as one of their favorite take-out foods. (Lesson 8-3)

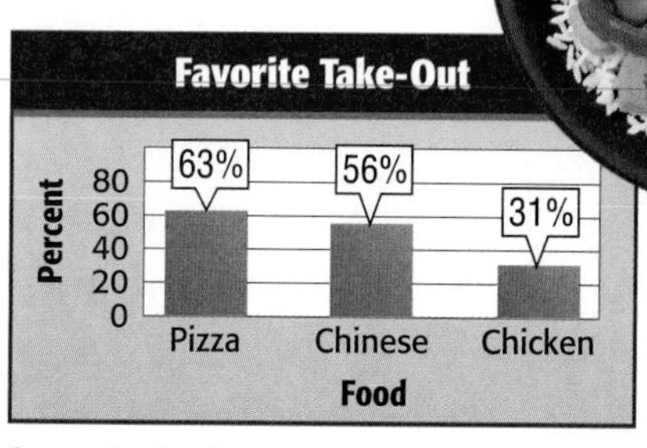

Source: *Bon Appetit*

GETTING READY FOR THE NEXT LESSON

PREREQUISITE SKILL Multiply. (Lesson 6-4)

39. $790 \cdot 0.185 \cdot 2$

40. $45 \cdot 0.04 \cdot 9$

41. $600 \cdot 0.03 \cdot 5\frac{1}{2}$

42. $162 \cdot 0.25 \cdot 3\frac{1}{2}$

8-6 Simple Interest

What You'll LEARN

Solve problems involving simple interest.

NEW Vocabulary

simple interest
principal

WHEN am I ever going to use this?

INVESTING Brooke plans to invest $1,000 in a certificate of deposit (CD). The graph shows CD rates for one year at various banks.

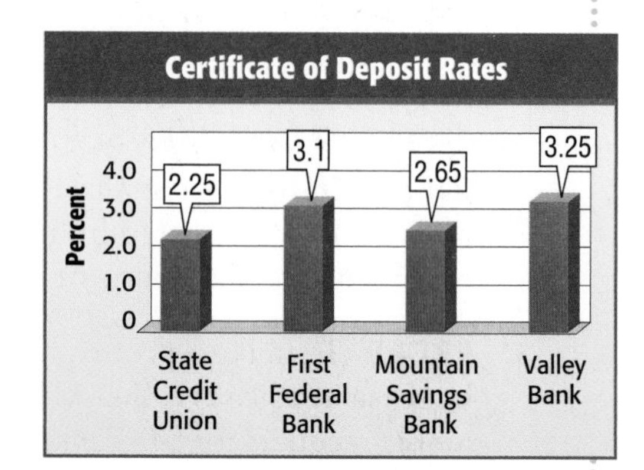

1. Calculate 2.25% of $1,000 to find the amount of money that Brooke can earn in one year for a CD at State Credit Union.
2. Find the amount of money that she can earn in one year at the other three banks.

When you deposit money in a CD, the amount that you earn is called interest. **Simple interest** is the amount paid or earned for the use of money. To find simple interest I, use the following formula.

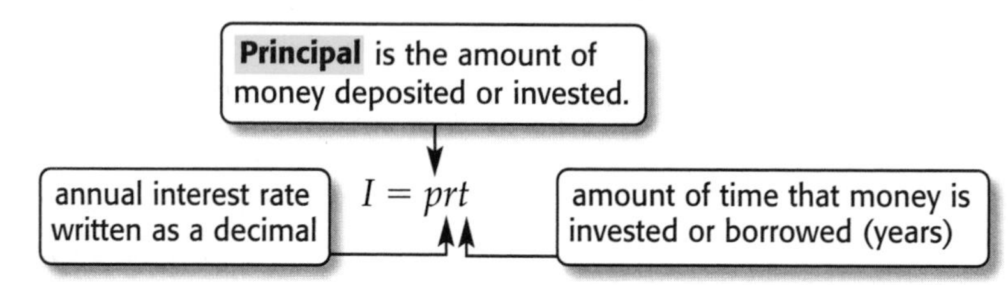

EXAMPLES Find Interest Earned

SAVINGS Raini has $750 in a savings account that pays 3% simple interest. How much interest will he earn in each amount of time?

1 **4 years**

$I = prt$	Formula for simple interest
$I = 750 \cdot 0.03 \cdot 4$	Replace p with $750, r with 0.03, and t with 4.
$I = 90$	Simplify.

Raini will earn $90 in interest in 4 years.

2 **9 months**

9 months $= \frac{9}{12}$ or 0.75 year	Write the time as years.
$I = prt$	Formula for simple interest
$I = 750 \cdot 0.03 \cdot 0.75$	$p = \$750, r = 0.03, t = 0.75$
$I \approx 16.88$	Simplify.

Raini will earn $16.88 in interest in 9 months.

READING Math

Formulas Read $I = prt$ as *Interest is equal to principal times rate times time.*

The formula $I = prt$ can also be used to find the interest owed when you borrow money. In this case, p is the amount of money borrowed, and t is the amount of time the money is borrowed.

EXAMPLE Find Interest Paid on a Loan

3 LOANS Emilio's father borrows $1,200 from the bank for a riding lawn mower. The interest rate is 8% per year. How much simple interest will he pay if he takes 2 years to repay the loan?

$I = prt$ — Formula for simple interest

$I = 1{,}200 \cdot 0.08 \cdot 2$ — Replace p with $1,200, r with 0.08, and t with 2.

$I = 192$ — Simplify.

Emilio's father will pay $192 in interest in 2 years.

EXAMPLE Find Total Paid on a Credit Card

4 CREDIT CARDS Cory charged a $600 TV on his credit card with an interest rate of 21%. If he has no other charges on the card, how much money will he owe after one month?

$I = prt$ — Formula for simple interest

$I = 600 \cdot 0.21 \cdot \frac{1}{12}$ — Replace p with $600, r with 0.19, and t with $\frac{1}{12}$.

$I = 10.5$ — Simplify.

The interest owed after one month is $10.50. So, the total amount owed would be $600 + $10.50 or $610.50.

Skill and Concept Check

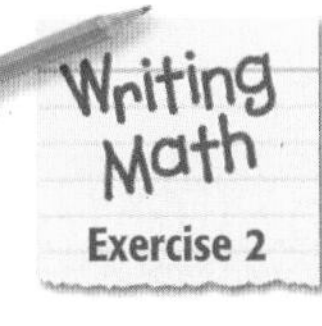

1. **List** the steps you would use to find the simple interest on a $500 loan at 6% interest rate for 18 months.

2. **OPEN ENDED** Suppose you earn 3% on a $1,200 deposit for 5 years. Investigate how the interest is affected if the rate or the time is increased.

GUIDED PRACTICE

Find the interest earned to the nearest cent for each principal, interest rate, and time.

3. $640, 3%, 2 years
4. $1,500, 4.25%, 4 years

Find the interest paid to the nearest cent for each loan balance, interest rate, and time.

5. $4,500, 9%, 3.5 years
6. $290, 12.5%, 6 months

7. **HISTORY** In 2002, a $5 bank note from 1886 was sold to a collector for $103,500. Suppose a person had deposited the $5 in a bank in 1886 with an interest rate of 4%. After 116 years, how much simple interest would have been earned on the account?

msmath2.net/extra_examples

Practice and Applications

HOMEWORK HELP	
For Exercises	See Examples
8–13	1, 2
14–19	3
20–22	4

Extra Practice
See pages 584, 603.

Find the interest earned to the nearest cent for each principal, interest rate, and time.

8. \$1,050, 4.6%, 2 years

9. \$250, 2.85%, 3 years

10. \$500, 3.75%, 1 year

11. \$3,000, 5.5%, 2 years

12. \$875, 6%, 4 months

13. \$98.50, $6\frac{1}{2}$%, 16 months

Find the interest paid to the nearest cent for each loan balance, interest rate, and time.

14. \$1,000, 7%, 2 years

15. \$725, 6.25%, 1 year

16. \$2,700, 8.2%, 3 years

17. \$175.80, 12%, 1.25 years

18. \$925, $19\frac{1}{2}$%, 3 months

19. \$800, 10.5%, 30 months

INVESTING For Exercises 20–22, use the following information.
Marcus has \$1,800 from his summer job to invest.

20. If he invests in a CD for 3 years at a rate of 5.25%, how much will the CD be worth after 3 years?

21. Suppose he invests the \$1,800 for 2 years and earns \$144. What was the rate of interest?

22. Marcus would like to have \$2,340 altogether. If he invests his money at 5% interest, in how many years will he have \$2,340?

23. **CRITICAL THINKING** Mrs. Williams deposits \$600 in an account that pays 4.5% annually. At the end of the year, the interest earned is added to the principal. Find the total amount in her account each year for 3 years.

Standardized Test Practice and Mixed Review

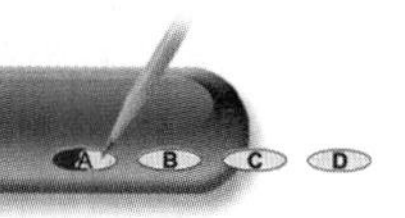

24. **MULTIPLE CHOICE** Antonia opened a savings account that pays 6.5% simple interest. How much money will be in Antonia's account after 3 years if she deposited \$250 at the beginning and never made any more deposits?

Ⓐ \$48.75 Ⓑ \$248.75 Ⓒ \$298.75 Ⓓ \$300.00

25. **MULTIPLE CHOICE** Mr. McMahon bought a \$562 freezer using a credit card that charges 18% annual interest. If he does not make any payments or any additional charges, how much will he owe after 1 month?

Ⓕ \$553.57 Ⓖ \$570.43 Ⓗ \$578.86 Ⓘ \$663.16

26. Find the total cost of a \$13.99 music CD if the tax rate is 7%. (Lesson 8-5)

Find each percent of change. Round to the nearest whole percent if necessary. State whether the percent of change is an *increase* or *decrease*. (Lesson 8-4)

27. original: 35
new: 45

28. original: 60
new: 38

29. original: \$2.75
new: \$1.80

msmath2.net/self_check_quiz

8-6b Spreadsheet Investigation

A Follow-Up of Lesson 8-6

What You'll LEARN

Use a spreadsheet to calculate simple interest.

Simple Interest

A computer spreadsheet is a useful tool for quickly calculating simple interest for different values of principal, rate, and time.

ACTIVITY

Max plans on opening a "Young Savers" account at his bank. The current rate on the account is 4%. He wants to see how different starting balances, rates, and times will affect his account balance. To find the balance at the end of 2 years for different principal amounts, he enters the values B2 = 4 and C2 = 2 into the spreadsheet below.

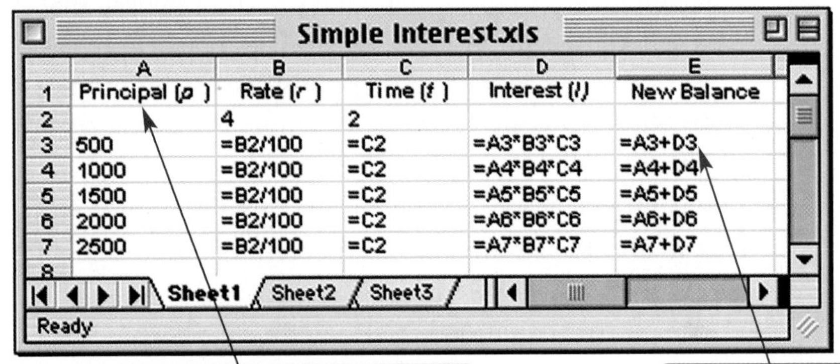

Simple Interest.xls

	A	B	C	D	E
1	Principal (*p*)	Rate (*r*)	Time (*t*)	Interest (*I*)	New Balance
2		4	2		
3	500	=B2/100	=C2	=A3*B3*C3	=A3+D3
4	1000	=B2/100	=C2	=A4*B4*C4	=A4+D4
5	1500	=B2/100	=C2	=A5*B5*C5	=A5+D5
6	2000	=B2/100	=C2	=A6*B6*C6	=A6+D6
7	2500	=B2/100	=C2	=A7*B7*C7	=A7+D7
8					

EXERCISES

1. Why is the rate in column B divided by 100?
2. What is the balance in Max's account after 2 years if the principal is $1,500 and the simple interest rate is 4%?
3. How much interest does Max earn in 2 years if his account has a principal of $2,000 and an interest rate of 4%?
4. Suppose you wanted to add a new row to the spreadsheet that represents a principal of $3,000. List each of the cell entries (A8, B8, C8, D8, and E8) that you would enter.
5. What entries for cells B2 and C2 would you use to calculate the simple interest on a principal of $1,500 at a rate of 7% for a 9-month period? What is the balance of this account at the end of the 9 months?
6. Explain how a spreadsheet is more useful than a pencil and paper when finding simple interest.

Study Guide and Review

Vocabulary and Concept Check

discount (p. 355)	percent of increase (p. 350)	sales tax (p. 354)
percent equation (p. 340)	population (p. 345)	simple interest (p. 358)
percent of change (p. 350)	principal (p. 358)	survey (p. 345)
percent of decrease (p. 350)	random sample (p. 345)	

State whether each sentence is *true* or *false*. If *false*, replace the underlined word, number, or equation to make a true sentence.

1. The sale price of an item is the regular price minus the discount.
2. When taking a survey, the total group of people that the survey is interested in is called the sample.
3. To find a percent of increase, compare the amount of the increase to the new amount.
4. The formula for simple interest is $I = prt$.
5. A sample is representative of the population if it is random.
6. A method for estimating the percent of a number is to find 21% of the number and then multiply.
7. The percent equation is part = percent · base.
8. A tax is the amount by which the regular price of an item is reduced.
9. The principal is the amount of money deposited or borrowed.
10. A sample is a question or set of questions designed to collect data about a specific group of people.

Lesson-by-Lesson Exercises and Examples

8-1 Percent and Estimation (pp. 334–337)

Estimate by using fractions.

11. 25% of 81
12. 33% of 122
13. 77% of 38
14. 19.5% of 96

Estimate by using 10%.

15. 12% of 77
16. 88% of 400
17. 52% of 1,000
18. 21% of 53
19. **PETS** About 12% of 291 households in a neighborhood have fish. Estimate how many households have fish.

Example 1 **Estimate 52% of 495.**

52% is about 50% or $\frac{1}{2}$, and 495 is about 500.

52% of 495 $\approx \frac{1}{2} \cdot 500$ or 250

So, 52% of 495 is about 250.

Example 2 **Estimate 68% of 80.**

10% of 80 $= 0.1 \cdot 80$ or 8 Find 10% of 80.

68% is about 70%.

$7 \cdot 8 = 56$ 70% of 80 $\approx 7 \cdot$ (10% of 80)

So, 68% of 80 is about 56.

msmath2.net/vocabulary_review

8-2 The Percent Equation (pp. 340–343)

Write an equation for each problem. Then solve. Round to the nearest tenth if necessary.

20. 32 is what percent of 50?
21. 65% of what number is 39?
22. Find 42% of 300.
23. 7% of 92 is what number?
24. 12% of what number is 108?

Example 3 **27 is what percent of 90?**

27 is the part and 90 is the base.
Let n represent the percent.

part = percent · base

$27 = n \cdot 90$ Write an equation.

$\frac{27}{90} = \frac{90n}{90}$ Divide each side by 90.

$0.3 = n$ The percent is 30%.

So, 27 is 30% of 90.

8-3 Using Statistics to Predict (pp. 345–347)

CAREERS In a university survey, 5.5% of the incoming freshmen chose elementary teacher as a career goal and 6.4% chose engineer.

25. Predict how many of the 3,775 freshmen would choose a career as an elementary teacher.
26. How many of the 3,775 freshmen would you expect to choose a career as an engineer?

Example 4 **In a survey of 150 students at McAuliffe Middle School, 12% said they have after-school jobs. Predict how many of the 644 students at the school have after-school jobs.**

Find 12% of 644.

$n = 0.12 \cdot 644$ Write an equation.

$= 77.28$ Multiply.

So, you could predict that about 77 students at McAuliffe Middle School have after-school jobs.

8-4 Percent of Change (pp. 350–353)

Find each percent of change. Round to the nearest whole percent if necessary. State whether the percent of change is an *increase* or *decrease*.

27. original: 172
    new: 254
28. original: $200
    new: $386
29. original: 75
    new: 60
30. original: $49.95
    new: $54.95
31. **GAMES** A computer game that sold for $24.95 last year is now priced at $27.95. Find the percent of change.

Example 5 **A magazine that originally cost $2.75 is now $3.55. Find the percent of change. Round to the nearest whole percent.**

The new price is greater than the original price, so this is a percent of increase.

amount of increase = 3.55 − 2.75 or 0.80

percent of increase = $\frac{\text{amount of increase}}{\text{original amount}}$

$= \frac{0.80}{2.75}$ Substitution

≈ 0.29 Simplify.

The percent of increase is about 29%.

8-5 Sales Tax and Discount (pp. 354–357)

Find the total cost or sale price to the nearest cent.

32. $25 backpack; 7% tax
33. $210 bicycle; 15% discount
34. $8,000 car; $5\frac{1}{2}$% tax
35. $40 sweater; 33% discount
36. $6.25 address book, 40% discount

Find the percent of discount to the nearest percent.

37. shirt: regular price: $42
    sale price: $36
38. snack: regular price, $2.50
    sale price: $1
39. boots: regular price, $78
    sale price: $70
40. DVD: regular price, $24.99
    sale price: $19.99

Example 6 **A new computer system is priced at $2,499. Find the total cost if the sales tax is 6.5%.**

First, find the sales tax.

$6.5\% \text{ of } \$2{,}499 = 0.065 \cdot 2{,}499$

≈ 162.44

Next, add the sales tax to the original price. The total cost is 162.44 + 2,499 or $2,661.44.

Example 7 **A pass at a water park is $58. At the end of the season, the same pass costs $46.40. What is the percent of discount?**

$58 - 46.40 = 11.60$ Find the amount of discount.

Next, find what percent of 58 is 11.60.

$11.60 = n \cdot 58$ Write an equation.

$0.2 = n$ Divide each side by 58.

The percent of discount is 20%.

8-6 Simple Interest (pp. 358–360)

Find the interest earned to the nearest cent for each principal, interest rate, and time.

41. $475, 5%, 2 years
42. $5,000, 10%, 3 years
43. $2,500, 11%, $1\frac{1}{2}$ years

Find the interest paid to the nearest cent for each loan balance, interest rate, and time.

44. $3,200, 8%, 4 years
45. $450, 13.5%, 2 years
46. $1,980, 21%, 9 months
47. **LOANS** Brian has a loan balance of $1,000. If he pays off the balance over 2 years at an annual simple interest rate of 18%, what is the total amount that he will pay?

Example 8 **Find the interest earned on $400 at 9% for 3 years.**

$I = prt$ Formula for simple interest

$I = 400 \cdot 0.09 \cdot 3$ $p = \$400, r = 0.09, t = 3$

$I = 108$ Simplify.

The interest earned is $108.

Example 9 **Elisa has a loan for $1,300. The interest rate is 7%. If she pays it off in 6 months, how much interest will she pay?**

$I = prt$ Formula for simple interest

$I = 1{,}300 \cdot 0.07 \cdot 0.5$ $p = \$1{,}300, r = 0.07, t = 0.5$

$I = 45.5$ Simplify.

The interest she will pay after 6 months is $45.50.

CHAPTER 8

Practice Test

Vocabulary and Concepts

1. **Describe** percent of change.
2. **State** the formula used to compute simple interest.

Skills and Applications

Estimate.

3. 18% of 246
4. 145% of 81
5. 71% of 324

Write an equation for each problem. Then solve. Round to the nearest tenth if necessary.

6. Find 14% of 65.
7. What number is 36% of 294?
8. 82% of what number is 73.8?
9. 75 is what percent of 50?

BOOKS For Exercises 10 and 11, refer to the table. It shows the results of a survey in which students at Haskell Middle School were asked to name their favorite types of fiction.

Type of Fiction	Percent
historical fiction	8%
mystery	24%
science fiction	38%
sports	30%

10. Predict how many of the 845 students at Haskell Middle School would choose science fiction as their favorite type.
11. How many of the 845 students would you expect to select mystery as their favorite type of fiction?

Find each percent of change. Round to the nearest whole percent if necessary. State whether the percent of change is an *increase* or *decrease*.

12. original: $60
new: $75
13. original: 145
new: 216
14. original: 48
new: 40

Find the total cost or sale price to the nearest cent.

15. $1,730 treadmill, $6\frac{1}{2}$% sales tax
16. $16 hat, 55% discount

Find the interest earned for each principal, interest rate, and time.

17. $3,000, 5.5%, 5 years
18. $2,600, 4%, 3 months

19. **LOANS** Leah borrows $2,200 to buy new furniture. Her loan has an annual interest rate of 16%. Find the simple interest that Leah will owe after 1 year.

Standardized Test Practice

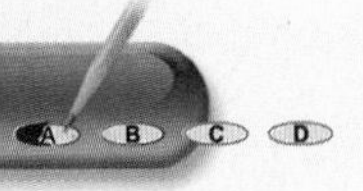

20. **MULTIPLE CHOICE** James earned $38 last week from mowing lawns in his neighborhood. This week, he earned $52. What was the percent of change?

Ⓐ 37% decrease　Ⓑ 27% decrease　Ⓒ 27% increase　Ⓓ 37% increase

CHAPTER 8

Standardized Test Practice

PART 1 Multiple Choice

Record your answers on the answer sheet provided by your teacher or on a sheet of paper.

1. Students at Karlon School are collecting canned foods for the local food pantry. The graph shows how many cans students in four of the grades collected in the first week.

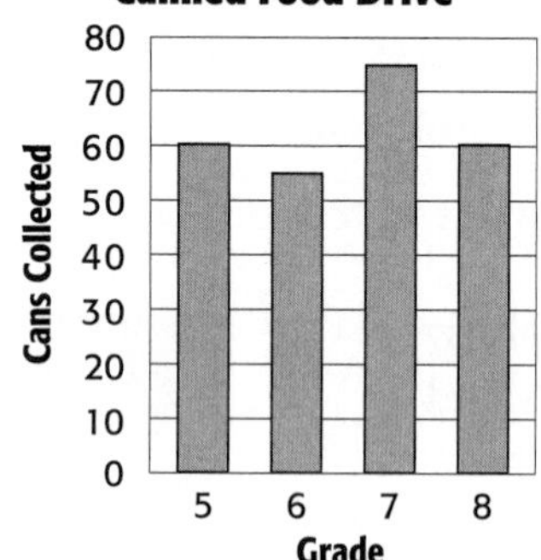

How many total cans did students in the 5th, 6th, 7th, and 8th grades collect during the first week? (Lesson 2-7)

Ⓐ 75 cans Ⓑ 120 cans
Ⓒ 240 cans Ⓓ 250 cans

2. Greenapple Books has 25 copies of this week's national best-selling book. On Friday, 12 copies of the book were sold. On Saturday, 7 copies of the book were returned. Which represents the copies of this book that Greenapple Books now has available? (Lesson 3-5)

Ⓕ −5 Ⓖ 5 Ⓗ 19 Ⓘ 20

3. Find the slope of the line graphed below. (Lesson 4-7)

Ⓐ −2

Ⓑ $-\frac{1}{2}$

Ⓒ $\frac{1}{2}$

Ⓓ 2

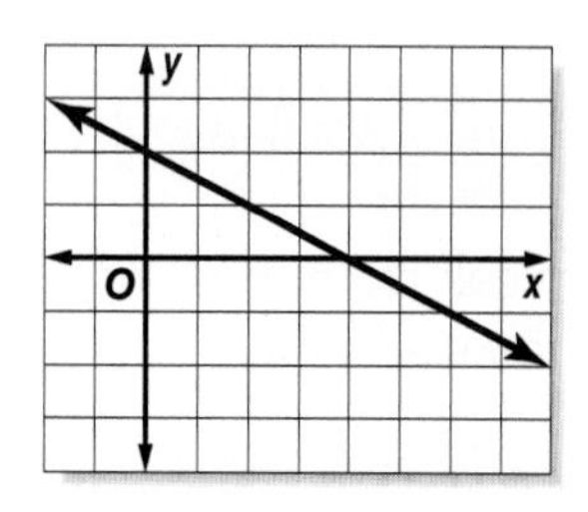

4. How far will you travel after 10 complete turns of the bicycle wheel shown at the right? Use 3.14 for π. (Lesson 6-9)

27 in.

Ⓕ 84.8 in.

Ⓖ 270 in.

Ⓗ 847.8 in.

Ⓘ 1,695.6 in.

5. The graph shows the results of an online poll to which 1,721 high school students responded.

Getting a Summer Job

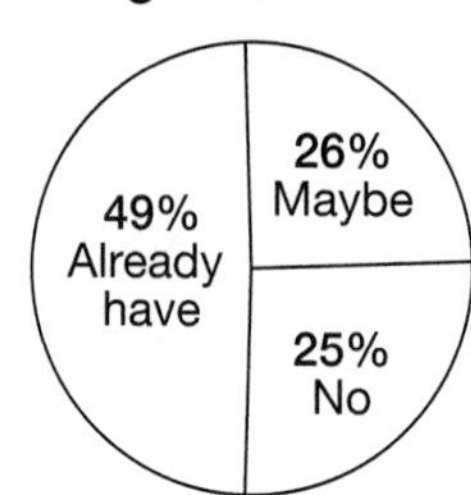

Source: Alloy online poll

Estimate how many of the students already had a summer job when they responded to the poll. (Lesson 8-1)

Ⓐ 50 Ⓑ 500 Ⓒ 850 Ⓓ 1,850

6. Zoe finished reading 85% of her book. If the book is 280 pages long, how many pages has she read? (Lesson 8-2)

Ⓕ 185 Ⓖ 208 Ⓗ 238 Ⓘ 265

7. A survey showed that 85% of youths who went to summer camp when they were 11 years old also attended summer camp when they were 12 years old. Out of 300 11-year-olds who went to summer camp this year, predict how many will go to summer camp next year. (Lesson 8-3)

Ⓐ 385 Ⓑ 255 Ⓒ 245 Ⓓ 45

PART 2 Short Response/Grid In

Record your answers on the answer sheet provided by your teacher or on a sheet of paper.

8. Jamal bought 2.4 meters of blue ribbon and 110 centimeters of white ribbon. How many centimeters longer is the blue ribbon than the white ribbon? (Lesson 1-8)

9. Write a percent to represent the shaded area of the model. (Lesson 5-5)

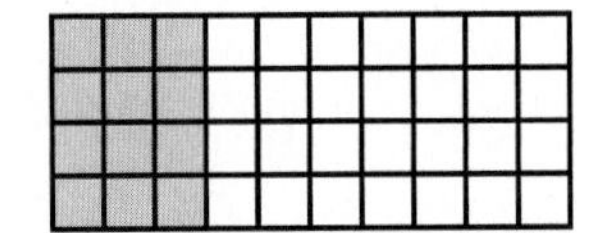

10. If a 5-pound bag of potatoes costs $3.75, how much does a 4-pound bag of the same potatoes cost? (Lesson 7-3)

11. A survey showed that 9 out of 10 teens expect to make a charitable contribution or volunteer during the holidays. Based on that survey, how many teenagers in a class of 400 expect to give money or volunteer the next holiday season? (Lesson 8-3)

12. What was the percent of increase in the price of first class stamps from 2001 to 2002? Round to the nearest whole percent. (Lesson 8-4)

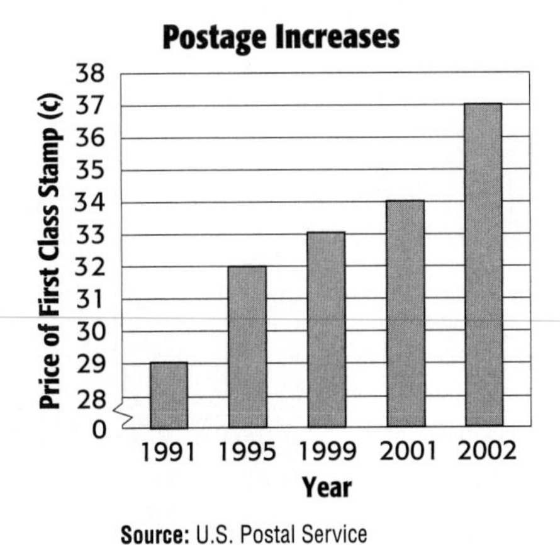

13. Midtown Veterinarians cared for 125 animals on Monday and 100 animals on Tuesday. What is the percent of decrease in number of animals? (Lesson 8-4)

14. Mr. Martinez wants to purchase a sound system that costs $245. If 6.2% sales tax is added, what will be the total cost of the system? (Lesson 8-5)

15. Kaitlyn and Camilia each have money in savings accounts, as shown below.

Money in the Bank		
Name	**Principal**	**Rate**
Kaitlyn	$400	3%
Camilia	$300	6%

Whose savings account will earn more interest after 1 year? (Lesson 8-6)

16. If a principal amount of $600 earned $75 in $2\frac{1}{2}$ years, find the interest rate. (Lesson 8-6)

PART 3 Extended Response

Record your answers on a sheet of paper. Show your work.

17. The Sybil family is working on its budget.
    a. Their rent increased from $750 to $907.50 a month. What is the percent of increase? (Lesson 8-4)
    b. Their landlord said that she would decrease the $907.50 monthly rent payment by 6% if the family swept the stairwell of the apartment building each week. If the Sybil family takes the landlord's offer, describe two ways to find how much rent they would pay. (Lesson 8-5)

TEST-TAKING TIP

Question 17 Remember to show all of your work. You may be able to get partial credit for your answers, even if they are not entirely correct.

CHAPTER 9

Probability

"What do games have to do with math?"

Rolling number cubes allows you to advance forward or backward in many board games. **Rolling doubles on a pair of number cubes happens only about 17% of the time.** In mathematics, you can use probability to help you determine how likely it is for events to take place.

You will solve problems about games in Lesson 9-1.

GETTING STARTED

Take this quiz to see whether you are ready to begin Chapter 9. Refer to the lesson or page number in parentheses if you need more review.

Vocabulary Review

Choose the correct term or number to complete each sentence.

1. A fraction is in (factored, simplest) form when the GCF of the numerator and denominator is 1. (Lesson 5-3)
2. The fraction $\frac{12}{38}$ in simplest form is $\left(\frac{6}{19}, \frac{3}{17}\right)$. (Lesson 5-3)

Prerequisite Skills

Multiply.

3. 7×15
4. 24×6
5. 13×4
6. 8×21
7. 5×32
8. 30×8
9. $6 \cdot 5 \cdot 4 \cdot 3$
10. $7 \cdot 6 \cdot 5$
11. $8 \cdot 7 \cdot 6$
12. $4 \cdot 3 \cdot 2 \cdot 1$
13. $10 \cdot 9 \cdot 8 \cdot 7$
14. $11 \cdot 10 \cdot 9$

Write each fraction in simplest form. Write *simplified* if the fraction is already in simplest form. (Lesson 5-3)

15. $\frac{8}{12}$
16. $\frac{3}{18}$
17. $\frac{4}{9}$
18. $\frac{5}{15}$

Find each value. (Lesson 5-3)

19. $\frac{6 \cdot 5}{3 \cdot 2}$
20. $\frac{9 \cdot 8 \cdot 7}{5 \cdot 4 \cdot 3}$
21. $\frac{4 \cdot 3 \cdot 2}{3 \cdot 2 \cdot 1}$
22. $\frac{7 \cdot 6 \cdot 5 \cdot 4}{4 \cdot 3 \cdot 2 \cdot 1}$

Probability Make this Foldable to help you organize the topics in this chapter. Begin with four sheets of $8\frac{1}{2}''$ by 11″ paper.

Stack Pages Place 4 sheets of paper $\frac{3}{4}$ inch apart.

STEP 2 **Roll Up Bottom Edges** All tabs should be the same size.

STEP 3 **Crease and Staple** Staple along fold.

STEP 4 **Label** Write the chapter title on the front. Label each tab with a lesson number and title.

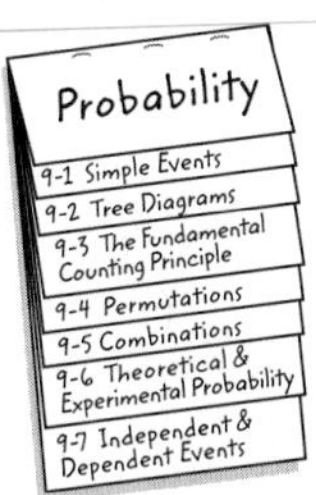

Reading and Writing As you read and study each lesson, write notes and examples under the corresponding tab.

9-1 Simple Events

What You'll LEARN

Find the probability of a simple event.

NEW Vocabulary

outcome
simple event
probability
random
complementary event

WHEN am I ever going to use this?

CANDY A box of saltwater taffy contains 48 pieces, six pieces of each flavor shown at the right.

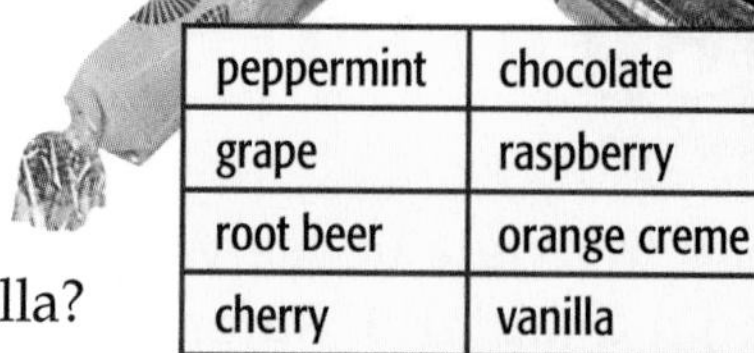

peppermint	chocolate
grape	raspberry
root beer	orange creme
cherry	vanilla

1. What fraction of the taffy is vanilla? Write in simplest form.
2. Suppose you take one piece of taffy from the box without looking. Are your chances of picking vanilla the same as picking root beer? Explain.

The 48 pieces of taffy in the box above are called **outcomes**. A **simple event** is one outcome or a collection of outcomes. For example, picking a piece of vanilla taffy is a simple event. The chance of that event happening is called **probability**.

Key Concept **Probability**

Words The probability of an event is a ratio that compares the number of favorable outcomes to the number of possible outcomes.

Symbols $P(\text{event}) = \frac{\text{number of favorable outcomes}}{\text{number of possible outcomes}}$

Outcomes occur at **random** if each outcome occurs by chance. A piece of vanilla taffy was selected at random in the activity above.

Read P(vanilla) as *the probability of picking a piece of vanilla taffy.*

$P(\text{vanilla}) = \frac{6}{48}$ ← number of pieces of vanilla taffy / total number of pieces of taffy

$= \frac{1}{8}$, or 12.5% Simplify.

EXAMPLE Find Probability

1 What is the probability of rolling an even number on a number cube marked with 1, 2, 3, 4, 5, and 6 on its faces?

$P(\text{even number}) = \frac{\text{even numbers possible}}{\text{total numbers possible}}$

$= \frac{3}{6}$ Three numbers are even: 2, 4, and 6.

$= \frac{1}{2}$ Simplify.

The probability of rolling an even number is $\frac{1}{2}$ or 50%.

EXAMPLE Find Probability

2 **PARCHEESI Jewel rolls two number cubes. She can move her game piece if a total of 5 is shown on the number cubes, or if a 5 is shown on at least one number cube. What is the probability that she can move her game piece on one roll of the number cubes?**

List all the possible outcomes. Then, find the pairs that total 5, or if a 5 is shown on at least one of the number cubes.

1, 1	1, 2	1, 3	1, 4	1, 5	1, 6
2, 1	2, 2	2, 3	2, 4	2, 5	2, 6
3, 1	3, 2	3, 3	3, 4	3, 5	3, 6
4, 1	4, 2	4, 3	4, 4	4, 5	4, 6
5, 1	5, 2	5, 3	5, 4	5, 5	5, 6
6, 1	6, 2	6, 3	6, 4	6, 5	6, 6

There are 36 possible outcomes and 15 of them are favorable. So, the probability that Jewel moves her game piece in one roll is $\frac{15}{36}$, or $\frac{5}{12}$.

REAL-LIFE MATH

PARCHEESI Parcheesi is derived from the Indian word *pacis*, which means 25.

Source: www.yesterdayland.com

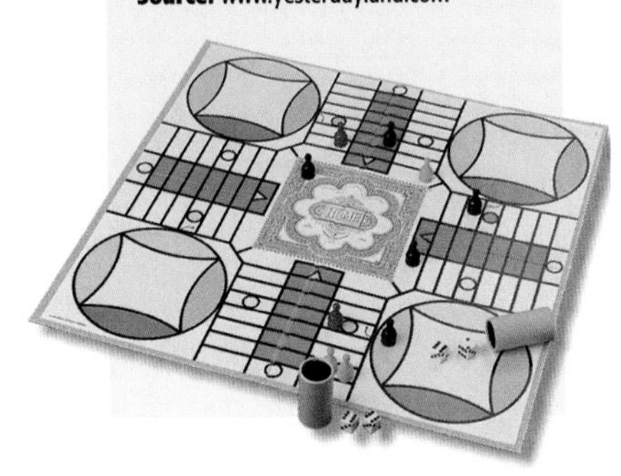

The probability that an event will happen is somewhere between 0 and 1. It can be shown on a number line.

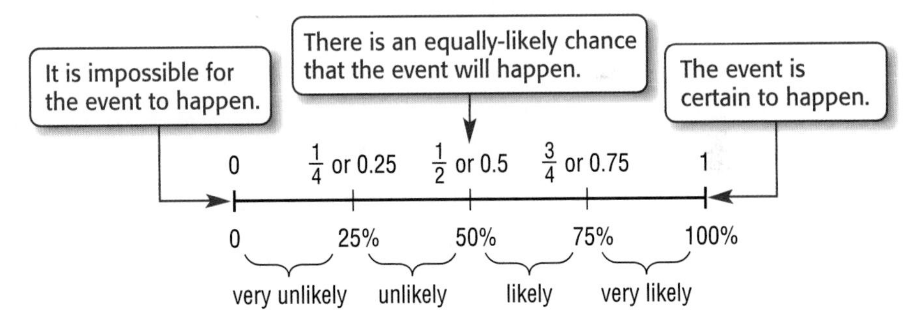

Either Jewel will be able to move her game piece or she will *not* be able to move it. Two events that are the only ones that can possibly happen are examples of **complementary events**.

EXAMPLE Find a Complementary Event

3 **PARCHEESI Refer to Example 2. Find the probability that Jewel *cannot* move her game piece on one roll of the number cubes.**

$$P(A) + P(\text{not } A) = 1$$

$$\frac{5}{12} + P(\text{not } A) = 1 \quad \text{Substitute } \frac{5}{12} \text{ for } P(A).$$

$$-\frac{5}{12} \qquad\qquad -\frac{5}{12} \quad \text{Subtract } \frac{5}{12} \text{ from each side.}$$

$$P(\text{not } A) = \frac{7}{12} \quad \text{Simplify.}$$

So, the probability that Jewel *cannot* move her game piece is $\frac{7}{12}$.

STUDY TIP

Check Your Answer You can check the answer by counting the number of outcomes that do not total 5 and that do not have a 5 on at least one number cube.

Skill and Concept Check

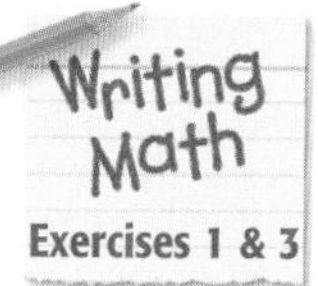

1. **Explain** why an event with a probability of 0.7 is likely to happen.
2. **OPEN ENDED** Describe an event that has a probability of $\frac{1}{5}$.
3. **Which One Doesn't Belong?** Identify the pair of probabilities that do not represent probabilities of complementary events. Explain.

$\frac{5}{8}, \frac{3}{8}$	0.65, 0.55	$\frac{4}{6}, \frac{1}{3}$	0.875, $\frac{1}{8}$

GUIDED PRACTICE

Use the spinner at the right to find each probability. Write as a fraction in simplest form.

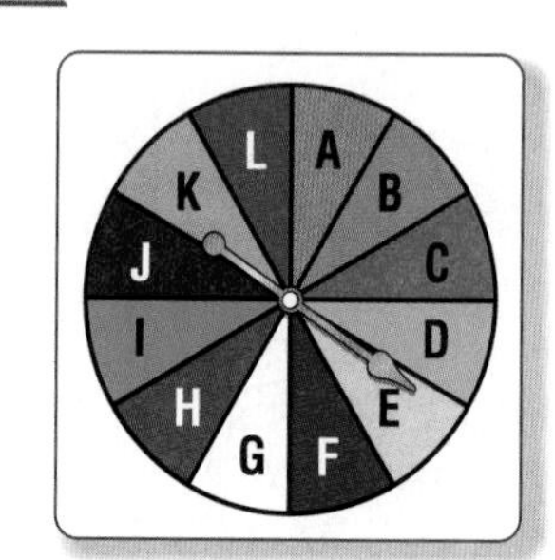

4. P(J)
5. P(vowel)

A bag contains 7 blue, 5 purple, 12 red, and 6 orange marbles. Find each probability if you draw one marble at random from the bag. Write as a fraction in simplest form.

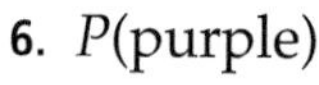

6. P(purple)
7. P(red or orange)
8. P(not blue)

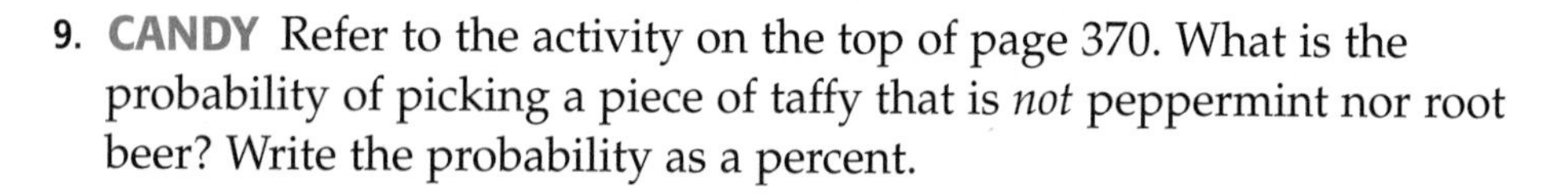

9. **CANDY** Refer to the activity on the top of page 370. What is the probability of picking a piece of taffy that is *not* peppermint nor root beer? Write the probability as a percent.

Practice and Applications

A set of 20 cards is numbered 1, 2, 3, ..., 20. Suppose you pick a card at random without looking. Find the probability of each event. Write as a fraction in simplest form.

HOMEWORK HELP	
For Exercises	See Examples
10–15	1
18, 20–23, 25–27	2
17, 19, 24, 28–29	3

Extra Practice
See pages 584, 604.

10. P(1)
11. P(not a factor of 10)
12. P(multiple of 3)
13. P(even number)
14. P(less than or equal to 20)
15. P(3 or 13)
16. How likely is it that an event with a probability of 0.28 will occur?
17. The forecast for tomorrow says that there is a 37% chance of rain. Describe the complementary event and its probability.

STUDENT COUNCIL The table shows the members of the Student Council. Suppose one student is randomly selected as the president. Find the probability of each event. Write as a fraction in simplest form.

Student Council	
girls	30
boys	20
8th graders	25
7th graders	15
6th graders	10

18. P(girl)
19. P(not 7th grader)
20. P(boy)
21. P(8th grader)
22. P(boy or girl)
23. P(6th or 8th grader)
24. P(not 6th grader)
25. P(5th grader)
26. Which event has a greater chance of happening: picking a president who is a girl or a president who is not an 8th grader? Explain.

27. **TECHNOLOGY** The graph shows the cost of 22 digital cameras. If one of the 22 cameras is chosen at random, what is the probability that it costs between $111 and $160?

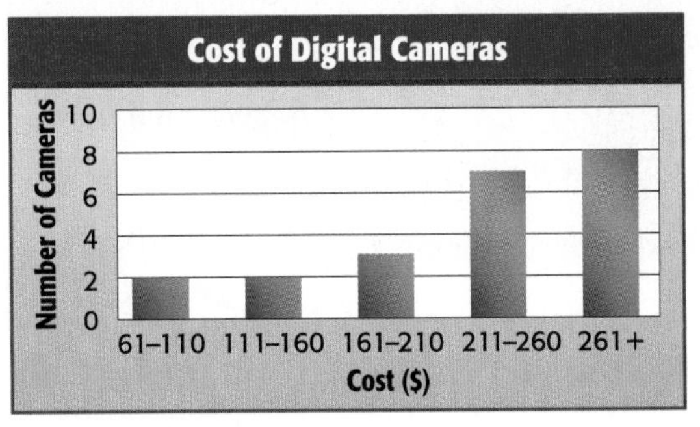

28. **MOVIES** The probability of buying a defective DVD is 0.002. What is the probability of buying a DVD that is *not* defective?

29. **MULTI STEP** The Jefferson Middle School Booster Club is selling raffle tickets for a new computer system. They sold 1,000 tickets at $2 each. Emilia's parents spent $200 on tickets. What is the probability that they will *not* win?

30. **CRITICAL THINKING** Melissa and Hakan are playing a game by rolling two number cubes. Hakan gets a point each time the sum of the number cubes is 2, 3, 4, 9, 10, 11, or 12. Melissa gets a point when the sum is 5, 6, 7, or 8. Does each player have an equal chance to win? Explain.

Standardized Test Practice and Mixed Review

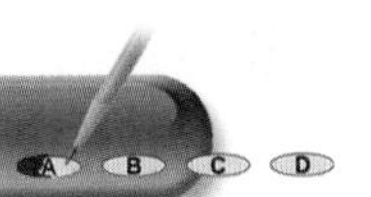

31. **MULTIPLE CHOICE** A bookshelf contains the books at the right. If you randomly choose a book from the shelf, what is the probability that it is science fiction?

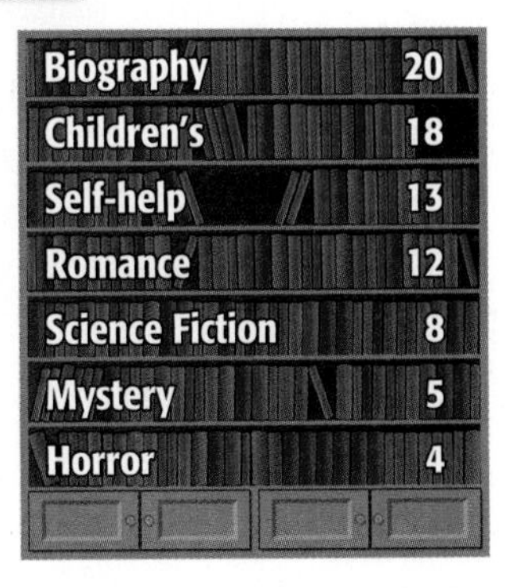

Ⓐ $\frac{1}{10}$ Ⓑ $\frac{3}{20}$ Ⓒ $\frac{9}{40}$ Ⓓ $\frac{1}{4}$

32. **MULTIPLE CHOICE** In a school raffle, one ticket will be drawn out of a total of 500 tickets. If the Hawkins family has 12 tickets, what is the probability that they will win?

Ⓕ 0.002 Ⓖ 2% Ⓗ $\frac{3}{125}$ Ⓘ $0.08\overline{3}$

Find the interest earned to the nearest cent for each principal, interest rate, and time. (Lesson 8-6)

33. $300, 10%, 2 years

34. $900, 5.5%, 4.5 years

35. **FOOD** The United States produced almost 11 billion pounds of apples in 2000. Use the information in the graph to find how many pounds of apples were used to make juice and cider. (Lesson 7-8)

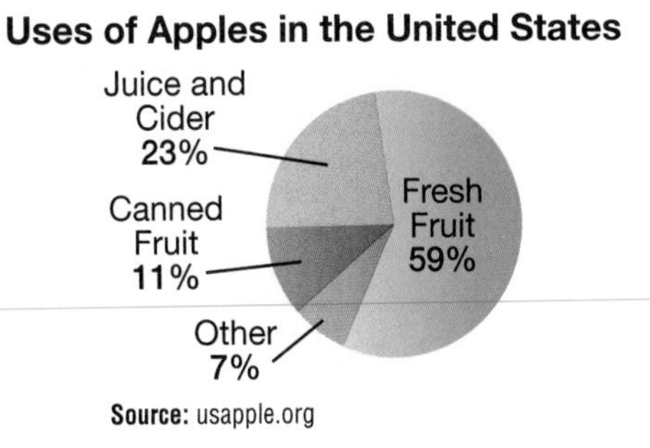

Source: usapple.org

Write each fraction or mixed number as a decimal. Use bar notation if necessary. (Lesson 5-4)

36. $\frac{3}{5}$

37. $\frac{9}{20}$

38. $6\frac{1}{8}$

39. $1\frac{7}{9}$

GETTING READY FOR THE NEXT LESSON

PREREQUISITE SKILL Write each fraction in simplest form. Write *simplified* if the fraction is already in simplest form. (Lesson 5-3)

40. $\frac{2}{6}$

41. $\frac{3}{8}$

42. $\frac{15}{30}$

43. $\frac{6}{16}$

44. $\frac{18}{32}$

9-2 Tree Diagrams

HANDS-ON Mini Lab

Materials
- 2 counters
- marker

Work with a partner.

Here is a probability game that you can play with two counters.

- Mark one side of the first counter A. Mark the other side B. Mark both sides of the second counter A.
- Player 1 tosses the counters. If both sides shown are the same, Player 1 wins a point. If the sides are different, Player 2 wins a point. Record your results.
- Player 2 then tosses the counters and the results are recorded. Continue alternating the tosses until each player has tossed the counters ten times. The player with the most points wins.

1. Before you play, make a conjecture. Do you think that each player has an equal chance of winning? Explain.
2. Now, play the game. Who won? What was the final score?
3. Collect the data from the entire class. What is the combined score for Player 1 versus Player 2?
4. Do you want to change the conjecture you made in Exercise 1? Explain.

What You'll LEARN

Use tree diagrams to count outcomes and find probabilities.

NEW Vocabulary

fair game
tree diagram
sample space

A game in which players of equal skill have an equal chance of winning is a **fair game**. One way you can analyze whether games are fair is by drawing a **tree diagram**. A tree diagram is used to show all of the possible outcomes, or **sample space**, in a probability experiment.

EXAMPLE Draw a Tree Diagram

1 **GAMES Refer to the Mini Lab above. Draw a tree diagram to show the sample space. Then determine whether the game is fair.**

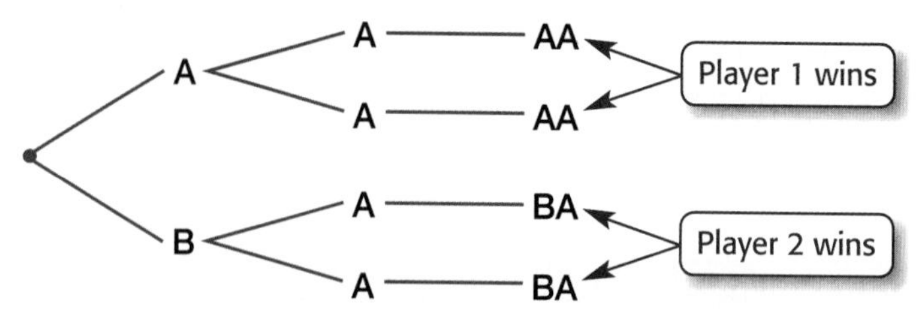

There are four equally-likely outcomes with two favoring each player. So, the probability that each player can win is $\frac{1}{2}$. Thus, the game is fair.

EXAMPLE Find the Number of Outcomes

REAL-LIFE MATH

SCOOTERS Kickboard scooters can travel approximately 4 miles per hour.

Source: www.factmonster.com

2 **SCOOTERS A certain type of kickboard scooter comes in silver, red, or purple with wheel sizes of 125 millimeters or 180 millimeters. Find the total number of color-wheel size combinations.**

Make a tree diagram to show the sample space.

Scooter Color	Wheel Size	Sample Space
silver	125 mm	silver, 125 mm
	180 mm	silver, 180 mm
red	125 mm	red, 125 mm
	180 mm	red, 180 mm
purple	125 mm	purple, 125 mm
	180 mm	purple, 180 mm

There are six different color and wheel size combinations.

Your Turn

a. How many outcomes are there if the manufacturer added an orange color and a 150-millimeter wheel size?

You can also use tree diagrams to help you find the probability of events.

EXAMPLE Find Probability Using Tree Diagrams

3 **COINS Suppose Pablo tosses three pennies. Find the probability that all three will show heads.**

Make a tree diagram to show the sample space. Then, find the probability of the three pennies showing heads.

Penny 1 Penny 2 Penny 3 Sample Space

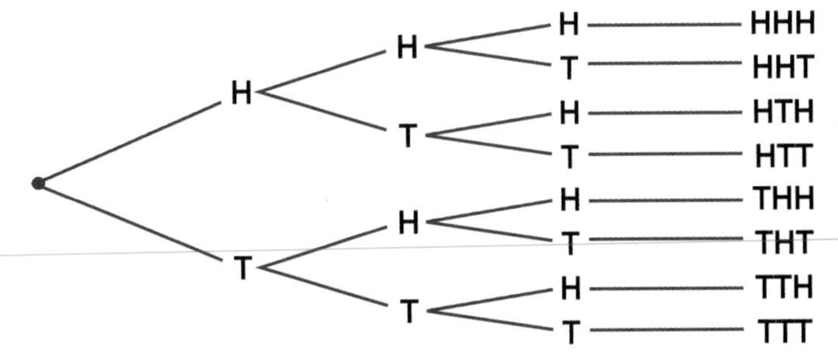

The sample space contains 8 possible outcomes. Only 1 outcome has all pennies showing heads. So, the probability of three pennies showing heads is $\frac{1}{8}$, or 12.5%.

Your Turn **Find each probability.**

b. P(3 tails) c. P(exactly 2 heads) d. P(at least 1 tail)

Skill and Concept Check

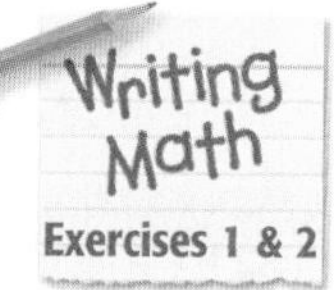

1. **Describe** a fair game between two players using one coin.

2. **OPEN ENDED** Give an example of a situation that has 8 outcomes.

GUIDED PRACTICE

The spinner is spun twice.

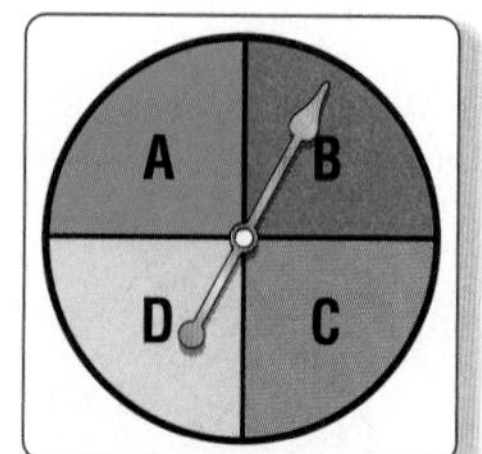

3. Draw a tree diagram to represent the situation.
4. How many outcomes are possible?
5. What is the probability of spinning two Cs?

6. **DELICATESSEN** A neighborhood deli sells sandwiches that can be made with ham, turkey, roast beef, salami, or bologna on rye, white, pumpernickel, or sourdough breads. Make a tree diagram to show all of the possible meat-bread choices.

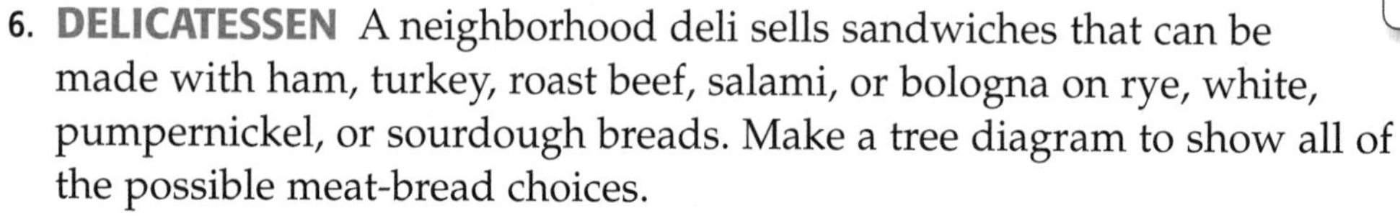

Practice and Applications

HOMEWORK HELP

For Exercises	See Examples
7–15, 18, 21	1, 2
16–17, 19	3

Extra Practice
See pages 585, 604.

For each situation, make a tree diagram to show the sample space. Then give the total number of outcomes.

7. tossing a coin and rolling a number cube
8. choosing black, blue, or brown socks with boots, gym shoes, or dress shoes
9. picking a number from 1 to 5 and choosing the color red, white, or blue
10. choosing a card with a shape and spinning the spinner from the choices at the right
11. choosing a letter from the word SPACE and choosing a consonant from the word MATH
12. choosing a purple, green, black, or silver mountain bike having 10, 18, 21, or 24 speeds
13. tossing a quarter, a dime, and a nickel
14. rolling a number cube, tossing a coin, and choosing a card from the cards marked A and B

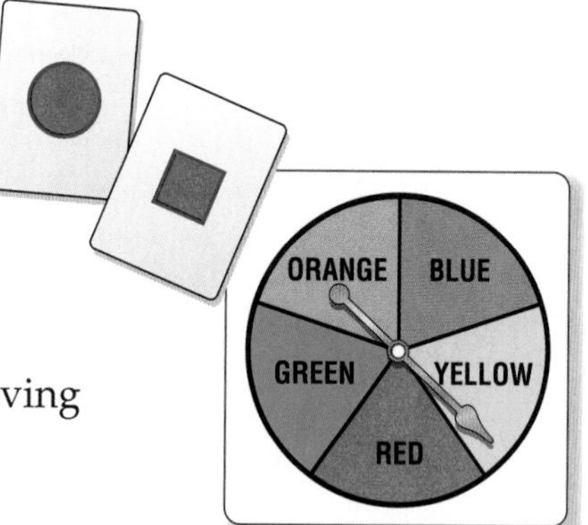

FAMILY For Exercises 15–17, use the information below.
Mr. and Mrs. Chen have three children. Suppose the chance of having either a boy or a girl is 50%.

15. Draw a tree diagram showing the possible arrangements of gender of the three children.
16. What is the probability of having 2 boys and 1 girl?
17. What is the probability of having the first two children boys and the last child a girl?

UNIFORMS For Exercises 18 and 19, use the information below and at the right.

Hunter is a big fan of the Houston Astros baseball team and wears a different jersey and cap every time he goes to a game. The table shows the number of different jerseys and caps Hunter owns.

	Home Jersey	Road Jersey	Practice Jersey	Cap
White	2	0	2	0
Black	1	0	0	1
Gray	0	1	0	0
Orange	1	1	0	2

18. How many jersey/cap combinations can Hunter wear when he goes to the baseball game to cheer on the Astros?

19. If Hunter picks a jersey/cap combination at random, what is the probability that he will wear a home jersey with a black cap?

20. **RESEARCH** Use the Internet or another resource to find the number of jerseys and caps your favorite major league baseball team has as part of its uniform. How many jersey/cap combinations are there for the team you picked?

21. **GAMES** Determine whether the following game for two players is fair. Explain.
    - Three counters are labeled according to the table at the right.
    - Toss the three counters.
    - If exactly 2 counters match, Player 1 scores a point. Otherwise, Player 2 scores a point.

	Side 1	Side 2
Counter 1	A	B
Counter 2	A	C
Counter 3	B	C

22. **CRITICAL THINKING** Refer to Exercise 21. Adjust the scoring of the game so that it is fair.

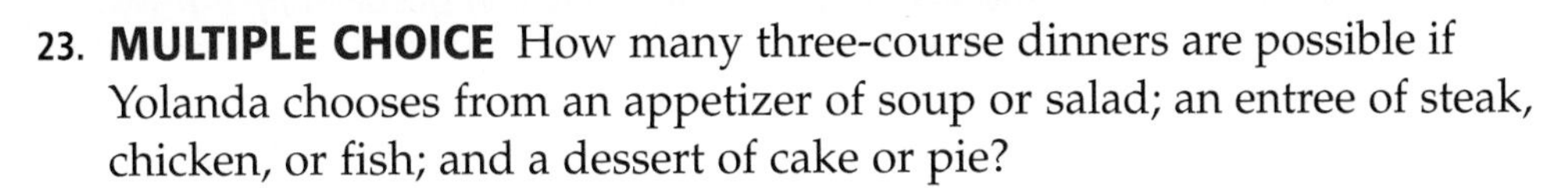

Standardized Test Practice and Mixed Review

23. **MULTIPLE CHOICE** How many three-course dinners are possible if Yolanda chooses from an appetizer of soup or salad; an entree of steak, chicken, or fish; and a dessert of cake or pie?

    (A) 7 (B) 10 (C) 12 (D) 15

24. **SHORT RESPONSE** Find the sample space of the tree diagram.

T-shirt	Color
short-sleeve	gray
	white
long-sleeve	gray
	white

PROBABILITY A spinner is equally likely to stop on each of its regions numbered 1 to 20. Find each probability as a fraction in simplest form. (Lesson 9-1)

25. a prime number 26. GCF(12, 18) 27. multiple of 2 or 3

28. **INTEREST** If Carlota invests $2,100 in a CD for 5 years at a simple interest rate of 4.75%, how much will the CD be worth after 5 years? (Lesson 8-7)

GETTING READY FOR THE NEXT LESSON

BASIC SKILL Multiply.

29. $7 \cdot 22$ 30. $5 \cdot 36$ 31. $11 \cdot 16$ 32. $23 \cdot 20$ 33. $131 \cdot 4$

msmath2.net/self_check_quiz

9-3 The Fundamental Counting Principle

What You'll LEARN

Use multiplication to count outcomes.

NEW Vocabulary

Fundamental Counting Principle

WHEN am I ever going to use this?

RETAIL SALES The Jean Factory sells juniors' jeans in different sizes and lengths. The table shows what they have available.

Sizes	Lengths
3	petite
5	regular
7	tall
9	
11	
13	
15	

1. According to the table, how many sizes of juniors' jeans are there?
2. How many lengths are there?
3. Find the product of the two numbers you found in Exercises 1 and 2.
4. Draw a tree diagram to help you find the number of different size and length combinations. How does the number of outcomes compare to the product you found above?

In the activity above, you discovered that multiplication, instead of a tree diagram, can be used to find the number of possible outcomes in a sample space. This is called the **Fundamental Counting Principle**.

Key Concept — **Fundamental Counting Principle**

If event M can occur in m ways and is followed by event N that can occur in n ways, then the event M followed by N can occur in $m \times n$ ways.

EXAMPLE Use the Fundamental Counting Principle

1 **FOOD A famous steak house allows customers to create their own steak dinners. The choices are shown at the right. How many different steak dinners are possible?**

451 Steakhouse

Steak	How Steaks Are Cooked	Potatoes
New York Strip	rare	mashed
Ribeye	medium	baked
Filet	well	twice baked
Porterhouse		au gratin
T-Bone		

$$\underbrace{5}_{\text{types of steaks}} \cdot \underbrace{3}_{\text{number of ways steaks can be cooked}} \cdot \underbrace{4}_{\text{types of potatoes}} = \underbrace{60}_{\text{total number of steak dinners}}$$

There are 60 different ways of choosing a steak dinner.

Check You can check your work by drawing a tree diagram and listing the 60 outcomes.

EXAMPLE Find Outcomes

Standardized Test Practice

2 **MULTIPLE-CHOICE TEST ITEM** There are 2 roads connecting Eastland and Harping, 3 roads connecting Harping and Sinclair, and 2 roads connecting Sinclair and Johnstown. Find the number of ways to drive from Eastland to Johnstown.

(A) 8 (B) 12 (C) 16 (D) 20

Test-Taking Tip *The Princeton Review*

Educated Guess Find out if there is a penalty for incorrect answers. If there is no penalty, making an educated guess can only increase your score, or at worst, leave your score the same.

Read the Test Item

To find the number of ways to drive from Eastland to Johnstown, multiply the number of roads from Eastland to Harping, from Harping to Sinclair, and from Sinclair to Johnstown.

Solve the Test Item

There are 2 roads from Eastland to Harping, 3 roads from Harping to Sinclair, and 2 roads from Sinclair to Johnstown. So, there are $2 \cdot 3 \cdot 2$, or 12 ways to drive from Eastland to Johnstown. So the answer is B.

Skill and Concept Check

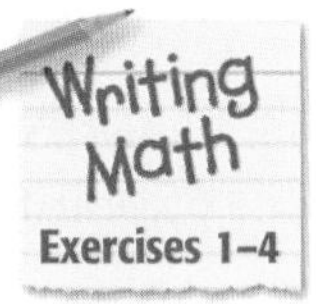

1. **Explain** how to use the Fundamental Counting Principle to count outcomes.
2. **OPEN ENDED** Give an example of a situation that has 48 outcomes using three events.
3. **NUMBER SENSE** Can a pair of events have a number of outcomes that is prime? Explain.
4. **Which One Doesn't Belong?** Identify the pair of events that do not have the same number of outcomes as the other three. Explain your reasoning.

12 meats, 8 breads	16 T-shirts, 6 colors
18 hats, 8 sizes	24 teams, 4 sports

GUIDED PRACTICE

Use the Fundamental Counting Principle to find the total number of outcomes in each situation.

5. tossing a quarter, a dime, and a nickel
6. choosing scrambled, sunny-side up, or poached eggs with bacon or sausage and milk, orange juice, or apple juice
7. choosing a card from twenty different cards and spinning the spinner at the right

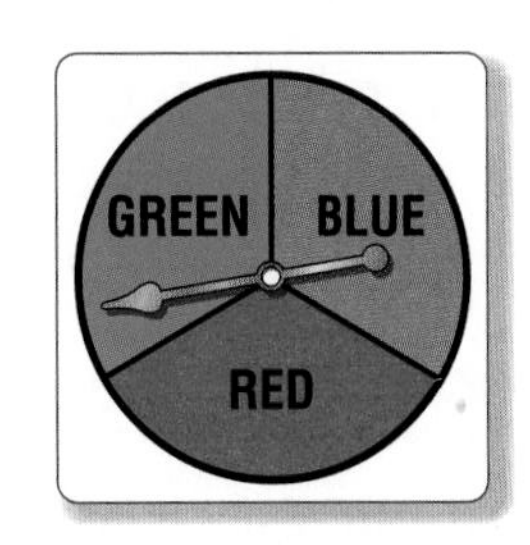

8. **STUDENT GOVERNMENT** If Opal, Jacob, Luanda, Tyrone, and Erica have an equal chance of being an officer of student council, what is the probability that Luanda will be an officer?

msmath2.net/extra_examples

Practice and Applications

HOMEWORK HELP	
For Exercises	See Examples
9–14	1
15–16	2

Extra Practice
See pages 585, 604.

Use the Fundamental Counting Principle to find the total number of outcomes in each situation.

9. rolling a number cube and tossing two coins
10. choosing a number from 1 to 20 and a color from 7 colors
11. choosing a plain, blueberry, garlic, or cinnamon-and-raisin bagel, with plain, chive, or sun-dried tomato cream cheese
12. choosing iced tea in regular, raspberry, lemon, or peach flavors; sweetened or unsweetened; and in a glass or in a plastic container
13. picking a date in the month of May and a day of the week
14. choosing a 4-digit Personal Identification Number (PIN) if the digits cannot be repeated

15. **ADVERTISING** The Wake-Up Restaurant advertises that you can have a different pancake breakfast every day of the year. It offers 25 different kinds of pancakes and 14 flavored syrups. If the restaurant is open every day of the year, is its claim valid? Explain.

16. **GAMES** A player rolls five number cubes to score the maximum number of points. Find the number of outcomes possible in one roll.

CRITICAL THINKING How many outcomes are possible if you toss each of the following?

17. one coin 18. two coins 19. three coins 20. n coins

Standardized Test Practice and Mixed Review

21. **MULTIPLE CHOICE** Elizabeth has 3 sweaters, 5 blouses, and 6 skirts that coordinate. How many different outfits can Elizabeth make?

    (A) 105 (B) 90 (C) 45 (D) 30

22. **GRID IN** WritePen makes 8 different styles of pens in several colors with 2 types of grips. If the company makes 112 kinds of pens, how many different colors do they make?

PROBABILITY What is the probability that the spinner shown at the right will stop on each of the following numbers? Write as a fraction in simplest form. (Lesson 9-1)

23. an even number 24. a multiple of 4 25. a number less than 10

26. **SCHOOL** Gustavo must choose from two geography classes, three history classes, and two statistics classes for next year. Make a tree diagram to show all of the possible schedules he can arrange. (Lesson 9-2)

GETTING READY FOR THE NEXT LESSON

BASIC SKILL Multiply.

27. $3 \cdot 2 \cdot 1$ 28. $9 \cdot 8 \cdot 7$ 29. $5 \cdot 4 \cdot 3 \cdot 2$ 30. $7 \cdot 6 \cdot 5 \cdot 4$

9-4 Permutations

HANDS-ON Mini Lab

Materials
- 3 index cards

Work with a partner.

How many different ways are there to arrange your first three classes if they are math, science, and language arts?

STEP 1 Write math, science, and language arts on the index cards.

STEP 2 Choose one of the subjects as the first class of the day. Choose one of the remaining two subjects for the second class. The third class is the card that remains.

MATH	SCIENCE	LANGUAGE ARTS

STEP 3 Record this arrangement of classes.

STEP 4 Change the order of the classes. Record this arrangement. Continue rearranging the cards until you have found all of the possible arrangements.

1. When you first started to make your list, how many choices did you have for your first class?
2. Once your first class was selected, how many choices did you have for the second class? Then, the third class?
3. Explain how you can use the Fundamental Counting Principle to find the number of arrangements.

What You'll LEARN

Find the number of permutations of a set of objects.

NEW Vocabulary

permutation
factorial

MATH Symbols

4! four factorial

A **permutation** is an arrangement, or listing, of objects in which order is important. You can use the Fundamental Counting Principle to find the number of possible arrangements.

There are **3** choices for the 1st class.
There are **2** choices that remain for the 2nd class.
There is **1** choice that remains for the 3rd class.

$3 \cdot 2 \cdot 1 = 6$ ← The number of arrangements of 3 classes.

The expression $3 \cdot 2 \cdot 1$ can be written as 3!, which is read *three factorial.*

Key Concept — **Factorial**

The expression n **factorial** (n!) is the product of all counting numbers beginning with n and counting backward to 1.

Technology You can find the factorial of a number on a calculator. To find 3!, press 3 PRB

  .

EXAMPLES Evaluate Factorials

Find the value of each expression.

1 **7!**

$7! = 7 \cdot 6 \cdot 5 \cdot 4 \cdot 3 \cdot 2 \cdot 1$ Definition of factorial

$= 5{,}040$ Simplify.

2 **$2! \cdot 3!$**

$2! \cdot 3! = 2 \cdot 1 \cdot 3 \cdot 2 \cdot 1$ Definition of factorial

$= 12$ Simplify.

EXAMPLE Find a Permutation

3 **VOLLEYBALL In how many ways can the starting six players of a volleyball team stand in a row for a picture?**

This is a permutation that can be written as 6!.

$6! = 6 \cdot 5 \cdot 4 \cdot 3 \cdot 2 \cdot 1$ Definition of factorial

$= 720$ Simplify.

So, there are 720 ways the six starting players can stand in a row.

REAL-LIFE MATH

SWIMMING In 1996, swimmer Amy Van Dyken was the first American woman to win 4 gold medals at one Olympic Games.

Source: sportsillustrated.cnn.com

Arrangements can include all or only part of a group.

EXAMPLE Find a Permutation

4 **SWIMMING The finals of the Middle School Appalachian League features 8 swimmers. In how many ways can the swimmers finish in first or second place?**

There are 8 choices for first place and 7 choices that remain for second place. So, there are $8 \cdot 7$ or 56 choices for first and second place.

Skill and Concept Check

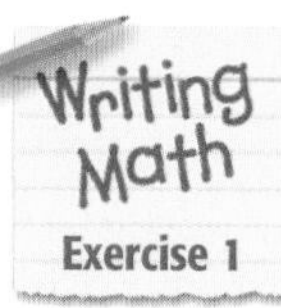

1. **Explain** the difference between 5! and $5 \cdot 4 \cdot 3$.
2. **OPEN ENDED** Describe a real-life situation that has 6 permutations.

GUIDED PRACTICE

Find the value of each expression.

3. 3!
4. $6! \cdot 2!$
5. In how many ways can you arrange the letters in the word *equal*?
6. **TRANSPORTATION** There are 7 students waiting at the bus stop. In how many ways can the students board the bus when it arrives?

msmath2.net/extra_examples

Practice and Applications

HOMEWORK HELP	
For Exercises	See Examples
7–14	1, 2
16	3
15, 17–18	4

Extra Practice
See pages 585, 604.

Find the value of each expression.

7. $5!$
8. $9!$
9. $4! \cdot 3!$
10. four factorial
11. $3! \cdot 6!$
12. $10 \cdot 9 \cdot 8$
13. $5! \cdot 4!$
14. $8! \cdot 2!$

15. In how many ways can a softball manager arrange the first four batters in a lineup of nine players?

16. How many different 5-digit zip codes are there if no digit is repeated?

17. **MUSIC** The chromatic scale has 12 notes. In how many ways can a song start with 4 different notes from that scale?

DOGS For Exercises 18 and 19, use the information below and at the right.
During the annual Westminster Dog Show, the best dog in each breed competes to win one of four top ribbons in the group.

18. In how many ways can a ribbon be awarded to a breed of dog in the Working group?

19. The top dog in each group competes against the other six group winners for Best of Show. If each dog has an equally-likely chance of winning Best of Show, what is the probability that a terrier will win?

2002 Westminster Dog Show

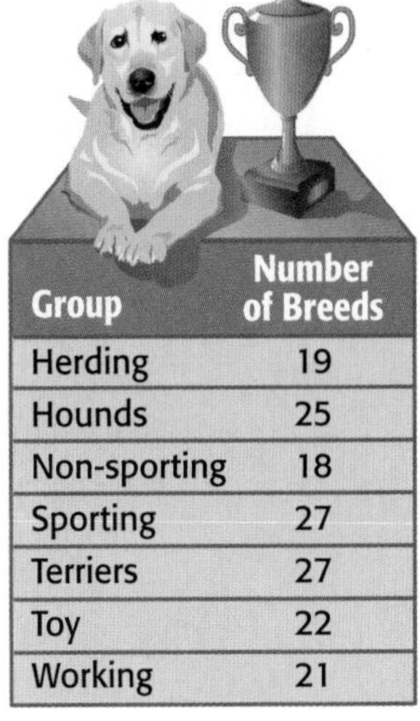

Group	Number of Breeds
Herding	19
Hounds	25
Non-sporting	18
Sporting	27
Terriers	27
Toy	22
Working	21

Source: westminsterkennelclub.org

Data Update Which breed of dog has won the most Best of Shows? Visit msmath2.net/data_update to learn more.

20. **CRITICAL THINKING** There are 1,320 ways for three students to win first, second, and third place during a debate match. How many students are there on the debate team?

Standardized Test Practice and Mixed Review

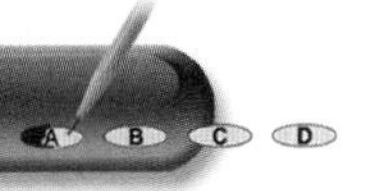

21. **SHORT RESPONSE** In a Battle of the Bands contest, how many ways can the four participating bands be ordered?

22. **GRID IN** How many different three-digit security codes can be made from the digits 1, 2, 3, 4, and 5 if no digit is repeated in a code?

23. **BREAKFAST** Find the total number of outcomes if you can choose from 8 kinds of muffins, 3 sizes, and 4 beverages. (Lesson 9-3)

24. **LUNCH** Make a tree diagram showing different ways to make a sandwich with turkey, ham, or salami and either cheddar or Swiss cheese. (Lesson 9-2)

GETTING READY FOR THE NEXT LESSON

PREREQUISITE SKILL Find each value. (Lesson 5-3)

25. $\frac{5 \cdot 4}{2 \cdot 1}$
26. $\frac{8 \cdot 7 \cdot 6}{3 \cdot 2 \cdot 1}$
27. $\frac{5 \cdot 4 \cdot 3}{4 \cdot 3 \cdot 2}$
28. $\frac{10 \cdot 9 \cdot 8 \cdot 7}{8 \cdot 7 \cdot 6 \cdot 5}$

msmath2.net/self_check_quiz

Mid-Chapter Practice Test

Vocabulary and Concepts

1. **Explain** what it means for an event to occur at random. (Lesson 9-1)
2. **State** the Fundamental Counting Principle. (Lesson 9-3)
3. **Define** *permutation.* (Lesson 9-4)

Skills and Applications

Two number cubes are rolled. Find each probability. (Lesson 9-1)

4. P(sum of 6)
5. P(sum less than 7)
6. P(sum of 7 or 11)

For each situation, make a tree diagram to show the sample space. Then give the total number of outcomes. (Lesson 9-2)

7. tossing a penny and tossing a dime
8. choosing cereal, French toast, or pancakes and choosing orange, apple, cranberry or grapefruit juice

Use the Fundamental Counting Principle to find the total number of outcomes in each situation. (Lesson 9-3)

Entrée	Salad	Dessert
ham	potato	ice cream
beef	tossed	pie
turkey	cole slaw	cake

9. choosing a dinner with one entrée, one salad, and one dessert
10. rolling a number cube and tossing three coins

Find the value of each expression. (Lesson 9-4)

11. 5!
12. 8!
13. $2! \cdot 7!$

14. **STUDENT COUNCIL** In how many ways can a president, treasurer, and a secretary be chosen from among 8 candidates? (Lesson 9-4)

Standardized Test Practice

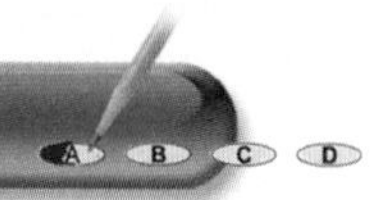

15. **GRID IN** Jeffrey has 2 action, 3 comedy, and 4 drama DVDs. If he randomly picks one DVD to watch, what is the probability that it will be a comedy? (Lesson 9-1)

16. **SHORT RESPONSE** Molly has three windows to display three of five best-selling books. How many different displays can she make if she puts one book in each window? (Lesson 9-4)

The Game Zone

A Place To Practice Your Math Skills

Math Skill
Probability

Cherokee Butterbean Game

GET READY!

Players: two, three, or four
Materials: 6 dry lima beans, bowl, marker

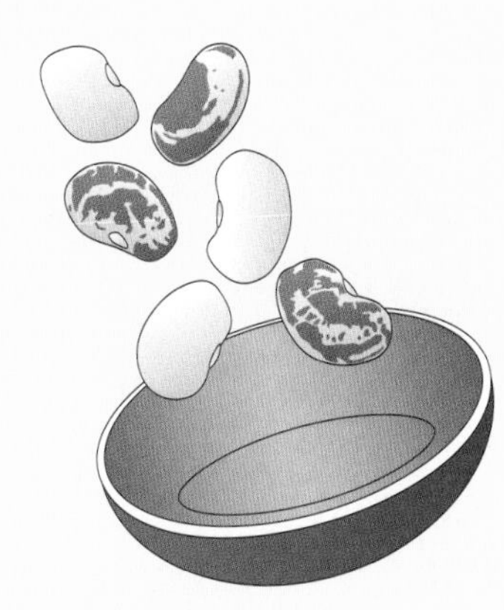

GET SET!

- This is a variation of a traditional Cherokee game.
- Color one side of each bean with the marker.

GO!

- The first player places the beans in the bowl, gently tosses the beans into the air, and catches them in the bowl. Points are scored as follows.
  - If all of the beans land with the marked or unmarked sides up, score 6 points.
  - If exactly one bean lands with the marked or unmarked side up, score 3 points.
  - If three beans are marked and three beans are unmarked, score 1 point.
- If a toss scores points, the player takes another turn. If a toss does not score any points, it is the next player's turn.
- **Who Wins?** The person with the most points after a given number of rounds is the winner.

9-5a HANDS-ON LAB

A Preview of Lesson 9-5

Exploring Combinations

What You'll LEARN

Find combinations.

Materials

- 5 index cards

INVESTIGATE *Work with a partner.*

The student safety club at Mahomet Middle School is planning to sell sundaes with two different toppings at the summer carnival. The choices are shown at the right.

Mahomet Middle School
Safety Club
Sundae Toppings:

hot fudge nuts
cherry caramel

STEP 1 Write the name of the four sundae toppings on the index cards.

STEP 2 To make a sundae, select any pair of cards. Make a list of all the different combinations that are possible. Note that the order of the toppings is not important.

Cherry and Nuts

is considered the same as

Nuts and Cherry

Writing Math

Work with a partner.

1. How many different combinations are possible with two toppings?
2. How many different sundaes could be made with two toppings if the order of the toppings *was* important? What is this type of arrangement called?
3. **Write** a fifth topping on an index card. Now find all two-topping sundae combinations. How many are there? How many sundaes can be made with two toppings if the order of the toppings was important?
4. **Compare and contrast** the way you can find a permutation with the way you find a combination.
5. **Make a conjecture** about how to find a combination given a permutation.

9-5 Combinations

WHEN am I ever going to use this?

BASKETBALL Coach Chávez wants to select co-captains for her basketball team. She will select two girls from the four oldest members on the team: Alita, Bailey, Charmaine, and Danielle.

1. Use the first letter of each name to list all of the permutations of co-captains. How many are there?
2. Cross out any arrangement that contains the same letters as another one in the list. How many are there now?
3. Explain the difference between the two lists above.

What You'll LEARN

Find the number of combinations of a set of objects.

NEW Vocabulary

combination

Link to READING

Everyday Meaning of combination: the result of putting together objects, as in a combination of ingredients

An arrangement, or listing, of objects in which order is *not* important is called a **combination**. For example, in the activity above, choosing Alita and Bailey is the same as choosing Bailey and Alita.

Permutations and combinations are related. You can find the number of combinations of objects by dividing the number of permutations of the entire set by the number of ways each smaller set can be arranged.

A permutation of 4 players, taken 2 at a time.

$$\frac{4 \cdot 3}{2!} = \frac{4 \cdot 3}{2 \cdot 1} = \frac{12}{2} = 6$$

There are 2! ways to arrange 2 players.

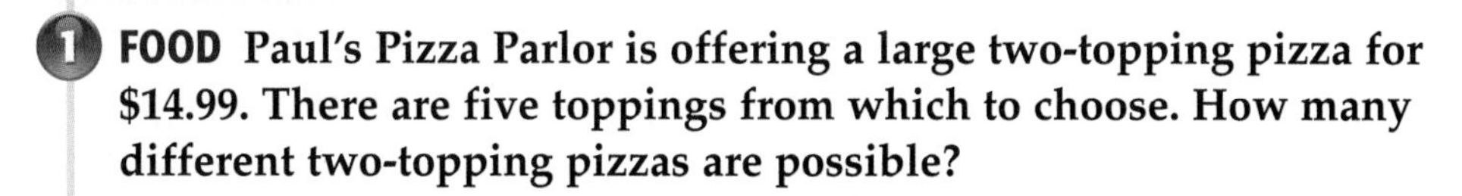

EXAMPLE Find the Number of Combinations

1 FOOD Paul's Pizza Parlor is offering a large two-topping pizza for $14.99. There are five toppings from which to choose. How many different two-topping pizzas are possible?

Method 1 Make a list.

The five toppings are labeled pepperoni (p), sausage (s), onions (o), mushrooms (m), and green pepper (g).

p, s	p, m	p, o	p, g	s, m
s, o	s, g	m, o	m, g	o, g

Method 2 Use a permutation.

There are $5 \cdot 4$ permutations of two toppings chosen from five.

There are 2! ways to arrange the two toppings.

$$\frac{5 \cdot 4}{2!} = \frac{20}{2} = 10$$

So, there are 10 different two-topping pizzas.

EXAMPLE Use a Combination to Solve a Problem

2 CHECKERS A checkers tournament features each of the top 8 regional players playing every opponent one time. The 2 players with the best records will then play in a final round to determine the champion. How many matches will be played if there are no ties?

Find the number of ways 2 players can be chosen from a group of 8.

$$\frac{8 \cdot 7}{2!} = \frac{56}{2} = 28$$

There are $8 \cdot 7$ ways to choose 2 people. → numerator
There are 2! ways to arrange 2 people. → denominator

There are 28 matches plus 1 final match to determine the champion. So, there will be 29 matches played.

Check Make a diagram in which each person is represented by a point. Draw line segments between two points to represent the games. There are 28 line segments. Then add the final-round match to make a total of 29 matches.

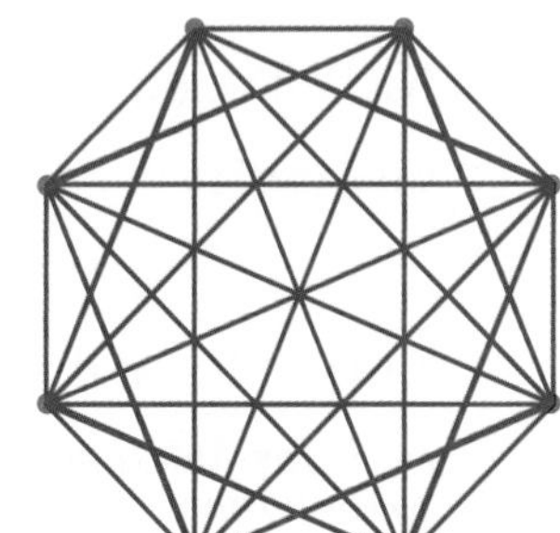

Your Turn

a. How many matches will be played if the top 16 players were invited to play?

REAL-LIFE MATH

CHECKERS Checkers is played on an 8-by-8 board in the U.S., England, Australia, and Ireland. People in eastern Europe play on a 10-by-10 board.

Source: www.about.com

The difference between permutations and combinations is that order is important in permutations, while order is *not* important in combinations.

EXAMPLES Identify Permutations and Combinations

Tell whether each situation represents a *permutation* or *combination*. Then solve the problem.

3 STUDENT GOVERNMENT The six students listed at the right are members of Student Council. How many ways can you choose a president, vice president, and treasurer from this group?

This is a permutation because the order of president, vice president, and treasurer is important. So, the number of ways you can choose the three officers is $6 \cdot 5 \cdot 4$, or 120 ways.

4 In how many ways can you choose a committee of three students from the six members in student council shown above?

This is a combination because the order of the students in the committee is not important.

$$\frac{6 \cdot 5 \cdot 4}{3!} = \frac{120}{6} = 20$$

There are $6 \cdot 5 \cdot 4$ ways to choose 3 people. → numerator
There are 3! ways to arrange 3 people. → denominator

So, there are 20 ways to choose the committee.

Skill and Concept Check

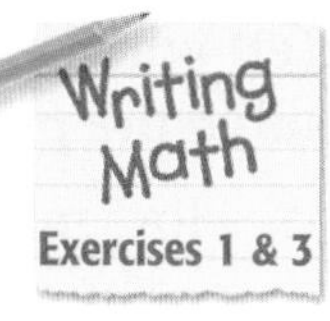

Exercises 1 & 3

1. **Explain** why a combination lock should be called a permutation lock.

2. **OPEN ENDED** Give an example of a permutation and combination.

3. **FIND THE ERROR** Allison and Francisca are calculating the number of ways that a 3-member committee can be chosen from a 7-member club. Who is correct? Explain.

Allison	Francisca
$7 \cdot 6 \cdot 5 = 210$ ways	$\frac{7 \cdot 6 \cdot 5}{3!} = \frac{210}{6} = 35$ ways

GUIDED PRACTICE

4. **VOLLEYBALL** Coach Malone has an 8-member volleyball team. He told his team that he would start six different players every game. How many games would it take to do this?

Tell whether each problem represents a *permutation* or *combination*. Then solve the problem.

5. How many ways can 10 students finish first, second, or third at the science fair?
6. How many ways can you pick 2 puppies from a litter of 7 puppies?

Practice and Applications

HOMEWORK HELP	
For Exercises	See Examples
7–10	1, 2
11–16	3, 4

Extra Practice
See pages 586, 604.

7. **FOOD** The International Club is selling hot dogs at the Spring Carnival. Customers can select three toppings from among chili, onions, cheese, mustard, or relish. How many combinations of three-topping hot dogs are there?

CIVICS For Exercises 8 and 9, use the information below and at the right.
If five of the nine Justices on the 2002 United States Supreme Court agree on a decision, they can issue a majority opinion.

8. How many different combinations of five Supreme Court Justices are there?
9. Before they take the Bench each day, the Justices engage in the "Conference handshake." Each Justice shakes hands with each of the other eight. How many handshakes take place?

U.S. Supreme Court, 2002	
Chief Justice, William H. Rehnquist	
Stephan G. Breyer	Antonin Scalia
Ruth Bader Ginsburg	David H. Souter
Anthony M. Kennedy	John Paul Stevens
Sandra Day O'Connor	Clarence Thomas

Source: www.washingtonpost.com

10. **SOCCER** There are 21 players trying out for 15 spots on the soccer team. How many ways does the coach have to create her team?

msmath2.net/self_check_quiz

Tell whether each problem represents a *permutation* or *combination*. Then solve the problem.

11. How many ways can you select four essay questions out of a total of 10 on the exam?
12. Six children remain in a game of musical chairs. If two chairs are removed, how many different groups of four students can remain?
13. How many ways can three flute players be seated in the first, second, or third seats in the orchestra?
14. In how many ways can four paintings be displayed from a collection of 15?
15. How many ways can seven students line up to buy concert tickets?
16. Given 12 Web sites, how many ways can you visit half of them?
17. **WRITE A PROBLEM** Write about a real-life situation that can be solved using a permutation and one that can be solved using a combination. Then solve both problems.
18. **CRITICAL THINKING** At a party, there were 105 handshakes. If each person shook hands exactly once with every other person, how many people were at the party?

Standardized Test Practice and Mixed Review

19. **MULTIPLE CHOICE** In how many ways can three gymnasts from a team of 10 be chosen to compete in a meet?

    Ⓐ 120 Ⓑ 180 Ⓒ 240 Ⓓ 720

20. **SHORT RESPONSE** Find $\frac{(n + 6)(n + 5)(n + 4)}{n!}$ if $n = 3$.
21. **SAILING** When Mr. Elms purchased his sailboat, it came with six different-colored flags to be used for sending signals. A specific signal depended on the order of the flags. How many different three-flag signals can he send? (Lesson 9-4)
22. **CARS** A certain brand of car has a choice of a 2.5, 3.1, or 4.0 liter engine; a radio with a cassette-player, CD-player, or CD-changer; cloth or leather seats; and silver, green, red, or white. How many different cars are possible? (Lesson 9-3)

Estimate. (Lesson 6-1)

23. $\frac{1}{10} + \frac{7}{8}$
24. $\frac{5}{12} - \frac{1}{9}$
25. $\frac{4}{9} \cdot 20$
26. $15\frac{7}{9} \div 3\frac{3}{5}$

GETTING READY FOR THE NEXT LESSON

PREREQUISITE SKILL Suppose you choose a card from a set of eight cards labeled 1–8. Write the probability of each event as a fraction in simplest form.
(Lesson 9-1)

27. P(even number)
28. P(number greater than 6)
29. P(multiple of 8)
30. P(number less than 9)
31. P(prime number)
32. P(*not* 2 or 5)

9-6a Problem-Solving Strategy

A Preview of Lesson 9-6

What You'll Learn
Solve problems by acting them out.

Act It Out

I wonder if tossing a coin would be a good way to answer a 5-question true-false quiz?

I'm not so sure, Whitney. Let's do an experiment with a coin and **act it out**!

Explore We know that there are five true-false questions on the quiz. We can carry out an experiment to test whether tossing a coin would be a good way to answer the questions and get a good grade.

Plan Let's toss a coin five times. If the coin shows tails, we will answer T. If the coin shows heads, we will answer F. Let's do four trials.

Solve Suppose the correct answers are T, F, F, T, F. Let's circle them.

Answers	T	F	F	T	F	Number Correct
Trial 1	Ⓣ	T	Ⓕ	F	T	2
Trial 2	F	Ⓕ	T	Ⓣ	Ⓕ	3
Trial 3	Ⓣ	Ⓕ	T	F	T	2
Trial 4	F	Ⓕ	T	F	Ⓕ	2

The experiment produces 2 or 3 correct answers.

Examine There is an equally likely chance that tossing a coin will produce a correct answer or a wrong answer for each question. Since the experiment produced about 2–3 correct answers on a 5-question quiz, it shows that tossing a coin to answer a true-false quiz is not the way to get a good grade.

Analyze the Strategy

1. **Explain** whether the results of the experiment would be the same if it were repeated.
2. **Explain** an advantage of using the act it out strategy to solve a problem.

Apply the Strategy

Solve. Use the act it out strategy.

3. **POP QUIZ** Determine whether using a spinner with four equal sections is a good way to answer a 5-question multiple-choice quiz. Each question has choices A, B, C, and D. Explain.

4. **PHOTOGRAPHS** Samuel is taking a picture of the Spanish Club's five officers. The club president will always stand on the left, and the vice president will always stand on the right. How many different ways can he arrange the officers for the picture?

Mixed Problem Solving

Solve. Use any strategy.

5. **MONEY MATTERS** Lola purchased a $35 book bag at a sale price of $27.50. What was the percent of decrease from the original price to the sales price?

6. **PICNIC** Examine the bar graph.

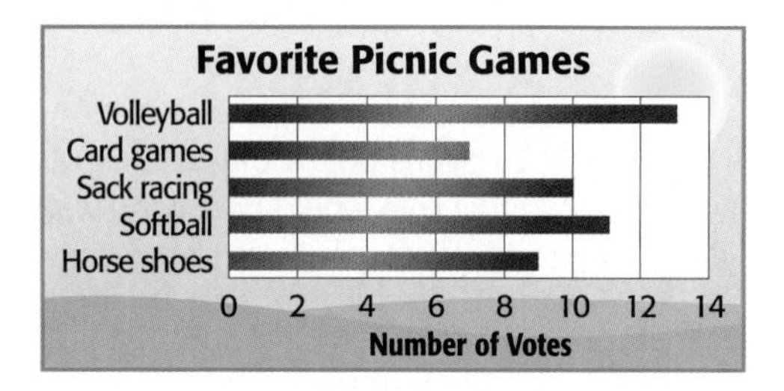

What is the probability that a student's favorite picnic game is sack racing?

7. **CROQUET** Sixteen teams are playing in a croquet tournament. If a team loses one game, it is eliminated. How many total games will be played in the tournament?

8. **EARTH SCIENCE** The Hawaiian mountain Mauna Kea is about 1.38×10^4 feet tall. Mt. Everest is about 29,000 feet. Which mountain is taller?

9. **BASEBALL** Each package of baseball cards contains a puzzle piece. If you collect all 6 different pieces, you win two tickets to a major league game. There is an equally-likely chance of getting a different puzzle piece each time. Explain how you would model this problem to find the answer.

10. **PATTERNS** The pattern below is known as Pascal's Triangle. Find the pattern and complete the 6th and 7th rows.

1st Row					1				
2nd Row				1		1			
3rd Row			1		2		1		
4th Row		1		3		3		1	
5th Row	1		4		6		4		1

11. **SPACE SCIENCE** A 21-kilogram sample of rocks from the moon is composed of about 40% oxygen and about 19.2% silicon. How much more is the oxygen mass of the rocks than the silicon mass?

12. **TRANSPORTATION** A taxi charges $2.35 for the first 0.4 mile and $0.75 for each additional 0.4 mile. Find the cost of a 4-mile taxi ride.

13. **STANDARDIZED TEST PRACTICE**
In a middle school of 900 students, about how many prefer the season of autumn?

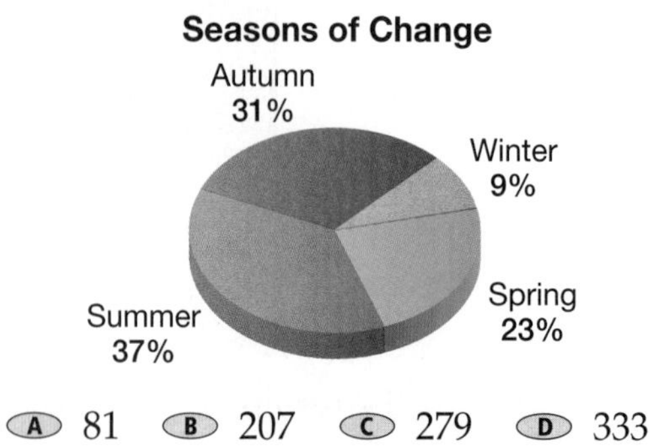

Ⓐ 81 Ⓑ 207 Ⓒ 279 Ⓓ 333

You will use the act it out strategy in the next lesson.

9-6 Theoretical and Experimental Probability

What You'll LEARN

Find and compare experimental and theoretical probabilities.

NEW Vocabulary

experimental probability
theoretical probability

REVIEW Vocabulary

sample space: a set of all possible outcomes **(Lesson 9-2)**

HANDS-ON Mini Lab

Materials

- 2 number cubes

Work with a partner.

STEP 1 Roll two number cubes 36 times. Record the number of times the sum of the number cubes is 7.

STEP 2 Use an addition table from 1 to 6 to help you find the expected number of times the sum of 7 should come up after rolling two number cubes 36 times. The top row represents one number cube, and the left column represents the other number cube. The table above has been started for you.

+	1	2	3
1	2	3	4
2	3	4	5

1. How many times did you roll a sum of 7? What is the probability of rolling a sum of 7?
2. How does your result compare to the results of other groups? Explain.
3. What is the expected probability of rolling a sum of 7?
4. How does your result compare to the expected probability of rolling a sum of 7? Explain any differences.

In activity above, you found the experimental probability of rolling a sum of 7 on two number cubes. **Experimental probability** is found using frequencies obtained in an experiment or game.

The expected probability of an event occurring is called **theoretical probability**. This is the probability that you have been using since Lesson 9-1. The theoretical probability of rolling a sum of 7 on two number cubes is $\frac{6}{36}$, or $\frac{1}{6}$.

EXAMPLE Experimental Probability

1 **Two number cubes are rolled seventy-five times and a sum of 9 is rolled ten times.**

What is the experimental probability of rolling a sum of 9?

$$P(9) = \frac{\text{number of times a sum of 9 occurs}}{\text{number of possible outcomes}}$$

$$= \frac{10}{75} \text{ or } \frac{2}{15}$$

The experimental probability of rolling a sum of 9 is $\frac{2}{15}$.

EXAMPLES Experimental and Theoretical Probability

2 The graph shows the results of an experiment in which a coin was tossed thirty times. Find the experimental probability of tossing tails for this experiment.

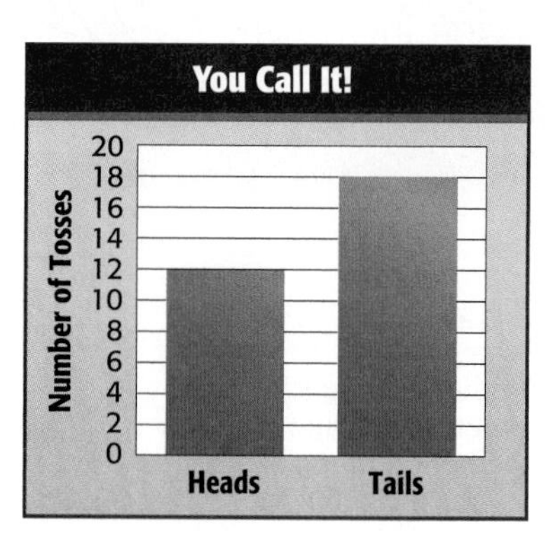

$$P(\text{tails}) = \frac{\text{number of times tails occurs}}{\text{number of possible outcomes}}$$
$$= \frac{18}{30} \text{ or } \frac{3}{5}$$

The experimental probability of tossing tails is $\frac{3}{5}$.

3 Compare the experimental probability you found in Example 2 to its theoretical probability.

The theoretical probability of tossing tails on a coin is $\frac{1}{2}$. So, the experimental probability is close to the theoretical probability.

Experimental probability can also be based on past performances and be used to make predictions on future events.

EXAMPLES Predict Future Events

REAL-LIFE CAREERS

How Does a Chef Use Math?

In addition to cooking, chefs order enough food and supplies based on customers' preferences and recent sales.

Online Research
For more information about a career as a chef, visit: msmath2.net/careers

4 FOOD In a survey, 100 people were asked to name their favorite Independence Day side dishes. What is the experimental probability of macaroni salad being someone's favorite dish?

Side Dish	Number of People
potato salad	55
green salad or vegetables	25
macaroni salad	12
coleslaw	8

There were 100 people surveyed and 12 chose macaroni salad. So, the experimental probability is $\frac{12}{100}$, or $\frac{3}{25}$.

5 Suppose 250 people attend the city's Independence Day barbecue. How many can be expected to choose macaroni salad as their favorite side dish?

$\frac{3}{25} = \frac{x}{250}$ Write a proportion.

$3 \cdot 250 = 25x$ Find the cross products.

$30 = x$ About 30 will choose macaroni salad.

Your Turn

a. What is the experimental probability of potato salad being someone's favorite dish?

b. About how many people can be expected to choose potato salad as their favorite dish if 250 attend the barbecue?

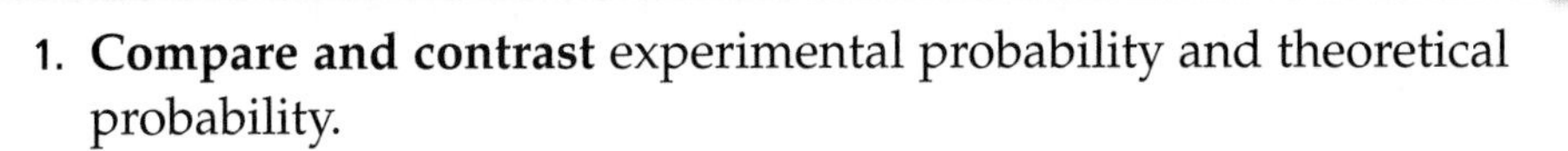

Skill and Concept Check

1. **Compare and contrast** experimental probability and theoretical probability.

2. **OPEN ENDED** Give an example of a situation that has a theoretical probability of $\frac{1}{10}$.

GUIDED PRACTICE

For Exercises 3 and 4, a coin is tossed 50 times, and it lands heads 28 times.

3. Find the experimental probability of the coin landing heads.
4. Find the theoretical probability of the coin landing heads.

5. **TRAFFIC** Between 3:00 P.M. and 4:00 P.M., 11 sedans, 15 trucks, and 6 sports cars drove through an intersection. Based on this information, what is the probability that a vehicle that drives through the intersection is a sports car?

Practice and Applications

HOMEWORK HELP

For Exercises	See Examples
6–9	1–3
10, 15	4
11–14, 16–17	5

Extra Practice
See pages 586, 604.

For Exercises 6–9, a number cube is tossed 20 times and lands on 1 two times and on 5 four times.

6. Find the experimental probability of landing on 5.
7. Find the theoretical probability of *not* landing on 5.
8. Find the theoretical probability of landing on 1.
9. Find the experimental probability of *not* landing on 1.

X GAMES For Exercises 10–12, use the graph of a survey of 50 students asked to name their favorite X Game sport.

10. What is the probability of inline being someone's favorite sport?
11. Suppose 500 people attend the X Games. How many can be expected to choose inline as their favorite sport?
12. Suppose 500 people attend the X Games. How many can be expected to choose speed climbing as their favorite sport?

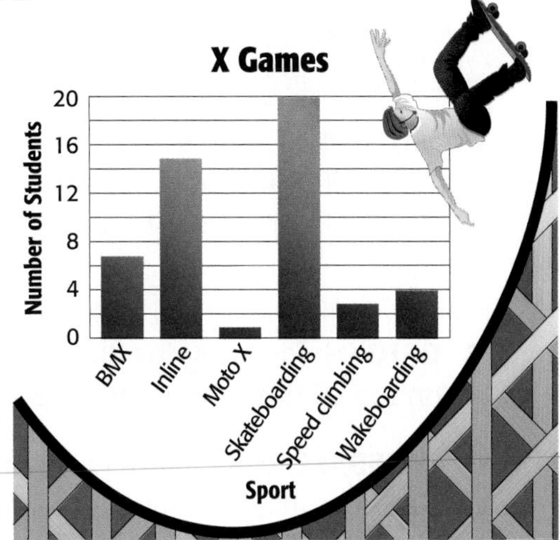

13. **REFRESHMENTS** In a survey taken at the beach, 47 people preferred cola, 28 preferred root beer, and 25 preferred ginger ale. If the manager of the Beach Hut is going to buy 50 cases of soda for the next day, about how many cases should be root beer?

14. **SPINNERS** A spinner marked with three sections A, B, and C was spun 100 times. The results are shown in the table. Make a drawing of the spinner based on its experimental probabilities.

Section	Frequency
A	24
B	50
C	26

GIFTS **For Exercises 15–17, use the results of the survey at the right.**

15. What is the probability that a mother will receive a gift of flowers? Write the probability as a fraction.
16. Out of 750 mothers that receive gifts, how many would you expect to receive flowers?
17. Out of 750 mothers that receive gifts, how many would you expect to receive jewelry?

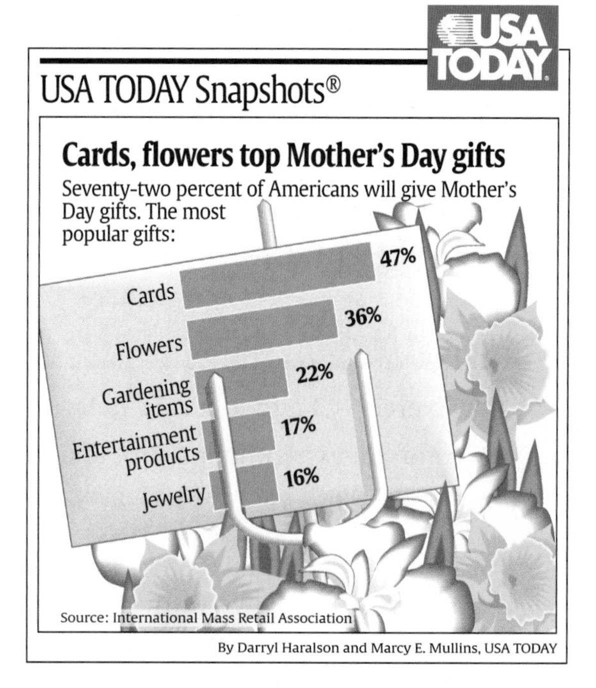

18. **CRITICAL THINKING** *Hot numbers* in a lottery are numbers that keep coming up. *Cold numbers* are numbers that have not "hit" in awhile. Some people select hot numbers or cold numbers when buying a ticket. Using probability, explain whether this makes sense.

EXTENDING THE LESSON

The *odds* of an event occurring is the ratio that compares the number of ways an event can occur (success) to the number of ways it cannot occur (failure).

Example Find the odds of rolling a number less than 5 on a number cube.

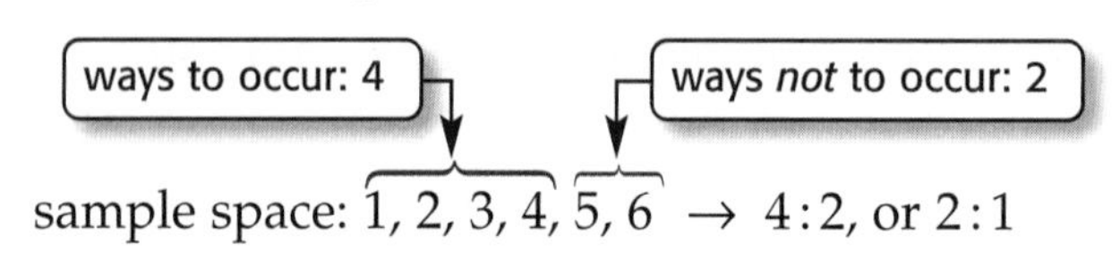

sample space: 1, 2, 3, 4, 5, 6 → 4 : 2, or 2 : 1

Find the odds of each outcome if a number cube is rolled.

19. 6
20. odd number
21. number greater than 4

Standardized Test Practice and Mixed Review

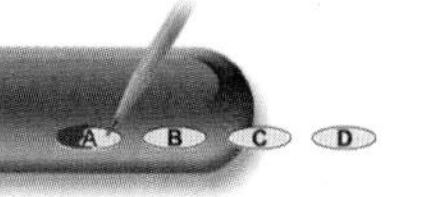

22. **MULTIPLE CHOICE** Myron has four dimes in his wallet with dates 1998, 1995, 2002, and 2000. If he randomly picks one dime from his wallet, what is the probability that it will have a date in the 1990s?

  Ⓐ $\frac{1}{4}$ Ⓑ $\frac{1}{2}$ Ⓒ $\frac{2}{3}$ Ⓓ $\frac{3}{4}$

23. **GRID IN** Rachel spins the spinner at the right 50 times, and it lands on red 15 times. What is the theoretical probability of spinning red?

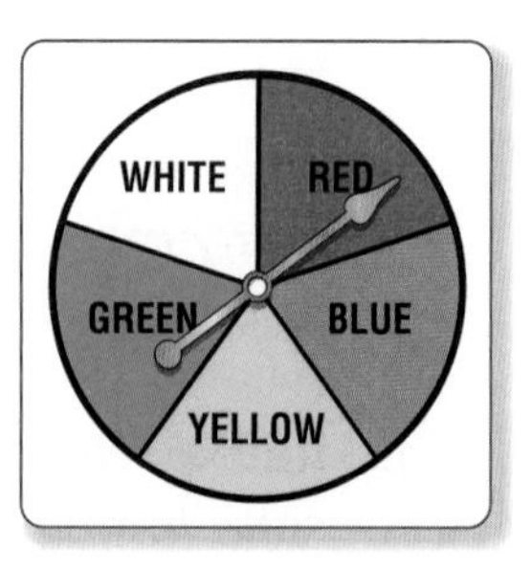

24. **BASEBALL** How many ways can a baseball coach select four starting pitchers from a pitching staff of eight? (Lesson 9-5)
25. **POLE VAULT** A pole vault competition has 10 people in it. In how many ways can first-, second- and third-place ribbons be awarded? (Lesson 9-4)

GETTING READY FOR THE NEXT LESSON

PREREQUISITE SKILL **Three coins are tossed. Find each probability.** (Lesson 9-1)

26. P(3 tails)
27. P(2 heads)
28. P(at least 2 heads)
29. P(at least 1 tail)

Experimental Probability

What You'll LEARN

Investigate experimental probability.

Materials

- paper
- scissors
- spinner arrows

INVESTIGATE *Work with a partner.*

Shannon is on the basketball team. She makes 75% of her free throws. In this lab, you will model, or *simulate* Shannon shooting free throws with a spinner and investigate the experimental probabilities of her being able to make a free throw.

- Create and cut out the spinner at the right.
- Spin the spinner 25 times to model Shannon shooting 25 free throws. Record your results in the table below. If the spinner lands on a line, spin again.

Outcome	Tally	Total
Made Free Throw		
Missed Free Throw		

- Combine your data with four other groups so you have 100 data points.

Writing Math

Work with a partner.

1. **Explain** why you combined your data with four other groups.
2. From your table, what is the experimental probability of making a free throw?
3. How does the experimental probability compare with Shannon's past performance?
4. What is the experimental probability of making two free throws in a row? **Explain** how you found your answer.
5. Suppose Shannon did shoot 100 free throws in a row. What factors would influence her making *or* not making 75 free throws? **Explain** your reasoning.
6. **Describe** the size of both sections of the spinner.
7. **Describe** another way to simulate Shannon shooting free throws.
8. **Draw** a spinner that simulates a free-throw shooter making 60% of her free throws.

9-7 Independent and Dependent Events

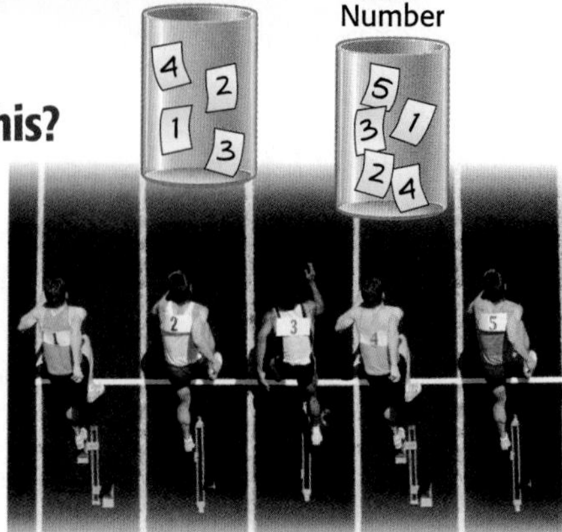

What You'll LEARN

Find the probability of independent and dependent events.

NEW Vocabulary

compound event
independent event
dependent event

WHEN am I ever going to use this?

TRACK AND FIELD The 100-meter dash features 20 runners competing in a preliminary round of 4 heats. The winner of each heat advances to the final race. Before the race, each runner chooses a number from jar 1 to determine the heat in which he runs and a number from jar 2 to determine one of five lanes he occupies. Omar is the first runner to choose from the jars.

1. What is the probability of Omar being in the second heat?
2. What is the probability of Omar being in lane 3?
3. Multiply your answers in Exercises 1 and 2 above. What does this number mean? Explain.

In the Mini Lab, choosing the heat and the lane is a compound event. A **compound event** consists of two or more simple events. Since choosing the heat number does not affect choosing the lane number, both events are called **independent events**.

Key Concept — Probability of Two Independent Events

Words The probability of two independent events can be found by multiplying the probability of the first event by the probability of the second event.

Symbols $P(A \text{ and } B) = P(A) \cdot P(B)$

EXAMPLE Independent Events

1 **A number cube is rolled, and the spinner at the right is spun. Find the probability of rolling a 2 and spinning a vowel.**

$P(2) = \frac{1}{6}$ $\quad$ $P(\text{vowel}) = \frac{2}{5}$

$P(2 \text{ and vowel}) = \frac{1}{\cancel{6}_{3}} \cdot \frac{\cancel{2}^{1}}{5}$ or $\frac{1}{15}$

So, the probability of rolling a 2 and spinning a vowel is $\frac{1}{15}$.

Check You can make a tree diagram to check your answer.

Probability Notation
P(A and B) is read the probability of A followed by B.

If the outcome of one event affects the outcome of a second event, the events are called **dependent events**.

EXAMPLE Dependent Events

2 SNACK BARS A box contains 2 oatmeal, 3 strawberry, and 6 cinnamon snack bars. Ruby reaches in the box and randomly takes two snack bars, one after the other. Find the probability that she will choose a cinnamon bar and then a strawberry bar.

$P(\text{cinnamon}) = \frac{6}{11}$ ← 11 snack bars, 6 are cinnamon

$P(\text{strawberry}) = \frac{3}{10}$ ← 10 snack bars after 1 cinnamon snack bar has been removed, 3 are strawberry

$P(\text{cinnamon, then strawberry}) = \frac{\overset{3}{\cancel{6}}}{11} \cdot \frac{3}{\underset{5}{\cancel{10}}} \text{ or } \frac{9}{55}$

So, the probability that Ruby will choose a cinnamon snack bar and then a strawberry snack bar is $\frac{9}{55}$, or about 16%.

STUDY TIP

Dependent Events The probability of *B* following *A* is the same as the probability of *A* following *B*.

Key Concept — Probability of Two Dependent Events

Words The probability of two dependent events is the probability of the first event times the probability that the second event occurs after the first.

Symbols $P(A \text{ and } B) = P(A) \cdot P(B \text{ following } A)$

Skill and Concept Check

Writing Math — Exercises 1 & 3

1. **Explain** the difference between independent and dependent events.
2. **OPEN ENDED** Describe two dependent events.
3. **FIND THE ERROR** Kimi and Shane are finding the probability of spinning an even number on two consecutive spins. Who is correct? Explain.

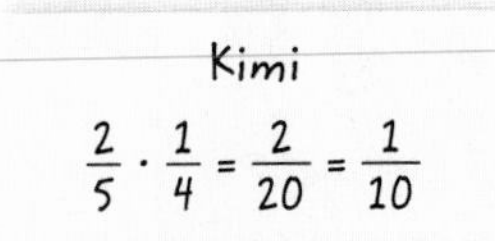

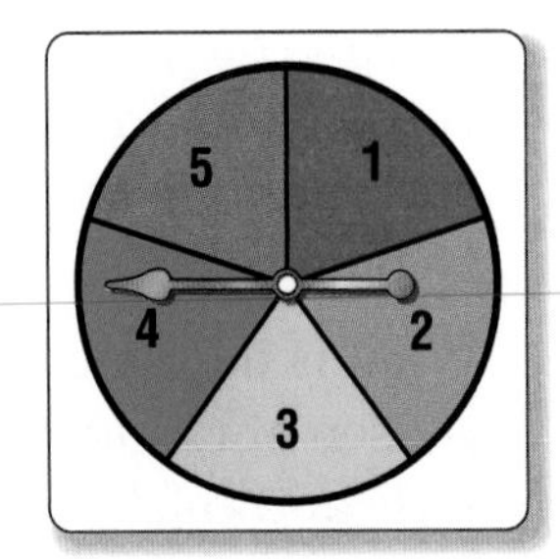

GUIDED PRACTICE

4. A coin is tossed, and a number cube is rolled. What is the probability of tossing heads and rolling a 3 or a 5?
5. A bag contains 5 red apples and 3 yellow apples. What is the probability of picking 2 red apples without the first being replaced?

Practice and Applications

HOMEWORK HELP	
For Exercises	See Examples
6–7, 17–19	1
8–9, 11–14	2

Extra Practice
See pages 586, 604.

6. A red and a blue number cube are rolled. Find the probability that an odd number is rolled on the red cube and a number greater than 1 is rolled on the blue cube.

7. Find the probability of heads on three consecutive tosses of a coin.

8. A cooler is filled with 12 colas and 9 diet colas. If Victor randomly chooses two without replacing the first, what is the probability that he will choose a cola and then a diet cola?

9. A deck of 30 cards is made up of the numbers 1–10 in three colors: red, purple, and green. Two cards are selected without either being replaced. Find the probability of choosing a purple 5 and then a red or green card.

10. Draw a Venn diagram to show the probability of two independent events *A* and *B*.

11. **CIVICS** In the 108th Congress, Tennessee had 4 Republicans and 5 Democrats serving in the House of Representatives. If a subcommittee of 2 representatives was formed to study Internet usage among middle school students, what is the probability that both would be Republicans?

CARDS For Exercises 12–14, use the information below.
A standard deck of playing cards contains 52 cards in four suits of 13 cards each. Two suits are red and two suits are black. Two cards are chosen from the deck one after another. Find each probability.

12. *P*(2 hearts)
13. *P*(red, black)
14. *P*(Ace, King)

Determine whether each event is *independent* or *dependent*.

15. choosing a student in the 7th grade and a student from the 8th grade
16. choosing a pair of shoes to try on, then choosing a smaller pair to try on

FASHION For Exercises 17 and 18, use the graphic at the right. Write your answers as decimals to the nearest hundredth.

17. What is the probability that a teen chosen at random said that friends and sports figures influenced her fashion choices?

18. What is the probability that a teen chosen at random said that personal style and celebrities influenced his fashion choices?

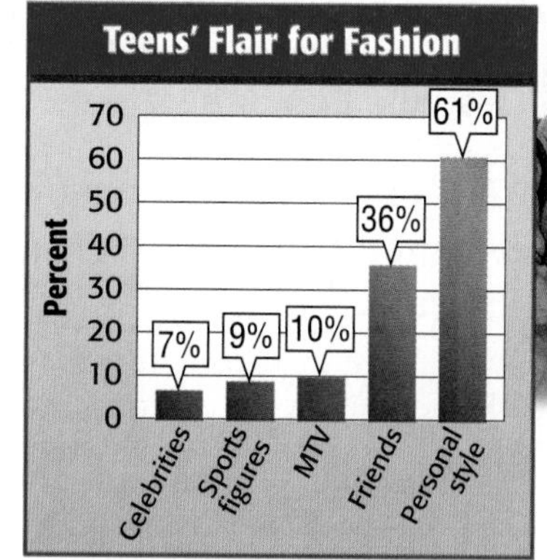

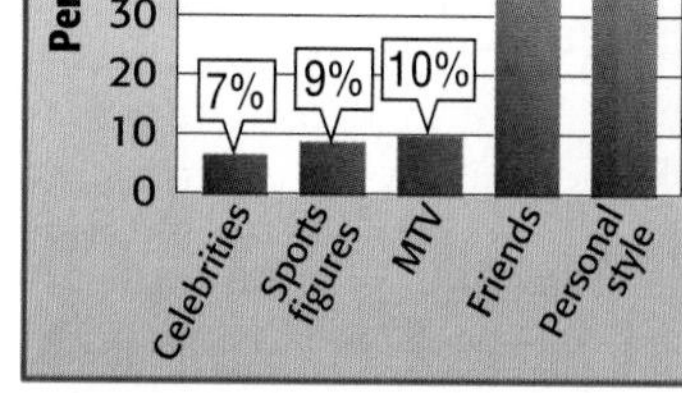

Source: American Express Retail Index

19. **TRAFFIC SIGNALS** There are three consecutive traffic signals on a street that operate independently of each other. The first is green 45% of the time, the second is green 60% of the time, and the third is green 50% of the time. What is the probability of a person driving down the street making all three green lights? Write as a percent to the nearest tenth.

CRITICAL THINKING For Exercises 20 and 21, use the spinner.

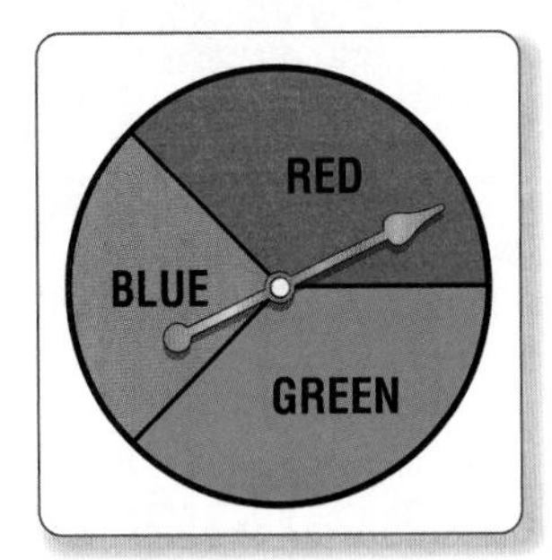

20. Make a tree diagram of all the possible outcomes of three successive spins of the spinner shown. How many paths in the tree diagram represent two red spins and one blue?

21. Suppose the spinner is designed so that for each spin there is a 40% probability of spinning red and a 20% chance of spinning blue. What is the probability of spinning two reds and one blue?

EXTENDING THE LESSON Events that cannot happen at the same time are *mutually exclusive.* The two events are connected by the word *or* and the probability is found by *adding* their individual probabilities.

Example Suppose a number cube is rolled. Find P(even or 5).

$$P(\text{even or } 5) = \frac{3}{6} + \frac{1}{6} = \frac{4}{6} \text{ or } \frac{2}{3}$$

Find each probability as a fraction in simplest form.

22. P(1 or 3)
23. P(5 or multiple of 3)
24. P(6 or less than 5)

Standardized Test Practice and Mixed Review

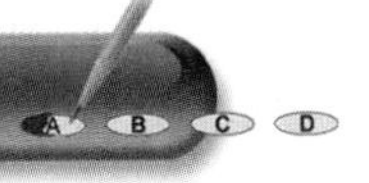

25. **MULTIPLE CHOICE** You guess the answers on a two-question multiple-choice quiz with five choices: a, b, c, d, and e. What is the probability that you will get both correct?

 (A) $\frac{1}{50}$ (B) $\frac{1}{25}$ (C) $\frac{2}{25}$ (D) $\frac{1}{5}$

26. **GRID IN** Find the probability of rolling an even number on a number cube, tossing heads on a coin, and choosing an ace in a standard deck of cards.

27. **PROBABILITY** Paz performed a probability experiment by spinning a spinner 20 times. The results are shown in the table. If the spinner is divided into four equal sections, how many sections would you expect to be blue? (Lesson 9-6)

Color	Frequency
red	~~||||~~
green	~~||||~~ ~~||||~~
blue	~~||||~~

28. **LOTTERY** Balls numbered 1 to 51 are dropped into a machine and mixed together. Six balls are then selected in any order to make up the winning numbers. How many different ways can the six numbers be chosen? (Lesson 9-5)

Evaluate each expression if $a = 6$, $b = -4$, and $c = -3$. (Lesson 3-6)

29. $9c$
30. $-8a$
31. $2bc$
32. $5b^2$

WebQuest Interdisciplinary Project

Step Right Up and Win a Prize

It's time to complete your project. Use the information and data you have gathered about carnival games to prepare a Web page or poster. Be sure to include a scale drawing of the game you design with your project.

msmath2.net/webquest

Study Guide and Review

Vocabulary and Concept Check

combination (p. 387)	fair game (p. 374)	random (p. 370)
complementary events (p. 371)	Fundamental Counting Principle (p. 378)	sample space (p. 374)
compound event (p. 398)	independent events (p. 398)	simple event (p. 370)
dependent events (p. 399)	outcome (p. 370)	theoretical probability (p. 393)
experimental probability (p. 393)	permutation (p. 381)	tree diagram (p. 374)
factorial (p. 381)	probability (p. 370)	

Choose the correct term to complete the sentence.

1. The set of all possible outcomes for an experiment is called the (sample space, probability).
2. The Fundamental Counting Principle counts the number of possible outcomes using the operation of (addition, multiplication).
3. The ratio of the number of times an event occurs to the number of trials completed is called the (theoretical, experimental) probability.
4. When the outcome of one event influences the outcome of a second event, the events are called (independent, dependent).
5. A (permutation, combination) is a listing of objects in which order is important.
6. A (complementary, compound) event consists of two or more simple events.
7. A (simple, random) event occurs by chance.
8. The probability of an event is the (product, ratio) of the number of ways an event can occur to the number of possible outcomes.
9. When using a combination, the order of the arrangement (is, is not) important.
10. The expression $n!$ is the (sum, product) of all counting numbers beginning with n and counting backward to 1.

Lesson-by-Lesson Exercises and Examples

9-1 Simple Events (pp. 370–373)

A bag contains 6 red, 3 pink, and 3 white bows. Suppose you draw a bow at random. Find the probability of each event. Write each fraction in simplest form.

11. P(red)
12. P(pink)
13. P(white)
14. P(red or white)
15. P(pink or white)
16. P(*not* white)

Example 1 **What is the probability of rolling an odd number on a number cube?**

$$P(\text{odd}) = \frac{\text{odd numbers possible}}{\text{total numbers possible}}$$

$$= \frac{3}{6} \quad \text{Three numbers are odd: 1, 3, and 5.}$$

$$= \frac{1}{2} \quad \text{Simplify.}$$

Therefore, $P(\text{odd}) = \frac{1}{2}$.

msmath2.net/vocabulary_review

9-2 Tree Diagrams (pp. 374–377)

For each situation, make a tree diagram to show the sample space. Then give the total number of outcomes.

17. rolling a number cube and then tossing a coin
18. choosing a red, blue, or white shirt with either black or gray lettering
19. tossing a coin and choosing a card from seven cards numbered from 1 to 7
20. choosing from white, wheat, or rye bread and turkey, ham, or salami to make a sandwich

Example 2 **Make a tree diagram to describe the possible outcomes for a family with two children.**

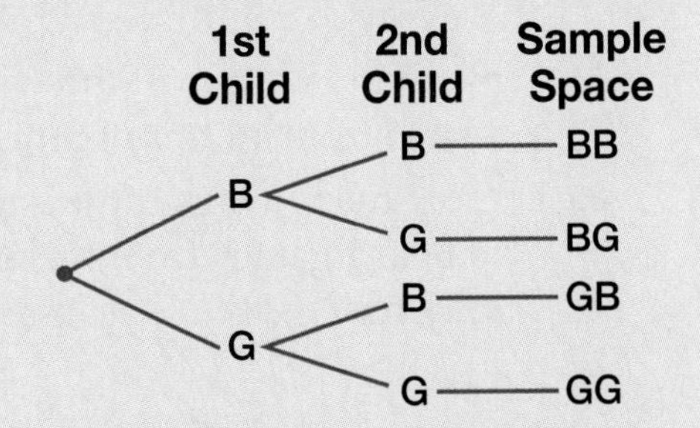

If a family has two children, there are 4 possible outcomes.

9-3 The Fundamental Counting Principle (pp. 378–380)

Use the Fundamental Counting Principle to find the total number of outcomes in each situation.

21. rolling two number cubes
22. selecting a car from 3 styles, 3 interior colors, and 3 exterior colors
23. making an ice cream sundae selecting from 5 flavors of ice cream and 4 different toppings
24. **SHOPPING** One catalog offers a jogging suit in two colors, gray and black. It comes in sizes S, M, L, XL, and XXL. How many possible jogging suits can be ordered?

Example 3 **Use the Fundamental Counting Principle to find the total number of outcomes for a family that has three children.**

There are 2 possible outcomes each time a family has a child, a boy or a girl. If a family has 3 children, there are $2 \cdot 2 \cdot 2$, or 8 outcomes.

9-4 Permutations (pp. 381–383)

Find the value of each expression.

25. 4!
26. 9!
27. $8 \cdot 7 \cdot 6 \cdot 5$
28. $3! \cdot 4!$
29. $5! \cdot 3!$
30. $7! \cdot 4!$
31. **BASKETBALL** In how many ways can five basketball players be placed in three positions?

Example 4 **In how many ways can a director and assistant director be chosen from among 4 candidates?**

There are 4 choices for director and then 3 choices remain for assistant director. So, there are $4 \cdot 3$, or 12, ways the director and assistant director can be chosen.

9-5 Combinations (pp. 387–390)

32. How many three-topping pizzas are possible given eight different toppings?
33. How many groups of five people are there from a committee of nine?
34. How many groups of three kittens are possible from a litter of nine?
35. In how many ways can Rondell select two board games from the ten that his family has?

Example 5 In how many ways can you choose two items from a menu with four items?

There are $4 \cdot 3$ permutations of two menu items chosen from four.

There are 2! ways to arrange the two menu items.

$\frac{4 \cdot 3}{2!} = \frac{12}{2}$ or 6

So, there are 6 ways to choose two items.

9-6 Theoretical and Experimental Probability (pp. 393–396)

The results of spinning a spinner labeled A–E fifty times are given. Find the experimental probability of each event.

Letter	Frequency
A	8
B	17
C	9
D	6
E	10

36. $P(\text{A})$ 37. $P(\text{D})$ 38. $P(\text{E})$
39. If the spinner is equally likely to land on each section, what is the theoretical probability of landing on B?

Example 6 A coin is tossed 75 times, and it lands on tails 55 times. What is the experimental probability of the coin landing on heads?

The coin landed on heads 20 times.

$$P(\text{heads}) = \frac{\text{number of times heads occurs}}{\text{number of possible outcomes}}$$

$$= \frac{20}{75} \text{ or } \frac{4}{15}$$

So, the experimental probability of the coin landing on heads is $\frac{4}{15}$.

9-7 Independent and Dependent Events (pp. 398–401)

For Exercises 40–43, a bag contains 6 green, 8 white, and 2 blue counters. Two counters are randomly drawn. Find each probability if the first counter is replaced before the second counter is drawn.

40. $P(\text{green, blue})$ 41. $P(\text{2 white})$

Find each probability if the first counter is *not* replaced before the second counter is drawn.

42. $P(\text{2 green})$ 43. $P(\text{blue, white})$

Example 7 A box contains 12 solid, 14 striped, and 10 spotted marbles. Suppose you reach in and grab two marbles. Find the probability of choosing a striped marble, replacing it, and then choosing a spotted marble.

$P(\text{striped}) = \frac{14}{36}$

$P(\text{spotted, after replacing the striped}) = \frac{10}{36}$

$\frac{14}{36} \cdot \frac{10}{36} = \frac{140}{1{,}296}$ or $\frac{35}{324}$

So, the probability of choosing a striped marble, replacing it, and then choosing a spotted marble is $\frac{35}{324}$, or about 11%.

CHAPTER 9

Practice Test

Vocabulary and Concepts

1. **Explain** the purpose of a tree diagram.
2. **Define** *combination*.

Skills and Applications

A spinner with sections labeled 1–8 has an equal chance of landing on each number. Find each probability.

3. P(odd number)
4. P(number greater than 1)
5. P(1 or 7)

For each situation, make a tree diagram to show the sample space. Then give the total number of outcomes.

6. tossing a coin three times
7. choosing a letter from the word MATH and then a digit from the number 123

Use the Fundamental Counting Principle to find the total number of outcomes in each situation.

8. choosing a 3-digit security code
9. rolling four number cubes

Find the value of each expression.

10. 8!
11. $10 \cdot 9 \cdot 8 \cdot 7$
12. $5! \cdot 4!$

13. **PARADES** If there are 50 floats in a parade, how many ways can a first place and a second place trophy be awarded?

14. **SURVEY** Two hundred fifty teenagers were asked what type of pet they owned. The results of the survey are in the table. What is the experimental probability that a teenager owns a pet? Write as a percent.

Pet	Number of Teenage Pet Owners
fish	26
cat	65
dog	86
bird	20
other	38
no pet	15

15. **YOGURT** A brand of yogurt has 15 different flavors. In how many ways can you choose three flavors?

Standardized Test Practice

16. **GRID IN** A variety bag of popcorn contains 10 pieces of regular popcorn, 6 pieces of cheese popcorn, and 4 pieces of caramel popcorn. Katie pulls out one piece of popcorn and eats it. Then she pulls out a second piece and eats it. What is the probability that her first piece was caramel and her second piece was regular?

msmath2.net/chapter_test

CHAPTER 9

Standardized Test Practice

PART 1 Multiple Choice

Record your answers on the answer sheet provided by your teacher or on a sheet of paper.

1. The ornithologist said there is a 0.6754 chance of the bird returning next year to its nest in a tree in the Glowinski family's backyard. What is the value of the 5 in this number? (Prerequisite Skill, p. 555)

   A five tenths

   B five hundredths

   C five thousandths

   D five ten thousandths

2. The line plot below shows the weight in grams of a single serving of different brands of cookies. Between which weights is there a cluster of data? (Lesson 2-3)

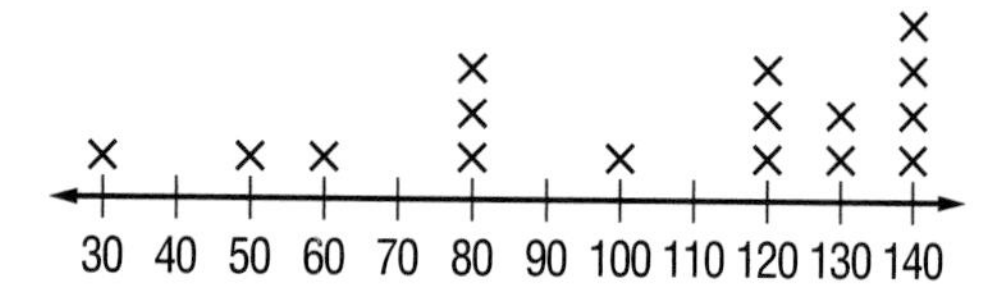

   F between 30 and 50

   G between 60 and 80

   H between 90 and 110

   I between 120 and 140

3. Park rangers measured the depth of a nearby river. It is 12 inches shallower than it was 4 years ago. Which formula best expresses the yearly change in the river? (Lesson 3-7)

   A $-12 \div 4$ B 12×4

   C -12×4 D $-12 \div -4$

4. Which of the following is the greatest common factor of 84 and 49? (Lesson 5-2)

   F 2 G 6

   H 7 I 12

5. If $792 was made in the sale of pens, what was the total amount of all sales to the nearest dollar? (Lesson 8-2)

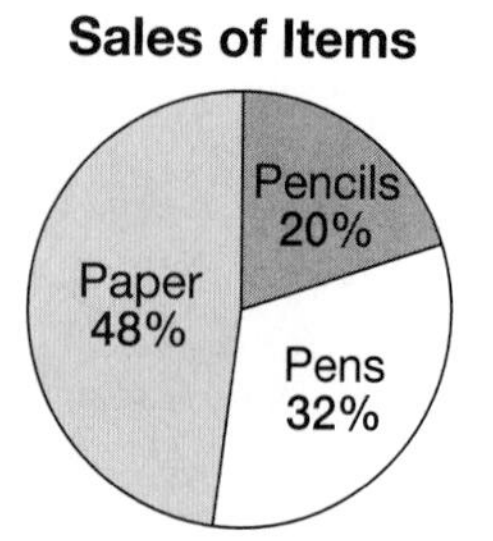

   A $2,534 B $2,475

   C $253 D $248

6. The tree diagram shows the number of ways a customer can order lunch in a diner. If all choices are equally likely, what is the probability that a customer will order meat, rice, and salad? (Lesson 9-2)

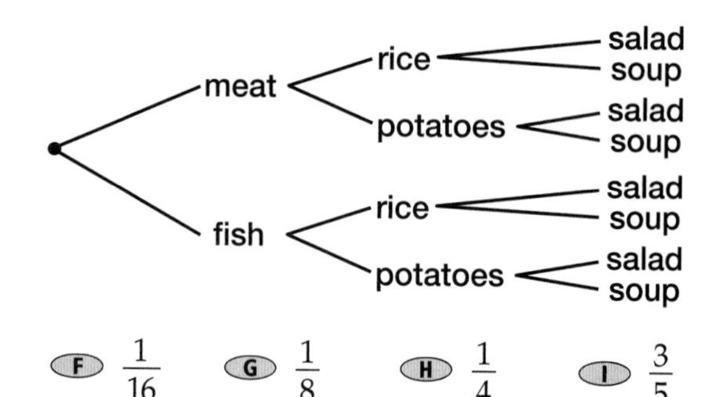

   F $\frac{1}{16}$ G $\frac{1}{8}$ H $\frac{1}{4}$ I $\frac{3}{5}$

7. Fernando has to wash carrots, celery, onions, and potatoes for dinner. Assuming that he washes each vegetable only once, in how many different orders can Fernando wash all of the vegetables? (Lesson 9-4)

   A 4 B 16 C 20 D 24

8. What is the theoretical probability of choosing a vowel from the word MATHEMATICS? (Lesson 9-6)

   F $\frac{4}{11}$ G $\frac{4}{9}$ H $\frac{4}{7}$ I $\frac{7}{11}$

TEST-TAKING TIP

Question 7 Find your own answer before looking at the answer choices. Doing so keeps you from being tempted by wrong answer choices that look correct, but are still wrong.

PART 2 Short Response/Grid In

Record your answers on the answer sheet provided by your teacher or on a sheet of paper.

9. Ebony shot −4, −2, 1, and −3 in four rounds of golf. The final score is the sum of the scores. What was her final score? (Lesson 3-4)

10. How much greater is the area of a professional basketball court than a college basketball court? (Lesson 6-8)

Basketball Court	Length (ft)	Width (ft)
college	84	50
professional	94	50

11. What is the percent of increase in two-vehicle households from 1990 to 2000? Write as a percent to the nearest tenth. (Lesson 8-4)

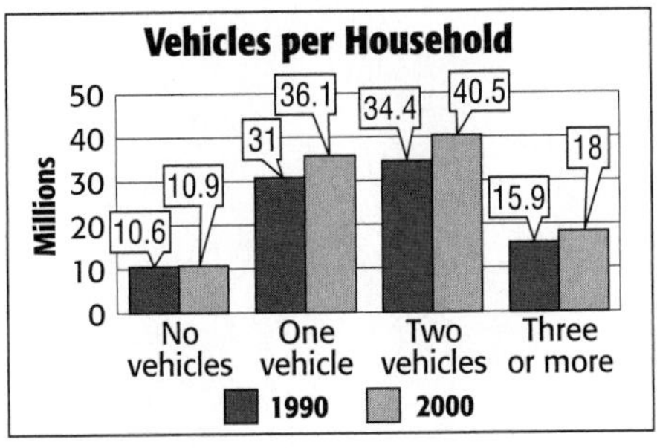

Source: Census Bureau

12. If two number cubes are rolled, what is the probability that their sum is 5 or 9? (Lesson 9-1)

13. Suppose you want to spend some time on the water. In how many ways can you choose a type of watercraft, the length of time, and the river? (Lesson 9-3)

Type of Watercraft	Time (hours)	River
paddleboat	1	Hamlin
canoe	2	Dustin
kayak	3	
water tricycle		

14. In how many ways can the Geiger family sit in a car if the two parents are in the front seat and the three children are in the back seat? (Lesson 9-4)

15. Twelve points are drawn on a circle. How many different line segments can be drawn between any two points? (Lesson 9-5)

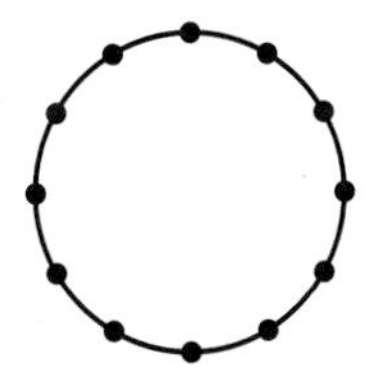

For Questions 16 and 17, use the information below and at the right.

A survey asked 400 students what type of materials they prefer to read. The table shows the results. (Lesson 9-6)

Type of Reading Materials	Number of Students
book	225
newspaper	35
magazine	140

16. What is the probability that a student prefers to read a magazine?

17. Suppose 1,000 students attend Franklin Middle School. About how many would prefer to read a book?

PART 3 Extended Response

Record your answers on a sheet of paper. Show your work.

18. Paula has five hobbies: jogging, painting, stamp collecting, quilting, and listening to music. She can enjoy her hobbies on either Friday night or Saturday night.
    a. Make a tree diagram that shows all of her possible outcomes. (Lesson 9-2)
    b. What is the probability that Paula will paint or jog on Friday night? (Lesson 9-2)
    c. Paula will not do the same hobby on both nights. What is the probability that Paula will quilt on Friday night and listen to music on Saturday night? (Lesson 9-7)

UNIT 5

Geometry and Measurement

In this unit, you will explore line and angle relationships, two- and three-dimensional figures, and the areas and volumes of these figures.

IT'S ALL GREEK TO ME

Are you ready for some time travel? You've been selected to join us on an adventure through the ages, back to the time of the ancient Greeks. Along the way, you'll research the life and mathematical discoveries of Pythagoras. You'll also explore many three-dimensional solids known to the ancient Greeks and construct one of your own. Our time machine will be leaving soon, so pack your geometry tool kit and prepare to meet a geometry giant!

Log on to msmath2.net/webquest to begin your WebQuest.

CHAPTER 10

Geometry

"What do video games have to do with math?"

The first video game programmers used quadrilaterals like rectangles, squares, and trapezoids in designing video games. Now they use these basic figures as a starting point to make the games more elaborate with real-life characters, breath-taking scenery, and silly animations.

You will solve problems about quadrilaterals in Lesson 10-5.

GETTING STARTED

Take this quiz to see whether you are ready to begin Chapter 10. Refer to the lesson or page number in parentheses if you need more review.

Vocabulary Review

Complete each sentence.

1. A comparison of two numbers by division is called a _?_. (Lesson 7-1)
2. The _?_ is written as part = percent · base. (Lesson 8-2)

Prerequisite Skills

Multiply or divide. Round to the nearest hundredth if necessary. (Pages 560 and 562)

3. $360 \cdot 0.85$
4. $48 \div 191$
5. $24 \div 156$
6. $0.37 \cdot 360$
7. $33 \div 307$
8. $0.69 \cdot 360$

Solve each equation. (Lessons 4-2 and 4-3)

9. $b + 36 = 89$
10. $74 = 22 + s$
11. $15 + r = 146$
12. $m + 78 = 93$
13. $153 = d + 61$
14. $6x = 360$
15. $180 = 2f$
16. $120 = 3c$
17. $5n = 270$
18. $15z = 90$

Solve each proportion. (Lesson 7-3)

19. $\frac{4}{a} = \frac{3}{9}$
20. $\frac{7}{16} = \frac{h}{32}$
21. $\frac{5}{8} = \frac{15}{y}$
22. $\frac{t}{42} = \frac{6}{7}$
23. $\frac{s}{12} = \frac{2}{3}$
24. $\frac{24}{p} = \frac{2}{3}$

Geometry Make this Foldable to help you organize your notes. Begin with a sheet of 11″ × 17″ paper and six index cards.

Fold
Fold lengthwise about 3″ from the bottom.

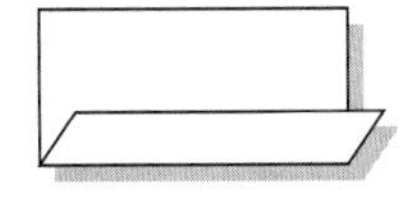

STEP 2 Fold Again
Fold the paper in thirds.

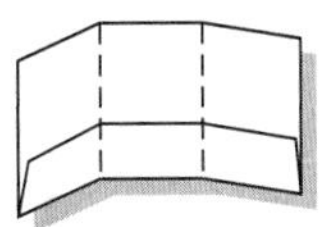

STEP 3 Open and Staple
Staple the edges on either side to form three pockets.

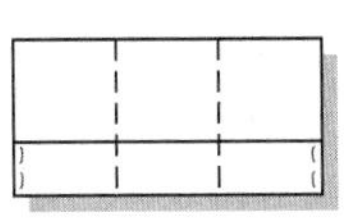

STEP 4 Label
Label the pockets as shown. Place two index cards in each pocket.

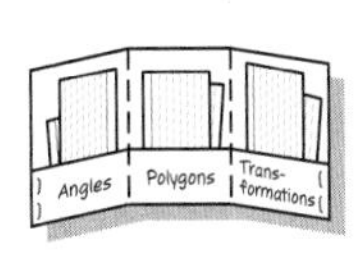

Reading and Writing As you read and study the chapter, write the properties and characteristics for angles, polygons, and transformations on the index cards.

10-1a HANDS-ON LAB

A Preview of Lesson 10-1

What You'll LEARN

Measure angles.

Materials

- protractor

Measuring Angles

To measure lines, you use a ruler. To measure angles, you use a *protractor*. Angles are measured in units called **degrees**. In this lab, you will learn how to measure angles using a protractor.

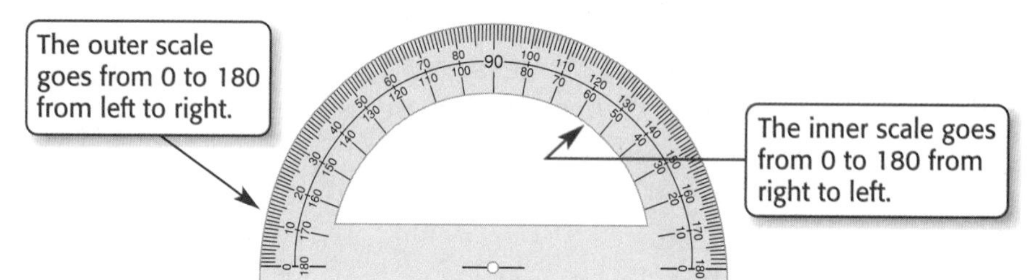

ACTIVITY *Work with a partner.*

Find the measure of the angle below.

STEP 1 Place the protractor on the angle so that the center is on the vertex of the angle and one side goes through 0° on the protractor.

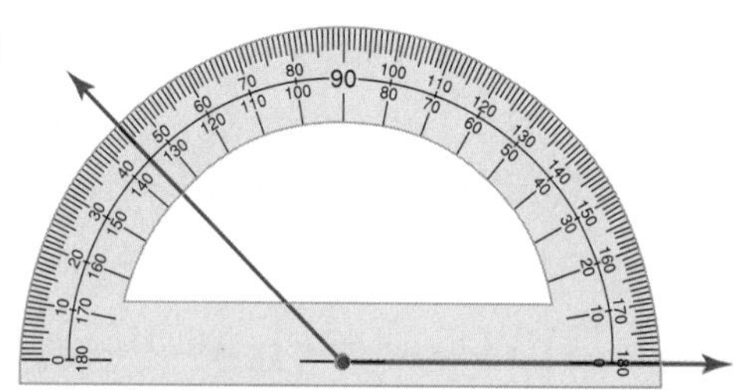

STEP 2 In this case, 0° is on the inner scale. So, follow the inner scale to the point where the other side of the angle meets the protractor. The inner number is the angle's degree measure.

So, the measure of this angle is 135°.

Your Turn **Find the measure of each angle.**

a.

b.

c.

Writing Math

1. **Explain** why there are two scales on the protractor.
2. Can you place either side of an angle through 0° and get the same angle measure? **Explain** your reasoning.
3. What angle has the same measure on both the inner and outer scales?
4. **Explain** how the two scales on the protractor are related.

10-1 Angles

WHEN am I ever going to use this?

CLOCKS The hour and minute hands of a clock form angles of different sizes.

3:10
less than 90°

3:00
90°

3:40
greater than 90°

1. Name other times in which the hands of a clock form an angle less than 90°, equal to 90°, and greater than 90°.
2. How many degrees is the angle that is formed by clock hands at 6:00?

What You'll LEARN

Classify and draw angles.

NEW Vocabulary

angle
degrees
vertex
acute angle
right angle
obtuse angle
straight angle

Link to READING

Everyday Meaning of Acute: characterized by sharpness, as in acute pain

An **angle** is made up of two rays with a common endpoint and is measured in units called **degrees**. If a circle were divided into 360 equal-sized parts, each part would have an angle measure of 1 degree.

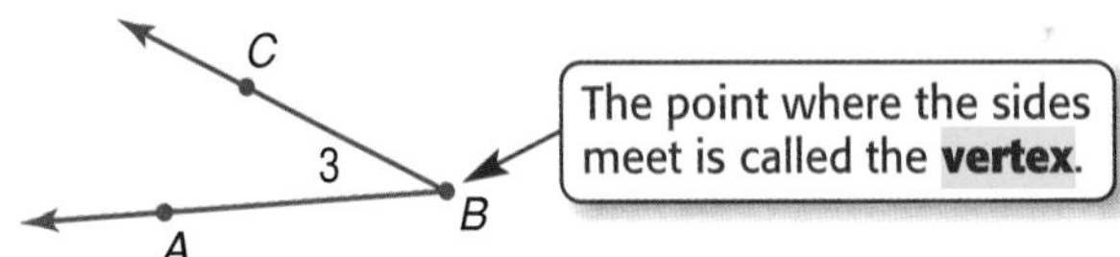

You can use the symbol for angle, ∠, to name an angle.

Use the vertex as the middle letter and a point from each side.	$\angle ABC$ or $\angle CBA$
Use the vertex only.	$\angle B$
Use a number.	$\angle 3$

Angles are classified according to their measure.

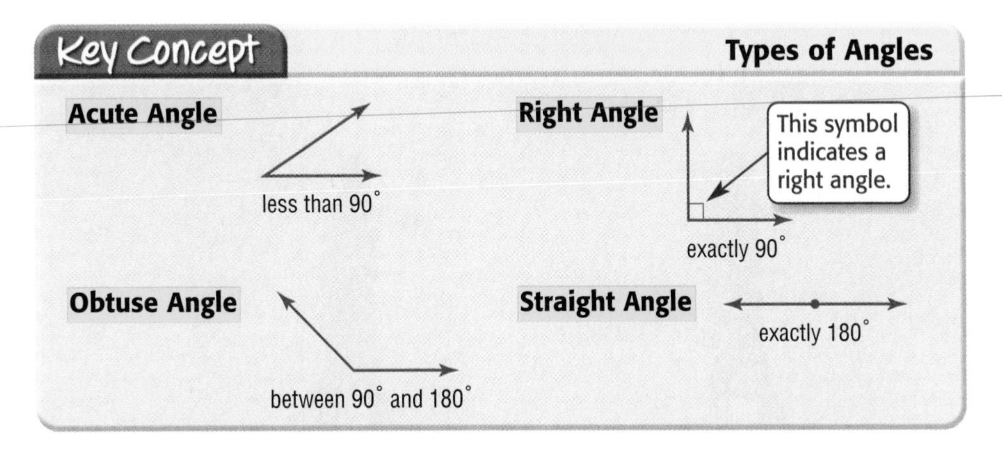

EXAMPLES Classify Angles

Classify each angle as *acute, obtuse, right,* or *straight.*

1 The angle is less than 90°, so it is an acute angle.

2 The angle is greater than 90°, so it is an obtuse angle.

EXAMPLE Draw an Angle

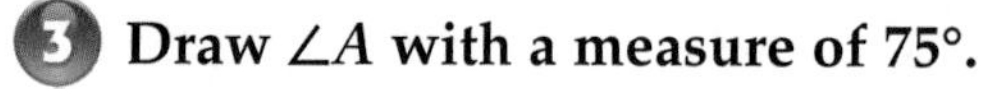

3 **Draw ∠*A* with a measure of 75°.**

- Draw a ray with endpoint *A*.
- Place the center point of a protractor on *A*. Align the ray with 0°.
- Use the scale that begins with 0°. Locate the mark labeled 75. Then draw the other side of the angle.

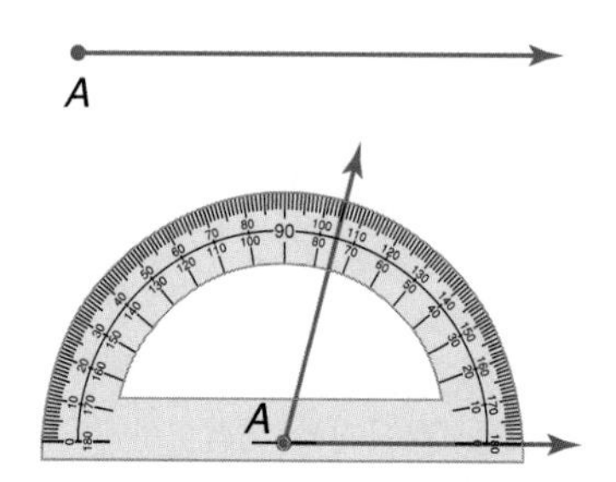

Skill and Concept Check

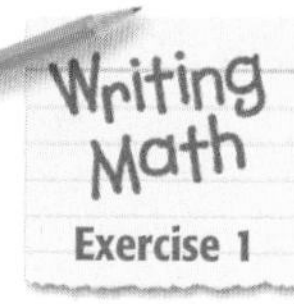

1. **Describe** a right angle.
2. **OPEN ENDED** Draw and label acute angle *XYZ*. Then find its measure.

GUIDED PRACTICE

Classify each angle as *acute, obtuse, right,* or *straight.*

3. 

4. 

Draw an angle having each measurement. Then classify the angle as *acute, obtuse, right,* or *straight.*

5. 90°
6. 170°
7. 11°

8. **BICYCLE SAFETY** A cyclist should be familiar with arm and hand signals indicating his or her movements. Describe the angles formed by the arm of the cyclist below.

Practice and Applications

HOMEWORK HELP	
For Exercises	See Examples
9–11, 18, 21	1, 2
12–17	3

Extra Practice
See pages 587, 605.

Classify each angle as *acute, obtuse, right,* or *straight*.

9. 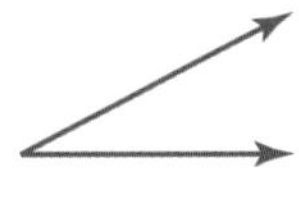

10. 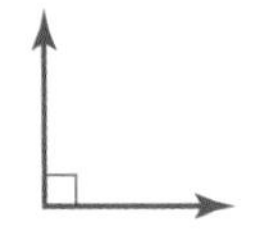

11. 

Draw an angle having each measurement. Then classify each angle as *acute, obtuse, right,* or *straight*.

12. 56°
13. 147°
14. 180°
15. 99°
16. 8°
17. 90°

For Exercises 18–20, use the figure at the right.

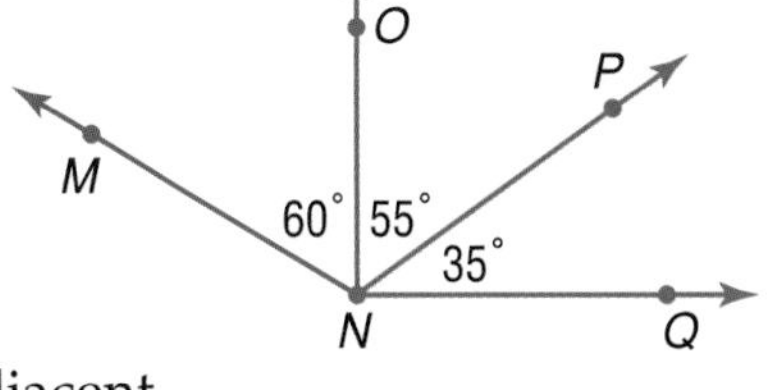

18. Name the angles that are obtuse.
19. The symbol $m\angle MNQ$ means the *measure of angle MNQ*. Find $m\angle MNQ$.
20. Angles are said to be *adjacent* if they have a common vertex and a common side between them. So, $\angle MNO$ is adjacent to $\angle ONP$. Name another angle adjacent to $\angle ONP$.

21. **SKI JUMPING** When skiers jump off a hill, they want to make the angle between their bodies and the front of their skis as small as possible. Describe where the ski jumper's legs and skis form acute and obtuse angles in the photograph.
22. **EARTH SCIENCE** Earth rotates 360° degrees in one day. Through how many degrees does it rotate in one hour?
23. **CRITICAL THINKING** How many times do the hands of a clock make a right angle in a 24-hour time period?

Standardized Test Practice and Mixed Review

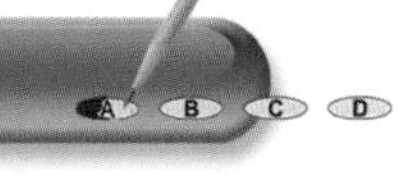

24. **SHORT RESPONSE** Name the obtuse angle in the figure at the right.

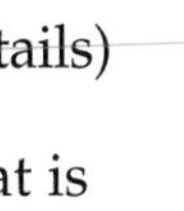

25. **MULTIPLE CHOICE** Which is the measure of an acute angle?
    (A) 87° (B) 95° (C) 120° (D) 180°

PROBABILITY A number cube is rolled, and a coin is tossed. Find each probability. (Lesson 9-7)

26. P(5 and heads)
27. P(odd and tails)
28. P(7 and heads or tails)
29. Anica spins a spinner fifty times, and it lands on 3 fifteen times. What is the experimental probability of *not* landing on 3? (Lesson 9-6)

GETTING READY FOR THE NEXT LESSON

PREREQUISITE SKILL Multiply or divide. Round to the nearest hundredth if necessary. (Pages 560 and 562)

30. $0.62 \cdot 360$
31. $360 \cdot 0.25$
32. $17 \div 146$
33. $63 \div 199$

10-1b Hands-On Lab

A Follow-Up of Lesson 10-1

Constructing and Bisecting Angles

What You'll LEARN

Construct and bisect angles.

Two angles that have the same measure are **congruent angles**. To **bisect** an angle means to divide it into two congruent angles. In this lab, you will learn how to construct congruent angles and bisect angles.

Materials

- straightedge
- compass

STUDY TIP

Symbols $\overrightarrow{LK}$ is read *ray LK*. A ray is a path that extends infinitely from one point in a certain direction.

ACTIVITY *Work with a partner.*

1 **Construct an angle congruent to ∠*ABC*.**

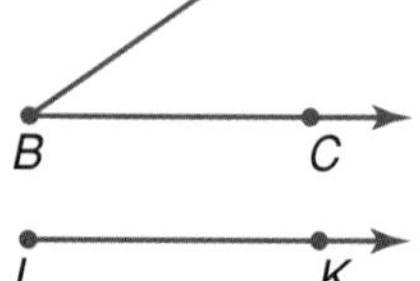

STEP 1 Use a straightedge to draw $\overrightarrow{LK}$.

STEP 2 With the compass at point *B*, draw an arc that intersects both sides of ∠*ABC*. Label the two points of intersection as *X* and *Y*.

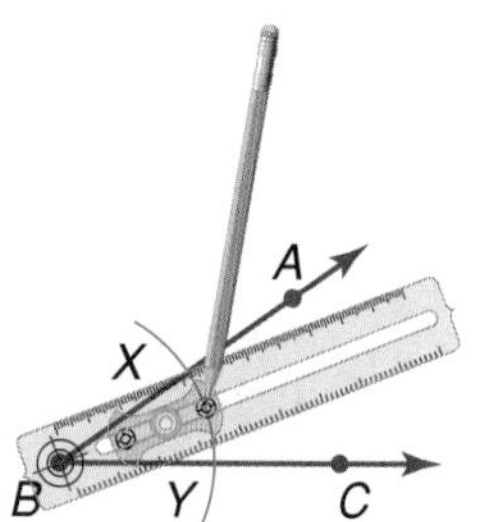

STEP 3 With the same setting on your compass, place your compass at point *L*. Draw another arc. Label the intersection *R*.

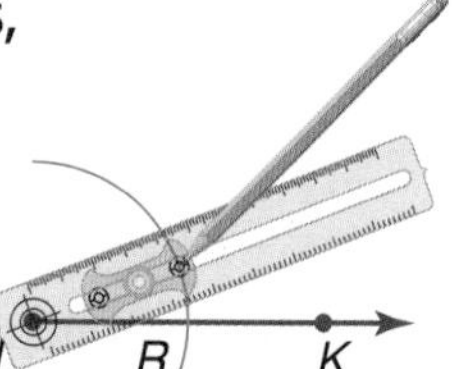

STEP 4 Open your compass to the same width as the distance between points *X* and *Y*. Then place the compass at point *R*. Draw an arc that intersects the arc you drew in Step 3. Label this point of intersection *S*.

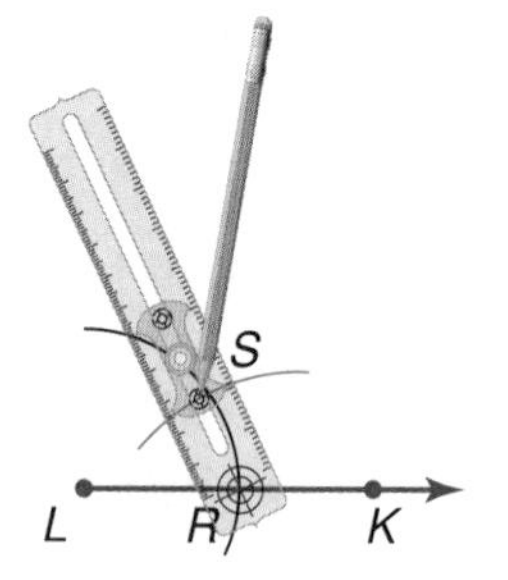

STEP 5 Draw $\overrightarrow{LM}$ through point *S*. Angle *MLK* is congruent to ∠*ABC*.

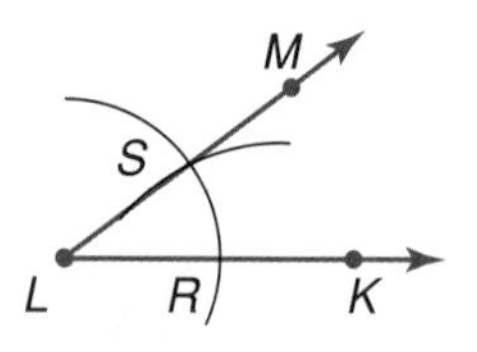

Your Turn

a. Draw an obtuse angle. Then construct an angle congruent to it.

ACTIVITY *Work with a partner.*

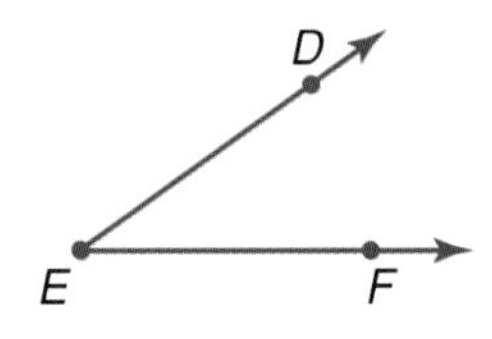

2 **Bisect ∠*DEF*.**

STEP 1 Place the compass at point *E* and draw an arc that intersects both sides of ∠*DEF*. Mark the two points of intersection as *H* and *I*.

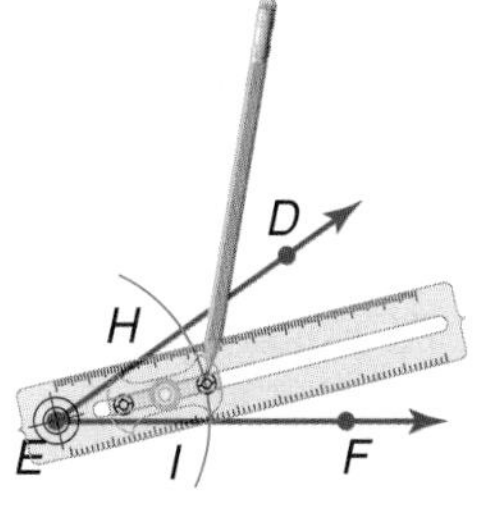

STEP 2 Place your compass at point *H*. Draw an arc inside ∠*DEF*. Using the same setting on your compass, place your compass at point *I*. Draw another arc inside ∠*DEF* intersecting the first arc. Label the intersection point *G*.

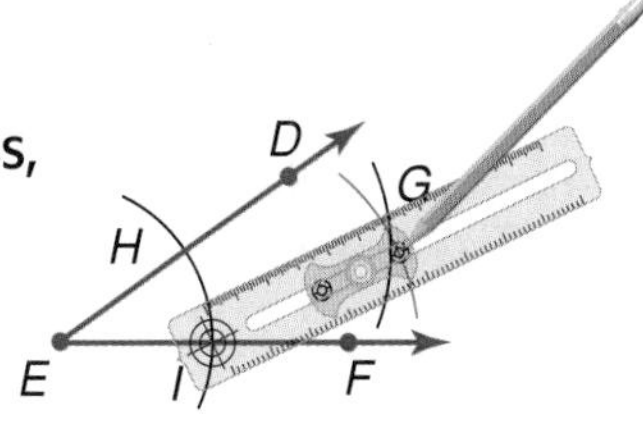

STEP 3 Draw $\overrightarrow{EG}$. $\overrightarrow{EG}$ is the bisector of ∠*DEF*.

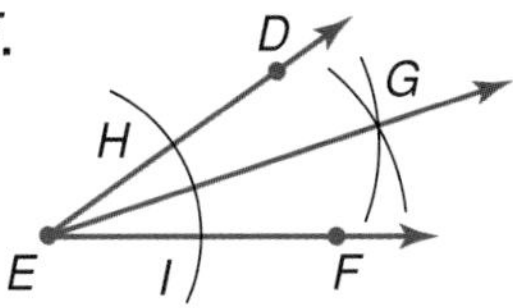

Your Turn

b. Draw a right angle. Then bisect it.

Writing Math

Work with a partner.

1. **Explain** how you could verify that ∠*ABC* and ∠*MLK* in Activity 1 are congruent.
2. **Explain** how you could verify that ∠*DEF* in Activity 2 is bisected.
3. A straight angle has a measure of 180°. What kind of angle is formed if you bisect a straight angle?
4. **Perpendicular lines** are lines that meet to form right angles. In the figure at the right, $\overline{PS}$ is perpendicular to $\overline{QR}$. Use a compass and a straightedge to construct perpendicular lines. (*Hint*: Use the procedure for bisecting angles.)

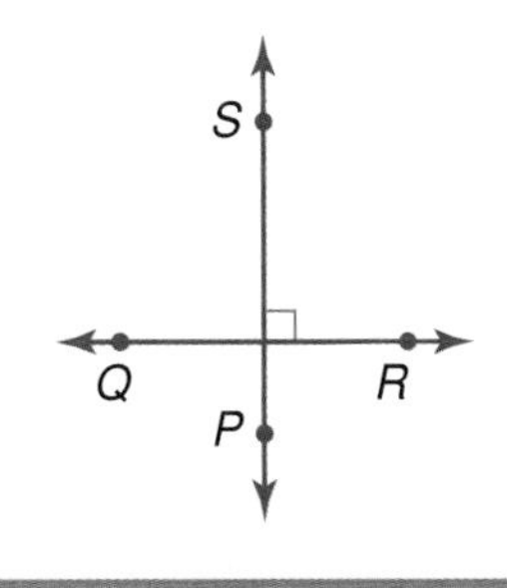

10-2 Making Circle Graphs

WHEN am I ever going to use this?

COLORS In a recent survey, people ages 13–20 were asked to choose their favorite shade of blue. The results are shown in the table.

1. Explain how you know that each person surveyed chose only one shade of blue.
2. If 500 people took part in the survey, how many preferred aquamarine?

Favorite Shades of Blue for People Ages 13–20	
Shade	**Percent**
navy	35%
sky/light blue	30%
aquamarine	17%
other	18%

Source: *American Demographics*

What You'll LEARN

Construct and interpret circle graphs.

NEW Vocabulary

circle graph

REVIEW Vocabulary

ratio: a comparison of two numbers by division **(Lesson 7-1)**

percent equation part = percent · base **(Lesson 8-2)**

A graph used to compare parts of a whole is called a **circle graph**. In a circle graph, the percents add up to 100.

EXAMPLE Construct a Circle Graph

1 COLORS Make a circle graph of the data in the table above.

- Find the degrees for each part. Round to the nearest whole degree.

  35% of 360° = 0.35 · 360° or 126°

  30% of 360° = 0.30 · 360° or 108°

  17% of 360° = 0.17 · 360° or about 61°

  18% of 360° = 0.18 · 360° or about 65°

- Use a compass to draw a circle with a radius as shown. Then use a protractor to draw the first angle, in this case 126°. Repeat this step for each section or *sector*.

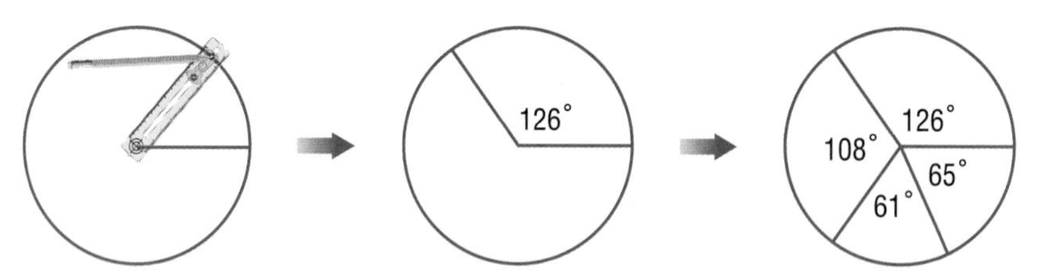

- Label each section of the graph with the category and percent. Give the graph a title.

Check To draw an accurate circle graph, make sure the sum of the angle measures equals 360°.

Favorite Shades of Blue for People Ages 13–20

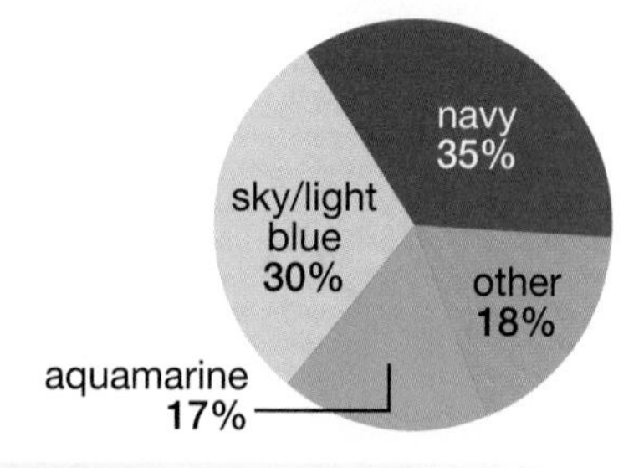

REAL-LIFE MATH

OLYMPICS The United States has won 2,117 Summer Olympic medals through 2000, the most by any country.

Source: infoplease.com

When constructing a circle graph, it is sometimes necessary to first convert the data to percents or decimals, and then to degrees.

EXAMPLE Construct a Circle Graph

2 **OLYMPICS The table shows the number of each type of medal won by the United States during the Summer Olympics from 1896 to 2000. Make a circle graph of the data.**

U.S. Summer Olympic Medals

Type	Number
gold	872
silver	658
bronze	587

Source: infoplease.com

- Find the total number of medals: 872 + 658 + 587 or 2,117.
- Find the ratio that compares each number with the total. Write the ratio as a decimal rounded to the nearest hundredth.

  gold: $\frac{872}{2{,}117} \approx 0.41$ silver: $\frac{658}{2{,}117} \approx 0.31$ bronze: $\frac{587}{2{,}117} \approx 0.28$

- Find the number of degrees for each section of the graph.

  gold: $0.41 \cdot 360° \approx 148°$

  silver: $0.31 \cdot 360° \approx 112°$

  bronze: $0.28 \cdot 360° \approx 101°$

  Because of rounding, the sum of the degrees is 361°.

- Draw the circle graph.

U.S. Summer Olympic Medals

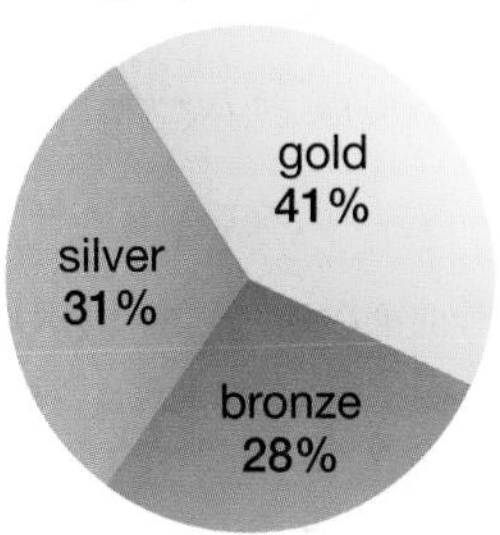

Check After drawing the first two sections, you can measure the last section of a circle graph to verify that the angles have the correct measures.

EXAMPLE Interpret a Circle Graph

3 **MONEY The circle graph shows the percent of Americans who favor, oppose, or don't know how they feel about a common currency for North America. Use the graph to describe the opinion of most Americans.**

Do Americans Favor Common North American Currency?

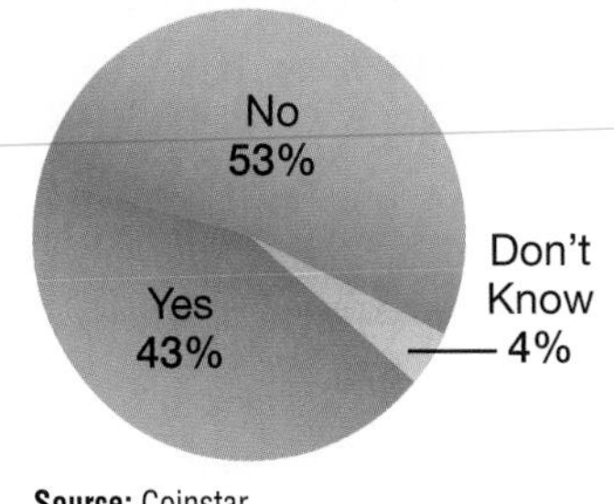

Source: Coinstar

The greatest percent of the circle graph is the section representing the "No" response.

So, most Americans do not favor a common North American currency.

Skill and Concept Check

Writing Math
Exercises 1 & 2

1. **OPEN ENDED** Draw a circle graph that has three parts. Label the parts and describe what they represent.

2. **DATA SENSE** The table shows the percent of people that use the sweeteners listed. Can the data be represented in a circle graph? Explain.

Type of Sweetener	Percent
sugar	74%
honey, molasses, syrup	46%
low calorie sweeteners	36%

Source: Yankelovich Partners

GUIDED PRACTICE

Make a circle graph of the data in each table.

3.

Favorite Shades of Blue for People Ages 21–34	
Shade	**Percent**
navy	48%
sky/light blue	23%
aquamarine	12%
other	18%

Source: *American Demographics*

4.

Speed Limit (mph)	Number of States
55	1
65	20
70	18
75	11

Source: *The World Almanac*

5. **EDUCATION** The circle graph shows the percent of students by grade level in U.S. schools. In which grades are most students?

6. Refer to the circle graph in Example 1 and the graph you drew in Exercise 3. What can you conclude about the favorite shades of blue for people in different age groups?

Grade Level of U.S. Students

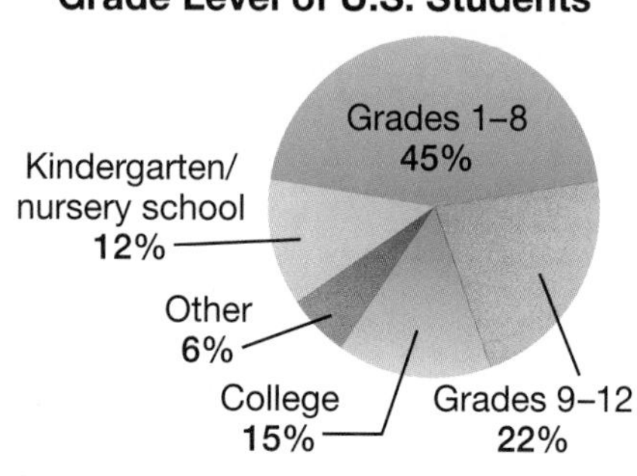

Source: U.S. Census Bureau

Practice and Applications

HOMEWORK HELP	
For Exercises	See Examples
7–8, 12–13	1
10–11	2
9, 14	3

Extra Practice
See pages 587, 605.

Make a circle graph of the data in each table.

7.

New York City Commuters	
Transportation	Percent
driving alone	24%
carpool	9%
public transit	53%
other	14%

Source: *Time Almanac*

8.

Los Angeles Commuters	
Transportation	Percent
driving alone	65%
carpool	15%
public transit	11%
other	9%

Source: *Time Almanac*

9. Compare and contrast the data from the circle graphs you drew in Exercises 7 and 8.

Make a circle graph of the data in each table.

10.

Endangered Species in U.S.	Number of Species
mammals	63
birds	78
reptiles	14
amphibians	10

Source: U.S. Fish and Wildlife Service

11.

U.S. Regions	Population (millions)
Northeast	54
Midwest	64
South	100
West	63

Source: *Time Almanac*

POLITICS For Exercises 12–14, refer to the graphic at the right.

12. How many 13- to 15-year olds participated in the survey?
13. Calculate the angle measure for each section.
14. Measure the angles for each section on the graph. Explain any differences.

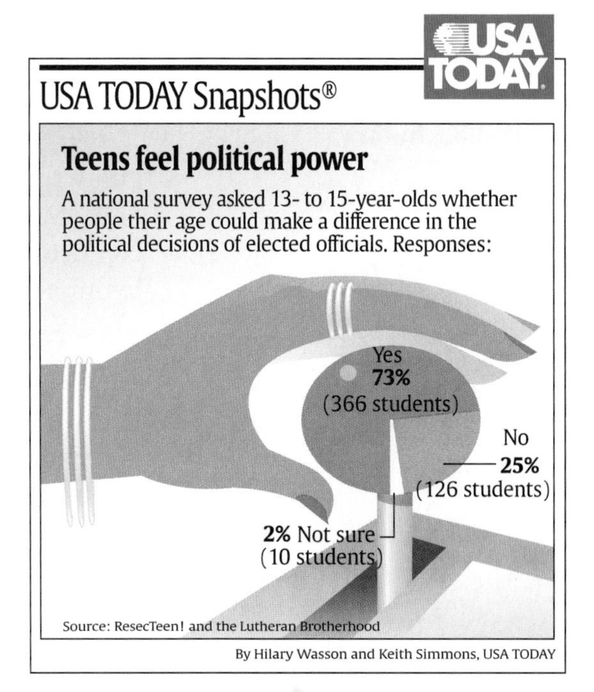

15. **RESEARCH** Use the Internet or another source to find data that add up to 100%. Make a circle graph of the data.
16. **CRITICAL THINKING** Line graphs are usually best for data that show change over time. When might it be more appropriate to display data in a circle graph? Give an example.

Standardized Test Practice and Mixed Review

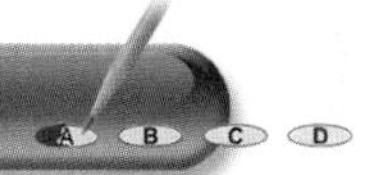

17. **MULTIPLE CHOICE** On a circle graph of the data in the table, which section would have an angle measure of about 38°?

   (A) Lake Erie (B) Lake Huron

   (C) Lake Michigan (D) Lake Ontario

Lake	Area (sq mi)
Erie	9,930
Huron	23,010
Michigan	22,400
Ontario	7,520
Superior	31,820

Source: www.infoplease.com

18. **SHORT RESPONSE** Find the angle measure of the Lake Michigan section in the circle graph referred to in Exercise 17. Round to the nearest whole degree.

Classify each angle as *acute, obtuse, right,* or *straight*. (Lesson 10-1)

19. 65° 20. 102° 21. 90°

22. **PROBABILITY** Corey has an after-school activity 15% of the time. Li-Cheng has one 20% of the time. Find the probability that they will both have an after-school activity on the same day. (Lesson 9-7)

GETTING READY FOR THE NEXT LESSON

PREREQUISITE SKILL Solve each equation. (Lesson 4-2)

23. $a + 18 = 90$ 24. $44 + x = 90$ 25. $180 = 39 + n$ 26. $123 + t = 180$

10-3 Angle Relationships

HANDS-ON Mini Lab

Materials

- protractor

What You'll LEARN

Identify and apply angle relationships.

NEW Vocabulary

vertical angles
congruent angles
supplementary angles
complementary angles

MATH Symbols

$\cong$ is congruent to

Work with a partner.

Draw two intersecting lines and label the angles as shown. Then measure each angle with your protractor and record the measurements.

1. Which angles have the same measure?
2. Draw two other pairs of intersecting lines. Measure their angles. **Make a conjecture** involving four angles created by intersecting lines.
3. What is the relationship between the measures of $\angle 1$ and $\angle 2$? $\angle 3$ and $\angle 4$?
4. Do other pairs of angles share the same relationship? Explain.

When two lines intersect, they form two pairs of opposite angles called **vertical angles**. In the Mini Lab, you found that vertical angles have the same measure. Angles with the same measure are **congruent angles**.

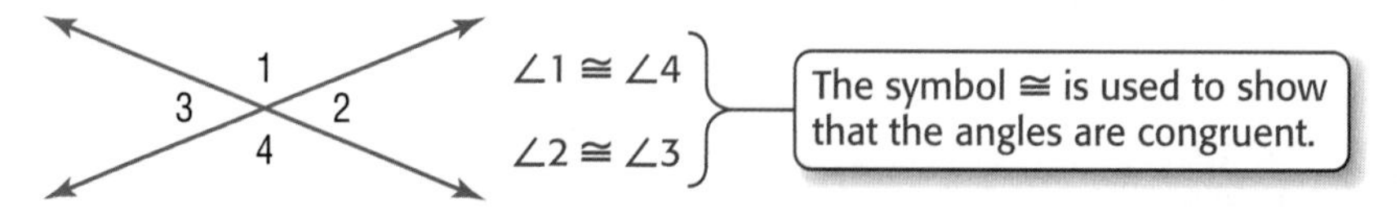

Pairs of angles can also have other relationships. In the Mini Lab, you found pairs of angles whose sum is 180°. Two angles are **supplementary** if the sum of their measures is 180°.

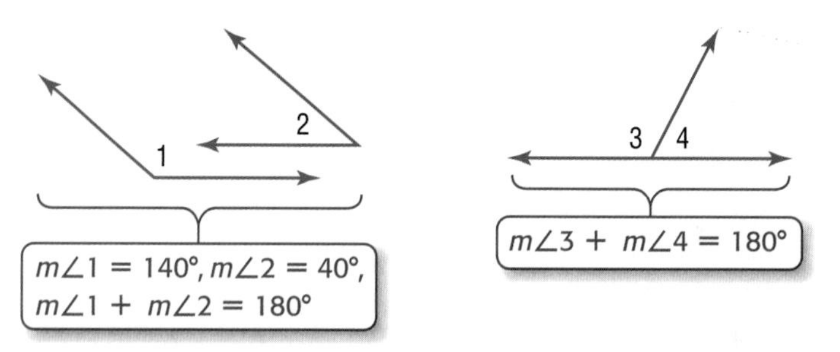

Two angles are **complementary** if the sum of their measures is 90°.

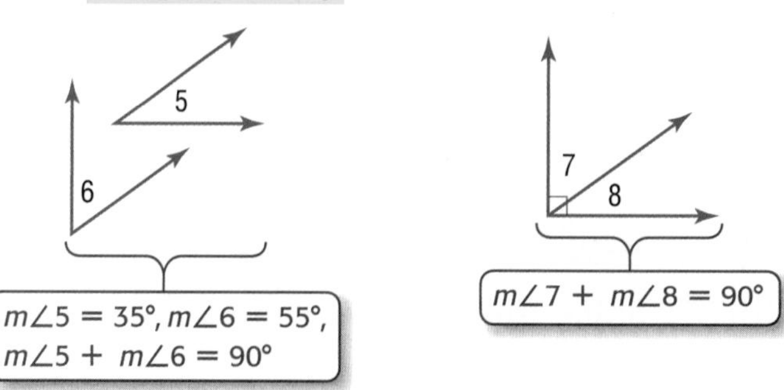

EXAMPLES Classify Angles

Classify each pair of angles as *complementary*, *supplementary*, or *neither*.

1

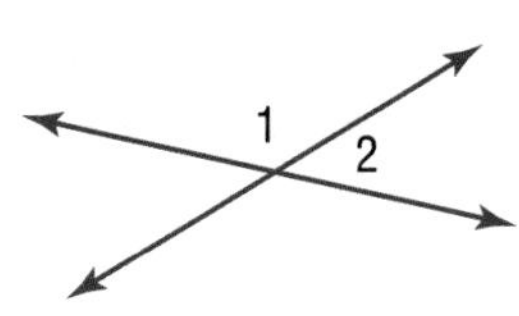

$\angle 1$ and $\angle 2$ form a straight line. So, the angles are supplementary.

2

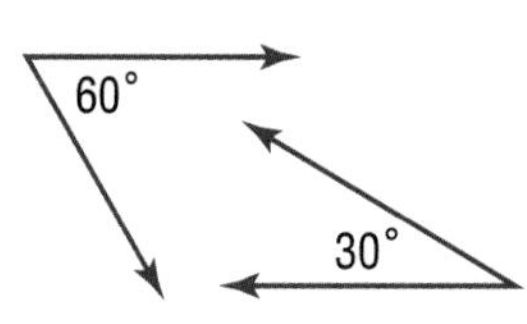

$60° + 30° = 90°$
The angles are complementary.

You can use angle relationships to find missing measures.

EXAMPLE Find a Missing Angle Measure

3 **Angles *ABC* and *CBD* are complementary. If $m\angle ABC = 28°$, find $m\angle CBD$.**

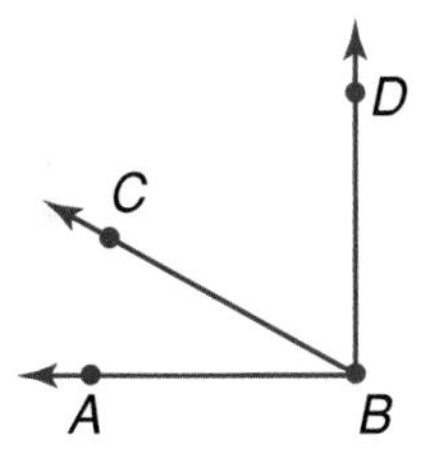

Since $\angle ABC$ and $\angle CBD$ are complementary, $m\angle ABC + m\angle CBD = 90°$.

$m\angle ABC + m\angle CBD = 90$		Write the equation.
$28 + m\angle CBD = 90$		Replace $m\angle ABC$ with 28.
$-28 \qquad\qquad -28$		Subtract 28 from each side.
$m\angle CBD = 62$		$90 - 28 = 62$

The measure of $\angle CBD$ is 62°.

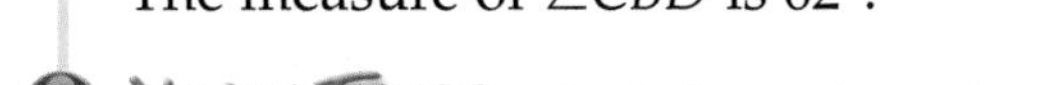

Your Turn **Find the value of *x* in each figure.**

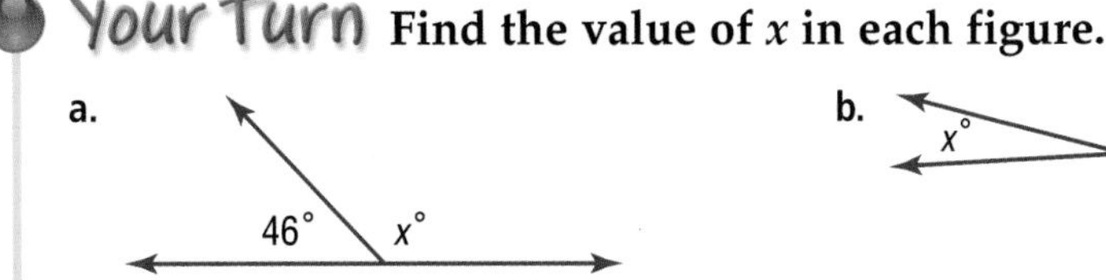

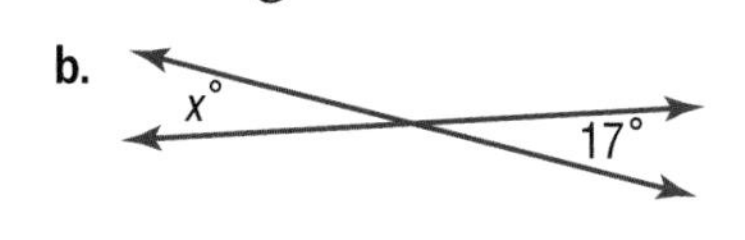

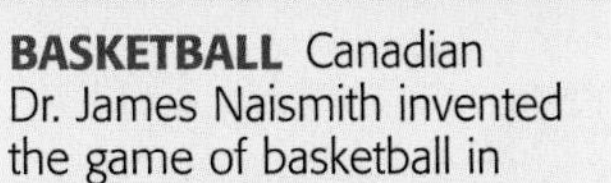

REAL-LIFE MATH

BASKETBALL Canadian Dr. James Naismith invented the game of basketball in 1891 using a ball the size of a soccer ball and two peach baskets.

Source: www.allsands.com

EXAMPLE Use Angles to Solve A Problem

4 **BASKETBALL Erin wants to make a bounce pass to Mackenzie. Find the value of *x* so that Erin's pass hits Mackenzie in the hands.**

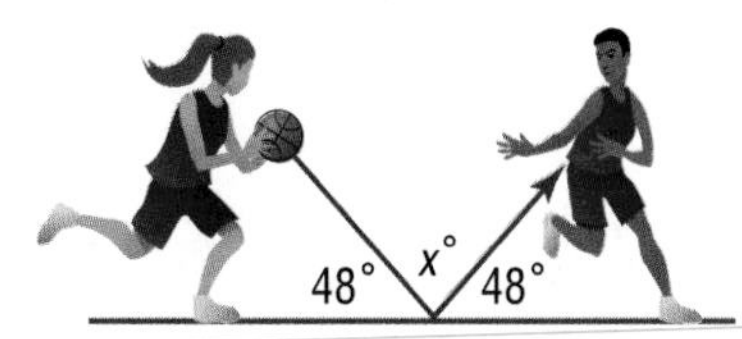

Since the sum of the three angles is 180°, $48° + x° + 48° = 180°$.

$48 + x + 48 = 180$	Write the equation.
$x + 96 = 180$	Simplify.
$-96 \qquad -96$	Subtract 96 from each side.
$x = 84$	$180 - 96 = 84$

So, the angle must be 84° for the ball to hit Mackenzie in the hands.

Skill and Concept Check

Writing Math
Exercise 2

1. **OPEN ENDED** Draw two angles that are complementary.

2. **NUMBER SENSE** Can a pair of angles be vertical and supplementary? Give an example or nonexample, with angle measurements, to support your answer.

GUIDED PRACTICE

Classify each pair of angles as *complementary, supplementary,* or *neither.*

3.
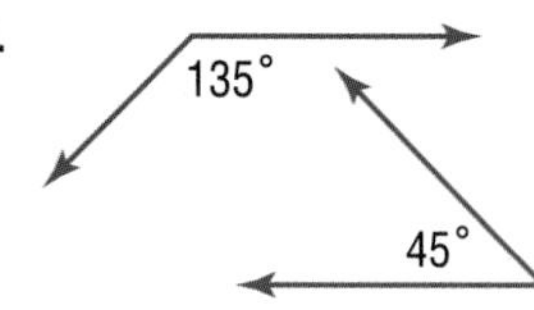

4.
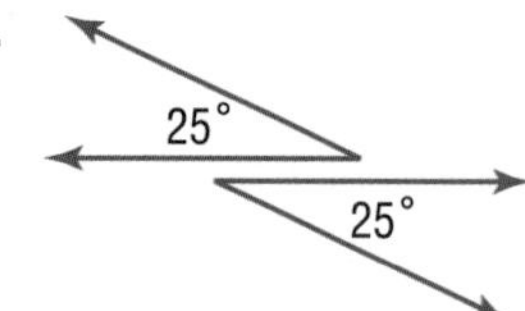

5.
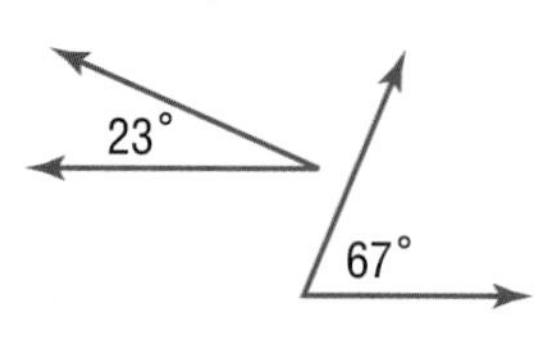

Find the value of x in each figure.

6.
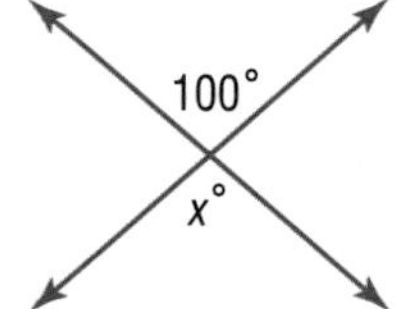

7.
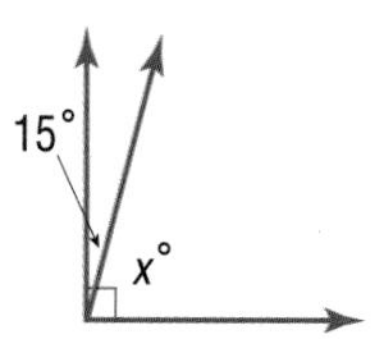

8.
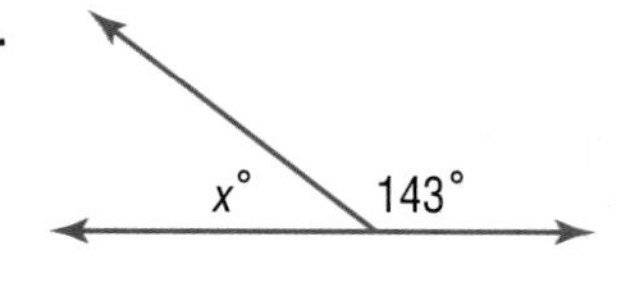

9. Suppose $\angle 1$ and $\angle 2$ are supplementary. If $m\angle 1 = 84°$, find $m\angle 2$.

Practice and Applications

HOMEWORK HELP	
For Exercises	See Examples
10–15, 22–23	1, 2
16–21	3, 4

Extra Practice
See pages 587, 605.

Classify each pair of angles as *complementary, supplementary,* or *neither.*

10.
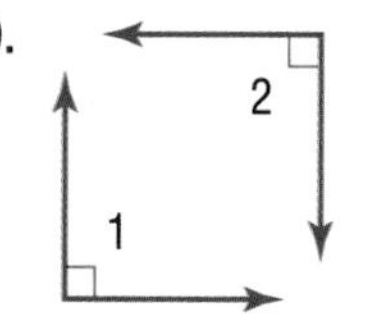

11.
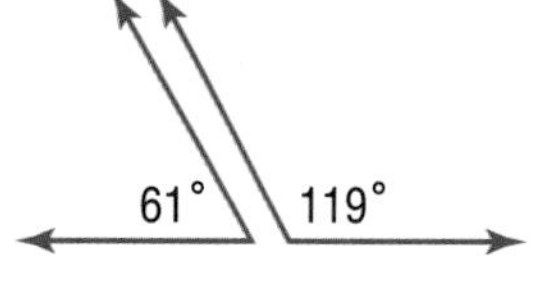

12.
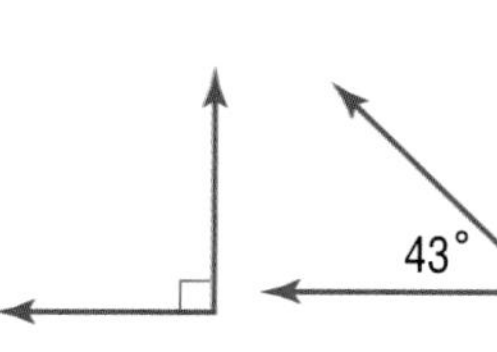

13. 1 2

14.
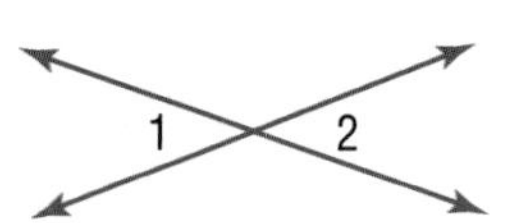

15.
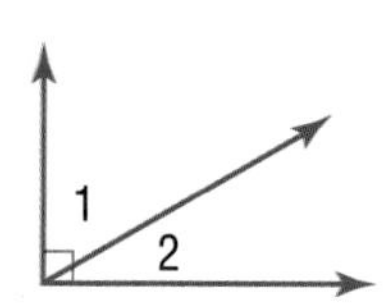

Find the value of x in each figure.

16.
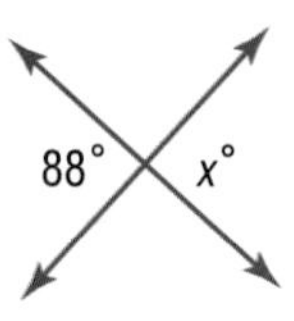

17.
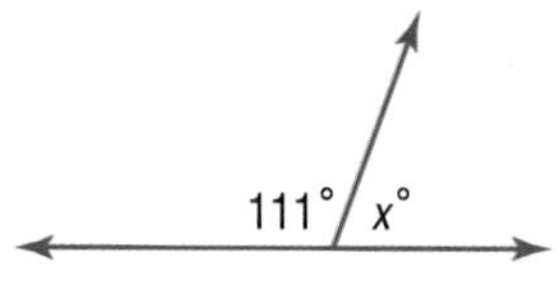

18.
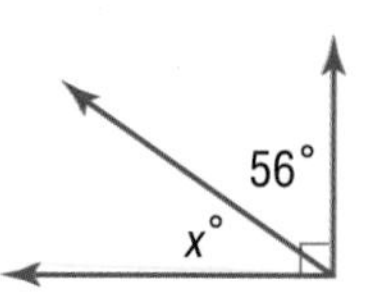

19.
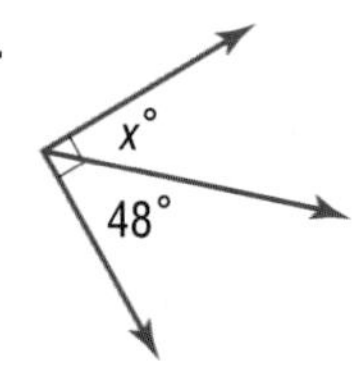

20.
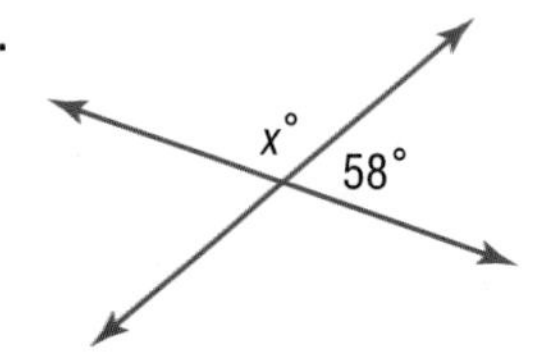

21.
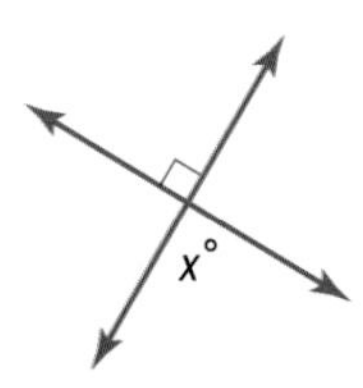

HEALTH For Exercises 22 and 23, use the graphic at the right.

22. Which two sections have angles that are complementary?

23. Which two sections have angles that are supplementary?

USA TODAY Snapshots®

Where the wild things are

Where Americans feel most at risk of getting sick from germs.

Public restroom **39%**

Restaurant **21%**

Airplane **20%**

Subway/train **11%**

Don't know/none of these **5%**

Movie theater seat **4%**

Source: Opinion Research Corporation

By In-Sung Yoo and Bob Laird, USA TODAY

Determine whether each statement is *sometimes*, *always*, or *never* true. Explain or give an example to support your answer.

24. Adjacent angles are complementary.

25. Two straight angles are supplementary.

26. Vertical angles have the same angle measure.

27. Adjacent angles are congruent.

28. **WRITE A PROBLEM** Write about a real-life object that has angles. Draw a diagram of it, measure all angles, and identify any complementary, supplementary, and vertical angles.

29. **CRITICAL THINKING** Angles E and F are complementary. If $m\angle E = x - 10$ and $m\angle F = x + 2$, find the measure of each angle.

Standardized Test Practice and Mixed Review

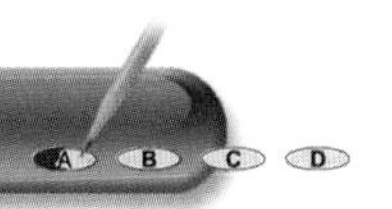

30. **SHORT RESPONSE** Find the measure of two angles that are vertical and complementary.

31. **MULTIPLE CHOICE** Find the value of a in the figure.

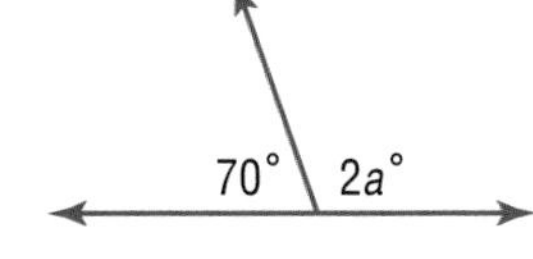

Ⓐ 90 Ⓑ 75 Ⓒ 55 Ⓓ 45

32. **STATISTICS** A company surveyed people about the type of crust they preferred on their pizza. Make a circle graph of the results shown at the right. (Lesson 10-2)

Type of Crust	Percent
regular thin	61%
thick	14%
deep dish	14%
extra thin	11%

Source: CREST

33. **TIME** Classify the angle the hands of a clock make at 3:30. (Lesson 10-1)

34. **REAL ESTATE** A house for sale has a rectangular lot with a length of 250 feet and a width of 120 feet. What is the area of the lot? (Lesson 6-8)

Write each percent as a decimal. (Lessons 5-5 and 7-6)

35. 53% 36. 78.5% 37. 431% 38. 0.23%

GETTING READY FOR THE NEXT LESSON

PREREQUISITE SKILL Solve each equation. (Lesson 4-2)

39. $x + 112 = 180$ 40. $50 + t = 180$ 41. $180 = 79 + y$ 42. $180 = h + 125$

msmath2.net/self_check_quiz

10-3b HANDS-ON LAB

A Follow-Up of Lesson 10-3

Constructing Parallel Lines

What You'll LEARN

Construct parallel lines, and discover angle relationships.

Materials

- straightedge
- compass
- protractor
- colored pencils
- notebook paper

Parallel lines are lines that do not intersect. The symbol ∥ means parallel. In this lab, you will learn how to construct parallel lines and discover angle relationships that are created from parallel lines.

ACTIVITY *Work with a partner.*

1 Construct a line parallel to line ℓ.

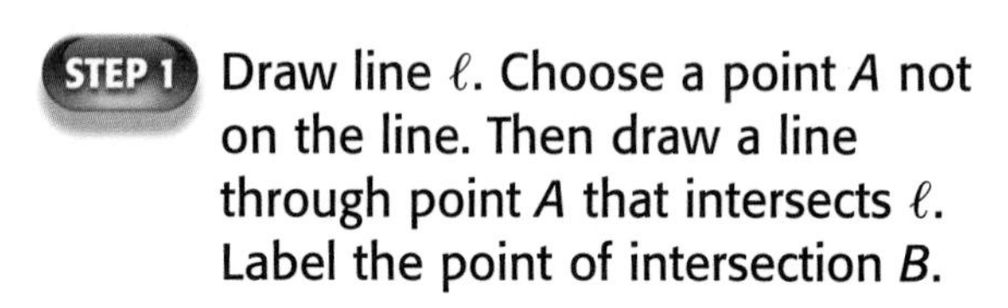

STEP 1 Draw line ℓ. Choose a point *A* not on the line. Then draw a line through point *A* that intersects ℓ. Label the point of intersection *B*.

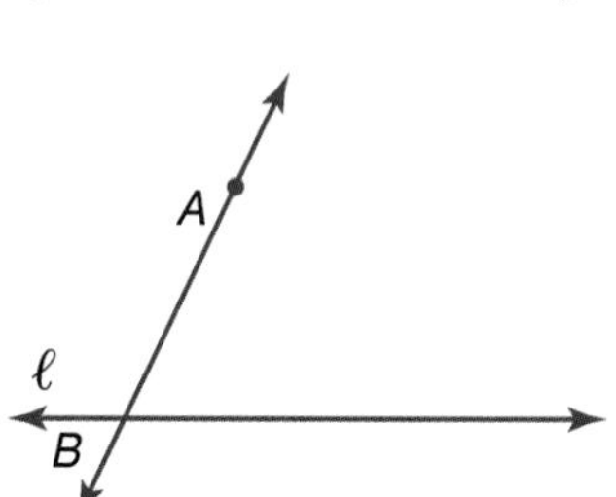

STEP 2 Place the compass at vertex *B* and draw an arc. Label the points of intersection *C* and *D*. With the same compass setting, place the compass at point *A* and draw an arc. Label the point of intersection *E*.

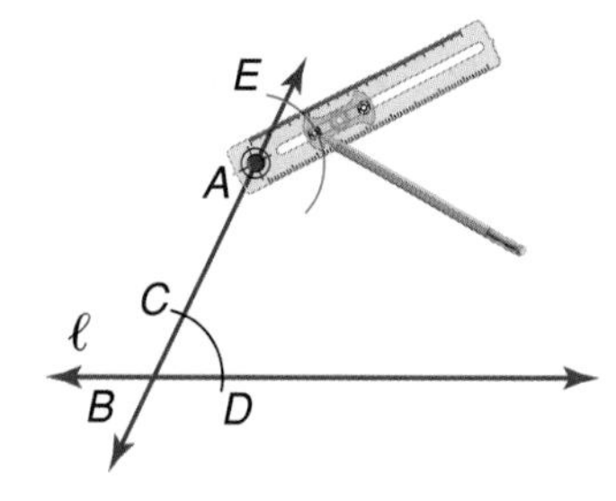

STEP 3 Open your compass to the same width as the distance between *C* and *D*. Then place the compass at point *E*. Draw an arc that intersects the arc you drew in Step 2. Label this point of intersection *F*.

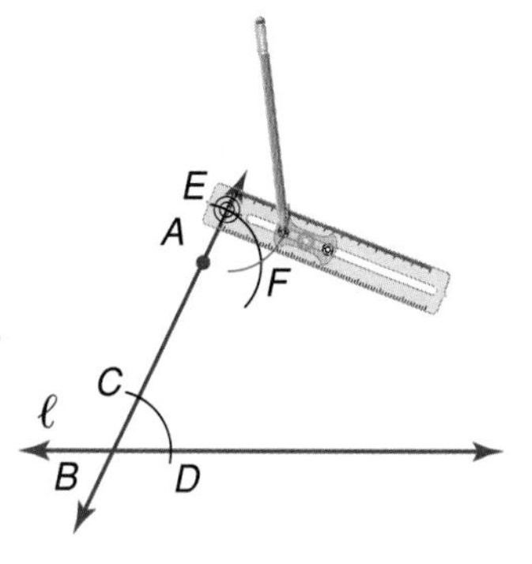

STEP 4 Draw a line through points *A* and *F*. Label the line *m*. Line *m* is parallel to line ℓ.

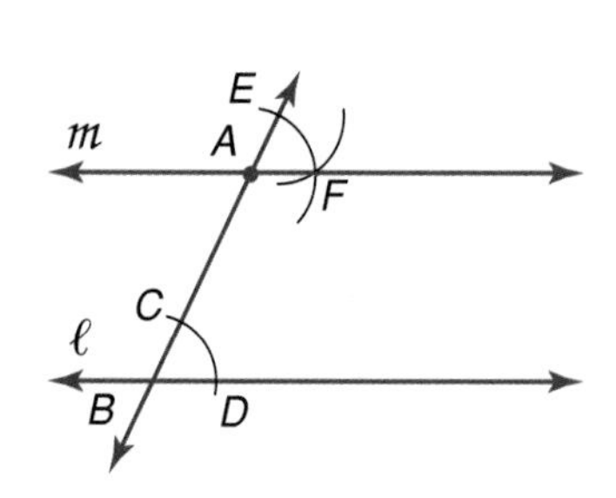

Your Turn

a. Draw a line. Then construct a line parallel to it.

A line that intersects parallel lines is called a **transversal**. Line t is a transversal for parallel lines ℓ and m. When a transversal cuts two parallel lines, there are several angle relationships formed.

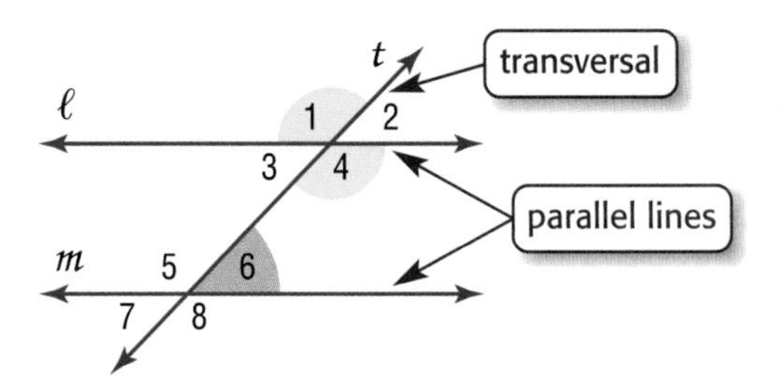

ACTIVITY *Work with a partner.*

2 **Find angle relationships in the figure above.**

STEP 1 Draw parallel lines and a transversal on notebook paper. Label the angles as shown above.

STEP 2 Measure each angle with your protractor. Record each measure.

STEP 3 Using colored pencils, shade any angles with the same measure one color. If there are angles with a different measure, shade them a second color. For example, $\angle 1$ and $\angle 4$ shown above are both shaded yellow because they have the same measure. Angle 6 is shaded blue because it has a different measure.

In the figure above, certain pairs of angles have special names. The angles that make up each pair have the same measure.

alternate interior angles	$\angle 3$ and $\angle 6$, $\angle 4$ and $\angle 5$
alternate exterior angles	$\angle 1$ and $\angle 8$, $\angle 2$ and $\angle 7$
corresponding angles	$\angle 1$ and $\angle 5$, $\angle 2$ and $\angle 6$, $\angle 3$ and $\angle 7$, $\angle 4$ and $\angle 8$
vertical angles	$\angle 1$ and $\angle 4$, $\angle 2$ and $\angle 3$, $\angle 5$ and $\angle 8$, $\angle 6$ and $\angle 7$

Writing Math

1. **Explain** why you think $\angle 1$, $\angle 2$, $\angle 7$, and $\angle 8$ are called exterior angles.
2. **Explain** why you think $\angle 3$, $\angle 4$, $\angle 5$, and $\angle 6$ are called interior angles.
3. What are $\angle 2$ and $\angle 4$ called?
4. If you know only one angle measure in the figure, **explain** how you can find the measures of the other angles without measuring.
5. Predict the measure of the other angles in the figure at the right using the 45° angle. Then copy the figure onto notebook paper and check by using a protractor.

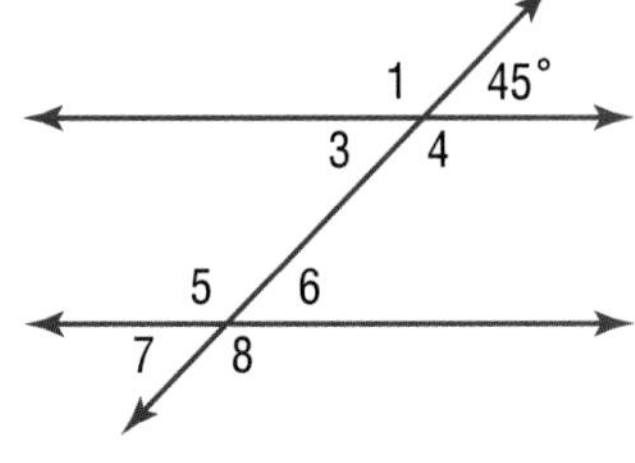

10-4 Triangles

HANDS-ON Mini Lab

Materials
- paper
- straightedge
- scissors

Work with a partner.

STEP 1 Use a straightedge to draw a triangle with three acute angles. Label the angles *A*, *B*, and *C*. Cut out the triangle.

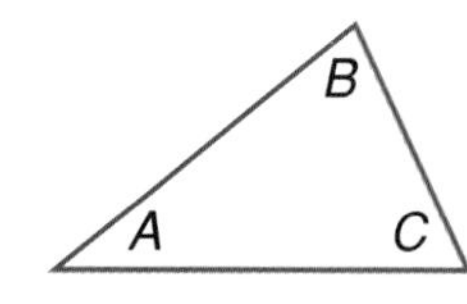

STEP 2 Fold $\angle B$ so that its vertex touches the line between angles *A* and *C*. The fold should be parallel to the base of the triangle.

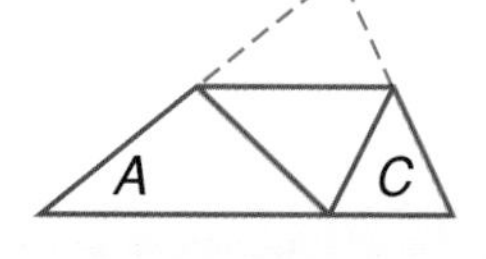

STEP 3 Fold $\angle A$ and $\angle C$ so the vertices meet.

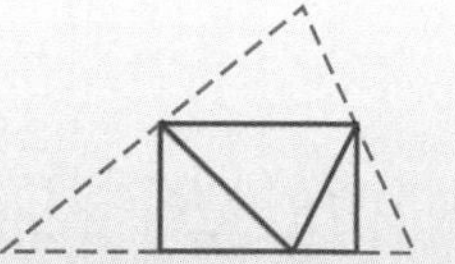

1. What kind of angle is formed where the three vertices meet?
2. Repeat the activity with another triangle. **Make a conjecture** about the sum of the measures of three angles of any triangle.

What You'll LEARN

Identify and classify triangles.

NEW Vocabulary

triangle
acute triangle
right triangle
obtuse triangle
congruent segments
scalene triangle
isosceles triangle
equilateral triangle

MATH Symbols

$m\angle 1$ measure of angle 1

A **triangle** is a figure with three sides and three angles. In the Mini Lab, you discovered a relationship among the three angles in a triangle.

Key Concept — **Angles of a Triangle**

Words The sum of the measures of the angles of a triangle is 180°.

Symbols $m\angle 1 + m\angle 2 + m\angle 3 = 180°$

Model

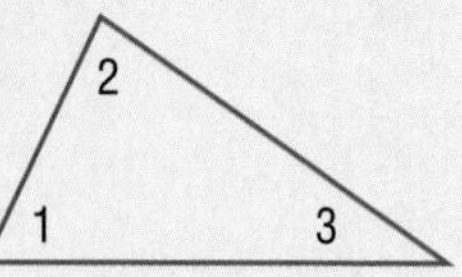

READING Math

Triangle The word *triangle* means *three angles*.

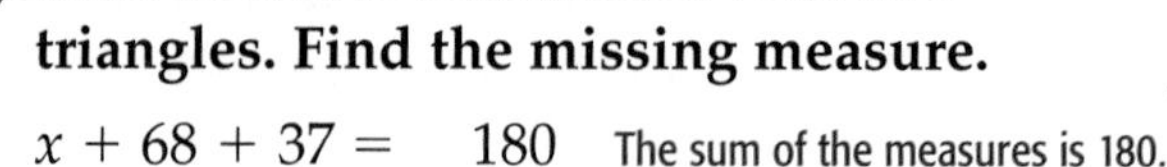

EXAMPLE Find Angle Measures of Triangles

1 KITES A kite is constructed with two triangles. Find the missing measure.

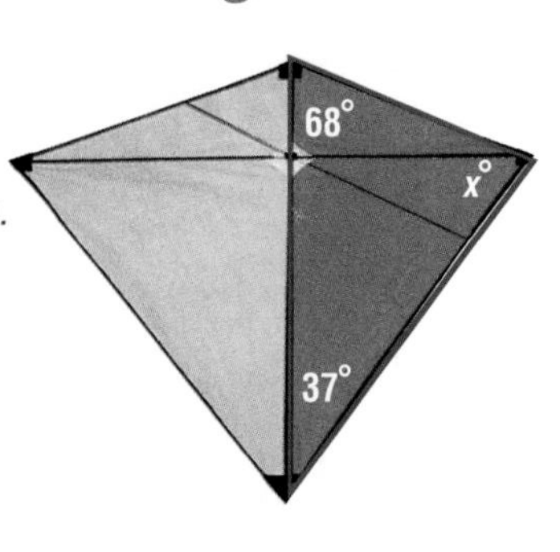

$x + 68 + 37 = 180$ The sum of the measures is 180.

$x + 105 = 180$ Simplify.

$-105 \quad -105$ Subtract 105 from each side.

$x = 75$

The missing measure is 75°.

EXAMPLE Find a Missing Measure

STUDY TIP

Check the Reasonableness of Results You can check whether you found the correct measure by adding 43 + 119 + 18. The sum should be 180.

2 **ALGEBRA Find $m\angle Z$ in $\triangle XYZ$ if $m\angle X = 43°$ and $m\angle Y = 119°$.**

Draw the triangle. Then write and solve an equation to find $m\angle Z$.

$m\angle X + m\angle Y + m\angle Z = 180$	The sum of the measures is 180.
$43 + 119 + m\angle Z = 180$	Substitute.
$162 + m\angle Z = 180$	Simplify.
$-162 \qquad -162$	Subtract 162 from each side.
$m\angle Z = 18$	

So, the measure of $\angle Z$ is 18°.

Since every triangle has at least two acute angles, one way you can classify a triangle is by using the third angle.

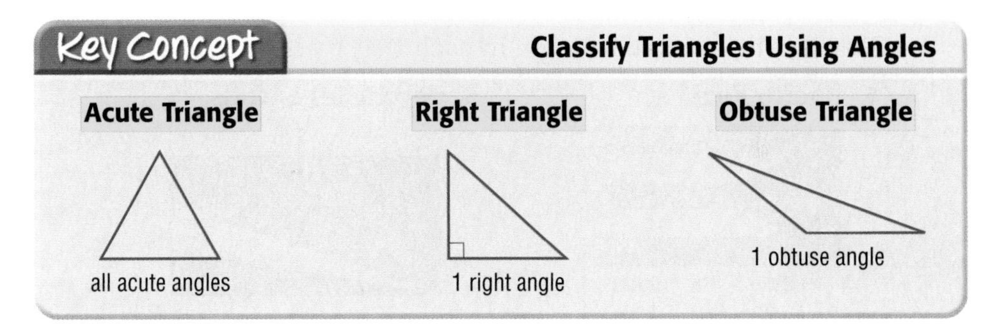

Another way to classify triangles is by their sides. Sides with the same length are **congruent segments**.

STUDY TIP

Congruent Segments The marks on the sides of the triangles indicate that those sides are congruent.

Key Concept **Classify Triangles Using Sides**

Scalene Triangle	Isosceles Triangle	Equilateral Triangle
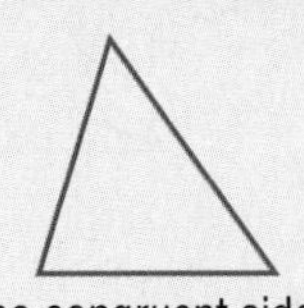	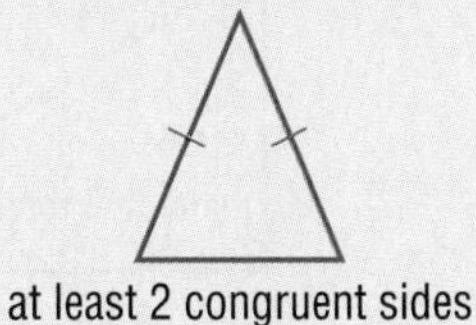	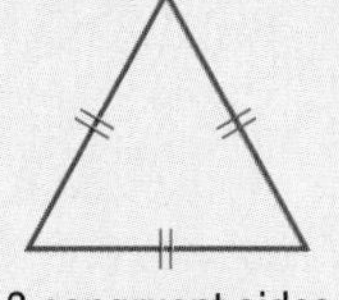
no congruent sides	at least 2 congruent sides	3 congruent sides

EXAMPLES Classify Triangles

Classify each triangle by its angles and by its sides.

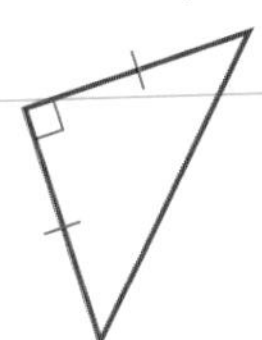

The triangle has a right angle and two congruent sides. So, it is a right, isosceles triangle.

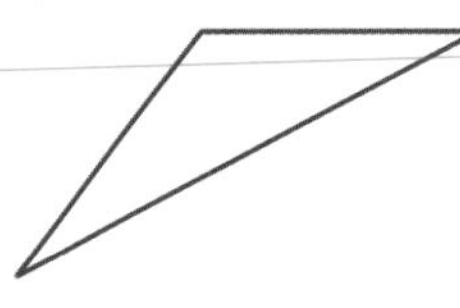

The triangle has one obtuse angle and no congruent sides. So, it is an obtuse, scalene triangle.

Skill and Concept Check

Writing Math
Exercise 1

1. **Describe** the angles in an obtuse triangle.

2. **OPEN ENDED** Draw an acute scalene triangle.

GUIDED PRACTICE

Find the missing measure in each triangle. Then classify the triangle as *acute, right,* or *obtuse.*

3. 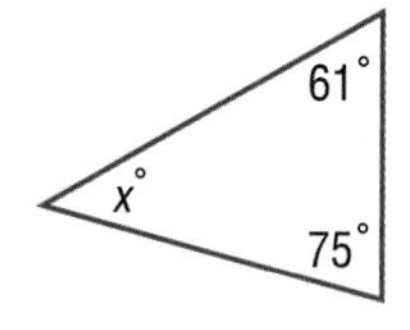

4. 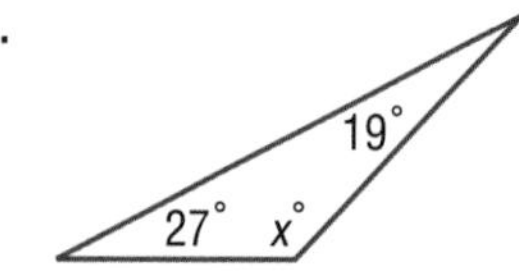

Classify the marked triangle in each flag by its angles and by its sides.

5. Puerto Rico

6. Seychelles Islands
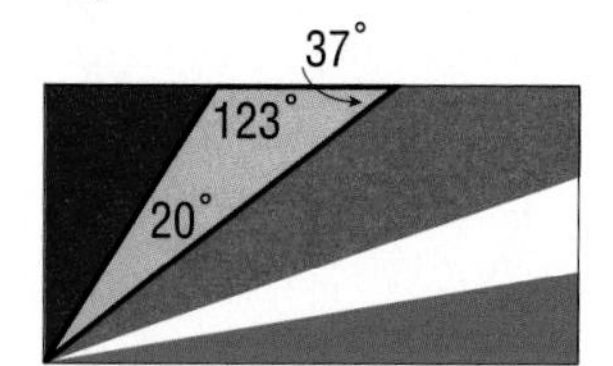

7. Bosnia-Herzegovina
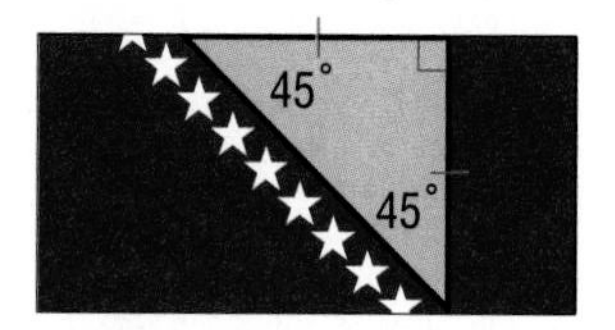

Practice and Applications

Find the missing measure in each triangle. Then classify the triangle as *acute, right,* or *obtuse.*

HOMEWORK HELP	
For Exercises	See Examples
8–14, 17, 24	1, 2
8–13, 15–16, 18–23	3, 4

Extra Practice
See pages 588, 605.

8. 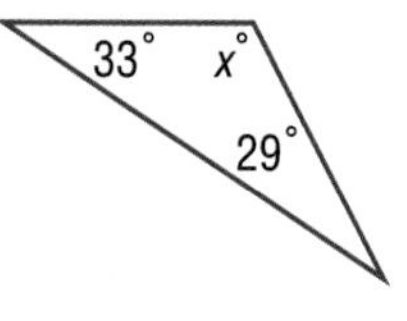

9. 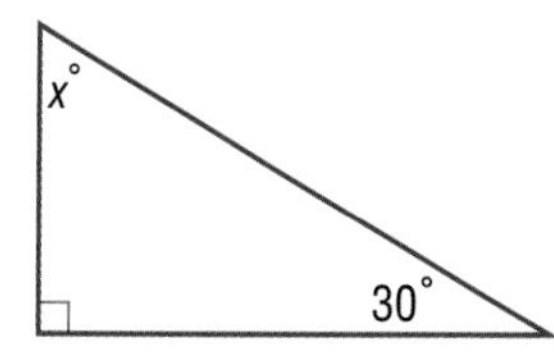

10. 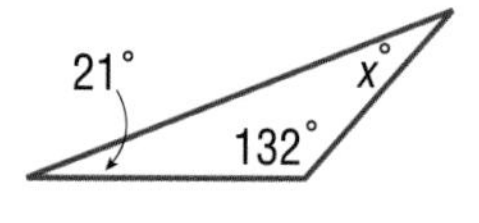

11. 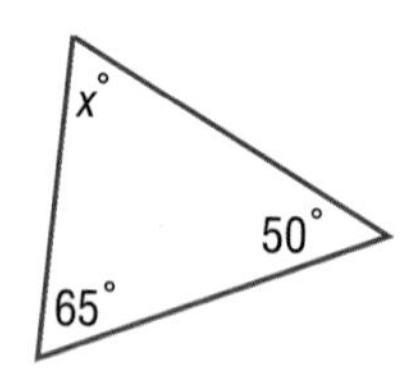

12. 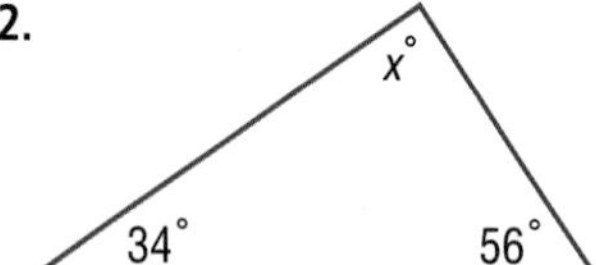

13. 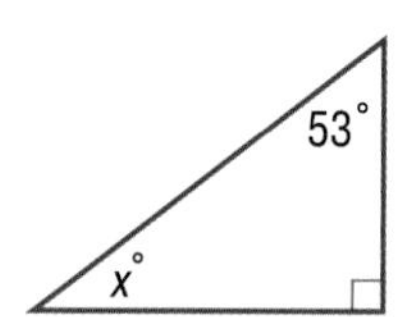

14. Find the third angle of a right triangle if the measure of one of the angles is 10°.

15. Three angles of a triangle measure 30°, 60°, and 90°. Classify the triangle by its angles.

16. Three sides of a triangle measure 5 meters, 8 meters, and 8 meters. Classify the triangle by its sides.

17. **ALGEBRA** Find $m\angle T$ in $\triangle RST$ if $m\angle R = 88°$ and $m\angle S = 75°$.

Classify each triangle by its angles and by its sides.

18. 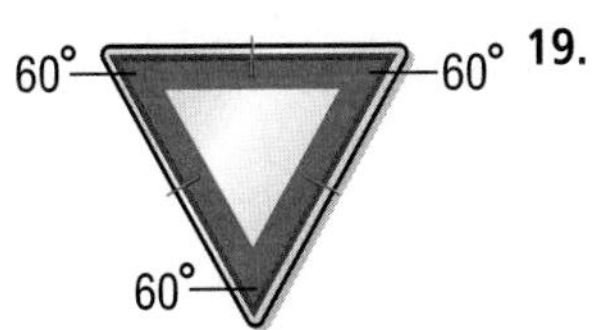

19. 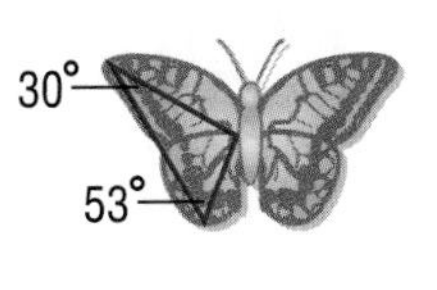

20. 45°

21. 

22. 

23. 

24. **GEOGRAPHY** A triangular area in the southern Atlantic Ocean where airplanes and boats have disappeared is known as the *Bermuda Triangle*. Find the missing angle measurement at Bermuda in the figure at the right.

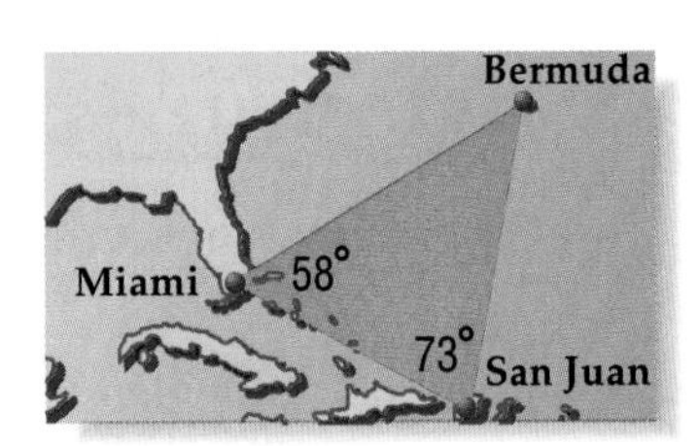

25. **WRITE A PROBLEM** Write about a real-life situation or object involving a triangle and its measures. Classify the triangle by its sides and its angles.

26. **CRITICAL THINKING** Find the missing angle measures in the figure.

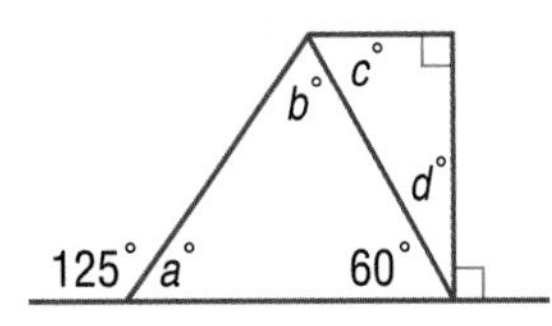

Standardized Test Practice and Mixed Review

27. **MULTIPLE CHOICE** How would you find $m\angle R$?

(A) Add 30° to 180°. (B) Subtract 60° from 180°.

(C) Subtract 30° from 90°. (D) Subtract 180° from 60°.

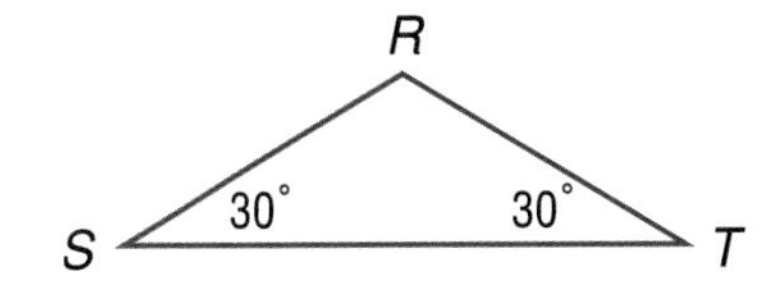

28. **SHORT RESPONSE** Find the measure of two congruent angles of a triangle if the third angle measure is 54°.

Classify each pair of angles as *complementary*, *supplementary*, or *neither*. (Lesson 10-3)

29. 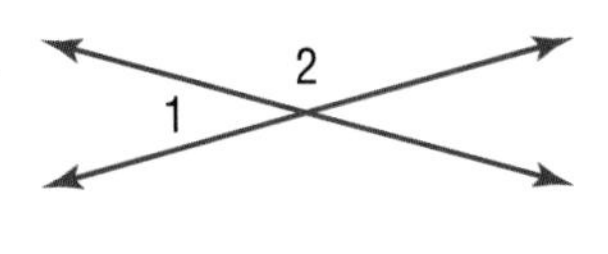

30. 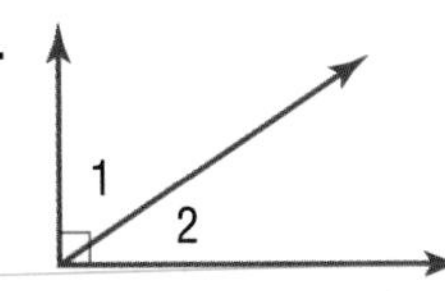

31. 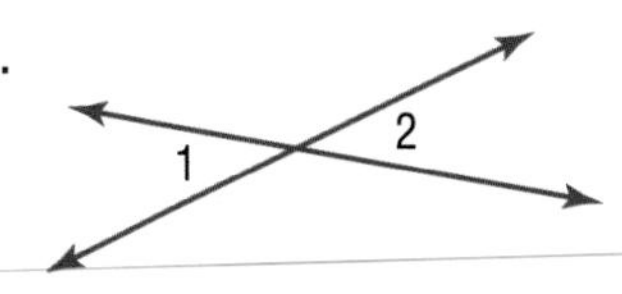

32. **BIRDS** A circle graph shows that 41% of people who bird watch live in the Northeast region of the United States. What is the measure of the angle of the Northeast section of the graph? (Lesson 10-2)

GETTING READY FOR THE NEXT LESSON

PREREQUISITE SKILL Solve each equation. (Lesson 4-2)

33. $x + 120 + 120 + 60 = 360$

34. $73 + 119 + x + 50 = 360$

10-4b HANDS-ON LAB

A Follow-Up of Lesson 10-4

Constructing Triangles

What You'll LEARN

Construct equilateral and isosceles triangles.

Materials

- straightedge
- compass
- protractor
- ruler

Equilateral triangles are triangles with three congruent sides. Isosceles triangles are triangles with at least two congruent sides. In this lab, you will learn how to construct both types of triangles.

ACTIVITY *Work with a partner.*

1 Construct an equilateral triangle.

STEP 1 Use your straightedge to draw line segment *AB*.

STEP 2 Open the compass to the same length as $\overline{AB}$. With the compass at point *A*, draw an arc above the line.

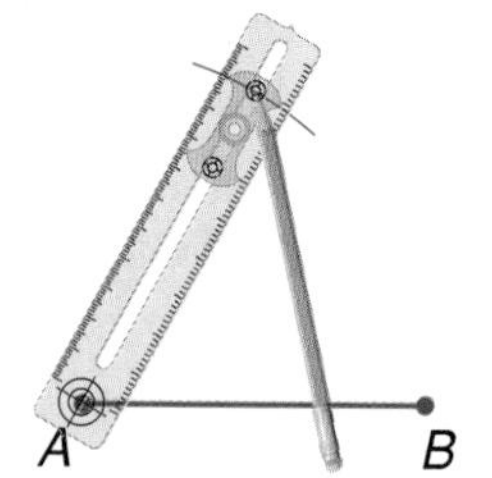

STEP 3 With the same compass setting, place the compass at point *B*. Draw an arc that intersects the arc you drew in Step 2. Label the intersection *C*.

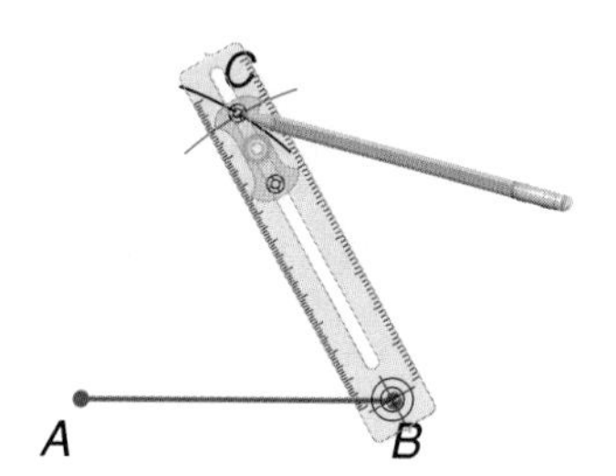

STEP 4 Connect the points to complete equilateral $\triangle ABC$.

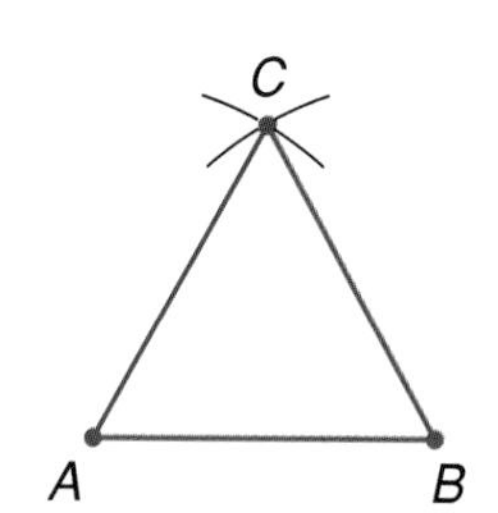

STUDY TIP

Check You can measure each side of the triangle to see if the sides are congruent.

Your Turn

a. Construct equilateral triangle *TUV* with sides measuring 3 inches.

In order to construct equilateral triangles in the activity above, you copied line segments using a compass. You constructed line segments with the same length by copying $\overline{AB}$ and making $\overline{AC}$ and $\overline{BC}$ congruent. So, a compass is also a useful tool to check whether line segments have the same length.

ACTIVITY *Work with a partner.*

2 Construct an isosceles triangle.

STEP 1 Use your straightedge to draw line segment DE.

D E

STEP 2 Open the compass to a length greater than $\overline{DE}$. With the compass at point D, draw an arc above the line.

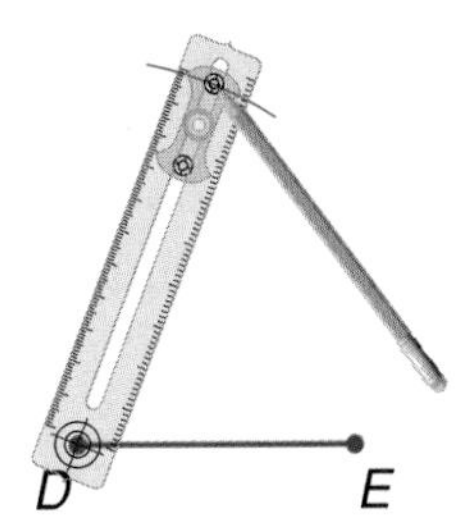

STEP 3 With the same compass setting, place the compass at point E. Draw an arc that intersects the arc you drew in Step 2. Label the intersection F.

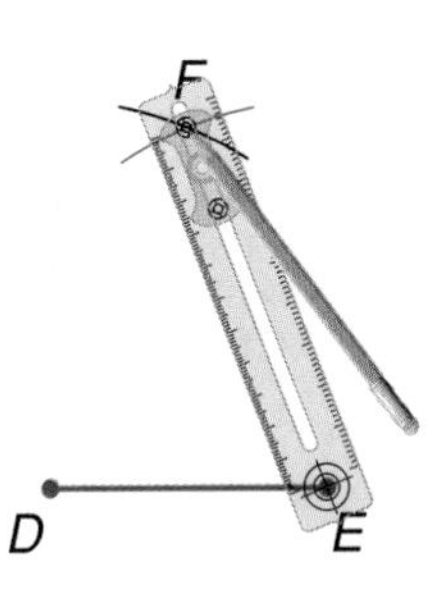

Check You can measure $\overline{DF}$ and $\overline{EF}$ to check if they are congruent.

STEP 4 Connect the points to complete isosceles $\triangle DEF$.

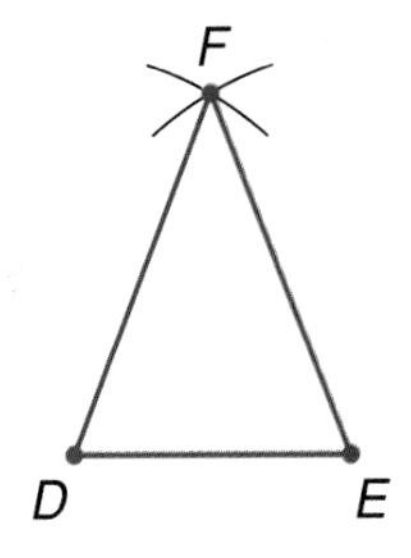

Your Turn

b. Construct isosceles triangle XYZ with sides measuring 3 centimeters, 5 centimeters, and 5 centimeters.

Writing Math

Work with a partner.

1. **Measure** the angles in the equilateral triangle you constructed with a protractor. Compare your measurements with other groups.
2. **Make a conjecture** about the measure of the angles of an equilateral triangle.
3. **Measure** the angles in the isosceles triangle you constructed with a protractor. Compare your measurements with other groups.
4. **Make a conjecture** about the measure of the angles of an isosceles triangle.

10-5 Quadrilaterals

What You'll LEARN

Identify and classify quadrilaterals.

NEW Vocabulary

quadrilateral
parallelogram
trapezoid
rhombus

REVIEW Vocabulary

parallel: lines that do not intersect **(Lesson 10-3b)**

WHEN am I ever going to use this?

BASKETBALL The photograph shows the free throw lane used in international basketball.

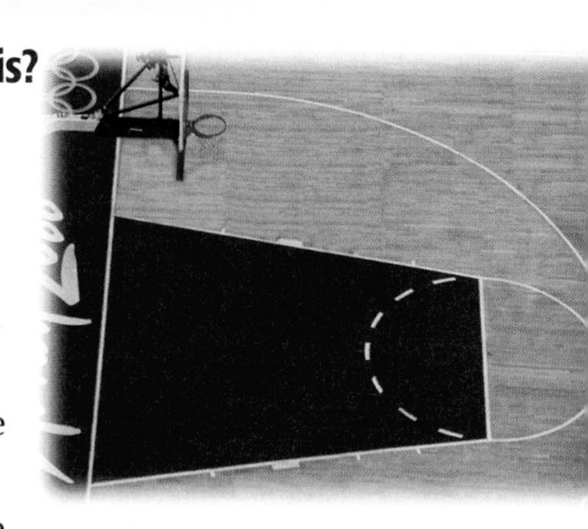

1. Describe the angles inside the 4-sided figure.
2. Which sides of the figure appear to be parallel?
3. Which sides of the figure appear to be congruent?

The shape of the free-throw lane above is called a trapezoid. Squares, rectangles, and trapezoids are examples of quadrilaterals. A **quadrilateral** is a closed figure with four sides and four angles. Quadrilaterals are named based on their sides and angles.

The diagram shows how quadrilaterals are related. Notice how it goes from the most general to the most specific.

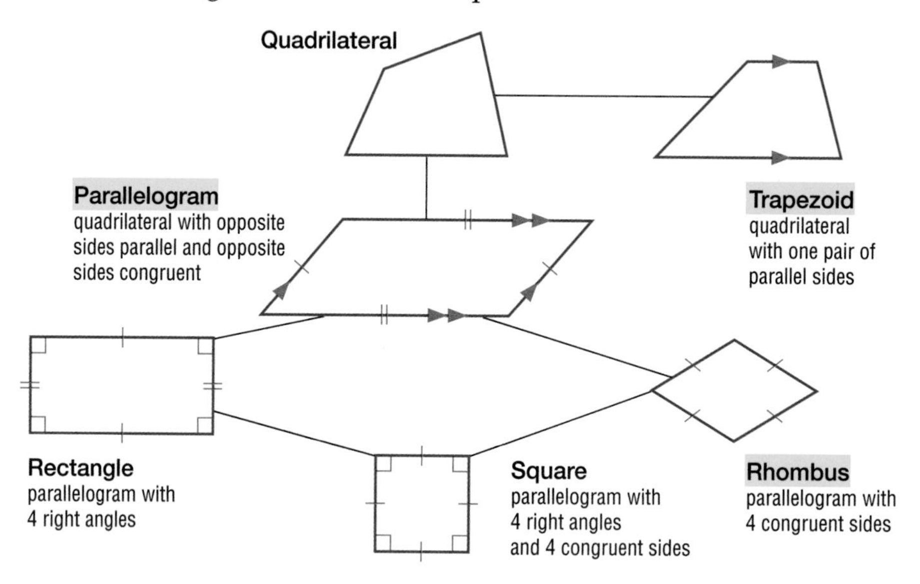

EXAMPLES Classify Quadrilaterals

READING Math

Parallel Lines The sides with matching arrows are parallel.

Classify the quadrilateral using the name that *best* describes it.

1 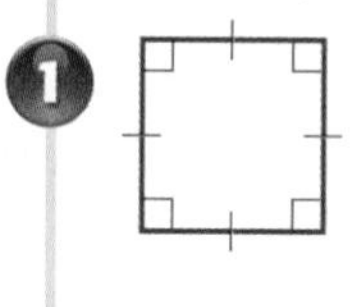

The quadrilateral has 4 right angles and 4 congruent sides. It is a square.

2 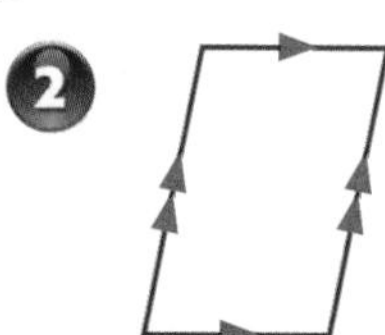

The quadrilateral has opposite sides parallel. It is a parallelogram.

A quadrilateral can be separated into two triangles, A and B. Since the sum of the angle measures of each triangle is 180°, the sum of the angle measures of both triangles is $2 \cdot 180$, or 360°.

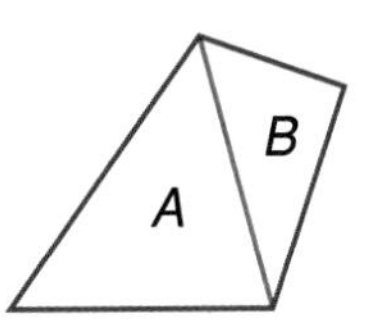

Key Concept **Angles of a Quadrilateral**

Words The sum of the measures of the angles of a quadrilateral is 360°.

Symbols $m\angle 1 + m\angle 2 + m\angle 3 + m\angle 4 = 360°$

Model

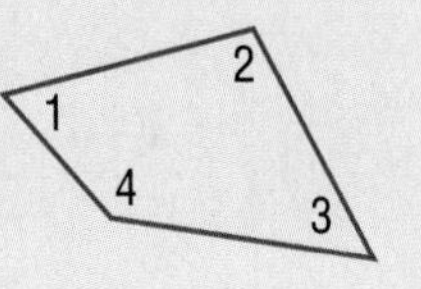

EXAMPLE Find a Missing Measure

3 Find the missing angle measure in the quadrilateral.

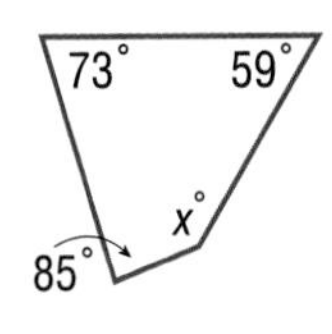

$85 + 73 + 59 + x = 360$ The sum of the measures is 360°.

$217 + x = 360$ Simplify.

$-217 \quad -217$ Subtract 217 from each side.

$x = 143$

So, the missing angle measure is 143°.

Skill and Concept Check

Writing Math Exercises 1 & 3

1. **Explain** why all squares are rectangles but not all rectangles are squares.
2. **OPEN ENDED** Describe a real-life example of a rhombus.
3. **FIND THE ERROR** Venus and Justin are describing a square. Who is correct? Explain.

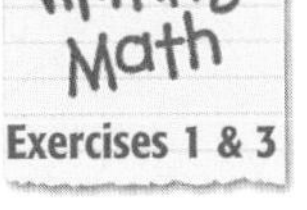

Venus	Justin
a parallelogram with 4 right angles	a rhombus with 4 right angles

GUIDED PRACTICE

Classify the quadrilateral using the name that *best* describes it.

4. 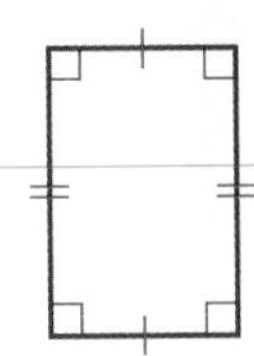

5. 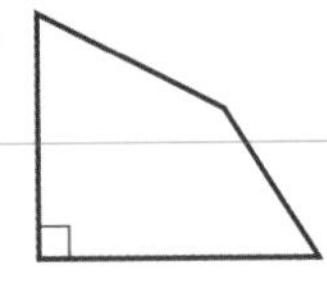

6. 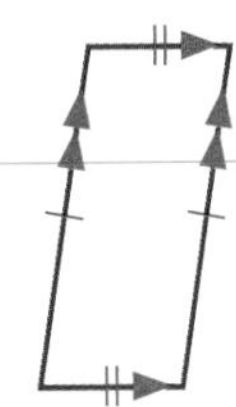

Find the missing angle measure in each quadrilateral.

7. 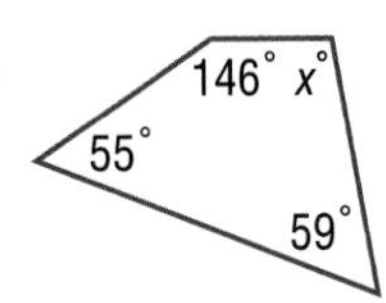

8. 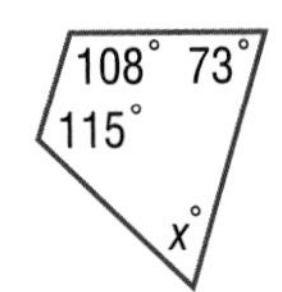

9. 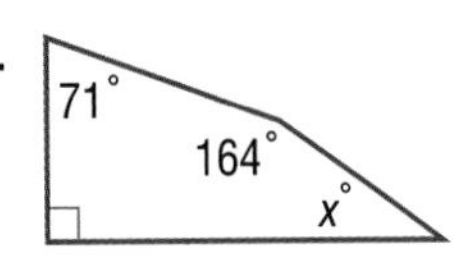

Practice and Applications

HOMEWORK HELP	
For Exercises	See Examples
10–15, 25, 27, 29–31	1, 2
19–24, 28	3

Extra Practice
See pages 588, 605.

Classify the quadrilateral using the name that *best* describes it.

10. 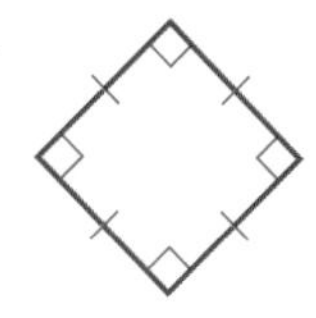

11. 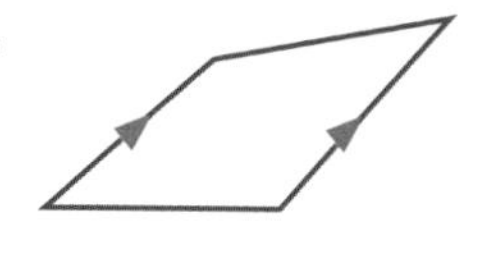

12. 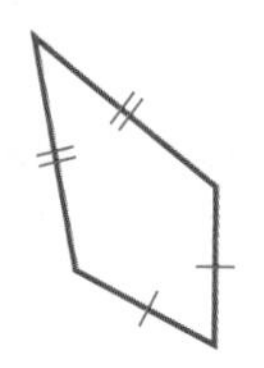

13. 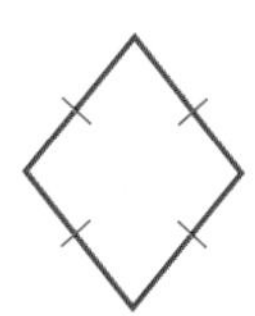

14. 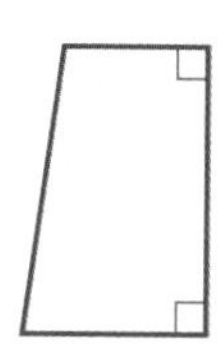

15. 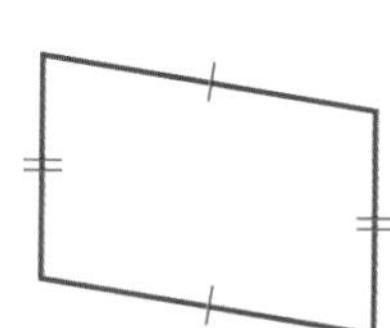

Determine whether each statement is *sometimes, always,* or *never* true. Explain.

16. A quadrilateral is a trapezoid.
17. A parallelogram is a rectangle.
18. A trapezoid is a parallelogram.

Find the missing angle measure in each quadrilateral.

19. 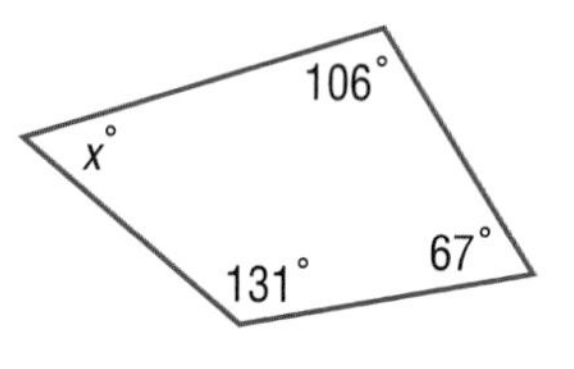

20. 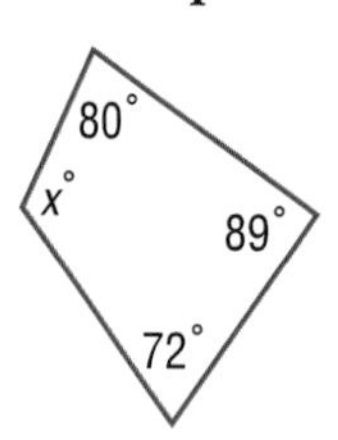

21. 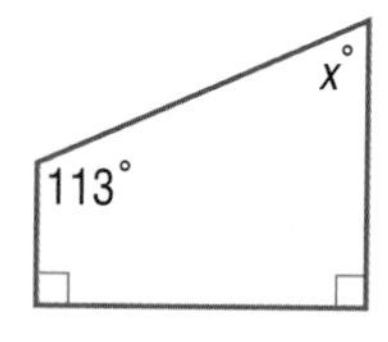

22. 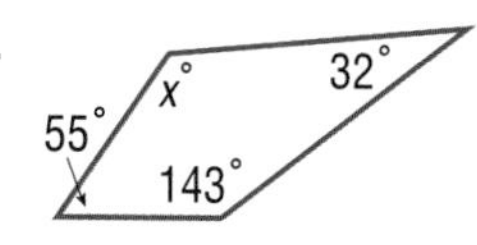

23. 

24. 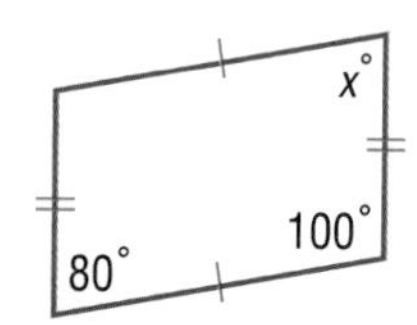

25. **VIDEO GAMES** The first video games used basic shapes like rectangles, squares, and circles in their interface. Name the quadrilaterals used in the video game shown at the right.

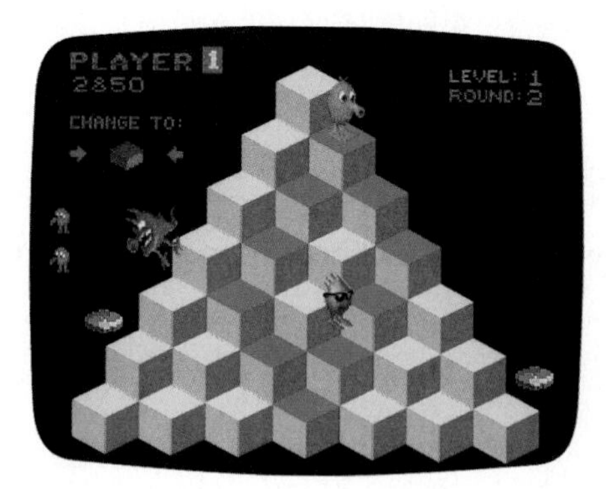

26. **RESEARCH** Use the Internet or another source to find other early video games that used different types of quadrilaterals as part of their interface. Classify the quadrilaterals in the games you found.

27. **ART** Design and draw a stained glass window that contains the following types of quadrilaterals: trapezoid, parallelogram, and rhombus.

28. **ALGEBRA** Find $m\angle B$ in quadrilateral $ABCD$ if $m\angle A = 87°$, $m\angle C = 135°$, and $m\angle D = 22°$.

Determine whether each figure described below can be drawn. Explain.

29. a quadrilateral that is both a rhombus and a rectangle

30. a trapezoid with 3 right angles

31. a trapezoid with two congruent sides

32. **CRITICAL THINKING** Find the value of x in the quadrilateral. Then find the measure of each angle.

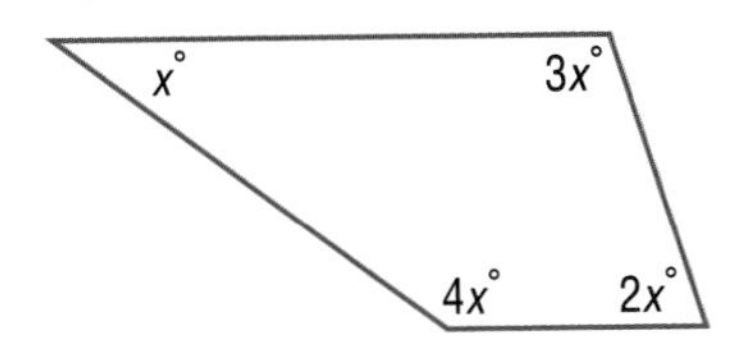

Standardized Test Practice and Mixed Review

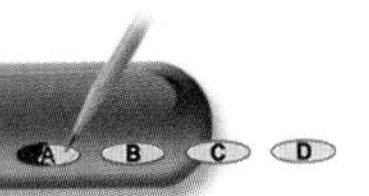

33. **MULTIPLE CHOICE** Which property is *not* characteristic of a rhombus?

Ⓐ opposite sides parallel
Ⓑ 4 right angles
Ⓒ 4 congruent sides
Ⓓ opposite sides congruent

34. **MULTIPLE CHOICE** Which quadrilateral does *not* have opposite sides congruent?

Ⓕ parallelogram
Ⓖ square
Ⓗ trapezoid
Ⓘ rectangle

Classify each triangle by its angles and by its sides. (Lesson 10-4)

35. 

36. 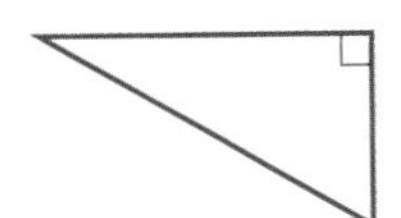

37. 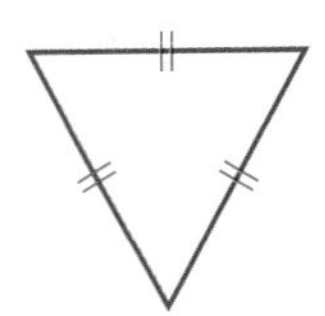

38. Suppose $\angle RST$ and $\angle TSU$ are supplementary angles. Find $m\angle TSU$ if $m\angle RST$ is 76°. (Lesson 10-3)

Find the sales tax or discount to the nearest cent. (Lesson 8-5)

39. \$54 jacket; 7% sales tax

40. \$23 hat; 15% discount

Find each number. Round to the nearest tenth if necessary. (Lesson 7-7)

41. 20% of what number is 17?

42. What number is 45% of 160?

43. 5 is 12% of what number?

44. 15 is what percent of 24?

45. **TRAVEL** On his summer vacation, Timothy drove 250 miles in 5 hours on the first day. He continued driving at the same rate the second day and drove for 8 hours. How many miles did Timothy drive the second day of his vacation? (Lesson 7-2)

GETTING READY FOR THE NEXT LESSON

PREREQUISITE SKILL Solve each proportion. (Lesson 7-3)

46. $\frac{3}{5} = \frac{x}{75}$

47. $\frac{a}{7} = \frac{18}{42}$

48. $\frac{7}{9} = \frac{28}{m}$

49. $\frac{3.5}{t} = \frac{16}{32}$

50. $\frac{3}{6} = \frac{c}{5}$

Mid-Chapter Practice Test

Vocabulary and Concepts

1. **Define** *supplementary angles.* (Lesson 10-3)
2. **Describe** the difference between a square and a rhombus. (Lesson 10-5)

Skills and Applications

Draw an angle having each measurement. Then classify each angle as *acute, obtuse, right,* or *straight.* (Lesson 10-1)

3. 134°
4. 90°
5. 180°
6. 17°

7. **SOCCER** Make a circle graph of total injuries of high school girls soccer players by position. (Lesson 10-2)

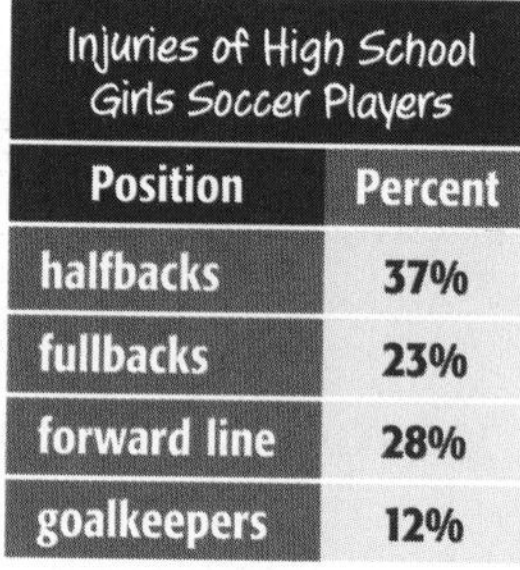

Injuries of High School Girls Soccer Players	
Position	**Percent**
halfbacks	37%
fullbacks	23%
forward line	28%
goalkeepers	12%

Source: National Athletic Trainers' Association

Find the value of x in each figure. (Lesson 10-3)

8. 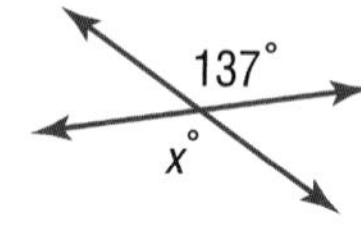

9. 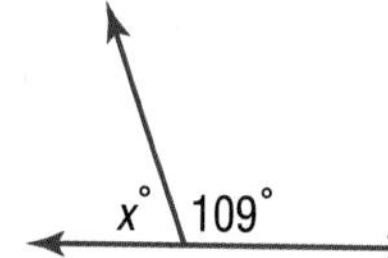

10. 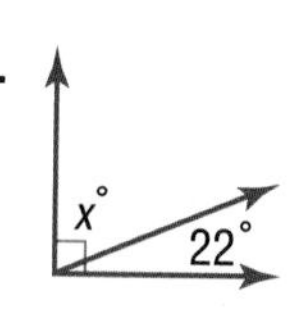

11. **ALGEBRA** Find $m\angle B$ in $\triangle ABC$ if $m\angle A = 62°$ and $m\angle C = 44°$. (Lesson 10-4)

Classify the quadrilateral using the name that *best* describes it. (Lesson 10-5)

12. 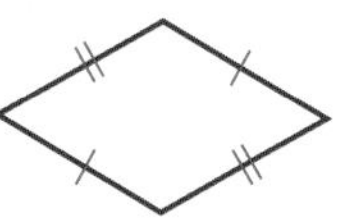

13. 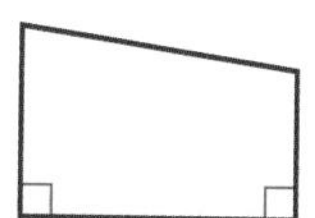

14. 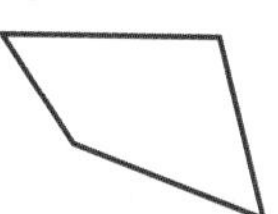

Standardized Test Practice

15. **MULTIPLE CHOICE** Which angle is complementary to $\angle CBD$? (Lesson 10-1)

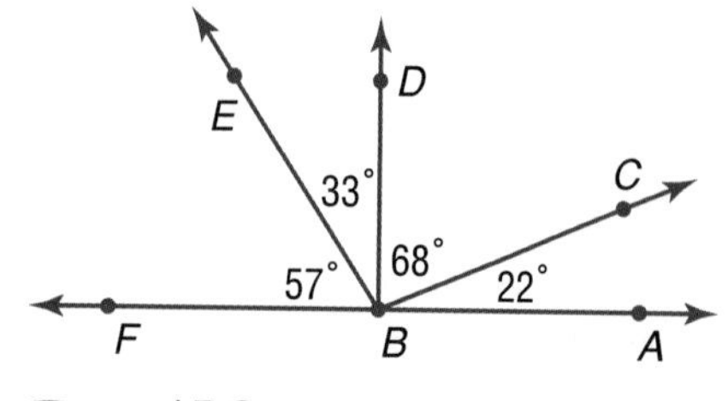

Ⓐ $\angle ABC$ Ⓑ $\angle DBE$

Ⓒ $\angle FBC$ Ⓓ $\angle EBF$

16. **SHORT RESPONSE** Which section of the circle graph has an angle measure of 63°? (Lesson 10-2)

What Do You Drink With Dinner?

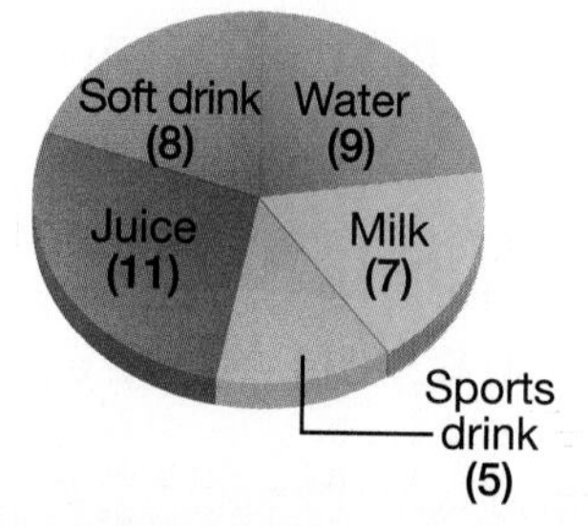

The Game Zone

A Place To Practice Your Math Skills

Math Skill
Finding Squares

Squares Everywhere!

GET READY!

Players: two
Materials: 8 red counters, 8 yellow counters, dot paper, straight edge

- Copy the game board onto dot paper.

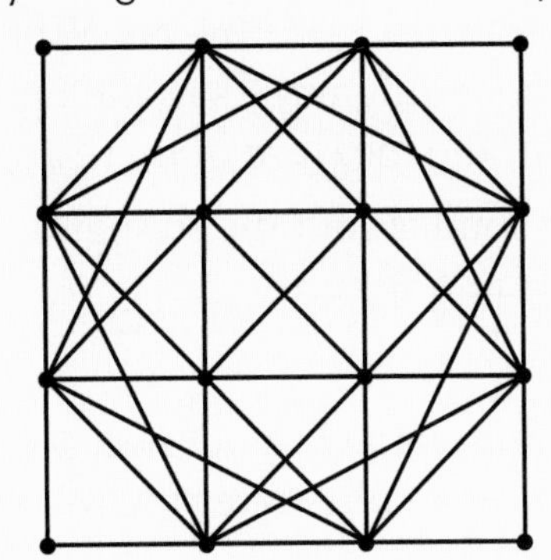

GO!

- The first player covers any dot with a counter. Then, players alternate turns.
- The object of the game is to cover the four vertices of a square with one set of counters. Note that the sides of a square don't have to be vertical or horizontal. The square can be "tilted" to one side.
- **Who Wins?** The first player to cover the four vertices of a square with his or her counters wins.
- Once you have played a game covering the four vertices of a square, change the game so you cover the four vertices of a rectangle or a parallelogram.

10-6 Similar Figures

HANDS-ON Mini Lab

Materials
- dot paper
- protractor

Work with a partner.

Copy each pair of rectangles and triangles onto dot paper. Then find the measure of each angle and the length of each side. Letters such as *AB* refer to the measure of the segment with those endpoints.

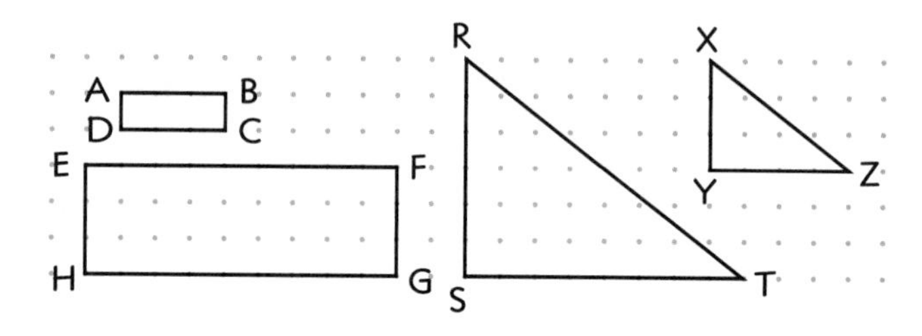

- Each pair of figures has the same shape but different sizes.
- Side *AB* "goes with" side *EF*. So, they are *corresponding sides*.

1. Write each fraction in simplest form.
   a. $\frac{AB}{EF}$; $\frac{BC}{FG}$; $\frac{DC}{HG}$; $\frac{AD}{EH}$ b. $\frac{RS}{XY}$; $\frac{ST}{YZ}$; $\frac{RT}{XZ}$
2. What do you notice about the ratios of corresponding sides?
3. Measure the corresponding angles in the figures above. What do you notice about the measure of these angles?
4. The rectangles are similar, and the triangles are similar. **Make a conjecture** about similar figures.

What You'll LEARN

Determine whether figures are similar and find a missing length in a pair of similar figures.

NEW Vocabulary

similar figures
indirect measurement

REVIEW Vocabulary

proportion: an equation that shows that two ratios are equivalent **(Lesson 7-3)**

MATH Symbols

~ is similar to

Figures that have the same shape but not necessarily the same size are **similar figures**. The Mini Lab illustrates the following definition.

Key Concept — **Similar Figures**

Words If two figures are similar, then
- the corresponding sides are proportional, and
- the corresponding angles are congruent.

Models

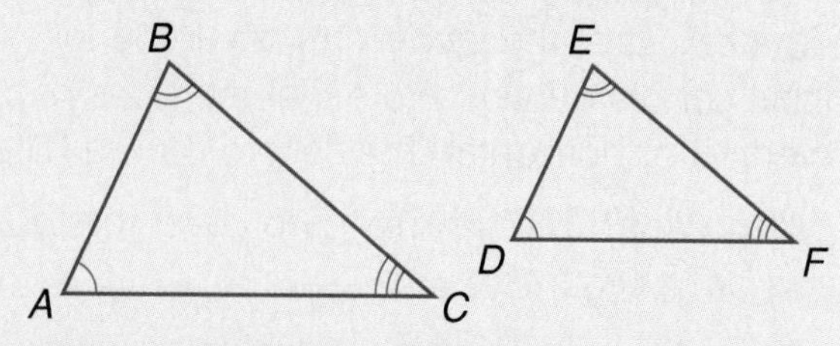

Symbols $\triangle ABC \sim \triangle DEF$ The symbol ~ means *is similar to.*

corresponding sides: $\frac{AB}{DE} = \frac{BC}{EF} = \frac{AC}{DF}$

corresponding angles: $\angle A \cong \angle D$; $\angle B \cong \angle E$; $\angle C \cong \angle F$

READING Math

Line Segments The symbol $\overline{XY}$ means the line segment XY. XY means the measure of $\overline{XY}$.

Look Back You can review **solving proportions** in Lesson 7-3.

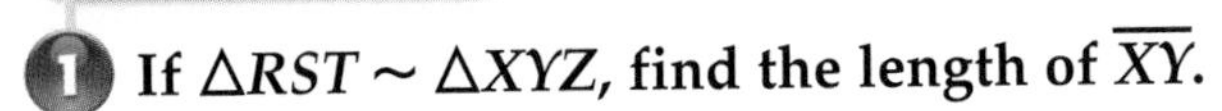

Find Side Measures of Similar Triangles

1 If $\triangle RST \sim \triangle XYZ$, find the length of $\overline{XY}$.

Since the two triangles are similar, the ratios of their corresponding sides are equal. So, you can write and solve a proportion to find XY.

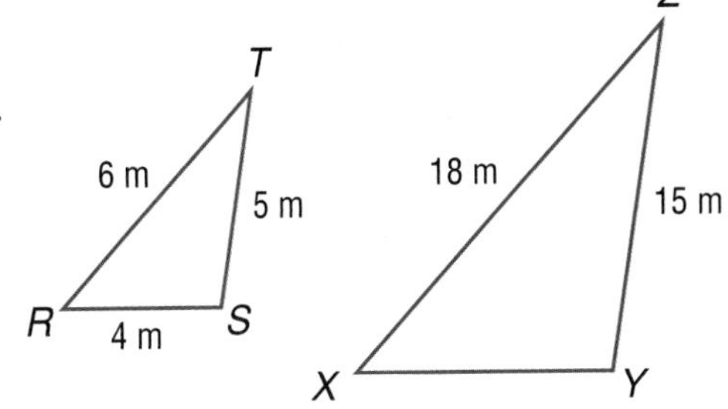

$\frac{RT}{XZ} = \frac{RS}{XY}$ Write a proportion.

$\frac{6}{18} = \frac{4}{n}$ Let n represent the length of $\overline{XY}$. Then substitute.

$6n = 18(4)$ Find the cross products.

$6n = 72$ Simplify.

$n = 12$ Divide each side by 6.

The length of $\overline{XY}$ is 12 meters.

You can use similar figures to find the length, width, or height of objects that are too difficult to measure directly. This type of measurement is called **indirect measurement**.

EXAMPLE Use Indirect Measurement

Standardized Test Practice

2 GRID-IN TEST ITEM Natalie wants to resize a 4-inch wide by 5-inch long photograph for the school newspaper. It is to fit in a space that is 2 inches wide. What is the length of the resized photograph in inches?

Read the Test Item

To find the length of the resized photograph, draw a picture and write a proportion.

Solve the Test Item

5 in.

4 in.

x

2 in.

$\frac{4}{2} = \frac{5}{x}$ Write a proportion.

$4x = 10$ Find the cross products.

$x = 2.5$ Divide each side by 4.

So, the length of the photograph is 2.5 inches.

Fill in the Grid

2	.	5		

Test-Taking Tip The Princeton Review

Grid In
When filling in the response grid, you can print your answer with the first digit in the left answer box, or with the last digit in the right answer box.

Skill and Concept Check

Writing Math
Exercise 1

1. **Describe** characteristics common to two similar triangles.

2. **OPEN ENDED** Draw two similar quadrilaterals and label the vertices. Then write equivalent ratios comparing all corresponding sides.

GUIDED PRACTICE

Find the value of x in each pair of similar figures.

3.

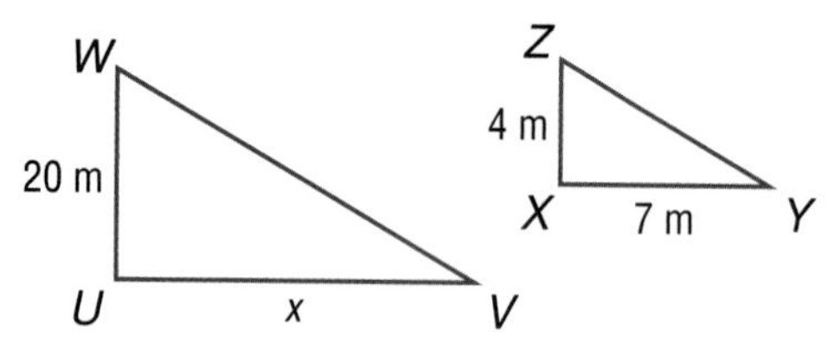

4.

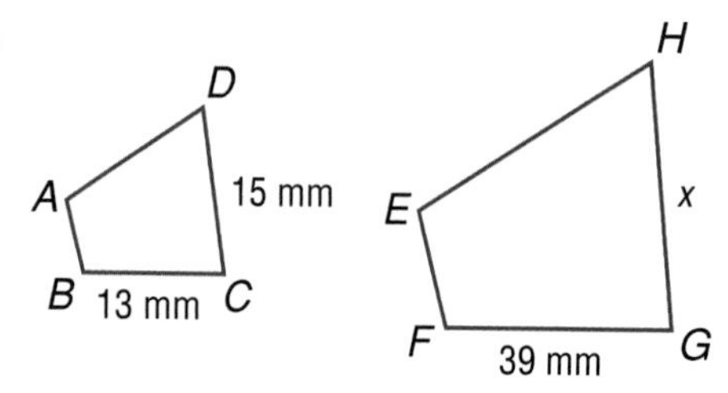

5. **MEASUREMENT** The height of an object and its shadow is proportional to the height of another object and its shadow. Suppose you are 6 feet tall and you cast a shadow 5 feet long. Find the height of the tree if it casts a shadow 40 feet long.

Practice and Applications

HOMEWORK HELP	
For Exercises	See Examples
6–9, 11–12	1
10, 13–15	2

Extra Practice
See pages 588, 605.

Find the value of x in each pair of similar figures.

6.

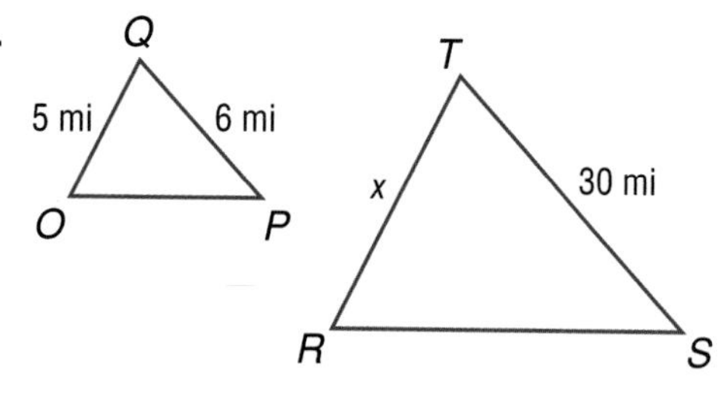

7.

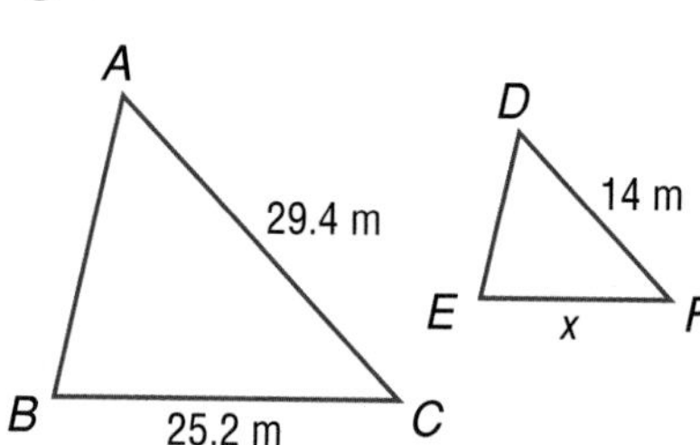

8.

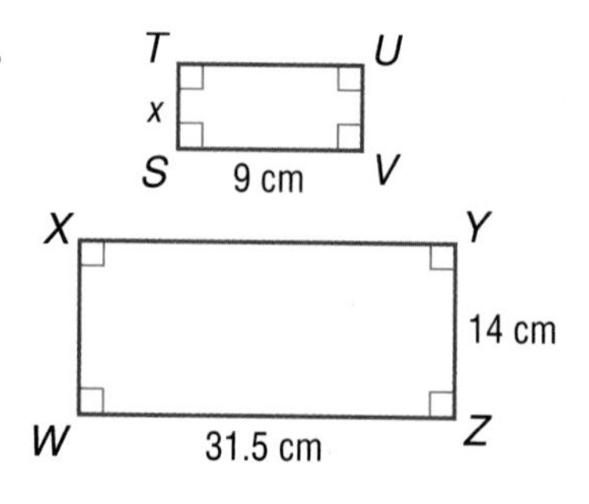

9.

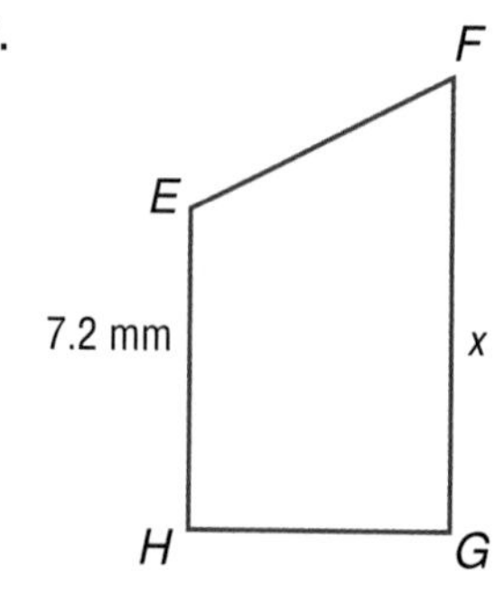

10. **FURNITURE** Kidco Furniture produces children's furniture that is similar in shape to full-sized furniture. The top of a full-sized desk measures 54 inches long by 36 inches wide. If the top of a child's desk is 24 inches wide, what is the length?

Determine whether each pair of figures is similar. Justify your answer.

11. 12.

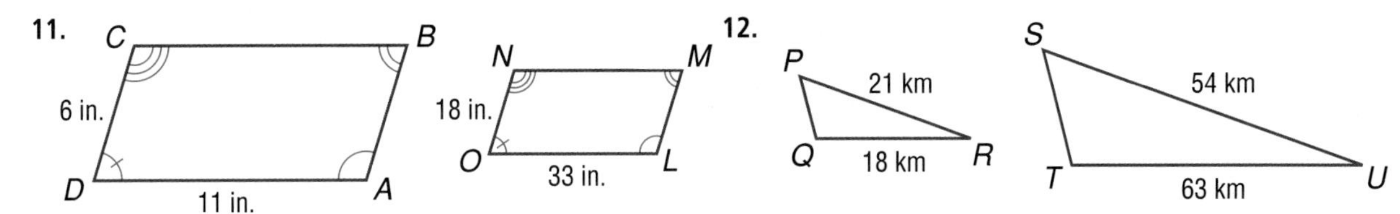

STATUES For Exercises 13 and 14, use the information below and at the right. Alyssa bought a miniature replica of the Statue of Liberty. The replica is 9 inches tall, and the length of the statue's right arm holding the torch is $1\frac{1}{4}$ inches.

13. **MULTI STEP** About how long is the Statue of Liberty's right arm?

14. **MULTI STEP** Alyssa's friend Garcia brought home a smaller replica. The length of the statue's right arm is $\frac{3}{4}$ inch. How tall is Garcia's statue?

15. **GEOMETRY** The ratio of square H's length to square I's length is 3:5. If the length of square H is 18 meters, what is the perimeter of square I?

CRITICAL THINKING Two rectangles are similar. The ratio of their corresponding sides is 1:4.

16. Find the ratio of their perimeters.

17. Find the ratio of their areas.

Standardized Test Practice and Mixed Review

18. **MULTIPLE CHOICE** Which of the following is *not* a true statement?

    (A) All squares are similar.
    (B) All rhombi are similar.
    (C) Some trapezoids are similar.
    (D) All equilateral triangles are similar.

19. **SHORT RESPONSE** Old Faithful in Yellowstone National Park shoots water 60 feet into the air and casts a shadow of 42 feet. What is the height of a nearby tree if it casts a shadow 63 feet long?

Classify the quadrilateral using the name that *best* describes it. (Lesson 10-5)

20. 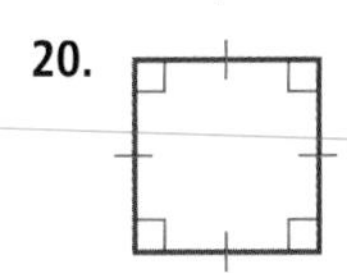

21. 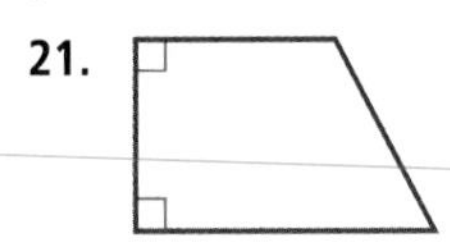

22. 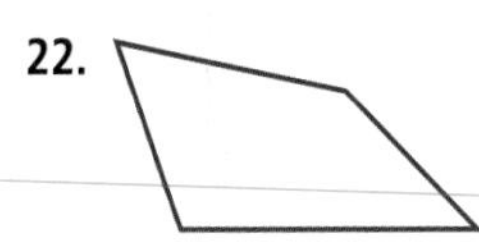

23. **SAILING** A triangular-shaped sail has angle measures of 44° and 67°. Find the measure of the third angle. (Lesson 10-4)

GETTING READY FOR THE NEXT LESSON

PREREQUISITE SKILL Solve each equation. (Lesson 4-3)

24. $5a = 120$
25. $360 = 4a$
26. $940 = 8n$
27. $6t = 720$

10-7a Problem-Solving Strategy

A Preview of Lesson 10-7

What You'll Learn
Solve problems using logical reasoning.

Use Logical Reasoning

Dion, how can we check that the garden we dug is in the shape of an equilateral triangle?

There is a relationship between the angles of an equilateral triangle that can help us. Nathaniel, let's **use logical reasoning** to find out!

Explore	Equilateral triangles have sides that are congruent. We need to find a relationship between the angles.
Plan	Let's draw several equilateral triangles and measure the angles.
Solve	sides = 1 cm sides = 2 cm sides = 3 cm sides = 4 cm Each angle of the triangles is 60°. So, if the angles of our triangular garden are 60°, then the garden is in the shape of an equilateral triangle.
Examine	Any triangle with angle measures of 60° is equilateral. If we try to draw a triangle with angle measures of 60° and different side lengths, the drawing would not be in the shape of a triangle. 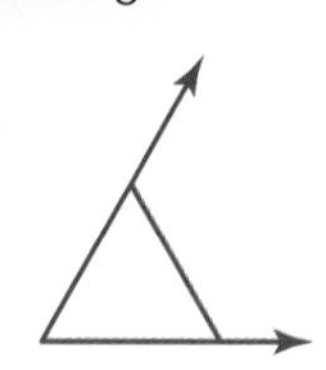

Analyze the Strategy

1. When you use *inductive reasoning,* you make a rule after seeing several examples. When you use *deductive reasoning,* you use a rule to make a decision. What type of reasoning did Nathaniel and Dion use to solve the problem? **Explain** your reasoning.
2. **Explain** how Nathaniel and Dion could have solved the problem using deductive reasoning.
3. **Explain** how the *look for a pattern* problem-solving strategy is similar to inductive reasoning.

Apply the Strategy

Solve. Use logical reasoning.

4. **GEOMETRY** Draw several isosceles triangles and measure their angles. What can you conclude about the measures of the angles of an isosceles triangle? See Exercise 1. Did you use inductive or deductive reasoning?

5. **TRAVEL** Use $d = rt$ where d is distance, r is rate, and t is time to find how far Mrs. Petricca drives if she drives 55 miles per hour for seven hours. See Exercise 1. Did you use inductive or deductive reasoning?

Mixed Problem Solving

Solve. Use any strategy.

6. **PHOTOCOPYING** Suppose you enlarge a drawing to 120% of its original size on the copy machine. If the drawing is 2 inches long and 3 inches wide, what are the dimensions of the copy?

7. **TOYS** Hannah was finding the relationship between the time it took a yo-yo to swing back and forth and its length. Predict the length of a yo-yo if it takes 5 seconds to swing back and forth.

Time (seconds)	Length (units)
1	1
2	4
3	9
4	16

GEOMETRY For Exercises 8 and 9, draw several rectangles and measure their diagonals.

8. Find a relationship between the diagonals of a rectangle.

9. See Exercise 1. Did you use inductive or deductive reasoning in Exercise 8?

10. **BASKETBALL** Placido, Dexter, and Scott play guard, forward, and center on a team, but not necessarily in that order. Placido and the center drove Scott to practice on Saturday. Placido does not play guard. Who is the guard?

11. **GARDENING** Shelby wants to fence in her garden with the dimensions shown. How much fencing will she need?

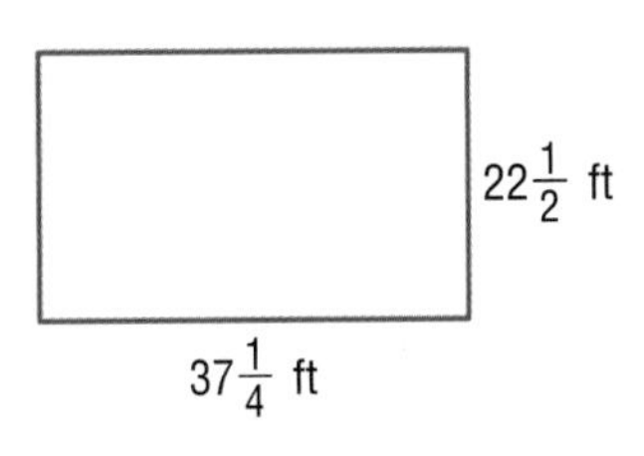

12. **CLOTHES** Careta put a $5 bill into one of seven pockets of her jacket. If she randomly reaches into one pocket, what is the probability that she will find the $5 bill?

13. **GEOMETRY** The large square has been divided into 9 squares. The lengths of the sides of two squares are given. Find the area of the entire square.

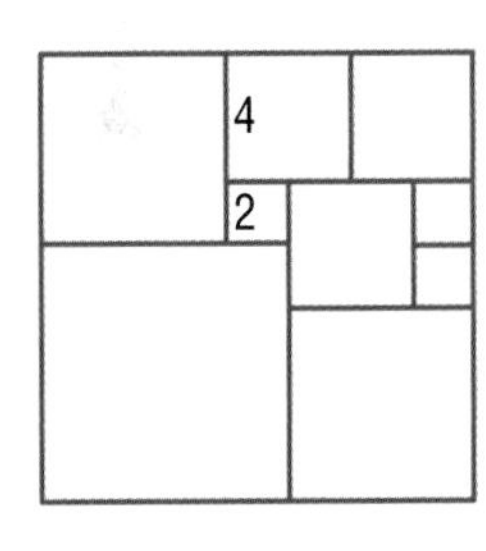

14. **STANDARDIZED TEST PRACTICE**
What can you conclude from the survey about favorite athletic shoes?

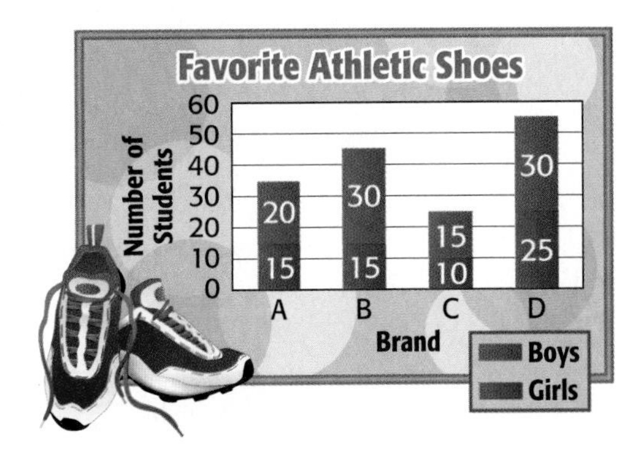

Ⓐ Brand A is the most popular among both boys and girls.

Ⓑ Brand B is two times more popular among boys than girls.

Ⓒ Brand C cost more than Brand D.

Ⓓ Brand A is less popular than Brand C.

You will use logical reasoning in the next lesson.

10-7 Polygons and Tessellations

What You'll LEARN

Classify polygons and determine which polygons can form a tessellation.

NEW Vocabulary

polygon
pentagon
hexagon
heptagon
octagon
nonagon
decagon
regular polygon
tessellation

WHEN am I ever going to use this?

GEOGRAPHY The size and shape of each state in the United States is different. Analyze the shapes of the states in both groups at the right.

Group 1	Group 2
Utah	Florida
Wyoming	Texas
New Mexico	Indiana
Colorado	North Carolina

1. Find the difference between the shapes of the states in Group 1 and the shapes of the states in Group 2.
2. Why do most states have boundaries that are not straight line segments?

The states listed in Group 1 above are examples of polygons. A **polygon** is a simple, closed figure formed by three or more straight lines. A simple figure does not have lines that cross each other. You have drawn a closed figure when your pencil ends up where it started.

Polygons	Not Polygons
• Line segments are called sides. • Sides meet only at their endpoints. • Points of intersection are called vertices.	• Figures whose sides overlap. • Figures that are open. • Figures that have curved sides.

Just as a triangle has three sides and a quadrilateral has four sides, other polygons can be classified by the number of sides they have.

Words	pentagon	hexagon	heptagon	octagon	nonagon	decagon
Number of Sides	5	6	7	8	9	10
Models						

READING Math

Polygons You can name polygons with more than 10 sides *n-gons*, with *n* representing the number of sides. For example, an 11-gon is a polygon with 11 sides.

A polygon that has all sides congruent and all angles congruent is called a **regular polygon**. Equilateral triangles and squares are regular polygons.

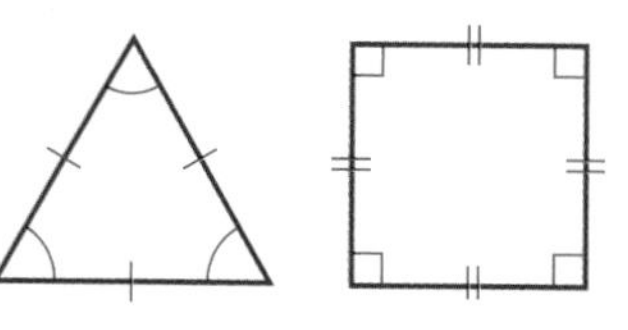

EXAMPLES Classify Polygons

Determine whether each figure is a polygon. If it is, classify the polygon and state whether it is regular. If it is *not* a polygon, explain why.

READING Math

Regular Polygons Since regular polygons have *equal-sized angles*, they are also called *equiangular*.

1

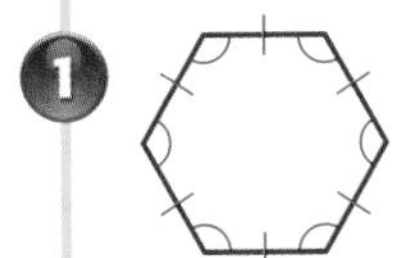

The figure has 6 congruent sides and 6 congruent angles. It is a regular hexagon.

2

The figure is not a polygon since it has a curved side.

Your Turn Determine whether each figure is a polygon. If it is, classify the polygon and state whether it is regular. If it is *not* a polygon, explain why.

a.

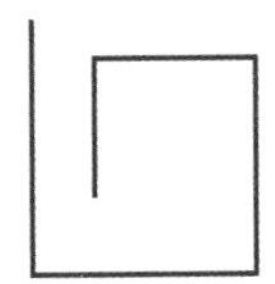

b.

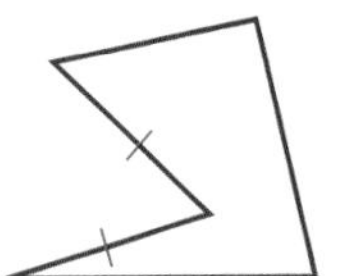

The sum of the measures of the angles of a triangle is 180°. You can use this relationship to find the measures of the angles of regular polygons.

EXAMPLE Angle Measures of a Polygon

STUDY TIP

Diagonal A diagonal of a polygon is a line segment that connects two vertices that are not next to each other.

3 **ALGEBRA Find the measure of each angle of a regular pentagon.**

- Draw all of the diagonals from one vertex as shown and count the number of triangles formed.
- Find the sum of the angle measures in the polygon.

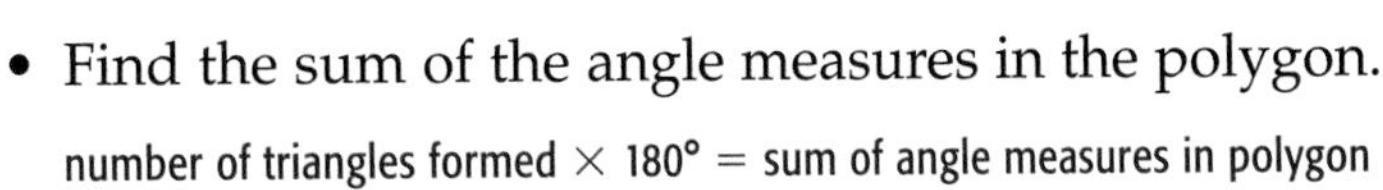

number of triangles formed $\times$ 180° = sum of angle measures in polygon

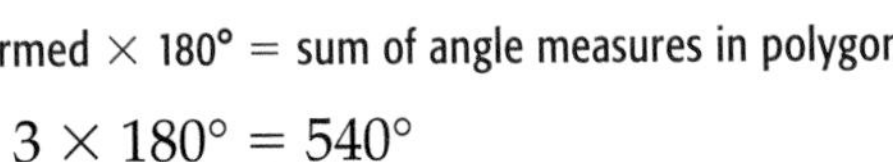

$$3 \times 180° = 540°$$

- Find the measure of each angle of the polygon. Let n represent the measure of one angle in the pentagon.

$5n = 540$ There are five congruent angles.

$n = 108$ Divide each side by 5.

The measure of each angle in a regular pentagon is 108°.

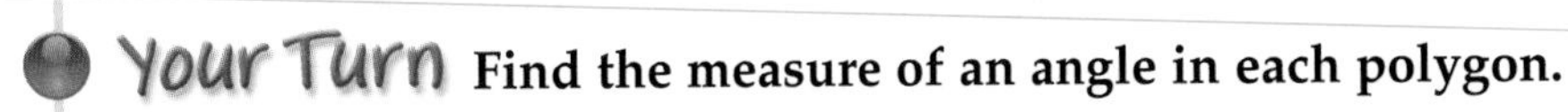

Your Turn Find the measure of an angle in each polygon.

c. regular octagon

d. equilateral triangle

A repetitive pattern of polygons that fit together with no overlaps or holes is called a **tessellation**. The surface of a chessboard is an example of a tessellation of squares.

msmath2.net/extra_examples

The sum of the measures of the angles where the vertices meet in a tessellation is 360°. The diagrams below show tessellations of equilateral triangles and squares.

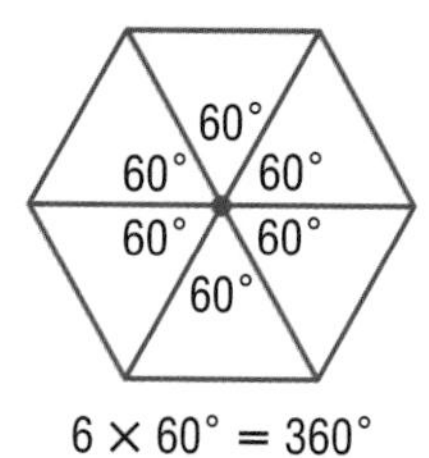

$6 \times 60° = 360°$

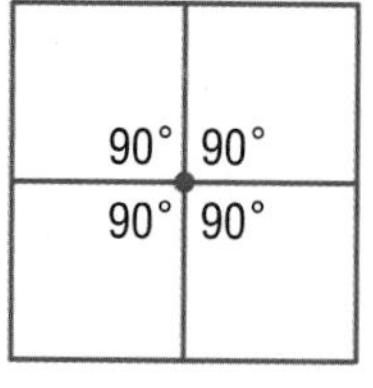

$4 \times 90° = 360°$

REAL-LIFE CAREERS

How Does a Landscape Architect Use Math?

Landscape architects multiply the length times the width of a garden to find the amount of mulch needed to cover the garden.

Online Research
For information about a career as a landscape architect, visit: msmath2.net/careers

EXAMPLE Tessellations

4 LANDSCAPING Mr. Brooks bought hexagonal-shaped stones to pave his patio. The stones are regular hexagons. Can Mr. Brooks tessellate his patio with the stones?

The measure of each angle in a regular hexagon is 120°.

The sum of the measures of the angles where the vertices meet must be 360°. So, solve $120n = 360$.

Since the solution, $n = 3$, is a whole number, a regular hexagon makes a tessellation.

Check You can check if your answer is correct by drawing a tessellation of regular hexagons.

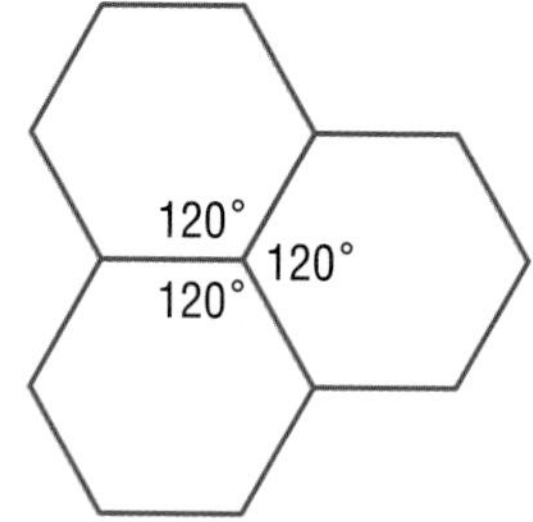

Skill and Concept Check

Writing Math Exercises 1 & 3

1. **Explain** how you know the parallelogram at the right can be used by itself to make a tessellation.

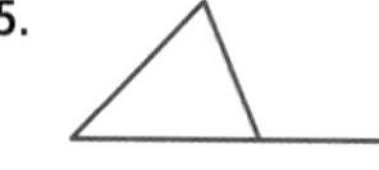

2. **OPEN ENDED** Draw examples of a pentagon, hexagon, heptagon, octagon, nonagon, and decagon.
3. **Explain** why a rhombus is not a regular polygon.

GUIDED PRACTICE

Determine whether each figure is a polygon. If it is, classify the polygon and state whether it is regular. If it is *not* a polygon, explain why.

4. 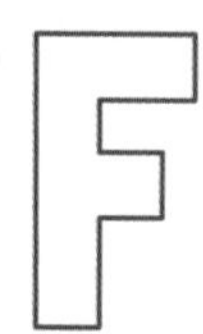

5. (figure)

6. Find the measure of each angle of a regular hexagon.
7. Can a regular polygon with an angle measure of 140° be used by itself to make a tessellation? Explain.

Practice and Applications

HOMEWORK HELP	
For Exercises	See Examples
8–13, 18, 20, 24–25, 27	1, 2
14–17, 19, 26	3
21–23, 28	4

Extra Practice See pages 589, 605.

Determine whether each figure is a polygon. If it is, classify the polygon and state whether it is regular. If it is *not* a polygon, explain why.

8. 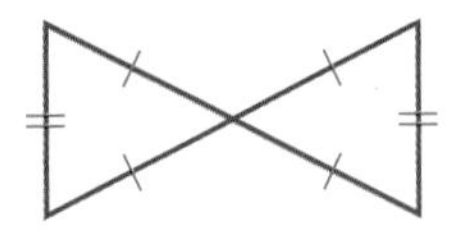

9. 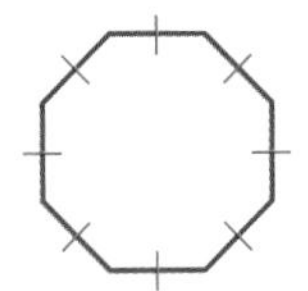

10. 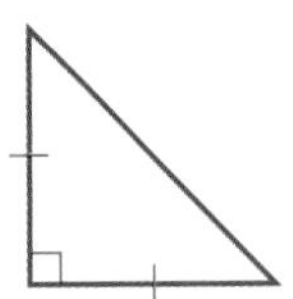

11.

12.

13.

Find the measure of an angle in each polygon if the polygon is regular. Round to the nearest tenth of a degree if necessary.

14. decagon **15.** nonagon **16.** 13-gon **17.** 20-gon

MANUFACTURING For Exercises 18–20, use the information below and the figure at the right.

A company designs cafeteria trays so that four people can place their trays around a square table without bumping corners. The trays are similar to the one at the right. The top and bottom sides of the tray are parallel.

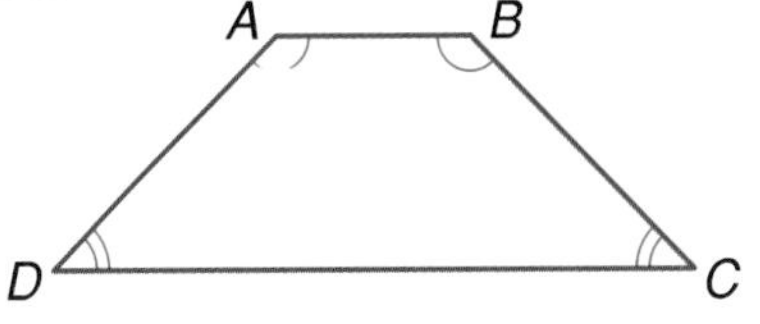

18. Classify the shape of the tray.

19. If $\angle A \cong \angle B$, $\angle C \cong \angle D$, and $m\angle A = 135°$, find $m\angle B$, $m\angle C$, and $m\angle D$.

20. Name the polygon formed by the outside of four trays when they are placed around the table with their sides touching.

Identify the polygons that are used to create each tessellation.

21. 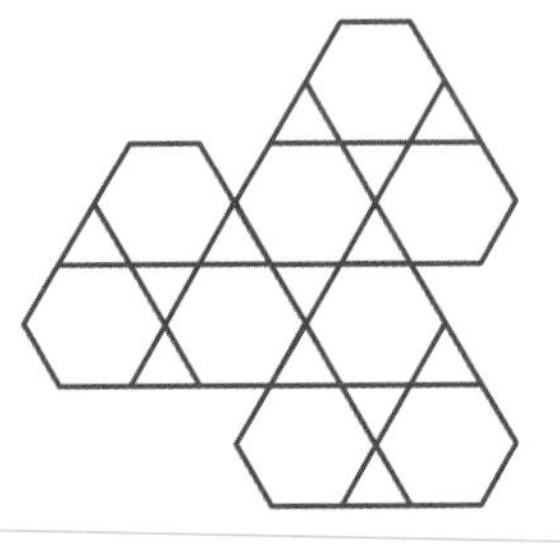

22. 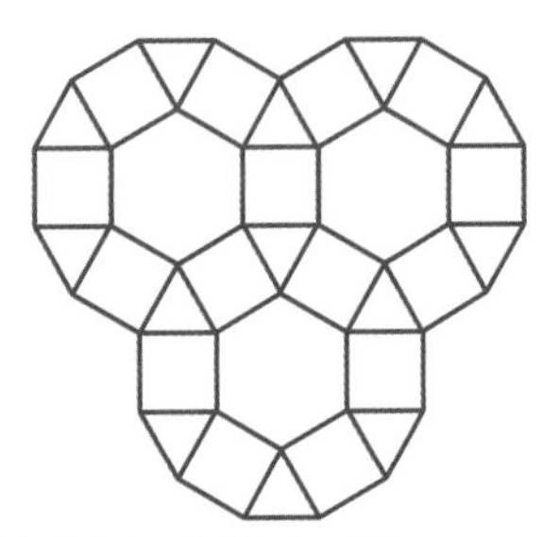

23. 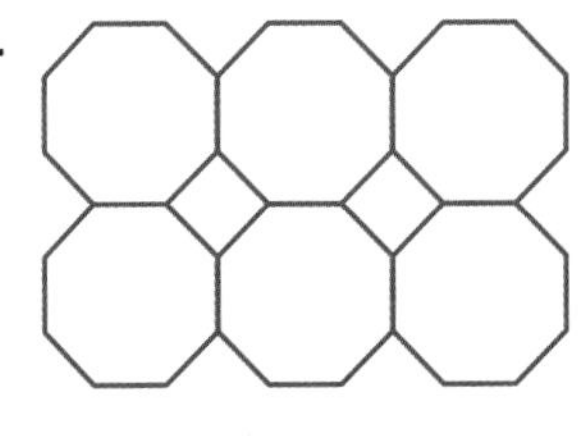

24. ART The mosaic at the right is decorated with handmade tiles from Pakistan. Name the polygons used in the tessellation.

25. What is the perimeter of a nonagon with sides 4.8 centimeters long?

26. Name the regular polygon if the sum of the measures of the polygon is 2,880° and the measure of one angle is 160°.

27. Find the perimeter of a pentagon having sides $7\frac{1}{2}$ yards long.

28. **SIGNS** Stop signs are made from large sheets of steel. Suppose nine stop signs can be cut from one sheet of steel. Can all nine signs be arranged on the sheet so that none of the steel goes to waste? Explain your reasoning.

29. **RESEARCH** Use the Internet or another source to find the shape of other traffic signs. Name the type of sign, its shape, and whether it is regular.

30. **WRITE A PROBLEM** Write about a real-life object or situation that contains a tessellation. Name the polygons used and draw a picture of the tessellation.

31. **CRITICAL THINKING** You can make a tessellation with equilateral triangles. Can you make a tessellation with any isosceles or scalene triangles? If so, explain your reasoning and make a drawing.

Standardized Test Practice and Mixed Review

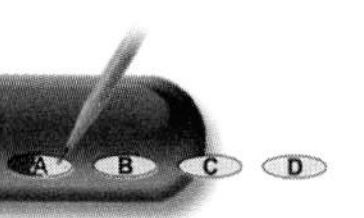

32. **MULTIPLE CHOICE** Which of the following flowchart symbols is *not* a polygon?

Ⓐ process box Ⓑ decision box Ⓒ stop box Ⓓ preparation box

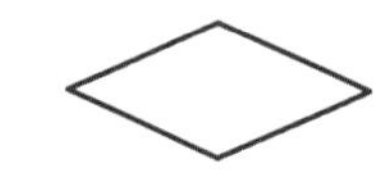

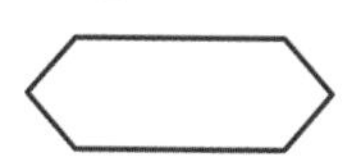

33. **MULTIPLE CHOICE** Which of the following regular shapes *cannot* be used by itself to make a tessellation?

Ⓕ hexagon Ⓖ triangle Ⓗ square Ⓘ pentagon

34. **SHORT RESPONSE** What is the sum of the measures of the angles of a quadrilateral?

For Exercises 35 and 36, use the figures at the right.

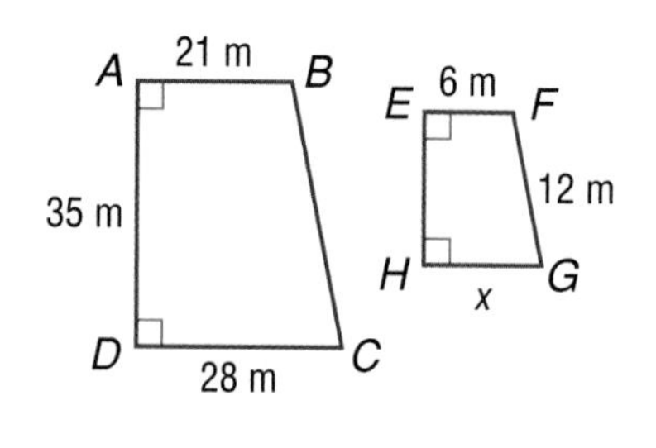

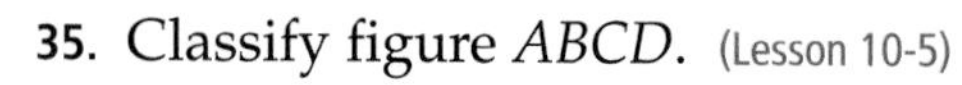

35. Classify figure *ABCD*. (Lesson 10-5)

36. The quadrilaterals are similar. Find the value of x. (Lesson 10-6)

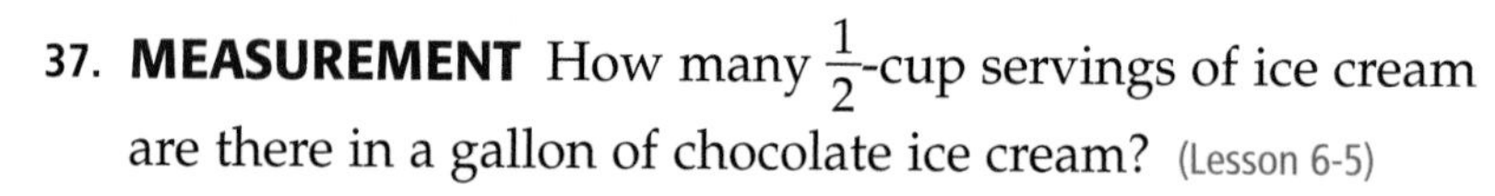

37. **MEASUREMENT** How many $\frac{1}{2}$-cup servings of ice cream are there in a gallon of chocolate ice cream? (Lesson 6-5)

Add or subtract. Write each sum or difference in simplest form. (Lesson 6-3)

38. $3\frac{2}{9} + 5\frac{4}{9}$

39. $5\frac{1}{3} - 2\frac{1}{6}$

40. $1\frac{3}{7} + 6\frac{1}{4}$

41. $9\frac{4}{5} - 4\frac{7}{8}$

GETTING READY FOR THE NEXT LESSON

PREREQUISITE SKILL On graph paper, draw a coordinate plane. Then graph and label each point. Connect the points in order. (Lesson 3-3)

42. $A(-2, 3)$

43. $B(4, 3)$

44. $C(2, -1)$

45. $D(-4, -1)$

10-8 Translations

HANDS-ON Mini Lab

Materials
- paper
- scissors

Work with a partner.

You can make changes in the polygons that tessellate to create new shapes that also tessellate.

STEP 1 Draw a square and cut a shape from it as shown.

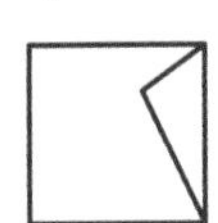

STEP 2 Slide the shape to the opposite side without turning it.

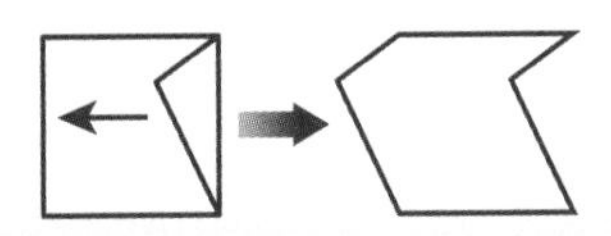

STEP 3 The new shape will make a tessellation.

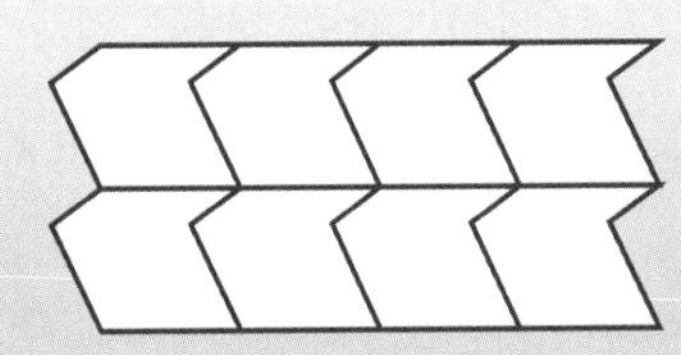

1. Make your own tessellation. Use different colors and shapes to make an interesting design.

What You'll LEARN

Graph translations of polygons on a coordinate plane.

NEW Vocabulary

transformation
translation

Link to READING

Everyday Meaning of Translation: a change from one language to another, as in an English-to-Spanish translation

Anytime you move a geometric figure, it is called a **transformation**. In the Mini Lab, you slid the polygon to a new position without turning it. This sliding motion is called a **translation**.

When translating a figure, every point of the original figure is moved the same distance and in the same direction.

EXAMPLE Graph a Translation

1 Translate trapezoid *HIJK* 3 units left and 5 units up.

- Move each vertex of the figure 3 units left and 5 units up. Label the new vertices H', I', J', and K'.
- Connect the vertices to draw the trapezoid. The coordinates of the vertices of the new figure are $H'(-2, 1)$, $I'(1, 1)$, $J'(1, 5)$, and $K'(-2, 4)$.

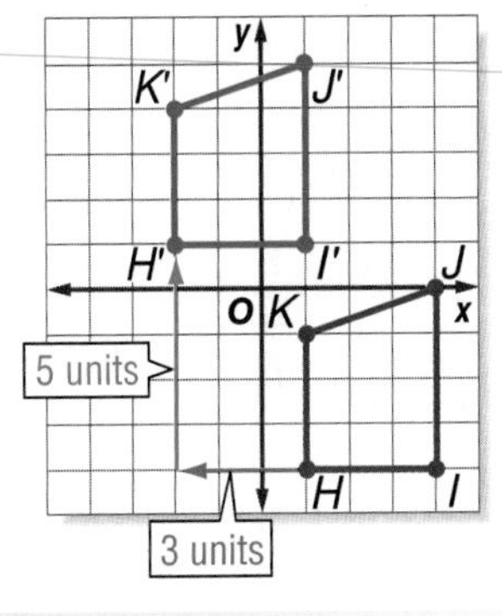

Transformations Whenever a figure is translated, use prime symbols for the vertices in the transformed image.
$A \rightarrow A'$
$B \rightarrow B'$
$C \rightarrow C'$

msmath2.net/extra_examples

When a figure has been translated, the original figure and the translated figure, or *image*, are congruent. In Example 1, trapezoid $HIJK$ is congruent to trapezoid $H'I'J'K'$.

You can also add to or subtract from the coordinates of the vertices of a figure to find the translated vertices.

EXAMPLE Find Coordinates of a Translation

2 **Triangle LMN has vertices $L(-1, -2)$, $M(6, -3)$, and $N(2, -5)$. Find the vertices of $\Delta L'M'N'$ after a translation of 6 units left and 4 units up. Then graph the figure and its translated image.**

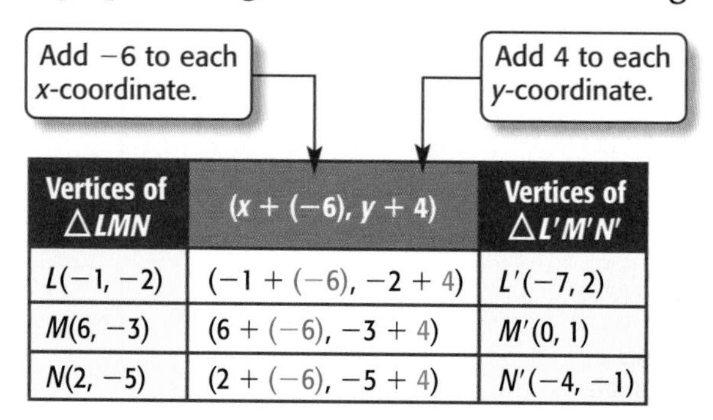

Add −6 to each x-coordinate.

Add 4 to each y-coordinate.

Vertices of ΔLMN	$(x + (-6), y + 4)$	Vertices of $\Delta L'M'N'$
$L(-1, -2)$	$(-1 + (-6), -2 + 4)$	$L'(-7, 2)$
$M(6, -3)$	$(6 + (-6), -3 + 4)$	$M'(0, 1)$
$N(2, -5)$	$(2 + (-6), -5 + 4)$	$N'(-4, -1)$

The coordinates of the vertices of $\Delta L'M'N'$ are $L'(-7, 2)$, $M'(0, 1)$, and $N'(-4, -1)$.

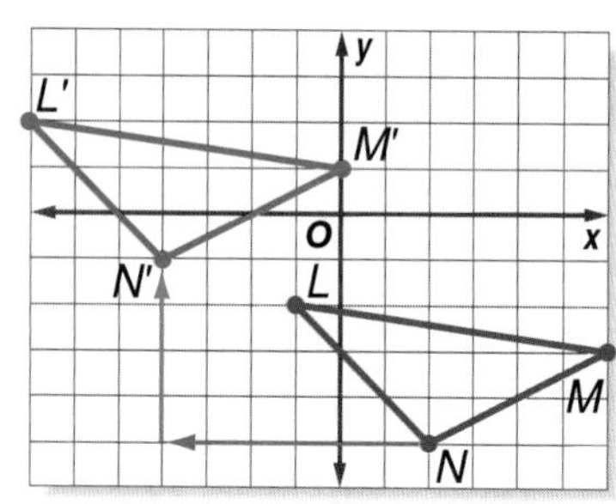

Your Turn **Triangle TUV has vertices $T(6, -3)$, $U(-2, 0)$, and $V(-1, 2)$. Find the vertices of $\Delta T'U'V'$ after each translation. Then graph the figure and its translated image.**

a. 3 units right and 4 units down **b.** 1 unit left and 6 units up

In Example 2, ΔLMN was translated 6 units left and 4 units up. This translation can be described using the ordered pair $(-6, 4)$.

EXAMPLE Naming Translations with Ordered Pairs

3 **GAMES When playing chess, the rook can only move vertically or horizontally across a chessboard. The chessboard at the right shows the movement of a rook after two turns. Describe this translation as an ordered pair.**

The rook moved 5 places left and 3 places up. So, the translation can be written as $(-5, 3)$.

STUDY TIP

Translation
A *positive* integer describes a translation right or up on a coordinate plane. A *negative* integer describes a translation left or down.

REAL-LIFE MATH

CHESS In 1977, Vlastimil Hort of Czechoslovakia played 550 opponents, 201 simultaneously, and lost only 10 games in just over 30 hours.

Source: www.research.ibm.com

Skill and Concept Check

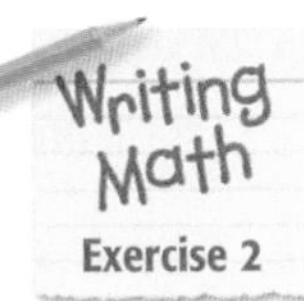

1. **OPEN ENDED** On a coordinate plane, draw a triangle and its translation 1 unit right and 4 units down.

2. **Which One Doesn't Belong?** Identify the translation that is not the same as the other three. Explain your reasoning.

(-5, -4)	(1, -3)	(6, -6)	(4, -2)

GUIDED PRACTICE

3. Translate $\triangle ABC$ 3 units left and 3 units down. Graph $\triangle A'B'C'$.

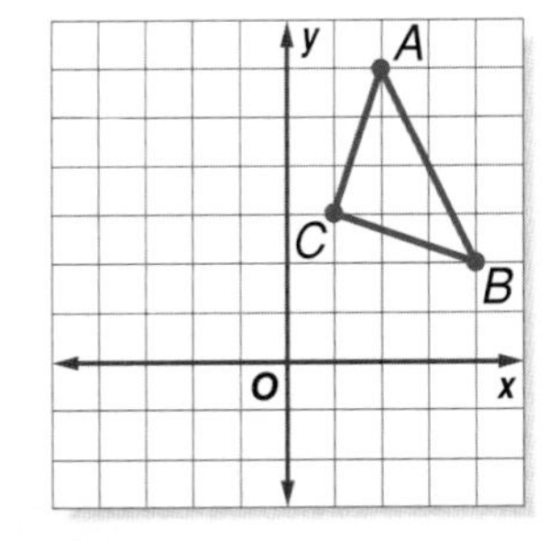

Quadrilateral *DEFG* has vertices *D*(1, 0), *E*(−2, −2), *F*(2, 4), and *G*(6, −3). Find the vertices of *D′E′F′G′* after each translation. Then graph the figure and its translated image.

4. 4 units right, 5 units down
5. 6 units right

Practice and Applications

6. Translate $\triangle HIJ$ 2 units right and 6 units down. Graph $\triangle H'I'J'$.

7. Translate rectangle *KLMN* 1 unit left and 3 units up. Graph rectangle *K′L′M′N′*.

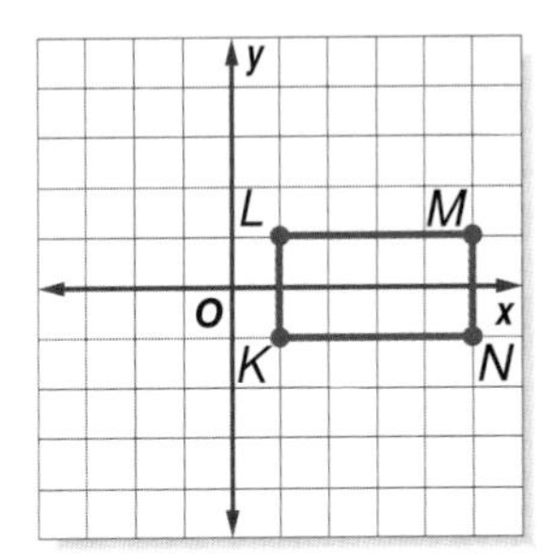

HOMEWORK HELP

For Exercises	See Examples
6–7	1
8–11, 15	2
12, 16	3

Extra Practice
See pages 589, 605.

Triangle *PQR* has vertices *P*(0, 0), *Q*(5, −2), and *R*(−3, 6). Find the vertices of *P′Q′R′* after each translation. Then graph the figure and its translated image.

8. 6 units right, 5 units up
9. 8 units left, 1 unit down
10. 3 units left
11. 9 units down

12. **MAPS** Payat lives at the corner of Wabash and Ohio. The school he attends is located at Huron and Dearborn. Describe Payat's walk from school to home as an ordered pair of the number of blocks.

13. **ART** Explain how translations and tessellations were used in *Horsemen*, created by M.C. Escher at the right.

Horsemen by M.C. Escher.

14. **RESEARCH** Use the Internet or another source to find other pieces of art that contain tessellations of translations. Describe how the artists incorporated both ideas into their work.

MULTI STEP For Exercises 15 and 16, use the following information.

$\triangle XYZ$ has vertices $X(7, 6)$, $Y(-3, 4)$, and $Z(1, -5)$. The translated figure $\triangle X'Y'Z'$ has vertices $X'(2, 4)$, $Y'(-8, 2)$, and $Z'(-4, -7)$.

15. Describe the translation.

16. Write the translation as an ordered pair.

17. **TESSELLATIONS** Use the figure at the right and patty paper to make a tessellation using translations.

18. **CRITICAL THINKING** Is it possible to make a tessellation with translations of equilateral triangles? Explain your reasoning.

Standardized Test Practice and Mixed Review

19. **MULTIPLE CHOICE** Which graph shows a translation of the letter Z?

(A)

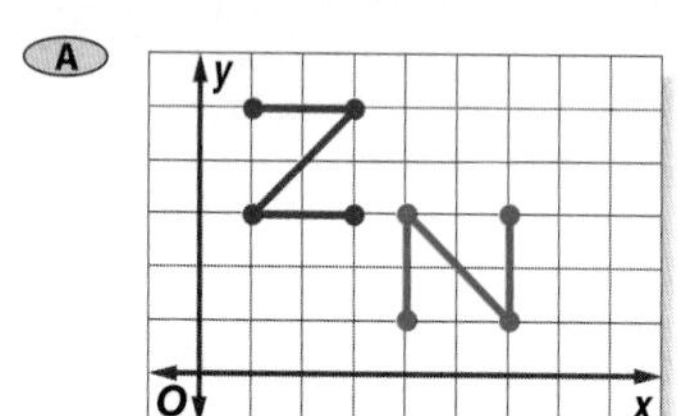

(B)

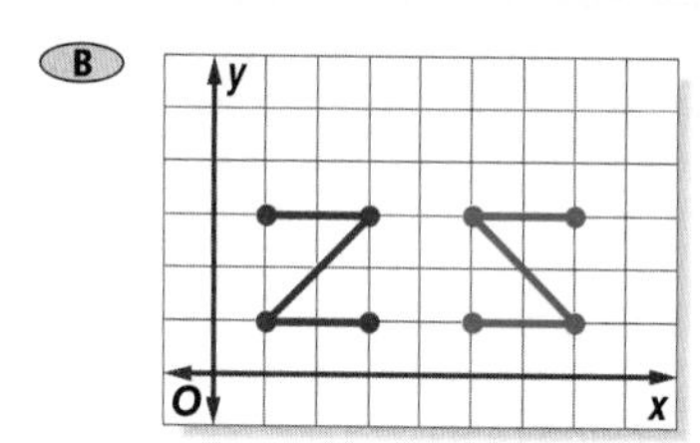

(C)

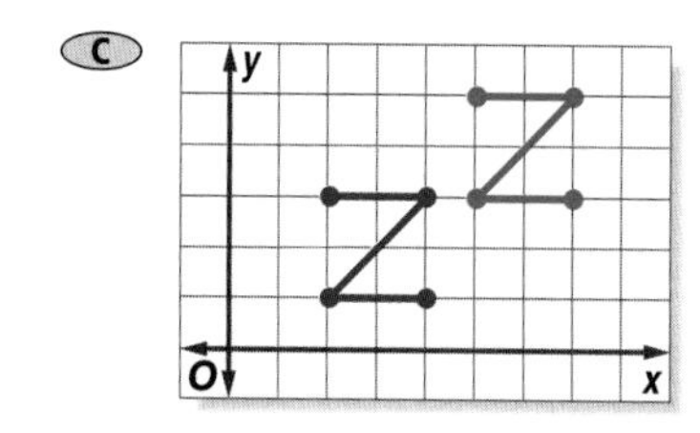

(D)

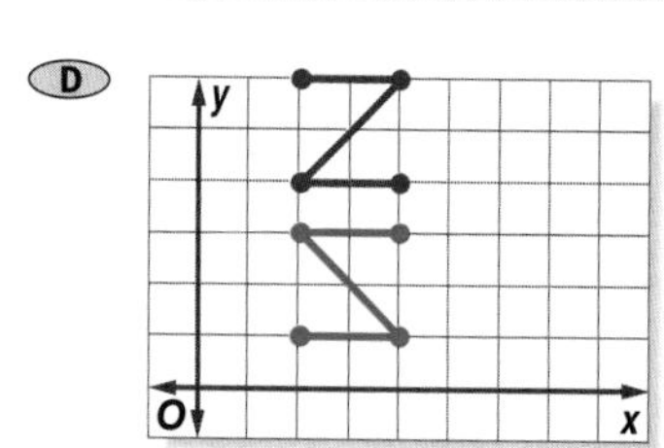

20. **SHORT RESPONSE** Triangle ABC with vertices $A(-5, 10)$, $B(5, 10)$, and $C(-10, 5)$ is translated by $(-10, 5)$. What are the coordinates of A'?

21. **GEOMETRY** What is the name of a polygon with eight sides? (Lesson 10-7)

22. **TOWERS** Horatio is 167 centimeters tall and casts a shadow 93 centimeters long. What is the approximate height of a cellphone tower if it casts a shadow 763 centimeters long? (Lesson 10-6)

GETTING READY FOR THE NEXT LESSON

BASIC SKILL Determine whether each figure can be folded in half so that one side matches the other. Write *yes* or *no*.

23. 24. 25. 26.

10-8b Spreadsheet Investigation

A Follow-Up of Lesson 10-8

What You'll LEARN

Use a spreadsheet to enlarge and reduce polygons.

Dilations

A computer spreadsheet is a useful tool for calculating the vertices of polygons. You can enlarge or reduce the polygons by using a spreadsheet to automatically calculate the new coordinates of the vertices. Enlarging or reducing a figure is a transformation called a **dilation**.

ACTIVITY

Emma has plotted a pentagon on graph paper. The coordinates of the vertices of the pentagon are (2, 2), (4, 2), (5, 4), (3, 6), and (1, 4). She wants to multiply the coordinates by 3 to enlarge the pentagon. She enters the coordinates on a spreadsheet as shown below.

Set up the spreadsheet like the one shown below.

Pentagon Dilation

	A	B	C	D
1	original		dilation	
2	x-coordinate	y-coordinate	x-coordinate	y-coordinate
3	2	2	=A3*3	=B3*3
4	4	2	=A4*3	=B4*3
5	5	4		
6	3	6		
7	1	4		
8				

Sheet1 Sheet2 Sheet3

Continue entering the formulas in columns C and D to complete the dilation.

EXERCISE

1. How will the formulas in columns C and D change the original pentagon? How do you know?
2. Graph the original pentagon and its dilation on graph paper.
3. What is the percent of increase of the original pentagon to its dilation?
4. Find the coordinates of the pentagon enlarged 5 times.
5. Find the coordinates of the pentagon reduced by one-half.
6. What type of dilation had occurred if the new coordinates of the pentagon are (5, 5), (10, 5), (12.5, 10), (7.5, 15), and (2.5, 10)? What is the scale factor?
7. Select another geometric figure and plot its points on graph paper. Set up a spreadsheet to find two dilations, one enlargement and one reduction of the same figure.

10-9 Reflections

HANDS-ON Mini Lab

Materials
- geomirror
- paper

Work with a partner.

- Write your first name in capital letters on a sheet of paper.
- Use the geomirror to trace the reflection of the letters in your name.
- Write your last name. Draw the reflection of the letters without using the geomirror.

1. Describe how you drew the reflection of your last name.
2. List the capital letters that look the same as their reflections.
3. Explain why the line where the geomirror and paper meet is called the *line of symmetry*.

What You'll LEARN

Identify figures with line symmetry and graph reflections on a coordinate plane.

NEW Vocabulary

line symmetry
line of symmetry
reflection

Figures that match exactly when folded in half have **line symmetry**. The figures below have line symmetry.

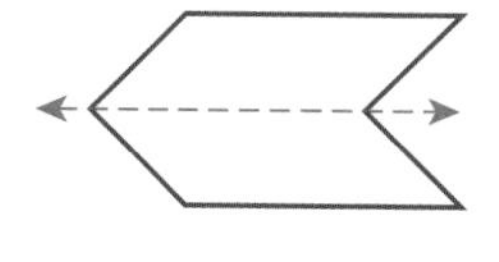

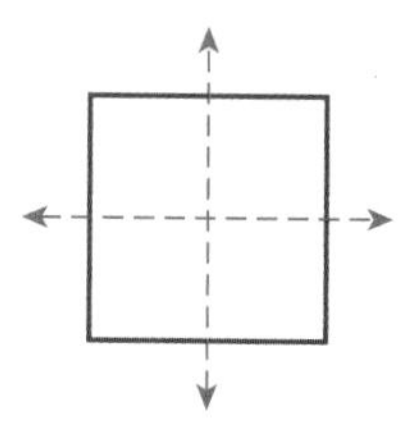

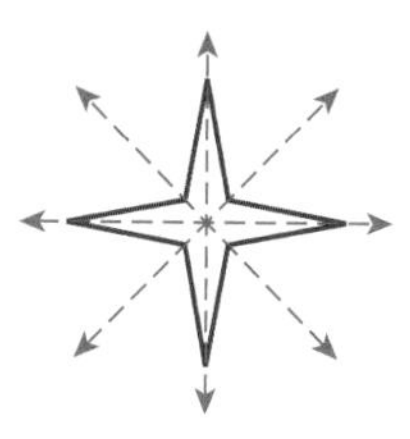

Each fold line is called a **line of symmetry**. Some figures can be folded in more than one way to show symmetry, so they have more than one line of symmetry.

EXAMPLES Identify Lines of Symmetry

Determine whether each figure has line symmetry. If so, copy the figure and draw all lines of symmetry.

1

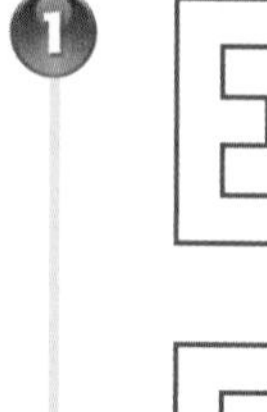

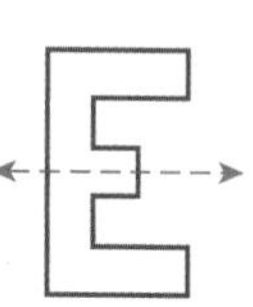

2

no symmetry

3

A type of transformation where a figure is flipped over a line of symmetry is a **reflection**. As with translations, the original figure and the reflected image are congruent.

STUDY TIP

x-axis When reflecting over the x-axis, change the y-coordinates to their opposites.

EXAMPLE Reflect a Figure Over the x-axis

4 **Triangle ABC has vertices A(5, 2), B(1, 3), and C(−1, 1). Find the coordinates of ABC after a reflection over the x-axis. Then graph the figure and its reflected image.**

Vertices of △ABC	Distance from x-axis	Vertices of △A′B′C′
A(5, 2)	2	A′(5, −2)
B(1, 3)	3	B′(1, −3)
C(−1, 1)	1	C′(−1, −1)

Plot the vertices and connect to form △ABC. The x-axis is the line of symmetry. So, the distance from each point on △ABC to the line of symmetry is the same as the distance from the line of symmetry to △A′B′C′.

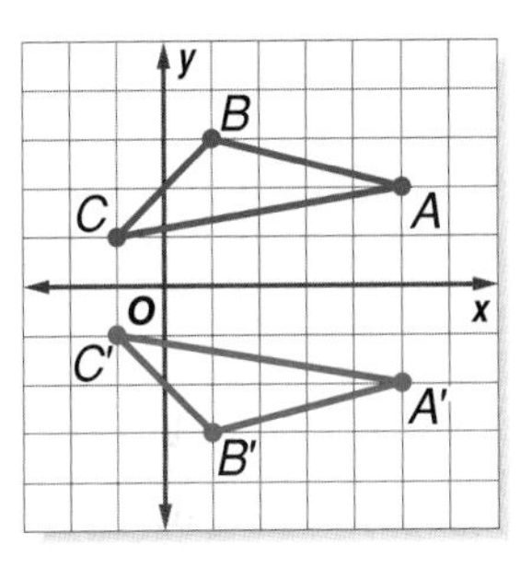

STUDY TIP

y-axis When reflecting over the y-axis, change the x-coordinates to their opposites.

EXAMPLE Reflect a Figure Over the y-axis

5 **Quadrilateral KLMN has vertices K(2, 3), L(5, 1), M(4, −2), and N(1, −1). Find the coordinates of KLMN after a reflection over the y-axis. Then graph the figure and its reflected image.**

Vertices of quad KLMN	Distance from y-axis	Vertices of quad K′L′M′N′
K(2, 3)	2	K′(−2, 3)
L(5, 1)	5	L′(−5, 1)
M(4, −2)	4	M′(−4, −2)
N(1, −1)	1	N′(−1, −1)

Plot the vertices and connect to form quadrilateral KLMN. The y-axis is the line of symmetry. So, the distance from each point on quadrilateral KLMN to the line of symmetry is the same as the distance from the line of symmetry to quadrilateral K′L′M′N′.

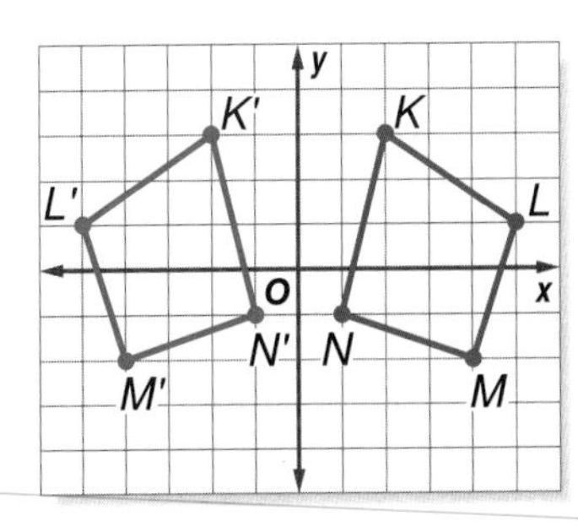

Your Turn

a. Find the coordinates of rectangle GHIJ with vertices G(3, −4), H(3, −1), I(−2, −1), and J(−2, −4) after a reflection over the x-axis. Then graph the figure and its reflected image.

b. Find the coordinates of triangle PQR with vertices P(1, 5), Q(3, 7), and R(5, −1) after a reflection over the y-axis. Then graph the figure and its reflected image.

Skill and Concept Check

Writing Math
Exercise 2

1. **OPEN ENDED** Draw a figure on a coordinate plane and its reflection over the y-axis.

2. **Which One Doesn't Belong?** Identify the transformation that is not the same as the other three. Explain your reasoning.

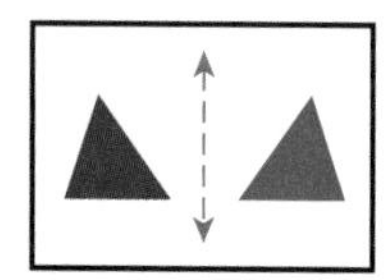 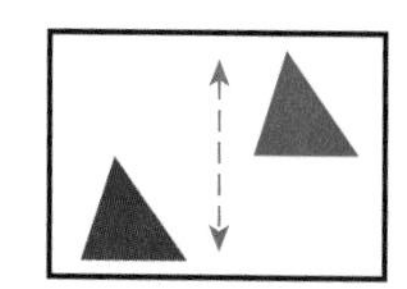 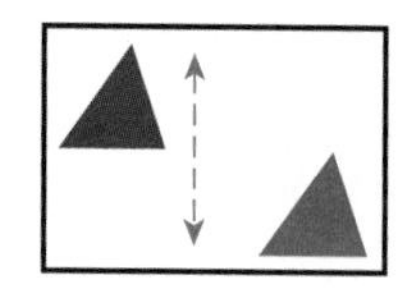 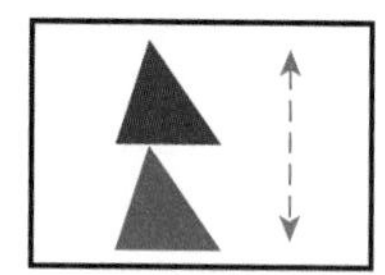

GUIDED PRACTICE

Determine which figures have line symmetry. Write *yes* or *no*. If *yes*, draw all lines of symmetry.

3. 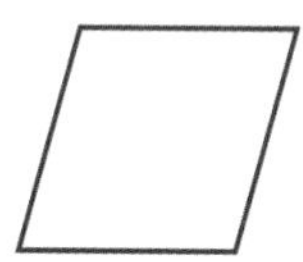

4. 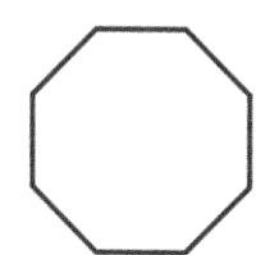

Find the coordinates of each figure after a reflection over the given axis. Then graph the figure and its reflected image.

5. $\triangle ABC$ with vertices $A(5, 8)$, $B(1, 2)$, and $C(6, 4)$; x-axis
6. parallelogram $WXYZ$ with vertices $W(-4, -2)$, $X(-4, 3)$, $Y(-2, 4)$, and $Z(-2, -1)$; y-axis

Practice and Applications

HOMEWORK HELP	
For Exercises	See Examples
7–10, 15–17	1–3
11–12	4
13–14	5
Extra Practice	

See pages 589, 605.

Determine which figures have line symmetry. Write *yes* or *no*. If *yes*, draw all lines of symmetry.

7. 

8. 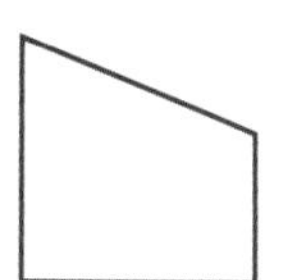

9. 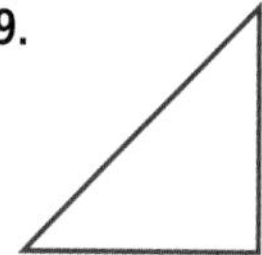

10. 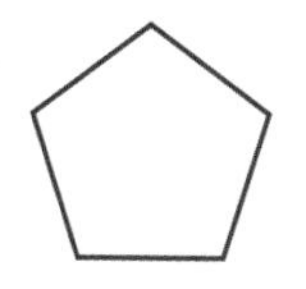

Find the coordinates of each figure after a reflection over the x-axis. Then graph the figure and its reflected image.

11. quadrilateral $DEFG$ with vertices $D(-3, 6)$, $E(-2, -3)$, $F(2, 2)$, and $G(4, 9)$
12. $\triangle TUV$ with vertices $T(-6, 1)$, $U(-2, -3)$, and $V(5, -4)$

Find the coordinates of each figure after a reflection over the y-axis. Then graph the figure and its reflected image.

13. $\triangle QRS$ with vertices $Q(2, -5)$, $R(4, -5)$, and $S(2, 3)$
14. parallelogram $HIJK$ with vertices $H(-1, 3)$, $I(-1, -1)$, $J(2, -2)$, and $K(2, 2)$

15. **BUILDINGS** Describe the location of the line(s) of symmetry in the photograph of the Taj Mahal.

16. **AMBULANCE** Explain why the word "AMBULANCE" is written backward and from right to left on the front of the emergency vehicle.

17. **SIGNAL FLAGS** International Marine Signal Flags are used by sailors to send messages at sea. A flag represents each letter of the alphabet. The flags below spell out the word *MATH*. Which flags have line symmetry? Draw all lines of symmetry.

M  A 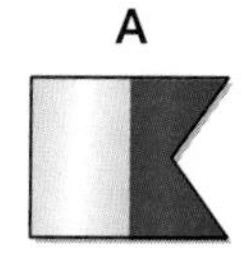 T 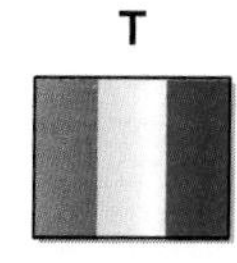 H 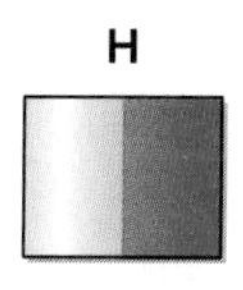

Data Update What other maritime signal flags have line symmetry? Visit msmath2.net/data_update for more information.

18. **TESSELLATIONS** Make a tessellation using a combination of translations and reflections of polygons. Explain how you created your tessellation.

19. **CRITICAL THINKING** Triangle JKL has vertices $J(-7, 4)$, $K(7, 1)$, and $L(2, -2)$. Without graphing, reflect the triangle over the x-axis and then over the y-axis. What are the new coordinates of the triangle after the double reflection?

Standardized Test Practice and Mixed Review

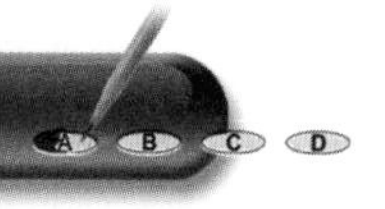

20. **MULTIPLE CHOICE** $\triangle ABC$ with vertices $A(-2, -5)$, $B(4, 1)$, and $C(3, -2)$ is reflected over the y-axis. Find the coordinates of the new figure.

   Ⓐ $A'(2, -5)$, $B'(-4, 1)$, $C'(-3, -2)$
   Ⓑ $A'(-2, 5)$, $B'(4, -1)$, $C'(3, 2)$
   Ⓒ $A'(2, 5)$, $B'(-4, -1)$, $C'(-3, 2)$
   Ⓓ $A'(2, 5)$, $B'(4, 1)$, $C'(3, 2)$

21. **SHORT RESPONSE** How many lines of symmetry, if any, are in the rhombus at the right?

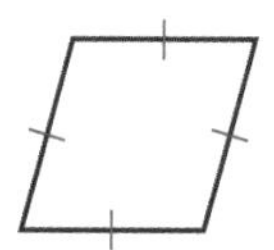

22. **GEOMETRY** Graph $\triangle FGH$ with vertices $F(-3, 7)$, $G(-1, 5)$, and $H(-2, 2)$ and its translation 4 units right and 1 unit down. Write the ordered pairs for the vertices of the new figure. (Lesson 10-8)

23. **ART** Aisha wishes to construct a tessellation for a wall hanging made only from regular decagons. Is this possible? Explain. (Lesson 10-7)

Estimate. (Lesson 6-1)

24. $\frac{4}{9} + 8\frac{1}{8}$

25. $\frac{1}{9} \times \frac{2}{5}$

26. $12\frac{1}{4} \div 5\frac{6}{7}$

27. Write an inequality for *six times a number is less than or equal to 18*. Then solve the inequality. (Lesson 4-5)

msmath2.net/self_check_quiz

Rotations

What You'll LEARN

Graph rotations on a coordinate plane.

Materials

- graph paper
- protractor
- ruler

Another type of transformation is a rotation. A **rotation** moves a figure around a central point. Another name for a rotation is a *turn*. In this lab, you will learn how to rotate a figure on a coordinate plane.

ACTIVITY *Work with a partner.*

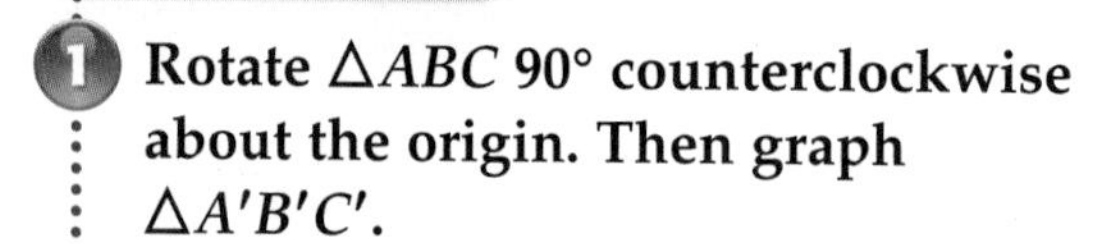

1 **Rotate $\triangle ABC$ 90° counterclockwise about the origin. Then graph $\triangle A'B'C'$.**

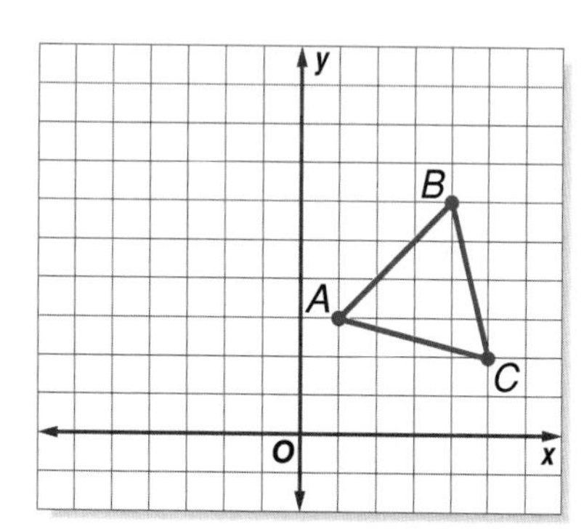

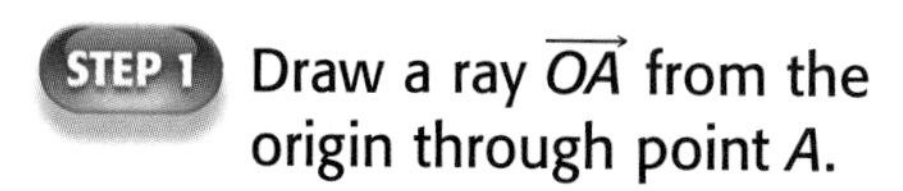

STEP 1 Draw a ray $\overrightarrow{OA}$ from the origin through point A.

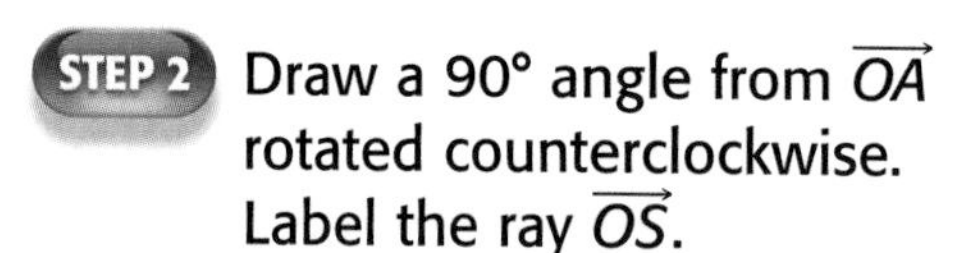

STEP 2 Draw a 90° angle from $\overrightarrow{OA}$ rotated counterclockwise. Label the ray $\overrightarrow{OS}$.

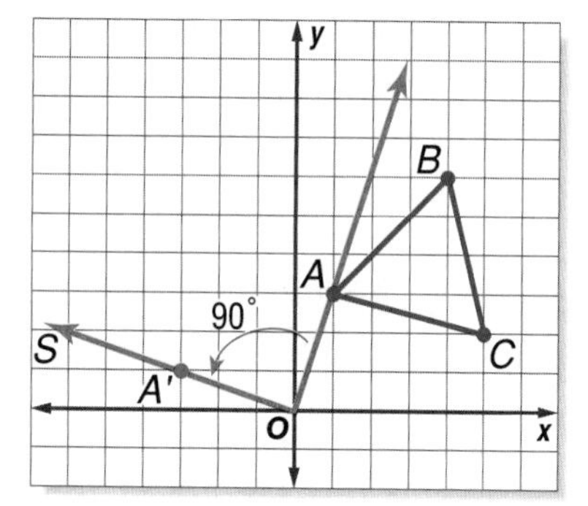

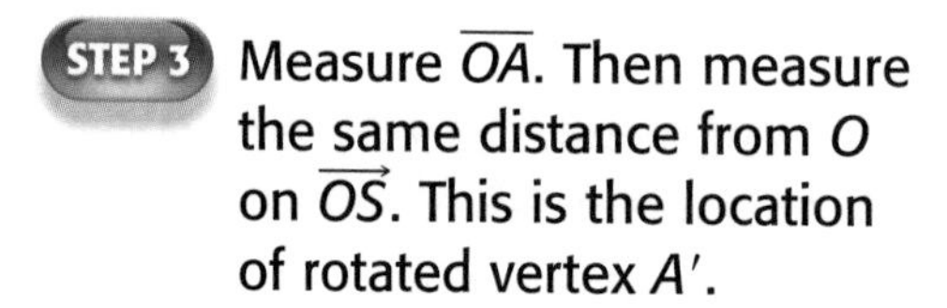

STEP 3 Measure $\overline{OA}$. Then measure the same distance from O on $\overrightarrow{OS}$. This is the location of rotated vertex A'.

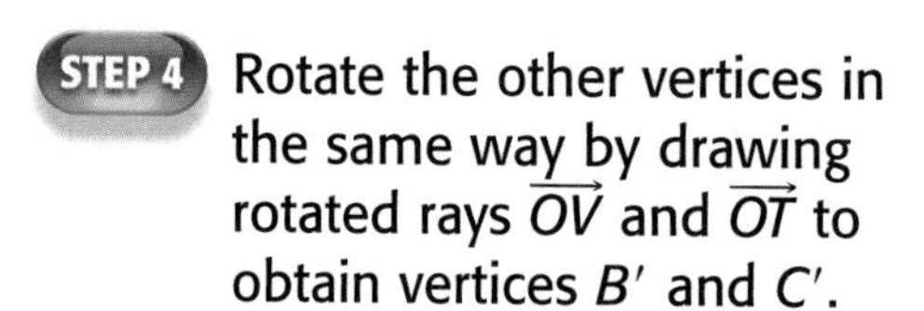

STEP 4 Rotate the other vertices in the same way by drawing rotated rays $\overrightarrow{OV}$ and $\overrightarrow{OT}$ to obtain vertices B' and C'.

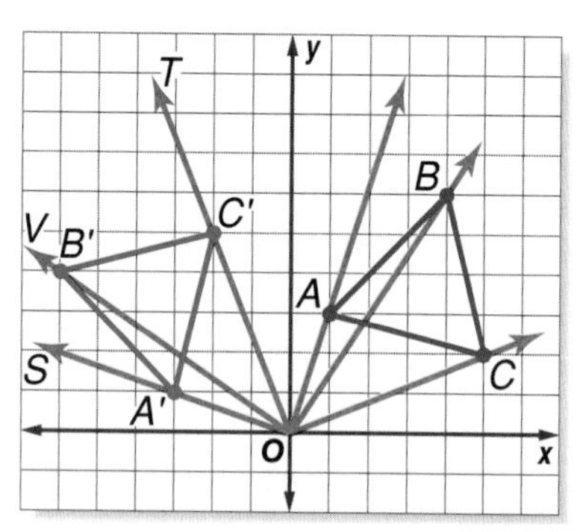

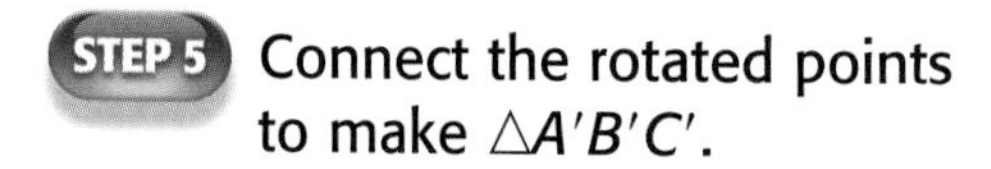

STEP 5 Connect the rotated points to make $\triangle A'B'C'$.

Rotated figures are congruent because the side lengths remain the same. So, $\triangle ABC \cong \triangle A'B'C'$.

READING Math

Rotations If you rotate an image once, the vertices contain a prime symbol ('). If you rotate a rotated image, the vertices contain a double prime symbol (").

Your Turn

a. Rotate $\triangle DEF$ 90° counterclockwise about the origin. Then graph $\triangle D'E'F'$.

b. Rotate $\triangle D'E'F'$ from Part a above 90° counterclockwise about the origin. Then graph $\triangle D''E''F''$.

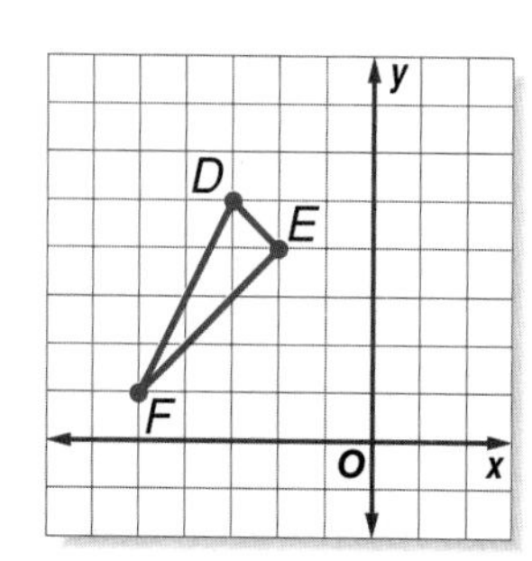

ACTIVITY *Work with a partner.*

2 **Identify the transformation as a *translation, reflection,* or *rotation.* The original figure is blue and the transformation is green.**

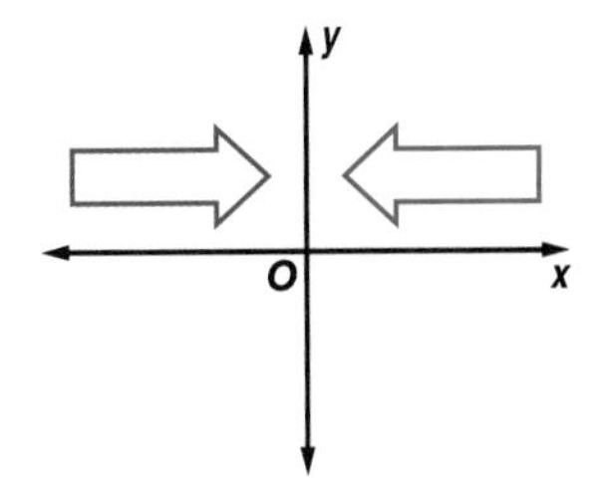

The distance from the point of the blue arrow to the line of symmetry is the same as the distance from the point of the green arrow to the line of symmetry. The y-axis is a line of symmetry. Therefore, the transformation is a reflection.

Your Turn **Identify the transformation as a *translation, reflection,* or *rotation.* The original figure is blue and the transformation is green.**

c.

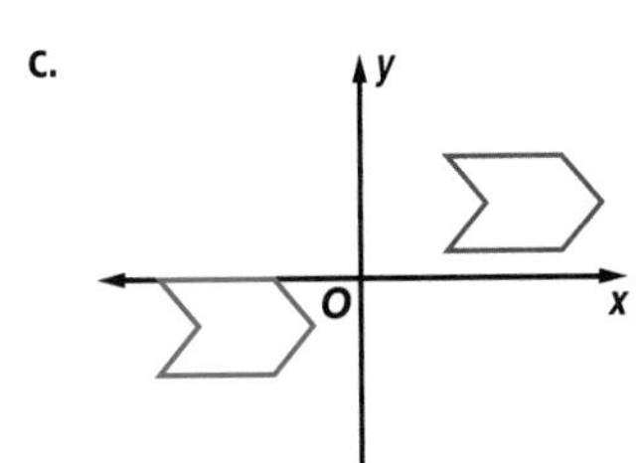

d.

e.

f.

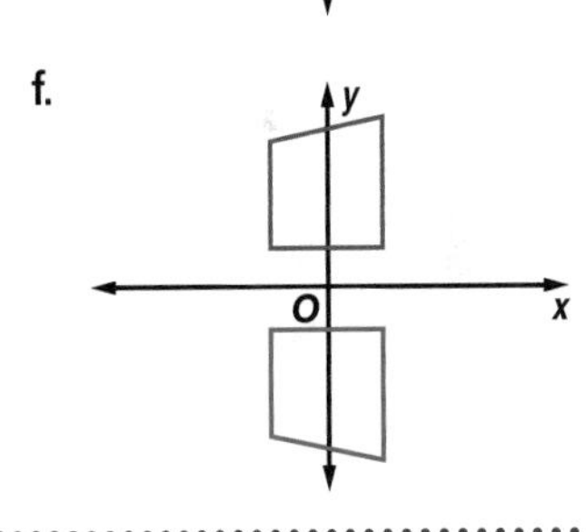

Writing Math

Work with a partner.

1. Rotate $\triangle DEF$ from Your Turn a 180° counterclockwise. Compare your rotated figure with your answer from Your Turn b.
2. **Make a conjecture** about rotating figures 180°.
3. Without rotating $\triangle DEF$ 360°, **make a conjecture** about rotating figures 360°. Describe the figure after a 360° rotation.
4. A 60° *counterclockwise* rotation is the same as a −300° *clockwise* rotation. Find an equivalent rotation of a 90° counterclockwise rotation.
5. Complete the following sentence. Since you can rotate figures in clockwise or counterclockwise directions, a 270° counterclockwise rotation produces the same result as a __?__ clockwise rotation.

Study Guide and Review

Vocabulary and Concept Check

acute angle (p. 413)
acute triangle (p. 429)
angle (p. 413)
circle graph (p. 418)
complementary angles (p. 422)
congruent angles (p. 422)
congruent segments (p. 429)
decagon (p. 446)
degrees (p. 413)
equilateral triangle (p. 429)
heptagon (p. 446)
hexagon (p. 446)
indirect measurement (p. 441)
isosceles triangle (p. 429)
line of symmetry (p. 456)
line symmetry (p. 456)
nonagon (p. 446)
obtuse angle (p. 413)
obtuse triangle (p. 429)
octagon (p. 446)
parallelogram (p. 434)
pentagon (p. 446)
polygon (p. 446)
quadrilateral (p. 434)
reflection (p. 457)
regular polygon (p. 446)
rhombus (p. 434)
right angle (p. 413)
right triangle (p. 429)
scalene triangle (p. 429)
similar figures (p. 440)
straight angle (p. 413)
supplementary angles (p. 422)
tessellation (p. 447)
transformation (p. 451)
translation (p. 451)
trapezoid (p. 434)
triangle (p. 428)
vertex (p. 413)
vertical angles (p. 422)

Choose the letter of the term that best matches each phrase.

1. the point where the sides of an angle meet
2. an angle whose measure is less than 90°
3. two angles whose measures add to 180°
4. a polygon with six sides
5. a triangle with no congruent sides

a. scalene triangle
b. complementary angles
c. vertex
d. hexagon
e. acute angle
f. supplementary angles

Lesson-by-Lesson Exercises and Examples

10-1 Angles (pp. 413–415)

Classify each angle as *acute, obtuse, right,* or *straight.*

6.
7.
8.
9.

10. Draw $\angle PQR$ with a measure of 128°.

Example 1 **Classify the angle as *acute, obtuse, right,* or *straight.***

The angle is an acute angle because its measure is less than 90°.

msmath2.net/vocabulary_review

10-2 Making Circle Graphs (pp. 418–421)

11. Make a circle graph of the data.

Favorite Soft Drink	Percent
Cola	36%
Diet Cola	28%
Root Beer	15%
Lemon Lime	7%
Other	14%

Example 2 Make a circle graph of the data of favorite season: 40% spring, 26% summer, 22% fall, and 12% winter.

First find the degrees for each part. Then construct the circle graph.

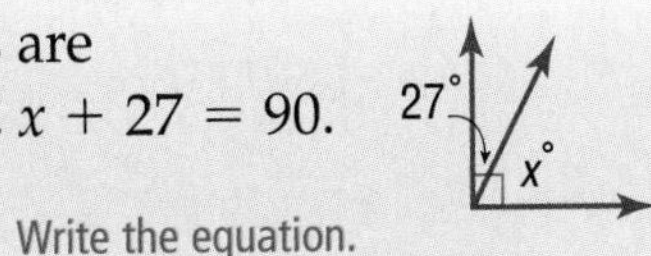

10-3 Angle Relationships (pp. 422–425)

Find the value of x in each figure.

12. 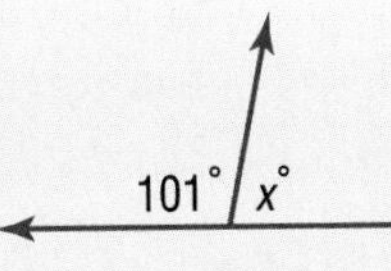

13.

14. If $\angle Y$ and $\angle Z$ are complementary angles and $m\angle Z = 35°$, find $m\angle Y$.

Example 3 Find the value of x.

Since the angles are complementary, $x + 27 = 90$.

$$\begin{aligned} x + 27 &= 90 && \text{Write the equation.} \\ -27 &\quad -27 && \text{Subtract 27 from each side.} \\ x &= 63 \end{aligned}$$

So, the value of x is 63.

10-4 Triangles (pp. 428–431)

Classify each triangle by its angles and by its sides.

15. 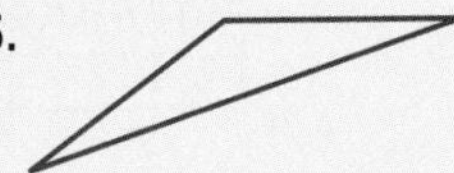

16. 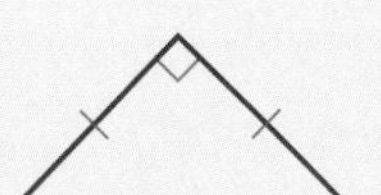

17. Find $m\angle S$ in $\triangle RST$ if $m\angle R = 28°$ and $m\angle T = 13°$.

Example 4 Classify the triangle by its angles and by its sides.

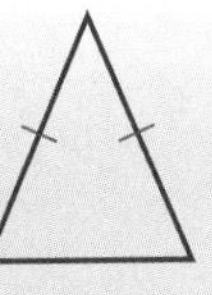

The triangle is acute since all three angles are acute. It is also isosceles because it has two congruent sides.

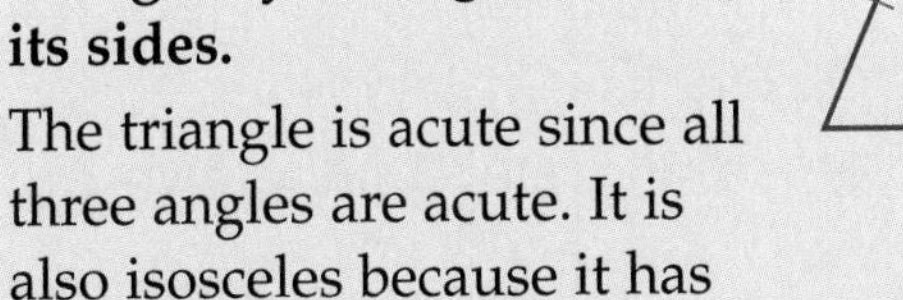

10-5 Quadrilaterals (pp. 434–437)

Classify the quadrilateral using the name that *best* describes it.

18. 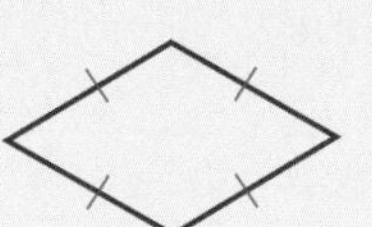

19. 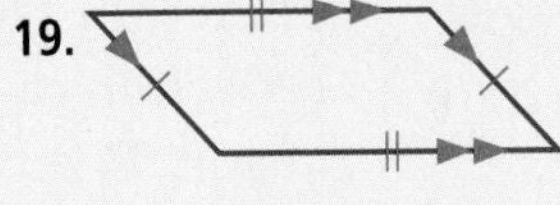

20. What type of quadrilateral may *not* have opposite sides congruent?

Example 5 Classify the quadrilateral using the name that *best* describes it.

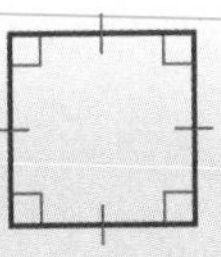

The quadrilateral is a parallelogram with 4 right angles and 4 congruent sides. It is a square.

10-6 Similar Figures (pp. 440–443)

Find the value of x in each pair of similar figures.

21. 5 cm, 8 cm, x, 16 cm

22. 4 m, 7 m, x, 6 m

Example 6 **Find the value of x in the pair of similar figures.**

x, 7 in., 16 in., 28 in.

$\frac{7}{28} = \frac{x}{16}$ Write a proportion.

$28x = 112$ Multiply.

$x = 4$ Divide.

10-7 Polygons and Tessellations (pp. 446–450)

Classify each polygon below and state whether it is regular.

23.

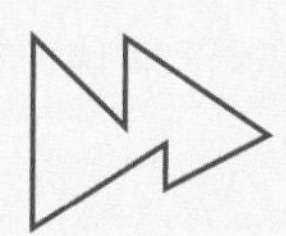

24.

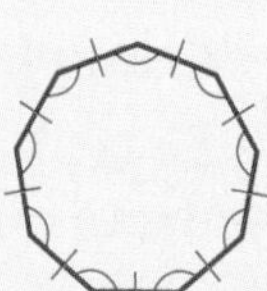

25. ALGEBRA Find the measure of each angle of a regular 15-gon.

Example 7 **Classify the polygon and state whether it is regular.**

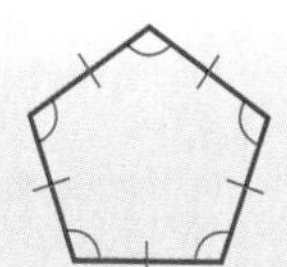

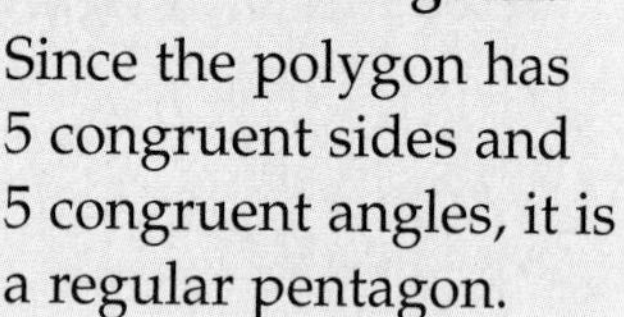

Since the polygon has 5 congruent sides and 5 congruent angles, it is a regular pentagon.

10-8 Translations (pp. 451–454)

Triangle PQR has coordinates $P(4, -2)$, $Q(-2, -3)$, and $R(-1, 6)$. Find the coordinates of $P'Q'R'$ after each translation. Then graph each translation.

26. 6 units left, 3 units up
27. 4 units right, 1 unit down
28. 3 units left
29. 7 units down

Example 8
Find the coordinates of $\triangle G'H'I'$ after a translation of 2 units left and 4 units up.

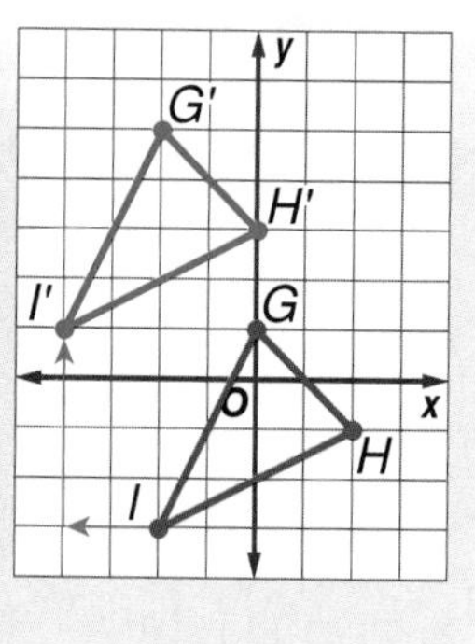

The vertices of $\triangle G'H'I'$ are $G'(-2, 5)$, $H'(0, 3)$, and $I'(-4, 1)$.

10-9 Reflections (pp. 456–459)

Find the coordinates of each figure after a reflection over the given axis. Then graph the figure and its reflected image.

30. $\triangle RST$ with coordinates $R(-1, 3)$, $S(2, 6)$, and $T(6, 1)$; x-axis
31. parallelogram $ABCD$ with coordinates $A(1, 3)$, $B(2, -1)$, $C(5, -1)$, and $D(4, 3)$; y-axis

Example 9
Find the coordinates of $\triangle C'D'E'$ after a reflection over the y-axis. Then graph its reflected image.

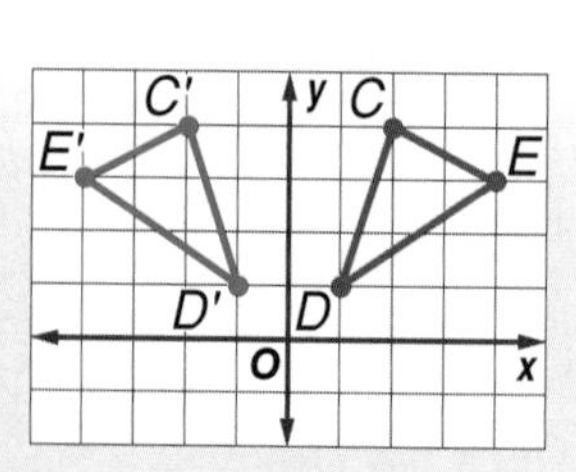

The vertices of $\triangle C'D'E'$ are $C'(-2, 4)$, $D'(-1, 1)$, and $E'(-4, 3)$.

CHAPTER 10 Practice Test

Vocabulary and Concepts

1. **List** the number of sides and number of angles in an octagon.
2. **Explain** how to translate a geometric figure.

Skills and Applications

Classify each angle as *acute, obtuse, right,* or *straight.*

3. 53°
4. 97°
5. 180°

Classify each pair of angles as *complementary, supplementary,* or *neither.*

6. 55° 125°

7. 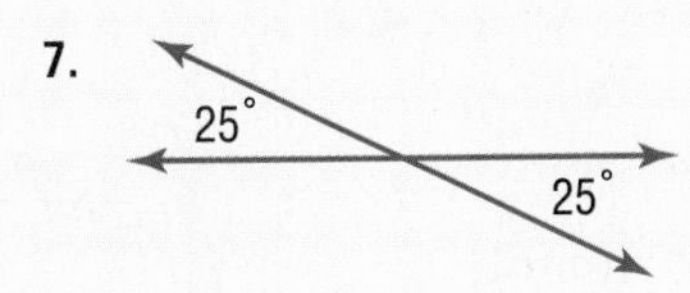

8. 45° 45°

9. **FOOD** Angelo's Pizza Parlor makes square pizzas. After baking, the pizzas are cut along one diagonal into two triangles. Describe the triangles.

Find the missing measure in each quadrilateral.

10. 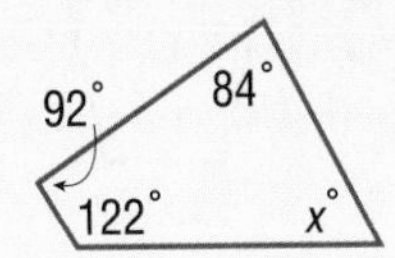

11. 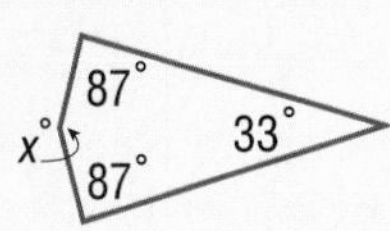

12. 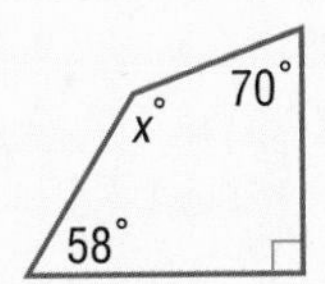

13. **GEOGRAPHY** Luz is drawing a map similar to one found in the atlas. The map in the atlas is 7 inches wide and 10 inches long. If she draws the width 17.5 inches, how long should she draw the map?

14. Can a regular heptagon, whose angle measures total 900°, be used by itself to make a tessellation? Explain your reasoning.

15. Draw a figure with exactly one line of symmetry.

Standardized Test Practice

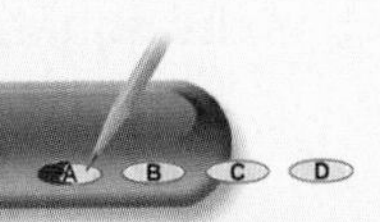

16. **MULTIPLE CHOICE** The table shows the results of a student survey of favorite type of book. Choose the statement that is *false*.

Favorite Type of Books	
Mystery	24
Science Fiction	8
Sports	26
Romance	30

(A) The science fiction section on a circle graph of the data at the right has an angle measure of about 33°.

(B) Romance books were the most favorite types of books.

(C) About 30% of students chose sports books as their favorite.

(D) The mystery and sports sections on a circle graph of the data at the right have supplementary angles.

CHAPTER 10

Standardized Test Practice

PART 1 Multiple Choice

Record your answers on the answer sheet provided by your teacher or on a sheet of paper.

1. Which inequality is false? (Prerequisite Skill, p. 556)

   Ⓐ $0.0059 > 0.0005$

   Ⓑ $6.1530 < 6.1532$

   Ⓒ $89.13 > 89.10$

   Ⓓ $5.06 < 5.006$

2. On Spring Street, there is an apartment building that houses four families, each with four members. One of the families moves away. Express this change in residents as an integer. (Lesson 3-1)

   Ⓕ 12 Ⓖ 4

   Ⓗ −4 Ⓘ −1

3. Find the algebraic expression for the following statement. *Blanca has five more than three times as many CDs as Garrett.* (Lesson 4-1)

   Ⓐ $B = \frac{3g}{5}$ Ⓑ $B = 3g - 5$

   Ⓒ $B = 3(g + 5)$ Ⓓ $B = 5 + 3g$

4. Which is the equation for the graphed line? (Lesson 4-7)

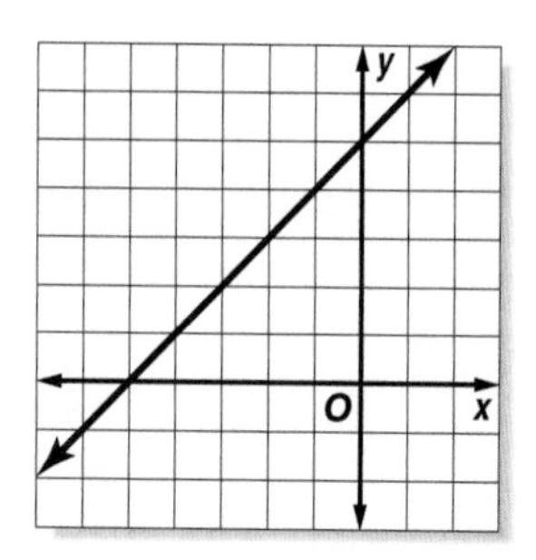

   Ⓕ $y = x + 5$ Ⓖ $y = x - 5$

   Ⓗ $y = 5x$ Ⓘ $y = 5x + 5$

5. The tree diagram shows the types of notebooks Esther can buy for school. If Esther selects one at random, what are the chances that she will select an orange notebook without lines? (Lesson 9-2)

   notebook with lines — orange, yellow, blue

   notebook without lines — orange, yellow, blue

   Ⓐ $\frac{1}{12}$ Ⓑ $\frac{1}{8}$ Ⓒ $\frac{1}{6}$ Ⓓ $\frac{1}{5}$

6. Which two of the figures below appear to be similar? (Lesson 10-6)

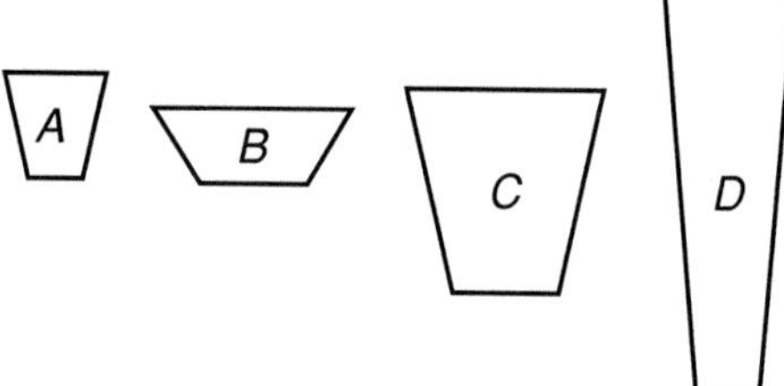

   Ⓕ figure A and figure C

   Ⓖ figure B and figure C

   Ⓗ figure A and figure D

   Ⓘ figure A and figure B

7. Kendrick chose a frame for a painting that looks like the polygon. Is this a regular polygon? (Lesson 10-7)

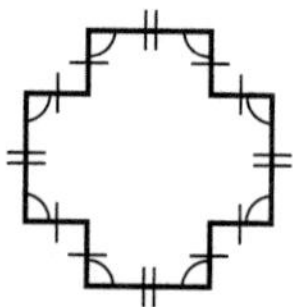

   Ⓐ Yes, because the angles are congruent.

   Ⓑ No, because the sides are not congruent.

   Ⓒ Yes, because the polygon is made up of line segments.

   Ⓓ Yes, because the angles and sides are congruent.

8. Which letter has two lines of symmetry? (Lesson 10-9)

   Ⓕ B Ⓖ E Ⓗ L Ⓘ X

PART 2 Short Response/Grid In

Record your answers on the answer sheet provided by your teacher or on a sheet of paper.

9. According to the box-and-whisker plot, what is the interquartile range? (Lesson 2-6)

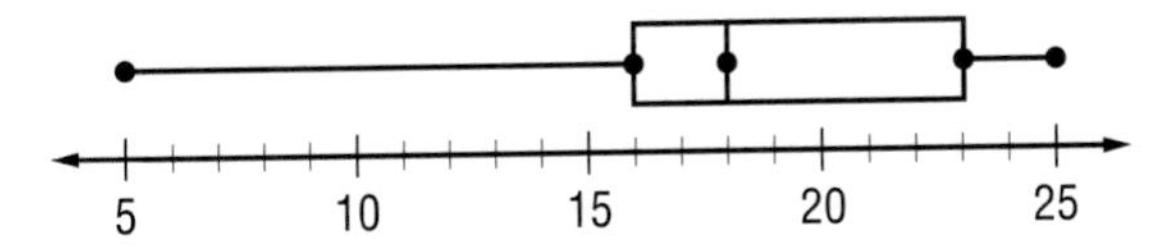

10. The table shows the values of p and q, where the values of p and q form a proportion. What are the values of Y and Z? (Lesson 7-3)

p	4	16	Z
q	7	Y	63

11. The veterinary hospital treated 162 animals last week. It treated 204 animals this week. What was the percent of increase? (Lesson 8-4)

12. Rebecca can choose seven different classes from five different class periods offered during the school day. How many possible ways can Rebecca arrange her class schedule? (Lesson 9-3)

13. How many degrees do the hands of a clock make at 9:00? (Lesson 10-1)

14. How long should the missing sides of the triangle be to make it isosceles with a perimeter of 15 centimeters? (Lesson 10-4)

3 cm

15. Find the sum of the angle measures of a hexagon. (Lesson 10-7)

16. How many lines of symmetry does the figure at the right have? (Lesson 10-9)

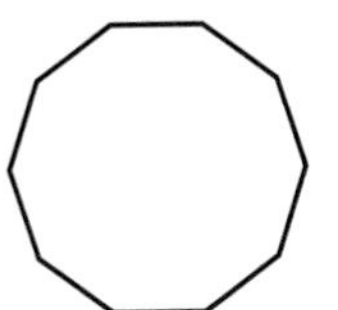

TEST-TAKING TIP

Question 17 Most standardized tests allow you to write in the test booklet. Use it to work out problems. For example, plot a figure and its translation on the coordinate graph given with the question in the booklet. Doing so will help keep you from making careless errors.

17. Triangle ABC with vertices $A(3, -7)$, $B(2, -3)$, and $C(8, -3)$ was translated to $A'B'C'$ with vertices $A'(-3, -3)$, $B'(-4, 1)$, and $C'(2, 1)$. Represent the translation as an ordered pair. (Lesson 10-8)

PART 3 Extended Response

Record your answers on a sheet of paper. Show your work.

18. Edmundo plotted polygon $JKLM$ on the coordinate plane below.

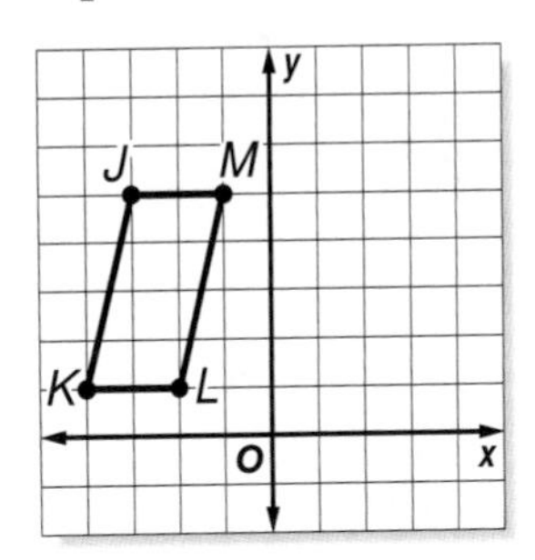

a. Classify $\angle J$. (Lesson 10-1)

b. Classify $\angle M$. (Lesson 10-1)

c. Classify polygon $JKLM$ using the name that *best* describes it. Be as specific as possible. (Lesson 10-5)

d. Can polygon $JKLM$ be used by itself to make a tessellation? Explain your reasoning. (Lesson 10-7)

e. If polygon $JKLM$ is translated 2 units right and 5 units down, what are the coordinates of the new figure? (Lesson 10-8)

f. If polygon $JKLM$ is reflected over the x-axis, what are the coordinates of the new figure? (Lesson 10-9)

Chapter 11 Geometry: Measuring Two-Dimensional Figures

"What does landscape design have to do with math?"

In designing a circular path, pool, or fountain, landscape architects calculate the area of the region. **They use the formula $A = \pi r^2$, where r is the radius of the circle.** There are many formulas involving two-dimensional figures that are useful in real life.

You will solve problems involving areas of circles in Lesson 11-6.

GETTING STARTED

Take this quiz to see whether you are ready to begin Chapter 11. Refer to the lesson or page number in parentheses if you need more review.

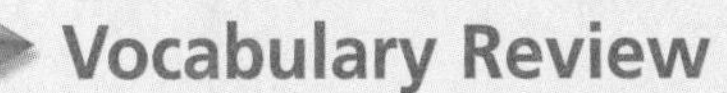

Vocabulary Review

State whether each sentence is *true* or *false*. If *false*, replace the underlined word or number to make a true sentence.

1. 0.5 is a <u>rational</u> number. (Lesson 5-4)
2. π is approximately equal to <u>3.14</u>. (Lesson 6-9)
3. A triangle with angle measures of 60°, 90°, and 30° is an <u>acute</u> triangle. (Lesson 10-4)

Prerequisite Skills

Replace each ● with <, >, or = to make a true sentence. (Page 556)

4. $67 \bullet 8.2 \cdot 8.2$
5. $11.1 \times 11.1 \bullet 123$
6. $5.9(5.9) \bullet 34.9$
7. $12.25 \bullet 3.5 \cdot 3.5$

Evaluate each expression. (Lesson 1-2)

8. 3^2
9. 8 squared
10. 5 to the third power
11. 6 to the second power

Find each value. (Lesson 1-3)

12. $\frac{1}{2}(5)(6)$
13. $\frac{1}{2}(4)(12 + 18)$
14. $9(3 + 3)$
15. $7(2)(8)$
16. $2(5)(4 + 1)$

Find the probability of rolling each number on a number cube. (Lesson 9-1)

17. $P(3)$
18. $P(6 \text{ or } 2)$
19. $P(\text{odd})$
20. $P(\text{greater than } 4)$

Measuring Figures Make this Foldable to help you organize your notes. Begin with a piece of 11″ by 17″ paper.

STEP 1 Fold Fold a 2″ tab along the long side of the paper.

Open and Fold Unfold the paper and fold in thirds widthwise.

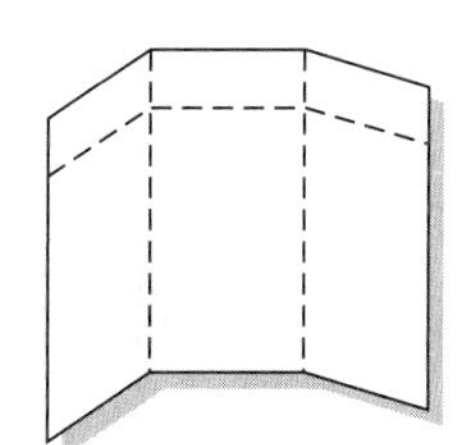

Open and Label Draw lines along the folds and label the head of each column as shown. Label the front of the folded table with the chapter title.

Squares and Square Roots	The Pythagorean Theorem	Finding Area

Reading and Writing As you read and study the chapter, write notes and examples in the appropriate columns of the table.

11-1 Squares and Square Roots

What You'll LEARN

Find squares of numbers and square roots of perfect squares.

NEW Vocabulary

square
perfect squares
square root
radical sign

HANDS-ON Mini Lab

Materials

- grid paper

Work with a partner.

The rectangle has a perimeter of 16 units and an area of 7 square units.

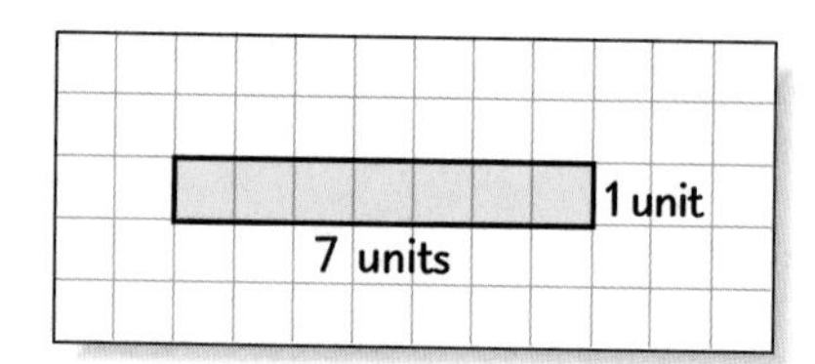

1. On grid paper, draw and label three other rectangles that have a perimeter of 16 units.
2. Summarize the dimensions and areas of the rectangles that you drew in a table like the one shown below.

Drawing	Dimensions (units)	Area (sq units)
▭▭▭▭▭▭▭	1 × 7	7

3. Draw three different rectangles that have a perimeter of 12 units and find their areas.
4. What do you notice about the rectangles with the greatest areas?

The area of the square at the right is 4 · 4 or 16 square units. The product of a number and itself is the **square** of the number. So, the square of 4 is 16.

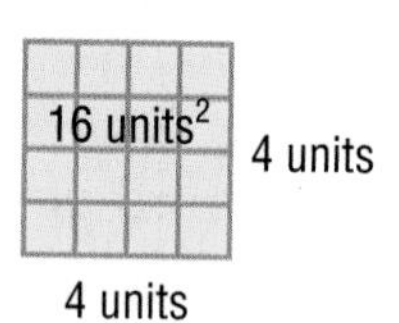

EXAMPLES Find Squares of Numbers

1 Find the square of 3.

$3 \cdot 3 = 9$

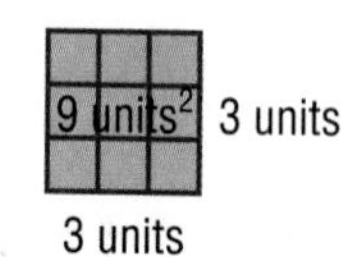

2 Find the square of 15.

15 [x^2] [ENTER] 225

Your Turn **Find the square of each number.**

a. 8 **b.** 12 **c.** 23

READING Math

Square a Number To *square a number* means to multiply that number by itself.

REAL-LIFE MATH

PHYSICAL SCIENCE The bungee bridge in Squaw Valley, California, is at an elevation of 8,200 feet, the highest in the world.

EXAMPLE Find a Square to Solve a Problem

3 PHYSICAL SCIENCE The falling distance of an object in feet d after t seconds is given by the formula $d = \frac{1}{2}(32)t^2$. If you went bungee jumping, how far would you fall 2.5 seconds after being released?

$d = \frac{1}{2}(32)t^2$ Write the formula.

$= \frac{1}{2}(32)(2.5)^2$ Replace t with 2.5.

$= \frac{1}{2}(32)(6.25)$ Use a calculator to square 2.5.

$= 100$ Simplify.

So, after 2.5 seconds, you would fall 100 feet.

Numbers like 9, 16, 225, and 6.25 are called **perfect squares** because they are squares of rational numbers. The factors multiplied to form perfect squares are called **square roots**.

Key Concept **Square Root**

Words A square root of a number is one of its two equal factors.

Symbols

Arithmetic	Algebra
$4 \cdot 4 = 16$, so 4 is a square root of 16.	If $x \cdot x$ or $x^2 = y$, then x is a square root of y.

Both $4 \cdot 4$ and $(-4)(-4)$ equal 16. So, 16 has two square roots, 4 and -4. A **radical sign**, $\sqrt{}$, is the symbol used to indicate the *positive* square root of a number. So, $\sqrt{16} = 4$.

READING Math

Square Roots Read $\sqrt{81} = 9$ as *the square root of 81 is 9.*

EXAMPLES Find Square Roots

4 Find $\sqrt{81}$.

$9 \cdot 9 = 81$, so $\sqrt{81} = 9$.

5 Find $\sqrt{196}$.

2nd [$\sqrt{}$] 196 ENTER 14

So, $\sqrt{196} = 14$.

6 SPORTS A boxing ring is a square with an area of 400 square feet. What are the dimensions of the ring?

2nd [$\sqrt{}$] 400 ENTER 20 Find the square root of 400.

So, a boxing ring measures 20 feet by 20 feet.

Your Turn Find each square root.

d. $\sqrt{25}$ e. $\sqrt{64}$ f. $\sqrt{289}$

Skill and Concept Check

Writing Math
Exercises 1 & 3

1. **Explain** how finding the square of a number is similar to finding the area of a square.
2. **OPEN ENDED** Write a number whose square is between 200 and 300.
3. **Which One Doesn't Belong?** Identify the number that is not a perfect square. Explain your reasoning.

256 | 121 | 529 | 116

GUIDED PRACTICE

Find the square of each number.

4. 6
5. 10
6. 17
7. 30

Find each square root.

8. $\sqrt{9}$
9. $\sqrt{121}$
10. $\sqrt{169}$
11. $\sqrt{529}$

12. **PHYSICAL SCIENCE** A model rocket is launched straight up into the air at an initial speed of 115 feet per second. The height of the rocket h after t seconds is given by the formula $h = -16t^2 + 115t$. What is the height of the rocket 3.5 seconds after it is launched?

Practice and Applications

HOMEWORK HELP

For Exercises	See Examples
13–20, 29–30, 34–35	1–3
21–28, 31–33, 36	4–6

Extra Practice
See pages 590, 606.

Find the square of each number.

13. 5
14. 1
15. 7
16. 11
17. 16
18. 20
19. 18
20. 34

Find each square root.

21. $\sqrt{4}$
22. $\sqrt{49}$
23. $\sqrt{144}$
24. $\sqrt{225}$
25. $\sqrt{729}$
26. $\sqrt{625}$
27. $\sqrt{1{,}225}$
28. $\sqrt{1{,}600}$

29. What is the square of -22?
30. Square 5.8.
31. Find both square roots of 100.
32. Find $-\sqrt{361}$.

33. **ALGEBRA** Evaluate $a^2 + \sqrt{b}$ if $a = 36$ and $b = 256$.

GEOGRAPHY For Exercises 34–36, refer to the squares in the diagram at the right. They represent the approximate areas of Texas, Michigan, and Florida.

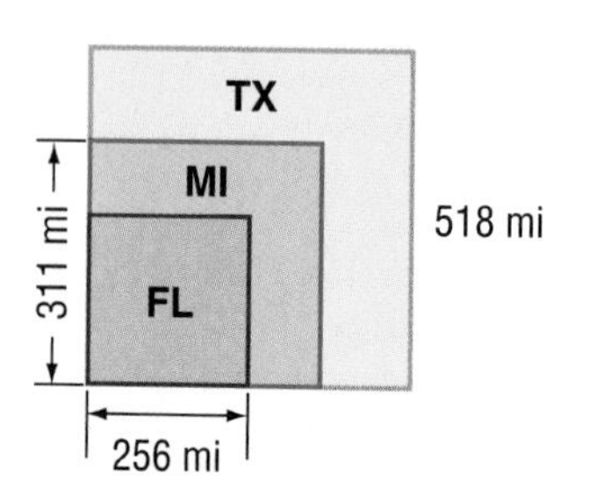

34. What is the area of Michigan?
35. How much larger is Texas than Florida?
36. The water areas of Texas, Michigan, and Florida are about 6,724 square miles, 40,000 square miles, and 11,664 square miles, respectively. Make a similar diagram comparing the water areas of these states. Label the squares.

MEASUREMENT For Exercises 37 and 38, refer to the garden, which is enclosed on all sides by a fence.

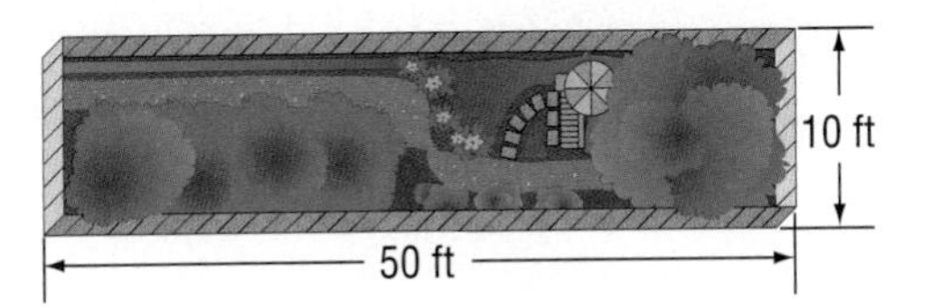

37. Could the garden area be made larger using the same fencing? Explain.

38. Describe the largest garden area possible using the same amount of fencing. How do the perimeter and area compare to the original garden?

39. **PROBABILITY** A set consists of all the perfect squares from 1 to 100. What is the probability that a number chosen at random from this set is divisible by 4 or 5?

ALGEBRA For Exercises 40–44, let the x-axis of a coordinate plane represent the side length of a square.

40. Let the y-axis represent the area. Graph the points for squares with sides 0, 1, 2, 3, 4, and 5 units long. Draw a line or curve that goes through each point.

41. On the same coordinate plane, let the y-axis represent the perimeter of a square. Graph the points for squares with sides 0, 1, 2, 3, 4, and 5 units long. Draw a line or curve that goes through each point.

42. Compare and contrast the two graphs.

43. For what side lengths is the value of the perimeter greater than the value of the area? When are the values equal?

44. Why do these graphs only make sense in the first quadrant?

45. **CRITICAL THINKING** The area of a square 8 meters by 8 meters is how much greater than the area of a square containing 9 square meters? Explain.

Standardized Test Practice and Mixed Review

46. **MULTIPLE CHOICE** A square plot of land has an area of 1,156 square feet. What is the perimeter of the plot?

Ⓐ 34 ft Ⓑ 102 ft Ⓒ 136 ft Ⓓ 289 ft

47. **SHORT RESPONSE** The perimeter of a square is 128 centimeters. Find its area.

For Exercises 48 and 49, refer to $\triangle ABC$ at the right. Find the vertices of $\triangle A'B'C'$ after each transformation. Then graph the triangle and its reflected or translated image.

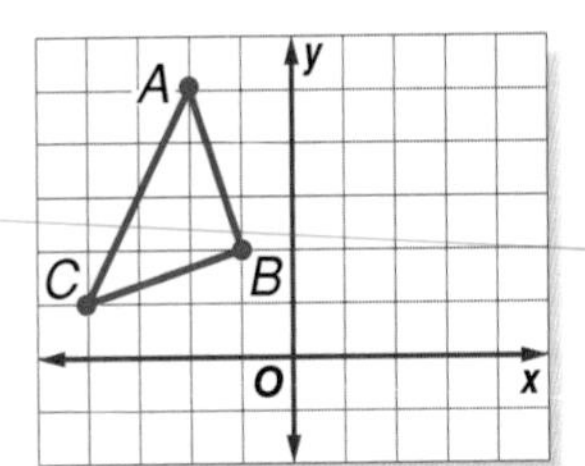

48. $\triangle ABC$ reflected over the y-axis (Lesson 10-9)

49. $\triangle ABC$ translated 5 units right and 1 unit up (Lesson 10-8)

GETTING READY FOR THE NEXT LESSON

PREREQUISITE SKILL Replace each ● with <, >, or = to make a true sentence. (Lesson 4-5)

50. $7 \bullet \sqrt{49}$ 51. $\sqrt{25} \bullet 4$ 52. $7.9 \bullet \sqrt{64}$ 53. $10.5 \bullet \sqrt{100}$

Study Skill

HOW TO...

Study Math Vocabulary

WEBS

If you've surfed the World Wide Web, you know it is a collection of documents linked together to form a huge electronic library. In mathematics, a web helps you understand how concepts are linked together.

A web can help you understand how math concepts are related to each other. To make a web, write the major topic in a box in the center of a sheet of paper. Then, draw "arms" from the center for as many categories as you need. You can label the arms to indicate the type of information that you are listing.

Here is a partial web for the major topic of *polygons*.

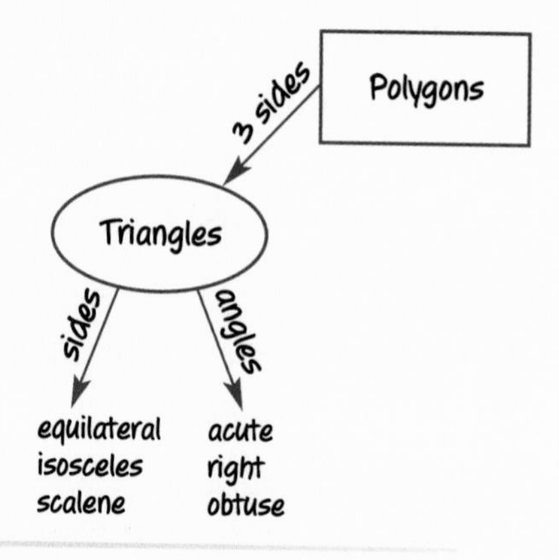

SKILL PRACTICE

1. Continue the web above by adding an arm for *quadrilaterals*.

2. In Lesson 5-8, you learned that the set of rational numbers contains fractions, terminating and repeating decimals, and integers. You also know that there are proper and improper fractions and that integers are whole numbers and their opposites. Complete the web below for *rational numbers*.

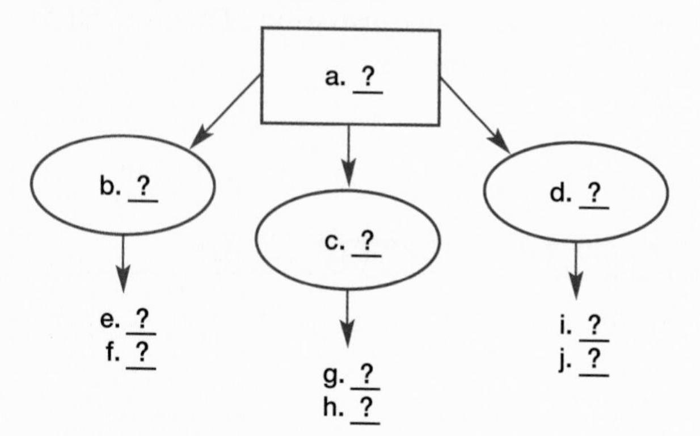

11-2 Estimating Square Roots

HANDS-ON Mini Lab

Materials

- **algebra tiles**

Work with a partner.

You can use algebra tiles to estimate the square root of 30.

- Arrange 30 tiles into the largest square possible. In this case, the largest possible square has 25 tiles, with 5 left over.

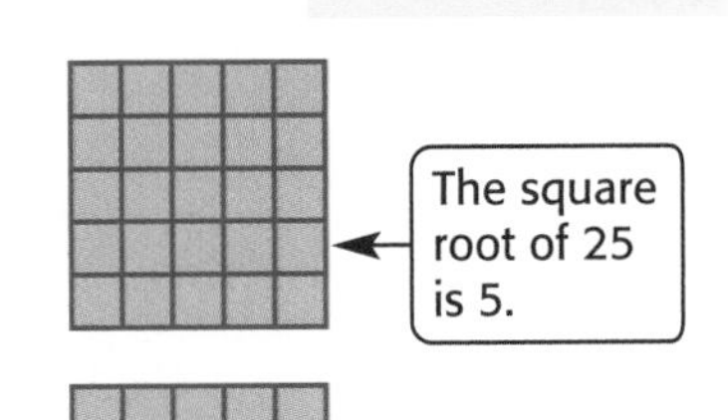

- Add tiles until you have the next larger square. So, add 6 tiles to make a square with 36 tiles.

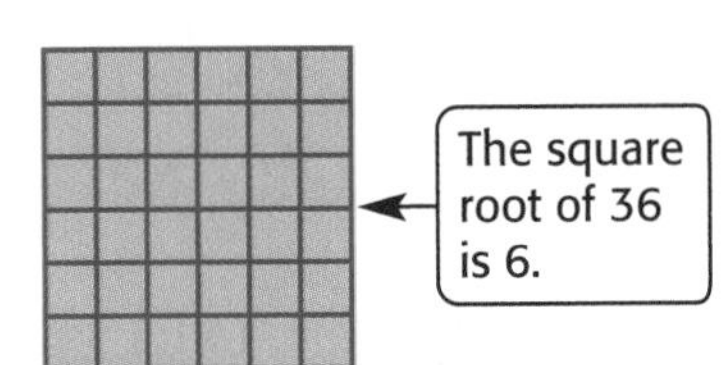

- The square root of 30 is between 5 and 6. $\sqrt{30}$ is closer to 5 because 30 is closer to 25 than to 36.

Use algebra tiles to estimate the square root of each number to the nearest whole number.

1. 40 **2.** 28 **3.** 85 **4.** 62

5. Describe another method that you could use to estimate the square root of a number.

What You'll LEARN

Estimate square roots.

NEW Vocabulary

irrational number

REVIEW Vocabulary

rational number: a number that can be written as a fraction **(Lesson 5-4)**

Recall that the square root of a perfect square is a rational number. You can find an estimate for the square root of a number that is *not* a perfect square.

EXAMPLE Estimate the Square Root

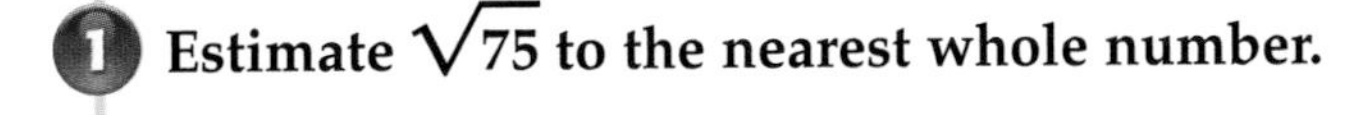

1 Estimate $\sqrt{75}$ to the nearest whole number.

List some perfect squares.

1, 4, 9, 16, 25, 36, 49, 64, 81, …

$64 < 75 < 81$	75 is between the perfect squares 64 and 81.
$\sqrt{64} < \sqrt{75} < \sqrt{81}$	Find the square root of each number.
$8 < \sqrt{75} < 9$	$\sqrt{64} = 8$ and $\sqrt{81} = 9$

So, $\sqrt{75}$ is between 8 and 9. Since 75 is closer to 81 than to 64, the best whole number estimate is 9. Verify with a calculator.

Irrational Numbers The square root of any number that is not a perfect square is an irrational number.

In Lesson 5-8, you learned that any number that can be written as a fraction is a rational number. These include integers as well as terminating and repeating decimals. A number that *cannot* be written as a fraction is an **irrational number**.

Rational Numbers $\sqrt{4}, 3\frac{1}{7}, 0.\overline{63}$

Irrational Numbers $\sqrt{2}, \pi, 0.636336333\ldots$

You can use a calculator to estimate square roots that are irrational numbers.

EXAMPLE Use a Calculator to Estimate

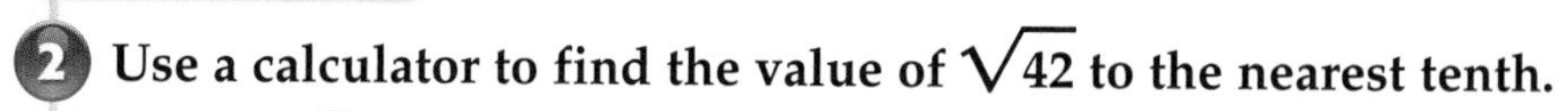

2 Use a calculator to find the value of $\sqrt{42}$ to the nearest tenth.

2nd [$\sqrt{\ }$] 42 ENTER 6.480740698

$\sqrt{42} \approx 6.5$

$\sqrt{42}$

1 2 3 4 5 6 7 8

Check $6^2 = 36$ and $7^2 = 49$. Since 42 is between 36 and 49, the answer, 6.5, is reasonable.

Your Turn **Use a calculator to find each square root to the nearest tenth.**

a. $\sqrt{6}$ b. $\sqrt{23}$ c. $\sqrt{309}$

Skill and Concept Check

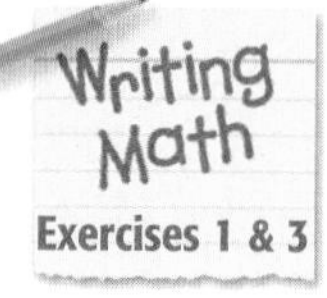

1. **Explain** why $\sqrt{30}$ is an irrational number.
2. **OPEN ENDED** List three numbers that have square roots between 4 and 5.
3. **NUMBER SENSE** Explain why 7 is the best whole number estimate for $\sqrt{51}$.

GUIDED PRACTICE

Estimate each square root to the nearest whole number.

4. $\sqrt{39}$ 5. $\sqrt{106}$ 6. $\sqrt{90}$ 7. $\sqrt{140}$

Use a calculator to find each square root to the nearest tenth.

8. $\sqrt{7}$ 9. $\sqrt{51}$ 10. $\sqrt{135}$ 11. $\sqrt{462}$

12. **GEOMETRY** Use a calculator to find the side length of the square at the right. Round to the nearest tenth.

Area = 95 cm²

x

x

Practice and Applications

HOMEWORK HELP

For Exercises	See Examples
13–20	1
21–32	2

Extra Practice
See pages 590, 606.

Estimate each square root to the nearest whole number.

13. $\sqrt{11}$
14. $\sqrt{20}$
15. $\sqrt{35}$
16. $\sqrt{65}$
17. $\sqrt{89}$
18. $\sqrt{116}$
19. $\sqrt{137}$
20. $\sqrt{409}$

Use a calculator to find each square root to the nearest tenth.

21. $\sqrt{15}$
22. $\sqrt{8}$
23. $\sqrt{44}$
24. $\sqrt{89}$
25. $\sqrt{160}$
26. $\sqrt{573}$
27. $\sqrt{645}$
28. $\sqrt{2{,}798}$

29. Order $\sqrt{87}$, 10, π, and $\frac{14}{3}$ from least to greatest.

30. Graph $\sqrt{34}$ and $\sqrt{92}$ on the same number line.

31. **ALGEBRA** Evaluate $\sqrt{a + b}$ if $a = 8$ and $b = 3.7$.

32. **DRIVING** Police officers can use a formula and skid marks to calculate the speed of a car. Use the formula at the right to estimate how fast a car was going if it left skid marks 83 feet long. Round to the nearest tenth.

$$s = \sqrt{39d}$$

- s = speed (mph)
- d = length of skid marks (ft)

33. **RESEARCH** In the 1990s, over 50 billion decimal places of pi had been computed. Use the Internet or another source to find the current number of decimal places of pi that have been computed.

34. **CRITICAL THINKING** You can use *Hero's formula* to find the area A of a triangle if you know the measures of its sides, a, b, and c. The formula is $A = \sqrt{s(s - a)(s - b)(s - c)}$, where s is half of the perimeter. Find the area of the triangle to the nearest tenth.

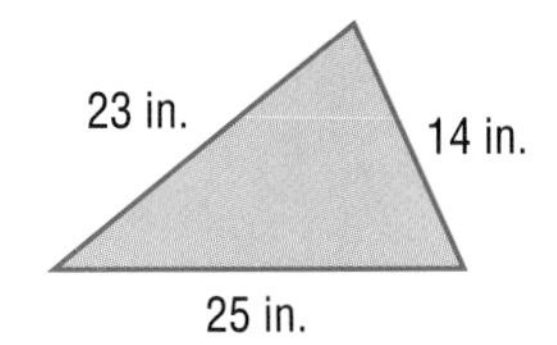

Standardized Test Practice and Mixed Review

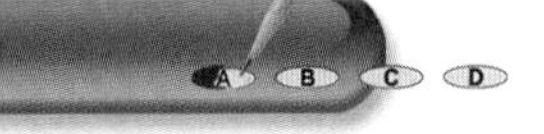

35. **MULTIPLE CHOICE** Identify the number that is irrational.

Ⓐ -4.1 Ⓑ 0 Ⓒ $\frac{3}{8}$ Ⓓ $\sqrt{6}$

36. **SHORT RESPONSE** Name the point that best represents the graph of $\sqrt{209}$.

Find each square root. (Lesson 11-1)

37. $\sqrt{169}$
38. $\sqrt{2{,}025}$
39. $\sqrt{784}$

40. Graph $\triangle JLK$ with vertices $J(-1, -4)$, $K(1, 1)$, and $L(3, -2)$ and its reflection over the x-axis. Write the ordered pairs for the vertices of the new figure. (Lesson 10-9)

GETTING READY FOR THE NEXT LESSON

PREREQUISITE SKILL **Solve each equation.** (Lesson 1-5)

41. $7^2 + 5^2 = c$
42. $4^2 + b = 36$
43. $3^2 + a = 25$
44. $9^2 + 2^2 = c$

11-3a HANDS-ON LAB

A Preview of Lesson 11-3

The Pythagorean Theorem

What You'll LEARN

Find the relationship among the sides of a right triangle.

Materials

- centimeter grid paper
- ruler
- scissors

INVESTIGATE *Work as a class.*

Four thousand years ago, the ancient Egyptians used mathematics to lay out their fields with square corners. They took a piece of rope and knotted it into 12 equal spaces. Taking three stakes, they stretched the rope around the stakes to form a right triangle. The sides of the triangle had lengths of 3, 4, and 5 units.

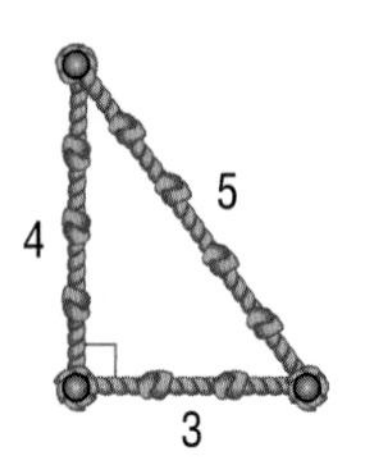

STEP 1 On grid paper, draw a segment that is 3 centimeters long. At one end of this segment, draw a perpendicular segment that is 4 centimeters long. Draw a third segment to form a triangle. Cut out the triangle.

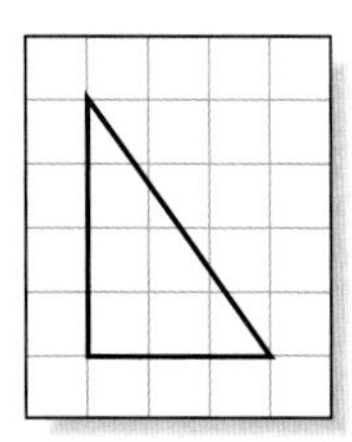

STEP 2 Measure the length of the longest side in centimeters. In this case, it is 5 centimeters.

STEP 3 Cut out three squares: one with 3 centimeters on a side, one with 4 centimeters on a side, and one with 5 centimeters on a side.

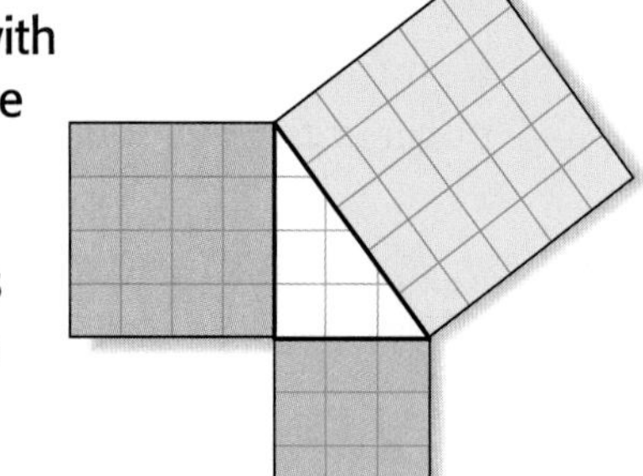

STEP 4 Place the edges of the squares against the corresponding sides of the right triangle.

STEP 5 Find the area of each square.

Writing Math

Work with a partner.

1. What relationship exists among the areas of the three squares?

Repeat the activity for each right triangle whose perpendicular sides have the following measures. Write an equation to show your findings.

2. 6 cm, 8 cm
3. 5 cm, 12 cm
4. **Write** a sentence or two summarizing your findings.
5. **MAKE A CONJECTURE** Determine the length of the third side of a right triangle if the perpendicular sides of the triangle are 9 inches and 12 inches long.

11-3 The Pythagorean Theorem

What You'll LEARN

Find length using the Pythagorean Theorem.

NEW Vocabulary

leg
hypotenuse
Pythagorean Theorem

REVIEW Vocabulary

right triangle: a triangle with exactly one angle that measures 90° **(Lesson 10-4)**

WHEN am I ever going to use this?

MOVING A square mirror 7 feet on each side must be delivered through the doorway.

1. Can the mirror fit through the doorway? Explain.
2. Make a scale drawing on grid paper to solve the problem.

The sides of a right triangle have special names, as shown below.

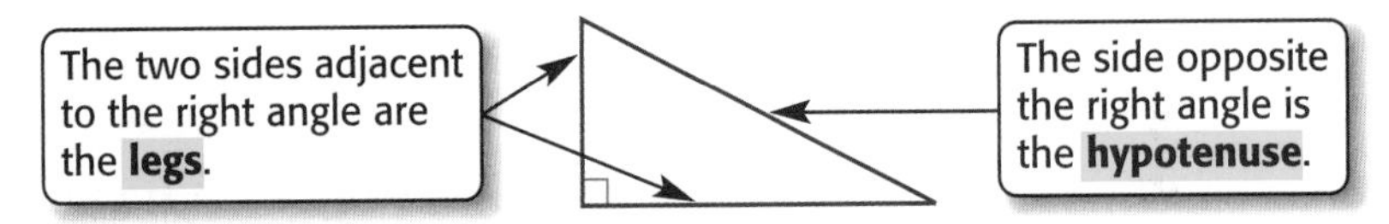

The **Pythagorean Theorem** describes the relationship between the length of the hypotenuse and the lengths of the legs.

Key Concept — Pythagorean Theorem

Words In a right triangle, the square of the length of the hypotenuse equals the sum of the squares of the lengths of the legs.

Model

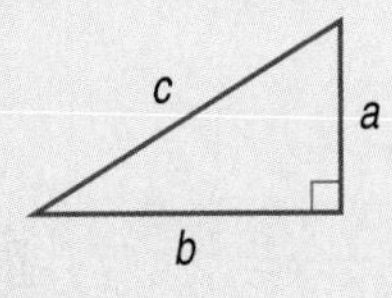

Symbols $c^2 = a^2 + b^2$

You can use the Pythagorean Theorem to find the length of the hypotenuse of a right triangle if the measures of both legs are known.

EXAMPLE Find the Length of the Hypotenuse

1 MOVING Determine whether a 7-foot square mirror will fit diagonally through the doorway shown at the right.

To solve, find the length of the hypotenuse c.

$c^2 = a^2 + b^2$	Pythagorean Theorem
$c^2 = 3^2 + 6.5^2$	Replace a with 3 and b with 6.5.
$c^2 = 9 + 42.25$	Evaluate 3^2 and 6.5^2.
$c^2 = 51.25$	Add.
$\sqrt{c^2} = \sqrt{51.25}$	Take the square root of each side.
$c \approx 7.2$	Simplify.

The length of the diagonal is about 7.2 feet. So, the mirror will fit through the doorway if it is turned diagonally.

msmath2.net/extra_examples

READING Math

Theorem A *theorem* is a statement in mathematics that can be justified by logical reasoning.

You can also use the Pythagorean Theorem to find the measure of a leg if the measure of the other leg and the hypotenuse are known.

EXAMPLE Find the Length of a Leg

2 **Find the missing measure of the triangle at the right.**

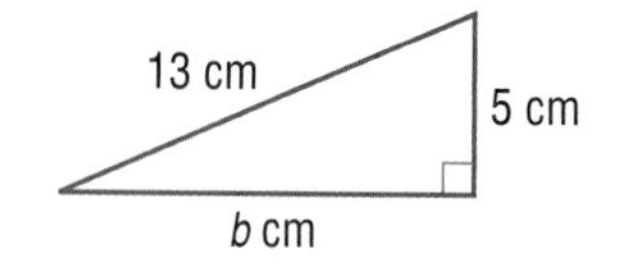

The missing measure is of a leg of the triangle.

$c^2 = a^2 + b^2$	Pythagorean Theorem
$13^2 = 5^2 + b^2$	Replace a with 5 and c with 13.
$169 = 25 + b^2$	Evaluate 13^2 and 5^2.
$169 - 25 = 25 + b^2 - 25$	Subtract 25 from each side.
$144 = b^2$	Simplify.
$\sqrt{144} = \sqrt{b^2}$	Take the square root of each side.
$12 = b$	Simplify.

The length of the leg is 12 centimeters.

Your Turn **Find the missing measure of each right triangle. Round to the nearest tenth if necessary.**

a.

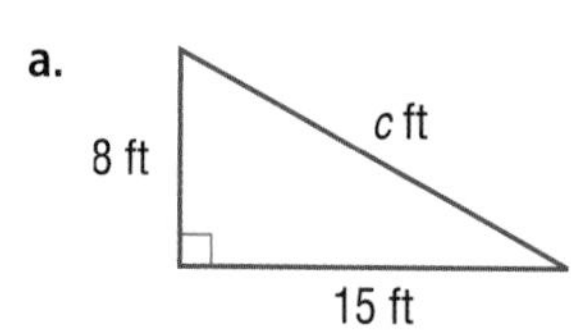

b.

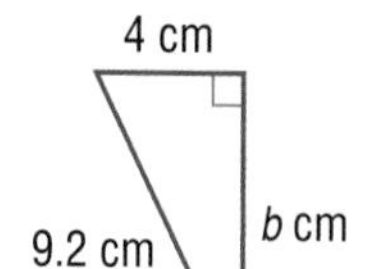

c. $b = 7$ in., $c = 25$ in.

REAL-LIFE CAREERS

How Does an Archaeologist Use Math? Before digging, archaeologists use the Pythagorean Theorem to calculate the diagonal of an excavation site to be sure that the area is a rectangle.

Online Research For information about a career as an archaeologist, visit: msmath2.net/careers

EXAMPLE Solve a Real-Life Problem

3 **ARCHAEOLOGY Archaeologists placed corner stakes to mark a rectangular excavation site, as shown at the right. If their stakes are placed correctly, what is the measure of the diagonal?**

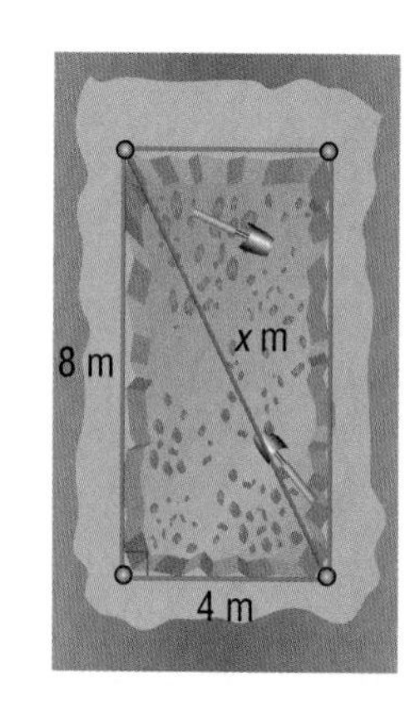

The diagonal of the rectangle is the hypotenuse of a right triangle. Write an equation to solve for x.

$c^2 = a^2 + b^2$	Pythagorean Theorem
$x^2 = 8^2 + 4^2$	Replace a with 8, b with 4, and c with x.
$x^2 = 64 + 16$	Evaluate 8^2 and 4^2.
$x^2 = 80$	Simplify.
$\sqrt{x^2} = \sqrt{80}$	Take the square root of each side.
$x \approx 8.9$	Simplify.

The length of the diagonal is about 8.9 meters.

You can determine whether a triangle is a right triangle by applying the Pythagorean Theorem.

EXAMPLES Identify Right Triangles

Determine whether a triangle with the given lengths is a right triangle.

STUDY TIP

Hypotenuse Remember that the hypotenuse is always the longest side.

4 **1.5 mm, 2 mm, 2.5 mm**

$c^2 = a^2 + b^2$

$2.5^2 \stackrel{?}{=} 1.5^2 + 2^2$

$6.25 \stackrel{?}{=} 2.25 + 4$

$6.25 = 6.25$ ✔

The triangle is a right triangle.

5 **4 ft, 6 ft, 8 ft**

$c^2 = a^2 + b^2$

$8^2 \stackrel{?}{=} 4^2 + 6^2$

$64 \stackrel{?}{=} 16 + 36$

$64 \neq 52$

The triangle is *not* a right triangle.

Your Turn **Determine whether each triangle with the given lengths is a right triangle. Write *yes* or *no*.**

d. 7.5 cm, 8 cm, 12 cm

e. 9 in., 40 in., 41 in.

Skill and Concept Check

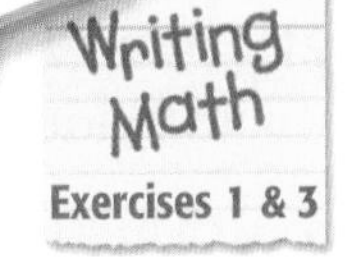

1. **Describe** the information that you need in order to find the missing measure of a right triangle.
2. **OPEN ENDED** Draw and label a right triangle that has one side measuring 14 units. Write the length of another side. Then find the length of the third side to the nearest tenth.
3. **FIND THE ERROR** Devin and Jamie are writing an equation to find the missing measure of the triangle at the right. Who is correct? Explain.

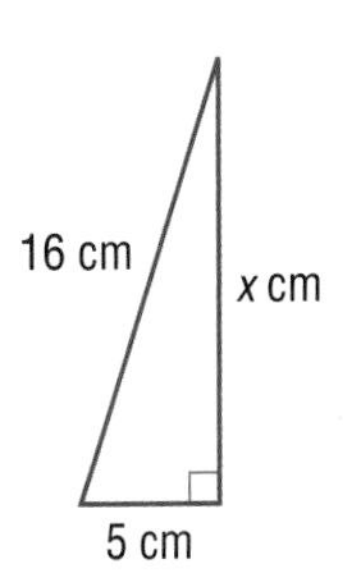

Devin
$16^2 = 5^2 + x^2$

Jamie
$x^2 = 16^2 + 5^2$

GUIDED PRACTICE

Find the missing measure of each right triangle. Round to the nearest tenth if necessary.

4.

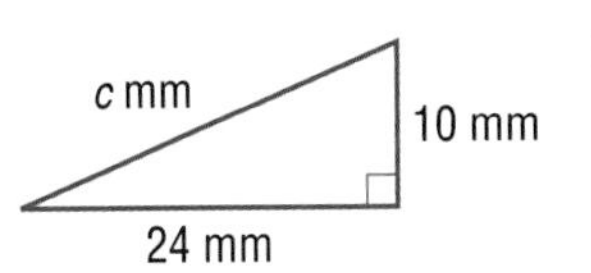

5.

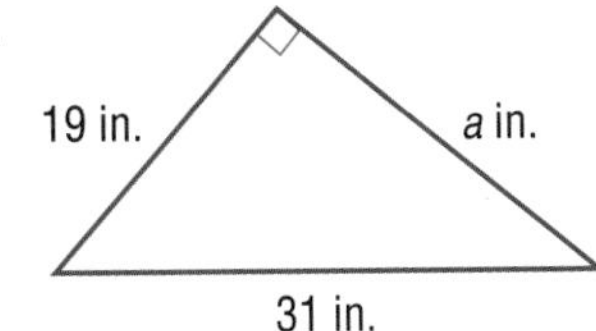

6. $b = 21$ cm, $c = 28$ cm

Determine whether a triangle with the given side lengths is a right triangle. Write *yes* or *no*.

7. 1.4 m, 4.8 m, 5 m
8. 21 ft, 24 ft, 30 ft

Practice and Applications

HOMEWORK HELP	
For Exercises	See Examples
9–16	1, 2
17–20	4, 5
21–22	3

Extra Practice
See pages 590, 606.

Find the missing measure of each right triangle. Round to the nearest tenth if necessary.

9.
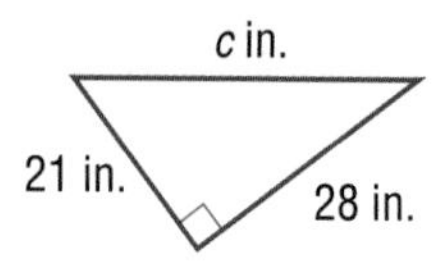

10.
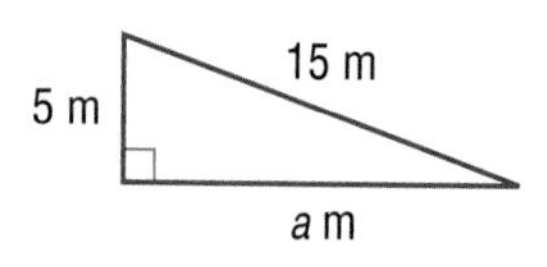

11.
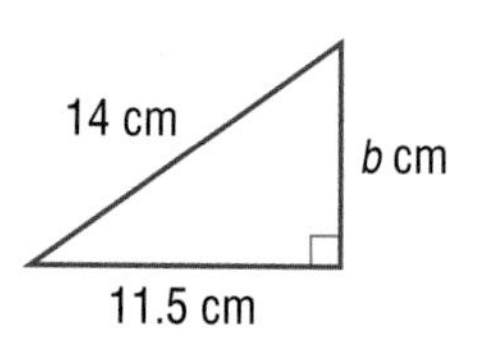

12.
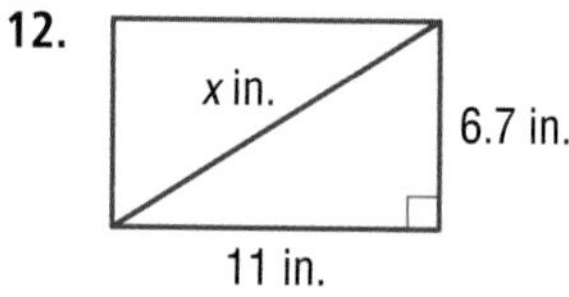

13. $a = 7$ in., $b = 24$ in.

14. $a = 13.5$ mm, $b = 18$ mm

15. $b = 13$ m, $c = 27$ m

16. $a = 2.4$ yd, $c = 3$ yd

Determine whether a triangle with the given side lengths is a right triangle. Write *yes* or *no*.

17. 12 cm, 16 cm, 20 cm

18. 8 m, 15 m, 17 m

19. 11 ft, 14 ft, 17 ft

20. 18 in., 18 in., 36 in.

21. SAFETY To the nearest tenth of a foot, how far up the wall x does the ladder shown at the right reach?

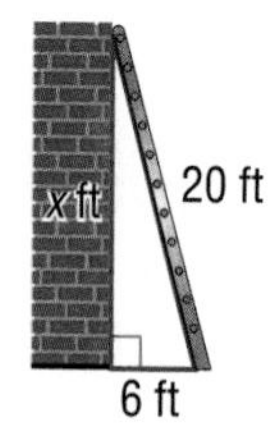

22. TRAVEL You drive 80 miles east, then 50 miles north, then 140 miles west, and finally 95 miles south. Make a drawing to find how far you are from your starting point.

23. CRITICAL THINKING What is the length of the diagonal of the cube shown at the right?

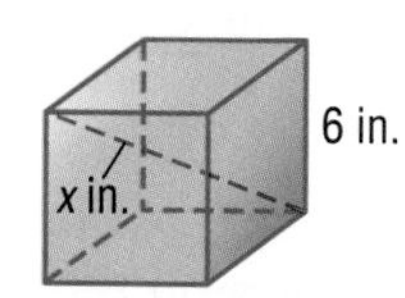

Standardized Test Practice and Mixed Review

24. MULTIPLE CHOICE Find the missing measure of a right triangle if $a = 20$ meters and $c = 52$ meters.

Ⓐ 24 m   Ⓑ 32 m   Ⓒ 48 m   Ⓓ 55.7 m

25. SHORT RESPONSE An isosceles right triangle has legs that are 8 inches long. Find the length of the hypotenuse to the nearest tenth.

Estimate each square root to the nearest whole number. (Lesson 11-2)

26. $\sqrt{61}$

27. $\sqrt{147}$

28. $\sqrt{40}$

29. $\sqrt{277}$

30. Find $\sqrt{256}$. (Lesson 11-1)

GETTING READY FOR THE NEXT LESSON

PREREQUISITE SKILL Multiply. (Lesson 6-4)

31. $17.8 \cdot 12$

32. $21.5 \cdot 27.1$

33. $3\frac{1}{2} \cdot 8$

34. $15\frac{1}{4} \cdot 18$

11-4 Area of Parallelograms

What You'll LEARN

Find the areas of parallelograms.

NEW Vocabulary

base
height

REVIEW Vocabulary

parallelogram: quadrilateral with opposite sides parallel and opposite sides congruent **(Lesson 10-5)**

HANDS-ON Mini Lab

Materials

- grid paper
- straightedge

Work with a partner.

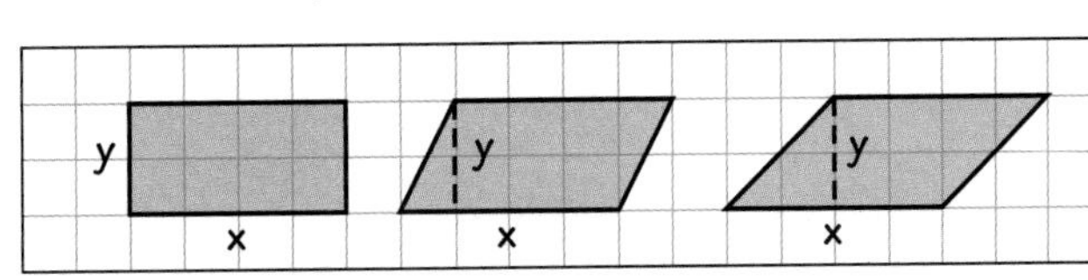

1. What is the value of x and y for each parallelogram?
2. Count the grid squares to find the area of each parallelogram.
3. On grid paper, draw three different parallelograms in which $x = 5$ units and $y = 4$ units. Find the area of each.
4. **Make a conjecture** about how to find the area of a parallelogram if you know the values of x and y.

In Lesson 10-5, you learned that a parallelogram is a special kind of quadrilateral. You can find the area of a parallelogram by using the values for the base and height, as described below.

The **base** is any side of a parallelogram.

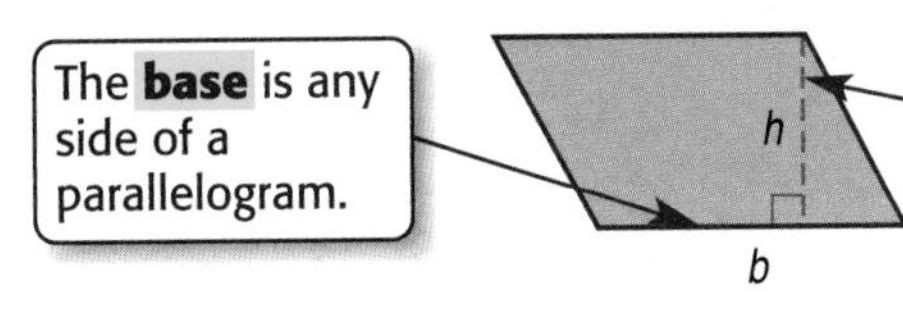

The **height** is the length of the segment perpendicular to the base with endpoints on opposite sides.

Key Concept — Area of a Parallelogram

Words The area A of a parallelogram equals the product of its base b and height h.

Symbols $A = bh$

Model

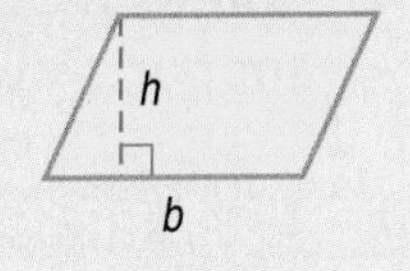

EXAMPLE Find the Area of a Parallelogram

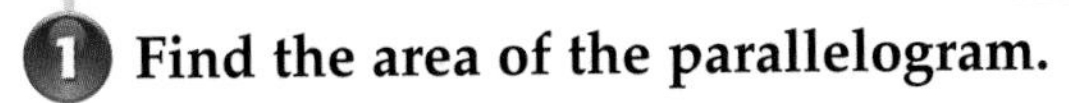

1 Find the area of the parallelogram.

5.8 cm
13 cm

Estimate $A = 13 \cdot 6$ or 78 cm^2

$A = bh$ Area of a parallelogram

$A = 13 \cdot 5.8$ Replace b with 13 and h with 5.8.

$A = 75.4$ Multiply.

The area of the parallelogram is 75.4 square centimeters. This is close to the estimate.

msmath2.net/extra_examples

STUDY TIP

Common Error When finding the area of a parallelogram, be sure to use the base and the height, not the base and a side.

EXAMPLE Find the Area of a Parallelogram

2 Find the area of the parallelogram at the right.

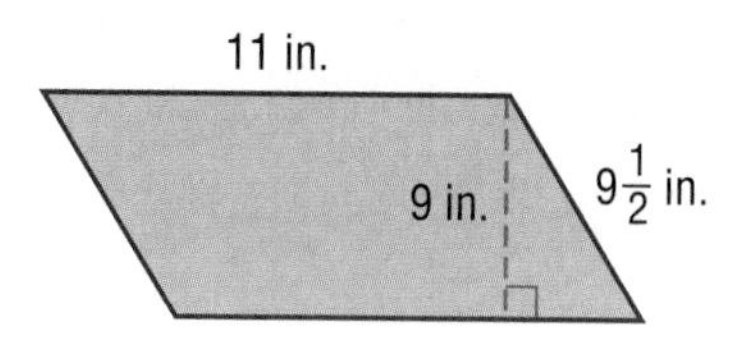

The base is 11 inches, and the height is 9 inches.

Estimate $A = 10 \cdot 10$ or 100 in^2

$A = bh$ Area of a parallelogram

$A = 11 \cdot 9$ Replace b with 11 and h with 9.

$A = 99$ Multiply.

The area of the parallelogram is 99 square inches. This is close to the estimate.

Your Turn **Find the area of each parallelogram.**

a. base = 5 cm
height = 8 cm

b. 10.2 ft
7 ft

Skill and Concept Check

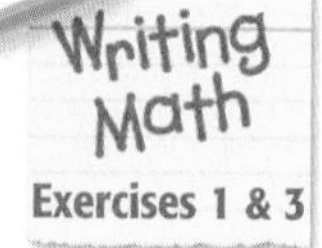

1. **Describe** what values you would substitute in the formula $A = bh$ to find the area of the parallelogram at the right.

2. **OPEN ENDED** Draw three different parallelograms, each with an area of 24 square units.

3. **NUMBER SENSE** *True* or *False*? The area of a parallelogram doubles if you double the base and the height. Explain or give a counterexample to support your answer.

GUIDED PRACTICE

Find the area of each parallelogram. Round to the nearest tenth if necessary.

4. 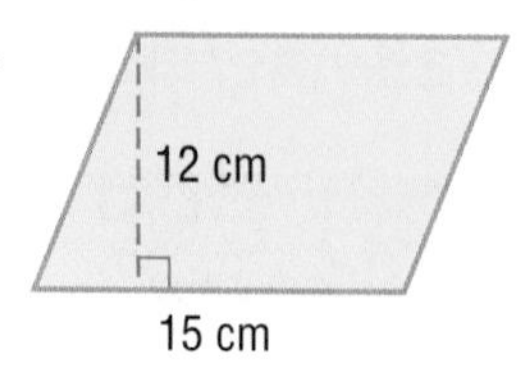

5. 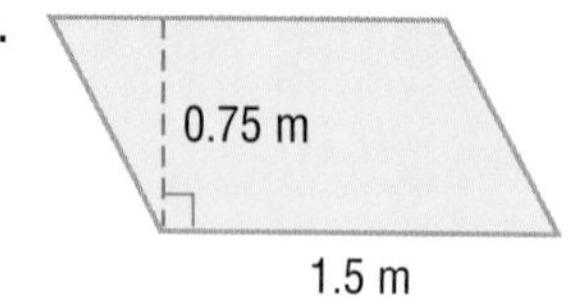

6. 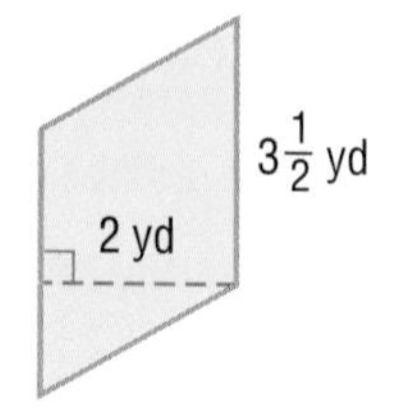

7. base = 16 in.
height = 4 in.

8. base = 3.5 m
height = 5 m

9. What is the area of a parallelogram with a base of 25 millimeters and a height that is half the base?

Practice and Applications

HOMEWORK HELP

For Exercises	See Examples
10–21	1, 2

Extra Practice
See pages 591, 606.

Find the area of each parallelogram. Round to the nearest tenth if necessary.

10.

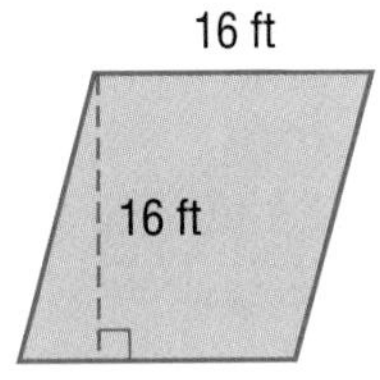

11.

21 mm

20.4 mm

12.

0.3 cm

0.5 cm

13.

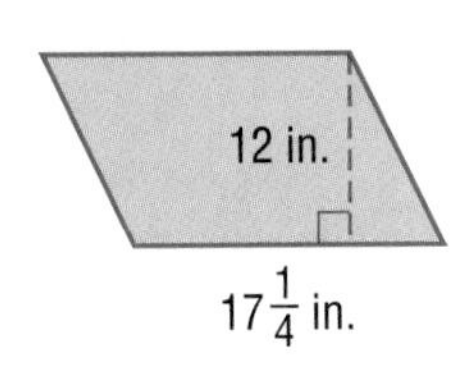

14.

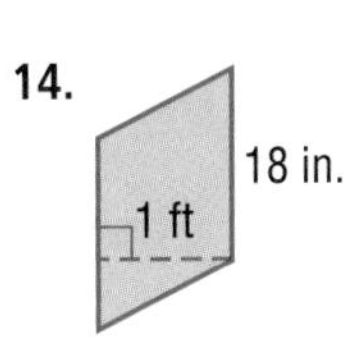

15.

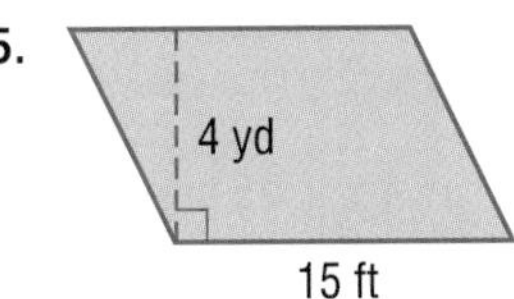

16. base = 13 mm
height = 6 mm

17. base = 45 yd
height = 35 yd

18. base = 8 in.
height = 12.5 in.

19. base = 7.9 cm
height = 7.2 cm

20. **MULTI STEP** A quilted block uses eight parallelogram-shaped pieces of cloth, each with a height of $3\frac{1}{3}$ inches and a base of $3\frac{3}{4}$ inches. How much fabric is needed to make the parallelogram pieces for 24 blocks? Write in square feet. (*Hint*: 144 in^2 = 1 ft^2)

21. What is the height of a parallelogram if the base is 24 inches and the area is 360 square inches?

22. **CRITICAL THINKING** Identify two possible measures of base and height for a parallelogram that has an area of 320 square inches.

Standardized Test Practice and Mixed Review

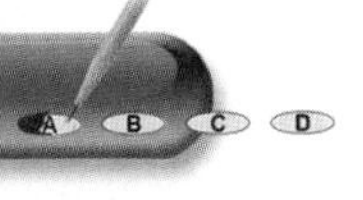

23. **MULTIPLE CHOICE** Find the area of the parallelogram.

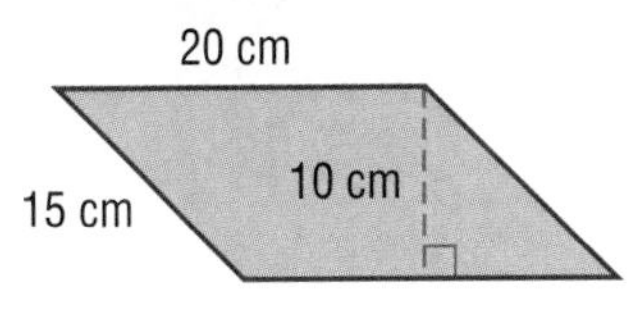

Ⓐ 75 cm^2 Ⓑ 150 cm^2

Ⓒ 200 cm^2 Ⓓ 300 cm^2

24. **SHORT RESPONSE** What is the base of a parallelogram if the height is 18.6 inches and the area is 279 square inches?

Determine whether a triangle with the given side lengths is a right triangle. Write *yes* or *no*. (Lesson 11-3)

25. 8 in., 10 in., 12 in. **26.** 12 ft, 16 ft, 20 ft **27.** 5 cm, 12 cm, 14 cm

28. Which is closer to $\sqrt{55}$, 7 or 8? (Lesson 11-2)

GETTING READY FOR THE NEXT LESSON

PREREQUISITE SKILL Find each value. (Lesson 1-3)

29. $6(4 + 10)$ **30.** $\frac{1}{2}(8)(8)$ **31.** $\frac{1}{2}(24 + 15)$ **32.** $\frac{1}{2}(5)(13 + 22)$

Mid-Chapter Practice Test

Vocabulary and Concepts

1. **Define** *square root.* (Lesson 11-1)
2. **State,** in words, the Pythagorean Theorem. (Lesson 11-3)
3. *True* or *False*? The area of any parallelogram equals the length times the width. Explain. (Lesson 11-4)

Skills and Applications

Find the square of each number. (Lesson 11-1)

4. 4
5. 12

Find each square root. (Lesson 11-1)

6. $\sqrt{64}$
7. $\sqrt{289}$

8. **LANDSCAPING** A bag of lawn fertilizer covers 2,500 square feet. Describe the largest square that one bag of fertilizer could cover. (Lesson 11-1)

Estimate each square root to the nearest whole number. (Lesson 11-2)

9. $\sqrt{32}$
10. $\sqrt{55}$

Find the missing measure of each right triangle. Round to the nearest tenth if necessary. (Lesson 11-3)

11. 7 m, 16.6 m, c m

12. $a = 8.2$ m
    $b = 15.6$ m

Find the area of each parallelogram. (Lesson 11-4)

13. base = 4.3 in.
    height = 9 in.

14. 7 mm, 6 mm, 12 mm

Standardized Test Practice

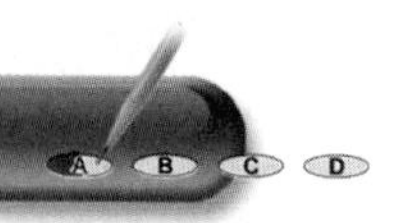

15. **MULTIPLE CHOICE** Which is the best estimate for $\sqrt{120}$? (Lesson 11-2)

    Ⓐ 10 Ⓑ 11 Ⓒ 12 Ⓓ 15

16. **SHORT RESPONSE** Miranda jogs 5 kilometers north and 5 kilometers west. To the nearest kilometer, how far is she from her starting point? (Lesson 11-3)

The Game Zone

A Place To Practice Your Math Skills

Math Skill
Finding Square Roots

Tic Tac Root

GET READY!

Players: two to four
Materials: index cards, construction paper

- Use 20 index cards. On each card, write one of the following square roots.

$\sqrt{1}$	$\sqrt{4}$	$\sqrt{9}$	$\sqrt{16}$
$\sqrt{25}$	$\sqrt{36}$	$\sqrt{49}$	$\sqrt{64}$
$\sqrt{81}$	$\sqrt{100}$	$\sqrt{121}$	$\sqrt{144}$
$\sqrt{169}$	$\sqrt{196}$	$\sqrt{225}$	$\sqrt{256}$
$\sqrt{289}$	$\sqrt{324}$	$\sqrt{361}$	$\sqrt{400}$

- Each player should draw a tic-tac-toe board on construction paper. In each square, place a number from 1 to 20, but do not use any number more than once. See the sample board at the right.

4	15	9
11	6	2
20	12	7

GO!

- The dealer shuffles the index cards and places them facedown on the table.
- The player to the left of the dealer chooses the top index card and places it faceup. Any player with the matching square root on his or her board places an X on the appropriate square.
- The next player chooses the top index card and places it faceup on the last card chosen. Players mark their boards accordingly.
- **Who Wins?** The first player to get three Xs in a row wins the game.

11-5a HANDS-ON LAB

A Preview of Lesson 11-5

Triangles and Trapezoids

What You'll LEARN

Find the areas of triangles and trapezoids using models.

Materials

- centimeter grid paper
- straightedge
- scissors
- tape

INVESTIGATE *Work as a class.*

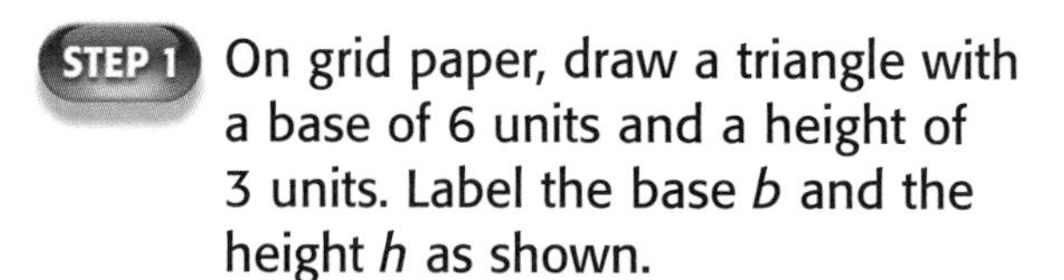

STEP 1 On grid paper, draw a triangle with a base of 6 units and a height of 3 units. Label the base b and the height h as shown.

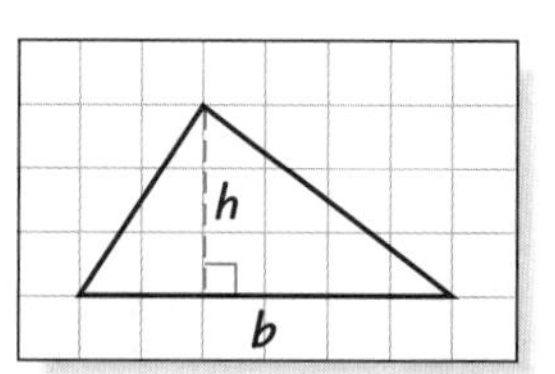

STEP 2 Fold the grid paper in half and cut out the triangle through both sheets so that you have two congruent triangles.

STEP 3 Turn the second triangle upside down and tape it to the first triangle.

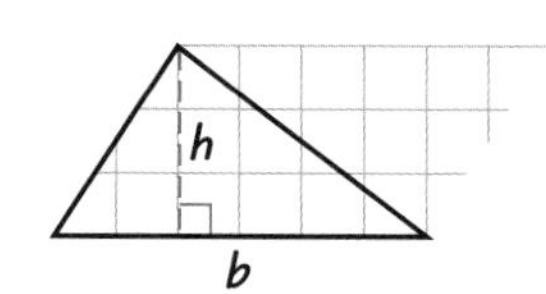

Writing Math

Work with a partner.

1. What figure is formed by the two triangles?
2. Write the formula for the area of the figure. Then find the area.
3. What is the area of each of the triangles? How do you know?
4. Repeat the activity above, drawing a different triangle in Step 1. Then find the area of each triangle.
5. Compare the area of a triangle to the area of a parallelogram with the same base and height.
6. **MAKE A CONJECTURE** Write a formula for the area of a triangle with base b and height h.

For Exercises 7–9, refer to the information below.

On grid paper, cut out two identical trapezoids. Label the bases b_1 and b_2, respectively, and label the heights h. Then turn one trapezoid upside down and tape it to the other trapezoid as shown.

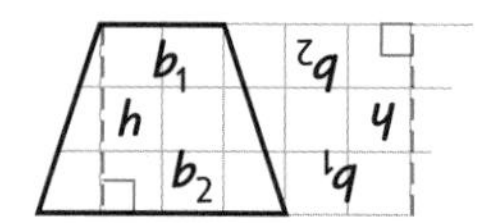

7. Write an expression to represent the base of the parallelogram.
8. Write a formula for the area A of the parallelogram using b_1, b_2, and h.
9. **MAKE A CONJECTURE** Write a formula for the area A of a trapezoid with bases b_1 and b_2, and height h.

11-5 Area of Triangles and Trapezoids

What You'll LEARN

Find the areas of triangles and trapezoids.

REVIEW Vocabulary

trapezoid: quadrilateral with one pair of parallel sides **(Lesson 10-5)**

HANDS-ON Mini Lab

Materials

- grid paper
- straightedge
- scissors

Work with a partner.

- Draw a parallelogram with a base of 6 units and a height of 4 units.
- Draw a diagonal as shown.
- Cut out the parallelogram.

1. What is the area of the parallelogram?
2. Cut along the diagonal. What is true about the triangles formed?
3. What is the area of each triangle?
4. If the area of a parallelogram is bh, then write an expression for the area A of each of the two congruent triangles that form the parallelogram.

Like parallelograms, you can find the area of a triangle by using the base and height.

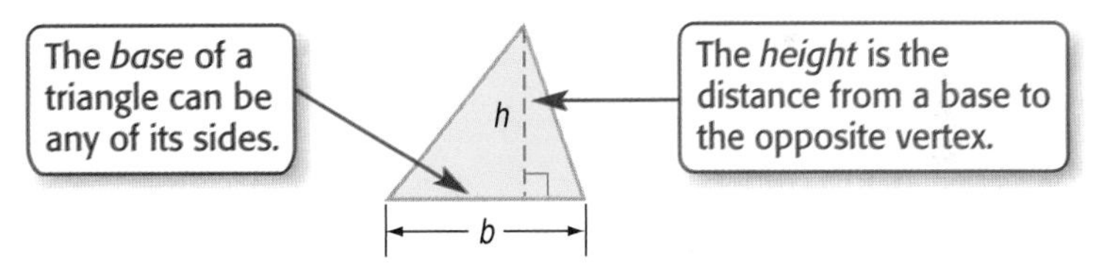

Key Concept — **Area of a Triangle**

Words The area A of a triangle equals half the product of its base b and height h.

Symbols $A = \frac{1}{2}bh$

Model

EXAMPLE Find the Area of a Triangle

1 Find the area of the triangle below. **Estimate** $\frac{1}{2}(10)(7) = 35$

$A = \frac{1}{2}bh$ Area of a triangle

$A = \frac{1}{2}(10)(6.5)$ Replace b with 10 and h with 6.5.

$A = 32.5$ Multiply.

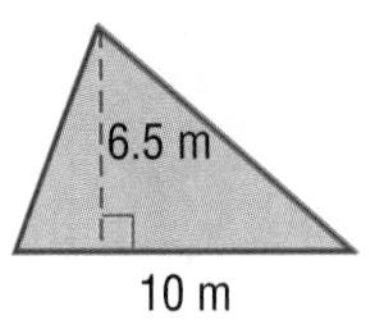

The area of the triangle is 32.5 square meters. This is close to the estimate.

A trapezoid has two bases, b_1 and b_2. The height of a trapezoid is the distance between the bases.

READING Math

Subscripts Read b_1 as *b sub 1*. Read b_2 as *b sub 2*. The subscripts mean that b_1 and b_2 represent different variables.

Key Concept — Area of a Trapezoid

Words The area A of a trapezoid equals half the product of the height h and the sum of the bases b_1 and b_2.

Symbols $A = \frac{1}{2}h(b_1 + b_2)$

Model

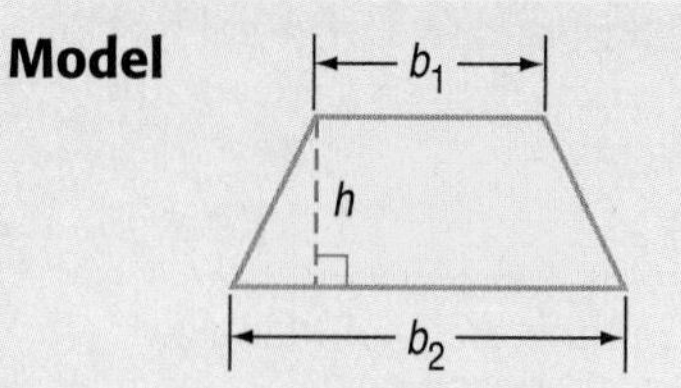

EXAMPLE Find the Area of a Trapezoid

2 Find the area of the trapezoid at the right.

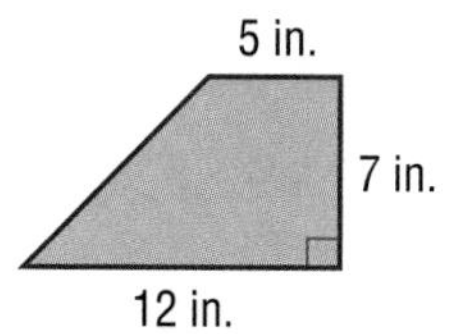

The bases are 5 inches and 12 inches. The height is 7 inches.

$A = \frac{1}{2}h(b_1 + b_2)$ Area of a trapezoid

$A = \frac{1}{2}(7)(5 + 12)$ Replace h with 7, b_1 with 5, and b_2 with 12.

$A = \frac{1}{2}(7)(17)$ Add 5 and 12.

$A = 59.5$ Multiply.

The area of the trapezoid is 59.5 square inches.

Your Turn **Find the area of each triangle or trapezoid. Round to the nearest tenth if necessary.**

a.

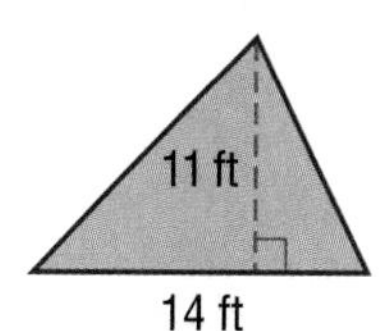

b.

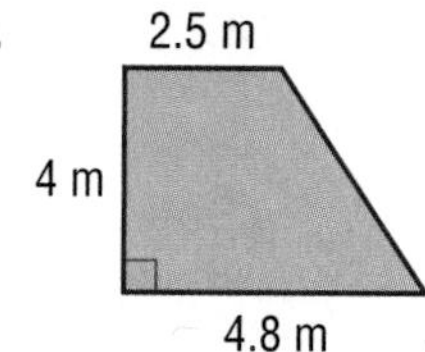

c.

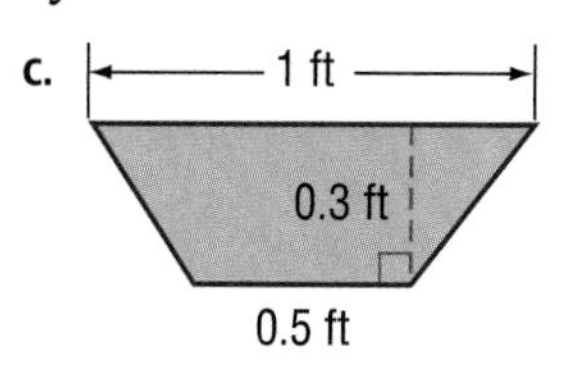

REAL-LIFE MATH

GEOGRAPHY The actual area of Tennessee is 42,244 square miles.

Source: *Merriam Webster's Collegiate Dictionary*

EXAMPLE Use a Formula to Estimate Area

3 GEOGRAPHY The shape of the state of Tennessee resembles a trapezoid. Estimate its area in square miles.

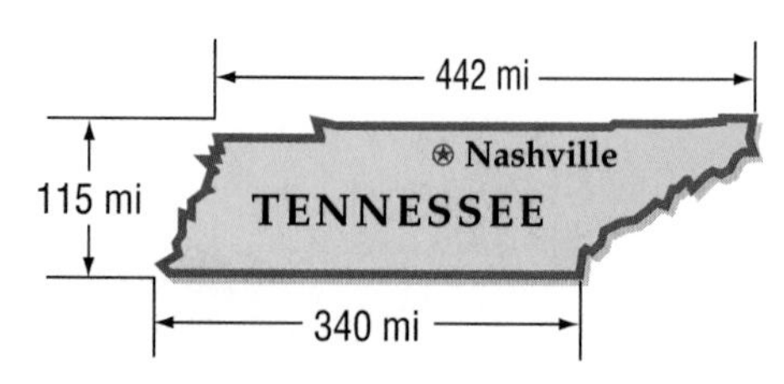

$A = \frac{1}{2}h(b_1 + b_2)$

$A = \frac{1}{2}(115)(442 + 340)$ Replace h with 115, b_1 with 442, and b_2 with 340.

$A = \frac{1}{2}(115)(782)$ Add 442 and 340.

$A = 44{,}965$ Multiply.

The area of Tennessee is about 44,965 square miles.

Skill and Concept Check

1. **Estimate** the area of the trapezoid at the right.

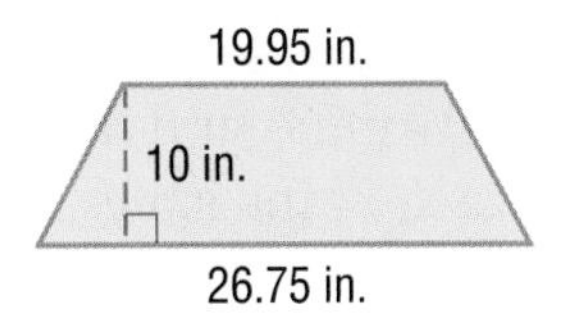

2. **OPEN ENDED** Draw a trapezoid and label the bases and the height. In your own words, explain how to find the area of the trapezoid.
3. **NUMBER SENSE** Describe the relationship between the area of a parallelogram and the area of a triangle with the same height and base.

GUIDED PRACTICE

Find the area of each figure. Round to the nearest tenth if necessary.

4.

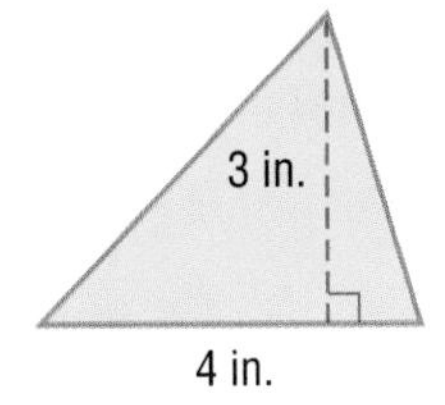

5.

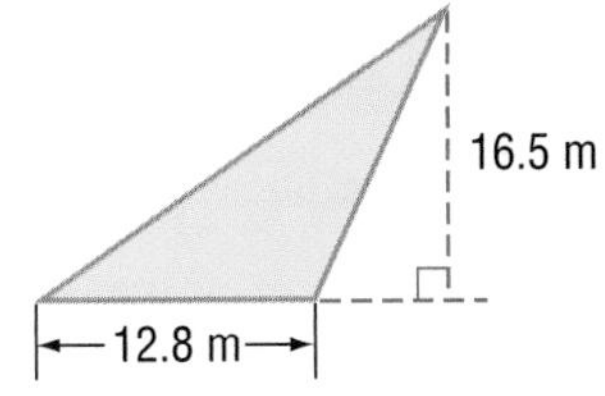

6.

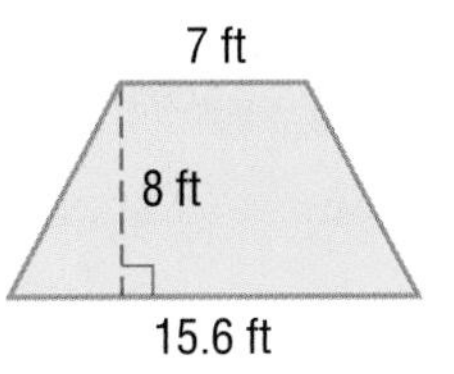

7. **MULTI STEP** The blueprints for a patio are shown at the right. If the cost of the patio is $4.50 per square foot, what will be the total cost of the patio?

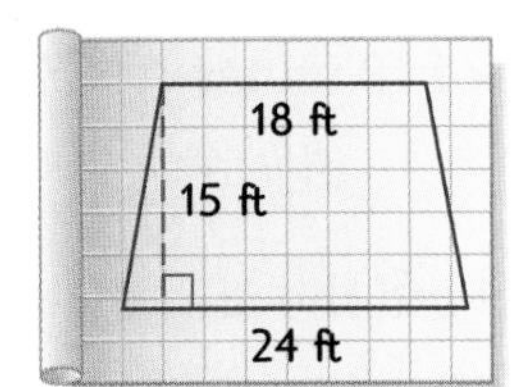

Practice and Applications

HOMEWORK HELP	
For Exercises	See Examples
8–9, 12, 14	1
10–11, 13, 15	2
20–21	3

Extra Practice
See pages 591, 606.

Find the area of each figure. Round to the nearest tenth if necessary.

8.

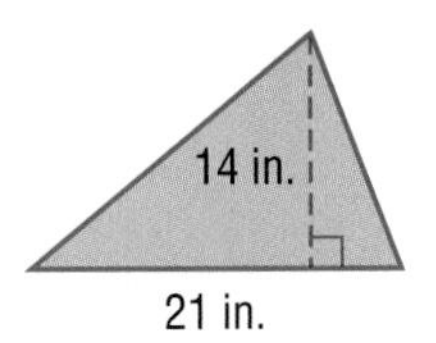

9.

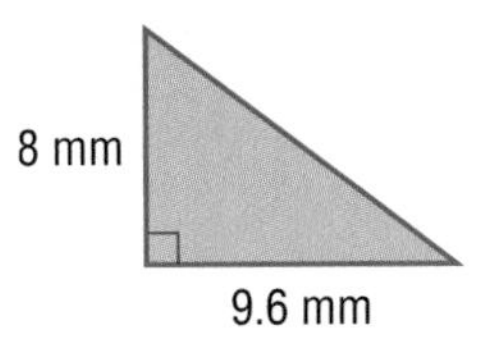

10.

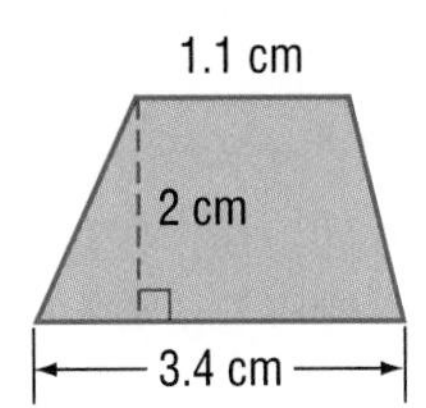

11. 17.75 m, 8 m, 10.25 m

12.

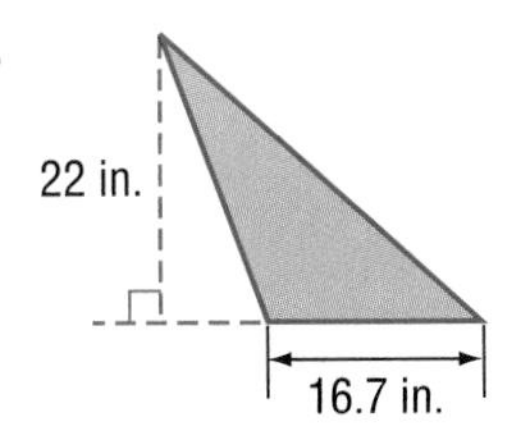

13.

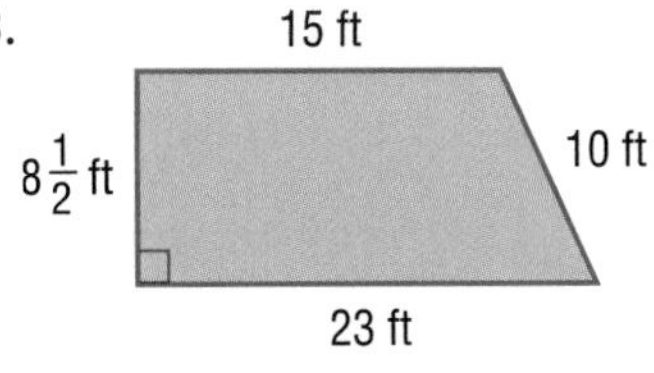

14. triangle: base = 4 cm, height = 7.5 cm
15. trapezoid: bases 13 in. and $1\frac{1}{4}$ ft, height 1 ft

Draw and label each figure on grid paper. Then find the area.

16. a triangle with no right angles
17. an isosceles triangle with a height greater than 6 units
18. a trapezoid with a right angle and an area of 40 square units
19. a trapezoid with no right angles and an area less than 25 square units

20. **GEOGRAPHY** Nevada has a shape that looks like a trapezoid, as shown at the right. Find the approximate area of the state.

21. **GEOGRAPHY** Delaware has a shape that is roughly triangular with a base of 39 miles and a height of 96 miles. Find the approximate area of the state.

Data Update How do the actual areas of Nevada and Delaware compare to your estimates? Visit msmath2.net/data_update to learn more.

22. **CRITICAL THINKING** A triangle has height h. Its base is 4. Find the area of the triangle. (*Hint*: Express your answer in terms of h.)

Standardized Test Practice and Mixed Review

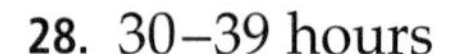

23. **SHORT RESPONSE** Find the area of the triangle at the right to the nearest tenth.

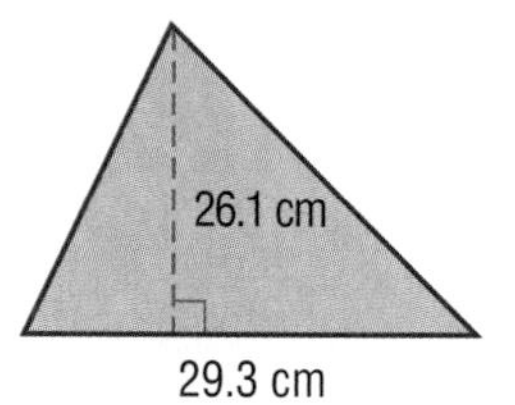

24. **MULTIPLE CHOICE** A trapezoid has bases of 15 meters and 18 meters and a height of 10 meters. What is the area of the trapezoid?

  Ⓐ 30 m^2 Ⓑ 60 m^2 Ⓒ 165 m^2 Ⓓ 330 m^2

25. **GEOMETRY** Find the area of a parallelogram having a base of 2.3 inches and a height of 1.6 inches. Round to the nearest tenth. (Lesson 11-4)

Find the missing measure of each right triangle. Round to the nearest tenth if necessary. (Lesson 11-3)

26. $a = 10$ m, $b = 14$ m

27. $a = 13$ ft, $c = 18$ ft

For Exercises 28–31, refer to the graphic at the right. Classify the angle that represents each category as *acute, obtuse, right,* or *straight.* (Lesson 10-1)

28. 30–39 hours

29. 1–29 hours

30. 40 hours

31. 41–50 hours

32. **MUSIC** Use the Fundamental Counting Principle to find the number of piano instruction books in a series if there are Levels 1, 2, 3, and 4 and each level contains five different books. (Lesson 9-3)

GETTING READY FOR THE NEXT LESSON

BASIC SKILL Use a calculator to find each product to the nearest tenth.

33. $\pi \cdot 13$ 34. $\pi \cdot 29$ 35. $\pi \cdot 16^2$ 36. $\pi \cdot 4.8^2$

11-6 Area of Circles

What You'll LEARN

Find the areas of circles.

REVIEW Vocabulary

pi (π): the Greek letter that represents an irrational number approximately equal to 3.14 **(Lesson 6-9)**

HANDS-ON Mini Lab

Materials

- large paper plate
- scissors

Work with a partner.

- Fold a paper plate in half four times to divide it into 16 equal-sized sections.
- Label the radius r as shown. Let C represent the circumference of the circle.
- Cut out each section; reassemble to form a parallelogram-shaped figure.

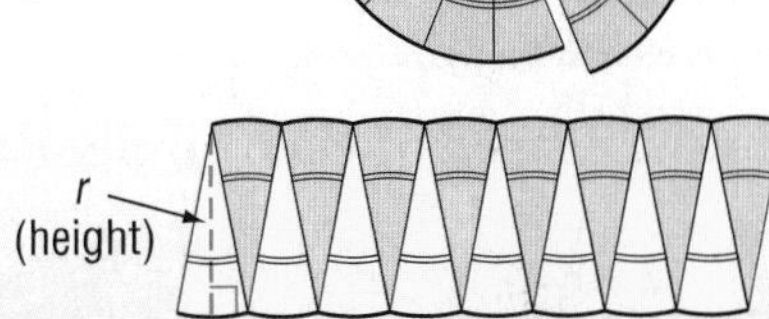

1. What is the measurement of the base and the height?
2. Substitute these values into the formula for the area of a parallelogram.
3. Replace C with the expression for the circumference of a circle, $2\pi r$. Simplify the equation and describe what it represents.

In the Mini Lab, the formula for the area of a parallelogram was used to develop a formula for the area of a circle.

Key Concept — **Area of a Circle**

Words The area A of a circle equals the product of pi (π) and the square of its radius r.

Symbols $A = \pi r^2$

Model

EXAMPLES Find the Areas of Circles

1 Find the area of the circle at the right.

$A = \pi r^2$ Area of a circle

$A = \pi \cdot 2^2$ Replace r with 2.

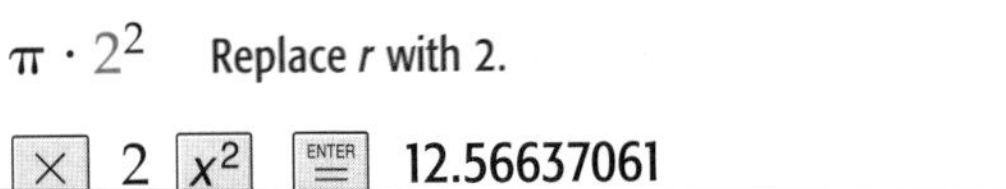

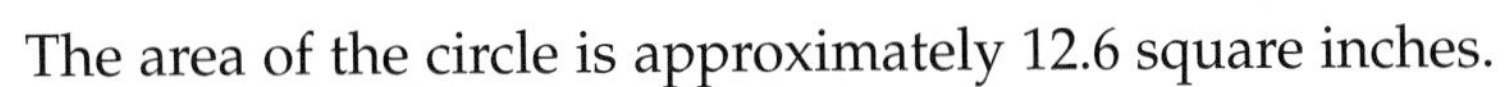

The area of the circle is approximately 12.6 square inches.

2 Find the area of a circle with a diameter of 15.2 centimeters.

$A = \pi r^2$ Area of a circle

$A = \pi \cdot 7.6^2$ Replace r with $15.2 \div 2$ or 7.6.

$A \approx 181.5$ Use a calculator.

The area of the circle is approximately 181.5 square centimeters.

Skill and Concept Check

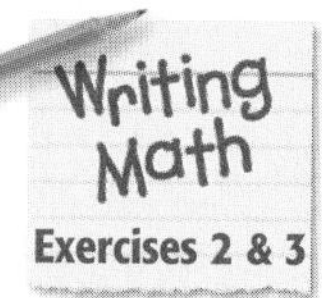

1. **OPEN ENDED** Draw and label the radius of a circle that has an area less than 10 square units.

2. **FIND THE ERROR** Carlos and Sean are finding the area of a circle that has a diameter of 12 centimeters. Who is correct? Explain.

Carlos	Sean
$A = \pi(12)^2$	$A = \pi(6)^2$
$\approx 452 \text{ cm}^2$	$\approx 113 \text{ cm}^2$

3. **NUMBER SENSE** Without using a calculator, determine which has the greatest value: 2π, $\sqrt{7}$, or 1.5^2. Explain.

GUIDED PRACTICE

Find the area of each circle. Round to the nearest tenth.

4. 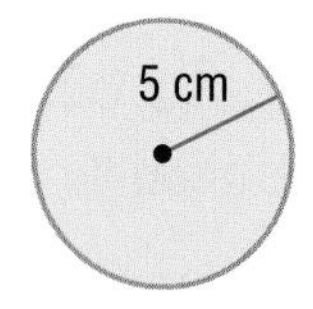

5. 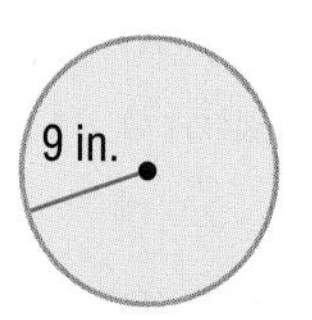

6. 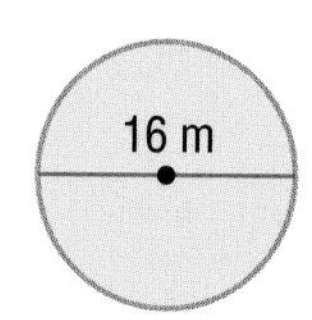

7. radius = 4.2 ft
8. diameter = 13 ft
9. diameter = 24 mm

10. **HISTORY** The Roman Pantheon is a circular-shaped structure that was completed about 126 A.D. Find the area of the floor if the diameter is 43 meters.

Practice and Applications

HOMEWORK HELP	
For Exercises	**See Examples**
11–16, 25	1
17–22, 24	2

Extra Practice
See pages 591, 606.

Find the area of each circle. Round to the nearest tenth.

11. 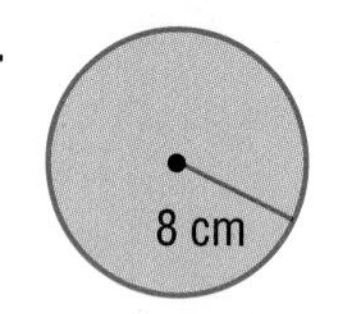

12. 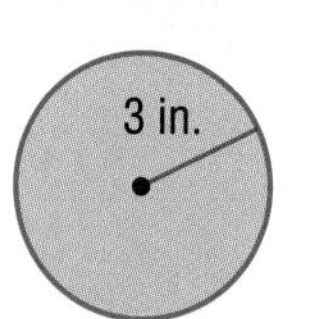

13. 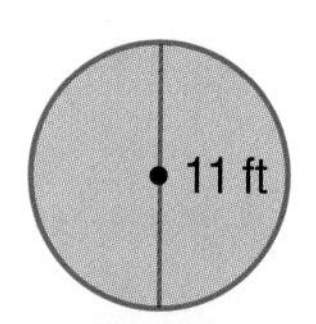

14. 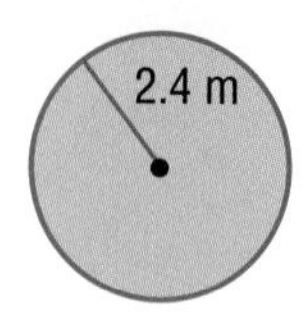

15. 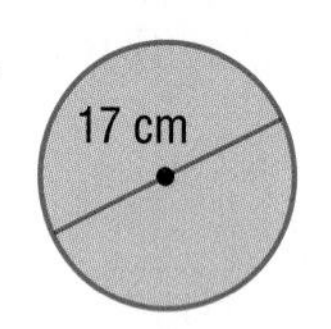

16. 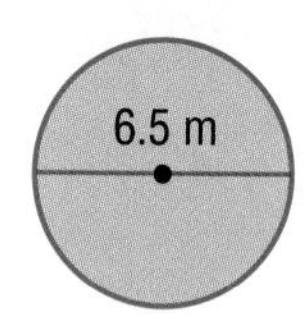

17. radius = 6 ft
18. diameter = 7 ft
19. diameter = 3 cm
20. radius = 10.5 mm
21. radius = $4\frac{1}{2}$ in.
22. diameter = $20\frac{3}{4}$ yd

23. A *semicircle* is half a circle. Find the area of the semicircle at the right to the nearest tenth.

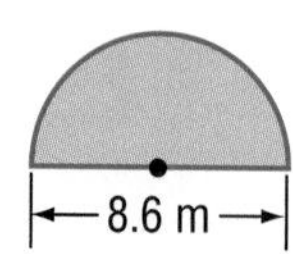

24. **MONEY** Find the area of the face of a Sacagawea $1 coin if the diameter is 26.5 millimeters. Round to the nearest tenth.

25. **LANDSCAPE DESIGN** A circular stone path is to be installed around a birdbath with radius 1.5 feet, as shown at the right. What is the area of the path? (*Hint*: Find the area of the large circle minus the area of the small circle.)

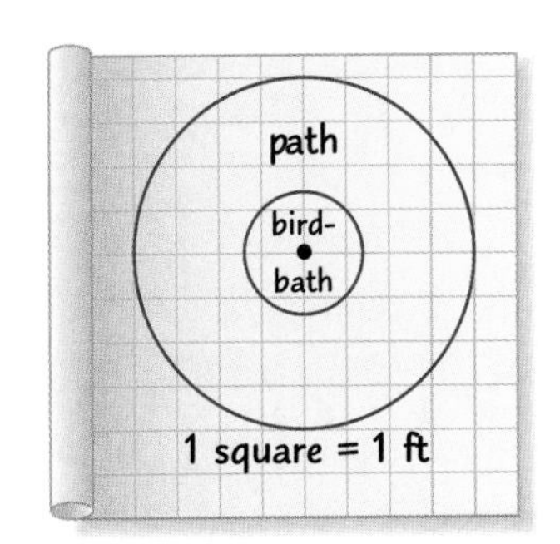

For Exercises 26 and 27, refer to the information below.
Let the x-axis of a coordinate plane represent the radius of a circle and the y-axis represent the area of a circle.

26. Graph the points that represent the circles with radii 0, 1, 2, and 3 units long. Draw a line or curve that goes through each point.

27. Consider a circle with radius of 1 unit and a circle with a radius of 2 units. Write a ratio comparing the radii. Write a ratio comparing the areas. Do these ratios form a proportion? Explain.

Find the area of the shaded region in each figure. Round to the nearest tenth.

28.

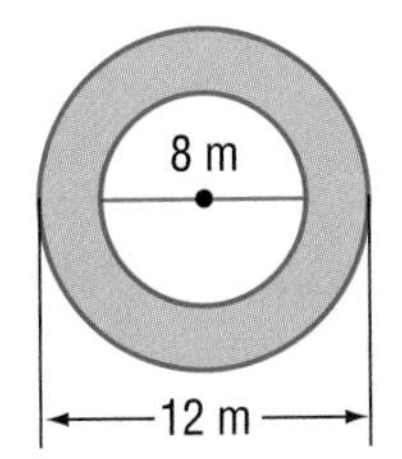

29.

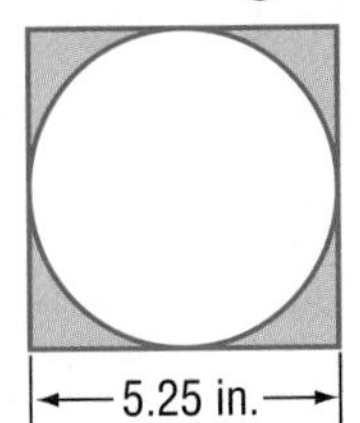

30.

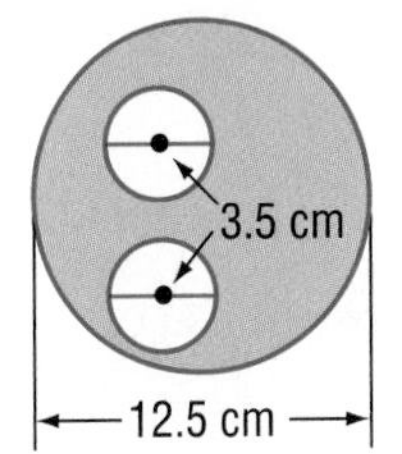

31. **CRITICAL THINKING** Determine whether the area of a circle is *sometimes, always,* or *never* doubled when the radius is doubled. Explain.

Standardized Test Practice and Mixed Review

32. **MULTIPLE CHOICE** A CD has a diameter of 12 centimeters. The hole in the middle of the CD has a diameter of 1.5 centimeters. Find the area of one side of the CD to the nearest tenth. Use 3.14 for π.

Ⓐ 111.3 cm^2 Ⓑ 113.0 cm^2 Ⓒ 349.4 cm^2 Ⓓ 445.1 cm^2

33. **SHORT RESPONSE** Find the radius of a circle that has an area of 42 square centimeters. Use 3.14 for π and round to the nearest tenth.

34. **GEOMETRY** Find the area of a triangle with a base of 21 meters and a height of 27 meters. (Lesson 11-5)

Find the area of each parallelogram. Round to the nearest tenth if necessary. (Lesson 11-4)

35.

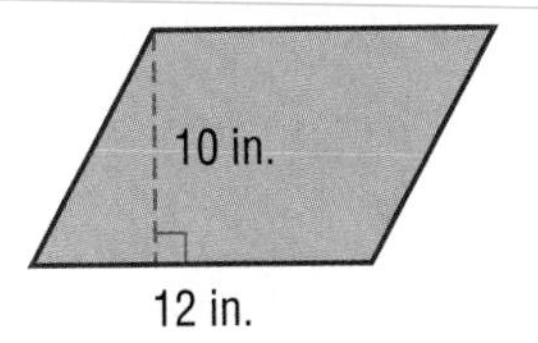

36.

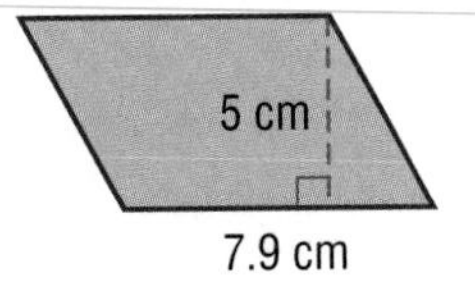

37.

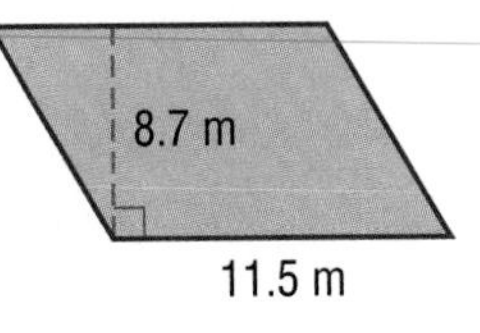

GETTING READY FOR THE NEXT LESSON

BASIC SKILL Simplify each expression. (Lessons 1-2 and 1-3)

38. 8.5^2

39. $3.14 \cdot 6^2$

40. $\frac{1}{2} \cdot 5.4^2 + 11$

41. $\frac{1}{2} \cdot 7^2 + (9)(14)$

11-7a Problem-Solving Strategy

A Preview of Lesson 11-7

What You'll Learn
Solve problems by solving a simpler problem.

Solve a Simpler Problem

The diagram shows the backdrop for our fall play. How much wallpaper will we need to cover the entire front?

We need to find the total area of the backdrop. Let's **solve a simpler problem** by breaking it down into separate geometric shapes.

Explore	We know that the backdrop is made of one large rectangle and two semicircles, which equal an entire circle.
Plan	We can find the areas of the rectangle and the circle, and then add.
Solve	area of rectangle: $A = \ell w$ $A = (8 + 8)7$ or 112 area of circle: $A = \pi r^2$ $A = \pi \cdot 4^2$ or about 50.3 total area: $112 + 50.3$ or 162.3 square feet So, we need at least 162.3 square feet of wallpaper.
Examine	Use estimation to check. The backdrop is 16 feet long and 11 feet high. However, it is less than a complete rectangle, so the area should be *less than* $16 \cdot 11$ or 176 feet. The area, 162.3 square feet, is less than 176 feet, so the answer is reasonable.

4 ft
7 ft
8 ft
8 ft

Analyze the Strategy

1. **Explain** why simplifying this problem is a good strategy to solve this problem.
2. **Describe** another way that the problem could have been solved.
3. **Write** a problem that can be solved by breaking it down into a simpler problem. Solve the problem and explain your answer.

Apply the Strategy

Solve. Use the solve a simpler problem strategy.

4. **LANDSCAPING** James is helping his father pour a circular sidewalk around a flower bed, as shown below. What is the area, in square feet, of the sidewalk? Use 3.14 for π.

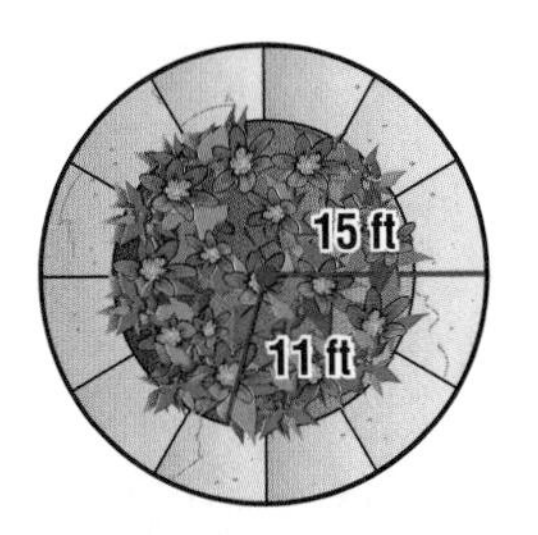

5. **COMMUNICATION** According to a recent report, one city has 2,945,000 phone lines assigned to three different area codes. How many of the phone lines are assigned to each area code?

Area Code	Percent
888	44.3%
777	23.7%
555	31.5%

Mixed Problem Solving

Solve. Use any strategy.

6. **EARTH SCIENCE** Earth's atmosphere exerts a pressure of 14.7 pounds per square inch at the ocean's surface. The pressure increases by 12.7 pounds per square inch for every 6 feet that you descend. Find the pressure at 18 feet below the surface.

7. **SALES** Deirdre is trying to sell $3,000 in ads for the school newspaper. The prices of the ads and the number of ads that she has sold are shown in the table. Which is the smallest ad she could sell in order to meet her quota?

Ad Size	Cost Per Ad	Number Sold
quarter-page	$75	15
half-page	$125	8
full-page	$175	4

8. **THEATER** Mr. Marquez is purchasing fabric for curtains for a theatrical company. The front of the stage is 15 yards wide and 5 yards high. The fabric is sold on bolts that are 60 inches wide and 20 yards long. How many bolts are needed to make the curtains?

9. **TELEVISION** The graph shows the results of a survey in which 365,750 people were asked to name their favorite television programs. Estimate how many people chose sitcoms as their favorite.

Favorite TV Shows

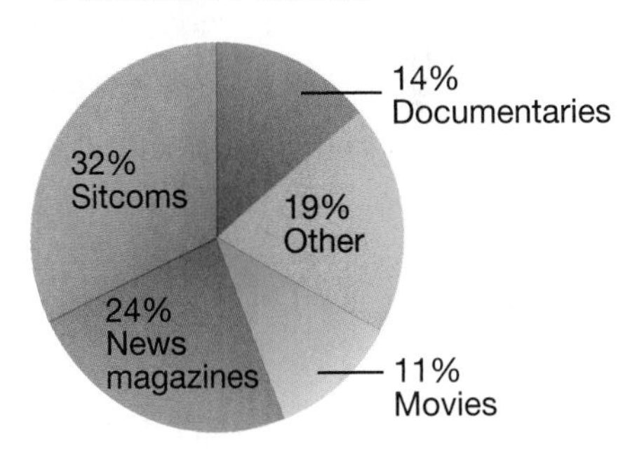

10. **STANDARDIZED TEST PRACTICE** Kara is painting one wall in her room, as shown by the shaded region below. What is the area that she is painting?

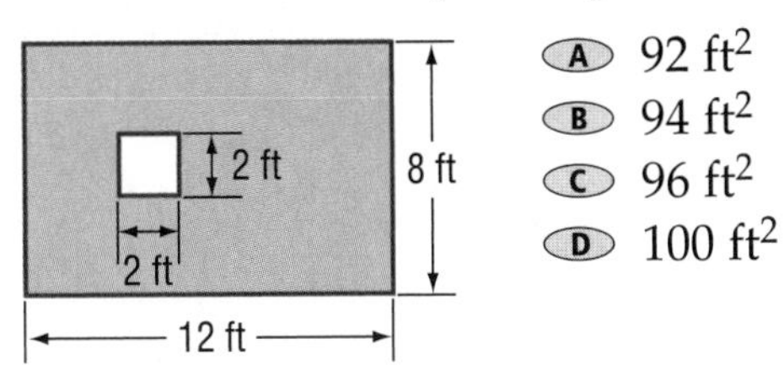

A. 92 ft^2
B. 94 ft^2
C. 96 ft^2
D. 100 ft^2

You will use the solve a simpler problem strategy in the next lesson.

11-7 Area of Complex Figures

What You'll LEARN

Find the areas of complex figures.

NEW Vocabulary

complex figure

Link to READING

Everyday Meaning of Complex: a whole made up of interrelated parts, as in a retail complex that is made up of many different stores

WHEN am I ever going to use this?

ARCHITECTURE Rooms in a house are not always square or rectangular, as shown in the diagram at the right.

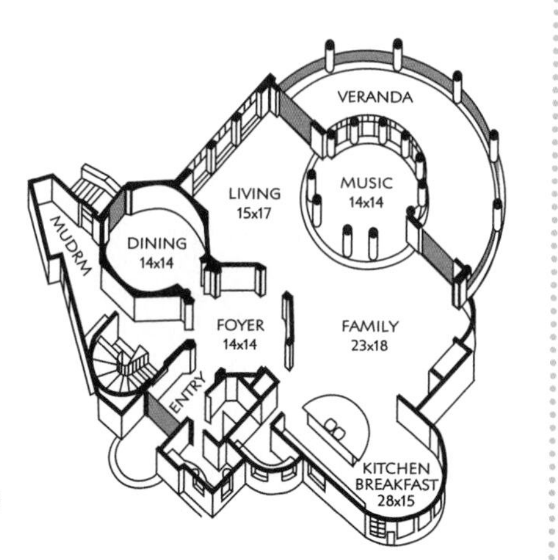

1. Describe the shape of the kitchen.
2. How could you determine the area of the kitchen?
3. How could you determine the total square footage of a house with rooms shaped like these?

A **complex figure** is made of circles, rectangles, squares, and other two-dimensional figures. To find the area of a complex figure, separate it into figures whose areas you know how to find, and then add the areas.

EXAMPLE Find the Area of an Irregular Room

1 **ARCHITECTURE** **Refer to the diagram of the house above. The kitchen is 28 feet by 15 feet, as shown at the right. Find the area of the kitchen. Round to the nearest tenth.**

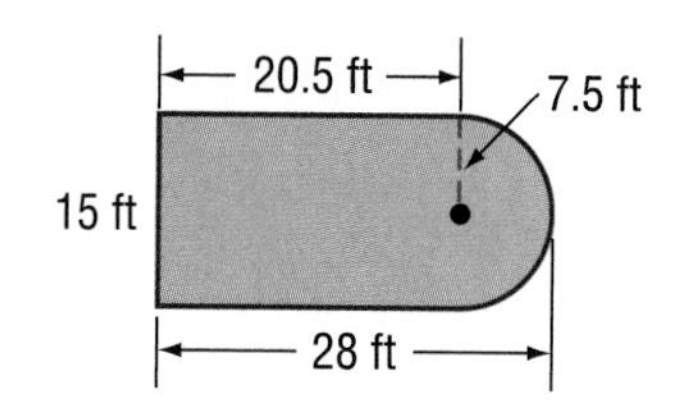

The figure can be separated into a rectangle and a semicircle.

Area of Rectangle

$A = \ell w$ Area of a rectangle

$A = 20.5 \cdot 15$ Replace ℓ with 20.5 and w with 15.

$A = 307.5$ Multiply.

Area of Semicircle

$A = \frac{1}{2}\pi r^2$ Area of a semicircle

$A = \frac{1}{2}\pi(7.5)^2$ Replace r with 7.5.

$A \approx 88.4$ Simplify.

The area of the kitchen is approximately 307.5 + 88.4 or 395.9 square feet.

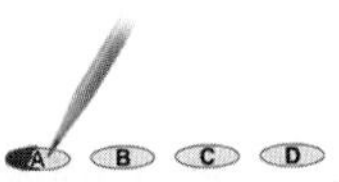

Standardized Test Practice

EXAMPLE Find the Area of a Complex Figure

2 **GRID-IN TEST ITEM** **Find the area of the figure at the right in square inches.**

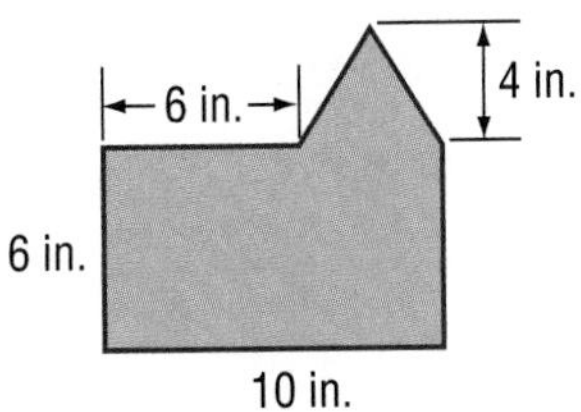

Read the Test Item

The figure can be separated into a rectangle and a triangle. Find the area of each.

Solve the Test Item

Area of Rectangle

$A = \ell w$	Area of a rectangle
$A = 10 \cdot 6$	Replace ℓ with 10 and w with 6.
$A = 60$	Multiply.

Area of Triangle

$A = \frac{1}{2}bh$	Area of a triangle
$A = \frac{1}{2}(4)(4)$	$b = 10 - 6$ or 4, $h = 4$
$A = 8$	Multiply.

The area is 60 + 8 or 68 square inches.

Fill in the Grid

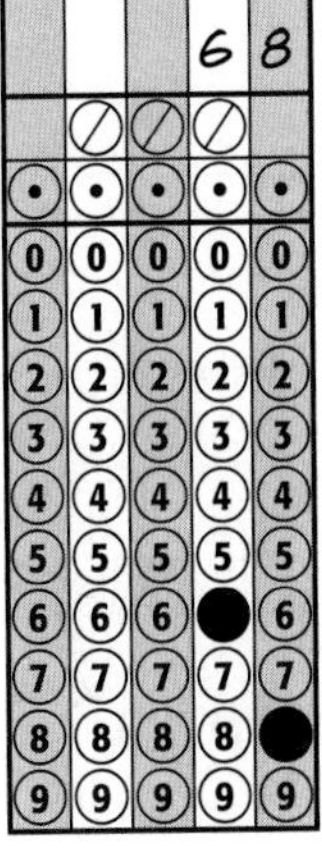

Test-Taking Tip

The Princeton Review

Drawings Be sure to add the areas of all the separate figures and not stop once you find the area of part of the figure.

Your Turn **Find the area of each figure. Round to the nearest tenth if necessary.**

a.

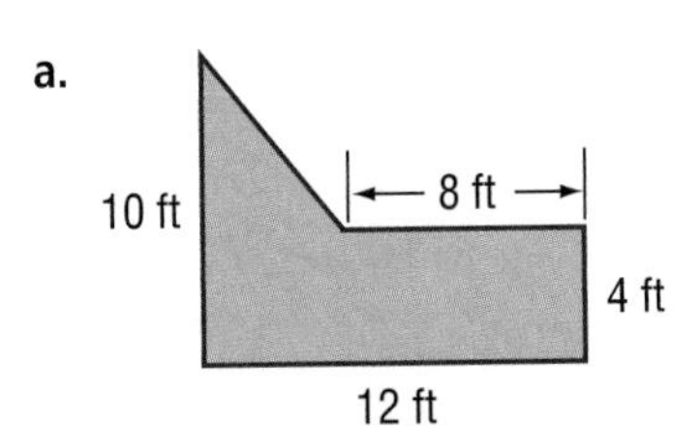

b.

2.5 cm

1.5 cm

9.2 cm

Skill and Concept Check

Writing Math Exercises 1 & 2

1. **Describe** how you would find the area of the figure at the right.

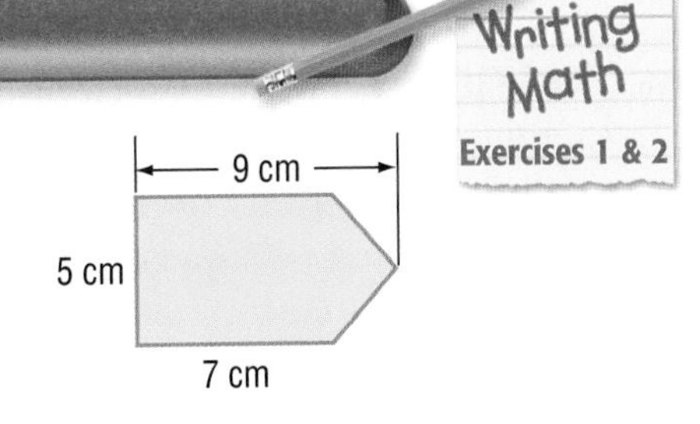

2. **OPEN ENDED** Sketch a complex figure and describe how you could find the area.

GUIDED PRACTICE

Find the area of each figure. Round to the nearest tenth if necessary.

3.

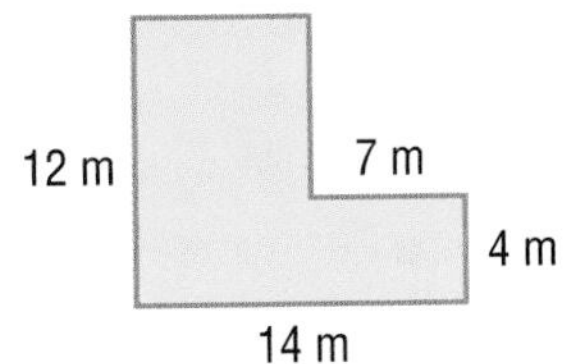

4.

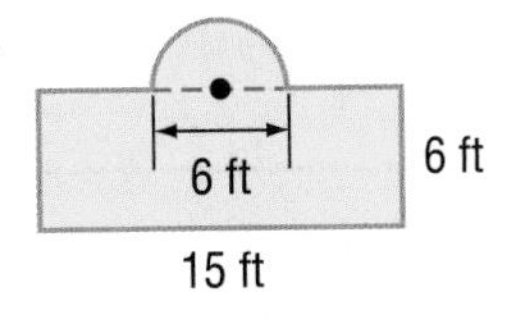

5.

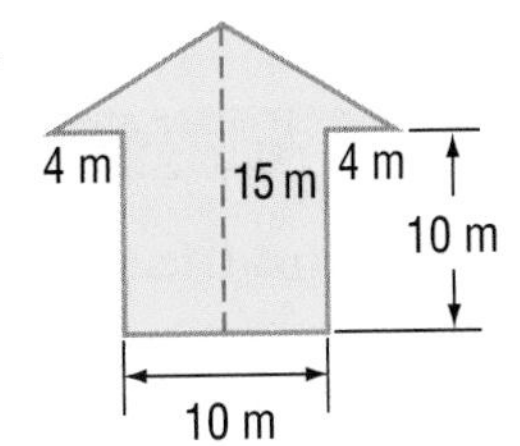

Practice and Applications

HOMEWORK HELP	
For Exercises	See Examples
6–12	1, 2

Extra Practice
See pages 592, 606.

Find the area of each figure. Round to the nearest tenth if necessary.

6. 15 cm, 7 cm, 10 cm

7. 5.3 in., 8 in., 4 in., 8 in.

8. 5 yd, 7 yd

9.
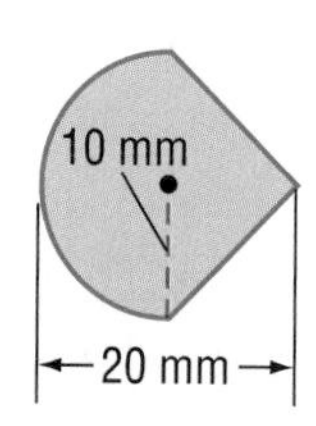

10. 12 in., 10 in., 15 in.

11.
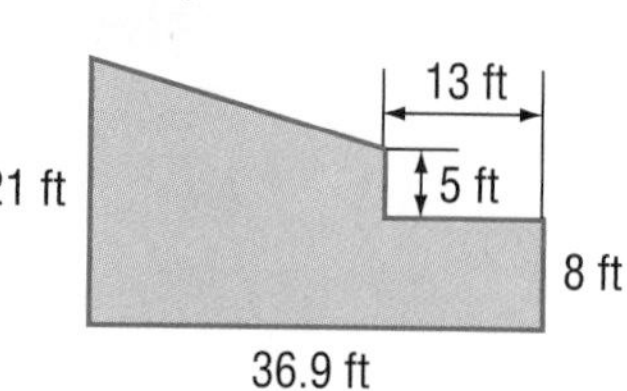

12. **INTERIOR DESIGN** The Eppicks' living room, shown at the right, has a bay window. They are planning to have the hardwood floors in the room refinished. What is the total area that needs to be refinished?

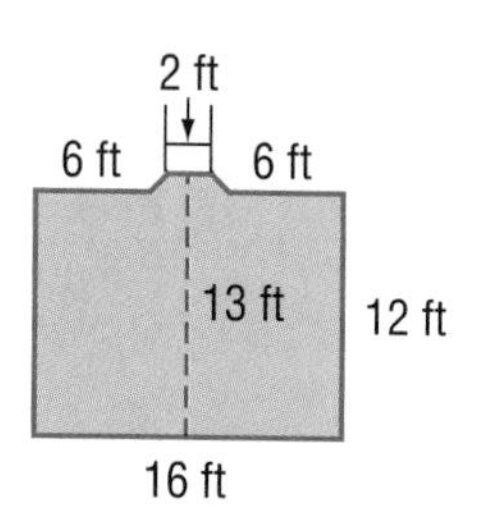

CRITICAL THINKING **Describe how you could estimate the area of each state.**

13.
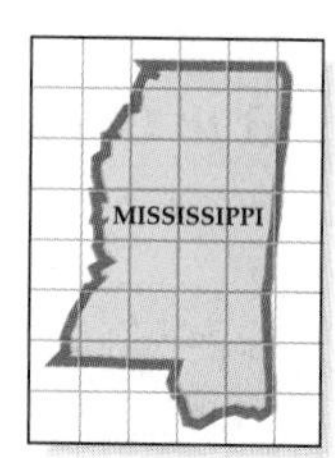

14.
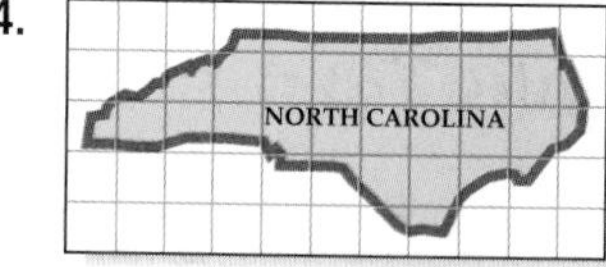

Standardized Test Practice and Mixed Review

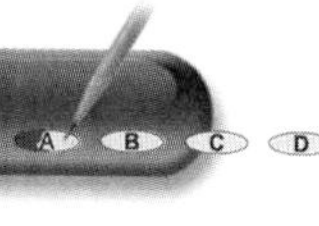

15. **SHORT RESPONSE** Find the area of the figure if each triangle has a height of 3.5 inches and the square has side lengths of 4 inches.

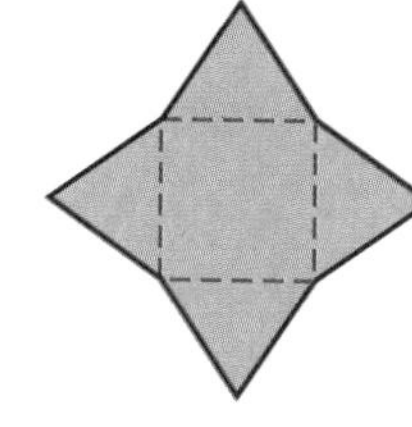

16. **MULTIPLE CHOICE** A rectangular room 14 feet by 12 feet has a semicircular sitting area attached with a diameter of 12 feet. What is the total area of the room and the sitting area?

Ⓐ 168 ft^2 Ⓑ 224.5 ft^2 Ⓒ 281.1 ft^2 Ⓓ 620.4 ft^2

Find the area of each circle. Round to the nearest tenth. (Lesson 11-6)

17. radius = 4.7 cm
18. radius = 12 in.
19. diameter = 15 in.

20. Find the area of a triangle that has a base of 3.8 meters and a height of 9 meters. (Lesson 11-5)

GETTING READY FOR THE NEXT LESSON

PREREQUISITE SKILL **Find the probability of rolling each number on a number cube.** (Lesson 9-1)

21. $P(2)$
22. $P(\text{even})$
23. $P(3 \text{ or } 4)$
24. $P(\text{less than } 5)$

11-8 Area Models and Probability

HANDS-ON Mini Lab

Materials

- two number cubes

What You'll LEARN

Find probability using area models.

REVIEW Vocabulary

probability: the ratio of the number of ways an event can occur to the number of possible outcomes **(Lesson 9-1)**

Work with a partner.

- Roll the number cubes and find the product of the numbers rolled.
- Repeat nine more times.
- Collect the data for the entire class. Organize the outcomes in a table.

1. Do certain products occur more often?
2. Make and complete a table like the one at the right to find all the possible outcomes.

×	1	2	3	4	5	6
1	1	2	3			
2	2	4	6			
3						

The grid at the right shows the possible products when two number cubes are rolled. The area of the grid is 36 square units. Notice that 6 and 12 make up $\frac{8}{36}$ of the area. So, the probability of rolling two numbers whose product is 6 or 12 is $\frac{8}{36}$.

1	2	3	4	5	6
2	4	6	8	10	12
3	6	9	12	15	18
4	8	12	16	20	24
5	10	15	20	25	30
6	12	18	24	30	36

The area of geometric shapes can be used to find probabilities.

EXAMPLE Use Area Models to Find Probability

1 PROBABILITY A randomly-dropped counter falls somewhere in the squares. Find the probability that it falls on the shaded squares.

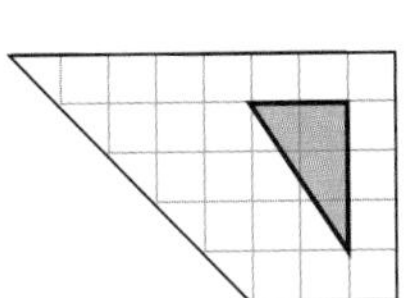

$$\text{probability} = \frac{\text{number of ways to land in shaded squares}}{\text{number of ways to land on squares}}$$

$$= \frac{\text{area of shaded squares}}{\text{area of all squares}}$$

Area of Shaded Squares

$A = \frac{1}{2}bh$ — Area of a triangle

$A = \frac{1}{2}(2)(3)$ — $b = 2$ and $h = 3$

$A = 3$ — Simplify.

Area of All Squares

$A = \frac{1}{2}h(b_1 + b_2)$ — Area of a trapezoid

$A = \frac{1}{2}(5)(8 + 3)$ — $h = 5, b_1 = 8, b_2 = 3$

$A = 27.5$ — Simplify.

So, the probability of a counter falling in the shaded squares is $\frac{3}{27.5}$ or about 10.9%.

EXAMPLE Find the Probability of Winning a Game

2 GAMES Suppose a dart is equally likely to hit any point on the board. What is the probability that it hits the white section?

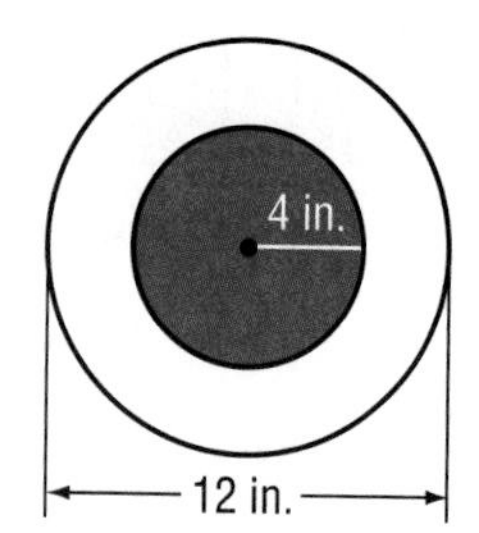

First, find the area of the white section. It equals the area of the large circle minus the area of the small circle.

Area of Large Circle

$A = \pi r^2$ Area of a circle

$A = \pi(6)^2$ Replace r with 6.

$A \approx 113.1$ Simplify.

Area of Small Circle

$A = \pi r^2$ Area of a circle

$A = \pi(4)^2$ Replace r with 4.

$A \approx 50.3$ Simplify.

Area of White Section $= 113.1 - 50.3$ large circle – small circle

$= 62.8$ Subtract.

$P(\text{white}) = \frac{62.8}{113.1}$ ← area of white section / ← area of entire model

≈ 0.5553 Use a calculator.

So, the probability of hitting the white section is about 55.6%.

REAL-LIFE MATH

GAMES The arrangement of numbers around a standard dartboard dates back over 100 years.

Source: mathpages.com

Skill and Concept Check

1. **Explain** how area models are used to solve probability problems.

2. **OPEN ENDED** Draw a spinner in which the probability of spinning and landing on a blue region is $\frac{1}{6}$. Explain your reasoning.

GUIDED PRACTICE

A randomly-dropped counter falls in the squares. Find the probability that it falls in the shaded regions. Write as a percent. Round to the nearest tenth if necessary.

3. 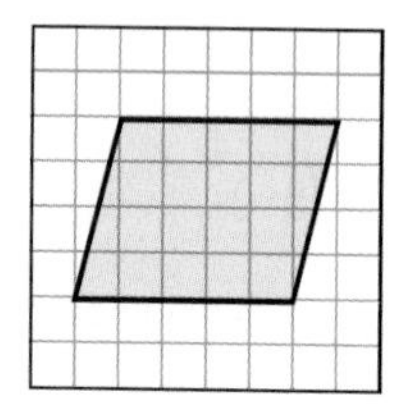

4. 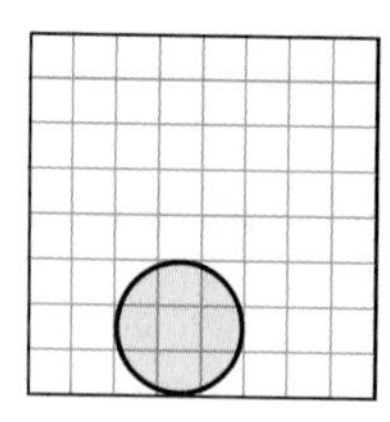

5. **GAMES** Suppose a dart is equally likely to hit any point on the dartboard at the right. What is the probability that it hits the red section?

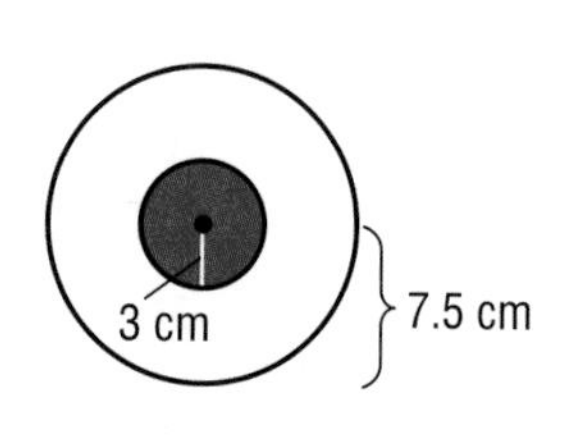

Practice and Applications

HOMEWORK HELP	
For Exercises	See Examples
6–11	1
12–14	2

Extra Practice
See pages 592, 602.

A randomly-dropped counter falls in the squares. Find the probability that it falls in the shaded regions. Write as a percent. Round to the nearest tenth if necessary.

6. 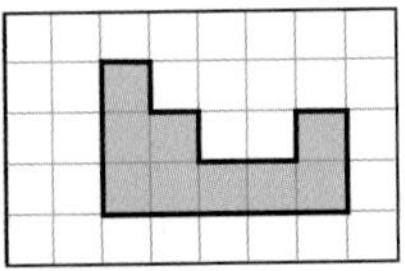

7. 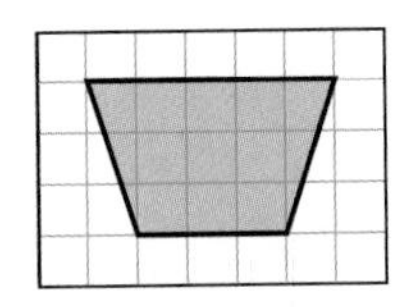

8. 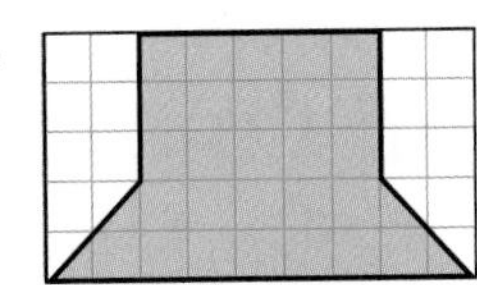

9. 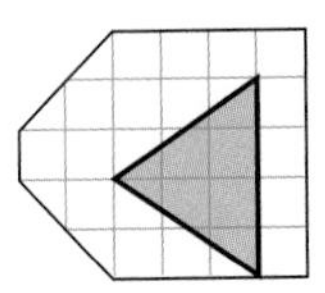

10. 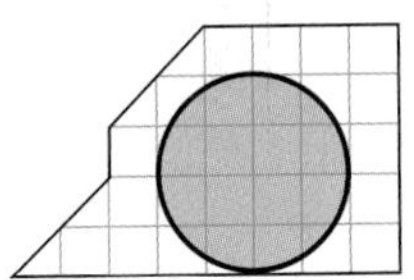

11. 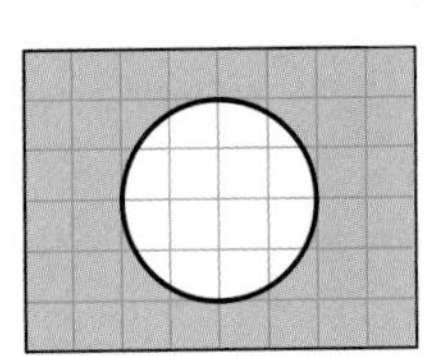

GAMES **Each figure represents a dartboard. If it is equally likely that a thrown dart will land anywhere on the dartboard, find the probability that it lands in the shaded region.**

12. 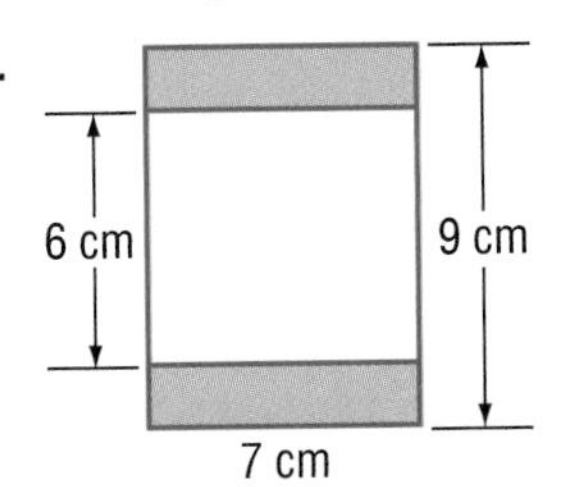

13. 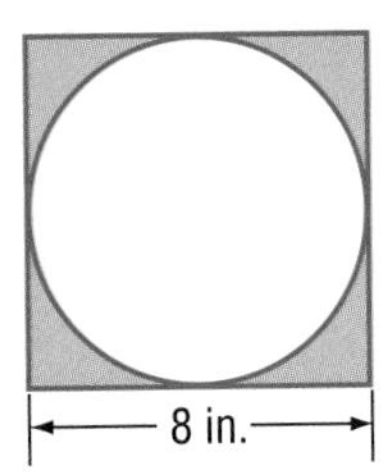

14. 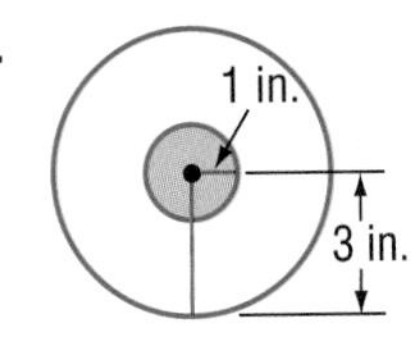

15. **CRITICAL THINKING** A quarter is randomly tossed on the grid board at the right. If a quarter has a radius of 12 millimeters, what is the probability that it does *not* touch a line when it lands? (*Hint*: Find the area where the center of the coin could land so that the edges do not touch a line.)

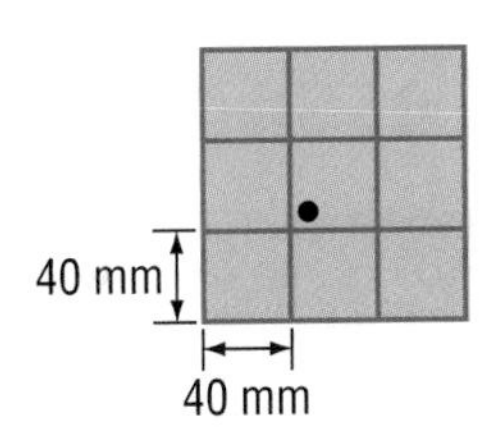

Standardized Test Practice and Mixed Review

16. **MULTIPLE CHOICE** At a carnival, a person wins a prize if their dart pops a balloon on the rectangular wall. If the radius of each circular balloon is 0.4 foot, approximately what is the probability that a person will win? Use 3.14 for π.

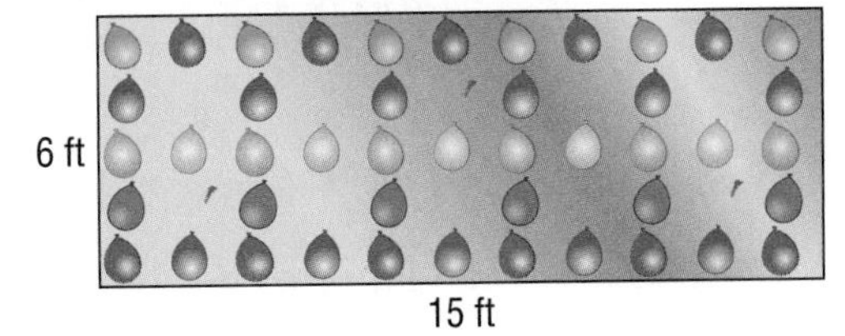

Ⓐ 25%  Ⓑ 29% 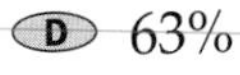 Ⓒ 30% Ⓓ 63%

17. **GRID IN** Find the probability that a randomly dropped counter will fall in the shaded region at the right. Write as a fraction.

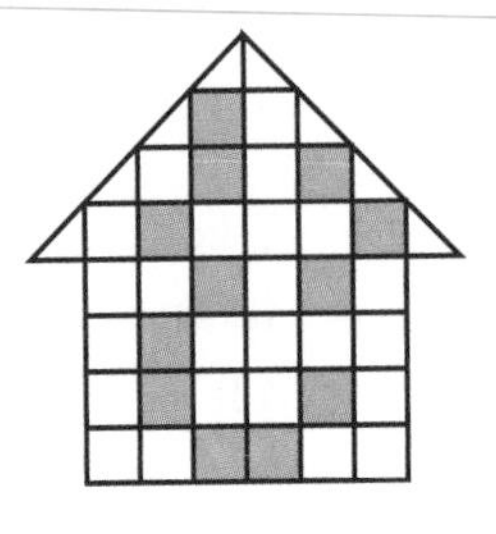

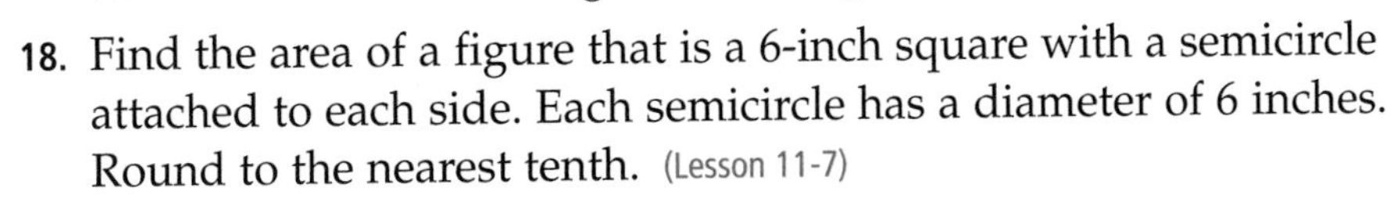

18. Find the area of a figure that is a 6-inch square with a semicircle attached to each side. Each semicircle has a diameter of 6 inches. Round to the nearest tenth. (Lesson 11-7)

19. Find the area of a circle with a radius of 5.7 meters. Round to the nearest tenth. (Lesson 11-6)

Study Guide and Review

Vocabulary and Concept Check

base (p. 483)
complex figure (p. 498)
height (p. 483)
hypotenuse (p. 479)
irrational number (p. 476)
leg (p. 479)
perfect square (p. 471)
Pythagorean Theorem (p. 479)
radical sign (p. 471)
square (p. 470)
square roots (p. 471)

Choose the correct term or number to complete each sentence.

1. In a right triangle, the square of the length of the hypotenuse is (equal to, less than) the sum of the squares of the lengths of the legs.
2. $A = \frac{1}{2}h(a + b)$ is the formula for the area of a (triangle, trapezoid).
3. The square of 25 is (625, 5).
4. A (square, square root) of 49 is 7.
5. The longest side of a right triangle is called the (leg, hypotenuse).
6. The $\sqrt{}$ symbol is called a (radical, perfect square) sign.

Lesson-by-Lesson Exercises and Examples

Squares and Square Roots (pp. 470–473)

Find the square of each number.

7. 6
8. 14
9. 23

Find each square root.

10. $\sqrt{16}$
11. $\sqrt{256}$
12. $\sqrt{900}$

Example 1 **Find the square of 9.**

$9^2 = 9 \cdot 9$ or 81

Example 2 **Find $\sqrt{121}$.**

Since $11 \cdot 11 = 121$, $\sqrt{121} = 11$.

11-2 Estimating Square Roots (pp. 475-477)

Estimate each square root to the nearest whole number.

13. $\sqrt{6}$
14. $\sqrt{99}$
15. $\sqrt{48}$
16. $\sqrt{76}$
17. $\sqrt{19}$
18. $\sqrt{52}$

Use a calculator to find each square root to the nearest tenth.

19. $\sqrt{61}$
20. $\sqrt{132}$
21. $\sqrt{444}$
22. $\sqrt{12}$

Example 3 **Estimate $\sqrt{29}$ to the nearest whole number.**

$25 < 29 < 36$ — 29 is between the perfect squares 25 and 36.

$\sqrt{25} < \sqrt{29} < \sqrt{36}$ — Find the square root of each number.

$5 < \sqrt{29} < 6$ — $\sqrt{25} = 5$ and $\sqrt{36} = 6$

So, $\sqrt{29}$ is between 5 and 6. Since 29 is closer to 25 than to 36, the best whole number estimate is 5.

msmath2.net/vocabulary_review

11-3 The Pythagorean Theorem (pp. 479–482)

Find the missing measure of each right triangle. Round to the nearest tenth if necessary.

23.

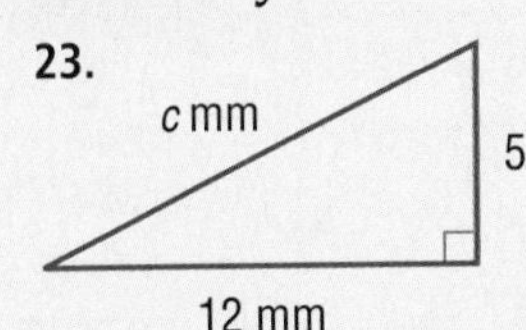

24.

21 in. b in. 34 in.

25. $a = 5$ ft, $b = 6$ ft

26. $b = 10$ yd, $c = 12$ yd

27. $a = 7$ m, $c = 15$ m

28. $a = 12$ in., $b = 4$ in.

Example 4 **Find the missing measure of the triangle. Round to the nearest tenth if necessary.**

4 cm c cm 12 cm

$c^2 = a^2 + b^2$ Pythagorean Theorem

$c^2 = 4^2 + 12^2$ $a = 4, b = 12$

$c^2 = 16 + 144$ Evaluate.

$c^2 = 160$ Add.

$c \approx 12.6$ Take the square root of each side.

The length of the hypotenuse is about 12.6 centimeters.

11-4 Area of Parallelograms (pp. 483–485)

Find the area of each parallelogram. Round to the nearest tenth if necessary.

29.

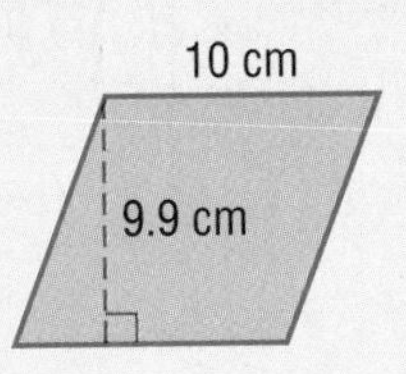

30.

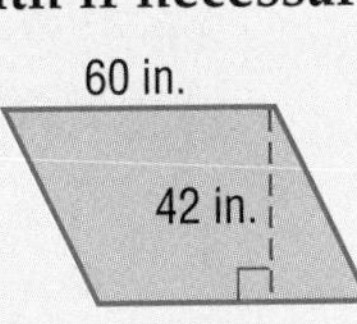

31. base = 9 cm, height = 15 cm

32. base = 24 m, height = 16.2 m

Example 5 **Find the area of a parallelogram if the base is 15 inches and the height is 8 inches.**

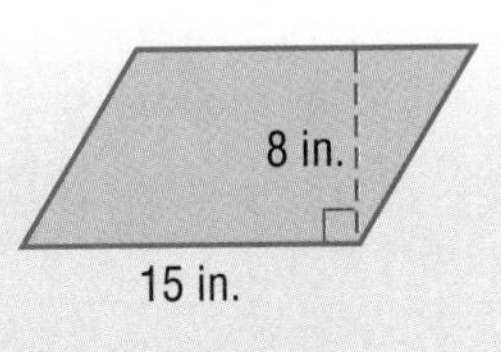

$A = bh$ Area of a parallelogram

$A = 15 \cdot 8$ Replace b with 15 and h with 8.

$A = 120 \text{ in}^2$ Multiply.

11-5 Area of Triangles and Trapezoids (pp. 489–492)

Find the area of each figure. Round to the nearest tenth if necessary.

33.

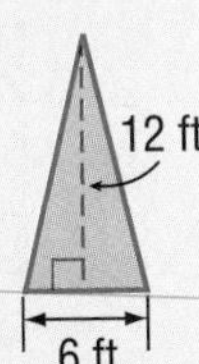

34.

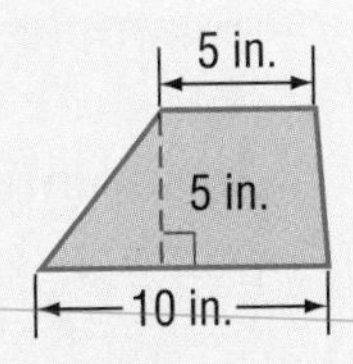

35. triangle: base = 24.7 cm, height = 15.2 cm

36. trapezoid: bases = 22 yd and 35 yd, height = 18.5 yd

Example 6 **Find the area of a triangle with a base of 8 meters and a height of 11.2 meters.**

$A = \frac{1}{2}bh$ Area of a triangle

$A = \frac{1}{2}(8)(11.2)$ or 44.8 m^2 $b = 8, h = 11.2$

Example 7 **Find the area of the trapezoid.**

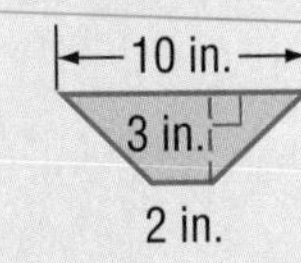

$A = \frac{1}{2}h(b_1 + b_2)$

$A = \frac{1}{2}(3)(2 + 10)$ $h = 3, b_1 = 2, b_2 = 10$

$A = \frac{1}{2}(3)(12)$ or 18 in^2 Simplify.

11-6 Area of Circles (pp. 493–495)

Find the area of each circle. Round to the nearest tenth.

37. radius = 11.4 in.

38. diameter = 44 cm

39. **GARDENING** A lawn sprinkler can water a circular area with a radius of 20 feet. Find the area that can be watered. Round to the nearest tenth.

Example 8 **Find the area of a circle with a radius of 5 inches.**

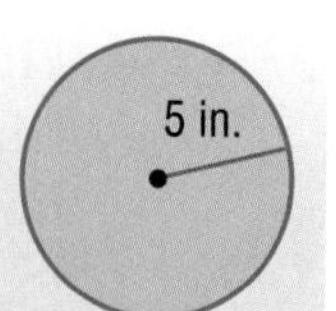

$A = \pi r^2$ Area of a circle

$A = \pi(5)^2$ Replace r with 5.

$A \approx 78.5$ Multiply.

The area of the circle is about 78.5 square inches.

11-7 Area of Complex Figures (pp. 498–500)

Find the area of each figure. Round to the nearest tenth if necessary.

40.

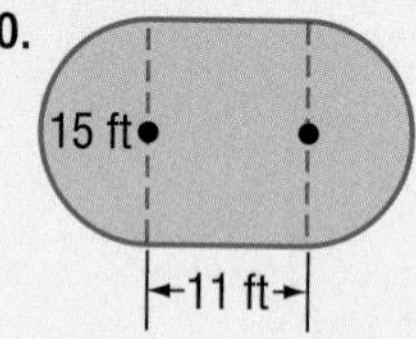

41.

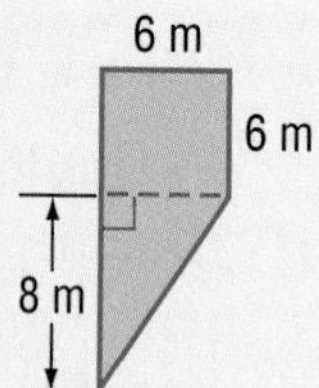

42.

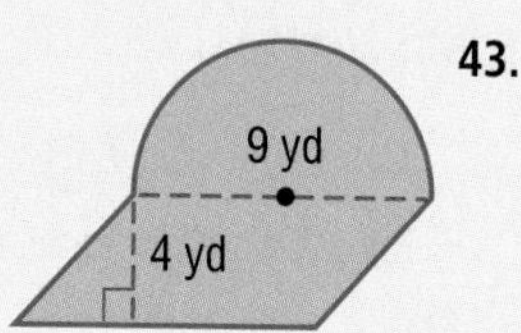

43.

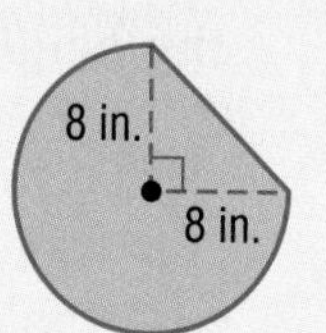

Example 9 **Find the area of the figure.**

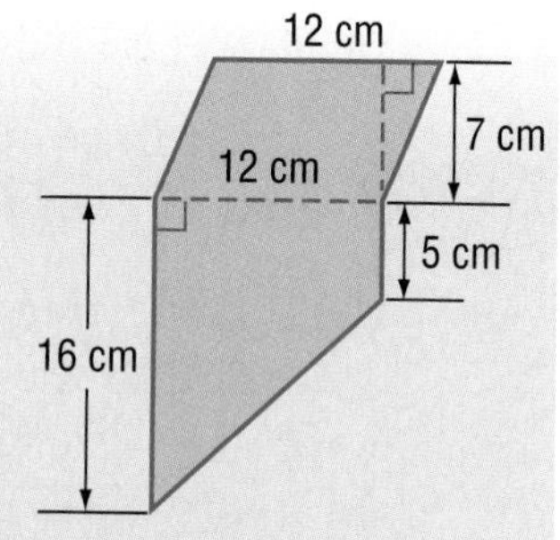

The figure can be separated into a parallelogram and a trapezoid.

parallelogram: $A = bh = (12)(7)$ or 84

trapezoid: $A = \frac{1}{2}h(b_1 + b_2)$

$= \frac{1}{2}(12)(16 + 5)$ or 126

The area of the figure is 84 + 126 or 210 square centimeters.

11-8 Area Models and Probability (pp. 501–503)

A randomly-dropped counter falls in the squares. Find the probability that it falls in the shaded regions. Write as a percent. Round to the nearest tenth if necessary.

44.

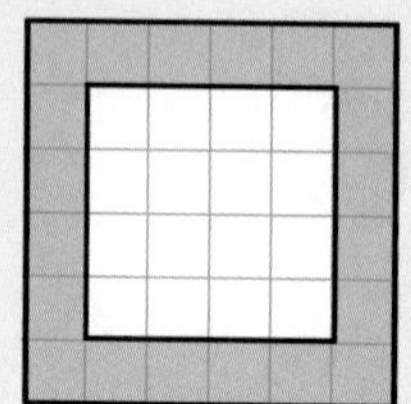

45.

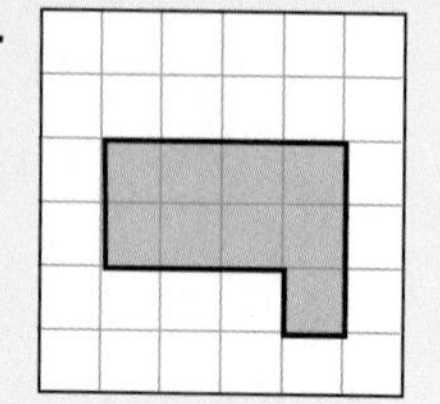

Example 10 **Find the probability that a randomly-dropped counter will land on a white square.**

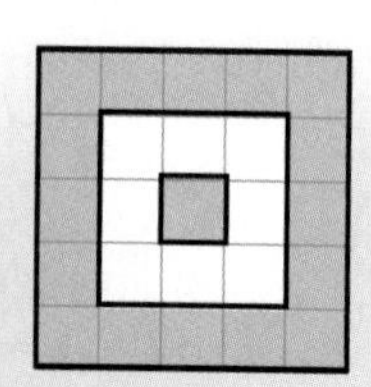

Probability of landing on a white square

$= \frac{\text{number of ways to land on white squares}}{\text{number of ways to land on squares}}$

$= \frac{\text{area of white squares}}{\text{total area}}$

$= \frac{8}{25}$ The probability is $\frac{8}{25}$ or 32%.

CHAPTER 11 Practice Test

Vocabulary and Concepts

1. **State** the measurements that you need in order to find the area of a parallelogram.
2. **Describe** how to find the area of a complex figure.

Skills and Applications

3. Find the square of 9.
4. Find $\sqrt{400}$.
5. **PHYSICAL FITNESS** This morning, Elisa walked 1 mile north, 0.5 mile west, and then walked straight back to her starting point. How far did Elisa walk? Round to the nearest tenth.
6. Estimate $\sqrt{23}$ to the nearest whole number.
7. Use a calculator to find $\sqrt{133}$ to the nearest tenth.

Find the missing measure of each right triangle. Round to the nearest tenth if necessary.

8. $a = 5$ m, $b = 4$ m
9. $b = 12$ in., $c = 14$ in.

Find the area of each figure. Round to the nearest tenth if necessary.

10. 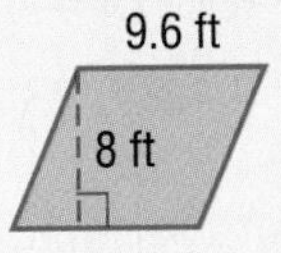

11. 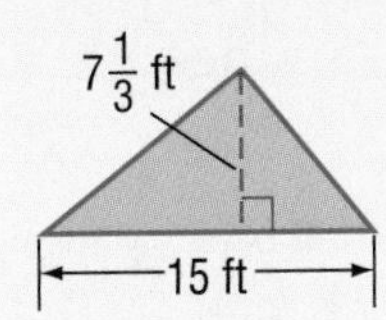

12. 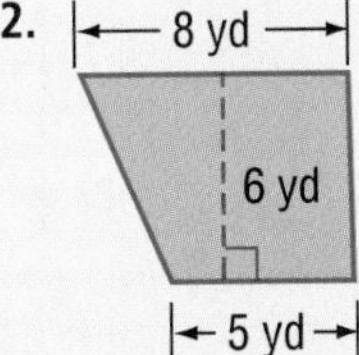

13. 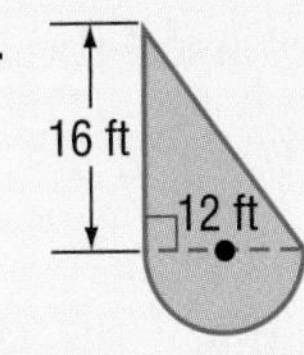

Find the area of each circle. Round to the nearest tenth.

14. radius = 9 ft
15. diameter = 5.2 cm

Standardized Test Practice

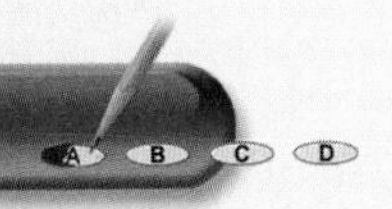

16. **MULTIPLE CHOICE** A randomly dropped counter falls in the squares. Find the probability that it falls in the shaded squares. Write as a percent. Round to the nearest tenth if necessary.

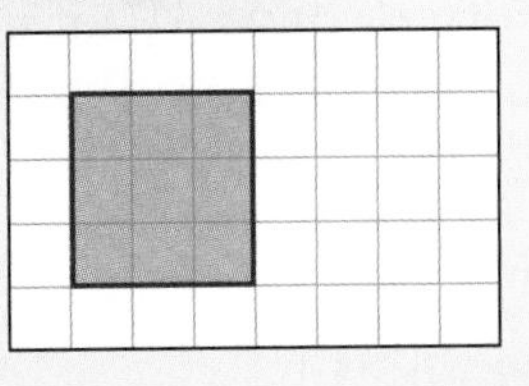

Ⓐ 9%　Ⓑ 14.1%　Ⓒ 22.5%　Ⓓ 40%

Standardized Test Practice

PART 1 Multiple Choice

Record your answers on the answer sheet provided by your teacher or on a sheet of paper.

1. Victoria bought 3 notebooks and 2 gel pens. Which expression represents her total cost if n is the cost of each notebook and p is the cost of each gel pen? (Lesson 1-4)

   (A) $n + p$ (B) $5(n + p)$
   (C) $5n \cdot 2p$ (D) $3n + 2p$

2. The table shows the weights of 13 dogs at a dog adoption center. Which measure of central tendency for these data is the least number? (Lesson 2-4)

Weight (lb)	Number of Dogs
20	|||
40	||
60	~~||||~~ ||
80	|

   (F) median (G) mean
   (H) mode (I) range

3. To make muffins, Desiree used $4\frac{1}{8}$ cups of flour, $4\frac{2}{3}$ cups of water, $4\frac{1}{4}$ cups of sugar, and $4\frac{1}{3}$ cups of milk. Desiree used the least amount of which ingredient? (Lesson 5-8)

   (A) flour (B) sugar
   (C) water (D) milk

4. Find $\frac{1}{7} \times \frac{2}{9}$. (Lesson 6-4)

   (F) $\frac{1}{63}$ (G) $\frac{2}{63}$ (H) $\frac{3}{16}$ (I) $\frac{9}{14}$

5. For his job, Marc drives 15,000 miles every 60 days. What is the average number of miles that Marc drives every day? (Lesson 7-2)

   (A) 250 mi (B) 300 mi
   (C) 900 mi (D) 1,500 mi

6. Trevor drives 45 miles per hour. Robin drives 54 miles per hour. What is the percent of increase from 45 miles per hour to 54 miles per hour? (Lesson 8-4)

   (F) 9% (G) 20% (H) 109% (I) 120%

7. Chi can take 3 different routes and 4 different modes of transportation to get to school, as shown below.

Route	Transportation
scenic	bike
quick	car
convenient	bus
	walking

   How many possible choices are there for Chi to use to get to school? (Lesson 9-3)

   (A) 7 (B) 9 (C) 12 (D) 16

8. Which is the square root of 441? (Lesson 11-1)

   (F) 21 (G) 22 (H) 23 (I) 24

9. Which is a reasonable *estimate* for the square root of 66? (Lesson 11-2)

   (A) 7.4 (B) 7.8 (C) 8.1 (D) 8.9

10. What is the value of x in the triangle? (Lesson 11-3)

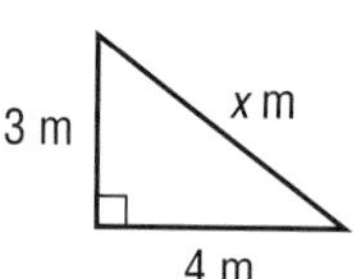

   (F) 2 (G) 5
   (H) 7 (I) 125

TEST-TAKING TIP

Question 8 You can use the answer choices to help you answer a multiple-choice question. For example, if you are asked to find the square root of a number, you can find the correct answer by calculating the square of the number in each answer choice.

PART 2 Short Response/Grid In

Record your answers on the answer sheet provided by your teacher or on a sheet of paper.

11. What value is represented by $(1 \times 10^5) + (2 \times 10^4) + (1 \times 10^3) + (1 \times 10^0)$? (Prerequisite Skill, page 555)

12. In the coordinate system, coordinates with a positive x value and a negative y value appear in what quadrant? (Lesson 3-3)

13. Write an equation to represent the following statement. (Lesson 4-1)

> The Palmas have 8 less than 2 times the number of trees in their yard as the Kandinskis have.

14. If you translate hexagon $LMNOPQ$ 4 units to the right, what are the new coordinates of point Q? (Lesson 10-8)

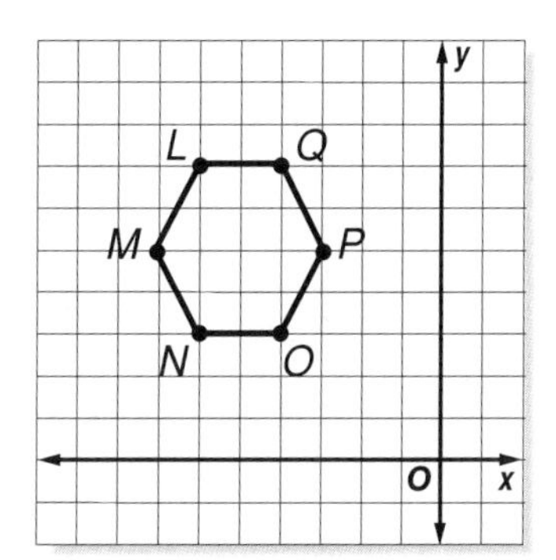

15. A window shade is being custom made to cover the triangular window shown at the right. What is the minimum area of the shade? (Lesson 11-5)

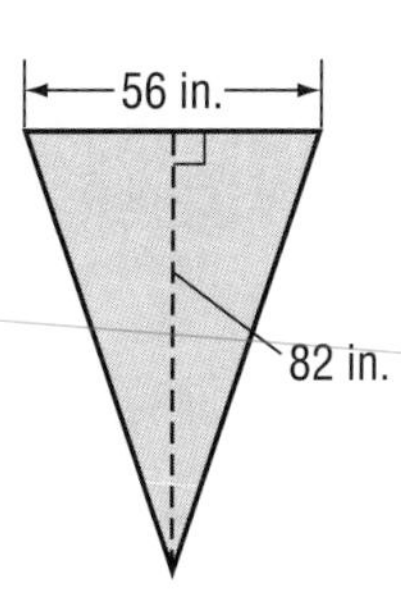

16. A kitchen chair has a circular seat that measures 14 inches across. What is the area of the seat on the kitchen chair? Use 3.14 for π and round to the nearest tenth. (Lesson 11-6)

17. The Connaught Centre building in Hong Kong has 1,748 circular windows. The diameter of each window is 2.4 meters. Find the total area of the glass in the windows. Round to the nearest tenth. (Lesson 11-6)

18. Find the area of the figure shown at the right. Use 3.14 for π. (Lesson 11-7)

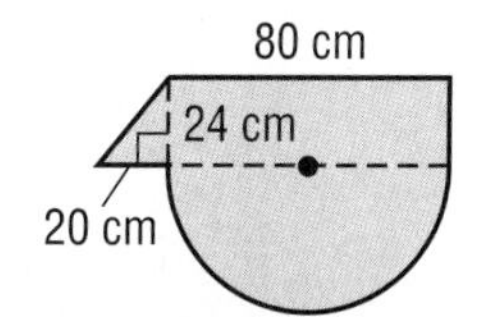

19. Amy plays hopscotch by throwing a small stone onto a numbered triangle. What is the probability that Amy's stone will land on a triangle numbered 5? Write as a fraction. (Lesson 11-8)

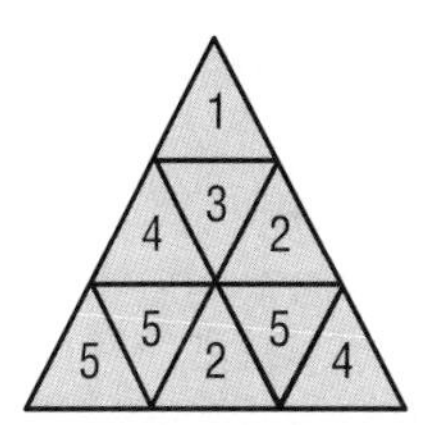

PART 3 Extended Response

Record your answers on a sheet of paper. Show your work.

20. Suppose you bought a new tent with the dimensions shown below.

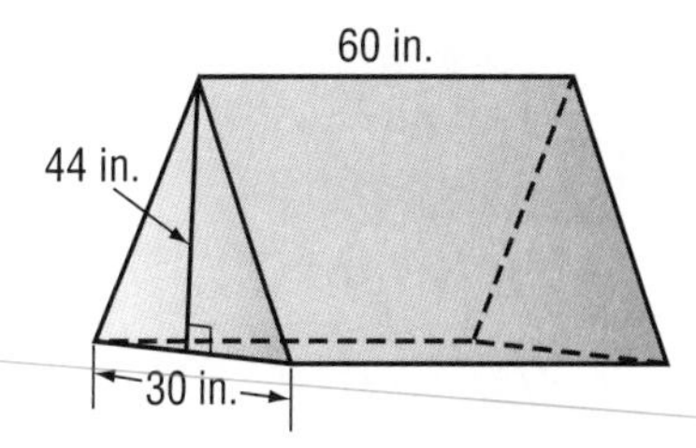

a. Is the area of the parallelogram-shaped side of the tent greater than or less than the area of the floor? Explain. (Lessons 11-3 and 11-4)

b. The front and back triangular regions are covered with screens. What is the total area of the screens? (Lesson 11-5)

Geometry: Measuring Three-Dimensional Figures

"What do paint cans have to do with math?"

Paint cans come in many different sizes, but they are all shaped like cylinders. **To find the volume V of a paint can, you can use the formula $V = \pi r^2 h$, where r is the radius of the lid and h is the height of the can.** A different formula can be used to find the surface area of a paint can.

You will solve problems involving volumes and surface areas of cylinders in Lessons 12-3 and 12-5.

GETTING STARTED

Take this quiz to see whether you are ready to begin Chapter 12. Refer to the lesson or page number in parentheses if you need more review.

Vocabulary Review

Choose the correct term to complete each sentence.

1. The distance around a circle is called (perimeter, circumference). (Lesson 6-9)
2. The expression πr^2 is used to find the (circumference, area) of a circle. (Lesson 11-6)

Prerequisite Skills

Estimate each product. (Page 558)

3. $9 \cdot 10.4$
4. $6.25 \cdot 3.8$
5. $7.03 \cdot 5.3$
6. $4.7 \cdot 2.15$

Evaluate each expression. Round to the nearest tenth if necessary. (Pages 559, 560)

7. $14.45 + 9.62$
8. $8.3 \cdot 6.4$
9. 36×5.2
10. $26.45 - 7.918$

Simplify. (Lesson 1-3)

11. $2 \cdot 6 + 4 \cdot 1 \cdot 3$
12. $1.2 \cdot 4 \cdot 4 + 3 \cdot 1.5$
13. $8 + 1 + 9 \cdot 2 + 5$
14. $7 + 3 \cdot 5 + 2 \cdot 6$

Multiply. (Lesson 6-4)

15. $4\frac{1}{2} \cdot 6$
16. $1\frac{1}{4} \cdot 5\frac{3}{5}$
17. $\frac{2}{7} \cdot 3\frac{3}{4}$
18. $3 \cdot 2\frac{2}{3}$

Find the area of each circle. Round to the nearest tenth. (Lesson 11-6)

19. diameter = 33 cm
20. radius = 3.8 yd
21. radius = 6 ft
22. diameter = 18 m

Surface Area and Volume Make this Foldable to help you organize information about solids. Begin with a piece of 11″ by 17″ paper.

Fold
Fold the paper in fourths lengthwise.

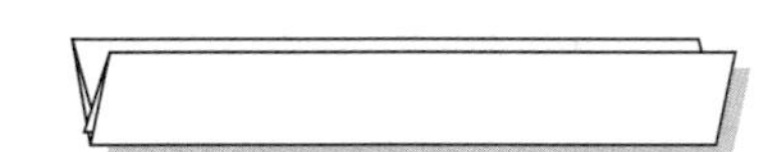

Open and Fold
Fold a 2″ tab along the short side. Then fold the rest in half.

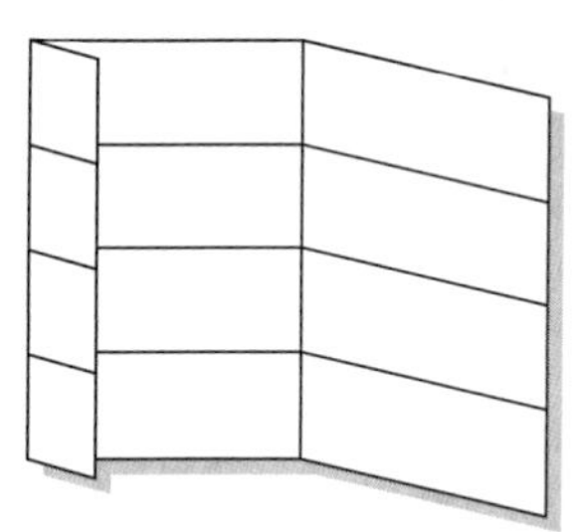

Label
Draw lines along folds and label as shown.

Ch. 12	Rectangular Prisms	Cylinders
Draw Examples		
Find Volume		
Find Surface Area		

Reading and Writing As you read and study the chapter, write the formulas for volume and surface area of each three-dimensional figure. Draw examples and list the characteristics of the figures.

12-1a HANDS-ON LAB

A Preview of Lesson 12-1

Building Three-Dimensional Figures

What You'll LEARN

Build three-dimensional figures given the top, side, and front views.

Materials

- centimeter cubes

Cubes are examples of three-dimensional figures because they have length, width, and depth. In this lab, you will use centimeter cubes to build other three-dimensional figures.

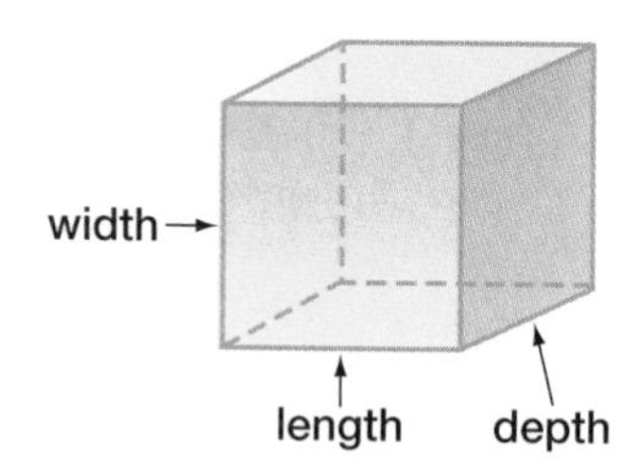

ACTIVITY *Work with a partner.*

The top view, side view, and front view of a three-dimensional figure are shown below. Use centimeter cubes to build the figure.

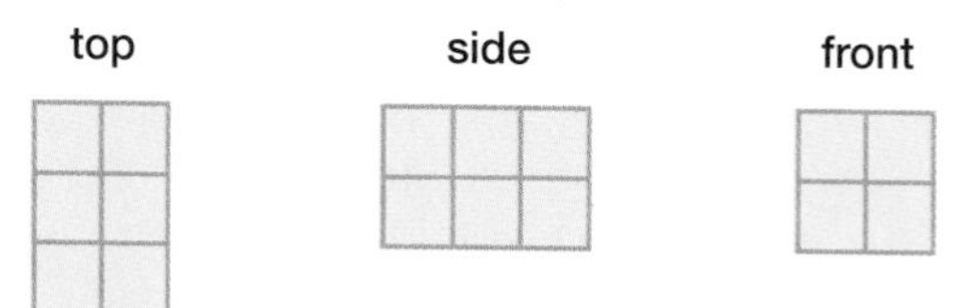

STEP 1 Use the top view to build the base of the figure. It is a 3-by-2 rectangle.

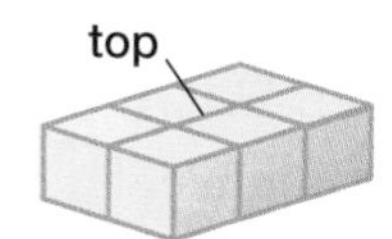

STEP 2 Use the side view to complete the figure. It is a 2-by-3 rectangle.

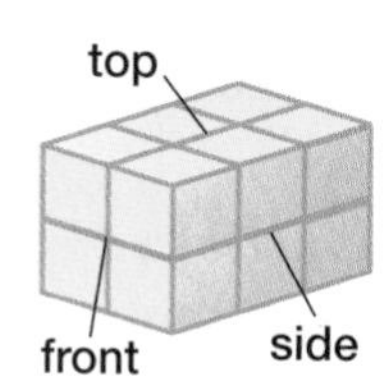

STEP 3 Use the front view to check the figure. It is a 2-by-2 square. So, the model is correct.

STUDY TIP

Plane A *plane* is a flat surface that extends in all directions. Each face of a cube represents a different plane.

Your Turn **The top view, side view, and front view of each three-dimensional figure are shown. Use centimeter cubes to build the figure. Then make a sketch of the figure.**

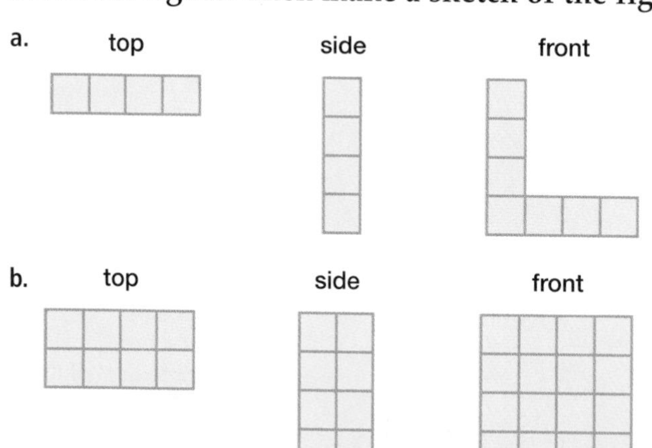

The top view, side view, and front view of each three-dimensional figure are shown. Use centimeter cubes to build the figure. Then make a sketch of the figure.

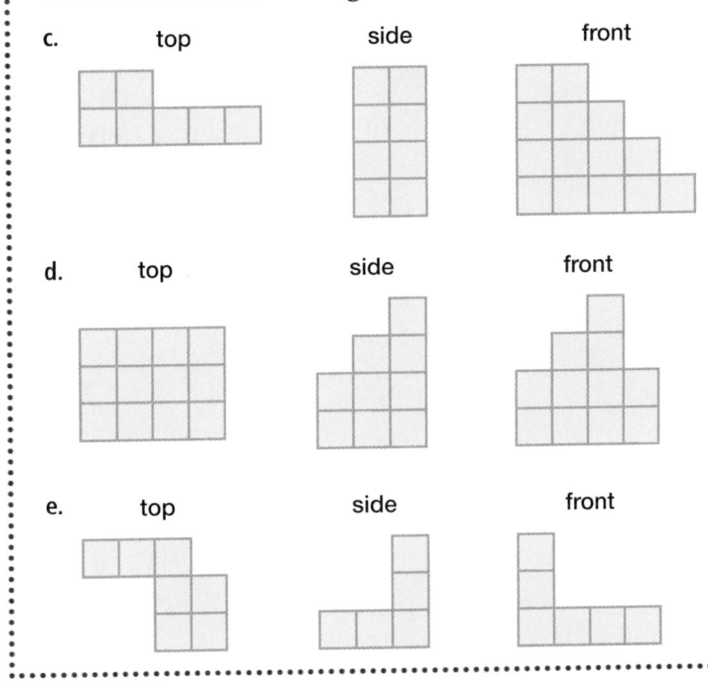

Writing Math

Work with a partner.

1. **Build** a model with cubes and draw the top, side, and front views. Give the drawing of the views to your partner and have him or her build the figure with cubes. Trade roles with your partner and repeat making the drawing and building the figure.
2. **Explain** how you began building the figures.
3. **Determine** whether there is more than one way to build each model. Explain your reasoning.
4. The figure at the right represents a building with a section that is 15 stories tall and another section that is 20 stories tall. Which view would you use to show the difference in height of each section?
5. **Build** two different models that would look the same from two views, but not the third view. Draw a top view, side view, and front view of each model.
6. **Describe** a real-life situation where it might be necessary to draw a top, side, and front view of a three-dimensional figure.

12-1 Drawing Three-Dimensional Figures

What You'll LEARN

Draw a three-dimensional figure given the top, side, and front views.

NEW Vocabulary

solid

WHEN am I ever going to use this?

COMICS For Exercises 1 and 2, refer to the comic below.

SHOE by Jeff MacNelly

1. Which view of the Washington Monument is shown in the comic?
2. Find a photograph of the Washington Monument and draw a side view.

STUDY TIP

Plane Figures In geometry, three-dimensional figures are *solids* and two-dimensional figures such as triangles, circles, and squares are *plane figures.*

A **solid** is a three-dimensional figure because it has length, width, and depth. You can draw different views of solids.

EXAMPLE Draw Different Views of a Solid

1 Draw a top, a side, and a front view of the figure at the right.

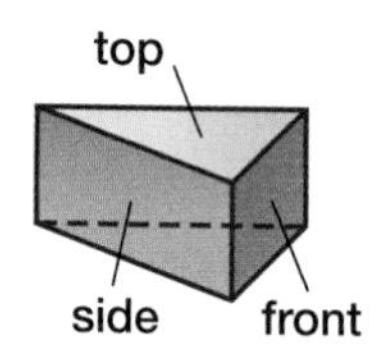

The top view is a triangle.
The side and front views are rectangles.

top

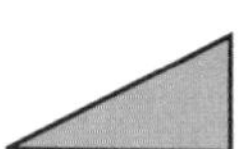

side

front

Your Turn Draw a top, a side, and a front view of each solid.

a. 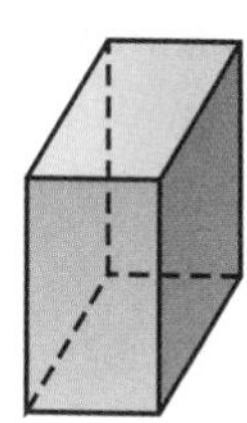

b. 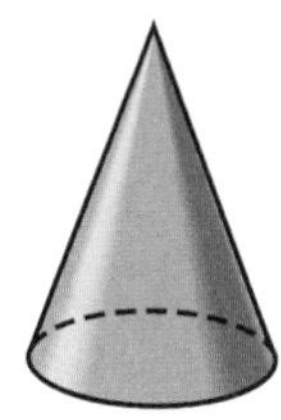

EXAMPLE Draw a Three-Dimensional Figure

2 **Draw the solid using the top, side, and front views shown at the right.**

top side front

Step 1 Use the top view to draw the base of the figure, a 1-by-3 rectangle.

Step 2 Add edges to make the base a solid figure.

Step 3 Use the side and front views to complete the figure.

STUDY TIP

Paper Use isometric dot paper for the drawings in this lesson, as shown at the right. This is different from rectangular dot paper.

Your Turn **Draw each solid using the top, side, and front views shown. Use isometric dot paper.**

c. top side front

d. top side front

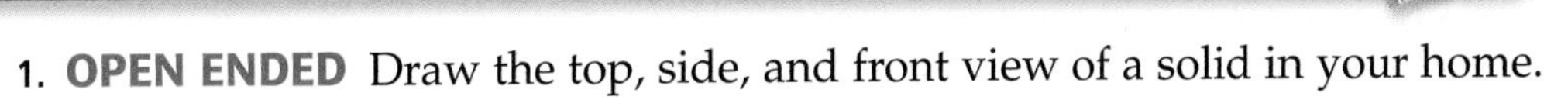

Skill and Concept Check

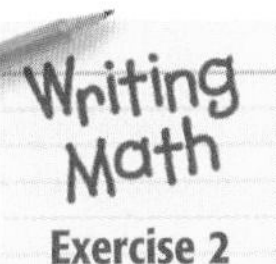

1. **OPEN ENDED** Draw the top, side, and front view of a solid in your home.

2. **Which One Doesn't Belong?** Identify the figure that does not have the same characteristic as the other three. Explain your reasoning.

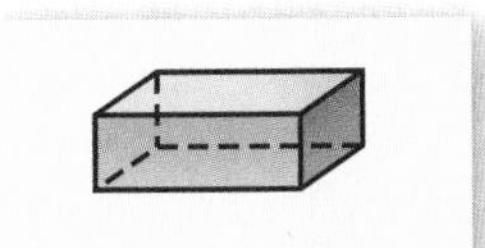

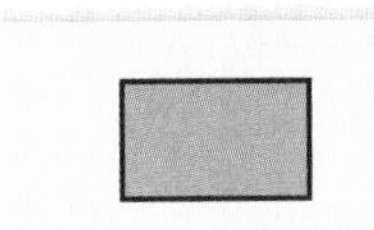

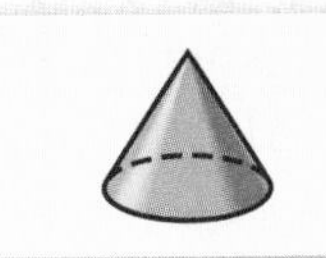

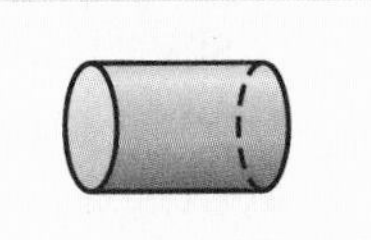

GUIDED PRACTICE

Draw a top, a side, and a front view of each solid.

3. 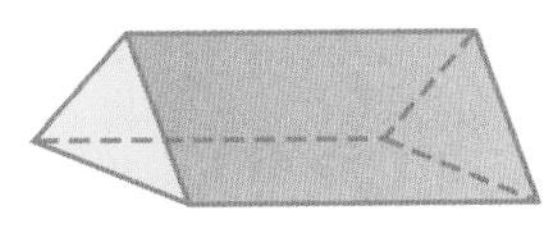

4. 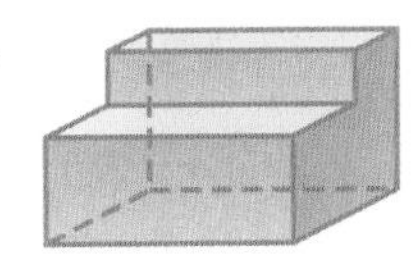

Draw each solid using the top, side, and front views shown. Use isometric dot paper.

5. top side front

6. top side front

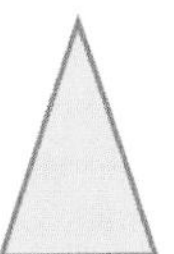

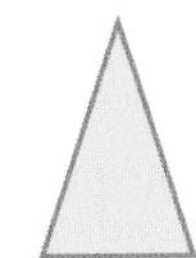

Practice and Applications

HOMEWORK HELP	
For Exercises	See Examples
7–12, 18	1
13–17, 19, 20	2

Extra Practice
See pages 593, 607.

Draw a top, a side, and a front view of each solid.

7. 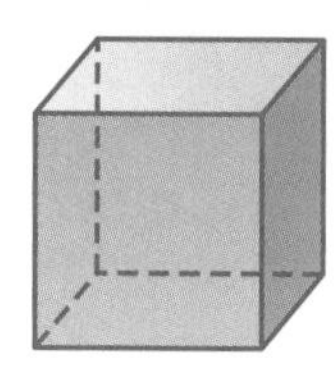

8. 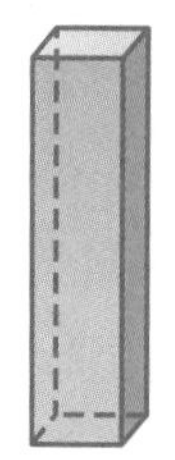

9. 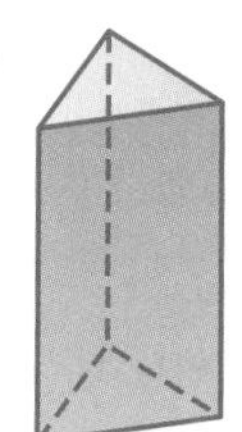

10. 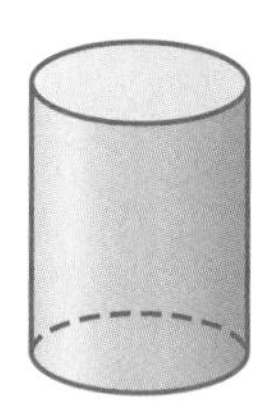

11. 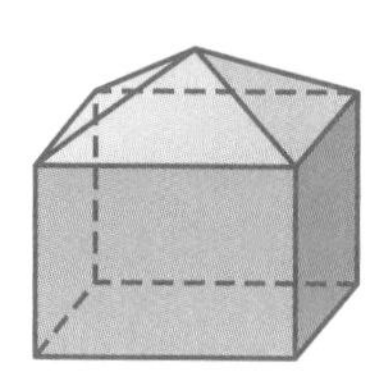

12. 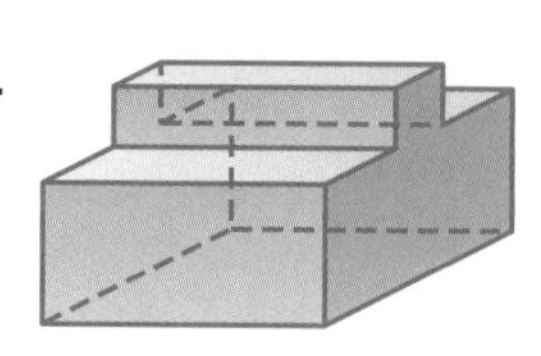

Draw each solid using the top, side, and front views shown. Use isometric dot paper.

13. top side front

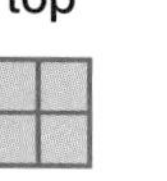
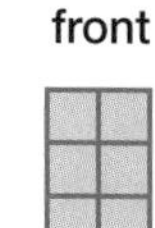

14. top side front

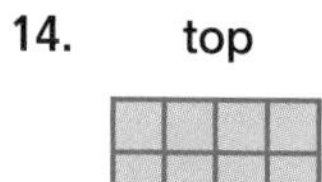
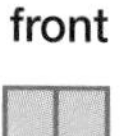

15. top side front

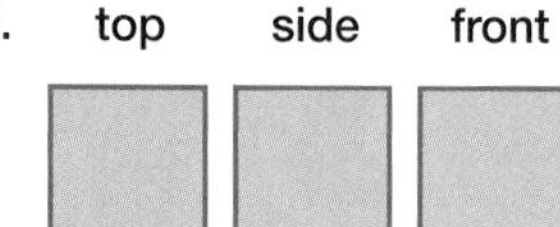

16. top side front

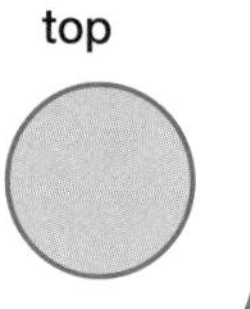
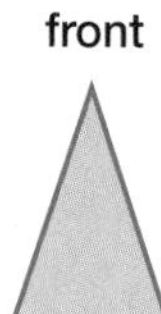

HISTORY For Exercises 17–20, use the information below.

The Lighthouse of Alexandria was the last of the seven Wonders of the Ancient World that disappeared. It contained three different-shaped levels.

top side front

17. The bottom level is the largest. Make a drawing of this level using the top, side, and front views shown at the right.

18. The middle level is a prism with a base that is a regular octagon. If the height and width of this level is approximately one third the height and width of the bottom level, draw the top, side, and front views.

19. The top level has the views shown at the right. What kind of solid is this?

top side front

20. Make a sketch of the Lighthouse of Alexandria that shows all three levels.

Data Update How does your drawing compare to the actual Lighthouse of Alexandria? Visit msmath2.net/data_update to learn more.

21. **RESEARCH** Use the Internet or another source to find a photograph of the only Wonder of the Ancient World existing today, the Great Pyramid of Giza. Draw a top view, a side view, and a front view of the pyramid.

22. **ARCHITECTURE** When a building is being designed, an architect provides a set of elevation drawings. These drawings show how the building appears from each side. Draw a set of elevation drawings for your home or school.

23. **CRITICAL THINKING** Draw a three-dimensional figure in which the front and top views each have a line of symmetry but the side view does not. (*Hint*: Refer to Lesson 10-9 to review lines of symmetry.)

Standardized Test Practice and Mixed Review

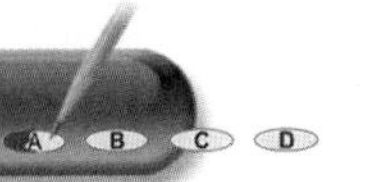

24. **MULTIPLE CHOICE** Which is the top view of the cylinder at the right?

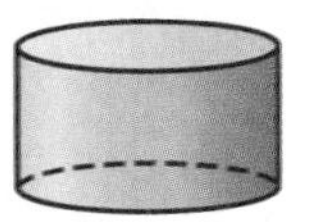

Ⓐ triangle　　Ⓑ rectangle

Ⓒ square　　Ⓓ circle

25. **MULTIPLE CHOICE** Which three-dimensional figure has the top, side, and front views shown at the right?

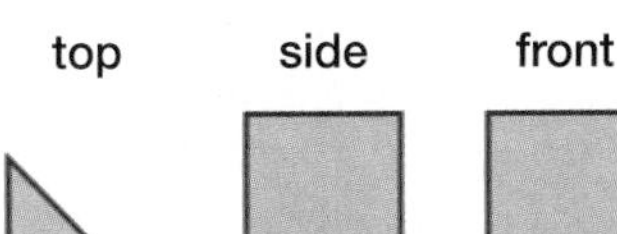

Ⓕ 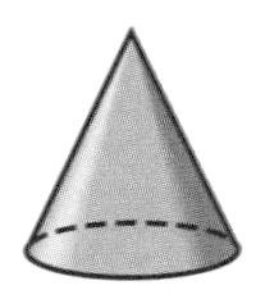

Ⓖ 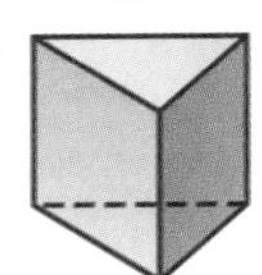

Ⓗ

Ⓘ 

26. Find the probability that a randomly dropped counter will fall in the shaded region. Write as a percent. (Lesson 11-8)

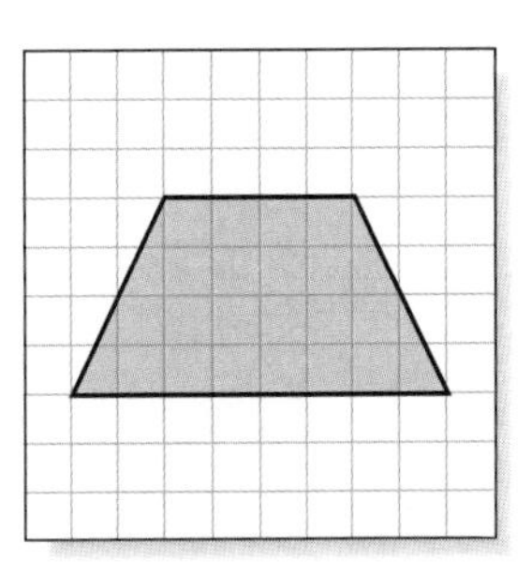

Find the area of each figure. Round to the nearest tenth if necessary. (Lesson 11-7)

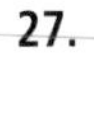

27. 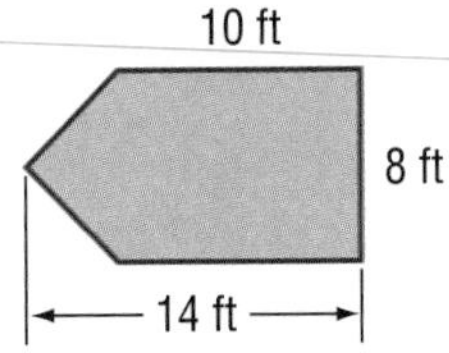

28. 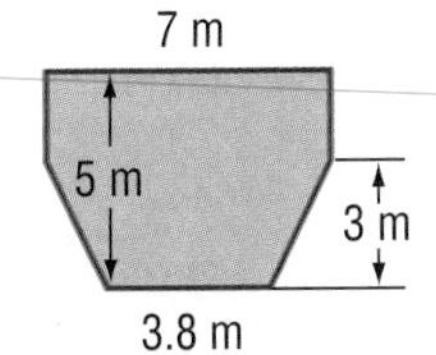

29. 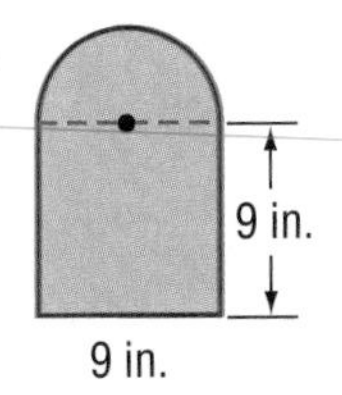

GETTING READY FOR THE NEXT LESSON

PREREQUISITE SKILL Multiply. (Lesson 6-4)

30. $7\frac{1}{2} \cdot 6$　　31. $8 \cdot 2\frac{3}{4}$　　32. $\frac{5}{6} \cdot 1\frac{4}{5}$　　33. $10\frac{1}{5} \cdot 6\frac{2}{3}$

12-1b Problem-Solving Strategy

A Follow-Up of Lesson 12-1

What You'll Learn
Solve problems by making a model.

Make a Model

I'm going to help my mom make 3-inch soft alphabet blocks for the children at her daycare center. About how much fabric do we need for one cube if there is a $\frac{1}{2}$-inch seam on each side?

We could **make a model** to find out.

Explore	We know that each cube is 3 inches long with $\frac{1}{2}$-inch seams.
Plan	We can make a cardboard model of a cube with sides 3 inches long. We could then cut the model into six squares and add $\frac{1}{2}$-inch paper extensions to each side as seams.
Solve	Make the cardboard model and unfold the cube. 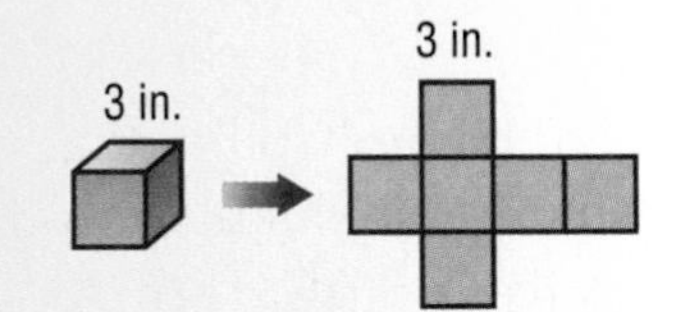 Cut the cube into six squares and tape $\frac{1}{2}$-inch seams on each side. Now each pattern piece is 4 inches by 4 inches or *about* 16 square inches. 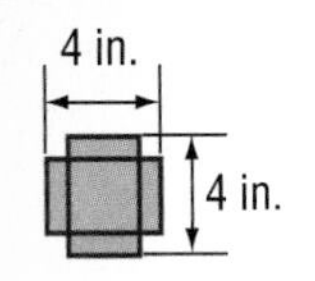 $16 \text{ in}^2 \times 6 \text{ squares} = 96 \text{ in}^2$ So, each cube needs about 96 square inches of fabric.
Examine	If the 4-inch-by-4-inch pattern pieces were laid as two columns and three rows, it would be 8 inches wide and 12 inches long or 96 square inches.

Analyze the Strategy

1. **Explain** when making a model is a better strategy than drawing a picture.
2. **Explain** why you think the students started with the three-dimensional model to make their pattern.
3. **Write** a problem that can be solved by making a model. Then solve the problem.

Apply the Strategy

Solve. Make a model.

4. **ART** Dominic is creating a layout of his bedroom for art class. The room measures 15 foot by 12 feet. If he uses a scale of 1 foot = $\frac{3}{4}$ inch, what are the dimensions of his bedroom on the model?

5. **BICYCLES** Eight customers lined up outside The Bike Shop with either a bicycle or a tricycle that needed repair. When the owner looked out the window, she counted 21 wheels outside the shop. How many tricycles and bicycles were there?

Mixed Problem Solving

Solve. Use any strategy.

6. **COMMUNITY SERVICE** There are four drop-off centers for the community food drive. Their total collections are shown in the table.

Center	Number of Cans
A	3,298
B	2,629
C	4,429
D	2,892

A newsletter reported that over 13,000 cans of food were collected. Is this estimate reasonable? Explain.

7. **SWIMMING** Yeti can swim one 20-meter lap in 1.25 minutes. How long will it take her to swim 100 meters at the same rate?

8. **TRAFFIC** At the four-way intersection shown below, the traffic lights change every 90 seconds. About 8 cars in one lane travel through the light in this amount of time. Determine the number of cars that travel through the intersection in 3 minutes.

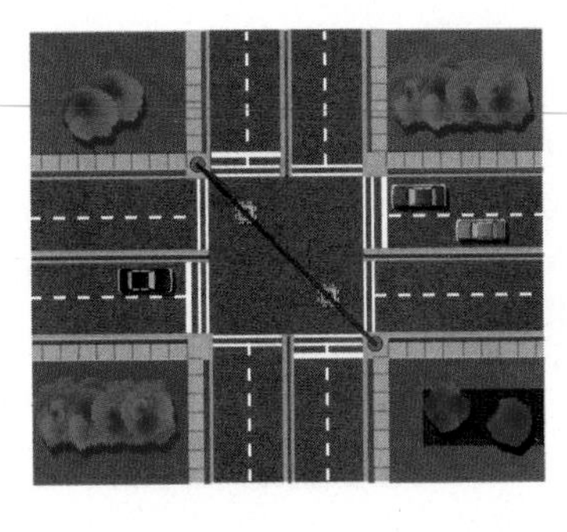

9. **MONEY** A top film actor made 16 films, which grossed over $\$5.09 \times 10^9$. On average, how much did each film make? Write in scientific notation. Round to the nearest tenth.

10. **DISPLAYS** Identical boxes are stacked in the corner of a store as shown below. How many boxes are *not* visible?

11. **REMODELING** How many square feet of wallpaper are needed to cover a wall that measures 9 feet by 16 feet?

12. **STANDARDIZED TEST PRACTICE**
Yolanda deposited \$450 in a new savings account in January, withdrew \$175 in February, and then began monthly deposits of \$75 from March through December. Which equation shows her balance b after her last deposit in December?

(A) $b = \$275 + 9(\$75)$

(B) $b = \$275 + 10(\$75)$

(C) $b = \$450 + 10(\$75)$

(D) $b = \$275 - 10(\$75)$

12-2 Volume of Rectangular Prisms

What You'll LEARN

Find the volumes of rectangular prisms.

NEW Vocabulary

volume
rectangular prism

Link to READING

Everyday Meaning of Volume: bulk or mass, as in shipping a large volume of merchandise

HANDS-ON Mini Lab

Materials

- centimeter grid paper
- scissors
- tape
- centimeter cubes

Work with a partner.

- On a piece of grid paper, cut out a square that is 10 centimeters on each side.
- Cut a 1-centimeter square from each corner. Fold the paper and tape the corners together to make a box.

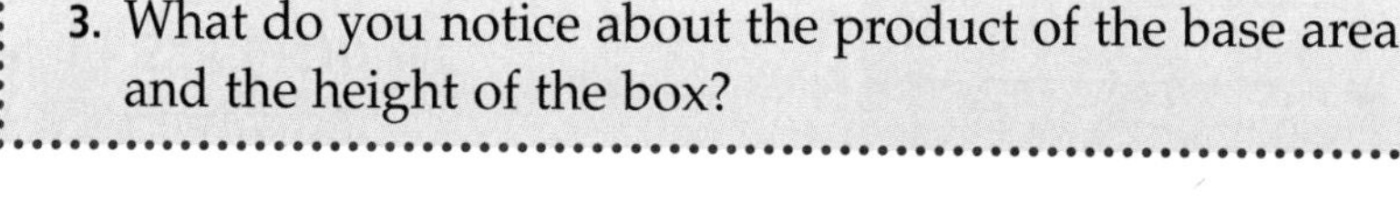

1. What is the area of the base, or bottom, of the box? What is the height of the box?
2. How many centimeter cubes fit in the box?
3. What do you notice about the product of the base area and the height of the box?

The **volume** of a solid is the measure of space occupied by it. It is measured in cubic units such as cubic centimeters (cm^3) or cubic inches (in^3). The volume of the figure at the right can be shown using cubes.

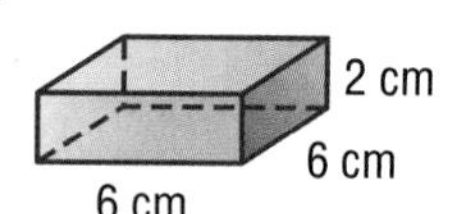

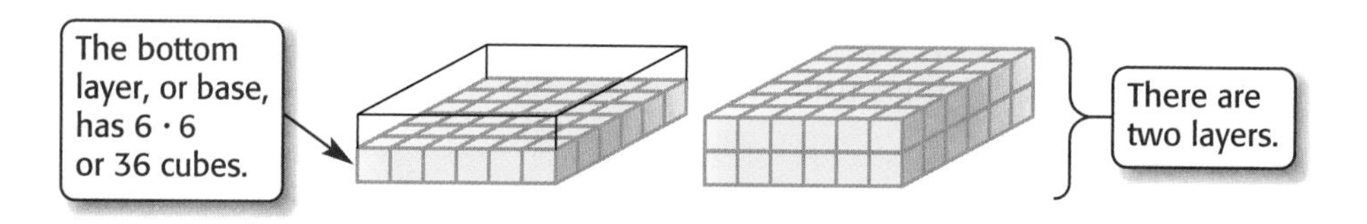

It takes 36 · 2 or 72 cubes to fill the box. So, the volume of the box is 72 cubic centimeters.

A **rectangular prism** is a solid figure that has two parallel and congruent sides, or bases, that are rectangles.

Key Concept — **Volume of a Rectangular Prism**

Words The volume V of a rectangular prism is the area of the base B times the height h. It is also the product of the length ℓ, the width w, and the height h.

Model

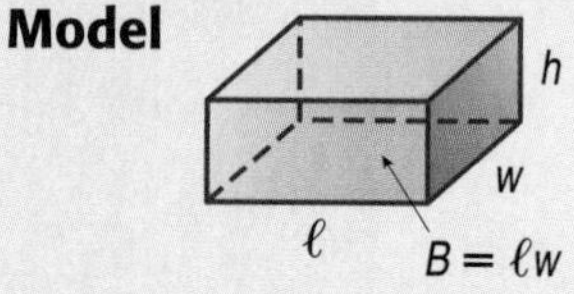

Symbols $V = Bh$ or $V = \ell wh$

EXAMPLES Find the Volumes of Prisms

1 Find the volume of the rectangular prism.

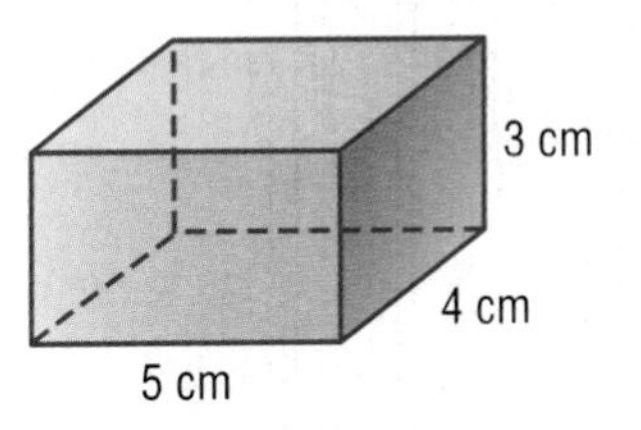

$V = \ell wh$ Volume of a rectangular prism

$V = 5 \cdot 4 \cdot 3$ Replace ℓ with 5, w with 4, and h with 3.

$V = 60$ Multiply.

The volume is 60 cubic centimeters.

STUDY TIP

Estimation It is helpful to first estimate the volume of a rectangular prism to determine whether your answer is reasonable.

2 CEREAL Find the volume of the cereal box.

Estimate $V \approx 8 \cdot 3 \cdot 10$ or 240

$V = \ell wh$ Volume of a rectangular prism

$V = 7.5 \cdot 2.5 \cdot 12$ Replace ℓ with 7.5, w with 2.5, and h with 12.

$V = 225$ Multiply.

The volume is 225 cubic inches.

Compare to the estimate.

Skill and Concept Check

1. **Write** the abbreviation for cubic yards using an exponent.

2. **OPEN ENDED** Draw and label a rectangular prism that has a volume between 400 and 600 cubic centimeters. State the volume of the prism.

3. **FIND THE ERROR** Cassandra and Ling are comparing the volumes of the two prisms at the right. Who is correct? Explain.

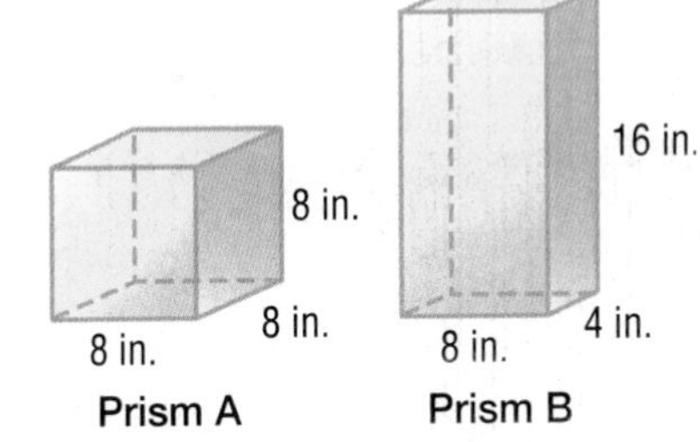

Cassandra Prism B has a greater volume because it is twice as tall as Prism A.	Ling Both prisms have the same volume.

GUIDED PRACTICE

Find the volume of each rectangular prism.

4. 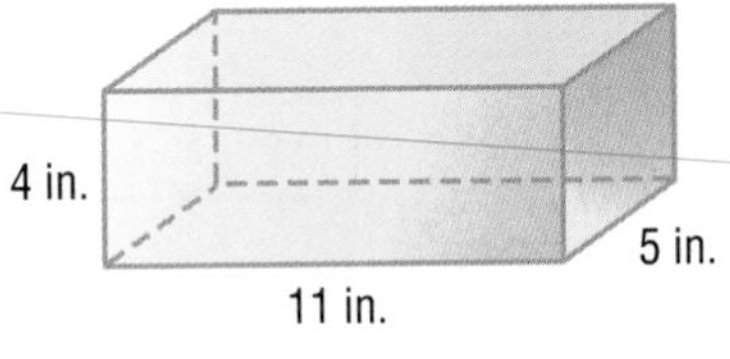

5. 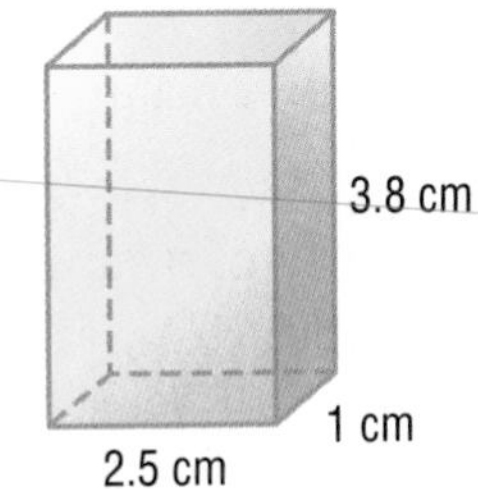

6. 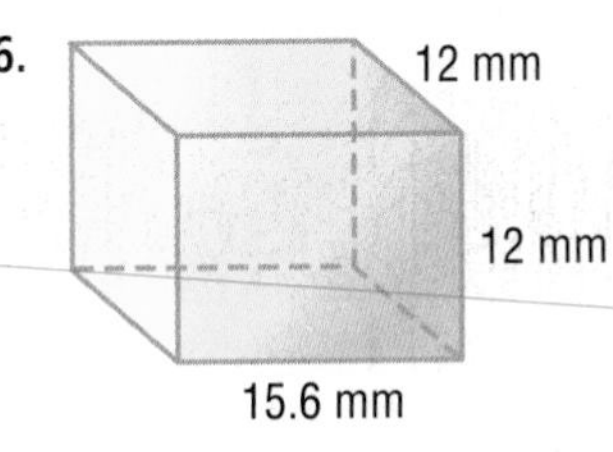

7. **STORAGE** A cabinet measures 20 inches by 30 inches by 60 inches. What is its volume?

Practice and Applications

HOMEWORK HELP	
For Exercises	See Examples
8–15	1, 2

Extra Practice
See pages 593, 607.

Find the volume of each rectangular prism. Round to the nearest tenth if necessary.

8.

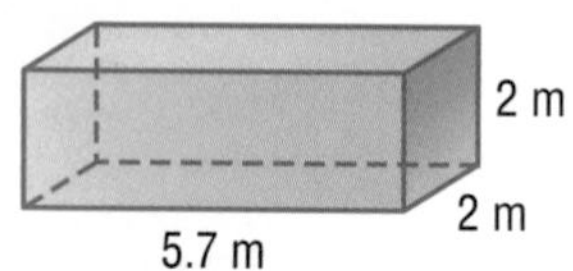

9.

6 in.
20 in.
8 in.

10.

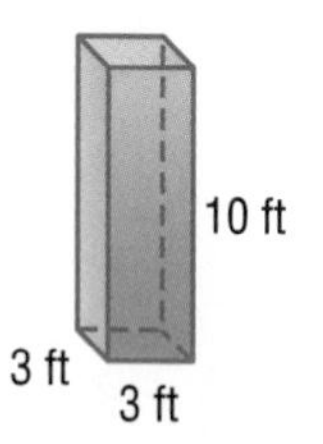

11.

12.5 cm
4.2 cm
4.5 cm

12.

$5\frac{1}{2}$ ft
3 ft
$2\frac{1}{4}$ ft

13. A cube has 3-centimeter edges. Find its volume.

14. The base of a rectangular prism has an area of 19.4 square meters. Find the height if the volume is 306.52 cubic meters.

15. **TRANSPORTATION** Find the approximate volume of the California trolley car shown at the right.

$8\frac{1}{4}$ ft
28 ft
$10\frac{1}{2}$ ft

16. **MEASUREMENT** How many cubic inches are in a cubic foot?

17. **MEASUREMENT** How many cubic centimeters are in a cubic meter?

18. **CRITICAL THINKING** A *triangular prism* is a prism that has bases that are triangles. Use $V = Bh$ to find the volume of the triangular prism at the right.

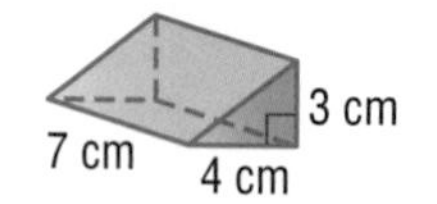

Standardized Test Practice and Mixed Review

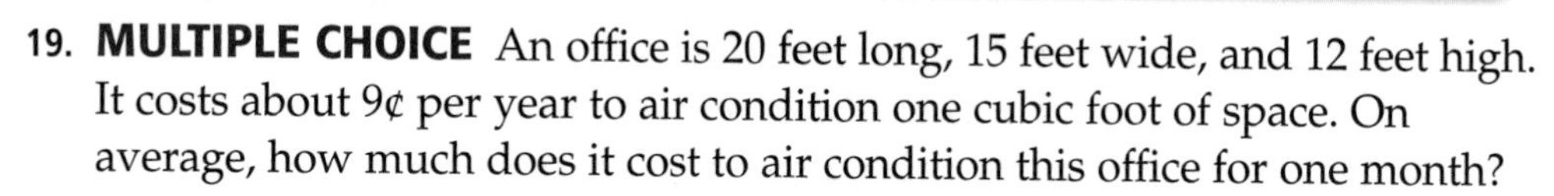

19. **MULTIPLE CHOICE** An office is 20 feet long, 15 feet wide, and 12 feet high. It costs about 9¢ per year to air condition one cubic foot of space. On average, how much does it cost to air condition this office for one month?

   (A) \$324   (B) \$300   (C) \$27   (D) \$2.25

20. **SHORT RESPONSE** A landscaper wants to cover a 40-foot-by-12-foot rectangular area with small stones. If she uses 120 cubic feet of stones, how deep will they be?

21. Draw a top, a side, and a front view of the solid at the right. (Lesson 12-1)

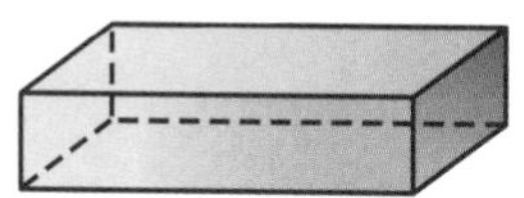

22. Draw an area model so that the probability of a randomly dropped counter falling in a triangular region is 25%. (Lesson 11-8)

GETTING READY FOR THE NEXT LESSON

PREREQUISITE SKILL Estimate. (Page 558)

23. $3.14 \cdot 6$ 24. $5 \cdot 2.7^2$ 25. $9.1 \cdot 8.3$ 26. $3.1 \cdot 1.75^2 \cdot 2$

msmath2.net/self_check_quiz

12-2b Spreadsheet Investigation

A Follow-Up of Lesson 12-2

What You'll LEARN

Use a spreadsheet to investigate the volumes of similar solids.

Similar Solids

A computer spreadsheet can help you calculate the sizes and volumes of similar rectangular prisms. You can enlarge or reduce dimensions of the prisms and the spreadsheet will automatically calculate the new volumes.

ACTIVITY

Mrs. Sanchez owns a box-making factory. She wishes to produce a set of similar boxes so that they will nest inside each other when assembled. She will need to show her customers the dimensions and volume of each type of box she plans to sell. This information can be placed into a spreadsheet.

Set up the spreadsheet like the one shown below.

Similar Boxes

	A	B	C	D	E
1		Dimensions (in.)			
2	Box	L	W	H	Volume (in³)
3	small	2.5	2.5	2.5	
4	medium	5	5	5	
5	large	7.5	7.5	7.5	
6					

Sheet1 / Sheet2 / Sheet3

EXERCISES

1. State the spreadsheet commands you used to find the volumes of the three boxes. Then find the volumes to the nearest tenth.
2. Expand the spreadsheet to calculate the volumes of the boxes when each of the dimensions is doubled. What happens to the volume of the small box when all of the dimensions of the box are doubled?
3. What happens to the volume of the small box when all of the dimensions of the box are tripled?
4. Suppose another box has a volume 216 times greater than the small box. What are the dimensions of this box? Use the spreadsheet to check your answer.
5. Extend the pattern in the volume column. Then express the pattern in exponential form.

12-3 Volume of Cylinders

HANDS-ON Mini Lab

Materials

- soup can
- centimeter grid paper
- scissors
- tape

Work with a partner.

Set a soup can on a piece of grid paper and trace around the base, as shown at the right.

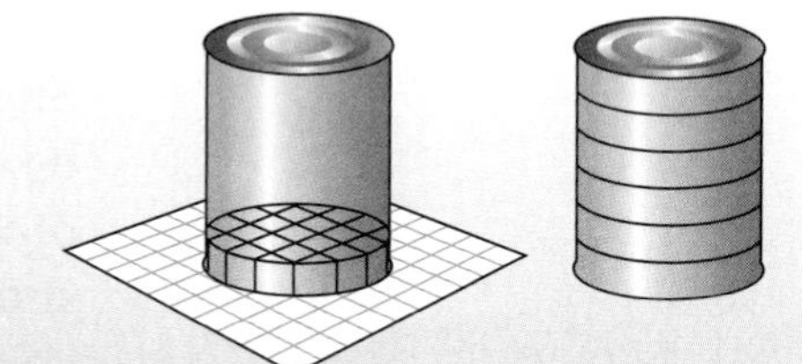

1. Estimate the number of centimeter cubes that would fit at the bottom of the can. Include parts of cubes.
2. How many layers would it take to fill the cylinder?
3. **Make a conjecture** about how you could find the volume of the soup can.

What You'll LEARN

Find the volumes of cylinders.

NEW Vocabulary

cylinder

A **cylinder** is a solid figure that has two congruent, parallel circles as its bases. As with prisms, the area of the base tells the number of cubic units in one layer. The height tells how many layers there are.

Key Concept — Volume of a Cylinder

Words The volume V of a cylinder with radius r is the area of the base B times the height h.

Symbols $V = Bh$ or $V = \pi r^2 h$, where $B = \pi r^2$

Model

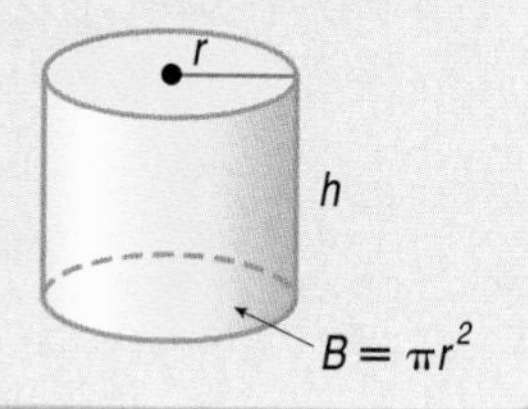

EXAMPLE Find the Volume of a Cylinder

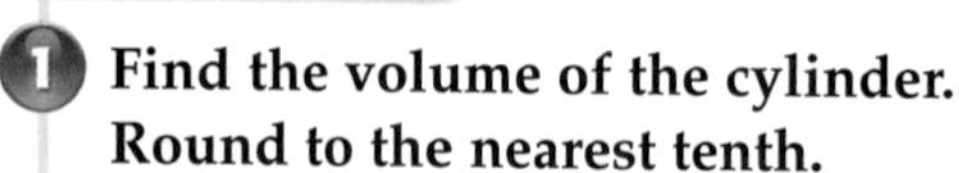

1 Find the volume of the cylinder. Round to the nearest tenth.

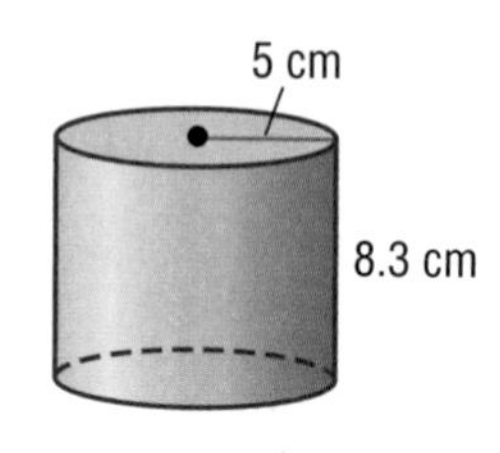

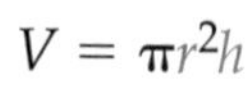

$V = \pi r^2 h$ Volume of a cylinder

$V = \pi(5)^2(8.3)$ Replace r with 5 and h with 8.3.

  π × 5 x^2 × 8.3 ENTER 651.8804756

The volume is about 651.9 cubic centimeters. Check by using estimation.

Standardized Test Practice

EXAMPLE Find the Volume of a Real-Life Object

2 **MULTIPLE-CHOICE TEST ITEM Find the volume of a cylinder-shaped paint can that has a diameter of 4 inches and a height of 5 inches.**

Ⓐ 20.0 in^3 Ⓑ 62.8 in^3 Ⓒ 125.7 in^3 Ⓓ 251.3 in^3

Read the Test Item

To find the volume, use the formula $V = \pi r^2 h$.

Solve the Test Item

$V = \pi r^2 h$ Volume of a cylinder

$V = \pi(2)^2(5)$ Replace r with 2 and h with 5.

$V \approx 62.8$ Simplify.

The answer is B.

Test-Taking Tip The Princeton Review

Reading
The radius of the can is $4 \div 2$ or 2 inches. Be sure to substitute that value into the formula. If you substitute 4 for r into the formula, you would get the incorrect answer D.

Skill and Concept Check

Writing Math Exercises 1 & 2

1. **OPEN ENDED** Draw and label a cylinder that has a larger radius, but less volume than the one shown at the right.

8 cm
16 cm

2. **Explain** how the formula for the volume of a cylinder is similar to the formula for the volume of a rectangular prism.

GUIDED PRACTICE

Find the volume of each cylinder. Round to the nearest tenth.

3. 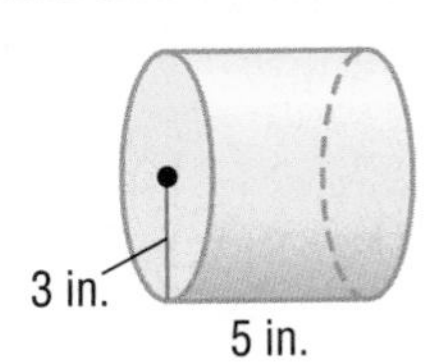

4. 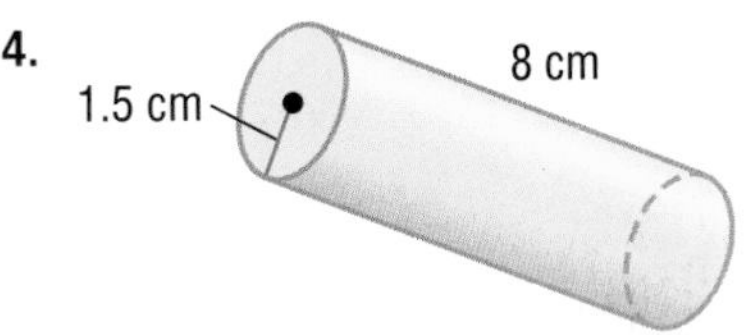

5. 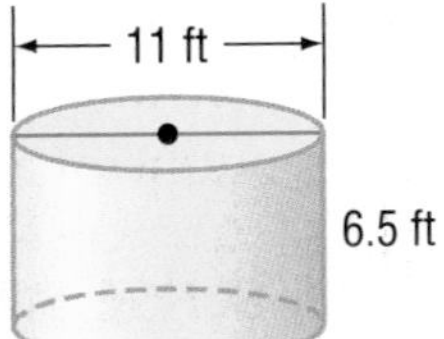

6. radius = 3 in.
   height = $5\frac{1}{2}$ in.

7. diameter = 12.4 m
   height = 2 m

8. **BAKING** Which will hold more cake batter, the rectangular pan, or the two round pans? Explain.

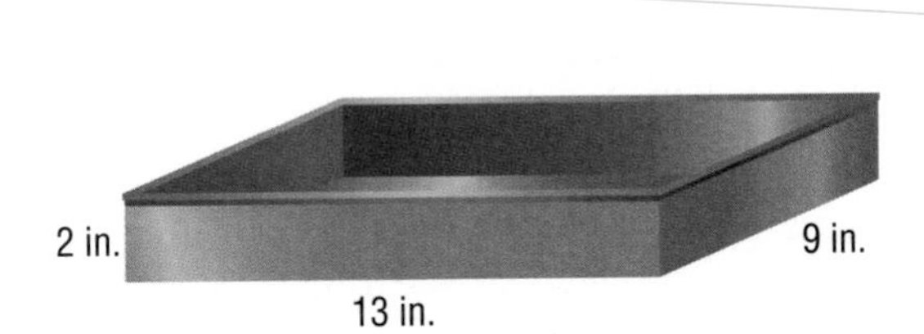

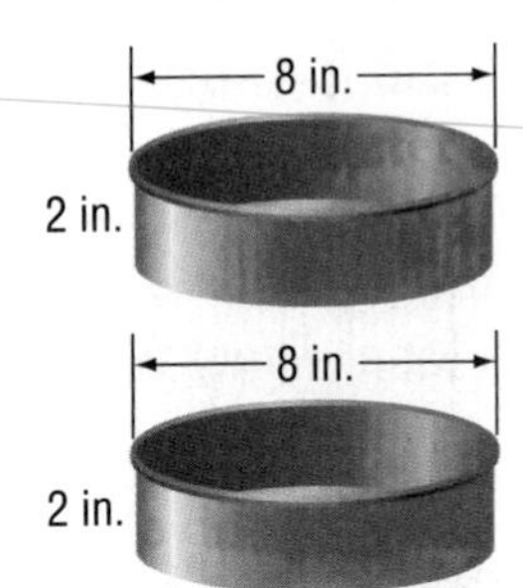

Practice and Applications

HOMEWORK HELP	
For Exercises	See Examples
9–18	1
19–22, 26–27	2

Extra Practice
See pages 594, 607.

Find the volume of each cylinder. Round to the nearest tenth.

9.
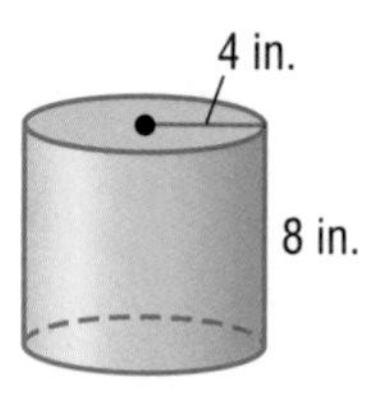

10.
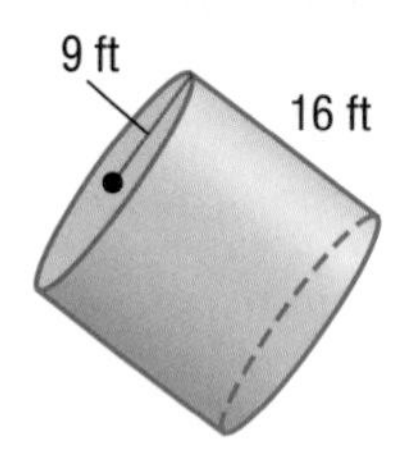

11.
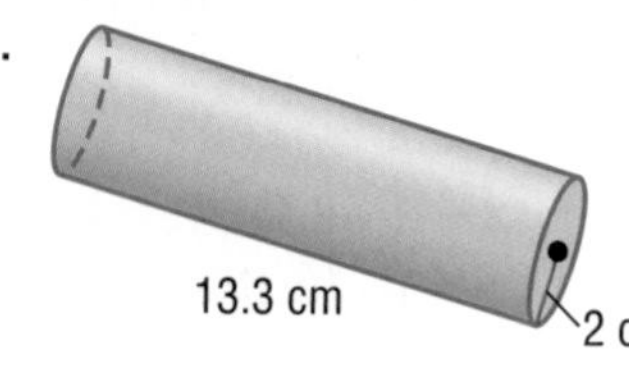

12.
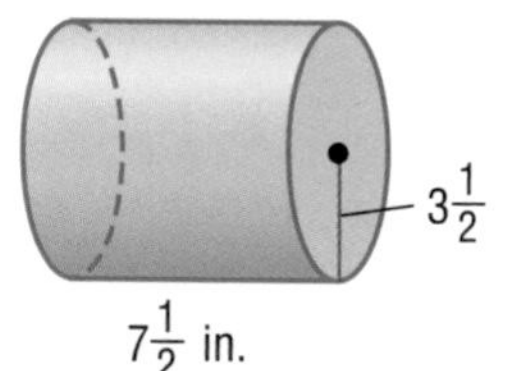

13.
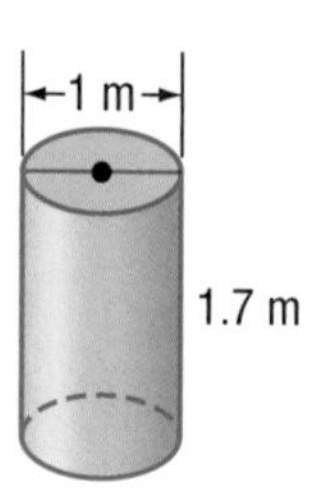

14.
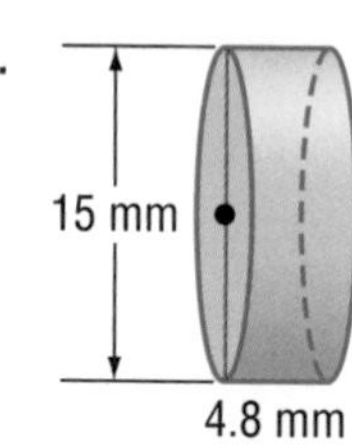

15. diameter = 21 mm
height = 8 mm

16. diameter = 4.5 m
height = 6.5 m

17. radius = 6 ft
height = $5\frac{1}{3}$ ft

18. radius = 12 ft
height = 11 yd

19. a soup can with a radius of 4 centimeters and a height of 12 centimeters

20. a hockey puck with a diameter of 3 inches and a height of 1 inch

21. a can of potato chips with a radius of $1\frac{1}{2}$ inches and a height of 8 inches

22. **SPACE SHUTTLE** The Hubble Space Telescope is cylinder-shaped, approximately the size of a school bus, as shown at the right. What is the volume to the nearest tenth?

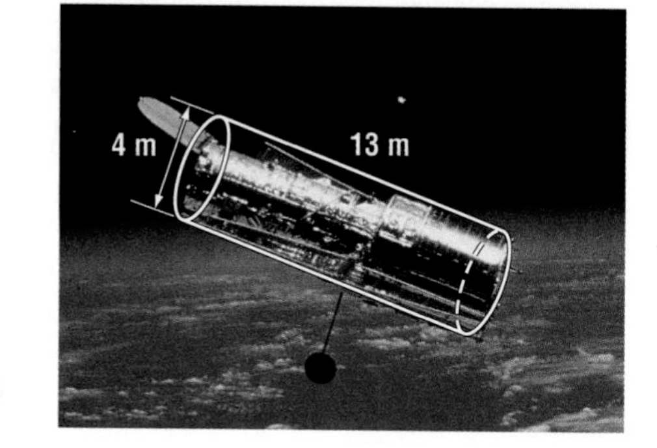

23. **WRITE A PROBLEM** Write a problem in which you find the volume of a cylinder.

24. **NUMBER SENSE** What is the ratio of the volume of a cylinder to the volume of a cylinder having twice the height but the same radius?

25. **NUMBER SENSE** What is the ratio of the volume of a cylinder to the volume of a cylinder having the same height but twice the radius?

26. **CONTAINERS** The two cans at the right have the same volume. Find h.

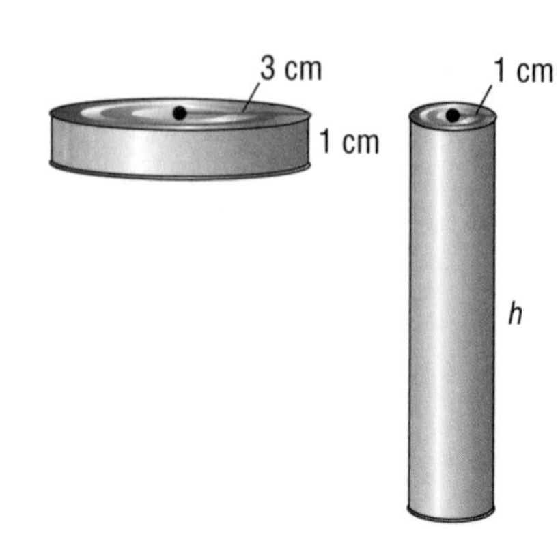

27. **MEASUREMENT** Firewood is usually sold by a unit of measure called a *cord*. A cord is a stack of wood that is 8 feet long, 4 feet wide, and 4 feet high. Suppose a tree has a diameter of 2 feet. Find the height of the tree trunk that would produce about 1 cord of firewood.

28. **CRITICAL THINKING** Two equal-sized sheets of paper are rolled along the length and along the width, as shown at the right. Which cylinder do you think has the greater volume? Explain.

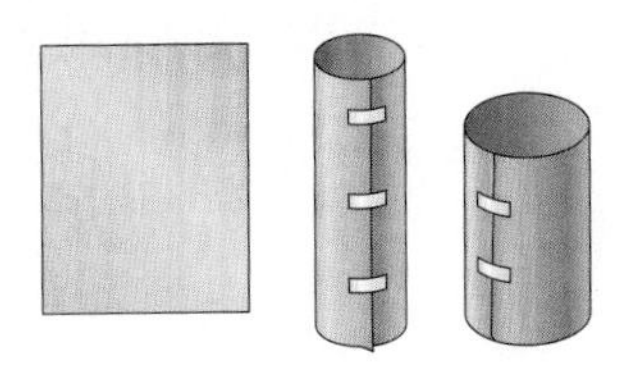

Standardized Test Practice and Mixed Review

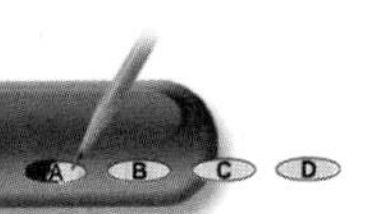

29. **MULTIPLE CHOICE** Which statement is true about the volumes of cylinders 1 and 2 shown below?

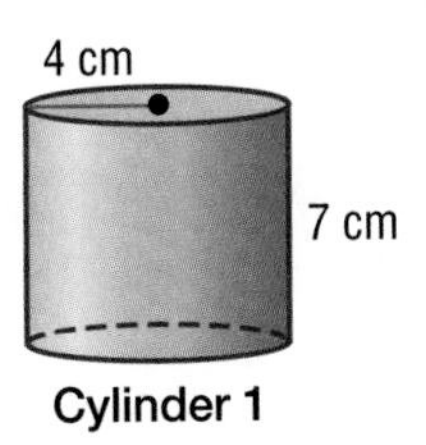

Cylinder 1

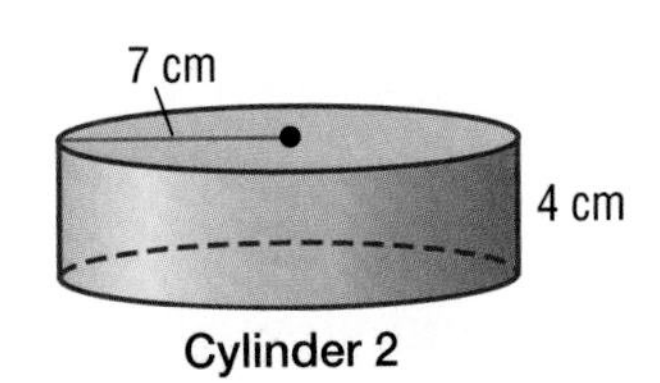

Cylinder 2

Ⓐ The volume of cylinder 1 is greater.

Ⓑ The volume of cylinder 2 is greater.

Ⓒ The volumes are equal.

Ⓓ cannot tell from the diagrams

30. **MULTIPLE CHOICE** A cylinder-shaped popcorn tin has a height of 1.5 feet and a diameter of 10 inches. Find the volume to the nearest cubic inch. Use 3.14 for π.

Ⓕ 118 in^3 Ⓖ 565 in^3 Ⓗ 1,413 in^3 Ⓘ 5,652 in^3

31. Find the volume of a rectangular prism with a length of 6 meters, a width of 4.9 meters, and a height of 5.2 meters. (Lesson 12-2)

Draw each solid using the top, side, and front views shown. Use isometric dot paper. (Lesson 12-1)

32. top side front

33. top side front

Find the area of each trapezoid. Round to the nearest tenth if necessary. (Lesson 11-5)

34.

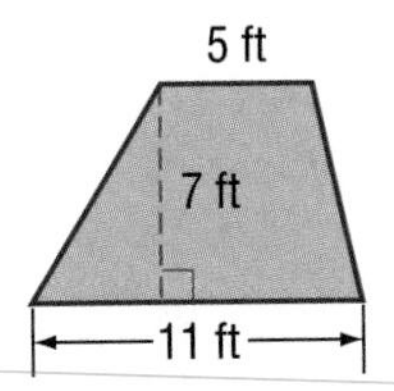

35.

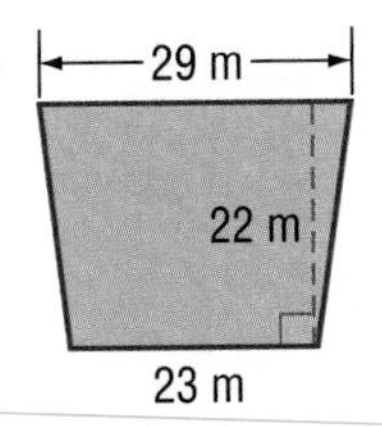

36.

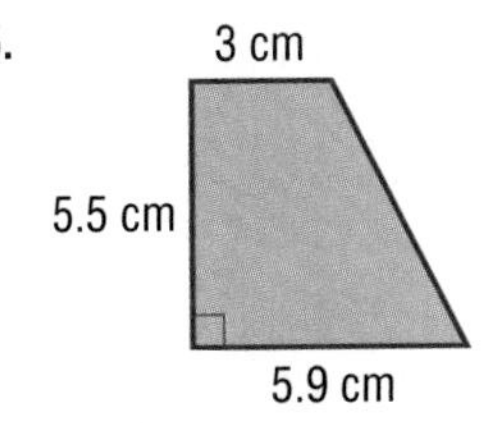

GETTING READY FOR THE NEXT LESSON

PREREQUISITE SKILL **Simplify.** (Lesson 1-3)

37. $3 \cdot 5 \cdot 8 + 2 \cdot 9 \cdot 3$

38. $7.6 \cdot 11 \cdot 2 + 2 \cdot 7.6 \cdot 3$

39. $2 \cdot 7 \cdot 1.5 + 2 \cdot 4 \cdot 1.5 + 1.5 \cdot 7$

40. $2 \cdot 2\frac{1}{4} \cdot 6 + 2 \cdot 9 \cdot 6 + 2 \cdot 2\frac{1}{4} \cdot 9$

Mid-Chapter Practice Test

Vocabulary and Concepts

1. **Define** *volume.* (Lesson 12-2)
2. **State**, in words, the formula for the volume of a cylinder. (Lesson 12-3)

Skills and Applications

Draw a top, a side, and a front view of each solid. (Lesson 12-1)

3. 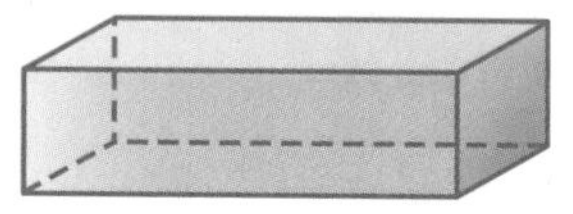
4. 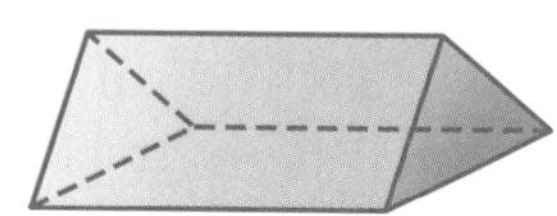
5. 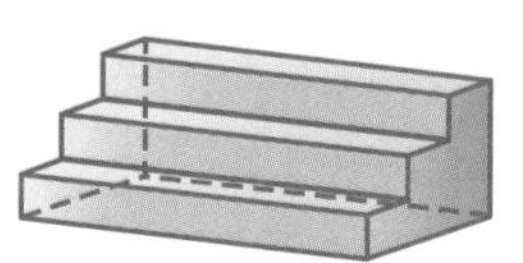

6. Draw the solid using the top, side, and front views shown at the right. Use isometric dot paper. (Lesson 12-1)

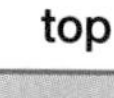
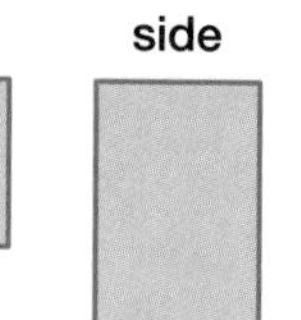

7. **GIFTS** A jewelry box measures 7 centimeters by 12 centimeters by 14 centimeters. What is its volume? (Lesson 12-2)

8. **TRUCKS** How tall is the trailer of a truck if it is 9 meters long, 7.2 meters wide and has a volume of 226.8 cubic meters? (Lesson 12-2)

9. **GARDENING** Jocelyn is buying potting soil to fill the window box. If one bag of potting soil contains 576 cubic inches, how many bags should she buy? (Lesson 12-2)

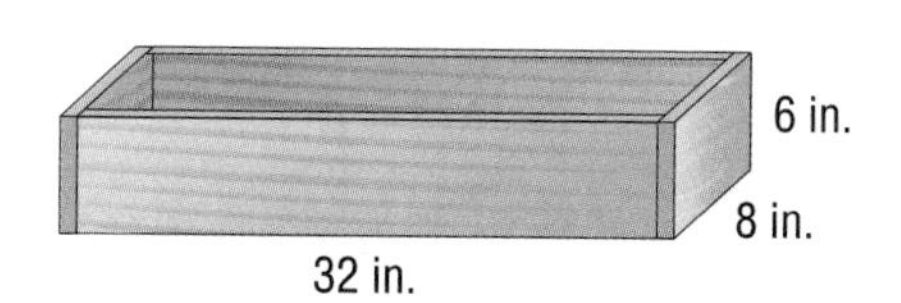

Find the volume of each cylinder. Round to the nearest tenth. (Lesson 12-3)

10. diameter = 25 ft
    height = 24 ft

11. radius = 2.2 cm
    height = 5 cm

12. radius = $1\frac{1}{2}$ in.
    height = $4\frac{1}{2}$ in.

Standardized Test Practice

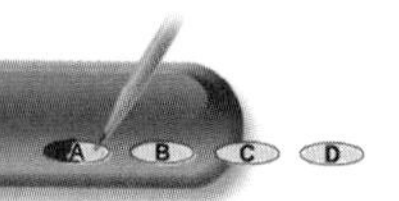

13. **GRID IN** A brick is 4 inches wide, 8 inches long, and 3 inches tall. What is the volume in cubic inches of a stack of 25 bricks? (Lesson 12-2)

14. **SHORT RESPONSE** A circular pond has a radius of 12 meters and it is 2.5 meters deep. What is its volume? Use 3.14 for π. (Lesson 12-3)

The Game Zone

A Place To Practice Your Math Skills

Math Skill
3-Dimensional Figures

Shape-Tac-Toe

GET READY!

Players: two
Materials: 1 index card, 2 large cubes, 10 two-color counters

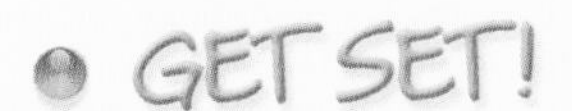

One player draws a game board on the index card like the one shown. The second player labels the six faces of each cube with the following words.

First Cube	Second Cube
prism	circular base
pyramid	triangle base
cylinder	square base
cone	2 parallel bases
flat surfaces	1 base
curved surface	congruent faces

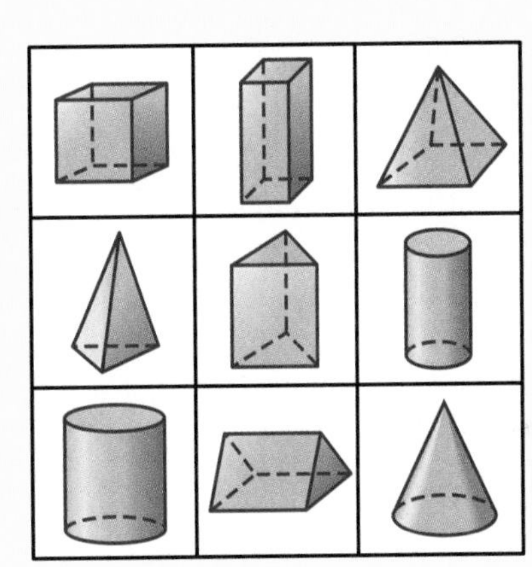

GO!

- The first player rolls both cubes and places a counter on any one shape that matches the conditions on the cubes. If it is impossible to cover a shape, the player loses his or her turn.
- Players alternate turns.
- **Who Wins?** The first player to cover three shapes in a row is the winner.

12-4a HANDS-ON LAB

A Preview of Lesson 12-4

Nets and Surface Area

What You'll LEARN

Use nets to find the surface area of rectangular prisms.

Materials

- rectangular dot paper
- scissors

Suppose you cut a cardboard box along its edges, open it up, and lay it flat.

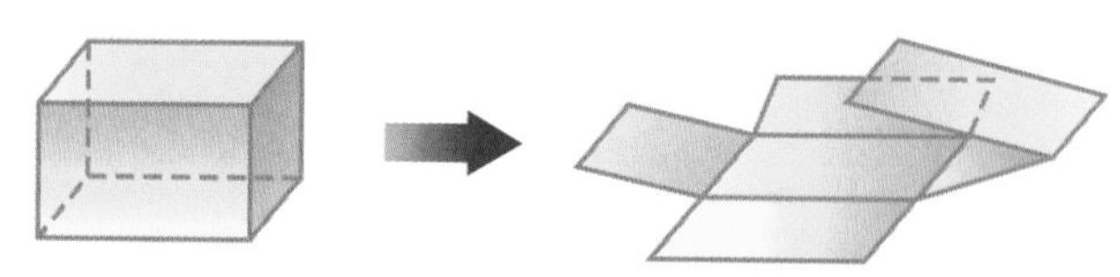

The result is a two-dimensional figure called a **net**. Nets can help you see the regions or faces that make up the surface of the figure.

ACTIVITY *Work with a partner.*

Make a net of the rectangular prism shown at the right.

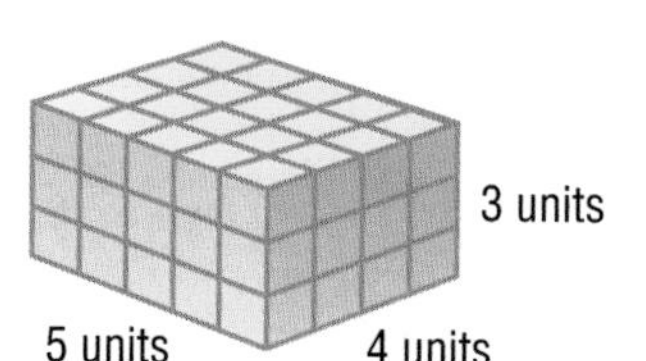

STEP 1 Begin by drawing the base of the prism. On the dot paper, draw a rectangle that is 5 units long and 4 units wide.

bottom

STEP 2 Visualize unfolding the prism along its edges. Draw the rectangles that represent the front, back, and sides of the prism.

back
side bottom side
front

STEP 3 Finally, draw the top of the prism.

This is only one of several possible nets that you could draw.

back
side bottom side top
front

Your Turn

Draw a net for each figure. Find the area of the net. Then cut out the net, fold it, and tape it together to form a three-dimensional figure.

a.

6 in.
4 in.
5 in.

b.

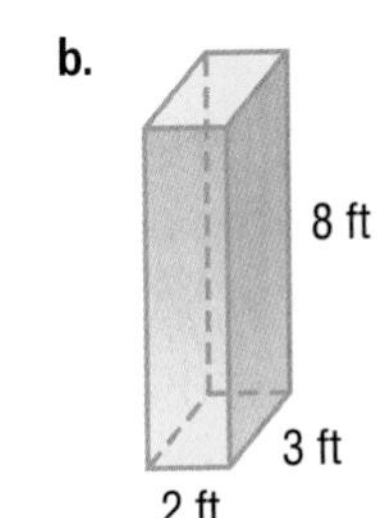

c.

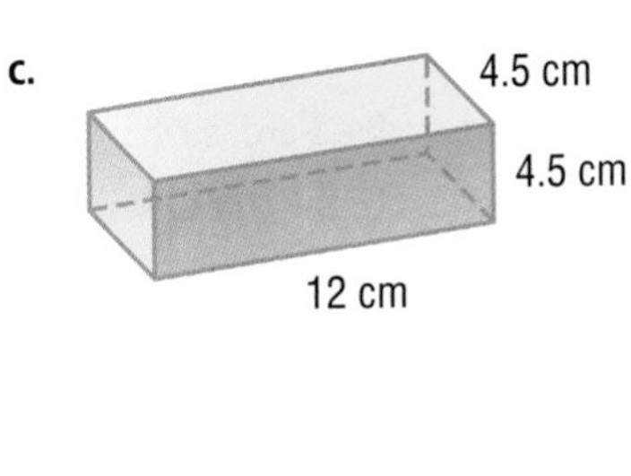

Writing Math

Work in groups of three.

1. The net shown on page 530 is made of rectangles. How many rectangles are in the net?
2. Explain how you can find the total area of the rectangles.
3. The *surface area* of a prism is the total area of its net. Write an equation that shows how to find the surface area of the prism at the right using the length ℓ, width w, and height h.

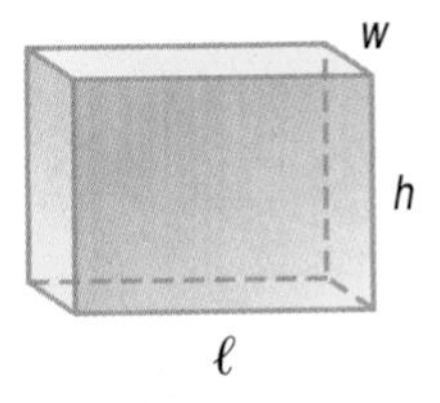

4. Find the surface areas of cubes whose edges are 1 unit, 2 units, and 3 units and graph the ordered pairs (side length, surface area) on a coordinate plane. Describe the graph.
5. **Describe** what happens to the surface area of a cube as its dimensions are doubled? tripled?
6. **Describe** how you would find the surface area of a square-based pyramid.

Draw a net for each figure.

7.

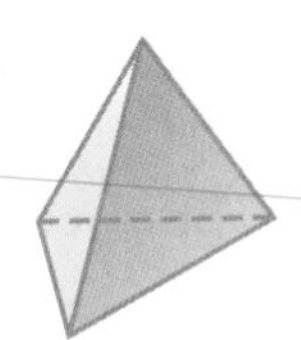

tetrahedron

8.

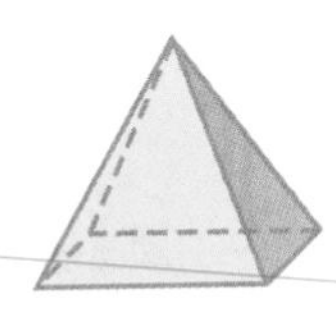

square-based pyramid

9. **Explain** how the formula for the surface area of a tetrahedron differs from the formula for the surface area of a square-based pyramid.

12-4 Surface Area of Rectangular Prisms

What You'll LEARN

Find the surface areas of rectangular prisms.

NEW Vocabulary

surface area

HANDS-ON Mini Lab

Materials

- 8 centimeter cubes

Work with a partner.

- Use the cubes to build a rectangular prism with a length of 8 centimeters.
- Count the number of squares on each face. The sum is the *surface area.*

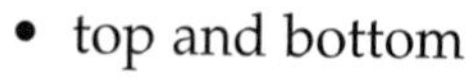

1. Record the dimensions, volume, and surface area in a table.
2. Build two more prisms using all of the cubes. For each, record the dimensions, volume, and surface area.
3. Describe the rectangular prisms with the greatest and least surface areas.

The sum of the areas of all of the surfaces, or faces, of a three-dimensional figure is the **surface area**.

EXAMPLE Use a Net to Find Surface Area

1 Find the surface area of the rectangular prism.

You can use a net of the rectangular prism to find its surface area. There are three pairs of congruent faces.

- top and bottom
- front and back
- two sides

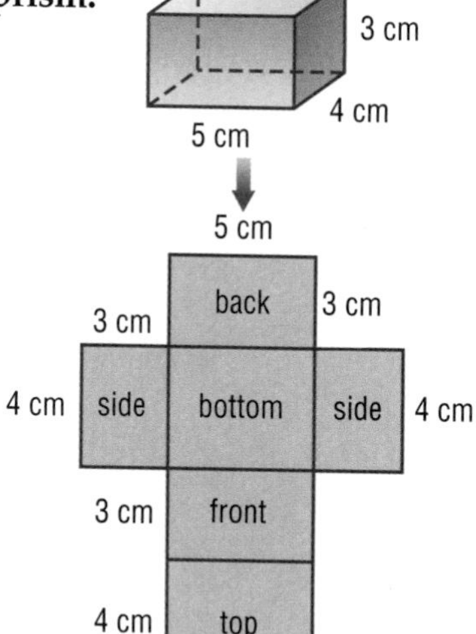

Faces	Area
top and bottom	$2(5 \cdot 4) = 40$
front and back	$2(5 \cdot 3) = 30$
two sides	$2(3 \cdot 4) = 24$
Sum of the areas	$40 + 30 + 24 = 94$

The surface area is 94 square centimeters.

Key Concept — Surface Area of a Rectangular Prism

Words The surface area S of a rectangular prism with length ℓ, width w, and height h is the sum of the areas of the faces.

Model

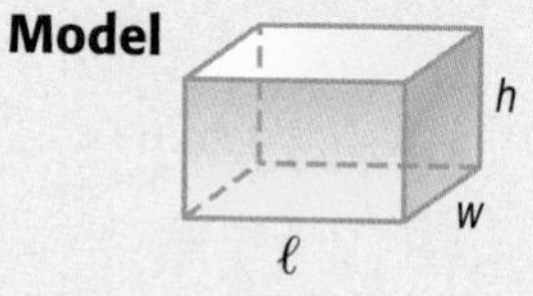

Symbols $S = 2\ell w + 2\ell h + 2wh$

EXAMPLE Use a Formula to Find Surface Area

2 **Find the surface area of the rectangular prism.**

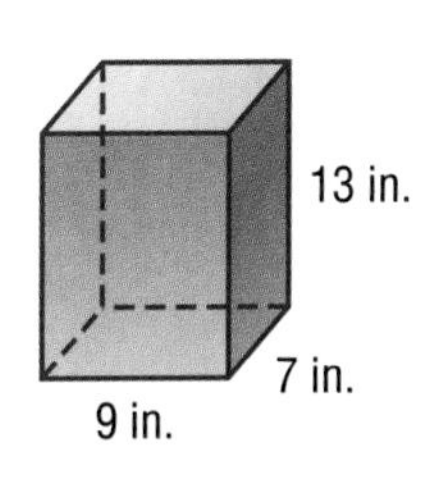

Replace ℓ with 9, w with 7, and h with 13.

surface area $= 2\ell w + 2\ell h + 2wh$

$= 2 \cdot 9 \cdot 7 + 2 \cdot 9 \cdot 13 + 2 \cdot 7 \cdot 13$

$= 126 + 234 + 182$ Multiply first. Then add.

$= 542$

The surface area of the prism is 542 square inches.

REAL-LIFE MATH

GIFTS Gifts have been wrapped since the invention of paper in 105 A.D. in China.

Source: www.hallmark.com

EXAMPLE Use Surface Area to Solve a Problem

3 **GIFTS Mario is wrapping a package 8 inches long, 2 inches wide, and 11 inches high. He bought a roll of wrapping paper that is 1 foot wide and 2 feet long. Did he buy enough to wrap the package?**

Step 1 Find the surface area of the package.

Replace ℓ with 8, w with 2, and h with 11.

surface area $= 2 \cdot 8 \cdot 2 + 2 \cdot 8 \cdot 11 + 2 \cdot 2 \cdot 11$

$= 252$ square inches

Step 2 Find the area of the wrapping paper.

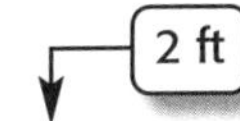

area $= 12$ in. $\cdot$ 24 in. or 288 in^2

Since 288 > 252, Mario bought enough wrapping paper.

Skill and Concept Check

Writing Math Exercises 1 & 3

1. **Explain** how to find the surface area of a rectangular prism.
2. **OPEN ENDED** Draw a rectangular prism and its net with the dimensions labeled on both figures. Then find the surface area.
3. **Explain** why surface area is measured in square units even though the figure is three-dimensional.

GUIDED PRACTICE

Find the surface area of each rectangular prism. Round to the nearest tenth if necessary.

4. 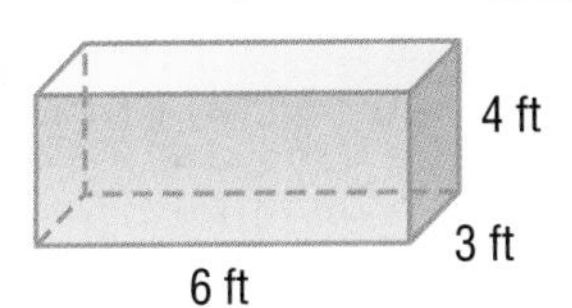

5. 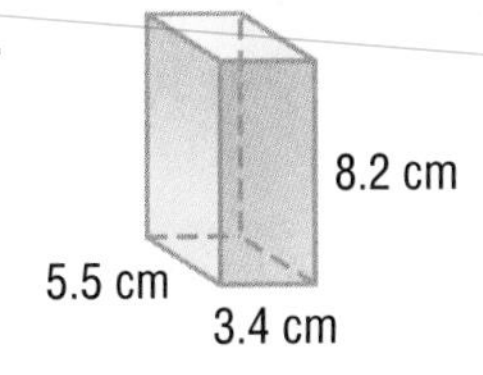

6. Find the surface area of a rectangular prism that has a length of 7 inches, a width of 11 inches, and a height of 9 inches.

msmath2.net/extra_examples

Practice and Applications

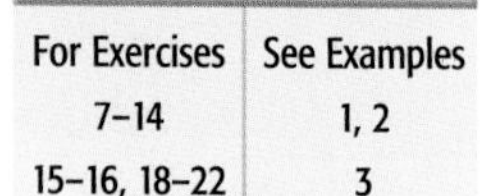

HOMEWORK HELP

For Exercises	See Examples
7–14	1, 2
15–16, 18–22	3

Extra Practice
See pages 594, 607.

Find the surface area of each rectangular prism. Round to the nearest tenth if necessary.

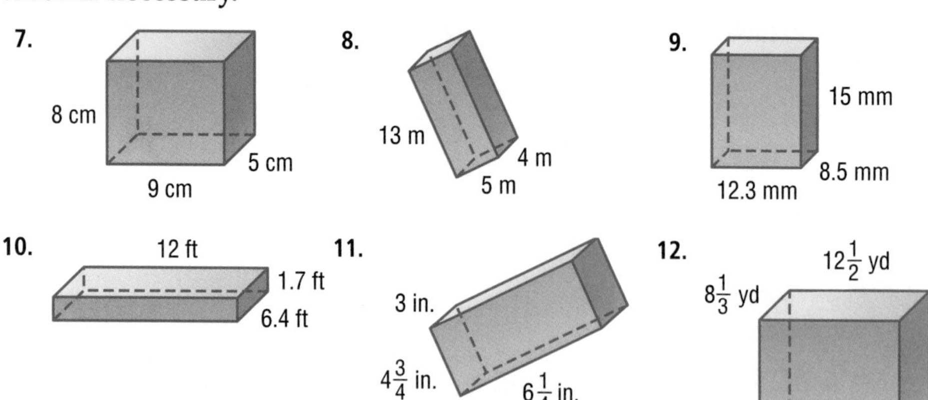

13. A cube has a surface area of 42 square meters. What is the area of one face?

14. Find the surface area of a rectangular prism that has a length of 5 feet, a width of 3 feet, and a height of 9 inches.

BOXES For Exercises 15 and 16, use the information below.
The largest corrugated cardboard box ever constructed was in Helmond, Netherlands. It measured 22.9 feet long, 8.5 feet high, and 7.87 feet wide.

15. Find the volume of the cardboard box to the nearest tenth.
16. Find the surface area of the cardboard box to the nearest tenth.

17. **ALGEBRA** Write a formula for the surface area of a cube in which each side measures x units.

PHYSICAL SCIENCE For Exercises 18–20, use the information below.
Granulated sugar dissolves faster in water than a sugar cube.

18. Suppose the length of each edge of a sugar cube is 1 centimeter. Find the surface area of the cube.
19. Imagine cutting the cube once in half horizontally and twice vertically. Find the total surface area of the eight cubes.
20. Make a conjecture as to why granulated sugar dissolves faster than a sugar cube.

CEREAL For the Exercises 21 and 22, use the following information.
Suppose you are designing a trial size cereal box that holds 100 cubic centimeters of cereal.

21. **MULTI STEP** Find the whole number dimensions of the box that would use the least amount of cardboard.
22. **MULTI STEP** If cardboard costs \$0.05 per 100 square centimeters, how much would it cost to make 100 boxes?

23. **WRITE A PROBLEM** Write about a real-life situation that involves finding the surface area of a rectangular prism. Then solve the problem.

24. **CRITICAL THINKING** A model is made by placing a cube with 12-centimeter sides on top of another cube with 15-centimeter sides. Find the surface area. (*Hint*: Do not include the area where the smaller cube covers the larger cube.)

Standardized Test Practice and Mixed Review

25. **MULTIPLE CHOICE** Find the surface area of the rectangular prism to the nearest tenth.

    (A) 101.5 cm^2 (B) 165.0 cm^2

    (C) 198.6 cm^2 (D) 203.0 cm^2

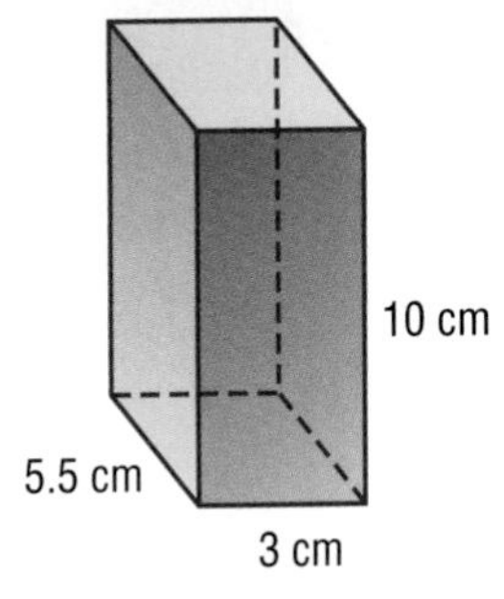

26. **MULTIPLE CHOICE** Find the amount of glass used for an aquarium that is 2.5 feet long, 1.6 feet wide, and 2 feet tall. (*Hint*: The top of the aquarium is open.)

    (F) 19.4 ft^2 (G) 20.4 ft^2 (H) 21.2 ft^2 (I) 24.4 ft^2

27. **GRID IN** The surface area of a cube is 294 square millimeters. What is the length of one side in millimeters?

Find the volume of each cylinder. Round to the nearest tenth. (Lesson 12-3)

28. 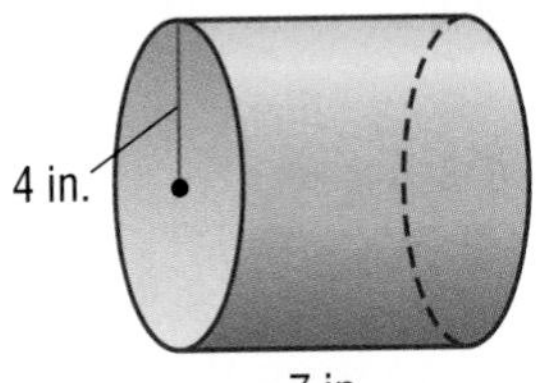

29. 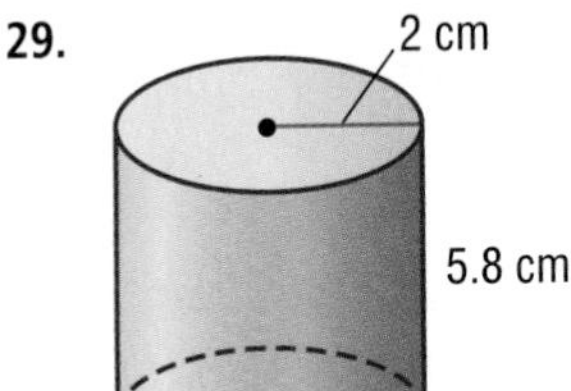

30. 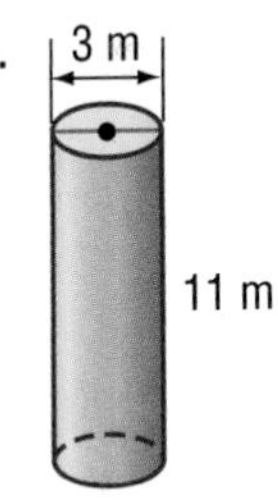

31. A rectangular prism is 14 inches long, 4.5 inches wide, and 1 inch high. Find its volume. (Lesson 12-2)

32. **CARPENTRY** The deck on a house is $25\frac{3}{4}$ feet long and $12\frac{1}{2}$ feet wide. The longer side of the deck is against the house. How many feet of wood does Jack need to buy to build a railing around the deck? (Lesson 6-6)

GETTING READY FOR THE NEXT LESSON

PREREQUISITE SKILL **Find the area of each circle. Round to the nearest tenth.** (Lesson 11-6)

33. 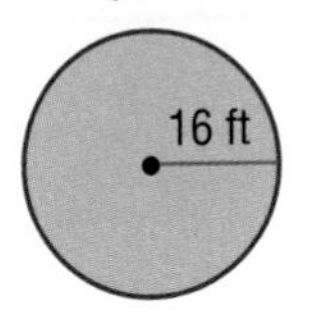

34. 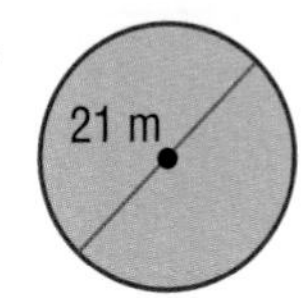

35. diameter = 13.6 yd

36. radius = 23 km

12-4b HANDS-ON LAB

A Follow-Up of Lesson 12-4

Changes in Volume and Surface Area

What You'll LEARN

Investigate changes in volume and surface area.

Materials

- isometric dot paper

Suppose you have a model of a rectangular prism and you are asked to create a similar model whose dimensions are twice as large. In this lab, you will investigate how changing the dimensions of a three-dimensional figure affects the surface area and volume.

ACTIVITY *Work with a partner.*

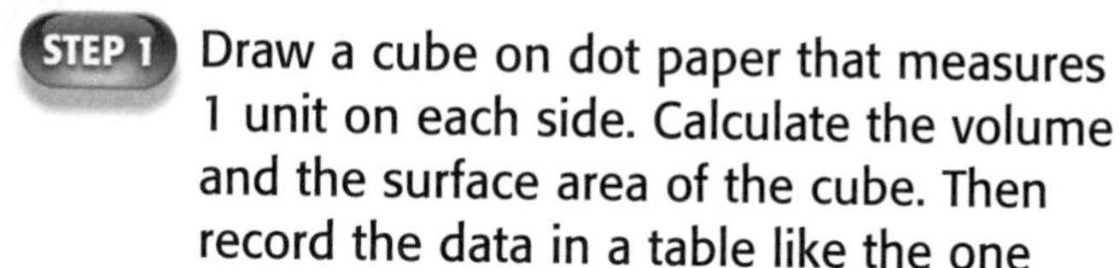

1

STEP 1 Draw a cube on dot paper that measures 1 unit on each side. Calculate the volume and the surface area of the cube. Then record the data in a table like the one shown below.

1 unit

STEP 2 Double the side lengths of the cube. Calculate the volume and the surface area of this cube. Record the data in your table.

2 units

STEP 3 Triple the side lengths of the original cube. Now each side measures 3 units long. Calculate the volume and the surface area of the cube and record the data.

3 units

STEP 4 For each cube, write a ratio comparing the side length and the volume. Then write a ratio comparing the side length and the surface area. The first one is done for you.

Side Length (units)	Volume (units^3)	Surface Area (units^2)	Ratio of Side Length to Volume	Ratio of Side Length to Surface Area
1	$1^3 = 1$	$6(1^2) = 6$	1:1	1:6
2				
3				
4				
5				
s				

Your Turn

a. Complete the table above.

ACTIVITY *Work with a partner.*

2 **STEP 1** Draw a cube on dot paper that measures 8 units on each side. Calculate the volume and the surface area of the cube. Record the data in a table like the one shown below.

STEP 2 Halve the side lengths of the cube in Step 1. Calculate the volume and the surface area of this cube and record the data.

STEP 3 Halve the side lengths of the cube in Step 2. Calculate the volume and the surface area of the cube and record the data.

STEP 4 For each cube, write a ratio comparing the side length and the volume and a ratio comparing the side length and the surface area. The first one is done for you.

Side Length (units)	Volume (units³)	Surface Area (units²)	Ratio of Side Length to Volume	Ratio of Side Length to Surface Area
8	$8^3 = 512$	$6(8^2) = 384$	8:512	8:384
4				
2				
s				

Your Turn

b. Complete the table above.

Writing Math

Work with a partner.

1. **Write** a formula for the volume V of a cube with side length s.
2. **Write** a formula for the surface area A of a cube with side length s.

Complete each sentence.

3. If the side length of a cube is doubled, the volume is __?__ times greater.
4. If the side length of a cube is doubled, the surface area is __?__ times greater.
5. If the side length of a cube is tripled, the volume increases by __?__ times and the surface area increases by __?__ times.
6. If the side length of a cube decreases by $\frac{1}{2}$, the surface area decreases by __?__.

12-5 Surface Area of Cylinders

HANDS-ON Mini Lab

What You'll LEARN

Find the surface areas of cylinders.

REVIEW Vocabulary

circumference: the distance around a circle **(Lesson 6-9)**

Materials

- soup can
- grid paper
- scissors

Work with a partner.

- Trace the top and bottom of the can on grid paper. Then cut out the shapes.
- Cut a long rectangle from the grid paper. The width of the rectangle should be the same as the height of the can. Wrap the rectangle around the side of the can. Cut off the excess paper so that the edges just meet.

1. Make a net of the cylinder.
2. Name the shapes in the net.
3. How is the length of the rectangle related to the circles?
4. Explain how to find the surface area of the cylinder.

The diagram below shows how you can put two circles and a rectangle together to make a cylinder.

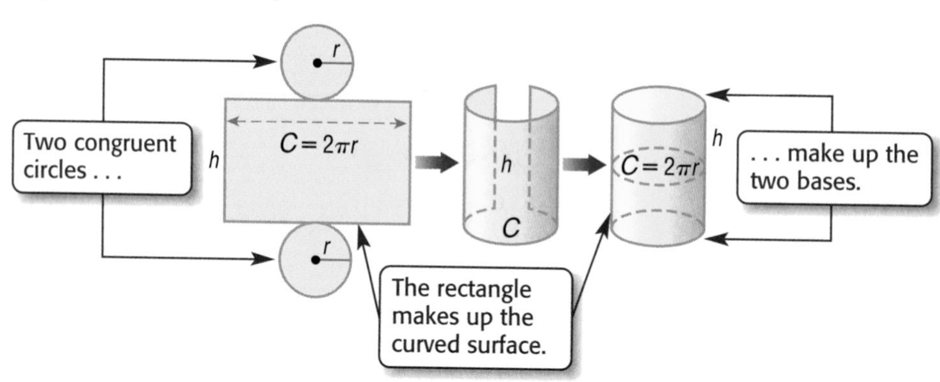

In the diagram above, the length of the rectangle is the same as the circumference of the circle. Also, the width of the rectangle is the same as the height of the cylinder.

$$\underbrace{S}_{\text{The surface area of a cylinder}} \underbrace{=}_{\text{equals}} \underbrace{2(\pi r^2)}_{\text{the area of two bases}} \underbrace{+}_{\text{plus}} \underbrace{(2\pi r)h}_{\text{the area of the curved surface.}}$$

Key Concept — Surface Area of a Cylinder

Words The surface area S of a cylinder with height h and radius r is the sum of the areas of the circular bases and the area of the curved surface.

Symbols $S = 2\pi r^2 + 2\pi rh$

Model

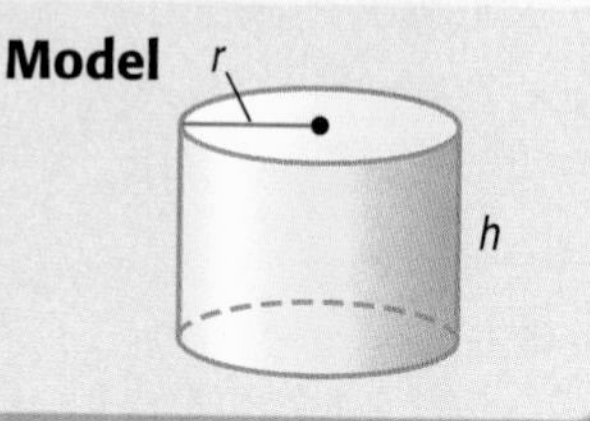

EXAMPLE Find the Surface Area of a Cylinder

1 Find the surface area of the cylinder. Round to the nearest tenth.

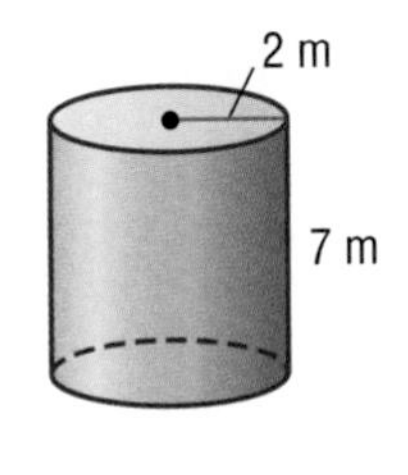

$S = 2\pi r^2 + 2\pi rh$	Surface area of a cylinder
$= 2\pi(2)^2 + 2\pi(2)(7)$	Replace r with 2 and h with 7.
≈ 113.1	Simplify.

The surface area is about 113.1 square meters.

Your Turn

a. Find the surface area of the cylinder. Round to the nearest tenth.

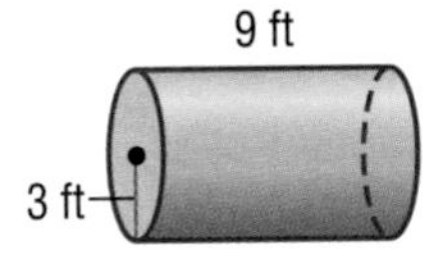

REAL-LIFE CAREERS

How Does a Graphic Designer Use Math?

A person who designs the packages of products must know the size of the package before making the design. This involves calculating the surface areas of packages that are rectangular prisms and cylinders.

Online Research
For information about a career as a graphic designer, visit: msmath2.net/careers

EXAMPLE Use Surface Area to Solve a Problem

2 DESIGN A can of soup is 5 inches high, and its base has a diameter of 4 inches. How much paper is needed to make the label on the can?

Since only the curved side of the can has a label, you do not need to include the areas of the top and bottom of the can.

$S = 2\pi rh$	Curved surface of a cylinder
$= 2\pi(2)(5)$	Replace r with 2 and h with 5.
≈ 62.8	Simplify.

So, about 62.8 square inches of paper is needed to make the label.

Skill and Concept Check

Writing Math
Exercise 2

1. **OPEN ENDED** Find the surface area of a cylinder found in your home.

2. **FIND THE ERROR** Matthew and Latonya have each drawn a cylinder with the dimensions shown below and insist that theirs has more surface area. Who is correct? Explain.

Matthew	Latonya
$r = 6$ cm, $h = 3$ cm	$r = 3$ cm, $h = 6$ cm

GUIDED PRACTICE

Find the surface area of each cylinder. Round to the nearest tenth.

3.

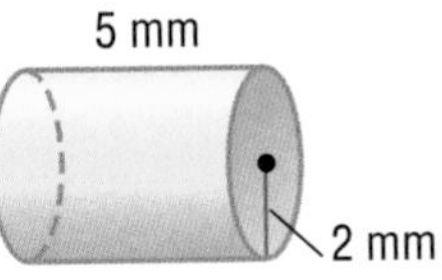

4.

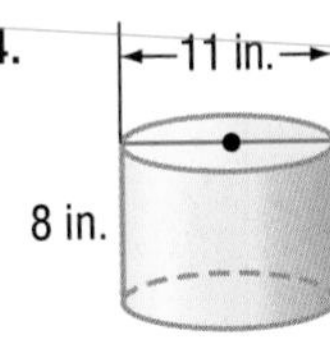

5. The radius of a cylinder is 8 meters, and its height is 5 meters. Find the surface area to the nearest tenth.

Practice and Applications

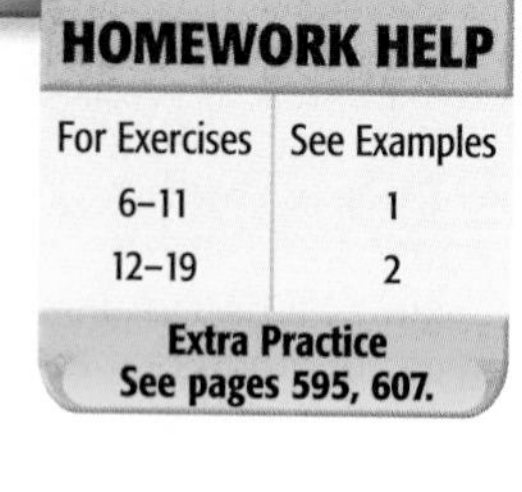

HOMEWORK HELP	
For Exercises	See Examples
6–11	1
12–19	2

Extra Practice
See pages 595, 607.

Find the surface area of each cylinder. Round to the nearest tenth.

6. 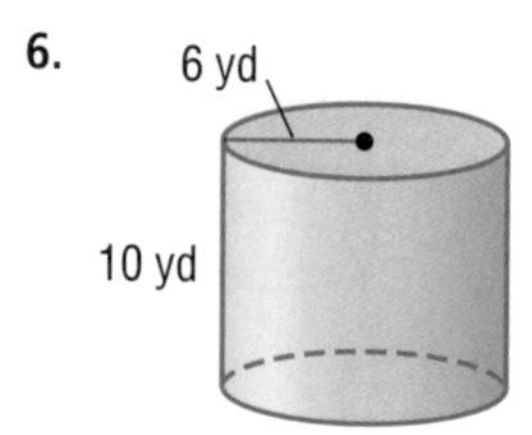

7. 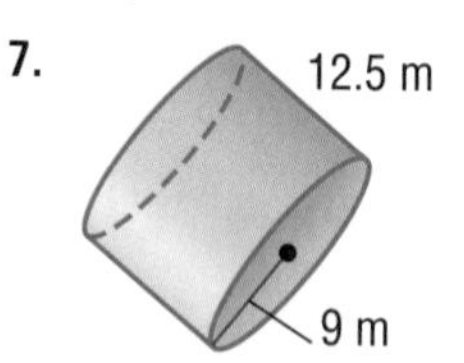

8. 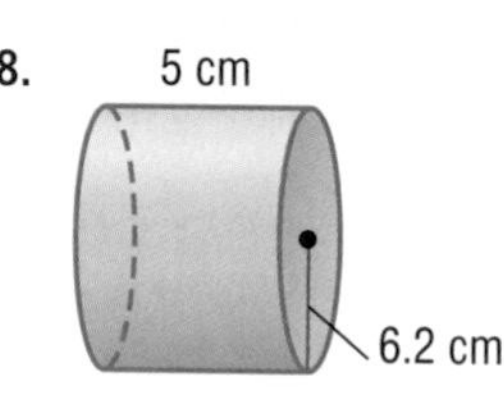

9. 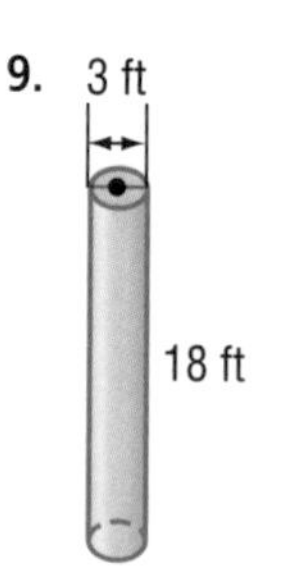

10. 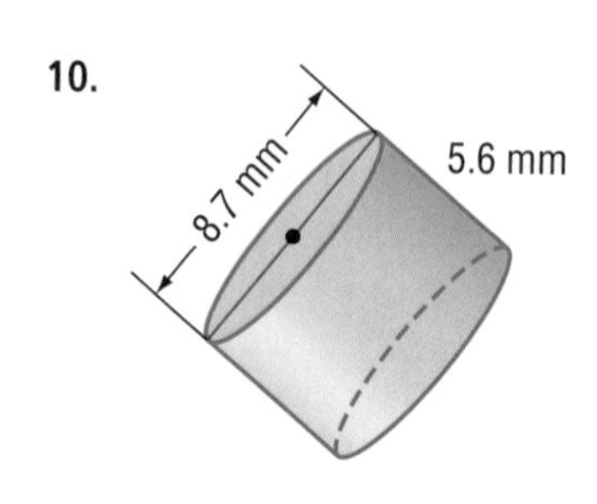

11. 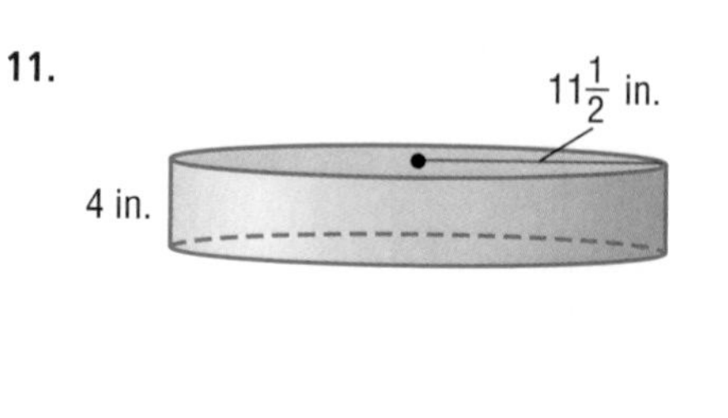

12. Find the area of the label on a can of tuna with a radius of 5.1 centimeters and a height of 2.9 centimeters.

13. The height of a water tank is 10 meters and has a diameter of 10 meters. What is the surface area of the tank?

14. An underground oil tank has a surface area of 2,915 square feet and a radius of 6 feet. What is the height of the oil tank?

BAKING **For Exercises 15 and 16, use the following information.** A *smash cake* is a smaller version of a birthday cake given to a child on his or her first birthday. Mrs. Jones baked a smash cake 3 inches high and 10 inches in diameter for her daughter when she turned 1 year old.

15. Mrs. Jones covers the top and sides of the cake with frosting. Find the area that the frosting covers to the nearest tenth.

16. Suppose Mrs. Jones first cuts the cake in two layers, frosts the top and sides of the bottom layer, puts the top layer on, and then frosts the rest of the cake. How much of the cake is covered with frosting now?

17. **MOVIES** The Student Council at Southwest Middle School is planning to sell popcorn in one of two open-top containers below at its "Movie Night." The cost depends on the amount of cardboard used to make each container. Which container should Student Council buy? Use volume and surface area measurements to explain your choice.

18. **CANDY** The largest piece of candy ever created was shaped like a cylinder 4.49 meters long and had a circumference of 1.14 meters. What was the surface area of the candy to the nearest tenth?

19. **MODELING** Suppose you took an $8\frac{1}{2}$-×-11-inch piece of paper and made two cylinders; one using the $8\frac{1}{2}$-inch side as the height (cylinder A) and the other using the 11-inch side as the height (cylinder B). Without calculating, which cylinder has the greater surface area? Explain.

20. **CRITICAL THINKING** If you double the height of a cylinder, will its surface area also double? Explain your reasoning.

21. **EXTENDING THE LESSON** In Lesson 12-1, you saw another solid with a curved surface, called a *cone*, as shown at the right. The curved surface is difficult to draw accurately in two dimensions since it is not a polygon. Make a cone with a piece of paper. Then draw the net for the cone using it as a pattern.

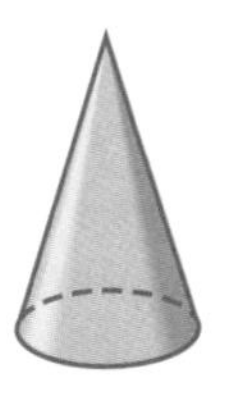

Standardized Test Practice and Mixed Review

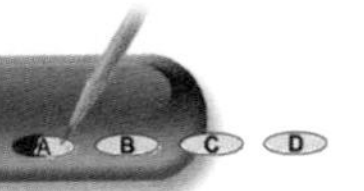

22. **MULTIPLE CHOICE** Gilberto completely covers a cylindrical can with construction paper. The can has a height of 15 inches and a radius of 4 inches. About how much construction paper does he use?

(A) 477.5 in² (B) 375.2 in² (B) 104.5 in² (D) 60.0 in²

23. **MULTIPLE CHOICE** The three metal containers below each hold about 1 liter of liquid. Which container has the greatest surface area?

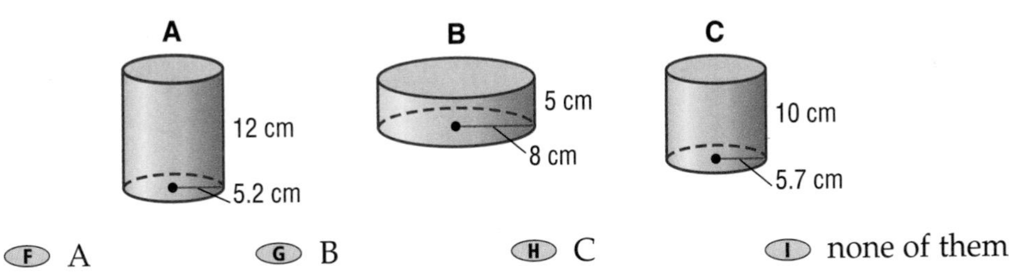

(F) A (G) B (H) C (I) none of them

Find the surface area of each rectangular prism. (Lesson 12-4)

24. 2 ft, 3 ft, 6 ft

25. 5 cm, 4 cm, 4 cm

26.

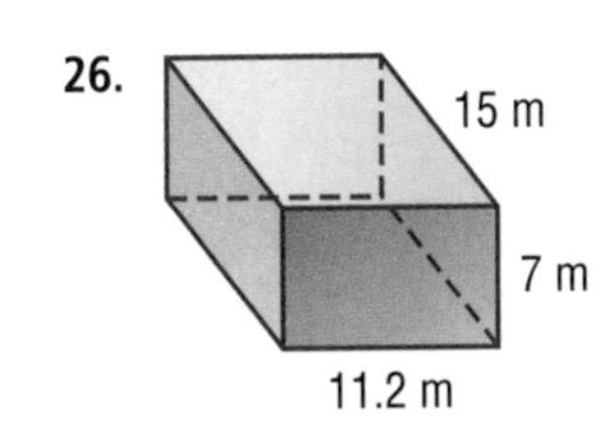

27. Find the volume of a cylinder having a radius of 4 inches and a height of 6.5 inches. (Lesson 12-3)

GETTING READY FOR THE NEXT LESSON

PREREQUISITE SKILL Find the value of each expression. Round to the nearest tenth. (Pages 559, 560)

28. $13.2 + 15.378$ 29. $23.7 - 9.691$ 30. $23 \cdot 7.1$ 31. $1.6(8.5)0.4$

12-6 Precision and Measurement

WHEN am I ever going to use this?

WORLD RECORDS In 2000, Joyce Samuels blew a bubble gum bubble that had a diameter of 27.94 centimeters.

1. What is the smallest unit of measure?
2. Explain whether a ruler whose smallest increment is 0.5 centimeter could have been used to measure the bubble.

What You'll LEARN

Determine and apply significant digits in a real-life context.

NEW Vocabulary

precision
precision unit
significant digits

The **precision** or *exactness* of a measurement depends on the unit of measure. The **precision unit** is the smallest unit on a measuring tool. As the units get smaller, the measurement gets "more precise."

EXAMPLES Identify Precision Units

Identify the precision unit of each measuring tool.

1

cm 1 2 3 4

The smallest unit is a half centimeter. So, the precision unit is 0.5 centimeter.

2

in. 1

The smallest unit is an eighth of an inch. So, the precision unit is $\frac{1}{8}$ inch.

Your Turn

a. Identify the precision unit of the scale at the right.

All measurements are approximate. You could estimate the measure on the scale below as 8 grams. A more precise method is to use significant digits. **Significant digits** include all of the digits of a measurement that you know for sure, plus one estimated digit.

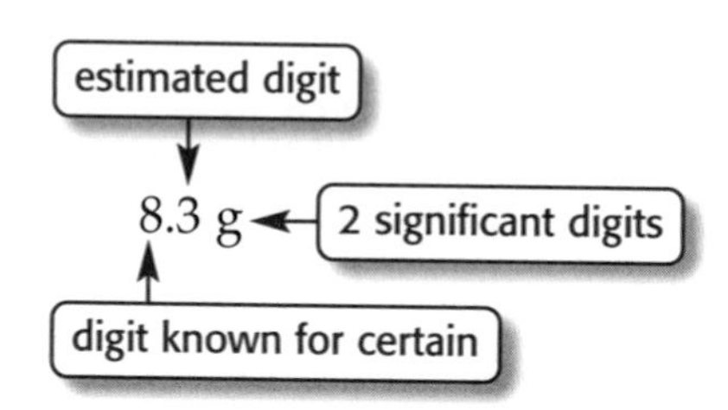

EXAMPLE Apply Significant Digits

3 **State the measure of the pencil using significant digits.**

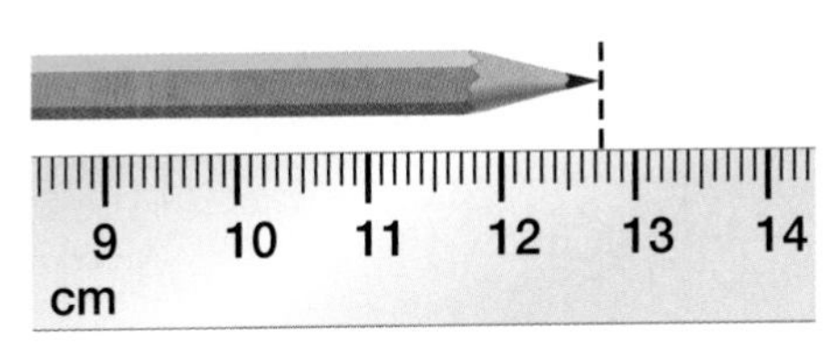

The precision unit is 0.1 centimeter. You know for certain that the length is between 12.7 and 12.8 centimeters. One estimate is 12.75 centimeters.

Your Turn

b. State the measure of the paper clip using significant digits.

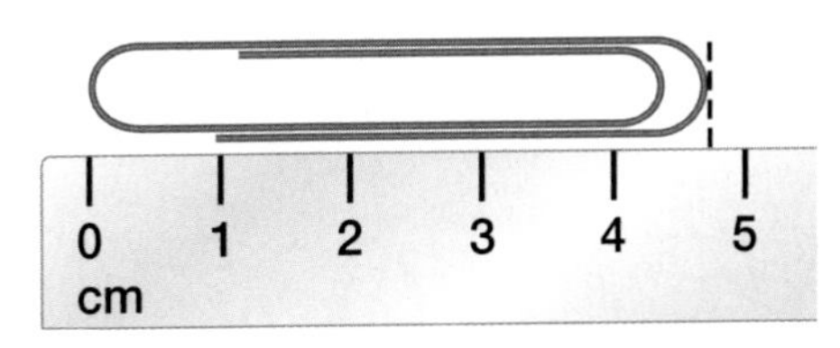

Skill and Concept Check

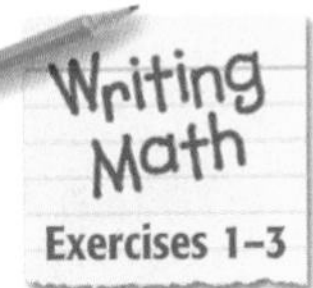

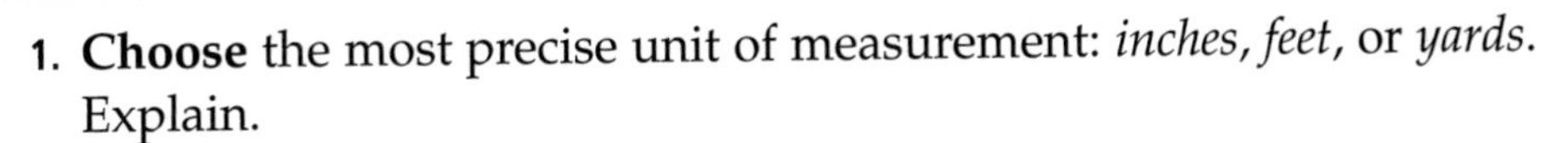

1. **Choose** the most precise unit of measurement: *inches*, *feet*, or *yards*. Explain.
2. **OPEN ENDED** Describe some items or situations in which an approximate measure is sufficient. Then describe some items or situations in which exactness in measurement is important and necessary.
3. **Which One Doesn't Belong?** Identify the measurement that does not have the same number of significant digits as the other three. Explain your reasoning.

8.6 mi	14.5 ft	3.22 cm	90.7 yd

GUIDED PRACTICE

4. **Identify** the precision unit of the ruler at the right.

State each measure using significant digits.

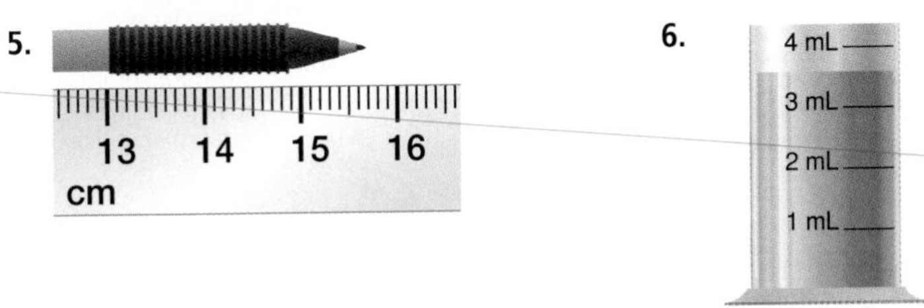

7. **OLYMPICS** Tom Dolan broke the Olympic Swimming World Record in 2000 for the 400-meter race. His time was 4 minutes, 11.76 seconds. Describe the precision of this measure.

Practice and Applications

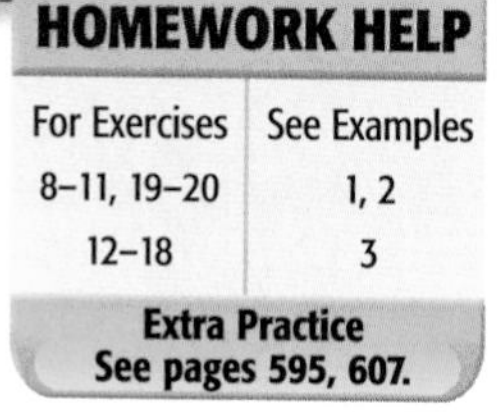

HOMEWORK HELP	
For Exercises	See Examples
8–11, 19–20	1, 2
12–18	3

Extra Practice
See pages 595, 607.

Identify the precision unit of each measuring tool.

8. 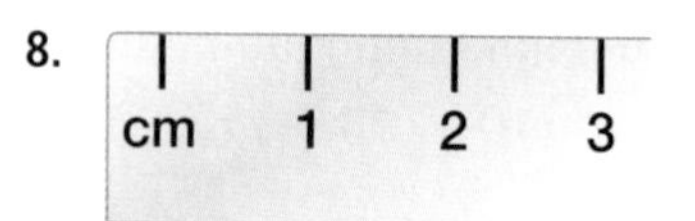

9. 14 15 16 17 18
cm

10. 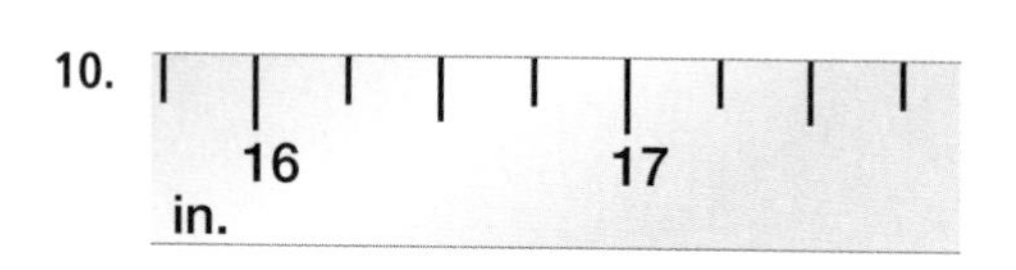

11. 110 10 100 20 90 30 80 lb 40 70 60 50

State each measure using significant digits.

12. 

13. 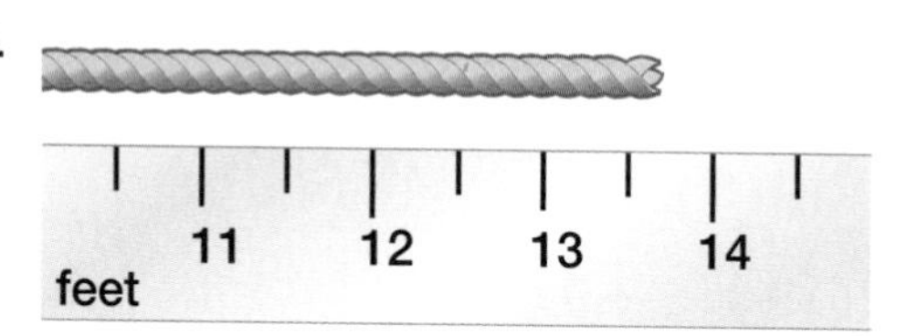

14. 31 32 33 34
cm

15. 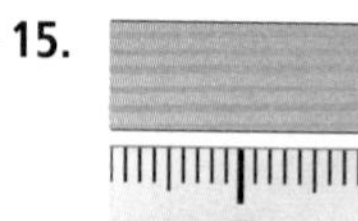 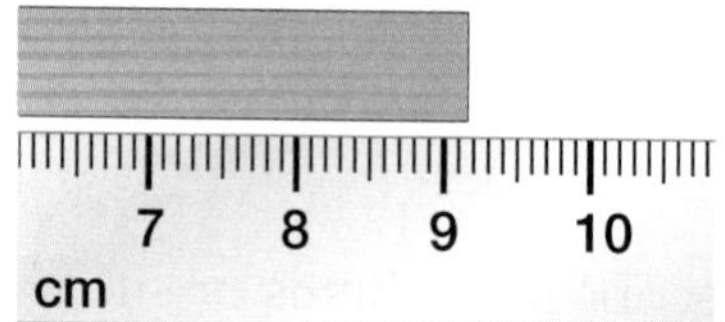

16. cm 1 2 3 4 5

17. 

TRIATHLONS **For Exercises 18–20, use the graphic at the right.**

18. How many significant digits are in the measurement describing the length of the cycling portion of the triathlon?

19. Which is a more precise measure: the number of miles cycled in a triathlon, or the number of miles run? Explain.

20. The running portion of the women's triathlon is equivalent to 5 kilometers. Which unit is more precise: a mile or a kilometer? Explain.

21. **RESEARCH** The silica tiles on space shuttles are cut accurately to 10,000th of an inch. Use the Internet or another source to investigate the technology that is used to make such precise measurements.

22. If you use a ruler with a precision unit of 1 millimeter, will your measurements *sometimes*, *always*, or *never* be exact? Explain.

23. **WRITE A PROBLEM** Write a measurement problem in which you use significant digits.

24. **CRITICAL THINKING** Choose the most precise measures to complete the sentence: The yarn at the right is between __?__ and __?__ inches long.

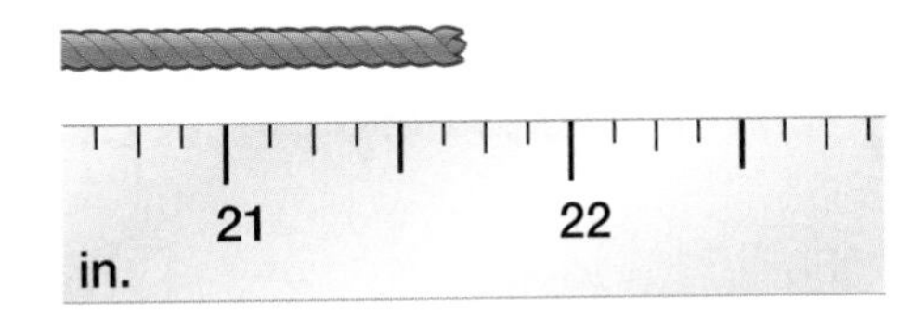

Standardized Test Practice and Mixed Review

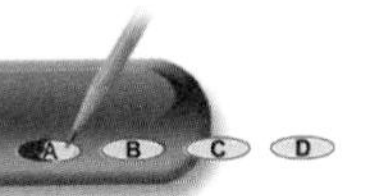

25. **MULTIPLE CHOICE** Choose the measuring tool that you would use to estimate the volume of a shoebox.

(A) scale (B) ruler (C) protractor (D) measuring cup

26. **MULTIPLE CHOICE** Choose the best precision unit for estimating the length of a gymnasium.

(F) 0.1 cm (G) 0.5 in. (H) 0.1 mi (I) 0.5 ft

Find the surface area of each cylinder. Round to the nearest tenth. (Lesson 12-5)

27.

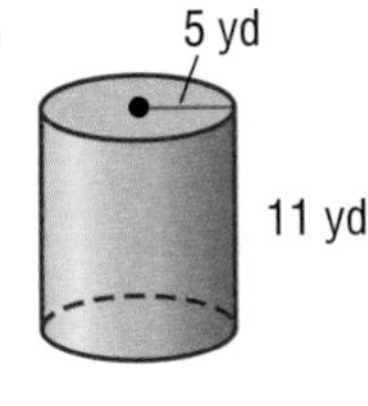

28.

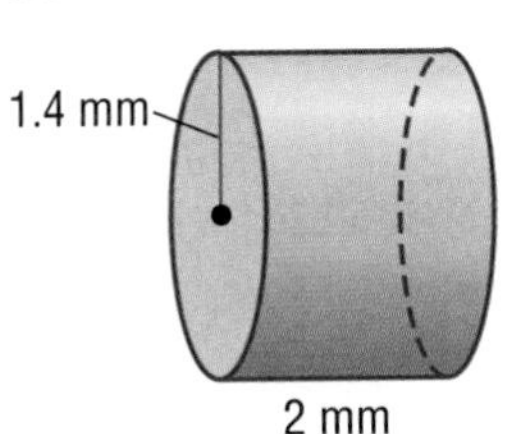

29.

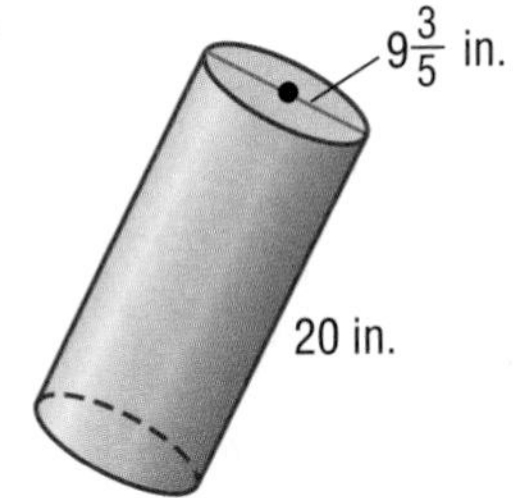

30. Find the surface area of a rectangular prism that has a length of 19 centimeters, a width of 8.8 centimeters, and a height of 13 centimeters. (Lesson 12-4)

ALGEBRA Solve each equation. Check your solution. (Lesson 6-5)

31. $\frac{n}{15} = 3$

32. $-26 = \frac{x}{2.4}$

33. $\frac{3}{8}y = 9$

WebQuest Interdisciplinary Project

It's All Greek to Me

It's time to complete your project. Use the information and data you have gathered about Pythagoras to prepare a Web page or poster. Be sure to include the three-dimensional solid you created with your project.

msmath2.net/webquest

CHAPTER 12

Study Guide and Review

Vocabulary and Concept Check

cylinder (p. 524)	precision unit (p. 542)	significant digits (p. 542)	surface area (p. 532)
precision (p. 542)	rectangular prism (p. 520)	solid (p. 514)	volume (p. 520)

Choose the correct term or number to complete each sentence.

1. A (rectangular prism, rectangle) is a solid figure that has three sets of parallel congruent sides.
2. The (volume, surface area) of a solid figure is the measure of the space occupied by it.
3. Volume is measured in (square, cubic) units.
4. The volume of a rectangular prism is found by (adding, multiplying) the length, the width, and the height.
5. A (cylinder, prism) is a solid that has two congruent, parallel circles as its bases.
6. Surface area is the sum of the (areas, volumes) of all of the outside surfaces of a three-dimensional figure.
7. The (smaller, larger) the precision unit, the more precise is the measurement.
8. Significant digits include all of the digits of a measurement that you know for sure, plus (one, two) estimated digit(s).

Lesson-by-Lesson Exercises and Examples

12-1 Drawing Three-Dimensional Figures (pp. 514–517)

Draw a top, a side, and a front view of each solid.

9. 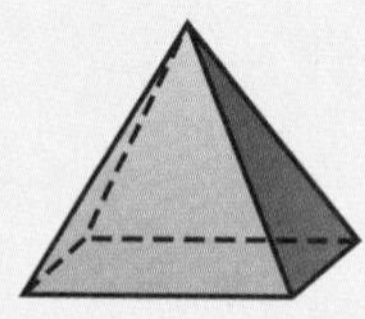

10. 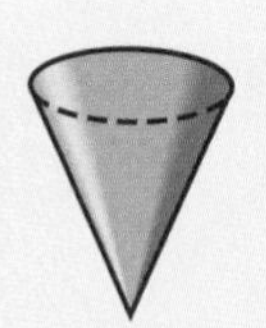

11. Draw a solid using the top, side, and front views shown. Use isometric dot paper.

top 

side 

front 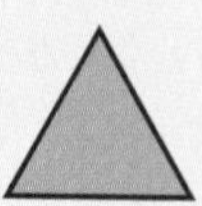

Example 1 **Draw the solid using the top, side, and front views shown.**

top side front

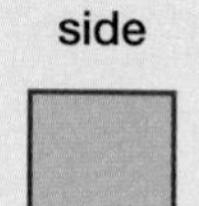

The side view is a square. The top and front views are rectangles.

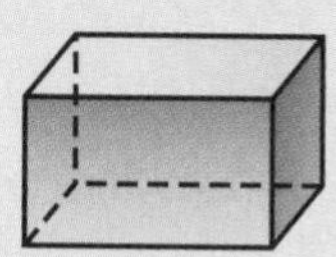

msmath2.net/vocabulary_review

12-2 Volume of Rectangular Prisms (pp. 520–522)

Find the volume of each rectangular prism. Round to the nearest tenth.

12.

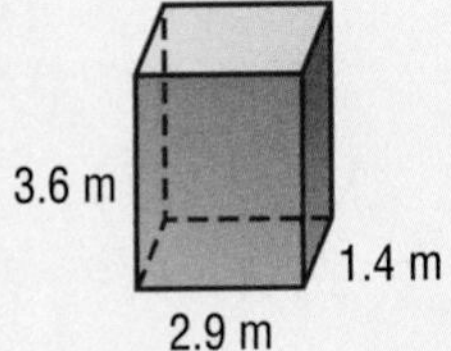

13.

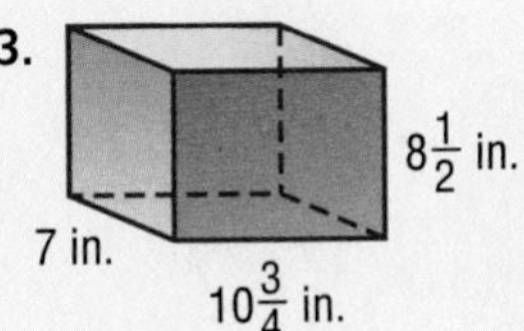

14. POOLS A swimming pool 25 yards long has 8 lanes that are each 3 yards wide. The water is 6 feet deep. Find the volume of water in the pool.

Example 2 **Find the volume of the rectangular prism.**

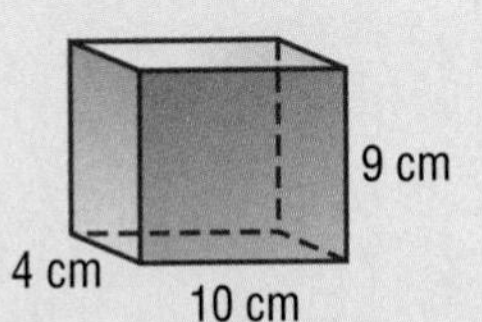

$V = \ell wh$ Volume of a rectangular prism

$V = (10)(4)(9)$ $\ell = 10$, $w = 4$, and $h = 9$.

$V = 360$ Multiply.

The volume is 360 cubic centimeters.

12-3 Volume of Cylinders (pp. 524–527)

Find the volume of each cylinder. Round to the nearest tenth.

15.

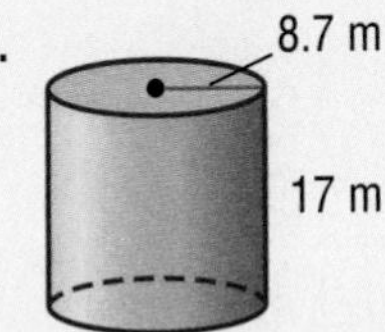

16.

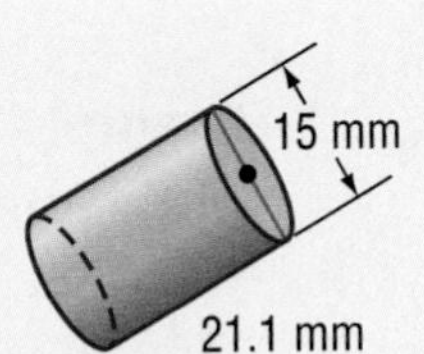

17. POTTERY In art class, Arturo made a vase in the shape of a cylinder. The diameter is 5 inches, and the height is 10 inches. Find the maximum volume of water the vase can hold.

Example 3 **Find the volume of the cylinder.**

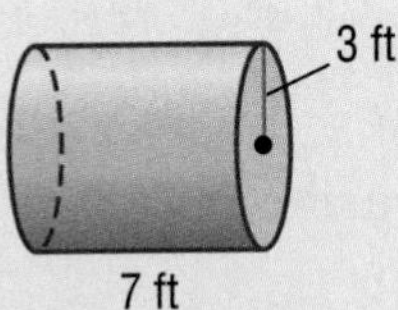

$V = \pi r^2 h$ Volume of a cylinder

$= \pi(3)^2(7)$ Replace r with 3 and h with 7.

≈ 197.9 Multiply.

The volume is about 197.9 cubic feet.

12-4 Surface Area of Rectangular Prisms (pp. 532–535)

Find the surface area of each rectangular prism. Round to the nearest tenth if necessary.

18.

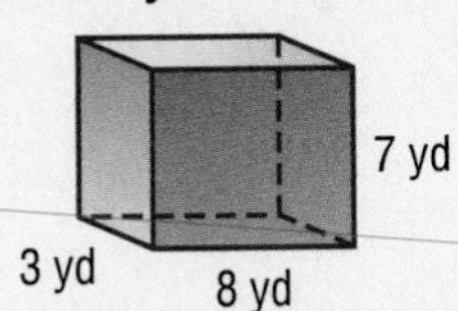

19.

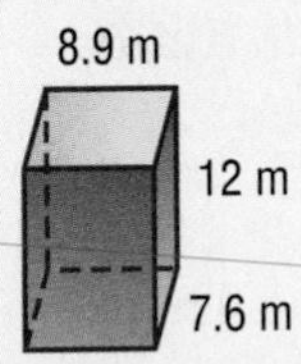

20. PETS A plastic pet carrier box is 2.5 feet long, 1 foot high, and 1.25 feet wide. How much plastic is used to make this carrier?

Example 4 **Find the surface area of the rectangular prism.**

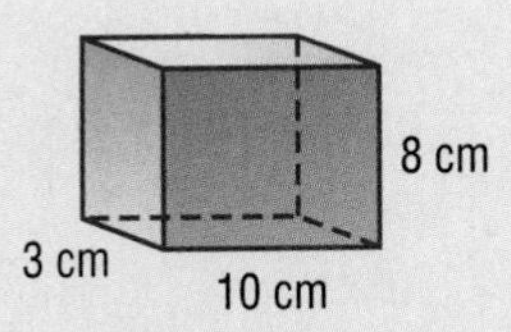

surface area $= 2\ell w + 2\ell h + 2wh$

$= 2(10)(3) + 2(10)(8) + 2(3)(8)$

$= 268$

The surface area is 268 square centimeters.

12-5 Surface Area of Cylinders (pp. 538–541)

Find the surface area of each cylinder. Round to the nearest tenth.

21. 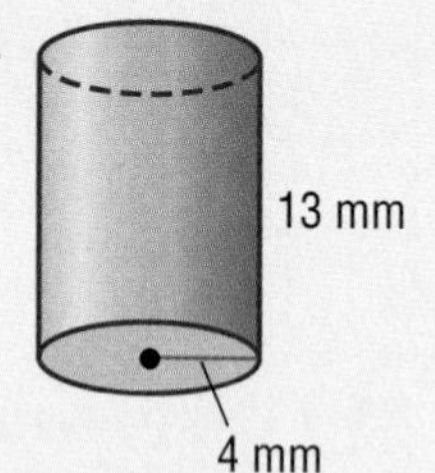

22. 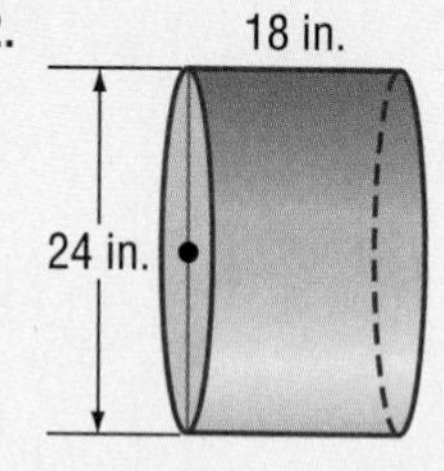

23. **DESIGN** A can of chicken noodle soup is $5\frac{1}{2}$ inches high, and its base has a radius of 2 inches. How much paper is needed to make the label on the can?

Example 5 **Find the surface area of the cylinder.**

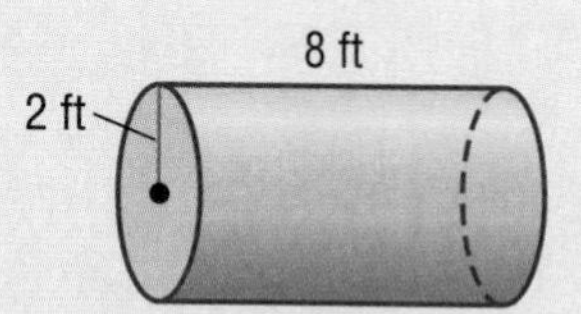

$$\text{surface area} = 2\pi r^2 + 2\pi rh$$
$$= 2\pi(2)^2 + 2\pi(2)(8)$$
$$\approx 125.7 \text{ ft}^2$$

The surface area is about 125.7 square feet.

12-6 Precision and Measurement (pp. 542–545)

Identify the precision unit of each measuring tool.

24. 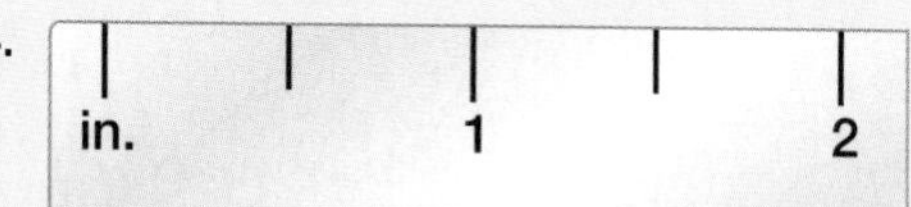

25. 

26. 

State each measure using significant digits.

27. 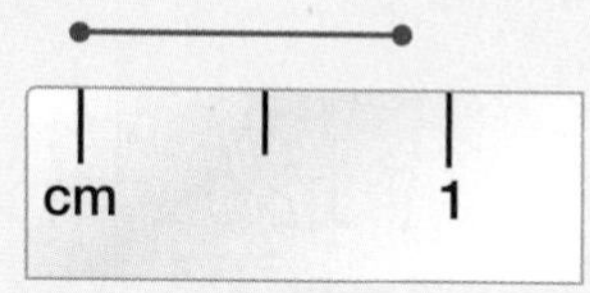

28. 

Example 6 **Identify the precision unit of the measuring tool.**

The smallest unit is one eighth of an inch. So, the precision unit is $\frac{1}{8}$ inch.

Example 7 **State the measure of the graduated cylinder using significant digits.**

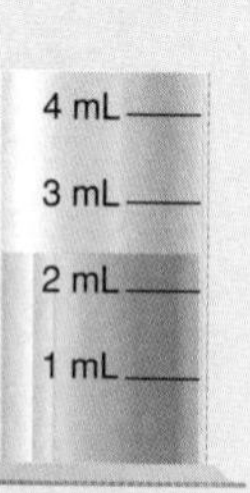

The precision unit is 1 milliliter. You know for certain that the volume is between 2 and 3 milliliters. One estimate is 2.4 milliliters.

CHAPTER 12 Practice Test

Vocabulary and Concepts

1. **Define** *surface area.*
2. **Explain** what a precision unit is.

Skills and Applications

Draw a top, a side, and a front view of each solid.

3. 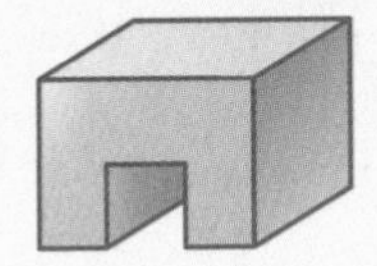

4. 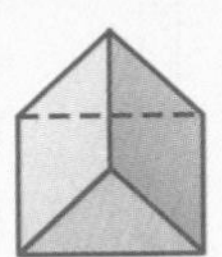

5. 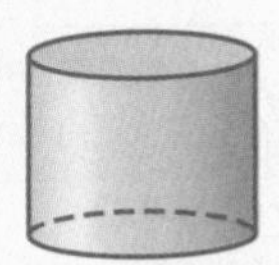

Find the volume and surface area of each rectangular prism and cylinder. Round to the nearest tenth if necessary.

6. 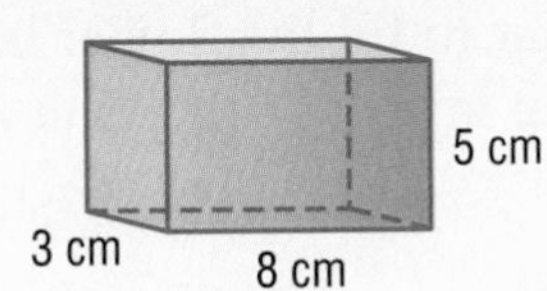

7. rectangular prism;
   length = 19.6 m
   width = 14 m
   height = 26.1 m

8. 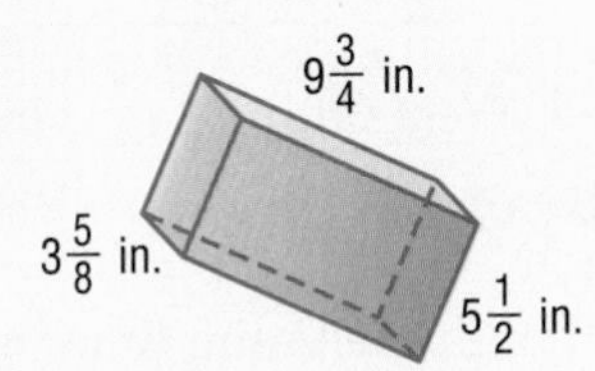

9. cylinder;
   radius = 6 ft
   height = 12 ft

10. 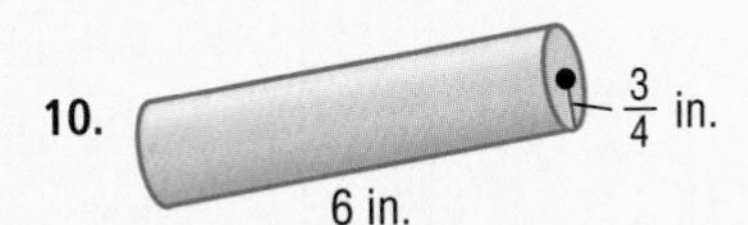

11. cylinder;
    diameter = 11.5 mm
    height = 20.7 mm

12. **DRINKING STRAWS** A drinking straw has a radius of $\frac{1}{8}$ inch and a height of $7\frac{3}{4}$ inches. How much liquid is in a straw that is half full?

13. **PACKAGING** What is the least amount of paper needed to wrap a box that is 9 inches by 18 inches by 4 inches?

Identify the precision unit of each measuring tool.

14. 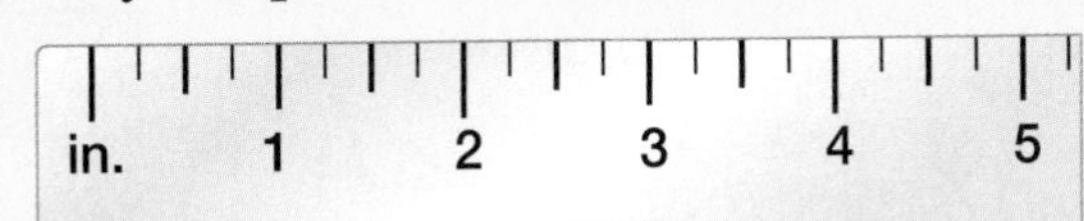

15. 

Standardized Test Practice

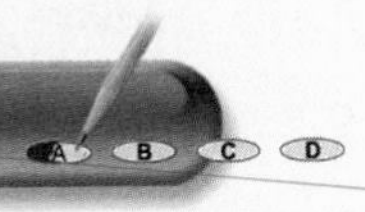

16. **MULTIPLE CHOICE** Find the best estimate of the measure of the paper clip using significant digits.

    Ⓐ 2 cm　　Ⓑ 2.25 cm

    Ⓒ 2.4 cm　　Ⓓ 2.75 cm

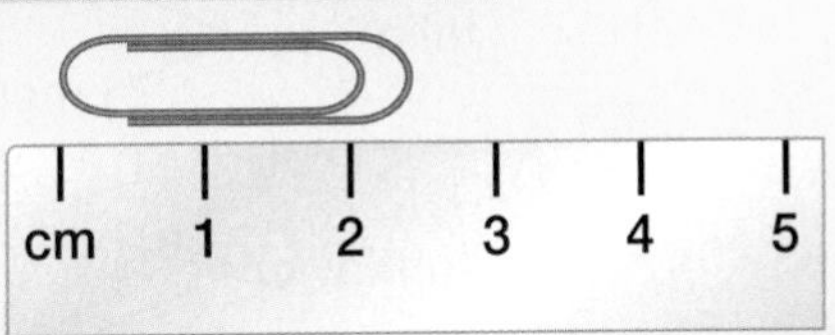

CHAPTER 12 Standardized Test Practice

PART 1 Multiple Choice

Record your answers on the answer sheet provided by your teacher or on a sheet of paper.

1. Which city in the table is the closest to sea level? (Lesson 3-2)

City	Altitude Compared to Sea Level (ft)
Ocean City	−25
Redmond	1,376
Reston	24
West Orange	−13

(A) West Orange (B) Reston

(C) Ocean City (D) Redmond

2. What is the value of w in the equation $6w + 12 = 30$? (Lesson 4-4)

(F) 3 (G) 12 (H) 18 (I) 30

3. Each inch of Tyrell's train set represents $4\frac{1}{2}$ feet of an actual train system. If the engine is 6 inches long, how long is an actual engine? (Lesson 7-4)

(A) $1\frac{1}{3}$ ft (B) 27 in.

(C) $10\frac{1}{2}$ ft (D) 27 ft

4. The tree diagram shows the choices of flowers and pots sold at Ivitz Gardens. How many different flower-pot choices are possible? (Lesson 9-2)

(F) 3

(G) 6

(H) 9

(I) 12

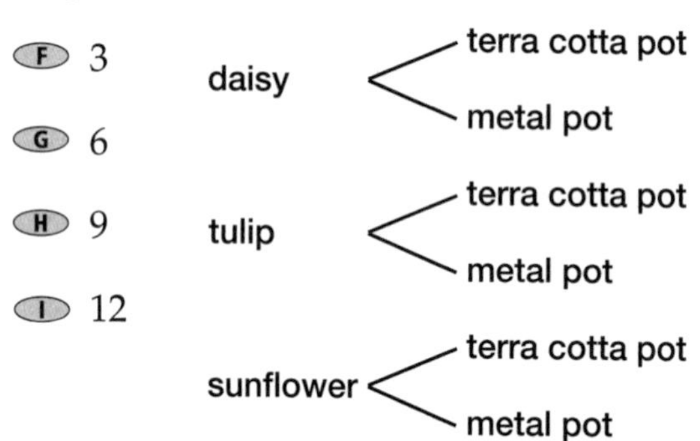

5. Which is the side view of the solid shown at the right? (Lesson 12-1)

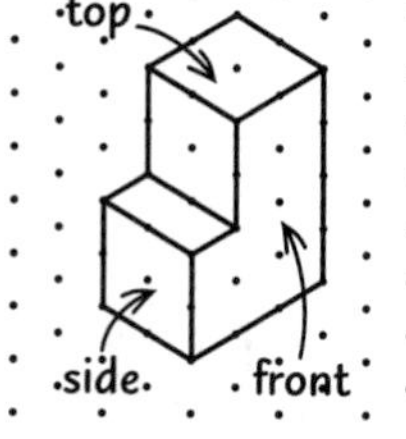

(A) (B)

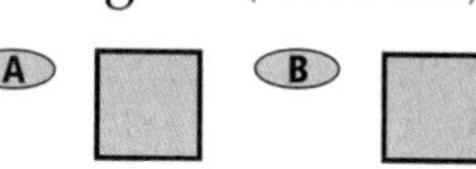

(C) (D)

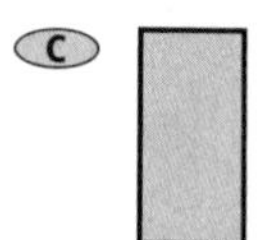

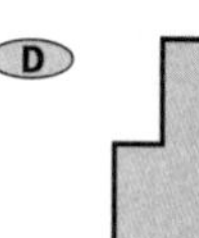

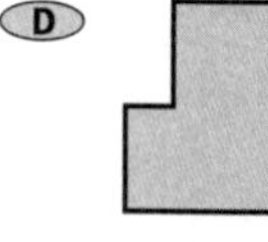

6. Gillian has a window box the shape of a rectangular prism. It is 36 inches wide and 6 inches long. How much soil should she buy in order to fill the flower box with a 4-inch-deep layer of soil? (Lesson 12-2)

(F) 46 in^3 (G) 576 in^3

(H) 864 in^3 (I) 1,296 in^3

7. Which is the volume of the cylinder? (Lesson 12-3)

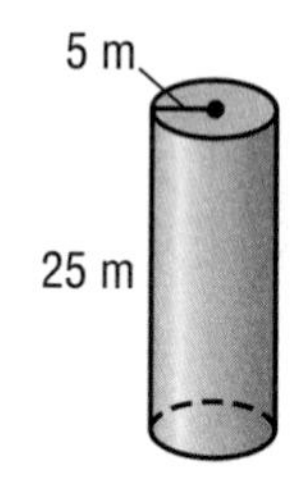

(A) 94.2 m^3

(B) 1,963.5 m^3

(C) 3,125.0 m^3

(D) 9,812.5 m^3

8. What is the surface area of the rectangular prism? (Lesson 12-4)

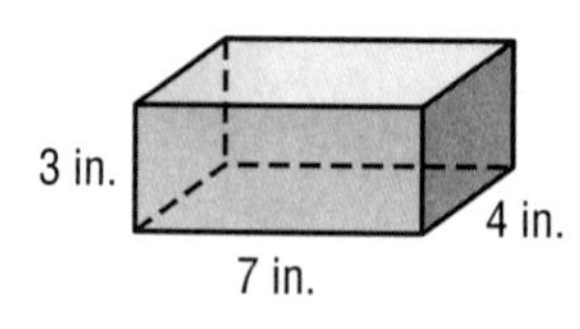

(F) 84 in^2

(G) 122 in^2

(H) 141 in^2

(I) 588 in^2

TEST-TAKING TIP

Questions 6–8 Most standardized tests include any necessary formulas in the test booklet. It helps to be familiar with formulas such as the volume or surface area of a cylinder, but use any formulas that are given to you.

PART 2 Short Response/Grid In

Record your answers on the answer sheet provided by your teacher or on a sheet of paper.

9. To find the perimeter of the rectangular tray shown below, Oi Ying wrote the equation $2 \times 8 + 2 \times 14$. What operation should Oi Ying do first? (Lesson 1-3)

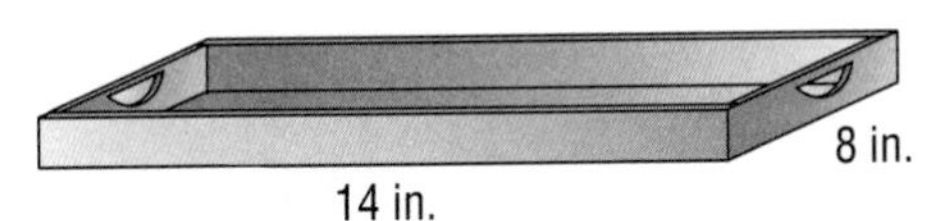

10. Marcos worked at a car wash during the summer. He worked with other employees to wipe the cars dry. Typically, Marcos wiped down 75% of the cars each day. What are two other ways that 75% can be expressed? (Lesson 5-6)

11. Find $5\frac{1}{2} \div \frac{3}{2}$. (Lesson 6-6)

12. At the Gilmour family reunion, 28 of the 70 family members ate potato salad. What percent of the Gilmours ate potato salad at the reunion? (Lesson 7-5)

13. At a sale, Shay finds a $125 coat marked down to $87.50. What percent of decrease is this? (Lesson 8-4)

14. Alisa wants a bicycle that usually costs $186. During a sale, the bicycle sells for 20% less than that. What is the price of the bicycle when it is on sale? (Lesson 8-5)

15. Triangle TUV is reflected over the x-axis. What are the new coordinates of point T? (Lesson 10-9)

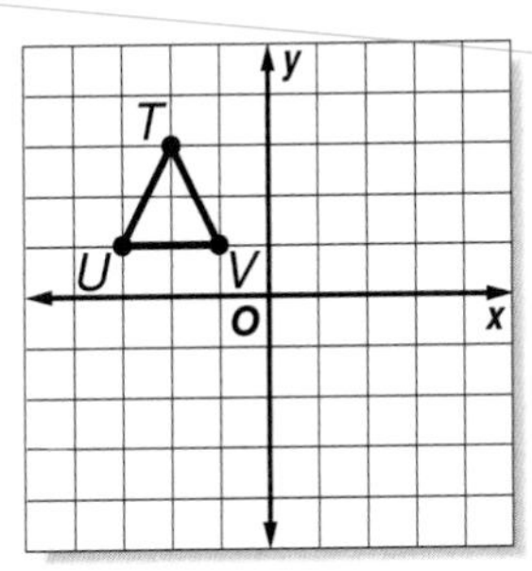

16. Find the total volume of the solid. (Lesson 12-2)

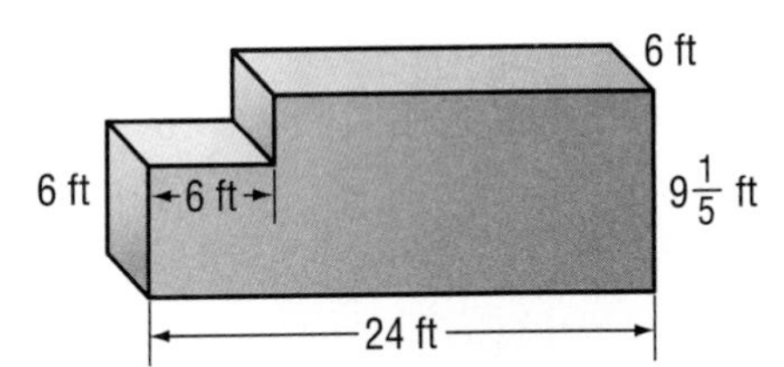

17. What is the surface area of a rectangular prism that is 14.2 centimeters long, 10.5 centimeters wide, and 3 centimeters high? (Lesson 12-4)

18. The cylindrical gasoline storage tank on Mrs. Simon's farm needs to be painted. The dimensions are shown below.

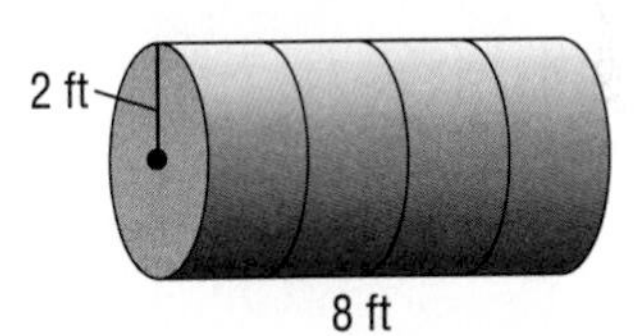

If one quart of paint covers 85 square feet, how many quarts does Mrs. Simon need for this job? (Lesson 12-5)

PART 3 Extended Response

Record your answers on a sheet of paper. Show your work.

19. Draw a top view, a front view, and a side view of the *tetrahedron* shown at the right. (Lesson 12-1)

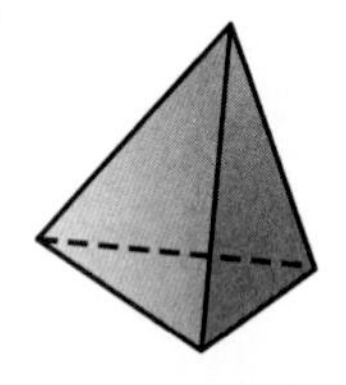

20. Determine whether the given precision unit is appropriate to use in estimating the length of each object. Explain your reasoning. If it is not, describe an appropriate precision unit. (Lesson 12-6)

 a. driveway: 1 inch

 b. compact disc: 1 meter

 c. calculator key: 0.1 centimeter

Skills

Reference

HOW TO...

USE THE STUDENT HANDBOOK

A Student Handbook is the additional skill and reference material found at the end of books. The Student Handbook can help answer these questions.

What If I Forget What I Learned Last Year?

Use the **Prerequisite Skills** section to refresh your memory about things you have learned in other math classes. Here's a list of the topics covered in your book.

- Divisibility Patterns
- Place Value and Decimals
- Comparing and Ordering Decimals
- Rounding Decimals
- Estimating with Decimals
- Adding and Subtracting Decimals
- Multiplying Decimals
- Powers of Ten
- Dividing Decimals
- Mixed Numbers and Improper Fractions

What If I Need More Practice?

You, or your teacher, may decide that working through some additional problems would be helpful. The **Extra Practice** section provides these problems for each lesson so you have ample opportunity to practice new skills.

What If I Have Trouble with Word Problems?

The **Mixed Problem Solving** portion of the book provides additional word problems that use the skills presented in each chapter. These problems give you real-life situations where the math can be applied.

What If I Forget a Vocabulary Word?

The **English-Spanish Glossary** provides a list of important, or difficult, words used throughout the textbook. It provides a definition in English and Spanish as well as the page number(s) where the word can be found.

What If I Need to Check a Homework Answer?

The answers to the odd-numbered problems are included in **Selected Answers**. Check your answers to make sure you understand how to solve all of the assigned problems.

What If I Need to Find Something Quickly?

The **Index** alphabetically lists the subjects covered throughout the entire textbook and the pages on which each subject can be found.

What If I Forget a Formula?

Inside the back cover of your math book is a list of **Formulas and Symbols** that are used in the book.

Need to Cover Your Book?
Inside the back cover are directions for a Foldable that you can use to cover your math book quickly and easily!

Prerequisite Skills

Divisibility Patterns

In $54 \div 6 = 9$, the quotient, 9, is a whole number. So, we say that 54 is **divisible** by 6. You can use the following rules to determine whether a number is divisible by 2, 3, 4, 5, 6, 9, and 10.

A number is divisible by:

- 2 if the ones digit is divisible by 2.
- 3 if the sum of the digits is divisible by 3.
- 4 if the number formed by the last two digits is divisible by 4.
- 5 if the ones digit is 0 or 5.
- 6 if the number is divisible by both 2 and 3.
- 9 if the sum of the digits is divisible by 9.
- 10 if the ones digit is 0.

EXAMPLE Use Divisibility Rules

Determine whether 972 is divisible by 2, 3, 4, 5, 6, 9, or 10.

2: Yes; the ones digit, 2, is divisible by 2.

3: Yes; the sum of the digits, $9 + 7 + 2 = 18$, is divisible by 3.

4: Yes; the number formed by the last two digits, 72, is divisible by 4.

5: No; the ones digit is not 0 or 5.

6: Yes; the number is divisible by 2 and 3.

9: Yes; the sum of the digits, 18, is divisible by 9.

10: No; the ones digit is not 0.

Exercises

Use divisibility rules to determine whether the first number is divisible by the second number.

1. 447; 3
2. 135; 6
3. 240; 4
4. 419; 3
5. 831; 3
6. 4,408; 4
7. 7,110; 5
8. 1,287; 9
9. 2,984; 9
10. 7,026; 6
11. 1,260; 10
12. 8,903; 6

Determine whether each number is divisible by 2, 3, 4, 5, 6, 9, or 10.

13. 712
14. 1,035
15. 8,901
16. 462
17. 270
18. 1,005
19. 32,221
20. 8,340
21. 920
22. 50,319
23. 64,042
24. 3,498

25. **MEASUREMENT** Jordan has 5,280 feet of rope. Can he cut the rope into 9-foot pieces and use all of the rope? Explain.

Place Value and Decimals

Our number system is based on units of ten. Numbers like 1.35, 0.8, and 25.09 are called decimals.

You can use a place-value chart like the one at the right to help you read decimals and write them in words. Use the word *and* to represent the decimal point.

ten thousands	thousands	hundreds	tens	ones	decimal point	tenths	hundredths	thousandths	ten-thousandths
10,000	1,000	100	10	1		0.1	0.01	0.001	0.0001
			2	0	•	0	9		

EXAMPLES Write Decimals in Words

Write each decimal in words.

1 **20.09**

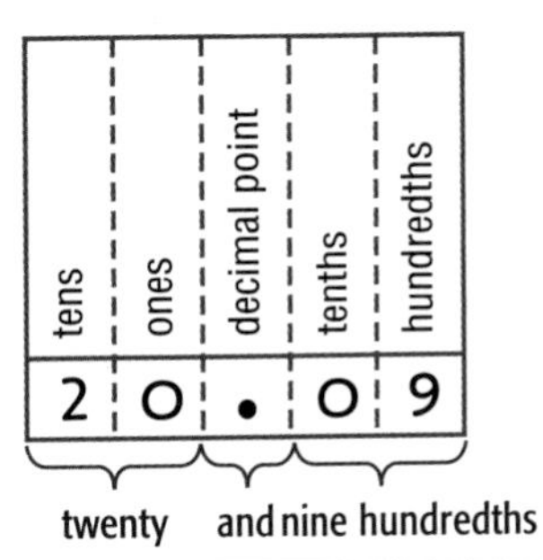

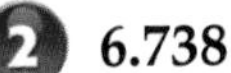

2 **6.738**

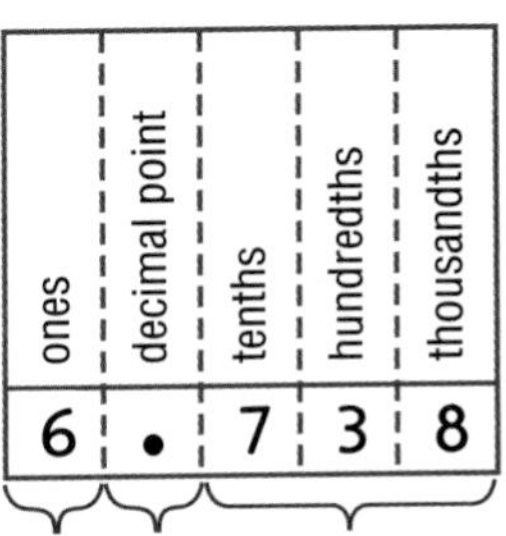

You can also write decimals in **expanded notation** using place value and decimals or their fraction equivalents as shown at the right.

Decimal	0.1	0.01	0.001	0.0001
Fraction	$\frac{1}{10}$	$\frac{1}{10^2}$	$\frac{1}{10^3}$	$\frac{1}{10^4}$

EXAMPLE Write a Decimal in Expanded Notation

3 **Write 2.814 in expanded notation using decimals and using fractions.**

Write the product of each digit and its place value.

$2.814 = (2 \times 1) + (8 \times 0.1) + (1 \times 0.01) + (4 \times 0.001)$

$2.814 = (2 \times 1) + \left(8 \times \frac{1}{10}\right) + \left(1 \times \frac{1}{10^2}\right) + \left(4 \times \frac{1}{10^3}\right)$ Replace the decimals with their fraction equivalents.

Exercises

Write each decimal in words.

1. 6.37 **2.** 13.7 **3.** 145.66 **4.** 92.03 **5.** 4,280.899

6. 0.6 **7.** 0.69 **8.** 0.0001 **9.** 7.005 **10.** 2000.0002

Write each decimal in expanded notation using decimals and using fractions.

11. 0.8 **12.** 5.3 **13.** 6.79 **14.** 9.132 **15.** 65.002

16. 0.625 **17.** 0.0072 **18.** 100.001 **19.** 548.845 **20.** 4,260.705

21. **BATS** The northern blossom bat is one of the world's smallest bats. It weighs just 0.53 ounce. Write its weight in expanded notation using fractions.

Comparing and Ordering Decimals

To determine which of two decimals is greater, you can compare the digits in each place-value position, or you can use a number line.

EXAMPLES Order Decimals

1 **Which is greater, 7.4 or 7.63?**

Method 1 Use place value.

7.4 Line up the decimal points.

7.63 Starting at the left, compare the digits in each place-value position.
↑

The digits in the tenths place are not the same.

Since 6 tenths > 4 tenths, 7.63 > 7.4.

Method 2 Use a number line.

Graph each decimal on a number line and compare.

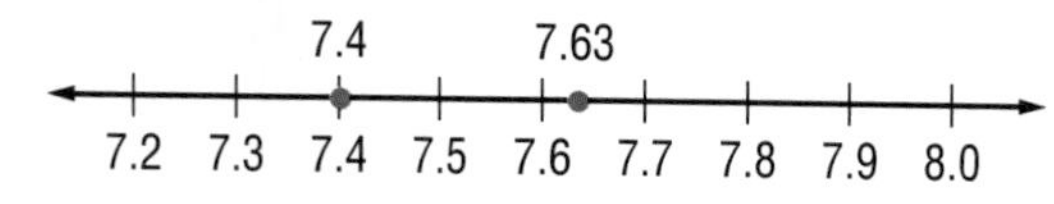

On a number line, numbers to the right are greater than numbers to the left.

7.63 is to the right of 7.4, so 7.63 > 7.4.

2 **Order 12.15, 12.08, and 12.103 from least to greatest.**

12.15 Line up the decimal points.

12.08 12.08 is less than 12.15 and 12.103 since 0 < 1 in the tenths place.

12.103 12.103 is less than 12.15 since 0 < 5 in the hundredths place.

So, the numbers from least to greatest are 12.08, 12.103, and 12.15.

Exercises

Replace each ● with <, >, or = to make a true sentence.

1. 1.22 ● 1.02
2. 0.97 ● 1.06
3. 7.90 ● 7.9
4. 1.3 ● 1.31
5. 4.03 ● 4.01
6. 0.77 ● 0.69
7. 0.8 ● 0.08
8. 0.68 ● 0.680
9. 3.28 ● 3.279
10. 25.23 ● 25.32
11. 77.55 ● 77.65
12. 1.29 ● 1.43
13. 310.36 ● 310.3600
14. 0.0034 ● 0.034
15. 9.09 ● 9
16. 1.67 ● 0.48
17. 738 ● 738.99
18. 1,000.01 ● 1,000

Order each group of decimals from least to greatest.

19. 5.13, 5.07, 5.009
20. 7.9, 7.088, 8.02, 7.98
21. 0.087, 0.901, 2, 1.001
22. 12.3, 12.008, 12.54, 12
23. 60.5, 60.05, 60.55, 60.505
24. 505.9, 505.09, 505.91, 505.99
25. 0.02, 0.2, 0.022, 0.002
26. 0.99, 0.90, 0.09, 0.999, 0.099

27. Place decimal points in 999, 463, 208, and 175 so that the resulting decimals are in order from least to greatest.
28. Order 3, 0.3, 3.33, 0.33, and 3.03 from least to greatest.

Rounding Decimals

When a number has more decimal places than you want or need, you can **round** it. Use the rules below to round a number to any place value.

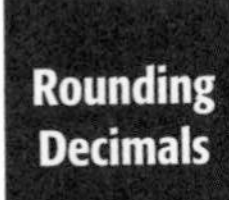

Rounding Decimals

Look at the digit to the right of the place being rounded.

- If the digit is 4 or less, the digit being rounded remains the same.
- If the digit is 5 or greater, the digit being rounded is rounded up.

EXAMPLES Round Decimals

1 Round 3.92 to the nearest tenth.

tenths place
↓
3.92 Look at the digit to the right of the tenths place.

Since 2 is less than 5, the digit in the tenths place stays the same.

3.92 rounded to the nearest tenth is 3.9.

2 Round 46.297 to the nearest hundredth.

hundredths place
↓
46.297 Look at the digit to the right of the hundredths place.

Since 7 is greater than 5, round 9 up.

46.297 rounded to the nearest hundredth is 46.30.

Exercises

Round each number to the given place value.

1. 0.315; tenth
2. 0.2456; hundredth
3. 17.499; tenth
4. 43.219; hundredth
5. 15.522; tenth
6. 9.6; ones
7. 7.0375; thousandth
8. 16.399; tenth
9. 6.95; tenth
10. 99.1283; thousandth
11. 1,000.37; tenth
12. 750.523; ones
13. 0.445; hundredth
14. 490.299; ones
15. 999.99; tenth
16. 476.835; hundredth
17. 682.596; tenth
18. 1,000.562; ones

Round each number to the underlined place-value position.

19. $0.7\underline{8}9$
20. $0.\underline{9}6$
21. $1.5\underline{7}246$
22. $2\underline{3}.48$
23. $1.7\underline{0}4$
24. $0.\underline{1}63$
25. $15.\underline{4}51$
26. $4.52\underline{9}88$
27. $0.\underline{7}87$
28. $3\underline{8}.56$
29. $5\underline{9}.61$
30. $\underline{0}.555$
31. $0.\underline{5}55$
32. $0.5\underline{5}5$
33. $10.\underline{9}29$
34. $10.9\underline{2}9$
35. $42\underline{5}.599$
36. $0.929\underline{2}9$

37. **HORSES** A trainer recorded a racehorse running at 47.54 miles per hour. Round this speed to the nearest mile per hour.

38. **POPULATION** In 2000, the population of Texas was 20.85 million. Round to the nearest million.

Prerequisite Skills

Estimating with Decimals

Estimation can be used to provide quick answers when an exact answer is not necessary. It is also an excellent way to check whether your answer is reasonable. One method of estimating is to use rounding. Round numbers to any place value that makes estimation easier.

EXAMPLES Estimate by Rounding

Estimate by rounding.

1 **23.485 − 9.757**

$$\begin{array}{rcr} 23.485 & \rightarrow & 23 \\ \underline{-9.757} & \rightarrow & \underline{-\ 10} \\ & & 13 \end{array}$$

Round to the nearest whole numbers.

The difference is about 13.

2 **43.9 × 37.5**

$$\begin{array}{rcr} 43.9 & \rightarrow & 40 \\ \underline{\times\ 37.5} & \rightarrow & \underline{\times\ 40} \\ & & 1{,}600 \end{array}$$

Round to the nearest ten.

The product is about 1,600.

3 **6.43 + 2.17 + 9.1 + 4.87**

$$\begin{array}{rcr} 6.43 & \rightarrow & 6 \\ 2.17 & \rightarrow & 2 \\ 9.1 & \rightarrow & 9 \\ \underline{+\ 4.87} & \rightarrow & \underline{+\ 5} \\ & & 22 \end{array}$$

Round to the nearest whole numbers.

The sum is about 22.

4 **432.87 ÷ 8.9**

8.9 → 9 Round the divisor.

432.87 → 450 Round the dividend to a multiple of 9.

432.87 ÷ 8.9 → 450 ÷ 9 = 50

The quotient is about 50.

Another way to estimate sums is to use **clustering**. This strategy is used when all the numbers are close to a common value.

EXAMPLE Estimate by Clustering

5 **Estimate 9.775 + 9.862 + 9.475 + 9.724 by clustering.**

All of the numbers are clustered around 10. There are four numbers.

So, the sum is about 4 × 10 or 40.

Exercises

Estimate by rounding.

1. 8.56 + 5.34
2. 34.84 − 17.69
3. 6.8 × 2.4
4. 40.79 ÷ 6.8
5. 6.9 + 5.2
6. 23.84 + 12.13
7. 34.3 − 18.9
8. 7.5 × 8.4
9. 65.48 ÷ 9.3
10. 26.3 × 9.7
11. 33.21 − 8.23
12. 67.86 − 24.35
13. 8.99 ÷ 2.6
14. 121.5 + 487.8
15. 32.5 × 81.4

Estimate by clustering.

16. 18.4 + 22.5 + 20.7
17. 56.9 + 63.2 + 59.3 + 61.1
18. 42.3 + 41.5 + 39.8 + 40.4
19. 77.8 + 75.6 + 81.2 + 79.9
20. 239.8 + 242.43 + 236.20 + 240.77
21. 9.9 + 10.0 + 10.3 + 11.1 + 9.8 + 11.2
22. 50.4 + 51.1 + 48.9 + 49.5 + 50.8
23. 100.5 + 97.8 + 101.6 + 100.2 + 99.3

Adding and Subtracting Decimals

To add or subtract decimals, write the numbers in a column and line up the decimal points. In some cases, you may want to *annex*, or place zeros at the end of the decimals, to help align the columns. Then add or subtract as with whole numbers and bring down the decimal point. Always estimate first to see whether your answer is reasonable.

EXAMPLES Add and Subtract Decimals

Find each sum or difference.

1 **12.6 + 4.5** **Estimate** 13 + 5 = 18

12.6	Line up the decimal points.
+ 4.5	Add.
17.1	The sum is close to the estimate.

2 **42.17 − 15.85** **Estimate** 42 − 16 = 26

42.17	Line up the decimal points.
− 15.85	Subtract.
26.32	The difference is close to the estimate.

3 **57.125 + 7.63** **Estimate** 57 + 8 = 65

57.125	
+ 7.630	Annex a zero to align the columns.
64.755	Compare to the estimate.

4 **25 − 15.25** **Estimate** 25 − 15 = 10

25.00	Annex two zeros to align the columns.
− 15.25	
9.75	Compare to the estimate.

Exercises

Find each sum or difference.

1. 0.132 − 0.021
2. 3.78 + 0.21
3. 13.2 + 12.8
4. 5.86 − 1.51
5. 42.07 − 38.78
6. 14.7 + 351.82
7. 42.3 + 0.81
8. 12.3 − 0.847
9. 342.9 − 0.18
10. 282.45 − 111.3
11. 100 − 0.48
12. 82.23 + 0.88
13. 128.01 − 39.117
14. 80.05 − 79.06
15. 104.98 − 0.12
16. 0.42 + 0.68
17. 0.48 + 2.901
18. 5.8 + 3.92
19. 38.63 + 38.63
20. 8 − 2.54
21. 16.354 − 0.2
22. 0.125 + 0.78
23. 8.2 − 6.9
24. 1.245 + 3.842
25. 3.2 + 1.23
26. 0.889 − 0.3
27. 22.22 + 1.475
28. 10 − 0.25
29. 33.16 − 0.08
30. 1.254 + 0.5
31. 44.698 − 14.903
32. 10 − 0.005
33. 722.86 + 0.024
34. 100.211 + 8.004
35. 86.124 + 32.822
36. 6.9 + 1.1
37. 75 − 0.24
38. 13 − 0.324
39. 0.8 + 1.2
40. What is the sum of 35.009 and 3.6?
41. Find 9.1 minus 5.625.
42. **AIR TRAVEL** The numbers of passengers, in millions, passing through four airports in a recent year were 80.2, 72.1, 68.5, and 60.7. Find the total number of passengers for the four airports.

Multiplying Decimals

To multiply decimals, multiply as with whole numbers. The product has the same number of decimal places as the sum of the decimal places of the factors. Use estimation to determine whether your answers are reasonable.

EXAMPLES Multiply Decimals

Multiply.

1.3×0.9 **Estimate** $1 \times 1 = 1$

$$\begin{array}{r l} 1.3 & \leftarrow \text{1 decimal place} \\ \underline{\times\, 0.9} & \leftarrow \text{1 decimal place} \\ 1.17 & \leftarrow \text{2 decimal places} \end{array}$$

The product is reasonable.

0.054×1.6 **Estimate** $0 \times 2 = 0$

$$\begin{array}{r l} 0.054 & \leftarrow \text{3 decimal places} \\ \underline{\times\, 1.6} & \leftarrow \text{1 decimal place} \\ 324 & \\ \underline{540} & \\ 0.0864 & \end{array}$$

Annex a zero on the left so the answer has four decimal places. Compare to the estimate.

Exercises

Place the decimal point in each product. Add zeros if necessary.

1. $1.32 \times 4 = 528$
2. $0.07 \times 1.1 = 77$
3. $0.4 \times 0.7 = 28$
4. $1.9 \times 0.6 = 114$
5. $1.4 \times 0.09 = 126$
6. $5.48 \times 3.6 = 19728$
7. $4.5 \times 0.34 = 153$
8. $0.45 \times 0.02 = 9$
9. $150.2 \times 32.75 = 4919050$

Multiply.

10. 0.2×6
11. 0.3×0.9
12. 0.45×0.12
13. 0.0023×32
14. 1.5×2.7
15. 10.1×9
16. 2×0.3
17. 6.78×1.3
18. 200×0.004
19. 0.0023×0.35
20. 15.8×11
21. 88×2.5
22. 33×0.03
23. 36×0.46
24. 0.003×482
25. 1.88×1.11
26. 0.6×2
27. 38.3×29.1
28. 0.7×18
29. 8×0.3
30. 12.2×12.4
31. 380×1.25
32. 42×0.17
33. 0.4×16
34. 0.23×0.2
35. 0.44×0.5
36. 0.44×55
37. 44×0.55

38. **JOBS** Antonia earns \$10.75 per hour. What are her total weekly earnings if she works 34.5 hours? Round to the nearest cent.

Powers of Ten

You can use a pattern to mentally find the product of any number and a power of 10. Count the number of zeros in the power of 10 or use the exponent. Then move the decimal point that number of places *to the right*.

Decimal	Power of 10	Product
19.7 ×	10^1 (or 10)	= 197
19.7 ×	10^2 (or 100)	= 1,970
19.7 ×	10^3 (or 1,000)	= 19,700
19.7 ×	10^4 (or 10,000)	= 197,000

EXAMPLES Use Mental Math to Multiply

Multiply mentally.

1 **12.562 × 100**

$12.562 \times 100 = 12.562$ — Move the decimal point two places to the right, since 100 has two zeros.

$= 1,256.2$

2 **0.59×10^4**

$0.59 \times 10^4 = 0.5900$ — Move the decimal point four places to the right, since the exponent is 4.

$= 5,900$

To mentally multiply by a power of ten that is less than 1, count the number of decimal places. Or, if the power is written as a fraction, use the exponent in the denominator. Then move the decimal point that number of places *to the left*.

Decimal	Power of 10	Product
19.7 ×	0.1 $\left(\text{or } \frac{1}{10^1}\right)$	= 1.97
19.7 ×	0.01 $\left(\text{or } \frac{1}{10^2}\right)$	= 0.197
19.7 ×	0.001 $\left(\text{or } \frac{1}{10^3}\right)$	= 0.0197

EXAMPLES Use Mental Math to Multiply

Multiply mentally.

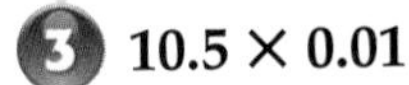

3 **10.5 × 0.01**

$10.5 \times 0.01 = 10.5$ — Move the decimal point two places to the left.

$= 0.105$

4 **5,284 × 0.00001**

$5,284 \times 0.00001 = 05284$ — Move the decimal point five places to the left.

$= 0.05284$

Exercises

Multiply mentally.

1. 12.53×10
2. 4.6×10^3
3. 78.4×0.01
4. 0.05×100
5. 4.527×10^0
6. $2.78 \times 1,000$
7. 13.58×0.01
8. 5.49×10^3
9. 0.1×0.8
10. 0.925×10
11. 99.44×10^2
12. 0.01×16
13. 1.32×10^3
14. $0.56 \times 10,000$
15. 1.4×0.001
16. 11.23×10^5
17. 68.94×0.01
18. 0.8×10^4
19. 28.1×0.01
20. 9.3×10^7
21. $625,799 \times 0.0001$

Prerequisite Skills

Dividing Decimals

To divide two decimals, use the following steps.

- If necessary, change the divisor to a whole number by moving the decimal point to the right. You are multiplying the divisor by a power of ten.
- Move the decimal point in the dividend the same number of places to the right. You are multiplying the dividend by the same power of ten.
- Divide as with whole numbers.

EXAMPLES Divide Decimals

Divide.

1 **25.8 ÷ 2** **Estimate** 26 ÷ 2 = 13

$$\begin{array}{r} 12.9 \\ 2\overline{)25.8} \\ \underline{-\,2} \\ 5 \\ \underline{-\,4} \\ 18 \\ \underline{-\,18} \\ 0 \end{array}$$

The divisor, 2, is already a whole number, so you do not need to move the decimal point.

Divide as with whole numbers. Then place the decimal point directly above the decimal point in the dividend.

Compared to the estimate, the quotient, 12.9, is reasonable.

2 **199.68 ÷ 9.6** **Estimate** 200 ÷ 10 = 20

$$\begin{array}{r} 20.8 \\ 9.6\overline{)199.68} \\ \underline{-\,192} \\ 7\,68 \\ \underline{-\,7\,68} \\ 0 \end{array}$$

Move each decimal point one place to the right.

Compare the answer to the estimate.

Exercises

Divide.

1. $0.3\overline{)9.81}$
2. $12\overline{)0.12}$
3. $3.2\overline{)5.76}$
4. $0.22\overline{)0.0132}$
5. $0.04\overline{)0.008}$
6. $3.18\overline{)0.636}$
7. $0.2\overline{)8.24}$
8. $82.3\overline{)823}$
9. $12.02\overline{)24.04}$
10. $0.5\overline{)85}$
11. $74.9\overline{)5.992}$
12. $19.2\overline{)4.416}$
13. $1.9\overline{)38.57}$
14. $13.8\overline{)131.1}$
15. $6.48\overline{)259.2}$
16. 812 ÷ 0.4
17. 0.34 ÷ 0.2
18. 14.4 ÷ 0.12
19. 90.175 ÷ 2.5
20. 39.95 ÷ 799
21. 88.8 ÷ 444
22. 613.8 ÷ 66
23. 2,445.3 ÷ 33
24. 20.24 ÷ 2.3
25. 45 ÷ 0.09
26. 2.475 ÷ 0.03
27. 4.6848 ÷ 0.366
28. 180 ÷ 0.36
29. 97.812 ÷ 1.1
30. 23 ÷ 0.023
31. 1,680.042 ÷ 44.2
32. **OLYMPICS** In the 2000 Olympics, Michael Johnson of the U.S. ran the 400-meter run in 43.84 seconds. To the nearest hundredth, find his speed in meters per second.
33. **PLANETS** It takes Pluto 247.69 Earth years to revolve once around the Sun. It takes Jupiter 11.86 Earth years to revolve once around the Sun. About how many times longer does it take Pluto than Jupiter to revolve once around the Sun?

Mixed Numbers and Improper Fractions

A **mixed number** is the sum of a whole number and a fraction. An **improper fraction** is a fraction with a numerator that is greater than or equal to the denominator.

mixed number: $8\frac{1}{2}$ improper fraction: $\frac{17}{2}$

To write a mixed number as an improper fraction, use the following steps.

- Multiply the whole number by the denominator.
- Add the numerator.
- Write the sum as the numerator of the improper fraction.

EXAMPLES Write Mixed Numbers as Improper Fractions

Write each mixed number as an improper fraction.

1 $6\frac{1}{2}$

$6\frac{1}{2} = \frac{(6 \cdot 2) + 1}{2}$ ← Multiply 6 by 2 and add 1. ← The denominator is 2.

$= \frac{12 + 1}{2}$ or $\frac{13}{2}$ Simplify.

2 $3\frac{4}{5}$

$3\frac{4}{5} = \frac{(3 \cdot 5) + 4}{5}$ ← Multiply 3 by 5 and add 4. ← The denominator is 5.

$= \frac{15 + 4}{5}$ or $\frac{19}{5}$ Simplify.

To write an improper fraction as a mixed number, divide the numerator by the denominator. Write the remainder as the numerator of the fraction.

EXAMPLE Write Improper Fractions as Mixed Numbers

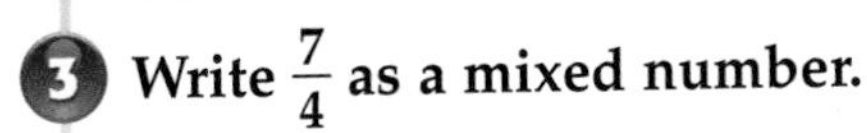

3 **Write $\frac{7}{4}$ as a mixed number.**

$7 \div 4 = 1 \text{ R } 3$ Divide the numerator by the denominator.

$= 1\frac{3}{4}$ Write the remainder as the numerator of the fraction.

So, $\frac{7}{4} = 1\frac{3}{4}$.

Exercises

Write each mixed number as an improper fraction.

1. $2\frac{1}{3}$	2. $1\frac{5}{8}$	3. $4\frac{5}{7}$	4. $3\frac{3}{4}$	5. $9\frac{1}{3}$
6. $1\frac{1}{10}$	7. $10\frac{2}{3}$	8. $12\frac{1}{4}$	9. $6\frac{3}{8}$	10. $11\frac{1}{2}$
11. $2\frac{3}{8}$	12. $20\frac{3}{5}$	13. $5\frac{7}{8}$	14. $27\frac{1}{3}$	15. $5\frac{2}{3}$

Write each improper fraction as a mixed number.

16. $\frac{15}{2}$	17. $\frac{9}{7}$	18. $\frac{15}{8}$	19. $\frac{100}{3}$	20. $\frac{28}{5}$
21. $\frac{3}{3}$	22. $\frac{19}{8}$	23. $\frac{17}{10}$	24. $\frac{51}{3}$	25. $\frac{17}{6}$
26. $\frac{5}{3}$	27. $\frac{99}{3}$	28. $\frac{25}{7}$	29. $\frac{46}{9}$	30. $\frac{27}{4}$

Extra Practice

Extra Practice

Lesson 1-1

(Pages 6–9)

Use the four-step plan to solve each problem.

1. The Reyes family rode their bicycles for 9 miles to the park. The ride back was along a different route for 14 miles. How many miles did they ride in all?
2. A farmer planted 389 acres of land with 78,967 corn plants. How many plants were planted per acre?
3. A group of 251 people is eating dinner at a school fund-raiser. If each person pays $8.00 for their meal, how much money is raised?
4. When Tamika calls home from college, she talks ten minutes per call for 3 calls per week. How many minutes does she call in a 15-week semester?
5. Darren runs at 6 feet per second and Kim runs at 7 feet per second. If they both start a race at the same time, how far apart are they after one minute?

Lesson 1-2

(Pages 10–13)

Write each power as a product of the same factor.

1. 13^4
2. 9^6
3. 1^7
4. 12^2
5. 5^8
6. 15^4

Evaluate each expression.

7. 5^6
8. 17^3
9. 2^{12}
10. 3^5
11. 1^4
12. 5^3
13. 10^2
14. 2^8
15. 8^2
16. 7^4
17. 20^3
18. 42^3

Write each product in exponential form.

19. $2 \cdot 2 \cdot 2 \cdot 2 \cdot 2$
20. $3 \cdot 3$
21. $1 \cdot 1 \cdot 1 \cdot 1 \cdot 1 \cdot 1$
22. $18 \cdot 18 \cdot 18 \cdot 18$
23. $9 \cdot 9 \cdot 9 \cdot 9 \cdot 9 \cdot 9 \cdot 9 \cdot 9$
24. $10 \cdot 10 \cdot 10 \cdot 10 \cdot 10 \cdot 10$

Lesson 1-3

(Pages 14–17)

Evaluate each expression.

1. $14 - (5 + 7)$
2. $(32 + 10) - 5 \times 6$
3. $(50 - 6) + (12 + 4)$
4. $12 - 2 \cdot 3$
5. $16 + 4 \times 5$
6. $(5 + 3) \times 4 - 7$
7. $2 \times 3 + 9 \times 2$
8. $6 \cdot (8 + 4) \div 2$
9. $7 \times 6 - 14$
10. $8 + (12 \times 4) \div 8$
11. $13 - 6 \cdot 2 + 1$
12. $(80 \div 10) \times 8$
13. $14 - 2 \cdot 7 + 0$
14. $156 - 6 \times 0$
15. $30 - 14 \cdot 2 + 8$
16. $3 \times 4 - 3^2$
17. $10^2 - 5$
18. $3 + (10 - 5 + 1)^2$
19. $(4 + 3)^2 \div 7$
20. 8×10^3
21. $10^4 \times 6$
22. 4.5×10^3
23. 1.8×10^2
24. $3 + 5(1.7 + 2.3)$
25. $4(3.6 + 5.4) - 9$
26. $10 + 3(6.1 + 3.7)$
27. $6(7.5 + 2.1) - 2.3$

Lesson 1-4

(Pages 18–21)

Evaluate each expression if $a = 3$, $b = 4$, $c = 12$, and $d = 1$.

1. $a + b$
2. $c - d$
3. $a + b + c$
4. $b - a$
5. $c - ab$
6. $a + 2d$
7. $b + 2c$
8. ab
9. $a + 3b$
10. $6a + c$
11. $\frac{c}{d}$
12. abc
13. $2(a + b)$
14. $\frac{2c}{b}$
15. $144 - abc$
16. $2ab$
17. $\frac{b}{2}$
18. a^2
19. $c^2 - 100$
20. $a^3 + 3$
21. $2b^2$
22. $b^3 + c$
23. $\frac{a^2}{d}$
24. $5a^2 + 2d^2$
25. $\frac{4d^2}{b}$
26. $\frac{15}{a}$
27. $3a^2$
28. $10d^3$
29. $\frac{ab}{c}$
30. $\frac{(a + b)}{d}$
31. $2.5b + c$
32. $\frac{10}{d}$
33. $\frac{(2c + b)}{b}$
34. $\frac{(b^2 + 2d)}{a}$
35. $\frac{(2c + ab)}{c}$
36. $\frac{(3.5c + 2)}{11}$

Lesson 1-5

(Pages 24–27)

Solve each equation mentally.

1. $b + 7 = 12$
2. $a + 3 = 15$
3. $s + 10 = 23$
4. $9 + n = 13$
5. $20 = 24 - n$
6. $4x = 36$
7. $2y = 10$
8. $15 = 5h$
9. $j \div 3 = 2$
10. $14 = w - 4$
11. $24 \div k = 6$
12. $b - 3 = 12$
13. $c \div 10 = 8$
14. $6 = t \div 5$
15. $14 + m = 24$
16. $3y = 39$
17. $\frac{f}{2} = 12$
18. $16 = 4v$
19. $81 = 80 + a$
20. $9 = \frac{72}{x}$
21. $66 = 22m$
22. $77 - 12 = a$
23. $9k = 81$
24. $95 + d = 100$
25. $b = \frac{72}{6}$
26. $z = 15 + 22$
27. $15b = 225$
28. $43 + s = 57$
29. $4w = 52$
30. $e - 10 = 0$
31. $62 - d = 12$
32. $14f = 14$
33. $48 \div n = 8$
34. $a - 82 = 95$
35. $\frac{x}{2} = 36$
36. $99 = c \div 2$

Lesson 1-6

(Pages 30–33)

Use the Distributive Property to write each expression as an equivalent expression. Then evaluate the expression.

1. $3(4 + 5)$
2. $(2 + 8)6$
3. $4(9 - 6)$
4. $8(6 - 3)$
5. $5(200 - 50)$
6. $20(3 + 6)$
7. $(20 - 5)8$
8. $50(8 + 2)$
9. $15(1{,}000 - 200)$
10. $3(2{,}000 + 400)$
11. $12(1{,}000 + 10)$
12. $7(1{,}000 - 50)$

Name the property shown by each statement.

13. $7(2 + 3) = 7(2) + 7(3)$
14. $7 \times 8 = 8 \times 7$
15. $(4 + 2) + 8 = 4 + (2 + 8)$
16. $92 + 3 = 3 + 92$
17. $4(12) + 4(3) = 4(12 + 3)$
18. $z \cdot 1 = z$
19. $(1 \times 5) \times 2 = 1 \times (5 \times 2)$
20. $72 + (8 + 6) = (72 + 8) + 6$
21. $4 \cdot (x \cdot y) = (4 \cdot x) \cdot y$
22. $a + 0 = a$
23. $a \times b = b \times a$
24. $43 + z = z + 43$

Lesson 1-7

(Pages 34–36)

Describe the pattern in each sequence and identify the sequence as *arithmetic, geometric,* or *neither.*

1. 5, 9, 13, 17, …
2. 3, 6, 12, 24, …
3. 10, 15, 25, 40, …
4. 90, 91, 94, 99, …
5. 8, 24, 72, 216, …
6. 4.5, 5.4, 6.3, 7.2, …
7. 0.3, 0.4, 0.5, …
8. 2.3, 3.4, 4.5, 5.6, …
9. 9.1, 8.4, 7.7, 6.0, …

Write the next three terms of each sequence.

10. 3, 11, 19, 27, …
11. 350, 300, 250, 200, …
12. 1,600, 800, 400, 200, …
13. 2, 6, 18, 54, …
14. 10, 17, 24, 31, …
15. 0, 7, 14, 21, …
16. 1, 5, 25, 125, …
17. 113, 107, 101, 95, …
18. 9, 81, 729, 6,561, …
19. 8,748, 2,916, 972, 324, …
20. 400, 1,000, 2,500, 6,250, …
21. 9.9, 8.8, 7.7, 6.6, …
22. 24, 22.5, 21, 19.5, …
23. 4,000, 2,000, 1,000, 500, …
24. 0.1, 1, 10, 100, …

Lesson 1-8

(Pages 38–41)

Complete.

1. 400 mm = __?__ cm
2. 4 km = __?__ m
3. 660 cm = __?__ m
4. 0.3 km = __?__ m
5. 30 mm = __?__ cm
6. 84.5 m = __?__ km
7. __?__ m = 54 cm
8. 18 km = __?__ cm
9. __?__ mm = 45 cm
10. 4 kg = __?__ g
11. 632 mg = __?__ g
12. 4,497 g = __?__ kg
13. __?__ mg = 21 g
14. 61.2 mg = __?__ g
15. 61 g = __?__ mg
16. __?__ mg = 0.51 kg
17. 0.63 kg = __?__ g
18. __?__ kg = 563 g
19. 662 m = __?__ km
20. 5,283 mL = __?__ L
21. 0.24 cm = __?__ mm
22. 380 kL = __?__ L
23. 10.8 g = __?__ mg
24. 83,000 mL = __?__ L
25. 17.8 m = __?__ cm
26. 2 km = __?__ m
27. 0.75 m = __?__ mm
28. 125 L = __?__ kL
29. 3 km = __?__ cm
30. 99 g = __?__ mg

Lesson 1-9

(Pages 43–45)

Write each number in standard form.

1. 7.6×10^2
2. 2.6×10^3
3. 4.7×10^5
4. 1.0×10^2
5. 6.4×10^6
6. 8.3×10^4
7. 2.76×10^3
8. 6.17×10^5
9. 8.70×10^5
10. 1.28×10^7
11. 4.32×10^9
12. 5.01×10^2
13. 1.89×10^6
14. 7.43×10^4
15. 3.03×10^9

Write each number in scientific notation.

16. 720
17. 7,560
18. 892
19. 1,400
20. 91,200
21. 51,000
22. 145,000
23. 90,100
24. 123,000,000,000
25. 4,500
26. 1,700,000
27. 10,000,000
28. 820,000,000
29. 680,000
30. 52,000
31. 60,600,000
32. 3,970,000
33. 574,000,000

Lesson 2-1

(Pages 54–57)

Make a frequency table of each set of data.

1.

Length of Time Walking (min)			
15	30	15	45
45	30	30	60
30	60	15	30
45	45	60	15

2.

Number of Raisins Eaten in Preschool			
40	49	45	49
42	41	45	41
45	41	41	40
41	43	40	41

3.

Favorite Type of TV Show				
M	S	N	S	M
M	S	S	S	S
G	M	N	S	M
S	G	G	G	G
N	S	M	G	G
S	S	N	N	N

M = movie, S = sitcom, N = news show, G = game show

4.

What Type of Transportation Did You Use on Your Last Family Vacation?					
C	A	C	T	C	A
C	C	C	C	S	A
B	B	C	T	A	A
T	S	A	C	A	A
A	A	A	A	A	S

C = car, T = train, S = ship, A = airplane, B = bus

Lesson 2-2

(Pages 60–63)

1. Rachel's quiz scores in science have been steadily going up since her parents hired a tutor for her. If the trend continues, predict Rachel's score on the next quiz.

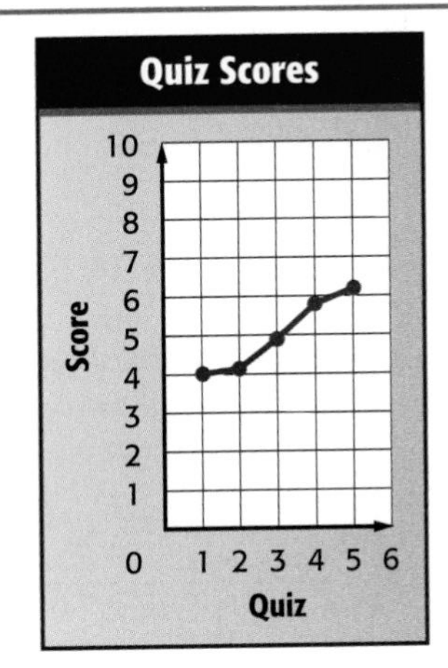

2. The table shows the average price paid to farmers per 100 pounds of sheep they sold.

Year	Price Per 100 Pounds ($)
1940	4
1950	12
1960	6
1970	8
1980	21
1990	23
2000	34

Source: *The World Almanac and Book of Facts*

a. Make a scatter plot of the data. Use the year on the horizontal axis and the price on the vertical axis.

b. Describe the relationship, if any, between the two sets of data.

c. Predict the price per 100 pounds for 2010. Explain.

Lesson 2-3

(Pages 64–68)

Make a line plot for each set of data. Identify any clusters, gaps, or outliers.

1.

Number of Pets in the Home				
0	1	3	4	0
2	1	0	1	1
10	0	1	5	2

2.

High Temperatures for 18 Days (°F)					
75	81	75	65	76	81
77	80	65	65	80	80
76	85	66	75	80	75

3.

Number of Stories for Buildings in Denver				
56	43	36	42	29
54	42	32	34	
52	40	32	32	

Source: *The World Almanac and Book of Facts*

4.

Ages of Children at Sunny Day Care (years)					
4	1	6	4	5	3
4	5	1	2	5	4
3	2	4	1	3	3

Extra Practice

Lesson 2-4

(Pages 69–72)

Find the mean, median, and mode for each set of data.

1. 1, 5, 9, 1, 2, 6, 8, 2
2. 2, 5, 8, 9, 7, 6, 3, 5, 1, 4
3. 1, 2, 1, 2, 2, 1, 2, 1
4. 12, 13, 15, 12, 12, 11
5. 82, 79, 93, 91, 95, 95, 81
6. 117, 103, 108, 120
7. 256, 265, 247, 256
8. 957, 562, 462, 847, 721
9. 47, 54, 66, 54, 46, 66
10. 81, 82, 83, 84, 85, 86, 87
11. 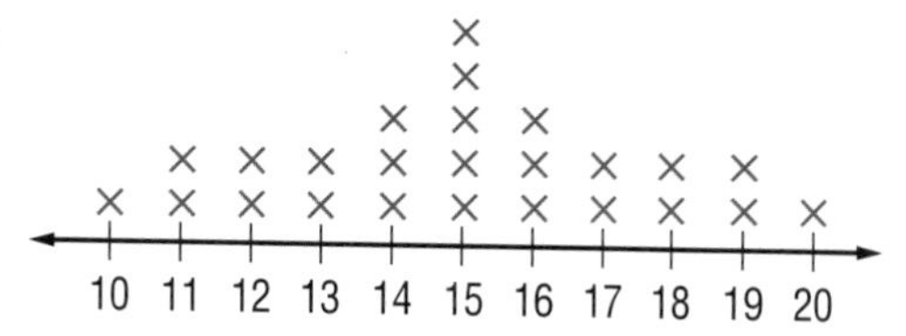

12.

Number of Absences	Tally	Frequency
0	IIII	4
1	~~IIII~~ IIII	9
2	~~IIII~~ I	6
3	~~IIII~~	5

Lesson 2-5

(Pages 76–79)

Make a stem-and-leaf plot for each set of data.

1. 23, 15, 39, 68, 57, 42, 51, 52, 41, 18, 29
2. 5, 14, 39, 28, 14, 6, 7, 18, 13, 28, 9, 14
3. 189, 182, 196, 184, 197, 183, 196, 194, 184
4. 71, 82, 84, 95, 76, 92, 83, 74, 81, 75, 96
5.

Average Monthly High Temperatures in Albany, NY (°F)			
21	46	72	50
24	58	70	40
34	67	61	27

Source: *The World Almanac and Book of Facts*

6.

Super Bowl Winning Scores 1985–2002					
38	42	20	30	35	23
46	20	37	49	31	34
39	55	52	27	34	20

Source: *The World Almanac and Book of Facts*

Lesson 2-6

(Pages 80–83)

For each set of data:

a. **Find the lower extreme, LQ, median, UQ, and upper extreme.**

b. **Draw a box-and-whisker plot of the data.**

1.

Price of Chicken Sandwiches at Various Restaurants			
\$1.00	\$3.70	\$3.50	\$3.90
\$2.50	\$1.70	\$2.20	\$2.10
\$1.20	\$1.90	\$4.20	

2.

Lengths of Principal Rivers in South America (mi)					
4,000	1,750	1,400	2,485	1,000	910
1,100	2,013	1,600	1,000	1,988	1,000
808	956	1,584	2,100	1,677	1,300

Source: *The World Almanac and Book of Facts*

3.

Height of World's Tallest Free-Standing Towers (ft)			
1,831	1,772	1,535	1,379
1,214	1,230	1,411	
1,815	1,362	1,369	

Source: *The World Almanac and Book of Facts*

4.

Mathematics Test Scores in Ms. Ruff's Class					
65	100	84	80	68	69
92	69	75	80	99	90
78	87	100	72	85	

Lesson 2-7

(Pages 85–89)

For each set of data, make a bar graph.

1.

Longest Snakes	
Snake Name	Length (ft)
Royal python	35
Anaconda	28
Indian python	25
Diamond python	21
King cobra	19
Boa constrictor	16

Source: *The Top 10 of Everything*

2.

Least Densely Populated States	
State	People per Square Mile
Alaska	1
Wyoming	5
Montana	6
North Dakota	9
South Dakota	10
New Mexico	15

Source: *The Top 10 of Everything*

For each set of data, make a histogram.

3.

Cost of a Movie Ticket at Selected Theaters			
\$5.25	\$6.50	\$3.50	\$3.75
\$7.50	\$9.25	\$10.40	\$4.75
\$10.00	\$4.50	\$8.75	\$7.25
\$3.50	\$6.75	\$4.20	\$7.50

4.

Highest Recorded Wind Speeds for Selected U.S. Cities (mph)					
52	55	81	46	73	57
75	54	58	76	46	58
60	91	53	53	51	56
80	60	73	46	49	47

Source: *The World Almanac and Book of Facts*

Lesson 2-8

(Pages 92–95)

Which graph could be misleading? Why?

1. Both graphs show pounds of grapes sold to Westview School in one week.

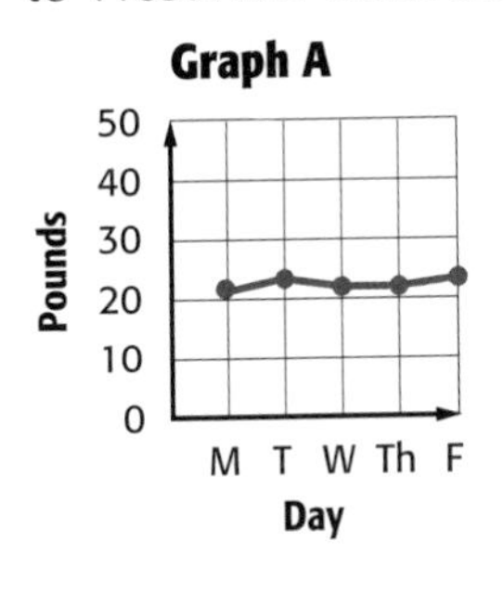

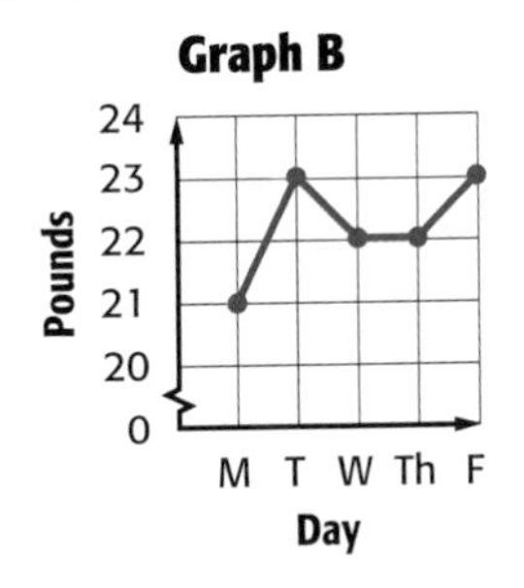

2. Both graphs show commissions made by Mr. Turner for a four-week pay period.

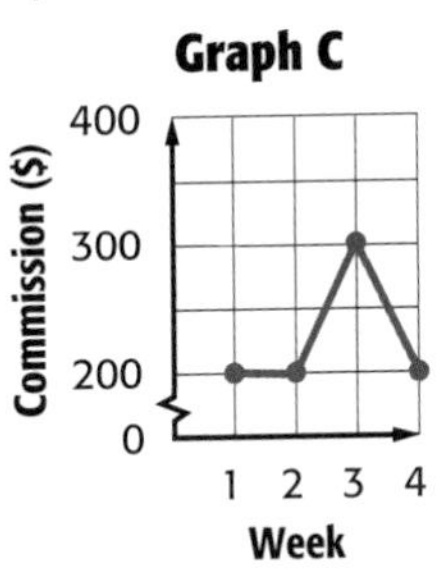

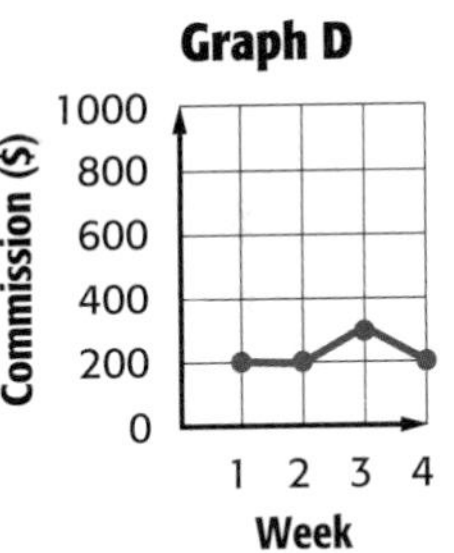

Lesson 3-1

(Pages 106–108)

Write an integer for each situation.

1. seven degrees below zero
2. a loss of 3 pounds
3. a loss of 20 yards
4. a profit of \$25
5. 112°F above 0
6. 2,830 feet above sea level

Evaluate each expression.

7. $|1|$
8. $|-8|$
9. $|0|$
10. $|-82|$
11. $|64|$
12. $|-128|$
13. $|-22| + 5$
14. $|-40| - 8$
15. $|-18| + |10|$
16. $|-7| + |-1|$
17. $|98| - |-5|$
18. $|-49| - |-10|$

Graph each set of integers on a number line.

19. $\{-2, 0, 2\}$
20. $\{1, 3, 5\}$
21. $\{-2, -5, 3\}$
22. $\{7, -1, 4\}$

Lesson 3-2

(Pages 109–111)

Replace each ● with < or > to make a true sentence.

1. 7 ● -7
2. -8 ● 4
3. -4 ● -9
4. -3 ● 0
5. 8 ● 10
6. -5 ● -4
7. 6 ● -7
8. -12 ● -13
9. 3 ● 1
10. -2 ● 2
11. 7 ● -1
12. -15 ● -20
13. -40 ● 30
14. 0 ● -3
15. -5 ● 0
16. 85 ● -17

Order the integers from least to greatest.

17. $-2, -8, 4, 10, -6, -12$
18. $19, -19, -21, 32, -14, 18$
19. $18, 23, 95, -95, -18, -23, 2$
20. $46, -48, -47, -52, -18, 12$
21. $0, -10, -6, -8, 12$
22. $-15, 18, -1, 0, 14, -20$

Lesson 3-3

(Pages 112–115)

Name the ordered pair for each point graphed at the right. Then identify the quadrant in which each point lies.

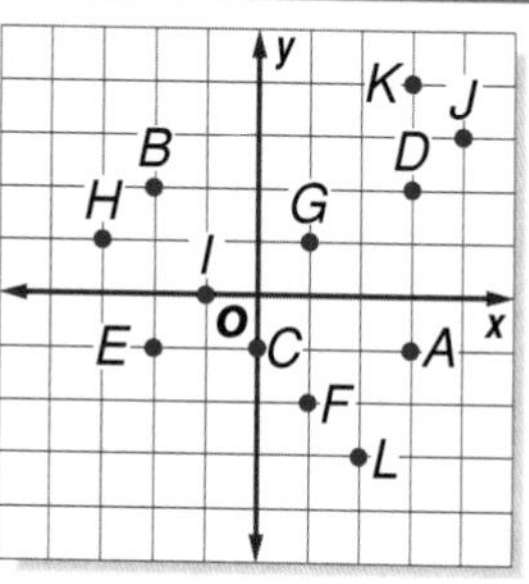

1. A
2. B
3. C
4. D
5. E
6. F
7. G
8. H
9. I
10. J
11. K
12. L

On graph paper, draw a coordinate plane. Then graph and label each point.

13. $N(-4, 3)$
14. $K(2, 5)$
15. $W(-6, -2)$
16. $X(5, 0)$
17. $Y(4, -4)$
18. $M(0, -3)$
19. $Z(-2, 0.5)$
20. $S(-1, -3)$
21. $A(0, 2)$
22. $C(-2, -2)$
23. $E(0, 1)$
24. $G(1, -1)$

Lesson 3-4

(Pages 120–124)

Add.

1. $-4 + 8$
2. $14 + 16$
3. $-7 + (-7)$
4. $-9 + (-6)$
5. $-18 + 11$
6. $-36 + 40$
7. $42 + (-18)$
8. $-42 + 29$
9. $18 + (-32)$
10. $12 + (-9)$
11. $-24 + 9$
12. $-7 + (-1)$

Evaluate each expression if $a = 6$, $b = -2$, $c = -6$, and $d = 3$.

13. $-96 + a$
14. $b + (-5)$
15. $c + (-32)$
16. $d + 98$
17. $-120 + b$
18. $-120 + c$
19. $5 + b$
20. $a + d$
21. $c + a$
22. $d + (-9)$
23. $b + c$
24. $d + c$

Simplify.

25. $x + 3 + (-8)$
26. $-7 + b + 3$
27. $9 + (-7) + f$
28. $10 + y + (-6)$
29. $m + (-11) + 11$
30. $-14 + a + (-19)$

Lesson 3-5

(Pages 128–131)

Subtract.

1. $3 - 7$
2. $-5 - 4$
3. $-6 - 2$
4. $8 - 13$
5. $6 - (-4)$
6. $12 - 9$
7. $-2 - 23$
8. $63 - 78$
9. $0 - (-14)$
10. $15 - 6$
11. $18 - 20$
12. $-5 - 8$
13. $21 - (-37)$
14. $-60 - 32$
15. $57 - 63$

Evaluate each expression if $k = -3$, $p = 6$, $n = 1$, and $d = -8$.

16. $55 - k$
17. $p - 7$
18. $d - 15$
19. $n - 12$
20. $-51 - d$
21. $k - 21$
22. $n - k$
23. $-99 - k$
24. $p - k$
25. $d - (-1)$
26. $k - d$
27. $n - d$

Lesson 3-6

(Pages 134–137)

Multiply.

1. $5(-2)$
2. $6(-4)$
3. $4(21)$
4. $-11(-5)$
5. $-6(5)$
6. $-50(0)$
7. $-5(-5)$
8. $-4(8)$
9. $3(-13)$
10. $12(-5)$
11. $-9(-12)$
12. $15(-8)$

Evaluate each expression if $a = -5$, $b = 2$, $c = -3$, and $d = 4$.

13. $-2d$
14. $6a$
15. $3ab$
16. $-12d$
17. $-4b^2$
18. $-5cd$
19. a^2
20. $13ab$

Simplify each expression.

21. $6(7b)$
22. $-8(2a)$
23. $-3(-5n)$
24. $5(-6x)$
25. $-9(8c)$
26. $14(-2m)$
27. $25(3d)$
28. $-10(17z)$

Lesson 3-7

(Pages 138–141)

Divide.

1. $4 \div (-2)$
2. $16 \div (-8)$
3. $-14 \div (-2)$
4. $\frac{32}{8}$
5. $18 \div (-3)$
6. $-18 \div 3$
7. $8 \div (-8)$
8. $0 \div (-1)$
9. $-25 \div 5$
10. $\frac{-14}{-7}$
11. $-32 \div 8$
12. $-56 \div (-8)$
13. $-81 \div 9$
14. $-42 \div (-7)$
15. $121 \div (-11)$
16. $-81 \div (-9)$
17. $18 \div (-2)$
18. $\frac{-55}{11}$
19. $\frac{25}{-5}$
20. $-21 \div 3$

Evaluate each expression if $a = -2$, $b = -7$, $x = 8$, and $y = -4$.

21. $-64 \div x$
22. $\frac{16}{y}$
23. $x \div 2$
24. $\frac{a}{2}$
25. $ax \div y$
26. $\frac{bx}{y}$
27. $2y \div 1$
28. $\frac{x}{ay}$
29. $-y \div a$
30. $x^2 \div y$
31. $\frac{ab}{1}$
32. $\frac{xy}{a}$

Lesson 4-1

(Pages 150–152)

Write each phrase as an algebraic expression.

1. six less than p
2. twenty more than c
3. the quotient of a and b
4. Juana's age plus 6
5. x increased by twelve
6. \$1,000 divided by z
7. 3 divided into y
8. the product of 7 and m
9. the difference of f and 9
10. twenty-six less q
11. 19 decreased by z
12. two less than x
13. thirty-five increased by a
14. four hundred divided into n
15. Joe's height minus nineteen
16. twice as many lemons
17. eight more than x
18. the product of m and n

Write each sentence as an algebraic equation.

19. Three times a number less four is 17.
20. The sum of a number and 6 is 5.
21. Twenty more than twice a number is -30.
22. The quotient of a number and -2 is -42.
23. Four plus three times a number is 18.
24. Five times a number minus 15 is 92.
25. Eight times a number plus twelve is 36.
26. The difference of a number and 24 is -30.
27. 42 more than five times a number is 156.
28. Thirteen and a number is 62.
29. The quotient of a number and -4 is -324.
30. The sum of a number and eleven is 41.

Lesson 4-2

(Pages 156–159)

Solve each equation. Check your solution.

1. $r - 3 = 14$
2. $t + 3 = 21$
3. $s + 10 = 23$
4. $7 + a = -10$
5. $14 + m = 24$
6. $-9 + n = 13$
7. $s - 2 = -6$
8. $6 + f = 71$
9. $x + 27 = 30$
10. $a - 7 = 23$
11. $-4 + b = -5$
12. $w + 18 = -4$
13. $k - 9 = -3$
14. $j + 12 = 11$
15. $-42 + v = -42$
16. $s + 1.3 = 18$
17. $t + 3.43 = 7.4$
18. $x + 7.4 = 23.5$
19. $p + 3.1 = 18$
20. $q - 2.17 = 21$
21. $w - 3.7 = 4.63$
22. $m - 4.8 = 7.4$
23. $x - 1.3 = 12$
24. $y + 3.4 = 18$
25. $0.013 + h = 4.0$
26. $7.2 + g = 9.1$
27. $z - 12.1 = 14$
28. $v - 18 = 13.7$
29. $w - 0.1 = 0.32$
30. $r + 6.7 = 1.2$

Lesson 4-3

(Pages 160–163)

Solve each equation. Check your solution.

1. $2m = 18$
2. $-42 = 6n$
3. $72 = 8k$
4. $-20r = 20$
5. $420 = 5s$
6. $325 = 25t$
7. $-14 = -2p$
8. $18q = 36$
9. $40 = 10a$
10. $100 = 20b$
11. $416 = 4c$
12. $45 = 9d$
13. $0.5m = 3.5$
14. $1.8 = 0.6x$
15. $0.4y = 2$
16. $1.86 = 6.2z$
17. $-8x = 24$
18. $8.34 = 2r$
19. $1.67t = 10.02$
20. $243 = 27a$
21. $0.9x = 4.5$
22. $4.08 = 1.2y$
23. $8d = 112$
24. $5f = 180.5$
25. $59.66 = 3.14m$
26. $98.4 = 8p$
27. $208 = 26k$

Lesson 4-4

(Pages 166–169)

Solve each equation. Check your solution.

1. $3x + 6 = 6$
2. $2r - 7 = -1$
3. $-10 + 2d = 8$
4. $2b + 4 = -8$
5. $5w - 12 = 3$
6. $5t - 4 = 6$
7. $2q - 6 = 4$
8. $2g - 3 = -9$
9. $15 = 6y + 3$
10. $3s - 4 = 8$
11. $18 - 7f = 4$
12. $13 + 3p = 7$
13. $7.5r + 2 = -28$
14. $4.2 + 7z = 2.8$
15. $-9m - 9 = 9$
16. $32 + 0.2c = 1$
17. $5t - 14 = -14$
18. $-0.25x + 0.5 = 4$
19. $5w - 4 = 8$
20. $4d - 3 = 9$
21. $2g - 16 = -9$
22. $4k + 13 = 20$
23. $7 = 5 - 2x$
24. $8z + 15 = -1$
25. $92 - 16b = 12$
26. $14e + 14 = 28$
27. $1.1j + 2 = 7.5$

Lesson 4-5

(Pages 172–175)

Solve each inequality. Graph the solution on a number line.

1. $x + 2 > -3$
2. $x + 2.9 \leq 9.1$
3. $8t \geq 24$
4. $v - 3 < -3$
5. $6y \geq -12$
6. $a + 3 \leq -2$
7. $k - 5 < -2$
8. $q - 3 \leq 14$
9. $c - 4 \leq -2$
10. $n + 2 > -5$
11. $j + 1.2 > 4.8$
12. $4x < 40$
13. $2y \leq 10$
14. $g + 8 < 10$
15. $2 + b < 4$

Write an inequality for each sentence. Then solve the inequality.

16. The product of a number and four is at least 32.
17. The sum of a number and -6 is at most 27.
18. Six times a number is greater than 48.
19. The difference between a number and 7 is less than 18.

Lesson 4-6

(Pages 177–181)

Copy and complete each function table. Identify the domain and range.

1.

x	$2x$	y
2		
1		
0		
−1		

2.

x	$3x + 1$	y
1		
0		
−1		
−2		

Graph each equation.

3. $y = 3x$
4. $y = 2x + 3$
5. $y = -x$
6. $y = 4x + 2$
7. $y = 0.5x + 2$
8. $y = -x + 3$
9. $y = 0.25x + 6$
10. $y = -3x + 6$
11. $y = 2x + 7$
12. $y = -5x + 1$
13. $y = 13 + x$
14. $y = 5 - 0.5x$

Make a function table for each sentence. Then write an equation using x to represent the first number and y to represent the second number.

15. The second number is four less than the first number.
16. The second number is twelve times the first number.
17. The second number is seven more than the first number.

Lesson 4-7

(Pages 182–185)

Find the slope of the line that passes through each pair of points.

1.

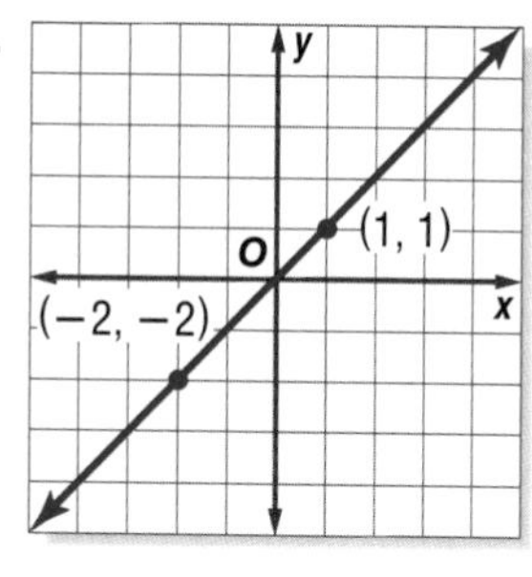

2.

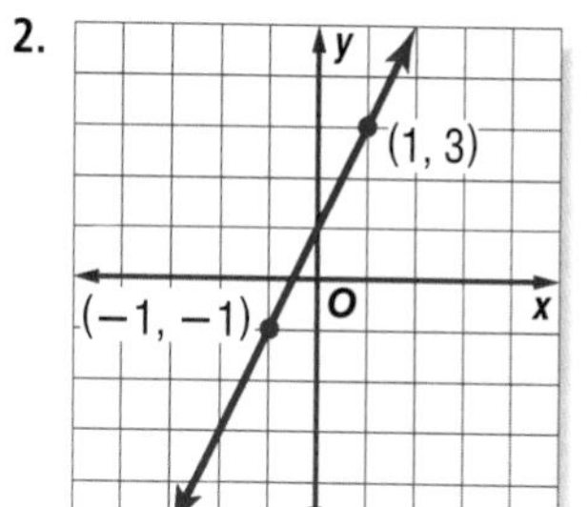

3.

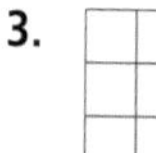
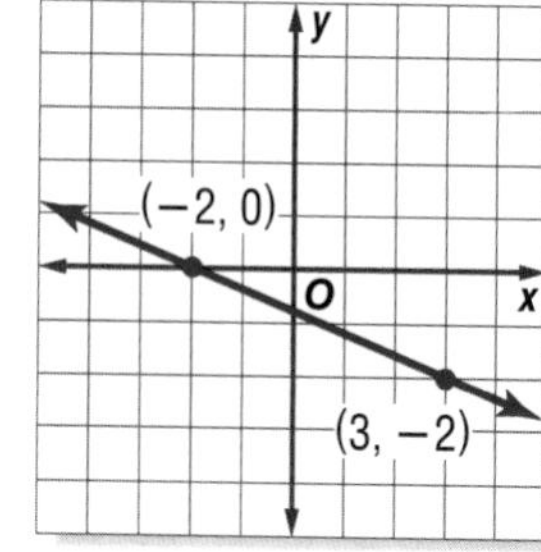

4. (3, 4), (−2, 0)
5. (9, −8), (−4, 1)
6. (−2, −3), (−3, −6)
7. (0, 0), (−2, −5)
8. (5, 2), (−1, 3)
9. (1, 5), (−2, 4)
10. (7, 2), (−4, 3)
11. (2, −6), (9, −5)
12. (4, −1), (0, −5)

Lesson 5-1

(Pages 197–200)

Determine whether each number is *prime* or *composite*.

1. 32
2. 41
3. 52
4. 21
5. 71
6. 102
7. 239
8. 93
9. 123
10. 251
11. 321
12. 423

Find the prime factorization of each number.

13. 81
14. 525
15. 245
16. 1,120
17. 750
18. 2,400
19. 914
20. 975
21. 423
22. 972
23. 144
24. 72

Write each expression as a product of its factors.

25. $35xy$
26. $14a^2$
27. $19bc$
28. $30n$
29. $27cd^2$
30. $37c^2$
31. $4s^2t^2$
32. $72abc$
33. $60p^2qr$

Lesson 5-2

(Pages 203–206)

Find the GCF of each set of numbers.

1. 12, 16
2. 63, 81
3. 225, 500
4. 37, 100
5. 32, 240
6. 412, 640
7. 36, 81
8. 140, 350
9. 72, 170
10. 51, 255
11. 48, 72
12. 86, 200
13. 12, 18, 42
14. 24, 56, 120
15. 48, 60, 84
16. 32, 80, 96
17. 14, 49, 70
18. 8, 10, 20

Find the GCF of each set of algebraic expressions.

19. $18b$, $24b$
20. $2a$, $3a$
21. $5n$, $5mn$
22. $12cd$, $24c$
23. $30x$, $50x^2$
24. $15az$, $25az$
25. $2c$, $4ac$, $8a$
26. d, $6c^2d$, $12d$
27. $10ab$, $15bc$, $20b^2$

Lesson 5-3

(Pages 207–209)

Write each fraction in simplest form.

1. $\frac{14}{28}$
2. $\frac{15}{25}$
3. $\frac{100}{300}$
4. $\frac{14}{35}$
5. $\frac{9}{51}$
6. $\frac{54}{56}$
7. $\frac{75}{90}$
8. $\frac{24}{40}$
9. $\frac{180}{270}$
10. $\frac{312}{390}$
11. $\frac{240}{448}$
12. $\frac{71}{82}$
13. $\frac{333}{900}$
14. $\frac{85}{255}$
15. $\frac{84}{128}$
16. $\frac{64}{96}$
17. $\frac{99}{99}$
18. $\frac{3}{99}$
19. $\frac{44}{55}$
20. $\frac{57}{69}$
21. $\frac{15}{37}$
22. $\frac{204}{408}$
23. $\frac{5}{125}$
24. $\frac{144}{216}$
25. $\frac{15}{75}$

Write two fractions that are equivalent to each fraction.

26. $\frac{2}{3}$
27. $\frac{1}{5}$
28. $\frac{2}{7}$
29. $\frac{3}{10}$
30. $\frac{7}{8}$
31. $\frac{5}{12}$
32. $\frac{3}{8}$
33. $\frac{8}{9}$
34. $\frac{1}{4}$
35. $\frac{2}{11}$

Lesson 5-4

(Pages 210–213)

Write each repeating decimal using bar notation.

1. 0.3333333...
2. 0.121212...
3. 4.3151515...
4. 7.023023023...
5. 0.544444...
6. 18.75484848...

Write each fraction or mixed number as a decimal. Use bar notation if the decimal is a repeating decimal.

7. $\frac{16}{20}$
8. $\frac{25}{100}$
9. $1\frac{7}{8}$
10. $\frac{1}{6}$
11. $\frac{11}{40}$
12. $5\frac{13}{50}$
13. $\frac{55}{300}$
14. $1\frac{1}{2}$
15. $\frac{5}{9}$
16. $2\frac{3}{4}$
17. $\frac{9}{11}$
18. $4\frac{1}{9}$

Write each decimal as a fraction in simplest form.

19. 0.26
20. 0.75
21. 0.4
22. 0.1
23. 4.48
24. 9.8
25. 0.91
26. 11.15

Lesson 5-5

(Pages 216–218)

Write each ratio as a percent.

1. 39 out of 100
2. $\frac{23}{100}$
3. 17 : 100
4. 72 per 100
5. 4 to 100
6. 98 in 100

Write each fraction as a percent.

7. $\frac{1}{2}$
8. $\frac{2}{5}$
9. $\frac{60}{100}$
10. $\frac{17}{20}$
11. $\frac{7}{25}$
12. $\frac{1}{20}$
13. $\frac{8}{100}$
14. $\frac{7}{7}$
15. $\frac{9}{10}$
16. $\frac{1}{100}$
17. $\frac{50}{50}$
18. $\frac{49}{50}$

Write each percent as a fraction in simplest form.

19. 12%
20. 23%
21. 1%
22. 94%
23. 36%
24. 4%
25. 72%
26. 100%
27. 65%
28. 47%
29. 15%
30. 48%

Lesson 5-6

(Pages 220–223)

Write each percent as a decimal.

1. 42%
2. 100%
3. 8%
4. 20%
5. 35%
6. 3%
7. 62%
8. 50%
9. 28%
10. 87%
11. 7.5%
12. 87.5%
13. 1.8%
14. 99.9%
15. $85\frac{1}{4}\%$
16. $24\frac{1}{2}\%$
17. $64\frac{4}{5}\%$
18. $36\frac{3}{4}\%$
19. $1\frac{1}{5}\%$
20. $2\frac{1}{2}\%$

Write each decimal as a percent.

21. 0.16
22. 0.1
23. 0.5
24. 0.98
25. 0.31
26. 0.76
27. 0.07
28. 0.8
29. 0.07
30. 0.10
31. 0.90
32. 1.00
33. 0.666
34. 0.725
35. 0.138
36. 0.899
37. 0.256
38. 0.038
39. 0.0525
40. 0.017

Lesson 5-7

(Pages 224–226)

Find the LCM of each set of numbers.

1. 4, 9
2. 6, 16
3. 24, 36
4. 48, 84
5. 8, 9
6. 49, 56
7. 42, 66
8. 15, 39
9. 56, 64
10. 24, 42
11. 80, 250
12. 16, 24
13. 13, 14
14. 36, 48
15. 10, 100
16. 25, 200
17. 1, 2, 5
18. 2, 3, 7
19. 1, 9, 27
20. 2, 24, 36
21. 7, 21, 35
22. 12, 18, 28
23. 32, 80, 96
24. 5, 18, 45
25. 11, 22, 33
26. 35, 70, 140
27. 25, 200, 400
28. 100, 200, 300

Lesson 5-8

(Pages 227–231)

Find the LCD for each pair of fractions.

1. $\frac{3}{8}, \frac{2}{3}$
2. $\frac{5}{9}, \frac{7}{12}$
3. $\frac{4}{9}, \frac{8}{15}$
4. $\frac{11}{24}, \frac{17}{42}$
5. $\frac{12}{36}, \frac{15}{42}$
6. $\frac{25}{27}, \frac{43}{81}$
7. $\frac{32}{64}, \frac{15}{48}$
8. $\frac{2}{6}, \frac{14}{15}$

Replace each ● with <, >, or = to make a true sentence.

9. $\frac{7}{9}$ ● $\frac{3}{5}$
10. $\frac{14}{25}$ ● $\frac{3}{4}$
11. $\frac{8}{24}$ ● $\frac{20}{60}$
12. $\frac{5}{12}$ ● $\frac{4}{9}$
13. $\frac{18}{24}$ ● $\frac{10}{18}$
14. $\frac{4}{6}$ ● $\frac{5}{9}$
15. $\frac{11}{49}$ ● $\frac{12}{42}$
16. $\frac{5}{14}$ ● $\frac{2}{6}$

Order each set of ratios from least to greatest.

17. 70%, 0.6, $\frac{2}{3}$
18. 0.8, $\frac{17}{20}$, 17%
19. $\frac{61}{100}$, 0.65, 61.5%
20. $0.\overline{42}$, $\frac{3}{7}$, 42%
21. 0.15, 10.5%, $\frac{7}{50}$
22. $\frac{1}{8}$, 0.81, 18%

Determine whether each number is rational. Write *yes* or *no*.

23. $4.\overline{28}$
24. $-\frac{5}{4}$
25. $\frac{0.5}{4}$
26. $\frac{1}{7}$
27. 0.414114111…
28. 10.987…
29. 7.22…
30. $3\frac{2}{5}$

Lesson 6-1

(Pages 240–243)

Estimate.

1. $\frac{3}{7} + \frac{6}{8}$
2. $\frac{3}{9} + \frac{7}{8}$
3. $\frac{1}{8} + \frac{8}{9}$
4. $3\frac{1}{8} + 7\frac{6}{7}$
5. $4\frac{2}{3} + 6\frac{7}{8}$
6. $3\frac{2}{3} \times 2\frac{1}{3}$
7. $\frac{4}{5} \cdot 3$
8. $9\frac{7}{8} - 6\frac{2}{3}$
9. $\frac{3}{7} - \frac{1}{15}$
10. $\frac{3}{4} \cdot \frac{7}{8}$
11. $7\frac{1}{4} \div \frac{2}{3}$
12. $\frac{5}{6} \div \frac{2}{3}$
13. $9\frac{3}{5} + 3\frac{1}{8}$
14. $5\frac{1}{3} - 2\frac{3}{4}$
15. $13\frac{7}{8} - 2\frac{1}{3}$
16. $\frac{13}{15} \cdot \frac{3}{8}$
17. $\frac{1}{9} \div 2$
18. $\frac{5}{8} - \frac{1}{16}$
19. $9\frac{2}{3} + 4\frac{7}{8}$
20. $\frac{1}{2} \cdot 25$
21. $35\frac{1}{3} \div 6\frac{3}{4}$
22. $100 \div 3\frac{3}{4}$
23. $\frac{1}{3} \cdot 38$
24. $\frac{15}{16} - \frac{3}{4}$
25. $\frac{1}{5} \cdot 16$
26. $47\frac{1}{2} \div 6\frac{5}{8}$
27. $\frac{1}{9} \cdot 79$

Lesson 6-2

(Pages 244–247)

Add or subtract. Write in simplest form.

1. $\frac{5}{11} + \frac{9}{11}$
2. $\frac{5}{8} - \frac{1}{8}$
3. $\frac{7}{10} + \frac{7}{10}$
4. $\frac{9}{12} - \frac{5}{12}$
5. $\frac{2}{9} + \frac{1}{3}$
6. $\frac{1}{2} + \frac{3}{4}$
7. $\frac{1}{4} - \frac{3}{12}$
8. $\frac{3}{7} + \frac{6}{14}$
9. $\frac{1}{4} + \frac{3}{5}$
10. $\frac{4}{9} + \frac{1}{2}$
11. $\frac{5}{7} - \frac{4}{6}$
12. $\frac{3}{4} - \frac{1}{6}$
13. $\frac{3}{5} + \frac{3}{4}$
14. $\frac{2}{3} - \frac{1}{8}$
15. $\frac{9}{10} + \frac{1}{3}$
16. $\frac{8}{15} + \frac{2}{9}$
17. $\frac{6}{7} + \frac{6}{9}$
18. $\frac{3}{7} + \frac{3}{4}$

Evaluate each expression if $a = \frac{2}{3}$ and $b = \frac{7}{12}$.

19. $\frac{1}{5} + a$
20. $a - \frac{1}{2}$
21. $b + \frac{7}{8}$
22. $\frac{7}{8} - a$
23. $a + b$
24. $a - b$

Lesson 6-3

(Pages 248–251)

Add or subtract. Write in simplest form.

1. $2\frac{1}{3} + 1\frac{1}{3}$
2. $5\frac{2}{7} - 2\frac{3}{7}$
3. $6\frac{3}{8} + 7\frac{1}{8}$
4. $2\frac{3}{4} - 1\frac{1}{4}$
5. $5\frac{1}{2} - 3\frac{1}{4}$
6. $2\frac{2}{3} + 4\frac{1}{9}$
7. $7\frac{4}{5} + 9\frac{3}{10}$
8. $3\frac{3}{4} + 5\frac{5}{8}$
9. $10\frac{2}{3} + 5\frac{6}{7}$
10. $17\frac{2}{9} - 12\frac{1}{3}$
11. $6\frac{5}{12} + 12\frac{5}{12}$
12. $7\frac{1}{4} + 15\frac{5}{6}$
13. $6\frac{1}{8} + 4\frac{2}{3}$
14. $7 - 6\frac{4}{9}$
15. $8\frac{1}{12} + 12\frac{6}{11}$
16. $7\frac{2}{3} + 8\frac{1}{4}$
17. $12\frac{3}{11} + 14\frac{3}{13}$
18. $21\frac{1}{3} + 15\frac{3}{8}$
19. $19\frac{1}{7} + 6\frac{1}{4}$
20. $9\frac{2}{5} - 8\frac{1}{3}$
21. $18\frac{1}{4} - 3\frac{3}{8}$
22. $1\frac{1}{8} + 2\frac{1}{12}$
23. $2\frac{1}{12} - 1\frac{1}{8}$
24. $10 - \frac{2}{3}$

Extra Practice

Lesson 6-4

(Pages 254–257)

Multiply. Write in simplest form.

1. $\frac{2}{3} \times \frac{3}{5}$
2. $\frac{1}{6} \times \frac{2}{5}$
3. $\frac{4}{9} \times \frac{3}{7}$
4. $\frac{5}{12} \times \frac{6}{11}$
5. $\frac{3}{8} \times \frac{8}{9}$
6. $\frac{3}{5} \times \frac{1}{12}$
7. $\frac{2}{5} \times \frac{5}{8}$
8. $\frac{7}{15} \times \frac{3}{21}$
9. $\frac{5}{6} \times \frac{15}{16}$
10. $\frac{6}{14} \times \frac{12}{18}$
11. $\frac{2}{3} \times \frac{3}{13}$
12. $\frac{4}{9} \times \frac{1}{6}$
13. $3 \times \frac{1}{9}$
14. $5 \times \frac{6}{7}$
15. $\frac{3}{5} \times 15$
16. $3\frac{1}{2} \times 4\frac{1}{3}$
17. $3\frac{5}{8} \times 4\frac{1}{2}$
18. $\frac{4}{5} \times 2\frac{3}{4}$
19. $6\frac{1}{8} \times 5\frac{1}{7}$
20. $2\frac{2}{3} \times 2\frac{1}{4}$
21. $\frac{7}{8} \times 16$
22. $5\frac{1}{5} \times 2\frac{1}{2}$
23. $7 \times \frac{1}{14}$
24. $22 \times \frac{3}{11}$
25. $8\frac{2}{3} \times 1\frac{1}{2}$
26. $4 \times 6\frac{1}{2}$
27. $\frac{1}{2} \times 10\frac{2}{3}$
28. $\frac{2}{3} \times 21\frac{1}{3}$
29. $\frac{7}{8} \times \frac{8}{7}$
30. $1\frac{2}{3} \times 1\frac{2}{3}$
31. $21 \times \frac{1}{2}$
32. $11 \times \frac{1}{4}$

Lesson 6-5

(Pages 258–261)

Find the multiplicative inverse of each number.

1. $\frac{2}{3}$
2. $-\frac{5}{4}$
3. 1
4. 10
5. -23
6. $-\frac{1}{7}$
7. $\frac{9}{16}$
8. 15
9. $1\frac{1}{3}$
10. $-3\frac{3}{4}$
11. $7\frac{3}{8}$
12. $-\frac{19}{10}$
13. $-\frac{3}{49}$
14. $6\frac{2}{5}$
15. $33\frac{1}{3}$
16. $-66\frac{2}{3}$

Solve each equation. Check your solution.

17. $\frac{a}{13} = 2$
18. $\frac{8}{9}x = 24$
19. $\frac{3}{8}r = 36$
20. $\frac{3}{4}t = \frac{1}{2}$
21. $16 = \frac{h}{4}$
22. $\frac{m}{8} = -12$
23. $\frac{5}{8}n = -45$
24. $10 = \frac{b}{10}$
25. $\frac{1}{7}x = 7$
26. $-5 = \frac{1}{5}y$
27. $\frac{4}{3}m = -28$
28. $\frac{2}{3}z = 20$
29. $\frac{c}{9} = 81$
30. $\frac{m}{9} = 9$
31. $-16 = \frac{4}{9}f$
32. $\frac{15}{8}x = 225$

Lesson 6-6

(Pages 264–266)

Divide. Write in simplest form.

1. $\frac{2}{3} \div \frac{3}{2}$
2. $\frac{3}{5} \div \frac{2}{5}$
3. $\frac{7}{10} \div \frac{3}{8}$
4. $\frac{5}{9} \div \frac{2}{5}$
5. $4 \div \frac{2}{3}$
6. $8 \div \frac{4}{5}$
7. $9 \div \frac{5}{9}$
8. $\frac{2}{7} \div 2$
9. $\frac{1}{14} \div 7$
10. $\frac{2}{13} \div \frac{5}{26}$
11. $\frac{4}{7} \div \frac{6}{7}$
12. $\frac{7}{8} \div \frac{1}{3}$
13. $15 \div \frac{3}{5}$
14. $\frac{9}{14} \div \frac{3}{4}$
15. $\frac{8}{9} \div \frac{5}{6}$
16. $2\frac{1}{2} \div 5$
17. $\frac{7}{8} \div 10$
18. $16 \div \frac{3}{4}$
19. $22 \div 3\frac{2}{3}$
20. $40\frac{5}{8} \div 4$
21. $\frac{3}{8} \div 2\frac{1}{2}$
22. $5\frac{1}{2} \div 2\frac{1}{2}$
23. $3\frac{1}{4} \div 5\frac{1}{2}$
24. $12\frac{5}{6} \div 2\frac{1}{6}$
25. $7\frac{1}{2} \div 3\frac{1}{2}$
26. $3\frac{1}{2} \div 7\frac{1}{2}$
27. $4\frac{2}{3} \div 2\frac{1}{3}$

Lesson 6-7

(Pages 267–269)

Complete.

1. 4,000 lb = __?__ T
2. 5 T = __?__ lb
3. 5 lb = __?__ oz
4. 12,000 lb = __?__ T
5. $\frac{1}{4}$ lb = __?__ oz
6. 12 pt = __?__ c
7. 3 gal = __?__ pt
8. 24 fl oz = __?__ c
9. 8 pt = __?__ c
10. 10 pt = __?__ qt
11. $2\frac{1}{4}$c = __?__ fl oz
12. 6 lb = __?__ oz
13. 10 gal = __?__ qt
14. 4 qt = __?__ fl oz
15. 4 pt = __?__ c
16. 13,200 ft = __?__ mi
17. 120 oz = __?__ lb
18. $9\frac{1}{4}$ gal = __?__ qt
19. 7,480 yd = __?__ mi
20. $12\frac{1}{2}$ lb = __?__ oz
21. $7\frac{1}{2}$ qt = __?__ pt
22. $3\frac{1}{8}$ c = __?__ fl oz
23. $2\frac{1}{4}$ mi = __?__ ft
24. $3\frac{2}{3}$ T = __?__ lb

Lesson 6-8

(Pages 270–273)

Find the perimeter and area of each rectangle.

1. 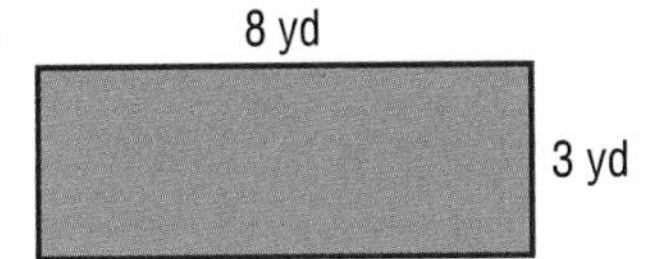

2. 

3. 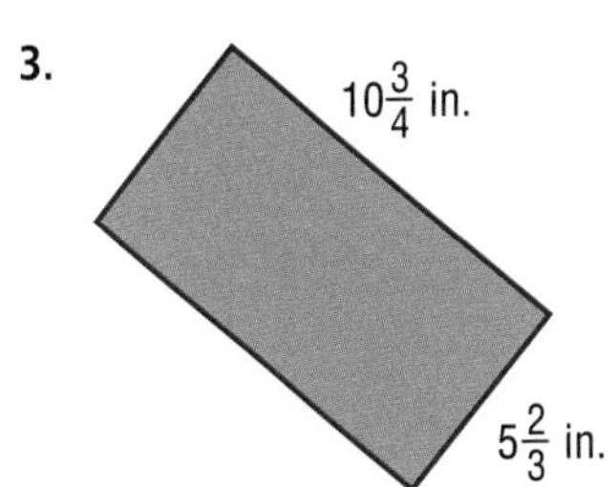

4. $\ell = 80$ yd, $w = 20$ yd
5. $\ell = 75$ cm, $w = 25$ cm
6. $\ell = 5.25$ km, $w = 1.5$ km
7. $\ell = 8.6$ cm, $w = 2.5$ cm
8. $\ell = 20.25$ m, $w = 4.75$ m
9. $\ell = 12$ ft, $w = 3$ ft
10. $\ell = 5\frac{1}{4}$ mi, $w = 2\frac{1}{2}$ mi
11. $\ell = 10\frac{2}{3}$ ft, $w = 5\frac{5}{6}$ ft
12. $\ell = 12\frac{1}{3}$ yd, $w = 5\frac{2}{3}$ yd
13. $\ell = 3\frac{7}{8}$ in., $w = 1\frac{1}{4}$ in.

Lesson 6-9

(Pages 275–277)

Find the circumference of each circle. Use 3.14 or $\frac{22}{7}$ for π. Round to the nearest tenth if necessary.

1. 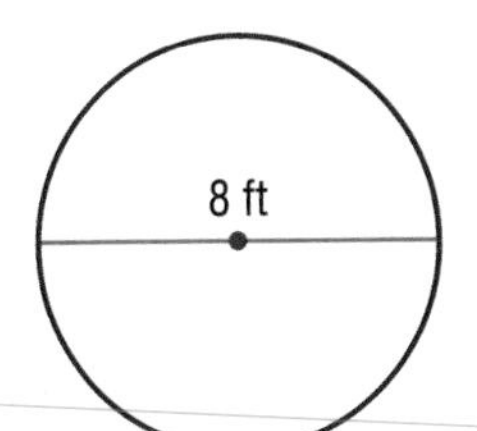

2. 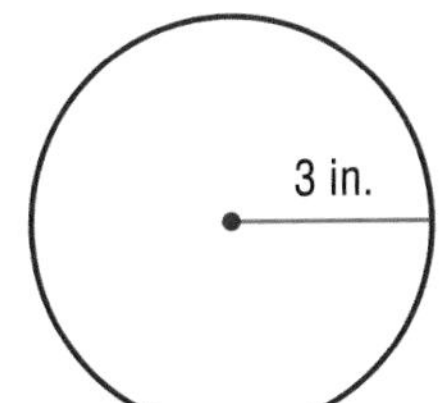
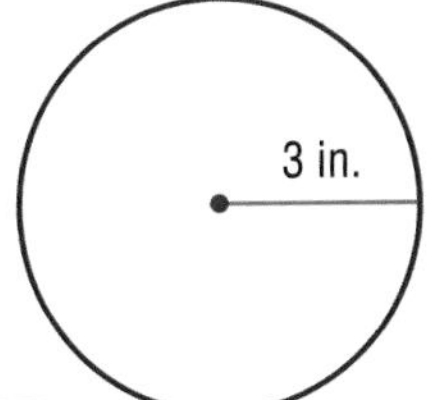

3. 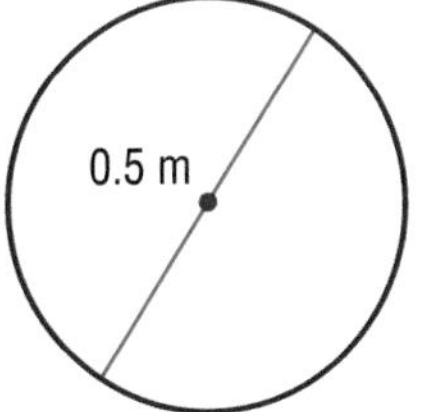
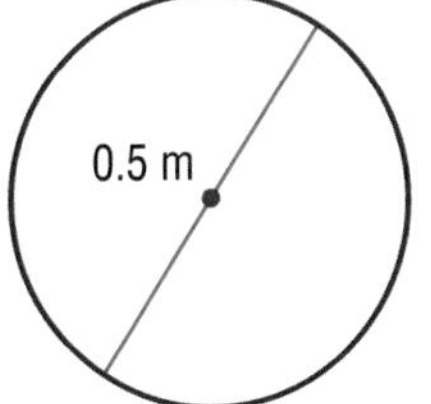

4. 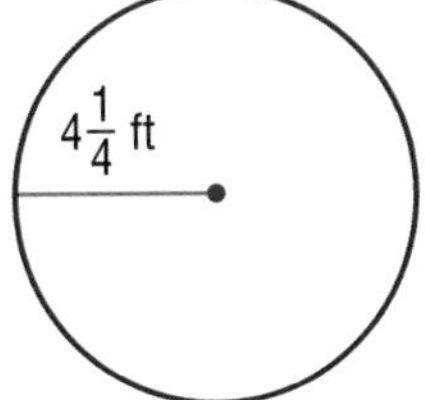
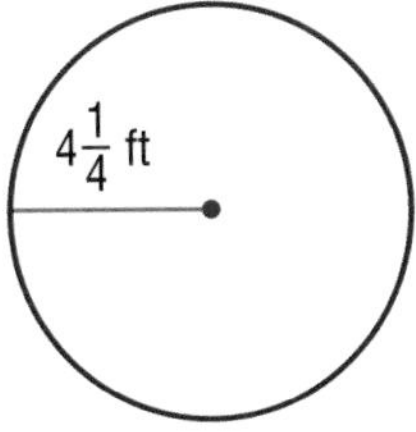

5. $r = 1$ m
6. $d = 2$ yd
7. $d = 5{,}280$ ft
8. $r = 350$ mi
9. $r = 1.5$ in.
10. $d = 1.5$ in.
11. $r = 0.5$ cm
12. $d = 6.4$ m
13. $r = 10.7$ km
14. $r = \frac{5}{8}$ in.
15. $d = 8\frac{1}{4}$ in.
16. $r = 2\frac{1}{3}$ yd
17. $d = \frac{3}{16}$ in.
18. $r = 5\frac{1}{2}$ mi
19. $d = 42\frac{3}{4}$ ft

Lesson 7-1

(Pages 288–291)

Write each ratio as a fraction in simplest form.

1. 45 to 15
2. 64 : 128
3. 14 to 49
4. 125 : 25
5. 18 to 81
6. 16 : 40
7. 120 to 180
8. 32 : 64
9. 90 to 100
10. 12 weeks out of 15
11. 33 min : 60 min
12. 10 ft to 8 yd
13. 25 pounds to 80 pounds
14. 15 minutes : 4 hours
15. 10 yards to 12 feet

Determine whether the ratios are equivalent. Explain.

16. 14 to 77 and 8 to 44
17. $\frac{48}{16}$ and $\frac{1}{3}$
18. 65 : 13 and 500 : 100
19. 72 to 90 and 20 to 16
20. 250 : 100 and 5 : 2
21. $\frac{32}{2}$ and $\frac{3}{48}$
22. 3 lb to 60 oz and 5 lb to 100 oz
23. 12 : 25 and 84 : 150
24. 28 in. : 5 ft and 40 in. : 8 ft
25. 8 hours to 5 days and 24 hours to 15 days

Lesson 7-2

(Pages 292–295)

Find each unit rate. Round to the nearest hundredth if necessary.

1. $240 for 4 days
2. 250 people in 5 buses
3. 500 miles in 10 hours
4. 18 cups for 24 pounds
5. 32 people in 8 cars
6. $4.50 for 3 dozen
7. 245 tickets in 5 days
8. 12 classes in 4 semesters
9. 60 people in 4 rows
10. 48 ounces in 3 pounds
11. 20 people in 4 groups
12. 1.5 pounds for $3.00
13. 45 miles in 60 minutes
14. $5.50 for 10 disks
15. 360 miles for 12 gallons
16. $8.50 for 5 yards
17. 24 cups for $1.20
18. 160 words in 4 minutes
19. $60 for 5 books
20. $24 for 6 hours

Lesson 7-3

(Pages 297–300)

Determine whether each pair of ratios forms a proportion.

1. $\frac{2}{5}$ and $\frac{5}{8}$
2. $\frac{125}{275}$ and $\frac{150}{300}$
3. $\frac{28}{42}$ and $\frac{70}{105}$
4. $\frac{55}{22}$ and $\frac{35}{14}$
5. $\frac{99}{100}$ and $\frac{100}{99}$
6. $\frac{85}{17}$ and $\frac{5}{1}$
7. $\frac{100\text{ ft}}{8\text{ s}}$ and $\frac{250\text{ ft}}{20\text{ s}}$
8. $\frac{\$15}{45\text{ lb}}$ and $\frac{\$20}{65\text{ lb}}$
9. $\frac{32\text{ mi}}{6\text{ h}}$ and $\frac{44\text{ mi}}{8.25\text{ h}}$

Solve each proportion.

10. $\frac{u}{72} = \frac{2}{4}$
11. $\frac{12}{m} = \frac{15}{10}$
12. $\frac{36}{90} = \frac{16}{t}$
13. $\frac{g}{32} = \frac{8}{64}$
14. $\frac{5}{14} = \frac{10}{a}$
15. $\frac{k}{18} = \frac{5}{3}$
16. $\frac{120}{150} = \frac{p}{20}$
17. $\frac{15}{w} = \frac{60}{4}$
18. $\frac{81}{90} = \frac{y}{20}$
19. $\frac{14}{s} = \frac{8}{4}$
20. $\frac{h}{3} = \frac{36}{9}$
21. $\frac{50}{8} = \frac{150}{t}$
22. $\frac{45}{8} = \frac{36}{d}$
23. $\frac{125}{v} = \frac{20}{5}$
24. $\frac{4}{5} = \frac{x}{3}$
25. $\frac{45}{75} = \frac{j}{3}$
26. $\frac{3}{7} = \frac{21}{d}$
27. $\frac{3}{10} = \frac{z}{36}$

Lesson 7-4

(Pages 304–308)

On a map, the scale is 1 inch = 50 miles. For each map distance, find the actual distance.

1. 5 inches
2. 12 inches
3. $3\frac{1}{2}$ inches
4. $2\frac{3}{8}$ inches
5. $\frac{4}{5}$ inch
6. $6\frac{3}{4}$ inches
7. $2\frac{5}{6}$ inches
8. 8 inches
9. 1.5 inches
10. 3.25 inches
11. 4.75 inches
12. 5.25 inches

On a scale drawing, the scale is $\frac{1}{2}$ inch = 2 feet. Find the dimensions of each room in the scale drawing.

13. 14 feet by 18 feet
14. 32 feet by 6 feet
15. 3 feet by 5 feet
16. 20 feet by 30 feet
17. 8 feet by 15 feet
18. 25 feet by 80 feet

Lesson 7-5

(Pages 312–315)

Write each percent as a fraction in simplest form.

1. 32%
2. 89%
3. 72%
4. 11%
5. 1%
6. 28%
7. 55%
8. 18.5%
9. 22.75%
10. 25.2%
11. 75.5%
12. 48.25%
13. 6.5%
14. 1.25%
15. 88.9%
16. $52\frac{1}{4}\%$

Write each fraction as a percent. Round to the nearest hundredth if necessary.

17. $\frac{14}{25}$
18. $\frac{28}{50}$
19. $\frac{14}{20}$
20. $\frac{7}{10}$
21. $\frac{17}{17}$
22. $\frac{80}{125}$
23. $\frac{9}{12}$
24. $\frac{4}{6}$
25. $\frac{11}{12}$
26. $\frac{9}{16}$
27. $\frac{8}{9}$
28. $\frac{3}{16}$
29. $\frac{5}{32}$
30. $\frac{1}{16}$
31. $\frac{8}{15}$
32. $\frac{9}{11}$

Lesson 7-6

(Pages 316–318)

Write each percent as a decimal and as a mixed number or fraction in simplest form.

1. 895%
2. 555%
3. 480%
4. 920%
5. 122%
6. 825%
7. 0.3%
8. 0.42%
9. 0.78%
10. 765%
11. 0.99%
12. 1,000%

Write each decimal as a percent.

13. 3.5
14. 12
15. 0.002
16. 6.78
17. 0.0056
18. 1.95
19. 0.0077
20. 0.0102
21. 14.0
22. 0.0064
23. 16.2
24. 44.3

Write each number as a percent.

25. $\frac{1}{250}$
26. $\frac{7}{400}$
27. $1\frac{1}{4}$
28. $7\frac{9}{10}$
29. $\frac{9}{10}$
30. $\frac{1}{500}$
31. $12\frac{1}{2}$
32. $\frac{1}{1,000}$
33. $4\frac{3}{4}$
34. 25
35. 900
36. $18\frac{2}{5}$

Extra Practice

Lesson 7-7

(Pages 319–321)

Find each number. Round to the nearest tenth if necessary.

1. 5% of 40 is what number?
2. What number is 10% of 120?
3. Find 12% of 150.
4. Find 12.5% of 40.
5. What number is 75% of 200?
6. Find 13% of 25.3.
7. 250% of 44 is what number?
8. What number is 0.5% of 13.7?
9. Find 600% of 7.
10. Find 1.5% of $25.
11. Find 81% of 134.
12. What number is 43% of 110?
13. What number is 61% of 524?
14. Find 100% of 3.5.
15. 20% of 58.5 is what number?
16. Find 45% of 125.5.
17. What number is 23% of 500?
18. Find 80% of 8.
19. 90% of 72 is what number?
20. What number is 32% of 54?

Lesson 7-8

(Pages 323–325)

Find each number. Round to the nearest tenth if necessary.

1. What number is 25% of 280?
2. 38 is what percent of 50?
3. 54 is 25% of what number?
4. 24.5% of what number is 15?
5. What number is 80% of 500?
6. 12% of 120 is what number?
7. Find 68% of 50.
8. What percent of 240 is 32?
9. 99 is what percent of 150?
10. Find 75% of 1.
11. What number is $33\frac{1}{3}\%$ of 66?
12. 50% of 350 is what number?
13. What percent of 450 is 50?
14. What number is $37\frac{1}{2}\%$ of 32?
15. 95% of 40 is what number?
16. Find 30% of 26.
17. 9 is what percent of 30?
18. 52% of what number is 109.2?
19. What number is 65% of 200?
20. What number is 15.5% of 45?

Lesson 8-1

(Pages 334–337)

Estimate by using fractions.

1. 28% of 48
2. 99% of 65
3. 445% of 20
4. 9% of 81
5. 73% of 240
6. 65.5% of 75
7. 48.2% of 93
8. 39.45% of 51
9. 287% of 122
10. 53% of 80
11. 414% of 72
12. 59% of 105

Estimate by using 10%.

13. 30% of 42
14. 70% of 104
15. 90% of 152
16. 67% of 70
17. 78% of 92
18. 12% of 183
19. 51% of 221
20. 23% of 504
21. 81% of 390
22. 41% of 60
23. 59% of 178
24. 22% of 450

Estimate.

25. 50% of 37
26. 18% of 90
27. 300% of 245
28. 1% of 48
29. 70% of 300
30. 35% of 35
31. 60.5% of 60
32. $5\frac{1}{2}\%$ of 100
33. 40.01% of 16
34. 80% of 62
35. 45% of 119
36. 14.81% of 986

Lesson 8-2

(Pages 340–343)

Write an equation for each problem. Then solve. Round to the nearest tenth if necessary.

1. Find 45% of 50.
2. 75 is what percent of 300?
3. 16% of what number is 2?
4. 75% of 80 is what number?
5. 5% of what number is 12?
6. Find 60% of 45.
7. Find 22% of 22.
8. 12% of what number is 50?
9. 38 is what percent of 62?
10. 80 is what percent of 90?
11. 90 is what percent of 95?
12. $28\frac{1}{2}$% of 64 is what number?
13. Find 46.5% of 75.
14. What number is 55.5% of 70?
15. 80.5% of what number is 80.5?
16. $66\frac{2}{3}$% of what number is 40?
17. Find 122.5% of 80.
18. 250% of what number is 75?

Lesson 8-3

(Pages 345–347)

1. The table shows the results of a survey of students' favorite cookies. Predict how many of the 424 students at Scobey High School prefer chocolate chip cookies.

Cookie	Number
chocolate chip	49
peanut butter	12
oatmeal	10
sugar	8
raisin	3

2. The circle graph shows the results of a survey of teens and where they would prefer to spend a family vacation. Predict how many of the 4,000 teens in the Central School District would prefer to go to an amusement park.

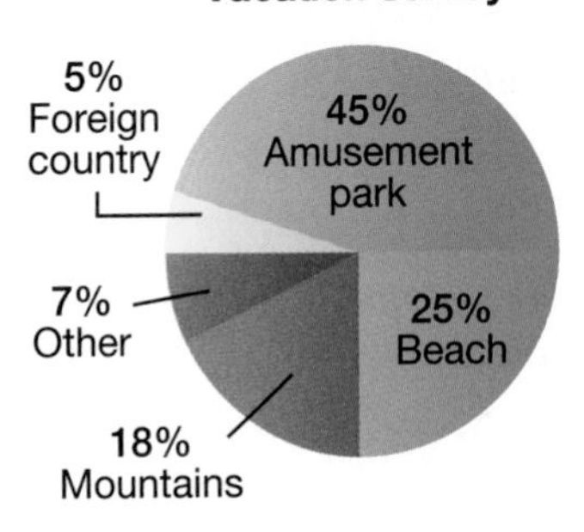

3. A survey showed that 7 out of 10 first graders received an allowance. Based on that survey, how many of the 800 first graders in the Midtown School District receive an allowance?
4. In 2000, about 29% of the foreign visitors to the U.S. were from Canada. If a particular hotel had 150,000 foreign guests in one year, how many would you predict were from Canada?

Lesson 8-4

(Pages 350–353)

Find each percent of change. Round to the nearest whole percent if necessary. State whether the percent of change is an *increase* or *decrease*.

1. original: 450
new: 675
2. original: $1,500
new: $1,200
3. original: 750
new: 700
4. original: 350
new: 420
5. original: 500
new: 100
6. original: $75
new: $50
7. original: 3.25
new: 2.95
8. original: 32.5
new: 44
9. original: 180
new: 160
10. original: $65
new: $75
11. original: 22
new: 88
12. original: 450
new: 445.5

Lesson 8-5

(Pages 354–357)

Find the total cost or sale price to the nearest cent.

1. $45 sweater; 6% tax
2. $18.99 CD; 15% discount
3. $199 ring; 10% discount
4. $29 shirt; 7% tax
5. $55 plant; 20% discount
6. $19 purse; 25% discount
7. $150 clock; 5% tax
8. $89 radio; 30% discount
9. $39 shoes; 5.5% tax
10. $145 coat; 6.25% tax
11. $300 table; 30% discount
12. $12 meal; 4.5% tax
13. $899 computer; 20% discount
14. $105 skateboard; $7\frac{1}{2}$% tax
15. $599 TV; 12% discount
16. $425 skis; 15% discount
17. $12,500 car; $3\frac{3}{4}$% tax
18. $49.95 gloves; $5\frac{1}{4}$% tax

Find the percent of discount to the nearest percent.

19. regular price: $72
    sale price: $60
20. regular price: $125
    sale price: $120
21. regular price: $360
    sale price: $280
22. regular price: $90
    sale price: $22.50
23. regular price: $25,000
    sale price: $22,000
24. regular price: $0.99
    sale price: $0.82

Lesson 8-6

(Pages 358–360)

Find the interest earned to the nearest cent for each principal, interest rate, and time.

1. $2,000, 8%, 5 years
2. $500, 10%, 8 months
3. $750, 5%, 1 year
4. $175.50, $6\frac{1}{2}$%, 18 months
5. $236.20, 9%, 16 months
6. $89, $7\frac{1}{2}$%, 6 months
7. $800, 5.75%, 3 years
8. $5,500, 7.2%, 4 years
9. $245, 6%, 13 months
10. $1,200, 3%, 45 months
11. $225, $1\frac{1}{2}$%, 2 years
12. $12,000, $4\frac{1}{2}$%, 40 months

Find the interest paid to the nearest cent for each loan balance, interest rate, and time.

13. $750, 18%, 2 years
14. $1,500, 19%, 16 months
15. $300, 9%, 1 year
16. $4,750, 19.5%, 30 months
17. $2,345, 17%, 9 months
18. $689, 12%, 2 years
19. $390, 18.75%, 15 months
20. $1,250, 22%, 8 months
21. $3,240, 18%, 14 months
22. $675, 15%, 2 years
23. $899, $10\frac{1}{2}$%, 18 months
24. $1,000, 8%, 1 year

Lesson 9-1

(Pages 370–373)

Use the spinner at the right to find each probability. Write as a fraction in simplest form.

1. *P*(even number)
2. *P*(prime number)
3. *P*(factor of 12)
4. *P*(composite number)
5. *P*(greater than 10)
6. *P*(neither prime nor composite)

A package of balloons contains 5 green, 3 yellow, 4 red, and 8 pink balloons. Suppose you reach in the package and choose one balloon at random. Find the probability of each event. Write as a fraction in simplest form.

7. *P*(red balloon)
8. *P*(yellow balloon)
9. *P*(pink balloon)
10. *P*(orange balloon)
11. *P*(red or yellow balloon)
12. *P*(*not* green balloon)

Lesson 9-2

(Pages 374–377)

For each situation, make a tree diagram to show the sample space. Then give the total number of outcomes.

1. rolling 2 number cubes
2. choosing an ice cream cone from waffle, plain, or sugar and a flavor of ice cream from chocolate, vanilla, or strawberry
3. making a sandwich from white, wheat, or rye bread, cheddar or Swiss cheese and ham, turkey, or roast beef
4. flipping a penny twice
5. choosing one math class from algebra and geometry and one foreign language class from French, Spanish, or Latin

Lesson 9-3

(Pages 378–380)

Use the Fundamental Counting Principle to find the total number of outcomes in each situation.

1. choosing a local phone number if the exchange is 398 and each of the four remaining digits is different
2. choosing a way to drive from Millville to Westwood if there are 5 roads that lead from Millville to Miamisburg, 3 roads that connect Miamisburg to Hathaway, and 4 highways that connect Hathaway to Westwood
3. tossing a quarter, rolling a number cube, and tossing a dime
4. spinning the spinners shown below

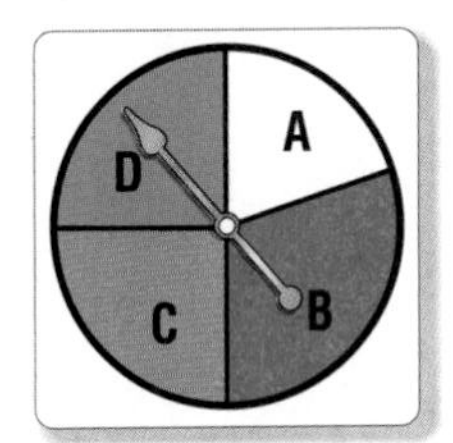

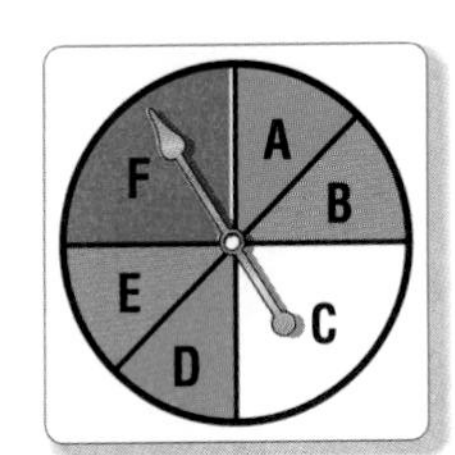

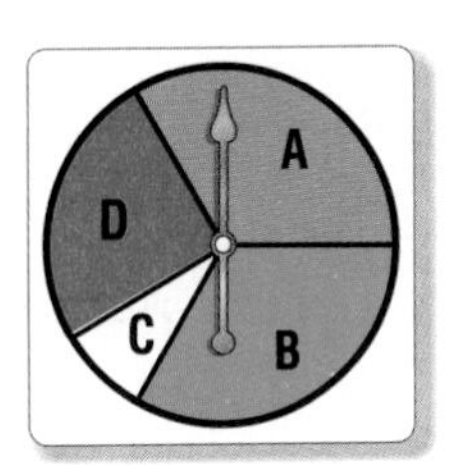

Lesson 9-4

(Pages 381–383)

Find the value of each expression.

1. $2!$	2. $7!$	3. $1!$	4. $8!$
5. $3! \cdot 5!$	6. $2! \cdot 5!$	7. 10 factorial	8. $1! \cdot 9!$
9. $9 \cdot 8 \cdot 7$	10. $6! \cdot 5!$	11. $7 \cdot 6 \cdot 5 \cdot 4$	12. $8! \cdot 3!$

Solve.

13. Eight runners are competing in a 100-meter sprint. In how many ways can the gold, silver, and bronze medals be awarded?
14. In a lottery for which 30 tickets were sold (all to different people), in how many ways can the grand prize, second prize, and third prizes be awarded?
15. Five-digit locker combinations are assigned using the digits 1–9. In how many ways can the combinations be formed if no digit can be repeated?
16. How many ways can you arrange the letters in the word *factor*?

Lesson 9-5

(Pages 387–390)

Tell whether each problem represents a *permutation* or *combination*. Then solve the problem.

1. How many ways can four students be placed in a line?
2. In how many ways can six books be displayed on a rack from a collection of 12?
3. In how many ways can you select three report topics from a total of 8 topics?
4. How many ways can nine students finish first, second, third, or fourth at a speech competition?
5. In how many ways can you ride five out of nine roller coasters if you don't care in what order you ride them?
6. In how many ways can you ride five out of nine roller coasters if you care in what order you ride them?

Lesson 9-6

(Pages 393–396)

The frequency table shows the results of a fair number cube rolled 40 times.

Face	Frequency
1	5
2	9
3	2
4	8
5	12
6	4

1. Find the experimental probability of rolling a 4.
2. Find the theoretical probability of *not* rolling a 4.
3. Find the theoretical probability of rolling a 2.
4. Find the experimental probability of *not* rolling a 6.
5. Suppose the number cube was rolled 500 times. About how many times would it land on 5?
6. Suppose the number cube was rolled 500 times. About how many times would it land on a prime number?

Lesson 9-7

(Pages 398–401)

Determine whether each event is *independent* or *dependent*.

1. selecting and eating a candy from a dish and then selecting another candy
2. tossing a coin and then rolling a number cube
3. choosing a card from deck 1 and then choosing a card from deck 2

Find each probability.

4. Two evenly balanced nickels are flipped. Find the probability that one head and one tail result.
5. A wallet contains four $5 bills, two $10 bills, and eight $1 bills. Two bills are selected without the first selection being replaced. Find *P*($5, then $5).
6. Two chips are selected from a box containing 6 blue chips, 4 red chips, and 3 green chips. The first chip selected is not replaced before the second is drawn. Find *P*(red, then green).
7. A bag contains 7 blue, 4 orange, 8 red, and 5 purple marbles. Suppose one marble is chosen and then replaced. A second marble is then chosen. Find *P*(purple, then red).

Lesson 10-1

(Pages 413–415)

Classify each angle as *acute, obtuse, right,* or *straight.*

1. 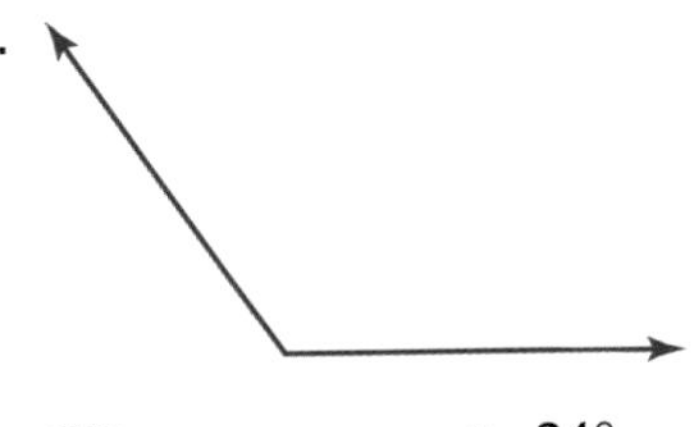

2. 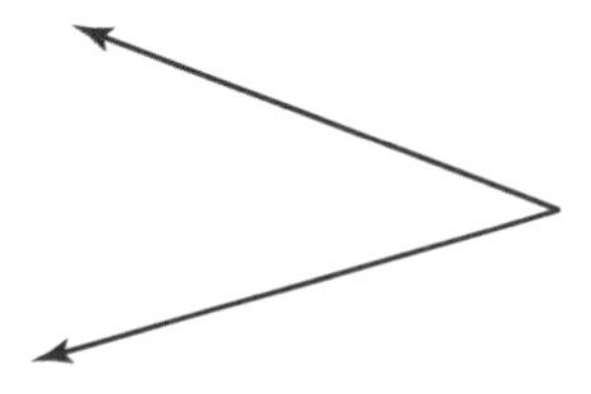

3. 

4. 65°
5. 24°
6. 110°
7. 95°
8. 90°

Draw an angle having each measurement. Then classify each angle as *acute, obtuse, right,* or *straight.*

9. 150°
10. 45°
11. 15°
12. 180°
13. 175°
14. 88°
15. 52°
16. 160°
17. 80°
18. 55°

Lesson 10-2

(Pages 418–421)

Make a circle graph of the data in each table.

1. Car Sales

Style	Percent
sedan	45%
SUV	22%
pickup truck	9%
sports car	13%
compact car	11%

2. Favorite Flavor of Ice Cream

Flavor	Number
vanilla	11
chocolate	15
strawberry	8
mint chip	5
cookie dough	3

Lesson 10-3

(Pages 422–425)

Classify each pair of angles as *complementary, supplementary,* or *neither.*

1. 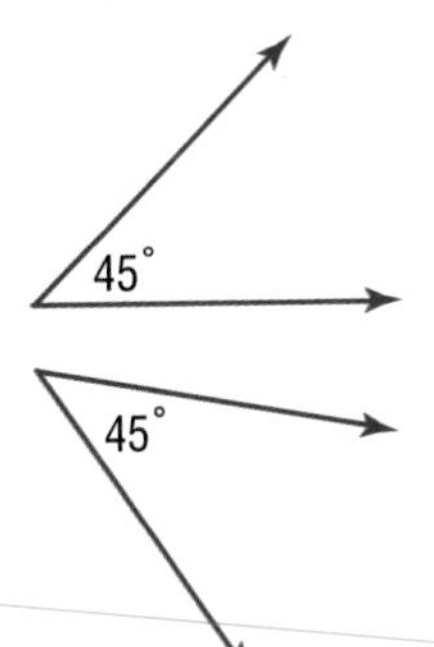

2. 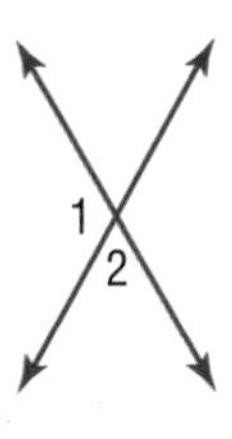

3. 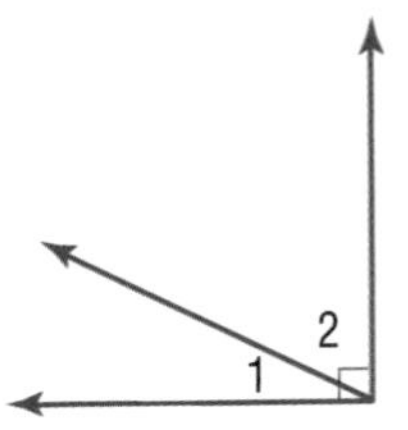

Find the value of x in each figure.

4. 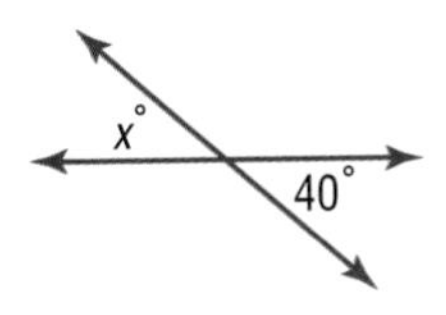

5. 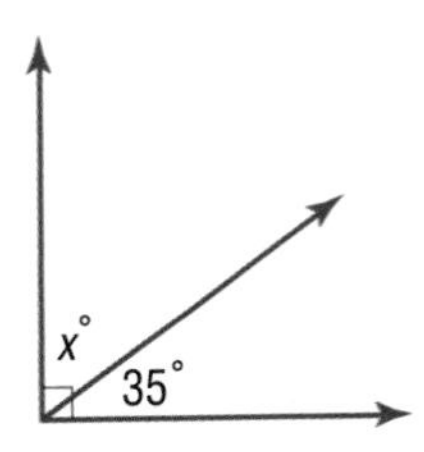

6. 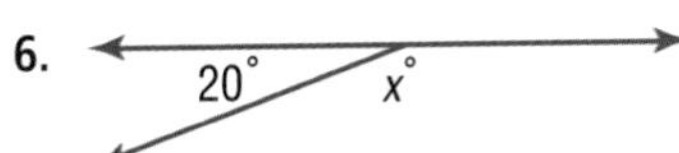

Lesson 10-4

(Pages 428–431)

Classify each triangle by its angles and by its sides.

1. 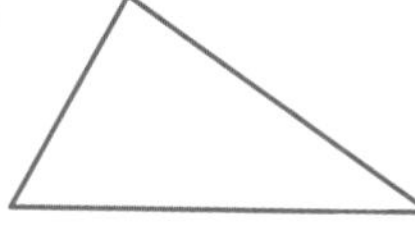

2. 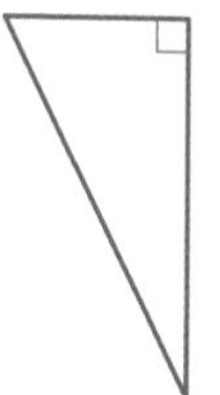

3. 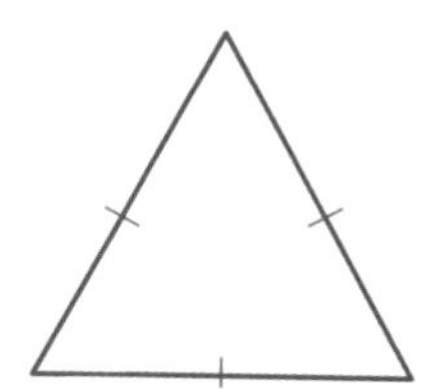

Find the missing measure in each triangle. Then classify the triangle as *acute*, *right*, or *obtuse*.

4. 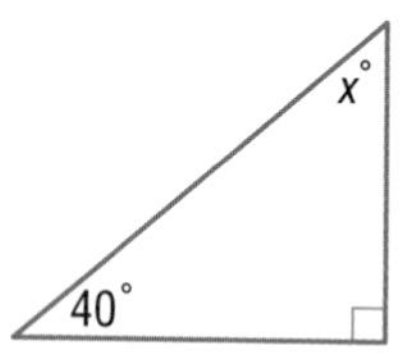

5. 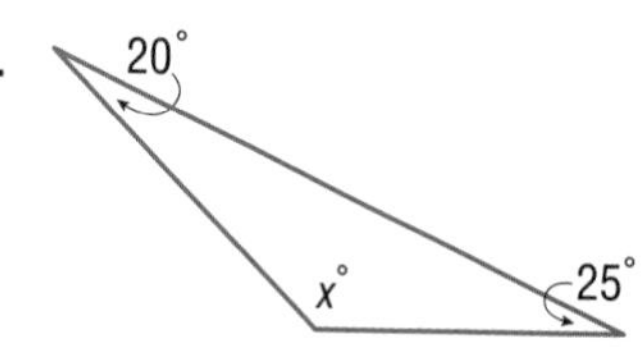

6. 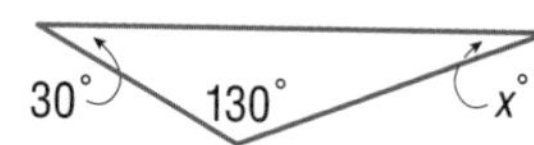

Lesson 10-5

(Pages 434–437)

Classify the quadrilateral using the name that *best* describes it.

1. 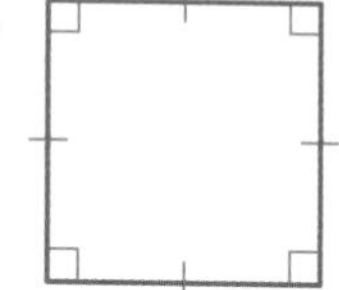

2. 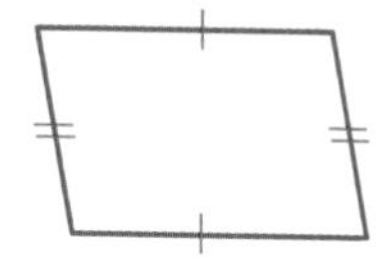

3. 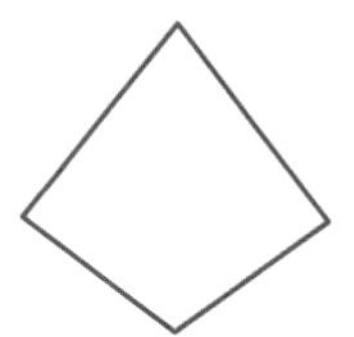

Find the missing measure in each quadrilateral.

4. 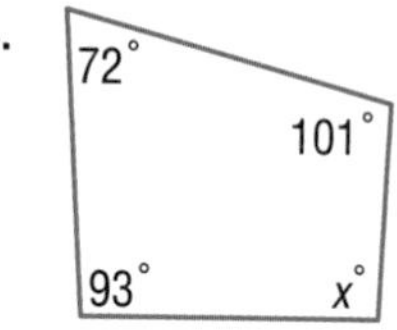

5. 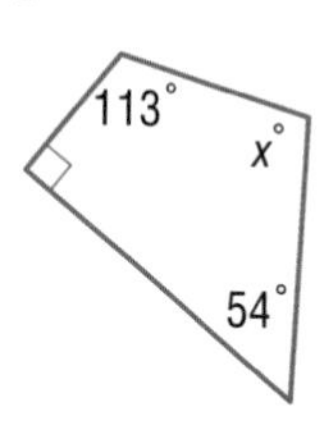

6. 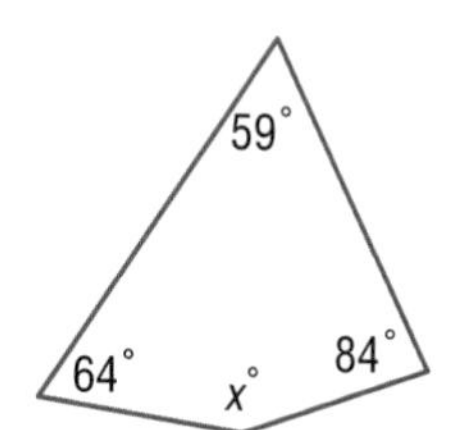

Lesson 10-6

(Pages 440–443)

Find the value of x in each pair of similar figures.

1. 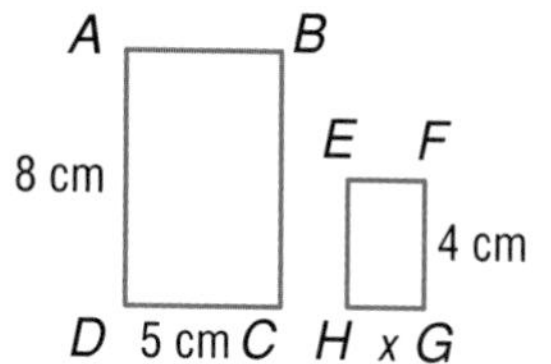

2. 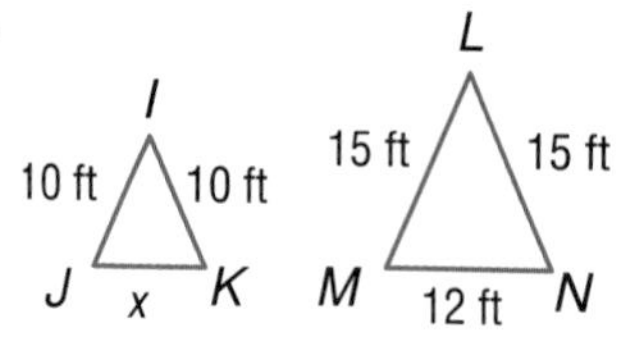

3. 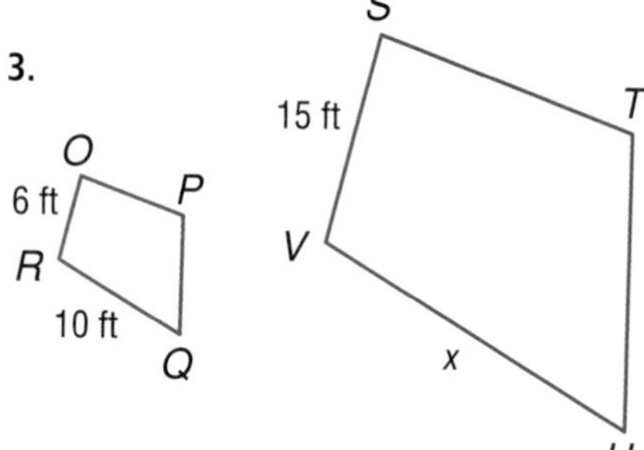

4. 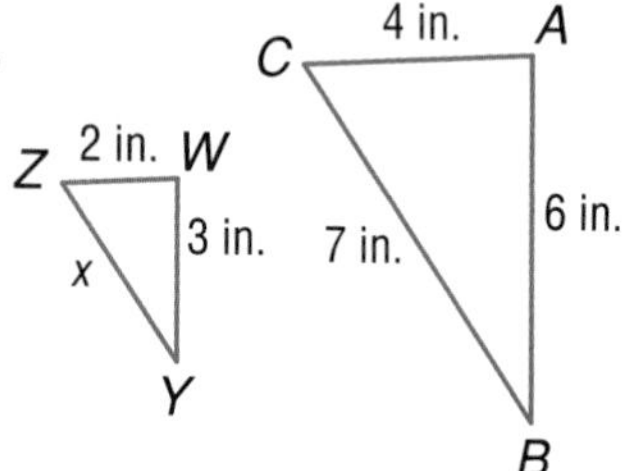

5. 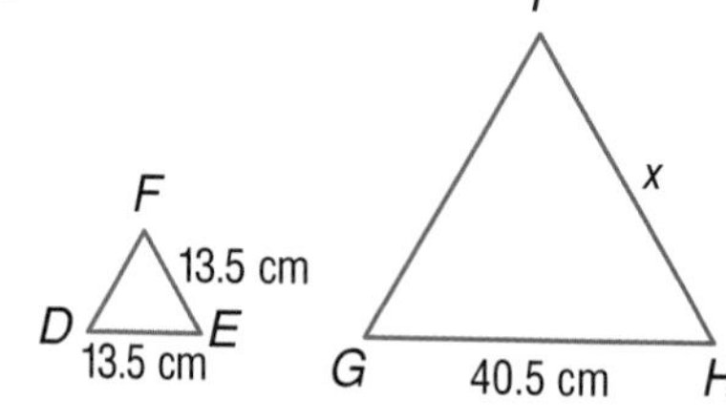

6. 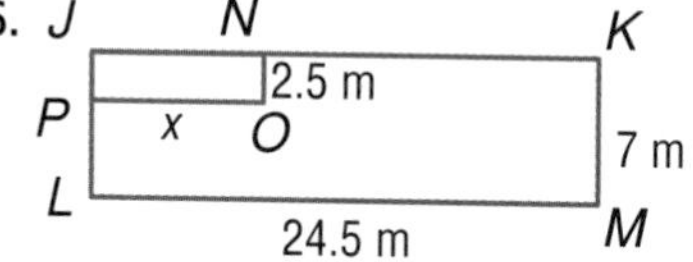

Lesson 10-7

(Pages 446–450)

Determine whether each figure is a polygon. If it is, classify the polygon and state whether it is regular. If it is *not* a polygon, explain why.

1. 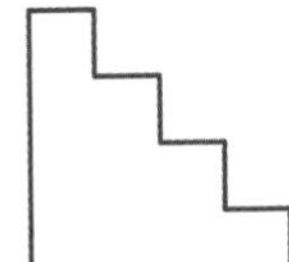

2. 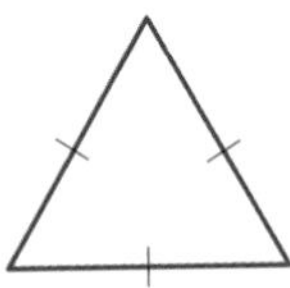

3. 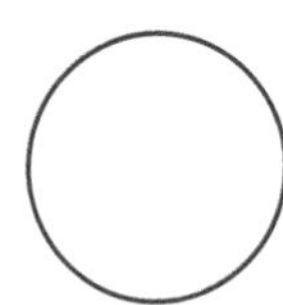

Find the measure of an angle in each regular polygon. Round to the nearest tenth degree, if necessary.

4. triangle
5. 30-gon
6. 18-gon
7. 14-gon

Lesson 10-8

(Pages 451–454)

1. Translate $\triangle ABC$ 2 units right and 1 unit down.

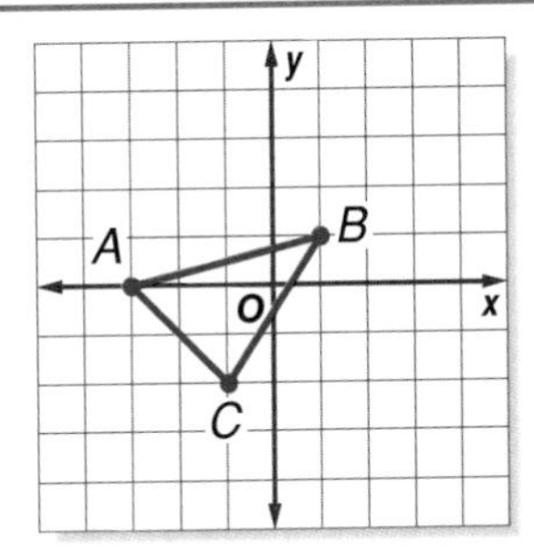

2. Translate quadrilateral *RSTU* 4 units left and 3 units down.

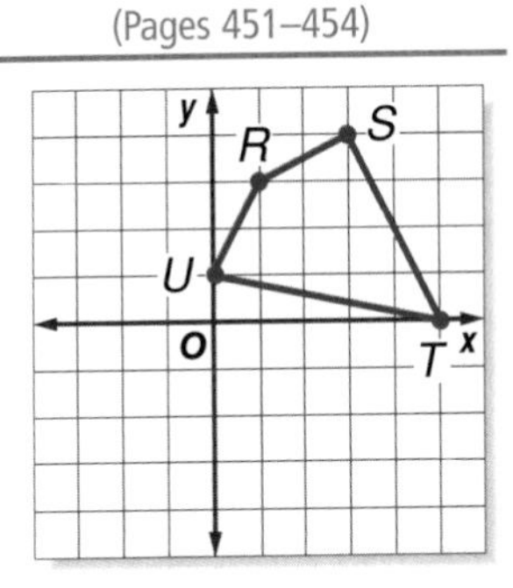

Triangle *TRI* has vertices $T(1, 1)$, $R(4, -2)$, and $I(-2, -1)$. Find the vertices of $T'R'I'$ after each translation. Then graph the figure and its translated image.

3. 2 units right, 1 unit down
4. 5 units left, 1 unit up
5. 3 units right
6. 2 units up

Lesson 10-9

(Pages 456–459)

Determine which figures have line symmetry. Write *yes* or *no*. If *yes*, draw all lines of symmetry.

1. 

2. 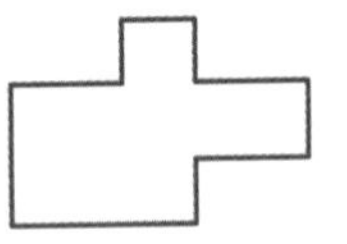

3. 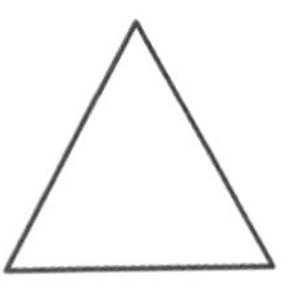

Find the coordinates of each figure after a reflection over the x-axis. Then graph the figure and its reflected image.

4. quadrilateral *QUAD* with vertices $Q(-1, 4)$, $U(2, 2)$, $A(1, 1)$, and $D(-2, 2)$
5. triangle *ABC* with vertices $A(0, -1)$, $B(4, -3)$, and $C(-4, -5)$

Find the coordinates of each figure after a reflection over the y-axis. Then graph the figure and its reflected image.

6. parallelogram *PARL* with vertices $P(3, 5)$, $A(5, 4)$, $R(5, 1)$, and $L(3, 2)$
7. pentagon *PENTA* with vertices $P(-1, 3)$, $E(1, 1)$, $N(0, -2)$, $T(-2, -2)$, and $A(-3, 1)$

Lesson 11-1

(Pages 470–473)

Find the square of each number.

1. 4
2. 19
3. 13
4. 25
5. 9
6. 2
7. 14
8. 24
9. 40
10. 50
11. 100
12. 250

Find each square root.

13. $\sqrt{324}$
14. $\sqrt{900}$
15. $\sqrt{2{,}500}$
16. $\sqrt{576}$
17. $\sqrt{8{,}100}$
18. $\sqrt{676}$
19. $\sqrt{100}$
20. $\sqrt{784}$
21. $\sqrt{1{,}024}$
22. $\sqrt{841}$
23. $\sqrt{2{,}304}$
24. $\sqrt{3{,}025}$

Lesson 11-2

(Pages 475–477)

Estimate each square root to the nearest whole number.

1. $\sqrt{27}$
2. $\sqrt{112}$
3. $\sqrt{249}$
4. $\sqrt{88}$
5. $\sqrt{1{,}500}$
6. $\sqrt{612}$
7. $\sqrt{340}$
8. $\sqrt{495}$
9. $\sqrt{264}$
10. $\sqrt{350}$
11. $\sqrt{834}$
12. $\sqrt{3{,}700}$
13. $\sqrt{298}$
14. $\sqrt{101}$
15. $\sqrt{800}$

Use a calculator to find each square root to the nearest tenth.

16. $\sqrt{58}$
17. $\sqrt{750}$
18. $\sqrt{1{,}200}$
19. $\sqrt{1{,}000}$
20. $\sqrt{5{,}900}$
21. $\sqrt{999}$
22. $\sqrt{374}$
23. $\sqrt{512}$
24. $\sqrt{3{,}750}$
25. $\sqrt{255}$
26. $\sqrt{83}$
27. $\sqrt{845}$
28. $\sqrt{200}$
29. $\sqrt{500}$
30. $\sqrt{10{,}001}$

Lesson 11-3

(Pages 479–483)

Find the missing measure of each right triangle. Round to the nearest tenth if necessary.

1.
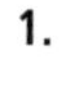
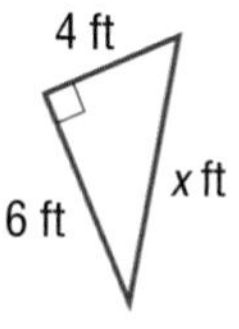

2.
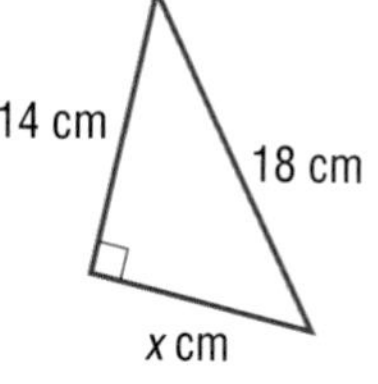

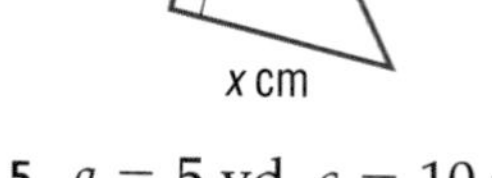

3.
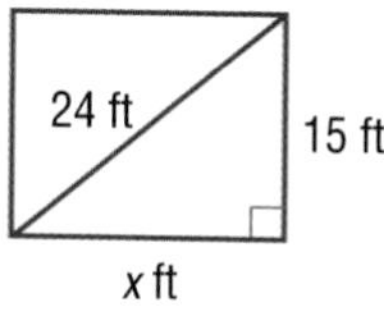
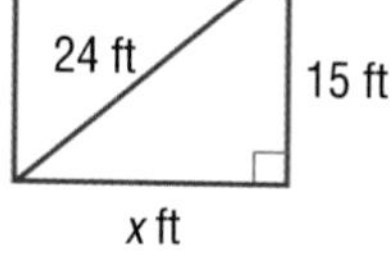

4. $a = 12$ cm, $b = 25$ cm
5. $a = 5$ yd, $c = 10$ yd
6. $b = 12$ mi, $c = 20$ mi
7. $a = 15$ yd, $b = 24$ yd
8. $a = 4$ m, $c = 12$ m
9. $a = 8$ mm, $b = 11$ mm
10. $a = 1$ mi, $c = 3$ mi
11. $a = 5$ yd, $b = 8$ yd
12. $b = 7$ in., $c = 19$ in.
13. $a = 50$ km, $c = 75$ km
14. $b = 82$ ft, $c = 100$ ft
15. $a = 100$ m, $b = 200$ m

Determine whether each triangle with the given side lengths is a right triangle. Write *yes* or *no*.

16. 6 mm, 8 mm, 10 mm
17. 12 ft, 15 ft, 20 ft
18. 300 m, 400 m, 500 m

Lesson 11-4

(Pages 483–485)

Find the area of each parallelogram. Round to the nearest tenth if necessary.

1.
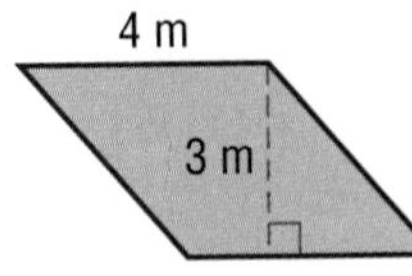

2.
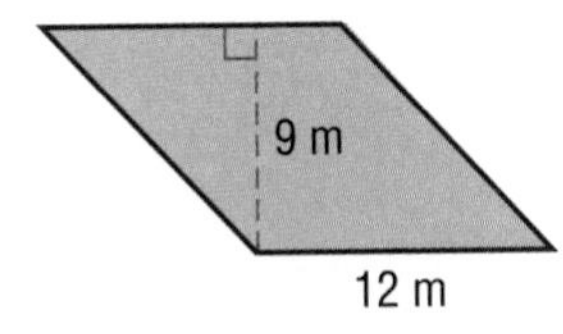

3.
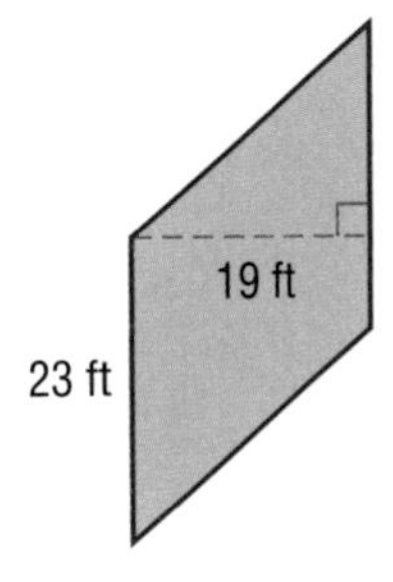

4. base = 19 m
   height = 6 m

5. base = 15 ft
   height = 12 ft

6. base = 25 yd
   height = 53 yd

7. base = 135 in.
   height = 15 in.

8. base = 8.2 cm
   height = 5.5 cm

9. base = 29.3 m
   height = 10.1 m

Lesson 11-5

(Pages 489–492)

Find the area of each figure. Round to the nearest tenth if necessary.

1.
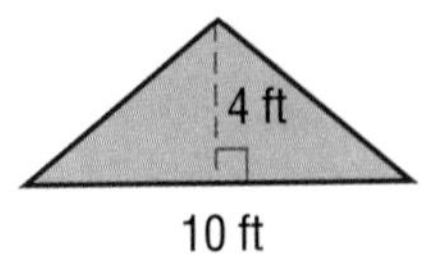

2.
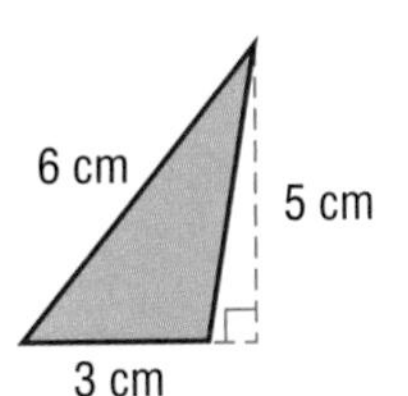

3.
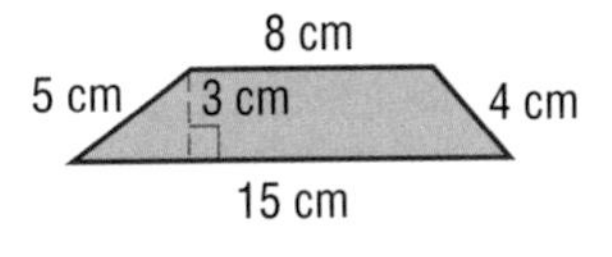

4. triangle: base = 5 in., height = 9 in.
5. trapezoid: bases = 3 cm and 8 cm, height = 12 cm
6. trapezoid: bases = 10 ft and 15 ft, height = 12 ft
7. triangle: base = 12 cm, height = 8 cm
8. trapezoid: bases = 82.6 cm and 72.2 cm, height = 44.5 cm
9. triangle: base = 500.5 ft, height = 254.5 ft

Lesson 11-6

(Pages 493–495)

Find the area of each circle. Round to the nearest tenth.

1.
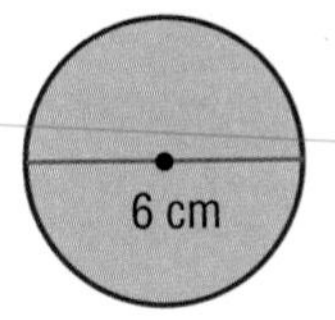

2.
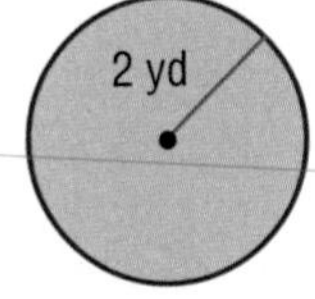
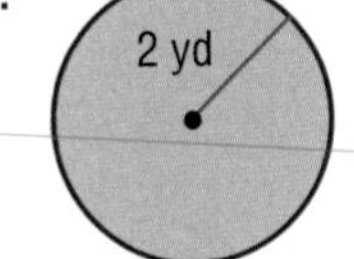

3.
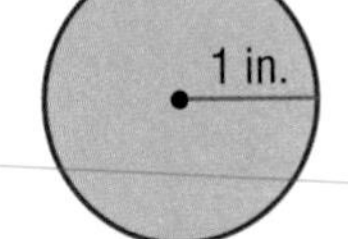
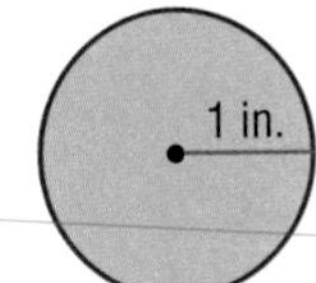

4. radius = 8 in.
5. diameter = 5 ft
6. radius = 24 cm
7. diameter = 2.3 m
8. diameter = 82 ft
9. radius = 68 cm
10. radius = 9.8 mi
11. diameter = 25.6 m
12. diameter = 6.75 in.
13. radius = $1\frac{1}{4}$ ft
14. diameter = $5\frac{2}{3}$ yd
15. diameter = $45\frac{1}{2}$ mi

Lesson 11-7

(Pages 498–500)

Find the area of each figure. Round to the nearest tenth if necessary.

1. 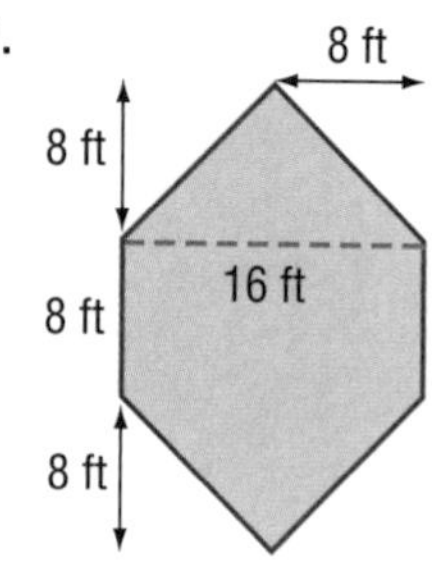

2. 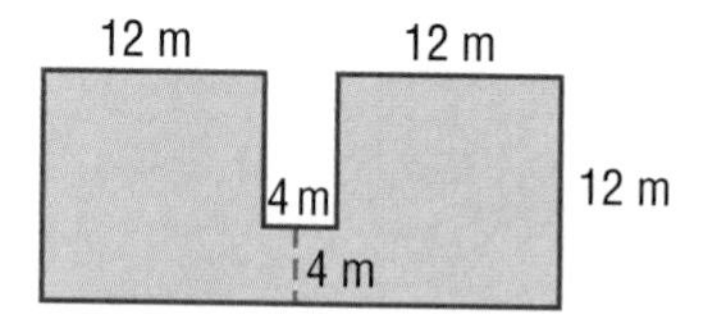

3. 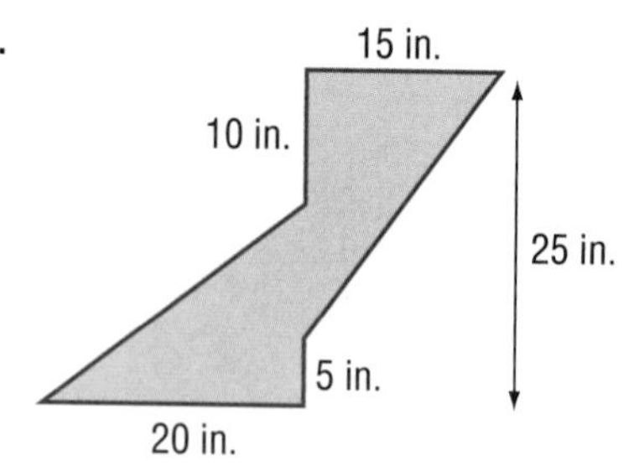

4. 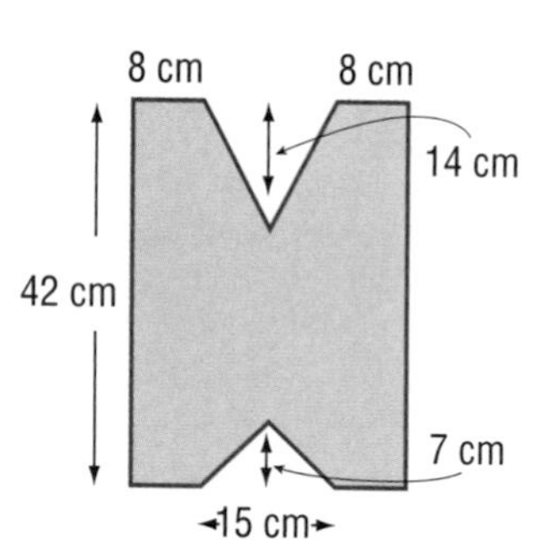

5. 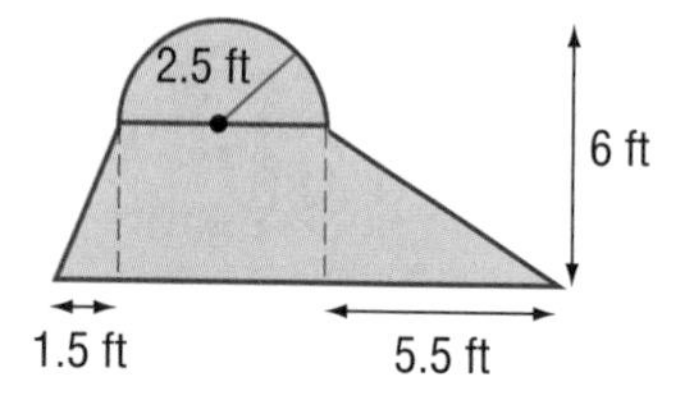

6. 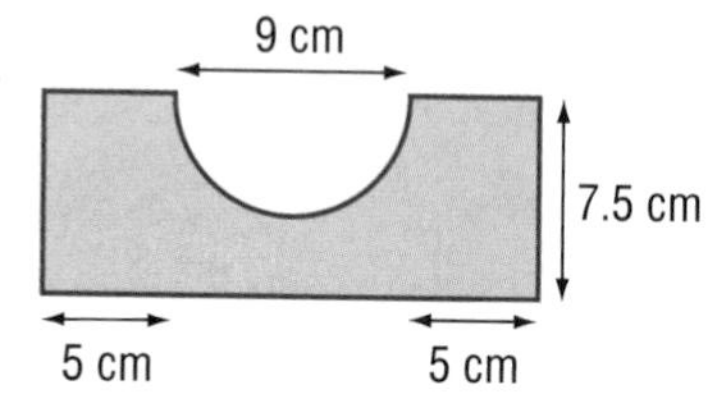

7. 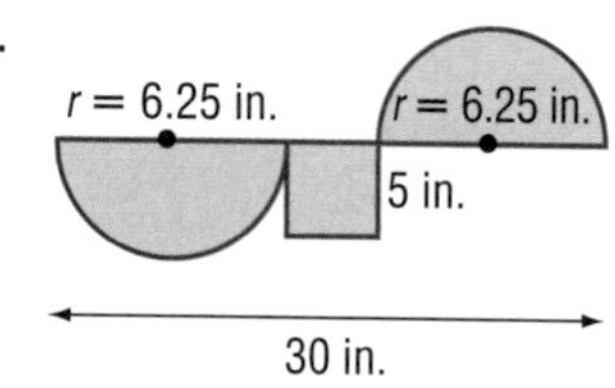

8. 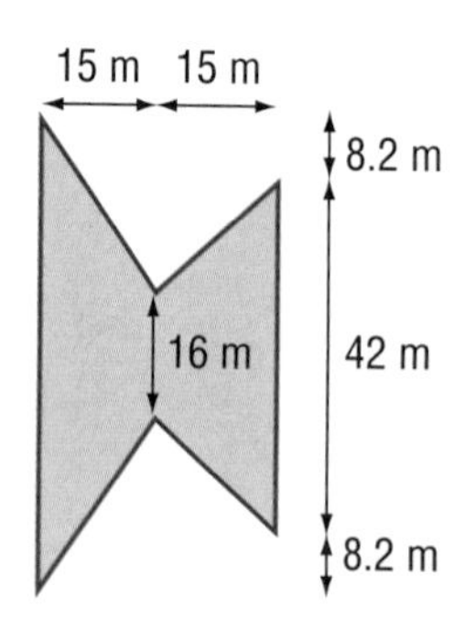

9. 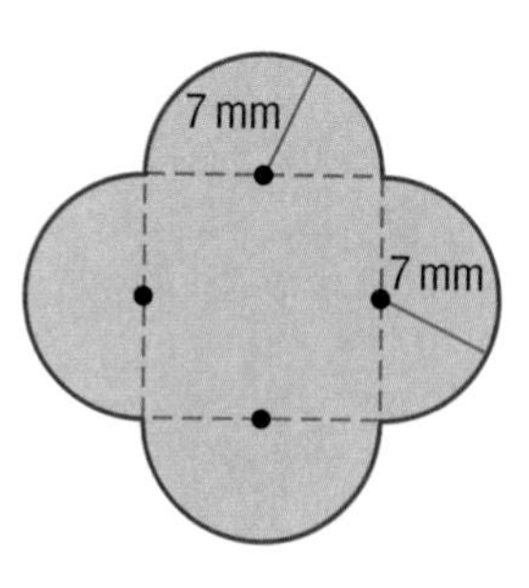

Lesson 11-8

(Pages 501–503)

A randomly dropped counter falls in the squares. Find the probability that it falls in the shaded squares. Write as a percent. Round to the nearest tenth if necessary.

1. 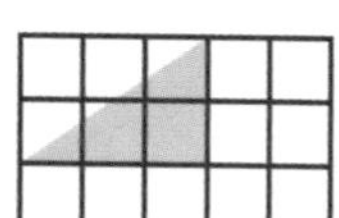

2. 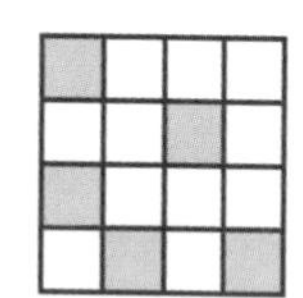

3. 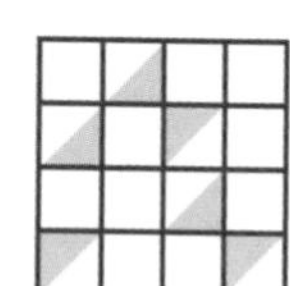

4. 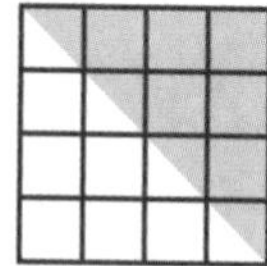 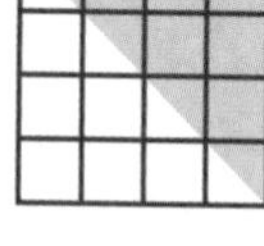

5. 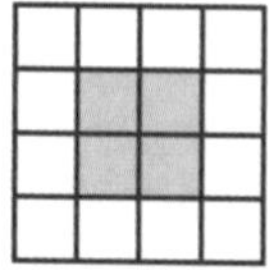 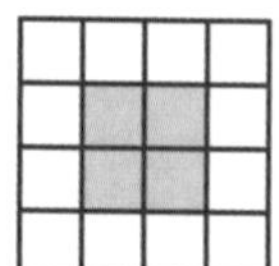

6. 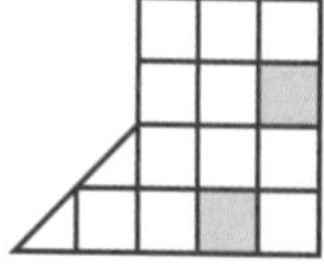 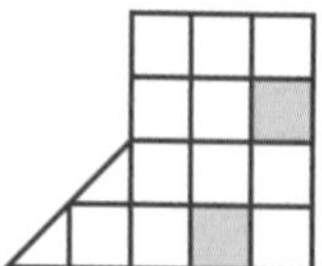

7. 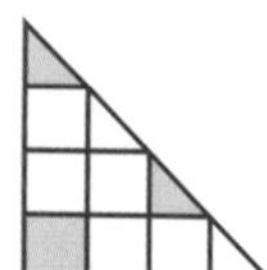 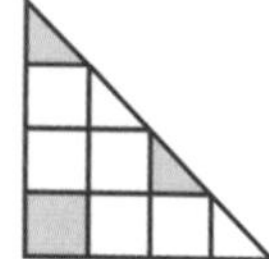

8. 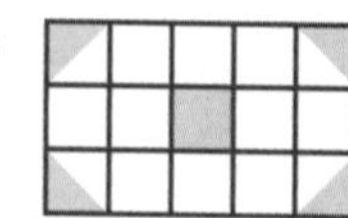

9. 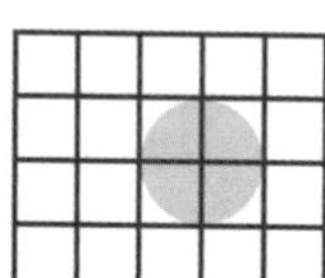

10. 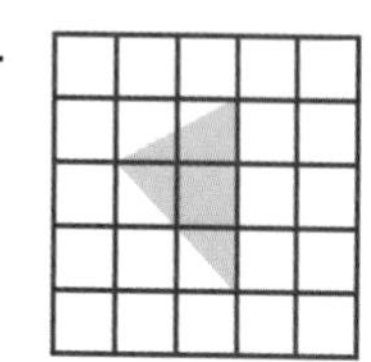

11. 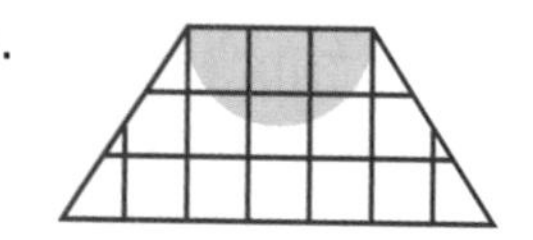

12. 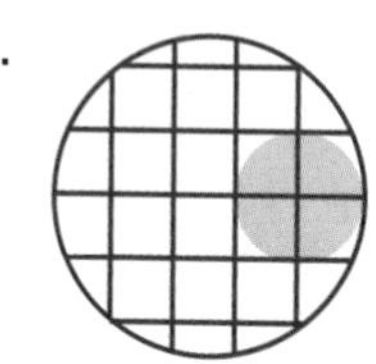

Lesson 12-1

(Pages 514–517)

Draw a top, a side, and a front view of each solid.

1. 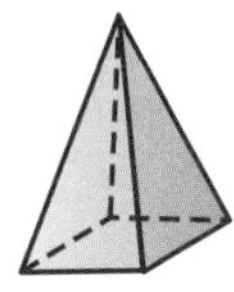

2. 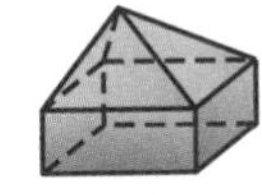

3. 

Draw each solid using the top, side, and front views shown. Use isometric dot paper.

4. top 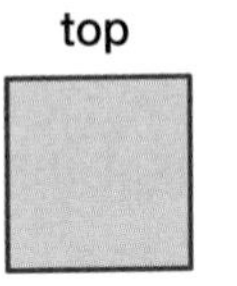 side 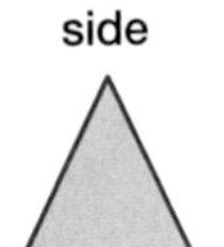 front 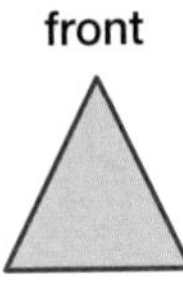

5. top 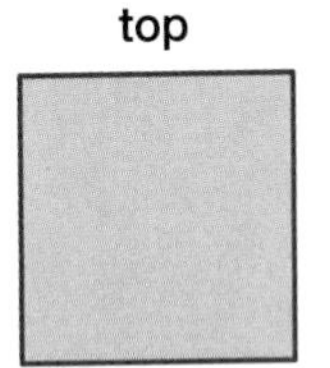 side 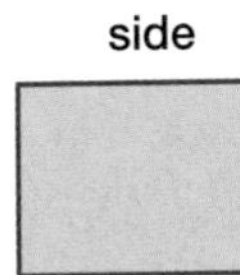 front 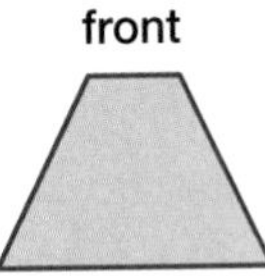

6. top  side 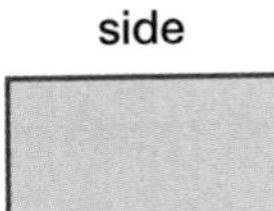 front 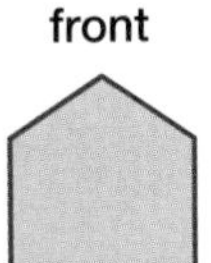

Lesson 12-2

(Pages 520–522)

Find the volume of each rectangular prism. Round to the nearest tenth if necessary.

1. 2. 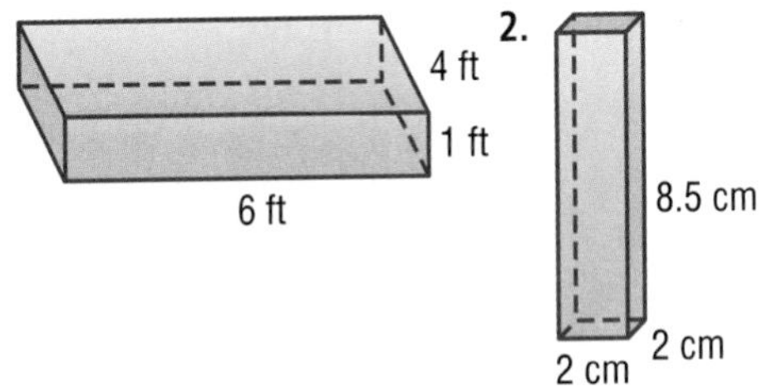

3. 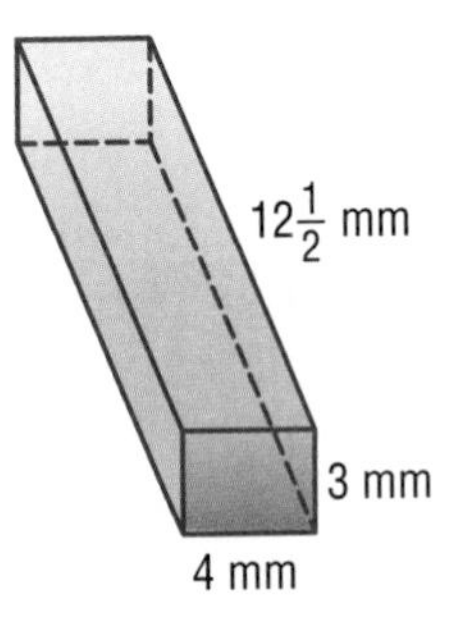

4. 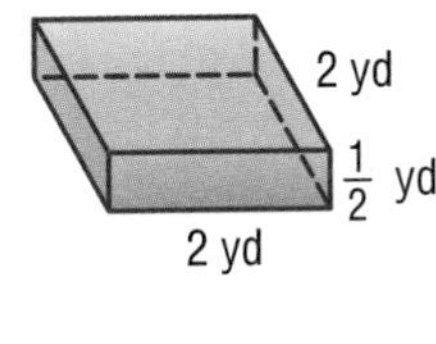

5. length = 3 ft
width = 10 ft
height = 2 ft

6. length = 18 cm
width = 23 cm
height = 15 cm

7. length = 25 mm
width = 32 mm
height = 10 mm

8. length = 1.5 in.
width = 3 in.
height = 6 in.

9. length = 4.5 cm
width = 6.75 cm
height = 2 cm

10. length = 16 mm
width = 0.7 mm
height = 12 mm

11. length = $3\frac{1}{2}$ ft
width = 10 ft
height = 6 ft

12. length = $5\frac{1}{2}$ in.
width = 12 in.
height = $3\frac{3}{8}$ in.

13. Find the volume of a rectangular prism with a length of 3 yards, a width of 5 feet, and a height of 12 feet.

Lesson 12-3

(Pages 524–527)

Find the volume of each cylinder. Round to the nearest tenth.

1. 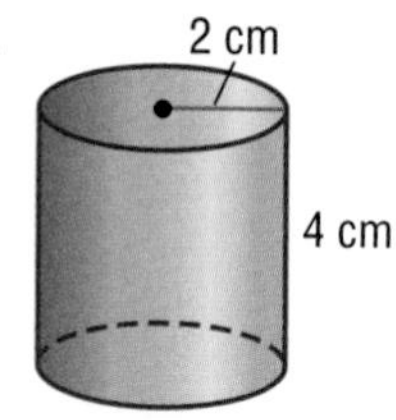

2. 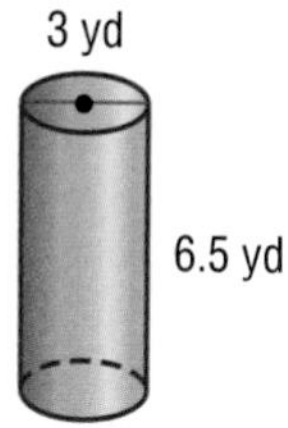

3. 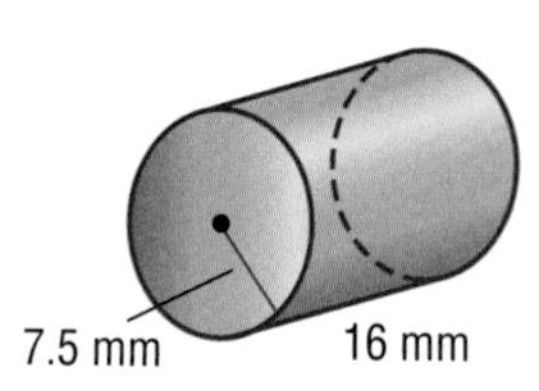

4. 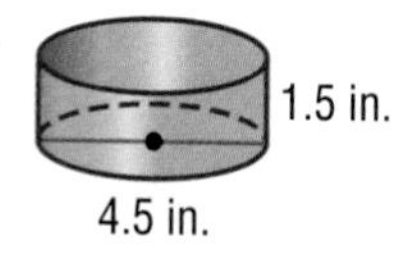

5. radius = 6 in.
   height = 3 in.
6. radius = 8 ft
   height = 10 ft
7. radius = 6 km
   height = 12 km
8. radius = 8.5 cm
   height = 3 cm
9. diameter = 16 yd
   height = 4.5 yd
10. diameter = 3.5 mm
    height = 2.5 mm
11. diameter = 12 m
    height = 4.75 m
12. diameter = $\frac{5}{8}$ in.
    height = 4 in.
13. diameter = 100 ft
    height = 35 ft
14. radius = 40.5 m
    height = 65.1 m
15. radius = 0.5 cm
    height = 1.6 cm
16. diameter = $8\frac{3}{4}$ in.
    height = $5\frac{1}{2}$ in.
17. Find the volume of a cylinder whose diameter is 6 inches and height is 2 feet.
18. How tall is a cylinder that has a volume of 2,123 cubic meters and a radius of 13 meters?

Lesson 12-4

(Pages 532–535)

Find the surface area of each rectangular prism. Round to the nearest tenth if necessary.

1. 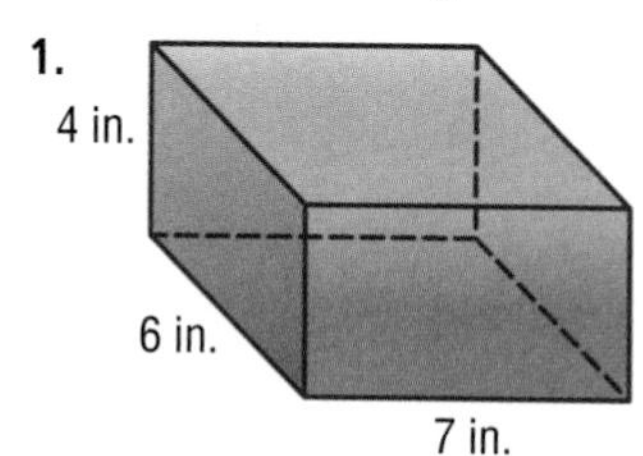

2. 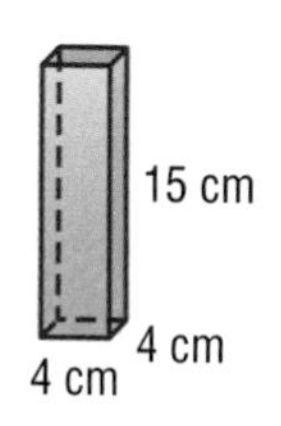

3. 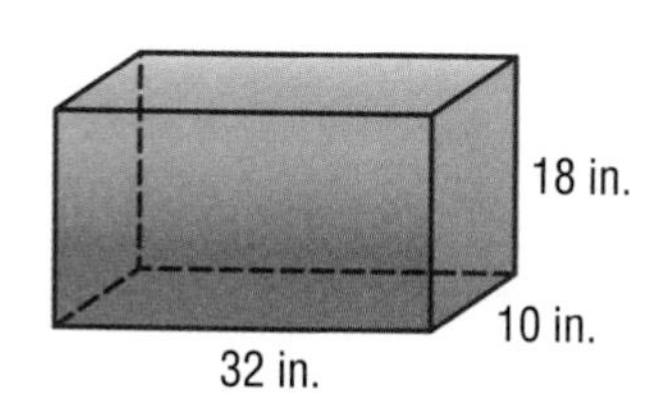

4. 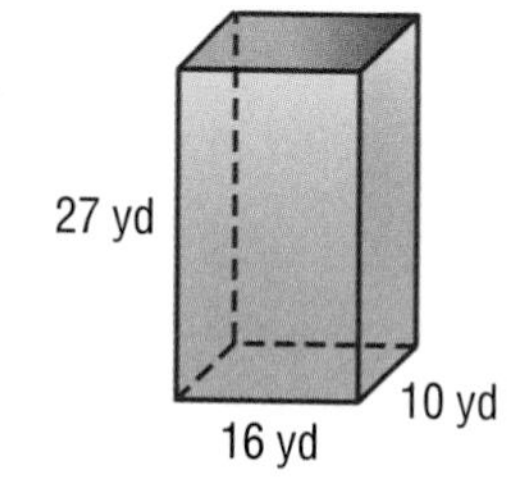

5. length = 10 m
   width = 6 m
   height = 7 m
6. length = 20 mm
   width = 15 mm
   height = 25 mm
7. length = 16 ft
   width = 20 ft
   height = 12 ft
8. length = 52 cm
   width = 48 cm
   height = 45 cm
9. length = 8 ft
   width = 6.5 ft
   height = 7 ft
10. length = 9.4 m
    width = 2 m
    height = 5.2 m
11. length = 20.4 cm
    width = 15.5 cm
    height = 8.8 cm
12. length = 8.5 mi
    width = 3 mi
    height = 5.8 mi
13. length = $7\frac{1}{4}$ ft
    width = 5 ft
    height = $6\frac{1}{2}$ ft
14. length = $15\frac{2}{3}$ yd
    width = $7\frac{1}{3}$ yd
    height = 9 yd
15. length = $4\frac{1}{2}$ in.
    width = 10 in.
    height = $8\frac{3}{4}$ in.
16. length = 12.2 mm
    width = 7.4 mm
    height = 7.4 mm
17. Find the surface area of an open-top box with a length of 18 yards, a width of 11 yards, and a height of 14 yards.
18. Find the surface area of a rectangular prism with a length of 1 yard, a width of 7 feet, and a height of 2 yards.

Lesson 12-5

(Pages 538–541)

Find the surface area of each cylinder. Round to the nearest tenth.

1. 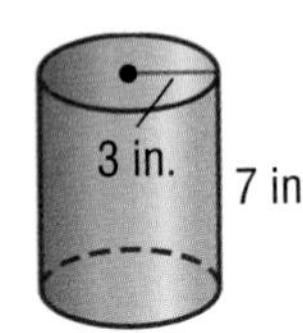

2. 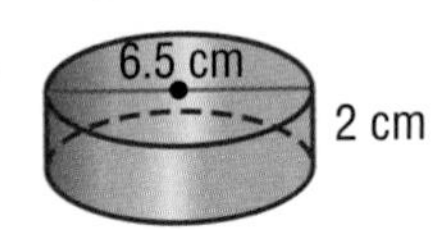

3. 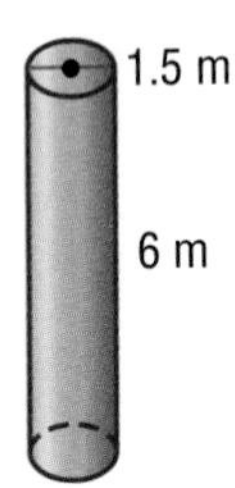

4. 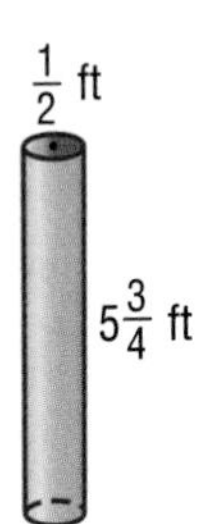

5. height = 6 cm
   radius = 3.5 cm

6. height = 16.5 mm
   diameter = 18 mm

7. height = 22 yd
   radius = 10.5 yd

8. height = 6 ft
   radius = 18.5 ft

9. height = 10.2 mi
   diameter = 4 mi

10. height = 8.6 cm
    diameter = 8.2 cm

11. height = 5.8 km
    diameter = 3.6 km

12. height = 32.7 m
    radius = 21.5 m

13. height = $2\frac{2}{3}$ yd
    diameter = 6 yd

14. height = $12\frac{3}{4}$ ft
    radius = $7\frac{1}{4}$ ft

15. height = $5\frac{1}{5}$ mi
    radius = $18\frac{1}{3}$ mi

16. height = $5\frac{1}{2}$ in
    diameter = 3 in

Lesson 12-6

(Pages 542–545)

Identify the precision unit of each measuring tool.

1. cm 1 2 3

2. 

3. 

4. 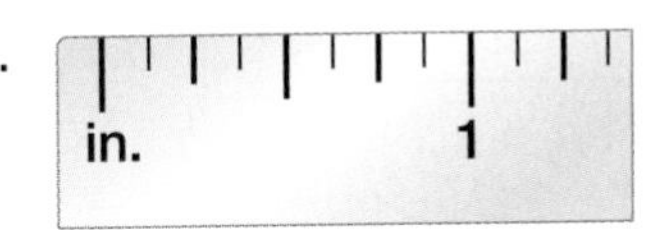

State each measure using significant digits.

5. 

6. 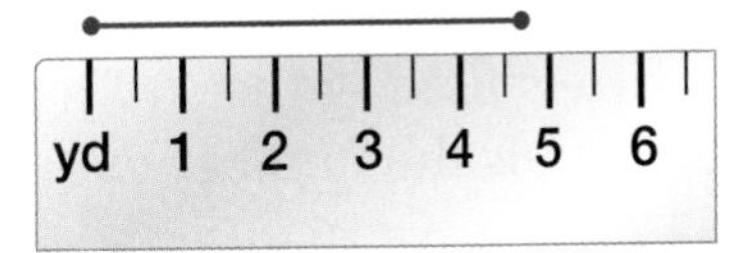

7. 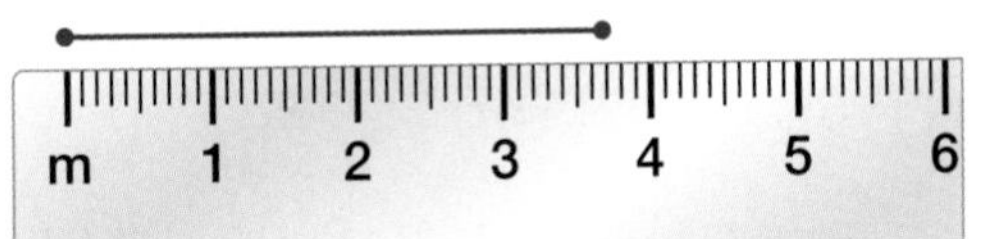

8. 

Mixed Problem Solving

Chapter 1 Decimal Patterns and Algebra

(pages 4–51)

HISTORY For Exercises 1 and 2, use the following information.
In 1932, Amelia Earhart flew 2,026 miles in 14 hours 56 minutes. (Lesson 1-1)

1. How many minutes did she fly?
2. To the nearest mile, what was her speed in miles per minute?

3. **TRANSPORTATION** An MD-80 aircraft burns 950 gallons of fuel per hour. How much fuel burns during a three-hour flight? (Lesson 1-1)

4. **LIGHT** The speed of light is about 67^3 kilometers per second. About how many kilometers per second is this? (Lesson 1-2)

Mixed Problem Solving

SALES For Exercises 5 and 6, use the following information.
A department store is having a back-to-school sale. The table shows the prices of three popular items.

Item	Price
jeans	$37.99
sweatshirt	$19.88
polo shirt	$22.50

Latonia wants to buy 2 pairs of jeans, 3 sweatshirts, and 1 polo shirt. (Lesson 1-3)

5. Write a numerical expression to represent the total cost of the six items.
6. Evaluate the expression to find the total cost.

7. **FITNESS** You can estimate how fast you walk in miles per hour by evaluating the expression $\frac{n}{30}$, where n is the number of steps you take in one minute. Find your speed if you take 96 steps in one minute. (Lesson 1-4)

8. **BASEBALL** Last year, Scott attended 13 Minnesota Twins baseball games. This year, he attended 24. Solve $13 + n = 24$ to find how many more games he attended this year than last. (Lesson 1-5)

9. **HOT AIR BALLOONS** Miyoki paid $140 for a four-hour hot air balloon ride over the Bridger Mountains. Solve $4h = 140$ to find the cost per hour of the ride. (Lesson 1-5)

ENTERTAINMENT For Exercises 10 and 11, use the following information.
The five members of the Wolff family went to an amusement park. They each purchased an all-day ride pass and a water park pass, as shown below. (Lesson 1-6)

Item	Price
all-day ride pass	$14.95
water park pass	$6.50

10. Use the Distributive Property to write two different expressions that represent the total cost for the family.
11. Find the total cost of the passes.

12. **NUMBER THEORY** Numbers that can be represented by a square arrangement of dots are called *square numbers*. The first four square numbers are shown below. (Lesson 1-7)

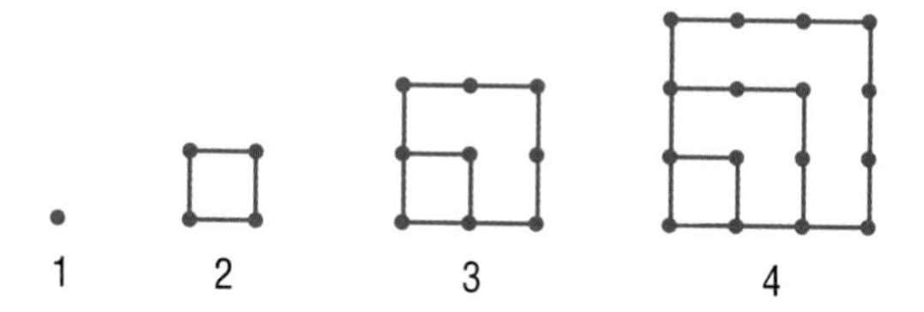

Write a sequence formed by the first eight square numbers.

13. **BIRDS** The world's smallest bird, the bee hummingbird, weighs only about 1.6 grams. How many milligrams is this? (Lesson 1-8)

14. **BRIDGES** The Akashi Kaikyo bridge in Japan is 2,003 meters long. What is the length of the bridge in kilometers? (Lesson 1-8)

15. **SPACE** The Andromeda Spiral is a galaxy 2,200,000 light years from Earth. Write this distance in scientific notation. (Lesson 1-9)

TRAVEL For Exercises 1 and 2, refer to the table below. It shows the number of passengers in twenty U.S. airports in one year. (Lesson 2-1)

Airport Passengers (millions)						
80	41	36	34	28	25	37
72	39	36	33	27	61	34
68	37	35	31	25	31	

Source: *The World Almanac*

1. Make a frequency table using the intervals 25–34, 35–44, 45–54, and so on.
2. Which interval has the greatest frequency? Explain what this means.
3. **DRIVING** The frequency table below shows the rural interstate speed limits for the fifty states. How many states have a speed limit of 75? (Lesson 2-1)

Speed Limit	Tally	Frequency
55	I	1
60		0
65	~~IIII~~ ~~IIII~~ ~~IIII~~ ~~IIII~~	20
70	~~IIII~~ ~~IIII~~ ~~IIII~~ III	18
75	?	?

Source: *The World Almanac*

SWIMMING For Exercises 4 and 5, refer to the table at the right. It shows the winning Olympic times for the Women's 4 × 100-meter Freestyle Relay in swimming. (Lesson 2-2)

Year	Time (s)
1976	225
1980	223
1984	224
1988	221
1992	220
1996	219
2000	217

Source: *ESPN Sports Almanac*

4. Make a line graph of the data.
5. Predict the winning time in this event in 2008.

NUTRITION For Exercises 6–9, use the data below. They are the grams of Carbohydrates in fifteen different energy bars. 16, 15, 20, 24, 16, 16, 16, 2, 20, 26, 14, 20, 20, 16, 16

6. Make a line plot of the data. (Lesson 2-3)
7. What is the range of the data? (Lesson 2-3)
8. Identify any clusters, gaps, or outliers and explain what they represent. (Lesson 2-3)
9. Find the mean, median, and mode. (Lesson 2-4)

BASKETBALL For Exercises 10–12, refer to the table below. It shows the number of games played by Michael Jordan each year from 1986–1987 to 2001–2002.

Number of Games Played							
82	82	81	82	82	80	78	0
17	82	82	82	0	0	0	60

Source: www.nba.com

10. Find the mean, median, and mode of the data. (Lesson 2-4)
11. Make a stem-and-leaf plot of the data. (Lesson 2-5)
12. Find the lower extreme, LQ, median, UQ, and upper extreme. (Lesson 2-6)

VACATION For Exercises 13–15, refer to the box-and-whisker plot below. It shows the number of vacation days for employees at a company. (Lesson 2-6)

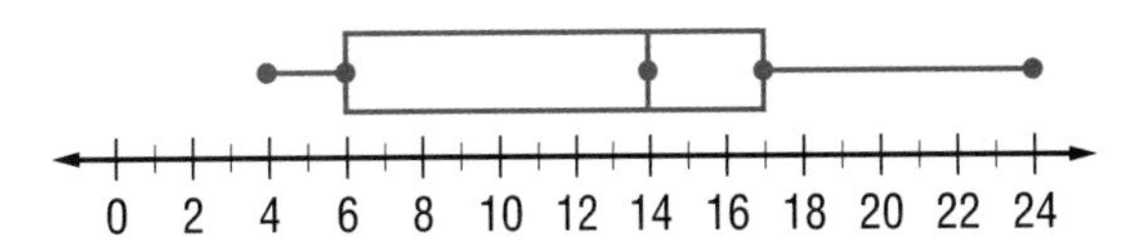

13. What fraction of the employees receive 6 to 14 vacation days?
14. What fraction of the employees receive 14 or more vacation days?
15. What is the range of the data?
16. **TOURISTS** The table shows the countries from which the most tourists in the United States came. Make a bar graph of the data. (Lesson 2-7)

Country	Number of Visitors (millions)
Canada	14.6
Mexico	10.3
Japan	5.0
United Kingdom	4.7

Source: Department of Commerce

TRAVEL For Exercises 17 and 18, refer to the data in Exercises 1 and 2.

17. Make a histogram of the data. (Lesson 2-7)
18. Would the mean or median best describe the data? Explain your reasoning. (Lesson 2-8)

Mixed Problem Solving

1. **AIR CONDITIONING** Jacob turned on the air conditioning, and the temperature in his apartment decreased 8 degrees. Write an integer to represent the drop in temperature. (Lesson 3-1)

EARTH SCIENCE For Exercises 2 and 3, use the table below. It describes the deepest land depressions in the world in feet below sea level.

Depth (ft)				
220	436	511	433	282
383	505	235	1,312	230

Source: *The Top 10 of Everything*

2. Write an integer to represent each depth. (Lesson 3-1)
3. Order the integers from greatest depth to least depth. (Lesson 3-2)

ENTERTAINMENT For Exercises 4–8, use the diagram below. It shows the locations of several rides at the Outlook Amusement Park. (Lesson 3-3)

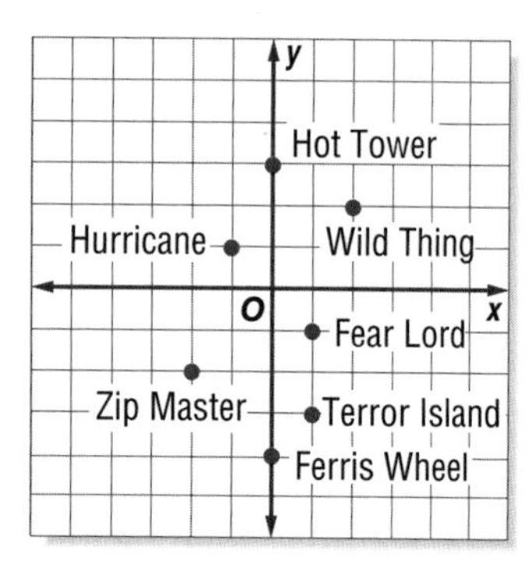

4. Which ride(s) is located in quadrant III?
5. Which ride(s) is located on the y-axis? Name the coordinates.
6. Which ride(s) has coordinates in which the x- and y-coordinates are equal?
7. In which quadrant is the Hurricane located?
8. A new ride is built with a location on the x-axis and 5 units left of the origin. Name the coordinates of this point.

9. **CAVERNS** Adriana is 52 feet underground touring the Lewis and Clark Caverns. She climbs a ladder up 15 feet. What is her new location? (Lesson 3-4)

10. **RECORDS** The lowest temperature recorded in Verkhoyansk, Russia, was about −90°F. The highest temperature was about 99°F. What is the difference between these temperatures? (Lesson 3-5)

11. **EARTH SCIENCE** The highest and lowest points in California are shown in the table. What is the difference in elevations? (Lesson 3-5)

Location	Elevation
Mount Whitney	14,494 ft above sea level
Death Valley	282 ft below sea level

Source: *The World Almanac of the U.S.A.*

12. **TEMPERATURE** Refer to the table below. How much colder is outer space than liquid helium? (Lesson 3-5)

Item	Temperature (°C)
dry ice	−108
outer space	−457
liquid helium	−452
bird's body	108

Source: *The Sizesaurus*

13. **RIDES** A glider ride over the Crazy Mountains has a maximum altitude of 12,000 feet, and it descends about 300 feet per minute. How long will it take the glider to reach an altitude of 6,000 feet? (Lesson 3-6)

14. **POOLS** A swimming pool holds 5,000 gallons of water. A hose can drain the pool at the rate of 225 gallons per hour. How much water will still be in the pool after it has been draining for 5 hours? (Lesson 3-6)

15. **TEMPERATURE** A temperature of −89°C was recorded in Antarctica. Use the expression $\frac{9C}{5} + 32$, where C is the temperature in degrees Celsius, to find the temperature in degrees Fahrenheit. (Lesson 3-7)

16. **DIVING** In two minutes, a scuba diver ascended from 40 feet below sea level to 8 feet below sea level. By how many feet did she ascend per minute? (Lesson 3-7)

1. **TOURISM** The Statue of Liberty in New York, New York, and the Eiffel Tower in Paris, France, were designed by the same person. The Statue of Liberty is 152 feet tall. It is 732 feet shorter than the Eiffel Tower, x. Write an equation that models this situation. (Lesson 4-1)

ELECTIONS For Exercises 2 and 3, use the table at the right and the following information.
New York has one more electoral vote than Texas. Pennsylvania has 9 fewer electoral votes than Texas.
(Lesson 4-2)

Number of Electoral Votes 2000	
California	54
New York	33
Texas	?
Florida	25
Pennsylvania	23

Source: *The World Almanac*

2. Write two different equations to find the number of electoral votes in Texas, n.
3. Find the number of electoral votes.

4. **ROLLER COASTERS** The track length of a popular roller coaster is 5,106 feet. The roller coaster has an average speed of about 2,000 feet per minute. At that speed, how long will it take to travel its length of 5,106 feet? Use the formula $d = rt$. (Lesson 4-3)

5. **BUSINESS** Carla's Catering charges a $25 fee to serve 15 or fewer people. In addition to that fee, they charge $10 per appetizer. You are having a party for 12 people and can spend a total of $85. How many appetizers can you order from Carla's Catering? (Lesson 4-4)

6. **MONEY** Koto works in a hardware store and earns $6 per hour. She plans on saving her money to buy a DVD player that costs $189. Write an inequality for the least number of hours x Koto needs to work in order to be able to purchase the DVD player. Then solve the inequality. (Lesson 4-5)

7. **CIVICS** The 26th amendment to the United States Constitution guarantees the right to vote to citizens who are eighteen years of age or older. Write an inequality showing the ages a of all voters. (Lesson 4-5)

8. **GEOMETRY** The formula for the perimeter of a square is $P = 4s$, where P is the perimeter and s is the length of a side. Graph the equation. (Lesson 4-6)

AGES For Exercises 9–12, use the table below. It shows how Jared's age and his sister Emily's age are related. (Lesson 4-6)

Jared's age (yr)	1	2	3	4	5
Emily's age (yr)	7	8	9	10	11

9. Write a verbal expression to describe how the ages are related.
10. Write an equation for the verbal expression. Let x represent Jared's age and y represent Emily's age.
11. Predict how old Emily will be when Jared is 10 years old.
12. Graph the equation.

13. **RAMPS** A ramp used to load cargo onto ships is modeled below. Is the slope of the ramp greater than or less than 1? Explain. (Lesson 4-7)

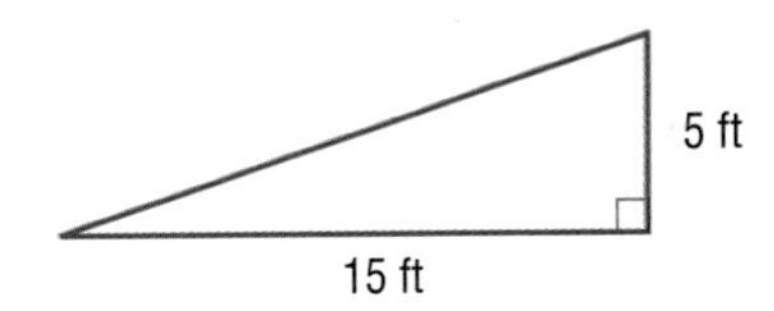

CYCLING For Exercises 14–16, use the table below. It shows the distances traveled by Jamal and Jasmine. (Lesson 4-7)

Hours Traveled	Cycling Distance (miles)	
	Jamal	Jasmine
1	8	11
2	16	22
3	24	33
4	32	44

14. Suppose each of these functions was graphed on a coordinate plane. Without calculating, which line is steeper? Explain.
15. Find the slope of each line.
16. What does the slope of each line represent?

Mixed Problem Solving

LAND For Exercises 1–3, use the information below.

A section of land is one mile long and one mile wide. (Lesson 5-1)

1. Write the prime factorization of 5,280.
2. Find the area of the section of land in square feet. (*Hint*: 1 mile = 5,280 feet)
3. Write the prime factorization of the area that you found in Exercise 2.

DECORATIONS For Exercises 4 and 5, use the information below.

Benito is cutting streamers from crepe paper to decorate for a party. He has a red roll of crepe paper 144 inches long, a white roll 192 inches long, and a blue roll 360 inches long. (Lesson 5-2)

4. If he wants to have all colors of streamers the same length, what is the longest length that he can cut?
5. How many total streamers can he cut?

OLYMPICS For Exercises 6 and 7, refer to the table below. It shows the medals won by the top three countries in the 2000 Summer Olympics.

Country	Medals		
	Gold	Silver	Bronze
United States	40	24	33
Russia	32	28	28
China	28	16	15

Source: *The World Almanac*

6. Write the number of gold medals that Russia won as a fraction of the total number that Russia won in simplest form. (Lesson 5-3)
7. Write the fraction that you wrote in Exercise 6 as a decimal. (Lesson 5-4)

8. **GEOGRAPHY** Alaska makes up about 0.16 of the total land area of the United States. Write this decimal as a fraction in simplest form. (Lesson 5-4)

9. **SPORTS** At Belgrade Intermediate School, 75 out of every 100 students participate in sports. What percent of students do *not* participate in sports? (Lesson 5-5)

ADVERTISING For Exercises 10–13, use the table below. It shows the results of a survey in which teens 13 to 17 years old were asked which types of advertising they pay attention to.

Type of Advertising	Percent of Teens
television	80%
magazine	62%
product in a movie	48%
ad in an E-mail	24%

Source: E-Poll

Write each percent as a fraction in simplest form. (Lesson 5-5)

10. television
11. magazine
12. product in a movie
13. ad in an E-mail

14. **POPULATION** In 2000, about 5.4% of the people in Wyoming lived in Laramie. Write this percent as a decimal. (Lesson 5-6)

GEOMETRY For Exercises 15–17, refer to the grid at the right. (Lesson 5-6)

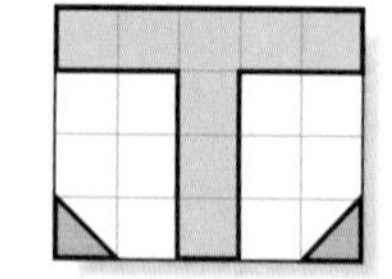

15. Write a decimal and a percent to represent the "T" shaded area.
16. Write a decimal and a percent to represent the area shaded pink.
17. What percent of the grid is *not* shaded?

18. **FLOWERS** Roses can be ordered in bunches of 6 and carnations, in bunches of 15. If Ingrid wants to have the same number of roses as carnations for parent night, what is the least number of each flower that she must order? (Lesson 5-7)

19. **WATER** The table at the right shows the fraction of each state that is water. Order the states from least to greatest fraction of water. (Lesson 5-8)

What Part is Water?	
State	Fraction
Alaska	$\frac{3}{41}$
Michigan	$\frac{40}{97}$
Wisconsin	$\frac{1}{6}$

Source: *The World Almanac of the U.S.A.*

1. **MEALS** A box of instant potatoes contains 20 cups of flakes. A family-sized bowl of potatoes uses $3\frac{2}{3}$ cups of the flakes. Estimate how many family-sized bowls can be made from one box. (Lesson 6-1)

RAINFALL For Exercises 2 and 3, use the table. It shows the average annual precipitation for three of the driest locations on Earth. (Lesson 6-2)

Location	Precipitation (in.)
Arica, Chile	$\frac{3}{100}$
Iquique, Chile	$\frac{1}{5}$
Callao, Peru	$\frac{12}{25}$

Source: *The Top 10 of Everything*

2. How much more rain does Iquique get per year than Arica?
3. How much more annual rain does Callao get than Iquique?

4. **CRAFTS** Kyle uses $2\frac{3}{4}$ yards of fabric for each craft item that he makes. If he uses a 25-yard fabric bolt and sells each item for $35, what is the greatest amount of money that he can earn? (Lesson 6-3)

5. **STARS** The star Sirius is about $8\frac{7}{10}$ light years from Earth. Alpha Centauri is half this distance from Earth. How far is Alpha Centauri from Earth? (Lesson 6-4)

LIFE SCIENCE For Exercises 6 and 7, use the table below. It shows the average growth per month of hair and fingernails. (Lesson 6-5)

Average Monthly Growth	
hair	$\frac{1}{2}$ in.
fingernails	$\frac{2}{25}$ in.

6. Solve $3 = \frac{1}{2}t$ to find how long it takes hair to grow 3 inches.
7. Solve $\frac{2}{25} = \frac{d}{12}$ to find how much fingernails grow in 1 year.

8. **SEWING** Jocelyn has nine yards of cloth to make table napkins for the senior citizens' center. She needs $\frac{3}{8}$ yard for each napkin. Use $\frac{3}{8}c = 9$ to find the number of napkins that she can make. (Lesson 6-5)

9. **WHALES** During the first year, a baby whale gains about $27\frac{3}{5}$ tons. What is the average weight gain per month? (Lesson 6-6)

10. **LIFE SCIENCE** An adult has about 5 quarts of blood. If a person donates 1 pint of blood, how many pints are left? (Lesson 6-7)

11. **COFFEE** In Switzerland, the average amount of coffee consumed per year is 1,089 cups per person. How many pints is this? (Lesson 6-7)

CHESS For Exercises 12–15, use the chess board below. (Lesson 6-8)

12. What is the perimeter of the chess board?
13. What is the area of the chess board?
14. What is the area of each small square?
15. A travel chess board has half the length and width of the board shown. What is the perimeter and area?

16. **LIFE SCIENCE** A nest built by bald eagles had a diameter of $9\frac{1}{2}$ feet. What was the circumference of the nest? (Lesson 6-9)

17. **EARTH SCIENCE** Earth has a diameter of 7,926 miles. Use the formula for the circumference of a circle to approximate the circumference of Earth at its equator. (Lesson 6-9)

Mixed Problem Solving

1. **SCHOOLS** In a recent year, Oregon had 924 public elementary schools and 264 public high schools. Write a ratio in simplest form comparing the number of public high schools to elementary schools. (Lesson 7-1)

2. **TEACHERS** The table shows the ratio of students to teachers for three states. Are any of the ratios equivalent? Explain. (Lesson 7-1)

State	Students:Teachers
Alabama	152:10
Kentucky	231:15
Montana	76:5

Source: *The World Almanac*

3. **EXERCISE** A person jumps rope 14 times in 10 seconds. What is the unit rate in jumps per second? (Lesson 7-2)

4. **FOOD** A 16-ounce box of cereal costs $3.95. Find the unit price to the nearest cent. (Lesson 7-2)

5. **PHOTOGRAPHS** Mandy is enlarging a photograph that is 3 inches wide and 4.5 inches long. If she wants the width of the enlargement to be 10 inches, what will be the length? (Lesson 7-3)

6. **GEOMETRY** If the ratio of length to width of a rectangle is about 1.6 to 1, then the rectangle is said to be a *golden rectangle*. Determine whether either figure below is golden rectangle. (Lesson 7-3)

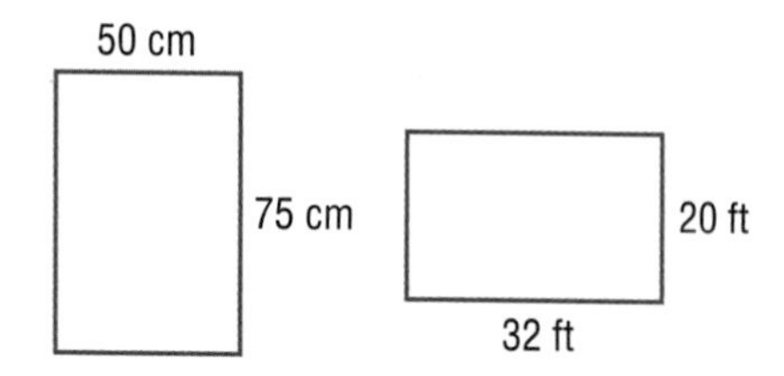

7. **MAPS** Washington, D.C., and Baltimore, Maryland, are $2\frac{7}{8}$ inches apart on a map. If the scale is $\frac{1}{2}$ inch:6 miles, what is the actual distance between the cities? (Lesson 7-4)

8. **MODELS** Ian is making a miniature bed for his daughter's doll house. The actual bed is $6\frac{3}{4}$ feet long. If he uses the scale $\frac{1}{2}$ inch $= 1\frac{1}{2}$ feet, what will be the length of the miniature bed? (Lesson 7-4)

9. **POPULATION** According to the U.S. Census Bureau, 6.6% of all people living in Florida are 10–14 years old. What fraction is this? (Lesson 7-5)

10. **INTERNET** In 2000, 41.5% of U.S. households had access to the Internet. Write this percent as a fraction in simplest form. (Lesson 7-5)

COINS For Exercises 11 and 12, use the table below. It shows the fraction of a quarter that is made up of the metals nickel and copper. Write each fraction as a percent. Round to the nearest hundredth if necessary. (Lesson 7-5)

Metal	Fraction of Quarter
nickel	$\frac{1}{12}$
copper	$\frac{11}{12}$

Source: *The Top 10 of Everything*

11. nickel

12. copper

13. **FARMS** Approximately 0.0009 of the land in Montana is farmland. Write this as a percent. (Lesson 7-6)

14. **POPULATION** In 2000, 1.5% of the U.S. population was 85 years and older. Write this percent as a fraction in simplest form. (Lesson 7-6)

15. **SEEDS** A packet of beans guarantees that 95% of its 200 seeds will germinate. How many seeds are expected to germinate? (Lesson 7-7)

16. **SKIS** Toshiro spent $520 on new twin-tip skis. This was 40% of the money he earned at his summer job. How much did he earn at his summer job? (Lesson 7-8)

1. **GEOGRAPHY** In Washington, about 5.7% of the total area is water. If the total area of Washington is 70,637 square miles, estimate the number of square miles of water by using 10%. (Lesson 8-1)

2. **POPULATION** According to the U.S. Census Bureau, 9.4% of the population in the Seattle-Bellevue metropolitan area is Asian. If this area has a population of 2,414,616, estimate the number of people who are Asian. (Lesson 8-1)

3. **FORESTS** In the United States, forests cover 24.7% of the land. If the total land area of the U.S. is 3,536,338 square miles, how many square miles are forests? Round to the nearest square mile. (Lesson 8-2)

4. **GOVERNMENT** Of the 435 members in the U.S. House of Representatives, 53 are from California and 13 are from North Carolina. To the nearest whole percent, what percent of the representatives are from California? from North Carolina? (Lesson 8-2)

FOOD For Exercises 5 and 6, use the graph below. It shows the results of a survey in which people were asked how they determine how long food has been in their freezer. (Lesson 8-3)

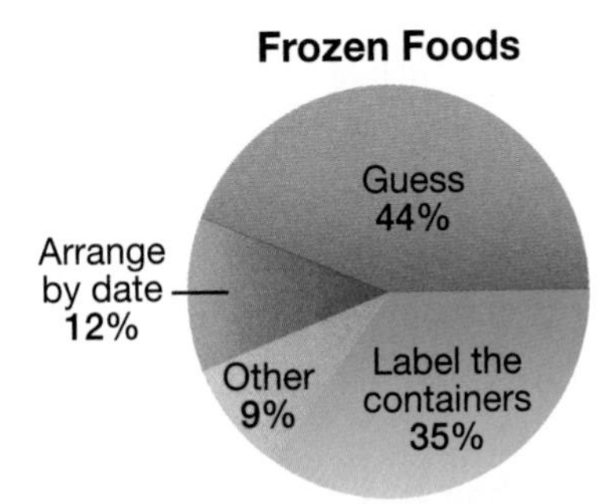

Source: Opinion Research Corp. for Tupperware

5. In a city of 56,000 people, predict how many of them guess to determine how long food has been in their freezer.

6. Suppose 420 people surveyed said they label their freezer containers. How many people took part in the survey?

SPORTS For Exercises 7–9, use the table below. It shows the number of participants ages 7 to 17 in the sports listed. (Lesson 8-4)

Sport	Number (millions)	
	1990	2000
in-line skating	3.6	21.8
snowboarding	1.5	4.3
roller hockey	1.5	2.2
golf	23.0	26.4

Source: National Sporting Goods Association

7. What is the percent of change in in-line skaters 7 to 17 years old from 1990 to 2000? Round to the nearest percent and state whether the percent of change is an *increase* or *decrease*.

8. Find the percent of change from 1990 to 2000 in the number of children and teens who played roller hockey. Round to the nearest percent.

9. Which sport had a greater percent of increase in participants from 1990 to 2000: golf or snowboarding? Explain.

10. **MUSIC** Find the total cost of a $15.50 CD if the sales tax is 8%. (Lesson 8-5)

COMPUTERS For Exercises 11 and 12, use the following information.
The Wares want to buy a new computer with a regular price of $1,049. (Lesson 8-5)

11. If the store is offering a 20% discount, what will be the sale price of the computer?

12. If the sales tax on the computer is 5.25%, what will be the total cost with the discount?

BANKING For Exercises 13–16, complete the table below. The interest earned is simple interest. (Lesson 8-6)

	Principal	Rate	Time (yr)	Interest Earned
13.	$1,525.00	5%	$2\frac{1}{2}$	?
14.	$2,250.00	4%	?	$337.50
15.	?	3.5%	4	$498.40
16.	$5,080.00	?	3	$952.50

1. **DENTISTS** A dental hygienist randomly chooses a toothbrush in a drawer containing 17 white, 12 green, and 5 blue toothbrushes. What is the probability that she chooses a green toothbrush? Write as a fraction in simplest form. (Lesson 9-1)

SURVEYS For Exercises 2 and 3, use the table below. It shows the results of a survey in which adults were asked how proud they were to be an American. (Lesson 9-1)

How Proud Are You?	
Response	**Number**
extremely	650
very	250
moderately	60
little/not at all	30
no opinion	10

Source: Gallup Poll

2. If one person participating in the survey is chosen at random, what is the probability that the person is extremely patriotic? Write as a fraction in simplest form.
3. If one person participating in the survey is chosen at random, what is the probability that he is *not* moderately patriotic? Write as a fraction in simplest form.

RANCHING For Exercises 4–6, use the following information.
For Roger to reach his cattle pasture, he must pass through three consecutive gates. Any of the three gates can be either *open* or *closed*. (Lesson 9-2)

4. Make a tree diagram to show all of the possible positions of the gates.
5. What is the probability that all three gates will be closed when Roger visits this pasture? Write as a fraction.
6. What is the probability that the first two gates are open? Write as a fraction in simplest form.
7. **HIKING** To *run the ridge* in the Bridger Mountains, you can choose two paths up the face of the first peak, four paths across the ridge, and three routes down the final peak. Find the number of ways you can *run the ridge*. (Lesson 9-3)
8. **SKATEBOARDS** World Sports makes skateboards with several different patterns on the deck. You can choose one of four deck lengths and one of six types of wheels. If they advertise that they have 120 different skateboards, how many deck patterns are there? (Lesson 9-3)
9. **READING** Mr. Steadman plans to read eight children's novels to his second graders during the school year. In how many ways can he arrange the books to be read? (Lesson 9-4)
10. **CRAFTS** Marina has print fabric in pink, blue, magenta, green, yellow, and tan. How many different stuffed bears can she make if each bear has only four different fabrics, and the order of the fabrics is not important? (Lesson 9-5)
11. **FOOD** The graph shows the results of a survey in which 7th graders at Plentywood Middle School were asked to name their favorite fruit.

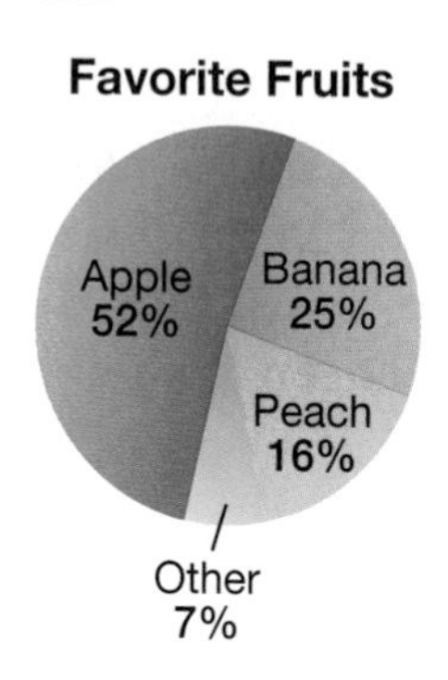

If a 7th grader at the school is randomly selected, what is the probability that they chose bananas as their favorite? Write as a fraction in simplest form. (Lesson 9-6)

12. **CANDY** A bag has 32 candies as shown in the table. What is the probability of randomly selecting one yellow candy and then one purple candy? Assume that the first candy is not replaced. (Lesson 9-7)

Color	Number
red	10
blue	6
purple	10
yellow	4
green	2

ART For Exercises 1 and 2, use the diagram of the Native American artifact. (Lesson 10-1)

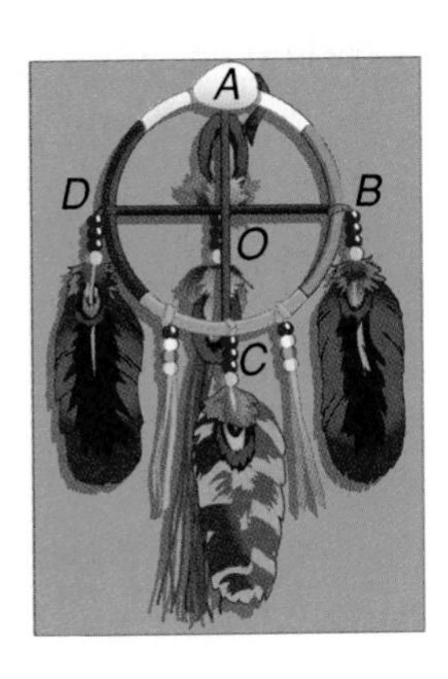

1. Name a right angle and a straight angle.
2. Suppose one side of an angle is $\overline{OA}$. Describe the location of point E so that $\angle AOE$ is acute.

TELEVISION For Exercises 3 and 4, use the table shown below. It shows the results of a survey in which families were asked how many T.V. channels they watched. (Lesson 10-2)

Channels that Families Watch	
Number	**Percent**
5 or fewer	30%
6–12	33%
13–25	19%
26 or more	14%

3. The fifth category in the survey is *No T.V. or no opinion*. What percent of the people surveyed were in this category?
4. Make a circle graph of the data.

PHYSICAL SCIENCE For Exercises 5 and 6, use the following information.
In physics, the *law of reflection* states that the angle of incidence of incoming light is equal in measure to the angle of reflection of outgoing light when light strikes a mirror, as shown in the diagram below. (Lesson 10-3)

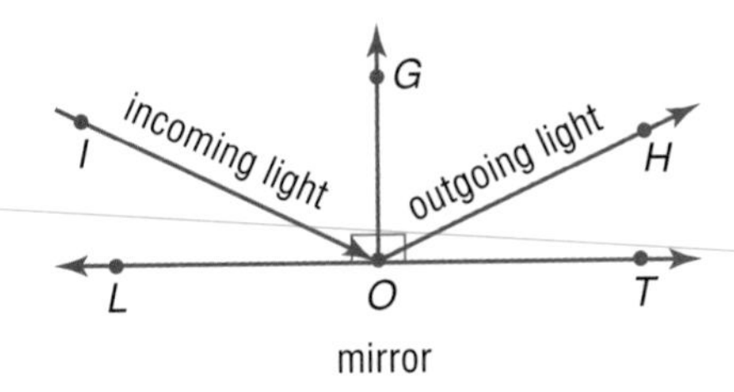

5. The angle of incidence is $\angle IOG$ and the angle of reflection is $\angle HOG$. If $m\angle IOG = 65°$, find $m\angle HOG$.
6. Find $m\angle LOI$. Name another angle with this measure.

For Exercises 7 and 8, use the figure below.

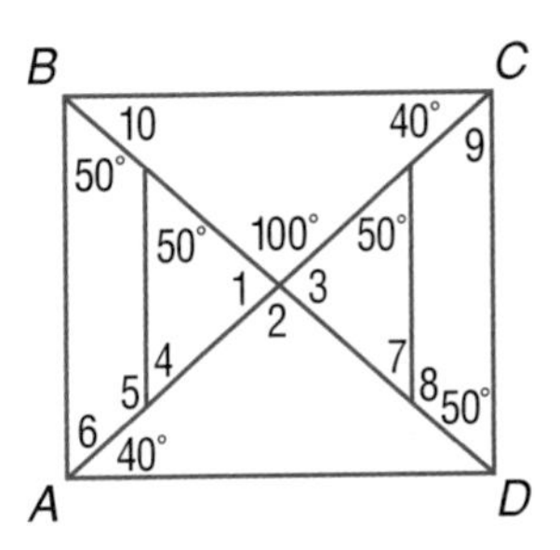

7. Find the measure of each angle numbered from 1–10. (Lesson 10-4)
8. Find the *best* name to classify quadrilateral $ABCD$. Explain your reasoning. (Lesson 10-5)
9. **CRAFTS** Priscilla makes porcelain dolls that are proportional to a real child. If Jody is $4\frac{2}{3}$ feet tall with a 23-inch waist, what should be the waist measure of a doll that is 13 inches tall? Round to the nearest inch. (Lesson 10-6)
10. **ART** Draw a tessellation using two of the polygons listed at the right. Identify the polygons and explain why the tessellation works. (Lesson 10-7)

regular triangles quadrilaterals pentagons hexagons octagons

For Exercises 11 and 12, use the quadrilateral *MOVE* shown below.

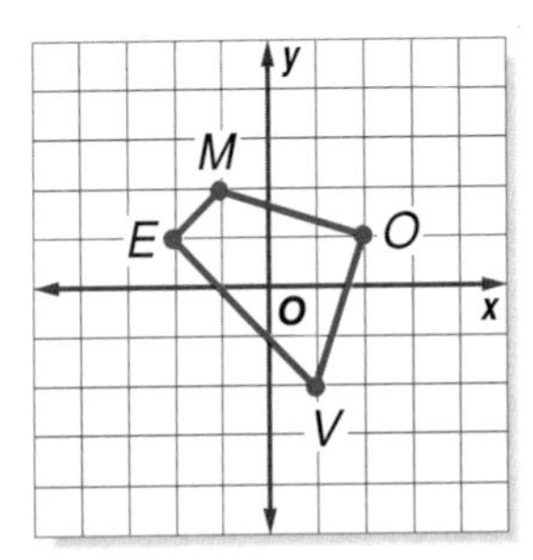

11. Describe the translation that will move M to the point at $(2, -2)$. Then graph quadrilateral $M'O'V'E'$ using this translation. (Lesson 10-8)
12. Find the coordinates of the vertices of quadrilateral $MOVE$ after a reflection over the y-axis. Then graph the reflection. (Lesson 10-9)

1. **PROPERTY** If a square section of land contains 1,742,400 square feet, what are the side dimensions? (Lesson 11-1)

2. **RECORDS** In Japan, a gigantic omelet was made with an area of 1,383 square feet. If the omelet was a square, what would be the side lengths? Round to the nearest tenth. (Lesson 11-2)

3. **SOFTBALL** A softball diamond is a square measuring 60 feet on each side.

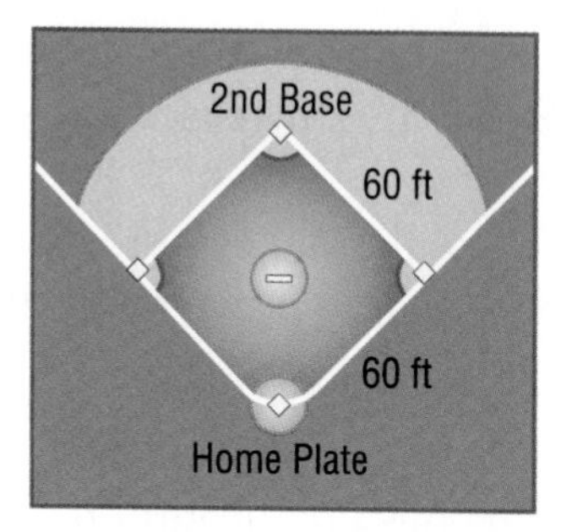

How far does a player on second base throw when she throws from second base to home? Round to the nearest tenth. (Lesson 11-3)

4. **BANDS** Mr. Garcia is planning a formation for the middle school band at a football game. The diagram shows the dimensions of the field.

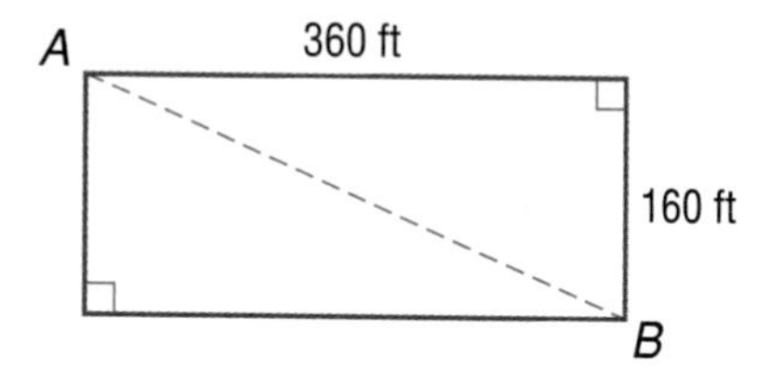

To the nearest foot, what is the distance from *A* to *B*? (Lesson 11-3)

5. **CRAFTS** A quilt pattern uses 25 parallelogram-shaped pieces of fabric, each with a base of 4 inches and a height of $2\frac{1}{2}$ inches. How much fabric is used to make the 25 pieces? (Lesson 11-4)

6. **FURNITURE** A corner table is in the shape of a right triangle. If the side lengths of the tabletop are 3.5 feet, 3.5 feet, and 4.9 feet, what is the area? Round to the nearest tenth if necessary. (Lesson 11-5)

7. **PUZZLES** A *tangram* is a square puzzle formed by the seven geometric figures shown at the right. Find the area of each small and large shaded triangle. (Lesson 11-5)

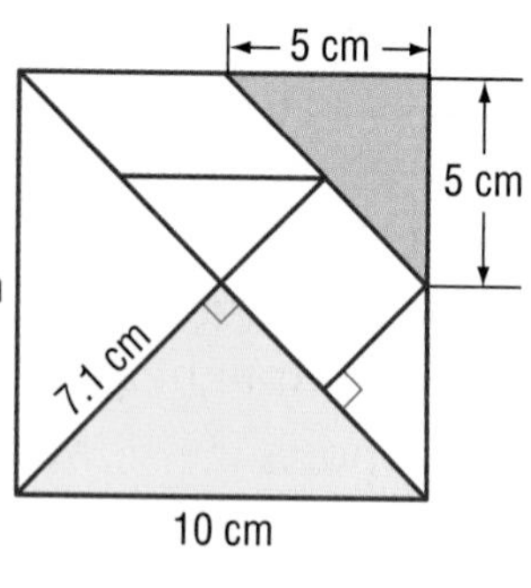

8. **COOKIES** In New Zealand, a giant circular chocolate chip cookie was baked with a diameter of 81 feet 8 inches. To the nearest square foot, what was the area of the cookie? (Lesson 11-6)

9. **LANDSCAPING** Find the area of the flower garden shown in the diagram at the right. Round to the nearest square foot. (Lesson 11-7)

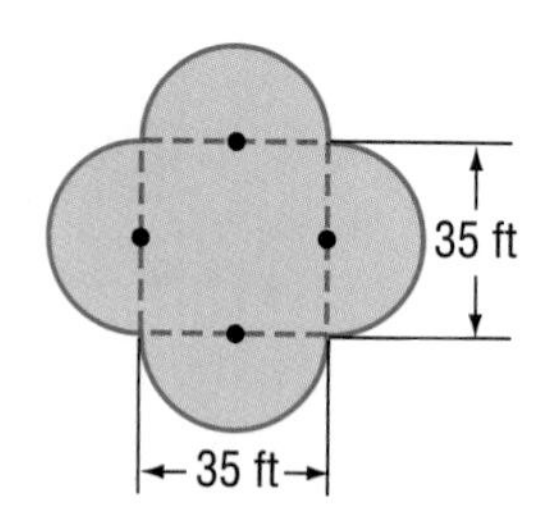

10. **CONSTRUCTION** A design for a deck is shown below. Find the area of the deck. (Lesson 11-7)

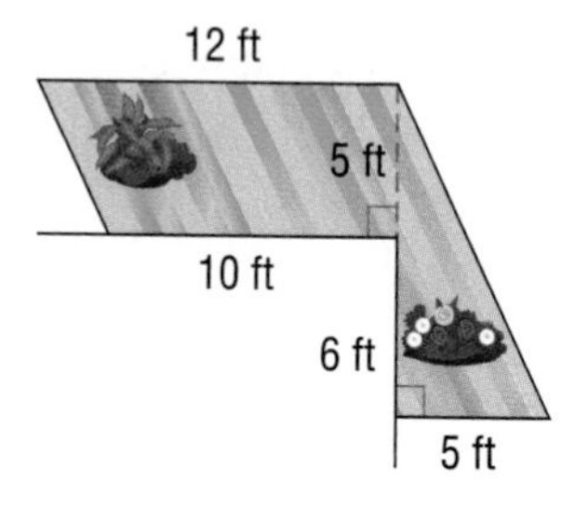

11. **GAMES** At the Moore Middle School carnival, one game has a dartboard as shown below.

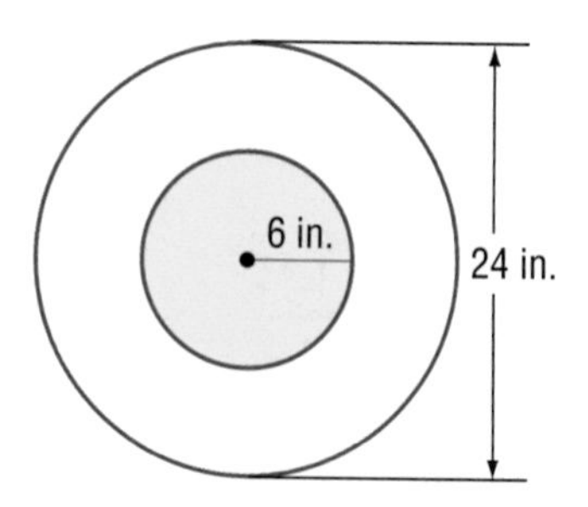

If it is equally likely that a thrown dart will land anywhere on the dartboard, find the probability that a dart lands in the shaded region. Write as a percent. (Lesson 11-8)

1. **RECORDS** According to the *Guinness Book of World Records*, the tallest hotel in the world is the 1,053-foot sail-shaped Burj Al Arab in Dubai, United Arab Emirates.

Draw possible sketches of the top, side, and front views of the hotel. (Lesson 12-1)

OCEANS For Exercises 2 and 3, use the following information.
The Atlantic Ocean has an area of about 33,420,000 square miles. Its average depth is 11,730 feet. (Lesson 12-2)

2. To the nearest hundredth, what is the average depth of the Atlantic Ocean in miles? (*Hint*: 1 mi = 5,280 ft)
3. What is the approximate volume of the Atlantic Ocean in cubic miles?

WATER For Exercises 4–6, use the cylinder-shaped water tank. (Lesson 12-3)

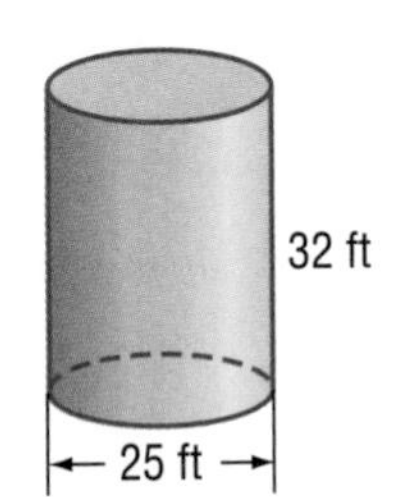

4. Find the volume of the tank. Round to the nearest cubic foot.
5. One gallon of water equals 231 cubic inches of water. Find the volume of the water tank in gallons.
6. A cylinder-shaped water tank holds 200,000 gallons of water. The height of the tank is 48 feet. To the nearest tenth of a foot, what is the radius of the tank?

For Exercises 7 and 8, use the following information.
Liz is designing some gift boxes. The small box is 6 inches long, 4 inches wide, and 2.5 inches high. The medium box has dimensions that are each 3 times the dimensions of the small box. (Lesson 12-4)

7. Find the surface area of the small box.
8. What are the dimensions of the medium box? Find the surface area of the medium box.

STORAGE For Exercises 9–11, use the following information.
The two canisters shown below each have a volume of about 628.3 cubic inches. (Lesson 12-5)

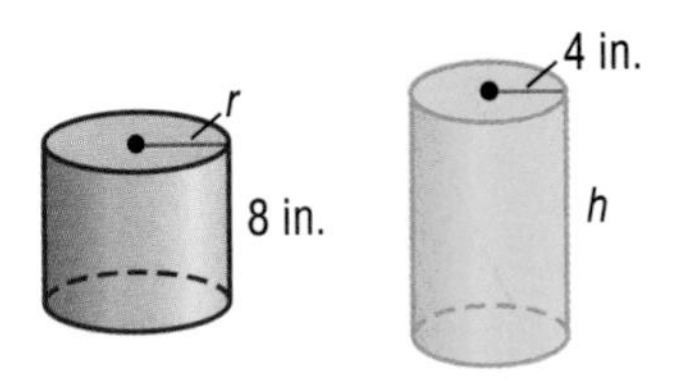

9. What is the radius of the blue canister? Round to the nearest tenth.
10. What is the height of the yellow canister? Round to the nearest tenth.
11. What is the difference between the surface area of the two canisters?

WATER For Exercises 12 and 13, refer to the water tank shown in Exercise 4. (Lesson 12-5)

12. What is the surface area of the water tank, including the bottom? Round to the nearest tenth.
13. A gallon of paint covers 400 square feet of surface. How many gallons are needed to cover the top and curved surface of the tank? Round to the nearest gallon.

For Exercises 14 and 15, use the following information.
In science class, Yvette and Tyrell use different rulers to measure the length of an earthworm. Using significant digits, Yvette records 13.2 centimeters, and Tyrell records 13.25 centimeters. (Lesson 12-6)

14. What is the most precise unit on Yvette's ruler? Explain.
15. What is the most precise unit on Tyrell's ruler? Explain.

Glossary/Glosario

Cómo usar el glosario en español:
1. Busca el término en inglés que desees encontrar.
2. El término en español, junto con la definición, se encuentran en la columna de la derecha.

English / Español

absolute value (p. 107) The distance the number is from zero on a number line.

valor absoluto Distancia a la que se encuentra un número de cero en la recta numérica.

acute angle (p. 413) An angle with a measure greater than 0° and less than 90°.

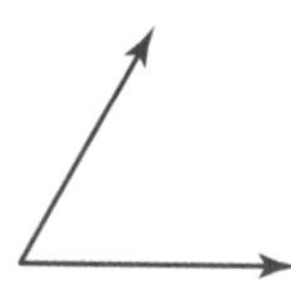

ángulo agudo Ángulo que mide más de 0° y menos de 90°.

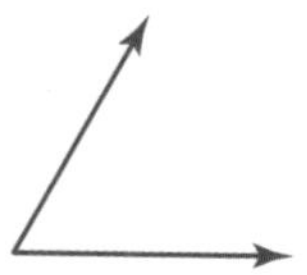

acute triangle (p. 429) A triangle having three acute angles.

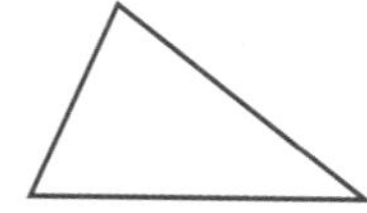

triángulo acutángulo Triángulo con tres ángulos agudos.

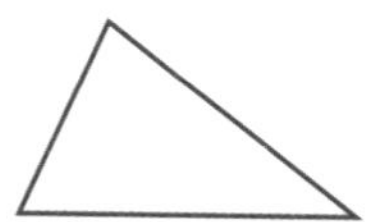

additive inverse (p. 121) The opposite of an integer. The sum of an integer and its additive inverse is zero.

inverso aditivo El opuesto de un entero. La suma de un entero y su inverso aditivo es cero.

algebra (p. 18) The branch of mathematics that involves expressions with variables.

álgebra Rama de las matemáticas que involucra expresiones con variables.

algebraic expression (p. 18) A combination of variables, numbers, and at least one operation.

expresión algebraica Combinación de variables, números y por lo menos una operación.

angle (p. 413) Two rays with a common endpoint form an angle. The rays and vertex are used to name the angle.

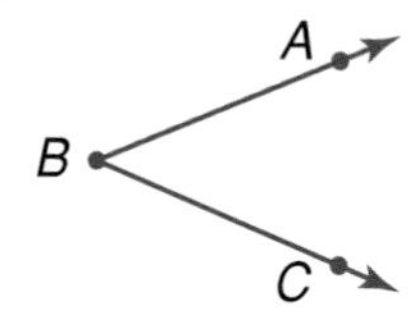

$\angle ABC$, $\angle CBA$, or $\angle B$

ángulo Dos rayos con un extremo común forman un ángulo. Los rayos y el vértice se usan para nombrar el ángulo.

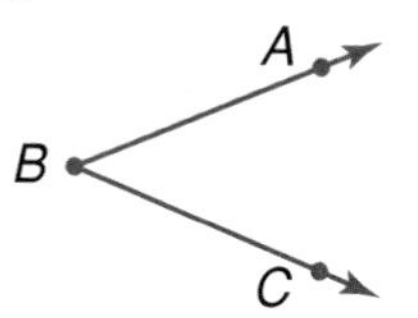

$\angle ABC$, $\angle CBA$, o $\angle B$

area (p. 271) The number of square units needed to cover a surface enclosed by a geometric figure.

área El número de unidades cuadradas necesarias para cubrir una superficie cerrada por una figura geométrica.

arithmetic sequence (p. 34) A sequence in which each term is found by adding the same number to the previous term.

sucesión aritmética Sucesión en que cada término se encuentra sumando el mismo número al término anterior.

Associative Property (p. 31) The way in which three numbers are grouped when they are added or multiplied does not change their sum or product.

propiedad asociativa La manera de agrupar tres números al sumarlos o multiplicarlos no cambia su suma o producto.

bar graph (p. 85) A graphic form using bars to make comparisons of statistics.

gráfica de barras Forma gráfica que usa barras para hacer comparaciones estadísticas.

bar notation (p. 211) In repeating decimals, the line or bar placed over the digits that repeat. For example, $2.\overline{63}$ indicates that the digits 63 repeat.

base (p. 10) In a power, the number used as a factor. In 10^3, the base is 10. That is, $10^3 = 10 \times 10 \times 10$.

base (p. 323) In a percent proportion, the whole quantity, or the number to which the part is being compared.

base (p. 483) The base of a parallelogram or triangle is any side of the figure. The bases of a trapezoid are the parallel sides.

box-and-whisker plot (p. 80) A diagram that summarizes data using the median, the upper and lower quartiles, and the extreme values. A box is drawn around the quartile values and whiskers extend from each quartile to the extreme data points.

notación de barra Línea o barra que se coloca sobre los dígitos que se repiten en decimales periódicos. Por ejemplo, $2.\overline{63}$ indica que los dígitos 63 se repiten.

base En una potencia, el número usado como factor. En 10^3, la base es 10. Es decir, $10^3 = 10 \times 10 \times 10$.

base En una proporción porcentual, la cantidad total o el número con el que se compara la parte.

base La base de un paralelogramo o triángulo es el lado de la figura. Las bases de un trapecio son los lados paralelos

diagrama de caja y patillas Diagrama que resume información usando la mediana, los cuartiles superior e inferior y los valores extremos. Se dibuja una caja alrededor de los cuartiles y se trazan patillas que los unan a los valores extremos respectivos.

C

center (p. 275) The given point from which all points on a circle or sphere are the same distance.

circle (p. 275) The set of all points in a plane that are the same distance from a given point called the center.

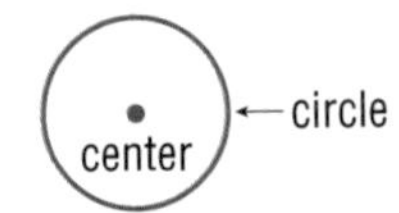

circle graph (p. 418) A type of statistical graph used to compare parts of a whole.

circumference (p. 275) The distance around a circle.

cluster (p. 65) Data that are grouped closely together.

coefficient (p. 19) The numerical factor of a term that contains a variable.

combination (p. 387) An arrangement, or listing, of objects in which order is not important.

common denominator (p. 227) A common multiple of the denominators of two or more fractions. 24 is a common denominator for $\frac{1}{3}$, $\frac{5}{8}$, and $\frac{3}{4}$ because 24 is the LCM of 3, 8, and 4.

Commutative Property (p. 31) The order in which two numbers are added or multiplied does not change their sum or product.

compatible numbers (p. 242) Numbers that are easy to compute mentally.

complementary angles (p. 422) Two angles are complementary if the sum of their measures is 90°.

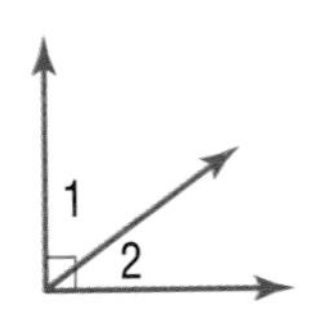

centro Un punto dado del cual equidistan todos los puntos de un círculo o de una esfera.

círculo Conjunto de todos los puntos en un plano que equidistan de un punto dado llamado centro.

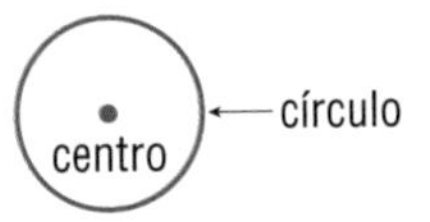

gráfica circular Tipo de gráfica estadística que se usa para comparar las partes de un todo.

circunferencia La distancia alrededor de un círculo.

agrupamiento Datos estrechamente agrupados.

coeficiente El factor numérico de un término que contiene una variable.

combinación Arreglo o lista de objetos donde el orden no es importante.

común denominador El múltiplo común de los denominadores de dos o más fracciones. 24 es un denominador común para $\frac{1}{3}$, $\frac{5}{8}$ y $\frac{3}{4}$ porque 24 es el mcm de 3, 8 y 4.

propiedad conmutativa El orden en que se suman o multiplican dos números no afecta su suma o producto.

números compatibles Números que son fáciles de computar mentalmente.

ángulos complementarios Dos ángulos son complementarios si la suma de sus medidas es 90°.

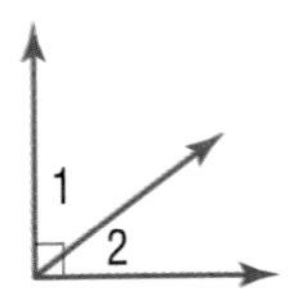

complementary events (p. 371) The events of one outcome happening and that outcome not happening are complementary events. The sum of the probabilities of complementary events is 1.

eventos complementarios Se dice de los eventos de un resultado que ocurren y el resultado que no ocurre. La suma de las probabilidades de eventos complementarios es 1.

complex figure (p. 498) A figure made of circles, rectangles, squares, and other two-dimensional figures.

figura compleja Una figura compuesta por círculos, rectángulos, cuadrados y otras dos figuras bidimencionales.

compound event (p. 398) An event consisting of two or more simple events.

evento compuesto Un evento que consiste en dos o más eventos simples.

composite number (p. 197) A whole number greater than 1 that has more than two factors.

número compuesto Un número entero mayor que 1 que tiene más de dos factores.

congruent angles (p. 422) Angles that have the same measure.

ángulos congruentes Ángulos que tienen la misma medida.

congruent segments (p. 429) Sides of a triangle having the same length.

segmentos congruentes Los lados de un triángulo que tienen la misma longitud.

constant (p. 19) A term that does not contain a variable.

constante Término que no contiene una variable.

coordinate grid (p. 112) A plane in which a horizontal number line and a vertical number line intersect at their zero points. Also called a coordinate plane.

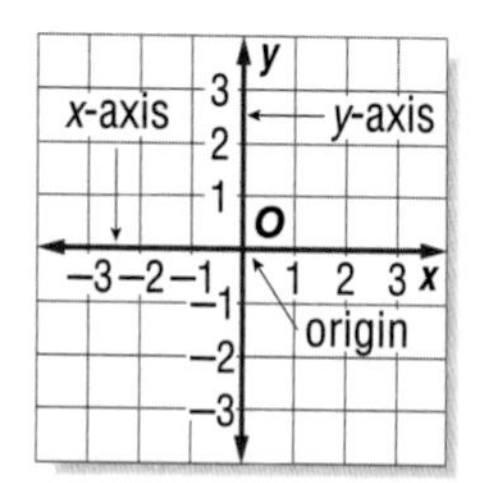

sistema de coordenadas Plano en el cual se han trazado dos rectas numéricas, una horizontal y una vertical, que se intersecan en sus puntos cero. También conocido como plano de coordenadas.

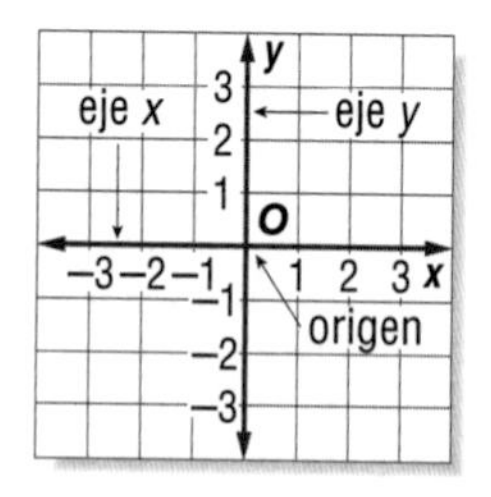

coordinate plane (p. 112) See coordinate grid.

plano de coordenadas Véase sistema de coordenadas.

cross product (p. 297) In a proportion, a cross product is the product of the numerator of one ratio and the denominator of the other ratio.

productos cruzados En una proporción, un producto cruzado es el producto del numerador de una razón y el denominador de la otra razón.

cubed (p. 10) The product in which a number is a factor three times. Two cubed is 8 because $2 \times 2 \times 2 = 8$.

al cubo El producto de un número por sí mismo, tres veces. Dos al cubo es 8 porque $2 \times 2 \times 2 = 8$.

cup (p. 267) A customary unit of capacity equal to 8 fluid ounces.

taza Unidad de capacidad del sistema inglés de medidas que equivale a 8 onzas líquidas.

cylinder (p. 524) A three-dimensional figure with two parallel congruent circular bases.

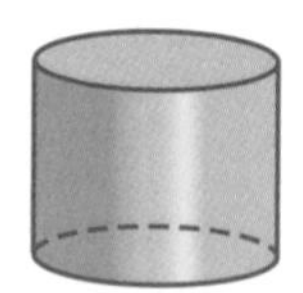

cilindro Figura tridimensional que tiene dos bases circulares congruentes y paralelas.

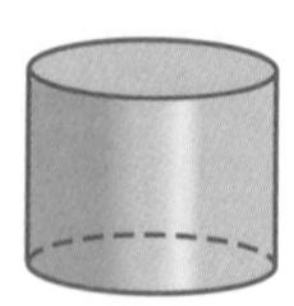

D

data (p. 54) Pieces of information, often numerical, which are gathered for statistical purposes.

datos Información, a menudo numérica, que se recoge con fines estadísticos.

decagon (p. 446) A polygon having ten sides.

defining the variable (p. 25) Choosing a variable to represent an unknown value in a problem, and using it to write an expression or equation to solve the problem.

degrees (p. 413) The most common unit of measure for angles. If a circle were divided into 360 equal-sized parts, each part would have an angle measure of 1 degree.

dependent events (p. 399) Two or more events in which the outcome of one event does affect the outcome of the other event(s).

diameter (p. 275) The distance across a circle through its center.

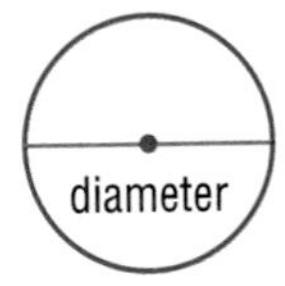

discount (p. 355) The amount by which the regular price of an item is reduced.

Distributive Property (p. 30) To multiply a sum by a number, multiply each addend of the sum by the number outside the parentheses.

domain (p. 177) The set of input values for a function.

decágono Un polígono con diez lados.

definir una variable El elegir una variable para representar un valor desconocido en un problema y usarla para escribir una expresión o ecuación para resolver el problema.

grados La unidad más común para medir ángulos. Si un círculo se divide en 360 partes iguales, cada parte tiene una medida angular de 1 grado.

eventos dependientes Dos o más eventos en que el resultado de uno de ellos afecta el resultado de los otros eventos.

diámetro La distancia a través de un círculo pasando por el centro.

descuento La cantidad de reducción del precio normal.

propiedad distributiva Para multiplicar una suma por un número, multiplica cada sumando de la suma por el número fuera del paréntesis.

dominio El conjunto de valores de entrada de una función.

equation (p. 24) A mathematical sentence that contains an equals sign, =.

equilateral triangle (p. 429) A triangle having three congruent sides.

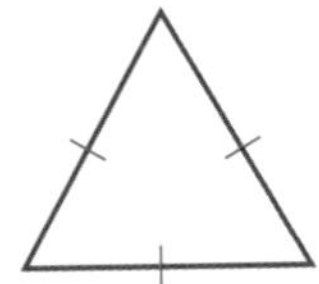

equivalent expressions (p. 30) Expressions that have the same value.

equivalent fractions (p. 207) Fractions that have the same value. $\frac{2}{3}$ and $\frac{4}{6}$ are equivalent fractions.

equivalent ratios (p. 289) Two ratios that have the same value.

evaluate (p. 11) To find the value of an expression.

experimental probability (p. 393) An estimated probability based on the relative frequency of positive outcomes occurring during an experiment.

exponent (p. 10) In a power, the number that tells how many times the base is used as a factor. In 5^3, the exponent is 3. That is, $5^3 = 5 \times 5 \times 5$.

exponential form (p. 11) Numbers written with exponents.

ecuación Enunciado matemático que contiene un signo de igualdad, =.

triángulo equilátero Triángulo con tres lados congruentes.

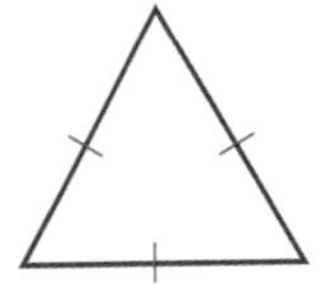

expresiones equivalentes Expresiones que tienen el mismo valor.

fracciones equivalentes Fracciones que tienen el mismo valor. $\frac{2}{3}$ y $\frac{4}{6}$ son fracciones equivalentes.

razones equivalentes Dos razones que tienen el mismo valor.

evaluar Calcular el valor de una expresión.

probabilidad experimental Estimado de una probabilidad que se basa en la frecuencia relativa de los resultados positivos que ocurren durante un experimento.

exponente En una potencia, el número que indica las veces que la base se usa como factor. En 5^3, el exponente es 3. Es decir, $5^3 = 5 \times 5 \times 5$.

forma exponencial Números escritos usando exponentes.

factorial (p. 381) The expression $n!$ is the product of all counting numbers beginning with n and counting backward to 1.

factors (p. 10) Two or more numbers that are multiplied together to form a product.

factor tree (p. 198) A diagram showing the prime factorization of a number. The factors branch out from the previous factors until all of the factors are prime numbers.

fair game (p. 374) A game in which players of equal skill have an equal chance of winning.

formula (p. 271) An equation that shows a relationship among certain quantities.

frequency table (p. 54) A table for organizing a set of data that shows the number of pieces of data that fall within given intervals or categories.

function (p. 177) A relation in which each element of the input is paired with exactly one element of the output according to a specified rule.

function table (p. 177) A table used to organize the input numbers, output numbers, and the function rule.

Fundamental Counting Principle (p. 378) Uses multiplication of the number of ways each event in an experiment can occur to find the number of possible outcomes in a sample space.

factorial La expresión $n!$ es el producto de todos los números naturales, comenzando con n y contando al revés hasta 1.

factores Dos o más números que se multiplican entre sí para formar un producto.

diagrama de árbol Diagrama que muestra la factorización prima de un número. Los factores se ramifican a partir de los factores previos hasta que todos los factores son números primos.

juego justo Juego en que jugadores con la misma habilidad tienen igual oportunidad de ganar.

fórmula Ecuación que muestra una relación entre ciertas cantidades.

tabla de frecuencias Tabla que se usa para organizar un conjunto de datos y que muestra cuántas piezas de datos caen dentro de intervalos o categorías dadas.

función Relación en que cada elemento de entrada es apareado con un único elemento de salida, según una regla específica.

tabla de funciones Tabla que organiza las entradas, la regla y las salidas de una función.

Principio Fundamental de Contar Este principio usa la multiplicación del número de veces que puede ocurrir cada evento en un experimento para calcular el número de posibles resultados en un espacio muestral.

G

gallon (p. 267) A customary unit of capacity equal to four quarts.

geometric sequence (p. 34) A sequence in which each term can be found by multiplying the previous term by the same number.

gram (p. 39) A unit of mass in the metric system equivalent to 0.001 kilogram.

graph (p. 106) The process of placing a point on a number line at its proper location.

greatest common factor (GCF) (p. 203) The greatest of the common factors of two or more numbers. The GCF of 18 and 24 is 6.

galón Unidad de capacidad del sistema inglés de medidas que equivale a 4 cuartos de galón.

sucesión geométrica Sucesión en que cada término se puede calcular multiplicando el término anterior por el mismo número.

gramo Unidad de masa del sistema métrico. Un gramo equivale a 0.001 de kilogramo.

graficar Proceso de dibujar o trazar un punto en una recta numérica en su ubicación correcta.

máximo común divisor (MCD) El mayor factor común de dos o más números. El MCD de 18 y 24 es 6.

H

height (p. 483) The length of the segment perpendicular to the base with endpoints on opposite sides. In a triangle, the distance from a base to the opposite vertex.

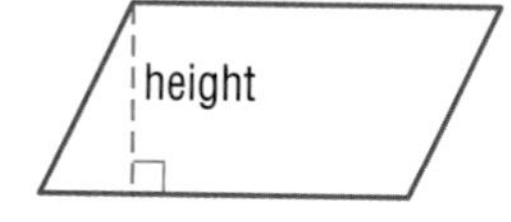

altura Longitud del segmento perpendicular a la base y con extremos en lados opuestos. En un triángulo, es la distancia desde una base al vértice opuesto.

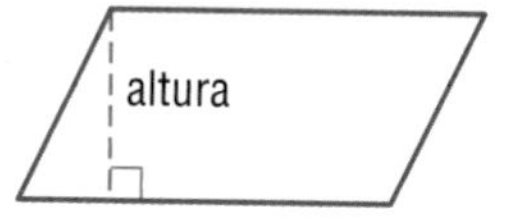

heptagon (p. 446) A polygon having seven sides.

hexagon (p. 446) A polygon having six sides.

histogram (p. 86) A special kind of bar graph in which the bars are used to represent the frequency of numerical data that have been organized in intervals.

hypotenuse (p. 479) The side opposite the right angle in a right triangle.

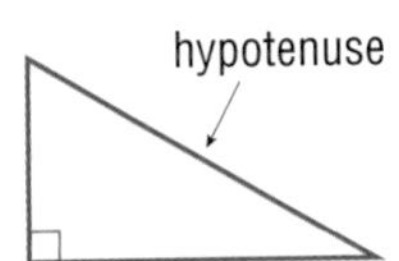

heptágono Polígono con siete lados.

hexágono Polígono con seis lados.

histograma Tipo especial de gráfica de barras que usa barras para representar la frecuencia de los datos numéricos, los cuales han sido organizados en intervalos iguales.

hipotenusa El lado opuesto al ángulo recto en un triángulo rectángulo.

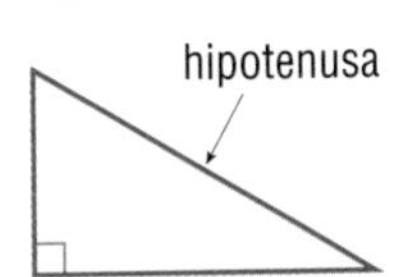

I

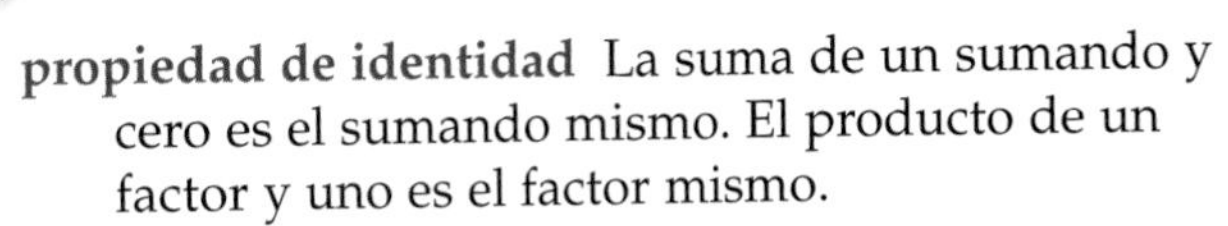

Identity Property (p. 31) The sum of an addend and zero is the addend. The product of a factor and one is the factor.

indirect measurement (p. 441) Finding a measurement by using similar triangles and writing a proportion.

integer (p. 106) The whole numbers and their opposites.
$\ldots, -3, -2, -1, 0, 1, 2, 3, \ldots$

independent events (p. 398) Two or more events in which the outcome of one event does not affect the outcome of the other event(s).

inequality (p. 172) A mathematical sentence that contains the symbols $<$, $>$, $\leq$, or $\geq$.

interquartile range (p. 81) The range of the middle half of a set of numbers.
Interquartile range = UQ − LQ

interval (p. 54) On a scale, the difference between the greatest and least values in each category.

inverse operations (p. 156) Operations that "undo" each other. Addition and subtraction are inverse operations.

irrational number (p. 476) A number that cannot be written as a fraction.

isosceles triangle (p. 429) A triangle having at least two congruent sides.

propiedad de identidad La suma de un sumando y cero es el sumando mismo. El producto de un factor y uno es el factor mismo.

medida indirecta Técnica que se usa para calcular una medida a partir de triángulos semejantes y proporciones.

entero Los números enteros y sus opuestos.
$\ldots, -3, -2, -1, 0, 1, 2, 3, \ldots$

eventos independientes Dos o más eventos en los cuales el resultado de uno de ellos no afecta el resultado de los otros eventos.

desigualdad Enunciado matemático que contiene los símbolos $<$, $>$, $\leq$ o $\geq$.

amplitud intercuartílica El rango de la mitad central de un conjunto de datos o números. La amplitud intercuartílica = CS − CI

intervalo En una escala, la diferencia entre el valor mayor y el menor en cada categoría.

operaciones inversas Operaciones que se "anulan" mutuamente. La adición y la sustracción son operaciones inversas.

número irracional Número que no se puede escribir como fracción.

triángulo isósceles Triángulo que tiene por lo menos dos lados congruentes.

K

kilogram (p. 39) The base unit of mass in the metric system equivalent to 1,000 grams.

kilogramo Unidad fundamental de masa del sistema métrico. Un kilogramo equivale a mil gramos.

leaf (p. 76) The second greatest place value of data in a stem-and-leaf plot.

least common denominator (LCD) (p. 227) The least common multiple of the denominators of two or more fractions.

least common multiple (LCM) (p. 224) The least of the common multiples of two or more numbers. The LCM of 2 and 3 is 6.

leg (p. 479) Either of the two sides that form the right angle of a right triangle.

linear equation (p. 178) An equation for which the graph is a straight line.

line graph (p. 60) A type of statistical graph using lines to show how values change over a period of time.

line of symmetry (p. 456) A line that divides a figure into two halves that are reflections of each other.

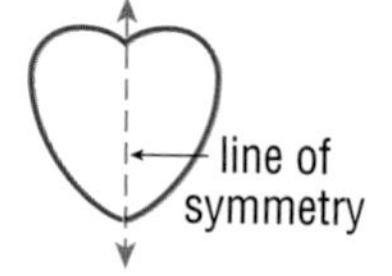

line plot (p. 64) A graph that uses an × above a number on a number line each time that number occurs in a set of data.

line symmetry (p. 456) Figures that match exactly when folded in half have line symmetry.

liter (p. 39) The base unit of capacity in the metric system. A liter is a little more than a quart.

lower extreme (p. 80) The least number of a set of data.

lower quartile (p. 80) The median of the lower half of a set of numbers, indicated by LQ.

hoja El segundo valor de posición mayor en un diagrama de tallo y hojas.

mínimo común denominador (mcd) El menor múltiplo común de los denominadores de dos o más fracciones.

mínimo común múltiplo (mcm) El menor múltiplo común de dos o más números. El mcm de 2 y 3 es 6.

cateto Cualquiera de los lados que forman el ángulo recto de un triángulo rectángulo.

ecuación lineal Ecuación cuya gráfica es una recta.

gráfica lineal Tipo de gráfica estadística que usa segmentos de recta para mostrar cómo cambian los valores durante un período de tiempo.

eje de simetría Recta que divide una figura en dos mitades que son reflexiones entre sí.

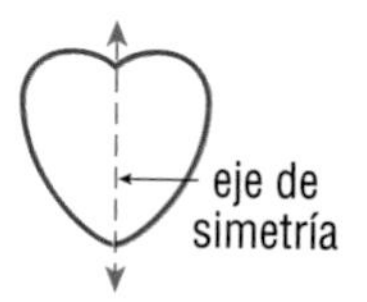

esquema lineal Gráfica que usa una recta numérica y una × sobre un número en la recta numérica, cada vez que el número aparece en un conjunto de datos.

simetría lineal Exhiben simetría lineal las figuras que coinciden exactamente al doblarse una sobre otra.

litro Unidad básica de capacidad del sistema métrico. Un litro es un poco más de un cuarto de galón.

extremo inferior El número menor de un conjunto de datos.

cuartil inferior La mediana de la mitad inferior de un conjunto de datos, la cual se denota por CI.

mean (p. 69) The sum of the data divided by the number of items in the data set.

measures of central tendency (p. 69) Numbers that are used to describe the center of a set of data. These measures include the mean, median, and mode.

median (p. 70) The middle number in a set of data when the data are arranged in numerical order. If the data has an even number, the median is the mean of the two middle numbers.

meter (p. 38) The base unit of length in the metric system.

metric system (p. 38) A base-ten system of measurement using the base units: meter for length, kilogram for mass, and liter for capacity.

media La suma de los datos dividida entre el número total de artículos en el conjunto de datos.

medidas de tendencia central Números que se usan para describir el centro de un conjunto de datos. Estas medidas incluyen la media, la mediana y la moda.

mediana Número central de un conjunto de datos, una vez que los datos han sido ordenados numéricamente. Si hay un número par de datos, la mediana es el promedio de los dos datos centrales.

metro Unidad fundamental de longitud del sistema métrico.

sistema métrico Sistema de medidas de base diez que usa las unidades fundamentales: metro para longitud, kilogramo para masa y litro para capacidad.

mode (p. 70) The number or numbers that appear most often in a set of data.

multiple (p. 224) The product of a number and any whole number.

multiplicative inverse (p. 258) The product of a number and its multiplicative inverse is 1. The multiplicative inverse of $\frac{2}{3}$ is $\frac{3}{2}$.

moda Número o números que aparece(n) más frecuentemente en un conjunto de datos.

múltiplo El producto de un número y cualquier número entero.

inverso multiplicativo El producto de un número y su inverso multiplicativo es 1. El inverso multiplicativo de $\frac{2}{3}$ es $\frac{3}{2}$.

negative integer (p. 106) An integer that is less than zero.

nonagon (p. 446) A polygon having nine sides.

numerical expression (p. 14) A combination of numbers and operations.

entero negativo Un entero menor que cero.

enágono Polígono que tiene nueve lados.

expresión numérica Combinación de números y operaciones.

obtuse angle (p. 413) Any angle that measures greater than 90° but less than 180°.

obtuse triangle (p. 429) A triangle having one obtuse angle.

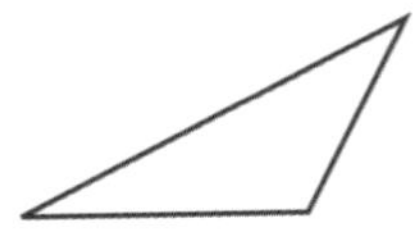

octagon (p. 446) A polygon having eight sides.

opposites (p. 121) Two integers are opposites if they are represented on the number line by points that are the same distance from zero, but on opposite sides of zero. The sum of two opposites is zero.

order of operations (p. 14) The rules to follow when more than one operation is used in a numerical expression.

1. Do all operations within grouping symbols first.
2. Evaluate all powers before other operations.
3. Multiply and divide in order from left to right.
4. Add and subtract in order from left to right.

ordered pair (p. 112) A pair of numbers used to locate a point in the coordinate plane. An ordered pair is written in the form (x-coordinate, y-coordinate).

origin (p. 112) The point at which the x-axis and the y-axis intersect in a coordinate plane.

ángulo obtuso Cualquier ángulo que mide más de 90° pero menos de 180°.

triángulo obtusángulo Triángulo que tiene un ángulo obtuso.

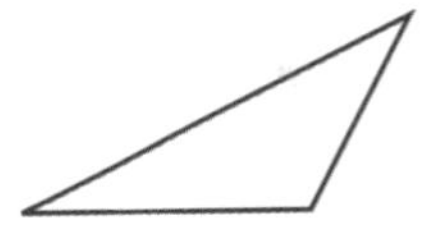

octágono Polígono que tiene ocho lados.

opuestos Dos enteros son opuestos si, en la recta numérica, están representados por puntos que equidistan de cero, pero en direcciones opuestas. La suma de dos opuestos es cero.

orden de operaciones Reglas a seguir cuando se usa más de una operación en una expresión numérica.

1. Primero ejecuta todas las operaciones dentro de los símbolos de agrupamiento.
2. Evalúa todas las potencias antes que las otras operaciones.
3. Multiplica y divide en orden de izquierda a derecha.
4. Suma y resta en orden de izquierda a derecha.

par ordenado Par de números que se utiliza para ubicar un punto en un plano de coordenadas. Se escribe de la siguiente forma: (coordenada x, coordenada y).

origen Punto en que el eje x y el eje y se intersecan en un plano de coordenadas.

ounce (p. 267) A customary unit of weight. 16 ounces equals 1 pound.

onza Unidad de peso del sistema inglés de medidas. 16 onzas equivalen a una libra.

outcome (p. 370) One possible result of a probability event. For example, 4 is an outcome when a number cube is rolled.

resultado Uno de los resultados posibles de un evento probabilístico. Por ejemplo, 4 es un resultado posible cuando se lanza un dado.

outlier (p. 65) A piece of data that is quite separated from the rest of the data. In a box-and-whisker plot, data that are more than 1.5 times the interquartile range from the quartiles.

valor atípico Dato que se encuentra muy separado del resto de los datos. En un diagrama de caja y patillas, los datos que distan de los cuartiles respectivos más de 1.5 veces la amplitud intercuartílica.

P

parallelogram (p. 434) A quadrilateral with opposite sides parallel and opposite sides congruent.

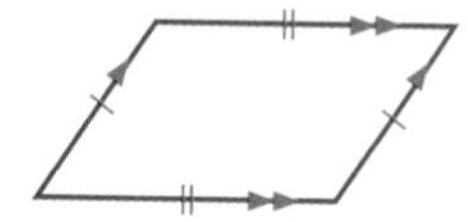

paralelogramo Cuadrilátero cuyos lados opuestos son paralelos y congruentes.

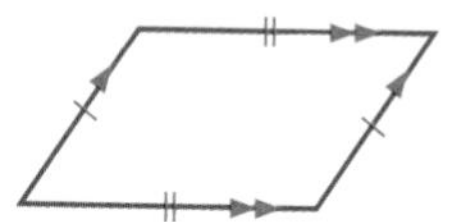

part (p. 323) In a percent proportion, the number that is compared to the whole quantity.

parte En una proporción porcentual, el número que se compara con la cantidad total.

pentagon (p. 446) A polygon having five sides.

pentágono Polígono que tiene cinco lados.

percent (p. 216) A ratio that compares a number to 100.

por ciento Razón que compara un número con 100.

percent equation (p. 340) An equation that describes the relationship between the part, base, and percent.
part = percent · base

ecuación porcentual Ecuación que describe la relación entre la parte, la base y el por ciento.
parte = por ciento · base

percent of change (p. 350) A ratio that compares the change in a quantity to the original amount.

porcentaje de cambio Razón que compara el cambio en una cantidad, con la cantidad original.

percent of decrease (p. 350) A percent of change when the original quantity decreased.

porcentaje de disminución Porcentaje de cambio cuando disminuye la cantidad original.

percent of increase (p. 350) A percent of change when the original quantity increased.

porcentaje de aumento Porcentaje de cambio cuando aumenta la cantidad original.

percent proportion (p. 323) Compares part of a quantity to the whole quantity using a percent.

$$\frac{\text{part}}{\text{base}} = \frac{\text{percent}}{100}$$

proporción porcentual Compara parte de una cantidad con la cantidad total mediante un por ciento.

$$\frac{\text{parte}}{\text{base}} = \frac{\text{por ciento}}{100}$$

perfect squares (p. 471) Numbers whose square root is a whole number. 25 is a perfect square because the square root of 25 is 5.

cuadrados perfectos Números cuya raíz cuadrada es un número entero. 25 es un cuadrado perfecto porque la raíz cuadrada de 25 es 5.

perimeter (p. 270) The distance around a closed geometric figure.

perímetro La distancia alrededor de una figura geométrica cerrada.

permutation (p. 381) An arrangement, or listing, of objects in which order is important.

permutación Arreglo o lista en que el orden es importante.

pint (p. 267) A customary unit of capacity equal to two cups.

pinta Unidad de capacidad del sistema inglés de medidas que equivale a dos tazas.

polygon (p. 446) A simple closed figure in a plane formed by three or more line segments.

polígono Figura simple cerrada en un plano, formada por tres o más segmentos de recta.

population (p. 345) The entire group of items or individuals from which the samples under consideration are taken.

población El grupo total de individuos o de artículos del cual se toman las muestras bajo estudio.

positive integer (p. 106) An integer that is greater than zero.

entero positivo Un entero mayor que cero.

pound (p. 267) A customary unit of weight equal to 16 ounces.

powers (p. 10) Numbers expressed using exponents. The power 3^2 is read *three to the second power,* or *three squared.*

precision (p. 542) The exactness of a measurement which depends on the unit of measure.

precision unit (p. 542) The smallest unit on a measuring tool.

prime factorization (p. 198) Expressing a composite number as a product of prime numbers. For example, the prime factorization of 63 is $3 \times 3 \times 7$.

prime number (p. 197) A whole number greater than 1 that has exactly two factors, 1 and itself.

principal (p. 358) The amount of an investment or a debt.

probability (p. 370) The chance that some event will happen. It is the ratio of the number of ways a certain event can occur to the number of possible outcomes.

properties (p. 31) Statements that are true for any number or variable.

proportion (p. 297) An equation that shows that two ratios are equivalent.

Pythagorean Theorem (p. 479) In a right triangle, the square of the length of the hypotenuse is equal to the sum of the squares of the lengths of the legs. $c^2 = a^2 + b^2$

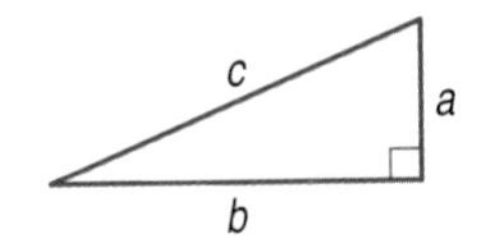

libra Unidad de peso del sistema inglés de medidas que equivale a 16 onzas.

potencias Números que se expresan usando exponentes. La potencia 3^2 se lee *tres a la segunda potencia* o *tres al cuadrado.*

precisión El grado de exactitud de una medida, lo cual depende de la unidad de medida.

unidad de precisión La unidad más pequeña de un instrumento de medición.

factorización prima Escritura de un número compuesto como el producto de números primos. La factorización prima de 63 es $3 \times 3 \times 7$.

número primo Número entero mayor que 1 que sólo tiene dos factores, 1 y sí mismo.

capital La cantidad de una inversión o una deuda.

probabilidad La posibilidad de que suceda un evento. Es la razón del número de maneras en que puede ocurrir un evento al número total de resultados posibles.

propiedades Enunciados que se cumplen para cualquier número o variable.

proporción Ecuación que afirma la igualdad de dos razones.

Teorema de Pitágoras En un triángulo rectángulo, el cuadrado de la longitud de la hipotenusa es igual a la suma de los cuadrados de las longitudes de los catetos. $c^2 = a^2 + b^2$

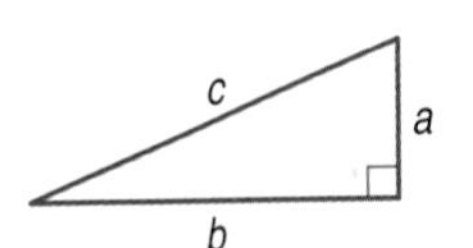

Q

quadrant (p. 113) One of the four regions into which the two perpendicular number lines of the coordinate plane separate the plane.

y-axis
Quadrant II | Quadrant I
O | x-axis
Quadrant III | Quadrant IV

quadrilateral (p. 434) A closed figure having four sides and four angles.

quart (p. 267) A customary unit of capacity equal to two pints.

cuadrante Una de las cuatro regiones en que dos rectas numéricas perpendiculares dividen el plano de coordenadas.

eje y
Cuadrante II | Cuadrante I
O | eje x
Cuadrante III | Cuadrante IV

cuadrilátero Figura cerrada que tiene cuatro lados y cuatro ángulos.

cuarto de galón Unidad de capacidad del sistema inglés de medidas que equivale a dos pintas.

R

radical sign (p. 471) The symbol used to indicate a nonnegative square root, $\sqrt{\ }$.

radius (p. 275) The distance from the center of a circle to any point on the circle.

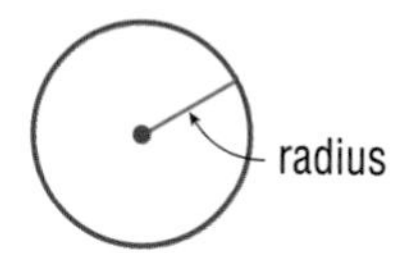

random (p. 370) Outcomes occur at random if each outcome is equally likely to occur.

random sample (p. 345) A sample is called random if the members of the sample are selected purely on the basis of chance.

range (p. 65) The difference between the greatest and least numbers in a data set.

range (p. 177) The set of output values for a function.

rate (p. 292) A ratio that compares two quantities with different kinds of units.

ratio (p. 216) A comparison of two numbers by division. The ratio of 2 to 3 can be written as 2 out of 3, 2 to 3, 2:3, or $\frac{2}{3}$.

rational numbers (p. 229) Numbers that can be written as fractions, including terminating and repeating decimals, and integers.

reciprocal (p. 258) The multiplicative inverse of a number.

rectangle (p. 434) A parallelogram having four right angles.

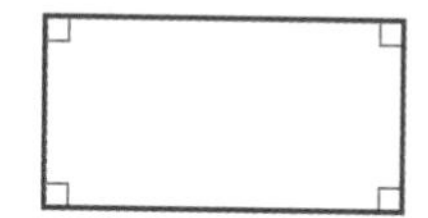

rectangular prism (p. 520) A solid figure that has two parallel and congruent bases that are rectangles.

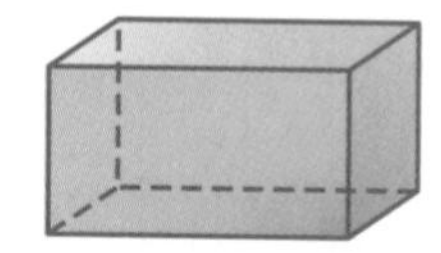

reflection (p. 457) A type of transformation where a figure is flipped over a line of symmetry.

regular polygon (p. 446) A polygon that has all sides congruent and all angles congruent.

repeating decimals (p. 211) A decimal whose digits repeat in groups of one or more. Examples are 0.181818… and 0.83333… .

signo radical Símbolo que se usa para indicar una raíz cuadrada no negativa, $\sqrt{\ }$.

radio Distancia desde el centro de un círculo hasta cualquier punto del mismo.

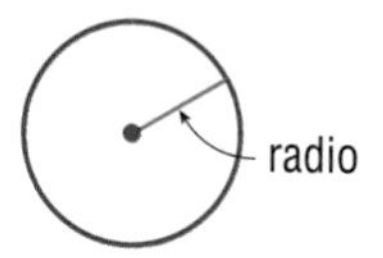

aleatorio Un resultado ocurre al azar si la posibilidad de ocurrir de cada resultado es equiprobable.

muestra aleatoria Se dice que una muestra es aleatoria si sus miembros sólo se eligen aleatoriamente.

rango La diferencia entre el número mayor y el menor en un conjunto de datos.

rango Conjunto de los valores de salida de una función.

tasa Razón que compara dos cantidades que tienen distintas unidades de medida.

razón Comparación de dos números mediante división. La razón de 2 a 3 puede escribirse como 2 de cada 3, 2 a 3, 2:3 ó $\frac{2}{3}$.

números racionales Números que pueden escribirse como fracciones, incluyendo decimales terminales y periódicos y enteros.

recíproco El inverso multiplicativo de un número.

rectángulo Paralelogramo con cuatro ángulos rectos.

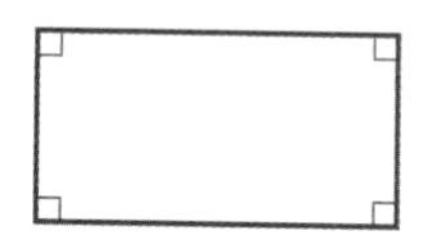

prisma rectangular Figura sólida con dos bases paralelas y congruentes que son rectángulos.

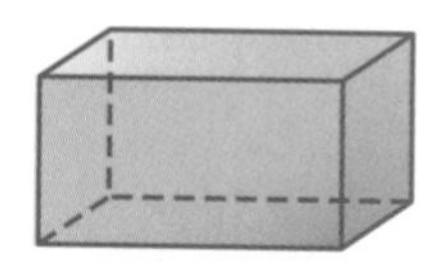

reflexión Transformación en que a una figura se le da vuelta de campana por encima de una recta.

polígono regular Polígono con todos los lados y todos los ángulos congruentes.

decimales periódicos Decimal cuyos dígitos se repiten en grupos de uno o más. Por ejemplo: 0.181818… y 0.83333… .

rhombus (p. 434) A parallelogram having four congruent sides.

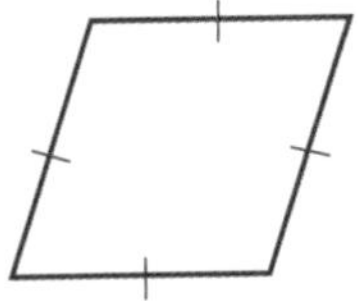

rombo Paralelogramo que tiene cuatro lados congruentes.

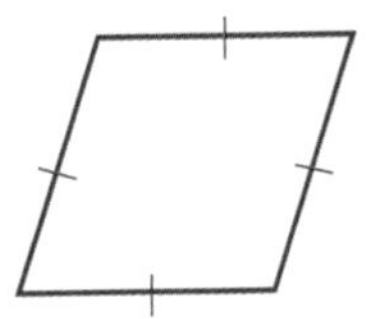

right angle (p. 413) An angle which measures exactly 90°.

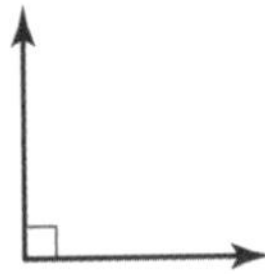

ángulo recto Ángulo que mide exactamente 90°.

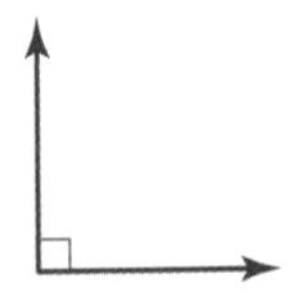

right triangle (p. 429) A triangle having one right angle.

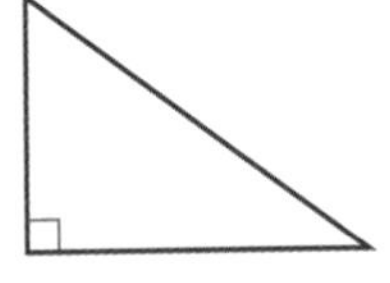

triángulo rectángulo Triángulo que tiene un ángulo recto.

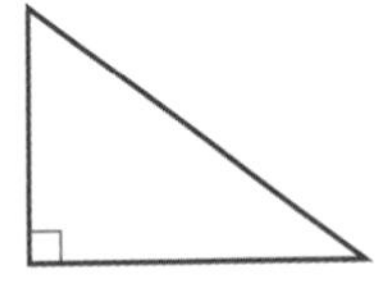

sales tax (p. 354) An additional amount of money charged on items that people buy.

impuesto sobre las ventas Cantidad adicional de dinero que se cobra por un artículo comprado.

sample space (p. 374) The set of all possible outcomes of a probability experiment.

espacio muestral Conjunto de todos los resultados posibles de un experimento probabilístico.

scale (p. 54) The set of all possible values of a given measurement, including the least and greatest numbers in the set, separated by equal intervals.

escala Conjunto de todos los posibles valores de una medida dada, el cual incluye los valores máximo y mínimo del conjunto, separados por intervalos iguales.

scale (p. 304) On a map, intervals used representing the ratio of distance on the map to the actual distance.

escala En un mapa, los intervalos que se usan para representar la razón de las distancias en el mapa a las distancias verdaderas.

scale drawing (p. 304) A drawing that is similar but either larger or smaller than the actual object.

dibujo a escala Dibujo que es semejante, pero más grande o más pequeño que el objeto real.

scale factor (p. 305) A scale written as a ratio in simplest form.

factor de escala Escala escrita como una tasa en forma reducida.

scale model (p. 306) A model used to represent something that is too large or too small for an actual-size model.

modelo a escala Réplica de un objeto real, el cual es demasiado grande o demasiado pequeño como para construirlo de tamaño natural.

scalene triangle (p. 429) A triangle having no congruent sides.

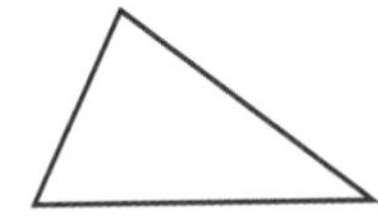

triángulo escaleno Triángulo sin lados congruentes.

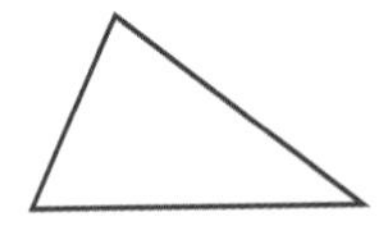

scatter plot (p. 61) In a scatter plot, two sets of related data are plotted as ordered pairs on the same graph.

diagrama de dispersión Diagrama en que dos conjuntos de datos relacionados aparecen graficados como pares ordenados en la misma gráfica.

scientific notation (p. 43) A number written as a product of a number that is at least 1 but less than 10, and a power of ten. For example, $687{,}000 = 6.87 \times 10^5$.

notación científica Números escritos como el producto de un número que es al menos igual a 1, pero menor que 10, por una potencia de diez. Por ejemplo, $687{,}000 = 6.87 \times 10^5$.

sequence (p. 34) A list of numbers in a certain order, such as 0, 1, 2, 3, or 2, 4, 6, 8.

sucesión Lista de números en cierto orden, tales como 0, 1, 2, 3 ó 2, 4, 6, 8.

significant digits (p. 542) All of the digits of a measurement that you know for sure, plus one estimated digit.

dígitos significativos Todos los dígitos de una medición que se sabe son exactos, más un dígito aproximado.

similar figures (p. 440) Figures that have the same shape but not necessarily the same size.

figuras semejantes Figuras que tienen la misma forma, pero no necesariamente el mismo tamaño.

simple event (p. 370) One outcome or a collection of outcomes.

eventos simples Un resultado o una colección de resultados.

simple interest (p. 358) The amount paid or earned for the use of money. The formula for simple interest is $I = prt$.

interés simple Cantidad que se paga o que se gana por el uso del dinero. La fórmula para calcular el interés simple es $I = prt$.

simplest form (p. 207) A fraction is in simplest form when the GCF of the numerator and the denominator is 1.

forma reducida Una fracción está escrita en forma reducida si el MCD de su numerador y denominador es 1.

slope (p. 182) A number that tells how steep a line is. Slope of a line is found by looking at the change in y with respect to the change in x.

pendiente Número que indica el grado de inclinación de una recta. La pendiente de una recta se calcula comparando el cambio en y con respecto al cambio en x.

solid (p. 514) A three-dimensional figure. Prisms, pyramids, cones, and cylinders are examples of solids.

sólido Una figura tridimensional. Los prismas, las pirámides, los conos y los cilindros son ejemplos de sólidos.

solution (p. 24) A value for the variable that makes an equation true. The solution of $12 = x + 7$ is 5.

solución Valor de la variable de una ecuación que hace verdadera la ecuación. La solución de $12 = x + 7$ es 5.

solving an equation (p. 24) The process of finding a solution to an equation.

resolver una ecuación Proceso de encontrar el número o números que satisfagan una ecuación.

square (p. 434) A parallelogram having four right angles and four congruent sides.

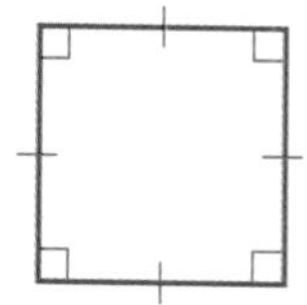

cuadrado Paralelogramo con cuatro ángulos rectos y cuatro lados congruentes.

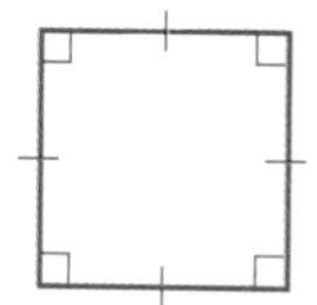

square (p. 470) The product of a number and itself. 36 is the square of 6.

cuadrado El producto de un número por sí mismo. 36 es el cuadrado de 6.

squared (p. 10) A number multiplied by itself; 4×4, or 4^2.

al cuadrado Número multiplicado por sí mismo; 4×4, ó 4^2.

square root (p. 471) One of the two equal factors of a number. The square root of 9 is 3.

raíz cuadrada Uno de dos factores iguales de un número. La raíz cuadrada de 9 es 3.

standard form (p. 11) Numbers written without exponents.

forma estándar Números escritos sin exponentes.

statistics (p. 54) The study of collecting, organizing, and interpreting data.

estadística Estudio de la recolección, organización e interpretación de datos.

stem (p. 76) The greatest place value common to all the data values is used for the stem of a stem-and-leaf plot.

tallo El mayor valor de posición común a todos los datos es el que se usa como tallo en un diagrama de tallo y hojas.

stem-and-leaf plot (p. 76) A system used to condense a set of data where the greatest place value of the data forms the stem and the next greatest place value forms the leaves.

straight angle (p. 413) An angle that measures exactly 180°.

supplementary angles (p. 422) Two angles are supplementary if the sum of their measures is 180°.

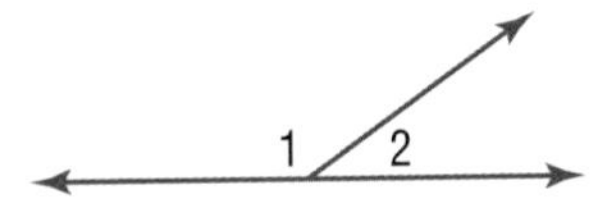

surface area (p. 532) The sum of the areas of all the surfaces (faces) of a three-dimensional figure.

survey (p. 345) A question or set of questions designed to collect data about a specific group of people.

diagrama de tallo y hojas Sistema que se usa para condensar un conjunto de datos y en el cual el mayor valor de posición de los datos forma el tallo y el segundo mayor valor de posición de los datos forma las hojas.

ángulo llano Ángulo que mide exactamente 180°.

ángulos suplementarios Dos ángulos son suplementarios si la suma de sus medidas es 180°.

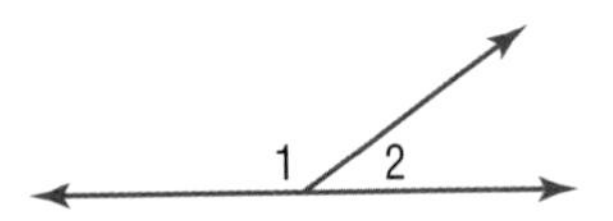

área de superficie La suma de las áreas de todas las superficies (caras) de una figura tridimensional.

encuesta Pregunta o conjunto de preguntas diseñadas para recoger datos sobre un grupo específico de personas.

term (p. 19) A number, a variable, or a product or quotient of numbers and variables.

term (p. 34) Each number in a sequence.

terminating decimals (p. 211) A decimal whose digits end. Every terminating decimal can be written as a fraction with a denominator of 10, 100, 1,000, and so on.

tessellation (p. 447) A repetitive pattern of polygons that fit together with no holes or gaps.

theoretical probability (p. 393) The ratio of the number of ways an event can occur to the number of possible outcomes.

ton (p. 267) A customary unit of weight equal to 2,000 pounds.

transformation (p. 451) A movement of a geometric figure.

translation (p. 451) One type of transformation where a geometric figure is slid horizontally, vertically, or both.

trapezoid (p. 434) A quadrilateral with one pair of parallel sides.

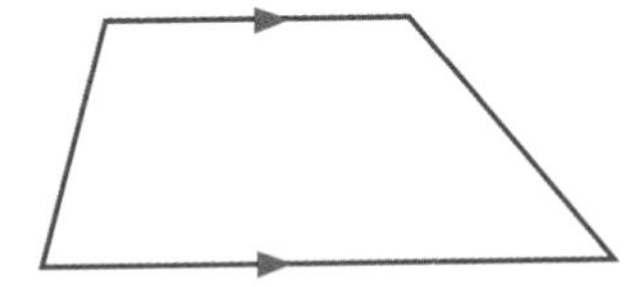

tree diagram (p. 374) A diagram used to show the total number of possible outcomes in a probability experiment.

término Un número, una variable o un producto o un cociente de números y variables.

término Cada número en una sucesión.

decimales terminales Decimal cuyos dígitos terminan. Todo decimal terminal puede escribirse como una fracción con un denominador de 10, 100, 1,000, etc.

teselado Un patrón repetitivo de polígonos que coinciden perfectamente, sin dejar huecos o espacios.

probabilidad teórica La razón del número de maneras en que puede ocurrir un evento al número total de resultados posibles.

tonelada Unidad de peso del sistema inglés que equivale a 2,000 libras.

transformación Movimientos de figuras geométricas

traslación Tipo de transformación en que una figura se desliza horizontal o verticalmente o de ambas maneras

trapecio Cuadrilátero con un único par de lados paralelos.

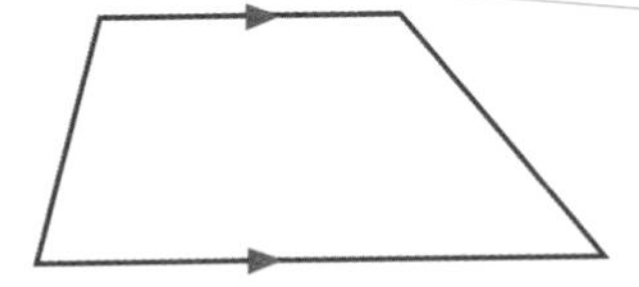

diagrama de árbol Diagrama que se usa para mostrar el número total de resultados posibles en experimento probabilístico.

triangle (p. 428) A polygon that has three sides and three angles.

two-step equation (p. 166) An equation having two different operations.

triángulo Polígono que posee tres lados y tres ángulos.

ecuación de dos pasos Ecuación que contiene dos operaciones distintas.

U

unit rate (p. 292) A rate with denominator of 1.

upper extreme (p. 80) The greatest number of a set of data.

upper quartile (p. 80) The median of the upper half of a set of numbers, indicated by UQ.

tasa unitaria Una tasa con un denominador de 1.

extremo superior El número máximo de un conjunto de datos o números.

cuartil superior La mediana de la mitad superior de un conjunto de números, denotada por CS.

variable (p. 18) A placeholder, usually a letter, used to represent an unspecified value in mathematical expressions or sentences. In $3 + a = 6$, a is a variable.

Venn diagram (p. 203) A diagram that uses circles to show how elements among sets of numbers or objects are related.

vertex (p. 413) A vertex of an angle is the common endpoint of the rays forming the angle.

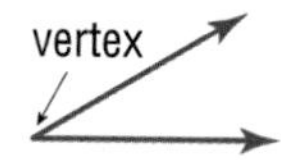

vertical angles (p. 422) Opposite angles formed by the intersection of two lines.

volume (p. 520) The number of cubic units needed to fill the space occupied by a solid.

variable Marcador de posición, por lo general, una letra, que se usa para representar un valor desconocido en expresiones o enunciados matemáticos. En $3 + a = 6$, a es una variable.

diagrama de Venn Diagrama que usa círculos para mostrar la relación entre los elementos en un conjunto de números u objetos.

vértice El vértice de un ángulo es el extremo común de los rayos que lo forman.

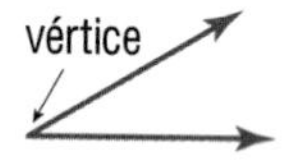

ángulos opuestos por el vértice Ángulos opuestos que se forman de la intersección de dos rectas.

volumen Número de unidades cúbicas que se requieren para llenar el espacio que ocupa un sólido.

***x*-axis** (p. 112) The horizontal number line in a coordinate plane.

***x*-coordinate** (p. 112) The first number of an ordered pair. It corresponds to a number on the x-axis.

eje *x* La recta numérica horizontal en el plano de coordenadas.

coordenada *x* El primer número de un par ordenado. Corresponde a un número en el eje x.

***y*-axis** (p. 112) The vertical number line in a coordinate plane.

***y*-coordinate** (p. 112) The second number of an ordered pair. It corresponds to a number on the y-axis.

eje *y* La recta numérica vertical en el plano de coordenadas.

coordenada *y* El segundo número de un par ordenado. Corresponde a un número en el eje y.

Selected Answers

Chapter 1 Decimal Patterns and Algebra

Page 5 Chapter 1 Getting Started
1. true **3.** 105.8 **5.** 60.64 **7.** 11.3 **9.** 27.69 **11.** 30.8 **13.** 17.01 **15.** 8.15 **17.** 1.1 **19.** 2,450 **21.** 61

Pages 8–9 Lesson 1-1
1. Sample answer: It helps to organize your thoughts and focus on how to approach solving the problem.
5. 13 E-mails **7.** 1,280,000
9.

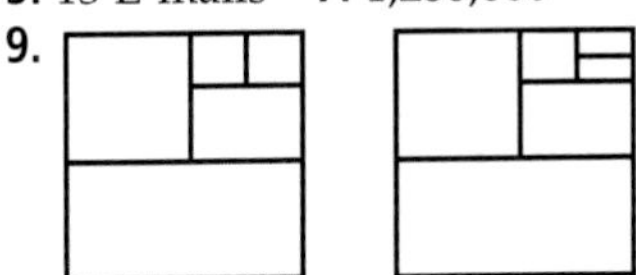

11. 2004 **13.** Sample answer: sweater, gloves, and chocolates **15.** 100 **17.** 625

Pages 11–13 Lesson 1-2
1. Five is used as a factor four times, or $5 \cdot 5 \cdot 5 \cdot 5 = 5^4$. **3.** Sample answer: exponential form: 4^2; standard form: 125 **5.** $6 \cdot 6$ **7.** $8 \cdot 8 \cdot 8 \cdot 8 \cdot 8$ **9.** 3,125 **11.** 5^6 **13.** 1,331 **15.** $4 \cdot 4$ **17.** $9 \cdot 9 \cdot 9$ **19.** $11 \cdot 11 \cdot 11$ **21.** 64 **23.** 625 **25.** 10 **27.** 4,096 **29.** 20,736 **31.** 125,000 **33.** 7^4 **35.** 6^5 **37.** 13^5 **39.** 2^8 **41.** Sample answer: A number taken to the third power is the same as the volume of a cube, or the amount of space inside a cube. **43.** false **45.** false **47.** false **49.** $6^3, 15^2, 3^5, 2^8$ **51.** $8^3, 3^6, 6^4, 5^5$ **53.** $11^3, 5^5, 9^4, 6^5$ **55.** Sample answer: It is easier and less error-prone to write 10^{15} than writing out 15 zeros. **57.** $5^4 \cdot 4^3$ **59.** Sample answer: $390{,}625 < 534{,}000 < 823{,}543$, or $5^8 < 534{,}000 < 7^7$ **61.** $5^1 \cdot 10^3$ **63.** C **65.** 3 **67.** false **69.** 3.4 **71.** 3.07 **73.** 33.98 **75.** 3.5

Pages 16–17 Lesson 1-3
1a. $14 - 7$ **1b.** $24 \div 3$ **3.** Cynthia; the first step is to do the division, $24 \div 6$. Yutaka incorrectly multiplied 6 and 2 first. **5.** 5 **7.** 27 **9.** 28 **11.** 53.5 **13.** 58 **15.** 27 **17.** 13 **19.** 13 **21.** 14.2 **23.** 5.8 **25.** 4 **27.** 10 **29.** 2 **31.** 88 **33.** 600 **35.** 195 **37.** 22.8 **39.** 19 **41.** Sample answer: $2(7) + 2(9.5) + 6$; 39 cm **43.** 450 **45.** 296,500 **47.** 57.9 **49.** $72 \div (9 + 27) - 2 = 0$ **51.** 16; The calculator followed the order of operations. Ruby incorrectly added $2 + 2$ first when she should have divided $12 \div 2$ first. **53.** D **55.** Sample answer: $30 - 20$, or about 10 days longer **57.** 121 **59.** 1,024 **61.** $8 \cdot 8 \cdot 8 \cdot 8 \cdot 8$ **63.** $4 \cdot 4 \cdot 4 \cdot 4 \cdot 4 \cdot 4$ **65.** 9 **67.** 18 **69.** 13

Pages 20–21 Lesson 1-4
1. coefficient: 2; constant: −6 **3.** $\frac{x}{y}$ is a division expression; the others are multiplication expressions. **5.** 20 **7.** 21 **9.** 5 **11.** 16 **13.** 5 **15.** 7 **17.** 8 **19.** 12 **21.** 48 **23.** 1.6 **25.** 3 **27.** 32 **29.** 65.2 **31.** 5 **33.** 18 **35.** 3 **37.** Always; ab means a is multiplied by b. **39.** $16n$ **41.** 2, 2.5, 3, 3.5 **43.** about 118 **45.** \$155; \$200; \$245 **47.** Sample answer: $x = 5, y = 2$ **49.** August **51.** 12 **53.** 4 **55.** true **57.** false

Pages 26–27 Lesson 1-5
1. Sample answer: Find the value of the variable that makes a true sentence. **3.** Ivan; $75 - 25 = 50$ is a true statement. **5.** 17 **7.** 54 **9.** 23 **11.** 4 **13.** 23 points **15.** 7 **17.** 38 **19.** 63 **21.** 5 **23.** 11 **25.** 20 **27.** 64 **29.** 20 **31.** 12

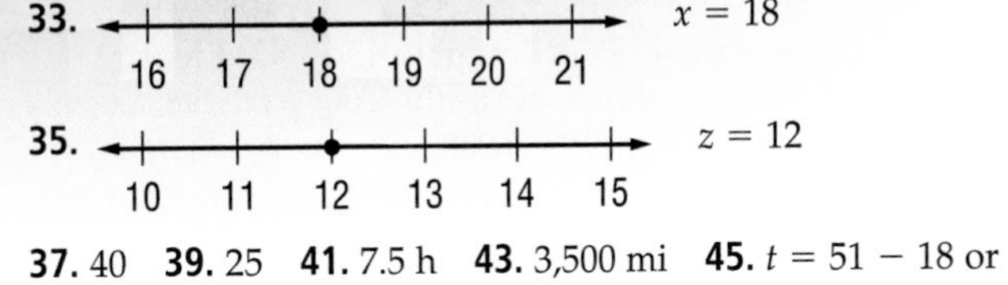

37. 40 **39.** 25 **41.** 7.5 h **43.** 3,500 mi **45.** $t = 51 - 18$ or $51 = t + 18$; 33 in. **47.** Sample answer: $b = 0$; a is any number **49.** 958 **51.** 31 **53.** 19 **55.** 28 **57.** 150

Pages 32–33 Lesson 1-6
1. $p(q + r)$ **3.** false; $(11 + 18) \times 5 = 145$; $11 + 18 \times 5 = 101$ **5.** $(10)2 + (8)2$; 36 **7.** Commutative Property of Addition **9.** Distributive Property **11.** $2(6) + 2(7)$; 26 **13.** $(11)8 + (3)8$; 112 **15.** $7(8) - 7(6)$; 14 **17.** Sample answer: The second expression, because it is easy to multiply $3 \cdot 400$, $3 \cdot 50$, and $3 \cdot 6$ mentally, and then add. **19.** Associative Property of Multiplication **21.** Commutative Property of Multiplication **23.** Associative Property of Addition **25.** Distributive Property **27.** $239 + 531 = 531 + 239$ **29.** $152 + 0 = 152$ **31.** $y + 5$ **33.** $32b$ **35.** $2x + 6$ **37.** $5c + 6$ **39.** $6(40 + 8) = 6(40) + 6(8) = 240 + 48 = 288$; Distributive Property **43.** A **45.** 8 **47.** 7.1 **49.** $7y$ **51.** 625 **53.** 54 **55.** 20 **57.** 32.4

Pages 35–36 Lesson 1-7
1. Sample answer: Each term in an arithmetic sequence is found by adding or subtracting the same number to or from the previous term. Each term in a geometric sequence is found by multiplying or dividing the previous term by the same number. **5.** Add 2 more than what was added to the previous term; neither. **7.** 324, 972, 2,916 **9.** 6.6, 7, 7.4 **11.** 70, 7, 0.7, 0.07 **13.** + 12; arith. **15.** × 3; geo. **17.** Multiply by 1 more than was multiplied by the previous term; neither. **19.** 20.6, 24.6, 28.6 **21.** 1.3, 1.6, 1.9 **23.** 0.2, 0.04, 0.008 **25.** 1, 3, 6, 10, 15, 21, 28, 36 **27.** 76 **29.** 100 **31.** Distributive **33.** 520 **35.** 3.2

Pages 40–41 Lesson 1-8
1. There are a greater number of smaller units. **3.** Hunter; Arturo multiplied 45.7 by 1,000; he should have divided by 1,000. **5.** 45,000 **7.** 370 **9.** 2,340 **11.** 7.2 **13.** 64,000 **15.** 0.925 **17.** 8.5 **19.** 9,100 **21.** 0.043 **23.** 0.997 **25.** 0.0821 **27.** 57 **29.** 0.169 km **31.** 0.02 km, 3,000 cm, 50 m **33.** 0.06 L, 660 mL, 6.6 kL **35.** 93,000 **37.** 0.04 **39.** 50 cm **41.** 5 cobblers **43.** 3,900,000,000 m **45.** C **47.** + 9; arithmetic **49.** + 1.1; arithmetic **51.** $5(9) + 5(7)$; 80 **53.** $8(7) - 8(2)$; 40 **55.** 800 **57.** 400,000 **59.** 6,500

Pages 44–45 Lesson 1-9
1. Sample answer: It is easier to keep track of the place value. **3.** 3.0×10^6; Sample answer: Chicago has a population of about 3 million people: $3.0 \times 10^6 = 3{,}000{,}000$. **5.** 100,000 **7.** 7.0×10^4 **9.** 1.75×10^7 **11.** 73,000 **13.** 338,000 **15.** 8,980,000 **17.** 798,000 **19.** 990,000,000 **21.** 80,600,000 **23.** 53,400, 5.03×10^5, 4.98×10^6, 4,980,100 **25.** 1.98×10^4 **27.** 9.97×10^6 **29.** 5.05×10^5 **31.** 2.36×10^7 **33.** 1.1×10^6 **35.** 6.08×10^{10} **37.** 54.3 quadrillion **39.** $>$ **41.** $<$ **43.** 2.94×10^{14} mi **45.** B **47.** 1,010 **49.** 2,330

Pages 46–48 Chapter 1 Study Guide and Review

1. b **3.** d **5.** f **7.** \$89.40 **9.** 40,353,607 **11.** 324 **13.** 10,000 **15.** 100 **17.** 25 **19.** 20 **21.** 33 **23.** 0 **25.** 11 **27.** 36 **29.** 56 **31.** \$965 **33.** 8 **35.** 7 **37.** 108 **39.** 67 **41.** Associative Property of Multiplication **43.** Identity Property of Multiplication **45.** arithmetic; +1.2; 6.8, 8.0, 9.2 **47.** \$125, \$62.50, \$31.25 **49.** 29,000 **51.** 0.007 **53.** 5.9×10^4 **55.** 3.24×10^5 **57.** 1.03×10^6 **59.** 9.1×10^6

Chapter 2 Statistics: Analyzing Data

Page 53 Chapter 2 Getting Started

1. hundredths **3.** greatest to least **5.** thousands **7.** ones **9.** 5.062, 5.16, 5.61 **11.** 0.398, 3.98, 39.8 **13.** 10.26, 1.26, 1.026 **15.** 69.79, 68.99, 68.9 **17.** 86 **19.** 555

Pages 56–57 Lesson 2-1

1. An advantage of using a frequency table is that it organizes data to provide a quick summary. It makes it easier to see quickly the number of times each item of data appears. A disadvantage is that a frequency table may not show the individual pieces of data. **3.** The interval in each category must be the same, 50.

5. Sample answer:

Number of Movie Rentals	Tally	Frequency
0–2	|||	3
3–5	卌	5
6–8	||	2
9–11	卌 |||	8
12–14	|||	3
15–17	||	2
18–20			1

7. cereal

9. Sample answer:

Year	Tally	Frequency
1900–1919	||	2
1920–1939	||||	4
1940–1959	||||	4
1960–1979	卌 ||||	9
1980–1999	卌 |||	8

11. Sample answer:

Chore	Tally	Frequency
wash dishes	卌 |||	8
vacuum	卌	5
clean room	卌 卌 ||	12
do laundry	卌		6
do yard work	||||	4

13. Sample answer:

Year	Tally	Frequency
1970–1979	||	2
1980–1989	|||	3
1990–1999	卌 |||	8
2000–2009	卌 ||	7

17. Sample answer: 91–110 Calories

19. Sample answer:

Price (\$)	Tally	Frequency
11–20	||	2
21–30	||||	4
31–40	||||	4
41–50	卌	5
51–60	卌	5
61–70	||	2
71–80	||||	4
81–90	|||	3
91–100	卌		6

Most of the scooter prices are spread out evenly among the intervals, with the \$91–\$100 category containing the greatest number. An advantage of using this scale and interval is that you can get a more detailed distribution. A disadvantage is that it is difficult to see clusters or trends.
21. F **23.** 533,000 **25.** 60,000,000 **27.** 78°F

Pages 61–63 Lesson 2-2

1. Graphs often show trends over time. If you continue the pattern, you can use it to make a prediction. **3.** Tally mark; the other three are ways to display data. **5.** Yes; in the 7th week he will be able to run 1 mile in about 7 minutes. **9.** Sample answer: 210 nations **11.** Sample answer: As the speed increases, the distance required to come to a complete stop increases. **13.** The data do not appear to fall along a straight line. So, birth month and height do not seem to be related. **15.** A **17.** Sample answer: 2; 18–47 **19.** 9^2 **21.** 7^3 **23.** 6.1, 6.2, 6.2, 6.2, 6.4, 6.5, 6.5, 6.7, 6.7, 6.8, 6.8, 6.8, 6.8, 7.0, 7.0

Pages 66–68 Lesson 2-3

1. Sample answer: the greatest and least values, clusters of data, gaps in data, how the data are spread out, the range, and the pieces of data that appear most often **3.** Darnell; 16 is the mode and 20 is the greatest value.

5.

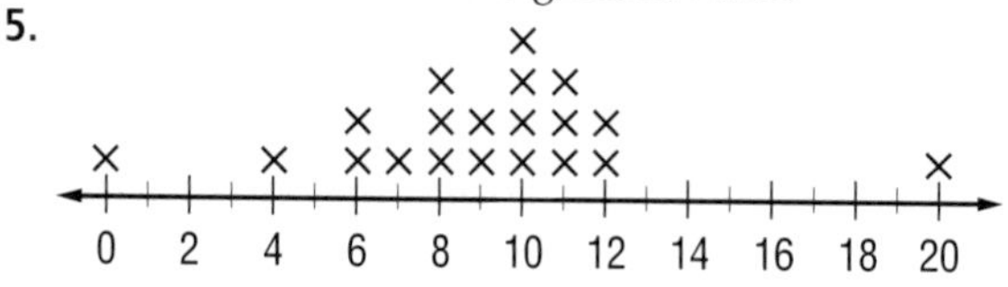

7. Sample answer: Most of the students in the class own 6–12 music CDs. **9.** 104°F **11.** The range would be 29 rather than 37.

13.

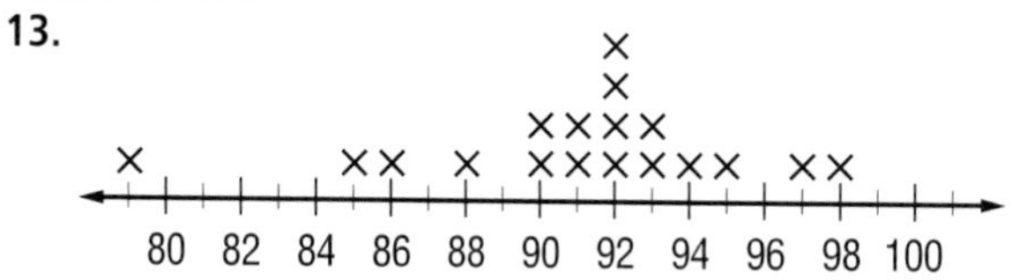

Sample answer: cluster 90–93; gap 79–85; outlier 79
15. Sometimes; clusters may appear anywhere on the line plot. **17.** 5

19.

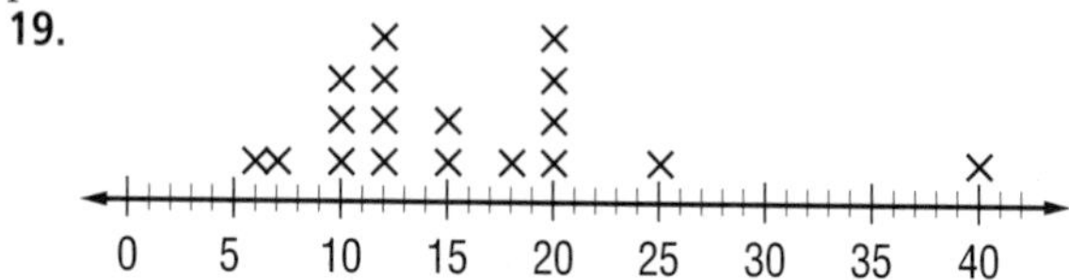

21. Sample answer: Most of the animals represented on the line plot have a life span of 10–20 years. **23.** Sample answer: Bowler B is more consistent because excluding the outlier of 90, the range of scores is 4 points, compared with a range of 10 for Bowler A. Bowler A has scores evenly spread out from 90 to 100. **25.** D **27.** Sample answer: The more it rains, the taller the grass becomes. **29.** 75 **31.** 29 **33.** 47 **35.** 16.1 **37.** 12.6 **39.** 36.4

Pages 71–72 Lesson 2-4

1. Sometimes; if there is an odd number of items, the median is the middle number. If there is an even number of items, the median is the mean of the two middle numbers. **3.** Jared; the median is the 11th term of the ordered data, or 2. The mode is the number that appears most often, or 3. **5.** 52.3; 57; 59 **7.** 8.2; 8; 7 **9.** Sample answer: Any of the three measures could be used to represent the data. The mean is slightly greater than most of the data items, and therefore is a less accurate description of the data. **11.** 8.7; 9.0; 8.0 and 9.3 **13.** \$12; \$9; \$6 **15.** \$7.94; \$8; \$7.50 **17.** Sample answer: The mean or median best represent the data because half of the items are less than these values and half are greater. Since there are four modes, the mode does not best represent the data. **19.** Sample answer: The mean would be most affected; the mode would be the least affected, since it would not change. **21.** 12 **23.** line graph **25.** ones **27.** tens

Pages 78–79 Lesson 2-5

1. Both a stem-and-leaf plot and frequency table show the frequency of data occurring. However, a stem-and-leaf plot shows individual data values and a frequency table shows only the number of data items occurring in each interval. **3.** 0, 1, 2, 3

5.

Stem	Leaf
1	6 9
2	1 3 5 5 9
3	1 3 4 5 9
4	2 7 8 *3|4 = 34*

7. 31 **9.** 32.9; 35; 40

11.

Stem	Leaf
1	3 3 5
2	4 8
4	0 1 2 2 5 6 8 8 8 *1|3 = 13*

13.

Stem	Leaf
40	3
41	1
42	
43	3 4 9
44	2 9
45	1 4 4 7 8 9 *43|9 = 439*

15. 26 **17.** \$110; \$115 and \$150

19.

Stem	Leaf
5	1 2 7
6	0 1 2 3 5 7 8 8
7	3 3 3 4 5 6 7 8 9
8	1 2 2 4 5 9
8	1 2 2 4 5 9
9	4 4 7
10	6 8 *9|4 = 94 points*

21. Sample answer: Most of the game point totals are clustered from 60 to 89 points. There do not appear to be any gaps or outliers.

23. Sample answer:

Price (\$)	Tally	Frequency
20–29	||||	4
30–39	~~||||~~ ~~||||~~	10
40–49	~~||||~~	5
50–59	|	1

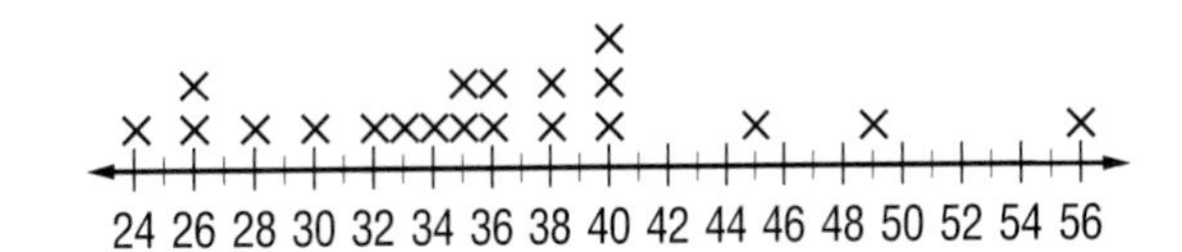

Stem	Leaf
2	4 6 6 8
3	0 2 3 4 5 5 6 6 8 8
4	0 0 0 5 9
5	6 *3|5 = \$35*

All three representations show the frequency of data occurring. The frequency table shows intervals of data and is useful in comparing price ranges. The line plot gives a good picture of the spread of the data. The stem-and-leaf plot shows individual prices as in the line plot, as well as intervals, such as \$20s, \$30s, and so on. **25.** 97 **27.** 4.3; 3.6; 3.6 and 2.4 **29.** 21 **31.** 13 **33.** 6, 7.9, 8, 8.4 **35.** 0.09, 0.14, 0.33, 0.7

Pages 82–83 Lesson 2-6

1. Sample answer: In a stem-and-leaf plot, you can see each data item and how the data are clustered. In a box-and-whisker plot, you can immediately see the median and how the data are dispersed. **3.** 270; 350; 380; 588; 1,221

5.

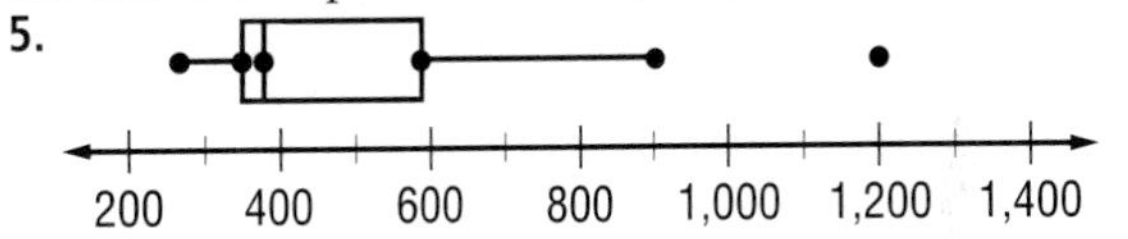

7. 15; 30.5; 45; 51.5; 58 **9.** one-fourth **11.** 0 and 4,600; no outliers **13.** one half **15.** A set in which an extreme value and a quartile are the same has no whisker on that side of the box-and-whisker plot. **17.** 20.5 **19.** 31.6; 32; none **21.** Sample answer: 5; 0–19

Pages 87–89 Lesson 2-7

1. Sample answer: Each interval represents a portion of the data set. The number of items in each interval is indicated by the frequency, typically shown along the vertical scale. By adding the frequencies for each interval, you can determine the number of values in the data set. **3.** World Cup Soccer titles of seven countries **5.** The horizontal axis represents the countries and the vertical axis represents the number of wins. **7.** the number of candies found in 50 snack-size packages

9. **Candies in Snack-Size Packages**

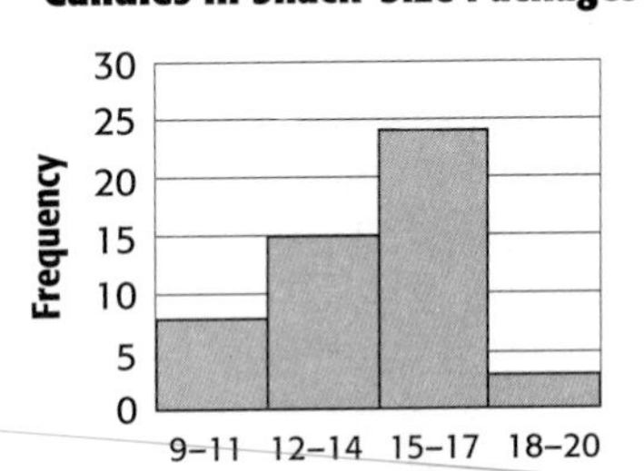

Candies in Package

11. Sample answer: The bars are horizontal; the categories are on the vertical axis and the scale is on the horizontal axis. **13.** Sample answer: 200,000

15. Sample answer: The histogram is strongly skewed toward the lower end, with a high concentration in the number of states with 10 or fewer representatives. The distribution spreads out from there, with fewer and fewer states as the number per state increases. **17.** Sample answer: Both histograms show that many players have salaries in the lowest interval of the histogram, less than

$1 million. Both histograms show one, two, or three players in several of the other intervals. The Diamondbacks' salaries show only three players' salaries of more than $7 million, while the Yankees have seven players in that same range. **19.** Sample answer: It is easier to compare two sets of data. **21.** histogram

23.

Stem	Leaf
0	2 5 9
1	0 2 3 6 7
2	3 5 5
3	1 *2∣3 = 23*

25. false

Pages 93–95 Lesson 2-8

1. Sample answer: Outliers may distort measures of central tendency; data shown in graphs may be exaggerated or minimized by manipulating scales and intervals. **3.** The cost is increasing over time. In about a year's time, the cost has more than tripled. **5.** Sample answer: The median, because it will not be affected by the high prices. **7.** Sample answer: Although both graphs show a decrease in network viewers, in Graph B, a change in the vertical scale makes it appear that the viewers are decreasing more rapidly. **9.** 488.3; none; 150 **11.** median **13.** Yes; all the other test scores were lower than 96. **15.** B **17.** the number of in-line skaters in New York, almost 2 million **19.** $7.57; $7.50; $7.50 and $8.00 **21.** 8,500 **23.** 0.643

Pages 96–98 Chapter 2 Study Guide and Review

1. true **3.** true **5.** false; outlier **7.** true

9. Sample answer:

Hours of Sleep	Tally	Frequency
5	I	1
6	III	3
7	~~IIII~~	5
8	IIII	4
9	III	3

11. Sample answer: 17 million

13.

Sample answer: cluster 7–9; gap 9–14; outlier 14

15.

Sample answer: cluster 19–23; gap 23–28; outlier 28

17. 24.4; 24; 18 and 31 **19.** 132.3, 130.5, 136

21.

Stem	Leaf
5	3
6	
7	5 8 8
8	3 5 7 7 9
9	1 2 *7∣5 = 75*

23.

Stem	Leaf
20	1 4
21	8 9
22	1 5 9
23	4 *23∣4 = 234*

25. 11.5

27.

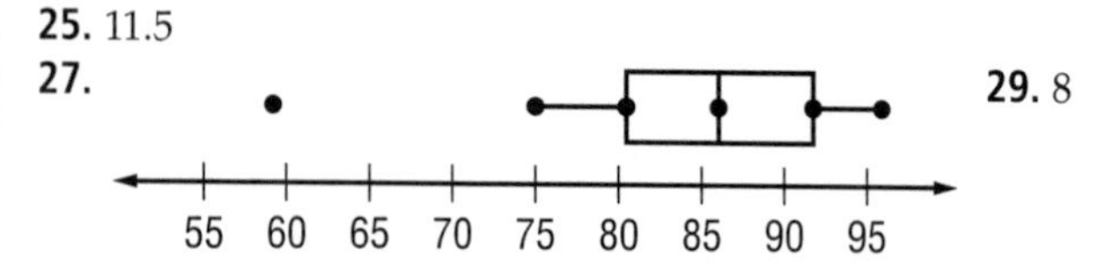

29. 8

Chapter 3 Algebra: Integers

Page 105 Chapter 3 Getting Started

1. false; median **3.** < **5.** < **7.** 123 **9.** 1,730 **11.** 30 **13.** 90 **15.** 7 **17.** 25 **19.** 14; 17 **21.** 100; 34

Pages 107–108 Lesson 3-1

1. Sample answers: temperature, golf scores, elevation, loss of money **3.** −3; The other expressions equal 3. **5.** −11 **7.** 7 **9.** 6 **11.** 9 **13.** 0 **15.** 2008 **17.** −50 **19.** 6 **21.** 9 **23.** 4 **25.** 14 **27.** 0 **29.** 2 protons: 2; 5 electrons: −5

31. **33.**

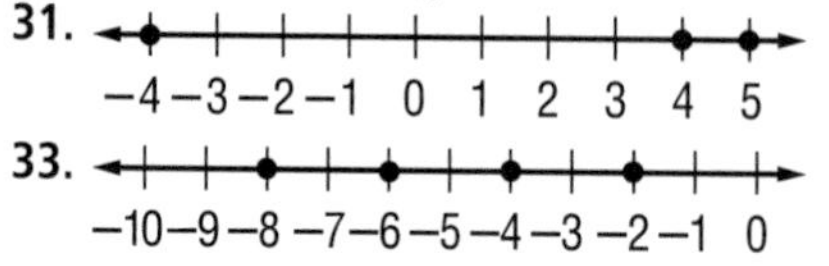

35. true **37.** B **39.** The very high salary increases the mean. **41.** > **43.** < **45.** >

Pages 110–111 Lesson 3-2

1.

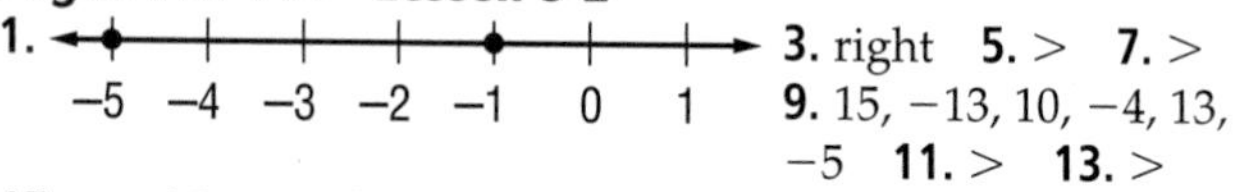

3. right **5.** > **7.** > **9.** 15, −13, 10, −4, 13, −5 **11.** > **13.** > **15.** > **17.** > **19.** > **21.** > **23.** false; Sample answer: 8 > 5 **25.** true **27.** true **29.** Mar. 2, Mar. 6, Mar. 1, Mar. 5, Mar. 3, Mar. 4, Mar. 7 **31.** −7, −6, −4, 1, 3, 5 **33.** They are negative integers. **35.** 7 **37.** 20

39. $x = 2$ **41.** $x = 3$

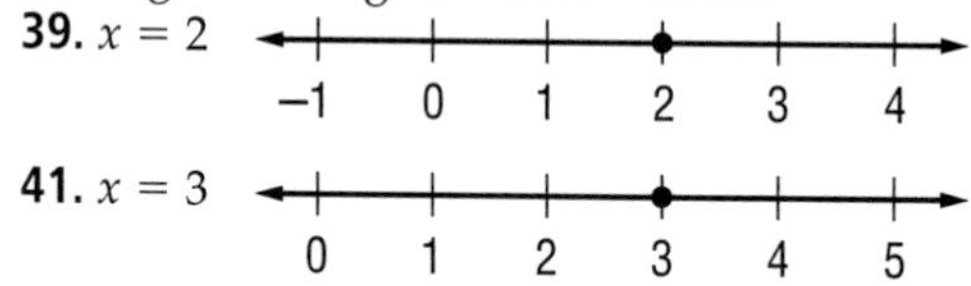

Pages 114–115 Lesson 3-3

1. Sample answer: Point *A* is 1 unit to the right and 2 units down from the origin, in quadrant IV. Point *B* is 2 units to the left and 1 unit up from the origin, in quadrant II. **3.** (−2, −4), III **5.** (0, 3), *y*-axis

6–8.

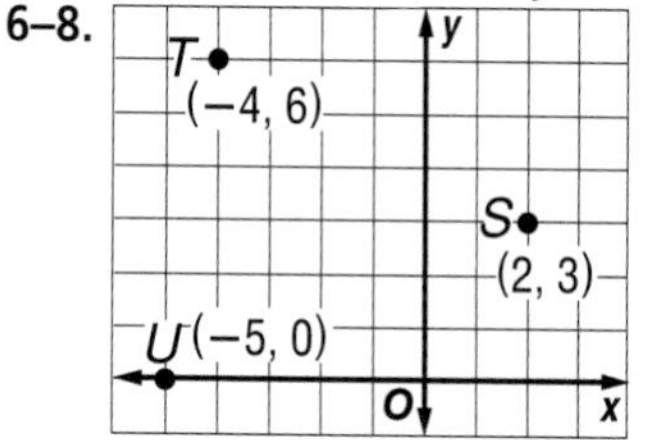

9. (−2, 2), II **11.** (−5, 0), *x*-axis **13.** (2, −2), IV **15.** (3, −5), IV **17.** (2, 2), I **19.** (0, −32)

20–31.

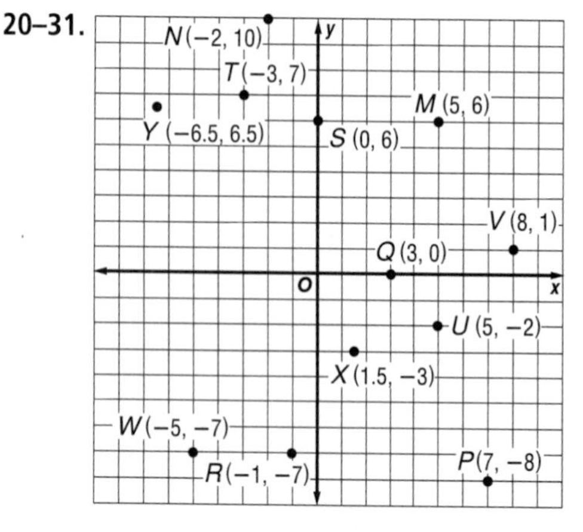

33. Sometimes; both (−2, 0) and (2, 0) lie on the *x*-axis. **35.** China **37.** Children's Zoo **39.** (−3, −2) **41.** Sample answer: On the coordinate plane, I would walk 1 unit right, then 4 units down. **43.** Quadrant I when both x- and y-coordinates are positive; quadrant III when both x- and y-coordinates are negative; the origin when both x- and y-coordinates are 0. **45.** G **47.** $<$ **49.** $>$ **51.** 101 **53.** 384 **55.** 7,470

Pages 122–124 Lesson 3-4

1. $2 + (-7) = -5$ Sample model:

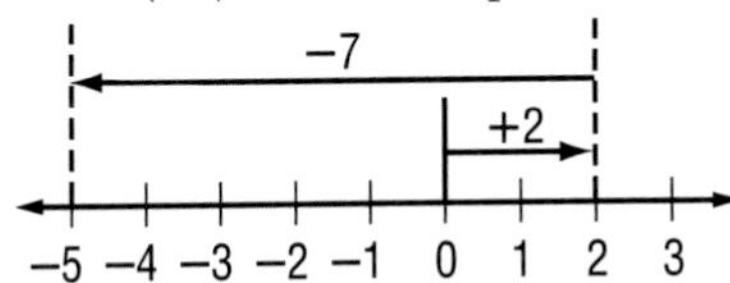

3. Brooke; Javier added correctly but did not apply the sign of the greater absolute value. **5.** -14 **7.** -4 **9.** less since $-\$42 + \$35 = -\$7$ **11.** 0 **13.** -6 **15.** -25 **17.** -14 **19.** 0 **21.** -38 **23.** 8 **25.** -30 **27.** $-3 + (-6)$; -9°F **29.** $-6 + 14$; 8 ft **31.** 4 **33.** -13 **35.** 12 **37.** 0 **39.** -3 **41.** -18 **43.** -12 **45.** Webb; $-14 < -12$ **47.** $y + 2$ **49.** n **51.** $f + (-8)$

53. Sample answer:

$$
\begin{aligned}
-6 + 9 + (-4) &= -6 + (-4) + 9 && \text{Comm. (+)}\\
&= (-6 + (-4)) + 9 && \text{Assoc. (+)}\\
&= -10 + 9 && \text{Simplify.}\\
&= -1
\end{aligned}
$$

55. Sample answer:

$$
\begin{aligned}
8 + 10 + (-8) &= 8 + (-8) + 10 && \text{Comm. (+)}\\
&= (8 + (-8)) + 10 && \text{Assoc. (+)}\\
&= 0 + 10 && \text{Additive Inv.}\\
&= 10
\end{aligned}
$$

57. Sample answer:

$$
\begin{aligned}
8 + (-9) + 9 &= 8 + ((-9) + 9) && \text{Assoc. (+)}\\
&= 8 + 0 && \text{Additive Inv.}\\
&= 8 && \text{Simplify.}
\end{aligned}
$$

61. -2; -2; No, addition is associative. **63.** -11 **65.** $(0, -2)$, y-axis **67.** $(1, 1)$, I

69.

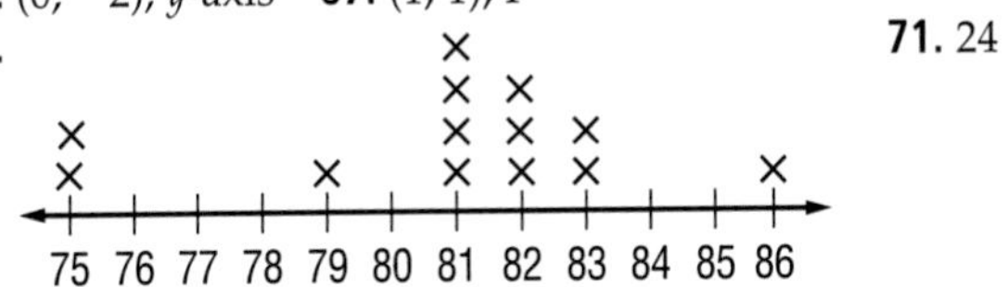

71. 24

Pages 130–131 Lesson 3-5

1. Sample answer: To subtract an integer, add its additive inverse. **3.** Mitsu; Bradley did not add the additive inverse of -19. **5.** -3 **7.** -2 **9.** -8 **11.** 265°F **13.** -10 **15.** 4 **17.** -20 **19.** 23 **21.** 29 **23.** 2 **25.** -6 **27.** -3 **29.** 1 **31.** -4 **33.** 18 **35.** 0 **37.** -12 **39.** 11 **41.** -3 **43.** 8 **45.** 2,139 m **47.** Tiberius; 79 yr **49.** always; $-2 - 3 = -5$ **51.** sometimes; $5 - 2 = 3$, $2 - 5 = -3$ **53.** true **55.** 300°C **57.** -11 **59.** -14 **61.** 70 **63.** 192

Pages 136–137 Lesson 3-6

1. $2(-3) = -6$

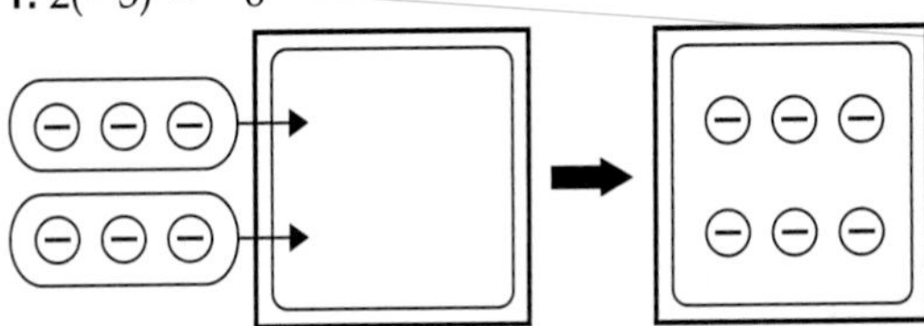

3. negative; Sample answer: $-2(-3)(-4) = -24$ **5.** 45 **7.** $-10a$ **9.** $45c$ **11.** 70 **13.** -875, or 875 feet below sea level **15.** 80 **17.** 170 **19.** 36 **21.** -450 **23.** -220 **25.** -64 **27.** 49 **29.** $-70d$ **31.** $-72f$ **33.** $24h$ **35.** $-10rs$ **37.** -40 **39.** -400 **41.** -300 **43.** -700 **45.** -32, 64; geometric: multiply the previous term by -2

47. Sample answer: new triangle $P'Q'R'$ is on the other side of the origin (in quadrant III) from original triangle PQR (which is in quadrant I).

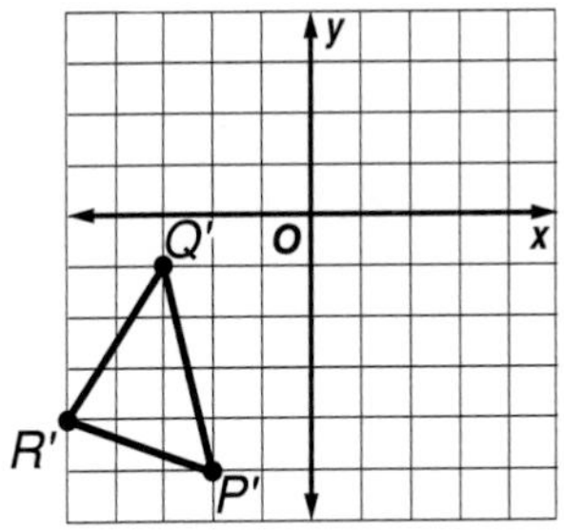

49. when n is an even number **51.** F **53.** 8 **55.** 0 **57.** 2 **59.** 20 and younger, 21–25, 26–30 **61.** 8 **63.** 4

Pages 140–141 Lesson 3-7

1. $-42 \div 7 = -6$, $-42 \div (-6) = 7$ **3.** $-18 \div (-9)$ has a positive quotient; the others have negative quotients. **5.** -8 **7.** -1 **9.** -4 **11.** -2 **13.** 5 **15.** -7 **17.** 4 **19.** 15 **21.** -1 **23.** -5 **25.** -1 **27.** -3 **29.** 2 **31.** 3 **33.** $p = -45 \div 3$; $p = -15$ or 15-yd penalty **35.** -3, 1; Divide the previous term by -3. **37.** -36 sea otters per year **39.** 1, 2, 4, 5, 10, 20, -1, -2, -4, -5, -10, -20 **41.** H **43.** -28 **45.** 60 **47.** 18 **49.** 0.003 g

Pages 142–144 Chapter 3 Study Guide and Review

1. negative **3.** 7 **5.** origin **7.** x-coordinate **9.** negative **11.** -150 **13.** 8 **15.** 11 **17.** 5 **19.** 25 **21.** $>$ **23.** $<$ **25.** $<$ **27.** -32, -23, -21, 14, 19, 25 **29.** 3°, 1°, 0°, -1°, -2° **31.** $(-3, -5)$, III **33.** $(0, 3)$, y-axis

34–37.

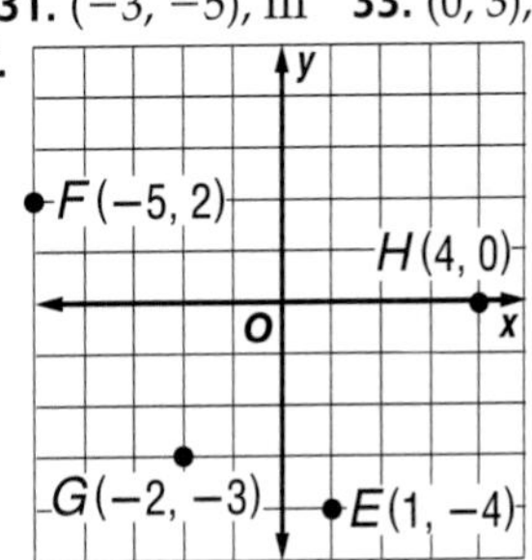

39. -13 **41.** 0 **43.** -13 **45.** 7 **47.** -12 **49.** 35 **51.** 28 **53.** -35 **55.** 5 **57.** -2 **59.** -50 ft per h

Chapter 4 Algebra: Linear Equations and Functions

Page 149 Chapter 4 Getting Started

1. solved **3.** 4 **5.** -11

7.

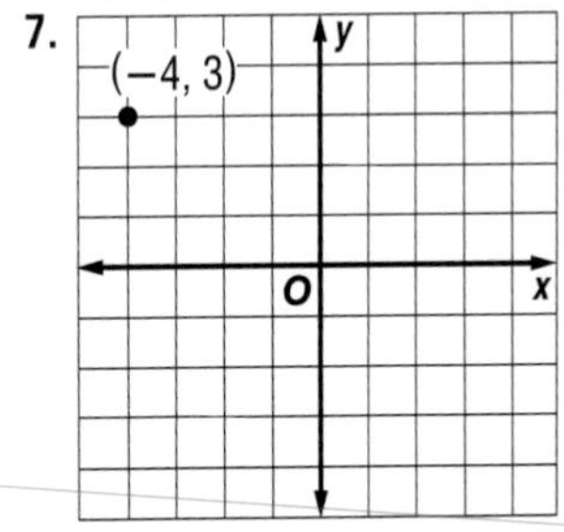

9. -8 **11.** 4 **13.** -11 **15.** 14 **17.** 2 **19.** -2

Pages 151–152 Lesson 4-1

1. Sample answer: 5 dollars more than my allowance is 8 dollars. **3.** $x + 8$ **5.** $10h$ **7.** $n + 4 = -9$ **9.** $a + 2.4 = 35.3$ **11.** $y + 5$ **13.** $n - 6$ **15.** $2f$ **17.** $\frac{z}{-12}$ **19.** $c + 2 = 4$ **21.** $10k = 280$ **23.** $n - 5 = 31$ **25.** $2z + 3 = 15$ **27.** $y + 7$ or $4.5y$ **29.** $x + 2$; $x - 2$ **31.** H **33.** -5 **35.** -8 **37.** -1 **39.** 0

Pages 158–159 Lesson 4-2

1. Addition Property of Equality **3.** $b + 5 = 4$; All of the other equations have solutions equal to 1. This one equals -1. **5.** -2 **7.** 11 **9.** -3 **11.** 7 **13.** -3 **15.** -9 **17.** 17 **19.** 7 **21.** -4 **23.** 61 **25.** -5 **27.** -12 **29.** -5 **31.** 18.4 **33.** 6.4 **35.** -0.68 **37.** $415 = 105 + x$; 310 ft **39.** The value of y must decrease by 3. **41.** 3 **43.** -5 **45.** 9 **47.** 1.2 **49.** 301

Pages 162–163 Lesson 4-3

1. No; $-3(-4)$ equals 12. **3.** Jesse; the variable is multiplied by -5. To solve for x, you need to divide each side of the equation by the entire coefficient, -5. **5.** 2 **7.** -4 **9.** 7 **11.** 6 **13.** -7 **15.** 3 **17.** 14 **19.** -5 **21.** -133 **23.** 10 **25.** 55 **27.** 40 **29.** 2.5 h **31.** 24 in. **33.** 1.5 **35.** 5.4 **37.** 0.8 **39.** 100 s **41.** *Blue Streak* 24.36 ft/s; *Corkscrew* 17.08 ft/s; *Magnum* 42.55 ft/s; *Mean Streak* 32.89 ft/s **43.** C **45.** -10 **47.** -3 **49.** $a + 5$ **51.** $2r$ **53.** -14 **55.** -21

Pages 168–169 Lesson 4-4

1. Sample answer: $3x + 10 = 4$ **3.** 2 **5.** 3 **7.** 6 **9.** 3 **11.** -4 **13.** 4 **15.** 9 **17.** 8 **19.** 11 **21.** -8 **23.** 0 **25.** 3 **27.** 2 **29.** 14 **31.** 2.25 **33.** 2.1 **35.** 28 **37.** 16 **39.** 3 h **41.** Less than; $-3°F$ is about $-19.4°C$. **43.** C **45.** 7 **47.** 41 **49.** 14 ft **51.** > **53.** <

Pages 174–175 Lesson 4-5

1.
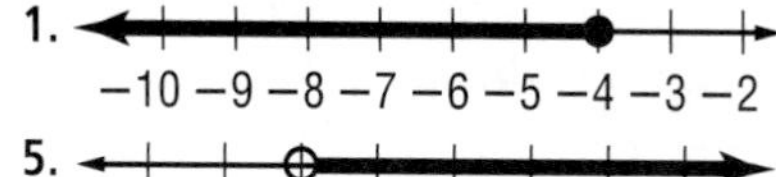

3. Sample answer: $2x < 4$ and $x + 1 < 3$

5.
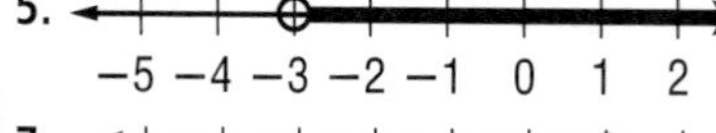

7. 1 2 3 4 5 6 7 8

9. $d \geq 4$

11.
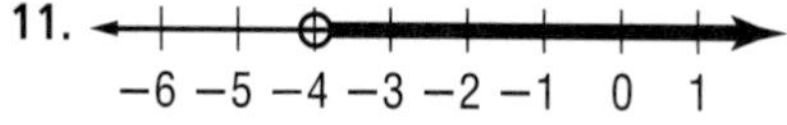

13.
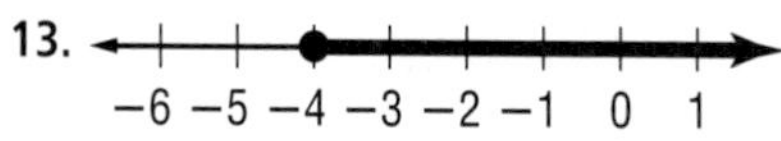

15. −3 −2 −1 0 1 2 3 4

17. −5 −4 −3 −2 −1 0 1 2

19. −10 −9 −8 −7 −6 −5 −4 −3

21. $x < -6$ **23.** $g > 7$ **25.** $u \geq 6$ **27.** $y \leq \frac{4}{3}$ **29.** $p \leq -1$ **31.** $d < 0$ **33.** $w + 1 \geq 5$; $w \geq 4$ **35.** $a \geq 16$ **37.** all except baseball-softball **39.** $h > 2{,}000{,}000$ **41.** $j \leq 2.14$, so 1 or 2 pairs of jeans **43.** B **45.** 8 **47.** -13

49.
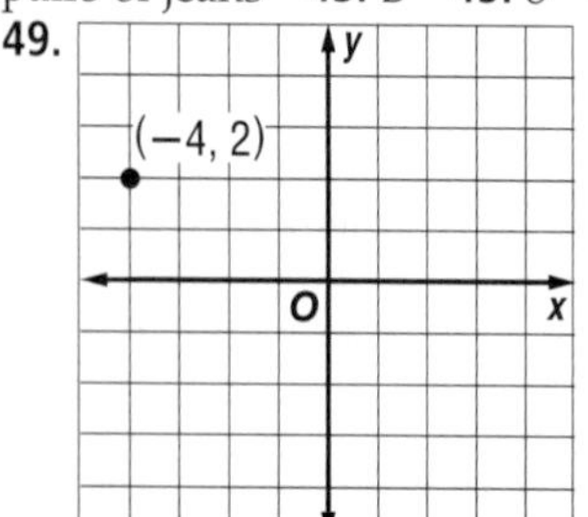

51.
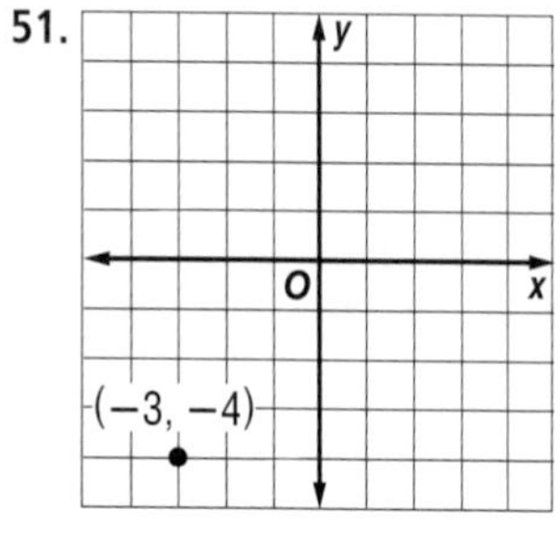

Pages 179–181 Lesson 4-6

1. Sample answer: $y = 2x$

3.

x	x − 2	y
1	1 − 2	−1
2	2 − 2	0
3	3 − 2	1
4	4 − 2	2

domain: {1, 2, 3, 4}; range: {−1, 0, 1, 2}

5.
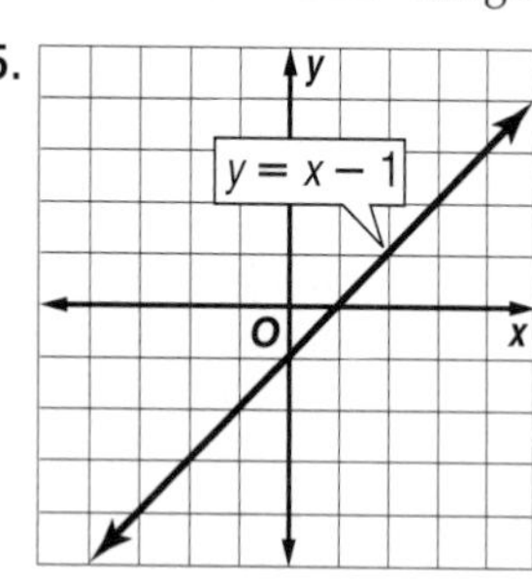

7.
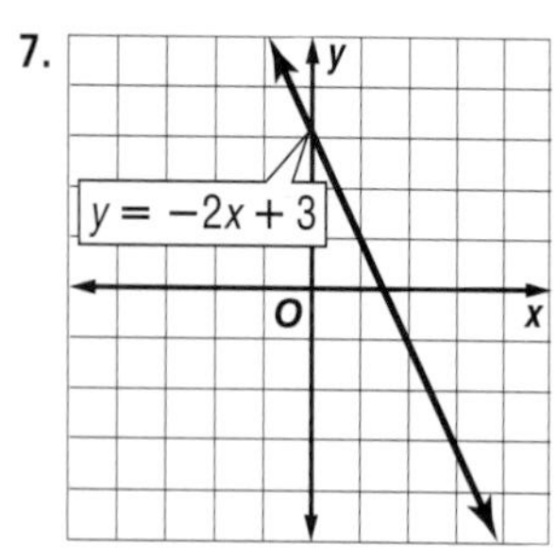

9.

x	x + 5	y
1	1 + 5	6
2	2 + 5	7
3	3 + 5	8
4	4 + 5	9

domain: {1, 2, 3, 4}; range: {6, 7, 8, 9}

11.

x	−6x	y
−1	−6(−1)	6
0	−6(0)	0
1	−6(1)	−6
2	−6(2)	−12

domain: {−1, 0, 1, 2}; range: {6, 0, −6, −12}

13.

x	−2x − 2	y
−1	−2(−1) − 2	0
0	−2(0) − 2	−2
1	−2(1) − 2	−4
2	−2(2) − 2	−6

domain: {−1, 0, 1, 2}; range: {0, −2, −4, −6}

15.
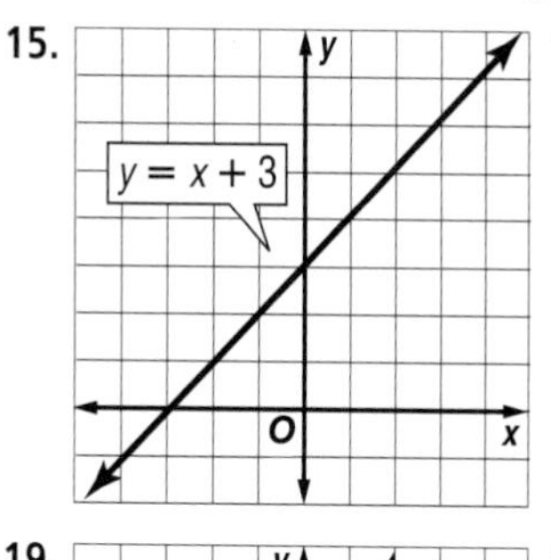

17.
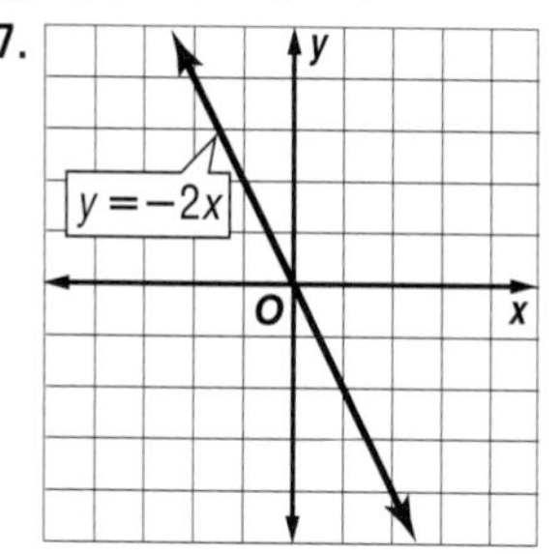

19.
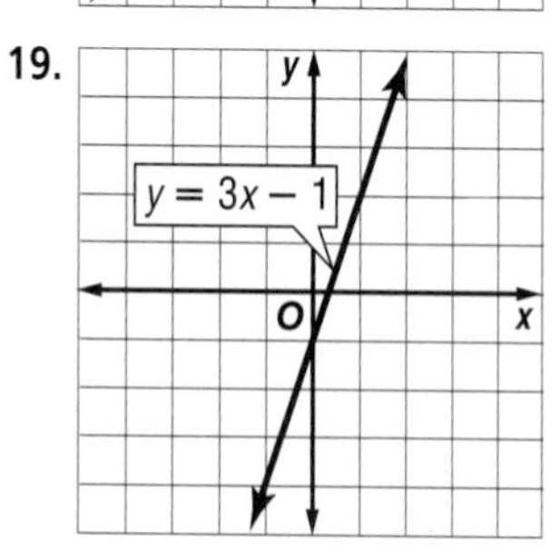

21.
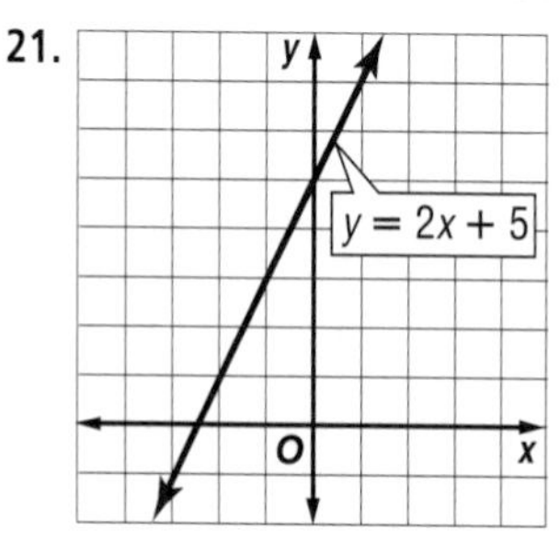

23.
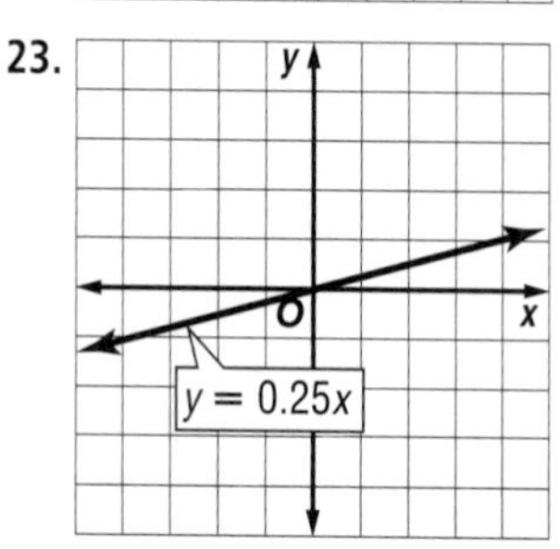

25.
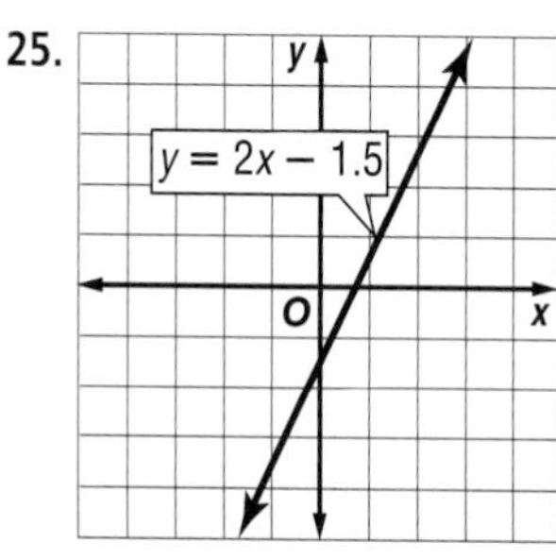

27. $y = x - 5$

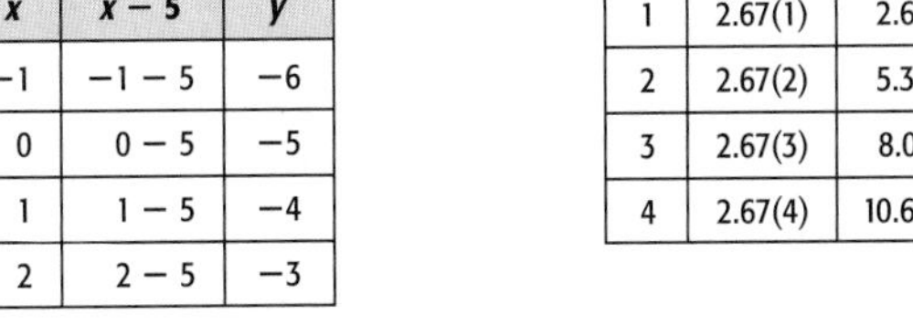

x	$x - 5$	y
−1	−1 − 5	−6
0	0 − 5	−5
1	1 − 5	−4
2	2 − 5	−3

29.

x	$2.67x$	y
1	2.67(1)	2.67
2	2.67(2)	5.34
3	2.67(3)	8.01
4	2.67(4)	10.68

31. **Average Defense Spending (per person)**

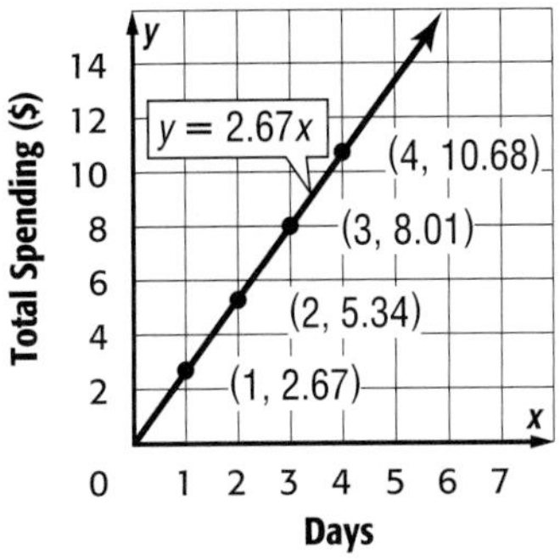

33. $y = 20x$

35. **Rectangles**

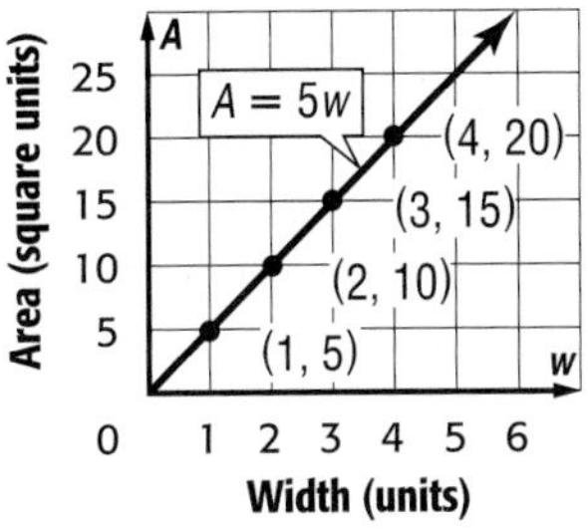

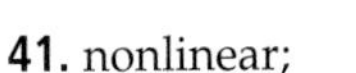

37. $y = x + 2$
39. $y = -4x$

41. nonlinear;

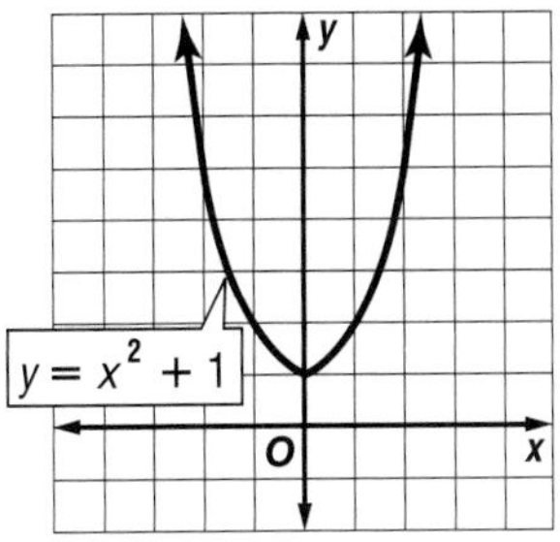

43. linear;

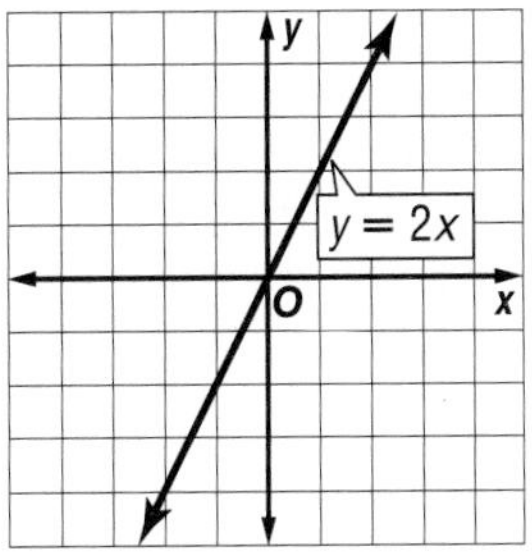

45. nonlinear;

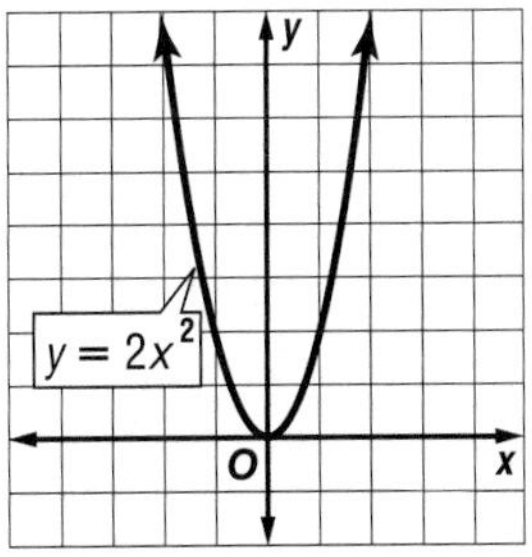

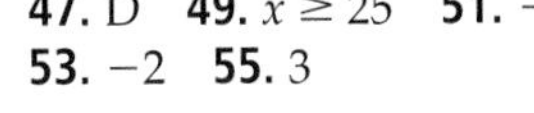

47. D 49. $x \geq 25$ 51. −13 53. −2 55. 3

Pages 184–185 Lesson 4-7

1. the steepness of a line from left to right
3. Sample answer:

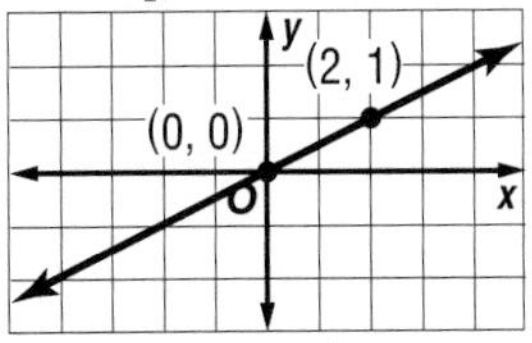

5. −1 7. 3 9. 1 11. −3 13. $\frac{1}{3}$ 15. $-\frac{1}{2}$ 17. Monica; her earnings are increasing faster.

19. The slope represents the amount of money earned with respect to the number of hours worked. 21. 1994–1995; The slope is steepest. 23. The sales remained constant; there is no change in the slope of the line. It is horizontal and the slope is equal to zero.
25. nonlinear

27.

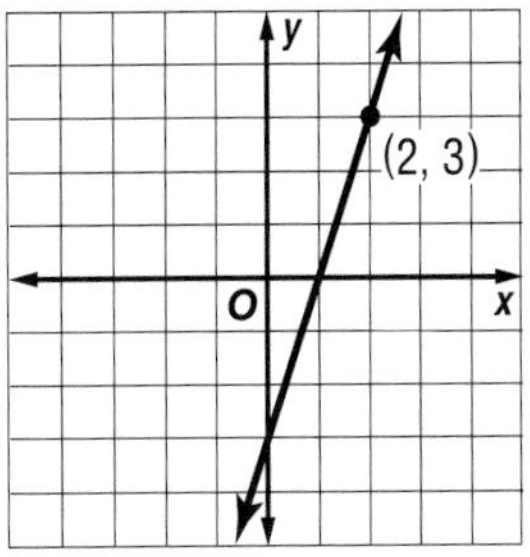

29.

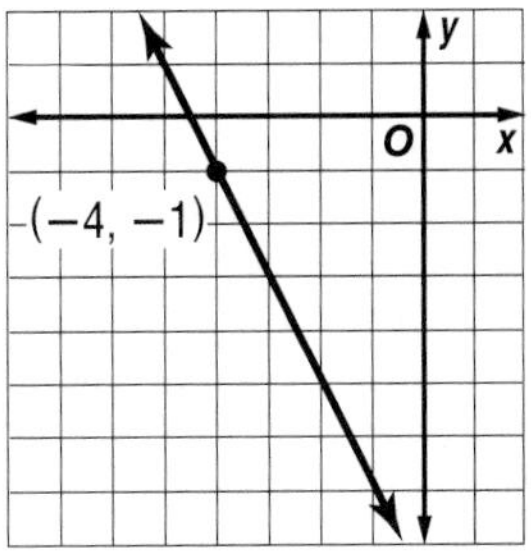

31. C

33.

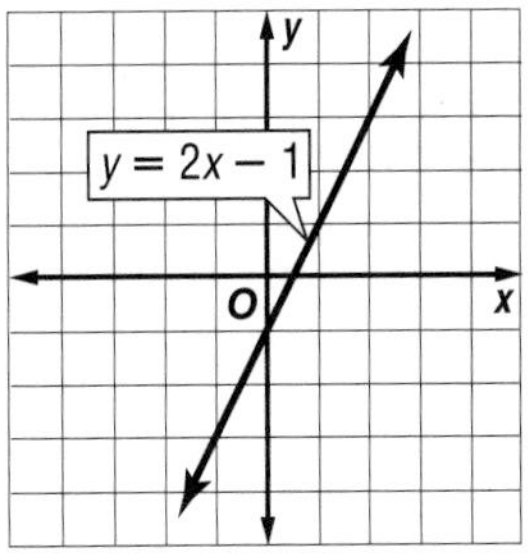

35.

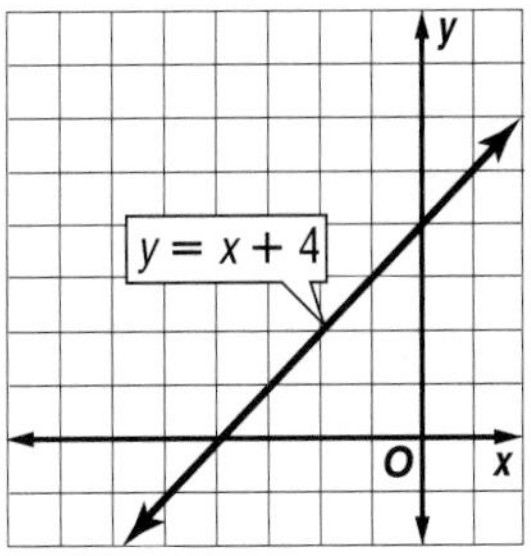

Pages 186–188 Chapter 4 Study Guide and Review

1. true 3. true 5. true 7. false; 7 9. false; −6
11. $n + 5$ 13. $2a$ 15. $a + 10 = 23$ 17. $\frac{56}{z} = 14$ 19. 3
21. −1 23. −13 25. 23 27. −2 29. 4 31. −12 33. −9
35. 11 37. 6 39. −1 41. 9.8 43. 3

45. $y > 3$

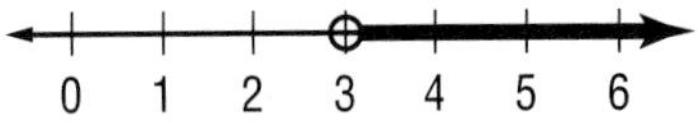

47. $d \leq 5$

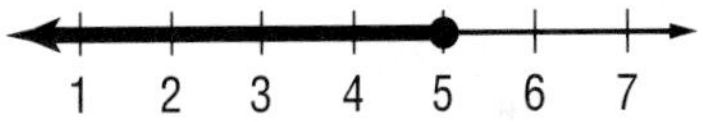

49. $s \leq 11$

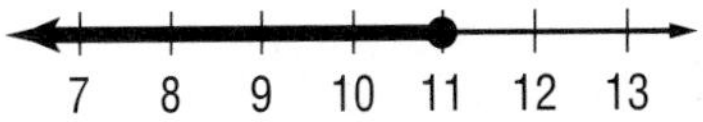

51. $m > -3$

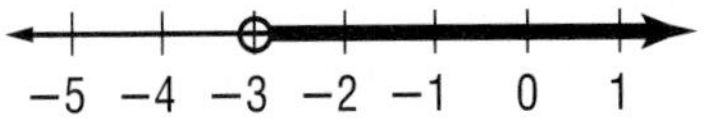

53. $t \geq 2$

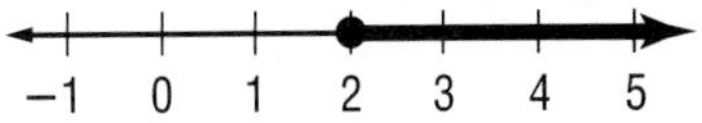

55.

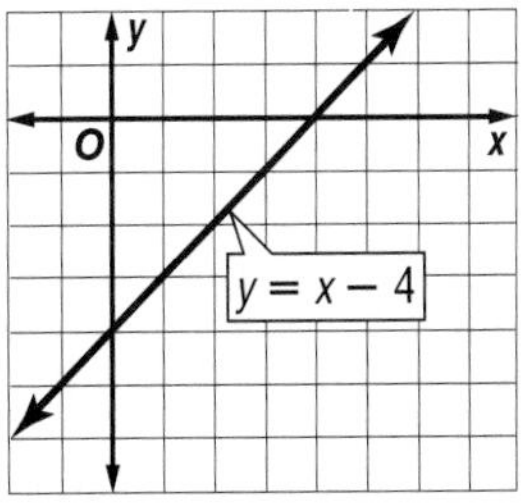

57.

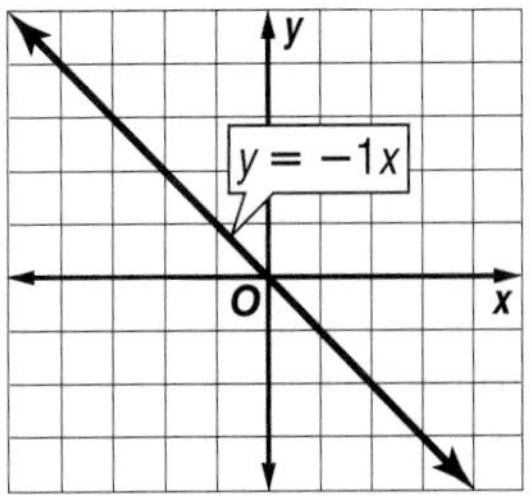

59.

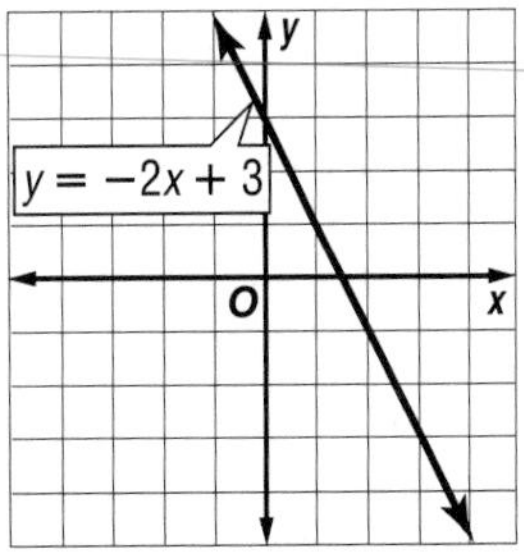

61. $y = 6x$ 63. 4 65. −1

Chapter 5 Fractions, Decimals, and Percents

Page 195 Chapter 5 Getting Started

1. greater **3.** 0.61 **5.** 0.33 **7.** 5 **9.** none **11.** 0.4 **13.** $2 \times 2 \times 2$ **15.** 7×7 **17.** 4^3 **19.** 1^5

Pages 199–200 Lesson 5-1

1. Sample answer: Both prime and composite numbers are whole numbers greater than 1. A prime number has exactly two factors, 1 and itself. Composite numbers have more than two factors. **3.** Sample answer: 9 is not prime. **5.** prime **7.** prime **9.** $2 \times 3 \times 5$ **11.** $2 \cdot 5 \cdot a \cdot c$ **13.** composite **15.** prime **17.** composite **19.** prime **21.** $2^2 \times 3^2$ **23.** $3^2 \times 11$ **25.** $2 \times 3 \times 5 \times 7$ **27.** $2 \times 3^2 \times 7$ **29.** composite **31.** $2 \cdot 2 \cdot 5 \cdot p \cdot q$ **33.** $7 \cdot 7 \cdot y \cdot y$ **35.** $2 \cdot 2 \cdot 2 \cdot 2 \cdot 3 \cdot a \cdot a \cdot b \cdot b$ **37.** 8 **39.** 7 **41.** 5^2 **43.** Sample answer: 5 and 7, 11 and 13, and 17 and 19 **45.** $2^3 \times 3 \times 5$ **47.** Sample answer: twenty 2-ft-by-3-ft tiles, or twelve 2-ft-by-5-ft tiles **49.** 36 **51.** composite **53.**

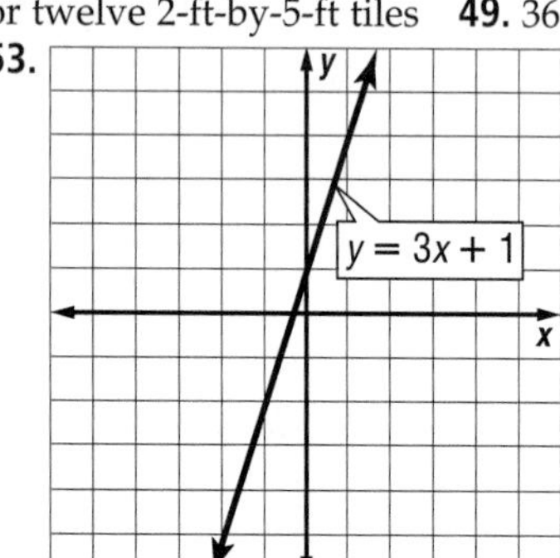

55. negative **57.** zero **59.** 2, 5, 10 **61.** 3, 9

Pages 205–206 Lesson 5-2

1. Sample answer: 32 and 48 **3.** Tiffany is correct; Charles included extra factors of 2 and 3. **5.** 15 **7.** 6 **9.** 6 **11.** 5 **13.** 24 **15.** 8 **17.** 6 **19.** 7 **21.** 5 **23.** never **25.** sometimes **27.** $9b$ **29.** $10x$ **31.** Sample answer: The first building is 9 in. high, 3 in. long, and 4 in. wide. The second building is 9 in. high, 6 in. long, and 5 in. wide. The third building is 9 in. high, 5 in. long, and 5 in. wide. **33.** No; Sample answer: There are a total of 16 pieces if the sandwiches are cut into 4-inch long pieces. **35.** True; their only common factor is 1. **37.** True; their only common factor is 1. **39.** G **41.** composite **43.** composite **45.** $\frac{1}{2}$ **47.** 0.02 L **49.** 24 **51.** 16

Pages 208–209 Lesson 5-3

1. Sample answer: A fraction is in simplest form if the GCF of the numerator and denominator is 1. **3.** Seki; Luther did not divide the numerator and denominator by the GCF. **5.** $\frac{3}{5}$ **7.** $\frac{4}{13}$ **9.** $\frac{5}{7}$ **11.** $\frac{7}{10}$ **13.** $\frac{4}{7}$ **15.** $\frac{6}{7}$ **17.** 1 **19.** $\frac{5}{6}$ **21.** $\frac{6}{10}, \frac{15}{25}$ **23.** $\frac{10}{16}, \frac{20}{32}$ **25.** $\frac{3}{5}$ **27.** $\frac{3}{4}$ **29.** No, because both the numerator and denominator can be divided by 2. **31.** Sample answer:

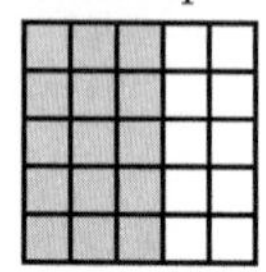

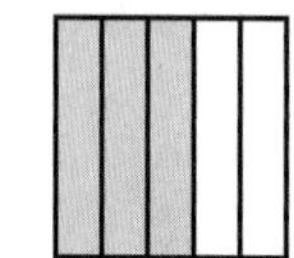

33. $2 \times 3^2 \times 5 \times 7$ **35.** 0.25 **37.** 0.375

Pages 212–213 Lesson 5-4

1. Sample answer: $\frac{1}{2} = 0.5$; $\frac{2}{9} = 0.\overline{2}$ **3.** $0.6\overline{3}$ **5.** $12.\overline{470}$ **7.** 3.625 **9.** $1.8\overline{3}$ **11.** $\frac{1}{10}$ **13.** 0.522 **15.** $5.92\overline{1}$ **17.** $13.1\overline{46}$ **19.** $30.6\overline{841}$ **21.** 0.8 **23.** $0.\overline{72}$ **25.** 0.15 **27.** 6.425 **29.** $0.\overline{714285}$ **31.** $4.\overline{203}$ **33.** $\frac{3}{4}$ **35.** $\frac{1}{5}$ **37.** $5\frac{24}{25}$ **39.** 13; $3\frac{12}{25}$ **41.** $8\frac{9}{10}$; $2\frac{3}{50}$ **43.** 2.3 h **45.** 3.160; Archimedes' **47.** B **49.** $\frac{5}{12}$ **51.** $\frac{2}{7}$ **53.** 12 **55.** $\frac{3}{9}$ **57.** $\frac{4}{12}$

Pages 218–219 Lesson 5-5

1. $\frac{9}{25}$; 36% **3.** 1%; the other ratios equal 10%. **5.** 29.2% **7.** 25% **9.** 80% **11.** $\frac{9}{10}$ **13.** $\frac{3}{4}$ **15.** 75% **17.** 12.2% **19.** $11\frac{2}{7}\%$ **21.** 99.9% **23.** 20% **25.** 80% **27.** 74% **29.** 26% **31.** 95% **33.** 12% **35.** 76% **37.** $\frac{1}{4}$ **39.** $\frac{9}{20}$ **41.** $\frac{1}{2}$ **43.** $\frac{31}{50}$ **45.** $\frac{17}{25}$ **47.** $\frac{11}{25}$ **49.** $\frac{1}{50}$ **51.** 36% **53.** 17 **55.** D **57.** $0.6\overline{5}$ **59.** $0.\overline{41}$ **61.** 162 **63.** 0.144

Pages 222–223 Lesson 5-6

1. Sample answer: 0.25, $\frac{1}{4}$, 25% **3.** 0.68 **5.** 0.276 **7.** 56% **9.** 8% **11.** 0.012 **13.** 0.7 **15.** 0.01 **17.** 0.022 **19.** 0.3025 **21.** 8% **23.** 60% **25.** 14.5% **27.** 70.25% **29.** 90.5% **31.** 80.8% **33.** 10% to 67% **35.** Z **37.** O **39.** 76.5% **41.** 99% **43.** $\frac{45}{80}$ **45.** G **47.** 72% **49.** 18% **51.** 9.375 **53.** 23 students **55.** 2×5^2 **57.** $2^2 \times 19$

Pages 225–226 Lesson 5-7

1. 3 and 4 are factors of 12; 12 is a multiple (and LCM) of 3 and 4. **3.** Sometimes; the LCM of 3 and 9 is 9; the LCM of 3 and 5 is 15. **5.** 42 **7.** 4 packages of hot dogs and 5 packages of buns **9.** 48 **11.** 72 **13.** 315 **15.** 72 **17.** 420 **19.** 144 **21.** $45n$, where $n = 5, 6, 7, \ldots$ **23.** 5 s; 3 s; 30 s **25.** $2^2 \times 5^2$, or 100 **27.** 2012 **29.** $\frac{17}{25}$ **31.** $\frac{3}{25}$ **33.** $<$ **35.** $<$

Pages 229–231 Lesson 5-8

1. Sample answer: $\frac{2}{15}, \frac{5}{6}$ **3.** 45 **5.** $>$ **7.** $>$ **9.** 0.59, 60%, $\frac{31}{50}$ **11.** yes **13.** 20 **15.** 70 **17.** 36 **19.** $>$ **21.** $<$ **23.** $=$ **25.** $<$ **27.** $>$ **29.** $<$ **31.** $<$ **33.** $>$ **35.** 3 saves out of 4; 3 out of 4 has an average of 0.75; 7 out of 11 has an average of about 0.64. **37.** 19%, $\frac{1}{5}$, 0.23 **39.** 0.615, 62%, $\frac{5}{8}$ **41.** 0.49, 0.5, $\frac{49}{50}$ **43.** Christopher **45.** no **47.** no **49.** yes **51.** Sample answer: $2\frac{3}{4}$ is the same as 2.75. Fill the cup to halfway between 2.7 and 2.8. **53.** $\frac{63}{32}$; the difference of $\frac{63}{32}$ and 2 is the least. **55.** McGwire: $\frac{70}{509} = 0.138$; Sosa: $\frac{66}{643} = 0.103$; McGwire **57.** Sample answer: When the numerators are the same, the fraction with the largest denominator is the least fraction. **59.** 47%, $\frac{4}{7}$, $47\frac{2}{5}$, 47.41, $47.\overline{4}$ **61.** 0.04 **63.** 0.14

Pages 232–234 Chapter 5 Study Guide and Review

1. c **3.** d **5.** a **7.** prime **9.** composite **11.** 2^7 **13.** 5×19 **15.** 9 **17.** 14 **19.** $18w$ **21.** $\frac{4}{5}$ **23.** $\frac{1}{11}$ **25.** $\frac{37}{45}$ **27.** $\frac{1}{20}$

29. 0.875 **31.** $4.\overline{3}$ **33.** $1.\overline{857142}$ **35.** $\frac{11}{25}$ **37.** $\frac{9}{50}$ **39.** $\frac{2}{25}$
41. 44% **43.** 40% **45.** $\frac{19}{20}$ **47.** $\frac{4}{25}$ **49.** 0.48 **51.** 0.125
53. 61% **55.** 19% **57.** 0.23 **59.** 8 **61.** 24 **63.** 120
65. = **67.** > **69.** =

Chapter 6 Applying Fractions

Page 239 Chapter 6 Getting Started

1. Property **3.** 35 **5.** 30 **7.** 21.6 **9.** 0.83 **11.** 7.5 **13.** $2\frac{6}{5}$
15. $5\frac{5}{4}$ **17.** $\frac{43}{4}$ **19.** $\frac{22}{5}$

Pages 242–243 Lesson 6-1

1. Sample answer: Models are useful to determine whether fractions are closer to 0, $\frac{1}{2}$, or 1, and to check estimates.
3a. Yes; the sum of $\frac{1}{2}$ and a fraction greater than $\frac{1}{2}$ is greater than 1. **3b.** Yes; both fractions are greater than $\frac{1}{2}$, so the sum is greater than $\frac{1}{2} + \frac{1}{2}$ or 1. **3c.** No; both fractions are less than $\frac{1}{2}$, so the sum is less than 1. **5–37. Sample answers are given.** **5.** $6 \times 3 = 18$ **7.** $1 - 0 = 1$ **9.** $22 \div 11 = 2$ **11.** $4 + 5 = 9$ **13.** $5 - 3 = 2$ **15.** $3 \cdot 6 = 18$
17. $1 + \frac{1}{2} = 1\frac{1}{2}$ **19.** $\frac{1}{2} - 0 = \frac{1}{2}$ **21.** $0 \times 1 = 0$ **23.** $1 \div 1 = 1$ **25.** $\frac{1}{2} \times 14 = 7$ **27.** $25 \div 5 = 5$ **29.** $12 \div 4 = 3$
31. $24 \div 3 = 8$ **33.** $-3 \times 9 = -27$ **35.** $-\frac{1}{6} \times (-66) = 11$
37. 3 − 1 or 2 c **39.** It will be greater than the actual quotient. **41.** H **43.** > **45.** < **47.** 12 **49.** 24

Pages 246–247 Lesson 6-2

1. Sample answer:

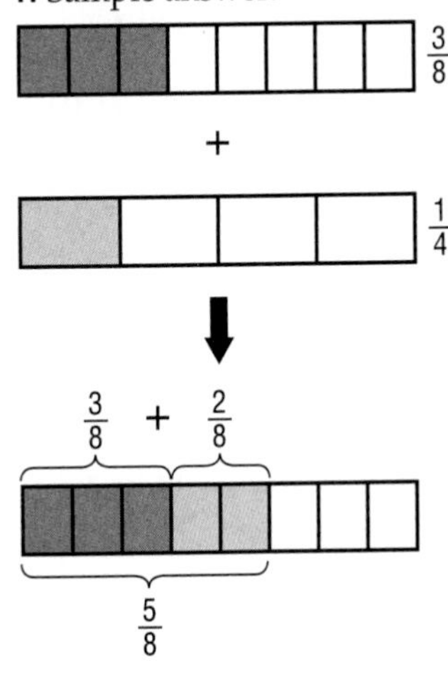

3. Jacinta; Marissa did not rename the fractions using the LCD. **5.** $1\frac{2}{7}$
7. $\frac{1}{4}$ **9.** $\frac{1}{4}$ **11.** $1\frac{5}{18}$
13. $\frac{8}{15}$ **15.** $\frac{4}{7}$ **17.** $\frac{2}{3}$
19. $\frac{2}{3}$ **21.** $\frac{19}{30}$ **23.** $\frac{1}{5}$
25. $1\frac{1}{4}$ **27.** $\frac{1}{24}$ **29.** $1\frac{13}{24}$
31. $\frac{1}{20}$ **33.** $\frac{11}{15}$ **35.** $\frac{5}{6}$
37. $\frac{1}{12}$ **39.** Jon; $\frac{1}{18}$ of the race **41.** Sierra; she has $\frac{1}{4}$ of her allowance left over; Jacob has $\frac{1}{15}$ left over.
43. C **45.** Sample answer: $1 - \frac{1}{2} = \frac{1}{2}$ **47.** Sample answer: $16 \div 8 = 2$ **49.** friend/family **51.** $4\frac{5}{3}$ **53.** $11\frac{6}{5}$

Pages 250–251 Lesson 6-3

1. Sample answer: A table is $8\frac{1}{2}$ feet long. A tablecloth is $2\frac{1}{3}$ feet longer. How long is the tablecloth? Answer: $8\frac{1}{2} + 2\frac{1}{3}$ or $10\frac{5}{6}$ ft **3.** less than; $7 + 1 < 2 + 7$ **5.** $4\frac{2}{3}$ **7.** $9\frac{13}{20}$ **9.** $8\frac{1}{4}$
11. $17\frac{5}{24}$ **13.** $4\frac{5}{8}$ **15.** $9\frac{5}{9}$ **17.** $8\frac{1}{3}$ **19.** $7\frac{5}{12}$ **21.** $7\frac{1}{2}$
23. $8\frac{1}{4}$ **25.** $5\frac{1}{15}$ **27.** $6\frac{5}{6}$ **29.** $19\frac{2}{3}$ **31.** $6\frac{3}{5}$ **33.** $4\frac{1}{3}$
35. $31\frac{3}{16}$ **37.** $38\frac{5}{8}$ **39.** $6\frac{6}{11}$ **41.** $13\frac{7}{18}$ **43.** 20 ft **45.** H
47. Sample answer: $1 \div 1 = 1$ **49.** Sample answer: $9 \times 7 = 63$ **51.** > **53.** < **55.** $\frac{19}{8}$ **57.** $\frac{51}{10}$

Pages 256–257 Lesson 6-4

1. Sample answer: $\frac{2}{3}$ and $\frac{4}{5}$ **3.** Less than; $\frac{4}{5}$ of 18 is a fraction of 18. **5.** $\frac{1}{5}$ **7.** $\frac{2}{3}$ **9.** $6\frac{3}{5}$ **11.** $\frac{3}{32}$ **13.** $\frac{1}{6}$ **15.** $\frac{1}{6}$
17. $\frac{1}{2}$ **19.** $\frac{5}{36}$ **21.** $4\frac{1}{2}$ **23.** 2,700 votes **25.** $2\frac{2}{3}$ **27.** 16
29. 42 **31.** $8\frac{2}{3}$ **33.** $\frac{1}{4}$ **35.** $\frac{7}{40}$ or 0.175 km **37.** Always; sample answer: the numerators of the fractions are less than the denominators, so the product of the numerators will be less than the product of the denominators, resulting in a fraction that is less than 1. **39.** 7/16 **41.** $\frac{6}{7}$ **43.** $1\frac{7}{18}$
45. 4 **47.** 3 **49.** −5°F **51.** 62.7 **53.** 2.88

Pages 260–261 Lesson 6-5

1. no; $\frac{8}{3} \neq 24$ **3.** $\frac{2}{3}$, 3; The other pairs of numbers are reciprocals, but these are not. **5.** $\frac{9}{2}$ or $4\frac{1}{2}$ **7.** $\frac{5}{29}$
9. −12 **11.** $1\frac{1}{2}$ **13.** 20.5 **15.** $\frac{2}{11}$ **17.** $-\frac{8}{3}$ or $-2\frac{2}{3}$
19. $\frac{1}{3}$ **21.** $\frac{5}{22}$ **23.** 36 **25.** 30 **27.** −24 **29.** $-\frac{4}{5}$
31. 42.12 **33.** 2.88 **35.** 137.5 mi **37.** Sample answer: Use the Multiplicative Property to multiply each side by 2. Then use the Division Property to divide each side by h. So, $b = \frac{2A}{h}$. **39.** 16 **41.** 9 **43.** $2\frac{2}{3}$ **45.** $\frac{6}{17}$ **47.** $\frac{5}{6}$
49. Sample answer: $24 \div 12 = 2$ **51.** Sample answer: $1 \div 1 = 1$

Page 266 Lesson 6-6

1. Sample answer: How many $\frac{1}{4}$-pound hamburgers can be made with 10 pounds of ground beef? **3.** $2\frac{2}{5}$ **5.** $\frac{1}{15}$
7. 24 boxes **9.** $\frac{2}{3}$ **11.** $1\frac{1}{6}$ **13.** $\frac{2}{9}$ **15.** $3\frac{3}{4}$ **17.** $\frac{1}{6}$ **19.** $2\frac{7}{15}$
21. greater than 1 because $7\frac{1}{6} > 3\frac{2}{3}$ **23.** $\frac{11}{16}$ lb **25.** $\frac{7}{6}$ or $1\frac{1}{6}$
27. $\frac{1}{8}$ **29.** $\frac{1}{16}$ **31.** 14 **33.** 16

Pages 268–269 Lesson 6-7

1. Sample answer: A recipe calls for 2 pints of strawberries. How many cups is this? Answer: 4 c **3.** 3 **5.** 3 **7.** 500
9. 100 T **11.** 6 **13.** 16 **15.** 9 **17.** 10,560 **19.** $4\frac{1}{2}$ **21.** $5\frac{1}{2}$
23. $1\frac{1}{3}$ **25.** $10\frac{3}{4}$ **27.** 82.8 **29.** $\frac{1}{2}$ **31.** $43\frac{3}{4}$ gal
33.

Pounds	Ounces
1	16
2	32
3	48
4	64

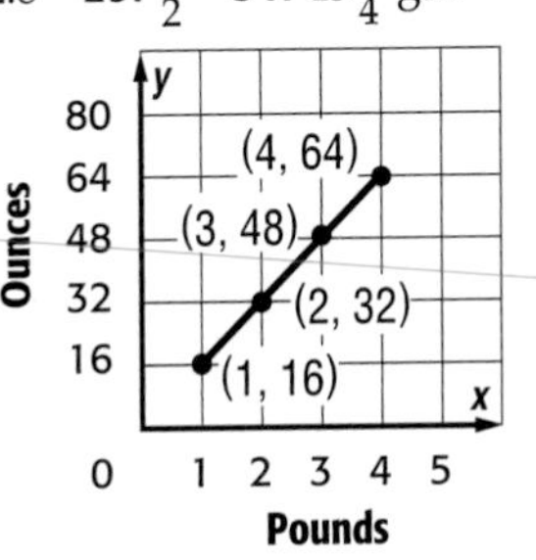

The graph is a straight line. For each x-value increase of 1, the y value increases by 16. **35.** A **37.** 28 **39.** −10.8
41. 34 **43.** 19

Pages 272–273 Lesson 6-8

1. Perimeter is the sum of measures, so the units are the same. Area is the product of measures, so the units are squared. **3.** 18 yd; 20 yd^2 **5.** 26 cm; 42 cm^2 **7.** $23\frac{1}{3}$ ft **9.** 82 in.; 414 in^2 **11.** 65.4 m; 183.6 m^2 **13.** $30\frac{1}{5}$ in.; $47\frac{7}{10}$ in^2 **15.** 60 cm; 216 cm^2 **17.** $15\frac{3}{20}$ in.; 14 in^2 **19.** $37\frac{1}{3}$ yd or 112 ft; $81\frac{2}{3}$ yd^2 or 735 ft^2 **21.** The perimeter is doubled, the area is quadrupled. **23.** $35\frac{13}{24}$ ft **25.** 7 boxes **27.** 1.5 acres; $65{,}010 \div 43{,}560$ **29.** The rectangles have the same perimeters; as the lengths and widths get closer, the areas increase. **31.** 36 **33.** A **35.** 10,500 **37.** $7\frac{1}{2}$ **39.** 33 **41.** 44 **43.** $4\frac{5}{7}$

Pages 276–277 Lesson 6-9

1. The circumference increases by 2π for every 1-unit increase in r. **3.** Mya; Aidan incorrectly applied the formula that uses the diameter. **5.** 44 m **7.** 73.5 cm **9.** 11 yd **11.** 50.2 ft **13.** 44 yd **15.** 47.4 m **17.** 12.6 km **19.** 285.7 m **21.** $16\frac{1}{2}$ in. **23.** 785 ft **25.** 2π or 6.28 **27.** 69.1 in. **29.** 17.6 cm; 19 cm^2 **31.** 8 yd

Pages 278–280 Chapter 6 Study Guide and Review

1. diameter **3.** Area **5.** reciprocal **7.** Sample answer: $3 \div 1 = 3$ **9.** Sample answer: $1 \times 0 = 0$ **11.** Sample answer: $\frac{1}{2} \cdot 26 = 13$ **13.** $\frac{1}{6}$ **15.** $\frac{1}{12}$ **17.** $1\frac{3}{5}$ **19.** $\frac{5}{24}$ **21.** $1\frac{1}{10}$ **23.** $7\frac{3}{5}$ **25.** $10\frac{1}{7}$ **27.** $2\frac{4}{15}$ **29.** $15\frac{1}{8}$ **31.** $\frac{6}{35}$ **33.** $\frac{2}{7}$ **35.** $1\frac{3}{4}$ **37.** $1\frac{11}{16}$ mi **39.** $\frac{1}{5}$ **41.** 16 **43.** -22.8 **45.** $\frac{7}{10}$ **47.** $6\frac{1}{2}$ **49.** $3\frac{3}{10}$ **51.** $\frac{3}{22}$ **53.** $\frac{3}{4}$ **55.** 24 **57.** 4 **59.** 3 **61.** $6\frac{1}{4}$ ft; $1\frac{5}{16}$ ft^2 **63.** 11 in.; $2\frac{1}{2}$ in^2 **65.** $13\frac{2}{3}$ ft; $10\frac{1}{2}$ ft^2 **67.** 25.1 yd **69.** 26.4 ft

Chapter 7 Ratios and Proportions

Page 287 Chapter 7 Getting Started

1. ratio **3.** 48.1 **5.** 7.4 **7.** $\frac{1}{5}$ **9.** $\frac{19}{23}$ **11.** $\frac{39}{50}$ **13.** $\frac{3}{50}$ **15.** 24 **17.** $5\frac{1}{2}$ **19.** 6 **21.** $1\frac{1}{2}$

Pages 290–291 Lesson 7-1

1. $\frac{8}{15}$, 8:15, 8 to 15 **3.** 3 to 2; The other ratios equal $\frac{2}{3}$. **5.** $\frac{4}{15}$ **7.** $\frac{11}{4}$ **9.** $\frac{4}{3}$ **11.** yes; $\frac{9}{15} = \frac{3}{5}$ **13.** $\frac{21}{23}$ **15.** $\frac{1}{5}$ **17.** $\frac{15}{7}$ **19.** $\frac{11}{30}$ **21.** $\frac{3}{16}$ **23.** $\frac{5}{1}$ **25.** $\frac{3}{8}$ **27.** yes; $\frac{9}{15} = \frac{3}{5}$ and $\frac{6}{10} = \frac{3}{5}$ **29.** yes; $8{:}21 = \frac{8}{21}$ and $16{:}42 = \frac{8}{21}$ **31.** yes; 6 lb:72 oz $= \frac{4}{3}$ and 2 lb:24 oz $= \frac{4}{3}$ **33.** $\frac{1}{13}$ **35.** no **37.** $\frac{2}{5}$; $\frac{7}{3}$; No, the ratio is greater for the dollar bill. **39.** 4; The ratios of successive terms increase by 1. **41.** G **43.** 26 ft; 36 ft^2 **45.** 19 cm; 15 cm^2 **47.** 4.9 **49.** \$0.31

Pages 294–295 Lesson 7-2

1. Sample answer: $\frac{30 \text{ mi}}{2 \text{ h}} = \frac{15 \text{ mi}}{\text{h}}$ **3.** a; Sample answer: $\frac{20 \text{ ft}}{1 \text{ min}} \rightarrow \frac{40 \text{ ft}}{1 \text{ min}}$ represents an increase in speed since 40 ft/min > 20 ft/min. If the same distance is covered in a longer period of time, then the rate is decreased. Consider $\frac{20 \text{ ft}}{1 \text{ min}} \rightarrow \frac{20 \text{ ft}}{2 \text{ min}}$. Since $\frac{20 \text{ ft}}{2 \text{ min}} = 10$ ft/min, the rate is decreased. **5.** 60 words per min **7.** \$0.98 per can **9.** Music Rox; \$14.95/CD < \$17.99/CD **11.** 152 customers per day **13.** \$1.48 per lb **15.** 270 C per serving **17.** 32 mi per gal **19.** 150 for \$11.99 **21.** 4,000 gal per min **23.** about 179 people per sq mi **25.** Sometimes; a ratio that compares two measurements with different units is a rate, such as $\frac{2 \text{ mi}}{10 \text{ min}}$. A ratio that compares two numbers or two measurements with like units is not a rate, such as $\frac{2 \text{ c}}{3 \text{ c}}$. **27.** human, 72 beats/min; elephant, 35 beats/min **29.** 32 tickets per h **31.** I **33.** $\frac{6}{1}$ **35.** $\frac{10}{7}$ **37.** 56 **39.** $1\frac{1}{3}$

Pages 299–300 Lesson 7-3

1. Compare the cross products. If they are equal, the ratios are equivalent. **3.** $\frac{2}{9}$; The other ratios equal $\frac{2}{3}$. **5.** no **7.** 2 **9.** 21 **11.** 9 **13.** 29.4 **15.** $6\frac{2}{3}$ or about 6.7 oz **17.** no **19.** no **21.** no **23.** 8 kittens **25.** 6 **27.** 6 **29.** 75 **31.** 2 **33.** 63 **35.** 22.5 **37.** 1,040 **39.** 13.5 **41.** $\frac{13}{15} = \frac{x}{100}$; 87% **43.** 4 c **45.** 210 words **47.** Sample answers: 1 and 36; 2 and 18; 4 and 9 **49.** 64 **51.** $\frac{5}{9}$ **53.** $\frac{7}{40}$ **55.** 56 **57.** $1\frac{3}{5}$ **59.** 52

Pages 306–308 Lesson 7-4

3. 200 km **5.** 3.5 cm **7.** $\frac{1}{90}$ **9.** 255 km **11.** 1,205 km **13.** 6 in.; $\frac{1}{240}$ **15.** 12 cm; $\frac{1}{300}$ **17.** 4 in. **19.** $\frac{3}{4}$ in. **21.** $\frac{1}{48}$ **23.** $12\frac{3}{16}$ mi **25.** $\frac{1}{12}$ **27.** 1 in. = 70 mi **29.** D **31.** 40 mi/h **33.** \$1.55 per lb **35.** $\frac{11}{40}$ **37.** $\frac{23}{75}$

Pages 314–315 Lesson 7-5

1. Sample answer: Write a proportion using the given fraction and $\frac{n}{100}$. Then solve for n and add a %. Or, write the fraction as a decimal. Then multiply by 100 and add a %. **3.** Sample answer: It is easier to compare percents than it is to compare fractions. **5.** $\frac{3}{16}$ **7.** $\frac{23}{40}$ **9.** 62.5% **11.** 36.36% **13.** $\frac{7}{40}$ **15.** $\frac{31}{500}$ **17.** $\frac{23}{80}$ **19.** $\frac{1}{3}$ **21.** $\frac{13}{16}$ **23.** $\frac{63}{80}$ **25.** 55% **27.** 37.5% **29.** 96.67% **31.** 71.43% **33.** 1.25% **35.** 41.67% **37.** 48% **39.** less than; $\frac{26}{125} = 20.8\%$ **41.** D **43.** 67.5 ft **45.** $1\frac{2}{3}$ **47.** 17.5 **49.** $\frac{9}{100}$ **51.** $\frac{42.5}{100}$

Pages 317–318 Lesson 7-6

1. 170%; 1.7 **3.** $63\frac{1}{2}$; It is the only number not equivalent to $63\frac{1}{2}\%$. **5.** 2; 2 **7.** 0.0015; $\frac{3}{2{,}000}$ **9.** 0.15% **11.** 0.05% **13.** 3.5; $3\frac{1}{2}$ **15.** 6; 6 **17.** 0.006; $\frac{3}{500}$ **19.** 0.0055; $\frac{11}{2{,}000}$ **21.** 2.6; $2\frac{3}{5}$ **23.** 0.0005; $\frac{1}{2{,}000}$ **25.** 850% **27.** 0.9% **29.** 264% **31.** 0.34% **33.** 350% **35.** 0.4% **37.** 162.5% **39.** 0.003; $\frac{3}{1{,}000}$ **41.** about 0.595% **43.** A **45.** $\frac{3}{40}$ **47.** $\frac{1}{16}$ **49.** 6 in. **51.** 0.065 **53.** 0.123

Pages 320–321 Lesson 7-7

1. $\frac{x}{22} = \frac{0.5}{100}$ **3.** Sample answer: Find the number of people surveyed who use a white erasable markerboard to record information during a meeting; 525 people. **5.** 110.5 **7.** 0.1 **9.** 65 **11.** \$160 **13.** 51.3 **15.** 12.5 **17.** 0.4 **19.** 0.6 **21.** $\frac{x}{1{,}746} = \frac{22}{100}$; about 384 students **23.** \$300.96 **25.** B **27.** 750% **29.** 0.04% **31.** 73.33% **33.** 20 **35.** 4.5

Page 325 Lesson 7-8
1. b **3.** 36% **5.** 0.7 **7.** 25% **9.** 8.6 **11.** 375 **13.** 7.5% **15.** 4.1 **17.** 30 **19.** 0.04 **21.** 60% **23.** 20% of 500, 20% of 100, 5% of 100; Sample answer: If the percent is the same but the base is greater, then the part is greater. If the base is the same but the percent is greater, then the part is greater. **25.** 65% **27.** 31.5

Pages 326–328 Chapter 7 Study Guide and Review
1. e **3.** a **5.** c **7.** g **9.** d **11.** $\frac{4}{3}$ **13.** $\frac{2}{3}$ **15.** yes; $\frac{3}{5} = \frac{21}{35}$ **17.** yes; 27:15 = $\frac{9}{5}$ and 9:5 = $\frac{9}{5}$ **19.** $4.75 per lb **21.** $9.50 per h **23.** $3\frac{1}{2}$ laps per min **25.** 6 **27.** 12.5 **29.** 9 **31.** 94 cm **33.** 27 cm **35.** 5.2 cm **37.** $\frac{27}{500}$ **39.** 12.5% **41.** 17.5% **43.** 0.0075; $\frac{3}{400}$ **45.** 0.005; $\frac{1}{200}$ **47.** 475% **49.** 0.95% **51.** 34.1 **53.** 0.6 **55.** 0.3 **57.** 130.2

Chapter 8 Applying Percent

Page 333 Chapter 8 Getting Started
1. true **3.** 48 **5.** 1,512 **7.** 1.75 **9.** 0.8 **11.** 130 **13.** 216.7 **15.** 0.4 **17.** 0.075 **19.** 39.5%

Pages 336–337 Lesson 8-1
1. Sample answer: Find $\frac{1}{5} \cdot 140$ or find 10% of 140 and multiply by 2; both estimates equal 28. **3.** Ian; Mandy incorrectly changed 1.5% to 1.5, which is 150%.
5–31. Sample answers are given. **5.** $\frac{1}{2} \cdot 160 = 80$ **7.** $\frac{3}{4} \cdot 20 = 15$ **9.** $0.1 \cdot 400 = 40$ and $2 \cdot 40 = 80$ **11.** $(1 \cdot 70) + \left(\frac{1}{2} \cdot 70\right) = 105$ **13.** $0.01 \cdot 80 = 0.8$ and $\frac{1}{2} \cdot 0.8 = 0.4$ **15.** $\frac{1}{4} \cdot 400 = 100$ **17.** $\frac{3}{4} \cdot 280 = 210$ **19.** $\frac{2}{3} \cdot 15 = 10$ **21.** $0.1 \cdot 40 = 4$ and $6 \cdot 4 = 24$ **23.** $0.1 \cdot 80 = 8$ and $8 \cdot 8 = 64$ **25.** $0.1 \cdot 120 = 12$ and $3 \cdot 12 = 36$ **27.** $(1 \cdot 54) + \left(\frac{1}{3} \cdot 54\right) = 72$ **29.** $4 \cdot 12 = 48$ **31.** $0.01 \cdot 400 = 4$ and $\frac{1}{2} \cdot 4 = 2$ **33.** Sample answer: $0.01 \cdot 3{,}000 = 30$ lb **35.** Sample answer: Find 1% of a number, then multiply by $\frac{3}{8}$. **37.** H **39.** 64.8 **41.** 157.1 **43.** 50 **45.** 357.1

Pages 342–343 Lesson 8-2
1. missing base **3.** $n = 0.88 \cdot 300$; 264 **5.** $3 = 0.12 \cdot n$; 25 **7.** 29.7%; 0.297 **9.** $n = 0.53 \cdot 470$; 249.1 **11.** $26 = n \cdot 96$; 27.1% **13.** $17 = 0.40 \cdot n$; 42.5 **15.** $30 = n \cdot 64$; 46.9% **17.** $1.45 = 0.33 \cdot n$; 4.4 **19.** $n = 0.752 \cdot 600$; 451.2 **21.** $14 = 0.028 \cdot n$; 500 **23.** $230 = n \cdot 200$; 115% **25.** 6% **27.** 42% **29.** 28% **31.** Sample answer: If the percent is less than 100%, then the part is less than the number; if the percent equals 100%, then the part equals the number; if the percent is greater than 100%, then the part is greater than the number. **33.** I **35.** Sample answer: $\frac{3}{10} \cdot 50 = 15$ **37.** 1,312.5 mg **39.** 18 **41.** 16.7% **43.** 31.6

Pages 346–347 Lesson 8-3
1. Sample answer: Randomly select a part of the group to get a sample. Find their preferences, and use the results to find the percent of the total group. **3.** 8.96 million **5.** 1,396 students **7.** about 110 owners **9.** 6,500 teens **11.** 800,000 people **13.** $54 = 0.72 \cdot n$; 75 **15.** 0.3 **17.** 4.5

Pages 352–353 Lesson 8-4
1. Sample answer: The sale price of a T-shirt is $14. If the shirt originally cost $25, what is the percent discount? Answer: 44% decrease **3.** Jada; Miranda did not write a ratio comparing the change to the original amount. **5.** 19% inc. **7.** 25% inc. **9.** 73% inc. **11.** 20% inc. **13.** 50% dec. **15.** 71% dec. **17.** 30% dec. **19.** 18% inc. **21.** 1% inc. **23.** 100% **25.** 21% inc. **27.** 1.2% increase **29.** 41% **31.** 16 to 17; 72% **33.** C **35.** 660 students **37.** $n = 0.21 \cdot 62$; 13.0 **39.** 0.055 **41.** 0.0675

Pages 356–357 Lesson 8-5
1. $6.86 **3.** Sample answer: The regular price of a CD is $17.95. The total cost is $18.98. **5.** $1,338.75 **7.** $47.33 **9.** 9% **11.** $46.40 **13.** $103.95 **15.** $2.58 **17.** $26.53 **19.** $26.07 **21.** 75% **23.** 23% **25.** $131.66 **27.** $600.88 **29.** $366.56 **31.** $57.60 **33.** B **35.** 50% inc. **37.** 73% inc. **39.** 292.3 **41.** 99

Pages 359–360 Lesson 8-6
1. Sample answer: Write the interest rate as a decimal and write the months in years. Then multiply $500 \cdot 0.06 \cdot 1.5$. **3.** $38.40 **5.** $1,417.50 **7.** $23.20 **9.** $21.38 **11.** $330 **13.** $8.54 **15.** $45.31 **17.** $26.37 **19.** $210 **21.** 4% **23.** $627, $655.22, $684.70 **25.** G **27.** 29% inc. **29.** 35% dec.

Pages 362–364 Chapter 8 Study Guide and Review
1. true **3.** false; original **5.** true **7.** true **9.** true **11–17. Sample answers are given.** **11.** $\frac{1}{4} \cdot 80 = 20$ **13.** $\frac{3}{4} \cdot 40 = 30$ **15.** $0.1 \cdot 80 = 8$ **17.** $0.1 \cdot 1{,}000 = 100$ and $5 \cdot 100 = 500$ **19.** Sample answer: $\frac{1}{10} \cdot 300 = 30$ households **21.** $39 = 0.65 \cdot n$; 60 **23.** $n = 0.07 \cdot 92$; 6.4 **25.** about 208 **27.** 48% inc. **29.** 20% dec. **31.** about 12% increase **33.** $178.50 **35.** $26.80 **37.** 14% **39.** 10% **41.** $47.50 **43.** $412.50 **45.** $121.50 **47.** $1,360

Chapter 9 Probability

Page 369 Chapter 9 Getting Started
1. simplest **3.** 105 **5.** 52 **7.** 160 **9.** 360 **11.** 336 **13.** 5,040 **15.** $\frac{2}{3}$ **17.** simplified **19.** 5 **21.** 4

Pages 372–373 Lesson 9-1
1. Sample answer: Since 0.7 is greater than 0.5 and less than 0.75, the event is likely to happen. **3.** 0.65, 0.55 are probabilities that are not complementary because $0.65 + 0.55 \neq 1$. The other pairs are complementary. **5.** $\frac{1}{4}$ **7.** $\frac{3}{5}$ **9.** 75% **11.** $\frac{4}{5}$ **13.** $\frac{1}{2}$ **15.** $\frac{1}{10}$ **17.** Sample answer: The complementary event is the chance of no rain. Its probability is 63%. **19.** $\frac{7}{10}$ **21.** $\frac{1}{2}$ **23.** $\frac{7}{10}$ **25.** 0 **27.** $\frac{1}{11}$ **29.** $\frac{9}{10}$ **31.** A **33.** $60 **35.** about 2.5 billion **37.** 0.45 **39.** $1.\overline{7}$ **41.** simplified **43.** $\frac{3}{8}$

Pages 376–377 Lesson 9-2
1. Sample game: Each player tosses a coin 10 times. If it comes up heads, player 1 receives 1 point. If it comes up tails, player 2 receives 1 point. The player with the most points at the end of 10 tosses wins.

3.
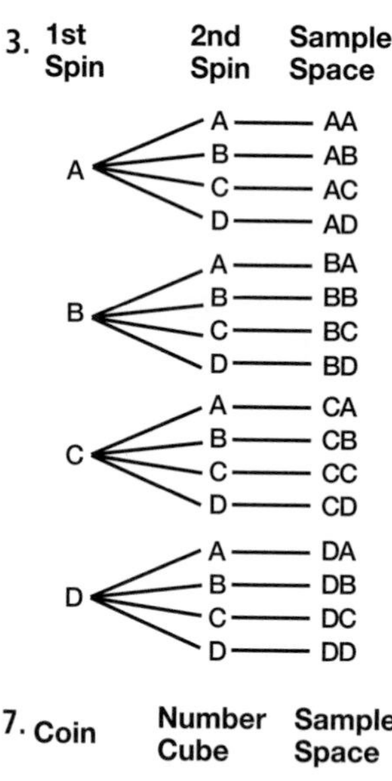

5. $\frac{1}{16}$

7.
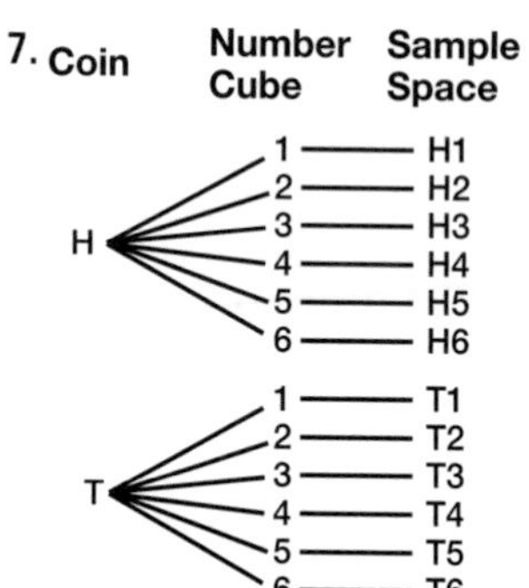

12 outcomes

9. Number Color Sample Space

Number	Color	Sample Space
1	red	1 red
	white	1 white
	blue	1 blue
2	red	2 red
	white	2 white
	blue	2 blue
3	red	3 red
	white	3 white
	blue	3 blue
4	red	4 red
	white	4 white
	blue	4 blue
5	red	5 red
	white	5 white
	blue	5 blue

15 outcomes

11.
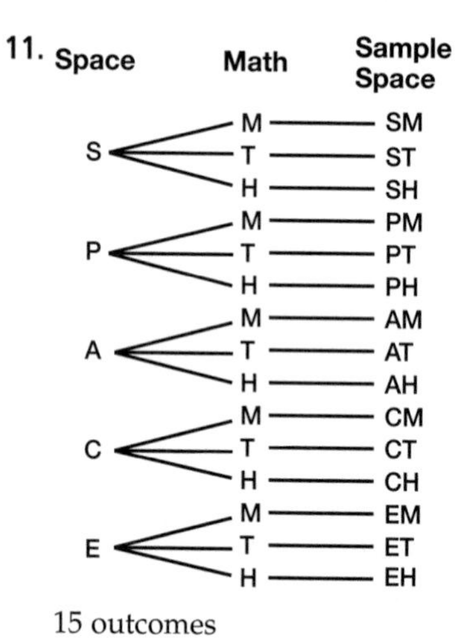

15 outcomes

13.
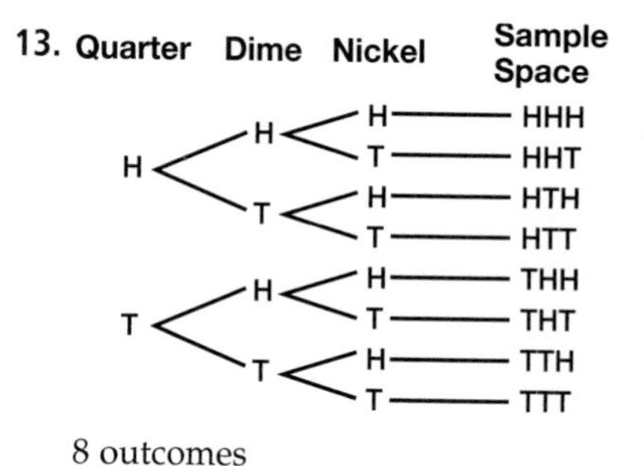

8 outcomes

15.
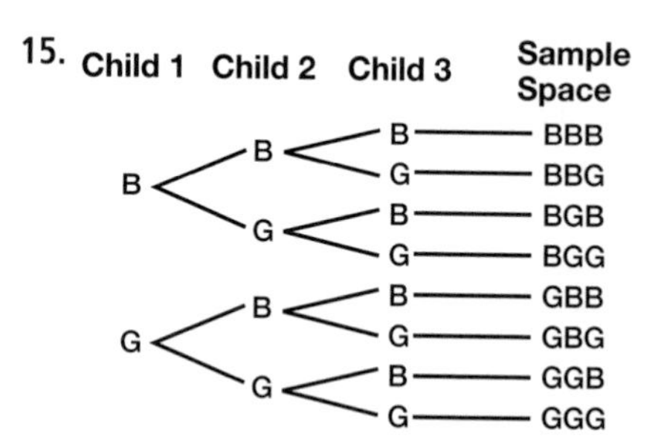

17. $\frac{1}{8}$ **19.** $\frac{4}{24}$ or $\frac{1}{6}$ **21.** $P(\text{Player 1}) = \frac{6}{8}$ or $\frac{3}{4}$; $P(\text{Player 2}) = \frac{2}{8}$ or $\frac{1}{4}$; Player 2 does not have an equal chance of winning. **23.** C **25.** $\frac{2}{5}$ **27.** $\frac{13}{20}$ **29.** 154 **31.** 176 **33.** 524

Pages 379–380 Lesson 9-3

1. Sample answer: Multiply the number of ways each event can occur to find the total number of outcomes. **3.** Sample answer: Yes, a pair of events can have a number of outcomes that is prime. For example, 11 different pairs of shoes can come in 1 color. $11 \cdot 1$, or 11 outcomes **5.** 8 **7.** 60 **9.** 24 **11.** 12 **13.** 217 **15.** No; the number of selections is 350, which is less than 365. **17.** 2 **19.** 8 **21.** B **23.** $\frac{1}{2}$ **25.** $\frac{3}{4}$ **27.** 6 **29.** 120

Pages 382–383 Lesson 9-4

1. Sample answer: Five factorial has extra factors of 2 and 1, $5! = 5 \cdot 4 \cdot 3 \cdot 2 \cdot 1$. **3.** 6 **5.** 120 **7.** 120 **9.** 144 **11.** 4,320 **13.** 2,880 **15.** 3,024 **17.** 11,880 **19.** $\frac{1}{7}$ **21.** 24 **23.** 96 **25.** 10 **27.** 2.5

Pages 389–390 Lesson 9-5

1. Sample answer: The order of the numbers is important in a lock. **3.** Francisca; Allison found the number of permutations of three people chosen from a group of seven. **5.** permutation; 720 ways **7.** 10 **9.** 36 **11.** combination; 210 **13.** permutation; 6 ways **15.** permutation; 5,040 **19.** A **21.** 120 signals **23.** Sample answer: $0 + 1 = 1$ **25.** Sample answer: $\frac{1}{2} \cdot 20 = 10$ **27.** $\frac{1}{2}$ **29.** $\frac{1}{8}$ **31.** $\frac{1}{2}$

Pages 395–396 Lesson 9-6

1. Sample answer: Both probabilities are ratios that compare the number of favorable outcomes to the total number of possible outcomes. Experimental probability is the result of an experiment. Theoretical probability is what is expected to happen. **3.** $\frac{14}{25}$ **5.** $\frac{3}{16}$ **7.** $\frac{5}{6}$ **9.** $\frac{9}{10}$ **11.** 150 **13.** 14 **15.** $\frac{9}{25}$ **17.** 120 **19.** 1:5 **21.** 2:4 or 1:2 **23.** 1/5 **25.** 720 **27.** $\frac{3}{8}$ **29.** $\frac{7}{8}$

Pages 399–401 *Lesson 9-7*

1. Sample answer: Events are independent when neither outcome affects the other. Events are dependent when the outcome of the first event affects the outcome of the second event. **3.** Shane; Kimi treated spinning the spinner twice as dependent events. **5.** $\frac{5}{14}$ **7.** $\frac{1}{8}$ **9.** $\frac{2}{87}$ **11.** $\frac{1}{6}$ **13.** $\frac{13}{51}$ **15.** independent **17.** 0.03 **19.** 13.5% **21.** 0.096 **23.** $\frac{1}{2}$ **25.** B **27.** 1 **29.** −27 **31.** 24

Pages 402–404 *Chapter 9* *Study Guide and Review*

1. sample space **3.** experimental **5.** permutation **7.** random **9.** is not **11.** $\frac{1}{2}$ **13.** $\frac{1}{4}$ **15.** $\frac{1}{2}$

17.

Number Cube	Coin	Sample Space
1	H	1H
	T	1T
2	H	2H
	T	2T
3	H	3H
	T	3T
4	H	4H
	T	4T
5	H	5H
	T	5T
6	H	6H
	T	6T

12 outcomes

19.

Coin	Card	Sample Space
H	1	H1
	2	H2
	3	H3
	4	H4
	5	H5
	6	H6
	7	H7
T	1	T1
	2	T2
	3	T3
	4	T4
	5	T5
	6	T6
	7	T7

14 outcomes

21. 36 **23.** 20 **25.** 24 **27.** 1,680 **29.** 720 **31.** 60 **33.** 126 groups **35.** 45 ways **37.** $\frac{3}{25}$ **39.** $\frac{1}{5}$ **41.** $\frac{1}{4}$ **43.** $\frac{1}{15}$

Chapter 10 Geometry

Page 411 *Chapter 10* *Getting Started*

1. ratio **3.** 306 **5.** 0.15 **7.** 0.11 **9.** 53 **11.** 131 **13.** 92 **15.** 90 **17.** 54 **19.** 12 **21.** 24 **23.** 8

Pages 414–415 *Lesson 10-1*

1. Sample answer: An angle with a measure of exactly 90°, which is indicated by a right angle symbol. **3.** obtuse **5.** right **7.** acute

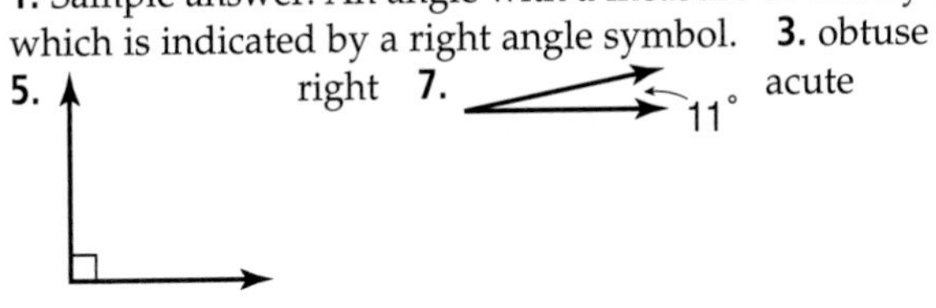

9. acute **11.** obtuse **13.** obtuse (147°) **15.** obtuse (99°) **17.** right

19. 150° **21.** Sample answer: acute angles: between their legs and the front of their skis; obtuse angles: between their legs and the back of their skis **23.** 48 times **25.** A **27.** $\frac{1}{4}$ **29.** $\frac{7}{10}$ **31.** 90 **33.** 0.32

Pages 420–421 *Lesson 10-2*

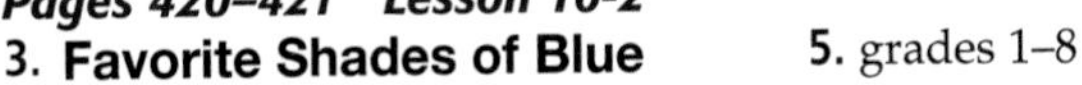

3. Favorite Shades of Blue

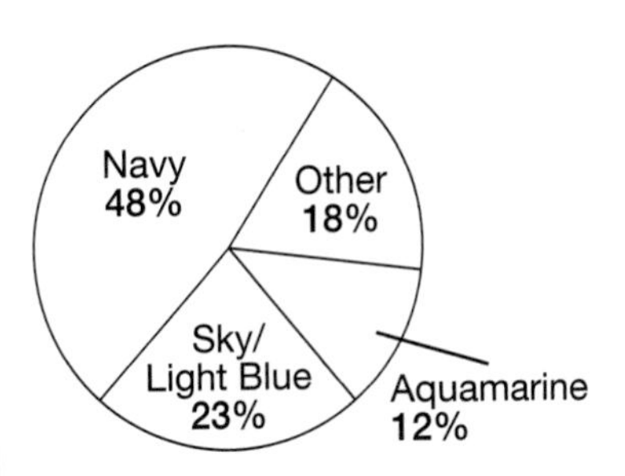

5. grades 1–8

7. NYC Commuters

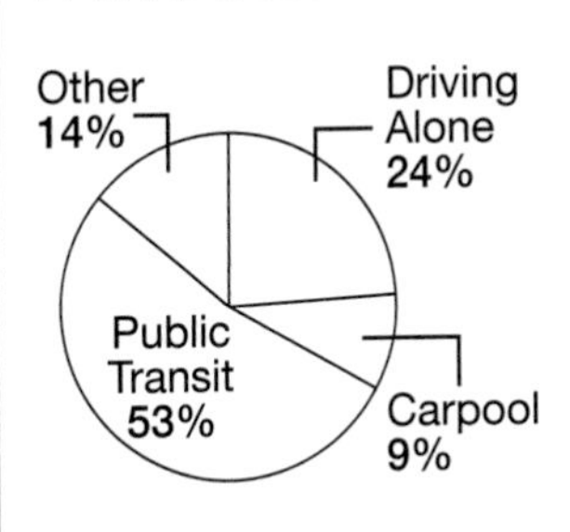

9. Sample answer: A greater percent of commuters in Los Angeles drive alone, while in New York, a greater percent use public transportation. A greater percent of people carpool in Los Angeles than in New York.

11. U.S. Regions

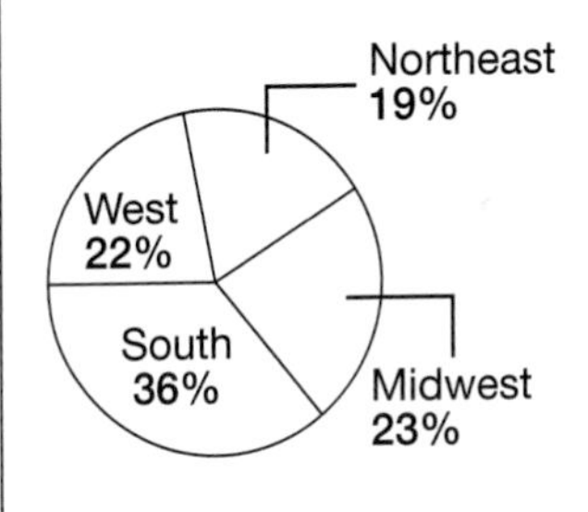

13. yes: 262.8°; no: 90°; not sure: 7.2° **17.** A **19.** acute **21.** right **23.** 72 **25.** 141

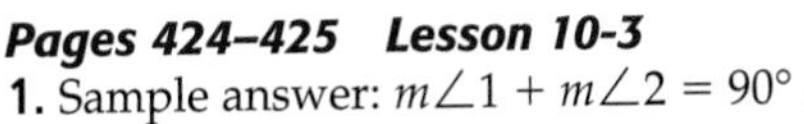

Pages 424–425 *Lesson 10-3*

1. Sample answer: $m\angle 1 + m\angle 2 = 90°$

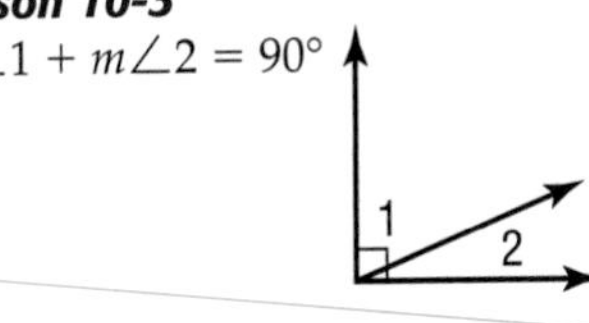

3. supplementary **5.** complementary **7.** 75 **9.** 96° **11.** supplementary **13.** supplementary **15.** complementary **17.** 69 **19.** 42 **21.** 90 **23.** Public Restroom and Subway/train **25.** never; Sample answer: One straight angle measures 180°. **27.** sometimes; Sample answer: Adjacent angles can have the same or different angle measure. **29.** $m\angle E = 39°$, $m\angle F = 51°$ **31.** C **33.** acute **35.** 0.53 **37.** 4.31 **39.** 68 **41.** 101

Pages 430–431 Lesson 10-4

1. An obtuse triangle has one obtuse angle and two acute angles. **3.** 44°; acute **5.** acute, equilateral **7.** right, isosceles **9.** 60°; right **11.** 65°; acute **13.** 37°; right **15.** right **17.** 17° **19.** obtuse, scalene **21.** right, scalene **23.** acute, isosceles **27.** B **29.** supplementary **31.** neither **33.** 60

Pages 435–437 Lesson 10-5

1. Sample answer: A square has all the properties of a rectangle. However, a rectangle does not have four congruent sides, which is a property of a square. **3.** Justin; Venus did not mention the 4 congruent sides of a square. **5.** quadrilateral **7.** 100° **9.** 35° **11.** trapezoid **13.** rhombus **15.** parallelogram **17.** sometimes; Sample answer: A parallelogram can also have angles with different degree measures. **19.** 56° **21.** 67° **23.** 90° **25.** rhombus, rectangle **29.** Yes; A square is a rhombus and a rectangle. **31.** Yes; One pair of opposite sides are parallel, while the other pair of opposite sides are congruent. This figure is called an isosceles trapezoid.

33. B **35.** obtuse, isosceles **37.** acute, equilateral **39.** \$3.78 **41.** 85 **43.** 41.7 **45.** 400 mi **47.** 3 **49.** 7

Pages 442–443 Lesson 10-6

1. Corresponding sides are proportional and corresponding angles are congruent. **3.** 35 m **5.** 48 ft **7.** 12 m **9.** 10.2 mm **11.** yes; $\frac{6}{18} = \frac{11}{33}$ **13.** about 42 ft **15.** 120 m **17.** 1:16 **19.** 90 ft **21.** trapezoid **23.** 69° **25.** 90 **27.** 120

Pages 448–450 Lesson 10-7

1. the point where vertices meet: 45° + 45° + 135° + 135° = 360° **3.** All of the angles of a rhombus do not necessarily have the same measure. **5.** not a polygon, it is not closed **7.** No; 140 does not go into 360 evenly. **9.** regular octagon **11.** not a polygon; the figure has a curved side **13.** regular decagon **15.** 140° **17.** 162° **19.** 135°, 45°, 45° **21.** hexagon and triangle **23.** octagons and squares **25.** 43.2 cm **27.** $37\frac{1}{2}$ yd **31.** Yes; the sum of the measures of angles of any triangle is 180°.

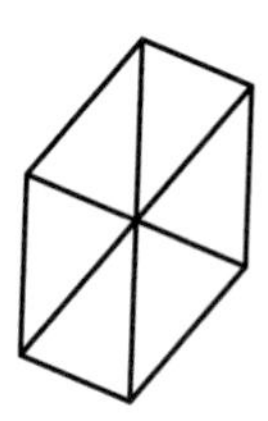

33. I **35.** trapezoid **37.** 32 servings **39.** $3\frac{1}{6}$ **41.** $4\frac{37}{40}$

42–45.

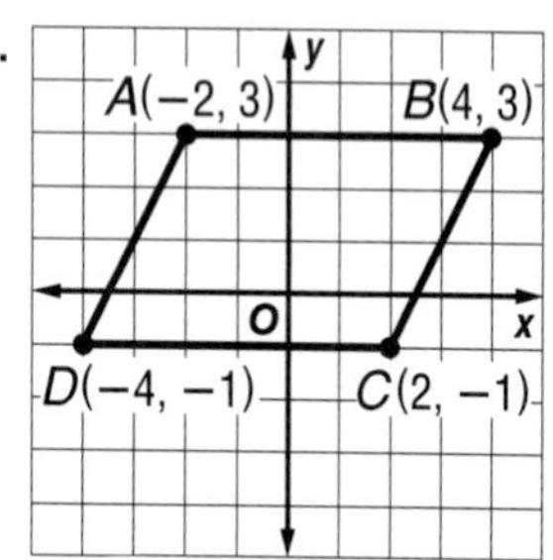

Pages 453–454 Lesson 10-8

1. Sample answer:

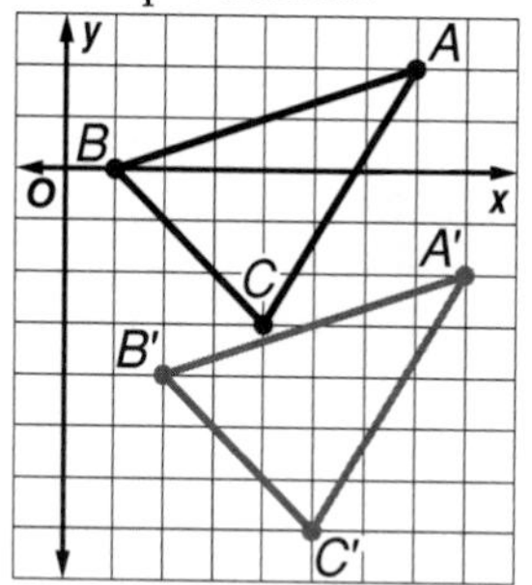

3.

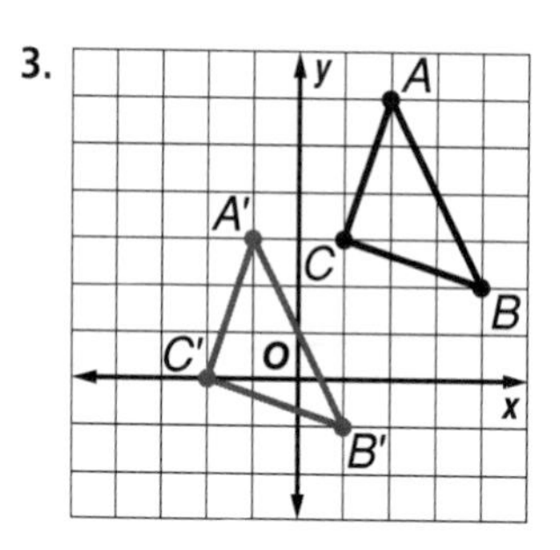

5. $D'(7, 0)$, $E'(4, -2)$, $F'(8, 4)$, $G'(12, -3)$

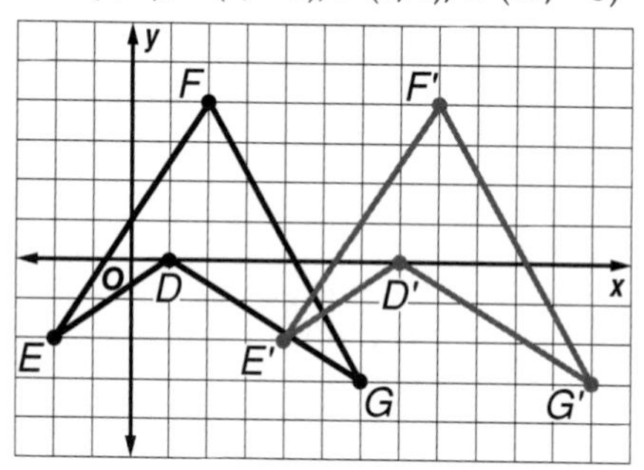

7.

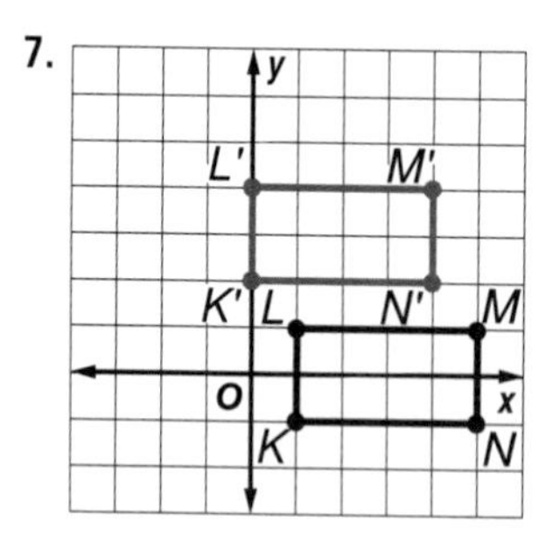

9. $P'(-8, -1)$, $Q'(-3, -3)$, and $R'(-11, 5)$

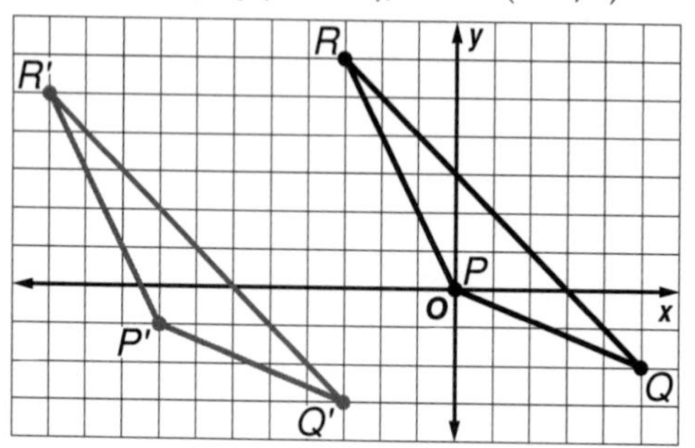

11. $P'(0, -9)$, $Q'(5, -11)$, and $R'(-3, -3)$

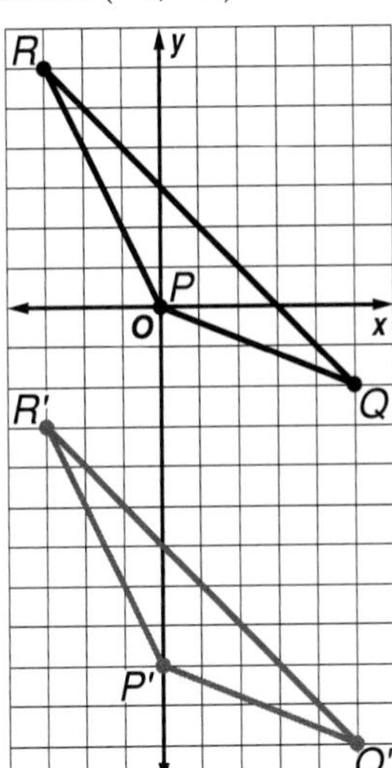

13. Sample answer: there are two main images, the brown horseman and the yellow horseman. Both main horsemen are translated to different parts of the picture. These translations allow for the tessellation of both horsemen. **15.** Sample answer: $\triangle X'Y'Z'$ was translated 5 units left and 2 units down.

17.

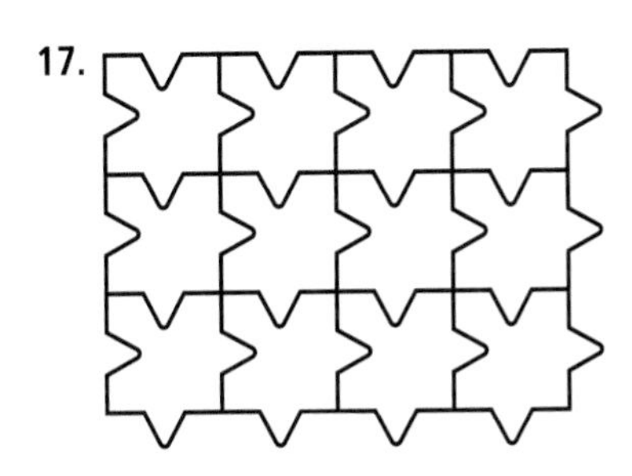

19. C **21.** octagon **23.** yes **25.** no

Pages 458–459 Lesson 10-9

1.

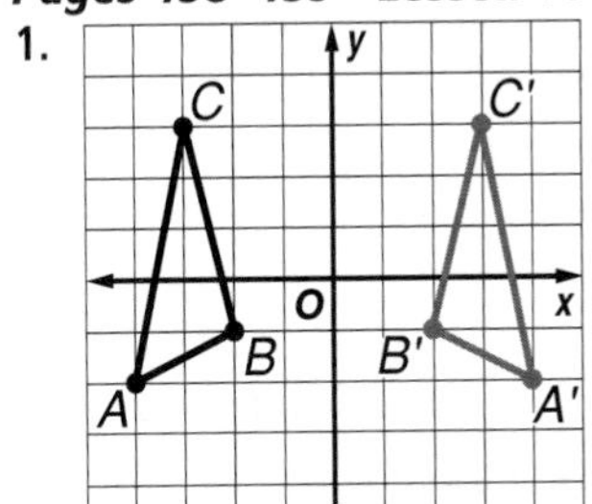

3. no

5. $A'(5, -8)$, $B'(1, -2)$, and $C'(6, -4)$

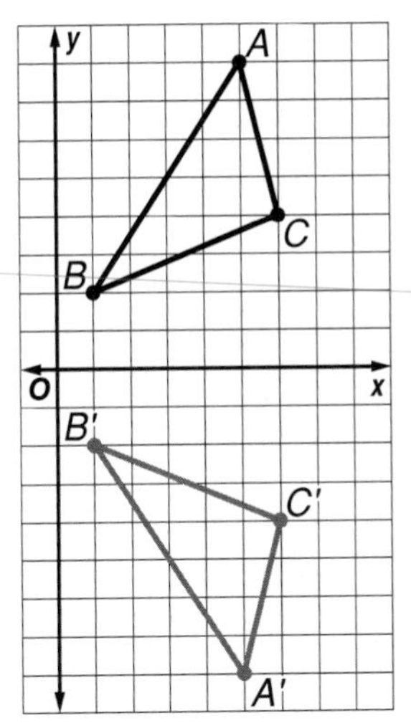

7. yes

9. yes

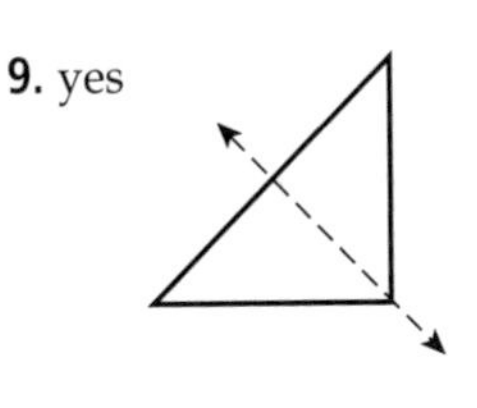

11. $D'(-3, -6)$, $E'(-2, 3)$, $F'(2, -2)$, and $G'(4, -9)$

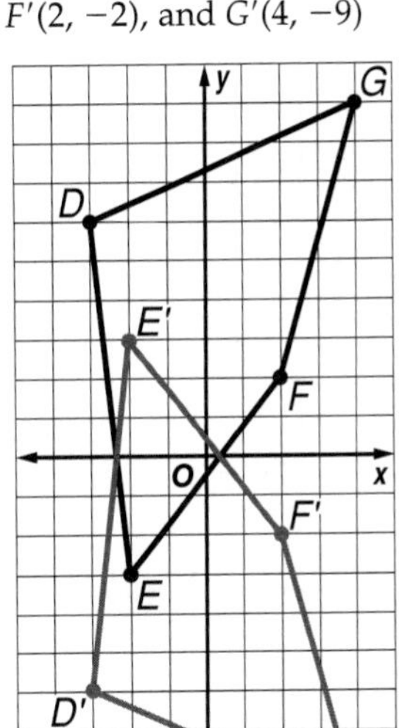

13. $Q'(-2, -5)$, $R'(-4, -5)$, and $S'(-2, 3)$

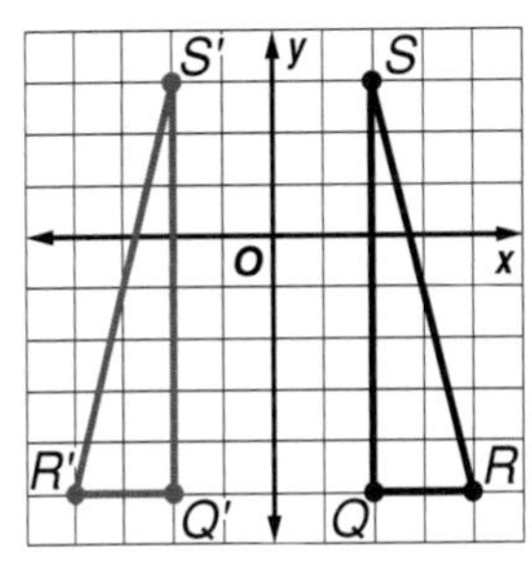

15. vertical line that divides the building in half and a line at the top of the water from the reflection in the water **17.** All four flags have line symmetry.

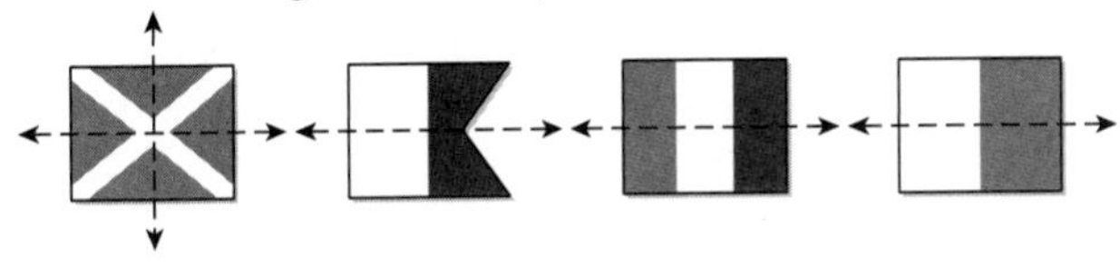

19. $J''(7, -4)$, $K''(-7, -1)$, and $L''(-2, 2)$ **21.** 2 **23.** No, the measure of one angle is 144°, which is not a factor of 360°. **25.** Sample answer: $0 \times \frac{1}{2} = 0$ **27.** $6n \le 18$; $n \le 3$

Pages 462–464 Chapter 10 Study Guide and Review

1. c **3.** f **5.** a **7.** right **9.** acute

11. Favorite Soft Drinks

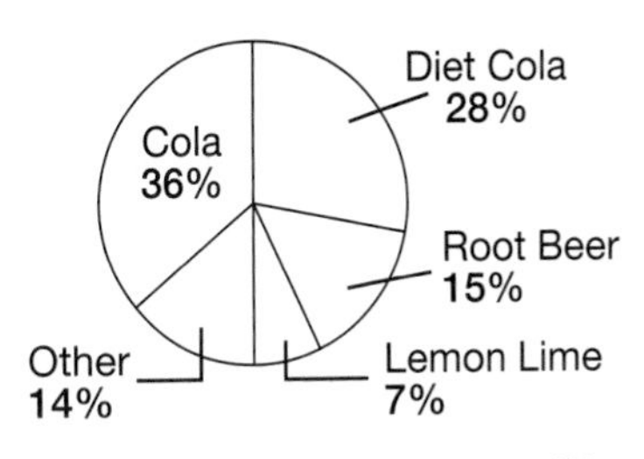

13. 79 **15.** obtuse, scalene **17.** 139° **19.** parallelogram **21.** 10 cm **23.** heptagon **25.** 156° **27.** $P'(8, -3)$, $Q'(2, -4)$, and $R'(3, 5)$

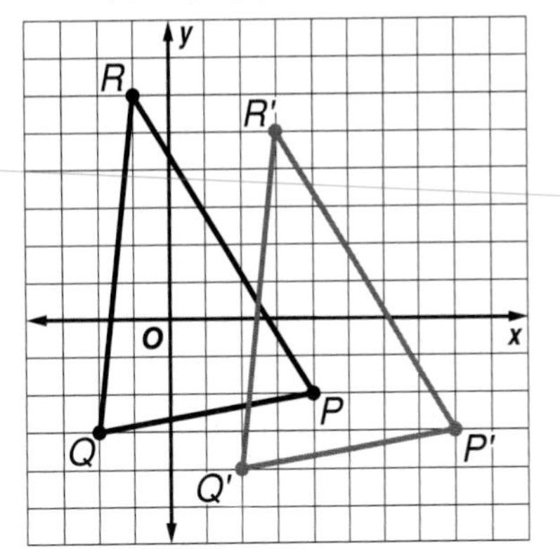

29. $P'(4, -9)$, $Q'(-2, -10)$, and $R'(-1, -1)$

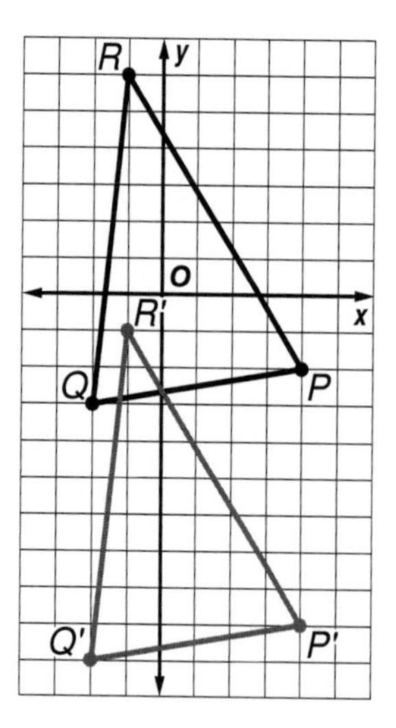

31. $A'(-1, 3)$, $B'(-2, -1)$, $C'(-5, -1)$, $D'(-4, 3)$

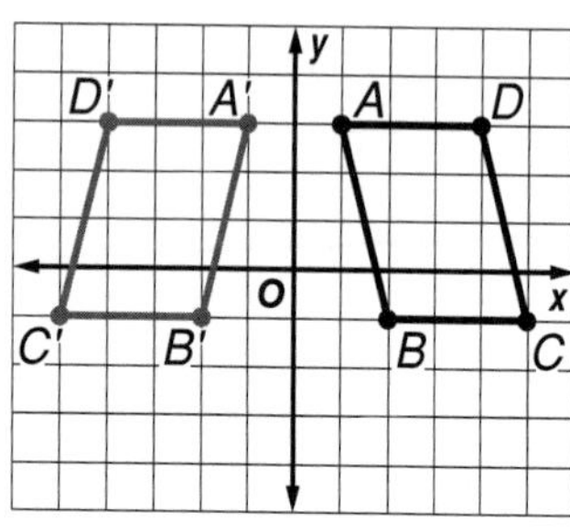

Chapter 11 Geometry: Measuring Two-Dimensional Figures

Page 469 Chapter 11 Getting Started

1. true **3.** false; right **5.** > **7.** = **9.** 64 **11.** 36 **13.** 60 **15.** 112 **17.** $\frac{1}{6}$ **19.** $\frac{1}{2}$

Pages 472–473 Lesson 11-1

1. Both involve multiplying a number by itself. **3.** 116; There are not two equal factors whose product is 116. **5.** 100 **7.** 900 **9.** 11 **11.** 23 **13.** 25 **15.** 49 **17.** 256 **19.** 324 **21.** 2 **23.** 12 **25.** 27 **27.** 35 **29.** 484 **31.** 10, −10 **33.** 1,312 **35.** 202,788 mi² **37.** Yes; for example, a garden that measures 30 ft by 30 ft has the same perimeter, but its area is 900 ft², which is greater than 500 ft². **39.** $\frac{3}{5}$

40–41. **Squares**

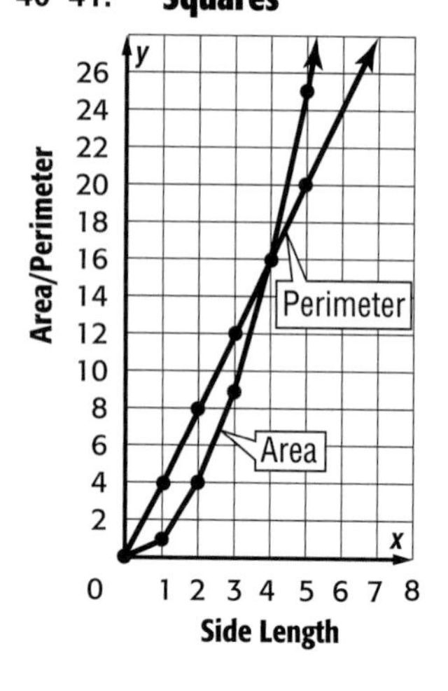

43. when the side length is between 0 and 4; when the side length is 0 or 4 **45.** 55 m² greater; The area of an 8-by-8-m square is 64 m². **47.** 1,024 cm²

49. $A'(3, 6)$, $B'(4, 3)$, $C'(1, 2)$

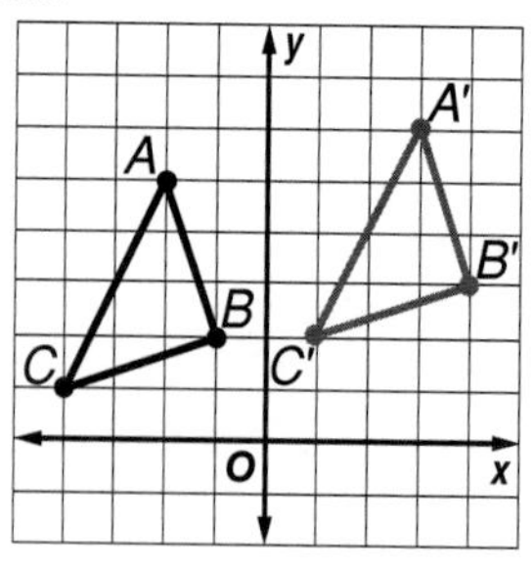

51. > **53.** >

Pages 476–477 Lesson 11-2

1. It cannot be written as a fraction. **3.** 51 is between the perfect squares 49 and 64. Since 51 is closer to 49 than to 64, $\sqrt{51}$ is closer to $\sqrt{49}$, or 7. **5.** 10 **7.** 12 **9.** 7.1 **11.** 21.5 **13.** 3 **15.** 6 **17.** 9 **19.** 12 **21.** 3.9 **23.** 6.6 **25.** 12.6 **27.** 25.4 **29.** π, $\frac{14}{3}$, $\sqrt{87}$, 10 **31.** 3.4 **35.** D **37.** 13 **39.** 28 **41.** 74 **43.** 16

Pages 481–482 Lesson 11-3

1. the lengths of both legs or the length of one leg and the hypotenuse **3.** Devin; the hypotenuse is 16 cm, not x cm as Jamie assumed. **5.** 24.5 in. **7.** yes **9.** 35 in. **11.** 8.0 cm **13.** 25 in. **15.** 23.7 m **17.** yes **19.** no **21.** 19.1 ft **23.** about 10.4 in. **25.** 11.3 in. **27.** 12 **29.** 17 **31.** 213.6 **33.** 28

Pages 484–485 Lesson 11-4

1. $b = 6$, $h = 3$ **3.** False; if the base and height are each doubled, then the area is $2b \cdot 2h = 4bh$, or 4 times greater. **5.** 1.1 m² **7.** 64 in² **9.** 312.5 mm² **11.** 428.4 mm² **13.** 207 in² **15.** 180 ft² or 20 yd² **17.** 1,575 yd² **19.** 56.9 cm² **21.** 15 in. **23.** C **25.** no **27.** no **29.** 84 **31.** 19.5

Pages 491–492 Lesson 11-5

1. Sample answer: $\frac{1}{2}(10)(20 + 30) = 250$ in² **3.** The area of a parallelogram is twice the area of a triangle with the same base and height. **5.** 105.6 m² **7.** $1,417.50 **9.** 38.4 mm² **11.** 112 m² **13.** 161.5 ft² **15.** 168 in² or 1.2 ft²

17. Sample answer: 28 in² **19.** Sample answer: 7 ft²

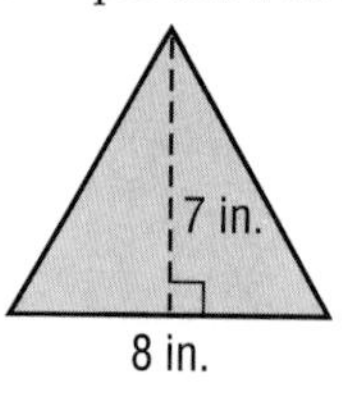

2 ft
2 ft
5 ft

21. about 1,872 mi² **23.** 382.4 cm² **25.** 3.7 in² **27.** 12.4 ft **29.** acute **31.** obtuse **33.** 40.8 **35.** 804.2

Pages 494–495 Lesson 11-6

1. Sample answer:

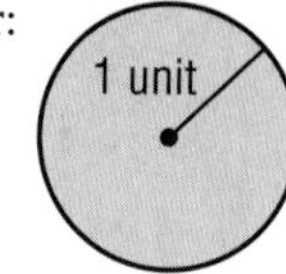

3. 2π; It is about $2 \cdot 3$, or 6; $\sqrt{7}$ is less than $\sqrt{9}$, or 3; 1.5^2 is less than 2^2, or 4. **5.** 254.5 in² **7.** 55.4 ft² **9.** 452.4 mm²

11. 201.1 cm^2 **13.** 95.0 ft^2 **15.** 227.0 cm^2 **17.** 113.1 ft^2 **19.** 7.1 cm^2 **21.** 63.6 in^2 **23.** 29.0 m^2 **25.** 43.2 ft^2 **27.** 1 : 2; 1 : 4; No, $\frac{1}{2} \neq \frac{1}{4}$. **29.** 5.9 in^2 **31.** Never; the area is quadrupled because $A = \pi(2r)^2$, or $A = 4\pi r^2$. **33.** 3.7 cm **35.** 120 in^2 **37.** 100.1 m^2 **39.** 113.04 **41.** 150.5

Pages 499–500 Lesson 11-7
1. Sample answer: Separate it into a rectangle and a triangle, find the area of each, and add. **3.** 112 m^2 **5.** 145 m^2 **7.** 58.6 in^2 **9.** 257.1 mm^2 **11.** 510.3 ft^2 **13.** Sample answer: Add the areas of a trapezoid and a small rectangle. **15.** 44 in^2 **17.** 69.4 cm^2 **19.** 176.7 in^2 **21.** $\frac{1}{6}$ **23.** $\frac{1}{3}$

Pages 502–503 Lesson 11-8
1. Sample answer: The probability of an event occurring is the ratio of the shaded area (the number of ways an event can occur) to the entire area (the number of possible outcomes). **3.** 31.3% **5.** 16% **7.** 34.3% **9.** 23.1% **11.** 73.8% **13.** 21.5% **15.** $\frac{4}{25}$ or 16% **17.** 3/10 **19.** 102.1 m^2

Pages 504–506 Chapter 11 Study Guide and Review
1. equal to **3.** 625 **5.** hypotenuse **7.** 36 **9.** 529 **11.** 16 **13.** 2 **15.** 7 **17.** 4 **19.** 7.8 **21.** 21.1 **23.** 13 mm **25.** 7.8 ft **27.** 13.3 m **29.** 99 cm^2 **31.** 135 cm^2 **33.** 36 ft^2 **35.** 187.7 cm^2 **37.** 408.3 in^2 **39.** 1,256.6 ft^2 **41.** 60 m^2 **43.** 182.8 in^2 **45.** 25%

Chapter 12 Geometry: Measuring Three-Dimensional Figures

Page 511 Chapter 12 Getting Started
1. circumference **3.** 90 **5.** 35 **7.** 24.1 **9.** 187.2 **11.** 24 **13.** 32 **15.** 27 **17.** $1\frac{1}{14}$ **19.** 855.3 cm^2 **21.** 113.1 ft^2

Pages 515–517 Lesson 12-1
3–17. Sample drawings are given.

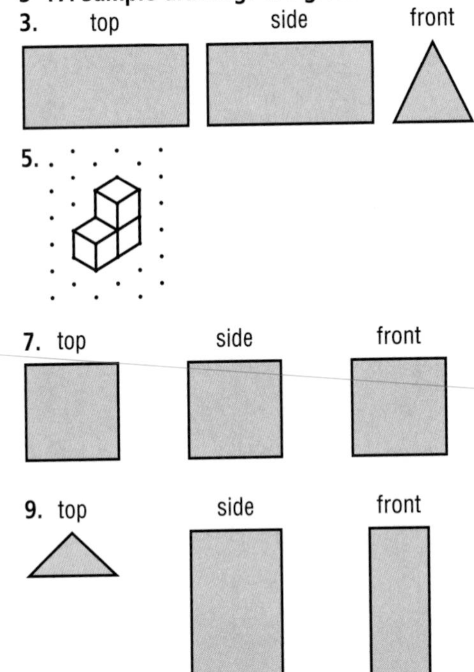

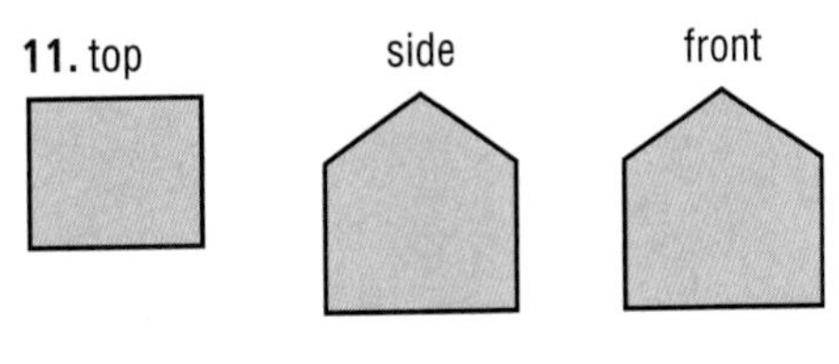

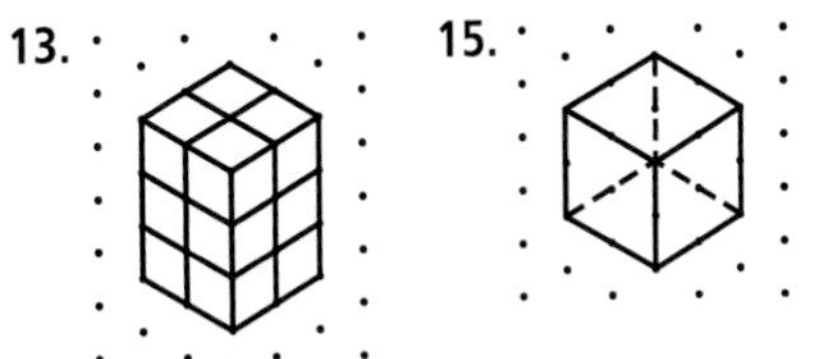

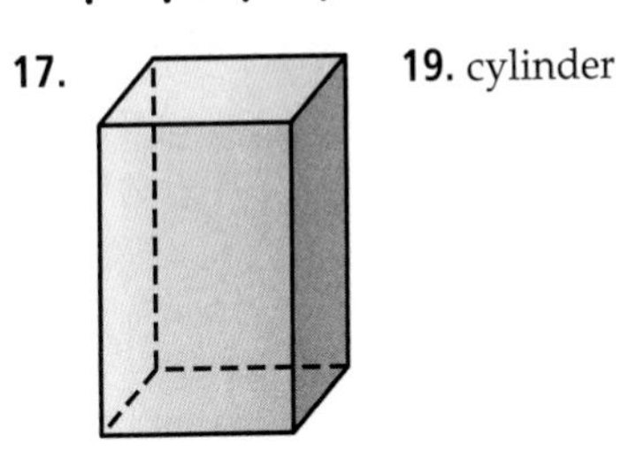

19. cylinder

21. Sample answer:

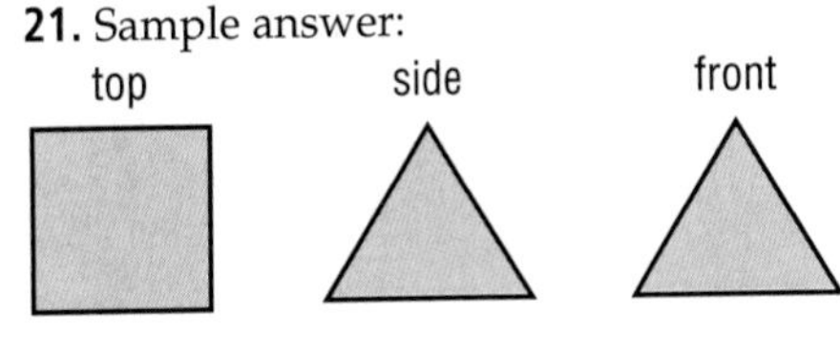

23. Sample answer:

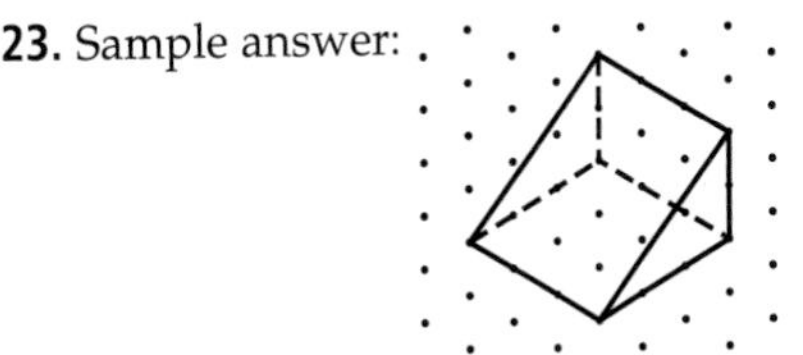

25. G **27.** 96 ft^2 **29.** 112.8 in^2 **31.** 22 **33.** 68

Pages 521–522 Lesson 12-2
1. yd^3 **3.** Ling; the width of Box B is half that of Box A. All three dimensions have to be considered when comparing volumes. **5.** 9.5 cm^3 **7.** 36,000 in^3 **9.** 960 in^3 **11.** 236.3 cm^3 **13.** 27 cm^3 **15.** 2,425.5 ft^3 **17.** 1,000,000 cm^3 **19.** C
21. Sample answer:

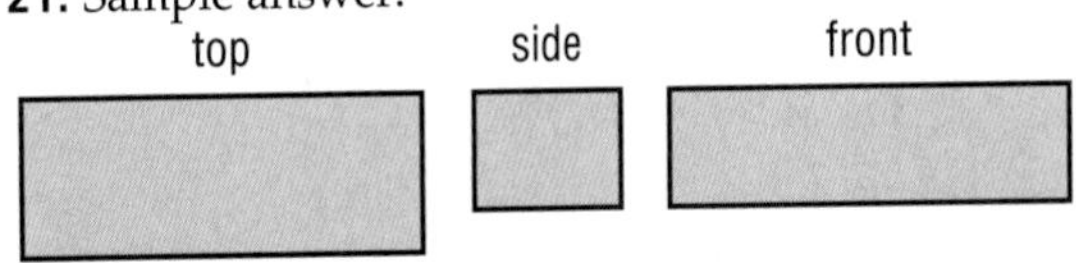

23. Sample answer: 18 **25.** Sample answer: 72

Pages 525–527 Lesson 12-3
1. Sample answer: 10 cm, 10 cm

3. 141.4 in^3 **5.** 617.7 ft^3 **7.** 241.5 m^3 **9.** 402.1 in^3 **11.** 167.1 cm^3 **13.** 1.3 m^3 **15.** 2,770.9 mm^3 **17.** 603.2 ft^3 **19.** 603.2 cm^3 **21.** 56.5 in^3 **25.** 1 to 4 **27.** about 41 ft **29.** B **31.** 152.88 m^3

33. Sample answer:

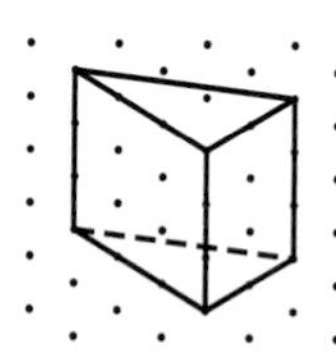

35. 572 m^2 **37.** 174 **39.** 43.5

Pages 533–535 Lesson 12-4

1. Find the sum of the areas of the faces. **3.** Surface area measures the area of the faces, and area is measured in square units. **5.** 183.4 cm^2 **7.** 314 cm^2 **9.** 833.1 mm^2 **11.** 125.4 in^2 or $125\frac{3}{8}$ in^2 **13.** 7 m^2 **15.** 1,531.9 ft^3 **17.** surface area = $6x^2$ **19.** 12 cm^2 **21.** 5 cm long, 4 cm wide, 5 cm high **25.** D **27.** 7 **29.** 72.9 cm^3 **31.** 63 in^3 **33.** 804.2 ft^2 **35.** 145.3 yd^2

Pages 539–541 Lesson 12-5

3. 88.0 mm^2 **5.** 653.5 m^2 **7.** 1,215.8 m^2 **9.** 183.8 ft^2 **11.** 1,120.0 in^2 or $1{,}119\frac{44}{45}$ in^2 **13.** 471.2 m^2 **15.** 172.8 in^2 **17.** Sample answer: prism: volume = 275 in^3, surface area = 237.5 in^2; cylinder: volume = 274.3 in^3, surface area = 211.1 in^2; Both containers hold about the same amount of popcorn. However, the cylinder uses less cardboard. So, the cylinder would be a more cost-effective container. **19.** Sample answer: The areas of both curved surfaces are the same. However, the areas of the top and bottom of cylinder A are greater than the areas of the top and bottom of cylinder B because it has a greater circumference (11 in.).
21. Sample answer:

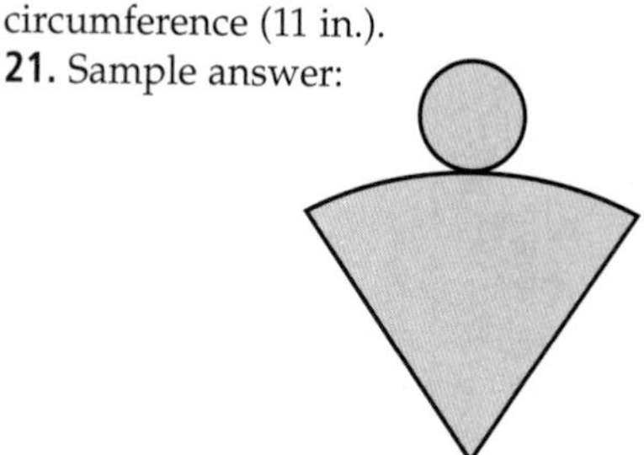

23. G **25.** 112 cm^2 **27.** 326.7 in^3 **29.** 14.0 **31.** 5.4

Pages 543–545 Lesson 12-6

1. Inches, because it is the smallest unit of measure. **3.** 8.6 mi, because it has 2 significant digits. The other measures have 3 significant digits. **5.** Sample answer: 15.65 cm **7.** Sample answer: The measure is precise to the nearest 0.01 second. **9.** 0.1 cm **11.** 2 lb **13.** Sample answer: 13.6 ft **15.** Sample answer: 9.15 cm **17.** Sample answer: 2.45 mL **19.** Neither; they are both given to the nearest tenth of a mile. **25.** B **27.** 502.7 yd^2 **29.** 748.0 in^2 **31.** 45 **33.** 24

Pages 546–548 Chapter 12 Study Guide and Review

1. rectangular prism **3.** cubic **5.** cylinder **7.** smaller
9. Sample answer:

top side front

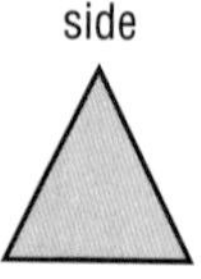

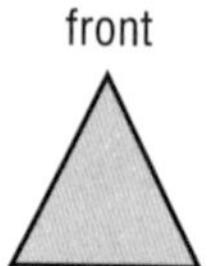

11. Sample answer:

13. 639.6 in^3 or $639\frac{5}{8}$ in^3 **15.** 4,042.4 m^3 **17.** 196.3 in^3 **19.** 531.3 m^2 **21.** 427.3 mm^2 **23.** 69.1 in^2 **25.** 5 min **27.** Sample answer: 0.8 cm

Photo Credits

Cover (tl b)Mark Gallup, (tr)PhotoDisc; **i** Mark Gallup; **iv v x** Aaron Haupt; **xi** Robert Laberge/Getty Images; **xii** Matthias Kulka/CORBIS; **xiii** Jeffrey L. Rotman/CORBIS; **xiv** Aaron Haupt; **xv** Haslin/Sygma/CORBIS; **xvi** PhotoDisc; **xvii** NASA; **xviii** CORBIS; **xix** DUOMO/CORBIS; **xx** CORBIS; **xxi** David Muench/CORBIS; **xxii** Kevin Fleming/CORBIS; **xxiii** John Evans; **xxiv** (t)PhotoDisc, (b)John Evans; **xxv** (t bl)PhotoDisc, (br)John Evans; **xxvi** John Evans; **2–3** Nicole Duplaix/Getty Images; **4–5** Jeff Greenberg/PhotoEdit; **7** CORBIS; **8** Dimitri Lundt/CORBIS; **13** Robert Laberge/Getty Images; **15** Chuck Savage/CORBIS; **19** Aaron Haupt; **22** (l)PhotoDisc, (r)Richard Hutchings/Photo Researchers; **24** Getty Images; **27** Tim Davis/Photo Researchers; **29** John Evans; **31** Brian A. Vikander/CORBIS; **32** Miles Ertman/Masterfile; **39** Aaron Haupt; **40** Phyllis Picardi/Stock Boston; **41** Bob Daemmrich/Stock Boston; **43** Owen Franken/Stock Boston; **52–53** Roger Ressmeyer/CORBIS; **54** Grant Symon/Getty Images; **55** Tim De Waele/TempSport/CORBIS; **57** George Clark, Hattiesburg American/AP/Wide World Photos; **58** John Evans; **59** Matthias Kulka/CORBIS; **60** Kelly–Mooney/CORBIS; **64** Joe Sohm/Photo Researchers; **67** CORBIS; **68** Craig Tuttle/CORBIS; **70** CORBIS; **73** First Image; **75** John Evans; **77 78** Matt Meadows; **79** Aaron Haupt; **80** John E. Kelly/Getty Images; **82** (tl tr)White House Historical Association, (b)Matt Meadows; **83** Jim Corwin/Photo Researchers; **85 86** CORBIS; **87** David Cannon/Getty Images; **94** Don Hebert/Getty Images; **95** Bob Daemmrich/Stock Boston; **102–103** Mike Powell/Getty Images; **104–105** Roger Ressmeyer/CORBIS; **106** Jed Jacobsohn/Getty Images; **109** Mark Tomalty/Masterfile; **111** (l)Timothy Fuller, (c)Thomas Veneklasen, (r)Crown Studios; **113** Geoff Butler; **117** John Evans; **120** Richard Kaylin/Getty Images; **122** Australian Picture Library/CORBIS; **123** Tony Feder/Getty Images; **129** (l)Jeffrey L. Rotman/CORBIS, (r)National Weather Service/Unisys Corp.; **131** Richard T. Nowitz/Getty Images; **132** John Evans; **133** Michelle Garrett/CORBIS; **137** Tim Wright/CORBIS; **139** NASA; **148–149** First Light Associated Photographers; **150** United Features Syndicate; **151 152** PhotoDisc; **158** (t)Michael Zito/SportsChrome, (c)CORBIS; **159** Six Flags, World of Adventure, Aurora, Ohio; **161 162** PhotoDisc; **163** Cedar Point; **164 165** PhotoDisc; **167** Tim Courlas; **169** PhotoDisc; **171** John Evans; **172** Aaron Haupt; **173** PhotoDisc; **176** CORBIS; **177** Aaron Haupt; **179** Brett Froomer/Getty Images; **183** CORBIS; **184** PhotoDisc; **192–193** ThinkStock/SuperStock; **194–195** Neil Rabinowitz/CORBIS; **199** Aaron Haupt; **201** John Evans; **204** Sandy Felsenthal/CORBIS; **206** KS Studios; **208** David Tardif/CORBIS; **210** Haslin/CORBIS SYGMA; **212** Carolyn A. McKeone/Photo Researchers; **215** John Evans; **216** Will & Deni McIntyre/Photo Researchers; **217** Doug Martin; **221** Ralph A. Clevenger/CORBIS; **223** Bob Krist/CORBIS; **227** Aaron Haupt; **228** Getty Images; **231** Ron J. Berard/DUOMO/CORBIS; **238–239** Roy Ooms/Masterfile; **243** Doug Martin; **245** PhotoDisc; **247** M. Angelo/CORBIS; **249** Aaron Haupt; **251** KS Studios; **252** John Evans; **253** Matt Meadows; **254** Richard Hamilton Smith/CORBIS; **257** (t)Joseph Nettis/Photo Researchers, (b)Latent Image; **263** John Evans; **265** Aaron Haupt; **268** Christopher Marona; **273** Charlotte Observer; **274** Matt Meadows; **275** Ben Wood/CORBIS; **277** Circlemakers; **284–285** James Richard/Stock Connection/PictureQuest; **286–287** Joseph Sohm/CORBIS; **289** Elaine Shay; **291** David Young-Wolff/PhotoEdit; **295** Johnny Johnson; **297** Doug Martin; **298** Project Exploration; **299** Aaron Haupt; **300** KS Studios; **301** Aaron Haupt; **302** (l)Ryan McVay/PhotoDisc, (r)KS Studios; **308** Underwood & Underwood/CORBIS; **311** John Evans; **312** Richard Lord/The Image Works; **315** Aaron Haupt; **317** Johnny Johnson; **319** Lynn M. Stone; **321** Matt Meadows; **323** NASA; **332–333** Richard Price/Getty Images; **334** CORBIS; **337** AFP/CORBIS; **338** (l)John Evans, (r)Matt Meadows; **341** Spencer Grant/PhotoEdit; **344** Mary Kate Denny/PhotoEdit; **346** Tim Fuller; **347** CORBIS; **349** John Evans; **351** Charles O'Rear/CORBIS; **352** David Young-Wolff/PhotoEdit; **354** Matt Meadows; **355** KS Studios; **356** Tony Freeman/PhotoEdit; **357** Morton & White; **359** Courtesy George Schweighofer/Currency Quest; **368–369** KS Studios; **370** Aaron Haupt; **371** Matt Meadows; **373** Laura Sifferlin; **375** Tim Fuller; **377** Aaron Haupt; **378** Geoff Butler; **382** DUOMO/CORBIS; **385** John Evans; **388** Geoff Butler; **389** The Supreme Court Historical Society; **391** Laura Sifferlin; **394** (t)KS Studios, (b)Michael Newman/PhotoEdit; **397** William Sallaz/DUOMO/CORBIS; **398** CORBIS; **400** Digital Vision/Getty Images; **408–409** Charles O'Rear/CORBIS; **410–411** Aaron Haupt; **414** Doug Martin; **415** Wally McNamee/CORBIS; **419** Jon Feingersh/CORBIS; **420** Tim Fuller; **423** Bettmann/CORBIS; **428** CORBIS; **434** Jed Jacobsohn/Getty Images; **439** John Evans; **441 443** PhotoDisc; **444** Laura Sifferlin; **447** Aaron Haupt; **448** (l)Tom Stack & Associates, (r)Aaron Haupt; **449** Christine Osborne/CORBIS; **450** EyeWire/Getty Images; **452** Wolfgang Rattay/Reuters/NewsCom; **454** Cordon Art-Baarn-Holland; **458** CORBIS; **459** Mug Shots/CORBIS; **468–469** John & Lisa Merrill/CORBIS; **471** Mark Gibson/H. Armstrong Roberts; **477** Stephen R. Swinburne/Stock Boston; **480** Collin Reid/AP/Wide World Photos; **487** John Evans; **490** David Muench/CORBIS; **496** (l)Laura Sifferlin, (r)Doug Martin; **497** Matt Meadows; **502** Doug Martin; **510–511** LWA-Dann Tardif/CORBIS; **514** Tribune Media Services, Inc.; **518** Laura Sifferlin; **522** Kevin Fleming/CORBIS; **523** Geoff Butler; **526** CORBIS; **529** John Evans; **533** Tim Fuller; **539** Alan Klehr/Getty Images; **540** (t)Ariel Skelley/CORBIS, (b)Geoff Butler; **542** Claudia Kunin/CORBIS; **552** Mark Gallup.

Index

Index

E

F

Index

Index

Symbols

Number and Operations

Symbol	Meaning
$+$	plus or positive
$-$	minus or negative
$a \cdot b$, $a \times b$, ab or $a(b)$	a times b
$\div$	divided by
$\pm$	plus or minus
$=$	is equal to
$\neq$	is not equal to
$>$	is greater than
$<$	is less than
$\geq$	is greater than or equal to
$\leq$	is less than or equal to
$\approx$	is approximately equal to
$\%$	percent
$a:b$	the ratio of a to b, or $\frac{a}{b}$
$0.7\overline{5}$	repeating decimal 0.75555...

Algebra and Functions

Symbol	Meaning
$-a$	opposite or additive inverse of a
a^n	a to the nth power
a^{-n}	$\frac{1}{a^n}$
$\lvert x \rvert$	absolute value of x
$\sqrt{x}$	principal (positive) square root of x
$f(n)$	function, f of n

Geometry and Measurement

Symbol	Meaning
$\cong$	is congruent to
$\sim$	is similar to
$^\circ$	degree(s)
$\overleftrightarrow{AB}$	line AB
$\overrightarrow{AB}$	ray AB
$\overline{AB}$	line segment AB
AB	length of $\overline{AB}$
∟	right angle
$\perp$	is perpendicular to
$\parallel$	is parallel to
$\angle A$	angle A
$m\angle A$	measure of angle A
$\triangle ABC$	triangle ABC
(a, b)	ordered pair with x-coordinate a and y-coordinate b
O	origin
π	pi $\left(\text{approximately } 3.14 \text{ or } \frac{22}{7}\right)$

Probability and Statistics

Symbol	Meaning
$P(A)$	probability of event A
$n!$	n factorial
$P(n, r)$	permutation of n things taken r at a time
$C(n, r)$	combination of n things taken r at a time

Formulas

Perimeter	square	$P = 4s$
	rectangle	$P = 2\ell + 2w$ or $P = 2(\ell + w)$
Circumference	circle	$C = 2\pi r$ or $C = \pi d$
Area	square	$A = s^2$
	rectangle	$A = \ell w$
	parallelogram	$A = bh$
	triangle	$A = \frac{1}{2}bh$
	trapezoid	$A = \frac{1}{2}h(b_1 + b_2)$
	circle	$A = \pi r^2$
Surface Area	cube	$S = 6s^2$
	rectangular prism	$S = 2\ell w + 2\ell h + 2wh$
	cylinder	$S = 2\pi rh + 2\pi r^2$
Volume	cube	$V = s^3$
	prism	$V = \ell wh$ or Bh
	cylinder	$V = \pi r^2 h$ or Bh
	pyramid	$V = \frac{1}{3}Bh$
	cone	$V = \frac{1}{3}\pi r^2 h$ or $\frac{1}{3}Bh$
Pythagorean Theorem	right triangle	$a^2 + b^2 = c^2$
Temperature	Fahrenheit to Celsius	$C = \frac{5}{9}(F - 32)$
	Celsius to Fahrenheit	$F = \frac{9}{5}C + 32$

Measurement Conversions

Length	1 kilometer (km) = 1,000 meters (m) 1 meter = 100 centimeters (cm) 1 centimeter = 10 millimeters (mm)	1 foot (ft) = 12 inches (in.) 1 yard (yd) = 3 feet or 36 inches 1 mile (mi) = 1,760 yards or 5,280 feet
Volume and Capacity	1 liter (L) = 1,000 milliliters (mL) 1 kiloliter (kL) = 1,000 liters	1 cup (c) = 8 fluid ounces (fl oz) 1 pint (pt) = 2 cups 1 quart (qt) = 2 pints 1 gallon (gal) = 4 quarts
Weight and Mass	1 kilogram (kg) = 1,000 grams (g) 1 gram = 1,000 milligrams (mg) 1 metric ton = 1,000 kilograms	1 pound (lb) = 16 ounces (oz) 1 ton (T) = 2,000 pounds
Time	1 minute (min) = 60 seconds (s) 1 hour (h) = 60 minutes 1 day (d) = 24 hours	1 week (wk) = 7 days 1 year (yr) = 12 months (mo) or 52 weeks or 365 days 1 leap year = 366 days
Metric to Customary	1 meter ≈ 39.37 inches 1 kilometer ≈ 0.62 mile 1 centimeter ≈ 0.39 inch	1 kilogram ≈ 2.2 pounds 1 gram ≈ 0.035 ounce 1 liter ≈ 1.057 quarts